Particulate Two-Phase Flow

BUTTERWORTH-HEINEMANN SERIES IN CHEMICAL ENGINEERING

SERIES TITLES

Bubble Wake Dynamics in Liquids and Liquid-Solid Suspensions *Liang-Shih Fan and Katsumi Tsuchiya*
Chemical Process Equipment: Selection and Design *Stanley M. Walas*
Chemical Process Structures and Information Flows *Richard S.H. Mah*
Computational Methods for Process Simulations *W. Fred Ramirez*
Constitutive Equations for Polymer Melts and Solutions *Ronald G. Larson*
Fluidization Engineering, 2nd ed. *Octave Levenspiel*
Fundamental Process Control *David M. Prett and Carlos E. Garcia*
Gas-Liquid-Solid Fluidization Engineering *Liang-Shih Fan*
Gas Separation by Adsorption Processes *Ralph T. Yang*
Granular Filtration of Aerosols and Hydrosols *Chi Tien*
Heterogeneous Reactor Design *Hong H. Lee*
Interfacial Transport Processes and Rheology *David Edwards, Howard Brenner, and Darsh T. Wasan*
Introductory Systems Analysis for Process Engineers *E. Bruce Nauman*
Laminar Flow and Convective Transport Processes: Scaling Principles and Asymptotic Analysis *L. Gary Leal*
Microhydrodynamics: Principles and Selected Applications *Sangtae Kim and Seppo J. Karrila*
Modeling With Differential Equations in Chemical Engineering *Stanley M. Walas*
Molecular Thermodynamics of Nonideal Fluids *Lloyd L. Lee*
Particulate Two-Phase Flow *M.C. Roco*
Phase Equilibria in Chemical Engineering *Stanley M. Walas*
Physicochemical Hydrodynamics: An Introduction *Ronald F. Probstein*
Slurry Flow: Principles and Practice *Clifton A. Shook and M.C. Roco*
Transport Processes in Chemically Reacting Flow Systems *Daniel E. Rosner*
Viscous Flow: The Practical Use of Theory *Stuart W. Churchill*

REPRINT TITLES

Advanced Process Control *W. Harmon Ray*
Applied Statistical Mechanics *Thomas M. Reed and Keith E. Gubbins*
Elementary Chemical Reactor Analysis *Rutherford Aris*
Kinetics of Chemical Processes *Michel Boudart*
Reaction Kinetics for Chemical Engineers *Stanley M. Walas*

Particulate Two-Phase Flow

Edited by

M.C. Roco

National Science Foundation and
University of Kentucky

With 50 contributing authors

Butterworth–Heinemann

Boston London Oxford Singapore Sydney Toronto Wellington

Recognizing the importance of preserving what has been written, it is the policy of Butterworth-Heinemann to have the books it publishes printed on acid-free paper, and we exert our best efforts to that end.

Library of Congress Cataloging-in-Publication Data

Particulate two-phase flow / edited by M.C. Roco, with 50 contributing authors.
p. cm. — (Butterworth-Heinemann series in chemical engineering)
Includes bibliographical references and index.
ISBN 0-7506-9275-8 (alk. paper)
1. Two-phase flow. 2. Particles. I. Roco, M. C. II. Series.
TA357.5.M84P37 1992
620.1'064—dc20 92-33076
CIP

British Library Cataloguing-in-Publication Data
A catalogue record for this book is available from the British Library.

Butterworth–Heinemann
80 Montvale Avenue
Stoneham, MA 02180

10 9 8 7 6 5 4 3 2 1

Printed in the United States of America

CONTENTS

CONTRIBUTORS

J. Abbott, Los Almos National Laboratory, Los Almos, NM 87545 (Chapter 1)

A. Acrivos, Chemical Engineering Department, CUNY City College, 138th St. at Convent Ave., New York, NY 10031 (Chapter 5)

R. Adrian, Department of Mechanics, University of Illinois, 216 Talbot Laboratory, 104 South Wright, Urbana, Illinois 61801 (Chapter 2)

H. Ahn, Division of Engineering and Applied Science, California Institute of Technology, Pasadena, CA 91125; Present address: The Aerospace Corporation, El Secundo, California (Chapter 7)

J.L. Binder, Department of Chemical Engineering, University of Illinois, Urbana, IL 61801 (Chapter 19)

J. Brady, Department of Chemical Engineering, California Institute of Technology, Room 210-41, Pasadena, CA 91125 (Chapter 26)

C.E. Brennen, Division of Engineering and Applied Science, California Institute of Technology, MS 104-44, Pasadena, CA 91125 (Chapter 7)

H. Brenner, Department of Chemical Engineering, MIT, Room 66-566, Cambridge, MA 02139 (Chapter 1)

T. Cader, Department of Mechanical Engineering, University of Kentucky, Lexington, KY 40506 (Chapter 13)

J.N. Chung, Department of Mechanical and Materials Engineering, Washington State University, Pullman, WA 99164 (Chapter 18)

H. Conrad, Department of Materials Science and Engineering, North Carolina State University, Box 7907, Raleigh, NC 27695-7907 (Chapter 12)

C. Crowe, Depatment of Mechanical and Materials Engineering, Washington State University, Pullman, WA 99164 (Chapter 18)

S.V. Dhodapkar, Department of Chemical Engineering, University of Pittsburgh, 323 BEH, Pittsburgh, PA 15261 ((Chapter 21)

D.A. Drew, Department of Mathematical Science, Rensselaer Polytechnic Institute, Troy, NY 12181 (Chapter 16)

R. Falco, Michigan State University, 231 Engineering Building, East Lansing, MI 48824 (Chapter 3)

L.-S. Fan, Department of Chemical Engineering, Ohio State University, 140 W. 19th Ave., Columbus, OH 43210-1180 (Chapter 23)

R. Flagan, Department of Chemical Engineering, California Institute of Technology, 210-41, Pasadena, CA 91125 (Chapter 14)

Y.O. Fuentes, Chemical Engineering Department, University of Wisconsin, 1415 Johnson Drive, Madison, WI 53706; Present address: Chemical Engineering Department, University of Colorado-Boulder, Campus Box 424, Boulder, CO 80309 (Chapter 27)

D. Gidaspow, Department of Chemical Engineering, Illinois Institute of Technology, IIT Center, Chicago, IL 60616 (Chapter 22)

A.L. Graham, Los Alamos National Laboratory, MEE-4, Los Alamos, NM 87545 (Chapter 1)

T.J. Hanratty, Department of Chemical Engineering, University of Illinois, Urbana, IL 61801 (Chapter 19)

G. Hetsroni, Department of Chemical and Nuclear Engineering, University of California-Santa Barbara, Santa Barbara, CA 93106 (Chapter 8)

R. Jackson, Department of Chemical Engineering, Princeton University, Princeton, NJ 08544 (Chapter 20)

D.D. Joseph, Department of Aerospace Engineering and Mechanics, University of Minnesota, 107 Akerman Hall, Minneapolis, MN 55455 (Chapter 10)

S. Kim, Chemical Engineering Department, University of Wisconsin, 1415 Johnson Drive, Madison, WI 53706 (Chapter 27)

G. Klinzing, Department of Chemical Engineering, University of Pittsburgh, 323 BEH, Pittsburgh, PA 15261 (Chapter 21)

H. Kytomaa, MIT, 77 Massachusetts Ave., Room 3-258a, Cambridge, MA 02139 (Chapter 24)

R.T. Lahey, Jr., Department of Nuclear Engineering, Rensselaer Polytechnic Institute, Troy, NY 12181 (Chapter 16)

D. Leighton, Department of Chemical Engineering, University of Notre Dame, Notre Dame, IN 46556 (Chapter 6)

Z.C. Liu, Department of Mechanics, University of Illinois, 216 Talbot Laboratory, 104 South Wright, Urbana, Illinois 61801 (Chapter 2)

J.E. Maneval, Department of Chemical Engineering, University of California, Davis, CA 95616 (Chapter 4)

O. Masbernat, Department of Mechanical Engineering, University of Kentucky, Lexington, KY 40506; presently at University of Toulouse, France (Chapter 13)

K.L. McCarthy, Department of Chemical Engineering, University of California, Davis, CA 95616 (Chapter 4)

M.J. McCarthy, Department of Chemical Engineering, University of California, Davis, CA 95616 (Chapter 4)

L.A. Mondy, Sandia National Laboratories, Division 1511, Albuquerque, NM 87185 (Chapter 1)

A.D. Nikolov, Illinois Institute of Technology, Chicago, IL 60616 (Chapter 11)

D. Nocera, Department of Chemistry, Michigan State University, East Lansing, MI 48824 (Chapter 3)

R.L. Powell, Department of Chemical Engineering, University of California, Davis, CA 95616 (Chapter 4)

A. Prosperetti, Department of Mechanical Engineering, The Johns Hopkins University, Baltimore, MD 21218 (Chapter 28)

I. Rampall, Department of Chemical Engineering, University of Notre Dame, Notre Dame, IN 46556 (Chapter 6)

M.C. Roco, Department of Mechanical Engineering, University of Kentucky, Lexington, KY 40506; and National Science Foundation, Engineering Directorate, Room 1115, Washington, D.C. 20550 (Chapter 13)

W.B. Russel, Department of Chemical Engineering, Princeton University, Princeton, NJ 08544-5263 (Chapter 17)

A.S. Sangani, Department of Chemical Engineering and Materials Science, Syracuse University, Syracuse, NY 13244 (Chapter 28)

S.L. Soo, Department of Mechanical and Industrial Engineering, University of Illinois, 1206 W. Green Street, Urbana, IL 61801 (Chapter 9)

T.R. Troutt, Department of Mechanical and Materials Engineering, Washington State University, Pullman, WA 99164 (Chapter 18)

K. Tsuchiya, Department of Chemical Engineering, Ohio State University, 140 W. 19th Ave., Columbus, OH 43210-1180 (Chapter 23)

O.R. Walton, Lawrence Livermore National Lab, P.O. Box 808, Mail Code L-207, Livermore, CA 94550 (Chapter 25)

D.T. Wasan, Department of Chemical Engineering, Illinois Institute of Technology, Chicago, IL 60616 (Chapter 11)

C. Zhu, Department of Mechanical and Industrial Engineering, University of Illinois, 1206 W. Green Street, Urbana, IL 61801 (Chapter 9)

C. Zukoski, Department of Chemical Engineering, University of Illinois, Urbana, IL 61801 (Chapter 15)

PREFACE

This book could have been entitled "Trends for Particulate Two-Phase Flow Research", since this was the guiding idea in preparing it. Substantial progress is noticeable in the last years in the areas of measurements of two-phase flow microstructure, production of new materials from ultrafine particles, micromechanical and probabilistic modeling, and numerical simulations on high performance computers. We have tried to identify contributions from those areas which will have a lasting impact on the field of fluid mechanics and particulate technology.

This publication brings together several of the most original and productive researchers in particulate and fluid flow research. Emphasis was placed on basic qualitative concepts, understanding flow phenomena, as well as new experimental, modeling and computational findings specific to particulate two-phase flows, at both low and large Reynolds numbers. Most of the chapters include a review section, one on industrial relevance, and another section on future research opportunities. The contributions use the same main notation, and when possible relate to each other. The book is a unique collection of 28 contributions. It offers a qualitative view of the basic research in the U.S. in the area of particulate two-phase flow, and is expected to be of interest to researchers all over the world.

Particulate flows are encountered in nature and a large variety of industrial settings such as chemical, metallurgical, plastics, pharmaceutical and food processing. In chemical plants, for instance, the production of about 2/3 of chemical products involves particulate two-phase flow. There are several emerging particulate technologies, such as nanoparticle engineering, electrorheology, clean room technology, materials processing based on metal clusters, and fabrication of optically nonlinear materials. Even a small improvement in the design of multiphase systems may lead to significant savings in industrial processes. This book provides basic concepts and theoretical background for design activities.

Particulate flow research is at the intersection of three disciplines: fluid mechanics, soil mechanics and solid mechanics. For the micron-sized particles, and particularly for nanoparticles under 50 nm, one may add physics and chemistry. The inroads made by the recently developed measuring techniques for two-phase flow microstructures (magnetic resonance imaging, chemical and radioactive tracers, tomography, etc.), micromechanical modeling (kinetic theory, multi-particle interaction models) and computational methods (direct numerical simulation, probabilistic simulation) are highlighted with examples in this book.

The text has been divided into two parts: experiments on particulate two-phase flow microstructure, and modeling and numerical simulation techniques. The chapters address the following generic topics:

- Description of multiphase flow microstructures, phenomena and interaction mechanisms: chapters 1 to 13, and 16 to 24.
- Transport phenomena of submicron particles, and nanoparticle engineering: chapters 11, 14, 15, as well as other chapters about colloids. Two chapters summarize particle production for development of new materials, via aerosols and colloids.
- Direct numerical simulation with local analytical and numerical solutions: chapters 25 to 28. The focus is on simulations which aim to reproduce the intrinsic physics of the flow.
- Inter-particle and particle-fluid interactions, including external fields, phase change and physico-chemical interactions: chapters 5, 6, 10 to 12, 17, 24 to 26.
- Particle-wall interactions: chapter 8.
- Fluid machinery for two-phase flow: chapter 13.
- Basic concepts for particle use and process control: chapters 12, 20 to 23.

The primary target of this book is researchers from universities and industry in chemical, mechanical, minerals, civil and pharmaceutical areas. Secondarily, the book will be useful to graduate students, as reference material for understanding phenomena, and as a guide to selecting advanced methods for engineers involved in design and production. This book summarizes research advances useful to both those entering the field and those planning for future research programs.

A message of the book is that the new experimental techniques and parallel processing are moving methods of investigation from the macroscopic to the microscopic. Growth is expected in research on particulate flow microstructure, direct numerical simulations and probabilistic techniques. Application to manufacturing, submicron particle use for materials with engineered microstructure, and particulate flow in environment are envisioned.

The scientific level of all contributions is high, but relevant to applications, and accessible to researchers and engineers in the field. We would like to thank all participants in the preparation of this publication. Have a useful and enjoyable reading.

PART I

EXPERIMENTS ON TWO-PHASE FLOW MICROSTRUCTURE

of this book] Maneval *et al.*, 1992). We have observed shear-induced microstructural changes in suspensions of large spheres (600 μm < diameter < 3.2 mm) using NMR imaging (Graham *et al.*, 1991; Abbott *et al.,* 1991). (The NMR imaging experiments discussed in this chapter were performed in collaboration with Drs. S. A. Altobelli and Eiichi Fukushima of Lovelace Medical Foundation.)

NMR imaging has been used to study flow-induced particle migration in concentrated suspensions (solids volume fraction $\alpha_S \geq 0.39$) since this technique can measure suspended-particle concentration distributions with good spatial resolution. When subjected to flow between rotating concentric cylinders separated by a wide gap (resulting in an *inhomogeneous* Couette flow), neutrally buoyant particles migrate away from the high shear-rate region near the inner rotating cylinder toward the lower shear-rate region near the outer fixed cylinder. This migration occurs in both monodisperse and bidisperse sphere suspensions. The monodisperse spheres migrate until they reach the region of lowest shear (existing near the outer wall), in which region the particles achieve (approximately) the maximum concentration of spheres possible for random packing. A similar migration occurs in bidisperse sphere suspensions. Furthermore, the larger particles in such suspensions segregate into concentric bands after being sheared. Visual observations of opaque particles in bands near the outer wall of the apparatus have shown that the larger spheres are arranged in two-dimensional hexagonal close-packed (hcp) sheets that rotate in concentric sheets relative to each other (Graham *et al.*, 1991; Abbott *et al.*, 1991).

Such ‘crystallization’ phenomena constitute a dramatic illustration of the shear-induced formation of structure in suspensions. If such phenomena are caused by shearing a suspension, then, in discussing the *intrinsic* rheological properties of a given (i.e., initially homogeneous) suspension, one must confront the need to determine the viscosity of the suspension without altering this property through the very act of *measuring* it. Conventional viscometers often employ inhomogeneous flow fields, which then, tend to influence the microstructure of the suspension. For example, suspended fibers, if not subject to strong, randomizing, rotary Brownian forces, will tend to align when undergoing shear flow; hence, the resulting rheological measurements are performed on suspensions possessing a flow-induced anisotropic structure rather than on the original isotropic (i.e., orientationally homogeneous) suspension.

We have shown (Mondy *et al.*, 1986; Milliken *et al.*, 1989b) that falling-ball rheometry can be used to determine the effective viscosity of a suspension without sensibly perturbing the original gross microstructure of that suspension. When the size of the probe (the falling ball) is of the order of the characteristic length of the suspended particles, the probe then perturbs only slightly the original microstructure of the (otherwise quiescent) suspension as it settles.

Also reviewed in this paper are experiments that illustrate the use of falling-ball rheometry in determining the viscosity of suspensions composed of both randomly-oriented rods and (approximately) aligned rods (Mondy *et al.*, 1990; Milliken *et al.*, 1989b). The effective, falling-ball, viscosity in the direction of the aligned rod axes is observed to be significantly lower than that observed with the rods randomly oriented. Furthermore, the viscosity of the oriented rod suspension agrees well with the viscosity measured with a (homogeneous flow) Couette rotational rheometer, implying that the strongly anisotropic rod alignment created within the quiescent suspension may mimic that induced by the shear flow.

The discrete, noncontinuum microstructure of the suspension also causes the settling ball to instantaneously deviate in both magnitude and direction from its mean, uniform-speed, rectilinear (vertical) path. Velocity data from the original falling-ball experiments (together with an additional series of experiments, wherein the full three-dimensional trace of each settling sphere through the suspensions was continuously monitored) allowed the dispersive component in the vertical, mean-settling direction of the ball to be determined (at various solid volume fractions α_S). Additionally, the detailed experiments showed the horizontal dispersive component to be an order-of-magnitude *smaller* than the comparable vertical component. Finally, at α_S = 50%, the dimensionless falling-ball dispersivity compared well with the dimensionless 'self-diffusivity' observed in the Couette devices. This suggests that the falling-ball fluctuations are caused by the same mechanism that gives rise to the shear-induced diffusivity undergone by a suspension subjected to a shear flow.

In the next section, we discuss the two distinct experimental protocols used (as well as their respective accuracies) in performing falling-ball experiments in concentrated suspensions. Section 1.3 discusses the accuracy and use of NMR imaging in determining local particle concentration and velocity profiles in concentrated suspensions. Section 1.4 addresses the falling-ball results for suspensions of both spheres and rods, examining the viscosity encountered by the settling ball, as well as the ball's dispersivity. Section 1.5 examines the behavior of suspensions of spheres in wide-gap Couette-flow devices, wherein the imposed flow field is strongly inhomogeneous.

1.2 FALLING-BALL TECHNIQUES

1.2.1 Real-Time X-Ray Radiography

Radiography, using penetrating radiation such as x rays, is a method useful in examining balls falling through *opaque* suspensions. If the imaging is done on a fluor, the image may be electronically amplified, observed by a video camera, and permanently recorded on videotape. This imaging technique allows the tracking of a radiographically opaque (high-atomic-Z-number) ball through the suspension.

The liquid used to buoyantly suspend the polymethyl methacrylate (PMMA) spheres was composed of a solution of practical grade 1,1,2,2 tetrabromoethane (TBE) from Eastman Kodak (12.5% by weight) and UCON oil (H-90,000), a polyalkylene glycol made by Union Carbide (87.5% by weight). A small amount (about 0.1% of the weight of TBE) of Tinuvin 328, manufactured by Ciba-Geigy, was dissolved in the TBE before mixing to prevent chemical breakdown of TBE because of ultraviolet radiation. Similar solutions, displaying a range of viscosities, can be created by using different UCON oils, which are commercially available over a wide range of viscosities.

The use of two complete x-ray systems focused on the same spatial point provides a stereo view of the tracer particles for use in three-dimensional tracking, as shown in Figure 1-1. The video system can support two cameras simultaneously. The use of split-screen recording ensures the synchronization of the two images. With this stereo view, we obtained four screen coordinates to determine the three spatial coordinates of the center of the falling ball. This constituted an overdetermined system, one that lent itself to linear regression analysis.

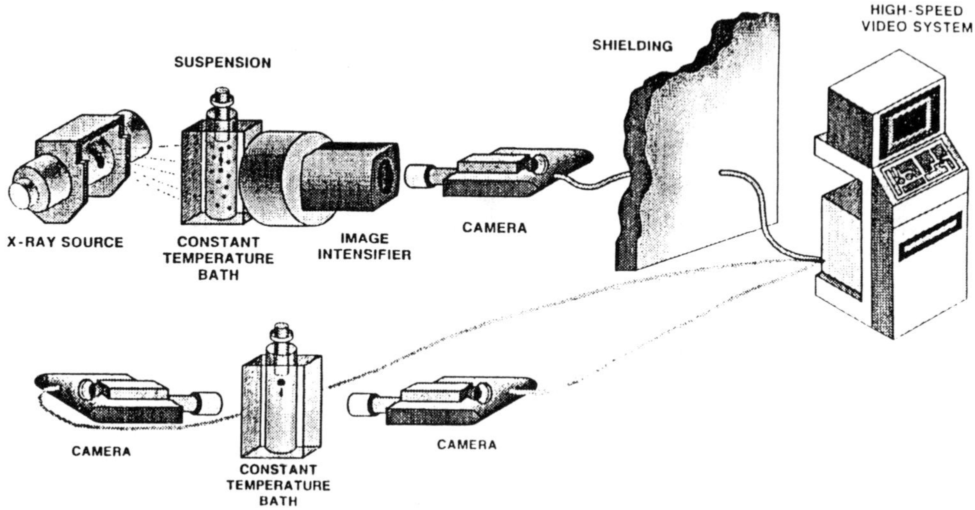

FIGURE 1-1 *Major pieces of equipment needed to perform both the real-time radiography and the visual falling-ball experiments. For simplicity, the two-camera setup is shown only for the visual experiments.*

The scheme outlined by Walton (1981), and extrapolated to handle nonlinear optics (Dingman *et al.*, 1992), was used to determine the laboratory-fixed coordinates from the camera coordinates.

Accuracy in locating the position of the falling ball depends upon the field of view, which, in turn, is governed by the needs of the particular experiment. In the experiments on model suspensions, discussed in the following subsections, an area about 150 × 75 mm was imaged on each split screen. This implied that the accuracy of the measured position of the particle center was within 0.5 mm. Note, however, that with an x-ray microfocus, a typical field of view is only about 10 mm × 10 mm, allowing very small balls to be tracked accurately. Steel balls with diameters of 432 μm have successfully been tracked (Graham *et al.*, 1987), both while settling in an otherwise quiescent propellant simulant, as well as while moving with the simulant in a capillary flow.

Measurements were made of the viscosities of suspensions of 5% to 55% by volume of uniform PMMA spheres in various density-matched Newtonian liquids (usually for the mixture of UCON oil and TBE described above) using suspended spheres 3.18 to 12.7 mm in diameter. A ball composed of a metal such as brass, nickel, or tungsten carbide that sufficiently attenuates the x-rays relative to the liquid, the suspended particles, and the container, produces an x-ray image that can be readily tracked as the ball falls through the suspension.

1.2.2 Refractive Index-Matched Suspensions

Although real-time radiography has the distinct advantage that it can be used with opaque suspensions, tracking of tracer particles can also be accomplished optically in *transparent* suspensions. We have developed a three-component, Newtonian liquid that matches *both* the refractive index and density of PMMA, allowing transparent suspensions of neutrally buoyant PMMA particles to be made. This suspending liquid was similar to that described in the previous section, except that it also contained Triton X-100, an alkylaryl polyether alcohol from J. T. Baker (50.27% by weight) in addition to the TBE (14.07% by weight) and the UCON oil (35.66% by weight).

Each interface shown in Figure 1-1 acts as a nonlinear lens because of Snell's Law. The first interface--between air and the PMMA tank holding cooling water--dominates the nonlinear behavior because it has the largest refractive index difference. An initial model, extending Walton's work by adding a simple plane interface at an arbitrary distance between the camera and the object being tracked, led to a highly nonlinear, implicit relationship between the camera coordinates and the object's real-space coordinates. However, by adding quadratic terms in the real-space coordinates to the linear relationship for the camera coordinates, errors in determining the positions of reference objects were rendered indistinguishable from random errors.

Balls of opaque plastic, glass, aluminum, and corundum, in addition to those utilized in the x-ray radiography experiments, were used in the transparent suspension cases. Typically, balls, approximately the same size as that of the suspended spheres themselves, were employed.

1.2.3 General Falling-Ball Techniques

On the length scale of the suspended spheres, the discrete, noncontinuum nature of the suspension is readily apparent in falling-ball experiments, as can be seen in Figure 1-2. Whereas a large ball appears to settle smoothly through a suspension of smaller particles, (i.e., its instantaneous velocity appears fairly constant, and its path strictly vertical), a ball of almost the same diameter as the suspended particles

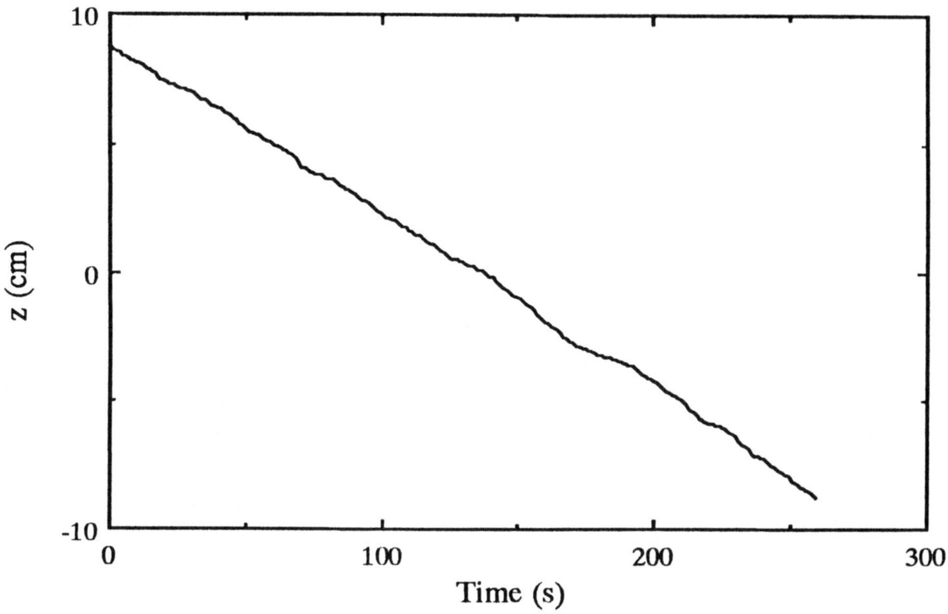

FIGURE 1-2 *Examples of experimentally measured ball traces: (a) Instantaneous vertical position (z) of a single 3.175-mm sphere settling in a suspension of 50-volume percent 3.175-mm neutrally buoyant spheres. Deviations from a uniform instantaneous settling speed are evidenced by the nonrectilinear nature of the trajectory.*

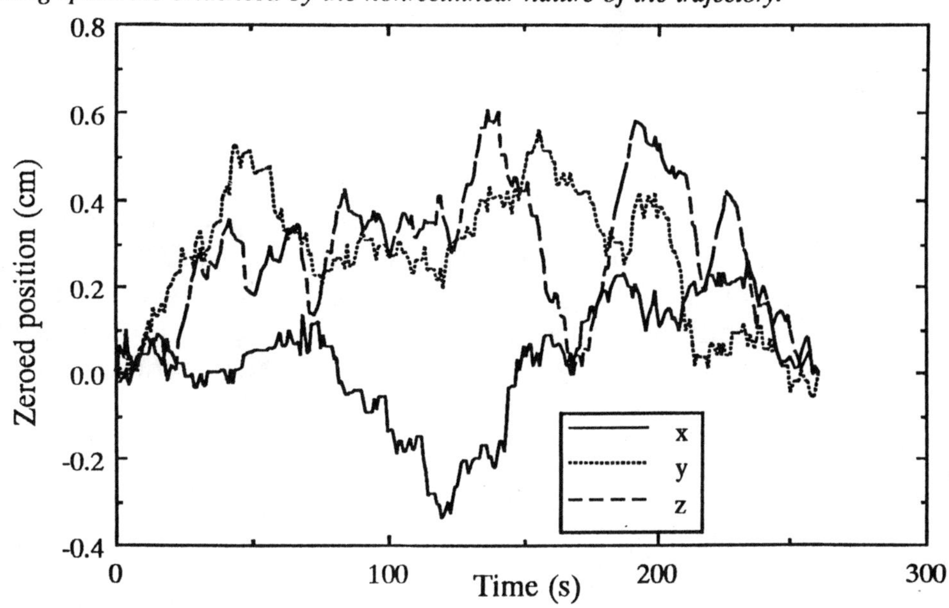

(b) By zeroing each direction, it is apparent that little or no correlation exists between the respective motions in each of the three directions. Observe that fluctuations in the vertical direction (z) display greater sweeps than in the horizontal (x,y) directions. Note that the z direction has been normalized such that the maximum variations in the vertical direction are of the same order as those in the x and y directions.

displays markedly erratic sedimentation behavior. Periods of almost no ball motion, as the falling-ball approaches and "rolls off" a suspended particle, alternates with periods of almost free fall through the interstitial fluid between suspended spheres. A statistical analysis reveals, however, that the *average* vertical terminal velocity of the ball, measured over a vertical distance (usually between 100 and 1000 suspended particle diameters), is accurately reproducible.

The chief advantage of falling-ball rheometry using relatively small balls, is that the effective viscosity of a suspension can be determined *while only slightly disturbing the respective initial positional and orientational distribution functions of the suspended particles.* Such rheological measurements furnish the so-called *zero-shear viscosity* of the suspension. Moreover, simultaneous observation of the dispersion of the falling ball about its mean settling mode furnishes useful information about the noncontinuum nature of the suspension.

In our experiments, the suspensions were confined within temperature-controlled circular columns. The suspensions were well stirred before each experiment to achieve a homogeneous spatial (and orientational) distribution of suspended particles.

1.3 NMR IMAGING OF SUSPENSIONS

The spatial distribution of suspended particles present in concentrated suspensions is difficult to measure because any observation of the interior of the suspension is obstructed by the high particle density. Indeed, most suspensions are opaque even at relatively low particle concentrations. However, the noninvasive technique of NMR imaging makes it possible to analyze the particle distributions existing in such suspensions. Background material on basic NMR phenomena can be found in standard textbooks (Morris, 1986; Abragam, 1961). Some details attendant upon the specific use of specific NMR imaging techniques are provided in the following paragraphs.

Techniques for imaging the annular domain between rotating concentric circular cylinders (i.e. wide-gap Couettes) have previously been described by Watkins and Fukushima (1988) and by Graham *et al.* (1991). The NMR imaging technique employed gives an imaged slice wherein the fluid furnishes a full-intensity signal and whereas the suspended particles furnish no signal whatsoever. The normalized value of the image intensity is proportional to the density of the liquid-phase protons in a volume element. As such, the NMR image, by allowing visualization of the individual particles themselves, furnishes quantitative information about the local spatial distribution of particles within the suspension.

To test the reproducibility of the technique, a pair of presumably identical experiments were performed two months apart. The results, expressed in terms of the variance S [cf. Eq. (1.4)], were accurately reproducible. Moreover, concentration images taken sequentially had errors of ±5%, pixel by pixel. The radially averaged concentration profiles were even more accurate.

A technique for measuring the tangential (angle θ) particle velocity field has been developed (Abbott *et al.*, 1991). Although, it is possible to obtain a full two-dimensional velocity profile from the image using a similar phase-shift technique, only axisymmetric bodies have thus far been studied. A recent review of this and other velocity measurements is provided by Caprihan and Fukushima (1990).

A long plastic connecting rod permitted the inner cylinder of the Couette apparatus to be turned remotely by a variable speed motor, as shown in Figure 1-3. This connecting rod was used to locate the motor sufficiently far from the magnet to permit the motor to function despite the strong magnetic field in its proximity. As discussed in the following section, *total* particle migration was established to be functionally dependent upon only the *total* strain (i.e., the number of rotations of the inner cylinder), independently of the shear rate. The number of revolutions contributing to the total strain were counted with a Radio Shack counter attached to the motor. The motor was stopped after a predetermined number of revolutions, and NMR images of the Couette gap were taken.

In particular, images were taken of a cross-sectional slice, 2.4 cm thick, perpendicular to the Couette axis, about midway along the apparatus axis. These images consisted of 128 rows, each with 128 elements, corresponding to an imaged area of 7.5 cm square. Data for successive rows were gathered 0.5 seconds apart, allowing the sample to

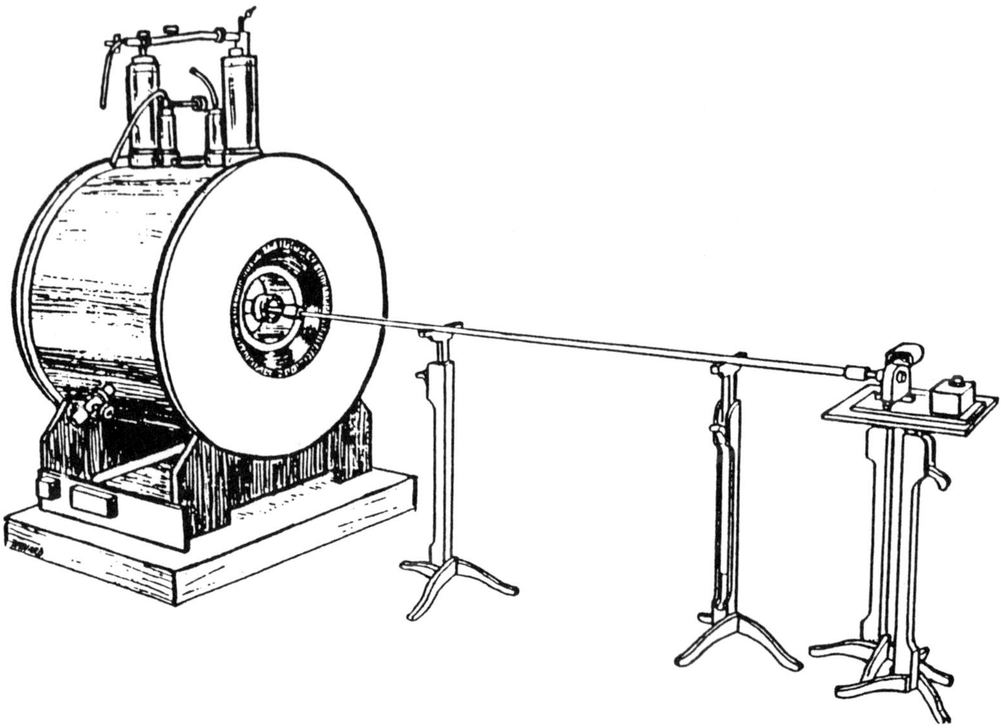

FIGURE 1-3 *Experimental NMR imaging apparatus for particle migration studies in wide-gap, inhomogeneous Couette flow device. Depicted are the horizontal bore magnet containing the Couette apparatus at left, the motor assembly at right, and the connecting rod, which allows the motor to be positioned far from the magnet.*

relax between NMR pulses. To reduce the noise in the signal, four iterations were summed to create each image. The time required for each image was ~8 min. Back-to-back repeated images indicated that no significant settling or particle redistribution occurred during this time interval. Increasing the resolution (i.e., taking more rows of data) or decreasing the signal-to-noise ratio (i.e., taking a larger number of iterations) altered the time required to record a single image in proportion to the square of the increase in resolution or the decrease in signal-to-noise ratio.

The volume fraction of liquid was determined in one of two ways from the images shown in Figure 1-4. In one method, a small syringe of pure liquid was taped to the outer wall to provide a reference. If the field was uniform within the measurement volume, the absolute fluid volume fractions could then be linearly scaled from the pure fluid measurement. Small variations near the edge of the magnetic field led

FIGURE 1-4 *NMR images of a cross section of a suspension (α_S = 50%) of 650 μm sieved beads between concentric cylinders. The image on the left represents the initial, well-dispersed state of the suspension. Note the lobes at the eight o'clock position near the outer wall. These regions were created when the suspension was poured down one side of the annular region and around the center cylinder. The image on the right was taken after rotating the inner cylinder until a steady state was achieved (4,000 revolutions). The dark area near the inner cylinder represents a larger fluid fraction, indicating that particles have migrated away from this high shear-rate region. The small deviations from perfect circular in the image on the right are caused by small asymmetries in the instrument.*

to larger errors when employing this method to establish the average concentration. In our experiments, the intensity values in the portion of the slice corresponding to the suspension were normalized such that the average intensity of the image matched the fluid fraction, which was known to a high degree of accuracy.

As shown in Figures 1-4 and 1-5, the initial spatial distribution of suspended particles after loading the Couette was essentially uniform, except near the edges. Three basic sources of nonuniformity in this region arose from: (i) particle exclusion from the region near the wall because of the nonzero particle size; (ii) mismatching between the imaging pixels and the physical edge of the Couette device; and (iii) rounding-off of the edges caused by measuring, in phase space, the step change in volumetric liquid concentration occurring in real space. This last effect, which is an inherent problem with spectral methods like NMR, appears to dominate the edge effects, although its effect is limited to a domain comprising but a few pixels near the edges.

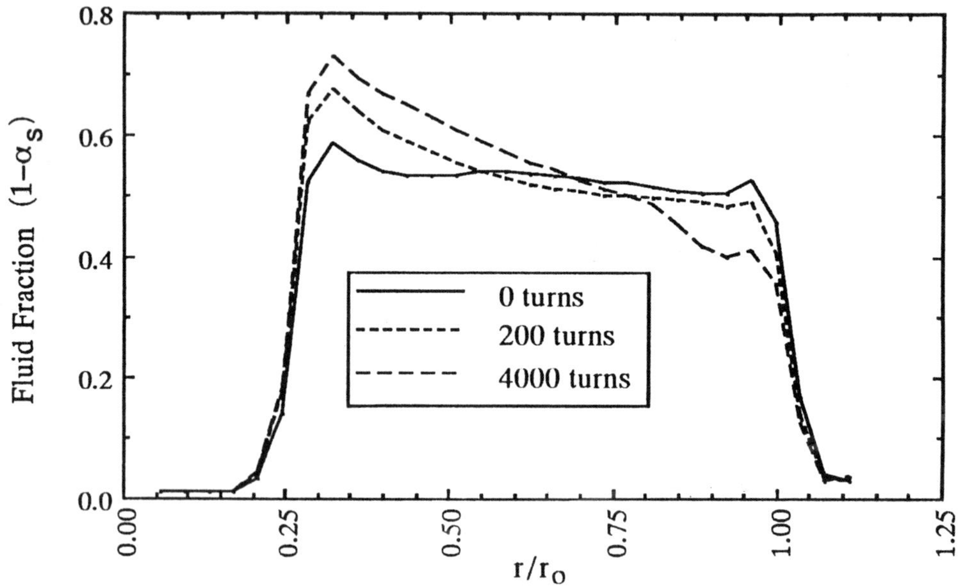

FIGURE 1-5 *The radial concentration profile is determined from the average value of the signal intensity as a function of radial position. The actual number of radial positions (r) at which data were taken in the annular space varied from 15 to 18, depending upon the NMR parameters selected for that particular experiment. Shown are data for a 50% suspension of 650-μm beads in its initial and final states and at one intermediate state. Note the 30% concentration difference that ultimately develops across the gap at steady state.*

1.4 FALLING-BALL EXPERIMENTS IN SUSPENSIONS

1.4.1 Measuring the Viscosity

Both optical and radiographic methods have been used to track the path of a dense ball settling slowly through a quiescent suspension. In a single-phase (i.e., *homogeneous*) Newtonian liquid, and at the small Reynolds numbers corresponding to our experiments, the terminal velocity of the sphere is directly related to the liquid viscosity via Stokes law (Bird *et al.*, 1960), corrected, if required, for the increased drag on the sphere occurring because of the presence of the container walls (Bohlin, 1960). In a suspension, an apparent viscosity can thus be comparably measured by supposing that the suspension is representable as a hypothetical homogeneous Newtonian liquid continuum (Mondy *et al.*, 1986) (and that, at the suspension scale, no slip occurs on the ball surface).

These measured terminal velocities (corrected for Newtonian wall effects), when translated into a suspension viscosity via Stokes law, yielded a viscosity which was independent of the size of the falling ball (relative to the diameter of the suspended spheres and of the container size) over a wide range of falling-ball sizes. (Anomalous behavior (Milliken *et al.*, 1989b; Mondy *et al.*, 1988) can occur with very small or very large balls.) For moderately concentrated suspensions (below about 30% solids), the average relative viscosity, μ_r (namely the viscosity of the suspension normalized with the suspending fluid viscosity), agrees with independent viscosity measurements taken in shear and capillary rheometers under flow conditions (Mondy *et al.*, 1986). These rheometers generally indicate that moderately concentrated suspensions of spheres behave as Newtonian liquids, without the anomalous strain-dependent results observed at higher particle concentrations. Consequently, falling balls experience the same average resistance to motion that they would if the combined effects of the interstitial fluid and suspended spheres were replaced by a hypothetical homogeneous Newtonian liquid characterized by a shear-independent viscosity.

At higher concentrations, time-dependent behavior of suspensions has been reported, along with the steady-state viscosity (Gadala-Maria and Acrivos, 1980; Leighton and Acrivos, 1987). As demonstrated by Surovec *et al.* (1989), the reproducibility of the steady-state viscosity for ideal suspensions in Couette devices is around 10%. Unfortunately, the initially measured viscosity is about 40% higher than the final viscosity. While the initially measured viscosity might correspond to that of a homogenous, random suspension, Surovec *et al.* (1989) also observed its dependence upon the loading technique used, suggesting that some flow-induced migration could have occurred during sample loading.

An illustration of the use of falling-ball rheometry as a tool with which to measure suspension viscosity, without inadvertently altering the microstructure of the suspension, can be found in recent work with suspensions of rodlike particles (Milliken *et al.*, 1989a; Powell *et al.*, 1989; Mondy *et al.*, 1990). For suspensions of non-Brownian rods in Newtonian fluids, viscometric and elongational flows (the usual flow fields employed by experimental rheologists) induce an alignment of the rods, thereby creating a nonrandom orientational particle distribution.

With falling-ball rheometry, however, the initial orientational distribution of the rods can be controlled; moreover, the rod-aligning effects of the falling-ball-induced flow field are minimal, at least for those portions of the suspensions distant from the falling-ball trace.

Falling-ball experiments have been performed to determine the viscosity of suspensions of randomly oriented rods (Milliken *et al.*, 1989a; Powell *et al.*, 1989). Suspensions of rods with an aspect ratio of 19.83 were observed to possess viscosities that varied linearly with a solids volume fraction, up to a volume fraction of about 0.125. Such apparently *linear* behavior thereby is obtained at volume fractions much higher than might be expected, based upon dilute suspension theories, and considerably higher than is found to occur with comparable suspensions of *spherical* particles.

Unlike the case for spherical particle suspensions, for suspensions of rods, one cannot rationally compare falling-ball measurements with comparable shear-flow viscosity measurements because the latter experiments cannot be performed on similar suspensions of *randomly oriented and spatially homogeneously distributed* rods. Such shear flows necessarily produce *nonrandom* orientational (and in the case of *inhomgeneous* shear flows, possibly spatial) distributions. Now, theoretical suspension viscosity predictions by Brenner (1974) exist for dilute homogeneous suspensions of rods subject to strong rotational Brownian motion. Furthermore, Haber and Brenner (1984) show that such Brownian rod suspensions behave rheologically in shear flow exactly as would suspensions with sustained random orientational distributions. It therefore appears legitimate to compare our (large rotary Peclet number) experiments with the theory for Brownian rods. The intrinsic viscosity predicted (Brenner, 1974) for rod suspensions of this aspect ratio is 29.2, whereas in the experimental value, it was 27.6. These values differ by only 6%. Further experiments with rods of other aspect ratios (Powell *et al.*, 1989; Morrison, 1989) show equally good agreement with this strong rotational Brownian motion theory.

During the falling-ball measurements performed with rod suspensions, the suspensions were stirred just prior to each experiment to assure orientational isotropy (as well as spatial homogeneity). In a subsequent series of experiments (Mondy *et al.*, 1990), the suspended rods were approximately aligned before each ball release, so that the

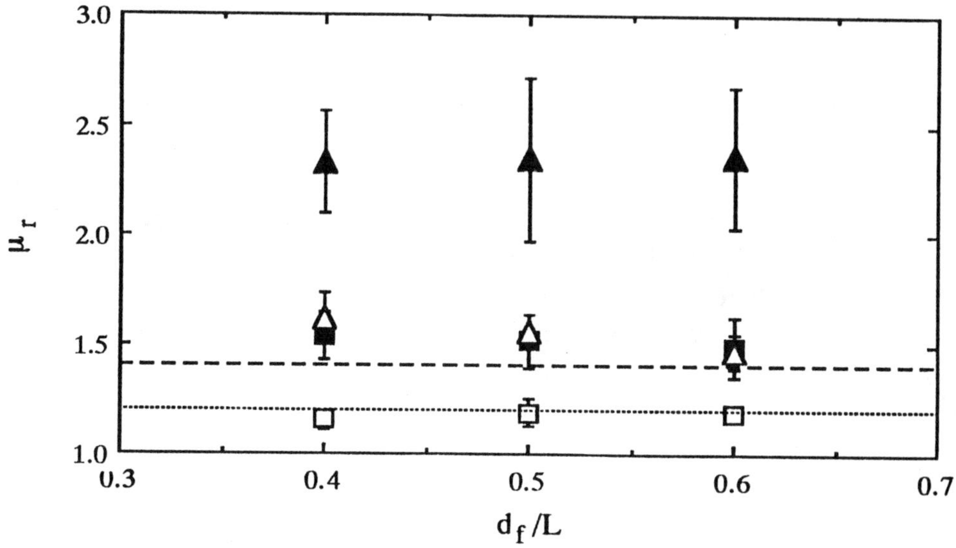

FIGURE 1-6 *Relative viscosity μ_r vs dimensionless ball diameter (diameter d_f of the falling ball divided by the length L of the suspended rods) for suspensions of aligned rods, as measured using falling-ball rheometry. Suspensions of various volume fractions are represented: $\alpha_S = 0.02$ (□), $\alpha_S = 0.05$ (Δ). Also shown are the relative viscosities of suspensions of randomly oriented rods: $\alpha_S = 0.02$ (■), $\alpha_S = 0.05$ (▲). The dotted and dashed lines represent data obtained (Ganani and Powell, 1986) during shearing flows for $\alpha_S = 0.02$ and $\alpha_S = 0.05$, respectively.*

suspensions were (initially) strongly anisotropic, as regards the distribution of rod orientations, with the orientational distribution approaching that of a Dirac delta function. Falling-ball measurements performed on these aligned suspensions yielded apparent viscosities (in a direction parallel to the rod axes) that were substantially less than those for suspensions having the same volume fraction of rods, but *randomly* oriented. Results are shown in Figure 1-6. In addition, viscosities measured for suspensions of aligned rods closely correlated with the viscosities of suspensions of short fibers (possessing similar rod aspect ratios and concentrations), as measured in shearing flows (Ganani and Powell, 1986). This suggests that such initial alignments may mimic the flow-induced orientation occurring in rotational rheometers. These observations demonstrate the possibility of using falling-ball rheometry to determine *configuration-specific*, anisotropic suspension viscosities possessing concomitant directional *tensorial* rheological properties; that is, in principle, such measurements afford the possibility of experimentally establishing the fourth-rank viscosity tensor (governing the complete stress tensor for a generalized Newtonian fluid) for an aligned rod suspension.

1.4.2 Measuring Ball Dispersivity During Falling-Ball Experiments

When instantaneously tracking a falling ball through a suspension, deviations from the average steady vertical settling path (due to "collisions", i.e., hydrodynamic interactions with the suspended particles) were observed to be relatively erratic. These fluctuations can be modelled in several ways, namely as a constrained deviation from an approximately periodic path, a confined fluctuation about the cylinder axis, or as an unconfined Fickian-like dispersion process.

To test the hypothesis that the deviations on a length-scale equivalent to several suspended sphere diameters might be random, the paths of several experiments were recorded at approximately the highest resolution in time (i.e., a time equivalent to the sphere settling one pixel), and, subsequently, transformed into frequency space, thus making it possible to examine the possibility of periodic ordering. The major supposition being tested was that the nearly maximum-packed structure arising in the 50% suspensions might lead to quasiperiodic microstructures. The fast Fourier transforms of the settling path revealed no significant frequency dependence, and the broad decay of the frequency response was akin to that known to exist for random walks.

For the observed deviations to be truly modelled as a random walk, such deviations need to be shown to be both independent and identically distributed. Upon testing the correlation existing between consecutive time intervals, the minimum time interval for uncorrelated or independent intervals was determined to be approximately the time needed for the ball to settle a vertical distance equivalent to about 6 to 10 suspended sphere diameters. Since a random walk is representable as a Fickian diffusion process, a Fickian-like diffusivity arising strictly from hydrodynamic interactions between the ball and the suspended particles was used to quantify the data.

To permit comparisons to be made between suspensions composed of different suspending fluids and suspending-sphere sizes, the measured dispersivities (D) were rendered dimensionless by using the diameter (d_S) of the suspended spheres and the mean falling-ball settling velocity (U) as characteristic scaling parameters. The dimensionless dispersivity (D^*) used in subsequent figures is therefore defined as

$$D^* = \frac{D}{U\, d_S} \quad . \tag{1.1}$$

In general, the dispersivity of a ball settling through a spatially homogeneous suspension need not be isotropic. The dispersivity ($D_{||}$) in the settling direction, parallel to gravity, may differ from that ($D_{\perp}$) in the horizontal direction, resulting in a transversely isotropic dispersivity dyadic characterized by these two scalar dispersivities.

The process of experimentally determining these dispersivities begins with establishing the minimum time necessary for the settling sphere to 'forget' its initial velocity, or for the deviation of its trajectory to appear random relative to its average settling motion. Because the corresponding time scale for Brownian fluctuations is of the order of the time between molecular collisions, all experiments were unquestionably performed on much longer time scales; hence, the question of molecular contributions to the process can be ruled out, even for short-time-dispersion coefficients. Because the ball's motion is highly deterministic and, thus, unlikely to be random durint the time it takes for the falling ball to pass around a suspended sphere, it was appropriate to establish the time interval (or, through the average velocity, the comparable vertical distance) required by the ball to become uncorrelated from its initial position. By correlating the displacements for successive time intervals of duration τ, the minimum time needed for the displacements to become uncorrelated was determined.

Once this minimum time was established, the dispersivity was simply obtained as being one-half the slope of the mean-squared displacement plotted as a linear function of time. Since more intervals are used in calculating some of the means (typically those for the shorter-time means), a weighted best-fit line was used to account for the unequal error limits. As can be seen by comparing Figure 1-7a with 1-7b, the vertical dispersivity component is about 20 times larger than the horizontal component, thereby rendering the dispersion dyadic strongly anisotropic. Also note that the line does not go through the origin.

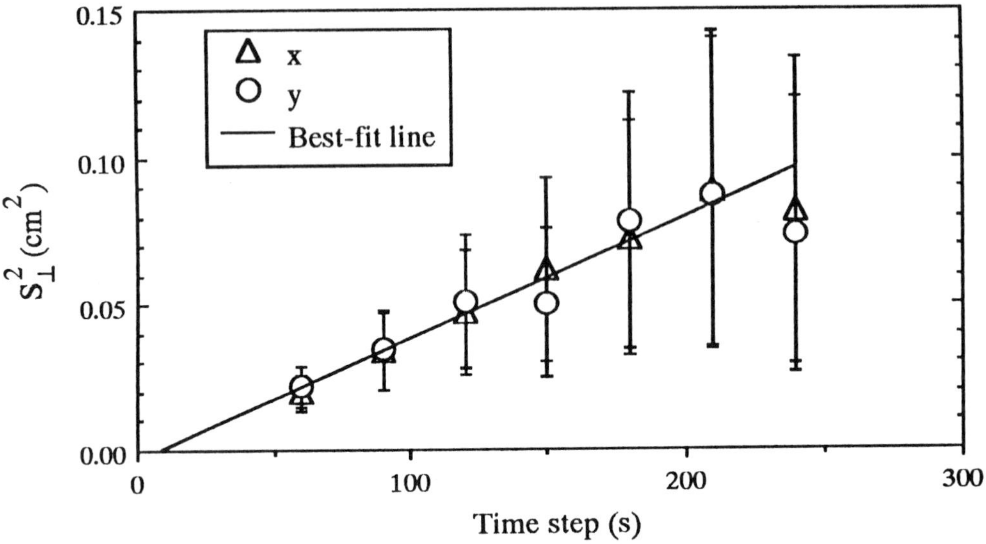

FIGURE 1-7 *Horizontal ($D_\perp$) and vertical ($D_{||}$) dispersivities: (a) The horizontal dispersivity data, shown above, is represented by one-half the slope of the mean-squared displacement ($S^2_\perp$) vs time data, using intervals greater than or equal to the minimum time interval. Information from each of the two mutually perpendicular, horizontal directions is combined to determine the horizontal dispersivity. A best-fit line, weighted with the confidence limits, is used to determine the slope.*

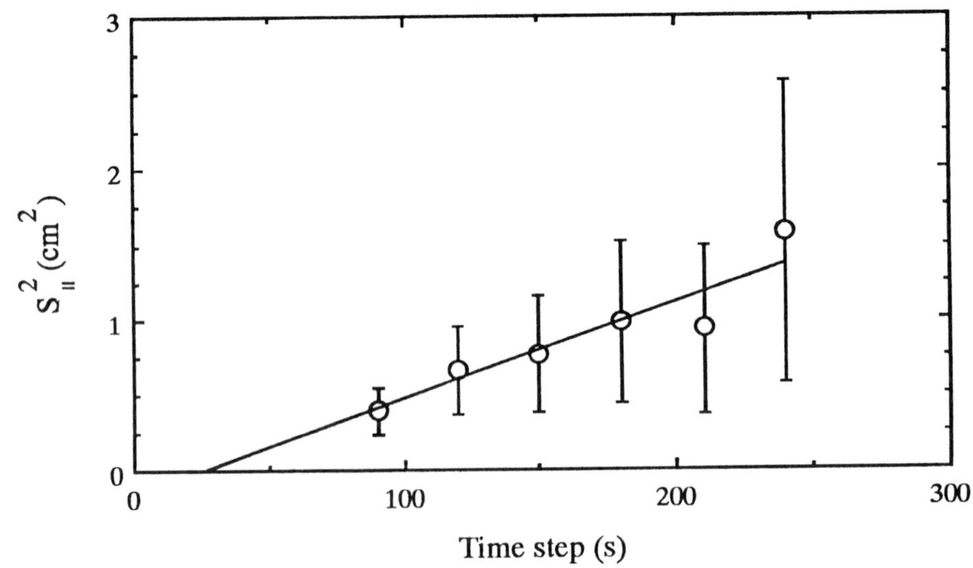

(b) The vertical dispersivity ($D_{||}$) is obtained by the same method as for the horizontal dispersivity in (a). Note that the vertical mean-squared displacements shown ($S^2_{||}$) are approximately 20 times larger than the horizontal displacements.

Since the dispersive model does not hold for times less than the minimum sampling interval, S^2 will grow quadratically near the origin instead of linearly. By assuming that the linear behavior then holds in the nonlinear region, an aphysical negative S^2 is predicted.

Recent efforts (Phillips *et al.*, 1991; Abbott *et al.*, 1991) aimed at calculating the collision-induced phenomenological dispersion coefficient governing the shear-induced lateral migration of suspended spheres, furnish a value for the above coefficient of about $2.4\alpha_S$. For a solids volume fraction of $\alpha_S = 0.50$, this predicts that the dimensionless Fickian diffusivity coefficient for such shear flows will be about unity, which is of the same order of magnitude as the dimensionless dispersivity of 0.5 observed for a ball (of the same size as the suspended spheres) settling through a disordered quiescent suspension. This suggests that the same basic physics governs both phenomena.

The vertical dispersivity component can, alternatively, be estimated by using mean settling velocity measurements gleaned from a series of experiments conducted over known settling distances (Mondy *et al.*,1986; Milliken *et al.*, 1989b). Use of the data from many previous experiments, together with our insights into the more complete information gathered at the 30 and 50% concentrations, enabled estimations of the dispersivity coefficients to be made over a wide range (5 to 55%) of concentrations.

In the dilute particle limit, the effect of particle concentration on the falling-ball settling velocity would be expected to be small, whereas the number of collisions per unit length is expected to be proportional to the concentration. This suggests that the dimensionless dispersivity should be proportional to the particle concentration in the dilute limit. At higher concentrations, an increase in the dispersivity would be caused by the increased number of interactions (all other things being equal). By plotting the dimensionless dispersivity versus the specific viscosity μ_r-1, as in Figure 1-8, an approximate functional relationship (dashed line) between these quantities is found to exist over two orders of magnitude in the specific viscosity:

$$D^*_{||} = 0.0579\ (\mu_r - 1)^{0.95} \qquad (spheres)\ . \qquad \textbf{(1.2)}$$

This near linear relation between dispersivity and specific viscosity suggests that multibody (suspended) particle effects contribute equally

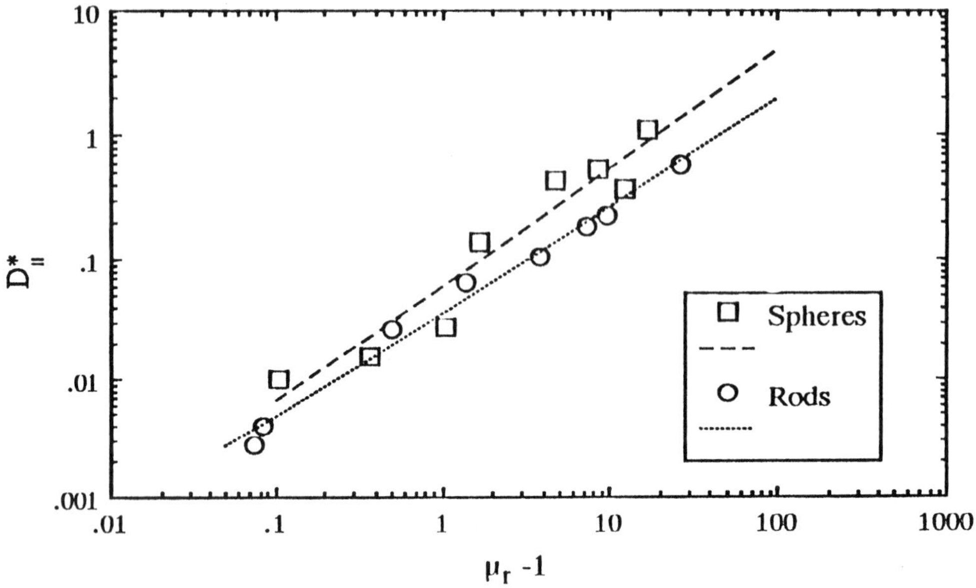

FIGURE 1-8 *Dimensionless vertical dispersivities for suspensions of spheres and suspensions of rods vs specific viscosity. The dispersivity for suspensions of spheres appears to be approximately linearly related (dashed line) to the specific viscosity (μ_r—1). This implies that the concentration dependence of the dispersivity is governed by the same mechanism as that which governs the viscosity of the suspension. For rod supension, the comparable dispersivity vs specific viscosity relationship shows a similar (dotted line) functional dependence to that for spheres. While the specific viscosity increases more rapidly in suspensions of the rods than in spheres at the same solids concentration, the dispersivity increases at about the same rate.*

to the observed increase in specific viscosity and dimensionless dispersivity.

Figure 1-8 also shows the dimensionless vertical dispersivity for a suspension of randomly oriented rods of aspect ratio 20 (using the length of the rods as the pertinent length scale for effecting the nondimensionalization), with the following equation giving the approximate functional dependence of the nondimensional dispersivity upon the specific viscosity:

$$D^*_{||} = 0.0349\ (\mu_r - 1)^{0.87} \quad (rods)\ . \tag{1.3}$$

Comparison with Eq. (1.2) shows that $D^*_{||}$ is greater for a suspension of rods than for a suspension of spheres at a given solids concentration. However, since the specific viscosity of a suspension of rods is greater than for a suspension of spheres at the same concentration, the dimensionless dipersivities in suspensions of rods and spheres are relatively the same at a given value of the specific viscosity. The coefficients of fit (R^2) of Eqs. (1.2) and (1.3) to the plotted data are 0.923 and 0.988 for the suspensions of sphere and rods, respectively.

1.5 NMR IMAGING OF THE FLOW BEHAVIOR OF SUSPENSIONS: WIDE-GAP, INHOMOGENEOUS COUETTE FLOW

A Newtonian fluid in a wide-gap Couette apparatus possesses perhaps the simplest flow field of any realizable *inhomogeneous* shear flow. As such, this device was used to study the effects of inhomogeneous shear on particle migration in highly concentrated suspensions. These suspensions consisted of polymethyl methacrylate (PMMA) particles suspended in a Newtonian oil. These particles were neutrally buoyant in the suspending liquids, and were sufficiently large to rule out the possibility that interparticle colloidal and/or Brownian forces could affect the suspension behavior. PMMA particles were obtained from three separate sources. The first set of particles, Lucite 47G (DuPont Chemical Company), possessed a mean diameter of 100 μm. The second set, Diakon MG102 (Imperial Chemical Industries), possessed a mean diameter of 675 μm. From visual examination, these particles appear to be spherical in shape. Unsieved (polydisperse) samples were used in the study exactly as received from the manufacturer. The standard deviations of particle diameters observed for these first two sets were approximately 35% and 15%, respectively. By sieving the particles (and taking the mean size to be the arithmetic average of the two respective mesh sizes of the sieve on which the particles were collected and that above it), sieved samples were produced with a standard deviation of about 5%. The third set of spheres consisted of individually ground, 3.175-mm diameter spheres (Clifton Plastics Company). Published tolerances of these large spheres were ±0.05 mm for the diameter and ±0.025 mm for the sphericity.

The same interstitial liquid used to create the optically transparent, neutrally buoyant suspensions employed in the falling-ball experiments was also used in the NMR imaging experiments. Although the temperature of each suspension was controlled only by the room air conditioner, no settling of the suspended particles was detected during the duration of the experiments. The liquid used also possessed the same refractive index as the PMMA, although this optical attribute is unnecessary when using the NMR technique. Nevertheless, this equality of indices proved useful in determining the extent and location of any air bubbles introduced during preparation of the suspensions. (Air bubbles, like the suspended particles, give no NMR signal.) Viscosity of the suspending liquid was 4.95 Pa·s at 21.5°C.

The Couette apparatus consisted of a solid rotating PMMA cylinder, concentrically positioned within a fixed outer PMMA circular tube capped with solid PMMA disks at each end. At one end the inner cylinder fitted into a recessed Teflon bushing, while at the other the shaft extended through the end cap sealed with an O-ring. The radius r_i of the inner cylinder was 0.64 cm, whereas the inner radius r_o of the outer cylinder was 2.38 cm. Length of the suspension-filled cavity was 25 cm. By imaging at both the center of the Couette, and near its end, it became apparent that any migration of particles in the axial direction was, necessarily, very slight, especially when compared with that occurring radially.

Rotation speeds of the inner cylinder varied between 17 and 117 rpm in this set of experiments, while being held constant during any one run. This gave rise to shear rates across the gap varying from about 1 to 25 sec^{-1} (assuming a Newtonian fluid response). Use of the calibration method discussed in section 1.3 permitted determination of the fluid fraction existing in each concentric annulus for each image examined via the following method: The centroid of each image was computed, corresponding to the Couette axis, and the average image intensity was subsequently, determined as a function of radial position measured from that centroid.

A further reduction of the data was performed to obtain a scalar quantification of the extent of particle migration. The variance S of the radial concentrations profiles (used to report the majority of these findings) was chosen as the most appropriate migration parameter because it depends upon the entire concentration profile and increases as the particles migrate away from the inner cylinder. The variance S uses the radial concentration profile weighted with the radial distance;

explicitly,

$$S(N) = \sum_{j=1}^{m} [\alpha_{S,r_j}(N) - \langle \alpha_S \rangle]^2 r_j \Big/ \sum_{j=1}^{m} r_j \quad , \qquad (1.4)$$

where α (with subscripts S,r_j) is the local volume fraction of solids in the jth concentric ring after N turns, $\langle \alpha_S \rangle$ is the average concentration of solids, and r_j is the radial distance from the center of the Couette device. The number m of discrete rings used in the computation varied (typically between 15 and 18) between experiments. Each ring was two pixels wide.

The experiments performed were aimed at determining the functional dependence of the particle migration rate upon the shear rate, viscosity of the suspending fluid, size of the suspended particles, etc. Total extent of migration was found to be dependent upon only the cumulative amount of shear (as well as upon the diameter and concentration of the suspended particles). These results are consistent with the hypothesis that the migration phenomenon is governed strictly by hydrodynamic particle interactions. The migration rate was found to be inversely proportional to the diameter of the suspended particles raised to a power lying between 2 and 3. Hence, particle size has a major effect on the migration, as predicted by Leighton and Acrivos (1987). As mentioned earlier, the coefficient governing the shear-induced migration determined from these experiments was approximately $2.4\alpha_S$ (Phillips *et al.*, 1991), which (within experimental uncertainty) is identical to the coefficient derived from the falling-ball experiments.

For this investigation the tangential velocity (v_θ) component as a function of radial position was determined in a narrow slice along the diameter. Thus, assuming θ symmetry, one obtains two velocity profiles (one from each side of the inner cylinder) for each experiment. Confirming velocity measurements performed on a particle-free viscous Newtonian fluid were observed to be in excellent agreement with the exact analytical solution for wide-gap Couette devices. Velocity measurements shown in Figure 1-9 were taken along the diameter of the Couette for the 50% suspension. This velocity profile differed significantly from that for the homogeneous Newtonian fluid case, as might be expected given the large particle concentration gradient existing

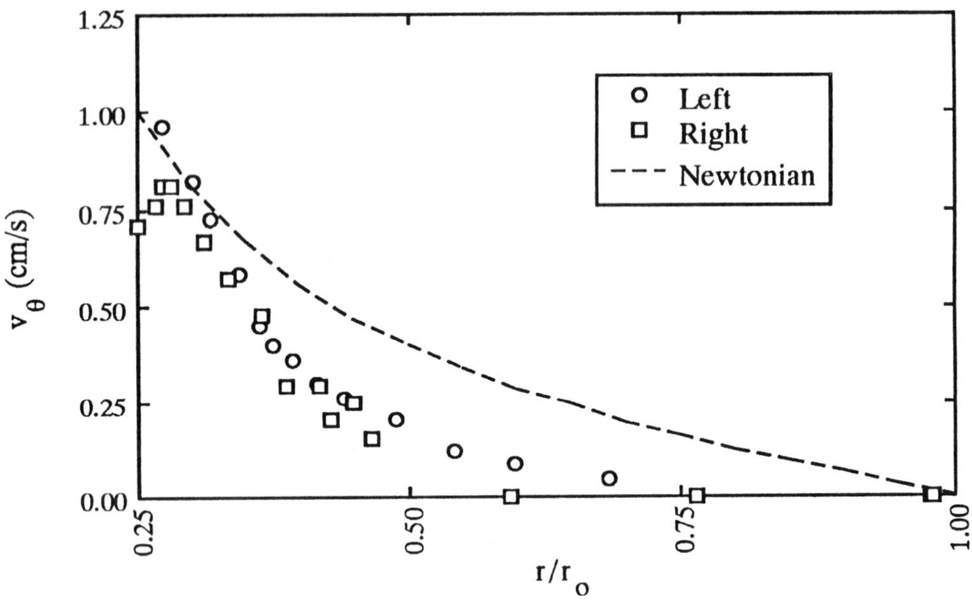

FIGURE 1-9 *Velocity profile for the steady-state image shown in Figure 1-4. The azimuthal velocity (v_θ) as a function of radial position (r) was measured along one diameter; by symmetry, independent measurements at radial positions respectively lying on the right- and left-hand sides of the center rod should yield equivalent values, as can be seen (approximately) in the above figure.*

across the gap at steady-state.

1.6 SUMMARY

Real-time radiography in opaque suspensions, as well as comparable visual techniques in transparent (refractive index-matched) suspensions, has been used to track the path of a single settling sphere (a 'falling ball') through an otherwise quiescent suspension of neutrally buoyant particles. The small settling sphere serves to probe the microstructure of a given suspension, without significantly disturbing the suspension's configuration. The average settling velocity of the ball can be translated via Stokes law (sans slip) into an effective viscosity of the suspension, just as if the suspension were a homogeneous Newtonian continuum. Detailed analysis of the trajectory of the falling ball shows significant vertical and horizontal "instantaneous" fluctuations about the mean position of the ball (based upon its *average* settling velocity), arising from its hydrodynamic interaction with the suspended particles. These dispersion data are shown to be quantified via an Einstein-like mean-square displacement law, linear in time for long times, whence the falling ball displays Fickian "diffusion" about its mean motion.

Effective continuum rheological properties have been measured via falling-ball experiments performed on both randomly oriented and aligned (orientationally ordered) systems of nonspherical particles. Viscosities derived from falling-ball experiments in aligned systems agree with Couette measurements in sheared suspensions, demonstrating that the falling-ball technique is capable of measuring configuration-specific rheological properties, corresponding to a specific geometric arrangement of the suspension particles.

NMR imaging has been used to noninvasively measure suspended particle velocity and concentration profiles in concentrated, nonquiescent, neutrally-buoyant suspensions undergoing macroscopically inhomogeneous shear. NMR imaging of a wide-gap Couette apparatus, with its concomitant radial shear rate gradients, allows direct observation of the changing microstructure of a suspension to be made when the suspension is subjected to a spatially inhomogeneous flow. Large concentration gradients were observed to arise during the flow of initially homogeneous suspensions, with the suspended particles migrating on average toward regions of low shear. These NMR experiments suggest that an accurate model of the suspension would require *a priori* knowledge of the details of the changing microstructure under the influence of the shear-induced particle migration.

Sequential use of falling-ball and NMR experiments performed on the same suspension allows the rheological properties of concentrated suspensions (as well as the microstructural changes that occur during flow) to be quantified and correlated.

1.7 CONCLUSIONS

NMR imaging of the spatial particle distribution reveals why the apparent rheological properties of an initially homogeneous suspension may depend upon the flow field used in its measurement. Our data, using NMR images of a concentrated suspension undergoing inhomogeneous shear in a wide-gap Couette apparatus, illustrate the existence of a flow-induced suspension microstructure. A simple model that supposes the suspension to possess a spatially uniform 'local' viscosity throughout the Couette apparatus would measure an apparent viscosity that appeared to depend upon the total strain and, hence, upon the duration of the experiment, at least until a steady-state mode of particle arrangement was achieved. Hence, a more detailed kinematical and dynamical model of the suspension is needed to account for the particle migration occurring as a consequence of the imposed macroscopic flow. In the model proposed by Leighton and Acrivos (1987), knowledge of the dependence of the local suspension viscosity upon the local properties of the suspension (i.e., concentration, orientation, alignment, etc.) is required in order to achieve a complete rheological theory. The ability of falling-ball rheometry to measure the dynamical configuration-specific properties of a suspension without significantly changing the suspension's microstructure therefore allows determination of such 'local' rheological properties, pertinent to experiments performed on permanently homogeneous suspensions.

In contrast with the complex rheological behavior displayed by inhomogeneously sheared suspensions, falling-ball measurements show that a spatially *homogeneous, orientationally isotropic* suspension often behaves as a simple Newtonian liquid. Such measurements do not sensibly perturb the structure of the suspension because the probe (namely the falling ball) size is fairly small compared with the size of the suspended particles. Experiments conducted on suspensions of rods illustrate the fact that falling-ball rheometry may prove a useful tool in isolating the effects of particle orientation upon suspension rheology, as well as in probing the relationship existing between the suspension microstructure and its macroscopic rheological properties (Mondy *et al.*, 1990). The dispersivity calculated from the detailed falling-ball settling paths was found to be accurately represented as a Fickian-like diffusion process (relative to the mean motion), albeit transversely isotropic rather than fully isotropic. The dispersivity component parallel to gravity was observed to more than an order-of-magnitude larger than the lateral component. Additionally, the magnitude of the vertical dispersivity arising from collisions between the falling ball and suspended particles appears to be the same as that which quantifies collision-induced migration in the Couette flows. Therefore, just as shear-induced migration phenomena demonstrate a fundamental distinction between a suspension and a truly homogeneous fluid, so too does the dispersion of a settling sphere about its average path equivalently arise owing to the noncontinuum nature of the suspension. In this sense the dispersion phenomenon may be likened to Brownian motion, the latter being a manifestation of the discrete (atomistic) nature of fluids, despite their seemingly continuous appearance at the macroscale.

1.8 FUTURE WORK

Currently, we are investigating the origin of flow-induced migration and the relationship between the continuum properties of a homogenous, random suspension and its time-dependent behavior in a wide-gap Couette device as measured by the torque reduction. The flow-induced migration depends upon an apparent irreversibility in the flow behavior of a suspension; otherwise, upon reversing the flow direction, the direction of migration would simply reverse. Because nonreversible diffusive behavior is observed, whereas the low-Reynolds number flow equations are reversible for smooth spheres, the questions arises as to what renders the flow irreversible, and over what extent of total strain does the motion remain reversible? In this context, we are examining the flow behavior of suspensions undergoing small amplitude oscillations. This should furnish insights into the origin of flow-induced lateral migration phenomena. Coupling of the long time behavior of suspensions in Couette devices with their homogenous viscosity and dispersivity properties should make the technique more accessible to rheologists without a large capital investment in equipment.

ACKNOWLEDGMENTS

Many colleagues not explicitly mentioned in the text contributed to various aspects of the work cited here. In chronological order they were, respectively, Ms. Julie Jensen, Mr. Larry Bryant, Jr., Ms. Laura Price, Professor Robert Powell, Professor Moshe Gottlieb, Dr. William Milliken, Mr. T. G. Morrison, Dr. Thomas S. Stephens, Lt. Wayne J. Tetlow, Lt. James E. Breck, Mr. Nicholas Tetlow and Mr. Raffy Mor.

This work was sponsored by the U. S. Department of Energy, at Los Alamos National Laboratory, under Contract W-7405-ENG-36 with the University of California and at Sandia National Laboratories, under Contract DE-AC04-76DP00789. All the authors gratefully acknowledge support for this work from Division of Engineering and Geosciences, Office of Basic Energy Sciences, the U.S. Department of Energy. Work at Los Alamos and Sandia National Laboratories was also partially supported by the Office of Naval Research.

REFERENCES

Abbott, J. R., Tetlow, N., Graham, A. L., Altobelli, S. A., Fukushima, E., Mondy, L. A., and Stephens, T. S., "Experimental Observations of Particle Migration in Concentrated Suspensions: Couette flow," *J. Rheol.*, **35**, 773-797 (1991).

Abragam, A., *The Principles of Nuclear Magnetism*, Clarendon Press, Oxford, (1961).

Ackerson, B. J. and Pusey, P. N., "Shear-Induced Order in Suspensions of Hard-Spheres," *Phys. Rev. Lett.*, **61**, 1033-1036 (1988).

Bird, R. B., Stewart, W. E., and Lightfoot, E. N., *Transport Phenomena*, John Wiley & Sons, New York (1960).

Bohlin, T., "On the Drag on a Rigid Sphere Moving in Viscous Liquid Inside a Cylindrical Tube," *Trans. Roy. Inst. Tech.* (Stockholm),**155**, 1-63 (1960).

Brenner, H., "Rheology of a Dilute Suspension of Axisymmetric Brownian Particles," *Int. J. Multiphase Flow*, **1**, 195-341 (1974).

Caprihan, A. and Fukushima, E., "Flow Measurements by NMR," *Phys. Reports*, **198**, 195-235 (1990).

Dingman, S., Ingber, M. S., Mondy, L. A., Abbott, J. R., and Brenner, H., "Particle Trackin gin Three-Dimensional Stokes Flow," *J. Rheol.* in press (1992)

Gadala-Maria, F. and Acrivos, A., "Shear-Induced Structure in a Concentrated Suspension of Solid Spheres," *J. Rheol.*, **24**, 799-811 (1980)

Ganani, E., and Powell, R. L., "Rheological Behavior of Rodlike Particles in Newtonian and Non-Newtonian Fluids," *J. Rheol.*, **30**, 995-1013 (1986).

Graham, A. L., Altobelli, S. A., Fukushima, E., Mondy, L. A., and Stephens, T. S., "NMR Imaging of Shear-induced Diffusion and Structure in Concentrated Suspensions Undergoing Couette Flow," *J. Rheol.*, **35**, 191-201 (1991).

Graham, A. L., Mondy, L. A., and Powell, R. L., "Applications of Real-time Radiography to Rheological Characterization and Processing Dynamics," presented at JANNAF Propellant Development Characterization Subcommittee Annual Meeting, Laurel, Maryland (1987).

Haber, S., and Brenner, H., "Rheological Properties of Dilute Centrally Symmetric Brownian Particles at Small Shear-Rates," *J. Colloid Interface Sci.*, **97**, 496-514 (1984).

Leighton, D., and Acrivos, A., "The Shear-induced Migration of Particles in Concentrated Suspensions," *J. Fluid Mech.*, **181**, 415-439 (1987).

Majors, P. D., Givler, R. C., and Fukushima, E., "Velocity and Concentration Measurements in Multiphase Flows by NMR," *J. Magn. Res.*, **85**, 235-243 (1989).

Maneval, J. E., Powell, R. L., McCarthy, M. J. and McCarthy, K. L., "Magnetic Resonance Imaging of Multiphase Systems," in *Particulate Two-Phase Flow* (M.C. Roco, Ed.), Chap. 4, Butterworth-Heinemann, Stoneham, Massachusetts (1992).

Milliken, W., Gottlieb, M., Graham, A. L., Mondy, L. A., and Powell, R. L. "The Viscosity-volume Fraction Relationship of a Randomly Oriented Suspension of Rods, *J. Fluid Mech.*, **202**, 217-232 (1989a).

Milliken, W., Mondy, L. A., Gottlieb, M., Graham, A. L., and Powell, R L., "The Effect of the Diameter of Falling Balls on the Apparent Viscosity of Suspension of Spheres and Rods," *PhysicoChem. Hydrodyn.*, **11**, 341-355 (1989b).

Mondy, L. A., Graham, A. L., and Gottlieb, M., "Microrheological Observations on the Onset of Non-Newtonian Behavior in Suspensions," presented at the Xth International Congress on Rheology, Sydney, Australia, August 1988.

Mondy, L. A., Graham, A. L., and Jensen, J. L., "Continuum Approximations and Particle Interactions in Concentrated Suspensions," *J. Rheol.*, **30**, 1031-1051 (1986).

Mondy, L. A., Morrison, T. G., Graham, A. L., and Powell, R. L., "Measurements of the Viscosities of Suspensions of Oriented Rods Using Falling Ball Rheometry, *Int. J. Multiphase Flow*, **14**, 651-662 (1990).

Morris, P. G., *Nuclear Magnetic Resonance Imaging in Medicine and Biology*, Clarendon Press, Oxford (1986).

Morrison, T. G., "Experimental Investigation of the Mechanics of Suspensions of Elongated Particles Using Falling-ball Rheometry," M.S. Thesis, University of California, Davis, California (1989).

Phillips, R. J., Armstrong, R. C., Brown, R. A., Graham, A. L. and Abbott, J. R., "A Constitutive Equation for Concentrated Suspensions that Accounts for Shear-Induced Particle Migration," in press *J. Rheol.* (1991).

Powell, R. L., Mondy, L. A., Stoker, G. G., Milliken, W., and Graham, A. L., "Development of a Falling-ball Rheometer with Applications to Opaque Systems: Measurements of the Rheology of Suspensions of Rods," *J. Rheol.*, **33**, 1173-1188 (1989).

Salem, A. J., and Fuller, G. G., "Small Angle Light Scattering as a Probe of Flow-induced Particle Orientation," *J. Colloid Interface Sci.*, **108**, 149-157 (1985).

Surovec, R. D., Husband, D. M., Stephens, T. S., and Graham, A. L., "Reproducibility in Measuring Rheology of Highly Concentrated Suspensions," presented at the JANNAF Propellant Development Characterization Subcommittee Annual Meeting, Laurel, Maryland (1989).

Walton, J. S., "Close-Range Cine-Photogrammetry: A Generalized Technique for Quantifying Gross Human Motion," Ph.D. Thesis, Pennsylvania State University (1981).

Watkins, J. C., and Fukushima, E., "High-pass Bird-cage Coil for Nuclear-magnetic Resonance," *Rev. Sci. Instr.*, **59**, 926-929 (1988).

CHAPTER 2

SIMULTANEOUS IMAGING OF THE VELOCITY FIELDS OF TWO PHASES

Z. C. Liu
R. Adrian

2.1 INTRODUCTION

Optical methods to visualize or to quantitatively measure the motion of particulates in multi-phase flows are valuable techniques that have a long history of applications. The development and wide spread availability of high quality imaging devices, in the form of CCD and CID arrays, and powerful image processing hardware and software, have had a significant impact upon these techniques, and they have made possible a new generation of instruments that are able to measure instantaneously the velocities at thousands of points in a flow field. These points may be fixed in space, in which case Eulerian fluid velocities are obtained on a large grid, thereby mapping out the entire flow field at one instant, or they may be the centers of suspended phase material such as particulates, bubbles or liquid droplets, in which case the Lagrangian velocities, and possibly trajectories, of the field of suspended phase objects are measured. These methods differ intrinsically from techniques such as laser Doppler velocimetry which provide time series measurements of velocity at a few points at most.

In addition to the velocity fields, optical imaging techniques are also capable of providing information about the concentration and size of the suspended phase and their distribution in space.

Modern optical techniques are variants of classical photography and cinematography whose basic principles differ relatively little from their predecessors. The primary differences lie in the vastly increased amount of optical image information that can be recorded and digitized, and the automated, computer-based processing that makes possible the analysis of large image arrays. In *particle tracking velocimetry* (PTV) one records multiple images of suspended phase particles at various times and locates the particle image centers to determine many simultaneous particle trajectories.[†] The number density of particles is relatively low enough to permit easy identification of the particle images and individual processing of them. PTV can also be applied to the measurement of instantaneous fluid velocity fields by means of using small particles that follow the flow, but the measured vectors are randomly located and somewhat sparse. These limitations are overcome in *particle image velocimetry* (PIV) by using higher concentrations of small marker particles to fill the fluid field and imaging the motion of small *groups* of these particles. The resulting measurements provide the instantaneous vector velocity field on a uniform grid of measurement sites. Particle imaging methods are complemented by the molecular marker technique described by Falco and Nocera elsewhere in this volume.

Many questions in the study of two-phase flow require knowledge of the motion of both the particulates and the fluid. Examples are studies of the complex velocity field of the liquid phase around a bubble or within a group of bubbles, and the slip velocity between particulates and the fluid phase. The subject of this chapter is the development of instrumentation capable of measuring simultaneously the instantaneous fluid velocity field and the trajectories of the particles. The approach is to combine PTV and PIV systems into a single instrument capable of discriminating between the two phases and matched to the optical and fluid characteristics of them.

Optical imaging methods are only one element in the array of instruments that are available for multiphase flow research. (Perhaps it would be more appropriate to say spectrum of instruments, because they can often be characterized by the wavelengths of the phenomena that are used to probe the fluid.) These include acoustic ultrasound, magnetic resonance imaging (see the chapter by Powell, McCarthy and McCarthy in this volume), and X-ray tomography (Zoltani, White and DiBianca, 1985), and neutron radiography (Cimbala et al., 1989) for example. It is clear that optical methods are restricted to media that are relatively transparent to optical wavelengths and this implies relatively low concentrations of suspended matter in a clear fluid, or refractive index matched suspensions of concentrated solids. While such considerations limit the range of application, there are, nonetheless many interesting and valuable experimental investigations that can be performed within these confines.

[†] If deformation and phase change do not cause significant changes in the shape of a droplet or a bubble, it is clear that imaging velocimetry can be applied to them as well as solid particles. For simplicity we shall refer to these masses as particles, regardless of their phase.

2.2 PARTICLE IMAGE VELOCIMETRY

Particle image velocimetry has been reviewed by Dudderar, Meynart and Simpkins (1988) and Adrian (1986, 1991). In its most common form PIV is a laser light sheet technique in which the sheet is double-pulsed and images of small particles are photographed as double exposures at right angles to the sheet, Figure 2-1. The laser is often a pulsed ruby or Nd:Yag laser which compresses energies of order one Joule into pulses of order 10*ns* in duration. High energy is necessary to obtain adequate scattering from marker particles which are very small—typically 1–2 micron liquid droplets or solid particles in a gaseous flow and 10–20 micron plastic or glass particles in a liquid flow. Double pulsed exposures are common because of laser capabilities, but three- and four-pulse laser systems are available and can be used to advantage. Argon ion lasers with swept beams can also be used in conjunction with somewhat larger particles that may be metal coated to improve scattering, and they can be chopped to form many pulses.

The particle concentration is of order 10 particles per mm^3, corresponding to volume fractions of order 10^{-8} to 10^{-5}. Thus, the fluid markers have negligible effect upon the rheology of the fluid, and they should not be considered to be part of the two-phase flow.

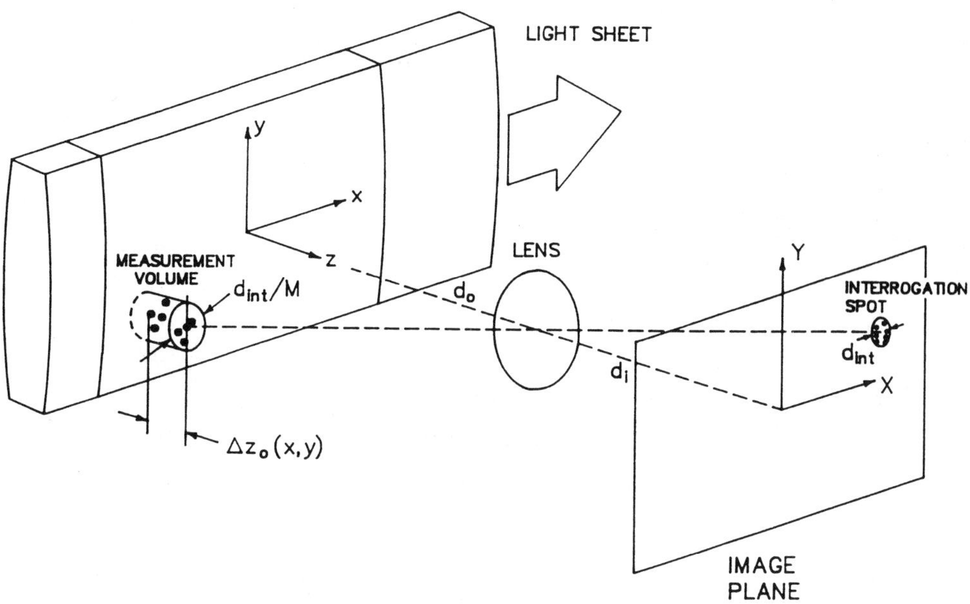

FIGURE 2–1 *Laser light sheet imaging.*

The velocity field in the plane of the light sheet is measured by recording a double or multiple exposure on a single photograph and extracting the in-plane vector displacements of the particles from the photograph on a pointwise basis by measuring the average vector displacement of the particles on a grid of "interrogation spots." The latter are defined by a He-Ne laser beam which illuminates a small spot on the photograph and a video camera which digitizes the spot. Alternatively, a lens placed after the spot can be used to Fourier transform the image field, and then the camera digitizes the resulting Young's fringe system pattern, Figure 2-2. Coherent illumination of the interrogation spot is necessary to form Young's fringes, but incoherent illumination can be used if the spot is digitized directly. In either case, the image is processed to extract the mean displacement of the particle image between successive exposures by forming the autocorrelation of the spot with itself and searching for the location of the peak correlation value that corresponds to each first image overlapping the next image.

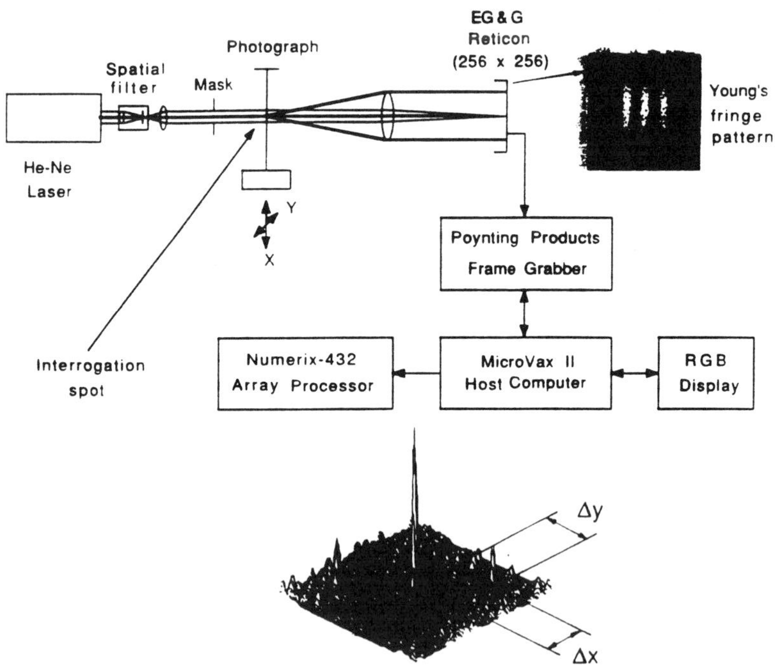

FIGURE 2–2 *Interrogation system for PIV photograph.*

When Young's fringes are used, a single 2-D FFT yields the correlation function, whereas direct digitization requires two FFT operations, the first one forming a numerical version of the Young's fringe system. Once the mean displacement has been found for a given spot, the image array data is discarded, the photograph is moved, and a new spot is digitized and analyzed. Variations on this process include cross-correlation analysis (Utami, Blackwelder and Ueno, 1991), use of a video array instead of a photograph (Willert and Gharib, 1991), and optical correlation methods as opposed to digital (Coupland and Halliwell, 1988).

The system used in our two-phase flow studies is a 2 Joule per pulse double-pulsed ruby laser, Apollo Model HD22, whose time interval between pulses can be adjusted from one microsecond to ten milliseconds. Photographs are taken with Kodak Technical Pan 4415, a Kodak TMax 100 or TMax 400 black and white film. Image analysis is performed by a Numerix

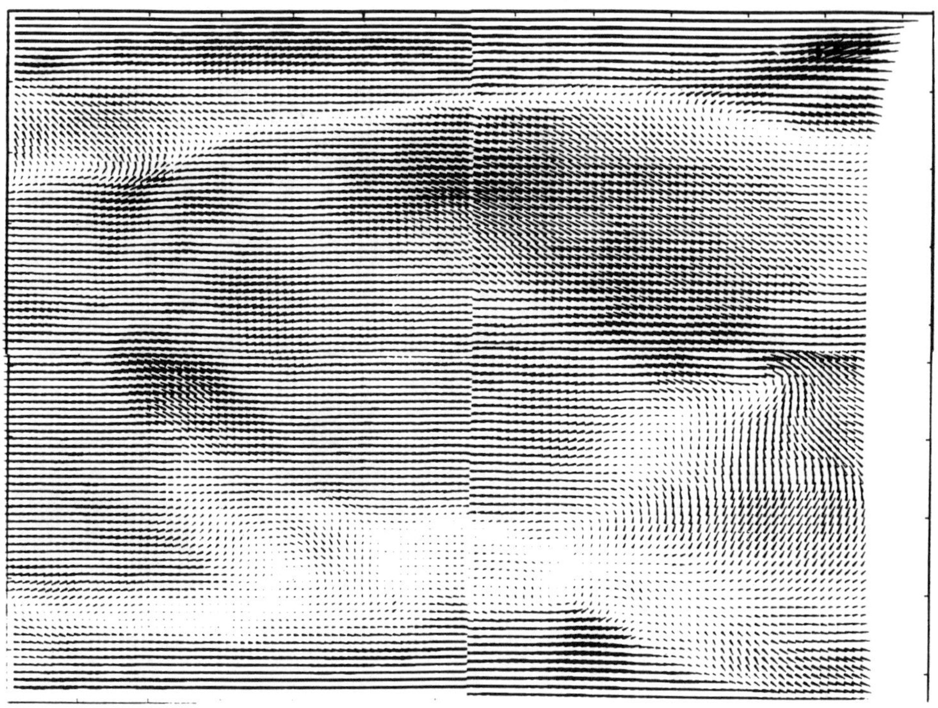

FIGURE 2–3 *Velocity vectors in turbulent channel flow. The measurement grid is 0.5mm and the mean bulk velocity has been subtracted. The data is from high image density PIV.*

432 array processor that operates at 30*M* flops and computes a 256×256 2-D complex FFT in 190*ms*. The image of the interrogation spot is acquired by a frame grabber and transferred through the Micro Vax II host to the array processor in a pipelined operation.

An example of a high image density PIV measurement from the work of (Liu et al., 1991) is shown in Figure 2-3. This is a vector field in a plane perpendicular to the flat wall of a 48.8*mm* high, 610*mm* wide channel containing water flow at 2872 Reynolds number. The bulk translational velocity has been subtracted from the measurements to render the field seen by an observer traveling with the bulk flow. The field consists of a grid of 115×100 vectors with 0.5*mm* spacing. The physical dimensions of each measuring volume were $1mm \times 0.3mm \times 0.8mm$ in the *x*-, *y*-, and *z*-directions, respectively. Five micron particles were used in a concentration of 35 per mm^3 to mark the fluid.

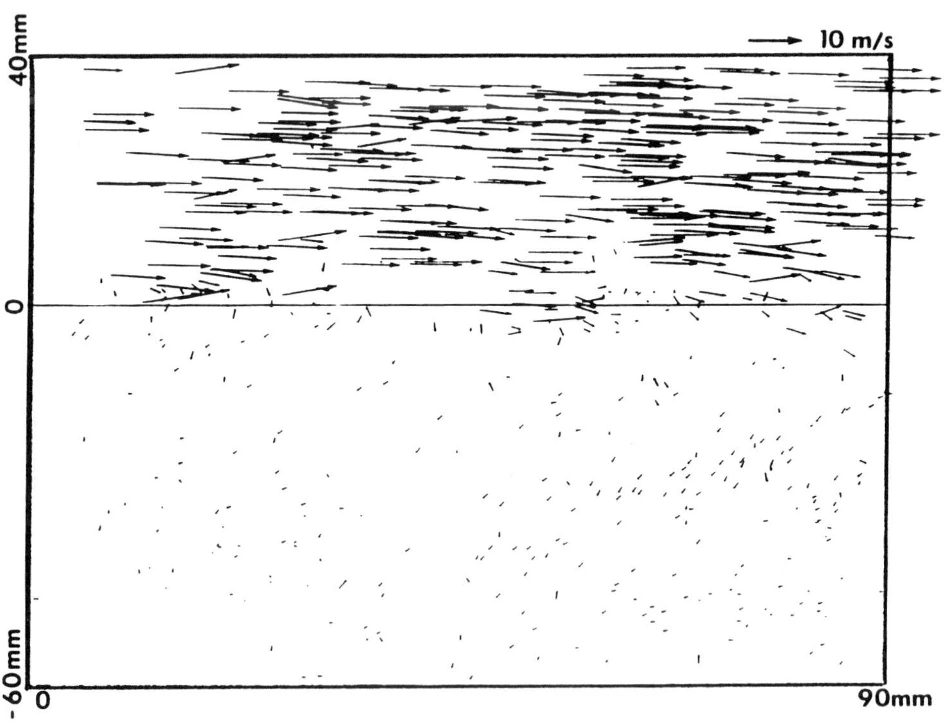

FIGURE 2–4 *Velocity vectors in a plane turbulent shear layer obtained by low image density PIV.*

The foregoing interrogation technique is particularly important to the process of correctly pairing images when there are many particles in each measuring volume (“high image density”). However, it can also be applied to low image density flows, as illustrated by the vector field in Figure 2-4. This measurement was made in a plane 2-D shear layer with $10 ms^{-1}$ velocity in the upper stream and $0\ ms^{-1}$ mean velocity below in the splitter plate. The fluid was water and the particles were 30 to 50 micron expanded plastic spheres having a density of $0.03 gm/cm^3$.

2.3 PARTICLE TRACKING VELOCIMETRY

2.3.1 Particulate Tracking

Particle tracking velocimetry is actually the low image density limit of PIV. At low concentrations, the spacing between particles is large compared to the displacements of the particles between exposures. Consequently, the pairs of images are easily recognized and one can analyze the motion of *individual* particles, as opposed to *groups* of particles. The

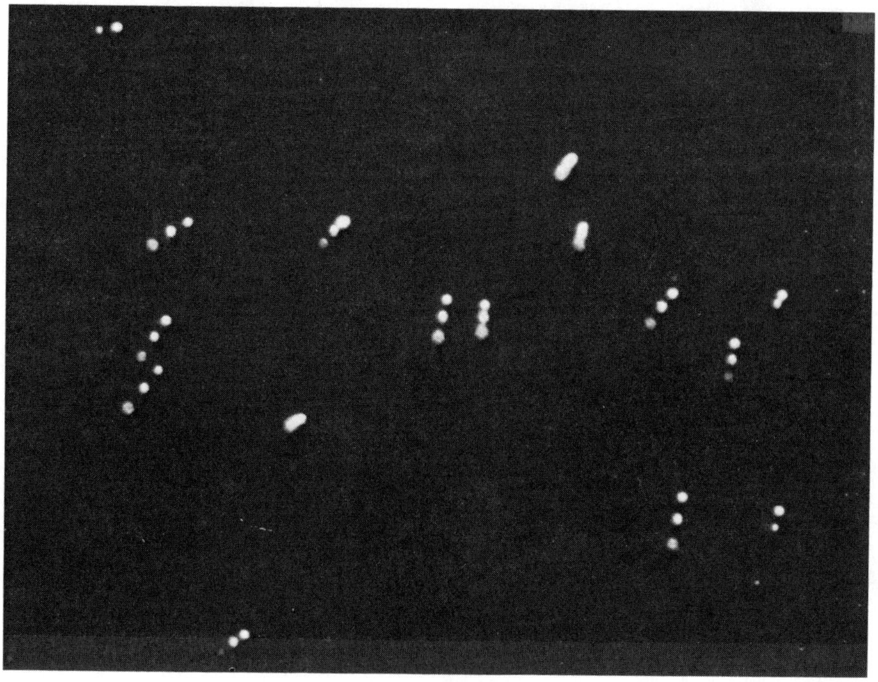

FIGURE 2–5 *Images of 90 micron water droplets from multiple pulse PTV in a pipe flow. [From Lee et al., 1989]*

inherent simplicity of identifying images in this case makes it possible to track the motion over long trajectories, as well as to measure Lagrangian velocities from shorter displacements. (See the chapter by Abbott, Brenner, Graham and Mondy in this volume.)

An example of PTV taken from the work of Lee et al. is shown in Figures 2-5 and 2-6. In their experiment 50 water droplets of various sizes were injected on the centerline of an air flow through a 50.8*mm* pipe at 52,000 Reynolds number, and the turbulent dispersion of the droplets was observed in a test section 2120*mm* downstream of injection. The particles were photographed from the axial direction by a camera located about 1000*mm* downstream from the test section. Within the region of the test section the particles were illuminated by a sequence of 10*mm* thick light sheets spaced 10*mm* apart in the axial direction and pulsed at a rate that

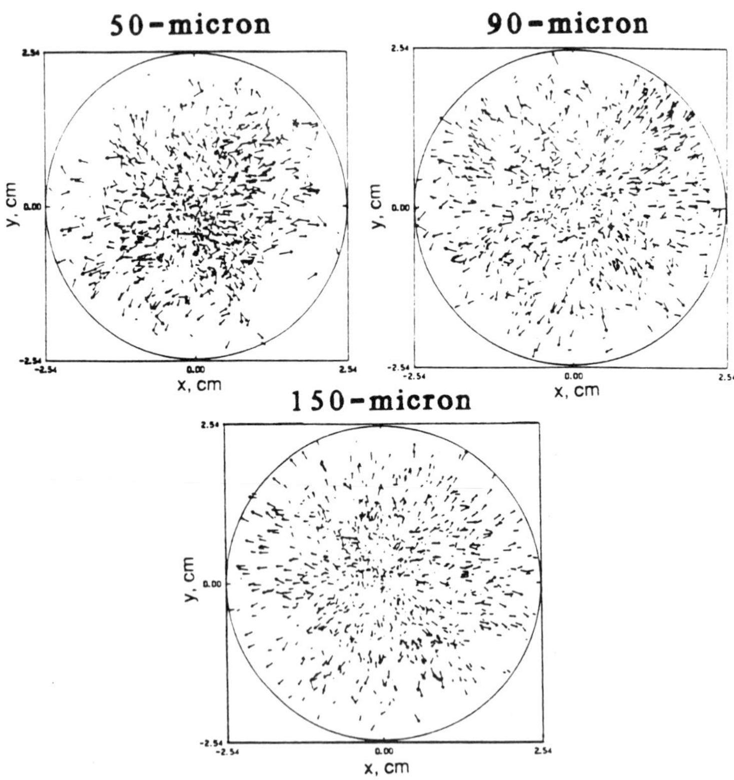

FIGURE 2–6 *Vector velocities of droplets dispersing in turbulent pipe flow. [From Lee et al., 1989]*

allowed the light sheets to follow the mean axial translation of the particles. In this way particles that were exposed in the first light sheet were kept within succeeding light sheets and the probability of losing images was minimized.

A magnified region of a typical three-pulse photograph is shown in Figure 2-5. The energy of the last pulse was reduced to provide a means of determining the direction of motion. Analysis of these images was performed by manually locating the cross hairs of a precision profile projector on the center of each image, digitizing the (x,y) coordinates and grouping the measurements into sets corresponding to a single particle. The results are shown in Figure 2-6 which contains an ensemble of measured velocity vectors from many individual photographs for each of three particle sizes, and at the approximate location (different for each size) where the droplets began to deposit on the wall. It is clear from the generally radial orientation of the vectors, especially for the 150 micron droplets, that the particle trajectories were still influenced by the location of the initial injection at this point. Further downstream the vectors become more isotropically oriented.

While semi-automated analysis is satisfactory for an experiment like the one just described, it is not feasible for experiments involving much larger amounts of data, and it is certainly unappealing intellectually. Consequently, many investigators have developed fully computerized image processing systems for the purpose of particle tracking (c.f. Perkins and Hunt, 1989; Nishino, Kasagi and Hirata, 1989; Racca and Dewey, 1988; Majumdar, Graham, Roco and Stroeve, 1987; Uemura, Yamamoto and Ohmi, 1989; Khalighi, 1989; and Majumdar, Stroeve, Fales, Algazi and Stewart, 1987, for example).

2.3.2 An Interrogation System for PTV

The PTV interrogation system that has been developed in the present work is typical of the current generation of equipment. It consists of four elements: an array camera, a host computer, digitization and image processing hardware and software, Figure 2-7. The camera that acquires the multi-exposed particle images is a 512×480 pixel TK-66 CCD array. The computer, an IBM/AT compatible 386/33 MHz, hosts a Data Translation DT-2861 frame grabber board which digitizes an image frame and stores it in one of 16 built-in buffers for further processing, and a DT-2858 frame processor board which performs image processing operations on the acquired frames. Camera data from the frame grabber is transferred to the frame processor via a fast bus, and it is also available to the AT bus. The frame processor uses a commercial image processing software IMAGEPRO II for generating bilevel images and other image enhancement. The PTV software was developed in-house. It identifies particulate images in the

The first stage of PTV interrogation involves two steps: pre-processing the image frame to remove noise and improve image quality, and generation of a bilevel image, Figure 2-8. The bilevel image replaces all grey levels above a prescribed threshold value by unity and all values below the multiple-exposure frames, calculates centroids, locates pairs and calculates displacements, from which velocities and accelerations are finally determined. The camera can acquire data directly, or from a photograph.

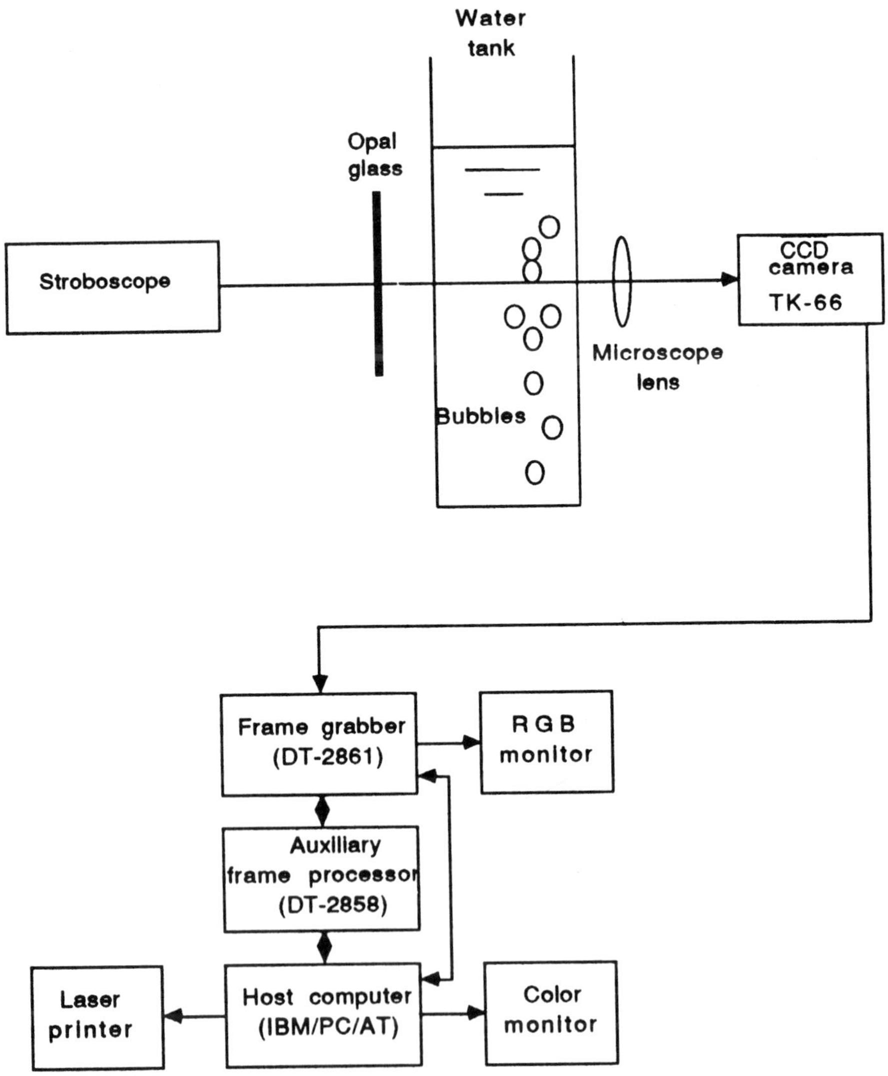

FIGURE 2–7 *Optical system and interrogation system for particle tracking velocimetry.*

threshold by zero. Image noise arises from nonuniform illumination, unfocused images, background light interference, grain in the photographic emulsion and electronic noise in the CCD camera. Digital low pass filtering attenuates the high frequency noise, and digital high pass filtering removes some of the background variation. The contrast of the images is enhanced by analyzing the distribution of grey levels and using this information to set the ranges in a histogram slide-and-stretch operation. The effect of this

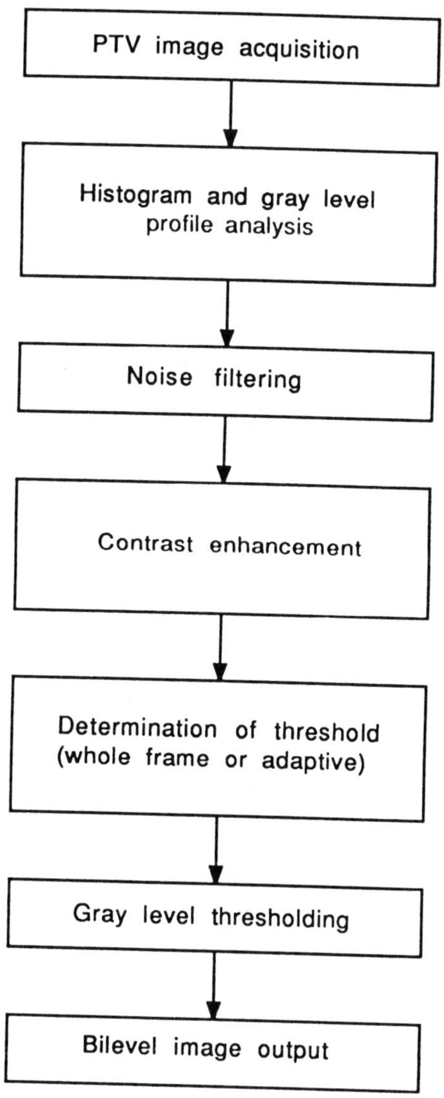

FIGURE 2–8 *Procedure for image enhancement, noise reduction and producing a bilevel image.*

operation is to amplify the brightest images up to full scale grey level. It is not strictly necessary when thresholding is performed, but it presents improved visual displays of the images. Nonuniformities in the background can also be removed by subtracting a reference background image. This is a particularly effective procedure if the background is steady. Thresholding the image performs two functions: image enhancement by setting the background to a uniform black level and setting each particle to a uniform

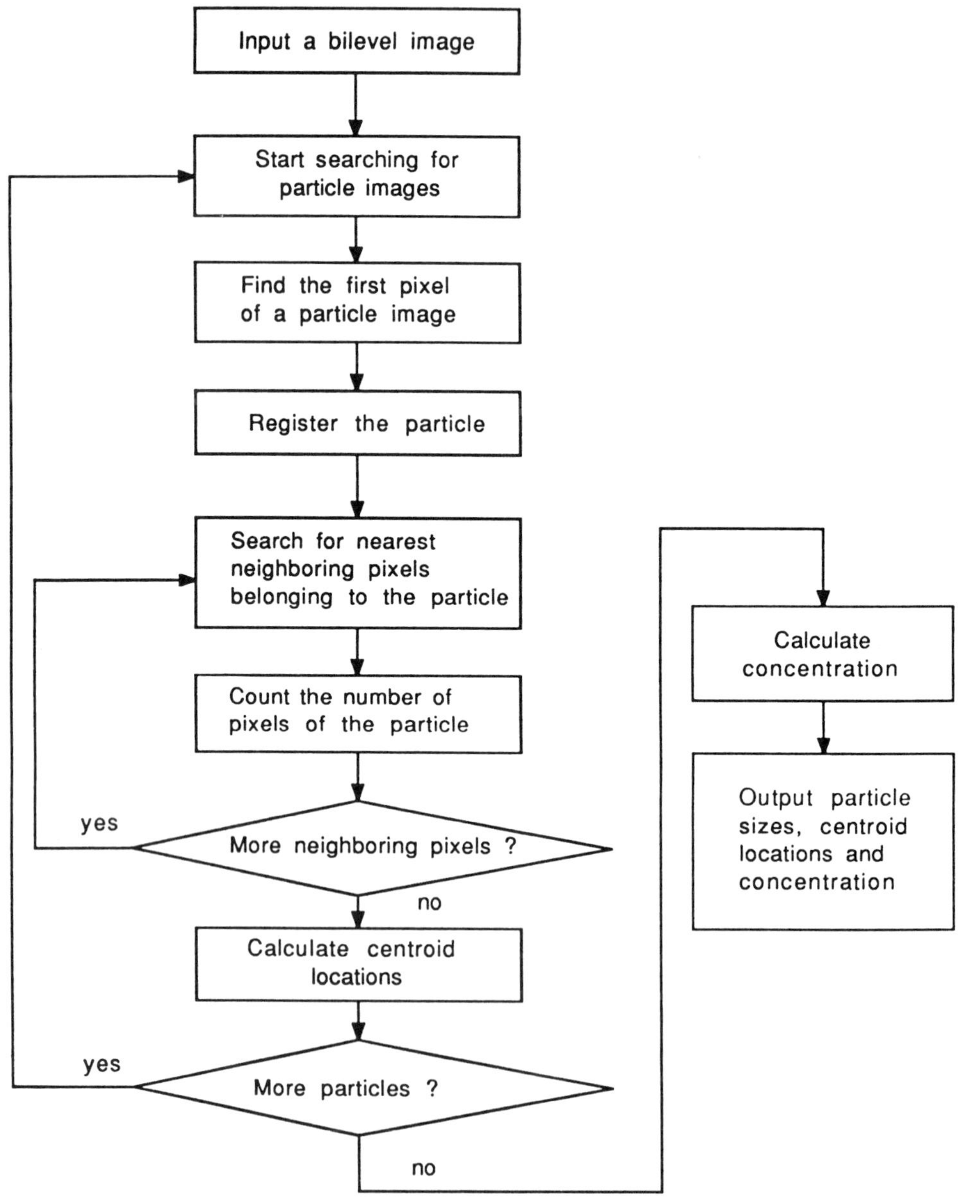

FIGURE 2–9 *Procedure to define particle images and locate their centroids.*

bright level, and image data simplification. In principle only the locations of the bright pixels need to be stored in memory, resulting in substantial image compression for sparse fields. However, the present system retains the full bilevel image field and postpones compression until the centroids of the images are located.

The next steps in the analysis are the identification of particle image, e.g. association of bright pixels with individual particles, and determination of the locations of each particle image, Figure 2-9.

The search for image pixels is performed by scanning the whole picture. The scanning is temporarily stopped when a pixel with value of 1 is found. A search process for nearest neighbors with values of 1 starts and the image size is grown. Once this operation is completed, the particulate image is registered. Scanning for the next particulate image resumes, and more particle images are defined, until the whole scanning process is finished. The centroids of the images are very simple to calculate based on the results of the scanning.

Having found the centroids of the images, one computes the displacement of successive images, and thus the velocities of particulates. An estimate of the particulate displacement is needed to start the search for the second image of an image pair or string of images. This estimate can be performed either from prior information about the particulate motion or by implementing an autocorrelation of the first searched sub-region. The determination of the pairing or the subsequent images is done by the procedure shown in the flow charts in Figures 2-10 and 2-11. The whole picture is scanned to find the centroids. Once the first centroid is found, the scanning is temporarily stopped. The estimated displacement is used to predict the location of the pairing centroid or the subsequent centroid of the multiple exposed images. A search is then performed around the predicted location over a small region defined by an assigned window. When a correctly paired image is found, the correct displacement is calculated, and it becomes the next estimated displacement for searching for the next image pair. Once an image pair is found, scanning for the first image of the next image pair continues, and the process is repeated until all the image pairs or image strings are finally found. The search process for the paired images is illustrated in Figure 2-12.

2.3.3 Measurement of Bubble Velocities in Water Flow

Experiments have been conducted in which velocities of bubbles rising in water were measured. Several approaches were tried to provide multiple images and better illumination of bubbles. Illumination from a stroboscope or a chopped argon ion laser were used because they proved to be best for bubbles. The incoherent strobe-light provided better bubble images because of the absence of coherent noise. Side- and back-illumination methods have been used for the CCD camera to take multi-pulsed images of bubbles. In side-illumination light sheet forming optics, including a cylindrical lens to expand the light beam in one dimension and a spherical lens to contract it

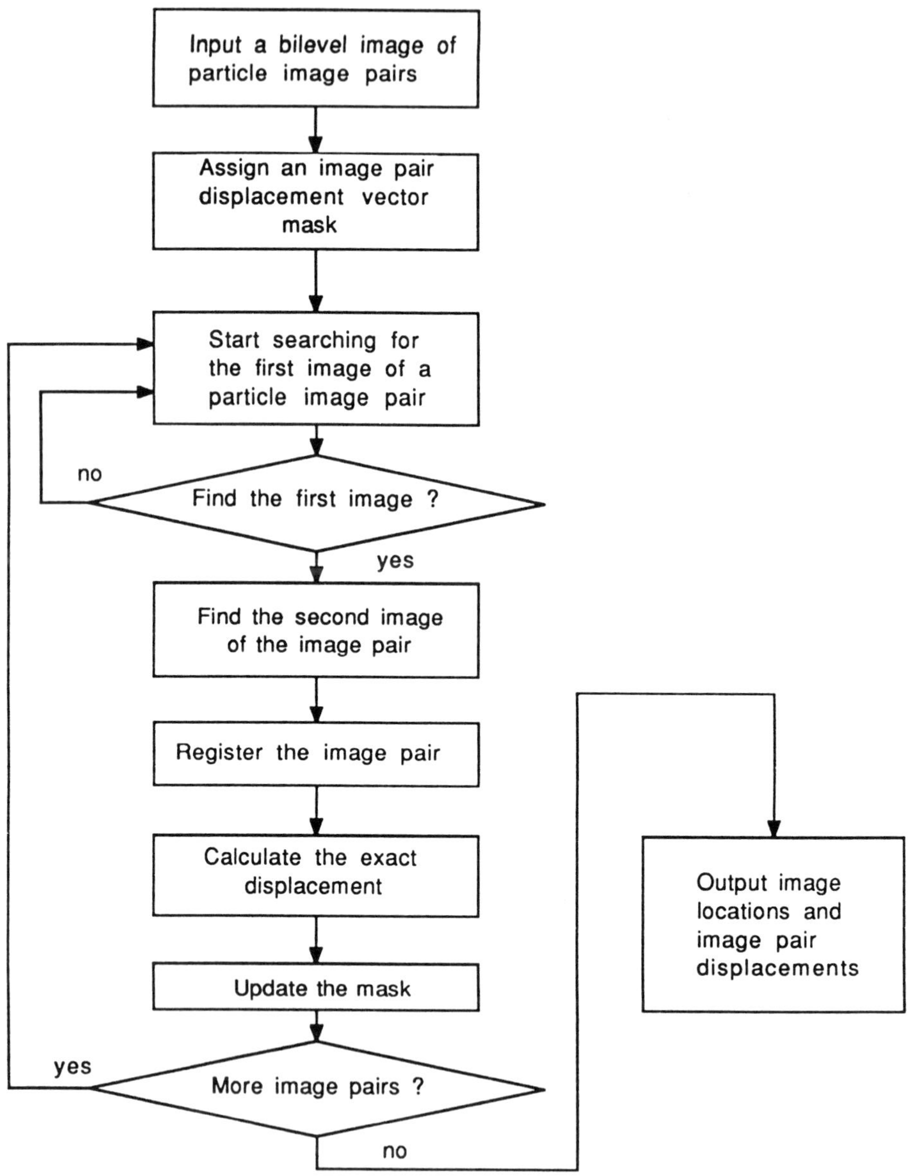

FIGURE 2–10 *Procedure to measure displacements of particle image pairs.*

in another dimension, were needed to generate a light sheet to illuminate a plane, when the bubble motion was investigated. The CCD camera received 90 degree scattered light. The image appeared as a bright spot that came from reflection on the outer surface of the bubble. Sometimes another less bright spot could be seen. That came from reflection on the opposite inner surface of the bubble (later experiments showed these spots quite clearly).

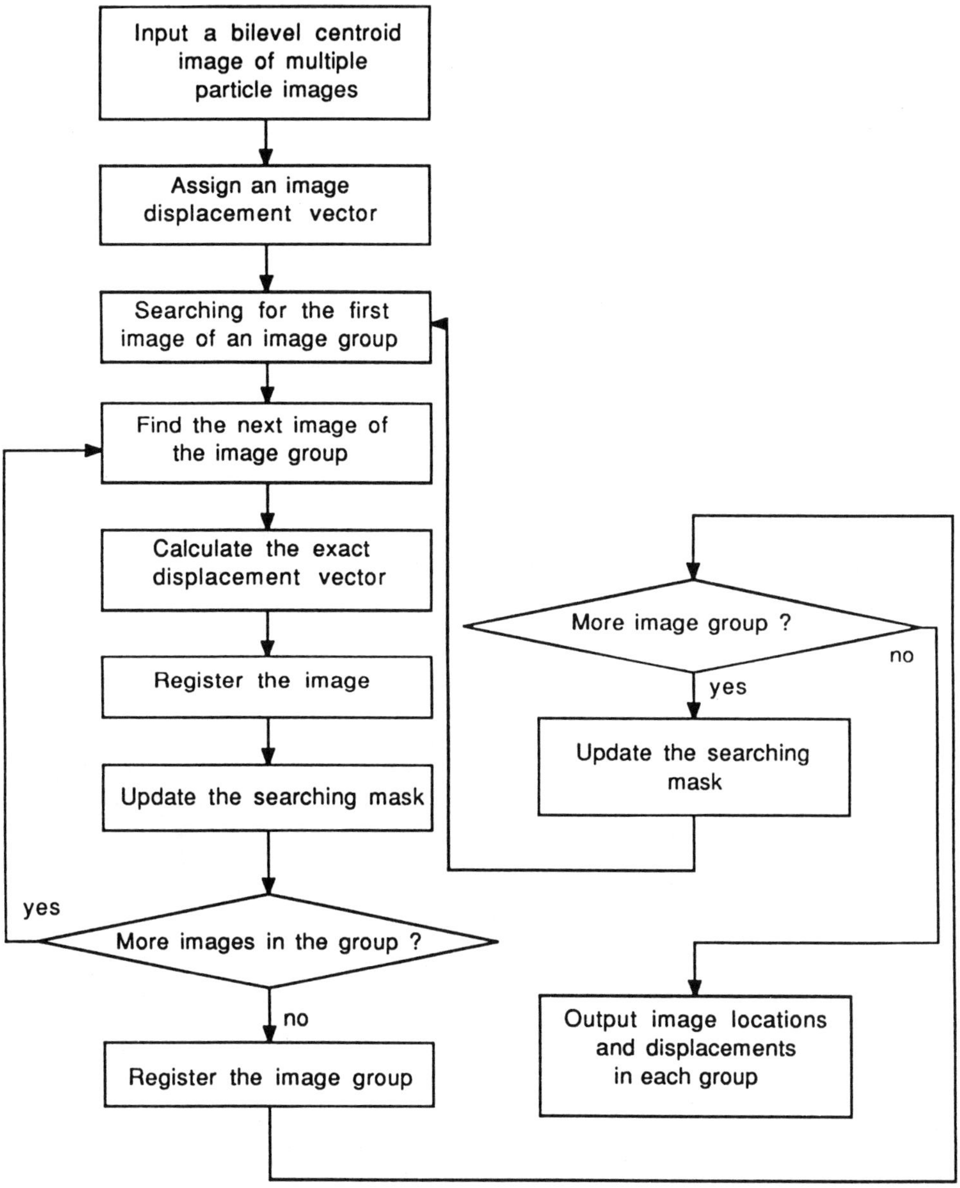

FIGURE 2–11 *Procedure for tracking multiple exposure images.*

For bigger bubbles this effect made it difficult to determine position accurately. Back illumination with an opal diffuser provided better bubble images with clear edges and background and made it easier and more accurate to determine their centroids. Because back illumination is actually a volume illumination, out-of-focus images also appear in the picture and need to be discriminated. If the number of bubbles in the flow is large, the background of the picture will be noisy. The selection of either illumination depends upon the particular experiment.

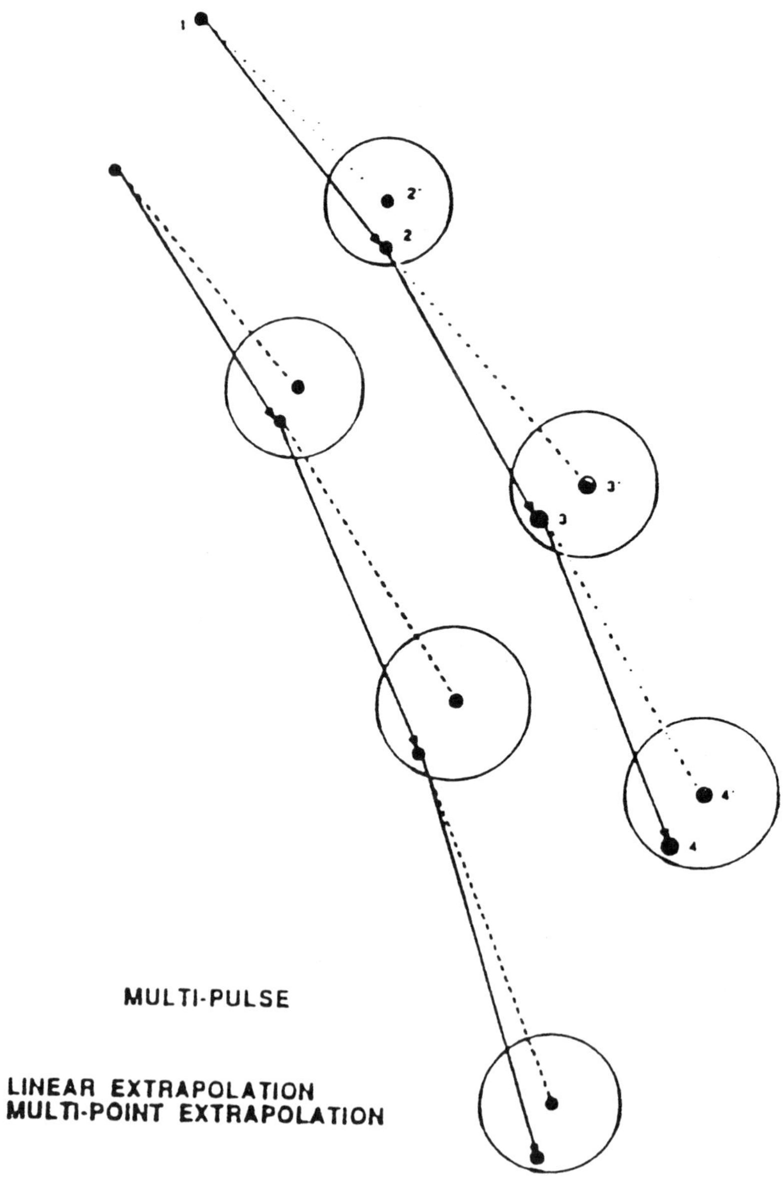

FIGURE 2–12 *Predictor/corrector search for successive images in a multiple exposure sequence.*

The use of a CCD camera facilitates the image acquisition, compared with photographic films. However, the CCD camera has limited spatial resolution, 512×512 pixels or higher up to 2k×2k. This limits the application of CCD cameras. It is desirable to seed the fluid with small particle tracers for PIV to measure the velocity field, and the CCD camera may not have enough spatial resolution to record these small particles. Therefore for simultaneous use of PIV and PTV to measure velocities of both fluid flow and bubble motion photographic film is a desirable recording material because of its high spatial resolution. The temporal resolution is not a problem for the photographic film, as it is determined by the pulsed light source.

The performance of our PTV interrogation system and software is illustrated by results from experiments using a double pulsed photograph of bubble rising in water, Figure 2-13, and a four pulsed CCD image of bubble motion moving with water in a square column, Figure 2-14.

2.4 SIMULTANEOUS VECTOR FIELDS OF TWO PHASES

It is clear from the foregoing discussions of PIV and PTV that it should be possible to combine these two techniques to achieve Lagrangian measurements of small suspend phase masses concurrently with Eulerian measurements of the fluid vector field. The feasibility of this effort is determined by the extent that a single instrument can be made to accommodate simultaneously the different fluid dynamic characteristics of the particle motion and the fluid flow, and the different light scattering characteristics of the particles and the fluid marker particles. Given the variability of these properties, there is probably no universal solution to the instrumentation problem, and the optimum apparatus will depend upon the two-phase system under consideration. In what follows we concentrate on a particular system, that of air bubbles in water.

The experimental flow consists of a submerged, horizontal, 10*cm/s* is water jet of 25*mm* diameter. Air bubbles approximately 100 to 500 microns in diameter were generated using a small syringe needle and injected into the jet. The bubble velocities were nearly the same as the water jet with an additional upward velocity due to the buoyancy effect. Spherical, 4 micron diameter polystyrene particles were seeded in the water flow and served as markers of fluid motion. The specific weight of the polystyrene particles was 1.02, and they followed the water accurately.

A double pulsed ruby laser was used to illuminate the markers in the water flow field. The energy level was 1 Joule per pulse. The pulse separation was adjusted from 1.2*ms* to 2.5*ms* to make the marker image pair displacement average around 0.2*mm* on the photograph. A light sheet with thickness of 1*mm* illuminated a vertical plane vertically from the bottom of the water tank through the centerline of the water jet.

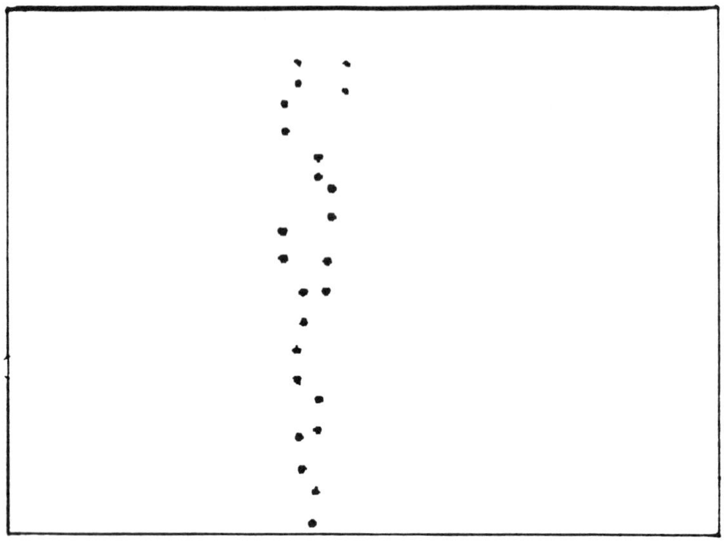

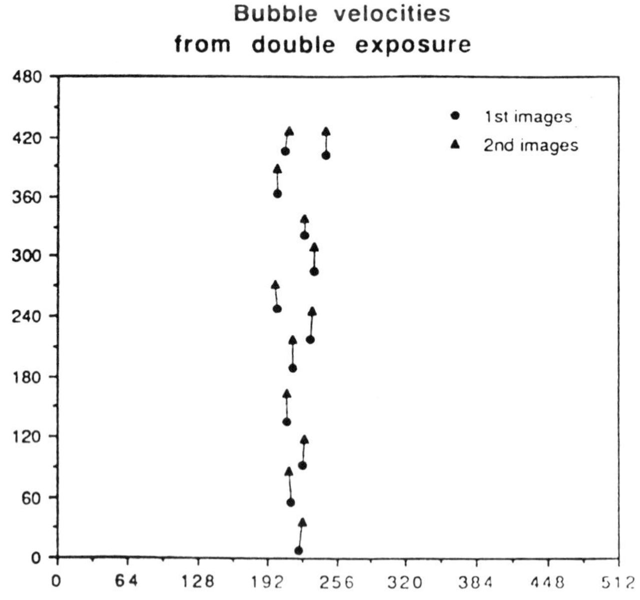

FIGURE 2–13 *(Top) Digitized bilevel images of a bubble column with double exposure; (Bottom) Vectors calculated by automatic algorithm.*

The two light sources were tried in the experiments to illuminate the bubbles: the first was a stroboscope whose output was condensed to a thin sheet of thickness 1.3 to 2*mm* and was directed horizontally to illuminate the same plane that the ruby laser illuminated. The second light source was an argon ion laser beam, also shaped into a sheet. The power of the argon laser was adjusted from 2 to 3 watts, and it provided enough light intensity to illuminate the bubbles in the flow. A rotating wheel chopped the continuous laser light into separate pulses. The frequency of the chopper could be tuned from 750 to 1500 Hz to provide 6 to 12 images for a single bubble.

A Nikon 35*mm* camera with a 2.8 macro lens served to record both water flow and bubble motion. The macro lens provided a magnification of unity when focused on the plane illuminated by the ruby laser and the

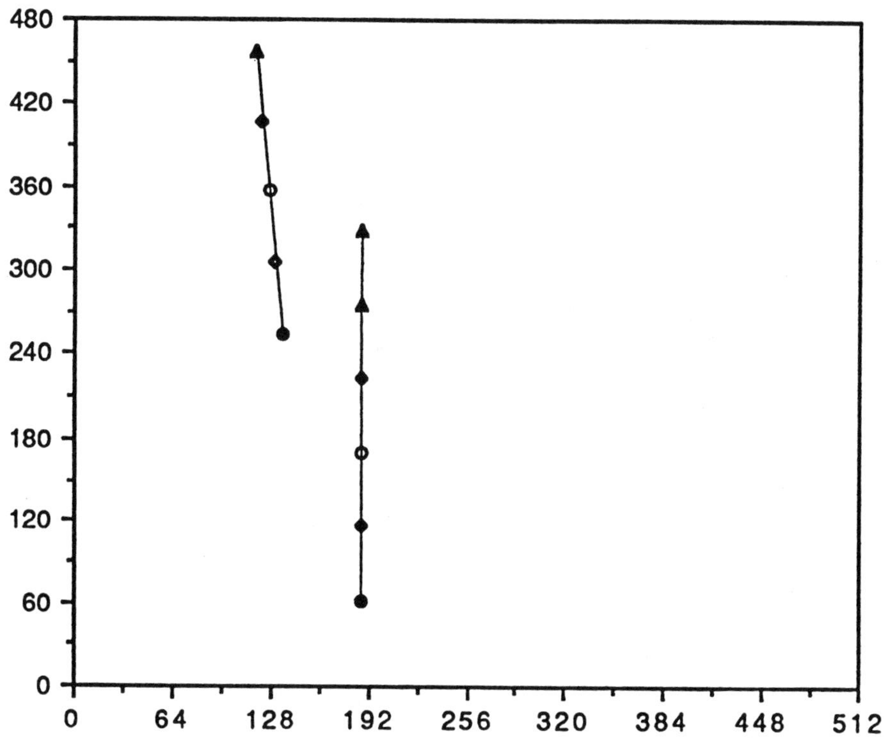

FIGURE 2–14 *Vectors obtained from automatic analysis of a four exposure PTV image obtained by direct video recording.*

chopped argon laser or stroboscope. A wide aperture (f number 2.8) resulted in small depth of field, 0.1*mm* by calculation. The shutter speed determined the number of images available for tracking a single bubble motion. A speed of 1/60*s* provided 12 images for a single bubble. A synchronization circuit was designed to synchronize the ruby laser pulses with the camera. Figure 2-15 shows the timing arrangement of the camera, ruby laser and the chopped argon laser or stroboscope.

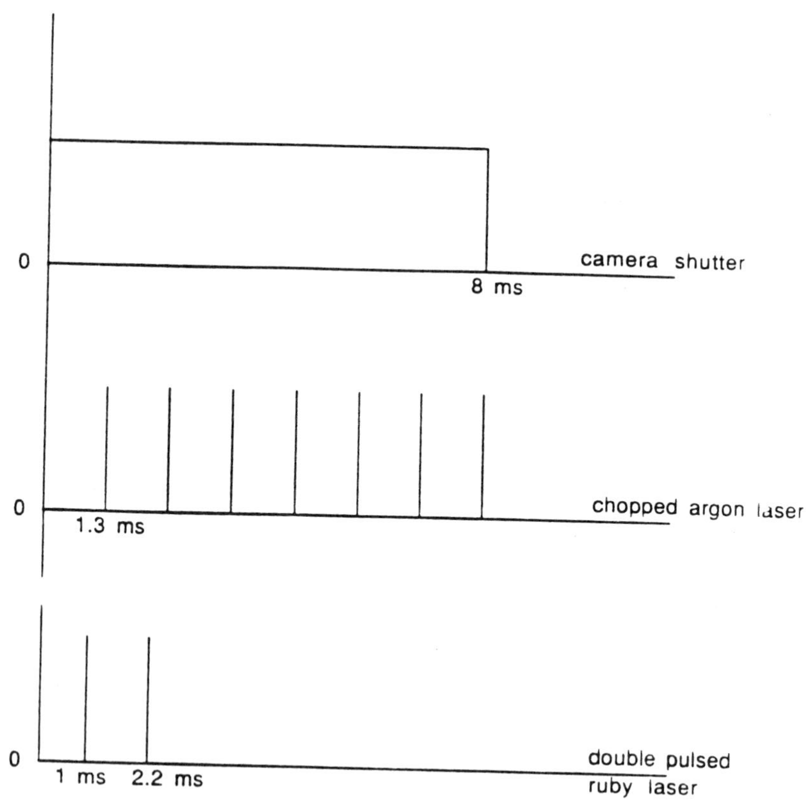

FIGURE 2–15 *Pulse timing for simultaneous PIV and PTV. The chopped argon laser can be replaced by a strobe light.*

The frequency response of the photographic film is quite important. Kodak Technical Pan 2415 film is sensitive to the ruby laser wavelength and it has high resolution. Unfortunately it is not sensitive enough for the low intensity strobe-light. TMZ p3200 high speed film is very fast and adequate for strobe-light, but it is not sensitive to the ruby laser wavelength. Figure 2-16 shows a recording using TMZ p3200. The bubble images illuminated by the stroboscope were clearly recorded, while only the larger fluid markers were recorded, and most of the marker images were not exposed. With the argon laser, TP 2415 film could be used to record both markers and bubble images.

Figure 2-17 shows a recording of marker image pairs illuminated by the double-pulsed ruby laser and bubble multiple-images illuminated by the chopped argon laser. The fluid velocity field shown by the markers and the bubble motion are clearly seen. There are 6 images for a single bubble from the argon laser. There also appeared two images from each bubble due to the ruby laser double-pulse. Some problems showed up with the bubble images. Firstly, the ruby laser is too intense for the bubbles, causing flare and obscuring the fluid marker images. Secondly, the directions of the illuminations of argon laser and the ruby laser were perpendicular, so their reflections by the bubble were from different spots on the bubble surface. The bubble positions determined from these different bright spots were different.

Several methods could be used to reduce the exposure of the bubble image. In Figure 2-18 color film has been used to distinguish the red marker images and green bubble images. The result is definitely positive: the red markers and the green bubbles are separated by different color. The green laser light is strong enough to exposure the bubble images, but not the marker particle images. The red laser light is strong enough to exposure both, and as in Figure 2-17 two large red exposures result from the double ruby pulse striking the bubbles. However, the green bubble tracks are easily distinguished in these recurs.

A further improvement to this system would be to use flourescent marker particles which would absorb the strong laser double pulse and re-emit a new wavelength different from that used to illuminate the bubbles. In this way the images of the bubbles and the markers could be completely separated by color at the time of recording by using either color film or color filters in front of two cameras. Flourescent markers have been used for this purpose in studies of flow through porous media (Northrup et al., 1991). When the suspended phase is solid or liquid, one can elect to make it flouresce and use elastic scattering from the marker particles. This would produce more intense images from the small particles.

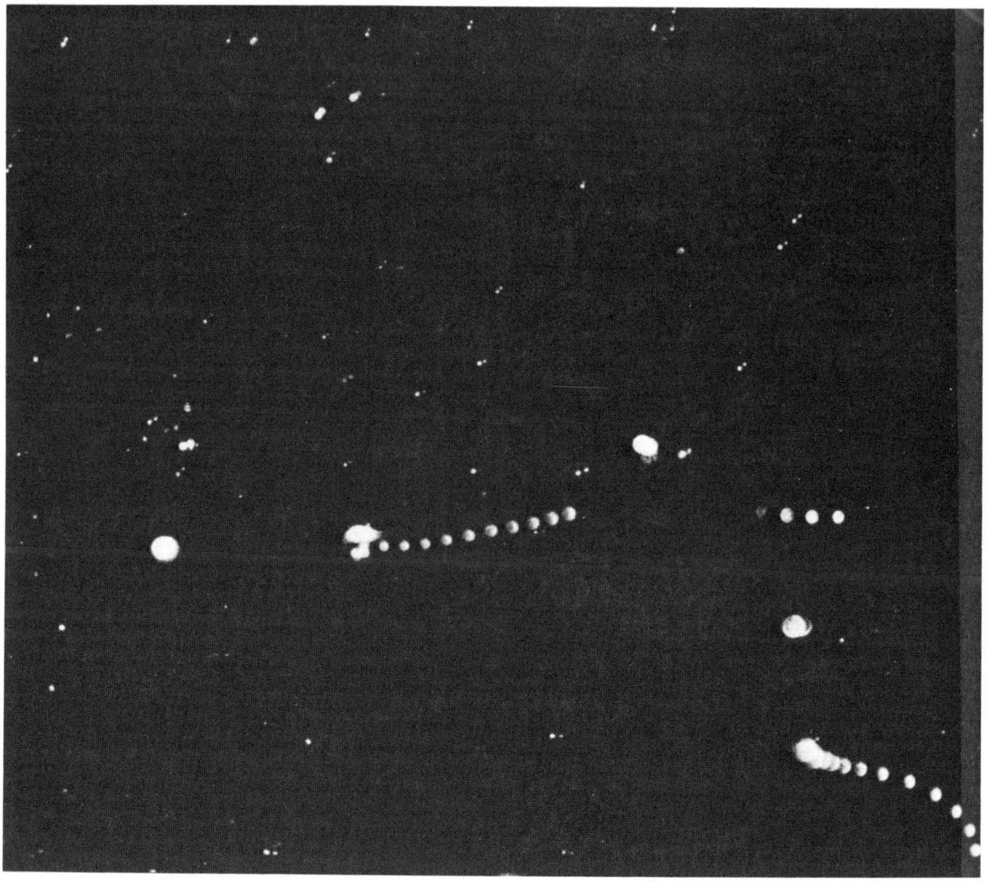

FIGURE 2–16 *Images of bubbles in water obtained from six stroboscopic pulses and two ruby laser pulses using film that is inadequately sensitive to red wavelengths.*

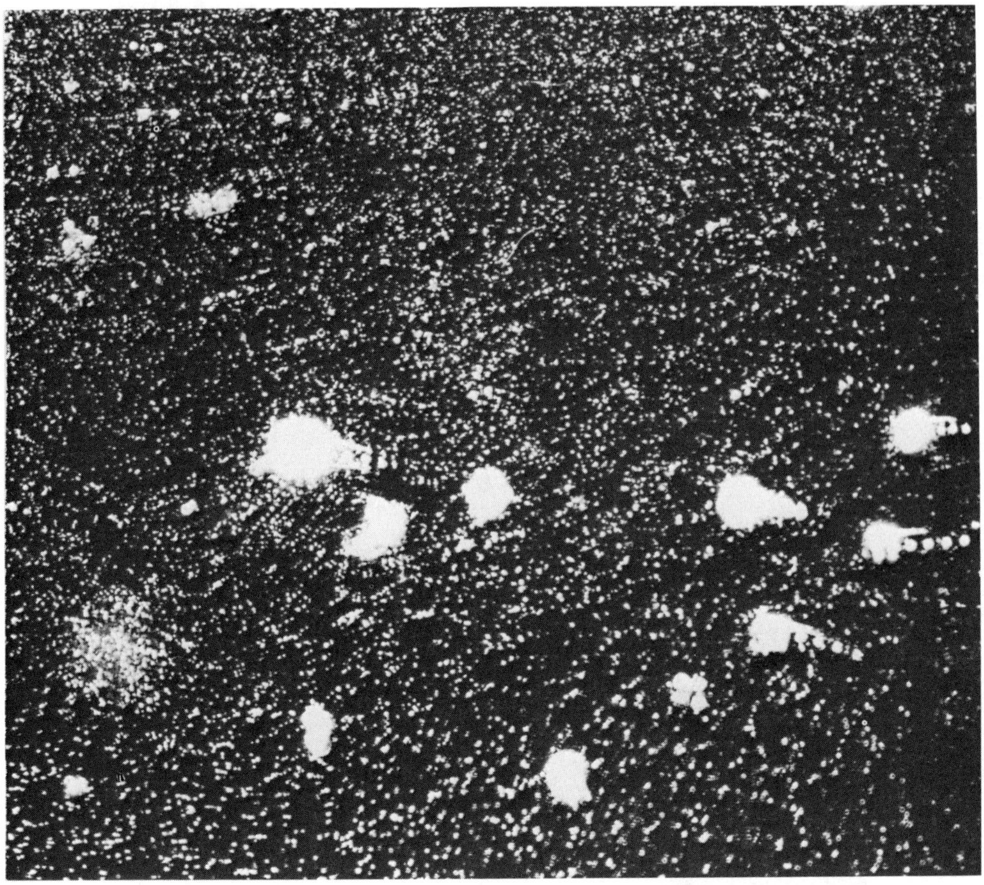

FIGURE 2–17 *Images of bubbles in water obtained from six chopped argon laser pulses and two ruby laser pulses using Technical Pan 4415 film. Note the overexposure due to the large bubbles scattering the light from the powerful ruby pulses.*

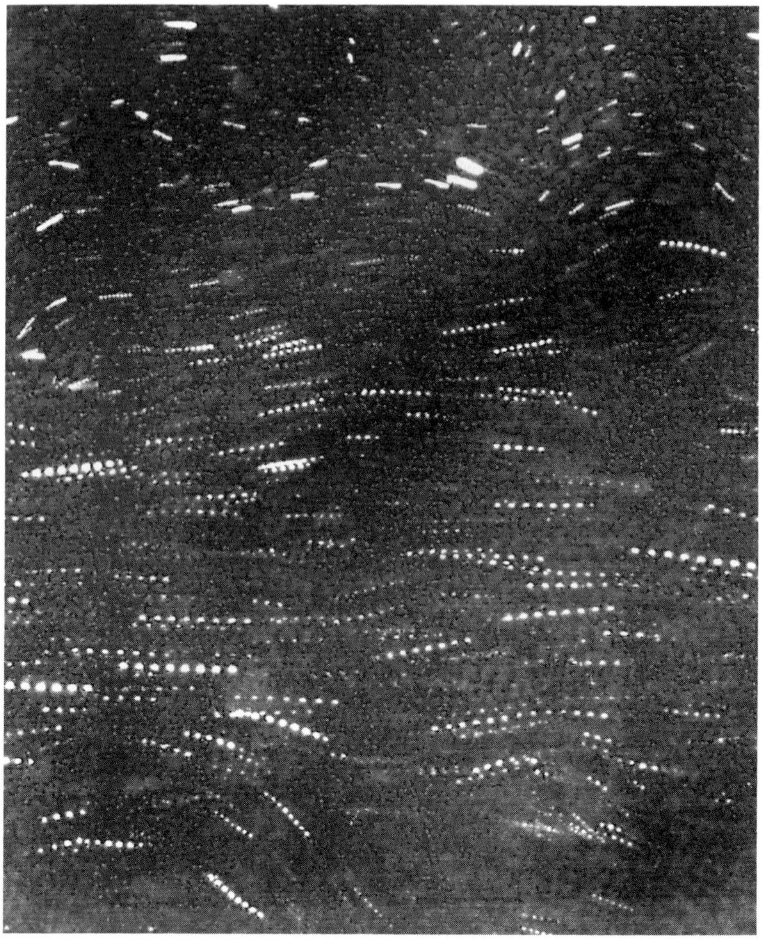

FIGURE 2–18 *Image field and conditions as in Figure 2-17, but using color film to discriminate between bubble images (green) and fluid marker images (red).*

Note: Green is seen here as white, and red as gray

2.5 SUMMARY

Particle image velocimetry and particle tracking velocimetry are useful new methods for the study of dilute or optically thin two phase flows. Under proper conditions they can be combined to measure simultaneously the Lagrangian particle velocities and the fluid velocity field in which the particles (or droplets or bubbles) are immersed.

Although the present discussion has focused on 2-D snapshots of the flow, extensions to 3-D and time varying flow are feasible and currently under development.

ACKNOWLEDGEMENT

This research was supported by a grant from the DOE Multiphase Flow Research Institute at Argonne National Laboratory under ANL Contract 82862403.

REFERENCES

Adrian, R. J., "Multi-point Optical Measurements of Simultaneous Vectors in Unsteady Flow—A Review," *Intl. J. Heat and Fluid Flow* **7**, 127–145 (1986).

Adrian, R. J., "Particle Imaging Techniques for Experimental Fluid Mechanics," *Annual Reviews of Fluid Mechanics* **23**, 261–304 (1991).

Cimbala, J. M., D. Sathianathan, and S. A. Cosgrove, "Flow Visualization Through Metal Enclosures with Neutron Radiography," *Flow Visualization*, FED-**85**, (B. Khalighi, M. J. Braun, and C. J. Freitas, eds.) 77–80, ASME: New York (1989).

Coupland, J. M. and N. A. Halliwell, "Particle Image Velocimetry: Rapid Transparency Analysis Using Optical Correlation," *Applied Optics* **27**, 1919–1921 (1988).

Dudderar, T. D., R. Meynart, and P. G. Simkins, "Full-Field Laser Metrology for Fluid Velocity Measurement," *Optics and Lasers in Engineering* **9**, 163–200 (1988).

Khalighi, B., "Study of the Intake Tumble Motion by Flow Visualization and Particle Tracking Velocimetry," *Experiments in Fluids* **10**, 230–236 (1991).

Lee, M. M., T. J. Hanratty, and R. J. Adrian, "An Axial Viewing Photographic Technique to Study Turbulence Characteristics of Particles," *J. Multiphase Flow* **15**, 787–802 (1989).

Liu, Z.-C., C. C. Landreth, R. J. Adrian, and T. J. Hanratty, ''High Resolution Measurement of Turbulent Structure in a Channel with Particle Image Velocimetry,'' *Experiments in Fluids* **10**, 301–312 (1991).

Majumdar, A., A. L. Graham, M. C. Roco, and P. Stroeve, ''Experimental Study on Solid Particle Dynamics in Shear Flow,'' *Powder Technology* **49**, 217–226 (1987).

Majumdar, A., P. Stroeve, J. L. Fales, V. R. Algazi, and J. E. Stewart, ''Teaching Computers to Track Particles. II. Auto Tracking Algorithm and Coordinate Transformation,'' *J. Imaging Science* **31**, 208–219 (1987).

Nishino, K., N. Kasagi, and M. Hirata, ''Three-Dimensional Particle Tracking Velocimetry Based on Automated Digital Image Processing,'' *J. Fluids Engineering* **3**, 384–391 (1989).

Northrup, M. A., T. J. Kulp, and S. M. Angel, ''Fluorescent Particle Image Velocimetry: Application to Flow Measurement in Refractive Index-Matched Porous Media,'' *Applied Optics* **30**, 3034–3040 (1991).

Perkins, R. J. and J. C. R. Hunt, ''Particle Tracking in Turbulent Flows,'' *Advances in Turbulence* **2**, 286–291, Springer-Verlag: Berlin (1989).

Racca, R. G. and J. M. Dewey, ''A Method for Automatic Particle Tracking in a Three-Dimensional Flow Field,'' *Experiments in Fluids* **6**, 25–32 (1988).

Utami, T., R. F. Blackwelder, and T. Ueno, ''A Cross-Correlation Technique for Velocity Field Extraction from Particulate Visualization,'' *Experiments in Fluids* **10**, 213–223 (1991).

Uemura, T., F. Yamamoto, and K. Ohmi, ''A High Speed Algorithm of Image Analysis for Real Time Measurement of Two-Dimensional Velocity Distribution,'' *Flow Visualization*, FED-**85** (B. Khalighi, M. J. Braun, and C. J. Freitas, eds.), 129–134, ASME: New York.

Willert, C. E. and M. Gharib, ''Digital Particle Image Velocimetry,'' *Experiments in Fluids* **10**, 181–193 (1991).

Zoltani, C. K., K. J. White, and F. A. DiBianca, ''Multichannel Flash X-ray Tomography of Microsecond Events,'' *Applied Optics* **24**, 4061–4063 (1985).

CHAPTER 3

QUANTITATIVE MULTIPOINT MEASUREMENTS AND VISUALIZATION OF DENSE LIQUID-SOLID FLOWS USING LASER INDUCED PHOTOCHEMICAL ANEMOMETRY (LIPA)

R. Falco
D. Nocera

3.1. INTRODUCTION

Our understanding of the physics of two-phase solid-liquid flows is improving rapidly, but is still in its infancy. Largely it consists of isolated experimental data in specially restricted laboratory flows, usually straight pipes. We have almost no information about the effects of area change, bends, dunes, bumps, geometry dislocations, secondary flows, rotating flows, etc., on the dynamics of the flowing solid-liquid mixtures. The paucity of data becomes apparent when, for example, one seeks the mean velocity profile in a solid-liquid flow with $\alpha_s > 15\%$ at speeds at which the flow condition is not laminar, for any particle size distribution and any pipe orientation. Little is known about the fluctuations in dilute solid-liquid flows, and essentially nothing in concentrated flows.

Almost all measurements in two-phase solid-liquid flow systems are difficult to make. Other than for flow rate, and in a number of cases pressure drop, special techniques are required to obtain measurements of the flow properties in the interior of the solid-liquid mixtures. The long standing problem has been that placement of probes into these mixtures is certain to change both the characteristics of the probe and the flow that is to be measured. Thus, optical measurement techniques, which are essentially non-intrusive, have evolved to enable velocity profiles and profiles of turbulence statistics to be obtained in mixtures of low to moderate concentration. The key to their use is to refractive index match the solid and liquid phases, allowing both flow visualization and quantitative measurement of a small number of particles seeded into the mixture that where not index matched. Cox and Mason (1971) provide a good review of early work in laminar flows. Majumdar et al. (1987a) have recently advanced visualization techniques significantly by using stereo photography to track a small quantity of opaque particles in a densely loaded solid-liquid turbulent flow. Most recently particle imaging velocimetry (PIV) has been proposed (see Liu and Adrian, this volume) as a means of obtaining two-dimensional results with selected marked particles in highly loaded mixtures of index matched particles; the number of opaque particles seeded in the mixture was somewhat larger than used before. The optical technique that has recently gained an important place in the measurement of two-phase systems is Laser Doppler Anemometry (LDA). By modifying the accepted bursts via special discrimination techniques (see Lee and Durst 1982) index matched mixtures can be studied. LDA has been used to obtain quantitative measurements of the kinematics and dynamics of the fluid, the particles and the combined mixture in turbulence flows. In contrast to particle imaging, LDA is sensitive to very small index of refraction mismatches, which can be used to detect the particles, and hence their motion. However, this sensitivity has, in practice, limited LDA to use in dilute solid-liquid mixtures. Very recently, the technique of NMR (Altobelli et al. 1991) has been applied to solid-liquid flows. Because of the time constants involved in switching the magnetics, it appears that NMR is limited to the measurement of mean velocity profiles in laminar flows.

Both particle imaging and LDA techniques are undergoing rapid development. However, aside from the issue of whether or not these techniques can provide reliable data in their respective regimes of application -- there are apparent systematic differences in the current data base as described below, that may be technique related -- a large gap still exists in our capability to measure the flow properties of solid-liquid mixtures. A technique is needed that can measure both single and multipoint fluid and solid particle statistics in heavily loaded flows. Furthermore, there is a strong need for the data of both phases to be obtained simultaneously. In addition, it is also important to gain an understanding of the three-dimensional motions of both the fluid and solid phases. A new technique is described which can meet these needs. It works by doping the liquid and/or solid phase with photoexcitable chemicals which upon excitation, mark specific points in the mixture, enabling measurement of the displacement of the selected points in the flow. The technique is non-intrusive and has the potential to provide reliable information over a wider range of concentrations than previously possible. The technique is called LIPA (pronounced lip - uh), an acronym for Laser Induced Photochemical Anemometry. We describe it in detail and present a number of possible implementations. Next the chemical underpinnings are reviewed, because this is a technique that depends on the presence of chemicals with long emission lifetimes, the development of which represents a field of chemistry that is seeing rapid progress. Some initial measurements in highly loaded solid-liquid mixtures flowing in a pipe are then presented. Finally, the outlook for the future is discussed. Throughout the chapter brief comparisons with other techniques are made.

3.1.1. Review and Discussion of Solid-Liquid Two-Phase Results

This brief review is intended to clarify the need for a new measurement technique to help better understand the physics of the flow of solid-liquid mixtures. We will emphasize existing measurements that help build a picture of the physics. From the data available, it is clear that interesting new physics is associated with solid-liquid mixtures (this is apparent even for dilute mixtures). In examining results, we will separate near-wall measurements from measurements in the core, because in single phase flows the turbulence is produced, and heat transfer is determined, by near-wall flow conditions. To date, as noted above, most measurements of mean and turbulence statistics in turbulent solid-liquid flows have been made by using LDA in mixtures that have been carefully refractive index matched. All of the data discussed below were taken in simulated, index matched, solid-liquid mixtures.

3.1.1.1. *Near-Wall Flow Studies*

The near-wall region is defined for a single phase flow as the region closer than 30 viscous units ($\nu/u_* < 30$). To our knowledge the only set of near-wall measurements in solid-liquid mixtures are those of Zisselmar and Molerus (1979) who used LDA in a 50 mm pipe with α_s up to 5.6%. By blocking out the signal from the particles to enable the fluctuations in the liquid to be measured, they found that the absolute value of u' and v' increased with increasing α_s for $\alpha_s < 3\%$, whereas u' decreased with v' continuing to increase for $\alpha_s > 3\%$. At the same time pressure measurements indicated that the pressure drop monotonically increased as the concentration was increased. Thus, when the turbulence is normalized by the friction velocity, u_*, u' will most likely continue to decrease for $\alpha_s >$ 3%. However, the normalized v' can go either way, and probably will continue to increase with some additional concentration increase. The heat transfer increased as the concentration was increased up to 3%, and then decreased. They also found that the absolute value of the Reynolds stress decreased monotonically, so that the wall friction normalized Reynolds stress values will decrease even faster.

To get some sense of the new physics, these changes may be compared with those found for dilute polymer solutions, a reference non-Newtonian flow that has been well characterized. The addition of small quantities of long chain polymers (1-100 ppm) to water strongly alters the turbulence, resulting in drag and heat transfer reduction. It is found that the near-wall streamwise fluctuations increase, but the normal fluctuations decrease when non-dimensionalized by the friction velocity, as the pressure drop decreases. The differences with the Zisselmar and Molerus findings suggest a very different mechanism of turbulence modification for dilute solid-liquid flows. Of particular interest is the fact that the decrease in the Reynolds stress is often associated with drag reduction. In dilute solid-liquid mixtures, the drag increases are associated with Reynolds stress reduction.

But the presence of particles in a liquid flow can also result in drag reduction. Dilute concentration of long fibers (see, for example, Metzner 1977) result in drag reductions of 5-20% for concentrations of 200-1000 ppm. This suggests that the particles have an important role to play in decoupling the phasing of the fluid turbulence, and that depending upon the shape of the particles either more or less drag will result. The information available is not enough to understand the physics. Clearly, more near-wall measurements are needed.

3.1.1.2. *Outer Region Studies*

Somewhat more information is available in the interior of the solid-liquid flows at speeds where we would expect turbulent motion. Dilute solid-liquid systems have been studied by Zisselmar and Molerus (1979), Lee and Durst (1982), Yianneskis et al. (1983), Nouri et al. (1987), Park et al. (1989) and Stein (1989).

Zisselmar and Molerus (1979) measured all three components and the Reynolds stress in the horizontal pipe flow study described above. The signal, from just the liquid, showed that the absolute values of u', v' and w' and <u'v'> monotonically decreased across the core as the particle concentration increased up to the maximum they used (5.6%). Nouri, et al. (1987) studied downward flow in vertical pipes with particles heavier than liquid (as in the case study we present below). They measured the streamwise component of the particle velocities only. Results for the axial turbulence intensity showed that it was less than the Newtonian values, in qualitative agreement with Zisselmar and Molerus in the core of their pipe flow. They suggested that a migration of the particles towards the walls occurred, consistent with the low Reynolds number results of Cox and Mason (1971), however, no independent confirmation of migration in their turbulent flow was made. It is likely that the classical Cox and Mason picture, where single particles heavier than the fluid migrate toward the wall in a downflow for dilute concentrations, and tube Reynolds numbers lower than 500, may not hold in turbulent flows. For example, Nouri et al. (1987) showed that when the concentration was increased to 14% α_S the mean velocity profile began to flatten out. The trend suggested that it might become flatter in the center of the pipe than observed in a Newtonian flow if they continued to add more particles, which suggests migration towards the center of the pipe. The differences that may be expected in turbulent flows are also suggested by the results of Cox and Mason (1971) in their oscillatory flow studies, which were made at low Reynolds number and with dilute particle concentrations. Under oscillation, they found that particles always migrated towards the center of the pipe.

In contradiction to the change in axial intensity found in the above two studies, Park et al. (1989), using a dual beam LDA system studied the flow of 1.5 μm particles in a 51 mm horizontal pipe with α_S also about 14%, and found that u' was greater than the Newtonian value across the outer region. Steen (1989) has reported a third picture. Also using LDA, he has made measurements of the mean flow and axial turbulence intensity of high aspect ratio fibers (10 μm x 1000 μm) at α_S = 1.2%, in an upward flowing vertical pipe of 25 mm diameter. These results were compared with single phase results. He found that for the same bulk velocity, the mixture had a flatter mean velocity profile in the center of the pipe, with higher velocities near the wall. Although he did not report measurements in the wall region, he suggested that there may be greater shear stress at the wall. The axial turbulence intensities of the mixture were generally lower in the center of the pipe, but higher than the single phase values closer to the wall (but still in the outer region).

We are not aware of any solid-liquid data for the radial turbulence intensity or the kinematic Reynolds stress <u'v'> to compare the Zisselmar and Molerus data with.

There is also considerable doubt about the form of the mean velocity profile in the region between the wall and the core, in high Reynolds number dilute-to-moderate concentration solid-liquid flows. This region has a logarithmic behavior in Newtonian pipe flows, with a universal slope (the Von Karman constant) and intercept when scaled on wall variables. Abbas and Crowe (1987) measured mean velocity profiles in an index matched solid-liquid system using LDA by scattering off the slight mismatch of the particles making up the solid phase. Their results showed that the profiles had a logarithmic behavior for mixtures up to 30% α_S, but they found a large change in the Von Karman constant as concentration changed. Earlier measurements of Mih (1979), using pitot tubes, did not find variations in the constant. Other authors (see Park et al. 1989) have preferred to fit a power law to moderate concentration solid-liquid flows.

This review has considered all of the turbulent solid-liquid flow profile data the authors are aware of as of Spring 1991. These results are sufficient to indicate the complexity of solid-liquid flows, and to clearly indicate the need for additional studies to understand their flow physics.

3.1.1.3. *Some Limitations of Available Techniques*

To an important extent, a more comprehensive picture is not available because existing measurement techniques have difficulty providing, or are unable to provide, the needed data. With the advent of LDA techniques and modern image acquisition and data reduction procedures, it would appear that eventually the need will be met, which has existed for several decades, for a database of quantitative information about the kinematics and dynamics of both phases and of the mixture in solid-liquid flows of arbitrarily high concentration. However, these optical techniques have limitations that may make this goal difficult to achieve.

Perhaps the most fundamental limitation is the need to receive scattered light from the particles making up the solid phase, or from the (usually smaller and more dilute concentrations of) 'seed' particles marking the liquid phase. If perfect index matching were possible, then a chosen number of opaque particles must be added to enable light to be scattered and subsequently imaged. Once opaque particles are added, they will disperse throughout the mixture and get between the measurement location and the receiving optics, thereby reducing the transmission of scattered light. Thus, they must be put into the flow in dilute quantities. This reduces the data acquisition rate of single point measurements severely. Furthermore, it even more severely reduces the possibility of obtaining multipoint correlations, spatial gradients, vorticity and strain-rate. When using LDA, the seeding of the opaque particles also increases the average index mismatch, increasing the error in penetrating both beams to a desired location in the mixture. This makes accurate beam crossing more difficult to achieve and adds unknown fluctuations in the beam crossing angle.

The near-wall measurements are especially difficult using scattered light techniques. For example, Steen (1989) reports that his efforts to make near-wall measurements using LDA were unsuccessful. He noted that although he took very careful measures to index match his mixtures, he could not match them closely enough to obtain reliable measurements. These difficulties in making near-wall measurements using LDA in solid-liquid flows are in addition to those encountered in single phase flows. In single phase flows, index matching has been used for the purpose of obtaining measurements closer to the walls in pipes. For example, Durst et al. (1979), who very carefully index matched their single phase fluid to the pipe wall material, reported positive Reynolds stresses near the wall in index matched Newtonian and dilute polymer solutions. This is not expected, and has not been found by others; see, for example, Niederschulte (1989), who obtained measurements even closer to the wall (in wall units) in a two-dimensional water channel. There is also the problem of stray scattered light from the wall itself, which can only be eliminated by perfect index matching of everything but the seed particles. When the need for matching within the interior of the flow is added to these concerns, the probability of accurate near-wall LDA measurements decreases further.

For measurements over the bulk of the flow, Yianneskis and Whitelaw (1983) suggested that it is necessary to keep the index of refraction differences to less than 0.0005 to obtain reliable LDA results. They accomplished this degree of matching by using temperature control. However, Steen (1989), in trying to meet this criterion, found, in attempts to make measurements in the core region, that solid particle -- liquid (which contained very small tracers) discrimination failed a larger percentage of the time because the overall signals were degraded as the laser beam crossing penetrated further and further into his 25 mm diameter pipe, which was loaded with only 1.2% α_s of an index matched solid. He made attempts to the use pedestal discrimination techniques of Lee and Durst (1982) to separately measure the solid particle velocity and the liquid velocity. Results for both his long and short fibers showed that the particle velocity was consistently higher than the liquid velocity. This is inconsistent with the extrapolation of results of Tsuji et al. (1984) for solid particles in an air flow, and Steen attributed the erroneous measurements to the difficulties resulting from excessive dropout of the liquid velocity signal. During dropout, the output voltage is fixed to the last reading but may decrease, giving a lower value to the liquid readings.

For measurements in simulated mixtures of practical interest, the situation is made even more complex because of the grossly non-homogeneous arrangements often assumed by two-phase systems, resulting in an uneven distribution of opaque particles and of the index of refraction variation. All of these difficulties are enhanced when attempts to measure three-dimensional flow field data are made.

3.1.1.4. *Additional Capabilities Needed in Solid-Liquid Flow Studies*

Although improvements in the LDA technique with respect to the issues of index matching sensitivity and discrimination ambiguity will be important for an understanding of the changes occurring in solid-liquid two-phase flows, other information not likely to be obtained with LDA is needed. In particular, data that can be used to validate ideas and computational studies will be most valuable. Many computational schemes rely on an understanding of the flow patterns. For each flow pattern the flow behavior is modeled. But this requires pattern detection, understanding of conditions which can change flow patterns, particle velocity information, and details like film thickness and interfacial shear and wave structure, for example. In general, single point techniques cannot give information about flow patterns along with the quantity measured. The situation is analogous to that in the study of turbulence in single phase flows, where the existence of coherent motions was not discovered even after decades of single point measurements. It was only through flow visualization, and by combining quantitative techniques with visualization, that the existence and importance of coherent motions were established. This provides even more reason to suspect that techniques which enable simultaneous flow pattern identification and quantitative time resolved measurement over an area will be particularly effective research tools for the study of multi-phase flows. LIPA can play an important role in these measurements.

In addition, most of the solid-liquid flows of interest are highly three-dimensional. Thus, a technique that is able to make three-dimensional measurements, especially in highly concentrated mixtures, is needed. It would be extremely helpful to be able to combine flow visualization with simultaneous quantitative three-dimensional flow field data. As we will indicate, LIPA can make three-dimensional measurements of all of the kinematic quantities.

The following list indicates the kind of information about solid-liquid flow systems needed to build a physical understanding of the behavior of these two-phase flows. A measurement technique is needed that can obtain the mean flow and turbulence quantities for:

- a systematic determination of the effects of particle loading;
- a systematic determination of the effects of particle size;
- a systematic determination of the effects of particle shape;
- a systematic determination of the effects of particle size distributions, including bi-modal distributions with widely differing modes;

all with special attention to near-wall measurements. It must also provide us with:

- drag measurements at the bounding walls;
- an understanding of the role particles play in the creation, diffusion and dissipation of vorticity, through local vorticity measurements.

To make these measurements, the technique must not be:

- very sensitive to index of refraction mismatch;
- sensitive to large spatial variations in particle loading;
- sensitive to the shape of the particles;
- sensitive to the addition of additives that are known to significantly change the pressure drop of solid-liquid flows.

It must be able to obtain:

- data in dense solid-liquid flows;
- the flow properties of the liquid and the solid phase simultaneously;
- the average flow properties of the mixture;
- measurements over a region of the flow, to enable measurement of pair distribution functions, spatial gradients and other spatial measures;
- three-dimensional data;

- data from experiments with complicated geometrical access, and perhaps with internal geometry changes.

The last point opens up the possibility of studying geometry induced non-equilibrium flows such as steps and bends, blade tip flows etc., about which essentially nothing is known. For example, we have essentially no idea how to model the effects of particles on separation.

It is clear that most of the above aspects have not been effectively addressed by current measurement techniques. Thus, we will examine a new technique that can potentially provide for many of these needs.

3.2. THE LIPA TECHNIQUE

In this section we introduce the LIPA technique. In the following section we will develop extensions of it for use in solid-liquid mixtures.

3.2.1. Introduction

LIPA was originally developed to meet an increasing need to obtain instantaneous flow field information over a region in strongly three-dimensional single phase flows of engineering interest. Examples include all aspects of turbulence studies, external and internal aerodynamics, studies of turbulence control, and mixing and combustion. In many cases there is a need to obtain more detailed information, including vorticity and strain-rate, in these as well as in considerably simpler flows.

A key objective in the development of LIPA has been to develop a technique that is easy to implement in a wide range of flows and over a wide range of speeds. Additionally, the analysis of its optical data will, ideally, be capable of total automation on PC level computers, and it should be robust, so as to minimize errors due to noise, etc., that automated data reduction is vulnerable to. LIPA has the possibility of achieving these goals.

3.2.2. Description of the LIPA Technique

Briefly, non-intrusive long lifetime photochromic and/or photoluminescent chemicals can be doped into liquids, solids or gases and selectively excited along laser lines. The marked regions can be imaged at two successive

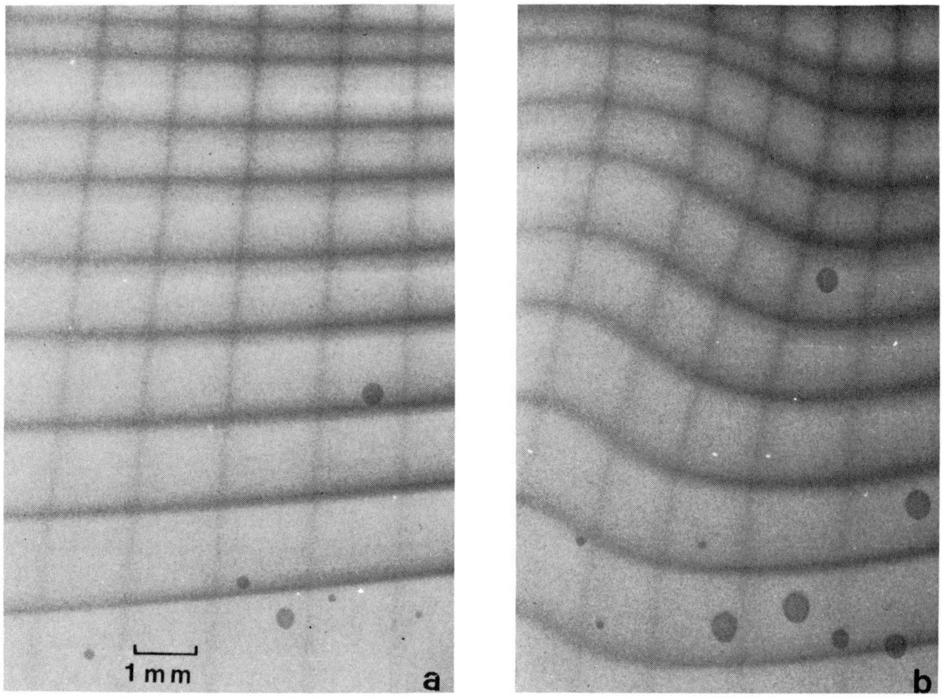

FIGURE 3-1. *LIPA grid used to measure streamwise vorticity; (a) shortly after the laser pulse, (b) 0.4 seconds later. A Kodak photochromic (see text) was dissolved in kerosene and excited by a 351 nm excimer laser. Line widths are the order of 100 microns.*

times, and from their displacement by the flow, one can obtain the kinematics of the flow. Measurements at many points over a region of space can be made by exciting the doped fluid along a grid of laser lines that is created in the interior of the fluid. Since the calibration requires only a clock (for the time scale) and a ruler (for the length scale), the technique is conceptually simple and direct. LIPA can be performed using the grid line intersections to define points in a liquid or a gas in which a photoactive chemical is dissolved or mixed. It can also be performed using photoactive solid particles seeded into a flow. This form results in an "extended PIV" technique, wherein a) we don't have reflected light problems, and b) we don't have problems with particles moving in and out of the light sheets. Furthermore, as we shall see in the next section, LIPA is relatively insensitive to refractive index mismatch.

Hummel and his group (see Popovich and Hummel (1967), and for recent work, Ojha et al. (1989) and references therein) were the first to use photochromic chemicals to study flows. They excited single lines, first using flash lamps, and later lasers. A single marked line enables the study of

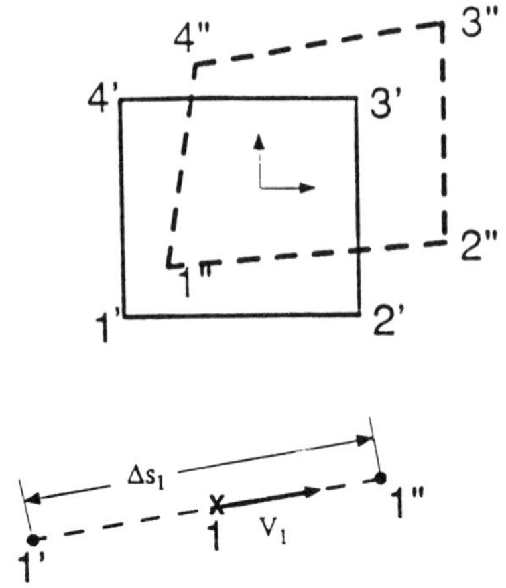

FIGURE 3-2. *Use of the distortion of one mesh of the grid to calculate velocities at four locations in the flow.*

essentially one-dimensional flows, where movement along the marked line was negligible, enabling the displacement of the time lines generated to yield streamwise velocity information. Falco and Chu (1987) and Chu (1987) also used photochromics in their work to advanced the technique, enabling it to get velocity, Reynolds stresses, vorticity and strain-rate components in two and three-dimensional flows, by using the idea of a grid of laser lines to isolate and follow specific marked points within a flow field

LIPA can also be performed by tracing the movement of laser induced long-time-fluorescence or phosphorescence in the same way as with laser induced color change. The photochromics are generally suited for use in organic solvents (kerosene is a relatively inexpensive choice). Recently, several chemicals have been developed (some of which are discussed below) which use long time fluorescence, or phosphorescence that is not collisionally quenched by oxygen or water molecules. These have enabled LIPA to be performed in water and in water droplet aerosols (because they are soluble in water), in particulate aerosols (for photoactive solids), or in gases (using photoluminescent gases).

Depending on the chemicals and data reduction capability, LIPA can be executed in several ways. Three which have seen use to our knowledge can be characterized as; Excite-Self-Decay mode, Excite-Forced-Decay mode, and the Sensitize-Delay-Excite mode. In the Excite-Self-Decay mode, a photoluminescent chemical, which either has a natural or a specially engineered long lifetime, is introduced into a fluid and generally

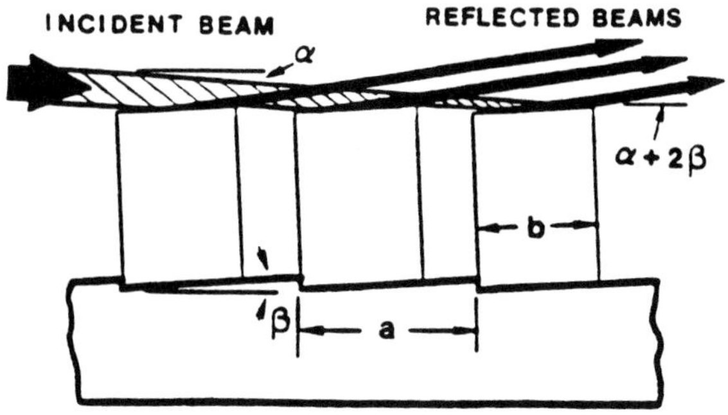

FIGURE 3-3. *Schematic of a beam divider, a device which produces an array of laser lines from a single incident beam.*

homogeneously dispersed. Ideally it is transparent to the observer before excitation. The doped fluid is then excited along predetermined multiple intersecting lines by a laser operating at a wavelength in the absorption region of the chemical. The laser excitation produces a grid of lines that can be observed because of the visible excitation of the chemical. At each intersection of the grid of lines, a 'fluid particle' is tagged. Once tagged, the LIPA technique works because the excitation time of the chemicals used is chosen to be long compared to flow time scales of interest. Thus, the fluid will remain tagged as it moves during the time needed for significant distortion of the tagged lines to occur. At least two images are recorded during this time. From the motion of an intersection point of the grid the local velocity in the plane of the grid can be measured. From the movement of two or more intersections spatial velocity gradients can be measured. From the movements of any mesh in the grid the vorticity can be obtained from circulation arguments. The details are explained below. This can be done for every intersection in the grid, yielding simultaneous information over an area. Since high-speed movies or high repetition rate electronic images can be recorded (synchronized in combination with a high repetition rate laser to excited the chemicals, for example), we can gain an understanding of the evolution of the dynamics of unsteady fluctuating flows.

Because the laser excitation is in a plane, our information will be very close to pre-selected planes in the flow, (the velocity vectors will, in general, move marked fluid out of the plane the grid, but we will need only

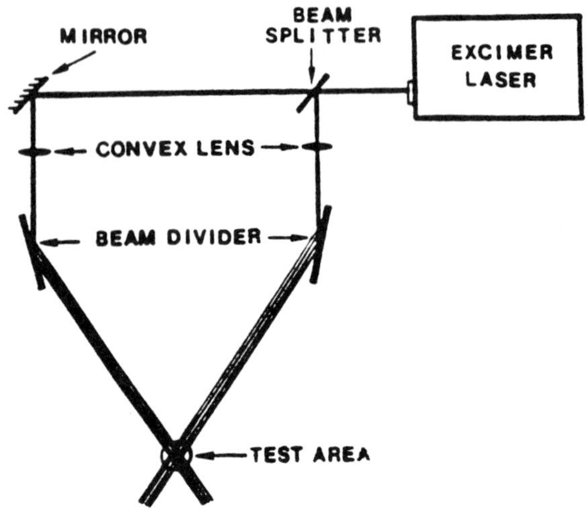

FIGURE 3-4. *A typical optical setup used to generate a grid of laser lines over a region of interest.*

small displacements to calculate the flow quantities, so the measurements will be in the region desired). This ability to choose to have information near the center-line or near the walls, for example, is extremely helpful. Both photochromic and photoluminescent chemicals are available that can be used in the Excite-Self-Decay mode. A key technical aspect of this mode of excitation is that the imaging system must provide a means of creating a short exposure of the distorted grid, at a pre-determined time during the decay period.

The use of LIPA in the Excite-Forced-Decay mode enables the use of a pulsed laser to act as the shutter. Here one pre-excites a molecule along the desired grid lines using laser radiation, and then de-excites it using a pulsed laser spread into a 'sheet' covering the area of the grid with a thickness sufficient to incorporate the maximum expected out of plane displacement, a predetermined time later. The emitted radiation only occurs during the duration of the de-exciting laser's pulse. Hence, the exposure is determined by the pulse duration of the de-exciting laser, which can be very short. This procedure usually requires quite precise laser wavelengths for both the excitation and de-excitation. It has been done with oxygen (see Miles et al. (1989), who call it RELIEF for Raman Excitation Plus Laser Induced Electronic Fluorescence).

A third approach is the Sensitize-Delay-Excite mode. Here, chemicals are used that are prevented from fluorescing because they have

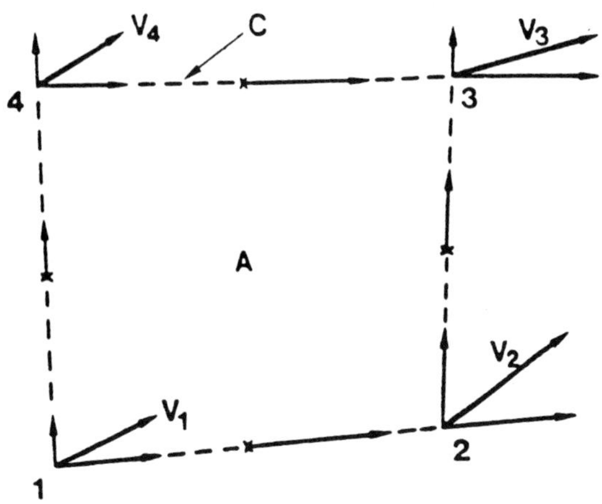

FIGURE 3-5 *Application of Gauss' theorem to a mesh of the grid (after conversion to velocities) for the purpose of estimating the vorticity from the circulation without the need to second difference.*

been specially engineered to have a protective shell built around them. LIPA is performed with them by removing the protective 'cage' so as to expose the dye along a grid of laser lines. After a pre-selected time delay, flood or sheet illumination at wavelengths that cause fluorescence is applied. The unshielded dye lines which have been distorted by the flow in the chosen time interval are then recorded (these dyes often fluoresce under visible radiation). In this case it is also possible to shutter by using a pulsed illumination source to cause fluorescence. The advantage of this approach is that the exciting laser beam pulse width essentially determines the exposure time. Alternatively, with these chemicals it is also possible to use a continuous fluorescence inducing light source for illumination of the modified chemical, and shutter using the imager. Unlike the Excite-Self-Decay mode, the chemicals needed don't return to original state, thus very slow flows can be studied. However, they are single use chemicals. This can result in significant expense, and has the additional disadvantage of having to clean out the flow facility periodically because of dye build up.

Figure 3-1 shows a grid and the distortion of that grid caused by the movement of the fluid. Note that the lines are ~ 100 μm wide and the mesh size is ~ 1 mm.

Figure 3-2 illustrates how the distortion of one grid box can simultaneously provide the projections of the velocities at four locations onto the plane of the imager which is set parallel to the grid box. With these velocities all the kinematic information obtainable from one 'fluid particle' may be calculated.

Figure 3-3 shows a schematic of the device used to produce an array of laser lines from a single laser beam. It has the advantages that no incident light is lost, and that the line widths and spacings can be specified by the facet design. It is similar in appearance to an oversized diffraction grating, but it gives specular reflection.

Figure 3-4 shows the setup, employing two of these devices, which has been used to create a grid of laser lines with a single laser pulse (typically 20 ns in duration). Ten simultaneous beams, for example, coming from each direction are easily created, enabling 100 different points to be tracked each time the laser fires.

Although second differencing can be used to obtain the desired spatial gradients, for the measurement of vorticity a more accurate procedure can be employed. The circulation, Γ, around each grid box, or any superset of grid boxes, can be calculated. It is defined as

$$\Gamma = \oint_{C} \underline{V} \bullet \underline{ds}$$

where the circuit, C, is any simply connected contour around an area A. Figure 3-5 indicates how Γ is estimated. Using Gauss' theorem to relate the line integral to an surface integral we obtain

$$\Gamma = \int_{A} \underline{\omega} \bullet \underline{n}dA$$

in which $\underline{\omega} \bullet \underline{n}$ is the vorticity component normal to the imaging plane containing A, a mesh of the grid. Thus, dividing Γ by the area of the mesh results in the average vorticity, which can be assigned to the centroid of the mesh.

The range of flow speeds over which this technique can be used is very large. Figure 3-6 shows flow velocity versus the lifetime of a photoluminescent or photochromic chemical required to distort the grid lines 1 mm.

This is an upper bound on the desired distortion, but serves as a simple rule of thumb. On the figure we have indicated some of the

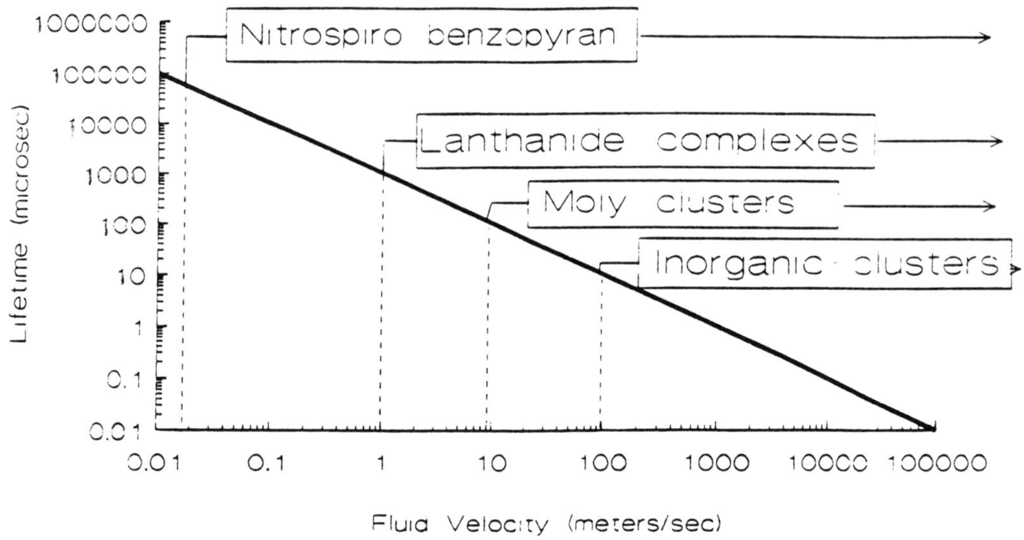

FIGURE 3-6. *Flow velocities required for a 1 mm movement of photochemically marked fluid during the lifetime of the chosen photoluminescent chemical. Thecompounds labeled with * have beendeveloped in our laboratories for these velocity regimes.*

chemicals that can be used in different flow regimes. The Kodak photochromic chemical 1',3',3',-trimethylspiro-8-nitro [2-H-1-benzopyran-2',2'-indoline] can be used in organic liquid flows; Flowmod's Flowlite I or II can be used in aqueous solutions; mixtures of phosphorescent gases have been used in high speed flows; and seeded phosphorescent particles have been used in moderate speed air and water flows. Table 3-1 and Figure 3-6 give additional details on these and other chemicals.

The capability of LIPA to obtain whole field measurements in gas flows makes possible detailed measurements in high speed flows. Most ultraviolet lasers pulse in nanoseconds, and the reactions leading to excitation occur over even shorter time scales. Thus, transonic, supersonic and hypersonic flow studies should be possible (the limitations will be determined by the capabilities of recording device). Miles et al. (1989) have pumped O_2 to an excited state, then deactivated it and traced a single laser line. If a grid were created using this technique, LIPA could be performed using excited-state oxygen.

The LIPA technique can be extended to provide three-dimensional information in the vicinity of a chosen plane in the flow. By imaging a single grid with two imaging systems (essentially stereo photography), we can obtain all three velocity components, and spatial gradients of the velocities formed in the grid plane. By creating two parallel grids, separated

CHEMICAL	TYPE	LIFETIME	EXCITATION	EMISSION	USE
Flowlite I*	luminescent	8.7μs	350-380	green	dissolves in water - high speed aerosol
Flowlite II*	luminescent	1.5ms	290-360	green	dissolves in water-not O_2 sensitive
Kodak (see text)	photo-chromic	10sec	350	blue	organic solvents
Crystalline phosphors	luminescent	μs-ms	250-450	visible spectrum	1-20 μm particles OK in air
SO_2	luminescent	1μs	308	yellow	in air at atm pressure
Biacetyl mixtures	luminescent	1ms	308 420	green	gas; no O_2

*Flowmod Corp., 301 Cowley, East Lansing MI 48823, USA

TABLE 3-1. *Representative chemicals that can be used for LIPA by doping them into air, water, organic liquids, and solids.*

in the third dimension by one mesh spacing, and viewing with two imaging systems each at 45 degrees to the plane of the laser lines, we can obtain all three components of vorticity, and all of the strain rate tensor components for every cube of the three-dimensional grid. By proper alignment of the grids and the cameras, a clear view to the intersections can be established, minimizing or eliminating the need for a third camera. In the next section we will amplify on these methods of data acquisition.

The LIPA technique is, furthermore, capable of a high degree of automation. Algorithms have been written which completely automate the process of finding the grid line intersections when the lines undergo small distortions (see Falco et al. 1989), and algorithms are currently being written that allow larger distortions. Because all of the information in the marked lines can be used to help find the intersection points, this technique is particularly robust, and it is especially suited for modern automated image analysis. In addition, the robustness is enhanced by the fact that the intersections from two *independent* successive images must be correlated. This reduces the probability of incorporating a spurious point (a result of noise or the algorithm being fooled) found on one data frame into the data set.

A final consideration is the data storage requirement. If the initial images are on film or video tape, each should be digitized to retain the maximum information stored on the original recording medium. However, immediately after finding the grid intersections, all of the digitized data can be discarded, except the coordinates of the grid intersections (the order of 100 - 1000 points per image) which are kept for subsequent analysis.

3.2.3. Accuracy of the Technique

LIPA is a non-intrusive measurement technique, which, in principle, requires only a clock and a ruler for calibration. Thus, it potentially can provide high accuracy results for flows over the entire range of speeds discussed above. An error analysis follows which indicates the expected error resulting from double differencing a grid to obtain vorticity.

Consider one mesh from the grid as shown in Figure 3-2. If the mesh width is ε and the maximum line displacement between the reference and data images is 1/2 of the mesh width, then after a time Δt the mesh will distort to the primed position. The spanwise vorticity component is defined:

$$\omega_z = \partial v / \partial x - \partial u / \partial y$$

$$= \left[\frac{\left(y_{1'} - y_{1''}\right) / \Delta t - \left(y_{2'} - y_{2''}\right) / \Delta t}{\left(x_{2'} - x_{2''}\right)} \right] - \left[\frac{\left(x_{1'} - x_{1''}\right) / \Delta t - \left(x_{4'} - x_{4''}\right) / \Delta t}{\left(y_{4'} - y_{4''}\right)} \right]$$

where,

$$\begin{aligned} y_{1'} - y_{1''} &= \Delta y_I = 1/2\varepsilon & x_{1'} - x_{1''} &= \Delta x_I = 1/2\varepsilon \\ y_{2'} - y_{2''} &= \Delta y_{II} = 1/2\varepsilon & x_{4'} - x_{4''} &= \Delta x_{II} = 1/2\varepsilon \\ y_{4'} - y_{1''} &= \Delta y_{III} = \varepsilon & x_{2'} - x_{1''} &= \Delta x_{III} = \varepsilon \\ t_2 - t_1 &= \Delta t \end{aligned}$$

the uncertainty in the measurement of vorticity $(\delta\omega_z)$ is thus,

$$\begin{aligned}(\delta\omega_z)^2 &= (\partial\omega_z / \partial\Delta y_I \cdot \delta\Delta y_I)^2 + (\partial\omega_z / \partial\Delta y_{II} \cdot \delta\Delta y_{II})^2 \\ &+ (\partial\omega_z / \partial\Delta y_{III} \cdot \delta\Delta y_{III})^2 + (\partial\omega_z / \partial\Delta x_I \cdot \delta\Delta x_I)^2 \\ &+ (\partial\omega_z / \partial\Delta x_{II} \cdot \delta\Delta x_{II})^2 + (\partial\omega_z / \partial\Delta x_{III} \cdot \delta\Delta x_{III})^2 \\ &+ (\partial\omega_z / \partial\Delta t \cdot \delta\Delta t)^2\end{aligned}$$

Estimates of the magnitudes of the terms are:

$$\begin{aligned}\partial\omega_z / \partial\Delta t &= -((1/2\varepsilon - 1/2\varepsilon)/\varepsilon)/(\Delta t)^2 \cong 0 \\ \partial\omega_z / \partial\Delta y_I &\cong 1/\Delta t \cdot \varepsilon \\ \partial\omega_z / \partial\Delta y_{II} &\cong 1/\Delta t \cdot \varepsilon \\ \partial\omega_z / \partial\Delta y_{III} &\cong (1/2\varepsilon - 1/2\varepsilon)/\Delta t \cdot \varepsilon^2 \cong 0 \\ \partial\omega_z / \partial\Delta x_I &\cong -1/\Delta t \cdot \varepsilon \\ \partial\omega_z / \partial\Delta x_{II} &\cong 1/\Delta t \cdot \varepsilon \\ \partial\omega_z / \partial\Delta x_{III} &\cong -(1/2\varepsilon - 1/2\varepsilon)/\Delta t \cdot \varepsilon^2 \cong 0\end{aligned}$$

Therefore,

$$\begin{aligned}(\delta\omega_z)^2 &= (\delta\varepsilon / 2\varepsilon\Delta t)^2 + (\delta\varepsilon / 2\varepsilon\Delta t)^2 \\ &+ (\delta\varepsilon / 2\varepsilon\Delta t)^2 + (\delta\varepsilon / 2\varepsilon\Delta t)^2 \\ &= 4(\delta\varepsilon / 2\varepsilon\Delta t)^2\end{aligned}$$

Finally,

$$\boxed{\delta\omega_z = (1/\Delta t)(\delta\varepsilon/\varepsilon)}$$

to the first order in ε. For example (see Figure 3-2) assume that the mesh spacing is 1000 μm, the line width is 100 μm, and that the image is digitized and analyzed in such a way that we have a 10% reading error associated with locating the center of the 100 μm line, i.e. δε = 10 μm, so δε/ε = 0.01. If Δt, the time between two frames is 0.01 sec, $\delta\omega_z = (1/\Delta t)(\delta\varepsilon/\varepsilon)$ = 1/sec.

The technique has been calibrated in an unsteady Stokes' layer flow by Falco and Chu (1987), for which the exact solution of the Navier-Stokes equations is known. These results show that strain-rate and vorticity can be measured to an accuracy of ± 1 sec^{-1}, which is consistent with the above analysis.

Although it may appear at first thought that the spatial resolution of LIPA is going to be less than that of particle imaging techniques, which can use very small particles, this is not necessarily the case. An important practical consideration is that the spatial resolution to which the location of the center of a single line can be found is not strongly coupled to the thickness of the laser lines or to the thickness of the ensuing photoluminescent lines (which are initially somewhat thinner). Because the marker is non-intrusive, a marked spot can be quite large compared to the point of measurement interest, as long as the center of the spot can be accurately determined from the image by image processing techniques. Although relatively thick lines may be used in LIPA, the motion of the field will not be constrained by the presence of the lines (irrespective of how large they are). This is, in general, not the case when a particle is used as a marker of the fluid, because the finite size of the marker distorts the flow it is in. During the time interval between marking and recording, the action of diffusion on the excited fluid lines is symmetric, simply spreading them about their centers. However, if the thickness of the lines is large enough that macroscopic flow field variations occur across them, the position of the centers will reflect an average that is biased towards the high velocity regions when imaged after the time delay. Thus, the laser line thickness and the time delay need to be optimized for the flow problem. As a result of these considerations, variations in the thickness of the laser lines across the measurement volume do not need to be considered separately. This lack of sensitivity to the thickness of the laser lines is important for solid-liquid flows because small index of refraction variations will focus and defocus the laser lines causing variations in their thickness along their length. Furthermore, the resolution of a vorticity measurement depends upon the mesh size of the grid, which can be adjusted independently of the grid line thickness. Finally, the field over which the measurements are made is also independent of these constraints, and depends only upon the number of grid lines created (here the key constraint is often the output energy of the laser).

When the lines are recorded on either film or video tape, the limiting resolution is dependent solely on the ratio of pixel size (or film grain) to the smallest length scales that need to be resolved in the flow. If the pixel size is small compared to the flow scales then the formation of relatively thick lines, on the order of 100 μm, imaged across ten pixels, results in greater accuracy than imaging thinner lines, on the order of 20 μm, across 2 pixels, when sub-pixel averaging is used to determine the centers. Furthermore, more advanced noise reduction techniques can be used on thicker grid lines without fear of breaking the individual lines into discontinuous segments. This is in contrast to PIV or holography where ~ 10 μm particles must be clearly defined in space, requiring enough pixels over these dimensions so that the particles are obviously separable from the noise in the image. Then enough particles must be averaged to allow correlation techniques to be implemented, which decreases the resolution to the order of a millimeter (if there are scales smaller than a millimeter, their contribution may not be evenly weighted by the particles imaged in that area). Thus, these considerations relax two important aspects of an experimental design for LIPA, the laser line width and the resolution of the imaging system.

3.2.4. Dynamic Range and Timing

One of the drawbacks of fluid or particle tracking techniques is that they can present limitations on dynamic range in flow situations where strong velocity gradients are present. In a quasi-uniform flow there is, of course, no problem because the time between frames can be chosen to obtain optimum displacement of the grid lines. If strong gradients are present, LIPA can avoid most of the limitations on dynamic range through a simple tactic. When used with a chemical that works in Excite-Self-Decay, or Sensitize-Delay-Excite modes, where the shuttering is done by the imaging system by using an imaging rate that will be appropriate for the highest velocities to be encountered in the field of view, the dynamic range is extended to be as wide as needed. To measure the slower velocities, the investigator simply skips an appropriate number of frames to allow the lower velocity regions to move the marked fluid over a distance predetermined to give the desired measurement accuracy. If the Excite-Forced-Decay mode is being used, the laser that forces the excited state to decay must have a repetition rate high enough to accomplish the task.

Figure 3-7 indicates an example of a typical timing sequence for experiments in the Excite-Self-Decay mode. In this sequence the high speed camera sends a signal every four revolutions to the laser, which pulses in nanoseconds, exciting the chemical in the fluid. Here, a chemical with a 1

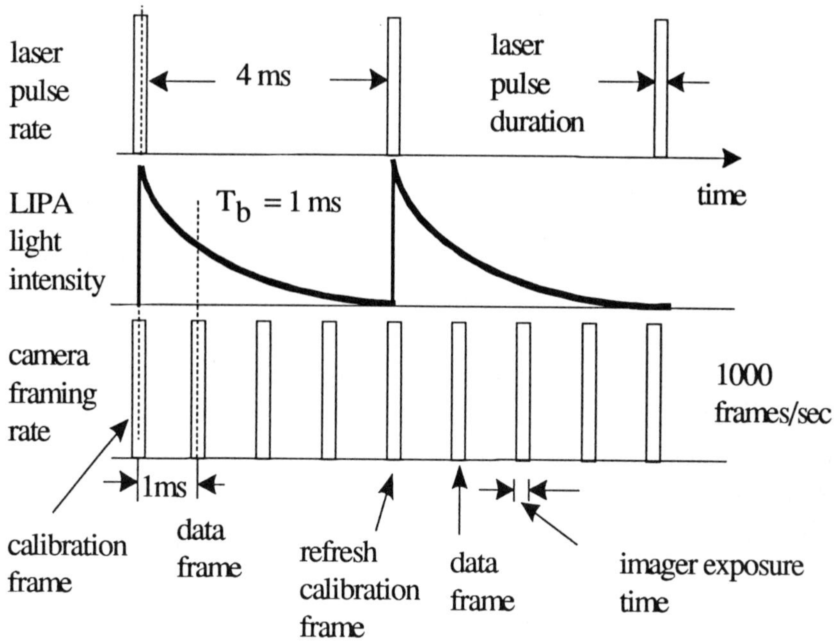

FIGURE 3-7. *A typical timing sequence for experiments of the Excite-Forced-Decay mode in which a high speed imager is used to record the results.*

ms lifetime is shown, which is suitable for a 1000 frame/sec imager (film camera or solid state), and a 250 Hz laser. In this timing arrangement, the first frame is the calibration (containing the laser energy), and the subsequent three frames contain data, with the highest quality data frame being the one after the laser is triggered. The overall decay of the chemical is indicated to be four framing intervals. The optimum framing rate is governed by the flow, as is the lifetime of the chemical chosen. If a high shear region exists, the dynamic range can be extended by simply increasing the framing rate, to have frames close enough to account for larger displacements per unit time in the high velocity regions.

3.2.5. Some Single Phase Applications

Important uses of LIPA in single phase flows focus on turbulence and the dynamics of unsteady flows. The LIPA technique (in the form first described Falco and Chu (1987), who used the Excite-Self-Decay mode) has been calibrated by measuring the velocity and vorticity in a Stokes layer, a flow which has an exact analytical solution. The results matched the exact solution to within the error suggested by the analysis described above.

Whole field velocity and vorticity distributions have been obtained for vortex rings by Falco and Chu (1988). The circulation in a starting vortex behind an impulsively-started airfoil was quantified by Falco et al. (1988). LIPA has been used to measure the skin friction on a moving compliant surface by Falco and Chu (1988), and the vortex induced motions near liquid/gas interfaces by Chu et al. (1991). It has also been used to measure streamwise vorticity in a study of the mechanism of turbulence production (Chu 1987). All of these experiments were recorded on film. They used photochromic chemicals dissolved in kerosene.

LIPA has also been performed in water, looking at elliptical turbulent jet flows (Falco, unpublished). For this study a photoluminescent chemical made at Michigan State University called Flowlite II (see Table 3-1) was dissolved in water to achieve a concentration of 10 ppm, and the results were imaged with an ITT intensified CID camera. Figure 3-8 is a still photo of the calibration grid which indicates the quality which can be achieved of grids which can be made in water/Flowlite II solutions.

It has furthermore been used in the gas phase to study scavenging in two-stroke engines (Hilbert and Falco 1991). They used biacetyl in a gaseous state, and nitrogen as the working gas. Images were made on an intensified CID array camera. LIPA is also well suited to very high speed flows. Bonnet and Falco (1989) studied the flow emanating from a 1 cm nozzle with exit Mach number equal to 3. These proof of concept experiments resulted in grids of the distributed ice particles in the combustor type flow field. The distorted grids captured on a Hadland image converter camera during 200 ns exposures at 1 μsec intervals were dense enough to validate the use of the technique for these flows.

3.2.6. Advantages of LIPA Measurements

The advantages of LIPA are summarized below. One aspect of the LIPA technique that should be singled out is the exact way in which it incorporates the effects of three-dimensionality of a flow field. When a single grid and a single imaging system with its imaging plane parallel to the initial grid plane is employed, we obtain the projections of the three-dimensional distortions of the initially planar grid onto the plane of the imager. This only requires that the distortion in the third dimension remain within the depth of field of the imaging system. Thus, LIPA does not have the inherent difficulty of laser sheet confined particle tracking techniques, such as PIV, where either the first or second image of a particle may not be recorded because its trajectory carries it through the boundaries of a light sheet due to three-dimensional motions.

In summary, the advantages of the LIPA technique are:

- Instantaneous measurements of velocity, velocity gradients, Reynolds stresses, vorticity and strain-rate made over an area,
- Direct wall shear stress measurements,
- Interpretation of all results is simple and direct,
- Simple calibration depends only on a clock and a ruler,
- Simple data acquisition,
- Not limited by interferometric quality access,
- Data reduction can be automated,
- Data storage is minimal,
- Can be set up to give highly accurate space-and-time-resolved measurements over a large range of velocities,
- Non-intrusive, and particles are not necessary,
- Can be used in Newtonian or non-Newtonian gases, liquids or solid / liquid mixtures,
- Three-dimensional measurements of the velocity field, the complete stress tensor and vorticity vector over an area are possible (as we describe below),
- Standard wavelengths of ultra-violet and blue lasers can be used to excite many chemicals with long lifetimes,
- Quantitative measurements can be combined with flow visualization of the doped chemicals,
- Mixing can be studied by using more than one chemical,
- Potentially can be combined with LIF (Laser Induced Fluorescence) to give concentration information along with the kinematic data.

The chief disadvantages of LIPA are:

- Specific chemicals must be found for flow velocities and fluids of interest,
- Cost of the chemicals,
- Need for ultraviolet light to excite many chemicals.

We will see that these attributes of LIPA form a foundation that can be enhanced to result in a technique that can enable systematic studies of solid-liquid two-phase flows as outlined above.

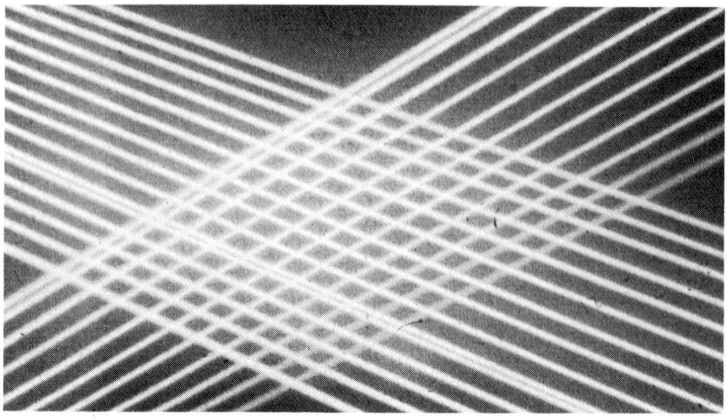

FIGURE 3-8. *A still photo of the calibration grid made in water. Ten ppm of Flowlite II in water produces a grid with lines that are of fairly uniform intensity, and 1.5 ms lifetime.*

3.3. LIPA FOR SOLID-LIQUID MIXTURES

LIPA can be extended to investigate the flow dynamics in the interior of refractive index matched solid-liquid flows. The extension is based upon two ideas. The first is that by index matching the doped two-phase mixtures, light can pass through the mixtures in straight lines, enabling essentially the same technique to be applied. The second idea is to dope the phases of the mixtures with photoluminescent chemicals that radiate different colors. Color can be used to distinguish between the solid and liquid phases. It can even be used to distinguish particles of different size distributions from the liquid they are in.

We will discuss the use of LIPA to measure the average flow properties of a solid-liquid mixture, and to separately but simultaneously measure the dynamics of the liquid and the solid phases. We then extend the discussion to procedures for three-dimensional measurements of the above cases. Although LIPA can be used for three-dimensional measurements in single phase flows, these techniques are described in this section because of the particular value of such measurements in solid-liquid flows. Next, techniques which enable simultaneous flow visualization and LIPA are discussed. This is followed by an outline of the measurement of wall shear stress using LIPA. LIPA imaging issues are then considered. In all of the above scenarios, once the flow is index matched and doped with photoluminescent chemicals, qualitative flow visualization of the mixture as a whole, or of either phase can be made, by putting a sheet of laser light into the flow, instead of a grid of laser lines.

3.3.1. Refractive Index Matching

The use of LIPA in solid-liquid flows requires refractive index matching. This constraint is also essential to the use of LDA or particle imaging techniques for measurements in solid-liquid flows. Majumdar et al. (1987a, b) in their work on stereo particle imaging discuss the precision with which index matching must be done for large particles. Nouri et al. (1987) have discussed the constraints put on LDA measurements, and have, so far, been able to make measurements in flows with concentrations up to 14% by volume. With careful matching it appears possible to use LDA at higher concentrations (Abbas and Crowe (1989) worked with concentrations to 30%). However, LIPA is inherently less sensitive to the consequences of index of refraction mismatch than LDA. Three important index matching issues are: the error incurred by variability in the degree to which beams coming through different regions of a flowing mixture cross to form intersections of the grid, the accuracy in locating a beam crossing position when there is variation, and the effect of refractive index induced beam width variations. In the following, we explore the crucial matter of the consequences of index of refraction variations in some detail.

We first note that LIPA does not require that the beams cross perfectly, or even with the same degree of error from instant to instant. Figure 3-9 shows a perfect beam crossing (I), and examples of possible alignment mismatches (II-V) that may occur from instant to instant in a somewhat imperfectly matched solid-liquid flow. The errors that these misalignments cause for LIPA are entirely due to the difference of the location from the hypothetical position of the grid plane. The index of refraction differences that caused the displacements of the laser lines do not affect the subsequent motion of the photoexcited lines by the fluid. As long as there are no scales in the flow of the order of the misalignment distance, there will be no additional relative displacement of the two lines about the intersection region. This error in the location of the grid intersections normal to the hypothetical grid plane is defined by the difference of the center of area of the locus of the circumferences of the intersecting laser lines, and thus is at a location outside of the hypothetical grid plane which is, at maximum, one half of a beam width -- if the beams have any common intersection. Since the photoluminescent molecules at the laser beam

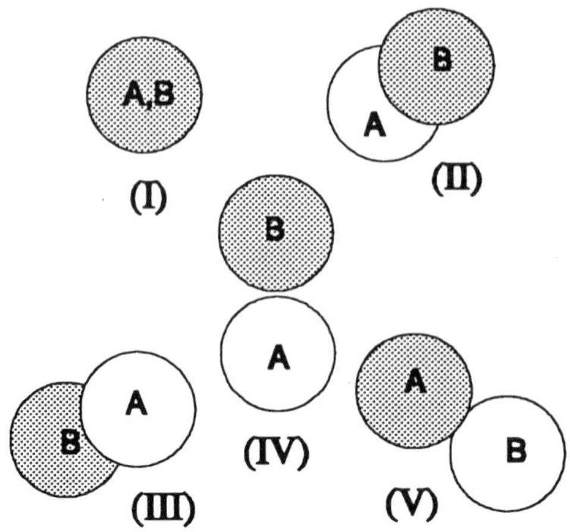

FIGURE 3-9. *Cross-sectional views of two crossing laser beams (drawn as circles) of the LIPA grid. (I) represents a perfect crossing, while (II-V) represent a range of probable changes in the crossings with time due to unavoidable refractive index variations that may exist in the flow.*

intersections receive more incident energy, they typically photoluminesce more brightly (as much as doubling the brightness), or undergo a stronger color change, so that a missed crossing can be detected in any data field. Thus, if desired, we could use the intensity of photoemission to detect whether beams have crossed in the third dimension, and establish a discriminator level, below which the data is not used. However, in general, even at a location where the beams do not cross (see Figure 3-9 IV, for example), in all but the most severe circumstances they will be close enough relative to the smallest scales in the flow (which would be the order of the grid's mesh size by design -- typically 10 times the line width), that we will obtain valid data, with the aforementioned decrease in spatial resolution in the third dimension. If the distortion were the order of the grid mesh size, this would be observed in the grid plane as well. Thus, in severe situations, the in-plane information could also serve as the basis of a criterion for rejecting a data point (or data set).

In general, there is no loss of resolution in the plane of the imager because the projections, although distorted by the (ever present, small) index mismatch, will appear to the imager to cross. Since it is only the difference in the positions of the crossing, over the allowed time interval, that is used to obtain LIPA data, in the plane of the imager there is no apparent problem. Thus, in the two-component data obtained from a three-dimensional flow field using the LIPA technique, the index of refraction variations show up only in the fact that the data is actually at a slightly different coordinate position in the third dimension. If we are ensembling the results of many realizations, then position errors similar to those created in the third direction become errors in positions of the intersections of the various members of the ensemble in the grid plane, because the laser lines are refreshed and sent into the fluid for each member of the ensemble, and the index of refraction fluctuations may cause differences between realizations. Of course, the index of refraction differences between realizations (which are typically only small fractions of a second apart) may be less than any differences within a given realization.

Since running a typical solid-liquid flow loop test requires significant time, it is difficult to keep the temperature constant over the course of a test, and one would expect small fluctuations in index of refraction from instant to instant. However, as we have suggested, LIPA can be used to make very useful measurements under these circumstances. It is important to note that LDA would result in loss of information, or bad information, for all options in Figure 3-9 except the perfect crossing. And, even if the beams do cross sufficiently well to get a good signal, one does not have information on the position change of any of the measurement volume coordinates, as is naturally available by using the LIPA technique. Furthermore, these variations will likely be accompanied by variations in the angles of beam crossing, and positions of the two beam waists, which are important to know in the LDA technique. Changes in the angle of beam intersection are not important in the LIPA technique because, once again, only the relative positions between time intervals matter. Comparative experiments to determine the relative insensitivity to index of refraction variations of LIPA with other techniques, particularly LDA, have not yet been performed.

An issue that is of particular importance for LIPA in index matched solid-liquid flows is that only one wavelength of light can be matched exactly, but at least two are needed for LIPA; the excitation laser wavelength, which is typically in the ultraviolet (see Table 3-1), and the visible emission from the chemicals. We propose the following procedure to reduce the distortions. Index match the solid-liquid mixture to optimize the transmission of the incident laser beam wavelength. This results in the best quality excitation. Immediately after excitation, the chemicals used will emit visible light, which will not be perfectly index matched on it passage out of the mixture to the imager. Since the index mismatch will result in the light appearing to be at slightly different position than the laser beam, the first image should be made immediately after the laser pulse. We should use this, rather than the laser lines as the reference grid. Then the grid is imaged again after the flow has moved it an appreciable amount. The displacements are calculated between these two images, and the results can be referred back to the geometric location of the crossing laser lines. Since the grid displacements are only a fraction of a grid mesh, the additional changes in the index mismatch due to grid motion by the flow will, in general, result in negligible additional error. Obviously, the development of chemicals that can be excited closer to the visible spectrum, or even within it, would reduce or eliminate the need for this procedure.

3.3.2. Data Acquisition

3.3.2.1. *Average Flow Properties of Either Phase, or of the Mixture*

To measure the average flow properties of the mixture, it is not necessary to distinguish between the liquid and solid phases. The procedures for the use of LIPA are thus identical to procedures in single phase LIPA discussed above, except that an index matched mixture must be created, which has an appropriate chemical doped into the liquid or into the solid phase or into both phases. Appropriate chemicals that enable us to index match the phases after doping are discussed below. Although both phases can be doped with the same chemical to obtain average flow properties, it may be convenient to dope each phase with different chemicals to enable a mixture to be used for other experiments in which the capability of distinguishing them is needed (see below). For the present purposes, one would simply not distinguish the phases either with the imaging system or during the data analysis. Under this doping arrangement, if it is desirable to measure the flow properties of either the liquid or of the solid, this can then be done by the simple addition of a filter that transmits either the wavelength of the chemical doped in the liquid or the solid, in front of the camera. A schematic of the optical arrangement is shown in Figure 3-10 (in this drawing and the following ones flat-sided index matched pipe enclosures, which would be helpful to reduce near-wall distortions of received light, are not shown). We will use the schematic convention of portraying the film size by the camera size. Typically, object-to-image magnification would be in the range of 5:1 to 0.5:1.

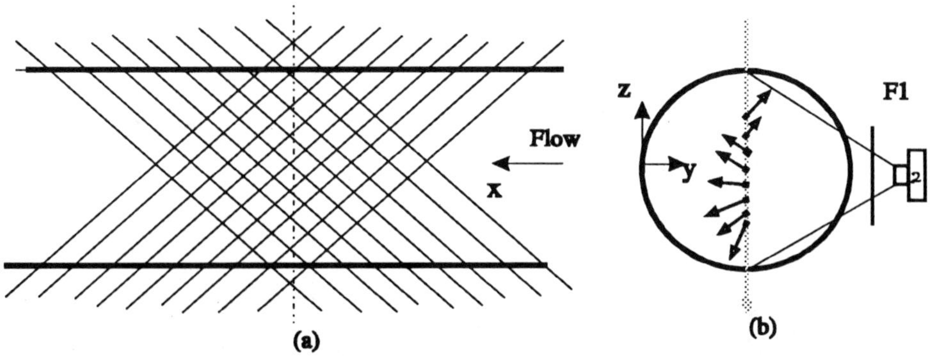

FIGURE 3-10. *LIPA arrangement to obtain average flow property measurements of a solid-liquid mixture or flow property measurements of each component separately. a) Axial slice (x coordinate direction) in the plane of the initial laser induced grid of lines, b) slice perpendicular to the pipe axis (y-z plane), showing possible out of plane motion that may occur, and the field of view of the camera. For measurement of the combined properties, both phases are marked and left indistinguishable in the analysis. With the addition of a filter (F1), the investigator could choose to measure only the response of the solid or the liquid phase. The shaded line is the end view of the laser grid. (In this drawing and the following ones flat-sided pipe enclosures, used to assist in index matching the pipe wall material so as to reduce near-wall distortions of received light, are not shown).*

3.3.2.2. *Simultaneous Measurement of Both Solid and Liquid Phases*

To enable separation of the dynamics of the solid and the liquid phases, or even a further separation of the dynamics of different size distributions of particles, the constituents to be distinguished need to be doped with chemicals that radiate different colors when excited. The correct choice of chemicals can enable excitation with the same laser radiation. Thus, only one grid need be impressed into the mixture. A discussion of the chemicals for doping is deferred to the next section. Here we shall assume, for ease of discussion, that the solid phase has been doped to radiate red and the liquid phase has been doped to radiate green.

Figure 3-11 shows an image acquisition arrangement that can give simultaneous solid and liquid phase dynamics. It consists of using a split field view for synchronization, to lower potential distortion (especially if film is used), and for economy. The color filters in front of each mirror, M1, will transmit light coming only from the solid or only from the liquid phase. Thus, on one half of each data frame we get the data for the liquid phase, and on the other half we get the data for the solid phase. Although we

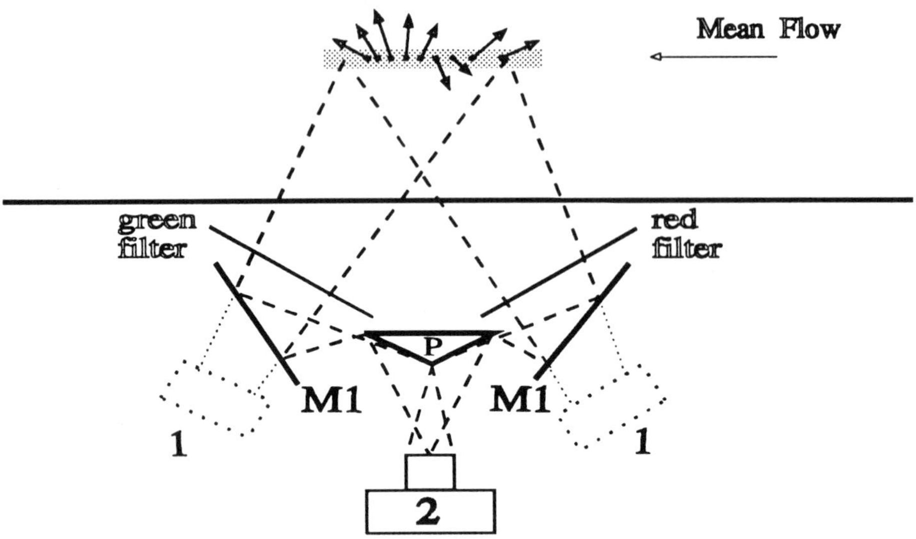

FIGURE 3-11. *Plan view of a data acquisition arrangement to enable simultaneous measurement of both liquid and solid phases; the grid is formed perpendicular to the page in the plane containing the shaded line, and would look like that in Figure 3-10 from the side. The mirrors eliminate the need for two imagers (indicated by dotted lines (1)). Arrows indicate distortions of the grid normal to its plane of formation, which determine the imaging system's depth of field. The shaded region width indicates the needed depth of field, but the arrows are drawn longer to make the effect clear to the reader. If an arrow is green (color emitted by the fluid, say), then its emitted light will pass through the green filter, left side mirror (M1) to the face silvered prism (P) and to the right portion of the real imager (2). The light emitted from the particles, which if red, will pass through the red filter and the left portion of the imager, will record the velocity of particles (which may be anywhere along the grid lines). Dashed lines indicate the field of view of the imaging system, the overlap of which provides simultaneous data.*

are using only one camera (either black and white film or electronic imaging -- or color film or color electronic imaging can be used), the optical setup is identical to an arrangement of two cameras (which are indicated by the dotted outlines at the positions where the apparent cameras would be). These cameras are both looking at the flow at a slightly different angle. (The angle is exaggerated in Figure 3-11). Thus the depth of field of the lens must accommodate the sine of the half angle of the image separation of the in-grid-plane field of view (for example, 5 degrees will result in a depth of field of 0.087 times the field of view). Furthermore, the components of the velocity recorded will be slightly different for the particles and the fluid using this arrangement. The differences can be kept small by reducing the

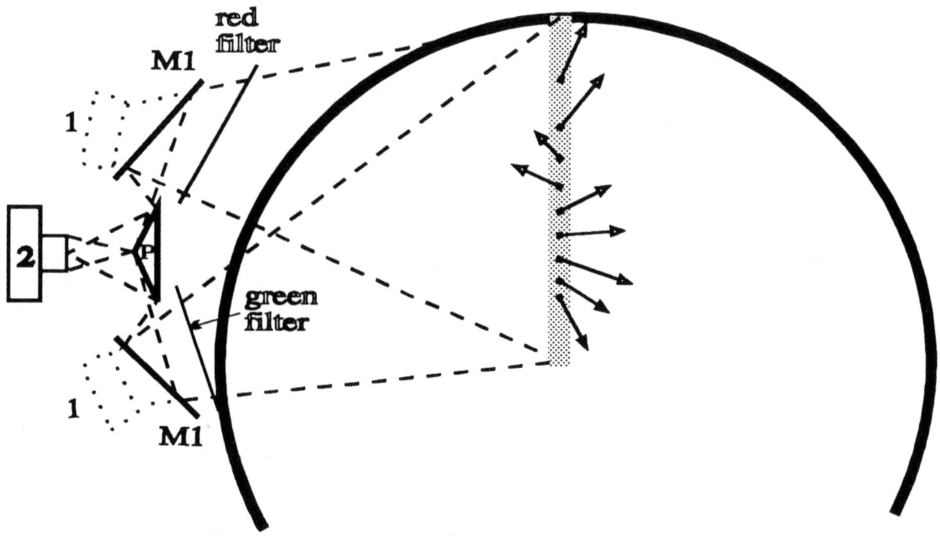

FIGURE 3-12. *Use of the above configuration to measure axial and radial components of the fluid and the particle velocity fields. This arrangement puts the left and right imaging differences in the same direction as the mean shear, increasing measurement error, but has the advantage of requiring minimal optical access. Nomenclature is the same as for previous figure.*

angle between the pseudo imagers to a minimum. From the point of view of flow dynamics, this arrangement is preferable to that shown in Figure 3-12 (which corresponds to rotating the assembly 90 degrees), because the differences in the relative accuracy of the left and right imaging are in the direction of the flow in which there are either no average differences or, at worst, slow changes (moving dunes for example). The above orientation also keeps the imaging optics and camera differences constant in the direction of mean shear, thus allowing the greatest accuracy.

The view of Figure 3-12 is one in which the flow is coming out of the page, and we are looking end-on at the grid (so we don't see the mesh). This configuration has the advantage of needing only a small ultraviolet transmitting access window for laser light access, and a disc (if a pipe flow is being investigated) of glass or plexiglass for recording the images, and thus it allows measurements to be more easily made in existing metal flow systems.

Another configuration, shown in Figure 3-13, can be used to eliminate the differences inherent in any left-right sided arrangement. It still only requires one imager. The view here is again one in which the flow is coming out of the page, and again we are looking end on at the grid (so we don't see the mesh). The left side mirror/filter arrangement will get the same view of fluid velocities as the right side mirror/filter arrangement gets of the solid particles, since the depth of field is exactly the same. This relaxes the constraints on the lens, enabling faster lenses to be used. The investigator would set the depth of field to include the maximum expected excursion of the fluid or particles from the initial plane of the grid lines (these excursions are indicated by the arrows). In this arrangement, both left and right side images have exactly the same precision at any position across the pipe, so a larger field of view is possible. Essentially the same visual access is required as for the configuration of Figure 3-12.

Of course two imagers can be used in an arrangement where they are on opposite sides of a pipe, for example. Figure 3-14 illustrates one possibility. The use of two imagers doubles the imaging resolution (the full imaging resolution can be used for each phase), and has the additional benefit of enabling the chosen delay time to be different for each imager. In a flow regime where a moving bed existed, it might be of interest to have a longer delay time for the solid phase measurements, since the solid would be moving at a different average speed than the liquid. The disadvantage is largely in the costs of the additional imager and associated synchronization electronics.

In general the use of filters as described above overcomes the need for carefully distinguishing different colors on a single piece of color film or on a color video camera. Because the phosphorescent chemicals used in LIPA emit at well defined frequencies, they can be sharply bandpass filtered. Without the filters, the excited lines would appear to have a brownish-orange color in general, because there is a mixture of both fluid and solid particles along each line and at the intersections of the grid, and thus a mixture of the two colors red and green, (using our color scheme). It is possible to filter after imaging, but the non-linear distortions of the imagers reduce the accuracy of this approach. Furthermore, filtering at acquisition time enables the use of black and white film, image intensifiers or image converter cameras, which don't distinguish color.

We note a few additional points. First, if the highest accuracy is required near the wall, the optical axis of the imaging optics should be located near the wall. For this purpose the symmetrical arrangements indicated in the above figures would be rearranged (in the sense that Figure 3-12 has been). Second, each planned limitation of the size of the field of view will enable higher resolution within the more limited field of view. Third, a simpler system, in which there is no spatial overlap of fluid and

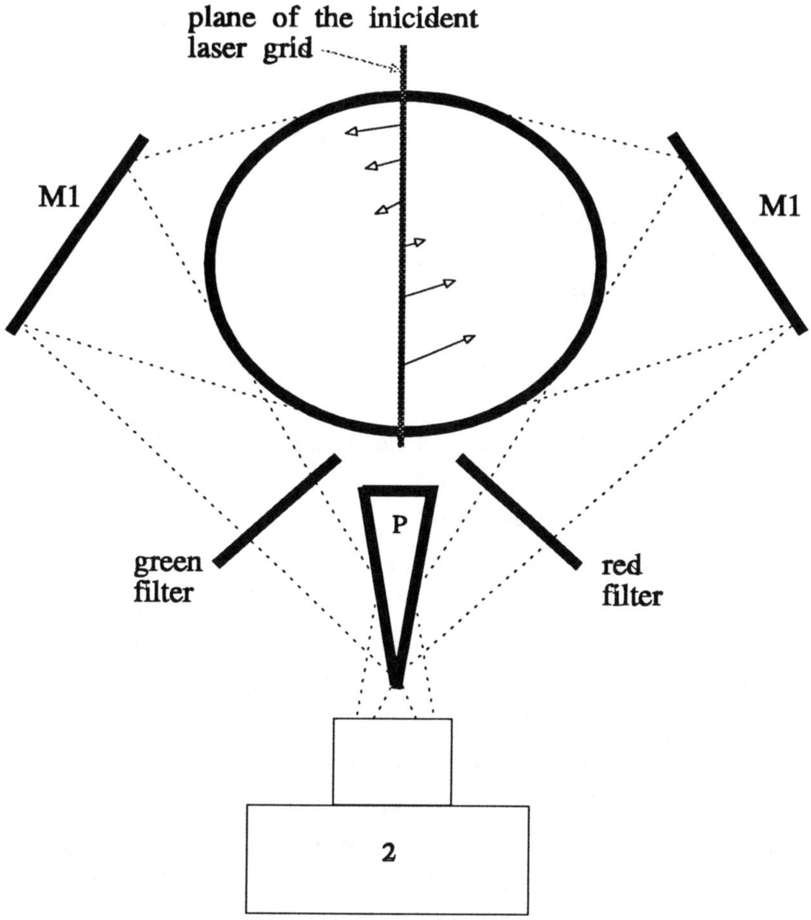

FIGURE 3-13. *Another data acquisition arrangement to enable simultaneous particle and fluid velocities using LIPA. Here the optical paths for both the fluid and the solid are the same, enabling a wider field of view without 'left/right' imaging biases. The apparatus identification numbers are the same as above.*

particulate information, i.e., the fluid velocities measured in the left half of the pipe and the particle velocities measured in the right half of the pipe, could simply use color filters to split the field and record particle and fluid velocities on different halves of the imager, without the use of the above optical systems. This arrangement could be used to obtain long time averaged statistics for each phase. Fourth, it should be emphasized that if the resolution chosen is sufficient to record the motion of individual particles, then particle velocity information will be available all along the paths of the excitation laser radiation, not only at the grid intersections. Finally, as indicated above, the use of color to separate phases, can be extended to include the separation of particles from different size distributions, as well as separation of the phases. If more than two colors are to be separated out using filters, it may be advisable to use multiple imagers.

Ensemble averaging the data obtained by these methods at each intersection point over a large number of images will result in the velocity, vorticity and stress fields of the particles of the liquid. Correlating data from two grid intersections of either marked liquid or marked solid particles can yield pair distribution functions. Because we are making multi-point measurements, we can also obtain three-point correlations, and other higher order statistics of the information of the components in the grid plane.

3.3.2.3. *Three-Dimensional Measurements Using LIPA*

Three-dimensional measurements of velocities and stresses of both solid and liquid phases, and of the strain-rates and vorticity of the fluid phase are inherently possible with the LIPA technique. The key enabling practical measurement of these quantities is that the data generated <u>is limited</u> to the spatial locations excited by the incident laser radiation (i.e. the grid of lines, which initially forms a plane). This provides an excellent compromise between the need to very sparsely load a field when doing three-dimensional stereo photography, and the need to have a high loading so particles (marking the fluid, say, as needed for holography, for example) are spaced close enough together so that, on average, the required spatial resolution is obtained. This capability to non-intrusively specify the three-dimensional localization of data that is desired, is a unique advantage of the LIPA technique. It results in manageable data acquisition and reduction, and is potentially fully automatable. Furthermore, high data rates can be obtained, which may prove to be more important for essentially three-dimensional dynamics than for two-dimensional flows. Reiterating, because the laser lines serve as a selector of a <u>subset</u> of the total number of fluid or solid particles in the flow field being measured, not as a delimiter of the space in which these particles may move, we can record a number that is easily reducible, while achieving high spatial resolution.

3.3.2.3.1. *Three-Dimensional Velocity Fields.* To measure all three velocity components simultaneously using LIPA, two views of the images are required as is the standard case for stereo photography. An important difference, from the use of two views to distinguish phases, or particles from different size distributions, is that the two views should be significantly further separated, optimally being at 90 degrees to each other, for best accuracy in the third dimension.

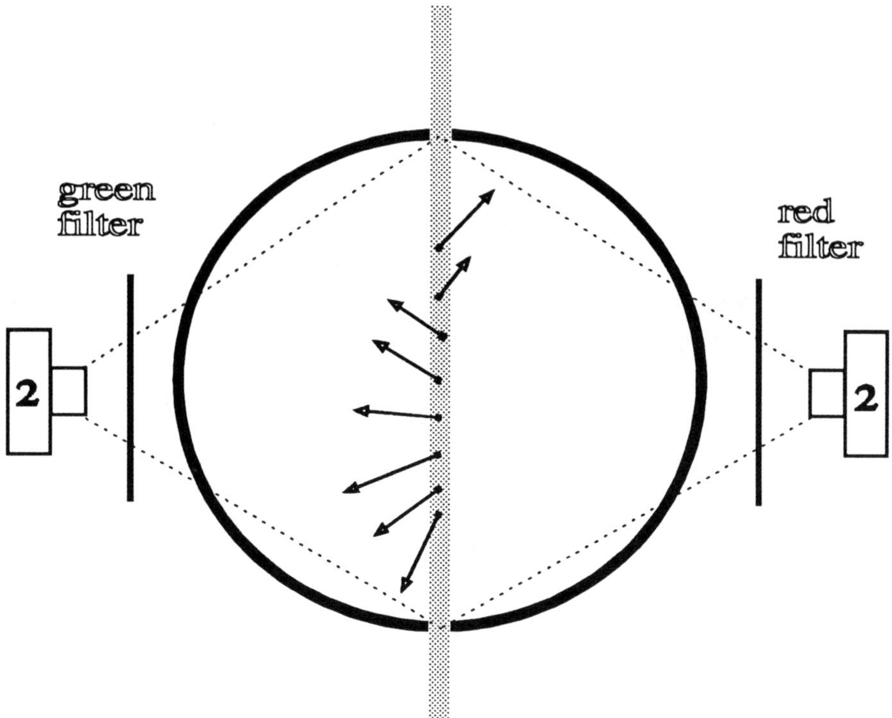

FIGURE 3-14. *The use of two imagers to obtain simultaneous solid and liquid velocities with LIPA.*

Figure 3-15 shows one possible arrangement, similar to that used by Falco (1972). It again uses face silvered mirrors and a prism to redirect the images to a single imager (instead of a pair of cameras indicated by dotted lines). As in several of the arrangements discussed above, this has the advantage of using a single frame on a roll of film, or a single solid state imager, which essentially eliminates synchronization and distortion problems. It also requires only one camera, which has important economic implications when the imager is a gated intensified CCD or CID array which, as we will see below, is likely to be needed.

An important constraint of stereo photography (as opposed to holography) is the limitation of depth of field, which is essentially no more than that of the optics on the receiving camera at the required magnification and lens aperture. For a given magnification the depth of field limitation can be overcome through the use of small aperture optics. This can work for LIPA largely because all of the information is confined to the region about the plane of the initial laser grid excitation, so the required depth of field will be proportional to the sine of the angle between the views times the extent of the field of view in the grid plane (see above). To compensate for the light loss, very sensitive film for slowly moving flows, or a gated image intensified camera for higher speed flows should be used. The additional compromise is between the need to have high magnification (to image individual grid intersections or particles) and the need to have large depth of field. Several approaches are possible:

- determine that the region of interesting three-dimensional information is local, say, proportional to the order of 100 particle diameters for 10 μm particles. This establishes essentially a "local" measurement technique, or,
- for examination of the liquid phase, use laser beam widths that are 10 times greater than the resolution of interest, say, so that spatial resolution of the camera system can be relaxed, then recover the resolution by subpixel averaging, or,
- relax the 90 degree separation angle of the two views, lowering accuracy of the third dimension.

The choice will depend upon the needs of the experiment. For example, for many experiments accurate three-dimensional measurements near a local region are adequate. In others less accurate information in the third dimension to establish a qualitative understanding is adequate.

3.3.2.3.2. *Three-Dimensional Vorticity Vectors, the Reynolds Stress and Strain-Rate Tensors.* To enable the measurement of these three-dimensional quantities with LIPA, a second grid must be placed behind the first one, displaced one mesh width in the third dimension. Figure 3-16 shows the arrangement in plan view (the y-z plane). Corresponding intersection points on the two grids combine to give the corners of a cube (it is, of course, not necessary to have the laser write lines in the third direction to construct the corners of the cubes). The left and right lines of sight are such that the grid lines on the rear grid are seen halfway between the forward grid lines when the mirrors are at 30 degrees to the axis of symmetry. It is furthermore necessary to angle the imaging apparatus a few degrees about the x axis to enable the lines parallel to the y direction of the rear grid to be viewed. Alternatively, the rear grid can be displaced up (or down) one-half a mesh width to prevent occultation. The same data acquisition optics as used for

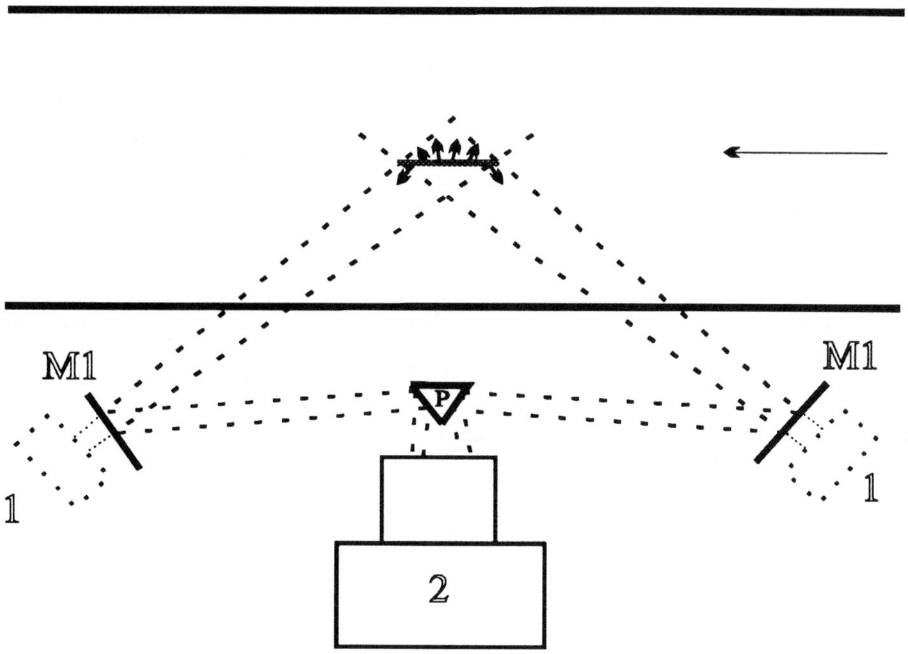

FIGURE 3-15. *Measurement of the three velocity components by using LIPA. A four mirror system (Falco 1972) is shown, which will result in two orthogonal views of the illuminated particles that are initially in the laser sheet. After excitation these particles will move both in the plane formed by the crossing laser lines and out of it (movement out of plane is indicated), and their three-dimensional images will be recorded. Movement out of the initial plane in which the grid was formed (in the third dimension) is indicated by the vectors. The arrangement results in a split field view on the recording device.*

the measurement of the three velocity components can be used for these measurements. Here the trade off is again in the direction of observing a more limited space to increase the measurement accuracy. Similar to the situation for the measurements of three velocity components, it appears that for this application only a few grid cubes would be appropriate. Thus, this becomes a 'point' probe capable of measuring all three vorticity components, all nine Reynolds stress tensor components, and all 27 strain-rate components. Data reduction will still require only one differentiation.

We have illustrated how both three-dimensional velocity measurements, and three-dimensional vorticity, stress and strain-rate measurements can be made in very dense solid-liquid systems by using LIPA. Because the mixtures must be index matched, and because we must have both the solid phase and the liquid phase doped with photoexcitable chemicals, we have an unparalleled opportunity to make three-dimensional measurements. It should be mentioned that a further extension is even possible. One could conceivably measure, separately and simultaneously, the particle and the fluid velocities in three dimensions. This will require two cameras, both of which are used in the split field mode, combined with color filtering.

3.3.2.4. *Simultaneous LIPA and LIF*

It is possible to do a void fraction analysis (or a concentration analysis of particles from a specific size distribution), and to obtain velocity fields at the same time, by simultaneously doing a LIPA and LIF experiment. LIF (Laser Induced Fluorescence), is a technique for measuring concentration, by recording the intensity of the emitted light from a doped fluid or solid. The intensity is essentially linearly related to the amount of dye at a given location. It is possible to excite a fluorescent dye with another laser over a sheet that is co-planar with the plane of the LIPA grid. A fluorescent dye can be mixed in the liquid along with the LIPA chemicals, and/or a fluorescent dye can also be doped into the solid phase, along with a longer radiating substance for LIPA. If properly chosen, a single laser can be used to excite all of the chemicals. We have used 308 nm radiation to excite both sodium fluorescein and Flowlite II (see Table 3-1). The laser beam was divided into a portion that produced a sheet for the LIF, and a portion that produced the grid lines.

The data acquisition configurations shown above (Figures 3-11 to 3-14) that can distinguish different phases can be used to distinguish the LIF and the LIPA data. In the case where we are interested in the average statistics of the flow, for example, the left camera can have a filter that only admits the LIF information and the right path can have a filter that only admits the LIPA information for the mixture. With a fast framing imager (the order of 1000 frames per sec), it is possible to use a ultraviolet filter and capture both LIF and LIPA data using a single imager. In this case, the LIF, because of its short lifetime can be captured on the frame synchronized to the laser firing (or immediately afterwards, within the lifetime of the dye fluorescence lifetime), while the LIPA data can be recorded on subsequent frames. Thus, by using fast framing imagers (for example, drum cameras, or Kodak's Ektapro 1000), it is possible to capture the LIF data and both fluid and solid phase velocity information simultaneously. Essentially all the data discussed in Section 3.2.2 would be obtained, but because of the more rapid imaging, additional frames would contain the LIF information in between

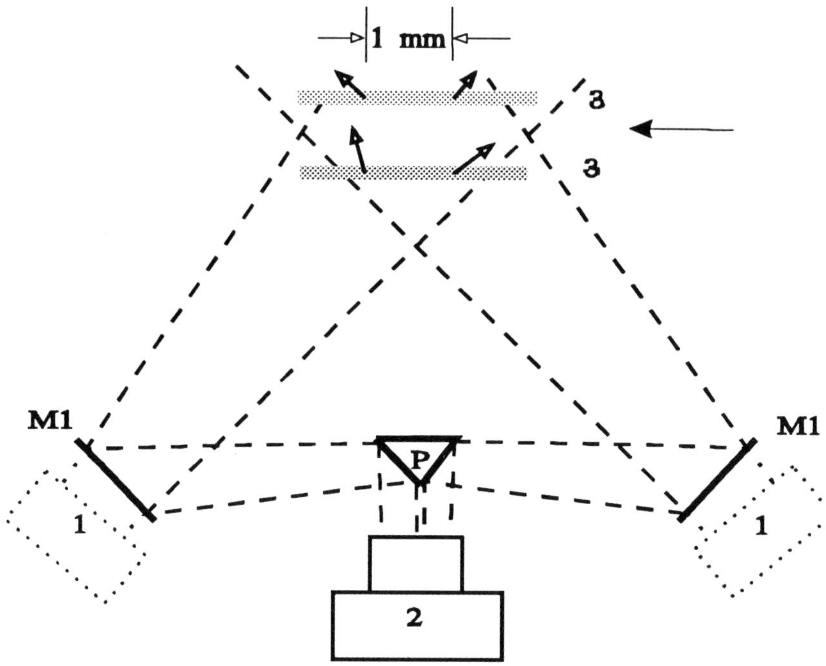

FIGURE 3-16. *Arrangement to measure the vorticity vector, the Reynolds stress tensor and the strain-rate tensor in solid-liquid systems. The volume covered is only a few grid cubes, making this essentially a local measurement. The shaded lines (3) represent the end views of the two parallel laser generated grids. The vectors are at positions of the grid line intersections.*

LIPA data frames. In this scenario, the LIF dyes used in the liquid and solid phases should pass through the same color filters that the LIPA dyes pass through. The high framing rates would, in general, result in many more frames than required for the LIPA analysis, and these would be discarded.

By relaxing the constraint of quantifying the concentration or determining macroscopic void fractions, the same approach will provide simultaneous flow visualization and LIPA data. The qualitative visualization data obtained with simultaneous velocity information will be very useful in understanding agglomeration dynamics of densely packed wet-solid mixtures, in examining flow regimes, and their formation mechanisms, etc., because of the ability to color separate the solid and liquid phases over the whole field.

3.3.2.5. *Wall Shear Stress Measurements Using LIPA*

LIPA can be used to make wall shear stress measurements. Essentially, the approach is to pass a line perpendicularly to the surface of interest, and measure the slope of the line after a time delay. Since the definition of the wall shear stress is $\tau_w = \mu du / dy|_w$ the displacement of the line at a fixed height in the neighborhood above the surface and the time interval is all that is needed, as long as there is a liquid layer near the wall.

3.3.2.6. *LIPA Imaging Considerations*

There is a close relationship between the resolution needed, the choice of laser linewidth and the field of view over which LIPA measurements can be made. Some examples of the constraints and tradeoffs are given below.

3.3.2.6.1. *Laser Linewidth and Particle Size.* The field of view of an imager and the chosen laser line width are intimately connected to the flow field resolution obtained from the imager. In general, it will be desirable to determine the solid and liquid flow fields to the same accuracy, and to resolve the smallest scales of motion in both of them. For the speeds and viscosities of most highly loaded solid-liquid flows (less than 1-2 meters/second, with viscosities commonly a factor of ten or more above that of water), experience suggests that there is unlikely to be much turbulent energy on scales smaller than 0.5 mm. Thus, a laser line width of 0.1 mm would be adequate to resolve the velocity field. This suggests that the laser beam widths should be on this order. The centers of these lines should be determined to 10% of their width.

The particles will, of course, vary in size depending on the flow being studied, but the laser beam widths used should be chosen both with the smallest scales of interest in mind and the particle distribution of the flow. Figure 3-17 shows three ranges of particle size to laser beam width. Although the flow is index matched, the information passing the color filters will produce images of the grid intersections that look similar to the shaded cross for the fluid, and to the white ovals for the solid on each of the separate solid and liquid phase exposures. If the particles are small with respect to the width of the laser beam, then the laser beam intersection will be easily definable (with the help of a hole filling algorithm) and the fluid velocity will be obtainable. As long as the imaging system has sufficient resolution to resolve the individual particles (say 50 μm in a 500 μm wide line) then the particles everywhere along the lines will provide particle velocity information. For flows where the particles become a sizable fraction of the line width, it will still be possible to determine the center of the intersection of the liquid by integrating information along the adjoining parts of the line. In cases where the particle loading is high, and the particle dimensions are the order of the laser line width, the fluid velocity will become more difficult to obtain, and it may be desirable to increase the laser linewidth to re-establish condition "b". However, the situation is somewhat better than suggested by Figure 3-17c because the lines have a depth equal to their width and this results in additional line illumination over portions of the line that the figure shows hidden by particles. However, our limited experience suggests a rule of thumb for concentrated solid-liquid mixtures -- that the width of the laser beam should be made approximately two times that of the mean of the solid particle distribution. Practically, this covers a wide range of particle sizes of interest, without losing the ability to resolve the small scales of motion. For very large particles the problems once again go away, and it is possible to get very detailed fluid velocities in the vicinity of the particles, as well as particle velocities. For very tiny particles, in multimodal distributions where the laser linewidth will be as much as 10-100 times the diameter of some of the particles, finding the centers of individual smaller particles would be difficult, but we can use the displacement of the intersections of the computer generated lines formed through groups of particles to measure the particle velocities (the idea is suggested by viewing Figure 3-17a but imagine many more particles, all of which would be smaller than the smallest particles shown).

Another point that is made clearer by viewing Figure 3-17a is the issue of the difficulty of separating the different colors on color film. The mixture (imagine red particles and green liquid) has the effect, especially when the particles are small, of mixing the colors and producing an brownish-orange color. This is an important reason for the use of color filters, and the reason a color imager is not desired.

3.3.2.6.2. *Field of View.* To enable magnification of the individual particles with the constraints of available imagers, we generally have to reduce our field of view as our resolution demands increase. We will consider our most limiting example, which is that of obtaining full stereo three-dimensional imaging enabling measurements of the vorticity vector, the Reynolds stress tensor and the strain-rate tensor, along with simultaneous solid-liquid concentration information from LIF. Using a contemporary high temporal resolution electronic camera like the Kodak Ektapro 1000, for example, we have a 192 x 239 pixel imaging area. To detect the center of a 50 μm particle to plus or minus 10%, we would have to image 5 μm onto each

(a) (b) (c)

FIGURE 3-17 *The effects of relative size of the particles and the laser lines on line identification.*

pixel (we are not including the additional precision available by pixel averaging in this discussion). The field of view is thus 5 μm x 192 pixels equal 960 μm, by 5μm x 239 pixels equal 1195 μm. If we are to include eight laser intersections in this area (to enable the cube to be constructed), we would make the laser beams 100 μm wide and space them 500 μm apart. This would allow the desired motion of the grid as a whole up to roughly 1/4 of a mesh size in any direction for all of the lines. It is obvious that to accomplish the task we have had to essentially confine ourselves to making a point measurement. Using a close range telescopic lens system there would be no problem with depth of field.

The other way to acquire this data is on photographic film. Using a 100 line pairs/mm film (400 ASA Kodak TMAX, for example), an 18 x 12 mm portion of the film -- which is the 'half frame' of a 35 mm high speed camera (18 x 24 mm format film) -- results in 1800 x 1200 equivalent pixels. Thus, our field of view can cover an area with dimensions of 5 μm x 1800 = 9 mm and 5 μm x 1200 = 6 mm. Keeping the same resolution, we can safely image 17 x 11 cubes. This gives us a significant amount of spatial data in three-dimensions. Because we are only penetrating one mesh deep, i.e., we need ~ 5 mm depth of field for each of the pseudo cameras, we can accomplish this using small apertures and image intensification. This field of view essentially covers the wall region in turbulent boundary layer and pipe flows, for example, or the shear layer region of a jet or combustor.

Relaxing the resolution constraint enables a wider field of view to be examined. For example, if we can accept 10% accuracy on a 1 mm dimension (the typical length of hot-wires, for example), we can, using the same reasoning, increase our field of view by 20 times that indicated above in the case of the electronic camera data acquisition.

3.3.3. Accessible Geometries and Access Issues

LIPA can be used in a wide range of flow studies. Because it is an optical technique it requires optical access to introduce the excitation laser energy, and it requires optical access to record the movement of the distorted flow. Although LIPA typically requires ultraviolet transmitting materials (such as fused silica, quartz, or ultraviolet transmitting glasses), it does not require interferometric optical access, or for that matter even optical access as "clean" as needed for LDA. This is a consequence of the fact that the technique uses the *difference* between two images to determine the state of motion. Thus, if the direction of the laser energy is changed upon entering the test chamber because of small refractive non-uniformities, the intersection points will be displaced (perhaps as in Figure 3-9), but the resulting measurements of the fluid movement of these lines will not be changed. The location of the measurement may change, but the intersections will tell the investigator where they are, as is the case with the movement due to index of refraction variations of the mixture itself. Thus, changes due to the passage through fixed geometric distortions are essentially automatically compensated for. This insensitivity makes it unnecessary, in general, to match the mixture to the quartz or glass walls of the device. Because the initial refraction at the walls will only move the laser beam crossings, these refractions can be compensated for during static setup and calibration, and the laser lines arranged as desired. The exception is, of course, measurements very near surfaces, where considerations discussed above may have to be implemented. Since LIPA does not require diffraction quality optical access, many more apparatus can be investigated. It is also important to note that since the emission wavelengths are in the visible, ordinary glass or plexiglass windows are all that is required to transmit the data to the image acquisition system.

Another degree of freedom allowed by LIPA is the ability to gain access using fiber optics. Single fibers brought to the laser head or bundles of quartz fibers that can withstand the energy of current commercial excimer lasers are now commercially available. The quality of connectors and the technology of fiber bundling is expected to rapidly improve.

3.4. CHEMICAL SYNTHESIS FOR LIPA

3.4.1. The Chemical Underpinning of LIPA

3.4.1.1. *Photophysical Properties Important to LIPA Imaging*

The underpinning of the LIPA technique for the fundamental study of basic flow properties of gas/liquid-solid flows is the design of photoluminescent compounds to enable the creation of a grid upon laser excitation with sufficient lifetimes to image the short time displacements of the solid and fluid phases. Photoluminescence is the fundamental physical phenomenon involving the emission of a photon from an electronically excited molecule or solid [Turro 1978]. The dissipation of excited state energy of an isolated molecule is described by the diagram in Figure 3-18. The luminescence process is characterized by a rate for photon emission, k_r, which is designated the radiative rate. Alternatively, reversion of the excited state to ground state can occur with the dissipation of energy in the form of heat. These nonluminescent decay pathways from electronic excited states are quantified by a nonradiative rate constant, k_{nr}.

The radiative and nonradiative decay pathways, which are directly competitive relaxation processes, are the crucial parameters to the LIPA technique because they define the lifetime and intensity of a photoluminescent molecule or solid. The observed lifetime (τ_o) of an electronic excited state is simply given by the reciprocal sum of the radiative and nonradiative rates (see Turro 1978),

$$\tau_o = \frac{1}{k_r + k_{nr}} \tag{3.1}$$

Photoluminescence intensity is directly dependent on the luminescence efficiency, usually called the quantum yield of luminescence ϕ_e, which is the ratio of the number of photons emitted per photons absorbed. The emission quantum yield is related to k_r and k_{nr} as follows,

$$\phi_e = \frac{k_r}{k_r + k_{nr}} = k_r \tau_o \tag{3.2}$$

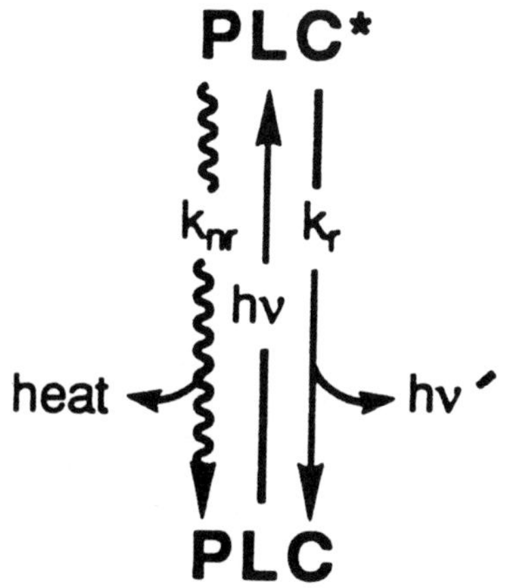

FIGURE 3-18. *Intrinsic luminescence (radiative) and thermal (nonradiative) decay pathways of an excited state molecule.*

From these expressions, we see that the design of successful imaging reagents requires control of the subtle interplay between k_r and k_{nr}. Because LIPA applications demand high luminescence intensities to facilitate detection of laser generated grids, successful imaging systems must display the intrinsic property of $k_r \gg k_{nr}$ for the emission quantum yield to approach its theoretical limiting value of unity. However, this property does not ensure the occurrence of lifetimes of sufficient duration to permit a significant displacement of a flow field to be observed. Flow velocities of 1 meter/sec require excited state lifetimes of at least 1 msec. Thus in addition to the design feature $k_r \gg k_{nr}$, the radiative rate must lie in the msec time regime for most solid-liquid flow studies.

Herein lies a paradox for LIPA measurements, because the radiative rate constant is related directly to the absorption cross-section of a molecule

$$k_r = 3x10^{-9} \nu_0^2 \int_{\nu_1}^{\nu_2} \varepsilon(\nu) d\nu \qquad (3.3)$$

where the argument of the integrand is the absorption band intensity with ν_0 representing the energy at the peak maximum. For strongly absorbing materials (such as organic molecules), k_r is on the order of 10^9 sec^{-1} (Turro 1978)· Thus even if $k_r \gg k_{nr}$, the lifetime (1 nsec) is too short for practical imaging. If the absorbance of the molecule is constrained to be weak, such

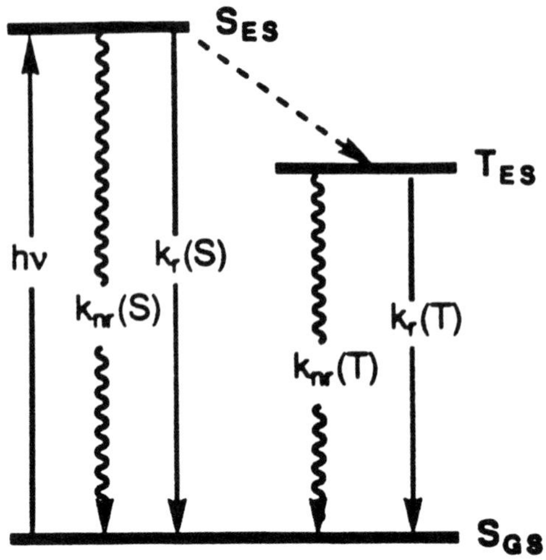

FIGURE 3-19. *Energy level diagram for a molecule possessing a photoluminescent singlet (S_{ES}) and triplet (T_{ES}) excited state. Photo-excitation designated by hν, and k_{nr} and k_r are the nonradiative and radiative rate constants, respectively, for excited state decay.*

that k_r approaches a msec time regime, the luminescence intensity will be correspondingly weak (despite $k_r \gg k_{nr}$) because not enough photons will be absorbed to produce the excited state (the emission quantum yield equals the photons emitted per photons absorbed). The concentration of a weakly absorbing species can be increased to ensure sufficient photon absorption [absorbance = (concentration) x (absorption cross-section) x (path length)] but often the concentrations are too large for practical engineering applications.

For these reasons the occurrence of long-lived intense photoluminescence is unusual for organic and inorganic molecules. Figure 3-19 shows a basic energy level diagram of a luminescent molecule. In the ground state virtually all molecules reside in states with all their electrons paired, a situation in which the molecule is said to be in a singlet ground state (S_{GS}). Quantum mechanical rules dictate that the most probable transitions to excited states will be to other singlet excited states (S_{ES}); hence the transitions are extremely intense (Porter 1983). From Equation 3.2, the rate for radiative relaxation $k_r(S)$ is so large that the nonradiative decay channels $k_{nr}(S)$ are generally not competitive, and intense emission is observed. But luminescence from singlet excited states, a process called fluorescence, is extremely short-lived owing to the large k_r's (see Equations 3.1 and 3.3). For every corresponding singlet excited state, a triplet excited state (in which two of the electrons of the molecule are unpaired) exists. The radiative decay rates from triplets, $k_r(T)$, can approach seconds, but typically $k_{nr}(T)$ will be large; and hence from equations 3.2 and 3.3 luminescence from the triplet, called phosphorescence, will be weak. Phosphorescence quantum yields for emission from most organic molecules are $< 10^{-5}$. Even if $k_{nr}(T)$ is slower, such that ϕ_e approaches unity, the absorption cross-section is typically so small that not enough photons can be absorbed by the sample to produce sufficient intensity for imaging.

To this end we have engineered novel molecular superstructures which contain a light harvesting center (LHC) and a remote photoluminescence center (PLC) (Yu et al. (1991), Pikramenou and Nocera (1991) and Dulebohn et al. (1990)). Figure 3-20 schematically represents the function and operation of these assemblies which we refer to as supramolecules owing to their multicomponent nature. Briefly, the LHC efficiently collects photons to produce its excited state. This excited state then channels its energy to the PLC, creating the ground state of the LHC, and the electronically excited PLC, which can subsequently relax by emission of photons at high efficiencies. With this approach, the aforementioned constraints on the radiative rate constant required for efficient light absorption are not imposed on luminescence. In this manner the optical excitation by the photon is efficiently pumped into a highly luminescent center irrespective of that center's absorption properties. The ability to synthesize these supramolecular assemblies and to define the paradigms that enable energy flow to be channeled to desired sites in the assembly constitutes a research field at the most fundamental level of chemistry, and is an area that is of special interest to the second author.

3.4.1.2. *Luminescence Quenching Pathways of Significance to LIPA*

The above discussion considers only the intrinsic properties of a molecule in the absence of external substituents. Because electronically excited molecules are highly energetic they are susceptible to physical or chemical interactions with their environment. Specifically, the excited state often reacts with external reagents (called quenchers, Q) by transferring its energy or an electron (Turro (1978), Scandola et al. (1983)),

$$PLC^* + Q \rightarrow PLC + Q^* \tag{3.4}$$

$$PLC^* + Q \rightarrow PLC^{+or-} + Q^{-or+} \tag{3.5}$$

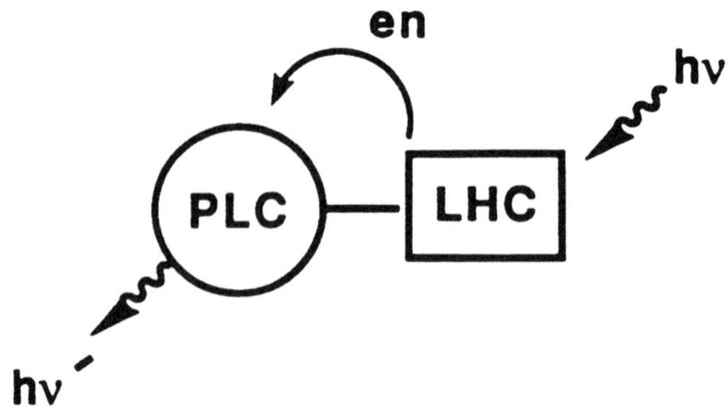

FIGURE 3-20. *Successful supramolecular design for LIPA imaging systems. Photons efficiently absorbed by a LHC and energy is then transferred to a PLC.*

Equations 3.4 and 3.5 are called energy and electron transfer quenching reactions, respectively, and are characterized by a quenching rate constant k_q. These pathways are nonradiative in nature because they introduce an additional mechanism for the dissipation of the excited state energy that leads to a diminishment (or quenching) of luminescence from the PLC. The introduction of a quenching pathway modifies the photoluminescence lifetime and intensity by adding k_q to the denominator of Equations 3.1 and 3.2. Hence the debilitating condition of quenching to the LIPA technique is manifested in a diminished lifetime and quantum efficiency. This decrease in luminescence lifetime and intensity is related directly to the concentration of quencher in the PLC's environment by the following expressions (Balzani et al. 1975),

$$\frac{I_o}{I} = 1 + k_q \tau_o [Q] \tag{3.6}$$

$$\frac{\tau_o}{\tau} = 1 + k_q \tau_o [Q] \tag{3.7}$$

where I_0 and τ_0 are the luminescence intensity and lifetime in the absence of a quencher, I and τ are the intensity and lifetime in the presence of a quencher of concentration [Q], and k_q is the rate of quenching. From these expressions we see that the lifetime and intensity decrease linearly with increasing concentration of quencher.

For engineering LIPA applications, quenching is an issue pervading most measurements. Water, oxygen, and residual metals in the environment are good quenchers of photoluminescence; and indeed the primary challenge in the design of any successful imaging system reduces to molecular designs that minimize these quenching pathways. Undeniably, the most problematic quencher in engineering applications is oxygen owing to is efficiency and prevalence in the environment. Oxygen quenches triplet excited states by the energy transfer mechanism described by Equation 3.4. The problems that oxygen brings to bear to the LIPA technique can be explicitly demonstrated by considering Equations 3.6 and 3.7. Typically the quenching rate for oxygen is diffusion-controlled (i.e. $k_q = 10^9\ M^{-1}sec^{-1}$), and its concentration in solution is 10^{-3} M and in air is 10^{-2} M. Thus for an imaging system possessing a 1 msec photoluminescence lifetime (permitting flow speeds of 0.5 meters/sec to be imaged), the lifetime and intensity of the imaging systems will be attenuated by 10^3 upon its introduction into oxygenated solutions and will be reduced by over 10^4 in air!

3.4.2. Previous Accomplishments in Two Phase LIPA Imaging

Two-phase flows of wet-solids, a term used for heavily loaded solid-liquid flows, have successfully been imaged in our laboratories by using the LIPA technique (Falco and Nocera 1991). In solids, quenching by oxygen is negligible because the concentration of oxygen in a solid is low, and its diffusion is slow. This feature permits a wide range of solids to be considered for LIPA applications. We have taken two approaches. The simplest is the introduction of 0.1% red phosphor into silica gel. The silica gel/phosphor solid admixture is insoluble in water and has a refractive index that can be adequately index of refraction matched with aqueous solutions of KSCN, thereby enabling the establishment of an optically transparent liquid-solid phase. The phosphor luminesces red with lifetimes approaching 10 ms, and thus relatively slow speed solid flows can be imaged. The drawback of this solid imaging system is that silica gel particles possess a negatively charged surface and hence will adsorb positively charged compounds (cations) from solution. Because our most successful imaging reagents for liquid solutions to date are cations, this solid phase presents technical difficulties in concurrent imaging of a two-phase wet-solid flow. To this end, a more attractive approach has been the synthesis of CaF_2 solid embedded with Eu^{3+} ion. Native CaF_2 has a refractive index which can be matched with aqueous solutions of KSCN, and thus is an ideal host solid for

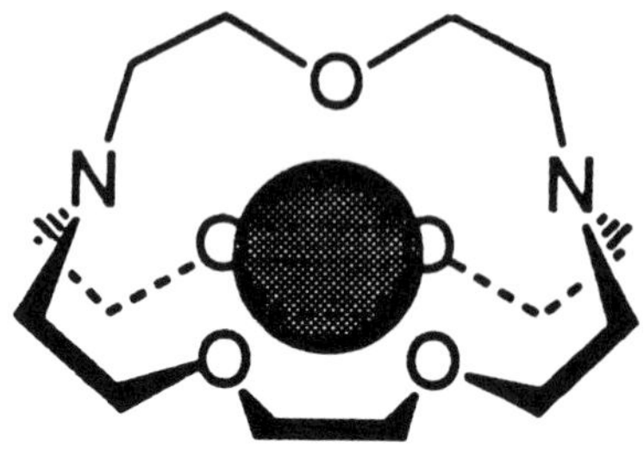

1

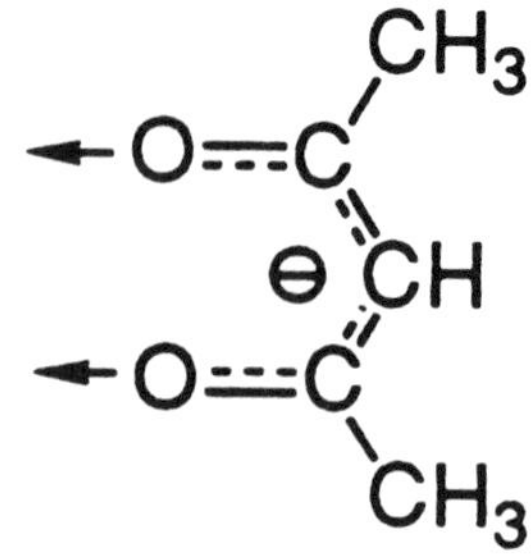

2

FIGURE 3-21. *The $Tb^{3+} \subset 2.2.1$ complex, and the light harvester acac. The arrows indicate the site of attachment of the acac to the encapsulated Tb^{3+} ion, which is indicated by the shaded ball.*

LIPA applications. The solid is prepared in our laboratories by precipitating the CaF_2 from solutions of varying concentrations of Eu^{3+} ion. Intense, long-lived (8 msec) red luminescence characteristic of Eu^{3+} lanthanide ion is observed from the Eu^{3+}/CaF_2 solid. We have found that a maximum loading of 5% of Eu^{3+} by weight can be achieved without dramatic changes in the refractive index of CaF_2. The solid compound does not interact with constituents in liquids thereby allowing the production of concomitant grid images for the solid and liquid in two-phase flows. We have successfully produced benchtop quantities of doped CaF_2. However, the scale-up to quantities necessary to perform normal laboratory engineering studies has not been tried.

The design of liquid imaging reagents is more complicated than that for solids owing to Equations 3.4 and 3.5. The long-lived excited states of organic compounds (the phosphorescent excited states) are subject to highly efficient oxygen quenching thereby obviating their direct use in LIPA studies of flow speeds that are less than hypersonic. Early in our studies we realized the inherent oxygen quenching problem, and for a variety of highly technical reasons turned our attention to inorganic compounds. Specifically we targeted compounds containing the lanthanide ion Tb^{3+}. Lanthanide ions in general exhibit intense luminescence from transitions involving electrons residing in orbitals buried deep within the ion [Brittain 1988]. For this reason, the luminescence excited state is shielded from the external environment. Although oxygen will physically contact the lanthanide ion upon collision, it cannot communicate with the luminescent excited state owing to poor electronic coupling (i.e. the physical overlap of the orbitals of the lanthanide ion with those orbitals of oxygen is poor) [Balzani et al. 1980]. Thus the energy transfer reaction between lanthanide ions and oxygen as described by Equation 3.4 is circumvented and oxygen quenching of lanthanide luminescence is not observed.

Yet water is an excellent quencher of lanthanide luminescence (Horrocks and Albin 1984). Water molecules can covalently bind to the lanthanide ion through their oxygen atom. The high frequency O-H oscillators of the coordinated water molecules result in large values of k_{nr} and therefore efficient nonradiative decay. Thus water has to be rigorously excluded from the lanthanide coordination sphere in the design of imaging reagents based on lanthanide ions.

The deleterious effects of O-H decay channels on the photoluminescence of Tb^{3+} ions can be suppressed by encapsulating the ion in molecular cages (Alpha et al. 1987) such as the cryptand ligands shown by structure 1 in Figure 3-21. The general class of molecular cages is called cryptands and the one displayed in structure 1 is given the name $\subset$2.2.1 cryptand. The two nitrogens and five oxygens comprising the three straps of the cryptand ligand occupy seven of the nine coordination sites of the Tb^{3+} ion (indicated by shaded ball) thereby shielding the lanthanide from seven of a possible nine waters. The presence of the two remaining coordination sites of the Tb^{3+} cryptand complex provides the opportunity to engineer highly efficient photoluminescent systems by establishing the process described in Figure 3-20. The $Tb^{3+}\subset$2.2.1 compounds are only weakly luminescent owing to the poor absorption cross-section of the lanthanide ions. By anchoring an LHC such as acetyl acetonate (called acac and represented by 2) at the two remaining coordination sites of the caged ion, we have been able to increase the luminescence efficiency of $Tb^{3+}\subset$2.2.1 complexes by upwards of three orders of magnitude (Yu et al. 1991, and Lessard et al. 1989). This chemical combination is called Flowlite II (see Table 3-1).

The red luminescence of the solid Eu^{3+}/CaF_2 or phosphor/silica solid phases is spectrally distinct from the green luminescence of the $Tb^{3+}\subset$2.2.1/acac in the liquid phase. Excitation of a two-phase flow with the 308-nm line of an excimer laser produces a superimposed green grid in the liquid phase and a red grid in the solid phase. The luminescence lifetimes of 1.8 msec of the $Tb^{3+}\subset$2.2.1/acac system and 10 msec of the

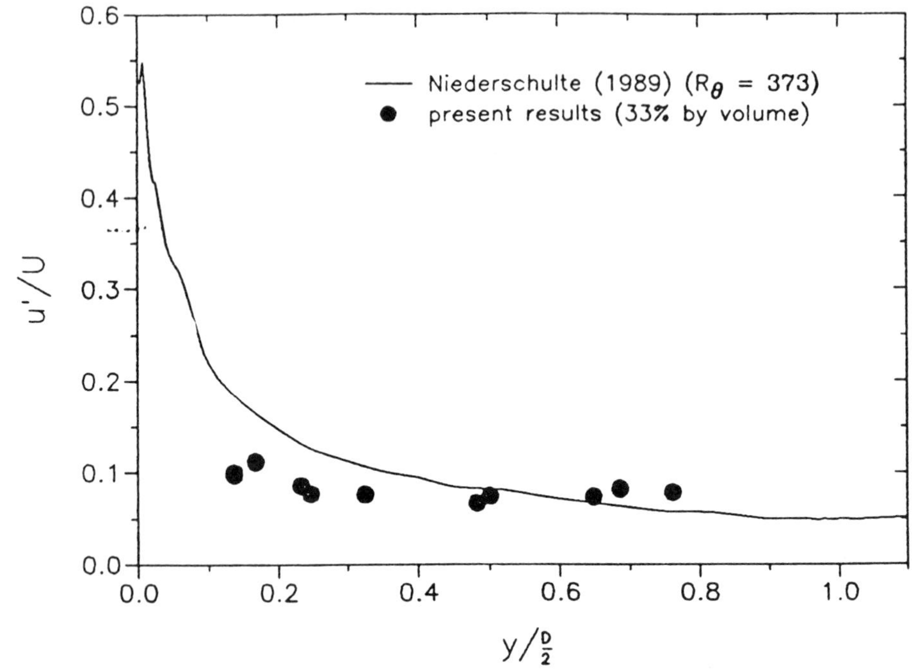

FIGURE 3-22. *Shows the streamwise fluctuations, u', at ten positions in the pipe with a liquid/solid mixture 33% by volume, compared with values for a low Reynolds number turbulent channel flow of a Newtonian fluid. We can see that the intensities are higher in the core of the pipe, but significantly lower near the walls.*

Eu^{3+}/CaF_2 or phosphor/silica solid systems is manifested in the creation of grids that persist for ~4 ms in the liquid phase and ~20 ms in the solid phase.

3.5. SOME NEW RESULTS FOR HEAVILY LOADED PIPE FLOW

As an example of the unique contribution that LIPA can make in particulate two-phase flows, we present results from a study that has been undertaken to understand the flow dynamics of heavily loaded solid-liquid two-phase pipe flow (known as a wet-solid). The focus has been to measure the flow properties of mixtures so dense that they exhibit yield stress and other highly non-Newtonian properties. LIPA was used to obtain the wall shear stress, turbulence intensities in the axial (u') and radial (v') directions, and the Reynolds stress <u'v'>. Apart from obtaining these statistics, the technique also enables us to focus on agglomeration dynamics of densely packed wet-solid mixtures. The focus of this first study of its kind was to obtain the average properties of the mixture. The results are preliminary and

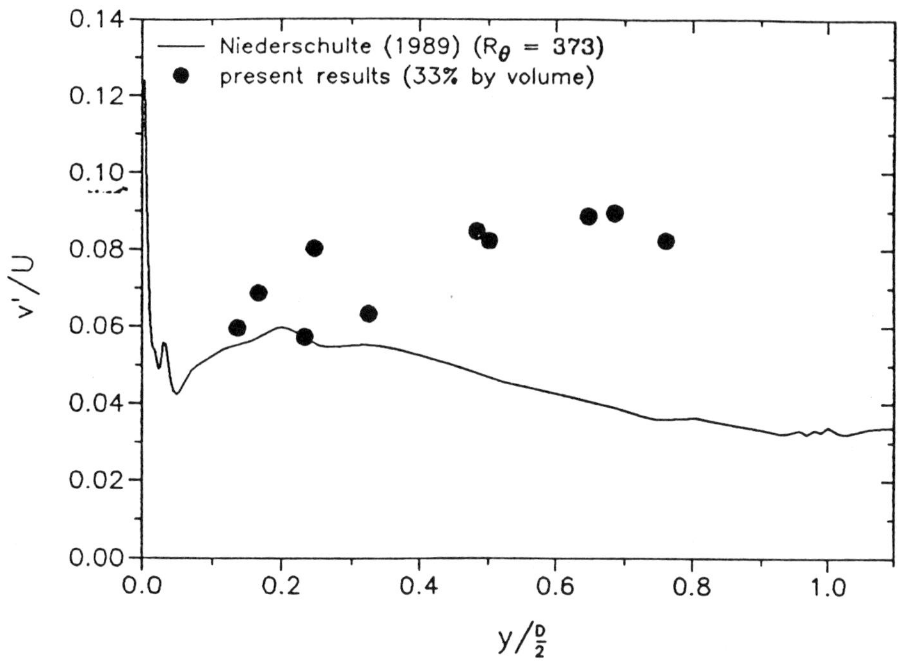

FIGURE 3-23. *Measurements of the radial component of the fluctuations of a liquid/solid mixture of 33% by volume. We see that v'/U is significantly higher in the core of the pipe, when compared to a low Reynolds number turbulent channel flow of a Newtonian fluid.*

are based on the simplest form of LIPA.

3.5.1. Experimental Details

We have used the index matched silica gel/phosphor admixture discussed in the previous section to mark the solid and Flowlite II to mark the liquid. The mixture, which ranged from 27% to 52% by weight of silica gel, was forced downward through a 2 cm vertical pipe flow by a piston pump. The average flow speed in the pipe was 1.8 m/s. Measurements were made 70 pipe diameters downstream from the contraction. The solid-liquid mixture was mixed by hand stirring to create a homogeneous system, and was poured into the cylinder of the pump immediately before each run. The resulting flow was in a state that did not resemble either a laminar or a turbulent flow. This will be described more fully below.

A gated image intensified CID camera, made by ITT, was triggered by a Lambda Physik excimer laser operating at 308 nm and 60 Hz. A time delay generator enabled each video exposure to be made at an optimum time

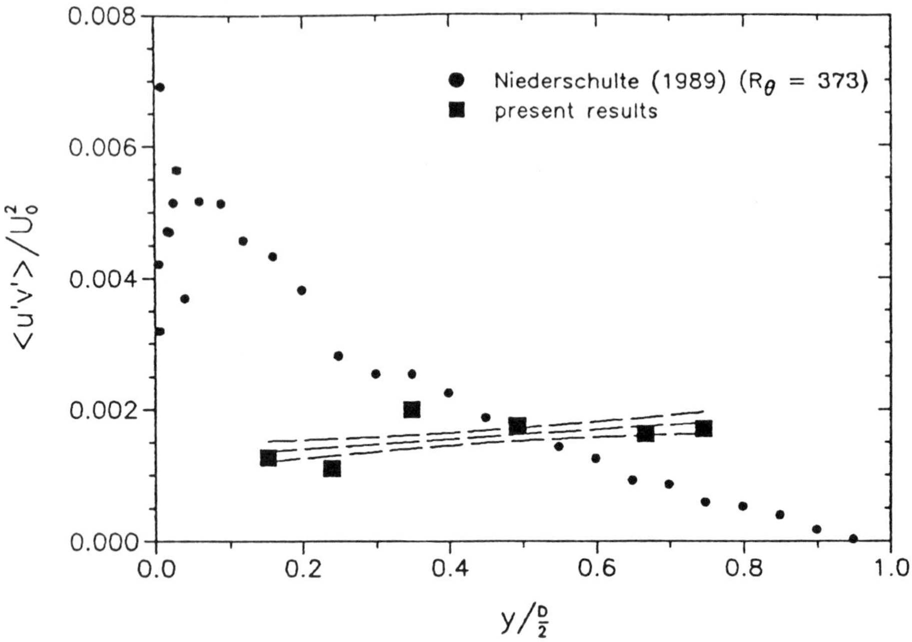

FIGURE 3-24. *The Reynolds stress <u'v'> for conditions of the previous figure. We see that <u'v'> is higher in the core of the pipe, and appears to decrease as we approach the wall region when compared to a low Reynolds number turbulent channel flow of a Newtonian fluid.*

delay for the flow. This was determined after repeated trials. Data reduction of the grid line and intersection positions was done by hand positioning a cursor on the data displayed on a CRT. The position of the point was then digitized by a Megavision 1024x image processor.

3.5.2. Results

3.5.2.1. *Turbulence Intensities and Reynolds Stress*

Results showing the distributions of velocity, turbulence intensity and Reynolds stress across a pipe are presented for a number of mixture concentrations ranging from 0 to 52% by volume. Figures 3-22 to 3-24 show the flow properties of our silica gel/water/thiocyanate mixture for the case where we have 33% solid by weight.

Figure 3-22 shows the rms axial turbulence intensity (u') normalized by the local mean velocity, and compares the distribution to the Newtonian situation. Note that the fluctuations are much lower in the wall region of the pipe, and become of equal or greater magnitude than the Newtonian values as we move towards the centerline.

Figure 3-23 shows the rms radial turbulence intensities (v') compared to the Newtonian values. Here we see that the radial intensities have increased dramatically in the center of the pipe, and are tending towards the Newtonian values as we approach the wall region.

Figure 3-24 show the Reynolds stresses (<u'v'>). They show a remarkable decrease from Newtonian values between the wall and half radius, and an increase in the center of the pipe. The increase in this correlation of the axial and radial velocity fluctuations, taken together with the u' and v' changes, strongly suggests that the flow in the core of the pipe has become more highly correlated than in a Newtonian pipe.

3.5.2.2. *Wall Shear Stress*

Figure 3-25 shows the probability distribution of the shear stress for a 27% α_S loaded pipe. The mean value of the wall shear stress, based on the assumption of water in the layer near the wall, is approximately 0.78 Pa. It should be noted that for a laminar flow in a 2 cm pipe with the same bulk averaged velocity, we would have a friction factor of 0.00064 vs. the 0.00036 found in our 27% solid loaded pipe flow. Experiments to obtain wall shear stress measurements were repeated three times. They confirmed one another and they support the possibility of laminar-like flow in the layer close to the wall. As the concentration was increased to 33%, we find very little change in the wall shear stress distribution, and a small increase in the magnitude to 0.845 Pa. If this were due to the increase in viscosity of a homogeneous mixture from 27% to 33% we would see an increase of the order of 100%, vs. the 8% measured (see Sengun and Probstein 1989).

3.5.2.3. *Core Annular Flow Indications*

It has been found that for concentrations ranging from 27% to 52%, a film of lower solid concentration than the average mixture concentration exists around the walls of the pipe at 70 diameters, even though a homogeneous mixture was introduced into the pipe. This low solids concentration layer was initially identified by the presence of very sharp changes in slope of the photoluminescing LIPA lines created in the mixture. It has subsequently

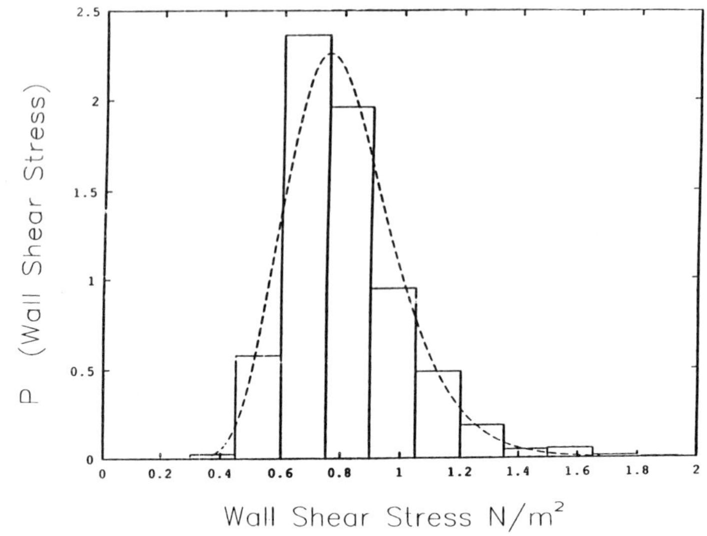

FIGURE 3-25. *The experimental probability distribution of the shear stress for a 27% α_s loaded pipe. Sample size is 1100. The mean wall shear stress is 0.78 N/m^2. The fitted lognormal distribution appears to adequately describe the experimental distribution.*

been confirmed and quantified by mechanical collection techniques (Buchner et al. 1991). The location of the sharp change in slope of the LIPA lines appears as a 'knee'. The slope of these lines, from the position of the 'knee' (the region of sharp change in slope) to the center of the pipe was much flatter than found in either Newtonian or dilute polymer solutions (for concentrations that give 60% drag reduction). The oscillations in the photoluminescent lines of the mixture were almost absent from the knee to the centerline, but were clearly observable between the wall and the knee. Figure 3-26 shows the average thickness of the low solids concentration layer, defined by the position of the 'knee', for four concentrations. It is approximately 1/10 of the pipe radius at 27% concentration by volume, decreasing to 0.025 in the mixture with 52% solids. The thickness of this wall layer of low concentration fluctuates with time for a fixed initial concentration, as well as changing with the initial concentration. The extent of the fluctuation of the height of the layer is indicated by the ±1 rms bars in Figure 3-26. An examination of the movies shows that this layer is fluctuating in thickness in a complicated manner that exhibits both long and short wavelength components. It appears that these phenomena may persist at higher concentrations.

With the understanding of the core of higher concentration mixture moving down the center of the pipe, the large changes in the fluctuations and Reynolds stress is interpretable. If a 'snaking' of mixture with higher

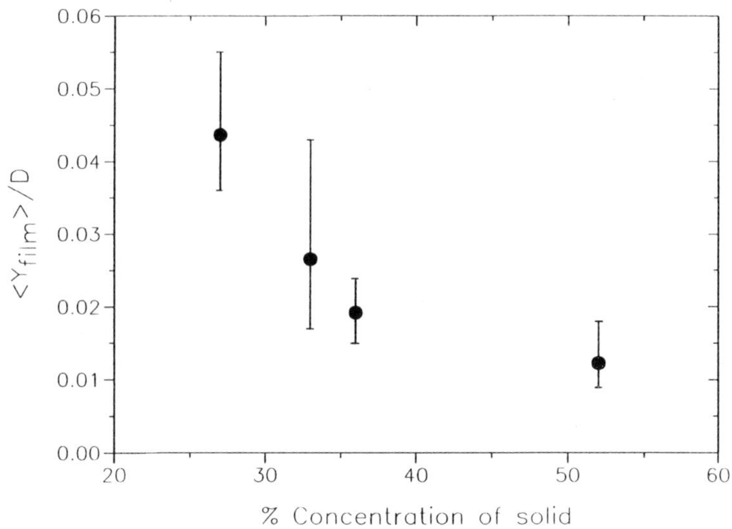

FIGURE 3-26. *The average extent of the boundary between the dense core and dilute wall layer, and the variation as a function of concentration.*

concentration than the average is imagined slithering down the pipe on a layer of lower concentration, we would expect it to be more highly correlated than a turbulent Newtonian liquid. Furthermore, from the observation that both the width and the precise location of this high concentration core is varying, we can explain why we would get the much higher 'v': a fixed observer sees the location of the core with respect to the centerline of the pipe constantly changing with time, creating large variations in the radial velocity component. Between the wall and the boundary of this high concentration region, the low concentration layer is more laminar-like, as is evidenced by the lower u' and the lower <u'v'>. This is consistent with our findings of lower shear stress at the wall. These results indicate a very different flow behavior than expected.

Because we find that the concentration gradient appears to be very steep, it can be thought of as occurring over a thin region that can be described as a boundary between the high concentration in the core and the low concentration in the region near the wall. Qualitatively, the flow over the wall region appears as a Couette flow with a moving wavy 'wall' of random amplitude and wavelength. However, it still has a pressure driven component. Overall, a sequence of favorable and adverse pressure gradients is impressed on the pipe wall, which may explain why the drag is lower than that of a steady laminar Newtonian flow.

We want to emphasize that we have measured a global

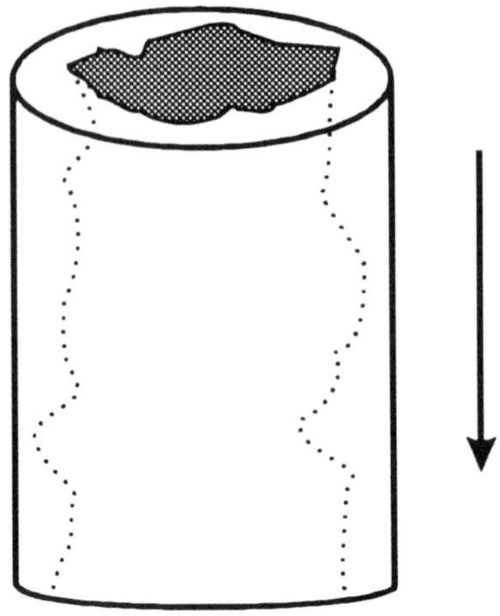

FIGURE 3-27. *A sketch of the annular segregation found in the flow of highly loaded solid-liquid initially homogeneous in a pipe. The shaded region contains the more highly loaded mixture. This core presents a wavy moving boundary over the more dilute layer of the mixture between it and the pipe wall.*

agglomeration effect for highly loaded solid-liquid mixtures. There may be an analogy with the separation of oil and water in pipe flows as found by Chen, Bai and Joseph (1990), and references contained therein. It may have potential positive value in a number of areas, including separation of the liquid, and drying of the solid phase in dense particulate two-phase solid-liquid mixtures. It may also indicate a need for extreme caution in interpreting results when some types of viscosimetry are used.

There is little evidence in the literature to compare these observations with. As noted above, Park et al. (1989) found that the u'/U_{cl} was lower in the 5.65% by weight solid loaded pipe flow (~14% by volume). Nouri et al. (1987) have shown that the intensity of the fluid fluctuations, and of the particle fluctuations, decreases with increasing concentration. Thus, it appears that at the loadings we investigated the situation is significantly different.

3.6. DISCUSSION

We want to emphasize the untested nature of many of the suggested approaches to LIPA for solid-liquid flow studies. We will briefly discuss some additional limitations and extensions of LIPA, and finish with a view of the future of the technique.

Resolution of the LIPA technique is limited by refraction and scattering due to mismatches in the index of refraction. In dense solid-liquid mixtures, slight mismatches cause degradation of the resolution through line broadening, which increases the signal to noise ratio, but does not fundamentally invalidate the data. Again, line position shifts due to index of refraction changes are less serious for LIPA than for Doppler techniques. In addition, absorption of the incident laser radiation is also a source of error, because of the decreased signal to noise ratio as the laser energy penetrates further into the flow apparatus. These phenomena also influence other optical techniques used to measure solid-liquid flows. The relative effects on each technique have yet to be quantitatively compared.

Other two-phase flow applications, with both liquid as well as solid particles, can be addressed using LIPA. Liquid-liquid flows are good candidates. For example, we can index of refraction match oil/water mixtures, dope the oil separately from the water, and perform LIPA in a manner similar to that described above for solid-liquid particulate flows. In addition, we can make LIPA measurements in relatively dilute mixtures of solid-liquid, gas-liquid or gas-solid or gas-solid-liquid flows that are not index matched. Doping chemicals are now available in our laboratories for all phases.

3.6.1. The Future of LIPA

As might be expected with a new technique a number of changes and improvements are expected in the near future. Perhaps the most important will be the use of solid state lasers. These lasers are characterized by low cost, high repetition rate, reliability, small size and light weight. Because they typically are low output power devices, their use in the LIPA technique would, for example, require one laser per line in a grid. The reason they are not useable now is that their radiation is in the red and infrared. However, the growing importance of optical mass storage of digitized data is ushering in the frequency doubled diode laser. Doubling an 810 nm laser, results in 405 nm, excellent for gases using biacetyl doping, for example. Doubling a 670 nm diode laser results in 335 nm which enables many of the other chemicals identified in Table 3-1 (as well as a host of others) to be excited. When this technology matures, the cost of doing LIPA will decrease, and the number of applications that can be addressed will increase.

Of almost equal importance is the need to manufacture doping chemicals that are less expensive, environmentally safe, environmentally insensitive, have high quantum efficiencies, and cover a broad range of lifetimes. Furthermore, chemicals that are excited closer to, or within, the visible range reduce the accuracy to which the mixtures must be index matched, as well as the compensation needed between excitation and received light. Finally, we do not yet have a good chemical to use in ordinary room temperature air, because of the rapid quenching due to oxygen and water vapor. New techniques in chemistry are rapidly changing this picture. The ability to engineer excited state properties at the molecular level is permitting us to create new chemicals for a variety of engineering applications as demand increases. One major breakthrough in this area is the design of supramolecular imaging systems that do not require the light harvesting center to be chemically attached to the photoluminescent center, but rather the LHC and PLC are mechanically juxtaposed by a large molecular cage. This avenue is currently under investigation in our laboratories, and is providing us with a new generation of LIPA imaging systems.

Although we have suggested LIPA is a technique that can provide quantitative information where others cannot, we can conceptually think beyond the above ideas to the day when X-ray lasers are commonplace, and LIPA is used, on location, to provide real time flow diagnostics on assembly lines, in processing plants, pipe lines, etc.

3.7. CONCLUSIONS

Laser Induced Photochemical Anemometry is a viable technique for making local flow field measurements in solid-liquid flows up to the maximum packing densities. Either photochromic or photoluminescent processes can be used. The accuracy of the technique is ultimately governed by length and time scale measurement.

We have indicated how LIPA can be used in solid-liquid flows to obtain:

- the flow properties of the liquid and the solid phases simultaneously;
- the average flow properties of the mixture;
- measurements over a region of the flow, to enable measurement of pair distribution functions, spatial gradients and other spatial

measures;
- wall friction data;
- three-dimensional data;
- data from experiments with complicated geometrical access, and/or with internal geometry changes.

Preliminary measurements indicate some of the rich new physics of high concentration particulate two-phase flows. In the dense solid-liquid flow, the mixture becomes redistributed so that a highly loaded solid core forms and slides down the center of the pipe on a more dilute mixture with the ease of an ice skater. The unsteady film that forms in the vertical pipe has two technologically important features. First, it enables the highly loaded flow to be pumped through the pipe using less energy. Second, if properly exploited, it can be used as a device for separating the liquid/solid mixture. This may enable transport of a lower solids concentration with an acceptable drying penalty at the delivery end.

The future of LIPA is bright. It is dependent upon the discovery and manufacture of less expensive photochemicals, and on advances in laser technology, image acquisition and processing. The chemical techniques are understood, although additional research is needed to provide a menu of imaging chemicals for the broad spectrum of LIPA applications. The rapid advances occurring in the other areas, which are driven by other market needs, are expected to reduce the cost and increase the potential arena of LIPA measurements dramatically.

ACKNOWLEDGMENTS

The experiments were performed with the assistance of Joe Klewicki and Rodney Hill. Most of the data was analyzed by Stephen Buchner and with help from Rodney Hill. Thanks are due to Al Folz for careful proof reading. This work was supported by DOE Contract DE-AC22-88PC88948 under the solids transport program. We wish to acknowledge Jim Bartis, Steve Passman, Shelby Rogers and Dave Wildman for their advice and encouragement.

NOTATION

d laser grid mesh width

I	intensity in the presence of a quencher
I_0	intensity in the absence of a quencher
k_{nr}	nonradiative rate constant
k_q	quenching rate constant
k_r	radiative rate constant
LHC	Light Harvesting Center
LIF	Laser Induced Fluorescence
LIPA	Laser Induced Photochemical Anemometry
NMR	Nuclear Magnetic Resonance
PIV	Particle Imaging Velocimetry
PLC	Photoluminescent Center
[Q]	concentration of a quencher
t	laser line width
U	Mean flow in the x direction
u'	rms of the velocity fluctuations in the x direction
u_*	friction velocity
v'	rms of the velocity fluctuations in the y direction
w'	rms of the velocity fluctuations in the z direction
x	coordinate in the axial direction
y	coordinate in the radial direction
z	coordinate in the azimuthal direction
α_s	concentration by volume
ϕ_e	quantum yield of luminescence
τ_o	observed lifetime of an electronic excited state
ν/u_*	viscous length scale
ν_o	energy at the peak maximum

REFERENCES

Abbas, M. A. and C. T. Crowe, "Experimental Study of the Flow Properties of a Homogeneous Slurry Near Transitional Reynolds Number," *Int J. of Multiphase Flow,* **13**, 357-364 (1987).

Alpha, B., V. Balzani, J-M Lehn, S. Perathoner, and N. Sabbatini, "Luminescence Probes: The Eu^{3+} and Tb^{3+} Cryptates of Polypyridine Macrobicyclic Ligands", *Angew. Chem. Int Ed. Engl*, **26**, 1266 (1987).

Altobelli, S. A. Givler, R. C. and Fukushima, E., 'Velocity and Concentration measurements of Suspensions by Nuclear Magnetic

Resonance Imaging', submitted for publication (1991).

Balzani, V., L. Moggi, M. F. Manfrin, and F. Bolletta, "Quenching and Sensitization Processes of Coordination Compounds", *Coord. Chem. Rev.* **15**, 321 (1975).

Balzani, V., F. Bolletta, and F. J. Scandola, "Vertical and Nonvertical Energy Transfer Processes: A General Classical Treatment", *J. Am. Chem. Soc.*, **102**, 2152 (1980).

Bonnet, J. P. and Falco, R. E. "Developpement de Techniques de Visualization et de Measure Non Intrusive de Vitesse en Ecoulement Turbulent" CERT Univ. Poitiers, Final Report DRET DS/SSR 88/1441 (1989).

Brittain, H. G., "Applications of Lanthanide Ion Luminescence from Inorganic Solids", in *Molecular Luminescence Spectroscopy Methods and Applications: Part 2: Chemical Analysis Series*; ed S. G. Schulman, John Wiley and Sons, New York, **77**, 401 (1988).

Buchner, S., R. E. Falco and D. G. Nocera, "Flow Field and Mixture Separation Measurements in a Dense Solid-Liquid Two-Phase Vertical Pipe Flow", *Bull. Am. Phy. Soc. Series II*, **36**, 2658 (1991).

Chen, K. Bai, R. and D. D. Joseph, "Lubricated pipelining. Part 3 Stability of core-annular flow in vertical pipes", *J. Fluid Mech.* **214**, 251-286 (1990).

Chu, C. C., "A Study of Turbulence Production and Modification in Boundary Layers Using a New Photochromic Visualization Technique," PhD Thesis, Dept. Mech. Engr., Michigan State University (1987).

Chu, C. C., C. S. Hsieh, S. T. Shi and C. T. Wang, "A Quantitative Study of the Flow Near Surfaces Induced by Vortex Motion," AIAA Paper 91-1816 (1991).

Cox, R. G. and S. G. Mason, "Suspended Particles in Fluid Flow Through Tubes", *Ann Rev. Fluid Mech*, **3**, 291-316 (1971).

Dulebohn, J. I., B. Van Vlierberge, K. A. Berglund, R. B. Lessard, J.-A, Yu, D. G. Nocera, "Light Harvesting Molecular Assemblies in the Design of Highly Luminescent Sol-Gel Derived Glasses", in *Proc. Sci. of Adv. Ceramics*, Mat. Res. Soc., Pittsburg, **180**, 733 (1990).

Durst, F., R. Kleine, and Th. Keck, "Turbulence Quantities and Reynolds Stress in Pipe Flow of Polymer Solutions Measured by Two-Channel Laser-Doppler Anemometry", Symposium on Turbulence, ed J. L. Zakin and G. K. Patterson, Univ of Miss-Rolla, 6-1 (1979).

Falco, R. E., "Measurement of the Intensity and Correlations from the

Initial to the Final Period of Decaying Grid Generated Turbulence", Ph.D. Thesis, Princeton University (1972).

Falco, R. E. and C. C. Chu, "Measurement of Two-Dimensional Fluid Dynamic Quantities Using a Photochromic Grid Tracing Technique," *Proc. SPIE,* **814**, 706-710 (1987).

Falco, R. E. and C. C. Chu, "Experimental Determination of Skin Friction Modifications Due to Elastic Compliant Surfaces Using Quantitative Visual Techniques", Report TSL-88-3, Dept. Mech. Engr., Michigan State University (1988).

Falco, R. E. C. C. Chu, M. H. Hetherington and C. P. Gendrich, "The circulation of an airfoil starting vortex obtained from instantaneous vorticity measurements over an area," AIAA Paper 88-3620 (1988).

Falco, R. E. and D. G. Nocera, "A Study of Flow Properties of Wet Solids Using Laser Induced Photochemical Anemometry (LIPA)", in *Liquid-Solid Flows-1991*, ed: M. C. Roco and T. Masuyama, ASME FED **118**, 143-145 (1991).

Hilbert, H. S. and R. E. Falco, "Measurements of Flows During Scavenging in a Two-Stroke Engine," SAE Technical Paper 910671. In *Two-Stroke Engine Design and Development SP-849,* SAE, Warrendale Pa. (1991).

Horrocks, W. DeW. Jr., and M. Albin, "Lanthanide Ion Luminescence in Coordination Chemistry and Biochemistry", *Prog Inorg Chem.* **31**, 1 (1987).

Lee, S. L. and F. Durst, "On the Motion of Particles in Turbulent Flows", *Int. J. Multiphase Flow,* **8**, 125-146 (1982).

Lessard, R. B., K. A. Berglund and D. G. Nocera, "Highly Emissive Lanthanide Compounds in Sol-Gel Derived Materials", *Proc. Sci. Adv. Ceramics*, ed. I. A. Aksay, D. R. Ulrich and G. L. MaVay, Materials Res. Soc. **155,** 109 (1989).

Majumdar, A., A. L. Graham, M. C. Roco and P. Stroeve, "Experimental Study on Solid Particle Dynamics in Shear Flow." *Powder Technology*, **49**, 217-226 (1987a).

Majumdar, A., P. Stroeve, J. L Fales, V. R. Algazi and J. E. Stewart, "Teaching Computers to Track Particles", *J. Imaging Science*, **31**, 208-219 (1987b).

Metzner, A. B., "Polymer Solution and Fiber Suspension Rheology and Their Relationship to Turbulent Drag Reduction," *Phy Fliuds,* **20**, S145-S149 (1977).

Mih, W., "Transporting Solid Particles in Smooth Pipelines", *ASCE Trans.*

Engr. J. **105**, 427-437 (1979).

Miles R. B., J. J. Connors, E. C. Markovitz, P. J. Howard, and G. Roth, "Instanstaneous Supersonic Velocity Profiles in an Underexpanded Sonic Air Jet by Oxygen Flow Tagging", *Phy. Fluids,* 389 (1989).

Niederschulte, M. A., "Turbulent Flow Through a Rectangular Channel," Ph.D. Thesis, University of Illinois, Urbana, Illinois (1989).

Nouri, J. M., J. H. Whitelaw, and M. Yianneskis, "Particle Motion and Turbulence in Dense Two-Phase Flows." *Int. J. Multiphase Flow,* **13**, 729 (1987).

Ojha, M., R.S.C. Cobbold, K. W. Johnston and R. L Hummel, "Pulsatile Flow Through Constricted Tubes: An Experimental Investigation Using Photochromic Tracer Methods", *J. Fluid Mech.* **203**, 173-197 (1989).

Park, J. T., R. J. Mannheimer, T. A. Grimley, and T. B. Morrow, "Pipe Flow Measurements of a Transparent Non-Newtonian Slurry", *J. Fluids Engineering* **111**, 331-336 (1989).

Popovich, A. T. and R. L. Hummel, "A New Method for Non-Disturbing Turbulent Flow Measurements Close to a Wall", *A.I.Ch.E. J.* **14**, 21-25 (1967).

Porter, G. B., "Introduction to Inorganic Photochemistry: Principles and Methods", *J. Chem. Ed,* **60**, 785 (1983).

Pikramenou, Z., and D. G. Nocera, "Luminescent Supramolecular Architectures: A Cyclodextrin Modified with a Europium(III) Crown Swing", *Inorg Chem.***31,** in press (1992)..

Scandola, F., and V. J. Balzani, "Energy-Transfer Processes of Excited States of Coordination Compounds", *J. Chem. Ed.,***60**, 814 (1983).

Sengun, M. Z. and R. F. Probstein, "Bimodal Model of Slurry Viscosity With Application to Coal-Slurries. Part 2. High Shear Limit Behavior." *Rheologica Acta,* **28**, 394-401 (1989).

Steen, M., "The Application of Refractive Index Matching for Two-Phase Flow Measurement in Turbulent Fiber Suspensions by Laser Doppler Anemometry", Norwegian Pulp and Paper Institue, Oslo (1989).

Turro, N. J. *Modern Photochemistry*, Benjamin/Cummings, Menlo Park, CA (1978).

Tsuji, Y., Y. Morikawa and H. Shiomi, "LDV Measurements of an Air-Solid Two-Phase Flow in a Vertical Pipe", *J. Fluid Mech.*, **139**, 417 (1984).

Yianneskis, M. and J. H. Whitelaw, "Velocity Characteristics of Pipe and

Jet Flows with high Particle Concentration", *Liquid-Solid Flows and Erosion Wear in Industrial Equipment.* ed M. C. Roco, ASME N.Y. 12-15 (1983).

Yu, J.-a,, R. B. Lessard and D. G. Nocera, "Direct Observations of Intramolecular Energy Transfer from a β-Diketonate to Terbium(III) Ion Encapsulated in a Cryptand",*Chem Phys Lett.* **187**, 263-268 (1991).

Zisselmar, R. and O. Molerus, "Investigation of Solid-Liquid Pipe Flow With Regard to Turbulence Modification", *The Chem. Engr. J.* **18,** 233-239 (1979).

CHAPTER 4

MAGNETIC RESONANCE IMAGING OF MULTIPHASE SYSTEMS

J. E. Maneval
R. L. Powell
M. J. McCarthy
K. L. McCarthy

4.1 INTRODUCTION

The purpose of this Chapter is to discuss the use of nuclear magnetic resonance (NMR) in the study of multiphase systems. In particular, we will emphasize recent efforts in applying NMR imaging techniques. NMR imaging (or MRI as it is commonly abbreviated) is an extension of classical NMR methods that allows for the spatial mapping of NMR-measureable quantities. As such, NMR imaging does not involve measurements of new physical quantities but rather adds a new dimension to an already flexible and useful technique.

The primary motivations for using NMR methods in the study of multiphase systems are NMR's non-invasive nature and its potential for making measurements in "real" systems. The discussion of applications (Section 4.5) should show that NMR can be used in a variety of complex and difficult-to-measure systems. For example, NMR is not limited to transparent materials, as an optical probe is and, in contrast to X-ray or gamma-ray methods, does not use ionizing radiation to generate the signals used in measurement. Because of these qualities, the development of NMR methods for medical imaging purposes has been intense. As will be shown, however, the qualities that make NMR a powerful imaging modality also make it a versatile probe for multiphase systems of scientific and engineering interest.

The discussion is set up as follows. First, the basic principles of NMR and their extension to NMR imaging are reviewed. The objective is to present a summary of the physical basis of NMR as well as define the vocabulary that is used to design and interpret experiments in static and flowing systems. With this foundation, specific applications of NMR in multiphase systems are illustrated. NMR imaging has been used to study a wide range of systems and the discussion will highlight the capabilities of NMR rather than focus on a detailed review. Finally, comments on the interpretation and use of NMR measurements in multiphase systems will be made.

4.2 BASIC CONCEPTS

Nuclear magnetic resonance methods are based on the magnetic properties of materials. Specifically, NMR experiments manipulate and measure the macroscopic magnetization of a sample of material. For almost all of the studies quoted in this chapter, the paramagnetic properties of hydrogen nuclei will be the most important quantity for measurement. Thus, materials with an appreciable quantity of hydrogen (e.g., water, oils, plastics or foods) are potentially NMR-observable and this accounts for the wide applicability of NMR methods. Restrictions will be pointed out in the discussions below.

4.2.1 NMR Physics

Materials on which NMR measurements are made are not magnetic in the sense that they exhibit a permanent magnetic moment. Rather, the sample magnetization is induced by the application of a strong (0.1 to 10 T) magnetic field. In the presence of this polarizing field, NMR-active nuclei (see Table 4.1, below) align with the field and produce a small, but measurable, magnetic moment of their own. The magnitude of the induced magnetization is quite small, usually on the order of 10^{-6} of the applied field. The relative weakness of the effect is the primary reason that NMR can be considered as being a non-invasive method. It also indicates that NMR is an inherently insensitive method of measurement and, generally, about 10^{19} nuclei (or spins) are required to give an adequate signal-to noise ratio in a measurement. This number of spins provides a lower limit to the concentration-resolution capabilities of NMR methods. A more in-depth discussion of the sensitivity of NMR imaging methods is provided by Brunner and Ernst (1979).

Most of the efforts using NMR imaging in multiphase systems have focussed on the hydrogen nucleus but there are other "NMR-active" nuclei that are of potential use in NMR studies. A list of some of the more common nuclei and their NMR characteristics is given in Table 4.1. Two measures of the detection sensitivity for a particular nucleus are provided in the Table. The first measure, the intrinsic sensitivity, indicates the strength of the NMR signal for the nucleus relative to the signal that may be obtained from a proton at a constant magnetic field strength. This measure is useful in estimating the relative feasibility of obtaining images based on the signal from a given nucleus under a given set of conditions. A low sensitivity implies that there will be difficulty in generating noise-free images.

The second measure is more indicative of the signal strengths that would be encountered in "natural settings" and could be termed an effective sensitivity. The effective sensitivity is calculated from the intrinsic sensitivity by multiplying the intrinsic sensitivity by the percent natural abundance of the particular isotope. For example, ^{13}C nuclei have an intrinsic sensitivity that is about 2% of that of an equivalent number of ^{1}H nuclei but, due to the relatively low natural abundance of the ^{13}C isotope (~ 1%), the signal that is available for image formation (say from the ^{13}C atoms in a plastic sample) is about 10^{-4} times that of the proton (under identical conditions). Unless enrichment techniques are used, creating images directly from ^{13}C will be difficult. Note also that the predominant form of carbon, ^{12}C, is not NMR-active and thus is unobservable by NMR methods.

The spin number given in Table 4.1 is an indication of the magnetic character of a nucleus. Spin-1/2 nuclei are the simplest NMR nuclei in terms of understanding their magnetic behavior. For non-interacting, spin-1/2 nuclei, a classical-mechanical model (see below) may be used to describe the spin dynamics while all other spin numbers require a full quantum-mechanical treatment. Subsequent discussion in this chapter will focus on the dynamics of spin-1/2 nuclei and on the dynamics of protons in particular.

Table 4.1
Some NMR-Active Nuclei and their Characteristics*

Nucleus	Spin Number	Gyromagnetic Ratio ($s^{-1}T^{-1}$)	Intrinsic Sensitivity (^1H = 1)	Natural Abundance Sensitivity (^1H = 1)
^1H	1/2	2.765×10^8	1.000	1.000
^2H	1	4.11×10^7	0.015	2.4×10^{-6}
^{13}C	1/2	6.73×10^8	0.023	2.5×10^{-4}
^{14}N	1	1.94×10^7	0.002	1.9×10^{-3}
^{15}N	1/2	-2.71×10^7	0.002	6.8×10^{-6}
^{17}O	5/2	-3.63×10^7	0.019	7.1×10^{-5}
^{19}F	1/2	2.52×10^8	0.846	0.846
^{23}Na	3/2	7.08×10^7	0.130	0.130
^{31}P	1/2	1.08×10^8	0.083	0.083

* Condensed from Morris (1986).

4.2.2 Pulse Sequences and the Bloch Equations

The NMR techniques discussed here are based on classical pulsed-NMR spectroscopic techniques. Pulsed-NMR spectroscopy uses various sequences of radio-frequency (RF) pulses to investigate the molecular state of an NMR-active nucleus. A schematic of a typical system used for an NMR experiment is given in Figure 4.1. The primary elements of the system are the magnet, a coil for transmitting and receiving RF energy (pulses and signals), an RF transmitter and a data acquisition system. For imaging applications, a system of coils for generating magnetic-field gradients is also included. The details of an NMR spectrometer or imager will vary considerably depending on the exact nature of the phenomena to be observed but Figure 4.1 gives the essence of what is involved in the measurement process. More information on the details of particular implementations can be found in Morris (1986) Mansfield and Morris (1982), Fukushima and Roeder (1981), and Gadian

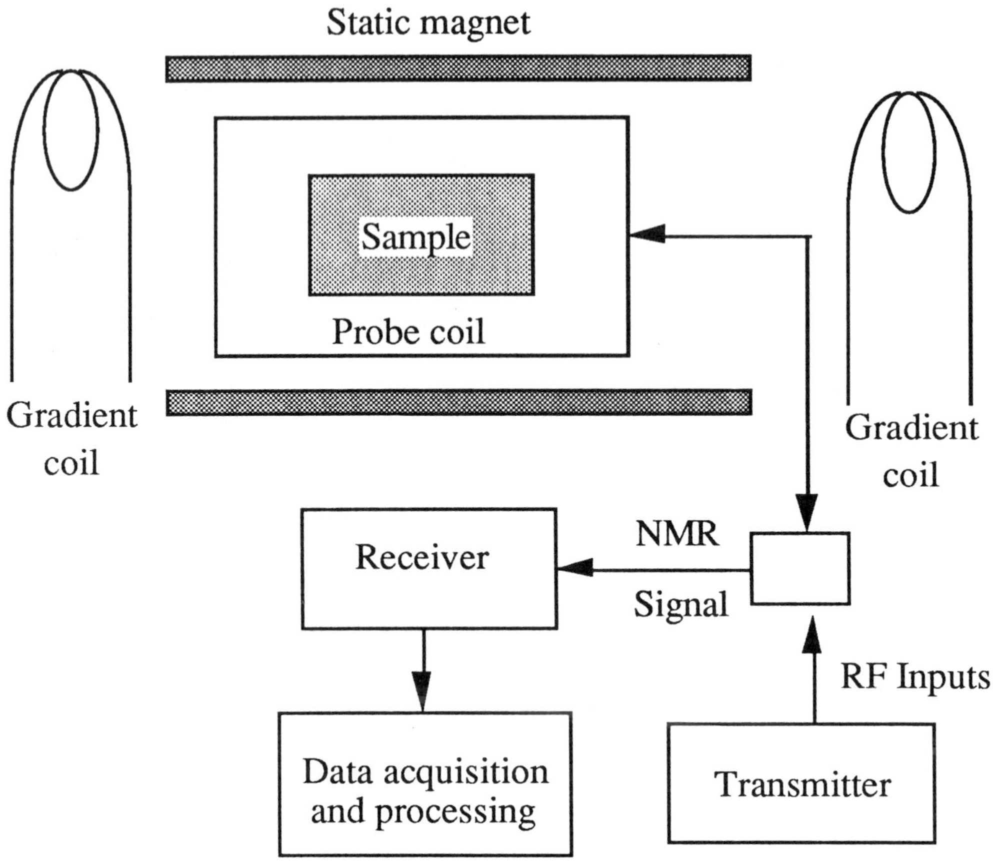

Figure 4.1. Schematic of a magnetic-resonance-imaging experimental set-up showing the primary apparatus and signal flows

(1982). Also, many of the references provided in Section 4.5 give details on the systems and techniques used in their measurements.

The interpretation of the results from an NMR experiment requires a model for the interaction of the spins in a sample with the applied RF pulse as well as with time-varying and static magnetic fields. For an assembly of non-interacting spin-1/2 nuclei, this model is given by (Abragam, 1961)

$$\frac{d\mathbf{M}}{dt} = \gamma\, \mathbf{M} \times \mathbf{B} \tag{4.1}$$

where **M** is the sample magnetization vector, **B** the external field vector and γ the gyromagnetic ratio for the nucleus (Table 4.1). These equations, with the addition of relaxation terms discussed below, are usually called the Bloch equations. They play a central role in understanding the dynamics of NMR experiments and relating NMR measurements to macroscopic quantities of interest. For a uniformly directed static field, $\mathbf{B} = B_0\mathbf{k}$, this set of equations yields a coupled set of ordinary differential equations for M_1 **i** and M_2 **j**. The solution to these equations predicts a precessional motion of the magnetization about the direction of the static field, much as a spinning top precesses in the earth's gravitational field. The frequency of this precession is

$$\omega_0 = \gamma B_0 \tag{4.2}$$

with clockwise rotations of the magnetization vector implied for positive ω_0. Equation 4.2 is referred to as the Larmor relation and is of fundamental importance to the interpretation of NMR results. The Larmor relation expresses the proportionality between the frequency at which a given nucleus absorbs and emits electromagnetic radiation and the magnetic field strength the nucleus finds itself in. In typical NMR systems, the Larmor frequency ranges from 1 to 500 MHz depending on the applied field strength. For example, at 2 Tesla, the proton frequency is 85.5 MHz.

NMR experiments usually deal with frequency information at or near the Larmor frequency. For this reason, it is useful to transform the Bloch equations to a coordinate frame which rotates at the Larmor frequency about the axis defined by the applied field direction (taken to be the z-axis). With ω_r as the frame rotation frequency (with counter-clockwise rotation of frame implied for positive rotations), the Bloch equations may be re-written as

$$\left(\frac{d\mathbf{M}}{dt}\right)_{rot} = \left(\frac{d\mathbf{M}}{dt}\right)_{lab} + \omega \times \mathbf{M} = \gamma \mathbf{M} \times \mathbf{B}_{eff} \tag{4.3}$$

where $\mathbf{B}_{eff} = \mathbf{B} + (\omega_r/\gamma)\mathbf{k}$ is the effective magnetic field in the rotating frame and **M** now stands for the magnetization in the rotating frame.

The effective field has several components that are of importance to the execution and analysis of an experiment. Choosing the frame-rotation frequency equal to the Larmor frequency ($\omega_r = -\omega_0$) focuses attention on frequencies around the Larmor frequency and the effective field in the rotating frame may then be expressed as

$$\mathbf{B}_{eff} = B_1 \mathbf{i} + \Delta B_0 \mathbf{k} \tag{4.4}$$

where ΔB_0 is a perturbation to the static field, B_0, and $\Delta B_0 << B_0$. A field perturbation can arise from instrumental and experimental sources. As will be evident from the discussion below, the nature and use of this field is the basis for understanding the extension of NMR to imaging and flow measurement. The B_1 field is a magnetic field applied at the Larmor frequency and represents the time-dependent RF pulses used to probe the spin system. The RF pulses are applied in a plane transverse to the direction of the static field and will be assumed to be directed along the x-axis of the rotating frame, as indicated in Eq. 4.4.

The Bloch equations as they stand do not describe all of the observed behavior of a system of spins. Specifically, there is no explicit means by which the spins can transfer the energy they receive from an RF pulse to their surroundings in order to return to their equilibrium state. To account for the observed relaxation processes, Bloch (1946) proposed the following modification to Eq. 4.3

$$\frac{d\mathbf{M}}{dt} = \gamma \mathbf{M} \times \mathbf{B}_{eff} - \frac{(M_x \mathbf{i} + M_y \mathbf{j})}{T_2} + \frac{(M_o - M_z)\mathbf{k}}{T_1} \tag{4.5}$$

where the quantities T_1 and T_2 are the so-called spin-lattice and spin-spin relaxation times, respectively and M_0 is the magnetization in equilibrium with the static field. Equation 4.5 assumes that the relaxation process are simple first-order process and this is found to be true for a wide range of liquids and liquid-like materials. The imaging techniques discussed in this chapter are assumed to have dynamics which follow Eq. 4.5.

The spin-lattice relaxation time, T_1, can be thought of as a time constant characteristic of an enthalpic transfer of energy between a sample and its surroundings (the lattice). Spin-lattice relaxation describes the return of the magnetization to its equilibrium alignment with the static field. The spin-spin relaxation time, T_2, on the other hand, can be thought of as a time constant characteristic of an entropic or randomizing process that degrades the coherence (ordered-ness) of an NMR signal. In all cases, $T_1 \geq T_2$. Relaxation times play an important role in the practical implementation of NMR experiments. Spin-lattice relaxation governs how rapidly a pulse sequence can be repeated while spin-spin relaxation control how long a signal will be available for observation. More detail on the nature of these relaxation processes can be found in Abragam (1961) and in Morris (1986).

4.2.3 Spin Echoes

An important NMR phenomenon that is used extensively in imaging practice is the spin echo. The spin echo was first observed by Hahn (1950) during attempts to improve measurement techniques for spin-spin relaxation rates. Spin echoes are best illustrated by examining the way in which they are generated. Consider the pulse sequence shown in Figure 4.2a. This spin-echo pulse sequence consists of two RF pulses separated by a time delay, τ. Using the Bloch equations (Eq. 4.1), it can be shown that an RF pulse of duration t_p rotates the magnetization vector by an angle of

$$\theta = \gamma B_1 t_p \tag{4.6}$$

about the axis of the B_1 pulse (the x-axis). As indicated on Figure 4.2b, the first pulse of the spin-echo sequence is designed so that it rotates the

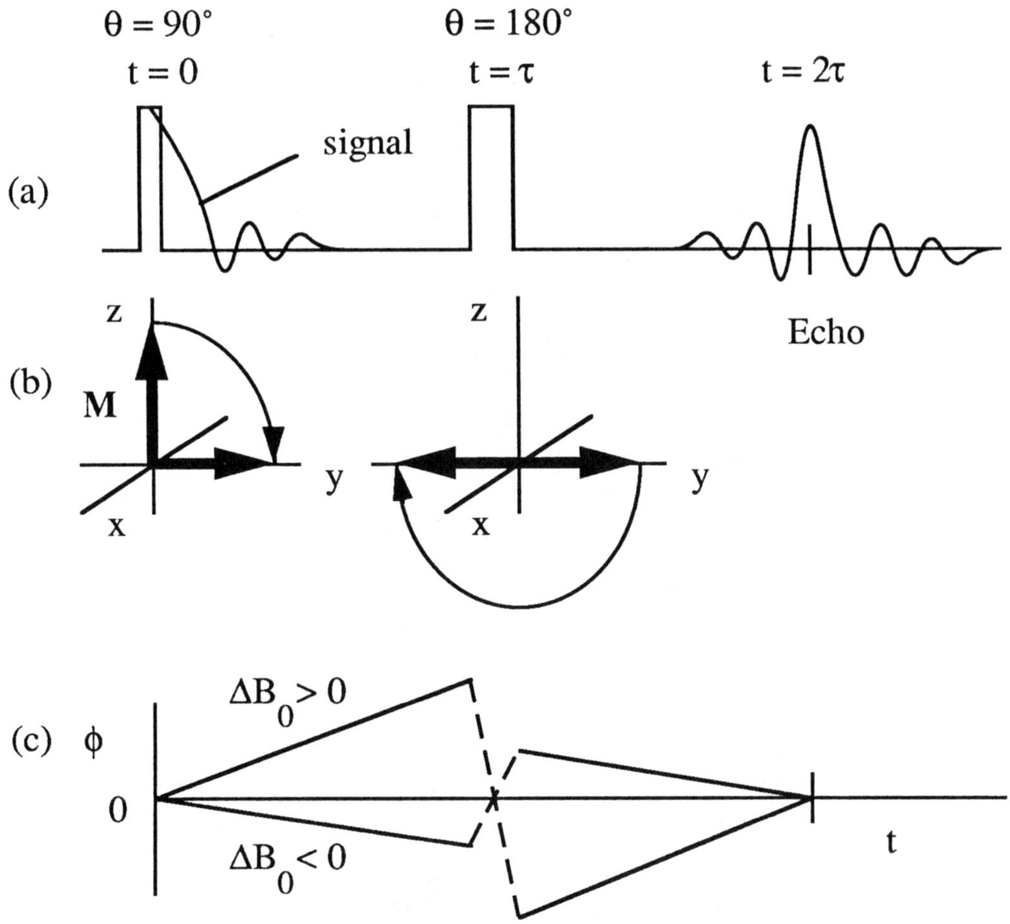

Figure 4.2. The Hahn spin-echo sequence. (a) Sequence of radio-frequency pulses and observed signal. (b) Rotation of magnetization vector (in the rotating frame) due to pulses. The magnetic field is assumed to be uniform. (c) Accumulation of phase angles for spins in regions where the field offset, ΔB_0, is non-zero. Acccumulations in only two such regions are shown, for clarity.

magnetization 90° from the z-axis and into the x-y plane. The second pulse rotates the magnetization an additional 180° about the x-axis.

Once the magnetization is in the x-y plane, it begins to precess at the Larmor frequency and produce an observable signal in the receiver coil. If there were no dispersive or relaxation processes, the net effect of the two pulses would be to re-orient the magnetization vector along the (-y)-axis (Figure 4.2b). However, if, for any reason, the static field is non-uniform (i.e., ΔB_0 is non-zero) different portions of the sample will be in slightly different magnetic fields. From the field-frequency equivalence expressed by the Larmor relation (Eq. 4.2), a distribution of fields implies a distribution of

Larmor frequencies within the sample and so there will be a distribution of phase angles in the sample as time progresses. A distribution of phase angles leads to an attenuation of the observed NMR signal from the sample.

A graphical representation of the effects of a non-uniform field on the signals observed in a spin-echo experiment is illustrated in Figure 4.2c. The phase angles accumulated by spins in regions of differing ΔB_0 are plotted as a function of elapsed time during the pulse sequence. By use of the Bloch equations, it can be shown that, during an interval, t, after the 90° pulse, a spin in a region with a particular value of ΔB_0 will accumulate a phase angle of

$$\phi = \gamma \Delta B_0 t \tag{4.7}$$

relative to a spin in a region where ΔB_0 is zero. The 180° pulse has the effect of reversing the phase angle that a spin acquires prior to the pulse (see Figure 4.2c) so that the net phase angle that accumulates *after* the 180° pulse is given by

$$\phi = \gamma \Delta B_0 (2t - \tau) \tag{4.8}$$

At a time $t = 2\tau$ after the 180° pulse (the "echo time"), the dispersion in phase angles is completely eliminated and a signal "reappears", producing a spin echo. For times $t > 2\tau$, the dephasing process starts again and the signal once more drops off to zero.

The re-appearance of the signal is sometimes called a refocussing and so 180° pulses are often called refocussing pulses. Refocussing pulses and echo formation permit a temporal separation between excitation (the 90° pulse) and detection (at the echo time) which has many advantages in designing experiments and equipment. Also, the amplitude of the echo at the echo time is governed by true spin-spin relaxation processes as well as the net displacement of a spin during the echo time. Hence, spin echoes provide a means of estimating the time constants for relaxation processes as well as for molecular and macroscopic motion. More information on measurement of motion will be given in Section 4.4, below.

4.3 SPIN DENSITY IMAGING

Classical NMR measurements of spin density provide no information on the spatial distribution of spins in a sample. In these measurements, observed NMR signals are proportional only to the total amount of material in a sample. NMR imaging, on the other hand, actually measures spatial variation of the spin density. The primary concepts for understanding the extension of NMR to NMR imaging are the Larmor relation and the magnetic field gradient.

The gradient of a magnetic field is, in general, a second-rank tensor. In NMR experiments, however, it suffices to consider only a portion of the components in the full gradient. Hence, NMR imaging deals with a magnetic-field gradient *vector*, **g**, defined by

$$\mathbf{g} = (\nabla\mathbf{B})\cdot\mathbf{k} = \nabla B_z \tag{4.9}$$

The field gradient is a generator of the offset field, ΔB_0. For a spin in a gradient, the offset field is given by

$$\Delta B_0 = \mathbf{r}\cdot\mathbf{g} = x g_x \tag{4.10}$$

where **r** is a position vector and the x-axis has been taken as the direction of the field gradient (i.e., $\mathbf{g} = g_x\,\mathbf{i}$) though other directions are possible.

The field gradient provides a means of generating a controlled dispersion in the Larmor frequencies within a sample. Using the Larmor relation, the precessional frequency for spins in the direction of the gradient is given by

$$\omega = \gamma\Delta B_0 = (\gamma g_x)x \tag{4.11}$$

which implies a proportionality between spatial differences in the direction of the gradient and frequency differences which are measured in an NMR experiment ($\Delta\omega \sim \Delta x$). The magnetic field gradient serves as the link between NMR frequency measurements and NMR imaging measurements and is the basis for almost all imaging strategies in current use.

The role of the gradient in NMR imaging is best illustrated by examining a commonly used pulse sequence, the so-called spin-warp pulse sequence (Edelstein, et al., 1980). The spin-warp sequence serves as the basis of most present-day NMR imaging methods (with exceptions as noted below) and contains many of the essential features of the use of gradients in NMR imaging. The pulse sequence is shown in Figure 4.3. The sequence consists of the basic spin-echo sequence from Figure 4.2 but includes several new features to allow for imaging. Creation of an NMR image via the spin-warp sequence takes place in three steps:

1. Selective excitation of a "slice" of finite thickness from an object by a selective 90° B_1 pulse
2. Phase-encoding of spatial information in one of the directions on the slice
3. Frequency-encoding of the spatial information in the orthogonal directionon the slice

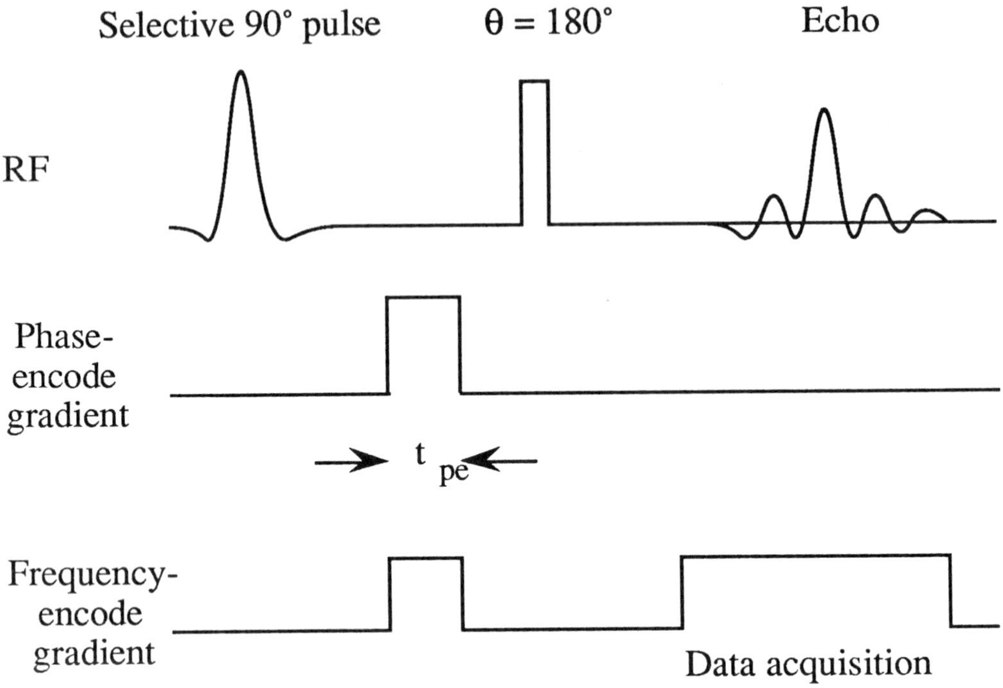

Figure 4.3. Spin-echo/spin-warp imaging sequence. Data for a full image is produced by repeating the entire sequence 2N+1 times using a different value of the phase-encode gradient on each repetition.

Each of these steps will be reviewed in turn. All results given may be derived from the Bloch equations (Eq. 4.5) in a straightforward manner and in the interest of clarity, many of the details have been omitted.

4.3.1 Selective Pulses

The selective pulse acts to physically restrict the portion of the sample that will ultimately produce the measured NMR signal. Proper design and analysis of the pulse can be quite complicated and the details would not add much to the basic concept of the process. A review of the selective-pulse process is provided by Warren and Silver (1989). The end result of a selective pulse is that a portion of spins in the sample is selectively excited by an RF pulse. Ideally, the excitation profile is given by

$$p_{sp}(z) = \begin{cases} 1 & |z - z_0| \le L_{sp}/2 \\ 0 & |z - z_0| > L_{sp}/2 \end{cases} \tag{4.12}$$

where L_{sp} is the slice thickness of the selective pulse, z_0 is the location of the slice center and **k** is taken as the unit vector normal to the slice. Actual selective pulses will produce approximations to this shape. The degree to which the ideal profile is approximated depends on interactions amongst the sample characteristics, the hardware used to generate the pulse and the time allowed for the pulse.

An important point to remember is that the use of a selective pulse implies that the resulting two-dimensional image actually arises from a three-dimensional region with a "hidden dimension" involving the slice thickness, L_{sp}. Proper interpretation requires a knowledge of the slice thickness as well as the in-plane resolution (which is controlled by the phase- and frequency-encoding processes discussed in Sections 4.3.2 and 4.3.3).

4.3.2 Phase Encoding

Phase-encoding is the process whereby information about the spin-density is encoded in the phase of the magnetization. During the phase-encode time, t_{pe}, a gradient is applied in order to create a distribution of phase angles across the sample in the direction defined by the phase-encode gradient. As noted in the discussion of spin echoes, this distribution of phase angles results in an attenuation of the signal when the echo is detected. However, since a gradient of known strength is applied, the amount and nature of this attenuation will be known and controllable.

Choosing the y-axis as the direction of the phase-encode gradient (i.e., $\mathbf{g}_{pe} = g_y\,\mathbf{j}$), an expression for the amplitude of the echo can be derived that reflects both the spin density as well as the dephasing effect of the gradient. The signal *at* the echo time can be written as

$$S(k_y) = \int \rho(y) \exp(-i2\pi k_y y)\, dy \tag{4.13}$$

where $\rho(y)$ is the spin density in the y-direction and k_y is a spatial frequency defined by

$$2\pi k_y = \gamma g_y t_{pe} \tag{4.14}$$

Equation 4.14 indicates that a single execution of the spin-warp sequence with a fixed value of the phase-encode gradient yields a single point in the spatial-frequency spectrum, $S(k_y)$. To obtain the full spectrum (i.e., to obtain the function $S(k_y)$ at many k_y values), the sequence must be repeated, each time using a different value of the phase-encode gradient. For example, if 2N+1 repetitions of the pulse sequence are used to estimate the spectrum, we have

$$S(k_y) = \int \rho(y) \exp(-i2\pi k_y y)\, dy \approx \sum_{n=-N}^{n=N} \rho(y) \exp(-i2\pi k_{y,n} y) \qquad (4.15)$$

where $k_{y,n}$ is the wavenumber in the phase-encode direction for the nth repetition of the sequence and is given by

$$2\pi k_{y,n} = \gamma g_{y,n} t_{pe} \qquad (4.16)$$

and $g_{y,n}$ is the value of the gradient during the repetition. The signal $S(k_y)$ in Eq. 4.15 is a discrete version of a Fourier transform of the spin density. An inverse-Fourier transformation will then yield the spin density, $\rho(y)$, in the direction of the phase encode gradient.

4.3.3 Frequency Encoding

As its name implies, frequency encoding encodes spatial information into the frequency of the NMR signal during detection. The basic idea is illustrated by considering the data acquisition portion of the pulse sequence shown in Figure 4.3. When the frequency-encode gradient, $\mathbf{g}_{fe} = g_x \mathbf{i}$, is applied, the signal from the sample *during* acquisition can be written as

$$S(k_x) = \int \rho(x) \exp(-i2\pi k_x x)\, dx \qquad (4.17)$$

where the spatial frequency in the direction of the frequency-encode gradient k_x is defined by

$$2\pi k_x = \gamma g_x (t - t_e) \qquad (4.18)$$

While the frequencies in the phase-encode direction, k_y, were acquired by changing the gradient magnitude over a series of pulse-sequence repetitions, the frequencies in the frequency-encode direction, k_x, are obtained all at once from the time-domain signal measured during signal acquisition. Putting these two encoding procedures together provides the raw data from which a spin-density image can be extracted.

The phase-encode process is inherently discrete and, because the frequency-encoded echo is acquired by digital methods, the net result of the spin-warp sequence is a two-dimensional array of numbers, $S(k_{x,m}, k_{y,n})$. Upon two-dimensional inverse-Fourier transformation, this array yields a two-dimensional array of intensities that correspond to the image of the spin density at a discrete set of positions, $\rho(x_m, y_n)$. A more detailed analysis of the pulse sequence (Maneval, et al., 1990) gives the result

$$S(k_x,k_y) = \iint \{\rho(x,y) \exp(-t_e/T_2)\} \exp[-2\pi k_x/(\gamma g_x T_2)] \times \exp(-i2\pi [k_x x + k_y y]) \, dxdy \qquad (4.19)$$

Equation 4.19 indicates that an inverse-Fourier transformation of the raw data will not directly produce the spin density image but rather will provide a relaxation-time weighted image. This image must be "corrected" for relaxation-time effects if the true image density is to be recovered. Correction is accomplished via a series of images (each acquired with a different echo time) and produces estimates the true local spin density as well as local values of the spin-spin relaxation time (Haacke, 1987; Haacke, et al., 1988).

4.3.4 Other Sequences

While the spin-warp sequence has wide applicability, it has weaknesses. The foremost shortcoming of the sequence is the speed at which an image can be acquired. Since the sequence must be repeated to fully sweep out the frequencies in the phase-encode direction, image times are usually on the order of 1 to 10 minutes. While this may be acceptable for many applications, it precludes the use of the sequence for the study of processes with time constants which are shorter than approximately five minutes or so.

In an effort to increase the image acquisition speed, Mansfield (1977) proposed the echo-planar method. In the echo-planar method, the entire image is acquired in a single execution of the pulse sequence. This feat is accomplished by creatively sampling the reciprocal space (*k*-space). Rather than sample *k*-space with a set of lines as the spin-warp sequence does (a line of k_x-values is acquired at fixed k_y for each repetition of the pulse sequence), the echo-planar technique varies the two gradients simultaneously during the pulse sequence to provide a more rapid and efficient sampling of *k*-space. Image times can be on the order of tens of milliseconds and are usually limited by the rate at which the gradients can be switched during pulse sequence execution. Recent reviews of the method (Cohen and Weisskopf, 1991; Guilfoyle, et al., 1991) give an indication of what is possible and the discussion in the applications section will show the utility of the method for the study of dynamic processes with short time-constants.

While the echo-planar method is a powerful technique, it does require that the spin-spin relaxation time be long enough to allow for extensive gradient switching. For many materials (oil cores, rocks and some plastics), these relaxation times are short and echo-planar methods are not easily applied. Two other approaches are available for these types of materials - the gradient-recalled echo method and projection reconstruction. In the gradient-recalled echo method, the 180° refocussing pulse is eliminated in favor of an "echo" generated by switching the direction of the applied field gradient

(Chang and Komoroski, 1989; Keuthe, 1989). By eliminating the refocussing pulse, an image may be formed can be produced in a relatively short time interval.

Projection reconstruction, like the echo-planar method, is an alternate approach to acquiring data in *k*-space (Brooks and DiChiro, 1976; Listerud, et al., 1989; Mansfield and Morris, 1982; Morris, 1986). While the method can be used with spin echoes (Callaghan, et al., 1988), projection reconstruction is generally used to produce images directly from the signal generated by the selective pulse. Since echoes are not needed, materials with short spin-spin relaxation times can be imaged in this way.

In summary, whatever the methods used, the net result of an NMR experiment is a two-dimensional frequency version of the desired spin density distribution, $\rho(x,y)$. The final NMR image is a suitable inverse transform of this data. The exact nature of the inverse may depend on the imaging method used. Additionally, interpretation of the final image depends on how the data was acquired (e.g., see Eq. 4.19). This issue will be elaborated on more fully in Section 4.6.1.

4.4 MEASUREMENT OF FLOW AND DIFFUSION

As with imaging, measurement of flow via NMR methods is vitally dependent on creative use of the magnetic field gradient. We will consider two principal methods used to measure flow velocity by NMR;

1. Time-of-Flight (TOF) - Two or more images are used to detect the displacement of spins during a fixed evolution time. The velocities are calculated from the measured displacements and the evolution time.

2. Phase encoding - The phase of magnetization is manipulated to provide information on velocities similar to the way that a phase-encode gradient is used to measure spin position.

A recent review of NMR methods for flow measurements (Caprihan and Fukushima, 1990) provides an excellent discussion of various pulse sequences that can be used to measure flow. Here, the basic concepts behind flow measurements will be presented to provide a foundation for understanding the range of options discussed in that review.

4.4.1 Time-of-Flight Methods

The TOF method is the more conceptually straightforward of the two methods. A time-of-flight measurement is a straightforward application of the Lagrangian description of motion. The method consists of three steps. First, a portion of the flowing material is "tagged" in order to distinguish it from the rest of the flow field and thereby focus attention on a material element of interest. This step defines a reference position for the element, $\mathbf{r}_0 = \mathbf{r}(t = 0)$, and sets the time origin for the experiment.

The next step is to allow the element to move with the velocity field for a set time interval, Δt, referred to as the evolution interval. Finally, at the end of the evolution interval, the new position of the element, **r**, is determined. With an assumption of steady flow during the evolution interval, the velocity field is calculated by dividing the displacement from the reference configuration by the duration of the evolution interval to give,

$$\mathbf{v} = \frac{(\mathbf{r} - \mathbf{r}_0)}{\Delta t} \tag{4.20}$$

where **v** is the velocity and **r** and $\mathbf{r}_0$ are the locations of the material element at times t = 0 and t = Δt, respectively.

Measurement of all components of the velocity at every point in an arbitrary flow field would be a formidable task. Restricting attention to rectilinear flows allows for considerable simplifications in analysis and experimentation while still covering many flows of practical interest. Rectilinear flows are those flows for which the velocity field is given by

$$\mathbf{v} = v_z(x,y)\mathbf{k} \tag{4.21}$$

where **k** is the unit vector in the direction of the flow (taken as the z-direction).

Since v_z is independent of the z-coordinate, the elements of any tagged region will follow streamlines which are parallel to the z-axis (Figure 4.4). Hence, the initial tagged region can be any planar element which is perpendicular to this axis. This characteristic of rectilinear flow is important to the application of NMR imaging to such flows because planar regions are easy to generate and because two-dimensional images of planar regions are straightforward to acquire (see, for example, the discussion of the spin-warp sequence in section 4.3).

There are several different means by which the displacement surface can be detected. Majors et al. (1989) developed a second-spin-echo technique to reconstruct the displacement surface while Frahm, et al. (1985) utilized a three-pulse stimulated-echo sequence to accomplish the same result. In either approach a three-dimensional data matrix is generated which is used to calculate the velocity field, $v_z(x,y)$. From Figure 4.4, the velocity at the point (x',y') is given by

$$v_z(x',y') = \frac{z(x',y';t = \Delta t) - z(x',y';t = 0)}{\Delta t} \tag{4.22}$$

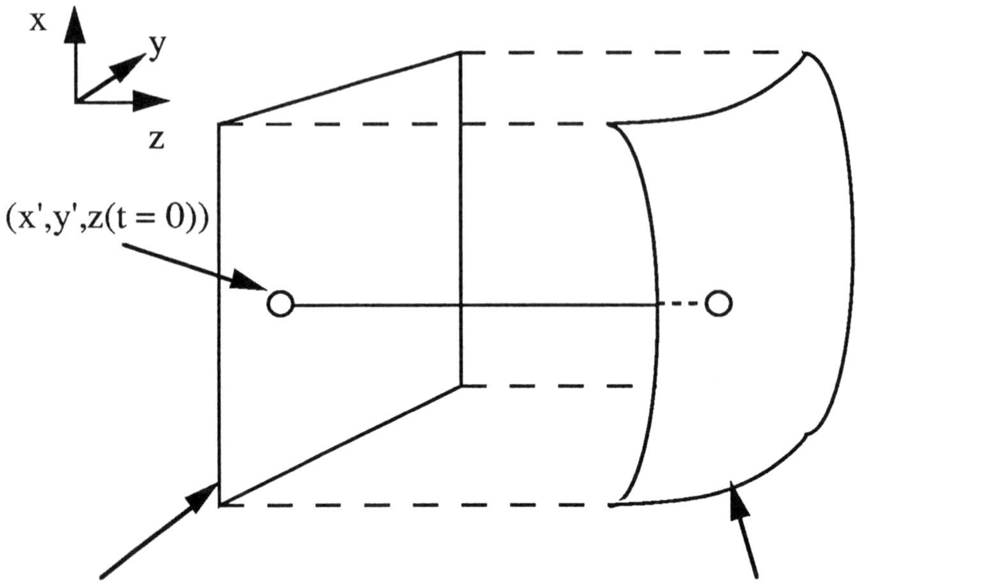

Figure 4.4. Illustration of displacement for a rectilinear flow. At t = 0, a tag plane is generated via a selective pulse. The flow field then deforms the tag for a time, Δt, at which the displacement surface, z(x',y';Δt) is detected.

4.4.2 Phase Methods

To understand the phase-encoding process for velocity measurements, we return to the Bloch equations and consider the motion of a spin in a magnetic field gradient. For coherent (bulk) motion of a spin, the displacement, **r**(t), is given by

$$\mathbf{r}(t) = \mathbf{r}_0 + \int_0^t \mathbf{v}(s)\,ds \tag{4.23}$$

If the spin moves at a constant velocity, v_z, in the direction of a field gradient it will acquire a phase angle in proportion to its motion. To see this, the Bloch equations along with Eq. 4.23, can be used to give

$$\phi = \gamma \int_0^t \mathbf{r}(s)\cdot\mathbf{g}(s)\,ds \tag{4.24a}$$

$$= \gamma z_0 \int_0^t g_z(s)\,ds + \gamma v_z \int_0^t s\,g_z(s)\,ds \tag{4.24b}$$

$$= \gamma (z_0 m_0 + v_z m_1) \quad (4.24c)$$

where m_0 and m_1 are the zeroth and first moments of the gradient waveform in the time domain . If the gradient is designed such that $m_0 = 0$ but $m_1 \neq 0$, this phase angle will be proportional to the velocity of the spin. Just as the phase of the magnetization was used to measure the spin density distribution, a properly designed gradient allows the phase to measure the distribution of velocities in a sample

Consider the pulse sequence given in Figure 4.5. The sequence is similar to the spin-warp sequence in Figure 4.3 with the exception that the phase-encode gradient shape has been changed so that $m_0 = 0$ but $m_1 \neq 0$. For a single execution of this pulse sequence, the amplitude of the spin echo will reflect the distribution of the velocities of the spins in the sample. Defining $f(v_z)$ as a velocity spectrum such that $f(v_z)dv_z$ is the fraction of spins in the sample with velocity between v_z and v_z+dv_z, we follow the development for phase encoding (Section 4.3) and use to the Bloch equations to find that the echo amplitude can be written as (compare to Eq. 4.13)

$$S(q_z) = \int f(v_z) \exp(-i2\pi q_z v_z)\, dv_z \quad (4.25)$$

where $2\pi q_z = \gamma\, g_z T^2$ is a reciprocal velocity which is the Fourier conjugate variable to the velocity and 2T is the duration of the velocity-encoding gradient. As was indicated in the discussion of phase encoding, repetition of the pulse sequence with different values of the gradient will "sweep out" the reciprocal domain and give the entire function, $S(q_z)$. An inverse-Fourier transformation of this function will then provide the velocity spectrum, $f(v_z)$.

The velocity spectrum is not the same as the velocity profile or distribution. The velocity profile gives the velocity as an explicit function of the spatial coordinates, $v_z(\mathbf{r})$. The velocity spectrum, on the other hand, is a measure of the velocities within a region without regard to the exact spatial distribution of the velocities. Hence, one could consider Eq. 4.25 as a definition of the function $f(v_z)$, a function that is measured by this NMR experiment. For well-ordered and simple flow fields (laminar flow in a tube, for example), there is a unique relation between the velocity spectrum and the velocity profile. In more complex settings, the relation is not unique as many flow fields could potentially generate the same spectrum.

Moran (1982) and Nalcioglu and co-workers (Nalcioglu and Cho, 1987; Xiang and Nalcioglu, 1988; 1989; 1990) have proposed and developed a velocity measurement approach based on phase-measurement concepts. In its full generality (Moran, 1982), the approach involves a six-dimensional joint position-velocity distribution, reminiscent of the phase-space methods in

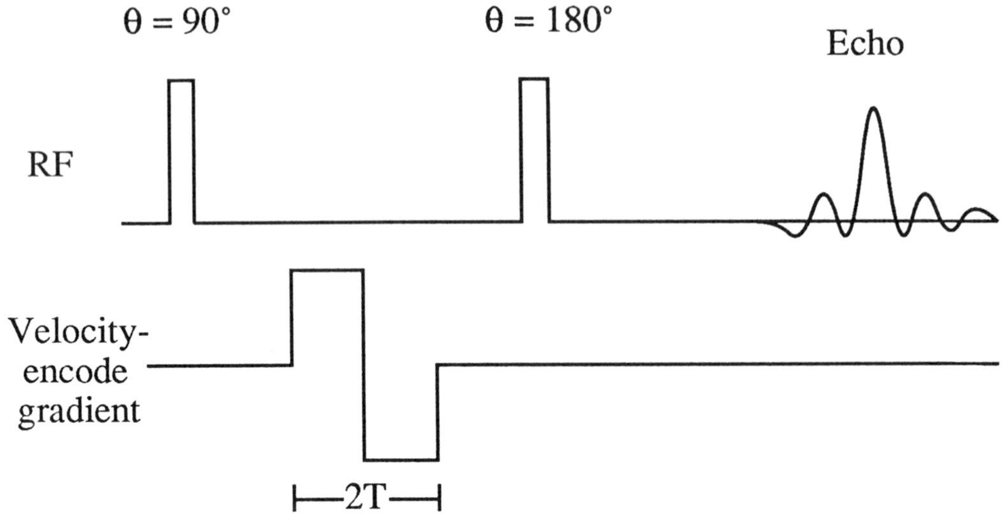

Figure 4.5. Velocity-spectrum measurement sequence. The height of the echo is proportional to the distribution of velocities (in the direction of the gradient) within the sample. The full velocity spectrum is obtained by repeating the sequence 2N+1 times, each time using the same shape but a different magnitude for the velocity-encode gradient.

statistical mechanics. In practical terms, the high-dimensionality of the method leads to long acquisition times and unmanageable data sets. However, lower-order "projections" (Xiang and Nalcioglu, 1990) as well as simple flow fields allow the method to be used as effectively as the time-of-flight method.

4.4.3 Diffusion

Bulk motion of a spin is a coherent process since the displacement of the spin is governed by a deterministic velocity field. In contrast, diffusive motion is a random process implying that the motion of a given spin can no longer be predicted with precision. This difference in the character of the motion requires a different approach to the description of the motion and produces different effects on NMR signals.

Measurement and analysis of diffusion effects have had a long history in the NMR literature, beginning with the spin-echo studies of Hahn (Hahn, 1950). More recent efforts have focussed of the use of the pulsed-field-gradient (PFG) techniques (Callaghan, 1984; Stilbs, 1987), first introduced by Stejskal and Tanner (1965). The basic PFG sequence is shown in Figure 4.6. It is similar in some respects to the sequence used to measure velocities (Figure 4.5) and indeed it may be used as such. For now however, the focus will be on its use to measure diffusive motion.

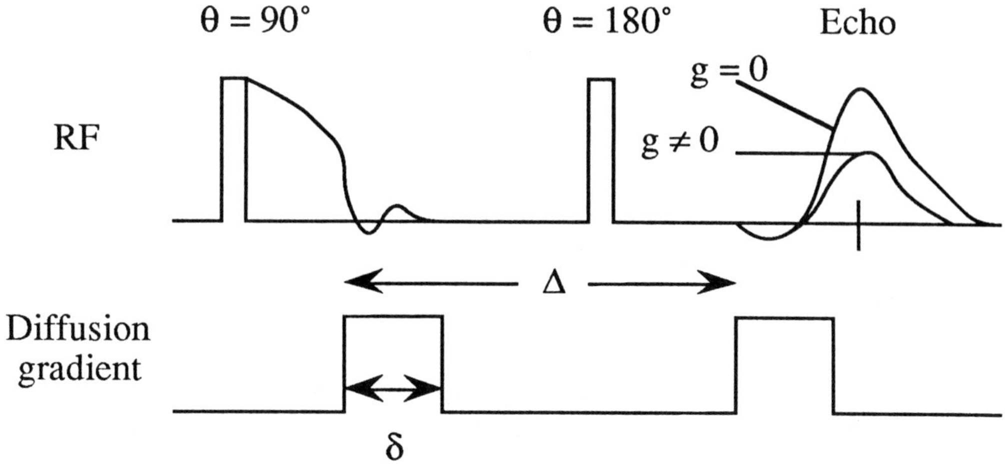

Figure 4.6. Basic pulsed-field-gradient (PFG) sequence. Measurement of the echo height as a function of the gradient strength, g, gradient duration, δ or spacing, Δ, provides a measure of the diffusion coefficient of a material.

As shown in Section 4.4.2, the effects of coherent motion manifest themselves in the phase of the NMR signal. In contrast, the effects of incoherent motion manifest themselves in an irreversible attenuation of the signal amplitude. If a spin does not move during the pulse sequence, there is no net effect of the gradient pulses in the PFG sequence ($m_0 = 0$ for this sequence) and the phase of the spin is unchanged at the echo time. However, if the spin makes a random jump to a new position during the interval between the gradient pulses, the dephasing effect of the first gradient pulse is imperfectly compensated by the second pulse and there results a net phase angle for the spin at the echo time. By considering an ensemble of spins moving via a random process (e.g., Brownian motion), it is clear that diffusive motion will result in a distribution of phase angles at the echo time which attenuates the echo amplitude. Since the motion is irreversible, diffusive motion manifests itself as an irreversible attenuation of the echo, much the same as intrinsic T_2-relaxation would.

More rigorous analyses of the effects of diffusion on NMR echo heights are available (Carr and Purcell, 1954; Douglass and McCall, 1958; Muller and Bloom, 1960; Woessner, 1961). For purposes of this Chapter, we simply quote the results. The extra attenuation brought about by diffusion in the PFG sequence is quantified by ratioing the echo height measured when $g \neq 0$ to that measured when $g = 0$. For liquids, this ratio is related to the parameters of the pulse sequence as

$$R = \exp\,[-\gamma^2 (\delta g)^2 D\,(\Delta - \delta/3)] \tag{4.26}$$

where g is the magnitude of the applied gradient, δ is its duration and Δ is the separation between the two gradient pulses (Figure 4.6). Measurement of R for several different gradient strengths or separations will give the diffusion coefficient from measurements of the slope of a semilogarithmic plot of R vs. Δ or g^2. The extension to spatially resolved measurements of local diffusion coefficients is discussed in Section 4.5.2.

4.5 APPLICATIONS

NMR and NMR imaging can be used in a variety of ways to study and characterize multiphase materials. For example, NMR has been used as an on-line monitor for bulk moisture content in foods (Rollwitz, 1963; Tollner and Rollwitz, 1987; Young, 1989) as well as measurement of the chemical components in a mixture (Abouelwafa and Kendall, 1980). Also, NMR-based methods have been developed that measure the flow of multicomponent and multiphase materials (Abouelwafa and Kendall, 1979; Jones and Child, 1976; Lynch and Segel, 1977). There is a long history of using bulk measurements of NMR relaxation times to probe the structure and dynamics of multiphase materials (Belton and Ratcliffe, 1985; Reeves, 1975) and recent efforts have extended the use of relaxation times to materials characterization (Halperin, et al., 1989) and estimation of permeability (Banavar and Schwartz, 1989). With the exception of some of the work by Halperin, et al. (1989), these efforts have focussed primarily on bulk property measurement of macroscopically homogeneous (though microscopically heterogeneous) materials and not on the spatial nature of the measurements.

The advance brought about by NMR imaging is that imaging allows one to make NMR measurements on a *spatially-resolved* basis. In principle, any classical NMR measurement or spectroscopic technique can be combined with the imaging methods outlined above to create a spatially resolved version of the measurements. In this way, NMR imaging becomes a powerful experimental method for the study of multiphase materials where the interest is in macro-scale heterogeneities.

We make distinctions among the various applications of NMR imaging on the basis of the method used for the particular measurement as well as the ultimate goal of the measurement. To this end, we consider the following categories;

1. Imaging of static and quasi-static systems - Such systems are investigated via a single image or a time sequence of images. It is usually assumed that the time constants of interest in the system are much longer than the time required to form an image. Motion and dynamics may be calculated from changes in sequences of images. The spin-warp, gradient-recalled echo and echo-planar sequences are typical of the methods used in these investigations.

2. Imaging of dynamic systems - The goal of these studies is usually to make spatially resolved measurements of bulk (coherent) flow or diffusive (incoherent) flow. Thus, the sensitivity of NMR to motion is directly exploited. The pulsed-field-gradient (PFG) methods are the usual basis of these methods.

Inevitably there will be overlap among these categories and some applications use more than one technique for the study of a system. However, since the NMR pulse sequence generally changes for each of the categories, this division has been found to be useful.

4.5.1 Imaging of Static and Quasi-Static Systems

A system is considered static if it does not change in any way for the duration of an NMR measurement. Quasi-static systems, on the other hand, will evolve in time but with time-constants that are much longer than the time required for image formation. As is evident, the definition of what is static depends on the system as well as the measurement technique.

4.5.1.1 *Static Systems*

The most straightforward application is to use imaging to see what is normally not visible. The internal structure of a biological specimen is typical of the samples studied. Indeed, the non-invasive study of internal human anatomy was (and still is) the primary driving force for development of NMR imaging methods. A good introduction to the application of NMR methods in these areas can be found in the books by Gadian (1982) and Morris (1986).

The use of NMR imaging in the characterization of oil-cores has been developed in recent years (Edelstein, et al., 1988; Rothwell and Vinegar, 1985; Vinegar, 1986). As a complementary technique to X-ray computed tomography (X-ray CT) and ultrasound, NMR imaging provides an important means of obtaining structural and chemical information about the fluids contained in rocks and oil-reservoirs. The information gained from such studies can be used to improve confidence and predictability of classical macroscopic measurements of porosity and saturation by providing information of the spatial distribution of these quantities in a sample.

NMR imaging has also been applied to study composite materials where a physical mixture of the component materials is important. The distribution of SiC whiskers in an alumina ceramic has been studied (Karunanithy, 1991) as well as the distribution of aluminum particles in a metal/rubber composite (Listerud et al., 1989). These studies provide the developer of a material with a rapid and non-invasive means of assessing the effects of processing on the quality of a final product.

NMR imaging has also been proposed as a key component of an on-line system for quality assessment of fruits and vegetables (Chen, et al.,

1989) as well as for lumber grading (Chang, et al., 1989). These schemes have as their goal improving the quality and utilization of natural products through the use of non-invasive monitoring methods. For example, the relative ripeness of fruits can be monitored by an imaging system so that only the ripest material is shipped. In the lumber-grading scheme, the NMR imaging would help improve utilization of the raw timber by identifying unseen defects before sawing commences.

4.5.1.2 *Quasi-Static Systems*

Typical two-dimensional NMR images acquired through the spin-warp sequence or its variants take from 1 to 10 minutes to acquire and so processes that may be observed by such techniques should have time constants on the order of hours or days. However, since echo-planar methods can acquire images in 30 to 100 msec, the concept of a quasi-static process does not necessarily exclude processes with small time constants.

Single-phase and multiphase flow of fluids in porous media has been an attractive area in which to apply NMR imaging. The time constants for macroscopic (NMR-observable) displacements is on the order of hours and so the normal imaging methods are easily used. Gaigalas, et al. (1989) observed single-phase flow by using water doped with a paramagnetic salt to replace pure water in saturated samples of clay, sand and rock. The dopant ($MnCl_2$) alters the relaxation-time behavior of the proton in water and creates effectively two types of water from the NMR point of view. In this way, water initially in the sample can be distinguished from water that is introduced so that displacement flow paths are readily imaged.

Graham, et al. (1991) and Abbott, et al. (1991) have used NMR imaging to show the evolution of the packing behavior of the solids in highly filled suspensions. They used a rotational Couette device and periodically stopped the rotation to look at the solids distribution. Solids migration and hexagonal close-packing were seen to develop from an initially random state.

The industrially important case of two-phase flow of oil and water has been observed by a number of authors. Chen, et al. (1988) and Baldwin and Yamanashi (1989) used paramagnetic salts to suppress the hydrogen signals from the water phase and thereby permit the NMR images to represent oil-saturation measurements. Time-course studies were then focussed on the geometry and efficiency of the displacement of oil by water. Majors, et al. (1990) used the chemical sensitivity of NMR spectroscopy to develop a method capable of observing oil and water saturations simultaneously during the displacement process.

Measurements of two-phase, air-water flow in building materials (plaster bars) were first carried out by Gummerson, et al. (1979). They were able to measure the time evolution of one-dimensional moisture profiles during imbibition into a dry specimen. Their results provided a means of estimating hydraulic diffusivities in such systems. Guillot, et al. (1989) and

Maneval, et al. (1991) have applied NMR imaging to the drying process to consider water removal from consolidated and unconsolidated samples.

Aqueous foams have been studied by NMR imaging to obtain mass-density profiles during drainage (Assink, et al., 1988; German and McCarthy, 1989). Such profiles are important in testing models of foam dynamics since in the past it has been extremely difficult to make such measurements. Also, the stability characteristics can be studied (Heil, et al., 1990).

Filtration of colloids has been studied (Horsfield, et al., 1989) demonstrating the ability of NMR to observe the build-up of filter cakes. Turney (1991) used NMR imaging to study sedimentation of suspensions of rods and spheres in Newtonian and non-Newtonian fluids. Kauten, et al (1991) studied the evolution of concentration profiles during separation of an emulsion using spatially resolved measurements of relaxation times.

Kose (1990) used echo-planar methods to produce a "snapshot" of transitional flow of water in a pipe and illustrated that the label of quasi-static is truly dependent on the imaging technique. Kose was able to observe the evolution of a "puff" turbulent vortex in tube flow at a Reynolds number of about 2200. He has recently extended the technique to the measurement of the full velocity field (Kose, 1991) and future extensions in this area can be expected.

4.5.2 Imaging of Diffusion and Flow

In this section, we review some applications of NMR imaging to the measurement of motion in multiphase systems. We include here motion by molecular diffusion, convection and dispersion. The examples presented here use NMR measurements to directly detect motion and are based on the pulsed-field-gradient methods developed by Tanner and Stejskal (1968).

4.5.2.1. *Diffusion Measurements*

If a sample has internal structure that can lead to spatial variations in the local diffusion coefficient in a sample, NMR imaging methods may be used to localize PFG measurements and thereby map out the diffusivities in a sample. The approach essentially grafts the PFG sequence onto the spin-warp imaging sequence to provide signals whose intensities are representative of local diffusive motion rather than local spin concentrations (Callaghan et al., 1988; Klammler and Kimmich, 1990; Taylor and Bushell, 1985). Examples include measurement of diffusion coefficients in various portions of intact chicken eggs (Klammler and Kimmich, 1990; Taylor and Bushell, 1985) and the study of non-Newtonian flow in micro-capillaries (Xia and Callaghan, 1990).

The measurements of (observation-) time dependent diffusivities in structured systems has been used as a means of non-destructive estimation of sample structural characteristics. These so-called restricted-diffusion measurements were presented with the introduction of the PFG methods (Stejskal, 1965; Tanner and Stejskal, 1968) and many of their applications and techniques have been reviewed recently (Callaghan, 1984; Kärger, et al., 1988). For example, Packer and Rees (1972) and Callaghan, et al. (1983) used restricted-diffusion methods to estimate droplet size distributions in emulsions. Comparison with optical and photographic methods showed that NMR produced satisfactory results.

More recent applications of the restricted-diffusion measurements include pore-space connectivity estimation (so-called *q*-space imaging) (Callaghan, et al., 1990; Packer and Zelaya, 1989), cell-size and compartmentation estimation (Cory and Garroway, 1990) and catalyst characterization (Cheng, et al., 1990). By use of the diffusive motion of spins, the restricted-diffusion methods hold the possibility of making less-than-image-resolution measurements on multiphase systems (Callaghan, et al., 1991; Cotts, 1991).

4.5.2.2. *Velocity Field Types*

As noted previously, NMR measurements of velocity use a PFG-like sequence but focus on the phase-dependence of the echo rather than the amplitude dependence to measure coherent motion. The range of flow phenomena that may be studied is quite broad and some degree of classification is needed to keep the discussion manageable. We have chosen to divide the systems to be studied into two general categories - those whose velocity characteristics that are "simple" and those whose velocity characteristics that are "complex". Some blurring of these categories is inevitable when the time-dependence of the flow field is considered and so this classification is considered useful but preliminary.

We define "simple" flow fields by two characteristics

1. The field to be measured is known in the sense that it may be calculated from a governing equation (e.g., the Navier-Stokes equations).

2. The length (or time) scale over which substantial changes in the flow field occur is larger than the length (or time) scale associated with the imaging measurement.

These characteristics imply that the NMR measurements are "point" measurements of the flow field where "point" is taken in a continuum mechanical sense.

"Complex" flow fields, by definition, have the opposite characteristics of simple fields. The primary characteristics of complex fields are

1. The field is unknowable in the sense that it is impossible (practically or otherwise) to predict the details from a governing equation.

2. The length (or time) scale for the flow field is on the same order of or smaller than the length (or time) scale of the imaging measurement.

Typical examples of complex flow fields include turbulent flow of fluids and flow through porous media. In these instances, the flow is described by its statistical characteristics both theoretically and experimentally. Temporal- and spatial-averaging of the underlying flow are important considerations when interpreting measurements in these systems.

4.5.2.3. *Measurements of Simple Flows*

A typical reason for measuring the velocity profile of a homogeneous fluid (water, glycerin) in a simple flow is to test the NMR methods used for the measurement. Hence, there are many examples of the measurement of the velocity profile for laminar flow in a tube and the review by Caprihan and Fukushima (1990) provides a sampling of such measurements. A simple geometry also provides a useful model for separating and understanding the processes that affect flow measurement by NMR. For example, Keuthe and Herfkens (1989) and Kose (1991) have studies the effects of rotational flow and so provide a good basis for understanding shear and turbulence effects in NMR images. Vera and Grutzner (1986) have studied the spectra of flow in a Taylor cell and have thereby provided a means for using NMR to study more complex hydrodynamic phenomena such as the transition to turbulence. Finally, Xia and Callaghan (1990) have indicated how fine a resolution is possible by their measurements of velocity profiles across 0.7mm-diameter tubes.

For multiphase flow systems (e.g., optically opaque suspensions), measurement of the velocity profile in a simple flow field often provides a direct means of testing or validating proposed rheological models. By comparing predicted profiles to experimentally determined profiles, model parameters can be estimated and used to improve modelling efforts. Therefore, the study of simple flow fields focuses attention on the physical phenomena rather than on the interpretation of the NMR signal.

The utility of NMR in this respect has been shown in the study of the flow of concentrated suspensions of solids. Unless index-matching techniques are used, concentrated suspensions (volume fractions above 5 or 10 volume percent) are generally optically opaque and so preclude the use of optical probes to measure velocity profiles. NMR time-of-flight methods have been shown capable of measuring flow profiles in highly concentrated suspensions of neutrally and non-neutrally buoyant materials (Majors et al., 1989; Sinton and Chow, 1991). Static imaging procedures have also been used to study these suspensions (Graham et al., 1991) (see above, Section 4.5.1.2). Fibrous suspensions are also readily imaged by time-of-flight methods (Maneval, et al., 1991) so that flow profiles can be measured in the optically opaque regions above 1% pulp by weight.

Examples of the abilities of NMR imaging are shown in Figures 4.7 and 4.8. In Figure 4.7a, the contours of the displacement surface from a time-of-flight measurement (see Section 4.4.1 and Figure 4.4) of the velocity profile for flow of an 6% tomato juice suspension. In Figure 4.7b, the profile derived from the data in Figure 4.7a is compared to the profile predicted by a power-law viscosity model. The parameters of the rheological model were measured independently of the flow measurements so that the comparison shown in Figure 4.7b does not involve any adjustable parameters. The aggreement is excellent and indicates the level of quantitative information available through NMR methods.

The range of velocities and concentrations that can be measured is illustrated in Figure 4.8 (Altobelli, et al., 1991). Each pair of images provides information on the solids concentration (upper image of each pair) and velocity profile (lower image of each pair) over the cross-section of a tube during flow of a neutrally-bouyant suspension of spheres. Moving left to right (over the columns), the solids volume fraction varies from 0 to 36 % while moving from the bottom to the top (over the rows), the average velocity varies from 2 to 22 cm/sec. The range of phenomena captured by the data is wide and shows that NMR imaging can provide data that is difficult to measure by any other means. For more details about this data, see the article by Altobelli, et al. (1991).

The flow of water and other homogeneous fluids in complex geometries also falls under the general heading of simple flows. In such instances, the non-invasive nature of NMR is used to maximum advantage. The flow of fluids in the extra-capillary space of a hollow-fiber bioreactor has been studied (Heath, et al., 1990) to check numerical predictions of the flow field though only in the well-defined case of a single fiber. Lee, et al. (1987) used NMR to measure velocity fields in a thermally stratified flow field. In this study, two of the velocity components were imaged and comparisons with laser-doppler methods (LDV) showed good agreement. Lucas, et al. (1977) used measurements of the velocity spectrum to study the thermally driven flow of cryogenic fluids. Kraft, et al. (1989) studied the effects of flow restrictions and development lengths in tube flow. They studied flow through a Venturi-like constriction (a stenosis) and were able to observe both the loss and recovery of laminar flow around the restriction. These studies were undertaken as a basis for the use of NMR to study flow in and around the human heart.

4.5.2.4. *Measurements in Complex Flow Fields*

As well as being complex in nature, complex flow fields are complex in their interpretation. Turbulent flows were the first flows of this type to be studied (De Gennes, 1969) and a series of studies by Fukuda and co-workers (Fukuda and Hirai, 1979; Fukuda, et al., 1985; Kawabe, et al., 1983) have illustrated the use of NMR to study statistical characteristics of laminar, transition and turbulent flow in tubes. While these studies were not imaging studies, the principles and methods that they developed could be extended to

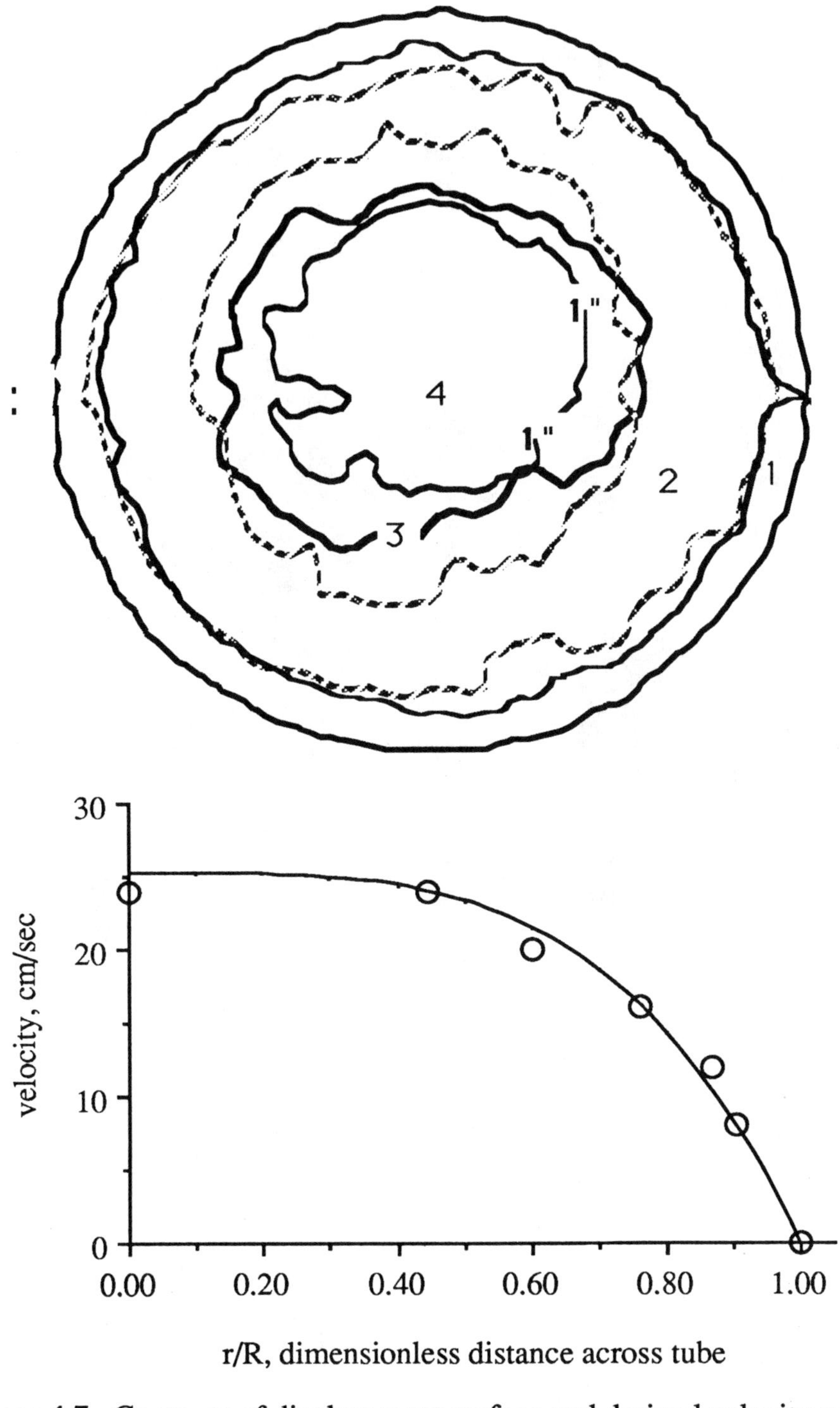

Figure 4.7. Contours of displacement surface and derived velocity profile for flow of tomato juice with a 6% solids fraction. A total of 7 contours were measured but only 4 are shown, for clarity. Prediction of a power-law viscosity model are also shown. Parameters for this model were determined independently of the flow measurement.

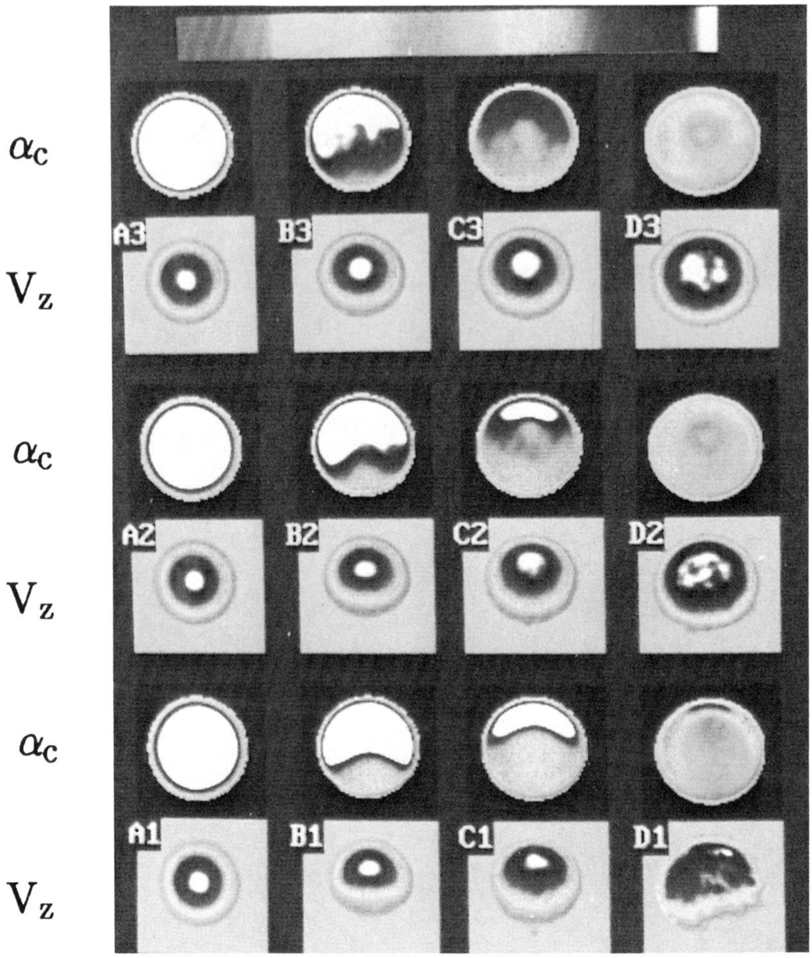

Figure 4.8. Velocity and particle-concentration distribution for flow of a suspension with a negatively bouyant solid phase. Solids fraction increases from left to right while velocity increases from bottom to top. From Altobelli, et al. (1991).
Notation: α_c = solids concentration; V_z = pipe axial velocity

imaging through the incorporation of a simple imaging sequence. Keuthe (1989) used a gradient-echo technique to image the intensity of turbulence (quantified by a turbulent diffusivity) created by the introduction of a jet into a quiescent fluid. His results for the dependence of the diffusivity on the mean velocity were in qualitative agreement with other measurement methods.

While the study of turbulent flows focuses on the temporal character of the flow field, the study of flow in porous systems focuses on the spatial character of the flow field and so NMR imaging should be a valuable tool in this area. The principal goal of the study of flow in permeable materials is to be able to quantitatively measure dispersive transport rates brought on by the spatially non-uniform character of the flow field. In medical research, such flows are typified by flow through capillary beds and there is a great effort to use NMR to measure "perfusion" in various organs in the body. There is also a wide range of other physical processes of engineering interest in which dispersive transport is important.

Grover and Singer (1971) provided some of the first results in this area by measuring the spectrum of blood-flow velocities in the human finger. More recent extensions have focussed on producing spatially-resolved measurements of velocity spectra (Moran, 1982) and the development of the so-called "IVIM" (*I*ntra-*V*oxel *I*ncoherent *M*otion) methods. IVIM methods are straightforward extensions of the PFG methods used to measure diffusion coefficients and discussions of the methods can be found in Caprihan and Fukushima (1990), LeBihan (1990) and Moonen, et al. (1990).

While NMR methods for measurement in complex flow fields are available, the interpretation of the results of the measurements is still under development. Measurements of first- and second-echo amplitude attenuations have been made for a range of materials such as chromatographic packings (Le Bihan, et al., 1988), glass beads (Snaar and Van As, 1990) and ion exchange resins (Duewell, et al., 1990). Typically, the results are interpreted in terms of an effective diffusivity and a mean convective flow but clear connections of these quantities with material characteristics is still under development (Maki, et al., 1991).

4.6 SUMMARY AND CONCLUSIONS

4.6.1 Static and Quasi-Static Systems

The systems discussed in Section 4.5.1 highlight the range of applications of NMR imaging in multiphase systems and demonstrate that it is a useful, non-invasive probe of materials an processes of engineering and scientific interest. Some issues remain, however. Correct interpretation and use of imaging data requires an understanding of the role of the pulse sequence (Maneval et al., 1990) in determining the length scale of the measurement relative to the length scale of the material. Knowledge of these length scales is important in defining what the terms "macroscopic" and "microscopic" actually mean for a given system. Additionally, since more than one pulse sequence can be used to create an image, a solid understanding of the dependence of the image intensities on the physics of NMR is required. For example, Chang and Komoroski (1989) have demonstrated that gradient-recalled-echo methods are more sensitive to voids and imperfections in a sample while spin-echo methods are more sensitive to the true spin-density distribution. Thus, the end use of the measurement may determine the way in which a measurement is made and interpreted.

Image acquisition time is an important issue, especially if the experiment is to be kept in the quasi-static regime. There are two primary factors that influence the length of time needed to form an image - signal level and data dimensionality. If the NMR signal from a sample is intrinsically weak due to low spin content, intrinsic insensitivity (see Table 4.1) or short relaxation times, multiple acquisitions are necessary to obtain satisfactory signal-to-noise ratios in the final image. For example, in the elastomer studies of Chang and Komoroski (1989), images required overnight acquisition times (8 to 10 hours). While echo-planar methods were developed to overcome long acquisition times, they are not currently applicable to systems with short relaxation times.

The dimensionality of the image also leads to relatively long acquisition times. For typical spin-echo methods, a one-dimensional image (signal as a function of one spatial dimension) can be acquired via frequency encoding in a second or less while a full two-dimensional, phase-encoded image might require several minutes. Gradient-recalled echo methods and echo-planar techniques can improve matters but may not be applicable to a given system. Addition of more dimensions to the data set (e.g., the three-dimensional chemical-shift images presented by Majors, et al. (1990)) can lead to long experiment times (one-half to on hour, typically). While many systems of engineering interest are not troubled by long data-acquisition times, there is ample reason to use data of low dimensionality to improve the flexibility and extend the range of application of NMR methods.

The ultimate spatial resolution of NMR imaging methods is a function of the equipment, the pulse sequence and the material. Microscopic images have been formed from a wide range of samples and currently available commercial machines can achieve a resolution of about 50 microns on the plane of the image (Ahn and Cho, 1989; Johnson, et al., 1986; Kuhn, 1990). The sample size is usually limited due to equipment constraints (gradient strength, primarily) and so the number of macroscopic systems that may be examined is restricted. Restricted-diffusion imaging methods (Section 4.5.1.2) can provide indirect means of "imaging" the microscale structure of a macroscopic material and may prove to be a useful alternative.

4.6.2 Flowing Systems

Flow measurements by NMR offer a sensitive and non-invasive means of observing flow in systems that have been experimentally intractable in the past. It appears that development of techniques for "simple" flow fields is sufficiently well advanced to make NMR a quantitative experimental tool in rheological research. For example, measurement of velocity spectra and profiles in rotational- and flow-viscometers (Graham et al., 1991; Sinton and Chow, 1991; Vera and Grutzner, 1986) can provide improved interpretation of rheological data from complex fluids and suspensions. Velocity spectra provide a rapid means of measurement when the material and flow field are well characterized and may also be used to probe transitional flows by observing changes in the spectrum with process parameters (Reynolds number, rotation rates). Velocity imaging can make accurate measurements of velocity profiles and thereby confirm or eliminate hypotheses about the flow field (e.g., the existence of shear layers and plug-flow regions).

The use of NMR for measurements in complex settings is still an area of active development. We restrict comment here to the spatial character of flow fields since that is primarily what NMR imaging will be used for in the study of multiphase flow. Comments on measurements in turbulent flows can be found in the references cited in Section 4.5.2.4.

The concepts of spatial averaging (Marle, 1982; Slattery, 1972; Whitaker, 1986) are central to the interpretation and application of NMR measurements in complex flow fields. The small-scale (continuum-scale) details of the flow are important only in terms of their average properties. It is then necessary to understand the nature and extent of averaging inherent in NMR measurement of such flows.

The spatial length-scale for flow measurement is set by the imaging method used to encode spatial position (e.g., spin-warp methods or echo-planar methods). A spatially resolved measurement of the flow field implies that one has information on the spectrum of velocities within a given set of volume elements (voxels). The spectrum can be used to calculate various quantities of interest for each voxel, such as the mean velocity

$$\bar{v}_z = \int v_z f(v_z)\, dv_z \tag{4.27}$$

and its dispersion,

$$\sigma_v^2 = \int (v_z - \bar{v}_z)^2 f(v_z)\, dv_z \tag{4.28}$$

It is usually anticipated that these quantities will be sufficient to characterize the spectrum and so be useful for measuring local mean flow rates and dispersivities.

Figure 4.9 illustrates two spectra that may be observed in a complex flow. The spectrum in Figure 4.9a shows a symmetrically shaped peak near a well defined mean velocity which would be representative of velocity measurements in a suspension flow. The mean velocity in this instance would be set by the velocity of the solid component of the suspension and, unless a macroscopic boundary is encountered, velocities near zero are not likely.

Figure 4.9b, on the other hand, shows a less well defined spectrum with a relatively broad range of velocities. Such a spectrum might be expected for measurements of flow through porous media where the no-slip

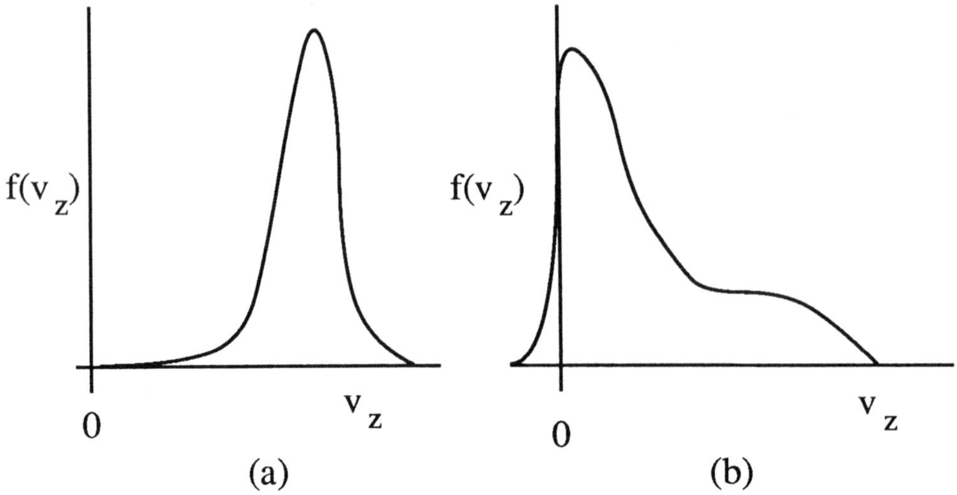

Figure 4.9. Schematics of velocity spectra in complex flow fields. (a) Unimodal spectrum with deviations which are small compared to the mean, typical of suspension flows. In this case, the mean and variance would be good descriptors of the spectrum. (b) Non-unimodal spectrum with large deviations from the mean. Such a spectrum might be typical of flow through porous media where the immobile solid phase creates a large peak in the spectrum at low velocities, due to no slip at the liquid-solid interface.

condition holds at the solid-liquid interfaces in a voxel. While the statistics defined in Eqs. 4.27 and 4.28 are still valid, it is not as likely that they can offer as complete a characterization of this spectrum as they could for the spectrum in Figure 4.9a. The issue is not necessarily resolved by improving the spatial resolution of the image. As spatial resolution becomes finer, the spectrum of any voxel will approach a delta-function, $\delta(v_Z - v_Z(\mathbf{r}))$, where $v_Z(\mathbf{r})$ is the continuum velocity at the point **r**. While this approach is valid, it defeats the goal of describing (and measuring) the flow in average terms.

There is also the issue of how the various moments of the velocity spectrum correspond to spatially averaged measurements of the the velocity field. The mean defined in Eq. 4.27 should correspond to a component of the spatially averaged velocity given by

$$\langle v_Z \rangle = \langle \mathbf{v} \rangle \cdot \mathbf{e}_Z = \frac{1}{V} \int_V \mathbf{v} \, dV \cdot \mathbf{e}_Z \qquad (4.29)$$

However, the variance of the velocity field is, in general, a tensor and so the variance defined in Eq. 4.28 would appear to provide only one component of a dispersion tensor. Also, the implicit dependence of any averaged quantity on the region of averaging (V in Eq. 4.29) requires some thought in making and interpreting the results of an NMR measurement.

Similar comments can be made regarding the IVIM techniques and their use in measuring "effective" diffusion coefficients. A goal of the IVIM methods is often to make "high resolution" maps of perfusion rates in body organs (Duewell et al., 1990) in terms of effective diffusivities and local flow rates. Since perfusion is an averaged quantity, dependent on the tortuosity of the capillary blood flow in the organ, the term "high resolution" must be qualified by reference to the length scale of the system and the measurement (Caprihan and Fukushima, 1990). Also, there is some doubt as to the relation of NMR-measured perfusion to classically measured perfusion (Henkelman, 1990). Future research will clear up these issues and perhaps show that NMR measurements can provide even more information about tissue viability in medical studies and dispersion mechanisms in general engineering studies of flow through porous systems.

4.6.3 Needs for Future Work

Research needs in the medical field have provided ample impetus for the development of sophisticated NMR imaging techniques and hardware. Thus, there are a range of options available for those interested in applying NMR to multiphase systems. From the engineering point of view, less costly and more flexible imaging systems would make NMR methods available to more research groups and thereby increase the utility of NMR in an engineering setting.

A large fraction of the cost of purchasing and operating an imaging system stems from the fact that most (but not all) of the static magnets used for imaging are super-conducting solenoidal magnets. Advances in high-temperature superconductor research would help to reduce operating costs (the magnet could be cooled by refrigerants less costly than liquid helium) and make higher field strengths available on a routine basis. Alternatively, magnet designs based on resistive or permanent magnets could be developed more extensively.

More flexible magnet designs would allow for a wider range of systems to be studied. For example, most imaging systems are designed for applications involving humans and are therefore horizontally oriented (the direction of the static field is perpendicular to the direction of gravity). Horizontal orientation works well for a range of systems of engineering interest such as fluid flow but in some cases (sedimentation or fluidization, for example), a vertical orientation (the magnet axis is parallel to gravity) would be preferable. Since the cost of an imaging system can be high, magnets designed to operate over a range of orientations would be welcomed.

The size of a sample to be imaged is not a particularly restrictive factor in the design of magnets since whole-body imaging systems are currently available for medical use. However, the size of the sample does limit the range and resolution of NMR measurements that can be made. This is due to the fact that the magnetic field gradient is central to the ability of NMR to form images and measure flow and diffusion. Production of a gradient system capable of delivering strong (1 to 10 T/m) and controllable pulses over a large sample volume (0.1 m^3) is a technical challenge that has yet to be met. While current and previous design efforts have yielded significant improvements in gradient design and construction, more work will be needed in this area.

ACKNOWLEDGEMENTS

This work was supported by NSF grants CTS-90-57660, CBT-87-18261 and CBT-88-12870 The authors would like to thank Mr. J. Seymour for providing the data shown in Figure 4.7.

REFERENCES

Abbott, J. R., N. Tetlow, A. L. Graham, S. A. Altobelli, E. Fukushima, L. A. Mondy and T. S. Stephens, "Experimental Observation of Particle Migration in Concentrated Suspensions: Couette Flow", *Journal of Rheology*, Vol. 35, 1991, pp. 773-797.

Abouelwafa, M. S. A. and E. J. M. Kendall, "Optimization of Continuous Wave Nuclear Magnetic Resonance to Determine in-situ Volume Fractions and Individual Flow Rates in Two-Component Mixtures", *Review of Scientific Instruments*, Vol. 50, 1979, pp. 1545-1549.

Abouelwafa, M. S. A. and E. J. M. Kendall, "In-situ Volume Fraction Determination in Multi-component Mixtures Using Pulsed Nuclear Magnetic Resonance", *Canadian Journal of Chemical Engineering*, Vol. 58, 1980, pp. 643-645.

Abragam, A., *Principles of Nuclear Magnetism,* Oxford University Press, Oxford, 1961.

Ahn, C. B. and Z. H. Cho, "A Generalized Formulation of Diffusion Effects in μm Resolution Nuclear Magnetic Resonance Imaging", *Medical Physics*, Vol. 16, 1989, pp. 22-28.

Altobelli, S. A., R. C. Givler and E. Fukushima, "Velocity and Concentration Measurements of Suspensions by Nuclear Magnetic Resonance Imaging", *Journal of Rheology*, Vol. 35, 1991, pp. 721-734.

Assink, R. A., A. Caprihan and E. Fukushima, "Density Profiles of a Draining Foam by Nuclear Magnetic Resonance Imaging", *AICHE Journal*, Vol. 34, 1988, pp. 2077-2081.

Baldwin, B. A. and W. S. Yamanashi, "Detecting Fluid Movement and Isolation in Reservoir Core with Medical NMR Imaging Techniques", *SPE Reservoir Engineering*, Vol. 4, 1989, pp. 207-212.

Banavar, J. R. and L. M. Schwartz, "Probing Porous Media With Nuclear Magnetic Resonance", in *Molecular Dynamics in Restricted Geometries*, J. Klafter and J. M. Drake ed., Wiley, New York, 1989.

Belton, P. S. and R. G. Ratcliffe, "NMR and Compartmentation in Biological Tissue", *Progress in NMR Spectroscopy*, Vol. 17, 1985, pp. 241-279.

Bloch, F., "Nuclear Induction", *Physical Review*, Vol. 70, 1946, pp. 460-474.

Brooks, R. A. and G. DiChiro, "Principles of Computer Assisted Tomography (CAT) in Radiographic and Radioisotopic Imaging", *Physics in Medicine and Biology*, Vol. 5, 1976, pp. 689-732.

Brunner, P. and R. R. Ernst, "Sensitivity and Performance Time in NMR Imaging", *Journal of Magnetic Resonance*, Vol. 33, 1979, pp. 83-106.

Callaghan, P. T., "Pulsed Field Gradient Nuclear Magnetic Resonance as a Probe of Liquid State Molecular Organization", *Australian Journal of Physics*, Vol. 37, 1984, pp. 359-387.

Callaghan, P. T., A. Coy, D. MacGowan, K. J. Packer and F. O. Zelaya, "Diffraction-Like Effects in NMR Diffusion Studies of Fluids in Porous Solids", *Nature*, Vol. 351, 1991, pp. 467-469.

Callaghan, P. T., C. D. Eccles and Y. Xia, "NMR Microscopy of Dynamic Displacements: *k*-Space and *q*-Space Imaging", *Journal of Physics E*, Vol. 21, 1988, pp. 820-822.

Callaghan, P. T., K. W. Jolley and R. S. Humphrey, "Diffusion of Fat and Water in Cheese as Studied by Pulsed Field Gradient Nuclear Magnetic Resonance", *Journal of Colloid and Interface Science*, Vol. 93, 1983, pp. 521-529.

Callaghan, P. T., D. MacGowan, K. J. Packer and F. O. Zelaya, "High-Resolution *q*-Space Imaging in Porous Structures", *Journal of Magnetic Resonance*, Vol. 90, 1990, pp. 177-182.

Caprihan, A. and E. Fukushima, "Flow Measurements by NMR", *Physics Reports*, Vol. 198, 1990, pp. 195-235.

Carr, H. Y. and E. M. Purcell, "Effects of Diffusion on Free Precession in Nuclear Magnetic Resonance Experiments", *Physical Review*, Vol. 94, 1954, pp. 630-638.

Chang, C. and R. A. Komoroski, "NMR Imaging of Elastomeric Materials", *Macromolecules*, Vol. 22, 1989, pp. 600-607.

Chang, S. J., J. R. Olson and P. C. Wang, "NMR Imaging of Internal Features in Wood", *Forest Products Journal*, Vol. 39, 1989, pp. 43-49.

Chen, J.-D., M. M. Dias, S. Patz and L. M. Schwartz, "Magnetic Resonance Imaging of Immiscible-Fluid Displacement in Porous Media", *Physical Review Letters*, Vol. 61, 1988, pp. 1489-1492.

Chen, P., M. J. McCarthy and R. Kauten, "NMR for Internal Quality Evaluation of Fruits and Vegetables", *Transactions of the American Society of Agricultural Engineers*, Vol. 32, 1989, pp. 1747-1753.

Cheng, W. C., N. P. Luthra and C. J. Pereira, "Study of Restricted Diffusion in Porous Catalysts by NMR", *AICHE Journal*, Vol. 36, 1990, pp. 559-568.

Cohen, M. S. and R. M. Weisskopf, "Ultra-Fast Imaging", *Magnetic Resonance Imaging*, Vol. 9, 1991, pp. 1-37.

Cory, D. G. and A. N. Garroway, "Measurement of Translational Displacement Probabilities by NMR: An Indicator of Compartmentation", *Magnetic Resonance in Medicine*, Vol. 14, 1990, pp. 435-444.

Cotts, R. M., "Diffusion and Diffraction", *Nature*, Vol. 351, 1991, pp. 443-444.

De Gennes, P. G., "Theory of Spin Echoes in a Turbulent Fluid", *Physics Letters*, Vol. 29A, 1969, pp. 20-21.

Douglass, D. C. and D. W. McCall, "Diffusion in Paraffin Hydrocarbons", *Journal of Physical Chemistry*, Vol. 62, 1958, pp. 1102-1107.

Duewell, S., R. Wüthrich, A. Buck, C. von Weymarn and G. K. von Schulthess, "MRI Signal Loss Due to Microcirculation: Phantom Studies", *Magnetic Resonance in Medicine*, Vol. 14, 1990, pp. 347-357.

Edelstein, W. A., J. M. S. Hutchinson, G. Johnston and T. Redpath, "Spin Warp Imaging and Applications to Human Whole-Body Imaging", *Physics in Medicine and Biology*, Vol. 25, 1980, pp. 751-756.

Edelstein, W. A., H. J. Vinegar, P. N. Tutunjian, P. B. Roemer and O. M. Mueller, "NMR Imaging for Core Analysis", 63rd Annual Technical Conference and Exhibition of the Society of Petroleum Engineers Proceedings, 101-112, Houston, TX, 1988.

Frahm, J., K.-D. Merboldt, W. Hänicke and A. Haase, "Stimulated Echo Imaging", *Journal of Magnetic Resonance*, Vol. 64, 1985, pp. 81-93.

Fukuda, K. and A. Hirai, "A Pulsed NMR Study on the Flow of Fluid", *Journal of the Physical Society of Japan*, Vol. 47, 1979, pp. 1999-2006.

Fukuda, k., A. Inouye, Y. Kawabe and A. Hirai, "Transition from Laminar to Turbulent Flow of Water in a Pipe Measured by a Pulsed NMR Method", *Journal of the Physical Society of Japan*, Vol. 54, 1985, pp. 4555-4560.

Fukushima, E. and S. B. W. Roeder, *Experimental Pulse NMR: A Nuts and Bolts Approach,* Addison-Wesley, New York, 1981.

Gadian, D. G., *Nuclear Magnetic Resonance and its Application to Living Systems,* Clarendon Press, Oxford, 1982.

Gaigalas, A. K., A. C. van Orden, B. Robertson, T. H. Mareci and L. A. Lewis, "Application of Magnetic Resonance Imaging to Visualization of Flow in Porous Media", *Nuclear Technology*, Vol. 84, 1989, pp. 113-118.

German, J. B. and M. J. McCarthy, "Stability of Aqueous Foams: Analysis Using Magnetic Resonance Imaging", *Journal of Agricultural and Food Chemistry*, Vol. 37, 1989, pp. 1321-1324.

Graham, A. L., S. A. Altobelli, E. Fukushima, L. A. Mondy and T. S. Stephens, "NMR Imaging of Shear-induced Diffusion and Structure in Concentrated Suspensions Undergoing Couette Flow", *Journal of Rheology*, Vol. 35, 1991, pp. 191-201.

Grover, T. and J. R. Singer, "NMR Spin-Echo Flow Measurements", *Journal of Applied Physics*, Vol. 42, 1971, pp. 938-940.

Guilfoyle, D. N., P. Gibbs, R. J. Ordidge and P. Mansfield, "Real-Time Flow Measurements Using Echo-Planar Imaging", *Magnetic Resonance in Medicine*, Vol. 18, 1991, pp. 1-8.

Guillot, G., A. Trokiner, L. Darasse and H. Saint-Jalmes, "Drying of a Porous Rock Monitored by NMR Imaging", *Journal of Physics D*, Vol. 22, 1989, pp. 1646-1649.

Gummerson, R. J., C. Hall, W. D. Hoff, R. Hawkes, G. N. Holland and W. S. Moore, "Unsaturated Water Flow Within Porous Materials Observed by NMR Imaging", *Nature*, Vol. 281, 1979, pp. 56-57.

Haacke, E. M., "Solving for Non-Ideal Conditions in Two-Dimensional Fourier Transform Magnetic Resonance Imaging Using a Generalized Inverse Transform", *Inverse Problems*, Vol. 3, 1987, pp. 421-435.

Haacke, E. M., Z.-P. Liang and J. A. Tkach, "T2-Deconvolution in MR Imaging and NMR Spectroscopy", *Journal of Magnetic Resonance*, Vol. 76, 1988, pp. 440-457.

Hahn, E. L., "Spin Echoes", *Physical Review*, Vol. 80, 1950, pp. 580-584.

Halperin, W. P., F. D'Orazio, S. Bhattacharja and J. C. Tarczon, "Magnetic Resonance Relaxation Analysis of Porous Media", in *Molecular Dynamics in Restricted Geometries*, J. Klafter and J. M. Drake ed., Wiley, New York, 1989.

Heath, C. A., G. Belfort, B. E. Hammer, S. D. Mirer and J. M. Pimbley, "Magnetic Resonance Imaging and Modeling of Flow in Hollow-Fiber Bioreactors", *AICHE Journal*, Vol. 36, 1990, pp. 547-557.

Heil, J. R., K. L. McCarthy, J. B. German, M. J. McCarthy and H. Patino, "Use of Magnetic Resonance Imaging for Evaluation of Beer Foam Characteristics", *American Society of Brewing Chemists Journal*, Vol. 48, 1990, pp. 119-122.

Henkelman, R. M., "Does IVIM Measure Classical Perfusion?", *Magnetic Resonance in Medicine*, Vol. 16, 1990, pp. 470-475.

Horsfield, M. A., E. J. Fordham, C. Hall and L. D. Hall, "1H NMR Imaging Studies of Filtration in Colloidal Suspensions", *Journal of Magnetic Resonance*, Vol. 81, 1989, pp. 593-596.

Johnson, G. A., M. B. Thompson, S. L. Gewalt and C. E. Hayes, "Nuclear Magnetic Resonance Imaging at Microscopic Resolution", *Journal of Magnetic Resonance*, Vol. 68, 1986, pp. 129-137.

Jones, D. W. and T. F. Child, "NMR in Flowing Systems", in *Advances in Magnetic Resonance*, J. S. Waugh ed., Academic, New York, 1976.

Kärger, J., H. Pfeifer and W. Heink, "Principles and Applications of Self-Diffusion Measurements by Nuclear Magnetic Resonance", in *Advances in Magnetic Resonance*, J. S. Waugh ed., Academic, New York, 1988.

Karunanithy, S., "NMR Imaging of Slipcast SiC Whisker-Reinforced Alumina", *Journal of Materials Science*, Vol. 26, 1991, pp. 2169-2172.

Kauten, R. J., J. E. Maneval and M. J. McCarthy, "Fast Determination of Spatially Localized Volume Fractions in Emulsions", *Journal of Food Science*, Vol. 56, 1991, pp. 799-801.

Kawabe, Y., K. Fukuda and A. Hirai, "A Note on the Signal Intensity of the Second Spin Echo in Carr-Purcell-Meiboom-Gill Sequence of Turbulent Flows", *Journal of the Physical Society of Japan*, Vol. 52, 1983, pp. 2308-2311.

Keuthe, D. O., "Measuring Distributions of Diffusivity in Turbulent Fluids with Magnetic-Resonance Imaging", *Physical Review A*, Vol. 40, 1989, pp. 4542-4551.

Keuthe, D. O. and R. J. Herfkens, "Fluid Shear and Spin-Echo Images", *Magnetic Resonance in Medicine*, Vol. 10, 1989, pp. 57-70.

Klammler, F. and R. Kimmich, "Volume-Selective and Spectroscopically Resolved NMR Investigation of Diffusion and Relaxation in Fertilized Hen Eggs", *Physics in Medicine and Biology*, Vol. 35, 1990, pp. 67-79.

Kose, K., "NMR Imaging of Turbulent Structure in a Transitional Pipe Flow", *Journal of Physics D*, Vol. 23, 1990, pp. 981-983.

Kose, K., "One-Shot Velocity Mapping Using Multiple Spin-Echo EPI and its Application to Turbulent Flow", *Journal of Magnetic Resonance*, Vol. 92, 1991, pp. 631-635.

Kraft, K. A., P. P. Fatouros, D. Y. Fei, S. E. Rittgers and P. R. S. Kishore, "MR Imaging of Model Fluid Velocity Profiles", *Magnetic Resonance Imaging*, Vol. 7, 1989, pp. 69-77.

Kuhn, W., "NMR Microscopy - Fundamentals, Limits and Possible Applications", *Angewandte Chemie International Edition in English*, Vol. 29, 1990, pp. 1-19.

Le Bihan, D., E. Breton, D. Lallemande, M.-L. Aubin, J. Vignaud and M. Laval-Jeantet, "Separation of Diffusion and Perfusion in Intravoxel Incoherent Motion MR Imaging", *Radiology*, Vol. 168, 1988, pp. 497-505.

LeBihan, D., "Magnetic Resonance Imaging of Perfusion", *Magnetic Resonance in Medicine*, Vol. 14, 1990, pp. 283-292.

Lee, S. J., M. K. Chung, C. W. Mun and Z. H. Cho, "Experimental Study of Thermally Stratified Unsteady Flow by NMR-CT", *Experiments in Fluids*, Vol. 5, 1987, pp. 273-281.

Listerud, J. M., S. W. Sinton and G. P. Drobny, "NMR Imaging of Materials", *Analytical Chemistry*, Vol. 61, 1989, pp. 23A-41A.

Lucas, P. G. J., D. A. Penman, A. Tyler and E. Vavasour, "Measurement of Fluid Flow Profiles using Pulsed NMR", *Journal of Physics E*, Vol. 10, 1977, pp. 1150-1152.

Lynch, G. F. and S. L. Segel, "Direct Measurement of the Void Fraction of a Two-Phase Fluid by Nuclear Magnetic Resonance", *International Journal of Heat and Mass Transfer*, Vol. 20, 1977, pp. 7-14.

Majors, P. D., R. C. Givler and E. Fukushima, "Velocity and Concentration Measurements in Multiphase Flow by NMR", *Journal of Magnetic Resonance*, Vol. 85, 1989, pp. 235-243.

Majors, P. D., J. L. Smith, F. S. Kovarik and E. F. Fukushima, "NMR Spectroscopic Imaging of Oil Displacement in Dolomite", *Journal of Magnetic Resonance*, Vol. 89, 1990, pp. 470-478.

Maki, J. H., J. R. MacFall and G. A. Johnson, "The Use of Gradient Flow Compensation to Separate Diffusion and Microcirculatory Flow in MRI", *Magnetic Resonance in Medicine*, Vol. 17, 1991, pp. 95-107.

Maneval, J. E., M. J. McCarthy and R. L. Powell, "NMR Measurements in Fibrous Suspensions", TAPPI Engineering Conference Proceedings, Nashville, 1991.

Maneval, J. E., M. J. McCarthy and S. Whitaker, "Use of Nuclear Magnetic Resonance as an Experimental Probe in Multiphase Systems", *Water Resources Research*, Vol. 26, 1990, pp. 2807-2816.

Maneval, J. E., M. J. McCarthy and S. Whitaker, "Studies of the Drying Process by NMR Imaging", in *Drying '91*, A. S. Majumdar and I. Filkova ed., Hemisphere, London, 1991.

Mansfield, P., "Multi-planar Image Formation Using NMR Spin Echoes", *Journal of Physics C*, Vol. 10, 1977, pp. L55-L58.

Mansfield, P. and P. G. Morris, *NMR Imaging in Biomedicine,* Academic Press, New York, 1982.

Marle, C. M., "On the Macroscopic Equations Governing Multiphase Flow with Diffusion and Chemical Reaction in Porous Media", *International Journal of Engineering Science*, Vol. 20, 1982, pp. 643-662.

Moonen, C. T. W., P. C. M. van Zijl, J. A. Frank, D. LeBihan and E. D. Becker, "Functional Magnetic Resonance Imaging in Medicine and Physiology", *Science*, Vol. 250, 1990, pp. 53-61.

Moran, P. R., "A Flow Velocity Zeugmatographic Interlace for NMR Imaging in Humans", *Magnetic Resonance in Medicine*, Vol. 1, 1982, pp. 197-203.

Morris, P. G., *NMR Imaging in Medicine and Biology,* Clarendon Press, Oxford, 1986.

Muller, B. and M. Bloom, "On the Study of Diffusion Using Spin Echoes", *Canadian Journal of Physics*, Vol. 38, 1960, pp. 1318-1327.

Nalcioglu, O. and Z. H. Cho, "Measurement of Bulk and Random Directional Velocity Fields by NMR Imaging", *IEEE Transactions on Medical Imaging*, Vol. MI-6, 1987, pp. 356-359.

Packer, K. J. and C. Rees, "Pulsed NMR Studies of Restricted Diffusion. I. Droplet Size Distributions in Emulsions", *Journal of Colloid and Interface Science*, Vol. 40, 1972, pp. 206-218.

Packer, K. J. and F. O. Zelaya, "Observation of Diffusion of Fluids in Porous Solids by Pulsed Field Gradient NMR", *Colloids and Surfaces*, Vol. 36, 1989, pp. 221-227.

Reeves, L. W., "Application of Non-Selective Pulsed NMR Experiments - Diffusion and Chemical Exchange", in *Dynamic Nuclear Magnetic Resonance Spectroscopy*, L. M. Jackman and F. A. Cotton ed., Academic Press, 1975.

Rollwitz, W. L., "Moisture Measurements in Various Hygroscopic Materials Using Nuclear Magnetic Resonance", in *Humidity and Moisture - Measurement and Control in Science and Industry*, P. N. Winn ed., Rheinhold, New York, 1963.

Rothwell, W. P. and H. J. Vinegar, "Petrophysical Applications of NMR Imaging", *Applied Optics*, Vol. 24, 1985, pp. 3969-3972.

Sinton, S. W. and A. W. Chow, "NMR Flow Imaging of Fluids and Solid Suspensions in Poiseuille Flow", *Journal of Rheology*, Vol. 35, 1991, pp. 735-772.

Slattery, J. C., *Momentum, Mass and Energy Transfer in Continua,* Kreiger, Malabar, Fla, 1972.

Snaar, J. E. M. and H. Van As, "Discrimination of Different Types of Motion by Modified Stimulated-Echo NMR", *Journal of Magnetic Resonance*, Vol. 87, 1990, pp. 132-140.

Stejskal, E. O., "Use of Spin Echoes in a Pulsed Magnetic-Field Gradient to Study Anisotropic, Restricted Diffusion and Flow", *Journal of Chemical Physics*, Vol. 43, 1965, pp. 3597-3603.

Stejskal, E. O. and J. E. Tanner, "Spin Diffusion Measurements: Spin Echoes in the Presence of a Time-Dependent Field Gradient", *Journal of Chemical Physics*, Vol. 42, 1965, pp. 288-292.

Stilbs, P., "Fourier Transform Pulsed-Filed Gradient Spin-Echo Studies of Molecular Diffusion", *Progress in NMR Spectroscopy*, Vol. 19, 1987, pp. 1-45.

Tanner, J. E. and E. O. Stejskal, "Restricted Self-Diffusion of Protons in Colloidal Systems by the Pulsed-Gradient, Spin-Echo Method", *Journal of Chemical Physics*, Vol. 49, 1968, pp. 1768-1777.

Taylor, D. G. and M.-C. Bushell, "The Spatial Mapping of Translational Diffusion Coefficients by the NMR Imaging Technique", *Physics in Medicine and Biology*, Vol. 30, 1985, pp. 345-349.

Tollner, E. W. and W. L. Rollwitz, "Nuclear Magnetic Resonance for Moisture Analysis", American Society of Agricultural Engineers 1987 Summer Meeting Proceedings, Paper No. 87-1008, Baltimore, 1987.

Turney, M. A., *Experimental Investigation of the Dynamics of Suspensions Using Magnetic Resonance Imaging*, Master of Science Thesis, University of California, Davis, 1991.

Vera, M. and J. B. Grutzner, "The Taylor Vortex: The Measurement of Viscosity in NMR Samples", *Journal of the American Chemical Society*, Vol. 108, 1986, pp. 1304-1306.

Vinegar, H. J., "X-Ray CT and NMR Imaging of Rocks", *Journal of Petroleum Technology*, Vol. 1986, pp. 257-259.

Warren, W. S. and M. S. Silver, "The Art of Pulse Crafting: Applications to Magnetic Resonance and Laser Spectroscopy", in *Advances in Magnetic Resonance*, J. S. Waugh ed., Academic, New York, 1989.

Whitaker, S., "Flow in Porous Media I. A Theoretical Derivation of Darcy's Law", *Transport in Porous Media*, Vol. 1, 1986, pp. 3-25.

Woessner, D. E., "Effects of Diffusion in Nuclear Magnetic Resonance Spin-Echo Experiments", *Journal of Chemical Physics*, Vol. 34, 1961, pp. 2057-2061.

Xia, Y. and P. T. Callaghan, "The Measurement of Diffusion and Flow of Polymer Solutions Using Dynamic NMR Microscopy", *Makromolecular Chemie Macromolecular Symposium*, Vol. 34, 1990, pp. 277-286.

Xiang, Q. S. and O. Nalcioglu, "Measurement of Mean and Variance of Velocity Fields Within Each Voxel by NMR Imaging", *IEEE Transactions on Medical Imaging*, Vol. 7, 1988, pp. 364-367.

Xiang, Q. S. and O. Nalcioglu, "Differential Flow Imaging by NMR", *Magnetic Resonance in Medicine*, Vol. 12, 1989, pp. 14-24.

Xiang, Q. S. and O. Nalcioglu, "Projection Images of the Position-Velocity Joint Spin Density Distribution", *Magnetic Resonance in Medicine*, Vol. 15, 1990, pp. 112-120.

Young, L., "Moisture Measurement Using Low Resolution Nuclear Magnetic Resonance", *Measurement and Control*, Vol. 22, 1989, pp. 54-55.

CHAPTER 5

THE RHEOLOGY OF CONCENTRATED SUSPENSIONS OF NON-COLLOIDAL PARTICLES

A. Acrivos

5.1 INTRODUCTION

Suspensions of solid particles in liquids are encountered in a variety of natural and industrial processes, e.g. in the flow of blood and in the transport of slurries in pipelines, and hence their rheology has attracted a great deal of attention from both the fundamental and practical points of view.

This overview will focus on the dynamics of such suspensions when the size of the particles, typically solid spheres, is beyond the colloidal range and when the particle Reynolds number is vanishingly small so that the creeping flow equations apply on a micro-scale. In addition, the discussion will focus on systems for which non-hydrodynamic forces are generally of secondary importance.

As is well-known, when the particle volume concentration α is small enough (typically less than about $3 - 4\%$), such suspensions behave like Newtonian fluids whose effective viscosity μ^* is given, for the case of solid spheres, by Einstein's formula

$$\mu^*/\mu = 1 + \frac{5}{2}\alpha + O(\alpha^2) \tag{5.1}$$

with μ being the viscosity of the ambient fluid. This expression, which is exact under the restrictions mentioned above, has formed the basis for much of the theoretical work on the subject since the turn of the century. For example, it has led to the commonly accepted view that such suspensions should behave as Newtonian fluids <u>for all values of the particle concentration</u> up to the maximum, and therefore numerous theoretical expressions have been proposed for extending the range of validity of Einstein's formula to higher concentrations. These are either exact expansions of eq. (5.1) to higher order in α, or approximate expressions that are largely empirical (having been obtained through the use of various types of cell models) although, in some cases, they have at least some theoretical basis. Among the the latter is the asymptotic expression (Frankel and Acrivos, 1967).

$$\mu^*/\mu \sim 1/[1 - \alpha/\alpha_m)^{\frac{1}{3}}] \quad as \quad \alpha \to \alpha_m \tag{5.2}$$

which was derived by noting that in a very concentrated suspension and provided that the spheres remain in relative motion (i.e. in the absence of any aggregation), the bulk of the viscous dissipation occurs within the thin lubricating layer separating two adjacent spheres.

The first rigorous extension of eq. (5.1) to $O(\alpha^2)$ was presented by Batchelor and Green (1972) whose analysis is particularly significant on at least three counts: (1) It describes a systematic procedure, subsequently extended to more complicated systems (Jeffrey and Onishi, 1984; Kim and Karrila, 1991), for solving creeping flow problems involving spheres, which, by taking advantage of the linearity and symmetry properties of such systems, greatly reduces the complexity of the ensuing analysis; (2) It further illustrates the application of a renormalization technique, developed earlier by Batchelor (1972), for recasting certain seemingly divergent expressions, for such bulk quantities as the particle stress, into expressions involving convergent integrals; and (3) It clearly demonstrates that such suspensions do <u>not</u> behave as effective Newtonian fluids, as had been generally supposed, in that the numerical value of k_2, the coefficient of the $O(\alpha^2)$ term in eq. (5.1), was shown to depend on the type of flow to which the suspension is being subjected. This is because, in contrast to the $O(\alpha)$ coefficient which is unaffected by how the particles are arranged, the calculation of k_2 requires knowledge of the pair-probability density function which differs in general from flow to flow and must be determined as part of the solution. Thus, for example, $k_2 = 5.00$, for the case of a hard-sphere equilibrium radial distribution function (Cichocki and Felderhof, 1988), while the corresponding coefficient of the $O(\phi^3)$ term has recently been reported (Thomas and Muthukumar, 1991) as being, approximately, 6.40 ± 0.40. On the other hand, for a suspension which is well-mixed on a macroscale, $k_2 = 7.6$ when the suspension undergoes an axisymmetric pure straining motion (Batchelor and Green, 1972).

This non-Newtonian feature uncovered by Batchelor and Green's rigorous calculation, is but another manifestation of the fact that, even under the idealized conditions referred to above, the rheological behavior of suspensions should be expected to differ in a significant way from that of a corresponding Newtonian fluid. Surprisingly, this fact was realized only recently, and may even not be widely appreciated; and yet, the experimental evidence was there all along (Jeffrey and Acrivos, 1976) for, as shown in Fig. 5.1,

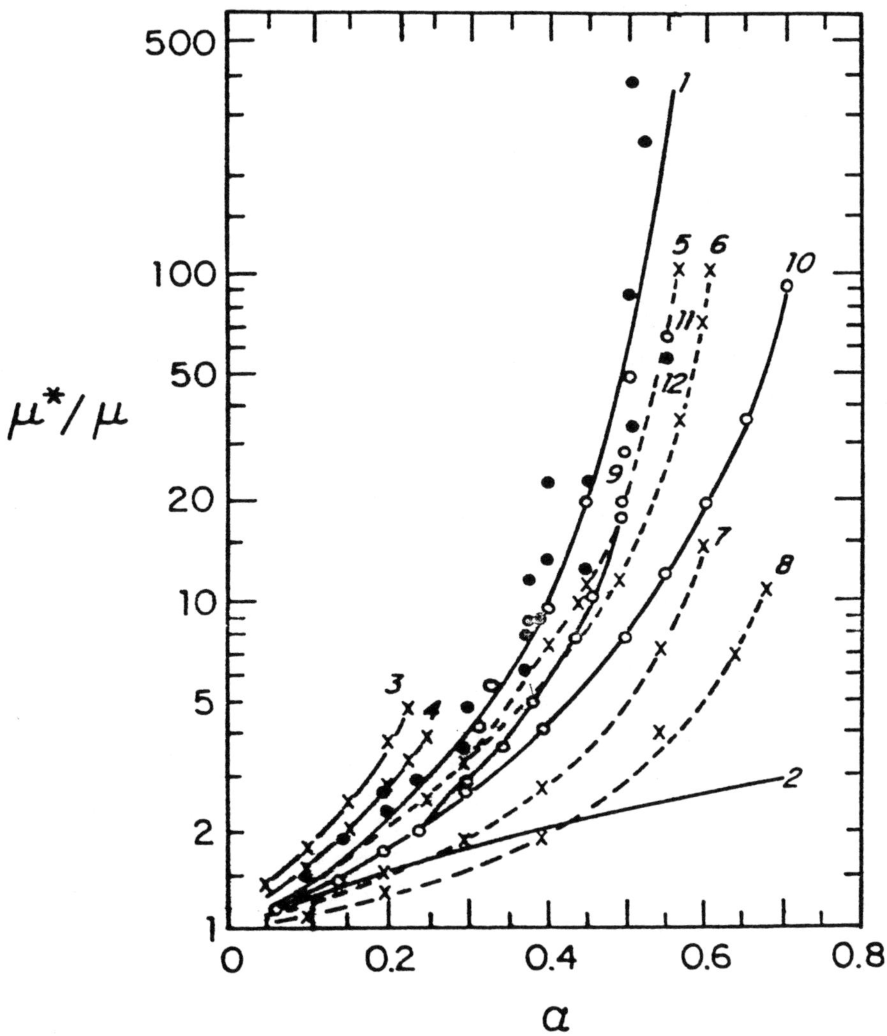

Figure 5.1 Data for the relative effective viscosity of suspensions collected by Rutgers (1962) and plotted against volume fraction. The explanation of the numbering of the curves can be found in the original paper.

it was recognized as far back as the early sixties, or even before, that viscometric measurements on suspensions using different devices, yield values for the effective viscosity as a function of the volume concentration α which can differ from one another by as much as an order of magnitude at the higher concentrations. This discrepancy among the various experimental sets pertaining to different investigations could be attributed only in small part to extraneous factors, such as the influence of weak colloidal forces, the presence of polydispersivity, small inertial effects, the use of slightly non-Newtonian suspending liquids, etc. Thus, more persuasive and plausible explanations had to be found.

It is the purpose of this overview to describe a number of important advances in this field, which came about as a result of some new and unexpected phenomena, that were discovered experimentally as well as through the development of new experimental methods and computational techniques, and which have led to a much better understanding of how such systems behave under flow than was the case even a decade ago. Special attention will be given those factors which, as is becoming increasingly clear, govern the rheology of concentrated suspensions in a major way.

5.2 PHENOMENOLOGY

This section summarizes several phenomena, associated with the flow of concentrated suspensions of non-colloidal particles in viscous fluids, which play an important role in determining their rheological behavior.

5.2.1 Shear-Induced Diffusion

When a concentrated (i.e. $\alpha > 0.30$) suspension of solid spheres of uniform concentration is made to flow under shear, an individual particle will experience a series of random displacements $\Delta\chi_i$ with zero mean relative to its mean trajectory as it interacts with its nearest neighbors whose instantaneous positions have also a random component. But, as is well known, in the presence of a concentration gradient, the random walk just described will induce a diffusive flux with the corresponding diffusion coefficient D_{ij}, in this case the self diffusion coefficient, having the familiar form

$$D_{ij} = \lim_{t \to \infty} \frac{1}{2} \frac{d}{dt} \langle (\Delta \chi_i)(\Delta \chi_j) \rangle \tag{5.3}$$

the brackets denoting an ensemble average. Under creeping flow conditions, this diffusion coefficient must be proportional to the frequency with which the test sphere interacts with its neighbors, and hence, from dimensional analysis we can conclude that D_{ij} must scale with γa^2 times a function of concentration, where γ is a characteristic shear rate and a is the particle radius.

To see more clearly how such a diffusive process can arise, consider, as an example, a dilute suspension of monodisperse solid spheres undergoing the simple shearing motion

$$u_i = \gamma \, \delta_{i1} \, \delta_{j2} \, x_j \tag{5.4}$$

where u_i is the fluid velocity, x_j is the position vector relative to a fixed origin and δ_{ij} is the Kronecker delta. Then, if χ_i denotes the position of the center of a given sphere, the so-called test sphere which is initially at the origin, it is well known that, on account of the reversibility of the creeping flow equations, this test sphere will return to its initial lateral position $\chi_2 = \chi_3 = 0$ at the end of an interaction with a second sphere originating at infinity. Thus, a net lateral displacement of the test sphere requires at least three-body interactions. On the other hand, $\Delta\chi_1$, the net longitudinal displacement of the test sphere, is in general non-zero following a two-sphere interaction, and therefore, as shown recently (Acrivos et al. 1991),

$$D_{11} = 1.27 \gamma a^2 \alpha ln \alpha^{-1} + O(\alpha) \tag{5.5}$$

whereas the remaining components of D_{ij} are $\gamma a^2 o(\alpha)$.

Eckstein, Bailey and Shapiro (1977) were, apparently, the first to identify shear-induced diffusion as an important mechanism for the migration of particles across streamlines in creeping flow and the first to determine experimentally the coefficient of lateral self-diffusion in the direction normal to the fluid velocity and the vorticity, i.e. D_{22}, by monitoring the random motion of a tagged spherical particle in a suspension being sheared in a Couette device. Their technique was later improved by Leighton and Acrivos (1987a) who covered a range of concentrations, shear rates and particle radii and verified the linear dependence of D_{22} on γa^2. These authors also found that $D_{22} \sim \frac{1}{2}\gamma a^2 \alpha^2$ as $\alpha \to 0$ which is consistent with the scaling predicted by the dilute analysis discussed above.

Self-diffusion is, however, not the only mechanism by means of which a marked particle can drift across the streamlines of the bulk flow for, as a result of interparticle interactions, such a diffusive flux can also take place from regions of high concentration to low as well as from regions of high shear to low (Leighton and Acrivos, 1987b). This was predicted theoretically using scaling arguments and was then confirmed experimentally in the following manner.

It had been noted a while back (Gadala-Maria and Acrivos, 1980) that, when a concentrated suspension was placed in the Couette device shown in Fig. 5.2 and then sheared over long periods of time at a constant shear rate, the measured torque and therefore the computed effective viscosity was found to decrease steadily with time and to attain an asymptotic value which, depending on the particle concentration, was significantly lower, i.e. by a factor of 2 or more, than that measured at the start of the experiment. In retrospect, it is easy to see that this decrease in the measured effective viscosity is consistent with the notion of shear-induced migration of particles from the suspension occupying the region of high shear, i.e. the narrow gap between the concentric cylinders of the Couette device, into the relatively stagnant pool at the bottom of the cup, thereby decreasing the particle concentration within that part of the suspension being sheared. This conjecture was in fact verified (Leighton and Acrivos, 1987b) by showing that the observed decrease in the measured torque could be eliminated by merely sealing the base of the gap with a layer of mercury and thereby preventing the particles from diffusing into the reservoir. In addition, by modeling the long-term viscosity decrease phenomenon, referred to above, as an unsteady-state one-dimensional diffusion process, it was found possible (Leighton and Acrivos, 1987b) to extract values for the shear-induced diffusion coefficient normal to the plane of shear, i.e. D_{33}, which again were found to scale with γa^2. An analogous technique was also used to infer values for the shear-induced diffusion coefficient along the plane of shear and normal to the flow direction, i.e. D_{22}, from the measured short-term viscosity increase associated with the procedure used in loading the sample in the viscometer (Leighton and Acrivos, 1987b).

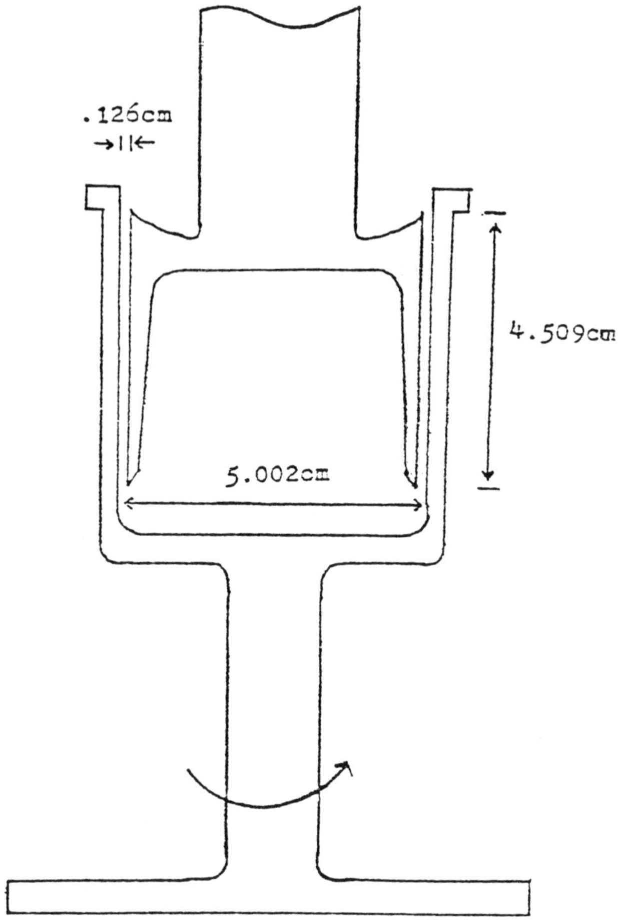

Figure 5.2 A typical Couette viscometer with a cup which contains the bulk of the suspension. The sample being sheared is in the narrow gap between the two cylinders.

The consequences of this shear-induced particle migration from regions of high shear to low and from regions of high concentration to low are far reaching and are believed to be responsible in large part, for the very large degree of scatter, depicted in Fig. 5.1, in the reported experimentally determined effective relative viscosities. This is so because, when α exceeds a certain threshold value of approximately 0.3, the slope of the effective relative viscosity vs. α curve begins to rise dramatically, so that a decrease in the particle concentation of a few percentage points, say from 45% to 42%, can lower the corresponding effective relative viscosity by a very significant amount - about 30% in this example. Thus, unless a viscometer is properly designed to eliminate shear-induced particle drifts, the non-uniformity in the particle concentration that results when the suspension is sheared, and which, of course, depends on the nature of the flow field thereby created and hence on the type of viscometer being used, will influence the bulk property being measured, e.g. the pressure drop in a capillary viscometer or the torque in a Couette device. Consequently, the effective relative viscosity thereby inferred from the measurements will differ significantly from its corresponding value when the concentration is uniform, and this difference will increase dramatically as the suspension becomes more and more concentrated.

To appreciate this fact more clearly, consider the laminar flow of a suspension in a tube or channel and suppose that the particle concentration is uniform at the inlet and that the flow is everywhere unidirectional. Then, as we proceed down the channel, the concentration profile will become non-uniform on account of the shear-induced migration of particles from the region of high shear, i.e. the wall, to the center, where the shear vanishes because of symmetry. But this particle flux will be opposed by a corresponding flux from regions of high concentration to low i.e. from the center to the wall, until a fully-developed concentration profile is established far from the entrance. It is not difficult to show (Phillips et al., 1991), moreover, that, according to the theoretical model currently in use, the centerline particle concentration of this fully developed profile will equal the maximum concentration $\alpha = \alpha_m$ where the effective viscosity becomes infinite and, hence, that the suspension would be expected to move as a plug in the central portion of the pipe or channel. In fact, this prediction applies irrespective of the magnitude of the inlet particle concentration! It is evident, therefore, that unless great care is taken to account for this non-uniformity in the particle concentration profile, any effective viscosities which are calculated from experimentally measured pressure drop vs. flow rate data and are then reported as corresponding to the inlet particle concentration could be seriously in error.

At this stage, it has not been established for sure whether shear-induced diffusion due to gradients in concentration or shear can arise in the presence of viscous forces alone or whether it is necessary to invoke the existence of irreversible forces, e.g. surface forces or those arising from the presence of a small amount of roughness on the surface of the particles. Although, to be sure, this is an important fundamental question which needs to be settled, the issue is rather moot from the practical standpoint because, in concentrated suspensions where the role of shear-induced particle diffusion is most pronounced, the particles are invariably so close to one another, that irreversible effects such as surface roughness are bound to exert an influence no matter what. Nevertheless, the precise role of these irreversible effects needs to be examined further.

5.2.2 Viscous Resuspension

A by-product of the shear-induced particle diffusion mechanism just described is the phenomenon of viscous resuspension wherein an initially settled bed of heavy particles in contact with a clear fluid above it can be resuspended under shear even under conditions of vanishingly small Reynolds numbers (Gadala-Maria, 1979; Leighton and Acrivos, 1986). As an example, consider the flow along a pipe or a channel of an initially well-mixed suspension of heavy particles in a viscous fluid. If the suspension is infinitely dilute, then the individual spheres will of course settle onto the bottom of the pipe or channel, because, in the absence of inertial forces, these spheres will not experience a hydrodynamic lift to counterbalance their sedimentation velocity along the direction of gravity. On the other hand, when particle-particle interactions become important, an individual sphere will experience shear-induced diffusion with the result that, as was explained in the last section, an upward flux of particles from regions of high concentration to low will be created which can eventually balance the downward particle flux due to gravity. Thus, as shown in Fig. 5.3, it is possible to set up a fully developed laminar flow downstream of the entrance which, in general will consist of three distinct regions (Leighton and Acrivos, 1986; Schaflinger et al., 1990):

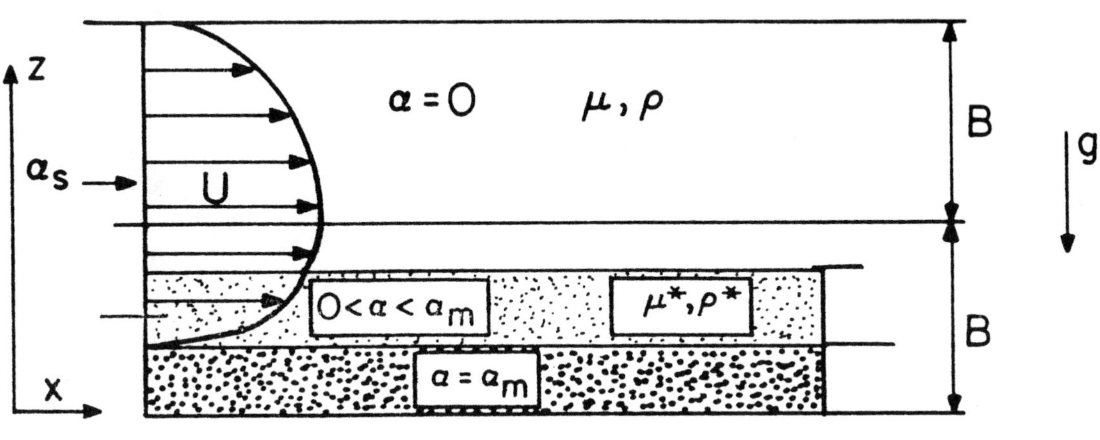

Figure 5.3 The fully-developed stratified flow of a suspension of heavy particles in a channel (Schaflinger et al., 1990).

a clear fluid layer, a flowing suspension within which the particle concentration α increases monotonically until it reaches its maximum value α_m where the effective viscosity vanishes, and a stationary sediment layer where $\alpha \geq \alpha_m$ throughout.

The dimensions of these three regions can then be determined theoretically by modeling the suspension as an effective fluid with a concentration dependent viscosity and then solving simultaneously the appropriate momentum equation and the particle mass balance equation. In the simple example of a uni-directional fully developed flow in a channel (Schaflinger et al., 1990), this particle mass balance equation reduces simply to

$$\frac{2}{9}\frac{a^2(\rho_s - \rho)}{\mu} f\alpha \; + \; \gamma a^2 \hat{D}\frac{d\alpha}{dz} \;\; = \;\; 0 \qquad (5.6)$$

where ρ_s and ρ are the solid and fluid densities, respectively, $f(\alpha)$ is the hindrance function, i.e. the ratio of the sedimentation velocity of a single sphere in the suspension to its value according to Stokes' law, $\gamma(z)$ is the local shear rate, $\hat{D}(\alpha) \equiv D/\gamma a^2$ is a dimensionless shear-induced diffusion coefficient, and z is the vertical coordinate asxis measured from the bottom of the channel. For a given dependence of μ^*/μ, f and $\hat{D}$ on α, it is then easy to show that the independent parameters of the system whose values completely determine the fully-developed velocity and concentration profiles, as well as the dimensions of the three regions referred to above, are (Schaflinger et al., 1990): (1) α_s, the particle concentration in the well-mixed suspension entering the channel; and (2) K, the modified Shields number

$$K \;\; = \;\; \frac{9}{16}\frac{\mu Q}{B^3 g(\rho_s - \rho_1)} \qquad (5.7)$$

where $2B$ is the spacing between the two plates of the channel and Q is the volume flux of clear fluid per unit depth in the well-mixed suspension. Several aspects of this model, such as the surprising prediction that the structure of the fully-developed stratified flow, e.g. the dimensions of the region occupied by the suspension, are independent of a, the particle radius, have already been confirmed experimentally (Leighton and Acrivos, 1986; Schaflinger et al., 1990). To-date though, the theoretical and experimental investigation of viscous resuspension has been confined to a few simple unidirectional (Leighton and Acrivos, 1986; Schaflinger et al., 1990) or quasi-unidirectional flows (Nir and Acrivos, 1990) in which the key variables, e.g. the longitudinal velocity component and the particle concentration, vary primarily in a direction normal to that of the flow. Clearly, it would be desirable to extend the theory to more general situations.

In addition to its practical importance in the design of systems for slurry transport, viscous resuspension can also play a useful role in the field of viscometry in that, by counteracting gravitational effects, it permits one to perform viscometric measurements on suspensions in which the particle and fluid densities need not necessarily be closely matched. This feature does not seem to have been exploited to-date.

5.2.3 Shear Induced Anisotropy

As was mentioned earlier in this overview, it is generally believed that, in the absence of non-hydrodynamic forces, suspensions of monodisperse solid spheres behave as effective fluids with a concentration dependent effective viscosity. Evidence that this cannot be so, however, is provided by an experiment (Gadala-Maria and Acrivos, 1980) that can best be explained with reference to Fig. 5.4. Specifically, after a concentrated suspension had been sheared steadily in the Couette device of Fig. 5.2, the cylinder which was being rotated, typically the outer cylinder, was suddenly stopped and, as seen in Fig. 5.4, the torque signal vanished almost instantaneously as would have been expected for a highly viscous fluid. However, when, after a pause of anywhere from a few minutes to a few hours, the direction of shear was reversed - by suddenly starting up the cylinder at the same speed as before, but in the opposite direction - the torque did not attain its expected value instantaneously, as would have been the case for a Newtonian fluid, but instead jumped to an intermediate value from which it gradually attained its final asymptote. On the other hand, when the rotating cylinder was stopped and then restarted, but in the same sense as before, the torque attained its earlier value instantaneously.

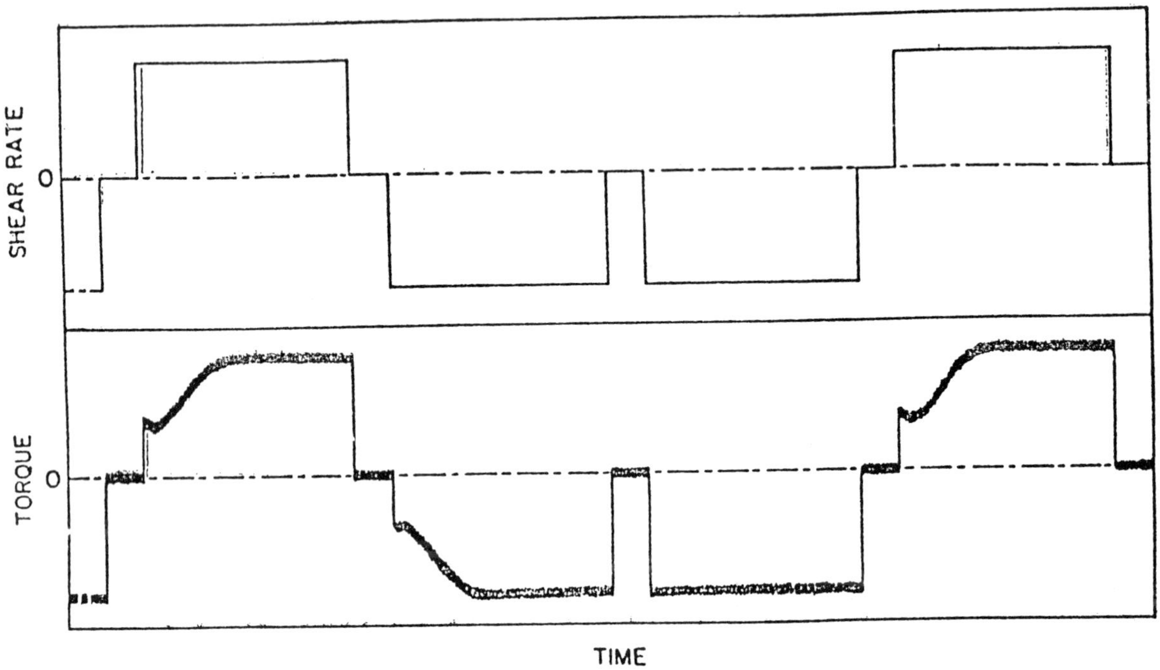

Figure 5.4 Schematic diagram showing the effect of a reversal in the direction of shear on the torque signal in the Couette device for a concentrated suspension (Gadala-Maria and Acrivos, 1980).

A similarly unexpected behavior was observed (Gadala-Maria and Acrivos, 1980) when a concentrated suspension, that had been sheared steadily for a while, was subjected to an oscillating shear, in that, as shown in Fig. 5.5,

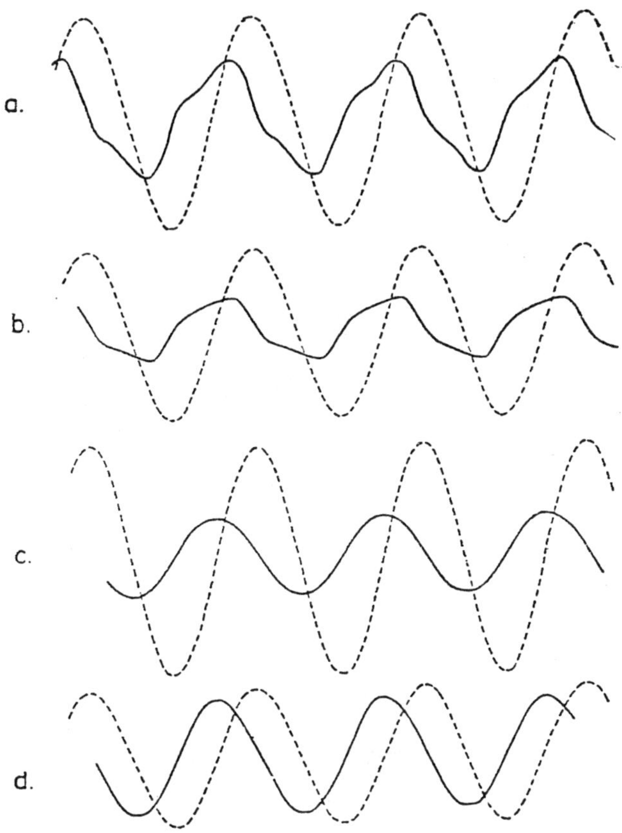

Figure 5.5 Shapes of the (_ _ _) input and, output (—) signals as a function of time from an oscillatory shear experiment (Gadala-Maria and Acrivos, 1980) on a $\alpha = 0.45$ suspension of polystyrene spheres (40 to $50 \mu m$ in diameter) in silicone oils between parallel plates at several input amplitudes. Strain amplitude at the edge (a) 1.05 (maximum angular displacement = 0.02 radians), (b) 0.84, (c) 0.45, and (d) 0.07. The input and output signals are on different scales. Oscillation frequency: 0.060 cycles/sec.

the response curve became sinusoidal only when the maximum amplitude of the input oscillation was reduced to values below $0.1°$ for a 45% suspension. Even so, the effective dynamic viscosity thereby calculated was found to be much lower (by a factor of approximately 2 for a 45% suspension) than the corresponding effective viscosity as determined from the steady shear experiment.

These observations clearly indicate that such suspensions can no longer be modeled as isotropic Newtonian fluids and that the shear creates a structure which renders them anisotropic. By structure it is meant that, whereas in an isotropic fluid, the pair distribution function $g(\vec{r})$, i.e. the probability of finding a sphere with its center at $\vec{r}$ given that there is a sphere with its center at the origin, has fore-and-aft symmetry, this is no longer the case when the suspension becomes anisotropic. In fact, the results of the transient experiments referred to above can be reconciled by supposing that, in a concentrated suspension, the asymmetry in $g(\vec{r})$ is such that the probability of finding a second sphere on the approach side is greater than that on the receding side. Thus, when the direction of the shear is suddenly reversed, the particles have to rearrange themselves into a mirror image of their original configuration thereby giving rise to the transient response noted in Fig. 5.4.

Owing to the reversibility of the creeping flow equations, the existence of such an asymmetry can only be due, of course, to the presence of non-hydrodynamic irreversible forces. These could be the usual surface forces which, although weak, begin to play an increasingly important role as the separation distance between two adjacent spheres decreases to zero. Alternatively, the irreversibility may result from a small degree of roughness which invariably exists on the surface of the spheres comprising the suspension. Clearly, though, the aspect of the subject deserves further study.

5.2.4 Wall Slip

It is now a well-accepted fact that, although the suspending fluid adheres to the walls of a viscometric device and satisfies the no slip boundary condition, this is not the case with the bulk velocity because, the solid particles can and do roll along these walls. A tangential bulk slip velocity is thereby induced proportional to the tangential component of the stress force, in which the slip coefficient is comparable in magnitude to the particle diameter d. The effect of this slip velocity on the measured apparent effective viscosity $\mu^*_{app.}$ in a plane Couette apparatus can be estimated, albeit rather crudely, by supposing that, adjacent to each of the two walls of the device, there exist a particle-free layer of thickness d. Then, since the shear stress is constant across the gap of the Couette, it is easy to show that

$$\mu^*_{app}/\mu^* = [1 + \frac{2d}{H}\frac{\mu^*}{\mu}]^{-1} < 1 \tag{5.8}$$

where μ^* is the effective viscosity of the suspension, which is now confined to the central portion of the Couette, and H is the spacing between the plates. Thus, as a result of slip, the apparent effective viscosity, as inferred from such viscometric measurements, will be lower than μ^* by an amount which depends on the ratio of the particle-diameter to the gap width. To be sure, this possible error in the interpretation of the data has been known for a long time and hence the implications of the above relation will not surprise anybody. What does not appear to have been widely appreciated, however, is how serious this error can become for concentrated suspensions even when d/H is as low as 0.01, since μ^*/μ can become very large indeed, i.e. $O(20)$ when $\alpha = 0.45$. It is thus essential that viscometric measurements on concentrated suspensions be corrected for the presence of slip, using techniques described in the literature (Yilmazer and Kalyon, 1989).

5.3 MICROMECHANICS

It is evident from what has been said thus far that the various phenomena discussed earlier on in this review owe their existence to the way in which the individual particles within a concentrated suspension interact both with one another as well as with the surrounding solvent. Thus, in order to fully understand the macroscopic properties of such systems, as well as to test the assumptions generally made in developing macroscopic models, it is often desirable to be able to study in some detail the corresponding micromechanics. Unfortunately, until very recently, this could be done theoretically or experimentally only for very dilute suspensions whose rheology differs only slightly from that of the suspending liquid. Thus, phenomena such as shear-induced diffusion or shear-induced anisotropy could be explained only by inferring, rather than by verifying through experiments or calculations, the behavior of the individual particles in a concentrated suspension under the action of shear. In the very recent past, however, a number of very powerful experimental and numerical techniques have been developed which have enabled us to probe the detailed micromechanics of such systems to a degree that was considered unthinkable even as far back as a few years ago. Since many of these advances will be reported in detail in other chapters of this book, the discussion which follows will be brief.

5.3.1 Novel Experimental Tools

In principle, it should be possible of course to observe the motion of individual particles using optical devices of sufficient intensity and magnification. The logistics of tracking and recording the paths of several hundred particles and then using this information to determine the statistical properties of flowing suspensions, e.g. the particle concentration and velocity profiles, can, however, be mind boggling. A case in point is the early work on the flow of concentrated suspensions in a tube by Karnis, Goldsmith and Mason (1966) who, using cinematography and then laboriously tracking and counting the particles in the projected images of the films, were able to show that the particle velocity profiles become increasingly blunted at the center as either the particle size or their bulk concentration is increased. They also reported that the particle concentration was uniform across the tube within experimental accuracy.

This procedure can now be greatly simplified, accelerated and automated, however, through the use of data acquisition systems consisting, for example, of a video camera, a VICOM computer and a VAX 11/780 computer or some similar accessories, which can identify the positions of each particle within the depth of field and then determine the relative positions of various interacting particles within a distance of several parameters (Husband and Gadala-Maria, 1987; Parsi and Gadala-Maria, 1987). The data, thereby obtained, can then be used to calculate not only bulk properties, e.g. the local particle concentration, but also statistical properties of fundamental importance such as the pair probability function $g(\vec{r})$ referred to earlier. Thus, it was verified experimentally (Husband and Gadala-Maria, 1987; Parsi and Gadala-Maria, 1987) that, indeed, in a concentrated suspension under shear, $g(\vec{r})$ loses its fore-and-aft symmetry and that a shear-induced anisotropic structure is thereby created in agreement with what had been inferred earlier from transient torque measurements in a viscometer following a reversal of the flow direction (Gadala-Maria and Acrivos, 1980). Other optical methods are outlined in Chapters 2 and 3.

Another important advance is the experimental adaptation of the well-known Laser Doppler Anemometer technique which was recently developed (Koh et al., 1991) for measuring the velocity and concentration profiles in flowing suspensions where the refractive indices of the solid and liquid phases were closely matched. The residual turbidity, due to small mismatches of the refractive indices, as well as impurities in the particles, allowed a Doppler signal to be detected when a particle passed a scattering volume. Although the measured concentration profiles showed the same trends as those computed on the basis of the shear-induced particle migration theory, significant quantitative differences between the two sets were found to exist. In addition, a large slip velocity between the particles and the fluid was observed which was found to increase with an increase in the particle concentration (Koh et al., 1991).

Another recently introduced experimental tool in the study of flowing suspensions is the nuclear magnetic flow imaging technique (Phillips et al., 1991; Altobelli et al., 1991; Sinton and Chow, 1991; Abbot et al., 1991), also discussed in Chapters 1 and 4, by means of which one can again measure both the velocity and concentration profiles. Although the method is to some extent still under development and requires, of course, the availability of some very elaborate and expensive equipment, it has already yielded valuable information regarding the non-uniformity of the particle concentration profiles which were found to be in good agreement generally with the theoretical predictions based on the shear-induced particle diffusion model.

It is evident therefore that remarkable progress has been achieved in the last few years in probing experimentally the detailed micromechanics of flowing concentrated suspensions and that new information is becoming available which should lead eventually to the development of improved theoretical models than the ones currently available.

5.3.2 Numerical Methods

When the particle Reynolds numbers is vanishingly small, the velocities of the individual particles comprising a suspension can be obtained, in principle, by solving the creeping flow equations within the space occupied by the solvent (assumed Newtonian) subject to the condition that the force and torque on each particle be specified (these being zero, of course, for neutrally buoyant particles). This is in addition to the macroscopic constraint which defines the flow field, e.g. a simple shear flow or a pure straining motion. Although, to be sure, the creeping flow equations are linear so that, as discussed in Chapter 27, their solution can be expressed in closed form in terms of a continuous distribution of singularities, the so-called single-layer and double-layer potentials, on the surface of each particle, the actual numerical evaluation of the particle velocities and thereby the evolution of the statistical configuration of the suspension was considered until recently to be a formidable problem owing to the presence of hydrodynamic intra-particle forces which are long-range. In addition, in order to properly account for the changing configuration, it is necessary to monitor the trajectories of a statistically significant number of particles which may be as large as several dozen.

The numerical technique of Stokesian Dynamics, developed by Brady and Bossis (Brady and Bossis, 1988; Brady et al., 1988) and described more fully in Chapter 26 of this book, has made this problem tractable, however, to the point where a number of fundamental phenomena, e.g. viscous resuspension, shear-induced particle diffusion, shear-induced anistropy, can now be investigated in detail and the influence of additional factors, e.g. non-hydrodynamic surface forces or Brownian motion effects, be evaluated quantitatively. This technique is exact in the sense that it can lead to results of arbitrarily high accuracy by increasing the number of particles that are being retained in the simulation, but unfortunately, it is also computationally intensive and therefore both time consuming and expensive.

Alternative methods have therefore been proposed which attempt to compute the macroscopic properties of a suspension, e.g. its effective viscosity, without requiring a detailed and exact simulation of the micromechanics. The technique of lattice-gas cellular automata (Ladd and Frenkel, 1990) is such a procedure which shows great promise even though it is not clear whether and under what conditions it can converge to exact results.

Numerical simulations are currently under very active development, of course, and it seems safe to predict that new, more powerful and computationally more efficient methods will be created which will make it possible to study more complicated systems, e.g. electrorheological fluids (Bonnecaze and Brady, 1991), and in more detail than has been found possible to-date.

5.4 SUMMARY

This overview has highlighted some of the more important recent developments pertaining to the flow of concentrated suspensions of non-colloidal particles. A few, surprising at first, experimentally observed phenomena were described and the role played by shear-induced particle diffusion, viscous resuspension, shear-induced anisotropy and partial slip along solid boundaries in the interpretation of rheological measurements has been emphasized. Aside from partial slip, the effects mentioned above have only recently been identified and their very considerably influence is only now beginning to be appreciated. In addition, new and novel experimental techniques as well as powerful and efficient numerical simulations are beginning to provide new insight regarding the detailed micromechanics of suspensions flowing under shear and thereby help to further identify the key parameters that affect their rheology in a significant way.

Clearly, much has happened to the field of supension rheology since the time of Einstein at the turn of the century, and especially so during the last ten to fifteen years when the major breakthroughs have occurred.

ACKNOWLEDGEMENTS

This work was supported in part by a grant from the U.S. Department of Energy (Grant #DE-FG02-20ER14139.A000) and by the State of New York under its Einstein Chair Program.

REFERENCES

Abbot, J.R., Tetlow, N., Graham ,A.L., Altobelli, S.A., Fukushima,E., Mondy, L.A. and Stephens, T.S., 1991 "Experimental Observations of Particle Migration in Concentrated Suspensions: Couette Flow", J.Rheol. 35, 773.

Acrivos, A., Batchelor, G.K., Hinch ,E.J. and Mauri, R., 1991 "Longitudinal Shear-Induced Diffusion of Spheres in a Dilute Suspension", J.Fluid Mech. (submitted).

Altobelli, S.A., Givler, R.C. and Fukushima, E., 1991 "Velocity and Concentration Measurements of Suspensions by Nuclear Magnetic Resonance Imaging", J. Rheol. 35, 721.

Batchelor, G.K. and Green, J.T., 1962 "The Hydrodynamic Interaction of Two Small Freely-Moving Spheres in a Linear Flow Field" J. Fluid Mech. 56, 375; ibid. "The Determination of the Bulk Stress in a Suspension of Spherical Particles to Order c^2", 56, 401.

Batchelor, G.K., 1972 "Sedimentation in a Dilute Dispersion of Spheres", J. Fluid Mech. 52, 245.

Bonnecaze, R.T. and Brady, J.F., 1991 "Dynamic Simulation of an Electrorheological Fluid", J. Chem. Phys. (submitted).

Brady, J.F. and Bossis, G., 1988 "Stokesian Dynamics", Ann. Rev. Fluid Mech. 20, 111.

Brady, J.R., Phillips, R.J., Lester, J.C. and Bossis, G., 1988 "Dynamic Simulations of Hydrodynamically Interacting Suspensions", J. Fluid Mech.195, 257.

Cichocki, B. and Felderhof, B.U., 1988 "Short-time Diffusion Coefficients and High Frequency Viscosity of Dilute Suspensions of Spherical Particles", J. Chem. Phys. 89, 149.

Eckstein, E.C., Bailey, D.G. and Shapiro, A.H., 1977 "Self-Diffusion of Particles in Shear Flow of a Suspension", J. Fluid Mech. 79, 191.

Frankel, N.A. and Acrivos, A.,1967 "On the Viscosity of a Concentrated Suspension of Solid Spheres", Chem. Eng. Sci. 22, 847.

Gadala-Maria, F., 1979, Ph.D. Thesis, Stanford University.

Gadala-Maria, F. and Acrivos, A., 1980 "Shear-Induced Structure in a Concentrated Suspension of Solid Spheres", J. Rheol. 24, 799.

Husband, D.M. and Gadala-Maria, F., 1987 "Anisotropic Particle Distribution in Dilute Suspensions of Solid Spheres in Cylindrical Couette Flow", J. Rheol. 31, 95.

Jeffrey, D.J. and Onishi, Y., 1984 "Calculation of the Resistance and Mobility Functions for Two Unequal Rigid Spheres in Low-Reynolds-Number Flow" J. Fluid Mech. 139, 261.

Jeffrey, D.J. and Acrivos, A.,1976 "The Rheological Properties of Suspensions of Rigid Particles", AIChE Journ. 22, 417.

Karnis, A., Goldsmith, H.L. and Mason, S.G. 1966 "The Kinetics of Flowing Dispersions, I. Concentrated Suspensions of Rigid Particles", J. Colloid Sci. 22, 531.

Kim, S. and Karrila, S.J., 1991 "Microhydrodynamics: Principles and Selected Applications", Butterworth-Heinemann, Stoneham, MA.

Koh, C.J., Hookham, P. and Leal, L.G., 1991 "An Experimental Investigation of Concentrated Suspension Flows in a Rectangular Channel", J. Fluid Mech. (submitted).

Ladd, A.J.C. and Frenkel, D., 1990 "Dissipative Hydrodynamic Interactions via Lattice-Gas Cellular Automata", Phys. Fluids A 2, 1921.

Leighton, D. and Acrivos, A., 1986 "Viscous Resuspension" Chem. Eng. Sci. 41, 1377.

Leighton, D. and Acrivos, A., 1987a "Measurement of Shear-Induced Self-Diffusion in Concentrated Suspensions of Spheres", J. Fluid Mech. 177, 109.

Leighton, D. and Acrivos, A., 1987b "The Shear-Induced Migration of Particles in Concentrated Suspensions", J. Fluid Mech. 181, 415.

Nir, A. and Acrivos, A., 1990 "Sedimentation and Sediment Flow on Inclined Surfaces", J. Fluid Mech. 212, 139.

Parsi, F. and Gadala-Maria, F., 1987 "Fore-and-Aft Asymmetry in a Concentrated Suspension of Solid Spheres", J. Rheol. 31, 725.

Phillips, R.J., Armstrong, R.C., Brown, R.A., Graham, A.L. and Abbot, J.R., 1991 "A Constitutive Equation for Concentrated Suspensions that Accounts for Shear-Induced Particle Migration", Phys. Fluids A (accepted).

Rutgers, J., 1962 "Relative Viscosity of Suspensions of Rigid Spheres in Newtonian Liquids", Rheol. Acta 2, 202.

Schaflinger, U., Acrivos, A. and Zhang, K., 1990 "Viscous Resuspension of a Sediment Within a Laminar and Stratified Flow", Int. J. Multiphase Flow 16, 567.

Sinton, S.W. and Chow, A.W., 1991 "NMR Flow Imaging of Fluids and Solid Suspensions in Poiseuille Flow", J. Rheol. 35, 735.

Thomas, C.U. and Muthukumar, M., 1991 "Three-Body Hydrodynamic Effects on Viscosity of Suspensions of Spheres", J. Chem. Phys. 94, 5810.

Yilmazer, U. and Kalyon, D.M., 1989 "Slip Effects in Capillary and Parallel Disk Torsional Flows of Highly Filled Suspensions", J. Rheol. 33, 1197.

CHAPTER 6

MEASUREMENT OF THE SHEAR-INDUCED MICROSTRUCTURE OF CONCENTRATED SUSPENSIONS OF NON-COLLOIDAL SPHERES

D. Leighton
I. Rampall

6.1 INTRODUCTION

The nature of interparticle interactions, in combination with the geometry of an applied flow field, together determine the local *microstructure* or local distribution of particles in a suspension. This microstructure strongly affects rheological properties of the suspension such as its apparent viscosity and possible normal stress differences. Knowledge of the microstructure also permits the calculation of effective medium properties such as the thermal and electrical conductivities and the dielectric constant, which are important in a variety of applications. Factors which influence the microstructure include van der Waals forces, Brownian motion, electroviscous effects, fluid inertia and any non-Newtonian properties of the suspending fluid. In this chapter we will consider suspensions of non-colloidal particles where Brownian motion is negligible and where surface forces such as van der Waals and electroviscous forces are unimportant on a length scale comparable to the diameter of the particles. We shall also restrict ourselves to sufficiently low Reynolds numbers that inertial effects are likewise negligible. Despite these restrictions, this class of suspensions contains many examples of industrial interest, and represents the high Peclet number / low Reynolds number limit of more general classes of suspensions.

Theoretical predictions of suspension microstructure have largely been limited to dilute suspensions in which only two particle interactions need be considered. The microstructure of a dilute suspension is most easily described by the pair-particle probability distribution function: the probability density of finding a second sphere at some position relative to a first sphere located at the origin. Batchelor and Green (1972) examined the pair distribution function of a dilute suspension of spheres undergoing purely hydrodynamic interactions in an arbitrary linear shear flow. The pair distribution function was determined by following the trajectories of pairs of spheres as they moved relative to each other in a shear flow. These authors demonstrated that the dilute pair distribution function was spherically symmetric in all regions in which the particle trajectories began at infinity, and was independent of the type of the linear shear flow. Of course, for simple shear flow (and indeed any planar shear flow for which there is a non-zero vorticity) there is a region over which the trajectories are closed. Since the trajectories in this region do not come from infinity, there is no way to determine the probability density function in the closed orbit region from two-particle interactions alone. As a consequence, the pair distribution function for a simple shear flow, for example, can only be determined if three-body interactions are taken into account, or if some non-hydrodynamic force such as Brownian motion or repulsive electroviscous forces is introduced to eliminate the closed orbit trajectories. Russel (Russel and Gast, 1986; Wagner and Russel, 1989) has developed a statistical approach for calculating equilibrium and non-equilibrium structures in colloidal suspensions where Brownian motion and other non-hydrodynamic forces are significant.

In concentrated suspensions the difficulties in determining the microstructure from exact theories are even greater since particle interactions in these suspensions intrinsically involve the interaction of many particles. Considerable progress in understanding the microstructure of a concentrated suspension has been made through numerical techniques pioneered by Bossis and Brady (Bossis and Brady, 1984; Durlofsky *et al.*, 1987; Brady and Bossis, 1988). This approach, termed Stokesian Dynamics, simulates the behavior of a suspension by following the trajectories of a discrete number of spheres in a cell which is periodically replicated to approximate an unbounded suspension. The hydrodynamic interaction between the spheres is modeled using the exact two-sphere mobility and resistance functions. This technique has also been used to simulate the behavior of bounded suspensions by taking into account the presence of plane walls (Durlofsky and Brady, 1989), suspensions in which Brownian motion, van der Waals, and electroviscous forces are present (Bossis and Brady, 1989; Bossis *et al.*, 1988), and also electrorheological fluids (Bonnecaze and Brady, 1990).

While this simulation procedure does an excellent job of capturing the purely hydrodynamic interaction between ideal particles, it is less clear that the simulation results correspond to the behavior of physical suspensions of particles. One problem is that the separation distance predicted by the simulation for interacting particles in the absence of non-hydrodynamic forces is exceedingly small. While two mathematically smooth spheres in a perfectly continuous fluid at zero Reynolds number will never come into contact due to the lubrication layer between the particles, the distance predicted from purely hydrodynamic interactions is much smaller than the typical surface roughness of the spheres used in rheological experiments (Smart and Leighton, 1989). As a consequence, real particles will come into contact in a sheared suspension, and it is as yet unresolved how to model contact forces in numerical simulations.

Recently Pakdel and Kim (1991) have developed numerical techniques to directly simulate the behavior of N particles interacting in a sheared suspension where the particles are not necessarily spherical. This simulation method can directly account for the presence of large scale surface roughness on spheres, as well as highly non-spherical particles such as flakes and rods. While computationally efficient, the technique employed requires large amounts of computer time. With the advent of fast parallel processors, however, it may provide a way of extending numerical simulation of suspensions to real particles.

The limitations of theoretical approaches to the problem of determining the microstructure of concentrated suspensions require us to turn to experimentation. A number of experimental investigations have been conducted. For our present purposes we shall group these into *indirect* and *direct* experimental techniques. We shall now describe these approaches in more detail.

6.2 INDIRECT EXPERIMENTS

6.2.1 Shear-Reversal Behavior

The macroscopic rheological behavior of a suspension is intimately related to the suspension microstructure, thus it is natural to attempt to use rheological measurements to draw inferences about the microstructure. The most compelling indirect evidence for the existence of anisotropic suspension microstructure in concentrated suspensions is found from the transient behavior in effective viscosity observed upon the reversal of the direction of an applied shear. In experiments with concentrated suspensions of 46μm diameter polystyrene spheres suspended in silicone fluids Gadala-Maria and Acrivos (1980) demonstrated that the apparent suspension viscosity would undergo a transient decrease upon reversal of the direction of shear. After subjecting the suspension to a steady shear for an extended period of time in a conventional Couette viscometer, the direction of the shear was reversed. The observed viscosity immediately after reversal of the direction of shear was found to be lower than the earlier steady state viscosity, and further decreased to some minimum value. After a total strain of about 3 in this new direction, the viscosity returned to its steady value. The decrease was associated only with reversal of the direction of motion since no such decrease was observed if the shearing were stopped and restarted again in the same direction. A schematic diagram of this shear reversal behavior observed for suspensions in a Couette viscometer is given in Figure 6-1.

The shear reversal phenomenon was further examined by Leighton (1985). In these experiments, also performed with 46μm spheres in silicone fluids, the minimum viscosity upon reversal of shear and the elapsed strain corresponding to this minimum were measured. The strain was found to be a strong function of concentration, varying from a dimensionless value of approximately 1.0 at 30% solids volume fraction to 0.17 at 50%. Intriguingly, the minimum viscosity upon reversal of shear was found to be a much weaker function of concentration than was the steady viscosity, as is shown in Figure 6-2. If both minimum and steady viscosities are fitted to the modified Eiler equation:

$$\mu = \left[1 + \frac{\frac{1}{2}[\eta]\,\alpha_s}{1 - \alpha/\alpha_{smax}} \right]^2 \tag{6.1}$$

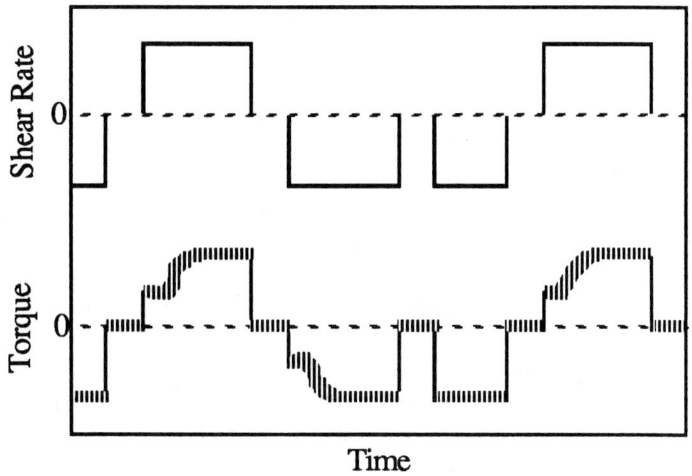

FIGURE 6-1 *Schematic diagram showing the torque response upon reversal of the direction of applied shear in a conventional Couette viscometer. Behavior shown is that observed for a 45% suspension of 46μm polystyrene spheres in silicone fluid. After Gadala-Maria (1979).*

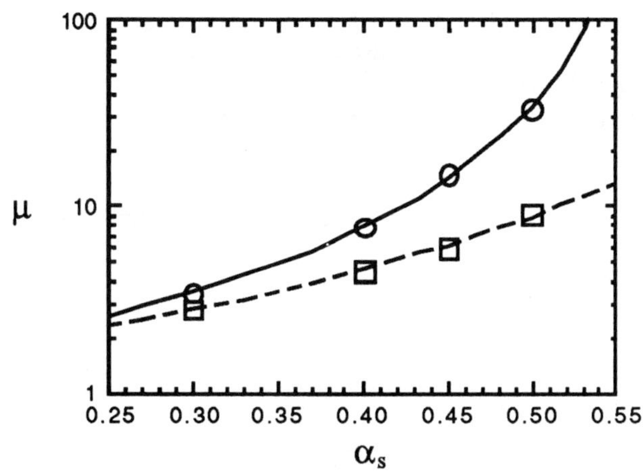

FIGURE 6-2 *Observed viscosity as a function of concentration. ○ - steady state viscosity, □ - minimum viscosity upon reversal of the direction of shear. Euler equation fitting parameters were: steady state - $\alpha_{s\,max}$ = .58, [η] = 2.76; minimum - $\alpha_{s\,max}$ = .78, [η] = 2.72. Data taken from Leighton (1985) for 46μm polystyrene spheres suspended in silicone fluids, sheared at 24s^{-1} in a Couette viscometer.*

where α_s is the solids concentration, [η] is the intrinsic viscosity, and $\alpha_{s\,max}$ is the maximum packing fraction, we find that the intrinsic viscosity of both sets of data are similar. The maximum packing fraction corresponding to the steady viscosity is 0.58, however, while that of the minimum viscosity is

0.78. While these values are very approximate measures of the maximum concentration at which the suspension will flow, particularly since they involve extrapolation from measurements made at concentrations only up to 50%, they do suggest that the instantaneous microstructure corresponding to the minimum in the torque signal is significantly different from the steady microstructure for simple shear.

The most straightforward explanation for the observed shear reversal behavior is that an anisotropic microstructure is acquired by a concentrated suspension undergoing steady simple shear flow, and that this microstructure flips over into its mirror image upon reversal of the direction of shear (Gadala-Maria and Acrivos, 1980). The decrease in the suspension viscosity corresponding to this reversal in shear direction can be most easily explained if the steady microstructure consists of elongated clusters aligned with the compression axis of the shear flow. Brady and Bossis (1988) have shown through numerical simulations that such structures lead to a large viscosity in much the same way as a suspension of rods has a much higher viscosity than a suspension of spheres with the same volume fraction. The microstructure corresponding to the minimum in the viscosity thus may be the result of the breakup of such structures, and the subsequent increase due to the formation of their mirror image in the reverse direction. The decrease in the strain corresponding to this minimum as the suspension concentration is increased is due to the smaller average interparticle separation distance at higher concentrations. The observation that the initial viscosity upon reversal of the direction of shear is less than the steady viscosity suggests the existence of some irreversible non-hydrodynamic force in the suspension. Brady and Bossis (1985) have shown via simulations that electroviscous forces can lead to qualitatively similar shear reversal behavior, however Leighton (1985) showed that such forces are negligible for the non-colloidal systems being considered here. Surface roughness and physical contact between particle surfaces has been proposed as the source of the irreversible force for these suspensions (Smart and Leighton, 1989), however this has yet to be conclusively demonstrated.

6.2.2 Normal Stress Differences

A second indirect observation consistent with anisotropic local microstructure is the existence of normal stress differences in viscometric flows. Normal stress differences are common in viscometric flows of viscoelastic materials such as high molecular weight polymers. The rod-climbing, or Weissenberg effect, is a manifestation of this phenomenon (Bird *et al.*, 1987). The imposed shear flow causes the conformation of the polymer to change to some anisotropic structure, and this structure further interacts with the shear flow to produce the normal stress differences. Because the deformation is proportional to the shear stress in the limit of small shear rates, the normal stress difference is proportional to the *square* of the shear rate for small shear rates. Quite a different behavior is observed in concentrated suspensions. First examined by Bagnold (1954) and more recently by Gadala-Maria (1979), the normal stress differences at low Reynolds numbers were found to be *linear* in the shear rate. This is the result of the structure acquired by the suspension being independent of the rate of shear, and only a function of the geometry of the applied shear. Any stresses (normal or shear) would thus be proportional to the shear rate. A plot of the normal stress difference function $\nu_1 - \nu_2 \equiv \tau_{11} + \tau_{22} - 2\tau_{33}$ for the simple shear flow $u_i = \dot{\gamma}\delta_{i1}x_3$ observed by Gadala-Maria for a suspension of 46μm polystyrene spheres in a parallel plate viscometer is depicted in Figure 6-3.

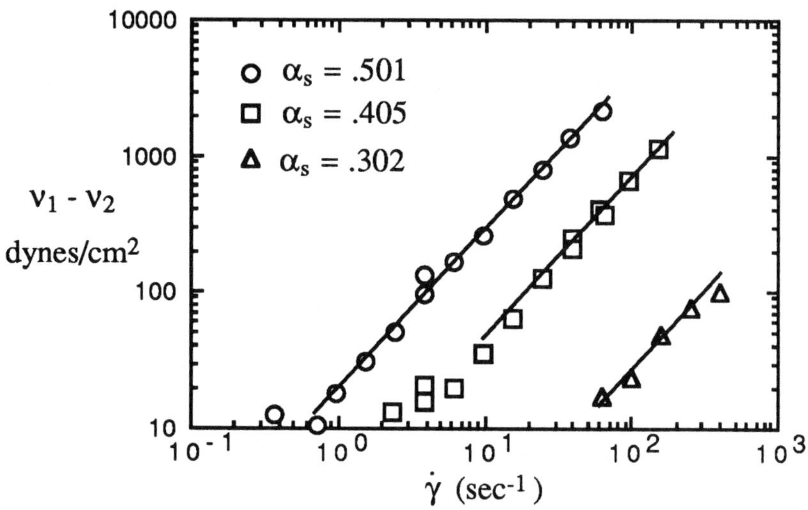

FIGURE 6-3 *Normal stress difference measurements in suspensions of 46μm polystyrene spheres in Ucon oil 75-H-450. After Gadala-Maria (1979).*

The normal stress differences were observed to be a strong function of concentration, as was the suspension viscosity. Leighton (1985) showed that the magnitude of the normal stress differences could be scaled via the anisotropy parameter A defined as:

$$A = \frac{\nu_1 - \nu_2}{\dot{\gamma}\,\mu\,\alpha_s} \tag{6.2}$$

where $\dot{\gamma}$ is the shear rate, μ is the suspension viscosity, and α_s is the suspension concentration. From the measurements by Gadala-Maria (1979) this parameter had a value of 0.04 at a concentration of 30% (where normal stress differences were first observable) and a value of .30 and .34 at a concentration of 40% and 50%, respectively. Note that between 40% and 50% the suspension viscosity increased by more than a factor of 4, while A was virtually unchanged.

The anisotropy parameter is intimately related to the suspension microstructure. For a dilute suspension of non-spherical particles, for example, A is a constant which depends only on the geometry of the shear flow, the shape of the particles, and their orientation. Thus, the observation that this parameter is constant for concentrated suspensions suggests that the structure of a 40% suspension is similar to that at 50%. Also, A can, in general, be positive or negative depending on the orientation and shape of the particles in suspension. For dilute suspensions of prolate spheroids, for example, A is positive when the major axis is aligned with the compression axis of a simple shear flow and negative when aligned with the extensional axis. As was first pointed out by Bagnold, the observed positive normal stress differences can be explained in terms of an anisotropic structure in which there is a reduction in the component of the velocity in the flow direction of a particle as it climbs over another particle during an interaction. Such a reduction in velocity would be the result of any irreversible repulsive force acting betweeen particles, such as electroviscous forces or, in the case of non-colloidal spheres, surface contact. This irreversible force leads to an imbalance in the particle distribution so that there is a relative excess of particles lined up along the compression axis of a shear flow (where their relative approach is slowed down) and a depletion along the extensional axis (where their recession is unhindered by the irreversible repulsive force), the same structure suggested by the shear reversal experiments.

6.3 DIRECT MEASUREMENTS OF SUSPENSION MICROSTRUCTURE

Because the suspension microstructure can only be inferred from the indirect measurements described above, it is desirable to examine the structure of the suspension directly. Several investigations of this type have been undertaken, all involving optical techniques. While optical techniques are quite useful for dilute suspensions, where the suspension itself is sufficiently transparent that the interior may be easily viewed, concentrated suspensions are much more challenging since they are, in general, optically opaque. In this section we will describe three recent experimental approaches.

6.3.1 Tracer Observation

One way in which optical techniques can be used in concentrated suspensions is through the use of refractive index matched suspensions - suspensions in which the suspending fluid is chosen to match the refractive index of the transparent particles. The behavior of the suspension is then determined by examining the behavior of a few opaque tracer particles. In this way, even a concentrated suspension is rendered optically dilute. Such techniques for concentrated suspensions were pioneered by Mason (Karnis, et. al, 1966; Karnis and Mason, 1967) many years ago, however one disadvantage of this technique is that it is generally only possible to obtain averaged information about the suspension such as the local concentration or the local particle velocity. Since only a small fraction of the spheres are opaque, it is difficult to acquire any information about how the particles are distributed relative to each other.

These difficulties were partially overcome by Graham (Graham and Bird, 1984; Graham and Steele, 1984) through the use of information theory. The purpose of these experiments was to determine the degree of particle aggregation which occured for moderately concentrated suspensions ($\alpha_s \leq 0.2$). A suspension of acrylic spheres was prepared using very large particles - 1/2 inch diameter ground spheres - and a suspending fluid which closely matched the refractive index of the acrylic. Because of the large size of the spheres, the suspending fluid was made very viscous (28.3p at 21.6^{0}C) so that the Reynolds number would still be small. Reynolds numbers for these experiments were on the order of 10^{-2} based on the shear rate and the particle radius. The suspensions were confined between two belts which were moved in opposite directions, resulting in a shear flow in the center. In order to minimize sedimentation effects, the device was oriented so that the motion was in the vertical direction.

Observations were performed by labelling tracer spheres with numbers so that each could be uniquely identified. Tracer particles were added to the suspension so that the concentration of the tracers was always 1%, sufficiently dilute that all of the particles could be seen. In operation, the suspension was observed using two synchronized video cameras oriented at right angles to each other, and a third video camera to track the motion of the belts driving the flow. From these images it was possible to directly calculate the absolute position of each of the tracers within the field of view. While such observations may be used to calculate concentration velocity distributions in the same manner as was done by Karnis, Graham also used the measurements to determine the degree of aggregation in the suspensions. A pair of tracer particles were taken to be aggregated if there was no measureable separation distance between the particles. For dilute suspensions where all of the particles are tracers, this directly yields the degree of aggregation. At 20% concentration, however, where only one out of every 20 particles are marked, Graham used information theory to infer the degree of aggregation. If a pair of tracer particles are observed to be touching in a concentrated suspension, it is likely that they are part of a much larger aggregate. By using statistics and probability theory it is possible to take observations of tracer spheres and infer the distribution of aggregate sizes. In their work, they found the aggregate size to be a rapidly increasing function of concentration, with about 6 spheres in an average cluster at 20% concentration. A plot of aggregate size vs. concentration is given in Figure 6-4. More recently, this work has been extended to include observations of fluctuating particle velocities at Reynolds numbers ranging from laminar flow up to and including turbulent conditions, although only at a concentration of 5% (Majumdar *et al.*, 1987). Measurements of particle angular velocities in turbulent shear flow have also been made for highly dilute ($\alpha_s = 0.05\%$) suspensions (Ye, *et al.* 1990).

6.3.2 Surface Observations

While the technique described above was very useful for acquiring aggregation information at moderate concentrations, no effort was made to determine the pair distribution function, or to extend the work to concentrations above 20% where interesting effects such as the shear-reversal behavior and non-zero normal stress differences occur. Indeed, structural information is very difficult to acquire at high concentrations from

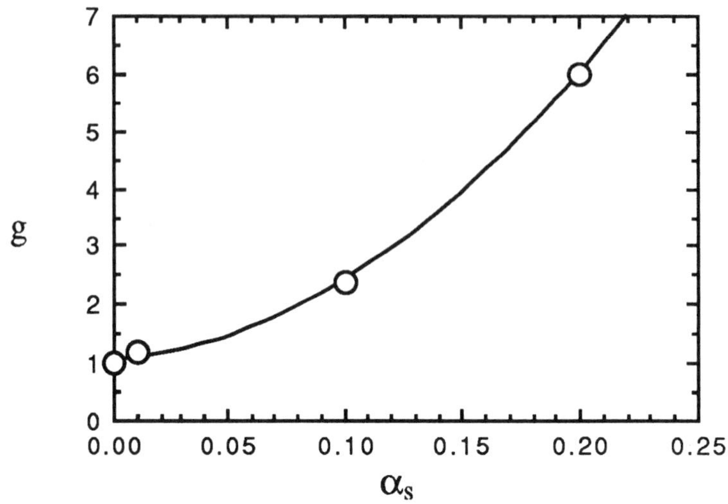

FIGURE 6-4 *Average number of spheres in a cluster as a function of concentration. Data were interpolated to a particle Reynolds number of 2.7 x 10^{-2}. After Graham and Steele (1984).*

tracer techniques because the percentage of marked spheres is small and the number of images required to obtain statistical significance increases accordingly. The first actual measurements of the pair distribution function for concentrated suspensions were made by Parsi and Gadala-Maria (1987) using a surface observation technique.

In these experiments, a concentrated suspension of 46μm polystyrene spheres was sheared in a conventional Couette viscometer, and the surface of the suspension was brightly illuminated. These investigators found that the light would be reflected off of the spheres in the layer closest to the top surface (there was no attempt at refractive index matching) and could be videotaped using a videomicroscope. The images consisted of a plane with bright dots in it, with these dots corresponding to the centers of the particles. By digitizing a large number of the videotaped images, it was possible to reconstruct the pair distribution function. The results these investigators obtained are reproduced in Figure 6-5.

The pair distribution function was found to have a maximum on the approach side of an interaction (along the compression axis) and a minimum on the recession side, in agreement with the interpretation of the indirect measurements described above. When the direction of applied shear was reversed, the pair distribution was found to evolve into its mirror image.

This approach offers many advantages over other optical techniques since it allows the use of non-refractive index matched suspensions and suspensions of much smaller spheres, however a number of significant difficulties were encountered. Since the suspension was opaque, only the surface layer of particles could be observed, and thus the results may have

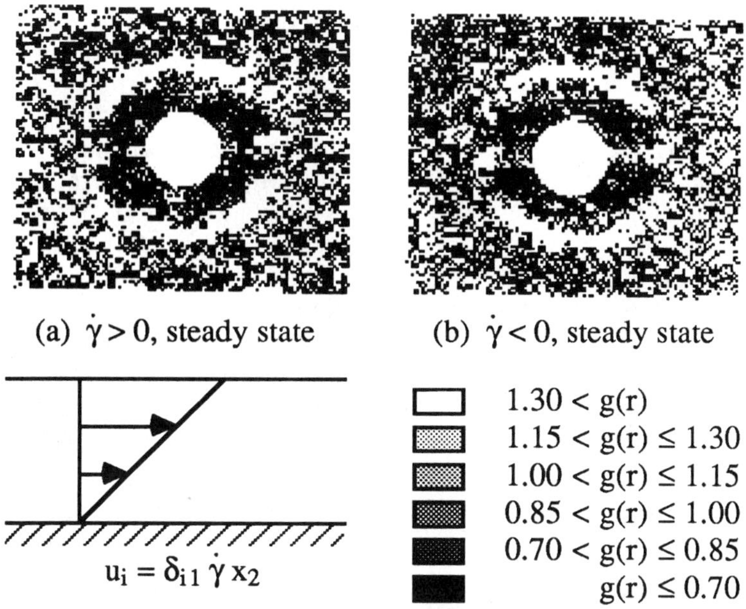

FIGURE 6-5 *Two dimensional pair distribution function for 46μm polystyrene spheres in silicone fluids. Suspension concentration was 40%. The white circle in each plot represents the reference sphere. Note the large maximum in the pair distribution function at r/a = 2 and the lack of fore-and-aft symmetry. After Parsi and Gadala-Maria (1987).*

been influenced by the proximity of the meniscus. It should be noted that this was not a problem with the corresponding measurements in a dilute suspension (Husband and Gadala-Maria, 1987), because for a dilute system the microscope could see deep into the suspension, with only the layer of particles in the plane of focus of the microscope being used in constructing the pair distribution. The calculated pair distribution function was also somewhat inaccurate because the spheres used in the calculation were not all in the same plane. As may be seen from Figure 6-5 this leads to some particles apparently being in the forbidden region less than two particle radii away from the sphere at the origin. To alleviate these difficulties somewhat, we have recently developed a new laser sheet imaging technique which will be described below.

6.3.3 Laser Sheet Imaging

The new imaging procedure which we have developed is similar to that used by Graham, *et al.* in that we employ index of refraction matching between particles and fluid, however instead of using tracer particles we illuminate a test section with a sheet of laser light. Due to the matching of the refractive index, the suspension is largely transparent to the laser, with a small amount of scattering occuring at the surfaces of the particles. If the laser sheet is thin compared to the radius of the particles (these experiments employed 1/8" and 1/4" diameter acrylic spheres), the surface scattering appears as a plane of bright circles when viewed normal to the plane of the laser sheet. The centers of the circles correspond to the projection of the centers of the spheres into the plane of the laser light, thus a single image provides the two-dimensional particle distribution function and multiple images of the system yield a two-dimensional velocity profile.

Experiments were conducted with suspension concentrations ranging from 40% to 50%. The acrylic spheres were suspended in a solution of 72.8% by weight Triton X-100, 14.0% zinc chloride, and 13.2% water. A small quantity of concentrated hydrochloric acid was added to the solution to prevent the formation of zinc hypochlorite precipitate. The fluid was found to match both the refractive index and density of the acrylic, and to be Newtonian with a viscosity of approximately 53p at 22.5^{0}C. The suspensions were confined inside a belt composed of 110mm photographic film from which the emulsion had been stripped, as depicted in Figure 6-6.

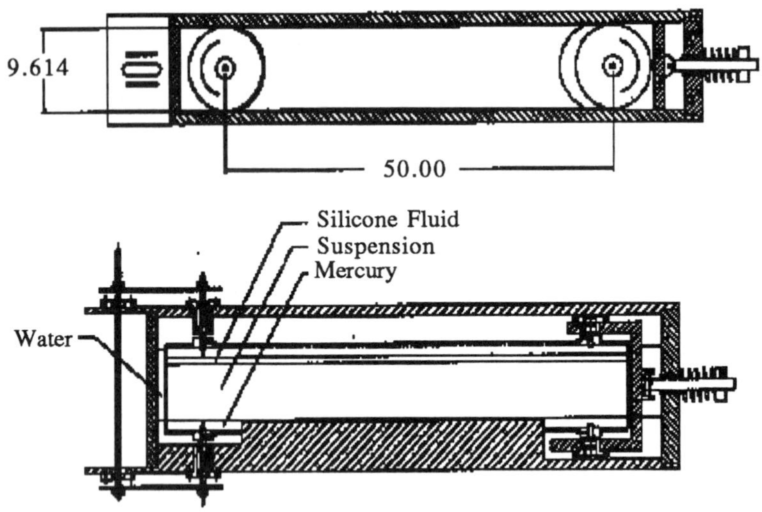

FIGURE 6-6 *Experimental geometry used in the flow visualization experiments. The suspension was confined in the interior of a rotating film loop. A stress-free lower boundary was provided by a layer of mercury, and the silicone fluid was added to eliminate distortion at the upper surface. All dimensions are in cm.*

A lower stress-free boundary was provided by a layer of mercury and an upper stress free boundary by a thin layer of immiscible silicone fluid with a viscosity of 1.2p. This upper layer was necessary to eliminate the distortion caused by particles being forced upward into the free surface by interactions in the shear flow. The suspensions were sheared at a shear rate of 0.07s^{-1}, yielding a particle Reynolds number of $1.6x10^{-4}$. The laser sheet was produced by a 5mW Helium-Neon laser using cylindrical lenses, and was projected through the film belt parallel to the plane of shear. The scattered light was observed through the upper free surface using a conventional video camera. A typical digitized image is reproduced in Figure 6-7.

In order to acquire useful information from these images it is necessary to determine the centers and radii of the circles in each image to a very high

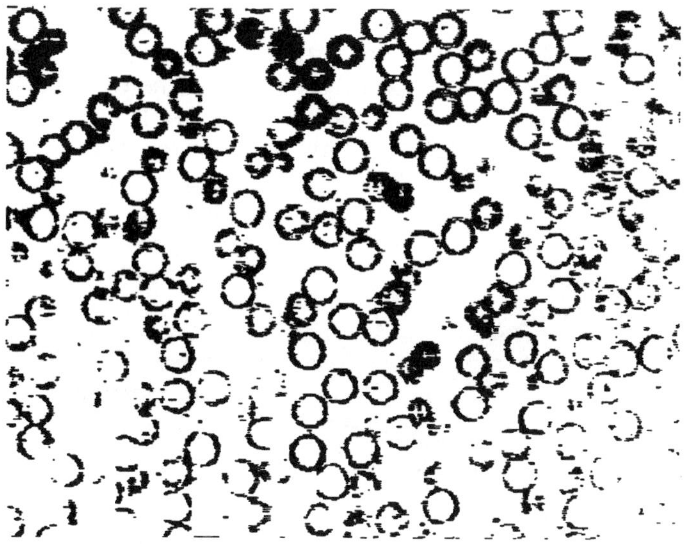

FIGURE 6-7 *Digitized image of a concentrated suspension. The suspension consisted of 1/8" acrylic spheres suspended in a mixture of Triton X-100, water and $ZnCl_2$ at 35% concentration. The shear direction was such that the velocity was to the left at the top of the picture and to the right at the bottom. Note the predominance of linear structures aligned with the compression axis.*

degree of accuracy. Conventional template matching techniques are capable of identifying the positions of circles of a specific radii, but cannot be used to locate circles of varying radii. In this visualization procedure circles of different radii are observed because the spheres (which are all ground to the same diameter) are not precisely in the plane of the laser sheet. We have extended match filtering techniques to overcome this difficulty. The new technique relies on convolving (using FFT routines) the images containing the circles with an image of a complex disk whose radius is that of the largest circle in the original image, with unit modulus, but whose phase varies radially so that the disk contains equal areas of all phases. This mask is given by:

$$M = \begin{cases} \exp\left(\dfrac{2\pi i r^2}{a^2}\right) & r \leq a \\ 0 & r \geq a \end{cases} \tag{6.3}$$

where a is the radius of the largest circle and $i = (-1)^{1/2}$. The convolution of this mask with an image containing circles results in a local maximum at the center of each circle, and the size of the circle in the original image may be calculated from the complex phase of the local maximum. The implementation of this algorithim for circle identification on a 512x512 digitized image takes approximately one minute on a Convex 220 computer, however it is considerably faster if specialized FFT accelerators are used.

From the location of circle sizes in each image it is possible to determine the pair-probability distribution of the particles in the suspension. Actually, because the sphere centers are not precisely in the plane of the laser sheet, the pair distribution calculated from the position of the circle centers is an integral of the underlying three-dimensional pair distribution function, as may be seen from Figure 6-8.

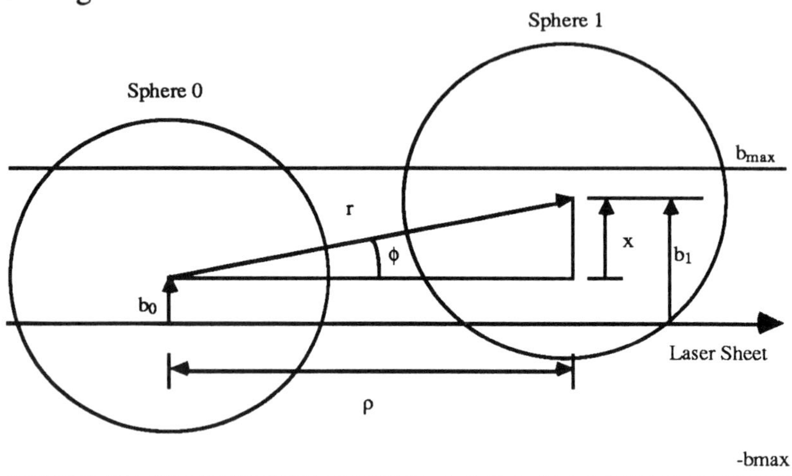

FIGURE 6-8 *Relationship between the observed pair distribution and the true pair distribution. All sphere centers were, in general, some distance b out of the plane of the laser sheet. Only spheres whose centers were within* $|b| < b_{max}$ *were used in constructing the pair distribution function.*

The relationship between the observed and actual distribution is given by the integral:

$$P_{obs}(\rho,\theta) = \frac{1}{2b_{max}} \int_{-b_{max}}^{b_{max}} \int_{-b-b_{max}}^{-b+b_{max}} P(r,\theta,\phi)\, dx\, db \tag{6.4}$$

where (ρ,θ) is the point in the plane of the laser light for which we are calculating the pair distribution function, and b_{max} is the maximum displacement of the sphere centers out of the plane. If we assume that the pair distribution is not a function of the angle ϕ out of the plane of shear, we may recast this integral into a convolution form:

$$P_{obs}(\rho,\theta) = \int_{-\infty}^{\infty} G(r^2 - \tau)\; P(\tau^{1/2},\theta)\, d\tau \tag{6.5}$$

where the convolution kernel G(t) is given by:

$$G(t) = \begin{cases} 0 & t < -4b_{max}^2 \\ \dfrac{1}{\sqrt{-t}} - \dfrac{1}{2b_{max}} & -4b_{max}^2 < t < 0 \\ 0 & t > 0 \end{cases} \tag{6.6}$$

which may then be deconvolved to obtain an approximation to the true pair distribution function in the plane of shear. The approximation that the pair distribution is independent of ϕ is reasonable as all of the particles used in constructing the pair distribution are close to the plane of shear and we know that $\partial P/\partial \phi = 0$ in the plane from symmetry. An example of the results of the deconvolution is given in Figure 6-9 where we have averaged over all angles in the plane of shear. Note that the deconvolution procedure significantly reduces the calculated magnitude of the pair distribution function in the forbidden region $r/a < 2$.

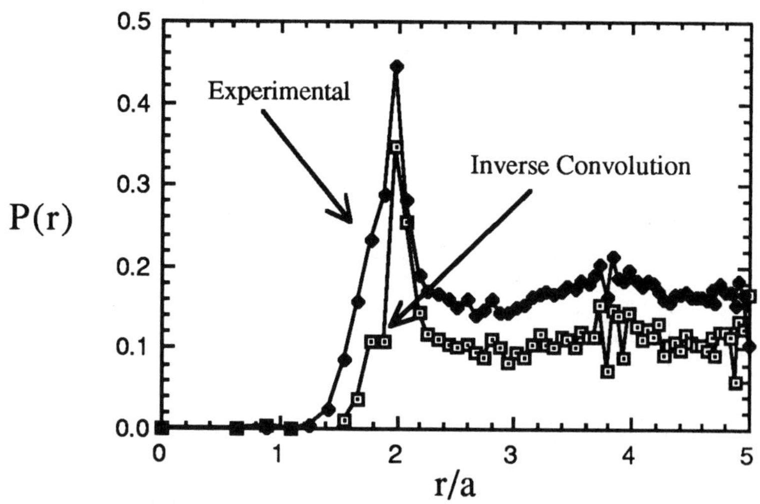

FIGURE 6-9 *Deconvolution of the radial pair distribution function. Data were for the pair distribution function averaged over all angles in the plane of shear. Note that the deconvolved pair distribution function has a much sharper singularity at r/a = 2. Data were for a 45% suspension.*

Plots of the radial component of the deconvolved pair distribution function for a 40% suspension are given in Figures 6-10 (the compression quadrant) and 6-11 (the extensional quadrant) for several angles. Rather than plotting the pair distribution directly, instead we have plotted its area average as a function of r/a. This greatly reduces the noise inherent in measurements of the pair distribution function. Note that the slope in Figure 6-10 is very sharp at a distance of two particle radii, and that the distribution is nearly independent of angle over the compression quadrant ($90^0 < \theta < 180^0$). A smaller secondary maximum at $r/a = 4$ may also be seen. In the extensional

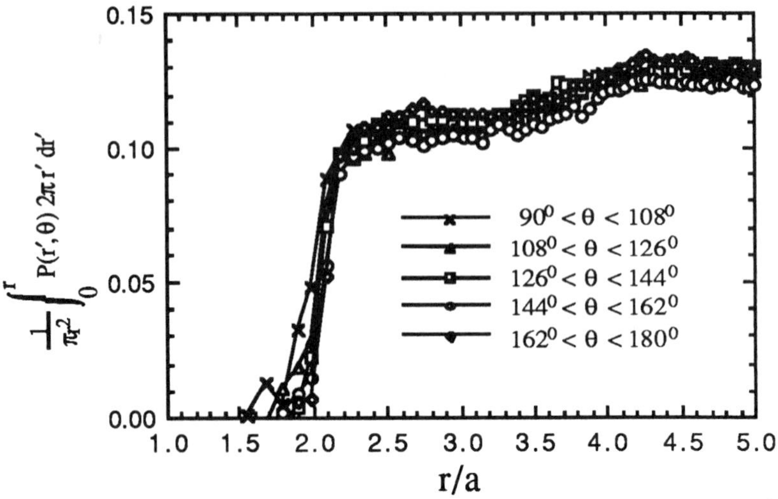

FIGURE 6-10 *Integrated pair distribution function for a 45% suspension in the compression quadrant. The pair distribution function was area averaged as a function of r to minimize noise. $\theta = 180°$ corresponds to the flow direction, $\theta = 90°$ is parallel to the velocity gradient. Note the lack of angular structure.*

quadrant ($0^0 < \theta < 90^0$) shown in Figure 6-11 a pronounced angular structure is apparent. The maximum at r=2a is sharply attenuated at $\theta = 27^0$, and the maximum is shifted out to slightly higher values of r. This may be seen more clearly in the corresponding plot of the angular distribution for several radial distances near r/a = 2 given in Figure 6-12. These observations are in agreement with the surface measurements of Parsi and Gadala-Maria (1987) discussed above, and also with the indirect rheological measurements.

6.4 CONCLUSIONS

From the indirect and direct measurements of suspension microstructure it is clear that a concentrated suspension in simple shear flow acquires a structure such that there is a depletion in the probability of finding a second sphere in the extensional quadrant relative to a first sphere located at the origin. This structure is consistent with the existence of irreversible contact forces between particles, and agrees well with the results of numerical simulation of colloidal particles which include repulsive electroviscous forces. The detailed structural information gained in the direct visualization experiments may be used to predict the rheological behavior of concentrated suspensions undergoing viscous flow, and information about particle rotational and translational velocities in turbulent flows can be used to better understand the interchange of kinetic energy between the fluid and particles in these systems. Measurements of the microstructure should also prove

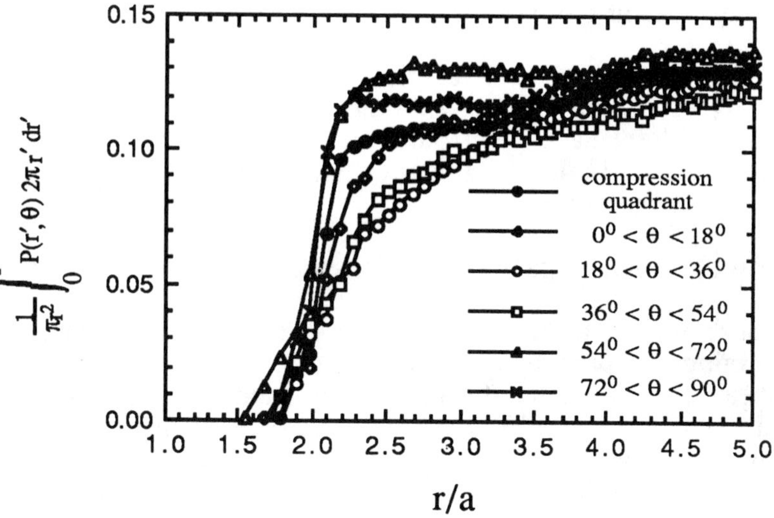

FIGURE 6-11 *Integrated pair distribution function for a 45% suspension in the extensional quadrant. The pair distribution function averaged over the compression quadrant is given by the filled circles for comparison. $\theta = 0°$ corresponds to the flow direction, $\theta = 90°$ is parallel to the velocity gradient. Note the pronounced angular structure in contrast to that in the compression quadrant. The average over the compression quadrant is also plotted for comparison.*

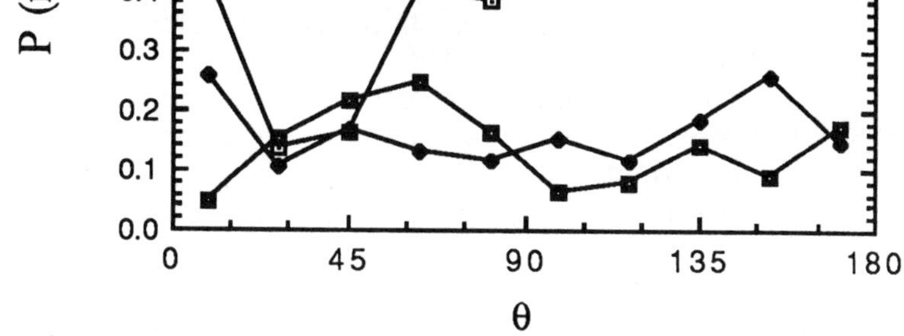

FIGURE 6-12 *Pair distribution function in the plane of shear as a function of angle. Data were for a 45% suspension. Note the absence of the r/a = 2 singularity in the extensional quadrant (0° < θ < 90°).*

useful in developing more realistic numerical simulations of concentrated suspensions.

While the direct and indirect measurements of the microstructure of concentrated suspensions of non-colloidal spheres undergoing simple shear flow described here have been quite successful, much work in this area remains to be done. In the remainder of this section we shall describe a few possible avenues for future investigation.

6.4.1 Particle Aggregation in Concentrated Suspensions

The measurement techniques already developed have proven successful in determining the pair distribution function of suspensions undergoing simple shear flow, however for a highly concentrated suspension this one function is insufficient to adequately describe the suspension microstructure. In particular, knowledge of the degree to which particles aggregate in the suspension is desirable: do particles flow as discrete units, or as parts of larger cluster of particles? It is difficult to extend the aggregation experiments of Graham to higher concentrations as at high concentration nearly all of the particles are in close proximity, yet they may not flow as a cluster. One way in which this problem may be addressed is through examination of the disturbance velocities of the particles in the suspensions - particles which belong to large clusters should have disturbance velocities which scale as the shear rate and the length scale of the cluster, rather than just the particle radius.

A related problem is the formation of percolation thresholds as a suspension is sheared. Such percolation phenomena have an important impact on transport properties such as electrical and thermal conductivity, particularly when the particles have a much higher conductivity than the suspending fluid. Numerical simulations suggest the existence of such a threshold (Brady and Bossis, 1985), and it would be useful to tie such simulations to actual observed microstructures.

6.4.2 Measurement of Particle Interactions

As was described earlier, it is uncertain how particles will interact when they come into actual physical contact. In order to develop a numerical simulation scheme which accurately reflects a physical suspension, it is important to know whether particles will "lock up" and rotate as a rigid pair of particles when in contact, or whether the surfaces will slide past one another. Such relative motion is likely to significantly affect the microstructure and degree of aggregation, particularly in concentrated suspensions. To address this question, it will be necessary to not only track particle motion, but to measure their rotational velocity as well. This is quite difficult as none of the techniques described above provide information on the angular position of the particles in concentrated suspensions. New measurement schemes will have to be developed to address this problem.

6.4.3 Extension to Other Flow Geometries and Systems

Thus far we have discussed the microstructure produced by simple shear flows only. While this is arguably the most important flow geometry (as well as being the easiest to work with) there are a host of other significant geometries. Important examples include sedimentation, with application to fluidized beds, and the extensional flow produced by a four roll mill. In addition, the structure produced by low amplitude oscillatory motion is of interest, as is the effect of finite Reynolds numbers, particle shape, and possible non-Newtonian characteristics of the suspending fluid on the structure. Recent experimental results suggest that when oscillatory shear is combined with a high degree of viscoelasticity, for example, a concentrated suspension acquires a highly ordered cubic structure (Kim and Gadala-Maria, 1990) which has important implications in composite material fabrication.

Most industrially occuring suspensions are polydisperse in particle size, thus it would be useful to examine the microstructure of bidisperse and polydisperse suspensions. This could have importance in calculating the burn rate of solid rocket fuel, for example, which is essentially a suspension of several different components, each of which have different size distributions. A suspension in which large particles are surrounded by small ones will burn differently than one in which large and small particles are clustered separately, for example. Measurements of the three-dimensional microstructure of suspensions also need to be performed.

In order to carry out these measurements it will be necessary to develop experimental techniques which go far beyond those described here. Optical techniques are largely limited to index matched suspensions, thus it is desirable to explore alternative approaches. One recently developed system which may be of use is high resolution magnetic resonance imaging. While this technique has been employed largely to measure particle concentration and velocity distributions, some information on microstructure for bidisperse suspensions was obtained by Abbot *et al.* (1991) using this technique. Other recently developed approaches to suspension imaging include acoustic imaging and x-ray tomography, although as yet these techniques lack sufficient resolution to determine suspension microstructure.

In conclusion, while much information has been obtained about the microstructure of concentrated suspensions and its relationship to the corresponding macroscopic rheological and effective medium properties, there still exist many avenues for further research in this area.

ACKNOWLEDGEMENTS

We thank Jeffrey Smart, Sean Beimfohr and Joseph Sladek for their assistance in preparing this document. This work was supported in part by the National Science Foundation.

REFERENCES

Abbot, J. R., N. Tetlow, A. L. Gragan, S. A. Altobelli, E. Fukushima, L. A. Mondy, and T. S. Stephens, Experimental Observations of Particle Migration in Concentrated Suspensions: Couette Flow, *J. Rheol.*, **5**, 773-795 (1991).

Bagnold, R. A., Experiments on a Gravity-Free Dispersion of Large Solid Spheres in a Newtonian Fluid Under Shear, *Proc. R. Soc. London, Ser. A.*, **225**, 49-63 (1954).

Batchelor, G. K. and J. T. Green, The Determination of the Bulk Stress in a Suspension of Spherical Particles to Order c^2, *J. Fluid Mech.*, **56**, 401-427 (1972).

Bird, R. B., R. C. Armstrong, and O. Hassager, *Dynamics of Polymeric Liquids*, Wiley, New York (1987).

Bonnecaze, R. T. and J. F. Brady, Dynamic Simulation of an Electrorheological Fluid, paper F15, 62nd Annual Meeting, The Society of Rheology, Santa Fe, New Mexico, October 21-25, 1990.

Bossis, G. and J. F. Brady, Dynamic Simulation of Sheared Suspensions. I. General Method., *J. Chem. Phys.*, **80**, 5141-5154, (1984).

Bossis, G. and J. F. Brady, Self-Diffusion of Brownian Particles in Concentrated Suspensions Under Shear, *J. Chem Phys.*, **87**, 5437-5448 (1987).

Bossis, G. and J. F. Brady, The Rheology of Brownian Suspensions, *J. Chem Phys.*, **91**, 1866-1874 (1989).

Brady, J. F. and G. Bossis, The Rheology of Concentrated Suspensions of Spheres in Simple Shear Flow by Numerical Simulation, *J. Fluid Mech.*, **155**, 105-129 (1985).

Bossis, G., J. F. Brady, and C. Mathis, Shear-Induced Structure in Colloidal Suspensions. I. Numerical Simulation, *J. Col. Interface Sci.* **126**, 1 (1988).

Brady, J. F. and G. Bossis, Stokesian Dynamics, *Ann. Rev. Fluid Mech.* **20**, 111-157 (1988).

Durlofsky, L. and J. F. Brady, Dynamic Simulation of Bounded Suspensions of Hydrodynamically Interacting Particles, *J. Fluid Mech.*, **200**, 39-67 (1989).

Durlofsky. L, J. F. Brady, and G. Bossis, Dynamic Simulation of Hydrodynamically Interacting Particles, *J. Fluid Mech.*, **180**, 21-49 (1987).

Gadala-Maria, F. A., *The Rheology of Concentrated Suspensions*, Ph.D. Thesis, Stanford University, 1979.

Gadala-Maria, F. and A. Acrivos, Shear-Induced Structure in a Concentrated Suspension of Solid Spheres, *J. Rheol.*, **24**, 799-814, (1980).

Graham, A. L. and R. B. Bird, Particle Clusters in Concentrated Suspensions. 1. Experimental Observations of Particle Clusters, *Ind. Eng. Chem. Fundam.*, **23**, 406-406, (1984).

Graham, A. L. and R. D. Steele, Particle Clusters in Concentrated Suspensions. 2. Information Theory and Particle Clusters, *Ind. Eng. Chem. Fundam.*, **23**, 411-420, (1984).

Husband, D. M. and F. Gadala-Maria, Anisotropic particle distribution in dilute suspensions of solid spheres in cylindrical Couette flow, *J. Rheol.*, **31**, 95-110, (1987).

Karnis, A., H. L. Goldsmith, and S. G. Mason, The Kinetics of Flowing Dispersions, I. Concentrated Suspensions of Rigid Particles. *J. Coll, Int. Sci.*, **22**, 531-553 (1966).

Karnis, A. and S. G. Mason, The Flow of Suspensions Through Tubes. VI. Meniscus Effects. *J. Coll. Int. Sci.*, **23**, 120-133 (1967).

Kim, C.-G. and F. Gadala-Maria, The Microstructure of Sphere-Filled Composites, paper A-19, 62nd Annual Meeting, The Society of Rheology, Santa Fe, New Mexico, October 21-25, 1990.

Leighton, D. T., *The Shear-Induced Migration of Particulates in Concentrated Suspensions*, Ph.D. Thesis, Stanford University, 1985.

Majumdar, A., A. L. Graham, M. C. Roco, and P. Stroeve, Experimental Study on Solid Particle Dynamics in Shear Flow, *Powder Technology*, **49**, 217-226 (1987).

Pakdel, P. and S. Kim, Mobility and stresslet functions of particles with rough surfaces in viscous fluids: A numerical study, *J. Rheol.*, **35**, 797-823 (1991).

Parsi, F. and F. Gadala-Maria, Fore-and-Aft Asymmetry in a Concentrated Suspension of Solid Spheres, *J. Rheol.*, **31**, 725-732 (1987).

Russel, W. and A. Gast, Non equilibrium statistical mechanics of concentrated colloidal dispersions: Hard spheres in weak flows, *J. Chem. Phys.*, **84**, 1815-1826 (1986).

Smart, J. R. and D. T. Leighton, Jr., Measurement of the Hydrodynamic Surface Roughness of Noncolloidal Spheres, *Phys. Fluids A*, **1**, 52-60 (1989).

Wagner, N. and W. Russel, Non equilibrium statistical mechanics of concentrated colloidal dispersions: Hard spheres in weak flows with many-body thermodynamic interactions, *Physica A*, **155**, 475-518 (1989).

Ye, J., Roco, M.C. and Staford, D., Particle rotation in Couette flow, ASME Vol. FED 98, 19–27 (1990), (also *Phys. Fluids A*, 1991).

CHAPTER 7

CHANNEL FLOWS OF GRANULAR MATERIALS AND THEIR RHEOLOGICAL IMPLICATIONS

H. Ahn
C. Brennen

7.1 BASIC FEATURES OF CHANNEL FLOWS OF DRY GRANULAR MATERIAL

7.1.1 Introduction

While the flow of a dry granular material down an inclined channel may seem at first sight to be a relatively simple flow, the experiments which have been conducted up to now suggest sufficient complexity to warrant further analysis and investigation. They also suggest complexities which may be present in all but the very simplest granular material flows; consequently it is important to our general understanding of granular material rheology that these experimental observations be fully understood. This review of the current knowledge of channel flows will focus on the basic mechanics of these flows and the contributions the observations have made to an understanding of the rheology. In order to make progress in this objective, it is necessary to avoid some of the complications which can occur in practice. Thus we shall focus only on those flows in which the interstitial fluid plays very little role in determining the rheology. In his classic paper, Bagnold (1954) was able to show that the regime in which the rheology was dominated by particle/particle or particle/wall interactions and in which the viscous stresses in the interstitial fluid played a negligible role could be defined by a single, Reynolds-number-like parameter. It transpires that the important component in this parameter is a number which we shall call the Bagnold number, Ba, defined by

$$Ba = \rho_s d^2 \delta / \mu_F$$

where ρ_s, μ_F are the particle density and interstitial fluid viscosity, d is the particle diameter and δ is the principal velocity gradient in the flow. In the shear flows explored by Bagnold δ is the shear rate. Bagnold (1954) found that when Ba was greater than about 450 the rheology was dominated by particle/particle and particle/wall collisions. On the other hand, for $Ba < 40$, the viscosity of the interstitial fluid played the dominant role. More recently Zeininger and Brennen (1985) showed that the same criteria were applicable to the extensional flows in hoppers provided the extensional velocity gradient was used for δ. This review will focus on the simpler flows at large Ba where the intestitial fluid effects are small.

Other important ancilliary effects can be caused by electrical charge separation between the particles or between the particles and the boundary walls. Such effects can be essential in some flows such as those in electrostatic copying machines. Most experimenters have observed electrical effects in granular material flows, particularly when metal components of the structure are not properly grounded. The effect of such electrical forces on the rheology of the flow is a largely unexplored area of research. The lack of discussion of these effects in this review should not be interpreted as a dismissal of their importance.

Apart from electrical and interstitial fluid effects, this review will also neglect the effects caused by non-uniformities in the size and shape of the particles. Thus, for the most part, we focus on flows of particles of spherical shape and uniform size. It is clear that while an understanding of all of these effects will be necessary in the long term, there remain some important issues which need to be resolved for even the simplest granular material flows.

7.1.2 Mass and Momentum Equations for Open Channel Flow

Perhaps the best starting point is the existing body of knowledge of open-channel flows for incompressible liquids. Thus Savage (1979) and Brennen *et al.*(1983) identify different types of granular material chute flow distinguished by a Froude number, Fr, defined in the present paper as

$$Fr = U/(gh)^{\frac{1}{2}} \tag{1}$$

where $U = u_m$ is the cross-sectionally-averaged velocity, h is the depth of flow and g is the acceleration due to gravity. The basic equations which govern the open-channel flow of any substance whether liquid or granular material are those of conservation of mass and a momentum equation. If the depth of the flow, h, varies slowly with position, z, along the channel then the flow can be represented by a velocity, u, parallel to the base and a solid fraction, α, which may be functions of the coordinate y measured perpendicular to the base (the origin of y is taken at the free surface). It is convenient to define profile shape parameters a, b and c as follows.

$$a = \int_0^1 \frac{\alpha}{\alpha_m} \frac{u}{U} d\left(\frac{y}{h}\right) \tag{2}$$

$$b = \int_0^1 \frac{\alpha}{\alpha_m} \left(\frac{u}{U}\right)^2 d\left(\frac{y}{h}\right) \tag{3}$$

$$c = \int_0^1 \frac{2p}{\rho_s \alpha_m g h \cos\theta} d\left(\frac{y}{h}\right) \tag{4}$$

where α_m is the cross-sectionally average solid fraction, ρ_s is the particle density and θ is the inclination of the channel to the horizontal. Note that if the solid fraction, α, and velocity, u, were uniform over the depth then $a = b = 1$. Moreover we define c so that if the pressure, p (normal stress in direction parallel to the base), varied hydrostatically over the depth then the third shape parameter, c, would also be unity.

If we assume for simplicity that the profiles of α, u and p are sufficiently self-similar so that a, b and c are simple constants then we can proceed with a conventional open-channel flow analysis in which, for steady flow, the conservation of mass requires that

$$d\left(\alpha_m h U\right)/dz = 0 \tag{5}$$

and, utilizing this, the equation of momentum can be written as

$$f = \frac{\tau P}{\rho_s \alpha_m g w h \cos\theta} = \tan\theta + \frac{h}{\alpha_m}\frac{d\alpha_m}{dz}\left(\frac{bFr^2}{\cos\theta} - \frac{c}{2}\right) + \frac{dh}{dz}\left(\frac{bFr^2}{\cos\theta} - c\right) \tag{6}$$

where τP is the shear force per unit channel length between the material and the walls and w is the width of the flow. Note that since $\rho_s \alpha_m g h$ is a typical normal stress at the channel base, the factor f is similar to a Coulomb friction factor. Note also that the momentum equation reduces to a more familiar form if the solid fraction is assumed constant ($d\alpha_m/dz = 0$) and the shape factors b and c are assumed to be unity (Patton *et al.*(1987)). The conventional behavior of an open-channel flow can be readily extracted from the above equation (6). In the simpler case of $d\alpha_m/dz = 0$, $b = c = 1$, it is clear that the evolution of the flow as it proceeds down the channel depends on whether the flow is subcritical ($Fr < 1$) or supercritical ($Fr > 1$). For example, if the angle θ is greater than a particular value given by f, if f remains constant as a Coulombic friction law might suggest and, if the flow is supercritical ($Fr > 1$), then the depth, h, will continuously decrease (since $dh/dz < 1$). Thus the flow will continuously accelerate and Fr will get larger still. In flows of liquid this process is terminated because f increases with velocity and hence the flow reaches an equilibrium in the sense that the depth and velocity no longer change with z. In the present paper the term "fully developed" is used to refer to such circumstances. In granular material flow the processes leading to fully developed flow are not well understood. We shall delay further discussion on this important issue until the basic concepts have been introduced.

Another consequence which is familiar in open channel flows or compressible flows is that, if one postulates a fully developed flow, then there exist two flows with the same flow rate (two "conjugate" states, one subcritical and the other supercritical) which satisfy the governing equations. This was confirmed analytically by Richman and Marciniec (1988) who searched for steady, fully developed solutions for the flow of granular material down "bumpy inclines." For a fixed flow rate, the range of inclinations for which fully-developed flow was possible was obtained. For certain combinations of inclination and flow rate, two conjugate solutions were found, one dilute and fast, the other dense and slow.

When an open channel flow is supercritical, kinematic waves cannot propagate upstream and so the evolution of the flow does not depend on downstream conditions or obstacles. What frequently happens under these circumstances is that a kinematic shock or "hydraulic jump" occurs in which the flow undergoes a transformation from a supercritical state ($Fr > 1$) to a subcritical state ($Fr < 1$). Hydraulic jumps have been observed in granular material flows and have been studied by Savage (1979), Morrison and Richmond (1976) and Brennen *et al.*(1983). If the upstream supercritical state and the downstream subcritical state are respectively denoted by subscripts 1 and 2 then application of the same mass and momentum equations leads to the following relation across the shock (Brennen *et al.*(1983)):

$$2Fr_1^2 \left\{ \left(\frac{b_2}{a_2^2} \right) \left(\frac{h_1}{h_2} \right) - \frac{b_1}{a_1^2} \right\} = c_1 - c_2 \frac{h_2^2}{h_1^2}. \tag{7}$$

The equations used by Savage (1979) and Morrison and Richmond (1976) are examples of this relation for special values of the shape parameters a, b and c; for example, Morrison and Richmond assume $a = b = 1$, $c_1 = (1 - \sin\phi)/(1 + \sin\phi)$, $c_2 = (1 + \sin\phi)/(1 - \sin\phi)$ where ϕ is the internal friction angle. Brennen *et al.*(1983) show that the experimental observations of hydraulic jumps in granular material agree with the form of equation (7).

Granular jumps are most commonly manifest in the following way. Most of the experimental facilities (Savage (1979), Ishida and Shirai (1979), Brennen *et al.* (1983), Johnson *et al.* (1990)) use a sluice-gate-like control valve to regulate the flow from a supply hopper to the open channel. At larger channel inclinations this generates a supercritical flow which continues to accelerate if unobstructed. However, when a sufficiently large obstruction is placed in the channel, it will create a granular jump which will propagate upstream like a hydraulic bore. If the obstruction is large enough the jump will propagate all the way to the sluice-gate and result in a substantial decrease in the flow rate. Lesser obstructions may cause the jump to find an equilibrium position some distance upstream of the obstruction (Brennen *et al.* (1983)). In either case, if the obstruction is removed, the flow will revert to its unobstructed state through an expansion wave which propagates upstream from the location of the obstruction and cancels out the hydraulic jump.

Before progressing further with a discussion of the flow of a granular material down an inclined chute it is necessary to examine two other relations which are needed to complement the mass and momentum equations. First we must discuss the concept of granular temperature (section 7.1.3) for it is through that concept that one can follow the constitutive laws (section 7.1.4) which relate the solid fraction, α, to the stresses and the velocity field. Secondly some discussion of the boundary conditions both at a solid wall and at the free surface (section 7.1.5) is necessary for these are clearly more complex than in conventional fluid mechanics.

7.1.3 Concept of Granular Temperature and its Conduction

One of the most important characteristics of rapid granular flow is the "granular temperature" which is a measure of the energy contained in the random or incoherent motions of the particles in the material. Clearly, the velocity of an individual particle can be decomposed into the mean or ensemble-averaged velocity and the fluctuation velocity associated with the departures from this mean. The mean-square value of these fluctuation velocities is commonly referred to as the granular temperature, and is analogous to the thermal motion of molecules in the kinetic theory of gases. As discussed in Campbell (1990), two mechanisms are responsible for the generation of granular temperature. The first mechanism is the collision between particles. The rapid flow of granular materials is characterized by high deformation rates and this rapid shearing motion of the flow causes collisions between particles. Since the directions of their resultant velocities after collision are dependent on the impact angle, their rotational velocity and particle properties, the collisions produce random velocity components. The other mechanism is a by-product of the random motions when these occur in a non-uniform flow field. A particle which moves in the direction of a velocity gradient yields departures from the local mean velocity. This "streaming" mechanism can only generate random velocity in the direction of the mean flow while the collisional mechanism yields components in all directions. Consequently, one would expect some anisotropy in the granular temperature especially at low densities where the streaming mechanism is more effective than the collisional mechanism as observed by Walton and Braun (1986a, b), Campbell and Gong (1986) and Campbell (1989).

Though many studies of granular materials draw on the analogy with the kinetic theory of gases, there are important differences between granular materials and gas molecules. One of the major differences is that the collisions between granular particles are inelastic. Thus fluctuation energy associated with the granular temperature is always dissipated by interparticle collisions. On the other hand, one of the similarities between granular materials and gas molecules is that granular temperature may be conducted by a temperature gradient. That is, fluctuation energy can be transferred from a region of high fluctuation energy to a region of low fluctuation energy through "granular conduction." Later we will discuss two mechanisms of granular conduction. For the present it is sufficient to observe that a local fluctuation energy equation is required to represent the balance between the generation of fluctuation energy due to shear motions, the dissipation due to inelastic collisions and the transport due to granular conduction. For example, when the fluctuation energy generated by shear motion is greater than the energy dissipation, the granular temperature of the system increases and/or excessive energy is conducted out of the system. In the opposite case, the granular temperature decreases and/or energy is conducted into the system from the outside.

For simple shear flow, no granular conduction is expected. In this case, the dissipation rate is uniform and equal to the production rate. Furthermore, the local magnitude of the granular temperature is influenced only by the local production of fluctuation energy, which is a direct result of the velocity gradient. Therefore, the granular temperature can be characterized by the velocity gradient alone. In channel flow, however, there are large gradients of granular temperature (Campbell and Brennen (1985) and Ahn *et al.* (1989b)). Therefore, the fluctuation energy is conducted along the gradient of granular temperature and the local granular temperature cannot be characterized by the local velocity gradient alone. In channel flow the direction of granular conduction plays a significant role in determining the profile of solid fraction and other characteristics of the flow (Ahn *et al.* (1989b)). If the fluctuation energy is being conducted from the bulk of flow to the channel base (that is, if granular temperature in the bulk is higher than near the channel base) then the solid fraction decreases monotonically with a distance from the base. On the other hand, if the energy is being conducted from the channel base to the bulk, the solid fraction is low at the channel base, reaches a maximum toward the center, and then decreases near the free surface. Ahn *et al.* (1989b) show that the direction of granular conduction in fully developed flow is determined by the channel inclination and the coefficient of restitution between particles.

In spite of its importance, the granular temperature had not been investigated experimentally until Ahn *et al.* (1988, 1989a) employed fiber optic probes to measure one of its components, namely the velocity fluctuation in the direction of the mean flow. Future developments will hopefully allow a more complete study of granular temperature in many different granular material flows.

7.1.4 Rheological Behavior of Granular Materials

Bagnold (1954), in his simple analyses and experiments of a cylindrical shear cell, distinguished three different regimes of flow behavior, namely, macroviscous, transitional, and grain inertia regimes, according to the importance of the role of the interstitial fluid in determining the dynamics of granular materials. In the grain inertia regime, where the interstitial fluid plays a minor role, the behavior of a granular flow is governed largely by direct interactions between particles. In this case, following Bagnold's argument, both the particle collisional rate and momentum transfer during the collision are proportional to the shear rate and, as a result, the stresses depend on the square of the shear rate.

Like Bagnold's experiments, most of stress measurements have been made in annular shear cells because the measurement technique in the shear cell is relatively simple compared to other experimental devices. The reader is referred to Savage (1978), Savage and McKeown (1983), Savage and Sayed (1984), Hanes and Inman (1985) and Craig *et al.* (1986). Savage and Sayed (1984) found that at the lower solid fractions and high shear rates both normal and shear stresses were proportional to the particle density and to the square of the shear rate and particle diameter. At higher solid fractions and lower shear rates, the stresses were found to be proportional to the shear rate raised to a power less than two. One of characteristics of the friction coefficient (the ratio of shear stress to normal stress) was that the friction coefficient was a weak function of the shear rate and that it increased with decreasing solid fraction. Campbell (1989) presented the results of computer simulations of an imposed simple shear flow with inelastic spheres. The stresses were normalized by the particle density and the square of the particle diameter and the shear rate, and were characterized by a "U" shape when presented as functions of solid fraction, with asymptotes toward large values as the solid fraction approached both zero and the shearable limit. The relative importance of the streaming (or kinetic) and collisional modes of momentum transfer was examined. The collisional mode dominated at high concentrations where collisions are frequent and the streaming mode dominated at low concentration where particles travel long distances between collisions. As a result, anisotropic behavior of the normal stresses was observed with decreasing concentration. The friction coefficient was also found to increase with decreasing solid fraction, mainly due to the streaming contribution.

For simple shear flow, all the theoretical analyses predict the same behavior which is a natural extension of that originally proposed by Bagnold, namely,

$$\tau_{ij} = \rho_s f_{ij}(\alpha) d^2 \left(\frac{du}{dy}\right)^2$$

where τ_{ij} is the stress tensor, ρ_s the particle density, f_{ij} a tensor function of the solid fraction α, d the particle diameter and du/dy the local mean shear rate. That the stress should be proportional to the square of the shear rate can be simply demonstrated by dimensional analysis. In the grain inertia regime where the interstitial fluid effect is negligible, the only available dimensional quantities are ρ_s, d, du/dy, the velocity fluctuations (or the square root of granular temperature) and the material properties of the particles. For simple shear flows in which the granular temperature is uniform, fluctuation energy is generated by the shearing motion and dissipated by the inelastic collisions. Thus, the magnitude of granular temperature is linearly proportional to the square of the shear rate. Savage and Jeffrey (1981) introduced the parameter S defined as

$$S = \frac{d\frac{du}{dy}}{T^{\frac{1}{2}}}$$

where T is the granular temperature. Lun *et al.* (1984) have evaluated S for a simple shear flow as a function of the solid fraction and coefficient of restitution between particles.

Channel flows are more complicated because the granular temperature may vary not only over the depth of the flow but also with position along the channel. The local granular temperature can not be described only by the local mean shear rate, the solid fraction, and the coefficient of restitution, but the fluxes of fluctuation energy (or granular temperature) in all directions have to be considered. Thus, the rheological behavior of the channel flow is not as simple as that of the simple shear flow and the above rheological model for simple shear flows does not hold for channel flows. Therefore, by studying the rheological behavior of the channel flow, more fundamental rheological models (not limited only to a simple case) can be examined.

7.1.5 Boundary Conditions at the Free Surface and the Wall

The conventional treatment of a solid boundary condition in fluid mechanics is the no-slip condition in which the fluid "particles" in contact with the wall cannot move relative to the wall. However in granular flow, there may be slip at a solid boundary. Furthermore, the boundary of the system generates granular temperature through shear work at a rate equal to the product of the slip velocity and the shear stress at the wall. At the same time, collisions of particles with the wall boundary dissipate fluctuation energy. Therefore, the boundary of the system may serve as either a source or a sink of fluctuation energy, depending on whether the generation or the dissipation dominates. These boundary conditions for granular flow can not be obtained independently of the flow field and studies of the slip velocity and the fluctuation energy flux at the boundary face many difficulties, some of which remain to be resolved (Hui *et al.* (1984), Jenkins and Richman (1986), Johnson and Jackson (1987), Richman and Chou (1988), Richman (1988) and Gutt and Haff (1988)).

The other type of boundary is a free surface and it is clear from the experimental observations that there are two types of free surface (Ahn *et al.* (1989a and b)). For dense and slow flows (perhaps all subcritical flows), the free surface is clearly defined. The material near the surface is being sheared either very slowly or not at all and particles are not thrown up out of the bulk. On the other hand for dilute and fast flows (perhaps all supercritical flows) particle saltation causes a substantial density gradient at the surface so that the precise location of the free surface is difficult to define. In a time-averaged view the solid fraction tends asymptotically approaches zero.

When the velocities at the base and free surface of a channel flow are compared, different base surface conditions yield quite different results. Various surface conditions have been explored by Bailard (1978) who used a surface onto which grains were glued, by Savage (1979) who attached roughened rubber sheets to the surface and by Ishida and Shirai (1979) who attached very rough sandpaper. In all these cases of rough base surface, the ratio of the velocity at the base, u_w, to the velocity at the free surface, u_s, was close to zero. Augenstein and Hogg (1978) obtained a spectrum of values for u_w/u_s

ranging from zero to values near one as the size of the particles glued to the base ranged from larger than the particles in the flow to much smaller than the particles in the flow. In the computer simulations of Campbell and Brennen (1985b) in which a smooth surface with high friction coefficient was employed, the ratio of u_w to u_s was about $0.4 \sim 0.5$. The high friction coefficient ensured a no-slip condition at the contact surfaces; however the particles at the wall still had non-zero slip velocities at their centers. Ahn *et al.* (1989a) have presented experimental measurements of the velocity ratio, u_w/u_s, as a function of the solid fraction for different surface conditions (see Figure 7-1).

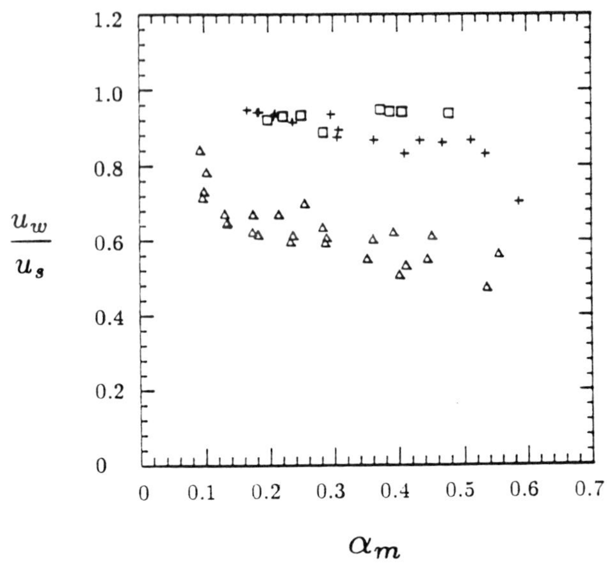

FIGURE 7-1. The ratio of velocity at the chute base to velocity at the free surface, u_w/u_s, presented as a function of the mean solid fraction, α_m. $\square$, the smooth surface; $+$, the moderately smooth surface; $\triangle$, the rubberized surface.

For a smooth aluminum surface, the ratio remains fairly constant and is independent of the solid fraction. For a surface coated with a thin rubber layer (the rubberized surface), the ratio gradually increases as the solid fraction decreases; an abrupt increase at a solid fraction of about 0.1 also occurred. Despite these data, the present state of knowledge does not allow prediction of the slip at the wall. Indeed the features of the surface or of the flow which determine the slip are not well understood.

The surface conditions also influence velocity fluctuations at the wall. If the friction between the particle surface and the wall surface is high, there will be no slip at the contact point during a collision and the particle will acquire substantial rotational velocity as a result of the collision with the wall. Subsequent collisions of this particle with other particles will cause some of this rotational energy to be converted to translational energy or granular temperature. On the other hand, if the friction at the collisional contact between the particle and the wall is small, slip will occur and relatively little rotational velocity will be imparted to the particle, resulting in lower granular temperatures. Therefore, the higher the friction, the larger the granular temperature near the wall. This is observed in the experiments of Ahn *et al.*(1989a), in which the measured velocity fluctuations were larger for the rubberized surface with a high Coulombic friction coefficient than for the smooth aluminum surface with a low Coulombic friction coefficient. Therefore, for rapid flows rough surfaces can result in flows of lower density and higher granular temperature than would be the case for smooth surfaces.

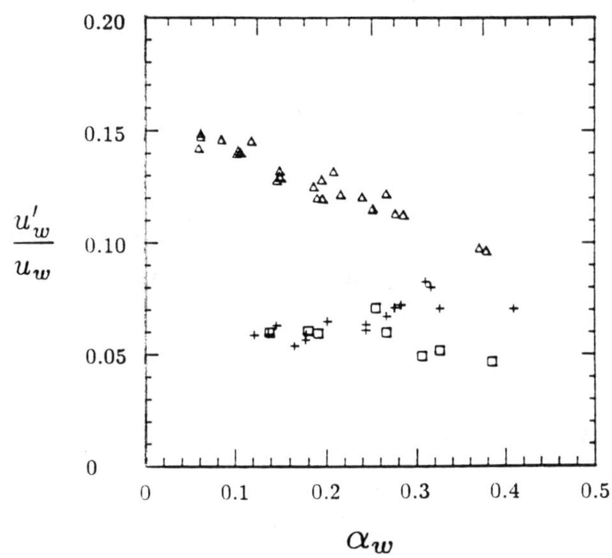

FIGURE 7-2. The longitudinal velocity fluctuation at the chute base normalized by the mean velocity, u'_w/u_w, against wall solid fraction, α_w. □, the smooth surface; +, the moderately smooth surface; △, the rubberized surface.

We will now turn to discussion of the fluctuation velocities themselves. Ahn *et al.*(1989a) present measurements of the velocity fluctuation near the base, u'_w, for various channel base surface conditions. This data is plotted against the solid fraction in Figure 7-2. For the smooth aluminum surface, the ratio of u'_w to the slip velocity, u_w is small and fairly constant. Note that u'_w itself increases with the decreasing solid fraction but the ratio, u'_w/u_w, appears to be constant. On the other hand u'_w/u_w for the rubberized surface is large and increases with decreasing solid fraction. These observations suggest the following explanation of why u_w/u_s for the rubberized surface abruptly increases at low solid fraction (Figure 7-1). The rubberized surface is characterized by large velocity fluctuations particularly at lower solid fractions as shown in Figure 7-2. The large fluctuations and the low solid fraction allow particles to move more freely from one location to another. One of the consequences is a decrease in the velocity gradient because when the particles move from a layer with low mean velocity to an adjacent layer with high velocity, the mean velocity of the adjacent layer is reduced. On the other hand when particles move from the upper layer with the higher mean velocity to the layer underneath they increase the mean velocity of the lower layer. This may help explain the large increase in u_w/u_s at low solid fractions exhibited in Figure 7-1.

7.1.6 Evolution of an Open Channel Flow of Granular Material

Having introduced and discussed some of the basic equations of open channel flows of granular material, it is appropriate at this point to pause in order to identify several key issues which remain unresolved. At the end of section 7.1.3 we briefly discussed the manner in which open-channel flows may become fully-developed. In the case of conventional open-channel flows of liquids, the flow becomes fully developed with the forces acting on an elemental length of the fluid reach equilibrium so that the depth and velocity no longer change with distance down the channel. For convenience we refer to this as a state of force-equilibrium. In liquids such force equilibria often occur because the shear stress, τ, increases with velocity. In granular material flow the variation of the shear stress with velocity, u, is less clear and consequnetly the circumstances under which force equilibrium may be reached are not obvious. If we assume that the wall friction, τ, is Coulombic so that f is a simple constant then equation

(6) would predict that equilibrium could only occur for one particular inclination, θ, given by $\tan^{-1} f$. Experiments (Bailard (1978), Ahn *et al.*(1989a and b)) seem to suggest that such a conclusion is too simplistic since τ depends on other factors such as the solid fraction, α.

But the evolution of a granular material flow depends on more than the momentum considerations of equation (6). Since the granular temperature affects the solid fraction and the stresses, it is also necessary that the flow reach "thermal" equilibrium as well as force equilibrium. In other words, in a fully developed flow the processes of production, conduction and dissipation of granular heat must also reach equilibrium before the stresses and the solid fraction become invariant with position, z. While the experimental evidence is not, as yet, conclusive, recent results (Johnson *et al.*(1990), Ahn *et al.*(1989a and b)) suggest that in a granular material flow the process of reaching thermal equilibrium is slower than the process of reaching force equilibrium so that the evolution is governed primarily by the former process.

It follows that one should observe different flows depending on how the granular material is supplied at the entrance to the channel for then the initial thermal state will be different. Johnson and Jackson (1990) have recently explored this feature in a valuable experimental study. They observed that for low mass flow rates the fully developed flow depends only on the mass flow rate with other conditions fixed and that it is independent of the entry conditions (how materials are fed into the channel). At higher mass flow rates it was unclear whether the flows in their channel were indeed fully developed. But unlike the low mass flow case, the flows far from the entry depended on both the flow rate and the entry conditions.

Johnson *et al.* (1990) also discuss how the flow develops as the granular material enters the channel. Because of difficulties in measuring the depth of flow, h, the nondimensionalized mass hold-up, m_T^*, defined as $m_T^* = \alpha_m h/d$ (instead of the solid fraction or velocity) was presented as a function of the nondimensionalized mass flow rate, $\dot{m}^*$. Since fully developed flow was desired, the channel was set at relatively small inclinations ($12° \sim 18°$); for larger inclinations the flow would accelerate over the length of the channel. Two methods of feeding granular material to the channel were investigated; "loose entry" and "dense entry." To achieve loose entry conditions, particles were dropped into the channel from a height, resulting in an energetic initial state at the channel entrance. Dense entry conditions were obtained by using a conventional sluice-gate which produced quite unenergetic initial conditions. For a given channel inclination and at low mass flow rates both loose and dense entry conditions yielded the same fully developed flow. With loose entry conditions, the energetic initial flow decelerated and the flow depth increased as the flow approached the fully developed state. On the other hand, with dense entry conditions, the unenergetic flow accelerated and the depth increased. However, as the mass flow rate increased, the two methods of entry yielded different results. In both cases the density and the depth of flow increased, but the increases in the case of dense entry were much larger than for loose entry. Likewise, the decrease in the velocity was much greater for dense entry. Further increase in the mass flow rate led to densities which were similar for both entry conditions, but the depth of flow for the dense entry was much larger, resulting in a lower velocity for the same mass flow rate. At the higher channel inclinations, dense entry flows were observed to be supercritical at low mass flow rate but subcritical at high mass flow rates. These results clearly suggest that channel flows of granular material can only be understood when the processes of production, dissipation and conduction of granular temperature are fully understood.

7.2 PROFILES OF CHANNEL FLOWS

7.2.1 Measurement of Profiles in Channel Flows

Few experimental studies of the detailed profiles of channel flows are reported in the literature because of the obvious difficulties involved in making point measurements of velocity, solid fraction and granular temperature within the interior of granular flows. Ridgway and Rupp (1970) used a horizontal knife-edge to split the channel flow into layers and, by measuring the mass flow rate in each layer, derived density profiles assuming uniform velocity. Augenstein and Hogg (1978) obtained vertical velocity distributions over the depth by examining the trajectories of sand particles leaving the channel exit. Bailard (1978) measured profiles in channen flows of sand with a layer of sand grains glued to the base of the channel. He used a horizontal splitter-collector assembly which was a combination of the techniques of both Ridgway and Rupp (1970) and Augenstein and Hogg (1974, 1978). By estimating the velocities from the trajectories of the particles at the end of the chute, the solid concentrations were calculated from the mass flux profiles. The concentration profiles showed a maximum at approximately mid-depth with reduced solid fractions both at the base and at the free surface. The velocity profiles were almost linear. More recently Drake and Shreve (1986) have taken high-speed motion pictures of nearly steady, uniform flows through the side walls of a channel. Their narrow channel was designed so that flows of a single particle width could be examined. Shirai *et al.* (1977) introduced a small probe made of three optical fibers of 0.2 mm in diameter to measure the particle velocity within the granular bed. Ishida and Shirai (1979) inserted fiber optic probes into the channel flow. In general, insertion of fiber optic probes causes interference with the flow, resulting in an unacceptable alteration of the flow (see Johnson and Jackson (1990)). The surface was covered with very rough sandpaper, and the velocity at the channel base was close to zero. Except for the region near the base, the velocity profile appeared to be linear, and the velocity gradient increased as the channel inclination became higher. Savage (1979) used fiber optic probes to measure the velocity of flow through glass side walls. Note, however, that the flow profile was probably affected by the presence of the side wall. Roughened rubber sheets were applied to the surface of the channel base and the velocity profile showed an inflexion point. Near the free surface, particle passages past the probes were infrequent, indicating the saltation of particles in the low density surface cloud. Johnson *et al.* (1990) also made similar efforts to obtain velocity profiles at the side walls using fiber optic probes. Both rough and smooth base surface conditions were investigated. Velocity profiles for both supercritical and subcritical flows were obtained for a smooth base. The rough base lined with a layer of sandpaper resulted in higher velocity gradients than the smooth base.

Ahn *et al.* (1989a) also employed fiber optic probes to obtain profiles of the velocity, solid fraction and one component of velocity fluctuation at the side walls. As shown in Figure 7-3(a), the velocity profiles were linear. It was also observed that the velocity gradient was higher for a rubberized base than for a smooth base. Profiles of the velocity fluctuation in the flow direction were presented for the first time. Figure 7-3(b) shows that the profile is fairly linear and that the fluctuations are larger at the free surface than at the channel base. These overall features are in contrast to the results obtained by Campbell and Brennen (1985b).

In their computer simulation, granular temperature near the solid

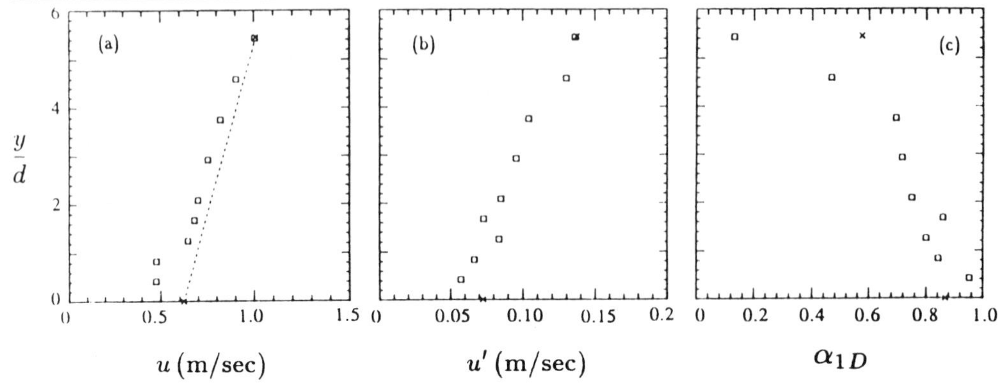

FIGURE 7-3. Vertical profiles at the side wall, □, for the (a) mean velocity, (b) velocity fluctuation, and (c) linear concentration as a function of coordinate y normalized by the particle diameter, d. Also shown is similar data at the center of the chute ($\times$). Dotted line is the assumed velocity profile at the center of the chute. Data were taken at $\theta = 17.8°$ on the rubberized surface; $\alpha_m = 0.30$, $h_o = 25.4$ mm, and $d = 3.04$ mm.

wall was substantially higher than near the free surface and the profile was far from linear. Ahn *et al.* also observed that the gradient of velocity fluctuations for the rubberized surface was larger than for the smooth surface. The profile of linear concentration which is a measure of the solid fraction is presented in Figure 7-3(c). In general, the linear concentration decreases monotonically except for the regions affected by the "corner effect." (The "corner effect" is that particles in the corner made of the channel base and the sidewall are arranged in a distinct line which has high solid fraction and low velocity.) The gradual decrease of the linear concentration near the free surface makes it difficult to define the location of the free surface. The shape of this density profile is quite different from those of Bailard (1978) and Campbell and Brennen (1985) where the density near the base is lower than in the center of the flow.

Compared to computer simulations of simple shear flows or Couette flows, relatively little work has been done on the computer simulation of gravity flows. Campbell and Brennen (1985b) simulated channel flows with two-dimensional disks, presenting the profiles of velocity, solid fraction, and granular temperature over the depth of flow, as shown in Figure 7-4. Their simulations employed periodic boundaries, which imply steady, fully developed flow. The results

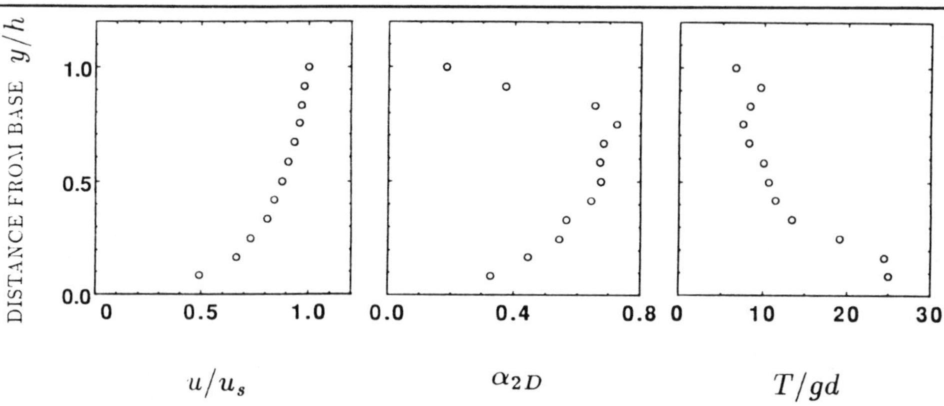

FIGURE 7-4. Profiles of the velocity nondimensionalized by the surface velocity, u_s, of the two-dimensional solid fraction α_{2D} and of the temperature T nondimensionalized by $gd/2$ from a computer simulation of two-dimensional discs flowing down a plane inclined at 30° ; $\epsilon_p = 0.6$ and $\epsilon_w = 0.8$ (Campbell and Brennen (1985b)).

showed that the granular temperature decreases substantially with the distance from the channel base, and that the solid fraction as observed in Bailard (1978) has its maximum value in the bulk of the flow and is smaller both at the free surface and at the base. Walton *et al.* (1988) used three-dimensional spheres to simulate gravity flow of particles through arrays of cylindrical horizontal rods and down inclined chutes. They employed a discrete particle model to simulate the chute flow tests conducted by Drake and Shreve (1986).

7.2.2 Types of Channel Flow Profile

As discussed in the previous section, the profiles of the granular temperature and solid fraction obtained by Ahn *et al.* (1989a) and Campbell and Brennen (1985b) are quite different. That is, in the experiments of Ahn *et al.*, the granular temperature increases and the solid fraction decreases monotonically with the distance from the channel base whereas, in the computer simulations of Campbell and Brennen, the temperature decreases with distance from the base and the solid fraction exhibits a maximum with lower values at the base and near the free surface. The analysis of Johnson and Jackson (1990) yielded both types of profile of the solid fraction but they do not present the corresponding profiles of granular temperature. For the rough surface with the inclination of 19°, the solid fraction decreased monotonically with distance from the base. A similar profile was manifest with a smooth surface at 16° inclination when the flow was dilute. However, at different type of profile closer to that of Campbell and Brennen occurred for very dense flow under the same conditions.

Ahn *et al.* (1989b) used the constitutive relations and governing equations of Lun *et al.* (1984) to generate solutions the fully developed chute flow of granular material. The solutions depend upon two parameters (the coefficient of restitution between particles, ϵ_p, and the channel inclination, θ) and three boundary values at the channel base (the values of solid fraction, granular temperature, and mean velocity). The results showed the significant role played by granular conduction in determining the profiles of granular temperature, solid fraction and mean velocity in chute flows. They also demonstrated that there exist two types of the fully developed flow depending on the choice of ϵ_p and $\tan\theta$. The two types are designated Type I and Type II. Type I flows were similar to those observed in the experiments of Ahn *et al.* (1989a); the granular temperature at the free surface is higher than at the channel base and fluctuation energy is conducted toward the base. In this type of flow the solid fraction monotonically decreases with the distance from the base and the free surface may not be well defined. On the other hand, Type II flow is characterized by higher granular temperature at the base than in the bulk and granular conduction transfers fluctuation energy from the base toward the bulk. Associated with these variations are a low solid fraction at the base and a more compact mass riding on this dilated layer. Thus the free surface of these Type II flows is well defined. The simulations of Campbell and Brennen produced Type II flows. Though Bailard (1978) did not measure granular temperature, his measured profiles of solid fraction are of Type II.

Ahn *et al.* (1989b) also discussed what determines the type of fully developed flow and Figure 7-5 presents a qualitative criterion. For a given inclination, fully developed flows of granular materials with high ϵ_p are of Type I, and particles with low ϵ_p yield fully developed flow of Type II. On the other hand, for a given granular material, fully developed flow with high inclination is of Type I, and the flow with low inclination is of Type II. (Here the borderline is a weak function of solid fraction, and near the borderline it is possible to have a flow of some of the characteristics of both Types I and II. Ahn *et al.* (1989b) referred to this flow as Type III.) It should be noted that the criterion in Figure 7-5 holds only if there exists a fully developed flow. The criterion is also consistent with physical intu-

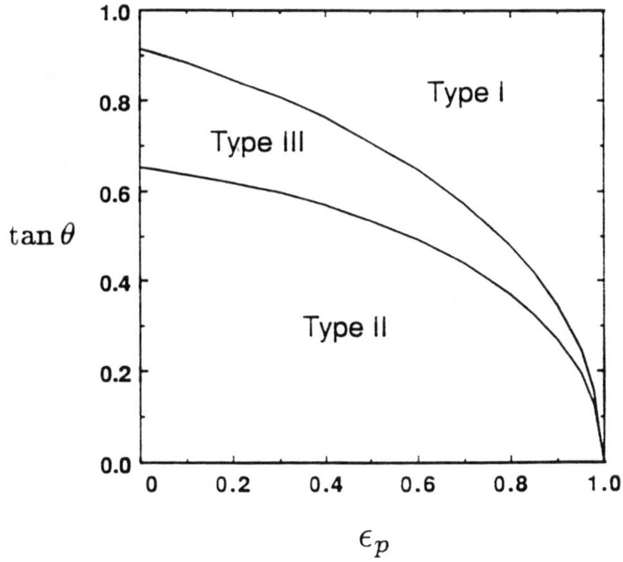

FIGURE 7-5. Flow Regimes in Channel Flow.

ition. The value $\tan\theta$ serves as a gravity force which generates the translational fluctuation energy through shear work. At larger values of $\tan\theta$, fluctuation energy production dominates. The dissipation of the fluctuation energy is a strong function of ϵ_p. For smaller ϵ_p, the fluctuation energy dissipation dominates. When the rate of generation of fluctuation energy is larger than the rate of dissipation, the channel base region has to absorb the excessive fluctuation energy. Otherwise, no fully developed flow exists for that $\tan\theta$ and ϵ_p. This implies that the coefficient of restitution between the base wall and the particle, ϵ_w, must be somewhat smaller than ϵ_p. (The role of the surface roughness is not clear.) In this case, the fluctuation energy is conducted from the bulk to the wall boundary and the granular temperature is expected to be smaller near the wall while larger temperatures occur at the free surface. On the other hand, when the rate of dissipation is larger than the rate of generation, the wall boundary will supply the fluctuation energy to the flow. Unless the rate of generation of fluctuation energy at the wall exceeds the rate of dissipation and thus supplies fluctuation energy to the bulk there can be no fully developed flow for that inclination and granular material. Therefore, ϵ_w is required to be somewhat larger than ϵ_p in order to maintain fully developed flow. In this case, the granular temperature near the wall is higher than near the free surface and the fluctuation energy is conducted from the wall to the free surface. (In the case where ϵ_w is smaller than ϵ_p for a given channel inclination so that dissipation dominates, fully developed flow could be established by external means, for example, by vibrating the channel base.) In summary, the surface boundary conditions of the channel base determine whether fully developed flow is possible or not for given $\tan\theta$ and ϵ_p. These parameters also determine whether the fully developed flow is of Type I or Type II.

This criterion is consistent with the results in the literature. Ahn *et al.* (1989a) used glass beads which have very high ϵ_p. (Lun and Savage (1986) estimated $\epsilon_p \simeq 0.95$ for glass beads.) The channel base was an aluminum with and without the rubber layer and ϵ_w was estimated to be about 0.7 and 0.5 respectively. For the case of Figure 7-3 where the channel inclination was 17.8°, Type I flow is expected and was observed in their experiments. In the computer simulations of Campbell and Brennen (1985), $\epsilon_p = 0.6$ and $\epsilon_w = 0.8$ with $\theta = 30°$. The criterion predicts the Type II flow which is consistent with the result of the simulation as shown in Figure 7-4. However, the results of Bailard (1978) can not be judged from this criterion because the channel inclination is high ($34° \sim 39°$) and ϵ_p has an unknown but probably small value.

Richman and Marciniec (1988) sought analytical solutions for channel flows by employing the constitutive theory of Jenkins and Richman (1985) for identical, smooth, nearly elastic spheres. They were concerned with a flat channel base wall to which identical smooth hemispherical particles were attached (see Jenkins and Richman (1986)). They provide the profiles of the granular temperature, solid fraction and velocity. Their results also show that there are two types of flow and that the profiles of two types of flow are consistent with those of Ahn *et al.* (1989b) (Type I and Type II flows). For this particular boundary condition and a given granular material, the analytical results indicate when fully developed flow is possible in terms of the inclination, mass flow rate, and solid fraction. For example, for Type I flow, fully developed flow can be established at high inclination for small mass flow rate and at low inclination for large mass flow rate. On the other hand, for Type II flow, higher mass flow rate can yield fully developed flow at higher inclination.

7.2.3 Transverse Profiles in Channel Flows

Transverse velocity profiles in channel flows have been investigated by Ishida and Shirai (1979), Savage (1979), and Ahn *et al.* (1989a). In Savage's experiments, the profiles at the free surface were obtained using marker particles and a movie camera. For smooth walls, the velocity profiles were quite uniform. For rough walls, the flows were very slow and the velocity profiles were roughly parabolic with much smaller velocities at the sidewalls than in the center. Using fiber optic probes, Ahn *et al.* obtained the velocity profiles both at the free surface and at the channel base. Comparison of the profiles on the free surface and on the channel base showed that the flow at the free surface was more uniform and less affected by the side wall than the flow at the base. It was also observed that the higher the velocity (higher channel inclination), the less significant the side wall effect. Indeed nonuniformity due to the side walls was significant only at the base and only at low velocities (low inclinations).

7.3 RHEOLOGICAL IMPLICATIONS OF THE CHANNEL FLOWS

7.3.1 Rheological Behavior in Channel Flows

The rheological behavior which is manifest in open channel flows of granular material has been examined by Patton *et al.* (1987). They used the momentum equation (6) to evaluate the shear stress, τ, at the channel base. The normal stress was calculated assuming a hydrostatic variation in pressure over the depth of flow. Both stresses were normalized by $\rho_s d^2 (U/h)^2$ and presented as functions of α (the shear rate was approximated as U/h). The normalized stresses showed fairly good correlations, and asymptotes at high solid fraction were very clear at high solid fraction. The friction coefficient was also plotted against the Froude number. It remained fairly constant over the wide range of the low Froude number, but as the Froude number became high the friction coefficient seemed to increase substantially though the data was quite scattered since, at these large Fr, the shear stress calculation using equation (6) is quite sensitive to $d\alpha_m/dz$ and dh/dz, quantities which are difficult to measure accurately. Since a high Froude number in the experiments of Patton *et al.* corresponds to high u and low α, the increasing friction coefficient with decreasing solid fraction is consistent with the trend observed in shear cell experiments and in the computer simulations.

Perhaps the most complete experimental investigation of the rheological behavior in the channel flow is found in Ahn *et al.* (1989a).

Two important instruments were used in their experiments; one is a gauge to measure shear stress, and the other is a set of two fiber optic probes to measure velocity, velocity fluctuations, and local solid fraction. In order to measure shear stress of flowing material at the channel base, a rectangular hole was cut into the channel base and replaced by a plate supported by strain-gauged flexures sensitive to the shearing force applied to the plate. The clearance between the plate and the rest of the channel base was adjusted to be about 0.2 mm, much smaller than the particle sizes. A system of fiber optic probes, similar to that devised by Savage (1979), was developed to measure particle velocities and their fluctuations at the channel base and at the free surface. Only one component of velocity fluctuations, u', was measured in a root-mean-square sense, and it should be some measure of the granular temperature. In addition to the mean solid fraction which was obtained by measuring the mean velocity, the depth of flow, and mass flow rate, local solid fraction was also estimated as follows. The number of particle passages per unit time detected by the probe was divided by the mean velocity to obtain the characteristic particle spacing, s, and in turn the linear concentration, α_{1D}, was calculated as $\alpha_{1D} = d/s$. An estimate of the local solid fraction near the wall, α_w, was calculated using $\alpha_w = \pi\alpha_{1D}^3/6$. Finally, the shear rate was approximated as $\Delta u/h$ where Δu is the velocity difference between at the free surface and at the wall. The normal stress was calculated in a manner similar to Patton *et al.* (1987).

Ahn *et al.* (1989a) presented the stresses with various normalizations, based on application of Lun *et al.* (1984) to the channel flow. They suggested that, for fully developed flow, the normal and shear stresses should be normalized by $\rho_s(d\Delta u/h)^2/\tan^2\theta$ and $\rho_s(d\Delta u/h)^2/\tan\theta$ respectively. For general flow (not necessarily simple shear flow or fully developed flow), it was suggested that the normal stress should be normalized by $\rho_s(u')^2$, and the shear stress by $\rho_s(d\Delta u/h)u'$. Some results are presented in Figures 7-6 through 7-8, and also are the theoretical results of Lun *et al.* (1984) included in the figures for comparison.

In Ahn *et al.* (1989a), the parameter S introduced by Savage and Jeffrey (1981) was examined with an approximation of $d(\Delta u/h)/u'$. The theoretical results of Lun *et al.* (1984) suggest that S or $S/\tan\theta$ should be a function only of α and ϵ_p for the case of simple shear flow or fully developed flow respectively. Figure 7-9 shows the results of some data which were close to fully developed flow, and the result

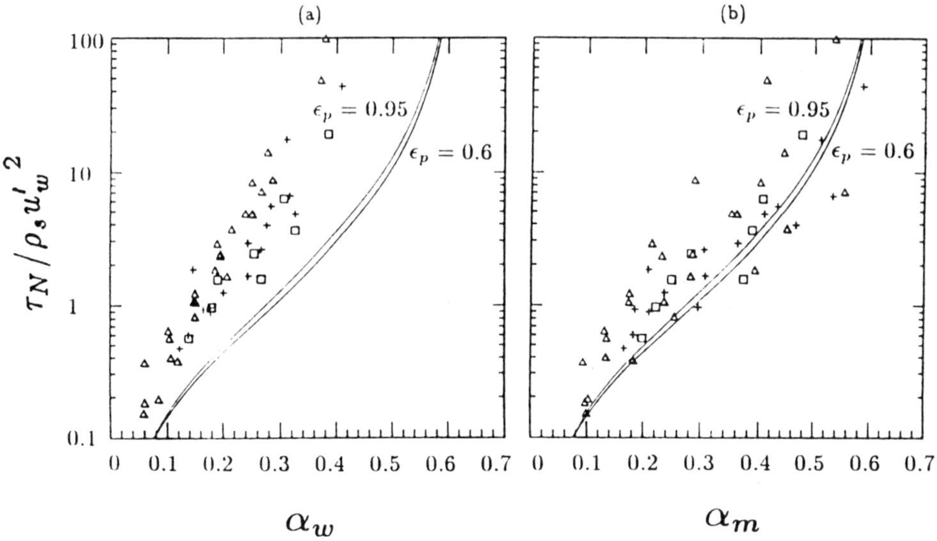

FIGURE 7-6. The normalized normal stress, $\tau_N/\rho_s u_w'^2$, presented as a function of (a) wall solid fraction, α_w and (b) mean solid fraction, α_m. □, the smooth surface; +, the moderately smooth surface; △, the rubberized surface. The solid lines are the results of Lun *et al.*[1984].

of Lun *et al.* (1984) is also plotted for comparison. The results show that the lower solid fraction, the lower ratio of the shear rate to the granular temperature. This is consistent with physical intuition that at low solid fraction particles move more freely, characterized by high granular temperature and thus by low shear rate as discussed in Section 7.1. This also implies that the ratio becomes zero at the free surface where no shear motion is expected.

Ahn *et al.* (1989a) also provided some data on friction coefficient for various surface conditions. First, the friction coefficient, f, was examined as a function of solid fraction, α. For the smooth aluminum surfaces, f is invariant with α. But for the rubberized surface which had higher Coulombic friction coefficient, f increased with decreasing α, and this is consistent with the results of other literature (for example, Savage and Sayed (1984) and Campbell (1989)). The friction coefficient was also plotted against u_w'/u_w where u_w and u_w' are the velocity and its fluctuation at the wall respectively. (See Figure 7-10.) For the smooth surfaces, all the data were clustered at one region. For the rubberized surface, f seemed to correlate quite well with u_w'/u_w; f increased with increasing u_w'/u_w. This phenomenon

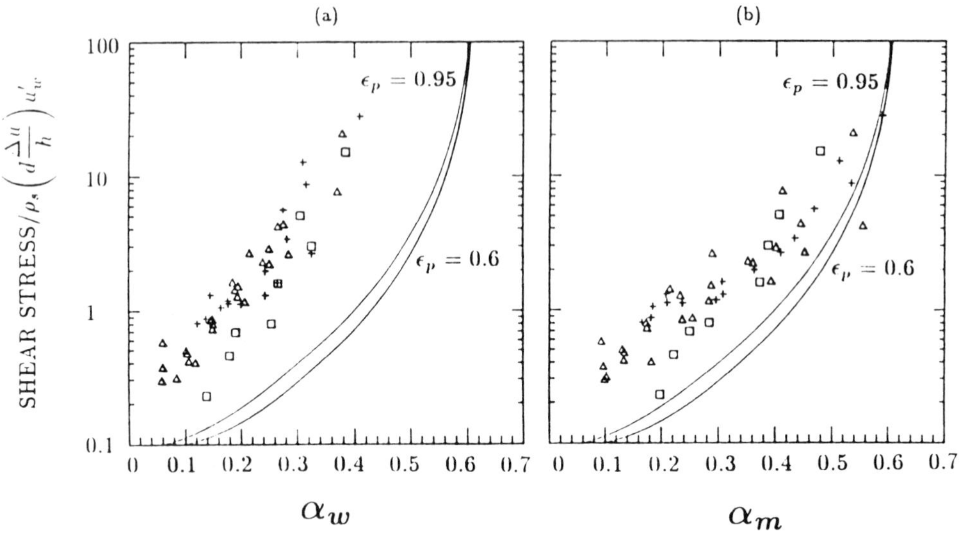

FIGURE 7-7. The normalized shear stress, $\tau_S/\rho_s d(\triangle u/h)u'_w$, presented as a function of (a) wall solid fraction, α_w, and (b) mean solid fraction, α_m. $\square$, the smooth surface; $+$, the moderately smooth surface; $\triangle$, the rubberized surface. The solid lines are the results of Lun *et al.*[1984].

was independent of particle size.

To explain these observations, they suggested the following. When a particle collides with a wall such that the shear stress at the contact point exceeds a shear stress limit which the surface can withstand for the given normal stress at the contact point, slip will occur. Then the ratio of the shear stress to the normal stress at the contact point is adjusted to the Coulombic friction coefficient of the surface, i.e. $f = \mu_c$. On the other hand, when the ratio at the impact does not exceed μ_c, there will be no slip between the contact surfaces of the particle and the wall. In this case, f is different from μ_c. For the smooth surface where μ_c was low, slip occurred at the contact between particles and the surface, and the slip condition resulted in the constant friction coefficient equal to μ_c. However, for the rubberized surface with high μ_c, a no-slip condition was applied, resulting in varying friction coefficient with solid fraction.

They also discussed why the friction coefficient for the rubberized surface varies with the solid fraction and u'_w/u_w. When there is no slip, the following relation results from a simple analysis of the oblique

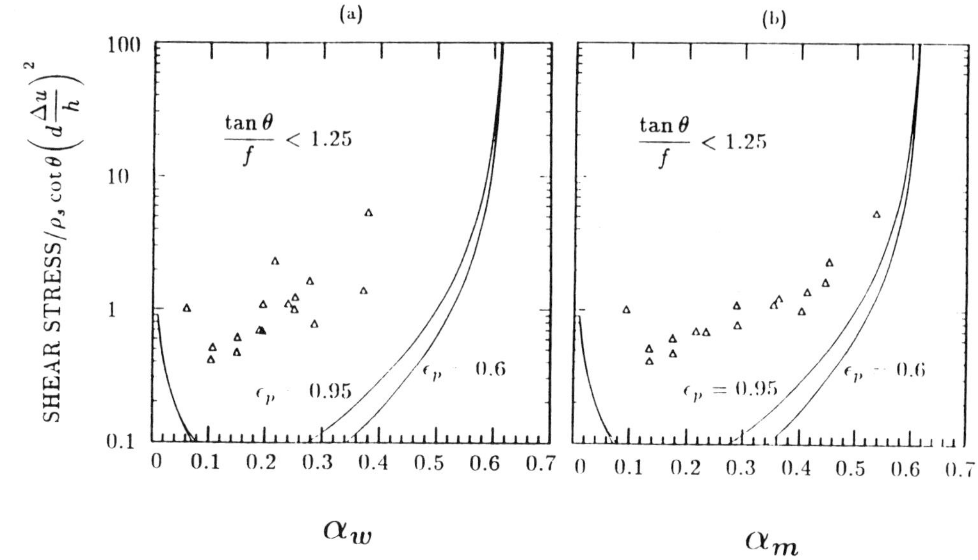

FIGURE 7-8. The normalized shear stress, $\tau_S \tan\theta/\rho_s(d\Delta u/h)^2$, presented as a function of (a) wall solid fraction, α_w, and (b) mean solid fraction, α_m. Data only for $\tan\theta/f < 1.25$. The solid lines are the results of Lun *et al.*[1984].

impact of a single sphere on the flat surface (see Ahn (1989)):

$$f = \frac{2}{7}\left(1 - \frac{\omega_1 d}{2u_1}\right)\frac{\tan\beta_1}{1+\epsilon_w}$$

where ω_1 is the rotational rate before impact, and u_1 is the velocity tangential to the wall before impact. The impact angle β_1 is defined by $\tan^{-1}(u_1/v_1)$ where v_1 is the velocity normal to the wall before impact and ϵ_w is the wall-particle coefficient of restitution. In this equation, the friction coefficient or the ratio of the shear stress to the normal stress at the surface depends on the ratio of rotational velocity to tangential velocity, $\omega_1 d/2u_1$, and on the impact angle, β_1. In their experiments, the value of β_1 could not be estimated. Another factor influencing f is $\omega_1 d/2u_1$. Campbell (1988) has shown that next to the wall ω is considerably larger than the mean value, but that with a small distance from the wall ω is slightly less than the mean value. Therefore, when a particle next to the wall with high ω hits the wall, the friction coefficient will be low, but if a particle at a distance from the wall with low ω comes down and collides with the wall, the friction coefficient will be relatively high.

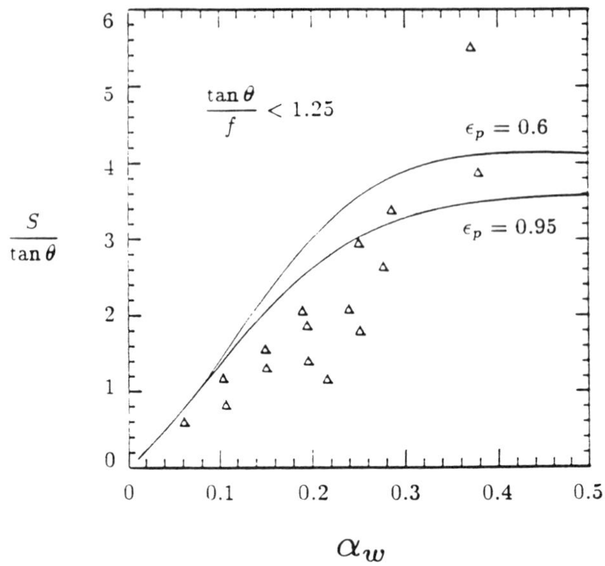

FIGURE 7-9. The parameter, $S/\tan\theta = d(\Delta u/h)/u'_w \tan\theta$, against wall solid fraction, α_w. Data only for $\tan\theta/f < 1.25$. The solid lines are the results of Lun *et al.*[1984].

These phenomena suggest a possible explanation for the increase in the friction coefficient as the solid fraction decreases. At low solid fraction, particles move more freely from one layer to another. Thus more particles in the upper layers with small values of $\omega d/2u$ move down to the boundary and collide with the wall. Because friction is measured in a statistical sense as a sum of frictions due to individual particles colliding with the wall, f is therefore high at low solid fraction. On the other hand, at high solid fraction and low granular temperature, very few particles in the upper layer with low $\omega d/2u$ penetrate to the wall. As a result, particles next to the wall with high rotational velocity will dominate collisions at the wall. Thus f would be smaller at high solid fraction. Furthermore, when higher u'_w/u_w exists, particles with low ω in the upper layer more easily move down to the boundary and collide with the wall. That is, as u'_w/u_w increases, the intrusion of particles with low ω from the upper layer into the boundary becomes more frequent, causing f to increase. As a result, the friction coefficient appears to be a fairly linear function of u'_w/u_w for the rubberized surface as shown in Figure 7-10.

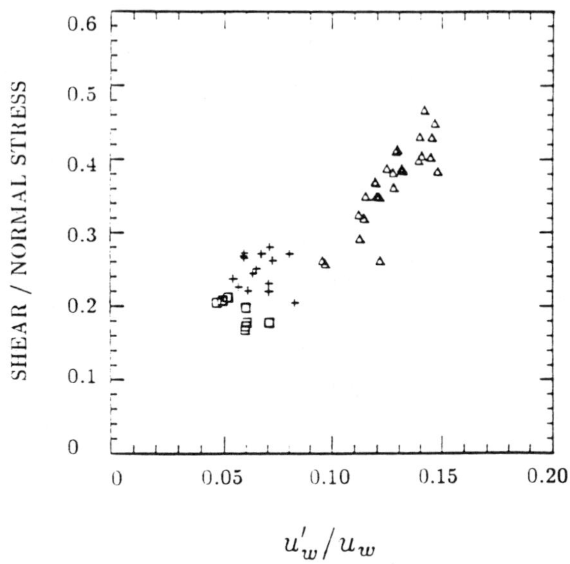

FIGURE 7-10. Friction coefficient at the wall, $f = \tau_S/\tau_N$, presented as a function of longitudinal velocity fluctuation at the wall normalized by mean velocity, u'_w/u_w. □, the smooth surface; +, the moderately smooth surface; △, the rubberized surface.

7.4. CONCLUDING REMARKS

One of the simplest granular material flows which can be generated in the laboratory is that down an inclined chute or channel. It therefore serves as a very convenient vehicle against which to test an existing knowledge of the rheology of granular material flows. In this review we have tried to demonstrate that much remains to be learned about the rheology from these flows which, when examined in detail, turn out to be far from simple. Perhaps the most important conclusion which should be drawn from this review is that we have reached a stage at which advancement in our understanding of the rheology of granular material flows will be led by our ability to devise methods to make detailed measurements of all of the fundamental properties of these flows, namely the mean velocity, solid fraction, fluctuating velocity and normal and shear stresses. In this review we have described some recent but incomplete attempts to experimentally document channel flows in this way. These attempts show that the kinetic theory models for granular material flow advanced by Savage (1984), Jenkins *et al.* (1985) and others have some validity but that more needs to be done before these models could be considered verified and validated. The other analytical approach, namely the computer simulation of granular material flows by Campbell and Brennen (1985a,b), Walton (1984) and others, also appears to have considerable value and yet many of the details necessary for these simulations (such as the collision mechanics) also require experimental validation.

Apart from the constitutive laws governing the mechanics of the flowing granular material there also remain some very important questions regarding the boundary conditions which should be applied to model solid walls or free surfaces which are the common boundaries of granular material flow. In this review we have described how the slip at a solid wall (channel base) is a function of the roughness of that wall and how that slip can effect the entire flow. The laws governing the slip are not at all well understood. Moreover, when one examines the situation more closely it is clear that a boundary condition on the granular temperature is also necessary and this, too, is a subject of considerable debate at the present time.

Another feature of channel flows which we have tried to highlight in this review is the lack of knowledge regarding the manner in which these flows approach equilibrium (if indeed they ever do) as they progress down the channel. Like open-channel flows of liquid they can approach a state in which the forces on an elemental length balance so that further acceleration or deceleration does not occur. But unlike open-channel flows of liquid, granular material flows will only reach equilibrium when the processes of production, dissipation and conduction of granular fluctuation energy have reached equilibrium. Though the work by Jackson has shed some very valuable light on this subject, much remains to be done before our understanding of this aspect of granular flow could be considered in any way complete.

ACKNOWLEDGEMENTS

The authors wish to thank Professor R. H. Sabersky and Professor M. L. Hunt for helpful discussions on the subject matter of this review.

NOTATION

a, b, c	Profile shape parameters
d	Particle diameter
f	Friction coefficient ($\tau P/\rho_s \alpha_m g w h$ or ratio of shear stress to normal stress)
Fr	Froude number $= U/(gh)^{\frac{1}{2}}$
g	Acceleration due to gravity
h	Depth of flow
p	Pressure
P	Perimeter of flow in contact with solid walls
s	Mean spacing between particle centers
S	Parameter $d\, du/dy/T^{\frac{1}{2}}$
T	Granular temperature, $(\langle u'^2\rangle + \langle v'^2\rangle)/2$
w	Channel width
u	Velocity of flow parallel with channel walls
u', v'	Fluctuating velocity components
U	Cross-sectionally averaged velocity $= u_m$
Δu	Velocity difference $= u_s - u_w$
v	Velocity of flow normal to channel base
y	Coordinate measured perpendicular to channel base
z	Coordinate measured parallel to channel walls
α	Solid fraction
α_{ID}	Linear concentration, d/s
β	Inclination of particle trajectory to wall normal
δ	Velocity gradient
θ	Channel inclination to horizontal
ρ_s	Particle density
φ	Internal friction angle
μ_c	Coulomb friction angle
μ_F	Liquid viscosity
τ	Shear stress at wall
τ_{ij}	Stress tensor
ϵ_p	Coefficient of restitution between particles
ϵ_w	Coefficient of restitution between particle and wall
ω	Particle rotational velocity

Subscripts

m	denotes quantity averaged over cross-section of channel flow
s	denotes quantity at the free surface
w	denotes quantity at the channel base

1 denotes quantity upstream of shock or before collision
2 denotes downstream of shock or after collision

Superscript

′ denotes fluctuating component

REFERENCES

Ahn, H. 1989. Experimental and Analytical Investigations of Granular Materials: Shear Flow and Convective Heat Transfer. Ph.D. Thesis, California Institute of Technology.

Ahn, H., Brennen, C.E. and Sabersky, R.H. 1988. Experiments on Chute Flows of Granular Materials. *Micromechanics of Granular Materials* (ed. M. Satake and J.T. Jenkins), Elsevier Pub. Co., Amsterdam, pp. 339-348.

Ahn, H., Brennen, C.E. and Sabersky, R.H. 1989a. Measurements of Velocity, Velocity Fluctuation, Density and Stresses in Chute Flows of Granular Materials. *J. Appl. Mech.*, to appear.

Ahn, H., Brennen, C.E. and Sabersky, R.H. 1989b. Analysis of the Fully Developed Chute Flow of Granular Materials. *J. Appl. Mech.*, to appear.

Augustein, D.A. and Hogg, R. 1978. An Experimental Study of the Flow of Dry Powders on Inclined Surfaces. *Powder Technology*, Vol. 19, pp. 205-215.

Bagnold, R.A. 1954. Experiments on a Gravity-free Dispersion of Large Solid Particles in a Newtonian Fluid Under Shear. *Proc. R. Soc. Lond.*, Vol. A225, pp. 49-63.

Bailard, J. 1978. An Experimental Study of Granular-Fluid Flow. Ph.D. Thesis, University of California, San Diego.

Campbell, C.S. 1986. The Effect of Microstructure Development on the Collisional Stress Tensor in a Granular Flow. *Acta Mechanica*, Vol. 63, pp. 61-72.

Campbell, C.S. 1988. Boundary Interactions for Two-dimensional Granular Flows: Asymmetric Stresses and Couple Stresses. *Micromechanics of Granular Materials* (ed. M. Satake and J.T. Jenkins), Elsevier Pub. Co., Amsterdam, pp. 163-174.

Campbell, C.S. 1989. The Stress Tensor for Simple Shear Flows of a Granular Material. *J. Fluid Mech.*, Vol. 203, pp. 449-473.

Campbell, C.S. 1990. Rapid Granular Flows. *Ann. Rev. of Fluid Mech.*, Vol. 22, pp. 57-92.

Campbell, C.S. and Brennen, C.E. 1985a. Computer Simulation of Granular Shear Flows. *J. Fluid Mech.*, Vol. 151, pp. 167-188.

Campbell, C.S. and Brennen, C.E. 1985b. Chute Flows of Granular Material: Some Computer Simulations. *J. Appl. Mech.*, Vol. 52, pp. 172-178.

Campbell, C.S. and Gong, A. 1986. The Stress Tensor in a Two-dimensional Granular Shear Flow. *J. Fluid Mech.*, Vol. 164, pp. 107-125.

Campbell, C.S. and Gong, A. 1987. Boundary Conditions for Two-dimensional Granular Flows. *Proceedings of the International Symposium on Multiphase Flows*, Hangzhou China, Vol. 1, pp. 278-283.

Campbell, C.S. and Wang, D.G. 1986. The Effective Conductivity of Shearing Particle Flows. *Proc. of 8th Int. Heat Transfer Conference*, Vol. 4, pp. 2567-2572.

Craig, K., Buckholz, R.H. and Domoto, G. 1986. An Experimental Study of the Rapid Flow of Dry Cohesionless Metal Powders. *J. Appl. Mech.*, Vol. 53, pp. 935-942.

Drake, T.G. and Shreve, R.L. 1986. High-speed Motion Pictures of Nearly Steady, Uniform, Two-dimensional, Inertial Flows of Granular Material. *J. Rheology*, Vol. 30, pp. 981-993.

Haff, P.K. 1983. Grain Flow as a Fluid-mechanical Phenomenon. *J. Fluid Mech.*, Vol. 134, pp. 401-430.

Hanes, D.M. and Inman, D.L. 1985. Observations of Rapidly Flowing Granular-Fluid Flow. *J. Fluid Mech.*, Vol. 150, pp. 357-380.

Hui, K., Haff, P.K., Ungar, J.E. and Jackson, R. 1984. Boundary Conditions for High-shear Grain Flows. *J. Fluid Mech.*, Vol. 145, pp. 223-233.

Ishida, M. and Shirai, T. 1979. Velocity Distributions in the Flow of Solid Particles in an Inclined Open Channel. *J. Chem. Eng. Jpn.*, Vol. 12, pp. 46-50.

Jenkins, J.T. and Richman, M.W. 1985. Kinetic Theory for Plane Flows of a Dense Gas of Identical, Rough, Inelastic, Circular Disks. *Phys. Fluid*, Vol. 28, pp. 3485-3494.

Jenkins, J.T. and Richman, M.W. 1986. Boundary Conditions for Plane Flows of Smooth, Nearly Elastic, Circular Disks. *J. Fluid Mech.*, Vol. 171, pp. 53-69.

Jenkins, J.T. and Richman, M.W. 1988. Plane Simple Shear of Smooth Inelastic Circular Disks: The Anisotropy of the Second Moment in the Dilute and Dense Limits. *J. Fluid Mech.*, Vol. 192, pp.

313-328.

Jenkins, J.T. and Savage, S.B. 1983. A Theory for the Rapid Flow of Identical, Smooth, Nearly Elastic Particles. *J. Fluid Mech.*, Vol. 130, pp. 187-202.

Johnson, P.C. and Jackson, R. 1987. Frictional-Collisional Constitutive Relations for Granular Materials, With Application to Plane Shearing. *J. Fluid Mech.*, Vol. 176, pp. 67-93.

Johnson, P.C., Nott, P. and Jackson, R. 1990. Frictional-Collisional Equations of Motion for Particulate Flows and their Application to Chutes. *J. Fluid Mech.*, Vol. 210, pp. 501-535.

Lun, C.K.K. and Savage, S.B. 1986. The Effects of an Impact Velocity Dependent Coefficient of Restitution on Stresses Developed by Sheared Granular Materials. *Acta Mechanica*, Vol. 63, pp. 15-44.

Lun, C.K.K., Savage, S.B., Jeffrey, D.J. and Chepurniy, N. 1984. Kinetic Theories for Granular Flow: Inelastic Particles in Couette Flow and Slightly Inelastic Particles in a General Flowfield. *J. Fluid Mech.*, Vol. 140, pp. 223-256.

Morrison, H.L. and Richmond, O. 1976. Application of Spencer's Ideal Soil Model to Granular Materials Flow. *J. Appl. Mech.*, Vol. 43, pp. 49-53.

Nakagawa, M. 1988. Kinetic Theory for Plane Flows of Rough, Inelastic Circular Discs. Ph.D. Dissertation, Cornell University, Ithaca, NY.

Ogawa, S. Umemura, A. and Oshima, N. 1980. On the Equations of Fully Fluidized Granular Materials. *J. Appl. Math. Phys. (ZAMP)*, Vol. 31, pp. 483-493.

Patton, J.S., Brennen, C.E. and Sabersky, R.H. 1987. Shear Flows of Rapidly Flowing Granular Materials. *J. Appl. Mech.*, Vol. 54, pp. 801-805.

Patton, J.S., Sabersky, R.H. and Brennen, C.E. 1986. Convective Heat Transfer to Rapidly Flowing Granular Materials. *Int. J. Heat Mass Transfer*, Vol. 29, No. 8, pp. 1263-1269.

Richman, M.W. 1988. Boundary Conditions Based Upon a Modified Maxwellian Velocity Distribution for Flows of Identical, Smooth, Nearly Elastic Spheres. *Acta Mech.*, Vol. 75, pp. 227-240.

Richman, M.W. and Chou, C.S. 1988. Boundary Effects on Granular Shear Flows of Smooth Disks. *ZAMP*, Vol. 39, pp. 885-901.

Richman, M.W. and Marciniec, R.P. 1988. Gravity-driven Granular Flows of Smooth Inelastic Spheres Down Bumpy Inclines. *J.*

Appl. Mech., to appear.

Ridgway, K. and Rupp, R. 1970. Flow of Granular Material Down Chutes. *Chem. Proc. Eng.*, Vol. 51, pp. 82-85.

Savage, S.B. 1979. Gravity Flow of Cohesionless Granular Materials in Chutes and Channels. *J. Fluid Mech.*, Vol. 92, pp. 53-96.

Savage, S.B. 1984. The Mechanics of Rapid Granular Flows. *Advances in Applied Mechanics*, Vol. 24 (ed. J. Hutchinson and T.Y. Wu), Academic Press, pp. 289-366.

Savage, S.B. and Jeffrey, D.J. 1981. The Stress Tensor in a Granular Flow at High Shear Rates. *J. Fluid Mech.*, Vol. 110, pp. 255-272.

Savage, S.B. and McKeown, S. 1983. Shear Stress Developed During Rapid Shear of Dense Concentrations of Large Spherical Particles Between Concentric Cylinders. *J. Fluid Mech.*, Vol. 127, pp. 453-472.

Savage, S.B. and Sayed, M. 1984. Stresses Developed by Dry Cohesionless Granular Materials in an Annular Shear Cell. *J. Fluid Mech.*, Vol. 142, pp. 391-430.

Sayed, M. and Savage, S.B. 1983. Rapid Gravity Flow of Cohesionless Granular Materials Down Inclined Chutes. *J. Appl. Math. Phys.*, Vol. 34, pp. 84-100.

Shirai, T., Ishida, M., Ito, Y., Inoue, N. and Kobayashi, S. 1977. Rotation of a Horizontal Disk Within and Aerated Particle Bed. *J. Chem. Eng. Jpn.*, Vol. 10, pp. 40-45.

Walton, O.R. 1984. Computer Simulation of Particulate Flow. Energy and Tech. Rev., Lawrence Livermore Lab., May 1984.

Walton, O.R. and Braun, R.L. 1986a. Viscosity, Granular-Temperature, and Stress Calculations for Shearing Assemblies of Inelastic, Frictional Disks. *J. of Rheology*, Vol. 30(5), pp. 949-980.

Walton, O.R. and Braun, R.L. 1986b. Stress Calculations for Assemblies of Inelastic Spheres in Uniform Shear. *Acta Mechanica*, Vol. 63, pp. 73-86.

Walton, O.R., Braun, R.L., Mallon, R.G. and Cervelli, D.M. 1988. Particle-dynamics Calculations of Gravity Flow of Inelastic, Frictional Spheres. *Micromechanics of Granular Materials* (ed. M. Satake and J.T. Jenkins), Elsevier Pub. Co., Amsterdam, pp. 153-162.

Zeininger, G. and Brennen, C.E. 1985. Interstitial Fluid Effects in Hopper Flows of Granular Materials. ASME Cavitation and Multiphase Flow Forum, June 1985, pp. 132-136.

CHAPTER 8

THE EFFECT OF PARTICLES ON THE TURBULENCE IN A BOUNDARY LAYER

G. Hetsroni

8.1 INTRODUCTION

In the relatively mature field of fluid dynamics there are two issues (or areas) which are poorly understood -- turbulence and two-phase flows. This review is on the interaction between these poorly understood areas, and therefore, there are gaps in our understanding of the physical phenomena and there are many difficulties in the modeling. There will be no mathematical modeling attempted here since it is premature to offer a complete model.

The problem of turbulence and particles interaction has two aspects: the dispersion of particles due to the turbulence of the carrier fluid, and the modulation of the turbulence due to the presence of the particles.

While these are closely interconnected, the former has been investigated quite thoroughly since the early 1920s, while there is paucity of data and analysis in the latter.

In this review only the effect of the particles on the turbulence in a boundary layer will be discussed in some depth. The experimental evidence in this area is relatively recent, because of the experimental difficulties. The experiments, reviewed below, were carried out with a hot wire anemometer (see Hetsroni & Sokolov 1971), in the earlier days. These data are sometimes open to questions since one can never be entirely sure how the hot wire responds to a two phase flow. In the 1980's the Laser Doppler Anemometer (LDA) was used and since then the data may be more reliable, but also more difficult to obtain. Many data have been obtained since then - but again - mostly on the motion of the particles. One of the difficulties in obtaining simultaneous data on the particles and the fluid, is to distinguish between the two signals. In other words, one needs to be sure whether the signal is due to a particle or to the tracers in the fluid. The discrimination between these is not trivial.

The various data sets are reviewed below, and some explanations are given on the possible physical phenomena. While these explanations seem plausible, there are many more data sets needed to make the underlying physics better understood.

Most of previous data sets, as well as reviews (e.g. Michaelides & Stock 1989, Parathasarathy & Fath 1990), were focused on the interaction between particles and the homogeneous or non-homogeneous turbulence of the mainstream. Here we focus our attention on the interaction between the particles and the coherent structures which form in a shear layer of a flow over a solid surface, such as in a flume.

The understanding of the underlying phenomena is of considerable practical interest since there is a direct correlation between the formation of ejections of the coherent structures at the wall, the turbulence in the mainstream, and the heat transfer from the wall. If the presence of particles may increase the frequency of ejections from the wall, and thereby enhance the turbulence of the mainstream, the heat transfer from the wall may also be increased. This may have a profound effect on the design and operation of heat transfer equipment. While this effect will not be reviewed here, the first steps to understanding the physical mechanisms are considered in detail.

8.2 COHERENT STRUCTURES

Observations by Kline et al. (1967) reveal that a boundary (whether wall or free shear) generates organized structures in the fluid, when mean shear is imposed on it. When the shear rate is increased at a boundary, alternating high-speed and low-speed regions are observed near the boundaries (figure 1). This flow pattern results from the counter rotating streamwise vortex pairs which occur as follows: It can be conjectured that due to the shear stress at the wall, a vortex filament is formed at the wall (figure 2a), in the spanwise direction. The filament's formation begins at a high-speed region, where its end is also attached to the wall, and is continuously elongated until two such filaments meet at the low-speed region. The filament is then stretched in the down-stream direction and attains a shape of a hairpin (figure 2b). Now it has the appearance of low-speed streaks which are formed between pairs of longitudinal counter rotating vortices (which are the legs of the hairpin). The center of this structure is lifted into the mainstream because the pressure under the vortex filament becomes higher than the pressure above it. The pairs of vortices are separated in the spanwise direction by $\Delta z^+ \approx 50$ and are inclined at an angle of about 20 - 25° to the boundary, from the location of attachments to the solid surface, to $y^+ < 20$. Here $z^+ = zu^*/\nu$. $y^+ = yu^*/\nu$, $u^* = \sqrt{\tau/\rho}$, where y^+ is the dimensionless distance from the wall and u^* is the wall shear velocity and ν is the kinematic viscosity. The streak spacing is measured at $y^+ = 5$ in units of $\lambda^+ = \lambda u^*/\nu$. The description in figure 2 is form Kline et al. 1967; Hinze 1975; Blackwelder & Eckelman 1979, Rashidi & Banerjee 1990. It has also been observed that the low-speed region, i.e. the center of the hairpin structure, is periodically broken off near the wall, and is ejected into the main stream. This process, the bursting process, is a random occurring event consisting of a gradual local lifting of the low-speed streaks, sudden oscillation and ejection of low-momentum fluid from the wall region into the mainstream. This is then followed by an insweep of high momentum fluid into the wall region, which most likely starts a new vortex filament, leading to another bursting. The spectacular phenomenon of bursting has been shown to be responsible for most of the turbulence energy production and to be the major contributor to the transport of "Reynolds stresses" (actually, these stresses do not exist. This is a mathematical description, resulting from certain separation of velocities, which appear in the Navier-Stokes equations).

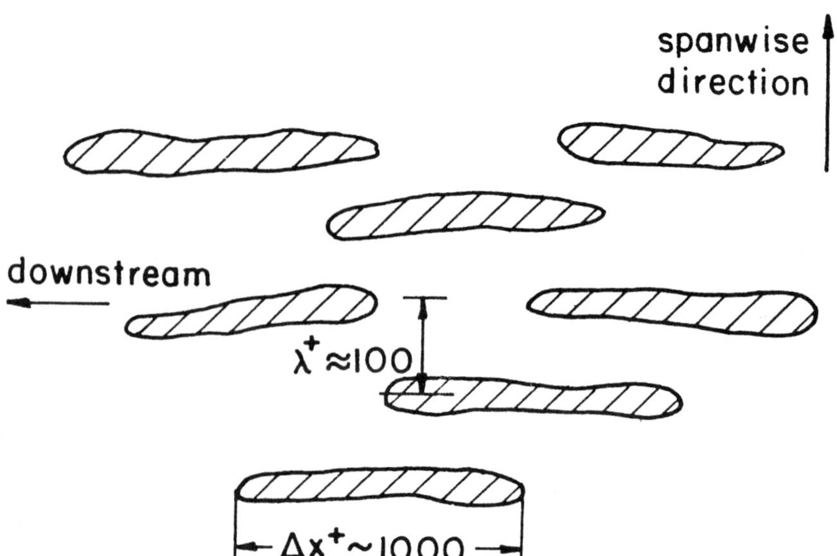

Figure 1. Low-speed, high-speed regions near the wall.

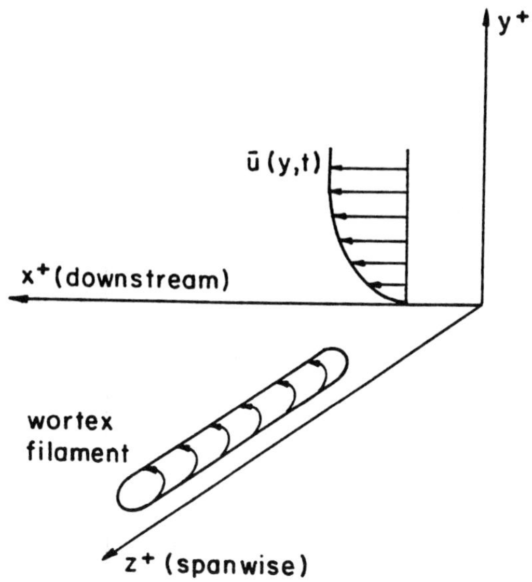

Figure 2a. Formation of the vortex filament.

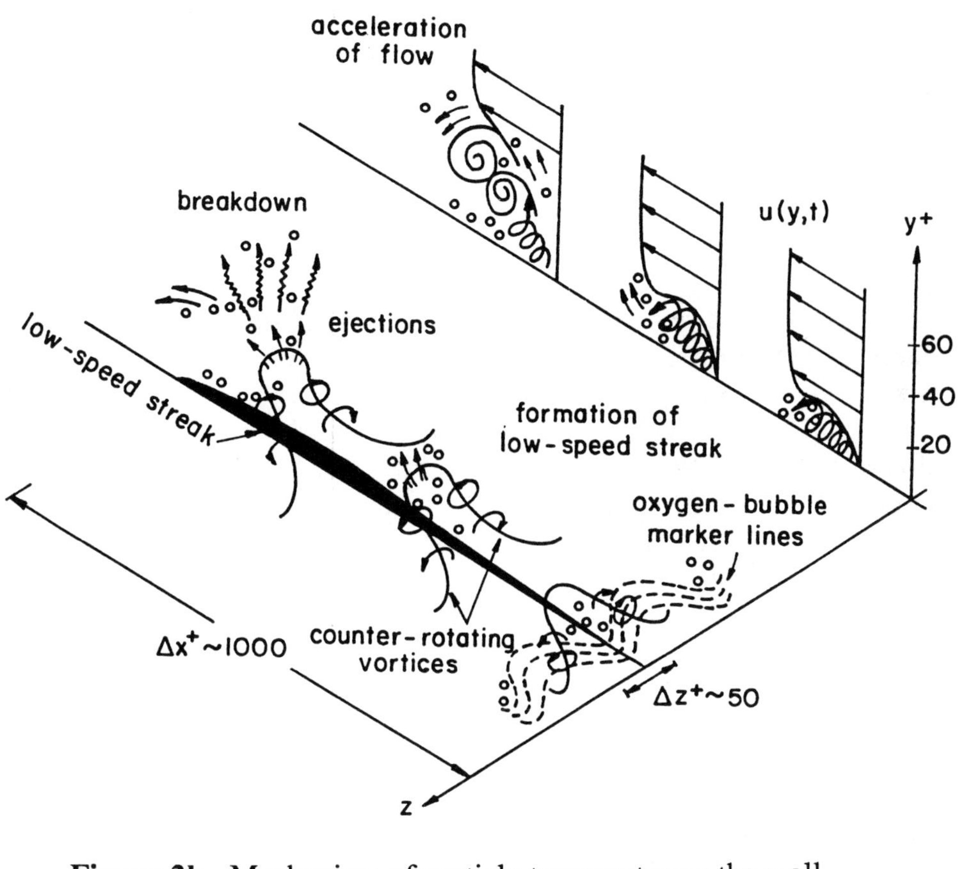

Figure 2b. Mechanism of particle transport near the wall.

8.3 PARTICLES IN TURBULENT FLOW

There have been a number of reviews on the interaction between turbulence and particles, e.g. Hetsroni 1989, Tsuji 1991. These will not be repeated here. It is worthwhile to repeat (from Hetsroni 1989) only the part on the experimental data - since there are very few good experimental data.

Early experiments (e.g. Hetsroni & Sokolov 1971; Serizawa 1974; Serizawa & Kataoka 1980; Serizawa et al. 1975) used hot-wire anemometry with special techniques to filter out the effect of the droplets (or bubbles) on the hot wire. These data have been criticized (Faeth 1983, 1987), claiming that only a relatively small fraction of the drops which impinged on the hot wire were accounted for, which tended to bias the results. Though it is, admittedly, possible that some bias existed, the results are still considered valid - in particular, the energy spectra, which could not have been biased in this matter. Later measurements were all conducted with a laser-Doppler anemometer - an experimental technique which is not without difficulties. Actually, there are only six sets of data of fluid - solid flow in a pipe, and most of them are not detailed enough for any definitive conclusions to be drawn. Other data were obtained using a free jet, a geometry which is experimentally much more convenient than a tube. In Table 1 (Gore & Crow 1989), all available experimental data are summarized. In the table, the geometry is given together with the ratio of the density of the particle ρ_p to the density of the fluid ρ_f, particle volume fraction c%, and the fluid's Re, i.e. the mainstream's Reynolds number. In the table there are sets of gas-solid, liquid-solid and gas-liquid data. The latter set was included because the very small liquid droplets in air behave very much as solid particles in air.

Other data sets (e.g. Snyder & Lumley 1971; Wells & Stock 1983) did not include measurements of turbulence as such (they measured only diffusivity), and are not included in Table 1.

It would be best now to plot the change in turbulence of the fluid, as a function of the local Re_p in order to get a clear indication of the effect of the particles on the turbulence of the mainstream. Unfortunately, not all of the data sets enables an exact calculation of the local Re_p, because the relative velocity is unknown. One could present the data as the change in turbulence due to the particles, as a function of the relative size of the particles, i.e. the ratio of the particle size to a characteristic length scale of the system. The characteristic length scale of the energy-containing eddies was taken as $0.2R$ at the pipe center and $0.05R$ or so near the wall (Hutchinson et al. 1971). For a single-phase jet, the data of Wygnanski & Fiedler (1969) that ℓ_e / x - 0.081, where x is the axial distance from the exit nozzle, was used. The percentage change in turbulence intensity is defined in figure 3 (Gore & Crowe 1989) as

$$\frac{\sigma_{TP} - \sigma_f}{\sigma_f} \times 100,$$

where $\sigma = \sqrt{u'^2} / u$ is the turbulence intensity, the subscript TP is for two-phase and f is for fluid.

Table 1. Experimental parameters

Reference	Geometry	$\frac{\rho_p}{\rho_r}$	C	Ref
Levy & Lockwood (1981)	Gas-solid downward jet	2000	6×10^{-4}	20,000
Hetsroni & Sokolov (1971)	Gas-liquid horizontal jet	775	2.5×10^{-6}	83,300
Tsuji et al. (1984)	Gas-solid upward pipe	850	5×10^{-3}	22,500
Modarress et al. (1984a)	Gas-solid downward jet	2500	2×10^{-4}	13,300
Tsuji & Morikawa (1984a)	Gas-solid horizontal pipe	833	4×10^{-3}	20,000
Shuen et al. (1985)	Gas-solid downward jet	2200	2×10^{-4}	19,000
Parthasarathy & Faeth (1987)	Liquid-solid downward jet	2.5	2.4-4.8	9,000
Modarress et al. (1984b)	Gas-solid downward pipe	2500	3.5×10^{-4}	17,000
Lee & Durst (1982)	Gas-solid	2080	1×10^{-3}	8,000
Zisselmar & Molerus (1979)	Liquid-solid horizontal pipe	2.5	4.0	100,000
Maeda et al. (1980)	Gas-solid upward pipe	7500	1×10^{-4}	20,000

In that 1989 paper some mechanisms were suggested to account for the increase in the turbulence of the carrier fluid in the main stream due to the presence of particles, (say in a jet or in a vertical pipe). The main mechanism is due to the relative velocity between the phases, which cause vortex shedding from the particles. These vortices are then found in the turbulent fluctuations in the main stream. The effect of particle rotation in the wall region (e.g. Ye et al. 1990) will also not be considered here.

In this review the stress will be placed on the interaction of particles and coherent structures, as discussed in the next section.

8.4 INTERACTION OF PARTICLES AND COHERENT STRUCTURES

This subject was reviewed by Rashidi et al. (1990). They state that "it is clear that particulate matter suspended in the boundary layer interacts with the coherent structures, and that this interaction may be quite complex. Grass (1974) recorded details of the suspension process in a turbulent boundary layer on a flat plate. He observed that sand particles were carried up from the bed region through virtually the total boundary layer thickness. Later, Summer & Deigaard (1981) made some measurements on the motion of 3.0 mm dia particles near a wall, using a photogrammetric system. They traced the motion of a single particle suspended in water in a horizontal channel. They found that, with a smooth bottom wall, the measured kinematical quantities characterizing the particle motion were in accord with available information on the bursting phenomena, as descried above, suggesting that this is perhaps a mechanism to cause particle suspension in the flow.

Dyer & Soulsby (1988) have similarly shown the importance of the bursting phenomenon on the transport of the particles on the ocean floor. The periodic ejection and insweep events associated with the bursting process have been captured in the instantaneous measurements of the Reynolds stress. This is seen as a sudden rise in the Reynolds stress magnitude of up to 30 times the mean Reynolds stress. In the ocean, these events are visible as swirls of sand that are lifted up above the ocean floor. Furthermore, it has been observed by Taylor & Dyer (1977), Itakura & Kishi (1980) and Soulsby & Wainwright (1987) that the particle movement affects the fluid velocity profile. Depending on he particle size and the particle velocity, the fluid velocity profiles may become concave or convex.

Rogers & Eaton (1989) measured the response of solid particles to a vertical turbulent boundary layer in air, using a low concentration of 50 and 90 μm glass particles. They found that the r.m.s. velocity fluctuations of both particle sizes nearly equaled the streamwise turbulence intensity of the flow, but the turbulence intensity of the particles in the normal direction was strongly attenuated, i.e. they were much lower than the fluid's. The power spectra showed that for the normal fluctuations the power spectrum shifts to higher frequencies relative to the streamwise fluctuations, therefore indicating that the particles did not closely follow the fluid fluctuations in the normal direction. Rogers & Eaton also used 70μm copper particles to create a greater particle mass loading of 20%. The presence of these particles tended to suppress turbulence, with the degree of turbulence suppression being a function of the distance downstream in the boundary layer. They attributed the turbulence suppression to an increase in the dissipation. However, they were unable to clarify their results in light of the observed dominant flow events near the wall.

Yung et al. (1989) investigated the interaction of the turbulent bursts and the deposited particles within the viscous sublayer near the wall. They carefully placed spherical particles of polystyrene and glass (50 μm dia) on the wall so that the particles were completely submerged within the viscous sublayer. Their flow visualization experiments showed that the bursting phenomenon has a insignificant effect on re-entrainment of the deposited particles within the viscous layer. It was shown that only about 1% of the particles are lifted up into the flow by the wall bursts. However, as the size of the particles increased, the number of the re-entrained particles also increased. Based on thee results they concluded that, in contrast to the previous hypothesis, the turbulent bursts are not effective to re-entrain the particles within the experimental range of $0.5 < d^+ < 1.3$, where d^+ is the non-dimensional particle diameter (du_*/ν). On the other hand, as d^+ increased, the importance of turbulent bursts in particle re-entrainment is substantially increased.

Cleaver & Yates (1973, 1975, 1976), in an attempt to model the particle transport in turbulent flows, suggested that the particles are re-entrained by the wall ejection events while they are deposited by the insweep events. Their model had some agreement with the limited experimental data in predicting the particle deposition and re-entrainment rates. Although their model and some recent theoretical attempts have been based on some characteristics of the bursting phenomenon, there is still not enough experimental evidence as to the details of the turbulence-particle interactions for any realistic modeling of this problem."

The work of Rashidi et al. is interesting in as much as the effect of particles on the frequency of bursting is clearly demonstrated. The experiments were carried out in a flume with water flow in a rectangular Plexiglass channel.

"The channel was made up of four equal sections and was 4.0 m long, 0.20 m wide and 0.15 m deep. It was designed so that, at the test section, flow was fully developed with an aspect ratio large enough to be free from wall effects at the center (figure 4).

The liquid was filtered continuously to remove solids > 5μm. Measurements of liquid temperature were performed at the entrance, the exit and the test section of the channel with thermocouples placed on the side walls and with a set of precision mercury thermometers placed in the channel. Great care was taken to eliminate wave formation at the inlet and reflections from the outlet. As a result, the flow was introduced into the inlet tank through several inlet holes in order to provide a setup that produced high flow rates with no significant interfacial waves. Efforts were also made to maintain a constant average flow and temperature throughout each run.

The particles (glass or polystyrene) were introduced in the flow at about 0.60 m upstream from the test section. This was done using a 3 mm dia tubing placed in the center of the channel and at the center of the flow depth. A known solution of each particle size was made in a separate mixing vessel, and was added into the flow by a variable-speed peristaltic pump, such that the particles were introduced in the flow with a known concentration and flow rate. The particle loading was controlled by changing the solution concentration or the pump's speed. The added particles were collected at the exit of the channel using a 50μm mesh filter. The effect of the tube presence, tube height location and particle acceleration on the mean flow velocity was later checked experimentally and theoretically to ensure a minimal disturbance of the flow. This was done theoretically be evaluating the appropriate length between the test section and the tube that provided negligible tube wake effect and enough time for particles to accelerate to the mean flow velocity. In addition, experiments were conducted with and without the tube presence to ensure that the tube wake effect had no influence on the visualized instantaneous velocity profiles. The effect of tube height location was also checked experimentally by changing the height of the tube with respect to the bottom wall and observing the effect on the measured wall ejection frequencies at the test section. This was done for three locations, $y/h = 0.1$, 0.5 and 0.9, and the effect was found to be insignificant for the present experimental conditions.

Measurements were made in liquid streams primarily by flow visualization. This was done through the use of very small (~ 20 Zμm dia) oxygen bubble tracers (Rashidi & Banerjee 1988) that were photographed using a high-speed video analyzer and alternatively a 35 mm camera with mechanically chopped flashes. The chopper had a constant speed of about 3700 rpm. It was designed so that several high-intensity beams of collimated light were produced by chopping a single flash. This resulted in well-spaced traces of both bubbles and particles in a film frame from which flow and particle velocities were found by image processing. The bubbles were produced from two platinum wires (25 μm dia) using a high-voltage pulse generator (20 - 200 pulse/s with a pulse duration of 1-5 ms at up to 300 V). One wire was placed horizontally parallel to the bottom of the channel and aligned in the spanwise direction (and could be moved up and down between the wall and interface), while the other wire was held vertically across the channel.

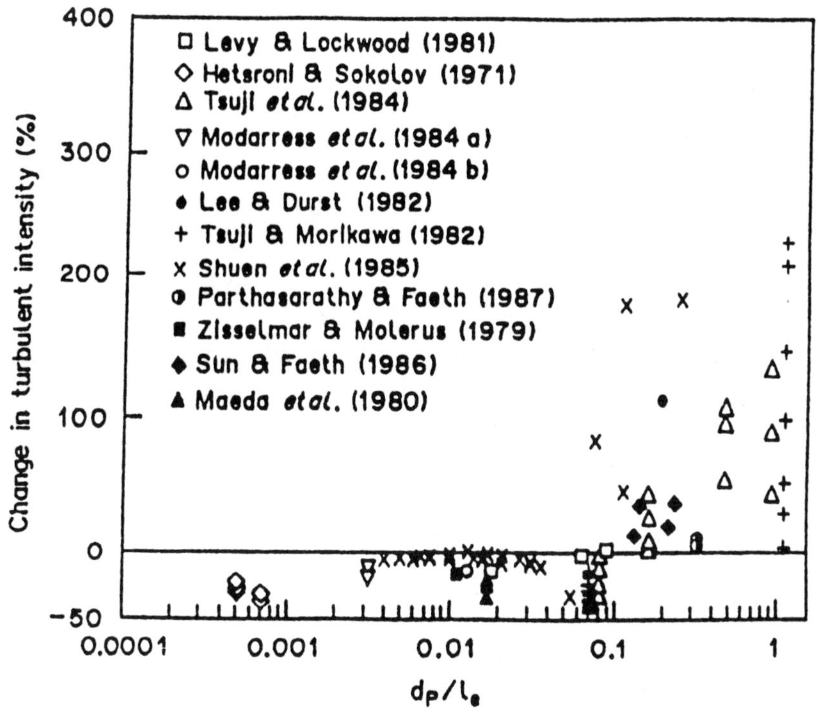

Figure 3. Change in turbulence intensity as a function of the length scale ratio (Gore & Crow, 1989).

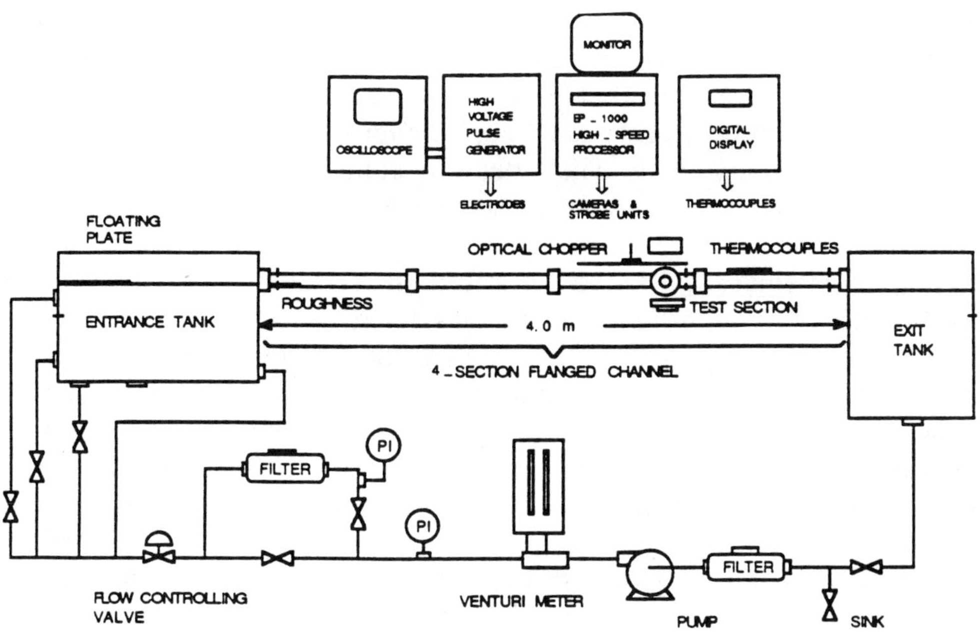

Figure 4. Experimental facility and arrangement of measurement devices.

The high-speed video recording system was a two-camera Kodak (Spin Physics) EP-1000 motion analyzer. It was used with both conventional and fiber-optic synchronized-strobe units. The system has a maximum full-screen capability of 1000 frame/s and a maximum split-screen capability of 6000 frame/s. In these experiments, data were generally recorded at 250 frame/s. The recorded data could be played back in slow motion as well as in single frames for detailed data analysis. The split-screen capability allowed one camera to be used to record the flow structures in the horizontal plane, while the other captured the structures in the vertical plane.

Experimental Conditions. The experimental runs were conducted in the described flow facilities. The flow conditions are summarized in Table 2. The experiments were done at four different Reynolds (and Froude) numbers under hydraulically subcritical conditions. Reynolds numbers based on flow depth and based on equivalent diameter are listed in Table 2. For the runs discussed here, it was possible to form a turbulent liquid flow with no significant interfacial waves. Therefore, particle-turbulence interaction was studied without the influence of interfacial waves. The values of the wall friction velocity, presented in Table 2, were evaluated from the measured velocity profiles and the Blasius formula. Both methods gave similar results.

Particle characteristics are listed in Table 3. Both spherical polystyrene and glass particles were used in this study. The polystyrene particles had sp. gr. 1.03 ± 0.02 and were employed in four different sizes of about 120, 220, 650 and 1100 μm dia. the glass particles had sp. gr. 2.50 ± 0.05 and were about 88 μm dia. Experiments were conducted at four different particle loadings, defined as Q_P/Q_F (were Q_P and Q_F are the particle and fluid volumetric flow rate, respectively); these were 0, 0.9×10^{-4}, 1.8×10^{-4} and 3.6×10^{-4}. The values of Q_P/Q_F were evaluated first from the measured values of particle and fluid flow rates, but later verified from the recorded video sequences of the flow field. The latter was done from the measurement of particle concentrations (from video sequences) averaged throughout each experimental run.

Experimental results. To investigate the dominant structures near the wall and their interactions with the particles, the high-speed video system was used. The system was used to view simultaneously the near-wall regions in the horizontal plane (upper portion of the screen) and the particle-burst interactions in the vertical plane (lower portion of the screen). Figure 5 illustrates a sequence of pictures obtained this way for Re = 5000 and 1100 μm polystyrene particles. As seen from these pictures and many other video sequences, the ejections originating at the wall clearly interact with the particles, and generally lift them up into the bulk flow. The ejected particles, after moving up a distance from the wall, lose their vertical momentum and return toward the wall. The cycle then repeats itself.

In order to study the details of turbulence-particle interaction, the influence of particles on the wall ejections was examined. For each run, a video sequence of at least 22,000 frames (~ 3 min) was recorded. From this sequence, about 100 wall ejections were measured. The measurements of wall ejection frequency were done from the EP-1000 recorded video sequences that provided the time and frame number of the events. In the video sequences, the wall ejections are characterized by the lift up and ejection of vortical structures in the upper portion of the screen (horizontal plane). This is seen in the lower screen (vertical plane) as a deceleration of instantaneous velocity profiles near the wall followed by a sudden ejection form the wall region. The ejections were measured at a fixed location 30 mm downstream from the vertical wire at $y^+ \approx 20$ to 30 (where

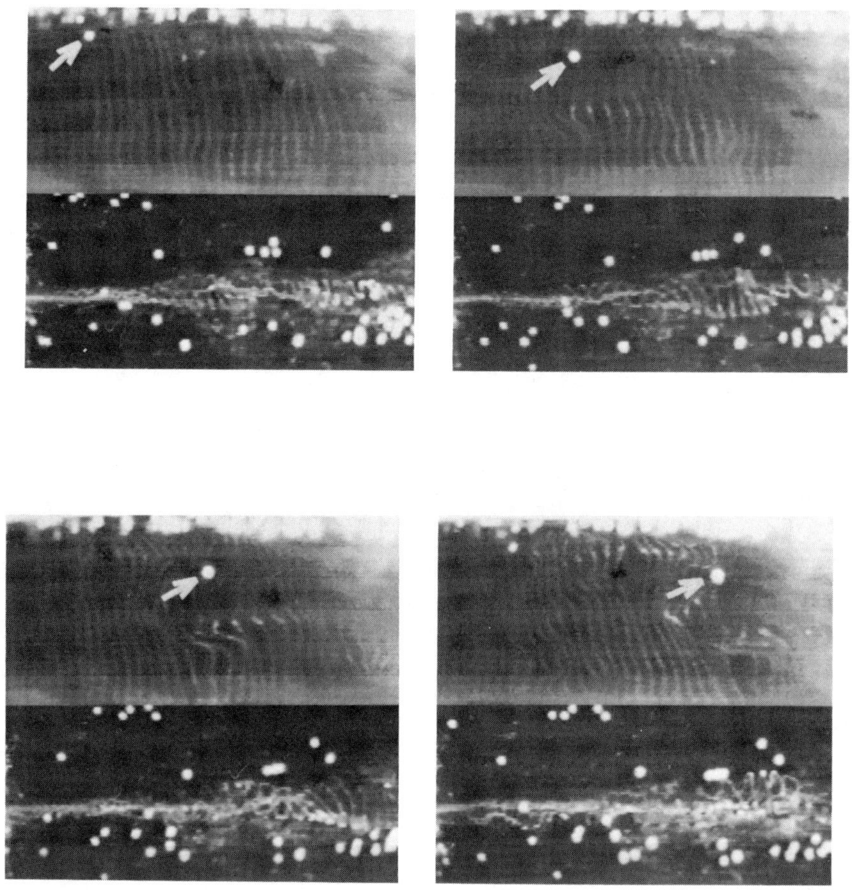

Figure 5. Sequential pictures (right to left) of wall burst interaction with the polystyrene particles of 1100 μm (indicated by the arrows) at Re = 5000. The time interval between the photographs is 0.12 s.

Table 2. Flow conditions

Reynolds number, Re ($u_M h/\nu$)	Flow depth h (cm)	Equivalent diameter D (cm)	Kinematic viscosity, $\nu \times 100$ (cm^2/s)	Mean velocity, u_M (cm/s)	Friction velocity, u_* (cm/s)	Froude number, Fr ($u_M/\sqrt{gh}$)	Reynolds number, Re_D ($u_M D/\nu$)
7500	2.75±0.05	8.63	0.848	23.6±2.3%	1.29	0.45	24,000
5000	2.75	8.63	0.83	15.6	0.90	0.30	16,000
3500	2.75	8.63	0.830	11.0	0.66	0.21	11,500
2500	2.75	8.63	0.821	7.7	0.46	0.15	8,000

Table 3. Particle characteristics

Particle type	Diameter (μm)	Specific gravity	Particle geometry	Loading $Q_P/Q_F \times 10^4$
Polystyrene	120	1.03±0.02	Spherical	0.9, 1.8, 3.6
Polystyrene	220	1.03	Spherical	1.8
Polystyrene	650	1.03	Spherical	1.8
Polystyrene	1100	1.03	Spherical	0.9, 1.8, 3.6
Glass	88	2.50±0.05	Spherical	1.8

Table 4. Ejection velocity

Run	Particle type	Particle type (μm)	d^+ ($u_* d/\nu$)	Particle loading, $Q_P/Q_F \times 10^4$	Mean dimensionless ejection	Re ($u_M D/\nu$)
1	Polystyrene	120	1.3	0.9	1.03	5000
2	Polystyrene	120	1.3	3.6	1.12	5000
3	Polystyrene	1100	11.8	0.9	0.99	5000
4	Polystyrene	1100	11.8	3.6	0.82	5000
5	Glass	88	0.9	1.8	1.00	5000

$y^+ = yu_*/\nu$ is the distance from the wall non-dimensionalized by the wall shear velocity, u_*, and kinematic viscosity, ν).

Figure 6 shows the effect of particle size on the average time between wall ejections or ejection periods (T_E) as a function of fluid Re. Here, the values of T_E are measured as described above and are normalized by the ejection periods obtained for the flow with no particles, $T_E^* = \left(T_E T_{E\ \text{no loading}}\right)$.

As seen from figure 6, the larger polystyrene particles of 1100 μm dia cause an increase in the number of ejections near the wall. The effect is slightly enhanced as Re decreases.

It is very clearly shown in figure 7 that at Re = 5,000 and with a constant particle loading, the frequency of ejections increases with particle size.

Particles with $d < 200$ μm cause actually a decrease in the number of ejections, while particles with $d > 200$ μm cause an increase in frequency, as compared to flow without particles.

The experimental conditions, listed in Table 2 and 3, indicate that the equivalent radius is $R = 0.043$ m. Using the data in figure 3 we note that particles with $d_p/l_e > 0.1$ increase the turbulence in the flow. With the experimental conditions of Rashidi et al. (1990) we find that $d_p > 215$ μm, the turbulence is enhanced when by the particles i.e. the data of Rashidi et al. (1990) fits extremely well with all other data in figure 3.

The effect of particle loading on the structures near the wall was also studied (Rashidi et al. 1990) with respect to the low-speed/high-speed streaks of the wall regions. Figure 8 illustrates the distribution of 1100 μm particles and their interactions with the low-speed/high-speed streaks that form near the wall. As seen from this sequence and others, particles generally accumulate in the low-speed regions of the streaks and then are taken away by the wall ejections into the bulk flow. Furthermore, it appears that the streaks have the same characteristics as those observed for the runs with no particles. Figure 9 (a,b) show the variation of the mean non-dimensional spanwise streak-spacing, $\lambda^+ = \lambda u_*/\nu$, as a function of particle loading for both 120 and 1100 μm particles at $y^+ = 5$. As before, for each run a video sequence of at least 22,000 frames (~ 3 min) was recorded and from this sequence 60-80 frames were analyzed. The reported mean streak-spacings correspond to at least 200 counts of the streaks. As seen from this figure, the change in particle loading does not influence the average spacing of the wall streaks. It appears that, for both 120 and 1100 μm particles, the mean spacing of the streaks non-dimensionalized with the wall shear velocity (u_*) and kinematic viscosity (ν) changes very little with increases in particle loading and exhibits consistent values of $\lambda^+ \approx 100$ at $y^+ = 5$. This is indeed a significant finding, since the particle introduction influences the frequency of wall ejections but has a negligible effect on the streaks' appearance and their spacing.

<u>Mechanism</u> - based on the experimental evidence we conclude that larger particles enhance the turbulence by increasing the frequency of ejections of coherent structures from the boundary layer. There could be a number of mechanisms which would explain this phenomenon e.g.:

The ends of the vortex filament, i.e. the ends of the horseshoe structure must remain attached to a solid surface. When a particle, of size comparable or larger than the diameter of the filament, passes by these ends, they may become detached from the surface of the wall and attach themselves to the particle,

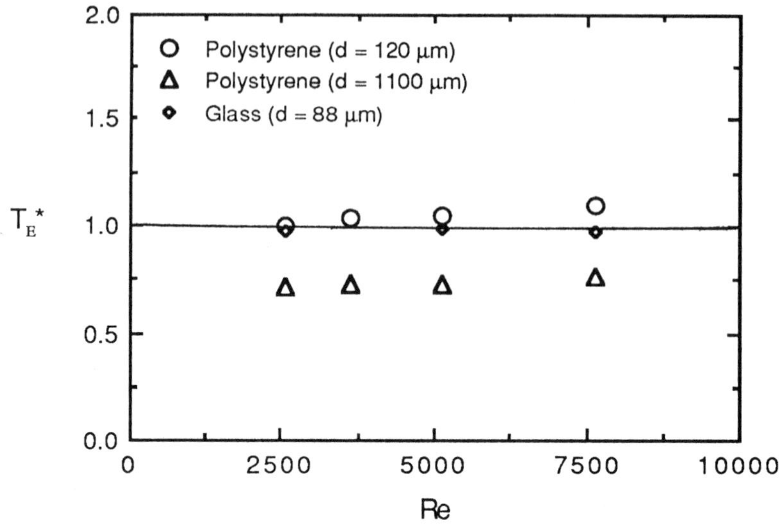

Figure 6. Effect of particles on the average, non-dimensional time between wall.

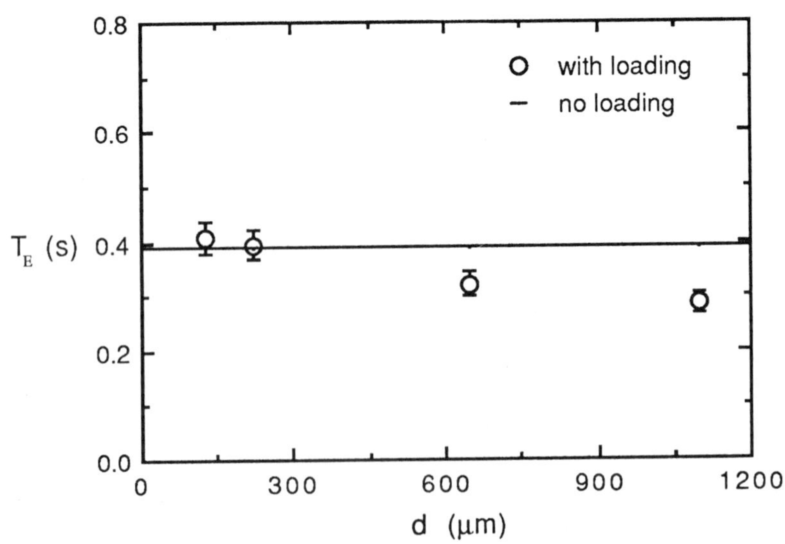

Figure 7. Effect of particle size on the average time between wall ejections, T_E, at $Re = 5000$, $Q_P/Q_F = 1.8 \times 10^{-4}$. The uncertainty in the values of T_E is $\pm 7\%$ at the 95% confidence level for 100 ejection measurements. (Rashidi et al. 1990).

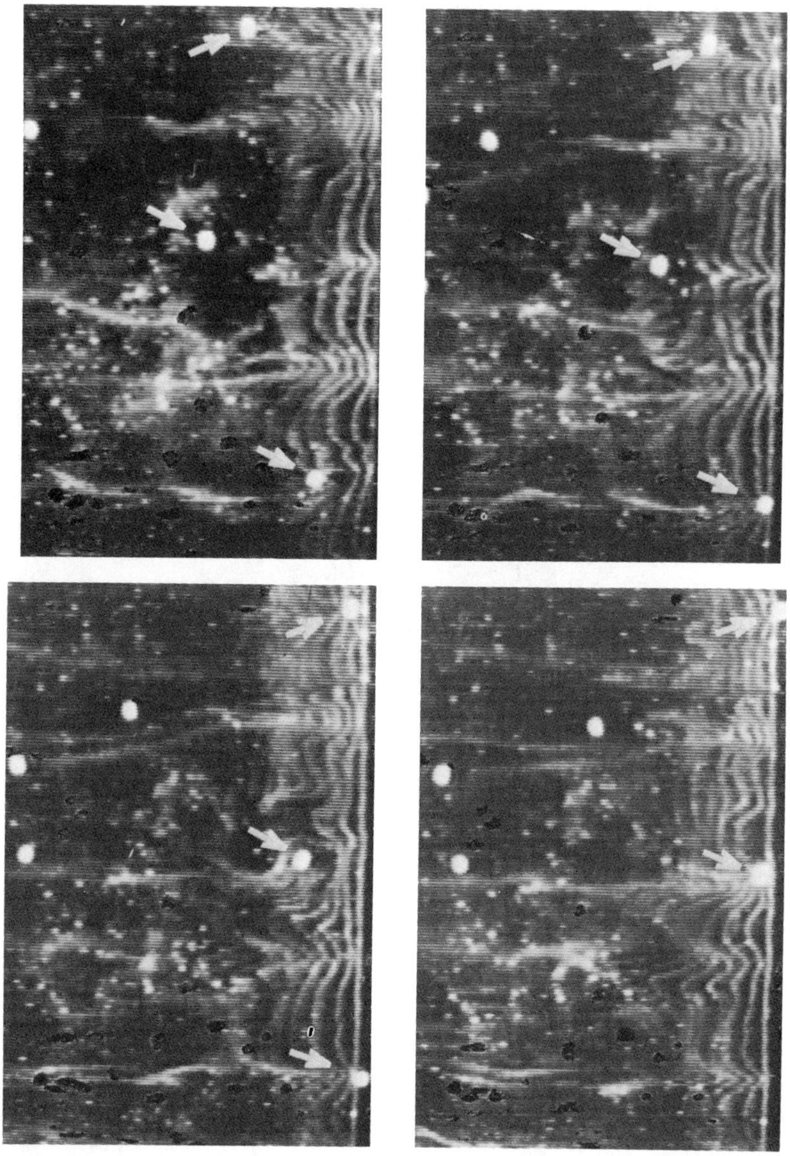

Figure 8. Illustration (right to left) of particle distributions with respect to the low-speed/high/speed streaks of the wall regions; d = 1100 μm particles. Re = 5000, $y^+ = 5\pm1$. The time interval between the photographs is 0.08s.

thereby freeing themselves from the wall and creating a "premature" burst (Haber 1991).

- Particles from the mainstream, moving at a velocity close to that of the fluid in the mainstream, settle on the wall. These particles have relatively high momentum, and could certainly be visualized as having a destabilizing effect in the boundary layer, causing premature bursts.

Currently, these are somewhat speculative calculations which are currently done and should shed further light on the phenomena.

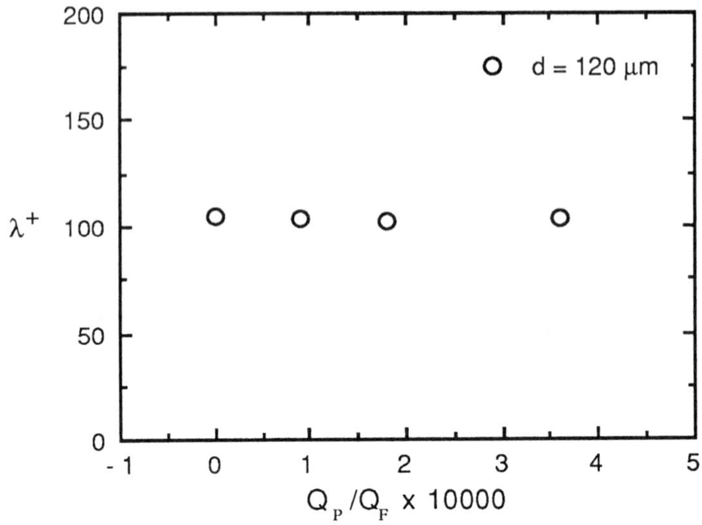

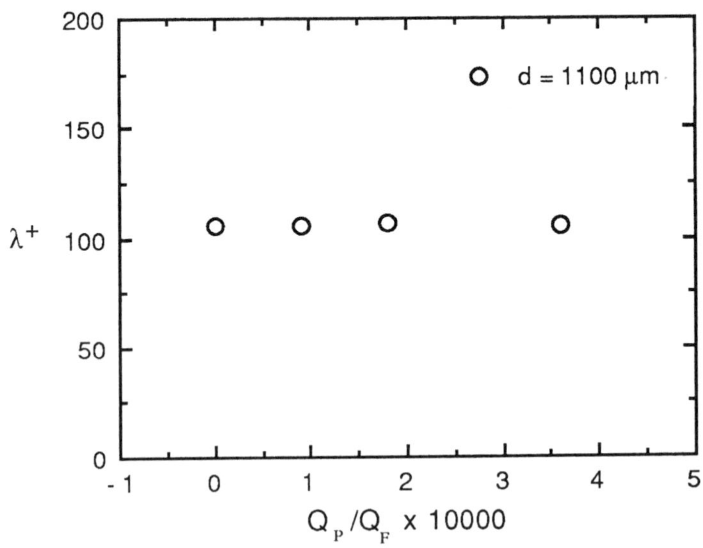

Figure 9. Variation of mean non-dimensional spanwise streak-spacing, $\lambda^+ = \lambda u_*/\nu$ as a function of particle loading: Q_P/Q_F at Re = 5000 and $y^+ = 5\pm1$. (a) 120 μm particles; (b) 1100 μm particles. The uncertainty in the values of λ^+ is about ± 7% at the 95% confidence level for 200 measurements.

8.5 CONCLUSIONS

From the experimental evidence we conclude that larger particles, of the order of 5 to 10 wall units (about 500 to 1000 μm in our case) enhance the turbulence of the fluid in a flume. Smaller particles suppress the turbulence.

Possible mechanisms for these phenomena could be: In the mean flow, i.e. not in the wall region, larger particles shed vortices. The energy is thus transferred from the mean flow to higher frequencies turbulent motions. The frequency of the vortex shedding depends on the relative velocity between the phases and on the size of the particles.

In the boundary layer, the larger particles increase the frequency of ejections of coherent structures by distabilizing these structures or by providing the vortex filament, which is the basis of the hairpin structure, with solid surfaces to which it can attach itself.

The mechanisms, which cause suppression of turbulence, are undoubtedly connected with increasing the number of velocity gradients in the flow, thereby causing additional energy dissipation.

Additional research is needed to confirm these postulates or to suggest other possible mechanisms.

8.6 REFERENCES:

Achenbach, E., "Vortex Shedding from Spheres," *J. Fluid Mech.* **62**, 209-221 (1974).

Al Taweel, A.M., and J. Landau, "Turbulence Modulation in Two-Phase Jets," *Int. J. Multiphase Flow*, **3**, 341-351 (1977).

Bagnold, R.A., "The Physics of Wind-Blown Sand and Desert Dunes," Methuen, London (1941).

Blackwelder, R.F., and H. Eckelmann, "Streamwise Vortices Associated with the Bursting Phenomenon," *J. Fluid Mechanics*, **94**, 577-594 (1979).

Besnard, D., F.H. Harlow, and R. Ravensahn, "Conservative and Transport Properties of Turbulence with Large Density Variations," Report LA-10911-MS (1987).

Cleaver, J.W. and B. Yates, "Mechanism of Detachment of Colloid Particles from a Flat Substrate in Turbulent Flow," *J. Colloid Interface Sci.* **44**, 464-473 (1973).

Cleaver, J.W. and B. Yates, "A Sublayer Model for Deposition of the Particles from Turbulent Flow," *Chem. Engng. Sci.* **30**, 983-992 (1975).

Cleaver, J.W., and B. Yates, "The Effect of Re-entrainment on Particle Deposition," *Chem. Engng. Sci.* **31**, 147-151 (1976).

Dyer, K.R., and R.L. Soulsby, "Sand Transport on the Continental Shelf," *A. Rev. Fluid Mech.* **210**, 295-324 (1988).

Faeth, G.M., "Recent Advances in Modeling Particle Transport Properties and Dispersion in Turbulent Flow," Proc. ASME-JSME Therm. Engng. Conference, **2**, 517-534 (1983).

Faeth, G.M., "Mixing Transport and Combustion in Sprays," *Prog. Energy Combust. Sci.* **13**, 293-345 (1987).

Gore, R., and C.T. Crowe, "Effects of Particle Size on Modulating Turbulent Intensity," *Int. J. Multiphase Flow* **15**, 279-285 (1989).

Grass, A.J., "Transport of Fine Sand on a Flat Bed: Turbulence and Suspension Mechanics," In Proc. Eurmech., Tech Univ. of Denmark, Copenhagen, Vol. 48, p. 33 (1974).

Haber, S., Personal communications (1991).

Hetsroni, G., "Particles-Turbulence Interaction," *Int. J. Multiphase Flow* **15**, 735-746 (1989).

Hetsroni, G., and M. Sokolov, "Distribution of Mass, Velocity and Intensity of Turbulence in a Two-Phase Turbulent Jet," *Trans. J. Appl. Mech.* **38**, 315-327 (1971).

Hinze, J.O., "Turbulence," 2nd edition, Mcgraw-Hill, New York (1975).

Hutchinson, P., G.F. Hewitt, and A.E. Dukler, "Deposition of Liquid or Solid Dispersion from Turbulent Gas Streams: A Stochastic Model," *Chem. Engng. Sci.* **26**, 419-439 (1971).

Itakura, T., and T. Kishi, "Open Channel Flow with Suspended Sediments," *J. Hydraul. Div.* ASCE **106** (8), 1325-1343 (1980).

Kim, H.T., S.J. Kline, and W. Reynolds, "The Production of Turbulence Near a Smooth Wall in Turbulent Boundary Layer," *J. Fluid Mech.* **50**, 133-160 (1971).

Kline, S.J., W.C. Reynolds, F.A. Schraub, P.W. Rundstaler, "The Structure of Turbulent Boundary Layer," *J. Fluid Mech.* **30**, 741-773 (1967).

Lee, S.L., and F. Durst, "On the Motion of Particles in Turbulent Flow," Report NUREG/CR-1554 (1980).

Lee, S.L., and F. Durst, "On the Motion of Particles in Turbulent Duct Flows," *Int. J. Multiphase Flow* **8**, 125-146 (1982).

Levy, Y., and F.C. Lockwood, "Velocity Measurements in a Particle Laden Turbulent Free Jet," *Combust. Flame* **40**, 333-339 (1981).

Maeda, M., K. Hishida, and T. Furutani, "Optical Measurements of Local Gas and Particle Velocity in an Upward Flowing Dilute Gas-Solids Suspension. In Polyphase Flow and Transport Tech., p. 211. Century 2-ETC, San Francisco, Calif. (1980).

Modarress, D., H. Tan, and S. Elghobashi, "Two-Component LDA Measurement in a Two-Phase Turbulent Jet," *AIAA J.* **22**, 624-630 (1884a).

Modarress, D., J. Wuerer, and S. Elghobashi, "An Experimental Study of a Turbulent Round Two-Phase Jet," *Chem. Engng. Commun.* **28**, 341-354 (1984b).

Nouri, J.M., J.H. Whitelaw, and M. Yianneskis, "Particle Motion and Turbulence in Dense Two-Phase Flows," *Int. J. Heat Mass Transfer* **13**, 729739 (1987).

Owen, P.R., "Pneumatic Transport," *J. Fluid Mech.* **39**, 407-432 (1969).

Pan, Y., G. Hetsroni, and S. Banerjee, "Direct Simulation of Particle Interaction in Shear Flows," AIChE meeting, Los Angeles, CA., Nov. 17-22 (1991).

Plachta, R., G. Hetsroni, and S. Banerjee, "Conditional Sampling in Coherent Sampling in Coherent Turbulent Structures with Combined 3-D Laser Anemometry and Flow Visualization," AIChE meeting, Los Angeles, CA., Nov.17-22 (1991).

Parthasarathy, R.N., and G.M. Faeth, "Structure of a Turbulent Particle-Laden Water Jet in Still Water," *Int . J. Multiphase Flow*, **13**, 699-716 (1987).

Parthasarathy, R.N., and G.M. Faeth, "Turbulence Modulation in Homogeneous Dilute Particle-Laden Flows," *J. Fluid Mech.* **220**, 485-537 (1990).

Rashidi, M., and S. Banerjee, "Turbulence Structure in Free-Surface Channel Flows," *Phys. Fluids* **31**, 2491-2503 (1988).

Rashidi, M., and S. Banerjee, "Streak Characteristics and Behavior Near Wall and Interface in Open Channel Flows," Trans. ASME *J. Fluids Engng.* **112**, 164-170 (1990).

Rashidi, M., G. Hetsroni, and S. Banerjee, "Particles-Turbulence Interaction in a Boundary Layer," *Int. J. Multiphase Flow* **16**, 935-949 (1990).

Rogers, C.B., and J.K. Eaton, "The Interaction Between Dispersed Particles and Fluid Turbulence in a Flat Plate Turbulent Boundary Layer in Air," Report MD-52, Stanford Univ., CA (1989).

Ruck. B., and B. Makiola, "Particle Dispersion in a Single-Sided Backward-Facing Step Flow," *Int. J. Multiphase Flow* **14**, 787-800 (1989).

Serizowa, A., "Fluid Dynamic Characteristics of Two-Phase Flow," Ph.D. Thesis, Kyoto Univ., Japan (1974).

Serizowa, A., and I. Kataoka, "Fundamental Aspects of the Drift Velocity in Turbulent Bubbly Flow," Tech. Rep. Inst. Atom. Energy, Kyoto Univ., No. 182 (1980).

Serizowa, A., I. Kataoka, and I. Michiyoshi, "Turbulence Structure of Air-Water Bubbly Flow," *Int. J. Multiphase Flow* **2**, 221 259 (1975).

Shuen, J.S., A.S.P. Solomon, Q-F. Zhang, and G.M. Faeth, "A Theoretical and Experimental Study of Turbulent Particle-Laden Jets," Report NASA CR 168293 (1983).

Shuen, J.S., A.S.P. Solomon, Q-F. Zhang, and G.M. Faeth, "Structure of Particle Laden Jets: Measurements and Predictions," *AIAA J.* **23**, 396-404 (1985).

Snyder, W.H., and J.L. Lumley, "Some Measurements of Particle Velocity Autocorrection Functions in a Turbulent Flow," *J. Fluid Mech.* **48**, 41-47 (1971).

Soo, S.L., H.K. Ihrig, and A.F. El Kough, "Experimental Determination of Statistical Properties of Two-Phase Turbulent Motion," Trans. ASME *J. Fluids Engng.* **D82**, 609-624 (1960).

Soulsby, R.L., and B.L.S.A. Wainwrite, "A Criterion for the Effect of Suspended Sediment on Near-Bottom Velocity Profiles," *J. Hydraul. Res.* **25**, 341-356 (1987).

Sumer, B.M., and R. Deigaard, "Particle Motions Near the Bottom in Turbulent Flow in an Open Channel: Part 2," *J. Fluid Mech.* **109**, 311-337 (1981).

Taylor, P.A., and K.R. Dyer, "Theoretical Models of Flow Near the Bed and Their Implications for Sediment Transport In The Sea," Vol. 6 (Edited by E.D. Goldberg, I.N. McCabe, J.J. O'Brian, and J.H. Steele), pp. 579-601, Wiley-Interscience, New York (1977).

Tsuji, Y., "Review: Turbulence Modifications in Fluid-Solid Flows," First Joint Conference, ASME-JSME Fluids Engineering (1991).

Tsuji, Y., and Y. Morikawa, "LDV Measurements of an Air-Solid Two-Phase Flow in a Horizontal Pipe," *J. Fluid Mech.*, **120**, 385-409 (1982).

Tsuji, Y., Y. Morikawa, and H. Shimi, "LDV Measurements of an Air-Solid Two-Phase Flow in a Vertical Pipe," *J. Fluid Mech.* **139**, 417-434 (1984).

Tsuji, Y., Y. Morikawa, T. Tanaka, K. Karmine, and S. Nishida, "Measurement of an Axisymmetric Jet Laden with Coarse Particles," *Int. J. Multiphase Flow* **14**, 565-574 (1988).

Wells, M.W., "The Effects of Crossing Trajectories on Diffusion of Particles in a Turbulent Fluid," (1982).

Wells, M.W., and D.E. Stock, "The Effects of Crossing Trajectories on the Dispension of Particles in Turbulent Fluids," (1983).

Wygnanski, I., and H. Fiedler, "Some Measurements in the Self-Preserving Jet," *J. Fluid Mech.* **38**, 577-612 (1969).

Ye J., M.K. Roco, and D. Stafford, "Particle Rotation in Couette Slurry Flow," ASME FED-98 (1990).

Yung, B.P.K., H. Merry, and T.R. Bott, "The Role of Turbulent Bursts in Particle Re-entrainment in Aqueous Systems," *Chem. Engng. Sci.* **44**, 873-882 (1989).

Zisselmar, R., and O. Molerus, "Investigation of Solid-Liquid Pipe Flow with Regard to Turbulence Modification," *Chem. Engng. J.* **18**, 233 (1979).

CHAPTER 9

UNSTEADY MOTION OF DENSE SUSPENSIONS AND RHEOLOGICAL BEHAVIOR

S. L. Soo
C. Zhu

9.1 INTRODUCTION

Modern manufacturing which involves the conveyance of particulate materials of high purity calls for dense phase pneumatic transport with minimum attrition of particulates and minimum erosion of the conduits. This is because attrition produces powdered waste which can become a health hazard if the dust is toxic, or may incur the hazard of dust explosion. Erosion of the piping system is to be avoided because of its waste and the eroded pipe material constitutes an impurity. The desirable system is therefore the one which conveys a dense suspension at a minimum speed and therefore also at a high energy efficiency. Many other applications of gas-solid systems are found in various industries (Soo, 1967). Many instrumentations and methods of measurement given in this chapter are applicable to industrial testing and production systems.

While this chapter emphasizes gas-solid suspensions, much of the fundamentals are applicable to liquid-solid suspensions or slurries. Comparisons to slurries will be made in the following under various topics. Other than the electrostatic effects, many similarities exist for their dilute suspensions. Steady horizontal pipe flow may exist only in a dense slurry; steady flow in a dense gas-solid suspension is not expected (Soo, 1987). Rheological properties of these two mixtures are especially different. Viscosity of gas-solid suspensions tends to be shear thickening, while slurries tends to be shear thinning (Brodkey, 1967; Cheremisinoff and Gupta, 1983).

9.1.1 Dilute Suspensions

The nature of a dilute suspension is well known (Soo, 1989). It is a case where the influence of particle-particle interaction is negligible in comparison to that of particle-fluid interactions. This leads to a number of simplifications in the formulation for predicting the velocity and density distributions of phases with validation by experiments. Prominent among the properties is the shear stress of the fluid phase in the mixture is attributable to the viscosity of the fluid phase μ_f (or μ_{ff} of fluid-fluid interaction) alone, while that of the particle phase is due to particle-fluid interaction μ_{sf}, where $\mu_{sf} \sim \alpha_s \bar{\rho}_s D_{sf}$, due to predominating fluid-particle interaction; D_{sf} being the diffusivity of the particle phase in the fluid and α_s is the volume fraction of the particle phase and $\bar{\rho}_s$ is the particle material density. However, much remains to be learned for the quantification of the motions of as the particle density increases. Decreasing energy was shown to occur in the low frequency part of the power spectrum and intensity of turbulence of the fluid phase even in dilute suspensions (Tsuji and Morikawa, 1982a). The nature of dilute suspensions is outlined in Table 9.1.

9.1.2 Dense Suspensions and Analogs

As the density increases, the effects of particle-particle interaction, the damping of fluid motion by particles (Baw and Peskin, 1971), and the interactions between the dense and the dilute phases (Soo, 1987) have to be treated by relating both averaged and statistical properties of the phases. A dense particulate suspension in a fluid at a low velocity, except in a very dense liquid slurry, is characterized by its density stratification and unsteady motions as has been described in a number of previous studies (Wen and Simon, 1959; Graf and Acaroglu, 1972; Takahashi et al., 1989; to cite a few). This is because, at high concentration of solid particles, instability leads to the

TABLE 9.1 *Comparison of dilute and dense suspensions.*

	Dilute suspension	Dense suspension
Relative motion between particles	Large	Small
Particle-particle interaction	Weak	Strong
Particle diffusivity, D_s	Large	Small
Apparent viscosity of particle phase	Due to particle-fluid interaction $\mu_{sf} = \alpha_s \bar{\rho}_s D_{sf}$	Due to particle-particle interaction $\mu_{ss} \propto 1/D_s$
Flow regime in application	Steady, turbulent	Pseudo-laminar to unsteady and stratified
Motion above minimum transport velocity	Stable	Stable in liquid slurry unstable in gaseous suspension
Analogy in molecular system	Rarefied gas flow	Molecular theory of liquids

formation of interacting phases of dilute and dense suspensions. A wavy interface between them was observed by Rossetti (1983). A clear delineation of various stages of development from gravitating pipe flow of a dilute suspension to a dense suspension with a wavy interface by Tsuji and Morikawa (1982) is shown in Figure 9-1.

Table 9.1 gives a comparison of the nature of a dense suspension to that of a dilute suspension. One notes that for the case of small relative motion between the phases in a dense suspension, strong interactions within the particle phase can be accounted for in formulating transport properties.

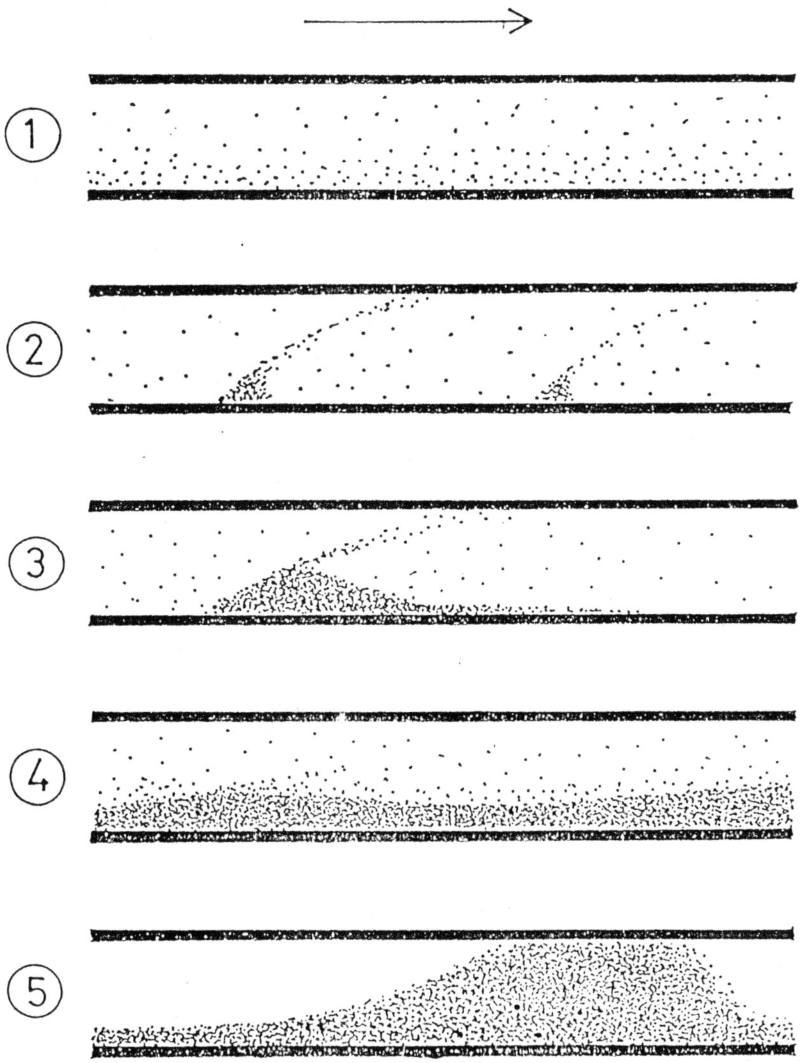

FIGURE 9-1 *Development from gravitating dilute suspension to flow with a dense phase with a wavy interface in a pipe (Tsuji and Morikawa, 1982; by permission of Pergamon Press PLC).*

In a dense gas-solid suspension in horizontal pipe flow, the statistical character of random motion of solids is notable; the correlations of velocity fluctuations and density fluctuations are measurable with a capacitive flowmeter and, for a given system, have been used as the basis for non-intrusive measurements (Bobis et al., 1986; Beck, 1981). Recognition of these fluctuations is also the basis of measurement of particle velocity in pneumatic transport of coal using the cross-correlation technique based on the triboelectric effect (Mathur and Klinzing, 1984). These studies have shown that fluctuation in density and velocity exist in dense suspensions of solids averaged over a pipe cross-section. In a preliminary experimental study, Soo and Mei (1987) determined the distributions of mass flows and their correlations and spectrum over a pipe section. The flow of a dense suspension of particles was shown by Konrad and Davidson (1984) to be analogous to gas-liquid flow, who concluded a corresponding empirical relation for bubble velocity as in a gas fluidized bed and a means for predicting the size of bubbles and slugs is needed. Where gravity effect is prominent, Roco and Shook (1984) showed that the vertical distributions in horizontal pipe flow are essentially two-dimensional; this was confirmed by Soo and Mei (1987) from radial traverses at various inclines with respect to the direction of gravity. Present chapter on recent studies demonstrates the feasibility of quantifying the nature of fluctuating motions in a dense suspension in horizontal flow using instrumentation and methods developed for such a purpose.

Transport properties of dense suspensions, that is, their viscosity and diffusivity were shown to have liquid (molecular) like behavior due to strong particle-particle interaction. The statistical nature of the flow of the particulate suspension is seen through the correlations of velocity and density leading to diffusion fluxes and Reynolds stresses. The viscosity due to collisional particle-particle interactions was shown to be given by (Soo, 1967, 1987):

$$\mu_{ss} \cong C_\mu \alpha_s^2 d^2 \bar{\rho}_s |\partial u_s / \partial r| \tag{9.1}$$

which is also validated by Ding and Gidaspow (1990) from solving the kinetic transport equation. The diffusivity is given by:

$$D_s \cong C_d (\bar{\rho}_f / \bar{\rho}_s)(d^2 / \alpha_s^3) |\partial u_s / \partial r| \tag{9.2}$$

where C_μ and C_d are empirical coefficients related to real particle behavior such as nonsphericity, roughness, surface contamination, etc., α_s is the volume fraction of particles, d is particle diameter, $\bar{\rho}_s$ is the density of the particle material, $|\partial u_s/\partial r|$ is the shear rate of particle motion, $\bar{\rho}_f$ is the density of the fluid. Therefore, different from the case of dilute suspensions, the dense suspension has $\mu_s \propto 1/D_s$; this has been demonstrated in the experimental results. One notes that most of the theoretical studies on cohesionless solid particles (Savage and McKeown, 1983; Shen and Ackerman, 1982) reached relations similar to Equation 9.1, that is, shear thickening characteristics due to collisional interaction, while frictional interactions at low shear rate leads to shear thinning characteristics (Roco et al., 1990). This type of relation shows that the viscosity of the mixture is primarily independent of the viscosity of the fluid phase and is applicable to a gas-solid suspension. When applied to liquid slurries, computer modeling (Ma and Roco, 1988; Brady, 1989) accounts for the shear thinning behaviors, with the high shear limit determined by Sengun and Probstein (1989).

Experimental approach to the study of particulate suspensions has been viewed from the triangular relation of cloud density (ρ_s), velocity (u_s) and mass flow ($\rho_s u_s$) of the particle phase (i.e., determination of a third quantity from measurement of any two of these quantities; Soo, 1982). Measurements of their local averages and fluctuations as well as their mean values have been made possible, together with air velocity and its distribution (Soo et al., 1989, 1989a, 1989b). Improved and newly developed instrumentation have made possible the measurements of phase velocities and instantaneous measurements of mass flows and densities of the particulate phase; computerized data system has made possible the determination of statistical correlations of properties. The latter enables us to quantify flow regimes of dense suspensions. From the measured fluctuations, autocorrelations and spectra, and cross-correlations have been determined.

9.2 BASIC CONSIDERATIONS

Theoretical considerations of horizontal pipe flow of a dense suspension include the condition of steady motion with sustained density gradient, the possibility of pure stratified flow and the origin of wavy stratified motion. These models, together with correlations of fluctuating motion and transport properties provide reference conditions for our experimental study.

9.2.1 Balance of gravity settling by shear lift and diffusion

In the case of horizontal pipe flow, the shear lift force acting on particles per unit mass is given by (Saffman, 1965):

$$f_L = [3(6.46)/4\pi](\bar{\rho}_f/\bar{\rho}_s)(U_o R/\nu)^{-1/2}(2R/d)(U_o/R)^{1/2}\Delta u|\partial u/\partial y|^{1/2} \qquad (9.3)$$

where R is the pipe radius, ν is the kinematic viscosity of the fluid, U_o is the mean or reference axial velocity of the fluid and $\Delta u = u_f - u_s$, the difference in the axial velocities; y is the coordinate through the vertical diameter. This, together with diffusion of particles given by Equation 9.2, act to maintain a steady density distribution of solid particles against the effect of gravity with the condition (Soo, 1987):

$$(dg/U_o^2)(U_o R/\nu)^{1/2} \leq 3.084(\bar{\rho}_f/\bar{\rho}_s)[1 - (\bar{\rho}_f/\bar{\rho}_s)]^{-1} R U_o^{-2}\Delta u|\partial u/\partial y| \qquad (9.4)$$

for gravitational acceleration g. It is seen that for a gas-solid suspension, $(\bar{\rho}_f/\bar{\rho}_s) \sim 10^{-3}$, large U_o is needed to attain steady flow even for very small particles. For a liquid-solid suspension, however, $\bar{\rho}_f/\bar{\rho}_s \sim 1$, for $d = 0.1$ mm, $R = 0.28$ m, $\nu = 10^{-6}$ m^2/s, the above condition is satisfied for $U_o = 1.6$ m/s, and $\Delta u/U_o = 0.1$ and $\partial(u/U_o)/\partial(y/R) = 1.5$; the two-dimensional analysis followed Roco and Shook (1987).

9.2.2 Steady stratified flow

When horizontal flow is stratified to an upper dilute layer and a bottom dense layer, a steady density distribution in the dilute phase is maintained by diffusion against gravity in the turbulent boundary layer of the dilute phase at the interface of the two regimes. The condition for steady flow was shown to be (Soo, 1987):

$$0.04\, U_o > [1 - (\bar{\rho}_f/\bar{\rho}_s)]g/F_{sf} \qquad (9.5)$$

where F_{sf} is the relaxation time constant for momentum transfer from fluid to particle, which, for the dilute layer is given by $F_{sf} = F^*18\mu/d^2[\,\bar{\rho}_s+(\bar{\rho}_f/2)]$, where F^* is the correction for deviation from Stokes flow. Note that $0.04U_o$ is nearly the turbulent intensity in the dilute phase. Again a nearly pure stratified flow is achieved in liquid-solid suspension but not a gas-solid suspension. For $U_o \sim 10$ m/s, F_{sf} has to be greater than 25, or $d < 100$ μm is required for pure stratified flow. This is an upper limit, experiments have shown that the above condition does not assure steady stratified flow and wave motion would occur.

9.2.3 Wave motions in Stratified Flow

Basic to our effort in quantifying unsteady motion in a dense suspension is the study of phase interactions. The theoretical study has extended from Soo (1987) on particle-particle interaction and flow stratification leading to a dense phase of solids along the bottom of the horizontal pipe with wave motions. The latter was validated, in part, by the theoretical study of de Crecy (1986) on stratified liquid-vapor flow.

The nature of fully developed stratified wave motion in pipe flow can be seen from a simplified analysis of fluctuations from mean plug flow where the average pressure gradient along the pipe is given by the average wall friction. The volume or time averaged continuity and momentum equations (Soo, 1989), when applied to the one-dimensional motion of phases of constant material density in a duct of depth D of infinite length can be expressed in terms of:

$$\partial\alpha_k/\partial t + (\partial/\partial z)\alpha_k u_k = 0 \qquad (9.6)$$

the continuity relation, and the momentum equation takes the form:

$$(\partial/\partial t)\alpha_k u_k + (\partial/\partial z)\alpha_k u_k^2 + (\varepsilon_k/\overline{p}_k)(\partial/\partial z)[\alpha(1-\alpha)(\Delta_g - \Delta_p)]$$

$$= -(\varepsilon_k/\overline{p}_k)\tau_i - (c_k \chi_f \tau_{wk}/A_k \overline{p}_k) \qquad (9.7)$$

where t is the time, α_k is the volume fraction of phase k of density $\overline{p}_k$ at velocity u_k in the z-direction along the duct with the volume fraction of the dense (subscript p) and the dilute phase (g) separated by an interface given by $y_i(z, t)$ as shown in Figure 9-2. The coefficients ε are given by:

$$\varepsilon_g = 1 \qquad \text{and} \qquad \varepsilon_p = -1$$

for α defined according to: (9.8)

$$\alpha_g = \alpha \qquad\qquad \alpha_p = 1 - \alpha$$

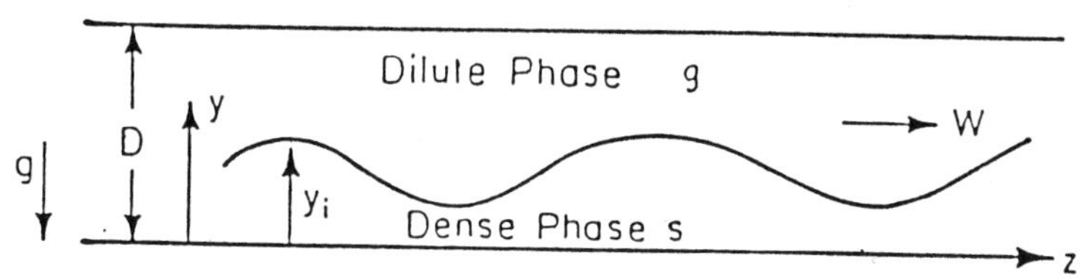

FIGURE 9-2 *Diagrammatic representation of phase interaction in duct flow.*

In addition, A_k is the cross-sectional area and c_k is a dimensionless coefficient of friction of phase k, χ_f is the frictional perimeter, τ_i is the interfacial shear stress per unit volume projected on the tube axis, τ_{wk} is the wall shear stress per unit volume projected on the tube axis for phase k, Δ's account for the field force and for the simple case in Figure 9-2,

$$\Delta_k = (1/D)\int_0^D (-p_k g)(y - y_i)\, dy \qquad (9.9)$$

For the dense phase,

$$p_k = \begin{cases} 0 & y > y_i \\ p_p & y \le y_i \end{cases}$$

where y_i is the position of the interface

$$\Delta_p = \bar{\rho}_p g y_i^2/2D \tag{9.10}$$

For the dilute phase,

$$\Delta_g = \rho_g g\,(D - y_i)^2/2D \tag{9.11}$$

When applied to the dilute phase, $k = g$ in Equations 9.6 and 9.7; now we drop the subscripts g; Equation 9.6 can be rearranged as:

$$\alpha\, \partial u/\partial t = -\int_{-\infty}^{\infty} (\partial^2 \alpha/\partial t^2) dz + F(t) - u(\partial\alpha/\partial t) \tag{9.12}$$

and Equation 9.7 then may take the form:

$$F(t) - \int_{-\infty}^{\infty} (\partial^2 \alpha/\partial t^2) dz - 2u(\partial\alpha/\partial t) - u^2(\partial\alpha/\partial z)$$

$$+ \bar{\rho}^{-1}(\partial/\partial z)[\alpha(1-\alpha)(\Delta_g - \Delta_p)] = -(\tau_i/\bar{\rho}) - (C_g \chi_f \tau_w/\bar{\rho}A) \tag{9.13}$$

For values of Δ_g and Δ_s in Equations 9.10 and 9.11, since $y_i\,/D = 1 - \alpha$, the term in Equation 9.13, namely,

$$(\partial/\partial z)(\Delta_g - \Delta_s)\alpha(1-\alpha) = (g/2D^3)[(4(\bar{\rho} + \bar{\rho}_p)y_i^3 - 3(3\bar{\rho} + \rho_p)Dy_i^2$$

$$+ 6\bar{\rho}D^2 y_i - \bar{\rho}D^3](\partial y_i/\partial z) \equiv F_1(y_i)(\partial y_i/\partial z) \tag{9.14}$$

Equation 9.13 can now be rearranged to:

$$(u/D^2)(y_i + D)(\partial y_i/\partial t) + [(u^2/D) - (F_1/\bar{\rho})](\partial y_i/\partial z)$$

$$= F(t) - \int_{-\infty}^{\infty} D^{-1}(\partial^2 y_i/\partial t^2)\, dz + (\tau_i/\bar{\rho}) + (C_g \chi_f \tau_w/\bar{\rho}A) \tag{9.15}$$

and the corresponding terms can be redefined to give:

$$C_t(\partial y_i/\partial t) + C_z(\partial y_i/\partial z) = C_c \tag{9.16}$$

with C_t, C_z, and C_c correspond to terms in Equation 9.15. Noting that only when $C_t = 0$, and C_z = constant, that steady flow is possible. In general, we have the following cases:

(a) $C_c = 0$ and y_i is in the form of wave motion, the integral term in Equation 9.15 vanishes, and $F(t)$ represents the average of friction of the wetted wall of the dilute phase and that at the interface;

(b) C_z/C_t = const. = b, Equation 9.16 represents the wave motion of the interface, with b as the wave velocity. The effect of the cylindrical geometry is readily accounted for.

9.2.4 Correlations of Velocity and Density of Particles

Expressing in simplest terms, we denote the particle cloud density ρ_s and velocity u_s in terms of local (time) averages(<> quantities) and fluctuations (' quantities):

$$\rho_s = \langle\rho_s\rangle + \rho_s' \qquad (9.17)$$

$$u_s = \langle u_s\rangle + u_s' \qquad (9.18)$$

Averaging theorems give, for mass flow of the particle phase:

$$\langle\rho_s u_s\rangle = \langle\rho_s\rangle\langle u_s\rangle + \langle\rho_s' u_s'\rangle \qquad (9.19)$$

in terms of product of averages and averages of fluctuation. The last term of Equation 9.19 is the covariance of ρ_s and u_s. It is seen that as long as there are fluctuations in density and velocity, this covariance should not be dropped. However, available instrumentation (Sec. 9.3.2) permits the determination of mass flow in terms of averages and fluctuations as:

$$\rho_s u_s = \langle\rho_s u_s\rangle + (\rho_s u_s)' \qquad (9.20)$$

to give the averages and correlations in the form:

$$\langle(\rho_s u_s)^2\rangle = \langle\rho_s u_s\rangle^2 + \langle(\rho_s u_s)'(\rho_s u_s)'\rangle \qquad (9.21)$$

and autocorrelation of density in Equation 9.17 is included in:

$$\langle\rho_s^2\rangle = \langle\rho_s\rangle^2 + \langle\rho_s'\rho_s'\rangle \qquad (9.22)$$

Substitution of Equations 9.17 and 9.18 into Equation 9.21 and dropping all triple correlations and higher, the average of particle velocity fluctuation can be expressed in the form:

$$\langle\rho_s\rangle^2\langle u_s' u_s'\rangle = \langle(\rho_s u_s)'(\rho_s u_s)'\rangle - \langle u_s\rangle^2\langle\rho_s'\rho_s'\rangle$$

$$- 2\langle\rho_s\rangle\langle u_s\rangle\langle\rho_s' u_s'\rangle + \langle\rho_s' u_s'\rangle^2 \qquad (9.23)$$

This equality provides us with a means of determining the intensity of particle motion in the axial direction and checking all the measurement of averages and fluctuations; the measurement of $\langle\rho_s' u_s'\rangle$ is needed. This has been accomplished by a recent development of simultaneous measurements of mass flow and density (Zhu et al., 1991). One notes that the normalized covariance $\langle\rho_s' u_s'\rangle/\langle\rho_s\rangle\langle u_s\rangle$ is the error of using isokinetic sampling (Soo et al., 1969) in the calibration of flow with fluctuation.

9.3 EXPERIMENTAL SYSTEM

Our experimental study has utilized the test loops of 51 mm (Figure 9-3) and 127 mm pipe diameters (illustrated in Soo et al.,1989a) which have been modified for study of dense suspensions in horizontal pipe flow. These test loops are unique in making possible testing of high flow rate of gas-solid suspensions with minimum investment of solids. For example, in the case of the 127 mm diameter pipe, to test with a solid flow rate of 6 kg/s (12.6 ton/hr) at 20 m/s for 10 minutes will require 2 bins of 4-ton capacity in a conventional system; our cyclone-standpipe recirculating system will only require 50 kg of solid particle in a standpipe of 3 m height for continuous testing without time limit. From the practical point of view of testing dense phase pneumatic transport of a specific material, such a system makes possible sustained tests with a limited amount of sample in a relatively large pipe.

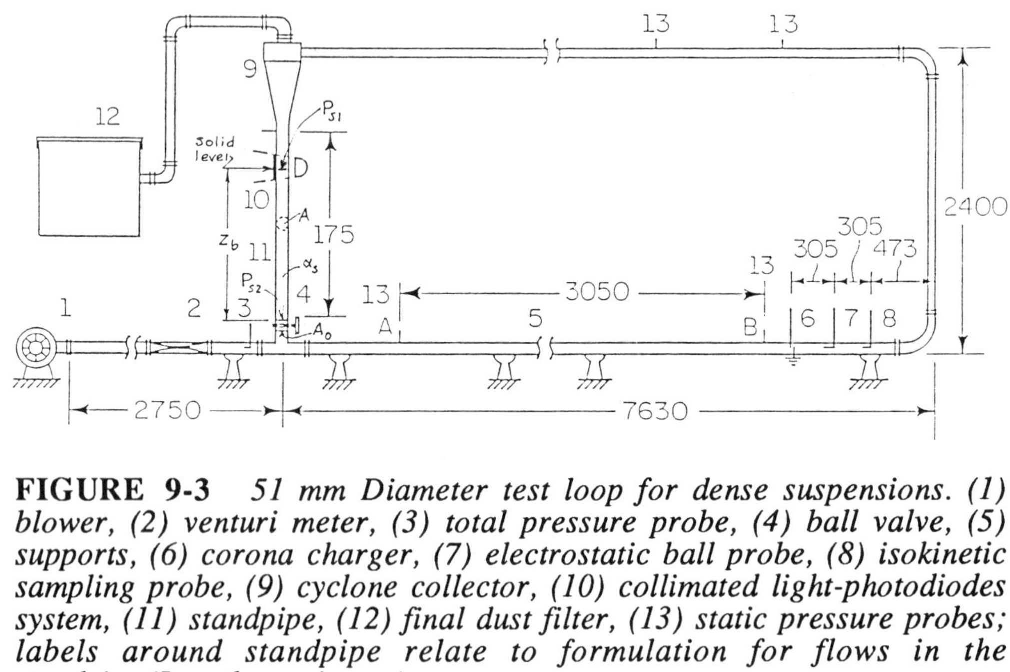

FIGURE 9-3 *51 mm Diameter test loop for dense suspensions. (1) blower, (2) venturi meter, (3) total pressure probe, (4) ball valve, (5) supports, (6) corona charger, (7) electrostatic ball probe, (8) isokinetic sampling probe, (9) cyclone collector, (10) collimated light-photodiodes system, (11) standpipe, (12) final dust filter, (13) static pressure probes; labels around standpipe relate to formulation for flows in the standpipe.(Lengths are in mm).*

9.3.1 Leakage flow of air in the standpipe

Satisfactory operation of the cyclone standpipe system depends on a small leakage flow up the standpipe. This leakage flow was taken into account in our design of the system. A delineation of the procedures at this point may benefit other users of this type of test system. Referring to the quantities identified around the standpipe in Figure 9-3, the relation of Li et al (1982) for flow in the standpipe can be expressed as, for non-cohesive particles:

$$\alpha_s \bar{\rho}_s U_s \doteq C_f (A_o/A)(2\bar{\rho}_s \alpha_{smf} \Delta P_s)^{1/2} = m^* \bar{\rho}_f U_o \qquad (9.24)$$

from the Bernoulli relation; where C_f is an orifice coefficient (~ 0.07), A_o is the area of the valve opening which can be varied to give various mass flow ratios m^* for mean air velocity U_o, average volume fraction solid α_{so} throughout the test section, A is the cross-sectional area of the standpipe, α_s is the volume fraction solid in the standpipe, α_{smf} is volume fraction of solids at minimum fluidization(~ 0.49). Equation 9.24 gives the solid volume fraction α_s in the standpipe as called for by m*. Chong et al. (1987) further proposed that the pressure head of solids ΔP_s as:

$$\Delta P_s \sim \alpha_s \rho_s g z_b \tag{9.25}$$

due to gravity in the standpipe with height z_b of solids. The pressure drop of the fluid in the standpipe: $\Delta P_f = P_{s2} - P_{s1}$; a pressure difference being measurable (Figure 9-3) and is determined by the flow condition and length of the flow loop. Based on the averaged momentum equation of the fluid in the standpipe (Soo, 1989):

$$0 = (1 - \alpha_s)\rho_f F_{fs}(U_s - U_f) - (dP_f/dz) + (1 - \alpha_s)\bar{\rho}_f g \tag{9.26}$$

where subscript f denotes fluid (air), F_{fs} is the relaxation time constant of momentum transfer from particles to fluid, and

$$(1 - \alpha_s)\rho_f F_{fs} = \alpha_s \bar{\rho}_s F_{sf} \tag{9.27}$$

with the time constant for momentum transfer from fluid to particle given by (Soo, 1967):

$$F_{sf} = 150(1 - \alpha_s)^2 \alpha_s^2 \mu / [d^2(\rho_s + \bar{\rho}_f/2)] \tag{9.28}$$

when the inertia force can be neglected. With pressures P_{si} measured at their respect stations, Equation 9.24 gives U_s and Equations 9.25 and 9.26 gives the interstitial air velocity U_f in the standpipe and the corresponding superficial velocity:

$$\begin{aligned} U_{fs} = (1-\alpha_s)U_f = &-(d^2/150\mu\alpha_s^3)(1-\alpha_s)^{-1}(\Delta P_f/z_b) \\ &+ (m^*\bar{\rho}_f U_o/\alpha_s \rho_s)(1-\alpha_s) \end{aligned} \tag{9.29}$$

subjecting to the condition that $|U_f|$ be less than the terminal velocity of the particles.

For 100 μm (88-105 μm) glass particles at mass flow ratios of 3, 6, 9 and 12, with a mean air velocity of 10 m/s, and $\bar{\rho}_s$ = 2500 kg/m^3, ρ_f = 1.3 kg/m^3, the above relations give the superficial air velocity in the standpipe U_{fs} in Table 9.2.

TABLE 9.2 *Typical operating condition of a standpipe.*

Mass flow ratio	z_b, m	α_s	ΔP_f, Pa	U_s, m/s (down)	U_{fs}, m/s (up)
3	1.5	0.6	2,770	0.026	0.06
6	1.45	0.55	4,450	0.057	0.10
9	1.3	0.5	5,550	0.094	0.14
12	1.05	0.47	6,300	0.140	0.22

When the level of the investment of solids in the standpipe is tall enough (to give large z_b), the air velocity in it could be downward with the solids. High level of z_b to give large m^*, however, demands increased blower power. In any case, the leakage flow of air up the standpipe can be determined and controlled.

9.3.2 Instrumentation

Central to the instrumentation and data systems was the use of analog-to-digital board linked to an IBM-AT computer with a sampling frequency of 4000 Hz. The following instruments were used in the measurements:

9.3.2.1 Overall flow rate of solids

The overall flow rate was given by the initial rate of rise of level of solids in the standpipe at the shut down of the feed valve (location 4 in Figure 9-3) into the pipe loop (Soo et al., 1989a). The photodiode array (EG&G Vactec VT-241 photodiodes) - light source system is shown in Figure 9-4. Since a change in resistance occurs in a photodiode when its incident light intensity changes, the output voltage from each of the 8 photodiode circuits gave the rate of change of the level of solids through a record as shown in Figure 9-5 and extrapolated to the initial rate. This measurement provides an overall check of local measurements of mass flow by isokinetic sampling (Soo et al., 1969) and integration over the pipe.

9.3.2.2 Electrostatic ball probe and local mass flow

Common to several of the local measurements was the electrostatic ball probe (Figure 9-6) (Cheng and Soo, 1970). The latter measures mass flow of particles in terms of its probe current produced by electrostatic charge separation from impaction and, ideally, single scattering. When applied to a dense suspension, the number of simultaneous impacts by particles on a ball probe is given by:

$$N_c = (\pi^{1/3}6^{2/3}/4)\alpha_s^{2/3}[(d_B/d)+1]^2 \quad = \quad n^{-1/3}/\lambda_B \qquad (9.30)$$

where d_B is the diameter of the ball probe, n is the number density of particles, λ_B is the mean free path of particle to probe collisions, assuming

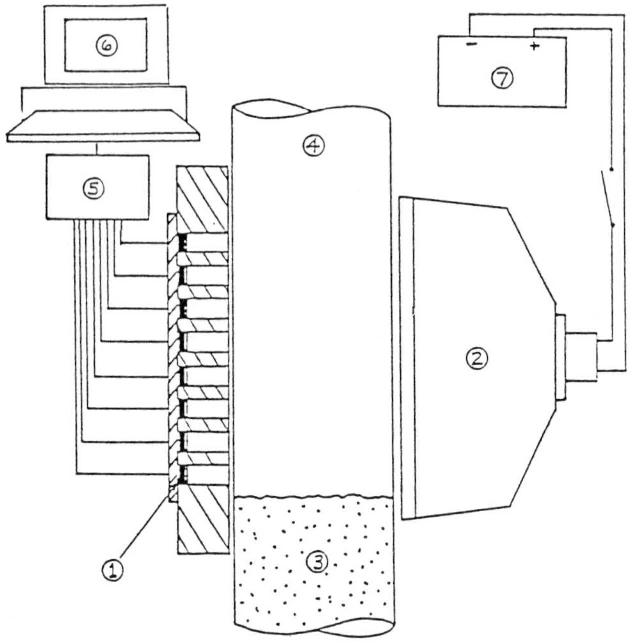

FIGURE 9-4 *Photodiodes-light source system for measuring bulk flow rate of particles. (1) Photodiode, (2) Collimated light source, (3) Glass particles, (4) Standpipe, (5) A/D board, (6) Computer, (7) 12-volt battery.*

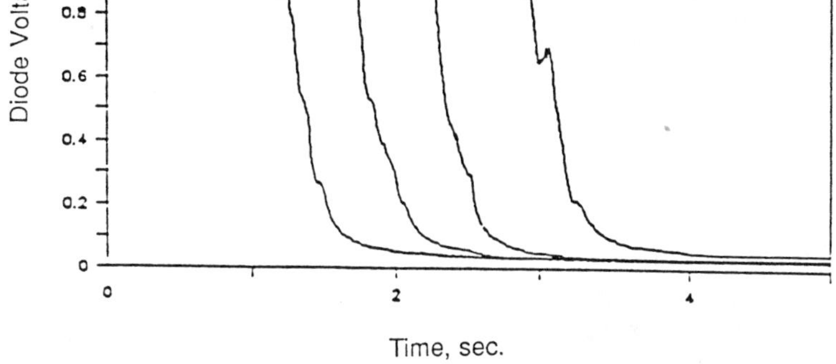

FIGURE 9-5 *Plot of first four photodiode signals used to obtain initial rate of particle level rise in the standpipe.*

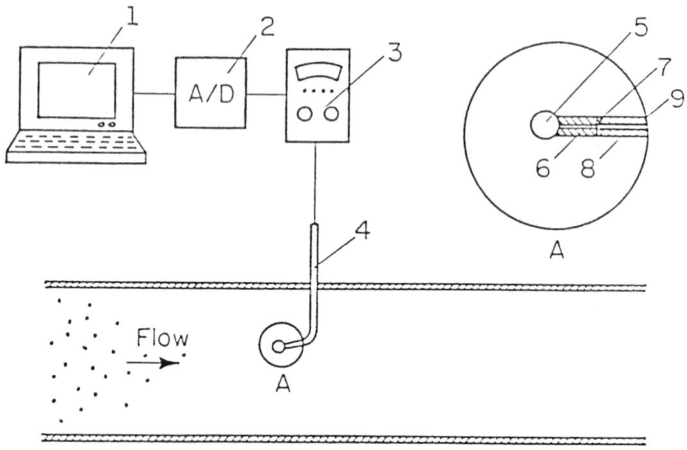

FIGURE 9-6 *Electrostatic ball probe system. (1) IBM PC-AT computer, (2) Analog-to-digital converter, (3) Electrometer, (4) Ball probe, (5) Bearing ball, (6) Epoxy, (7) Wire, (8) Stainless steel tubing, (9) Insulator.*

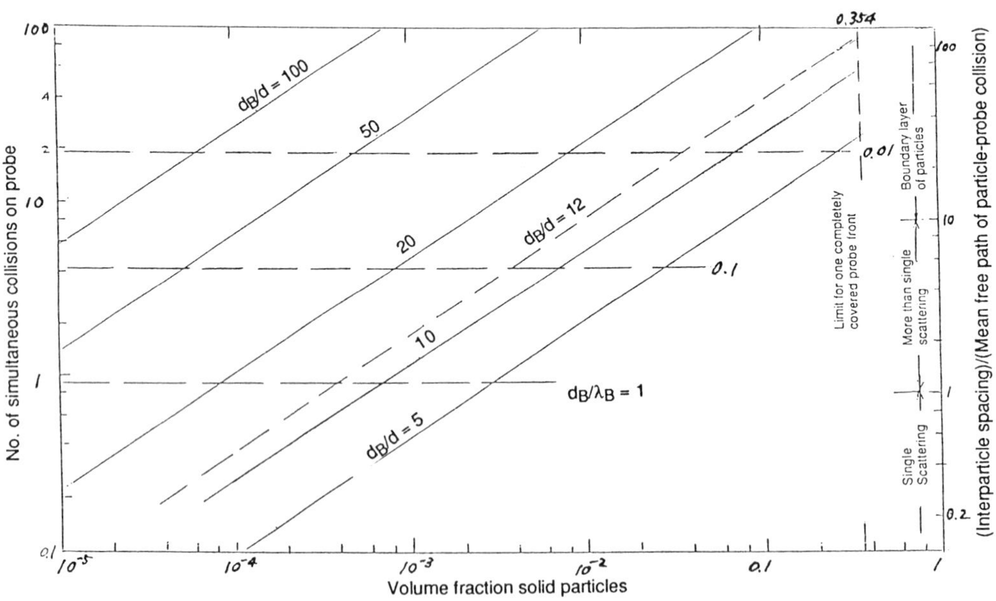

FIGURE 9-7 *Particle flow ranges over an electrostatic probe.*

monodispersed particle diameter d. Figure 9-7 shows flow conditions ranging from single scattering to "boundary layer flow" of particles. For our tests with 100 and 200 μm particles, probe diameters of 2.4 mm and smaller were used to assure single scattering over a wide range. An improved probe theory for the three ranges in Figure 9-7 accounts for local velocity effect, probe current saturation due to multiple scattering and sliding of particles over the spherical surface in the dense range. The basic relation in Soo (1989) is now modified to give a semi-empirical relation:

$$i_B = C_{B1} u_s^{3/5} [\exp(-C_{B2} \rho_s / u_s)] (\rho_s u_s) \qquad (9.31)$$

where $\rho_s u_s$ is the mass flow of solids and the probe constants C_{B1} is given by:

$$C_{B1} = 18.55 \eta_{Bs} h_{eBs} (k_B + k_s)^{4/5} (1 + r^*)^{4/5} \bar{\rho}_s^{-1/5} (V_{so} - V_B) d_B^2 \qquad (9.32)$$

and C_{B2} is an empirical constant accounting for the effect of particle concentration, velocity, material properties of particles (s) and probe (B) and their diameters d and d_B. Other terms in Equation 9.32 are: fraction impacted η_{Bs} (nearly 1), charge transfer coefficient h_{eBs}, elastic parameter $k = (1-\nu)/\pi E$, ν the Poisson ratio, E the modulus of elasticity of material, r^* the coefficient of restitution of impact, V_{so} is the reference potential of the particles, and V_B the potential of the probe (nearly zero). Since C_{B1} is given by collision of ideal spheres, its determination is, in practice, by calibration with isokinetic sampling and is treated as an empirical constant.

Figure 9-8 is a plot of (Probe output)/$u_s^{2.6}$ vs. (mass flow)/u_s^2, thus making the velocity effect implicit, for glass particles of various sizes on a stainless steel ball probe of 2.4 mm diameter; the effect of particle concentration is readily seen. The probe output is in mV for probe current of the magnitude of 10^{-8} A through an equivalent resistance to ground of 1 megohm. With this conversion, the magnitude of the probe constants corresponding to those in Equation 9.31 are: $C_{B1}' = 0.022$; $C_{B2} = 0.05$, 0.06, and 0.075; for 100, 200, and 400 μm glass particles. The data scatter is a result of calibration based on instantaneous probe outputs by averages in isokinetic sampling.

Local mass flow of particles and its fluctuation were measured with the electrostatic ball probe. This gives $\langle \rho_s u_s \rangle$ and $\langle (\rho_s' u_s')(\rho_s' u_s') \rangle^{1/2}$, the root-mean-square fluctuations, correlations and power spectrum and integral time scale (Soo et al., 1989b).

9.3.2.3 Local average particle velocity distribution

Cross-correlation of a propagation of a disturbance in the form of electric output of two electrostatic ball probes at identical radial locations over a given distance along the pipe (Figure 9-9) gave the local average particle velocity (Lucht and Soo, 1990). The electrical signals x and y at these locations arose through charge separation at the ball probes and are correlated according to (Bendat and Piersol, 1986):

$$R_{xy}(\tau) = \lim_{T \to \infty} \frac{1}{T} \int_0^T x(t) y(t+\tau) dt \qquad (9.33)$$

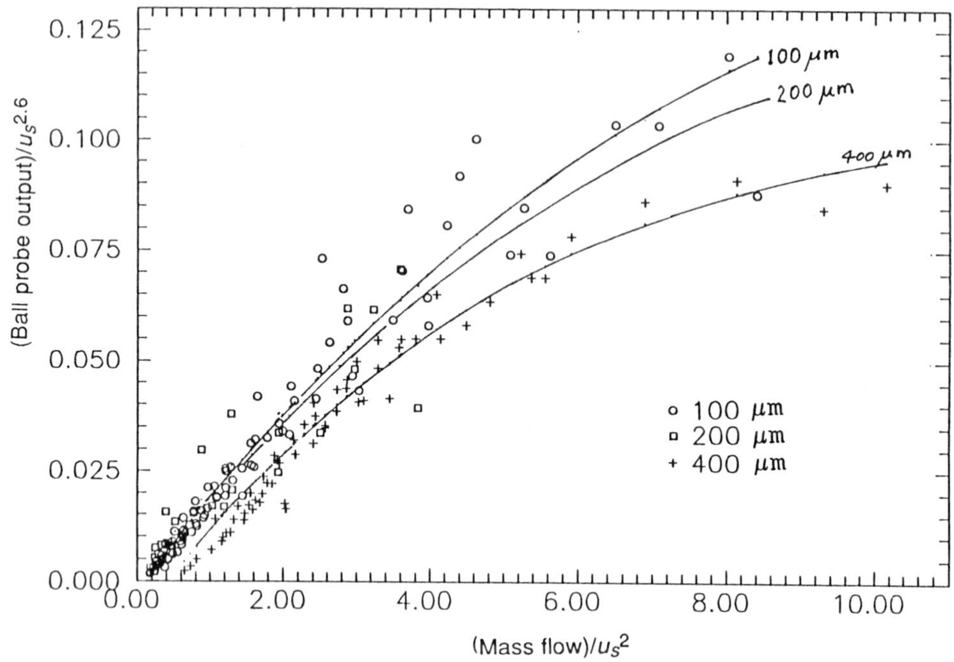

FIGURE 9-8 *Typical calibration curves for an air suspension of glass particles on an electrostatic probe formed by a 2.4 mm sphere.*

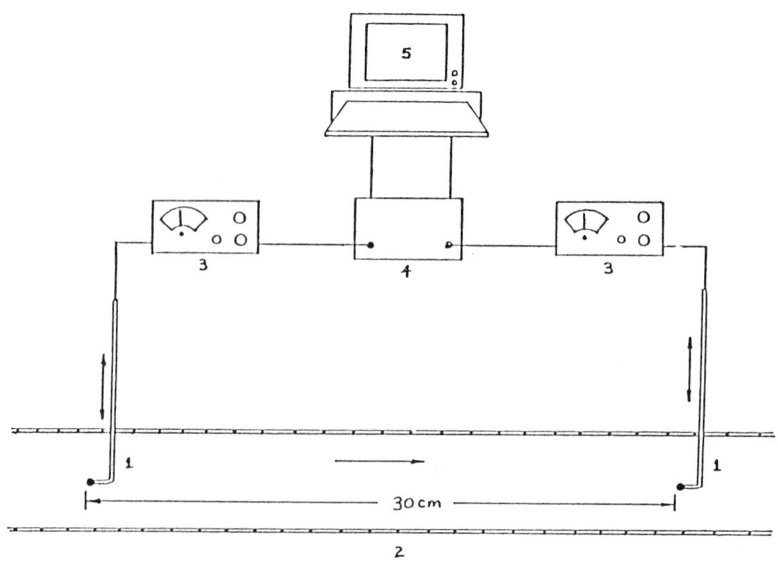

FIGURE 9-9 *Electrostatic ball probes positioned for measuring local mean particle velocity by cross-correlation. (1) 1/8" diameter electrostatic ball probe, (2) Test section, (3) Electrometer, (4) A/D board, (5) Computer.*

An example of the output plot of R_{xy} vs. lag time is shown in Figure 9-10, with local average velocity $<u_s>$ of particles given by the distance between the probes and the lag time at maximum R_{xy}. An alternate method based on the same statistical principle was through correlating attenuated signals from photocell responses to dual laser beams produced by a beam splitter (Zhu et al., 1991); improved consistency was achieved and development continues.

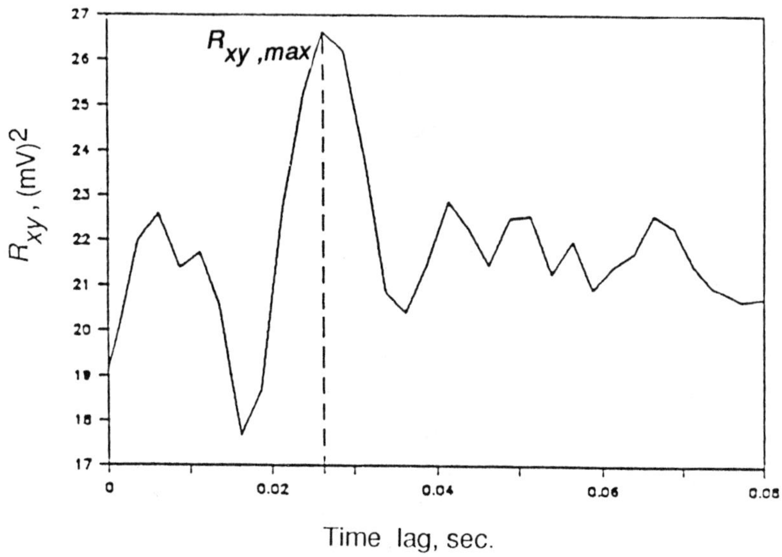

FIGURE 9-10 *Plot of cross-correlation vs. lag time for determining local mean particle velocity.*

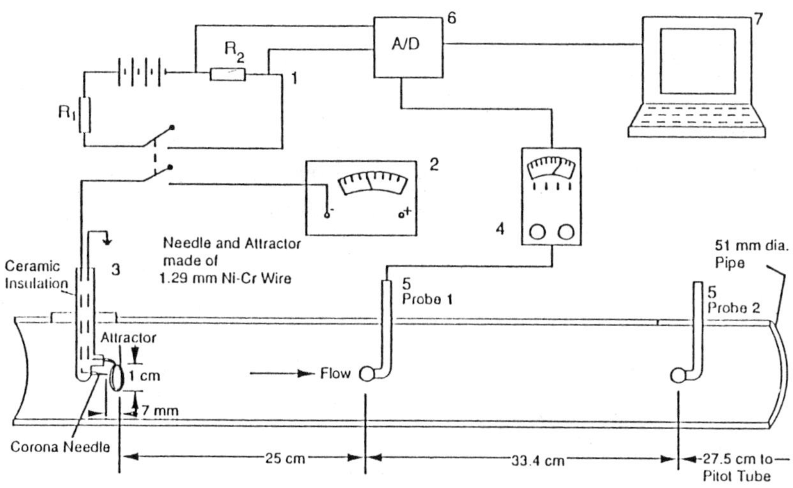

FIGURE 9-11 *Corona discharge and probe system installed in a pipe. (1) Synchronizing circuit, (2) High voltage power supply, (3) Corona charger assembly, (4) Electrometer, (5) 3.16 mm dia. electrostatic ball probe, (6) A/D converter, (7) IBM PC-AT computer.*

9.3.2.4 Local velocities of phases

Local phase velocities were given by the corona charger-ball probe system. With the corona charger and the ball probe at identical locations over a given distance along the pipe (Figure 9-11), the different times of arrival of gas molecules (Δt_a) and particles (Δt_s) carrying different charges from an impulse corona discharge gave the velocities of the phases (Figure 9-12) (Soo et al., 1989). This provides a mutual check with $<u_s>$ from cross-correlation. This measurement for the velocity of the gas phase is needed because the Shapiro probe (Cheng et al., 1970) became erratic when applied to a dense suspension. The disadvantage of this system is that it gives instantaneous velocities, switching of high corona voltage (nearly 7 kV) to obtain average and fluctuations is not readily accomplished with the present system.

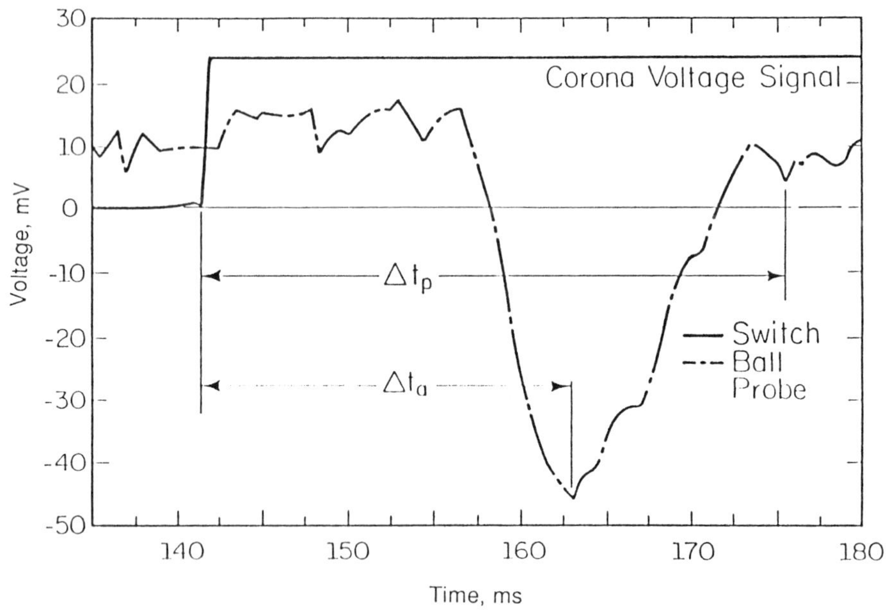

FIGURE 9-12 *Measurement of velocity of phases along center line of pipe (sampling frequency: 4 kHz). Δt_a is the time of arrival of ionized air; Δt_p is the time of arrival of particle cloud with modified charge.*

9.3.2.5 Local density of particles and its fluctuation

Local densities of particle clouds in a dense suspension were measured by attenuation of a laser beam (Lucht and Soo, 1990). This simple laser (10 mW HeNe) system has an output fluctuation of ±5% is useful for developing this technique. This development was necessary because current laser-Doppler velocimetry and laser holography have only been developed for the measurement of velocity and density of particles in a dilute suspension (Tsuji and Morikawa, 1982a, Schaefer et al., 1987). Figure 9-13 shows the alignments of components for traversing a light beam in a vertical cross section normal to the direction of gravity. This is valid inasmuch as pipe flow

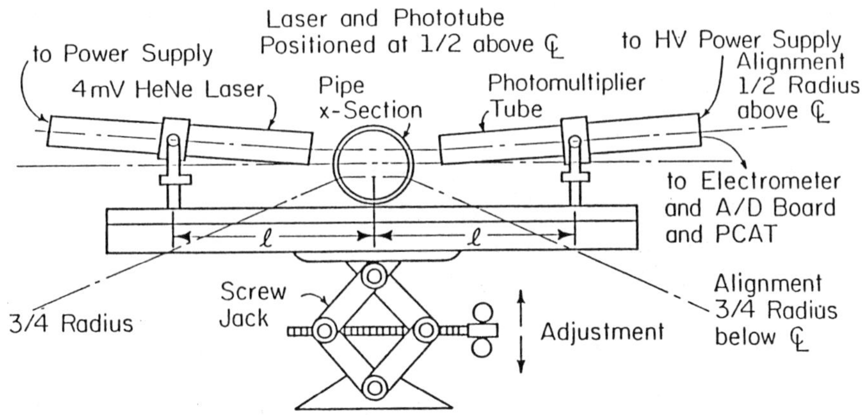

FIGURE 9-13 *Laser probe for density of particles aligned for measurements at 1/2 radius above center line of pipe and alignments for other heights over the pipe cross-section.*

of a particulate suspension under the influence of gravity has a nearly two-dimensional density distribution from theoretical considerations (Roco and Shook, 1984, 1987) and measurements of mass flow by radial traverses at various angles with respect to the direction of gravity (Soo and Mei, 1987). For given number density N_s of monodispersed particles, and negligible scattering of light by air, the laser beam intensity decreases from I from I_o

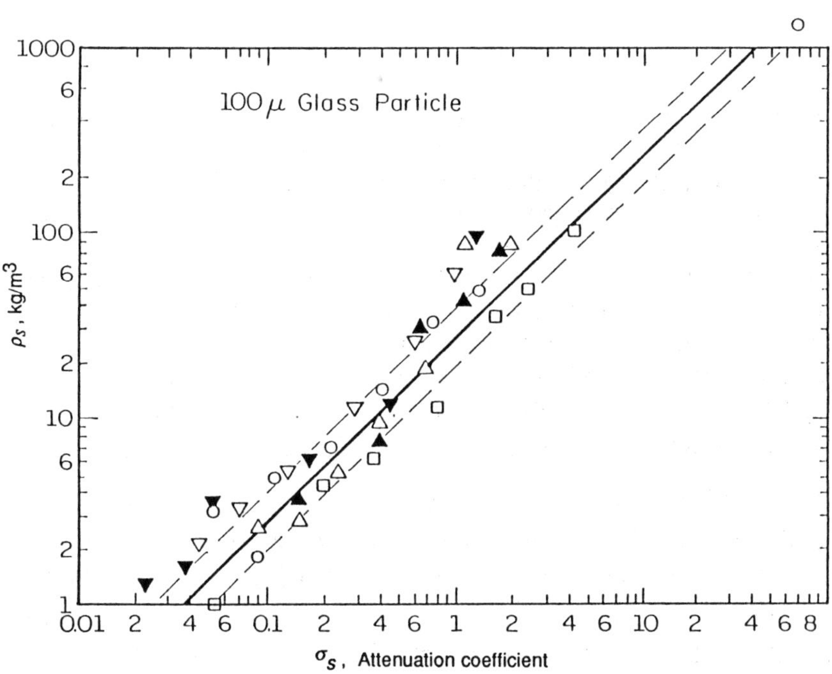

FIGURE 9-14 *Calibration plot of particle density vs. attenuation coefficient.*

over a distance L (beam length inside the cylindrical pipe) according to:

$$I(L) = I_o \ \exp(-\sigma_s N_s L) \tag{9.34}$$

and the attenuation coefficient σ_s is given by the product of particle scattering cross-section σ_s and N_s. The latter is related to density ρ_s of the particle phase by the particle mass m. With the linear relationship between the light intensity and the PMT voltage, Equation 9.34 can be expressed as:

$$\sigma_s = (-1/L) \ln [V(L)/V_o] \tag{9.35}$$

for voltage V_o at intensity I_o. A calibration plot of the log of particle density vs. the log of attenuation coefficient validated the assumption in Eq. (9.34), giving $\rho_s = 23.3 \pm 5.2\ \sigma_s$ as the best fit. A typical calibration based on mass flow of particles by isokinetic sampling and local average particle velocity based on cross-correlation is as shown in Figure 9-14; the highest particle density is obtained with a packed bed. The spread of data in Figure 9-14 was due largely to the calibration with isokinetic sampling which was based on long time average of fluctuating density. Once calibrated, however, its read-out maintained high degree of consistency. The laser output gives $<\rho_s>$ and $<\rho_s'\rho_s'>^{1/2}$, the root-mean-square fluctuations, correlations, and power spectrum and integral time scale.

9.3.2.6 Local phase velocities and particle density

Measurements were made by our Lexel laser-Doppler velocimeter and Aerometrics receiving system (Liljegren and Vlachos, 1990). This gives phase velocities, particle density and their fluctuations for the dilute phase, but more work is needed in modifying it for dense suspensions.

The primary standard is still the isokinetic sampling technique (Soo et al., 1969) which provides the calibration for the above electrostatic ball probe and the density probe based on laser attenuation, losing the last term of average mass flow fluctuation in Equation 9.20. The availability of advanced system such as the Aerometrics laser receiving optics does not improve the situation: For a collecting volume v and count of N passes of particles over time Δt, knowing the average velocity $<u_s>$, the local density is given by $N/(\Delta t)<u_s>A = NL/(\Delta t)<u_s>v$, and neither the effective area A in v normal to $<u_s>$, nor the characteristic path length L is clearly known for an independent determination of particle density ρ_s without basing the calibration on mass flow. Obtaining an independent primary standard for the measurement of local instantaneous density remains a challenge.

Measurement of normalized covariance identified in Section 9.2.4 was made by Zhu et al. (1991) by using the ball probe for instantaneous mass flow and the laser attenuation for particle density at close proximity. Results show that the normalized covariance of mass flow is always negative and tends to be small in the dense phase (below 1%) and large in the dilute phase (up to 5%); it is also small at high particle velocities.

One further notes that, in spite of the seemingly elaborate instrumentation schemes, we have barely accounted for all the variables in the flow of dense suspensions.

9.4 EXPERIMENTAL RESULTS

The experimental results and their deduction were illustrated with 100 μm (88 to 105 μm) particles. Tests were conducted at 7, 10, and 15 m/s mean air velocities at various mass flow ratios. Using the instrumentation in Section 9.3.2, and the data of local averages of mass flow, density, and velocities and fluctuations of mass flow and density (Soo et al., 1989, 1989a, 1989b; Lucht & Soo, 1990) were obtained as shown in Figures 9.15, 9.16 and 9.17 for 10 m/s mean air velocity. An autocorrelation is given by (Bendat and Piersol, 1986):

$$R_{xx}(\tau) = \frac{1}{T}\int_0^T x(t)x(t+\tau)dt \tag{9.36}$$

which, for discrete data, is given by computation of the product of sample data values:

$$R_{xx}(r\Delta t) = (1/N\text{-}r)\sum_{n=1}^{N-r} x_n x_{n+r} \tag{9.37}$$

The root mean square values are given by the root of $R_{xx}(0)$.

The power spectral density is given by the Fourier transform of the autocorrelation function:

$$S_{xx}(f) = 2\int_0^\infty [R_{xx}(\tau)\cos(2\pi f\tau)]d\tau \tag{9.38}$$

which, was calculated based on the fast Fourier transform algorithms according to, for $k = 0, 1, 2, \ldots, N/2$:

$$S_{xx}(f_k) = (2/N\Delta\tau)|X(f_k)|^2 \tag{9.39}$$

where

$$X(f_k) = \Delta t \sum_{n=0}^{N-1} x(n)\exp[-j2\pi kn/N] \tag{9.40}$$

The power spectral density also gives the dominating frequency and the maximum lag number M. The integral time scale (ITS) is given by:

$$\text{ITS} = \sum_{r=0}^{M} [R(r\Delta t)/R(0)]\Delta\tau \tag{9.41}$$

Figure 9.18 illustrates the autocorrelations and spectra of density across the pipe height and is representative. Figure 9.19 illustrates the integral time scales from correlations of both density and mass flow.

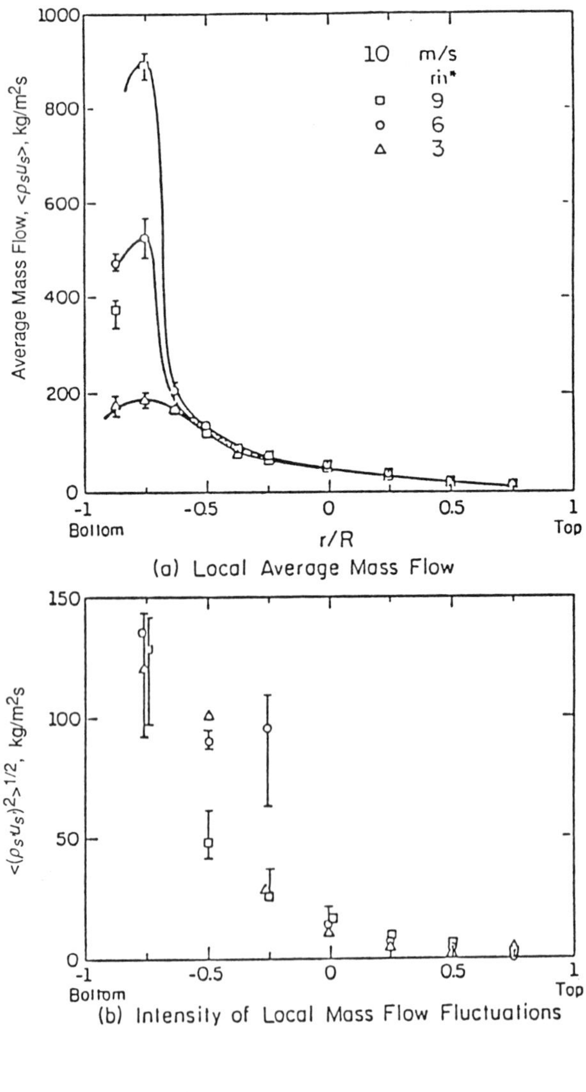

FIGURE 9-15 *Distribution of local mass flow and its fluctuation at 10 m/s mean air velocity and mass flow ratios of 3, 6, and 9 .*

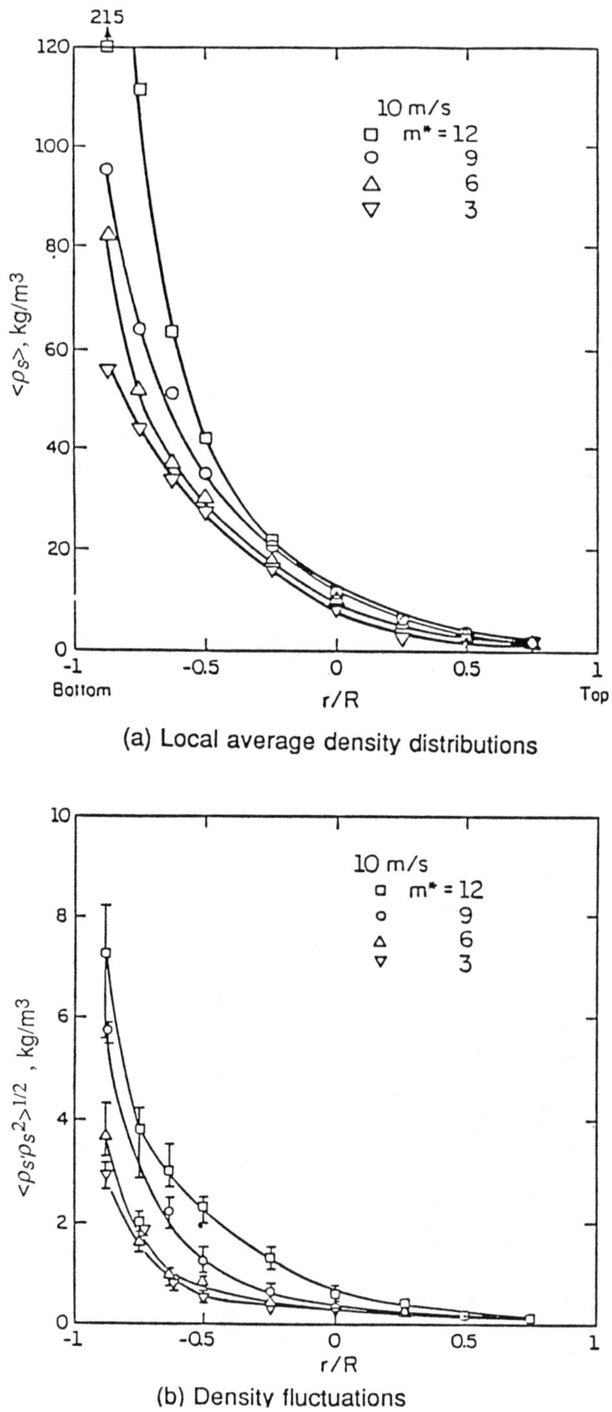

FIGURE 9-16 *Distribution of local density and its fluctuation at 10 m/s mean air velocity and various mass flow ratios.*

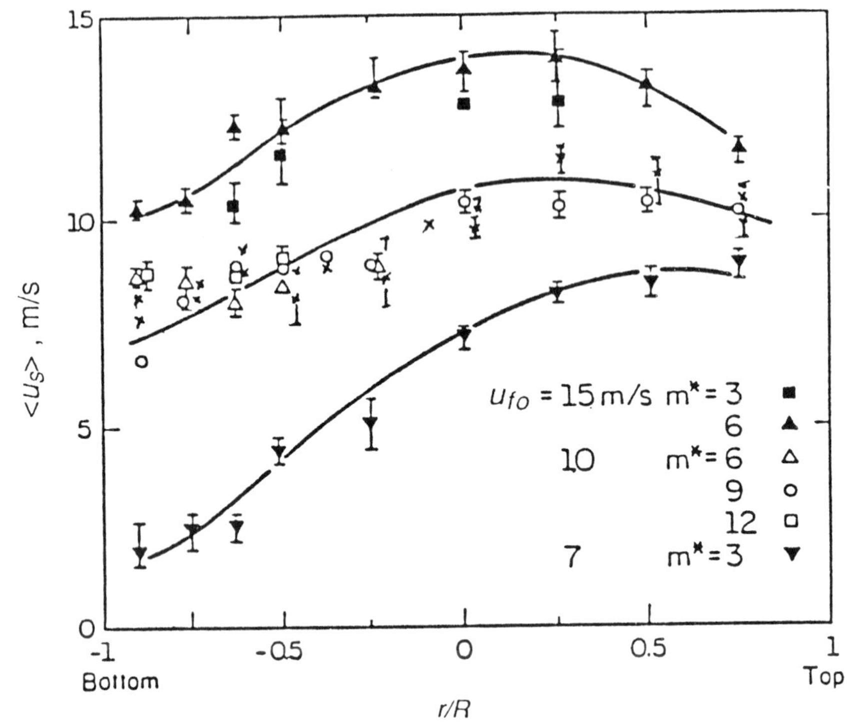

FIGURE 9-17 *Average particle velocity at mean velocities of 7, 10, and 15 m/s at various mean mass flow ratios. (x 's give range of measurement by corona charger-electrostatic probe system)*

9.4.1 Intensity and Stress of the Particle Phase

Using the data in Figures 9.15, 9.16 and 9.17, we were able to determine the Reynolds stress term $<\rho_s><u_s'u_s'>$ for small $<\rho_s'u_s'>$ from Equation 9.23. This approximation was necessary while a technique for simultaneous measurement of ρ_s and $\rho_s u_s$ is being developed. This is because of the inherent nature of calibration for the measurement of local density from local mass flow and local average velocity. Figure 9.20 illustrates a typical trend for the case of 100 micron particles with 10 m/s air velocity at a mass flow ratio of 6 for intensity and stress of random motion of particles which can be understood from the nature of stratification. The range of intensity was 10 to 20% of local average velocity. We note that this Reynolds stress is limited to the axial component; the other components would be more useful.

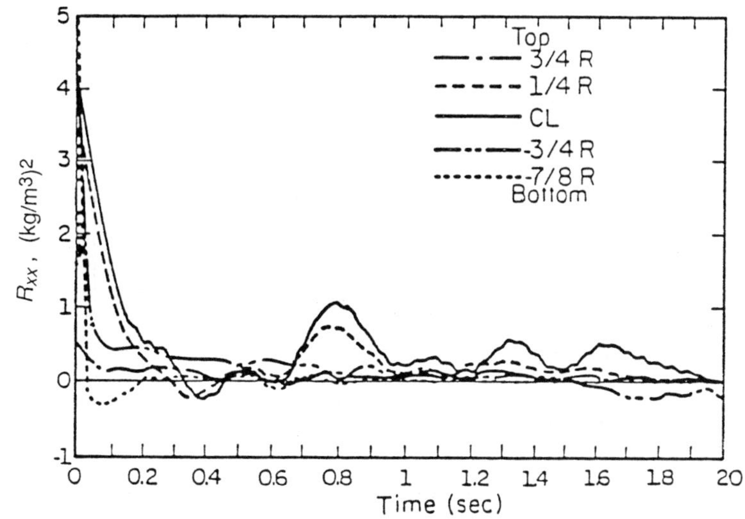

(a) Density Autocorrelation

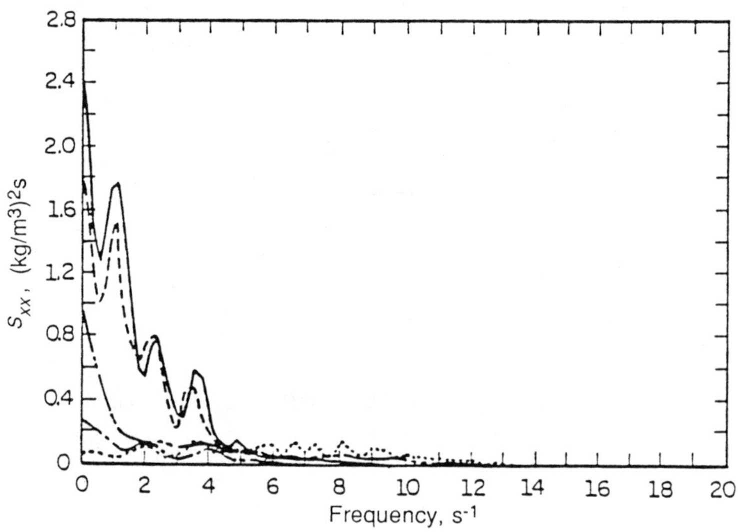

(b) Density Autospectrum

FIGURE 9-18 *Autocorrelation and spectrum of particle density at mass flow ratio of 9 and mean air velocity of 10 m/s.*

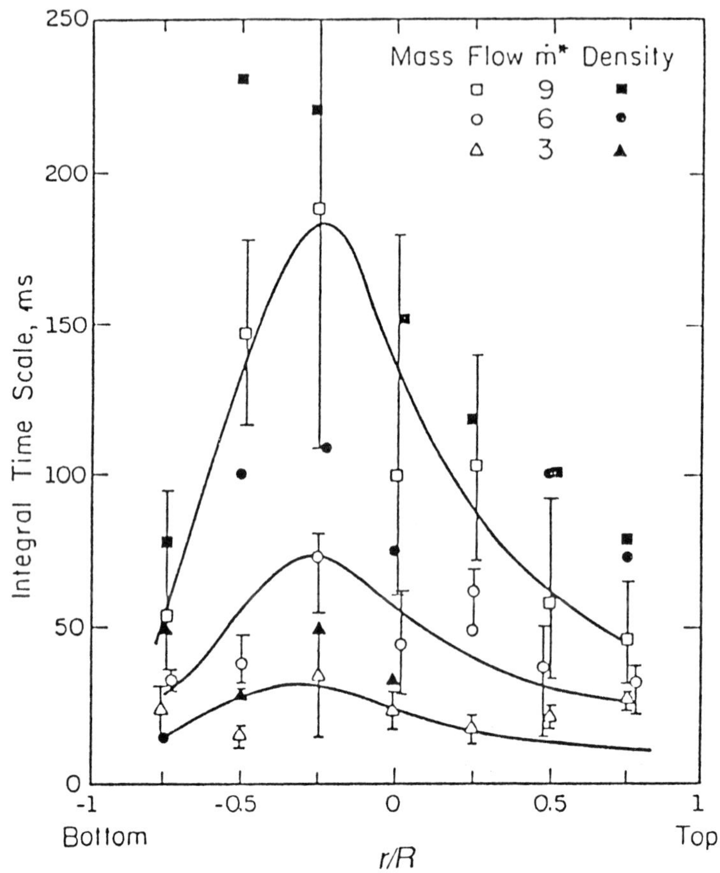

FIGURE 9-19 *Integral time scale of fluctuation. Mean air velocity of 10 m/s, mass flow ratios of 3, 6, and 9.*

9.4.2 Viscosity of the Dense Phase

The magnitude of viscosity was demonstrated by computation applying Equation 9.1 to our recent experimental results of particle velocities from cross-correlation of data from two electrostatic ball probes and densities from laser beam attenuation (Lucht and Soo, 1990) in Figures 9.16(a) and 9.17. This was performed by extending the relations on stratified flow (Soo,1987)

to account for variable particle density $\alpha_s \rho_s$. Referring to Figure 9.21 for geometry and notations, and assuming an infinitely long duct,

$$-(dP/dz) + (\partial \tau_s / \partial y) = 0 \quad (9.42)$$

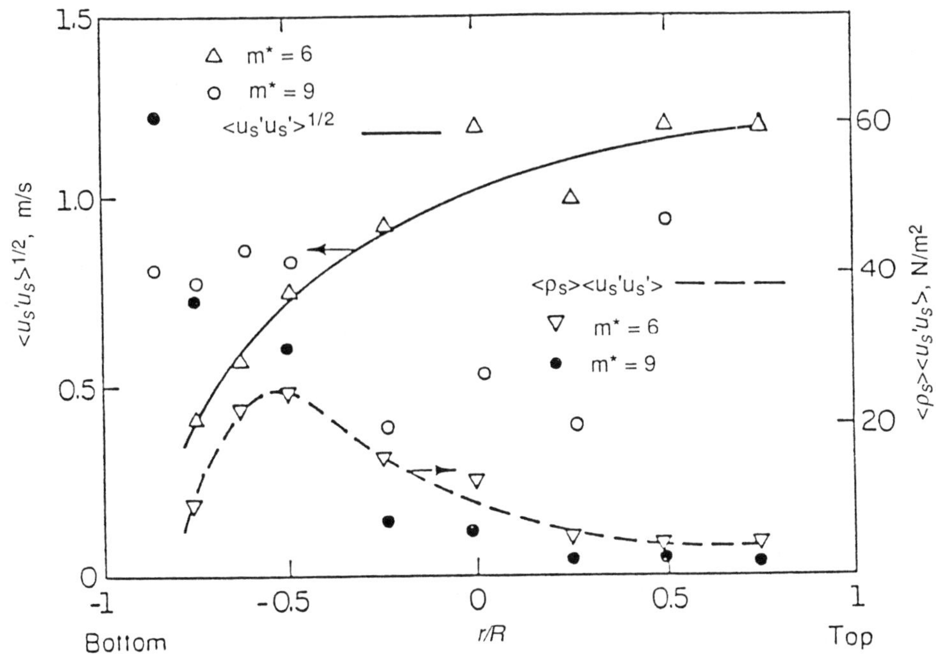

FIGURE 9-20 *Variation of intensity of fluctuating particle (100 μm glass) motion and corresponding stress in the vertical plane of a pipe. m* is mass flow ratio of solid to air. (States correspond to Figures 9-15, 9-16)*

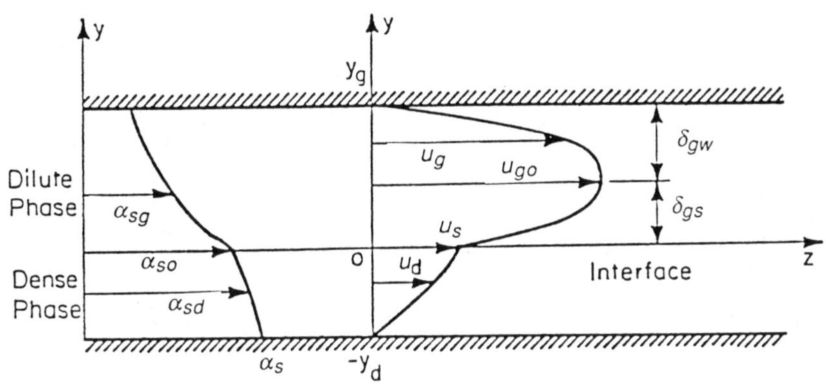

FIGURE 9-21 *Idealized stratified flow.*

where the shear stress in the dense phase

$$\tau_s|_y = -C_\mu \alpha_s^2 d^2 \rho_s (\partial u_s/\partial y)^2 \tag{9.43}$$

while at the interface the shear stress is equal to that in the dilute phase:

$$\tau_s|_{y=0} = \tau_o = 0.0225(\nu_g u_o/\delta_{gw})^{1/4} \rho_g u_o^2 \tag{9.44}$$

Denoting $C = -dP/dz$, it is readily shown that the shear rate is given by:

$$\partial u_s/\partial z = -\alpha_s^{-1}[(C_\mu d^2 \rho_s/4)^{-1}(Cy + \tau_o)]^{1/2} \tag{9.45}$$

Integration gives the relation to the velocity of the interface and C_μ is given by:

$$C_\mu = [\int_0^y \alpha_s^{-1}(Cy + \tau_o)^{1/2}\, dy]^2 4/d^2 \rho_s u_s^2 \tag{9.46}$$

For the above cases, the two-dimensional approximation gives, for a mean air velocity of 10 m/s, y_d = -1/4 duct height, so that integration covers the dense layer. The results are given in Table 9.3.

Table 9.3 *Transport properties of stratified gas-solid suspension.*

m^*	6	9	12
$-dP/dz$, Pa/m	80	140	320
C_μ	39	39	36
u_s, m/s	4.5	4.8	5.0
μ_{ss}, kg/m s	1.2×10^{-3}	1.3×10^{-3}	5.3×10^{-2}
μ_o, kg/m s	4×10^{-5}	5×10^{-5}	5.3×10^{-5}
D_s/C_d, m^2/s	2.4×10^{-5}	9.4×10^{-6}	2.2×10^{-7}

With the viscosities of the dense suspension of Schügerl (1971) up to 0.44 kg/m s and air as the upper and lower limits. The corresponding diffusivities due to particle-particle interaction are given in terms of D_s/C_d; C_d is of order 10^{-1}.

9.5. DISCUSSIONS

Our measurements have so far been made with 200 micron (210 to 230 mm) and 100 micron (88-105 mm) glass particles. As expected, the small particles tend to have a dense layer of smaller particle cloud density and a dilute layer of larger cloud density of particles than in the case of large particles. These results demonstrate that the quantitative description of the motion of a dense suspension in terms of local averages of density, velocity and mass flow, and their fluctuations, autocorrelations, spectra, and integral time scales can be made with some confidence. Simultaneous measurements using the ball probe and the laser attenuation system have been made to determine $<\rho_s'u_s'>$. Though seemingly elaborate, it is recognized that the present results are in the nature of a preliminary study.

Measurements have shown the existence of low frequency components of fluctuation in all cases, confirming wave motion of a dilute-dense interface. A wide range of low velocity dense phase transport was shown to be possible. The nature of wave motion is seen in the case of m^* =6 and 10 m/s mean air velocity. Figure 9.19 gives an integral time scale of 75 ms along the center line and Figure 9.20 gives an intensity of 1.1 m/s of particle motion. These data give a length scale of 8.25 cm (or 4 times of the pipe diameter) for the wave motion; the length scale at one-half radius from bottom was 3.75 cm. This further explains the 2-dimensional nature of flow distributions in horizontal pipe flow. With the terms in Equation 9.16 converted those corresponding to a cylindrical conduit, Zhu (1991) determined the wave velocity of the interface to be 4.2, 6.1, and 8.6 m/s for cases of mean air velocities of 8, 10, and 15 m/s.

In practical applications, there is a need for predicting pressure drop in pipe flow of a dense suspension. While various empirical correlations for specific designs have been introduced continuously, a basic approach calls for the determination of viscosity of these suspensions. General interest in transport properties of suspensions suggests an overall consideration at this stage. The formulation of Batchelor (1970) for the stress system of a suspension and Batchelor and Green (1972) for dilute simple shear flow extends from Einstein (1906) to account for a simple shear flow with vanishing particle (subscript k) Reynolds number to give a total particle-presence stress of the mixture in the form:

$$\tau_m = (1- \alpha_s)\tau_f^{(i)} + \alpha_s\tau_s^{(i)} = 2\mu_f(1 + 2.5\alpha_s)\Delta_m \tag{9.47}$$

where μ_f is the fluid viscosity, superscript (i) denotes intrinsic stresses within each phase, and Δ_m is the bulk strain-rate of the mixture. However, general two-phase motion includes bifurcated velocities of the phases and often the interest is to determine the individual motions of the phases. Therefore, a knowledge of individual expression of shear stress of each phase in the mixture is needed. Hwang and Shen (1989) suggested that, for the fluid:

$$\tau_{fm} = \tau_f^{(v)} + \tau_f^{(t)} \tag{9.48}$$

where superscripts (v) and (t) denote viscous and turbulent stresses of the fluid phase in the mixture respectively, and for the particle phase:

$$\tau_{sm} = \tau_s^{(c)} + \tau_s^{(k)} + \tau_s^{(s)} \qquad (9.49)$$

where superscripts (c), (k), and (s) denote collisional, kinetic, and particle-presence stresses of the particle phase in the mixture.

Among the above stresses:

(a) $\tau_f^{(v)}$ and $\tau_s^{(s)}$ together constitutes τ_m in Equation 9.1 but the split is not obvious. For a dilute suspension, $\tau_s^{(s)}$ from particle-fluid interaction arises from the resistance to the transfer of collisionless gradient of momentum of particle by a resistance to the momentum gradient with a "viscosity" $\rho_s D_{sm}$, where D_{sm} is the diffusivity of the particle phase k through the mixture and is approximated by the diffusivity of k through the fluid f (Soo, 1989).

(b) $\tau_s^{(c)}$ due to particle-particle interaction in shear motion in a dense suspension was shown to follow Equation 9.1.

(c) $\tau_f^{(t)}$ and $\tau_s^{(k)}$ are readily obtained from time-averaging of either the volume-averaged or time-averaged equations in the form of a Reynolds analysis to give Reynolds stresses typically in the form of $<\rho_s u_s' u_s'>$ and $<\rho_f u_f' u_f'>$.

It is seen that the stress system of general phase configurations is incomplete at this time except for very dilute or very dense suspensions of spheres. Present study is an attempt to determine those in the intermediate range.

9.6 SUMMARY AND CONCLUSIONS

Unsteady stratified motions have been observed in dense gas-solid suspensions at moderate and low velocities but means for prediction is lacking. Quantifcation of unsteady motion in a dense suspension has been made through the study of phase interactions in horizontal stratified flow. Fluctuations in mass flow and density of solid phase have been correlated in terms of frequency spectrums and integral time scales.Transport properties of dense suspensions, that is, their viscosity and diffusivity were shown to have liquid (molecular) like behavior due to strong particle-particle interaction. The statistical nature of the flow of the particulate suspension is seen through the correlations of velocity and density leading to diffusion fluxes and Reynolds stresses.

The needed instrumentation and facility for the study of a dense suspension, while improvements will continue, are now complete for further extensive measurements. Measurement of local density remains challenging. In addition to the determination in terms of averages, fluctuations, correlations, spectra, and integral time scale of a dense suspension, our effort has been extended to determining the intensities and corresponding Reynolds stresses. Estimation of viscosity and diffusivity of the dense and intermediate range of concentrations of particles confirms the basic kinetic nature of a dense suspension, although much work is needed to cover the intermediate range.

ACKNOWLEDGEMENTS

Studies leading to the present chapter have been supported by the National Science Foundation under Grant CBT-8709974 of the Particulate and Multiphase Processing Program directed by Drs. M. S. Ojalvo and S. L. Goren, and under Grant CTS 9014914 of the Particulate and Hydraulic System Program directed by Dr. M. C. Roco. The discussions with Professor B. T. Chao are appreciated. Other participants in this continuing research were D. A. Baker, T. E. Lucht, L. M. Liljegren, and M. C. Slaughter. Their contributions are identified in reference citations.

NOTATION

A	cross-sectional area
c_k	dimensionless coefficient of friction
C	coefficients defined by Equations 9.15 and 9.16, or empirical coefficients in Equations 9.1, 9.2, and 9.31
C_f	friction coefficient in standpipe
d	diameter or particle diameter
D	height of channel
D_s	particle diffusivity
E	modulus of elasticity of material
g	gravitational acceleration
f_L	lift force per unit mass
F	relaxation time constant, or a function defined by Equation 9.13
h_e	charge transfer coefficient
i	current
k	elastic parameter
m^*	mass flow ratio
n	number density
N	number of impacts
r	radial coordinate
R	pipe radius or correlation according to subscripts
S	power spectral density function
t	time
T	averaging time
u	local axial velocity
$\mathbf{u}$	vectorial velocity
U	mean velocity in the pipe or standpipe as designated
u_k	axial component of velocity of phase k
x, y	functions
y	vertical coordinate
z	axial coordinate

Greek symbols

α_k	volume fraction of phase k
δ	thickness as defined
Δ	area averaged weight
Δ	deformation tensor
ε	coefficient of Δ terms
λ	mean free path
μ	viscosity of the fluid
ν	kinematic viscosity or Poisson ratio of material
χ_f	frictional perimeter
ρ	density of a phase or cloud density
$\overline{\rho}$	material density
σ	attenuation coefficient
τ	time duration
τ_i	interfacial shear stress
τ_{wk}	wall shear stress
τ	stress tensor

Subscripts

b	bed height
B	ball probe
d	diffusivity or dense phase
g	for dilute phase, gas
i	interface
f	fluid or air (or no subscript)
fs	transfer from s to f, or superficial velocity
k	general phase k
km	phase k in the mixture
m	mixture
mf	minimum fluidization
o	reference value or valve opening
p	for dense phase
s	for particle phase
sd	for dense phase in stratified flow
sf	from fluid to particle
sg	for dilute phase in stratified flow
so	for interface
ss	particle-particle interaction
xx	autocorrelation of quantity x
xy	cross correlation of quantities x and y
w	wall
μ	viscosity

Superscripts

'	fluctuating quantities
(c)	collisional stress
(k)	kinetic stress
(s)	particle presence stress
(t)	turbulent stress
(v)	viscous stress

REFERENCES

Batchelor, G. K., "The Stress System in a Suspension of Force-free Particles," *J. Fluid Mech.,* **41**, 545-570 (1970).

Batchelor, G. K., and J. T. Green, "The Determination of the Bulk Stress in a Suspension of Spherical Particles to Order α_k^2," *J. Fluid Mech.*, **56**, 401-472 (1972).

Baw, P. S., and R. L. Peskin, "Some Aspects of Gas-solid Suspension Turbulence," *Trans. ASME, D., J. of Basic Eng.*, **93**, 631-636 (1971).

Beck, M. S., "Correlation in Instruments: Cross Correlation Flowmeters," *Instruments Science and Technology, J. Phys. E: Sci. Instrum. (Great Britain),* **14**, 7-19 (1981).

Bendat, J. S., and A. G. Piersol, *Random Data,* Second Ed., John Wiley, NY, 11, 122, 146, 385, 406 (1986).

Bobis, J. P., K. G. A. Porges, and A. C. Raptis, "Particle Velocity and Solid Volume Fraction Measurements with a New Capacitive Flowmeter at Solid-Gas Flow Test Facility," *Gas-Solid Flows-1986, ASME FED -* **35**, 79-86 (1986).

Brady, V. F., "Modeling Particulate Flows by Stokesian Dynamics Simulation," Proc. Joint DOE/NSF Workshop on Fluid-solid Transport, PETC, USDOE, 134-135 (1989).

Brodkey, R. S., The Phenomena of Fluid Motions, Addison Wesley, Reading, MA (1967).

Cheremisinoff, N. and R. Gupta, *Handbook of Fluids in Motion*, Ann Arbor Science, MI (1983).

Cheng, L., and S. L. Soo, "Charging of Dust Particles by Impact," *J. Appl. Phys.,* **41**, 585-591 (1970).

Cheng, L., S. K. Tung, and S. L. Soo, "Electrical Measurement of Flow Rate of Pulverized Coal Suspensions," *Trans. ASME, J. Eng. for Power,* **92** (A-2), 135-149 (1970).

Chong, Y. O., C. S. Teo, and L. S. Leung, "Recent Advances in Standpipe Flow," *J. Pipelines,* **6**, 121-132 (1987).

Ding, J., and D. Gidaspow, "A Bubbling Fluidization Model Using Kinetic Theory of Granular Flow," *AIChE J.,* **36**(4), 523-538 (1990).

de Crecy, F., "Modeling of Stratified Two-phase Flow in Pipes, Pumps, and other Devices," *Int. J. Multiphase Flow,* **12**, 307-323 (1986).

Einstein, A. (1906), "Eine Neue Bestimmung der Moleküldimension," *Annln. Phys.*, **19**, 289 and **34**, 591 (1906).

Graf, W. H., and E. R. Acaroglu, "Sediment Transport in Conveyance Systems. Part I. A Physical Model for Sediment Transport in Conveyance Systems," *Bull. Intern. Assoc. Sci. Hydrol.,* **23**(2), 20-39; "Part II. The Modes of Sediment Transport and Their Related Bed Forms in Conveyance Systems," *ibid.,* **23**(3), 123-135 (1972).

Hwang, G. J., and H. H. Shen, "Modeling the Solid Phase Stress in a Fluid-solid Mixture," *Int. J. Multiphase Flow,* **15**(2), 257-268 (1989).

Konrad, K., and J. F. Davidson, "The Gas-Liquid Analogy in Horizontal Dense-phase Pneumatic Conveying," *J. Powder Tech.,* **39**, 191-198 (1984).

Li, X. G., D. L. Liu, and M. Kwauk, "Pneumatically Controlled Multistage Fluidized Beds - II," *Proc. Joint Meeting of Chem. Eng.. SIESC and AIChE, Chem. Industry Press,* Beijing, 382-391 (1982).

Liljegren, L. M., and N. S. Vlachos, "Laser Velocimetry Measurements in a Horizontal Gas-Solid Pipe Flow," *Experiments in Fluids,* **9**, 205-212 (1990).

Lucht, T. R., and S. L. Soo, "Density and its Fluctuation in Stratified Flow of a Dense Particulate Suspension," *Advanced Powder Technol.* **2** (2), pp. 1-11 (1991).

Ma, D. N., and M. C. Roco, "Probabilistic Three-Dimensional Model for slow Shearing Particulate Flow: Wet Friction," *Proc. Third International Symposium on Liquid-solid Flows,* ASME FED **75**, 53-63 (1988).

Mathur, M. P., and G. E. Klinzing, "Measurement of Particle Velocity in Pneumatic Transport of Coal Using Cross-Correlation Technique," *Particulate Sci. and Tech.,* **2**, 223-235 (1984).

Roco, M. C., and C. A. Shook, "Computational Method for Coal Slurry Pipelines with Heterogeneous Size Distribution," *J. Powder Tech.,* **39**, 159-176 (1984).

Roco, M. C., and C. A. Shook, "New Approach to Predict Concentration Distribution in Fine Slurry Flow," *Physico-Chemical Hydrodynamics J.,* **8**(1), 43-60 (1987).

Roco, M. C., V. Makhijani, and D. N. Ma, "Probabilistic Micromechanical Model for Two-Dimensional Slow Particulate Flow: Dry Friction," *Powder Tech.,* **60**, 223-243 (1990).

Rossetti, S. J., "Concepts and Criteria for Gas-Solid Flow," *Handbook of Fluids in Motion* (N. P. Cheremisinoff and R. Gupta, Eds.), Ann Arbor Science, MI, Ch. 24, 638-652 (1983).

Savage, S. B., and S. McKeown, "Shear Stress Developed During Rapid Shear of Dense Concentrations of Large Spherical Particles Between Concentric Cylinders," *J. Fluid Mechanics,* **127**, 453-472 (1983).

Schaefer, M., M. Bottlinger, F. Loeffler, and H. Umbauer, "Realization of a Concept for the Complete Evaluation of Double Pulse Holograms of Particle Phases in Flow," *Particle Characterization,* **4**, 166-174 (1987).

Schügerl, K., "Rheological Behavior of Fluidized Systems," *Fluidization* (J. F. Davidson an D. Harrison, Eds.) Academic Press, London, 261-292 (1971).

Sengun, M. Z., and R. F. Probstein, "Bimodal Model of Slurry Viscosity with Application to Coal-slurries. Part 2. High Shear Limit Behavior," *Rheol. Acta* , **28**, 382-393 (1989).

Shen, H. and N. L. Ackerman, "Constitutive Relationships for Fluid-Solid Mixtures," *J. Engineering Mechanics Div.*, ASCE, **108** (EMS), 748-763 (1982).

Soo, S. L., *Fluid Dynamics of Multiphase Systems*, Ginn-Blaisdell, (1967); revised edition as *Multiphase Fluid Dynamics*, Science Press-Gower Aldershot, Hampshire, England (1991).

Soo, S. L., J. J. Stukel, and J. M. Hughes, "Measurement of Mass Flow and Density of Aerosols in Transfer," *Environmental Sci. and Tech., Ind. Engr. Chem.*, **3**(4), 386-393 (1969).

Soo, S. L. , "State of Multiphase Instrumentation," *Developments in Theoretical and Applied Mechanics,* **11**, (University of Alabama, Huntsville) 563-576 (1982).

Soo, S. L., "Pipe Flow of a Dense Suspension," *J. Pipelines,* **6**, 193-203 (1987).

Soo, S. L., and Mei R., "Dense Suspensions and Fluidization," *Proc., Powder and Bulk Solids Conf.*, Rosemont, IL, Session 3, 134-143 (1987).

Soo, S. L., *Particulates and Continuum: Multiphase Fluid Dynamics,* Hemisphere Pub. Corp., NY (1989).

Soo, S. L., Baker, D. A., Lucht, T. E., and Zhu, C., "A Corona Discharge Probe System for Measuring Phase Velocities in a Dense Suspension," *Rev. Sci. Instrum.*, **60**, 3475-3477 (1989).

Soo, S. L., T. R. Lucht, and C. Zhu, "Measurement of Average Flow Rate of Solids in a Recirculating Pneumatic Pipe Flow Loop," 14th Powder & Bulk Solids Conferences, Rosemont, IL, Session B, 350-354 (1989a).

Soo, S. L., C. Zhu, T. R. Lucht, "Unsteady Pipeflow of Dense Suspensions," Proc. DOE/NSF Workshop on Fluid-Solids Transport, Pleasanton, CA, 83-97 (1989b).

Takahashi, H., T. Masuyama, and K. Noda, "Unstable Flow of a Solid-Liquid Mixture in a Horizontal Pipe," *International J. Multiphase Flow,* **15**, 831-841 (1989).

Tsuji, Y., and Y. Morikawa, "Flow Pattern and Pressure Fluctuation in Air-Solid Two-phase Flow in a Pipe at Low Air Velocities," *International J. Multiphase Flow,* **8**(4), 329-341 (1982).

Tsuji, Y., and Y. Morikawa, "LDV Measurements of an Air-solid Two-phase Flow in a Horizontal Pipe," *J. Fluid Mech.,* **120**, 385-409 (1982a).

Wen, C. Y., and H. P. Simons, "Flow Characteristics in Horizontal Fluidized Solids Transport," *AIChE J.,* **5**, 263-267 (1959).

Zhu, C., M. C. Slaughter, and S. L. Soo, "Measurement of Velocity of Particles in a Dense Suspension by Cross-correlation of Dual Laser Beams," *Rev. Sci. Instrum.* **62**, 2036-2037 (1991).

Zhu, C., M. C. Slaughter, and S. L. Soo, "Covariance of Density and Velocity Fields of a Gas-solid Suspension," *Rev. Sci. Instrum.* **62**, (11), 2835-2836 (1991a).

Zhu, C., Dynamic Behavior of Unsteady Turbulent Motion in Pipe Flows of Dense Gas-Solid Suspensions, Ph. D. Thesis, University of Illinois at Urbana-Champaign, Urbana, IL 61801 (1991).

CHAPTER 10

FINITE SIZE EFFECTS IN FLUIDIZED SUSPENSION EXPERIMENTS

D. Joseph

10.1 INTRODUCTION

"Two fluid" equations for fluidized suspensions of solid particles can be rigorously formulated as ensemble averages (Joseph and Lundgren 1990). Even though these equations have a rigorous foundation they are not useful unless the interaction forces are correctly modeled. Modeling, the framing of constitutive hypotheses, is basically a guess about the solutions of the equations of motion. A constitutive hypothesis is when you guess what the solution of a problem of dynamics might be without actually obtaining the solution.

Conventional theories of fluidized suspensions are generally formulated as one-dimensional approximations in which quantities perpendicular to the direction of flow are averaged over cross sections. One-dimensional theories describe the evolution in time t of averaged quantities in the direction z of flow. A one-dimensional, two-phase flow (Wallis 1969) is a system of four "two-fluid" equations for the volume fraction α of solids, the volume fraction $\varepsilon = 1 - \alpha$ of fluids, the average particle velocity $u_p = u$ and fluid velocity u_f. In one-dimensional, one-phase flows (Foscolo and Gibilaro 1984, 1987; Batchelor 1988, Joseph 1990) the forces which the fluid exert on the solids are modeled, and the decoupled equations for α and u are of interest. The modeling of forces in two-phase and one-phase flows is fraught with difficulties. It is not even clear what the variables should be as our discussion in §10.3 of finite size effects will reveal. In one-phase theories it is not even possible to guarantee that the range α (z,t) $\in$ [0,1] will be protected (Singh and Joseph 1990). Moreover, there are three effects of fundamental importance which are omitted entirely in conventional one-dimensional theories: finite size effects §10.3); wakes §10.4); and turning couples on long bodies leading to stable fluidized structures which we call wake architectures §10.5). In this essay we shall focus our attention on these omitted finite size and nonlinear effects.

10.2 ONE-DIMENSIONAL, ONE-PHASE FLOW

Conventional theories are framed in terms of conservation equations for the number density $N = \rho\alpha/m$ of particles of mass m and density ρ whose mass centers are on the plane z. Since ρ and m are constant,

$$\alpha_t + (\alpha\, u)_x = 0. \qquad (10.2.1)$$

The momentum equation of motion may be written in the first order approximation as

$$\rho\alpha\,(U_t + uu_z) = \mathfrak{F}\,[\alpha,u] - \Pi_z\,[\alpha,u] \tag{10.2.2}$$

where $\Pi\,[\alpha,u]$ is the particle phase pressure, (10.2.1) and (10.2.2) are the equations for the particle phase and

$$\Pi_z = \Pi_\alpha\,\alpha_z + \Pi_u\,U_z\,. \tag{10.2.3}$$

Various kinds of modeling assumptions about $\mathfrak{F}$ and Π have been made by the authors named in §10.1. There are two points of view, Joseph (1990) argues that in a first order system of equations for α and u, one ought to assume say that Π_u and Π_α are not zero, unless an argument which says one or both are zero is given. In the other view, adopted by Batchelor (1988), it is necessary to give the reasons why Π_α or Π_u is not zero, from the point of view of physics, and he reasons that $\Pi_\alpha \neq 0$ but presents no argument to justify putting $\Pi_u=0$. Foscolo and Gibilaro (1987) also try to obtain Π from a physical argument and they also get a Π that doesn't depend on u. Joseph (1990) noticed that you could write their Π_z as $\frac{2}{3}D\,\mathfrak{F}_\alpha\alpha_z$ with their $\mathfrak{F}$=NF (α,u) where N is the number density, D is the particle diameter and

$$F(u,\alpha) = m\tilde{g}\left\{-(1-\alpha) + \left[\frac{u_c-u}{U(0)}\right]^{\frac{4.8}{n}}(1-\alpha)^{-3.8}\right\} \tag{10.2.4}$$

on a single particle. Here m is the mass of a single sphere of radius R, $\tilde{g}=(\rho-\rho_f)g/\rho$ is reduced gravity, ρ is the density of the sphere, ρ_f is the density of the fluid, $U(\alpha)$ is the steady fall velocity under gravity of a sphere in a uniform dispersion of spheres of solids fraction α. In a uniform dispersion u=0 and α is independent of z. The fluidization velocity $u_c=u_f(1\text{-}\alpha)-u\alpha$, where u_f is the fluid velocity, is in general solenoidal, independent of z, and when u=0, u_c satisfies the Richardson and Zaki correlation

$$u_c = U(\alpha) = U(0)\,(1-\alpha)^n\,. \tag{10.2.5}$$

U(0) is the velocity of one sphere in a pure liquid and it can be expressed in terms of the Reynolds number using various empirical correlations and n(Re) is the Richardson and Zaki exponent; it lies between 4.8 for small Reynolds numbers $\mathrm{Re} = \frac{u_c 2R}{\nu}$ and 2.4 for large Re. The ingenious force law (10.2.4) was invented by Foscolo and Gibilaro but it cannot be regarded as established. Their argument about the form of Π_z does not exclude the idea that $\Pi = N_A F$ where $N_A = 4\alpha/\pi D^2$ is the number density per unit area, and $N_A/N = \frac{2}{3}D$. Then

$$\Pi_z = \frac{2}{3} D \left(\mathfrak{F}_\alpha \alpha_z + \mathfrak{F}_u U_z\right) \tag{10.2.6}$$

where $\mathfrak{F}_u \neq 0$.

In the early days it was thought that zeroth order approximation in which Π_z and higher derivatives are zero might suffice. In fact, zeroth order theories have been discredited partly because they give rise to Hadamard instability of the state of uniform fluidization.

The state of uniform fluidization, u=0 and $\alpha = \alpha_o$ where α_o is a constant, is a solution of (10.2.1) and (10.2.2) because $\mathfrak{F}[\alpha_o, 0] = 0$ (see §10.5) and $\Pi_z = 0$. To study stability we write $\alpha = \alpha_o + \psi$ linearize for small ψ and u where ψ and u satisfy

$$\psi_t + \alpha_o u_z = 0, \tag{10.2.7}$$

$$u_t = \hat{a}u - \hat{b}\psi - \hat{c}\psi_z - \hat{d}u_z. \tag{10.2.8}$$

The values

$$[\hat{a}, \hat{b}, \hat{c}, \hat{d}] =$$

$$\left[\frac{4.8\mathfrak{g}}{nU(0)}\left(1 - \alpha_o\right)^n, 4.8\mathfrak{g}, -\frac{2}{3}D\hat{b}, -\frac{2}{3}D\hat{a}\right] \tag{10.2.9}$$

used by Foscolo and Gibilaro (1987) with $\hat{d}=0$ and by Joseph (1990) with $\hat{d} \neq 0$ are representative for all the conventional theories so far proposed. The dispersion relation

$$\sigma^2 + \sigma(\hat{a} - i\beta\hat{d}) + \alpha_o(\beta^2\hat{c} + i\beta\hat{b}) = 0 \tag{10.2.10}$$

governs the growth rate σ to waves of length $2\pi/\beta$ of disturbance proportional to $\exp(\sigma t + i\beta x)$. When the particle phase pressure gradient $\Pi_z = 0$ so that $\hat{c} = \hat{d} = 0$, we get

$$\text{re}\,\sigma \to \sqrt{\hat{b}\alpha_o}\ \sqrt{\beta} \tag{10.2.11}$$

for large β. This is a catastrophic instability with an unbounded growth rate for small waves, a Hadamard instability with square root growth. The particle phase pressure will regularize this catastrophic instability, but the system with coefficients (10.2.6) is unstable with positive growth rates re σ (β) uniformly bounded in β, not Hadamard unstable. On the other hand, if $\hat{d}=0$ but the other coefficients are as in (10.2.6), Foscolo and Gibilaro (1984, 1987) have shown that stability is possible and the criterion for instability is in good agreement with the formation of bubbling gas beds for some beds.

The considerations of the previous paragraph explain why so much attention has been given to the particle phase pressure in conventional theories of fluidized suspensions. Batchelor (1988) argues that the α_z term represents fluctuation induced diffusion against a gradient in which particles drift into empty places by random walks as in Brownian motions. The fluctuations here would have to be supplied by hydrodynamics, vortex shedding, unsteady particle motions or other causes whose magnitude areas are yet unknown.

Hadamard instabilities can also be regularized by higher order gradients; for example, by viscosity which is associated with a second derivative of u. Viscosity, like other gradient regularizers, will kill the short waves but will not stabilize an otherwise unstable system. The regularization of Hadamard instability is only one property among many which could be expected of a physically realistic mathematical theory. In the two-dimensional beds of large spherical particles studied by Singh and Joseph (1991a,b) there is a structured response in which certain discrete waves do not grow. These blocked wave lengths cannot be predicted from conventional gradient theories, but they are predicted by a global theory which accounts for the fact that the portion of the plane at z blocked by particles is determined also by spheres whose centers are not at z.

10.3 FINITE SIZE EFFECTS (Singh and Joseph 1990,1991a,b)

To account for the finite size of spheres we are going to introduce a third variable, the fractional area α^a, which is related to the volume fraction α by a convolution generated by the following construction. Consider a plane at z. Locate the origin of a coordinate x on z. Spheres at a distance $|x|<R$ from z intersect z and their area of intersection is $\pi(R^2-x^2)$. If N(x+z,t) is the number density of spheres with centers at x+z, then N(x+z,t)Adx is the number of such spheres in an infinitesimal volume Adx, where A is the area of, say, some square $A=L^2$ with a very large L. The area A_s on the plane which is cut out by spheres is obtained by summing all of the areas of intersections coming from infinitesimal volumes centered on x+z as x varies from -R to R,

$$A_s = \int_{-R}^{R} N(x+z,t)\, \pi(R^2 - x^2)\, A dx. \qquad (10.3.1)$$

After writing $A_s/A = \alpha^a$, using $N = \rho\alpha/m = 3\alpha/4\pi R^3$, we get

$$\alpha^a = \frac{3}{4R^3} \int_{-R}^{R} \alpha(x+z,t)\, (R^2 - x^2)\, dx\ . \qquad (10.3.2)$$

When α is independent of x, then $\alpha^a(z,t) = \alpha(z,t)$. Since α is independent of x when R is small, our theory reduces, but not uniformly (see (10.3.12)), to the conventional one for small particles. The criterion for smallness is that the variation in number density of an ensemble average is small relative to the size of a particle.

10.3.1 Three-Variable Theory

We construct a three-variable theory by replacing the volume fraction in force law (10.2.4) and (10.2.5) with the area fraction α^a. A nonlinear three-variable theory without gradients is given by

$$\alpha_t + (u\alpha)_z = 0\ , \qquad (10.3.3)$$

$$\alpha^a = \frac{3}{4R^3} \int_{-R}^{R} \alpha(x+z,t)(R^2 - x^2)\, dx\ , \qquad (10.3.4)$$

$$u_t + uu_z = \tilde{g}\left\{ -\varepsilon + \left[\frac{u_c - u}{U(0)}\right]^{\frac{4.8}{n}} \varepsilon^{-3.8} \right\} \qquad (10.3.5)$$

where $\varepsilon = 1 - \alpha^a$ and the composite velocity u_c, which is the superficial velocity at the inlet of the bed is given in

$$u_c = U(0)\varepsilon_o^n. \qquad (10.3.6)$$

Uniform fluidization, $u = 0$ and $\alpha = \alpha_o$ is a solution of this system.

In our three-variable model we think of $\alpha(z,t)$ and $u(z,t)$ as an ensemble of averaged quantities. It is necessary to think of ensemble averages rather than volume averages because we shall be looking at fields with rapid variation over the length of the microstructure. The number density is then taken as $N = \rho/m\alpha$. The fractional area α^a is the convolution of an ensemble average and is not itself an ensemble average. In our experiments we actually measure α^a as simple averages over horizontal lines of pixels. The main idea of our model is to replace the volume fraction α with fractional area α^a in the force law, where α^a and α are related by the simple construction just given. We have constructed our model as a hypothesis which seems reasonable but which we cannot prove. We wish to draw out the predictions of our model and see how they compare with experiments. We should not be deterred by bogeyman mathematicians because we don't know the exact words to scare away the bogeyman.

10.3.2 Linearized Stability

Equations (10.3.3), (10.3.4) and (10.3.5) may be linearized around uniform fluidization; $\alpha = \alpha o + \psi$ where ψ is small. Then

$$\psi_t + \alpha_o u_z = 0,$$

$$\alpha_1^a = -\varepsilon_1 = \frac{3}{4R^3} \int_{-R}^{R} \psi(z+x,t)\,(R^2-x^2)\,dx, \qquad (10.3.7)$$

$$u_t = \hat{a}u - \hat{b}\varepsilon_1$$

where $\hat{a}$, $\hat{b}$ are given by (10.2.6), and after eliminating ε_1 and u_1, we find

$$\psi_{tt} = -\hat{a}\,\psi_t - \hat{b}\,\frac{3\alpha_o}{4R^3}\frac{\partial}{\partial z} \int_{-R}^{R} \psi(z+x,t)\,(R^2-x^2)dx\,. \qquad (10.3.8)$$

Writing $\psi = \psi_o\, e^{\sigma t}\, e^{i\beta z}$ we find a complex dispersion relation of the form

$$\sigma^2 + \sigma\,\hat{a} + i\beta\,\hat{b}\,\alpha_o\Theta(\beta R) = 0. \qquad (10.3.9)$$

where

$$\Theta(\beta R) = 3\left[\frac{\sin \beta R}{(\beta R)^3} - \frac{\cos \beta R}{(\beta R)^2}\right] \qquad (10.3.10)$$

is the blockage function (figure 10.3.1)

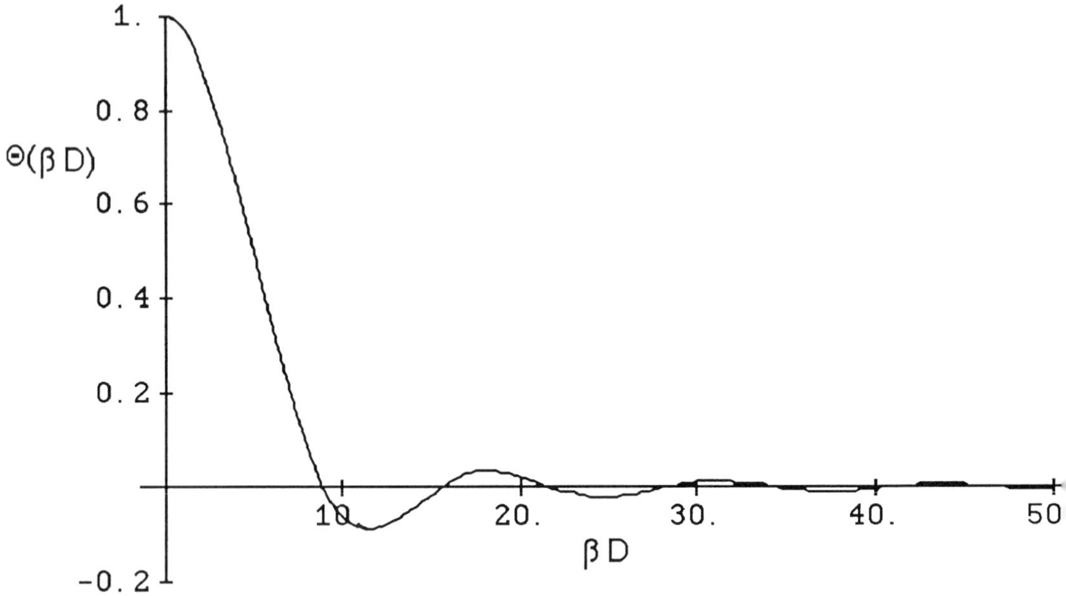

Figure 10.3.1 Θ(βD) is plotted as a function of (βD), D=2R.

We may solve the quadratic equation for

$$\sigma = -\frac{\hat{a}}{2} \pm \frac{\hat{a}}{2}\sqrt{1 - i\,\Sigma}, \tag{10.3.11}$$

$$\Sigma = -\frac{4\beta\hat{b}\,\alpha_o\Theta(\beta R)}{\hat{a}^2}.$$

The most dangerous root is associated with the plus sign in (10.3.10) and with the first zero 2.6/D of d(βΘ/dβ. At this most dangerous β,

$$\Sigma = \Sigma_m = \frac{2.17\, n^2\, \alpha_o u_c^2}{\varepsilon_o\, D\tilde{g}}. \tag{10.3.12}$$

Singh and Joseph (1991a,b) have argued that Σ_m is a basic parameter distinguishing the response of different beds. The growth rate and frequency vanish, $\sigma = 0$, at the zeros of $\beta\Theta(\beta R)$, at $\beta = 0$, $\beta R = 4.493$, 7.7253, 10.904, etc., and $\beta\Theta(\beta R) \to 0$ for a fixed value of R, as $\beta \to \infty$. It follows that re $\sigma \to 0$ as $\beta \to \infty$ and our zeroth order model with blockage is always unstable but not Hadamard unstable. On the other hand, if $\beta<1.3/R$ is large and fixed, then $\Theta(\beta R) \to 1$ as $R \to 0$ and our model reduces, nonuniformly, to the conventional one which is effectively but not actually Hadamard unstable. The value $\beta= 1.3/R$ is a cutoff for Hadamard instability where the growth rate re σ is a maximum.

10.3.3 Nonlinear Theory (Singh and Joseph 1991b)

Singh and Joseph (1991b) solved the initial value problem for the three-variable, one-dimensional theory by numerical methods for periodic boundary and different initial conditions. No solutions of permanent form with discrete spectral peaks were found. Instead, they found chaotic solutions with nearly stationary levels of the power when a certain dimensionless "growth rate" parameter is below a critical value and unstable solutions whose power levels increase without bound when this value is above the critical one. The power spectrum of the bounded solutions is continuous and such that power levels are very low for wave numbers in the blocked set which are mariginally stable in the linear theory.

10.3.4 Comparison With Experiments

Singh and Joseph (1990) made video recordings of experiments on beds of fluidized spheres confined to move in two dimensions between glass plates. A typical configuration seen in the experiments look like the photograph shown in figure 10.3.2. Analysis of the digitized signals gives the fractional

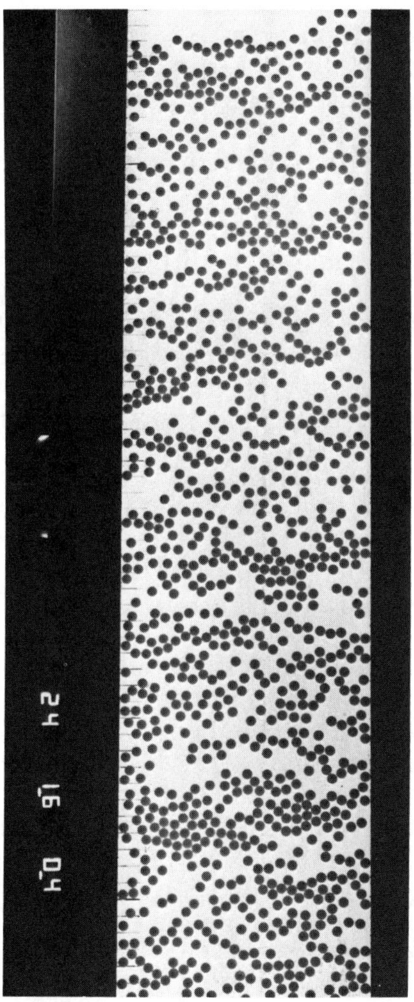

Figure 10.3.2 Fully expanded fluidized bed of plastic beads, Re=300, α=0.88. Most spheres line up in horizontal arrays, vertical pairs are drafting.

area as a function of space and time at a discrete set of points. The amplitude of the measured spectra is a strong function of frequency and wave number with deep minima at the blocked wave numbers predicted by stability theory and the numerical integration of the initial value problem. At low Reynolds numbers, the measured spectrum looks like the growth rate function from linear theory. As the Reynolds number increases, the graph of the power spectrum gets higher energy at smaller wave numbers, consistent with the increasing importance of wake effects.

In figure 10.3.3 we have compared plots of the logarithm of the growth rate log ($Re\sigma$) from linear theory as a function of βD with the magnitude $\log\alpha$ of the power spectrum from nonlinear theory and experiments. The blocked values of βD are evident in all cases. Many more and different comparisons of theory and experiments are discussed by Singh and Joseph (1990) with similar agreements. We think that the agreements between the predictions of the simple theory with finite size but no gradient are astonishing, since there are no disposable constants and nothing has been adjusted. On the other hand, the predictions of the gradient theory do not agree with the experiments on two-dimensional fluidizations of large particles. Using the Foscolo-Gibilaro theory with $\hat{d} = 0$ we find from (10.2.7) that the beds we used in our experiments are operating in regions of stable uniform fluidization with

$$Re=300,\ u_c=.044\ \text{m/s},\ \alpha\approx 0.8,\ n\approx 3.0,\ D=0.0063\ \text{m},\ \frac{\rho}{\rho_f}=1.12,\ \tilde{g}=1.05\ \text{m/s}^2.$$

It is difficult to use Batchelor's theory to check the stability since some of the coefficients are not known, but he says "...$\frac{\rho}{\rho_f}$ is seldom less than 2 for a marginally stable bed." These theories disagree with our experiments and those Volpicelli, Massimilla and Zenz (1966) in which all the flows were decidedly not uniform.

We have shown that finite size effects are effectively lost for sufficiently small particles where conventional theories with gradients may be appropriate. However, the Hadamard instability will never arise in the case of a small particle limit in the three-variable theory because the growth rate is a maximum at $\beta=1.3/R$, so that waves shorter than $2\pi R/1.3$ are regularized. In the limiting theory β is not totally independent of R and a limit $\beta \to \infty$ for a fixed R is not useful.

10.4 WAKES AND TURNING COUPLES

It is not clear why the one-dimensional theory with finite size effects works so well. Perhaps one reason is that one-dimensional dynamics are generated by nonlinear effects associated with drafting, kissing and tumbling. This kind of one dimensional dynamics is evident in the photographs of two-dimensional beds shown in figure 10.3.2, which are like the ones studied by Singh and Joseph (1990) and in all the other photographs of two-dimensional fluidization shown in the papers of Volpicelli, et al. (1966), Fortes, Joseph and Lundgren (1987) and Joseph, Fortes, Lundgren and Singh (1987). It may not be a bad idea to base a force law for such a motion on a reliable empirical correlation, like the Richardson and Zaki (1954) correlations used by Foscolo and Gibilaro in deriving (10.2.4) which is valid under the very nonlinear conditions prevailing in our experiments. It is perhaps appropriate here to note that Volpicelli, et al. (1966) showed that the Richardson-Zaki correlations do hold in two-dimensional beds of spheres after one accounts for the effects of walls in shifting the curves of correlation.

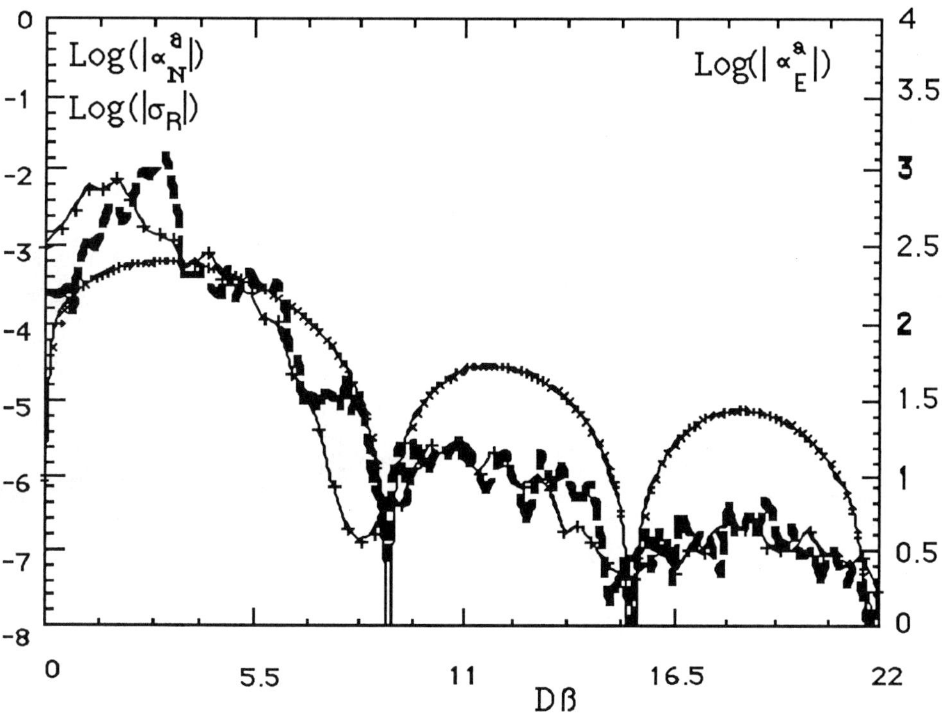

Figure 10.3.3 Comparison of log$|\sigma_R|$ ( ), log $|\alpha^a_N|$ ( ) and log$|\alpha^a_E|$ ( ), where σ_R is the real part of the growth rate of the linear theory of stability of uniform fluidization, log $|\alpha^a_N|$ is the amplitude of the power spectrum computed numerically for the three variable nonlinear theory and log$|\alpha^a_E|$ where α^a_E is the power spectrum measured in the experiments. The value of the parameter Σ_m which we estimated for the experiments is 1.79. For the nonlinear numerical solution Σ_m is 0.0958. The amplitude is a strong function of Σ_m, but the blocked wave numbers are robust, independent of sphere size or Reynolds number.

We may explain the dynamic generation of effective one-dimensional motions at modest, but not too small, Reynolds numbers larger than some critical value which may lie between 0.5 (Happel and Pfeffer 1960) and 8 (figure 10.4.2), as a pairwise coupling of spheres whose line of centers is parallel to the stream. We describe the dynamic scenario associated with this pairwise coupling as drafting, kissing and tumbling. The second sphere is drafted into the wake of the first sphere; they kiss, then tumble (figures 10.4.1, 10.4.2). The falling motion of kissing spheres whose line of centers is parallel to the stream is unstable to couples of the type which turn streamlined bodies broadside on, stagnation point pressures of the type that cause an airship to stall (figure 10.4.3). Neighboring spheres with centers aligned cross stream are a characteristic feature easily observed in all photographs of two-dimensional fluidized beds (figures 10.3.2, 10.4.6, 10.4.7, 10.4.8, 10.5.1 and 10.5.2).

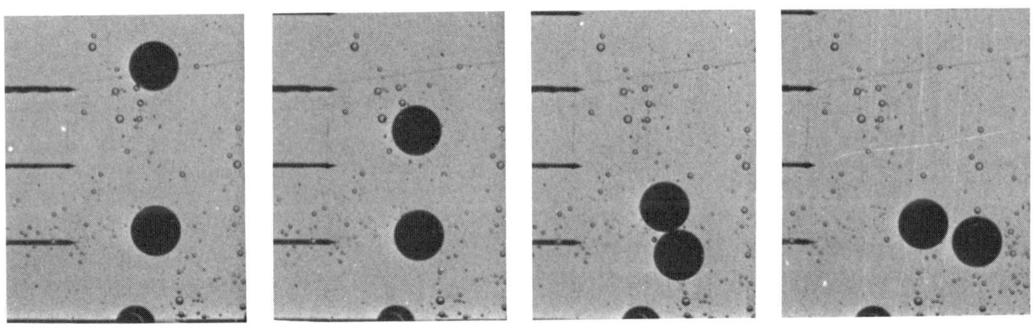

(a) (b) (c) (d)

Figure 10.4.1 Drafting, kissing of plastic spheres in water, Re=700: (a) the line of centers lies along the stream; (b) both spheres are falling but the second is falling faster than the first; (c) kissing after drafting; and (d) tumbling after kissing. Tumbling is an instability, well-known of long bodies which like to orient themselves broadside to the flow.

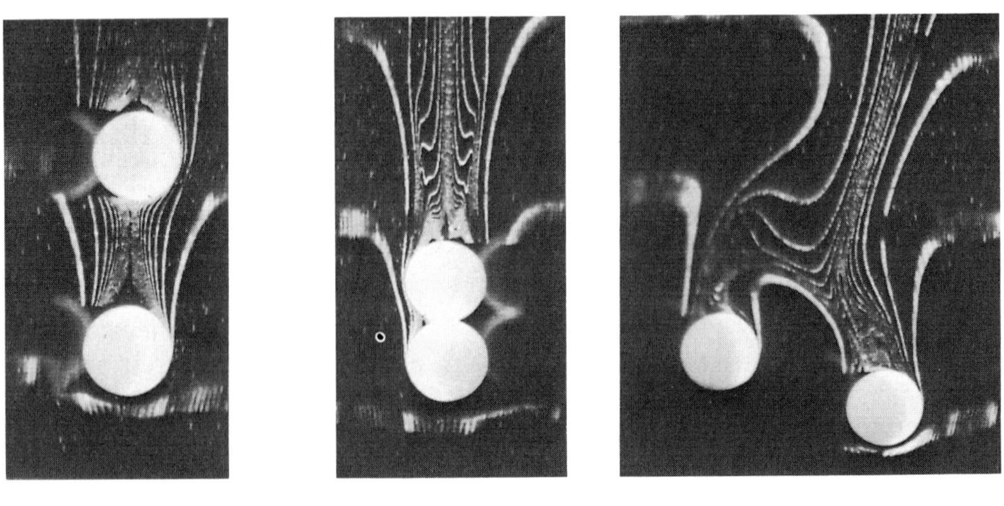

Figure 10.4.2 Interaction of two Delrin spheres 11.113 mm diameter at Re=8.5, showing drafting, kissing and tumbling.

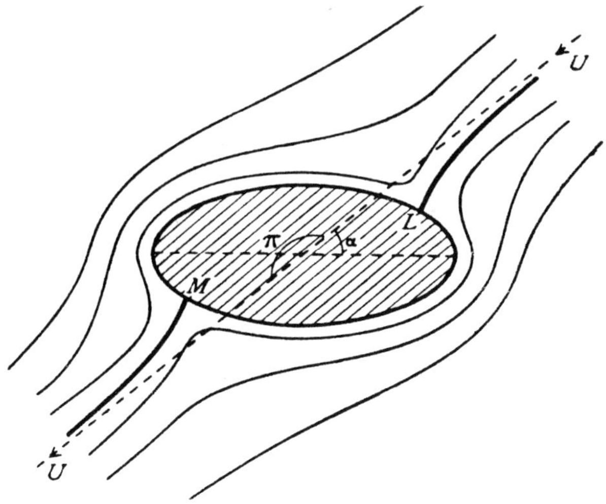

Figure 10.4.3 Potential flow solution for uniform flow past a slender body. The high pressure at the points of stagnation give rise to a couple causing the body to turn broadside-on. In real flows this mechanism operates only on the front side, with a wake on the back.

The robust stability of the cross stream alignments are particularly dramatic in a two-dimensional bed of a single line of particles (see figures 10.4.8 and 10.5.1). It is perhaps a subtle feature of nature that the same hydrodynamic forces which cause long bodies to float broadside-on endow beds of tumbling spheres with an anisotropic structure of horizontal arrays which may lend itself to study in a one-dimensional approximation.

It must be noted that the drafting of spheres is a rearrangement mechanism which sucks spheres together, an effect of an entirely different natutre than the Brownian type of motions which cause spheres to diffuse into empty places by the mechanism which is invoked to explain the stabilizing action of the particle phase pressure (Batchelor 1988). Wakes lead to aggregation rather than dispersion. The photographs shown in figures 10.4.4 to 10.4.9 show how aggregates may form and collapse under the action of wakes. The formation and collapse of aggregates appears to be closely related to the propogation of voidage waves in a manner to be inferred by comparing figures 10.4.4 with 10.4.6 and 10.4.5 with 10.4.8. In the typical case voidage shock waves propogated by a mechanism in which particles fall out of the aggregate at the roof of the void and collect at the top of the aggregate at the floor of the void as in our figures 10.4.6 and 10.4.8, and in figure 10 of Fortes, et al. (1987). This type of shockwave of voidage is called slugging in three-dimensional fluidized beds and is well-known.

The reader should wonder how in the world the aggregates at the top of the beds shown in figures 10.4.4, 10.4.5 and 10.4.7 ever formed. They are stationary aggregates of close-packed spheres, porous media, held together

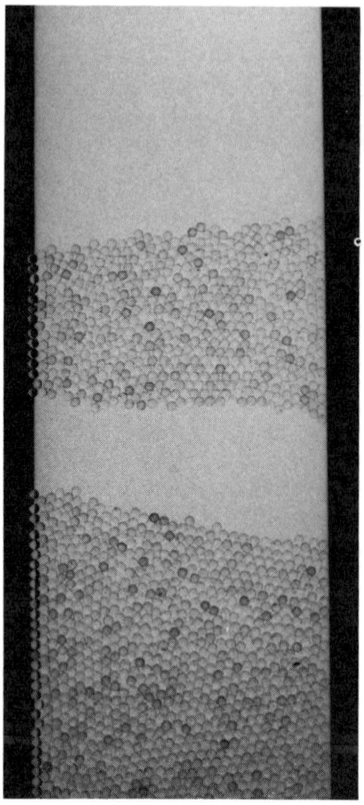

Figure 10.4.4 A stationary void and stationary aggre-gates of 0.231 diameter glass beads in a 6 inch bed. Re=1400 is based on the sphere diameter and fluidi-zation velocity. This is a stationary voidage shock, a two-dimensional "slug." The bed is slightly inclined from the vertical (20°) and wall friction helps to stabilize the shock wave.

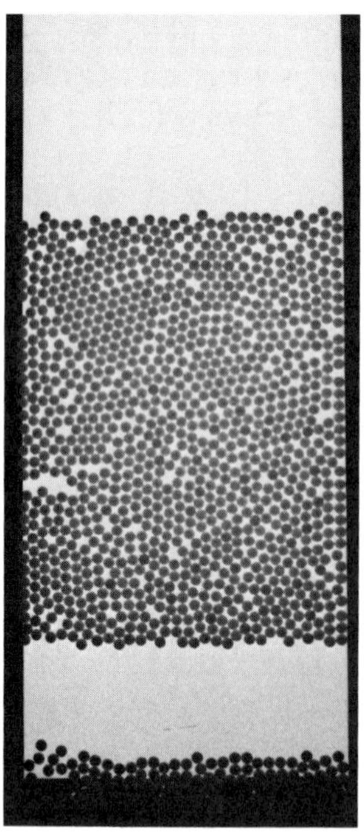

Figure 10.4.5 A stationary void and stationary aggre-gates of a $\frac{1}{4}$ inch plastic spheres in a 6 inch bed, Re=380.

(a) (b)

Figure 10.4.6 Propagation of voidage waves in beds of 0.231 in glass beads in water. (a) A shock wave of voidage propagates by particle release from aggregate at the roof of the void. (b) A bed has expanded but close-packed clusters are still in evidence. The clusters appear to be held together by wake forces. The spheres are either in clusters, in cross stream arrays or they are drafting, kissing and tumbling.

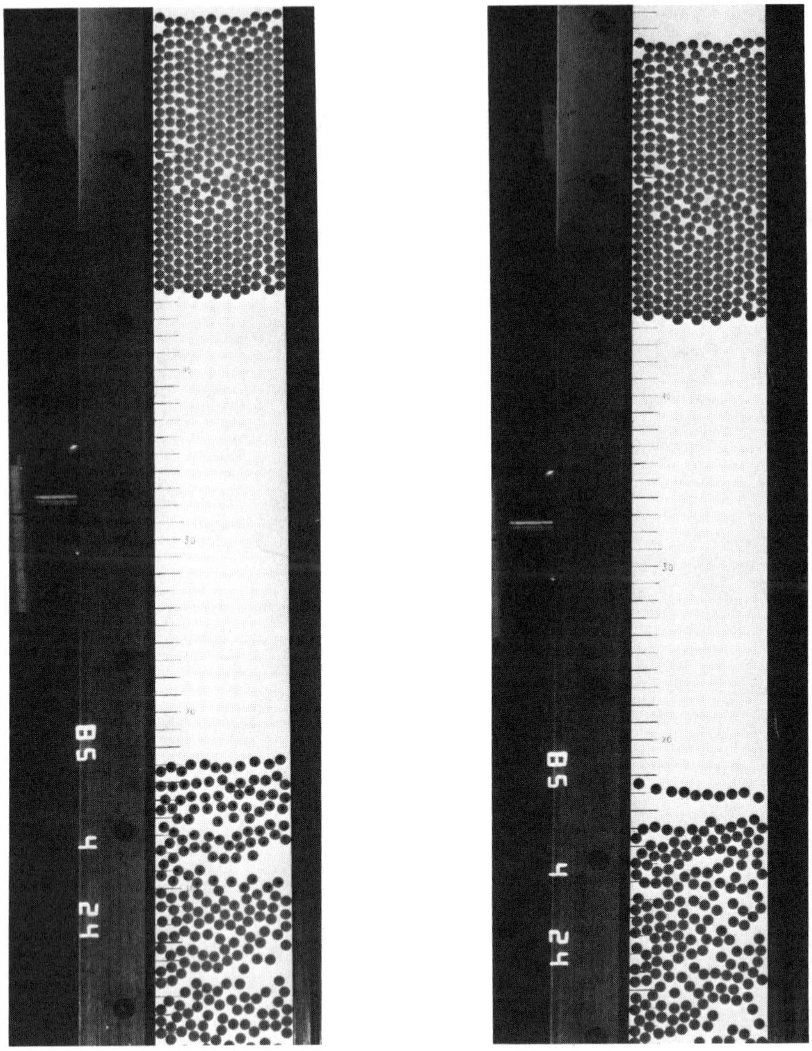

(a) (b)

Figure 10.4.7 A stationary shock wave, closed-packed, $\frac{1}{4}$ inch diameter plastic spheres at the top of the void and propagating "one dimensional" voidage waves at the bottom of a 3 inch bed, Re=390. There is nothing to restrain the aggregated beads at their top or bottom, other than hydrodynamic forces. Earlier, the aggregate formed by sucking spheres into the bottom row of the packed bed.

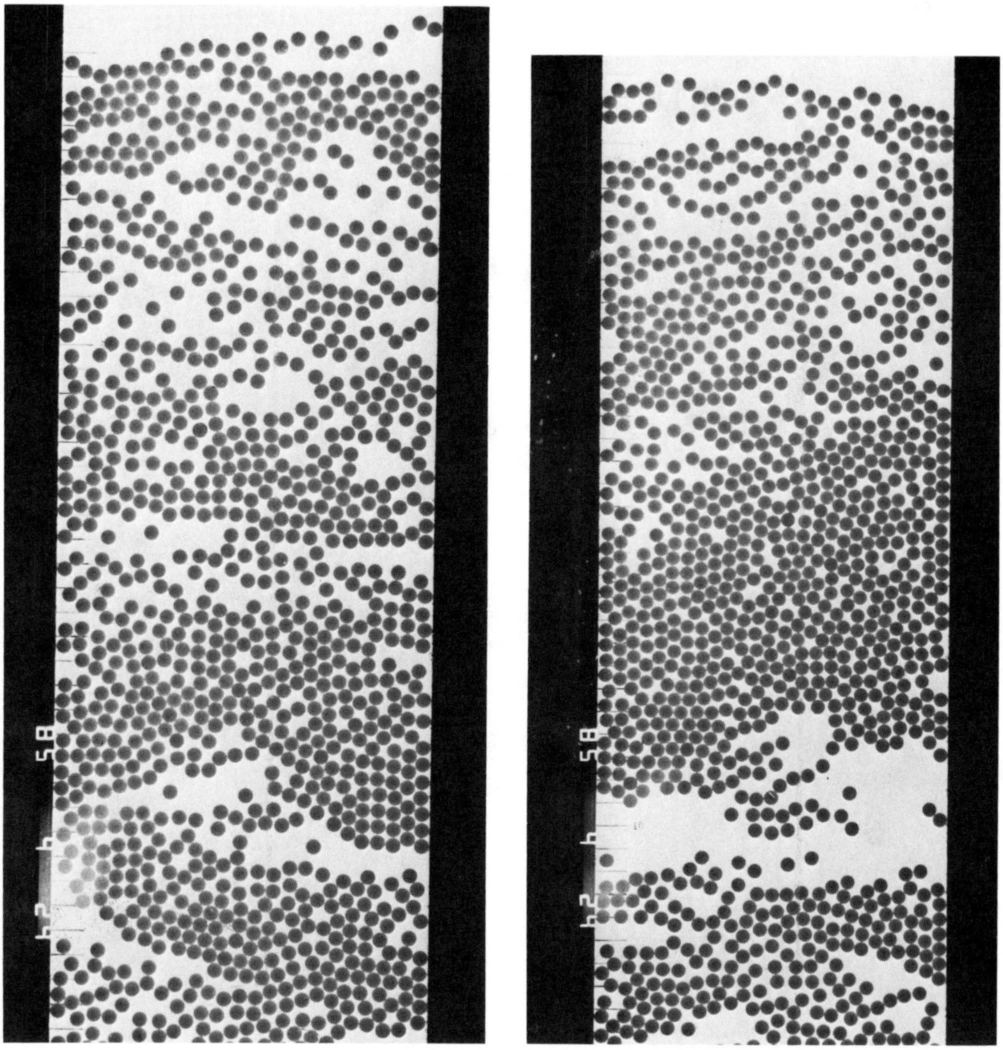

(a) (b)

Figure 10.4.8 Fully fluidized bed of $\frac{1}{4}$ inch plastic spheres in a six inch bed, Re= 370. It appears as if there are clusters of spheres which are not individually fluidized. We think that these clusters are held together by mutual interactions of wakes. The spheres are either in clusters, in cross stream arrays or are drafting, kissing and tumbling.

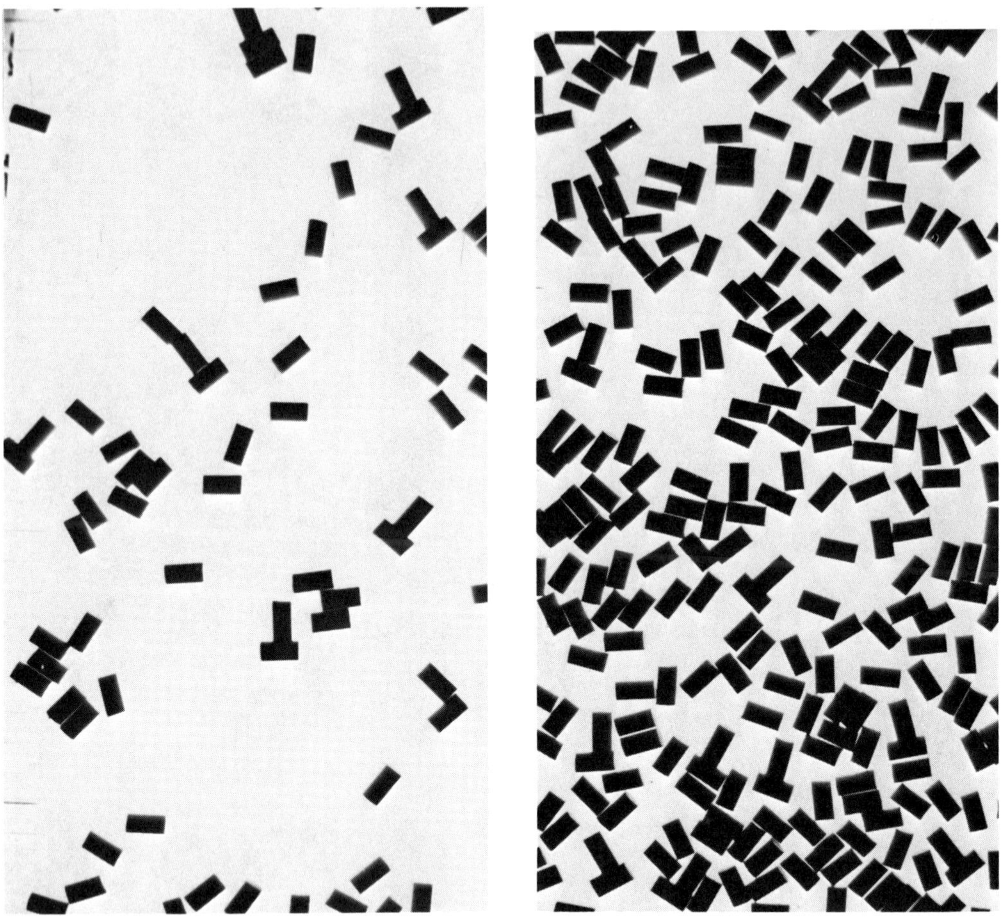

Figure 10.4.9 Fluidization of round cylinders with flat ends, l/D≈2. Individual cylinders rock violently from vortices shed at the sharp corner at a Reynolds number $0(10^3)$. They oscillate around the horizontal. Wake interactions are evident in the piles of cylinders which contact along their long sides, and in the characteristic inverted T in which the top cylinder stands end on in the wake of the bottom. The formations are examples of wake architectures. Most of the cylinders are glued by wakes in small clusters dispersed over the bed. We may entertain the idea that the clusters defined by range of influence of the wakes in a given architecture are dispersed in the flow, rather than the individual particles.

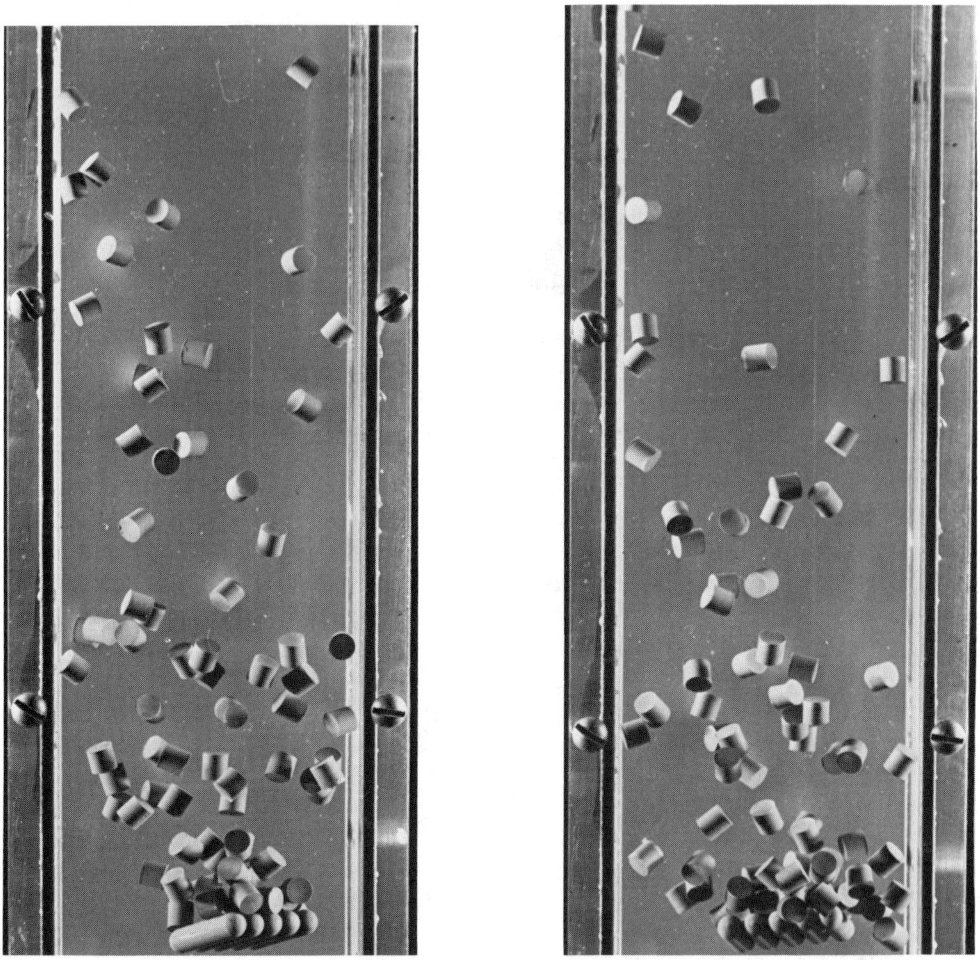

Figure 10.4.10 Short cylinders are sucked into the wake of a raft made by glueing five long cylinders. The raft is fluidized in water in a three-dimensional bed and the drag of short cylinders and the raft is about the same. The raft always floats broadside on and the violent fluctuations around the mean due to vortex shedding are suppressed by the short cylinder.

by hydrodynamic forces only, we think by the wakes behind each and every sphere in the array. We have seen how particles drift into these aggregates when they are close to the roof of the void and other conditions which we have not identified are satisfied. They are pulled in by pressure forces which we think are generated in the wakes behind the aggregated spheres, clearly a three-dimensional effect for close-packed spheres in a two-dimensional bed. This formation of aggregates is exactly the inverse of mechanism which causes propogation of voids because the particles are sucked into, rather than released from, the packed aggregate at the roof of the void. The whole matter could stand more study.

It is well-known, but I think less well known than it ought to be, that the average orientation of a long body moving relative to a stream is robustly stable when the long side is across the stream.

> Thus a ship has to be kept on course by the helmsman, an elongated airship requires similar attention. A sailing ship will not sail permanently before the wind with the helm lashed, but tends to set itself at right angles to the wind. A body sinking in liquid tends to sink with its longest dimension horizontal (Milne-Thompson 1960).

The tendency for long bodies to float broadside on in a two-dimensional flow is the same as the tendency of flat or broad body to float with its broad side facing the flow in a three-dimensional setting. This, together with the well-known power of wakes behind blunt bodies is exhibited in figure 10.4.10 where we show small cylinders bombarding a fluidized raft floating broadside on. The raft was made by glueing five long cylinders together and when fluidized without the short cylinders, it rocked with a frequency which we suppose to be associated with vortex shedding around an equilibrium position perpendicular to the stream. We matched the drag on the raft with the drag on a small cylinder of length/diameter = 0.25in./0.0255in. and fluidized many such cylinders with the raft. The cylinders are sucked into the wake of the raft like debris behind a fast-moving truck. The composite body, raft plus entrained cylinders, is very stable; the oscillations are greatly suppressed.

10.5 STABLE WAKE ARCHITECTURES

Many stable configurations of spherical and long particles have been found in our experiments in two-dimensional beds. Some of them were described in the papers by Fortes, et al. (1987) and Joseph, et al. (1987) and some will be described in a forthcoming paper by Fortes and Joseph (1991). These stable configurations appear to be held in place by wake forces and turning couples in a manner of whose details have yet to be determined. The stable configurations found so far are

(1) Stable horizontal line of spheres (figure 10.5.1)
(2) Stable arch of aluminim spheres (figure 10.5.2)
(3) Nested wake of two spheres (figure 10.5.3)
(4) Nested wake of three spheres (figure 10.5.3)
(5) Nested wake of four spheres (figure 10.5.3)
(6) Stable doublets of long particles (figure 10.5.4)
(7) Stable triplets of long particles (figure 10.5.4)
(8) Stable quadruplets of long particles (figure 10.5.4)

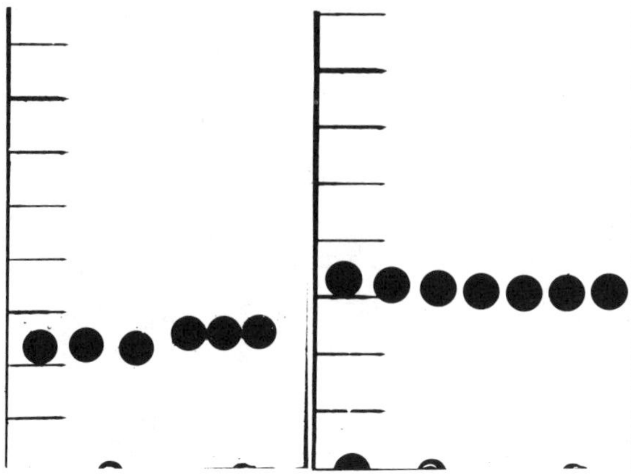

(a) (b)

Figure 10.5.1 Long bodies float broadside on and spheres line up broadside on. (a) Three cemented spheres on the right are robustly stable when their line of centers is perpendicular to the stream. Three spheres on the left are not cemented. They line up broad-side on because otherwise they would interact in the wakes of other pairs. (b) A line of seven individual spheres fluidized with their line of centers perpendicular to the stream.

Figure 10.5.2 Stable arch of aluminum spheres.

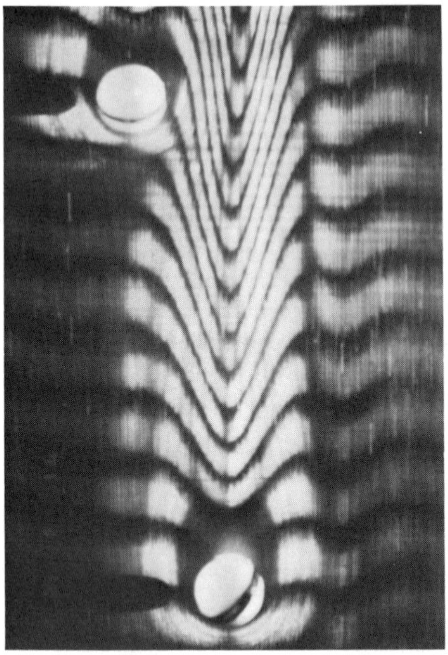

(a)

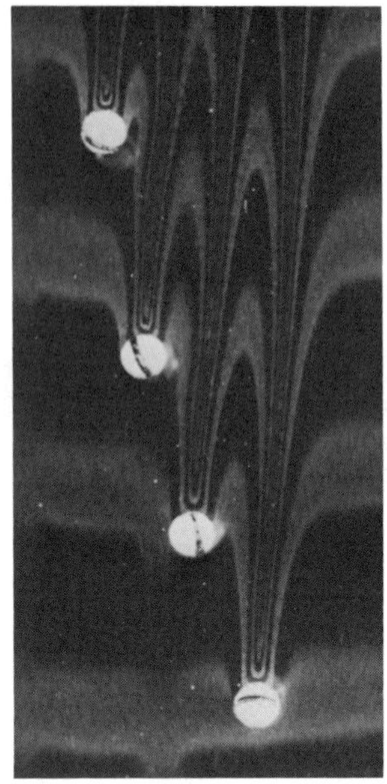

(c)

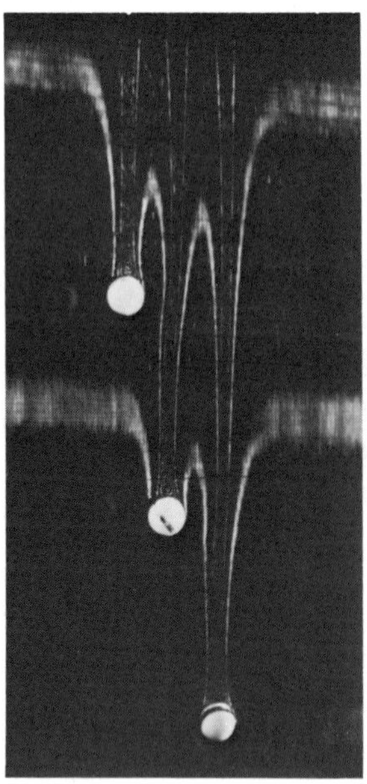

(b)

Figure 10.5.3 (Fortes and Joseph, 1991) Fluidization of teflon spheres of 6.3 mm diameter in aqueous glycerin in perfect equilibrium between weight and drag. The spheres are locked together in a steady flow and are stable in a range of Reynolds numbers $22<\mathrm{Re}<43$. The velocity profiles are visualized by pulsed hydrogen bubbles illuminated in a laser sheet: (a) nested wake of two spheres, Re=22.48; (b) three spheres, Re=22.00; and (c) four spheres, Re=28.4.

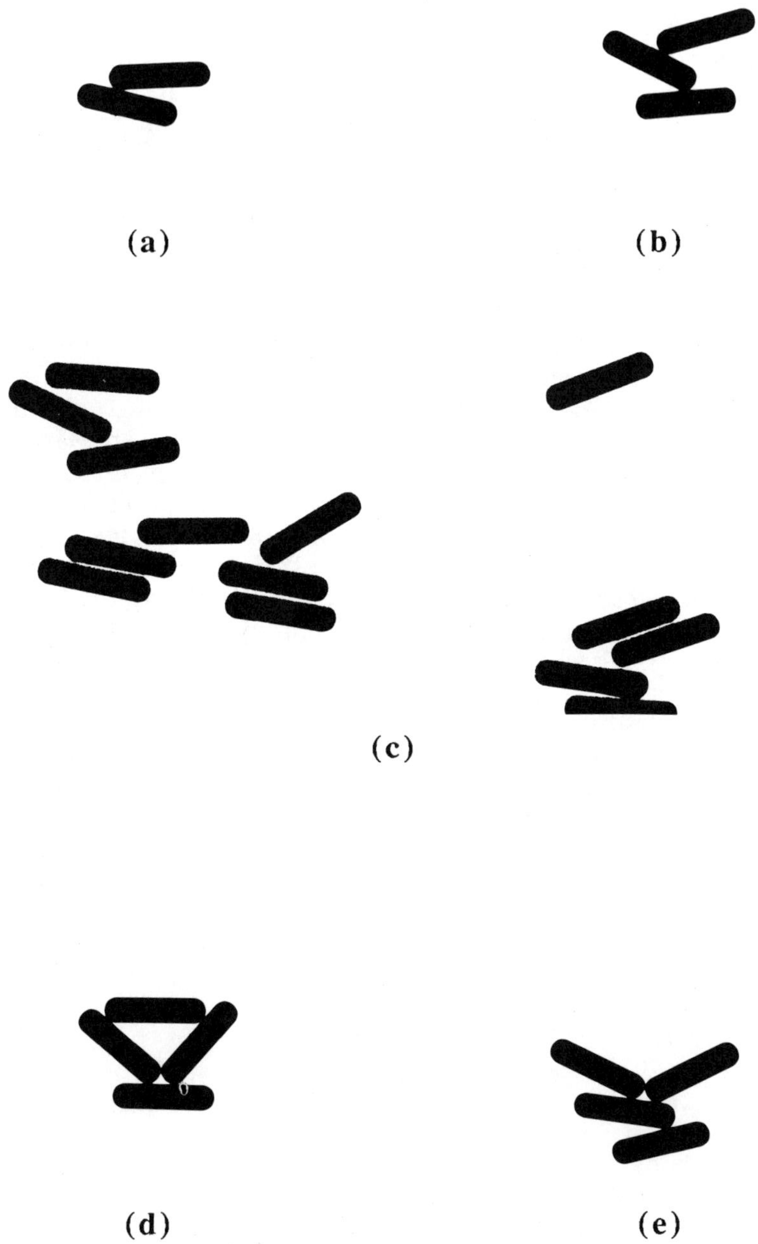

Figure 10.5.4 Stable doublets, triplets and quadruplets. Long bodies float broadside on. They generate strong wakes. If several long bodies are present, they interact to form stable wake architecture: (a) stable doublet, (b) stable triplet of the first kind, (c) stable triplets of the first and second kinds form on the left, (d) stable quadruplet, and (e) stable quadruplet of the second kind.

We are going to confine our explanations of the remarkable stability of these configurations to captions on the figures. It is our belief that a good theoretical understanding of stable wake architectures can be best achieved by comparing experiments with direct numerical simulation of the Navier-Stokes equation and the particles equation of motion. The first paper to move in this direction, by Singh, Caussignac, Fortes, Joseph and Lundgren (1989) who tested the stability of such a single line of particles (as in figure 10.5.1a) in a direct, but strictly two-dimensional simulation of the Navier-Stokes equations for steady flow across circular cylinders Re<100. They perturbed the periodically spaced horizontal line of circular cross sections of cylinders all equally spaced in various ways and computed the steady flow and the hydrodynamic forces on the cylinder to see if the forces would tend to restore stability. They found that the two-dimensional array was stable to small disturbances with greater stability for high Reynolds numbers. A further step was taken recently by Hu, Crochet and Joseph (1991) who computed the trajectories of "spheres" dropped from rest in a channel. They found that a single cylinder sedimenting in a channel at small Reynolds numbers will drift to the center of the channel. The channel center appears to be an equilibrium position at Reynolds numbers below the critical one for vortex shedding. At higher Reynolds numbers the cylinder drifts off the center of the channel. It rotates due to uneven shear and oscillates due to vortex shedding.

The numerical simulation for two sedimenting cylinders gave rise to the kind of drafting, kissing and tumbling which is observed in the sedimentation (and in the fluidization) of two spheres. However, unlike the spheres the cylinders do not actually touch and appear to repel each other at a distance at large Reynolds numbers apparently due to lubrication forces or to a failure of the numerical scheme for touching cylinders at small Reynolds numbers.

A detailed comparison of the two-dimensional computation with the two-dimensional motion of sedimenting spheres has yet to be made. There is a huge body of information which can be exracted from the two-dimensional simulations but eventually a full three-dimensional will be needed. The two-dimensional bed is an excellent laboratory for comparing computations with experiments because everything can be seen.

10.6 SUMMARY

We have considered three effects which are neglected in conventional theories of fluidized suspensions: finite size effects, wakes and turning couples on long bodies and on spherical bodies in momentary contact.

We developed a three variable, one-dimensional, one-phase theory to account for the finite size of particles. The theory uses the Foscolo-Gibilaro force law except that the force that the fluid exerts on the solid is assumed to depend on the fractional area rather than the volume fraction of solids and the gradient of the volume fraction which expresses the particle phase pressure is put to zero. The area and volume fraction are related by a simple geometrical construction in the case of fluidization of a monodisperse suspension of spheres of radius R, taking into account the fact that the portion of the plane at Z blocked by particles is determined also by spheres whose centers are not at Z. Our one-dimensional theory then has three unknowns, the volume fraction, the area fraction and particle velocity, rather that two. These equations admit uniform fluidization as a solution which is Hadamard unstable in the two-variable theory but simply unstable in the three-variable theory, with a distinguished set of marginally stable modes belonging to a countable set of blocked wave numbers generated by the relation between the volume and area fraction. The blockage function regularizes the ill-posed problem and generates a structured dynamical response with a set of discrete stationary marginal modes. The finite size, three-variable theory reduces to the conventional two-variable theory as the particle radius R tends to zero, but not uniformly. The limiting problem is never ill-posed, the maximum growth rate Σ_m is at an approximate wave number of 1.3/R, and is therefore bounded for any finite R no matter how small. The initial value problem for our three-variable, one-dimensional theory is solved by numerical methods for periodic boundary and different initial conditions. Solutions are bounded with nearly stationary power when Σ_m is small, and no such solutions exist when Σ_m is large. The power spectrum of bounded solutions contains low power for the blocked wave numbers and is in astonishing agreement with the values of the power spectrum measured in the experiments of Singh and Joseph (1991 a, b) with fluidized spheres constrained to move in two dimensions (see figure 3.3).

The nonlinear effects, which are neglected in conventional treatments of fluidized suspensions, are associated with wakes and the moment distributions of hydrodynamic forces on long bodies (and spheres in near proximity), which tend to turn the bodies so that the long side (or line of centers) is perpendicular to the stream. These mechanisms are active in the balance that determines the main features of fluidization at all but the smallest Reynolds numbers and they are of practical importance. Nonlinear fluidization of spheres is dominated by mechanisms associated with drafting, kissing and tumbling which are illustrated in figures 4.1 and 4.2. The drafting of spheres is a wake effect which sucks particles together; exactly opposite to the effects of hydrodynamic dispersion in which fluctuations drive particles apart. Many robust effects of aggregation can be explained by wake effects which are neglected in conventional theories. The tumbling of spheres into positions with lines of center perpendicular to the stream, gives an anisotropic structure to the bed in which cross-stream alignments are favored. The nonlinear mechanisms underway in the fluidization when there are only a few particles, are accessible to rigorous treatment by numerical simulations using the Navier-Stokes equations and the particles equations of motion.

ACKNOWLEDGEMENTS

This work was supported by the National Science Foundation and the Department of Energy. Numerical results were obtained under grants from the Minnesota Super Computer Institute and AHPCRC.

REFERENCES

Batchelor, G.K., A new theory of the instability of a uniform fluidized bed. *J. Fluid Mech.*, **193**, pp. 75-110 (1988).

Fortes, A., D.D. Joseph, and T. Lundgren, Nonlinear mechanics of fluidization of beds of spherical particles. *J. Fluid Mech.*, **177**, pp. 467-483 (1987).

Fortes, A., and D.D. Joseph, Wake architectures in two-dimensional fluidization of spheres. *AHPCRC* preprint 91-68 (1991).

Foscolo, P.V., and L.G. Gibilaro, A fully predictive criterion for transition between particulate and aggregate fluidization. *Chem. Eng. Sci.,* **39**, pp. 1667-1674 (1984).

Foscolo, P.V., and L.G. Gibilaro, Fluid dynamic stability of fluidized suspensions. The particle bed model. *Chem. Eng. Sci.*, **42**, pp. 1489-1500 (1987).

Hu, H., M.J. Crochet, and D.D. Joseph, Direct simulation of fluid particle motion. AHRCRC Report 91-45. Submitted to *Theor. Computat. Fluid Dynamics* (1991).

Happel, J., and R. Pfeffer, The motion of two spheres following each other in a viscous fluid. *A. I. Ch. E. Journal*, **6,1**, pp. 129-133 (1960).

Jackson, R., The mechanics of fluidized beds: Part I: The stability of the state of uniform fluidization. *Trans. Instn. Chem. Engrs*, **41**, pp. 13-21 (1963).

Jones, A.V., and A. Prosperetti, On the suitability of first-order differential models for two-phase flow prediction. *Int. J. Multiphase Flow*, **11**, pp. 133-148 (1985).

Joseph, D.D., A. Fortes, T. Lundgren, and P. Singh, Nonlinear mechanics of fluidization of beds of spheres, cylinders and disks in water. Advances in multiphase flow and related problems. Ed., G. Papanicolau, *SIAM*, pp. 101-122 (1987).

Joseph, D.D., Generalization of Foscolo-Gibilaro analysis of dynamic waves. *Chem. Eng. Sci.*, **45**, pp. 411-414 (1990).

Joseph, D.D., and T.S. Lundgren, Ensemble averaged and mixture theory equations for incompressible fluid-particle suspensions. *Int. J. Multiphase Flow*, **16**, pp. 35-42 (1990).

Milne-Thompson, L.M., Theoretical hydrodynamics. *4th Ed. Macmillan*, pp. 530 (1960).

Prosperetti, A., and A.V. Jones, The linear stability of general two-phase flows models-II. *Int. J. Multiphase Flow*, **13**, pp. 161-171 (1987).

Prosperetti, A., J.V. Satrape, Two phase flows in fluidized beds, sedimentation and granular flows. IMA volumes in Mathematics and its Applications 26, *Springer-Verlag*, pp. 98-117 (1989).

Richardson, J.F., and W.N. Zaki, Sedimentation and fluidization: Part I. *Trans. Instn. Chem. Engrs*, **32**, pp. 35-53 (1954).

Singh, P., P. Caussignac, A. Fortes, D.D. Joseph, and T. Lundgren, Stability of periodic arrays of cylinders across the stream by direct simulation. *J. Fluid Mech.*, **205**, pp. 553-571 (1989).

Singh P., and D.D. Joseph, One-dimensional, particle bed models of fluidized suspensions, two phase flows in fluidized beds, sedimentation and granular flows. IMA volumes in Mathematics and its Applications 26, *Springer-Verlag*, pp. 130-149 (1990).

Singh, P., and D.D., Joseph, Dynamics of fluidized suspensions of finite size. AHRCRC Report 91-60. Submitted to *J. Fluid Mech.* (as a revision of the 1989 paper "Chaos and structure in two-dimensional beds of spheres fluidized by water") (1991a).

Singh, P., and D.D. Joseph, Finite size effects in fluidized beds in liquid-solid flows. FED 118, Eds. M.C. Roco and T. Mayasuma, *American Society of Mech. Engrs.*, pp. 77-86 (1991b).

Volpicelli, G., L. Massimilla, and F.A. Zenz, Non-homogeneties in solid-liquid fluidization. *Chem. Eng. Symp. Series*, **67**, pp. 62, 42-50 (1966).

CHAPTER 11

DISPERSION STABILITY: A THIN FILM MODEL FOR CONCENTRATED COLLOIDAL DISPERSIONS

D. T. Wasan
A. D. Nikolov

11.1 INTRODUCTION

Numerous products such as paints, inks, coatings, emulsions, foams, gels and polymer latexes, and processes such as coagulation, flocculation, sedimentation, lubrication, and paper making depend to a large extent on the dispersion quality and stability of particle suspensions. The last decade has seen an intensified interest in understanding the nature of interactions in such concentrated suspension or colloidal dispersions. A number of experimental techniques including scanning electron microscopy, tunneling microscopy, dielectric spectroscopy, x-rays, dynamic light scattering and rheometry have been used to probe particular particle interactions in concentrated colloidal systems.

A new era of research on understanding the nature of interactions within superamolecular fluids such as concentrated suspensions of particles, surfactant micellar solutions, and micro-emulsions was opened when we discovered that the thinning films formed from aqueous, concentrated dispersions of latex particles changed thickness with regular step-wise jump transitions during which the films exhibited a number of metastable equilibria until a stable final state is reached. The height of the steps is approximately constant and nearly corresponds to the effective diameter of the particles. Such a step-wise thinning of the film (Nikolov and Wasan 1989) is due to the formation of a long-range crystal-like structure within the film and a layer by layer thinning of such an ordered structure. This ordering occurs because highly charged Brownian particles interact via repulsive forces and are forced into the restricted volume of the film.

One way to demonstrate the presence of ordered structure inside a stratifying film is to form a large film (2.5cm diameter) in a vertical frame (see Figure 11-1a) inside of a glass cylinder with films formed from 44 vol% latex particle suspensions and a diameter of 156nm. One observes a series of horizontal stripes of different, uniform colors at the upper part of the film. The different colors are due to interference of the common (polychromatic) light reflected by the surfaces of the different, uniform thickness stripes. The boundaries between the stripes are very sharp - a consequence of the step-wise profile of the film surface in this region - and the liquid meniscus below the film appears as a region with gradually changing colors. When observed in reflected light, the next lower stripes have the following colors: white, yellow, blue, red and green-yellow. Following this, a sequence of alternating green and red stripes of different thickness are observed. After these stripes is the order/disorder transition region. A sketch of the structure of the film containing uniform colloidal particles is shown in Figure 11-1b.

The different thicknesses of the stripes as determined by the difference in reflectivity are marked on this figure. According to the colloid crystal-like model, the different color stripes contain different numbers of particle layers. More specifically, the black stripe (very thin film) does not contain particles; the white stripe contains one layer of particles, the yellow-two layers, the blue stripe - three layers, the red stripe - four layers, and so on. Figure 11-1c shows a similar photograph for the 20 vol.% silica particles having a diameter of 19nm. In this case, one observes more stripes because the particles are smaller and, therefore, it is to be expected that more ordering should exist. These photographs also show the almost circular spots in the order/disorder region. The colored spots of lesser thickness than the surrounding film move upwards in the lower stripes and, eventually, fuse with the corresponding colored stripe. By measuring the velocities and size (radii) of these spots, one can estimate the effective dynamic viscosity of the ordered structure inside the film. This point is discussed in a later section.

We observed similar sharply defined stripes with films formed from micellar solutions of nonionic surfactant (e.g. ethoxylated alcohol, with 30 ethoxy groups and 12-15 carbon chains) with a micellar diameter of about 10nm. (see Figure 11-1d). However, all stripes were gray in color, though with different intensities, because the diameter of the micelles is small.

In 1973, Japanese scientists found that latex particles in an aqueous bulk solution formed ordered structures (solid-like arrange-

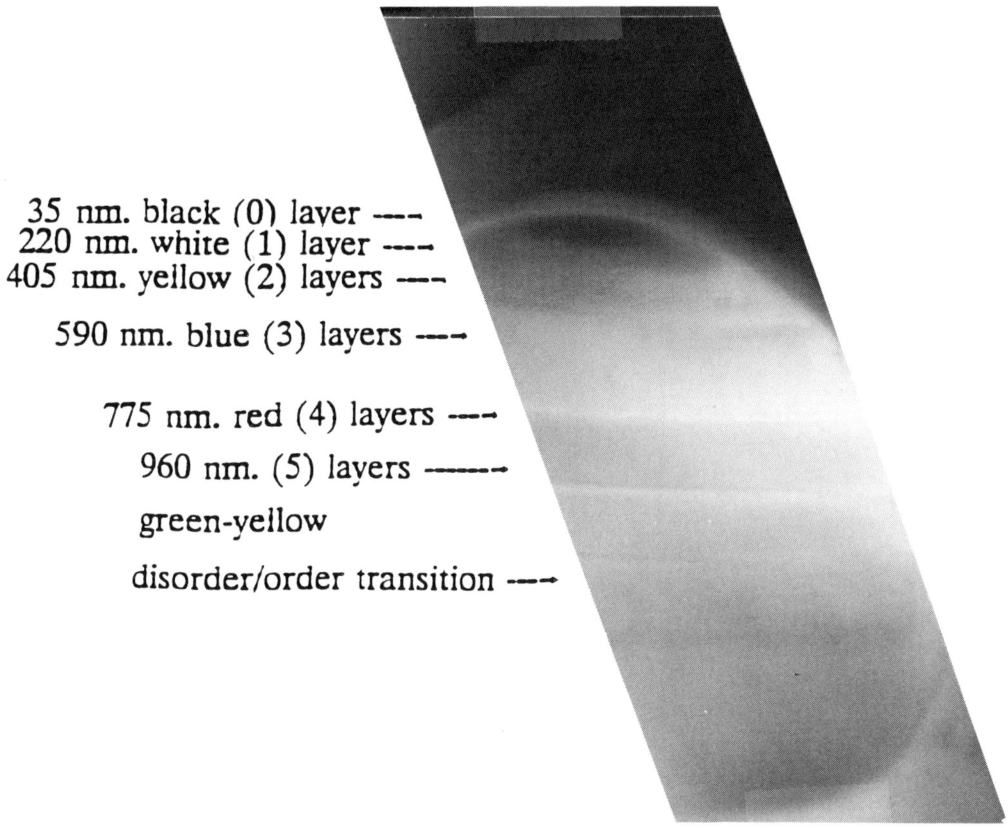

Figure 11-1a. *Interference stripes in a vertical stratifying film formed from a 44 V% latex suspension with a particle diameter of 156nm. Each stripe represents a different number of particle layers inside the thinning film.*

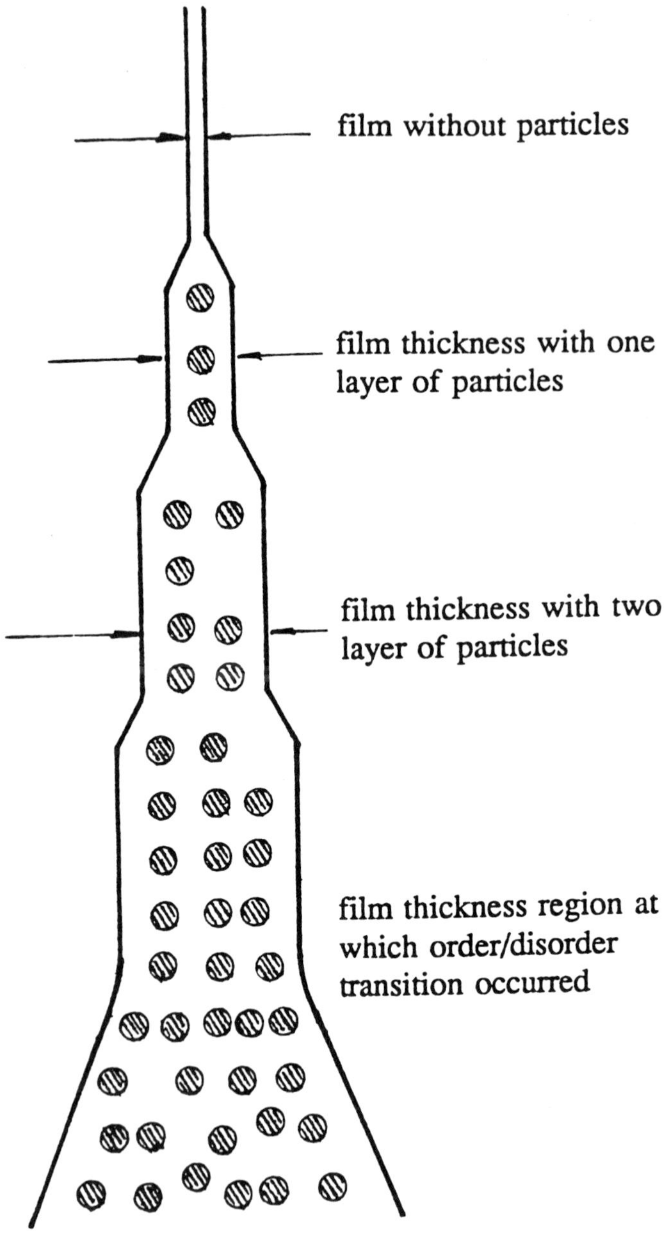

Figure 11-1b. *Sketch of the structure of vertical film containing uniform colloidal particles. The film thicknesses are marked with arrows.*

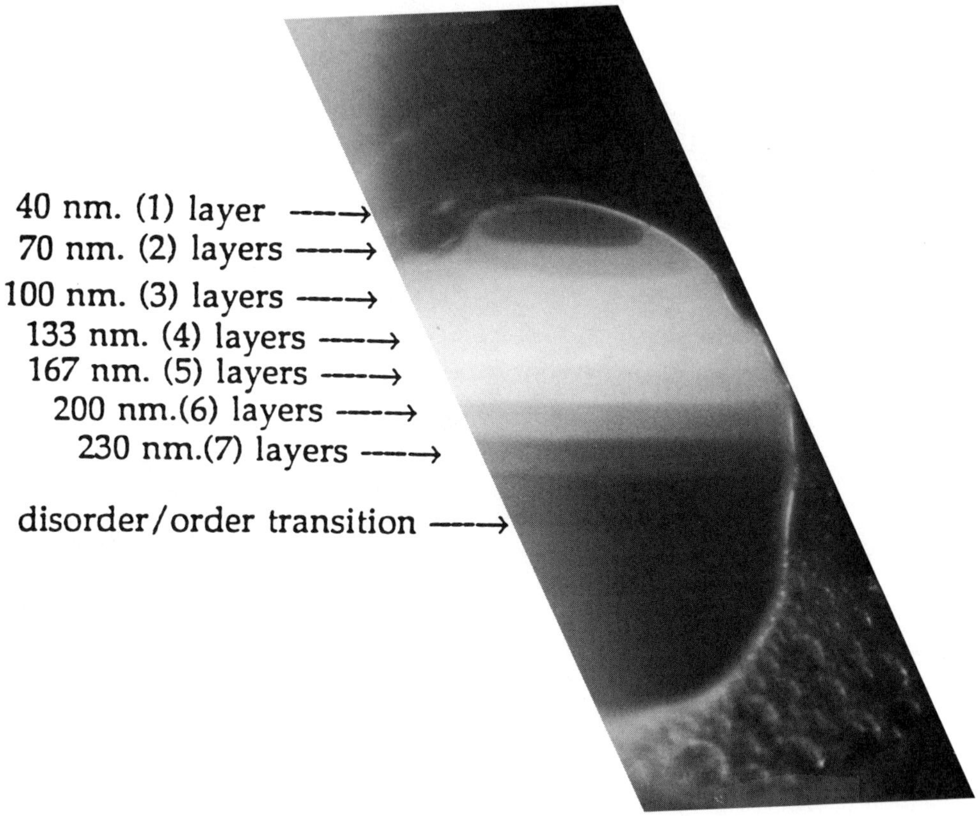

Figure 11-lc. *Interference stripes in a vertical stratifying film formed from 20 V% silica suspensions with particle diameters of 19nm. Each color stripe represents a different number of particle layers inside the thinning film.*

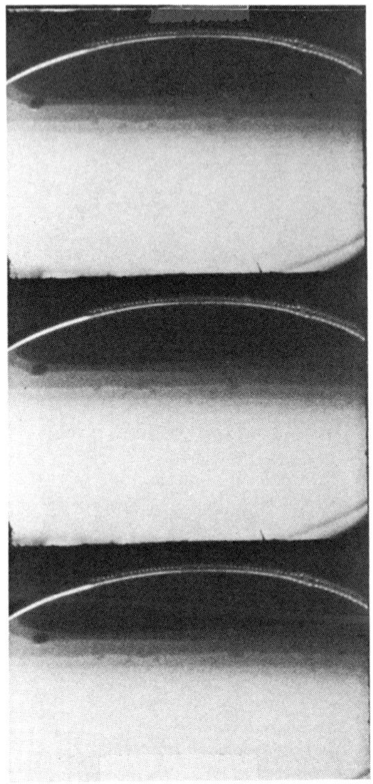

Figure 11-1d. *Interference patterns in a vertical stratifying film formed from 5.2 x 10^{-2}mol/l ENORDET AE 1215-30 solution with micellar diameter of about 10nm.*

ments of latex particles) and this finding aroused considerable interest (Hachisu et al., 1973). The classical DLVO theory explained order in such systems as a balance of van der Waals forces and electrostatic forces, however, relative to the size of the particles, the interparticle distances were too large for van der Waals forces to be significant. The ordering, then, had to be entirely due the electrostatic interactions.

In recent years, much attention has been devoted to experimental characterization and theoretical prediction of the structural and rheological properties of bulk concentrated suspensions (Castillo et al., 1984; Russell and Benzing, 1983; Pieranski, 1983; Reiner and Radke, 1991). For a comprehensive summary of these studies, the reader is referred elsewhere (Ise and Sogami, 1988; Russell et al., 1989). Suffice it to say that the dynamics of structure formation are strongly dependent on the Brownian motion of the particles and the nature of their interactions. In the time-average scale, this relates directly to osmotic compressibility and, thereby, to the structure factor and radial-distribution functions.

Following, is our study of the thin films of self-organizing microstructures, such as colloidal particles as well as surfactant micelles, which assume ever increasing importance in current, as well as in new and emerging technologies. The study of thin films of self-organizing microstructures has applications to such diverse areas as ceramics processing, paints and inks, coatings, magnetic tapes and disks, and emulsion and foam systems. The thickness of these films is typically on the order of the long-range intermolecular forces ($<0.1\mu m$ or 1000Å).

The dynamic process of stratification, or ordered microstructuring in submicron thin liquid films is shown to be important tool for probing the long-range structural or interaction forces in concentrated particle suspensions and colloidal dispersions. It is important to understand these forces in order to control the overall stability of films and thereby, colloidal dispersions.

11.2 EXPERIMENTAL INVESTIGATION OF STRATIFYING FILMS

In the experimental method we developed for investigating thin film behavior (Nikolov and Wasan 1989), the first step is to form a biconcave drop of the liquid under study in a cylindrical capillary (of radius $R=1.5x10^{-1}cm$) with hydrophilic inner walls. Next, liquid is slowly sucked out of the drop through an orifice in the capillary wall, creating a horizontal flat film (of radius about 10^{-2} cm) encircled by a biconcave liquid meniscus. This type of specially designed cell is similar to the one introduced first by Scheludko (1957). The glass assembly is enclosed in a thermostat and a constant temperature is maintained (±0.1°C). The entire assembly is placed on the stage of a metallograph microscope which is mounted on a vibration free table to keep any external dis-turbances from affecting the film thinning process (Figure 11-2). Mono-chromatic light (wavelength 546nm) from the bottom of the glass cell is incident on the films surface and the light reflected from a small portion of the film area is conducted through a fiber optic probe to a photomultiplier. Here the optical signal is converted to an electrical signal, amplified through an electrometer and, finally, recorded on a strip chart recorder which represents photocurrent as a function of time. As the film thins the thickness changes, producing interference patterns. The film thickness, h, at any instant of time, can be estimated from the intensity of the photocurrent using the expression derived by Lyklema *et al.* (1965).

$$h=\frac{\lambda}{2\pi n_0}\left(l\pi \pm \arcsin\sqrt{\Delta/[1+(4\beta/(1-\beta))(1-\Delta)]}\right) \tag{11.1}$$

where

$$\beta=\frac{(n-1)(n-n_0)}{(n_0+1)(n+n_0)} \tag{11.2}$$

λ = wavelength of monochromatic light (546nm)
l = order of interferences
n, n_0 = refractive index of film liquid and dispensed phase respectively

$$\Delta=\frac{I-I_{\min}}{I_{\max}-I_{\min}} \tag{11.3}$$

I = instantaneous value of the photocurrent
I_{min}, I_{max} = value of the photocurrent at the last interferential minimum and maximum, respectively.

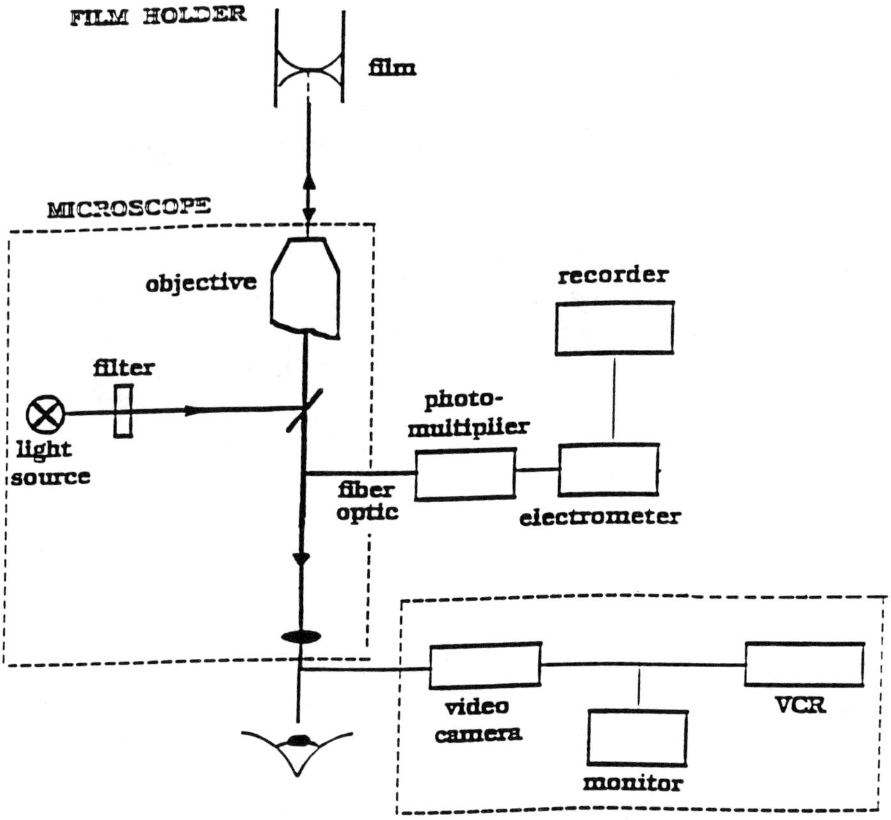

Figure 11-2. *Sketch of the experimental set-up for measuring film thickness by light interferometry.*

The video in conjunction with a monitor and VCR records, in detail, the process of film thinning. The method described here is a general one, appropriate for studying films formed, not only from micellar surfactant solution, but also from suspensions of latex, silica particles and other colloidal dispersions.

Using a film formed from a micellar surfactant solution as an example, we observed the following phenomena: As soon as the film forms, it starts to decrease in thickness. After it is thinner than 104nm, i.e. 1040Å (the last interferential maximum) the film thickness changes in steps (see Figure 11-3). The film first rests for a few seconds in a metastable state with uniform thickness. Then, one or several spots (with smaller thickness than the remaining part of the film) appear and gradually increase in size (see Figure 11-4a). Soon, the spots cover the entire film and the film rests for several seconds in a new metastable state. Then, even darker spots appear and, after their expansion, a subsequent metastable state ensues. This process continues until the film finally reaches a stable state with no more stepwise changes. The metastable state of the film appears in the interferogram as a step whose width is proportional to the lifetime of the respective state. The calculated height of the steps is also shown in Figure 11-3 and the magnitude is approximately constant for all steps (about 10.6 nm), which corresponds to micelle diameter, about 10nm.

Film formed from polystyrene latex suspensions (see Figure 11-4b) stratify in a similar fashion to these formed from micellar solutions, with one observable difference. Because the particles are much larger (in our case 91nm) than surfactant micelles, with a diameter exceeding the thickness of the last interferential maximum, there can be constructive as well as obstructive interferences, and the thinner spots sometimes appear brighter rather than darker than the remaining thicker film, as shown in Figure 11-4b. The average height of the step in these cases is about 91nm, and the diameter of the latex particle is 91nm.

In every case, the height of the steps is approximately constant, corresponding closely to the effective diameter of the surfactant micelles or the latex particles. For ionic surfactants (Figure 11-4a), the effective micelle diameter includes the Debye diameter of the surrounding electric double layer.

The experimental data for stratifying films formed from latex suspensions or from micellar solutions of anionic and nonionic surfactant (Nikolov and Wasan, 1989; Nikolov et al., 1990), and our theoretical analysis of this data (Nikolov et al., 1989; Kralchevsky et al., 1990; Nikolov et al., 1988), show that stratification is a universal

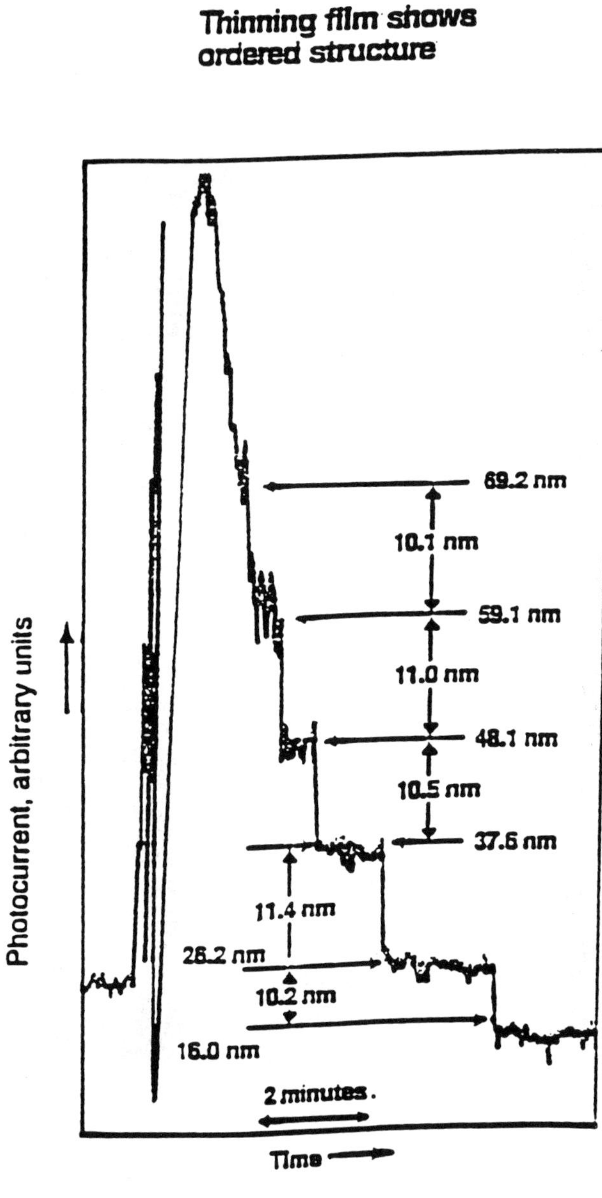

Figure 11-3. *Interferogram of film formed from solution of nonionic detergent (Enordet AE 1215-30, 0.052 mol/l). As film thins, less light is reflected. Formation of metastable states of uniform thickness is revealed by "steps." Height of step (black arrow) corresponds to thickness of film. Vertical distance between steps corresponds to micelle diameter, about 10nm. Width of steps is proportional to lifetimes of respective metastable states.*

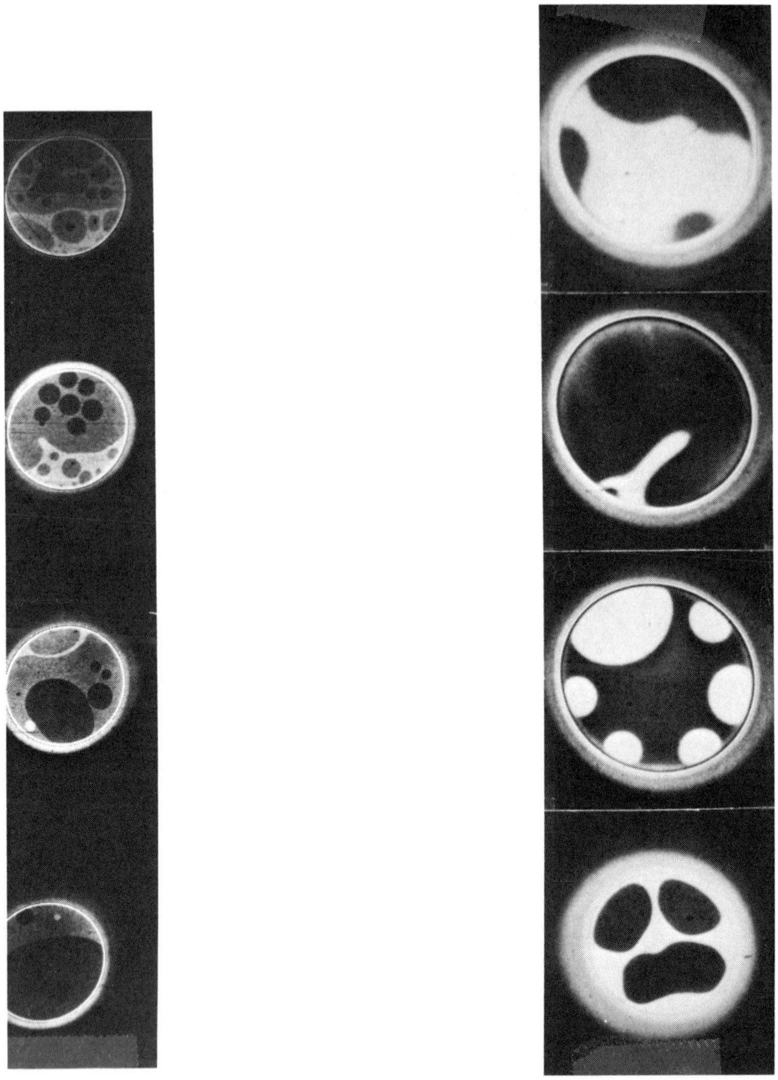

a.

b.

Figure 11-4. *Stratification of films:*
a. 0.1 mol/l sodium dodecylsulfate
b. 30wt% latex suspension with a particle diameter of 91nm

phenomenon which occurs when ordered structures of latex particles or micelles form inside the film and that the stepwise transitions are a layer-by-layer thinning of that structure.

A direct proof of the presence of microstructures within thin films is of paramount importance. We have conducted some preliminary experiments using the low-angle transmitted light diffraction technique with a laser wavelength of 632.8nm. A vertical microscopic film of diameter 2.5cm was formed from a 40wt% latex suspension on a vertical plate frame inside of a glass cell which was especially designed to prevent film thinning caused by mechanical disturbances and evaporation. The monodispersed charged latex beads had a diameter of about 150nm and a charge of ca. 5μcolum/cm^2. Figure 11-5 shows the diffraction pattern obtained. The pattern symmetry indicates the presence of a colloid crystal-like structure.

Latex suspensions have been known to undergo phase transitions and to form ordered structures called colloid crystals in the bulk phase. This is due to the interaction between particles which are strongly repulsive. When the repulsive force is electrostatic (as in latexes and micellar solutions of ionic surfactants), adding salt to the mixture suppresses stratification as seen in Figure 11-6, which is a plot of the effective volume fraction of micelles of sodium dodecyl sulfate anionic surfactant (SDS) versus NaCl concentration. The solid curve which represents the threshold concentration, separates the regions with and without stratification. The dashed line on this same phase diagram shows the results of (Hachisu et al., 1973; Ohtsuki et al., 1978) the order-disorder phase transition in bulk latex suspensions (the ordered phase region is on the left of the dashed line). The close resemblance of the two plots bears out our explanation of the stratification as being a restricted volume, induced phase transition in the film.

We have also recently shown that when the repulsion is the result of steric forces (the case with micellar solutions of nonionic surfactants), stratification is temperature sensitive (Nikolov et al. 1990).

11.3 DISJOINING PRESSURE DUE TO ORDERED MICROSTRUCTURE

As discussed in our recent papers (Nikolov et al., 1990; Nikolov et al., 1989; Kralchevsky et al., 1990; Nikolov et al., 1988), the excess (disjoining) pressure Π in the stratifying film has the

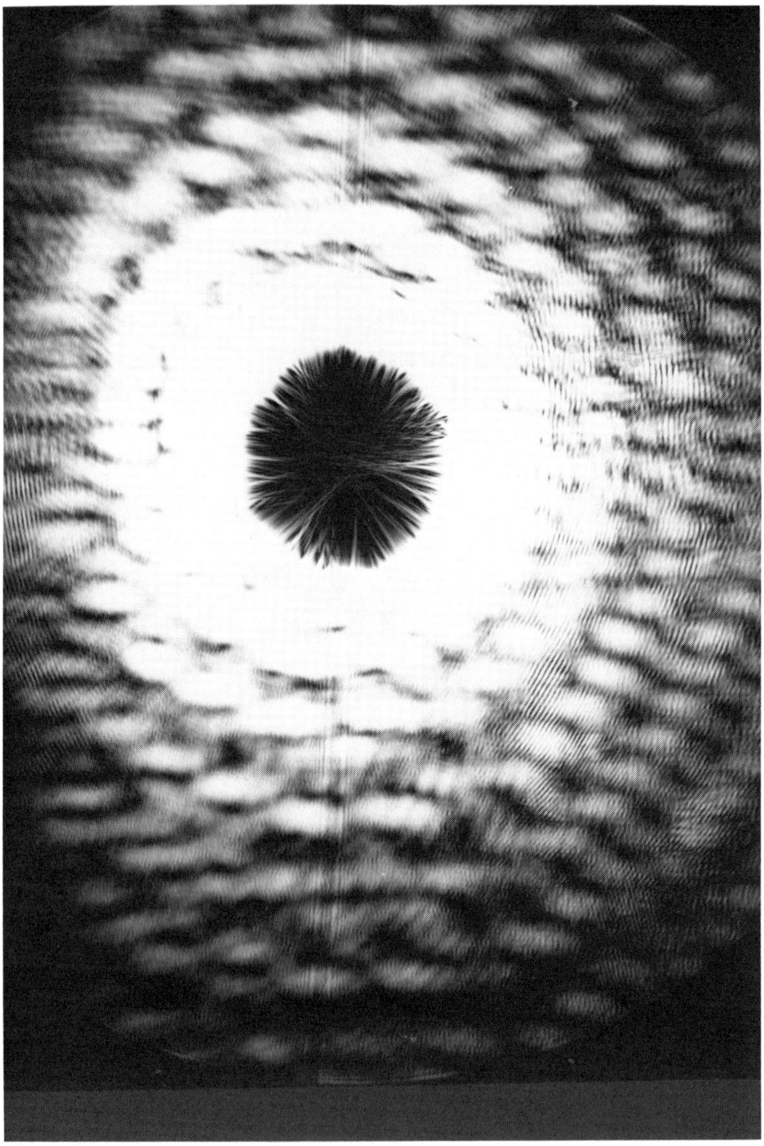

Figure 11-5. *Diffraction patterns from a vertical foam film formed from 40wt% latex suspension, with particle diameter of 156nm.*

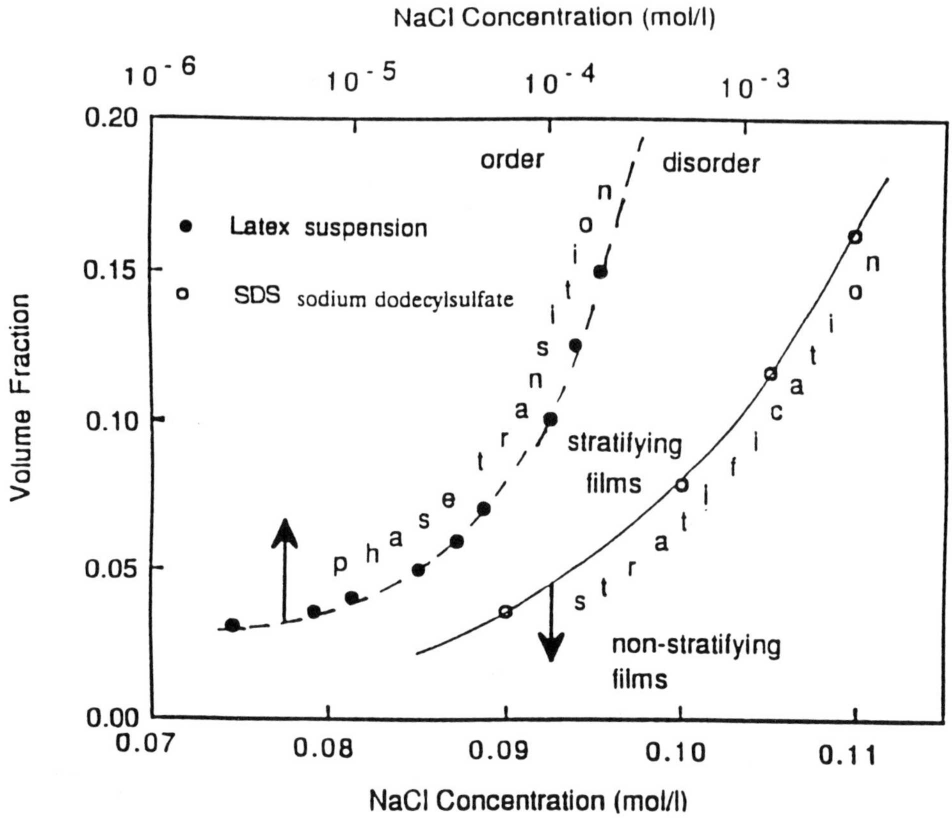

Figure 11-6. *Volume fraction versus concentration of added NaCl for latex solution and micelles. The dashed line is the experimental curve of Hachisu et al., (Ref. 2) for the added electrolyte order-disorder phase diagram for latex solutions. The results for stratifying-non-stratifying films are also plotted as a phase diagram.*

additional component Π^{ms} because of the ordered microstructure inside the film.

By using the effective potential of the interparticle repulsion $U^{eff} = (U_o/\delta) \exp(-\kappa\delta)$, δ is the distance between the centers of the colloid particles along normal to the film, κ and U_o are constants, (Beresford Smith et al., 1985), we can derive the following expression for the structural component of the disjoining pressure, Π^{ms} as

$$\Pi^{ms} = \frac{U_0}{\delta_l^2}\left[(\kappa+\frac{1}{\delta})\frac{e^{-\kappa\delta}}{\delta} - (\kappa+\frac{1}{\delta_l})\frac{e^{-\kappa\delta_l}}{\delta_l}\right] \tag{11.4}$$

where δ_l is the distance between the two particles in the bulk solution.

During the process of thinning $\delta < \delta_l$, and then $\Pi^{ms} \neq o$.

The total disjoining pressure Π_n, for a film with *n* particle layers in it also contains terms accounting for the electrostate repulsion and van der Waals attraction between the film surfaces. By integration of Π_n with respect to the film thickness, *h*, we derived an expression for the excess free energy, ω_n, per unit area of the film (Nikolov et al. 1989). The curves ω_n (*h*), *n* = 0,1,2.. (see Figure 11-7) exhibit minima, which corresponds to the metastable states (*n* = 1,2,3,-.) and to the final stable state (*n* = 0) of the film. A stepwise film thickness transition can be interpreted as a transition from a given metastable state to the next one. The experimental values of the film thickness are in good agreement with the ones theoretically calculated (Nikolov et al. 1989). The shape of the excess energy curves, ω_n (*h*), also properly reflects the phenomenon of stratification, the energy of the metastable state decreases with the decrease in film thickness. These findings for the stratification phenomenon in thin liquid films are consistent with the known fact that in a concentrated suspension where the interactions between neighboring particles are strongly repulsive, the suspension will tend to form an ordered lattice to minimize its potential energy (Hirtzel et al., 1985).

Recently, we measured the contact angle and, thereby, the film tension and excess film energy by forming small lenses in the stratifying film (Kralchevsky et al., 1989). Table 1 lists the values of the contact angles, as well as the changes in the film tension ($\Delta\gamma$) as a function of film thickness and micellar layers (*n*) inside the film for 0.1M sodium dodecylsulfate surfactant solution. This table clearly shows that the change in excess film energy increases as the film becomes thinner, (i.e., the energy barrier to jumpwise transitions

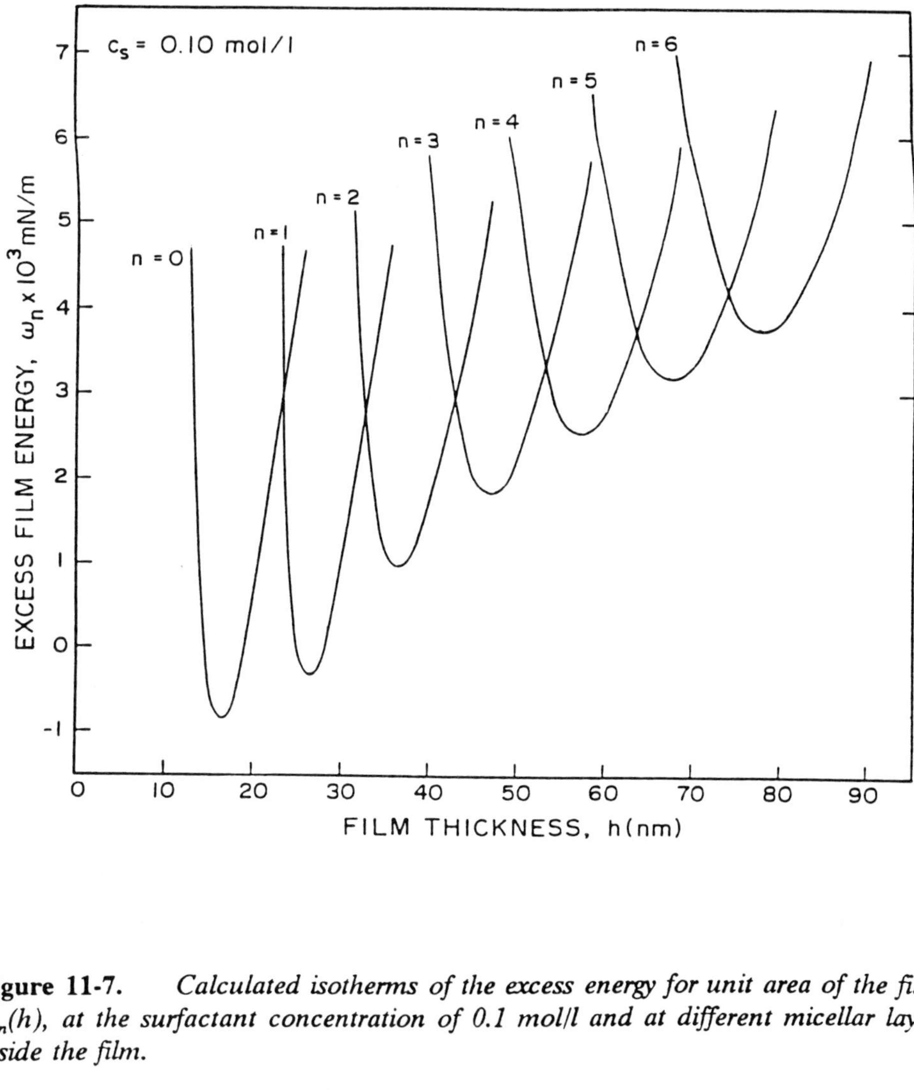

Figure 11-7. *Calculated isotherms of the excess energy for unit area of the film, $\omega_n(h)$, at the surfactant concentration of 0.1 mol/l and at different micellar layers inside the film.*

Table 1. Measured Values of the Film Thickness $h_{n,}$ and Contact Angle $\theta_{n,}$ of Stratifying Film from 0.1 mol/l SDS

Number of Micellar Layers n	*Film Thickness* $h_{n,}$ *nm*	*Contact Angle* $\theta_{n,}$ *deg*	*Change in the Film Tension* $\Delta\gamma \times 10^3$, *mN/m*
0	16.2	1.26	
1	26.1	0.906	9.05
2	35.9	0.790	2.69
3	47.0	0.685	2.29
4	57.9	0.665	0.85

increases with decreasing number of layers). This is consistent with our calculations as depicted in Figure 11-7.

It is noteworthy that the curves in Figure 11-7 resemble, by shape, the oscillation of the force vs. distance curves (Israelachivili and Pashley, 1983) which are due to the short range solvation forces; in our case, the phenomenon of stratification is governed by long-range screened electro-static repulsion between colloidal particles. Indeed, at a most recent symposium on surface chemistry, Kekicheff (1991) presented the results on the oscillatory structural effects in thin films of micellar solutions contained between two solid mica sheets using their surface force apparatus. Their force vs. distance measurements for micellar films where the electrostatic interaction dominate, have independently confirmed our results (See Figure 11-7).

Hirtzel and Rajagopalan (1985) have shown that osmotic pressures of even moderately-interacting dispersions can be many times over the pressure of uncharged dispersion of the same size. Furthermore, the osmotic pressure of ordered suspensions is known to be strongly dependent upon the effective volume fraction of interacting particles.

Therefore, we report, here, our data on the osmotic pressure versus the effective volume fraction of the micelles for the nonionic surfactant solution (Figure 11-8). These data were obtained by using the well known light scattering technique and based on the calculation

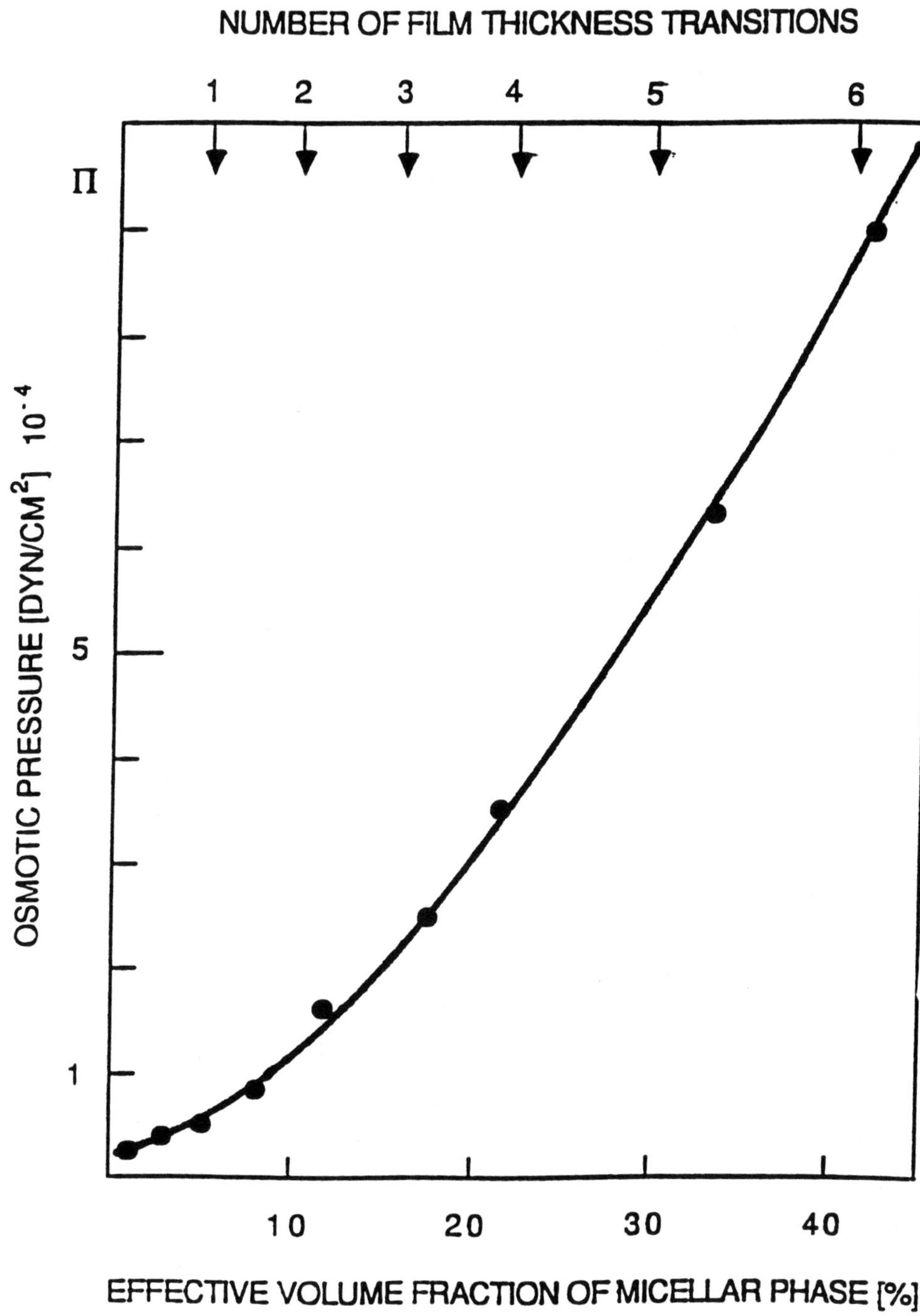

Figure 11-8. *Osmotic pressure versus effective volume fraction of micellar phase of ENORDET AE 1215-30 nonionic surfactant solution.*

of the second viral coefficient. Also shown on this graph are the film thickness transitions determined by using our microinterferometric method. We found that the osmotic pressure increases with increasing the volume function of the micellar phase. Also the number of transition increases when the micellar volume fraction is increased. The osmotic pressure also increases with increasing the number of transitions.

11.4 SPOT FORMATION

The dimension of a stratifying film in the normal direction is much less than the dimension of the film in a lateral direction. For this reason, the equilibrium, with respect to the forces acting normal to film surfaces is achieved faster than the equilibrium between the film and the bulk solution in the meniscus. When the lateral equilibrium is reached the process of stepwise thinning of the film stops. Depending on the experimental conditions (capillary pressure, film size, electrolyte concentration, particle size and polydispersity, etc.), the film in its final state can contain $n = 0,1,2$, or more particle layers.

During the process of stepwise thinning, the presence of normal stability at the metastable states (the minimum in Figure 11-7) is accompanied with lateral instability. The driving force of the thinning is the difference between the chemical potentials of the particles inside the film and inside the meniscus. Under the action of the gradient of the chemical potential, i.e. the osmotic pressure gradient, particles or micelles leave the film and vacancies appear in their places. Because the particles diffuse, the vacancies (initially created at the film periphery) move throughout the entire film area. For example, particles at 20 vol% concentration and with diameter of 20nm can move through the film diameter of 3×10^{-2} cm in about 10^{-2} second. Phase transitions occur at sufficient large concentrations of vacancies and, as a result spots are formed. After a spot has been formed, it gradually increase its area. The vacancies created at the film periphery "condense" at the spot thus increasing its area. At a given concentration of the vacancies, ($\theta = \theta_{eq}$) equilibrium between the film and the solution in the meniscus is reached, which leads to the arrest of the stepwise thinning. This diffusive-osmotic mechanism of the spot formation is analyzed in a recent paper (Kralchevsky et al., 1990); this mechanism is applicable both to films containing micelles and to films containing latex or other colloid particles. Our model predicts the appearance of spots above a given critical concentration, θ_{cr} of vacancies in the film. Spots will not appear in the film if $\theta_{cr} > \theta_{eq}$, and stratification will not occur. The model also predicts a pronounced increase of this critical concentration of vacancies with the decrease of the film area. In this way, a film with a given thickness exhibits stepwise thinning when its area is large but, when its area is small, it stays in the equilibrium state. This very important feature of the structuring film can be used to control the stability of the dispersion system.

By increasing the film area, we increase the vacancy flux and this increases the probability of spot formation and accelerates the film thinning process. By decreasing the film area, we decrease the vacancy flux and the chance of spot formation. Therefore, we can stop the thinning process. This film in its stable state can contain several layers of particles.

Let J be the total influx of vacancies in the film and let r be the radius of the spot. If J is constant during the spot expansion, we can derive the simple relationship between the size of the spot and time:

$$r^2 = \frac{\delta_l^2}{\pi} Jt + constant \tag{11.5}$$

Where δ_1^2 is the area of a vacancy, that is r^2 should be a linear function of time. And, in fact, this was observed experimentally (Kralchevsky et al., 1990).

It is evident from the preceding discussion that the vacancy flux controls the film stratification process and is dependent, not only on the film size but also on the particle characteristics, such as size, concentration and polydispersity, as well as on surfactant concentration. Some of these factors affecting this thin film stratification phenomenon are discussed below.

11.5 EFFECT OF PARTICLE CONCENTRATION

We have previously mentioned that increasing the surfactant micellar concentration enhanced film stratification. This is, indeed, the result of the increased effective volume fraction of the micellar phase. We now report our results on the effect of solid particle concentration on the stratification phenomenon.

Figure 11-9 shows the successive movie frame photographs depicting the thinning behavior of the film formed from an aqueous

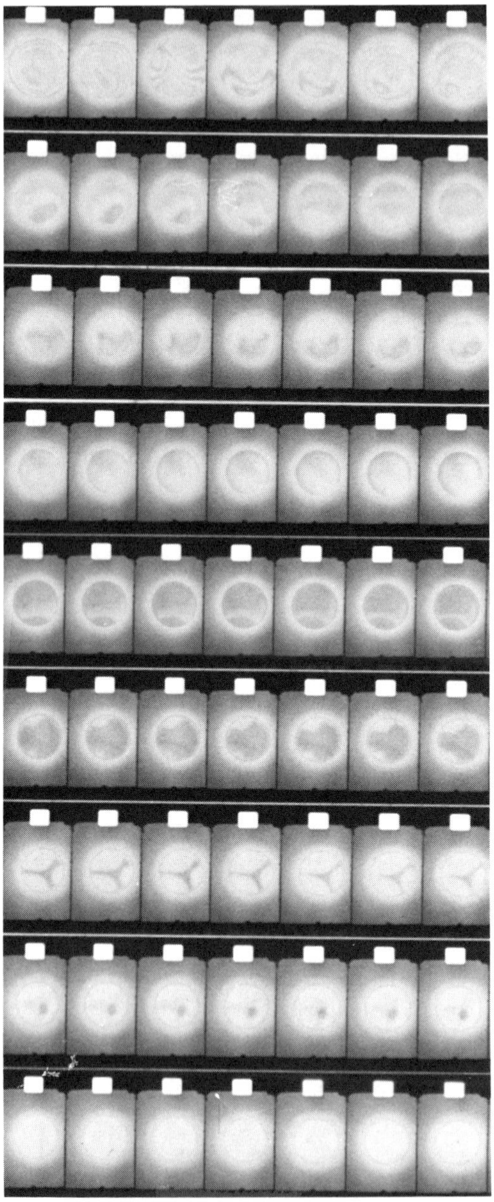

Figure 11-9. *Successive movie frame photographs show the thinning of a horizontal flat film with diameter $6x10^{-2}$cm formed from 20 V% water suspension of monodisperse silica.*

dispersion of 20 V% of silica particles with a diameter of about 19nm. The thinning behavior of this film is similar to that of the surfactant micellar film and also to that of the film containing latex particles (Figure 11-4b). Initially, when the film is formed, interference patterns (circles) appear representing the dimple formation due to the hydrodynamic interactions between the approaching surface of the film. The dimples disappear after a film thickness of 400nm is reached. The interferogram is shown in Figure 11-10. The film thins through several maxima and minima in the intensity of light (the successive maximum and minima represents a change in thickness of about 100nm). After the last maximum (at about 300nm), the first step-transition at a film thickness of about 230nm, is observed, followed by several others at 200, 167 and 133nm, respectively. The film stops thinning after about 2 minutes, at about 100nm. In this particular case, during the last stage of film thinning, the light intensity increases by steps which indicates that the stratification phenomena is occurring at high values of film thickness (i.e., before the last light intensity maximum). The mean height of the step is about 30nm. It is noteworthy that a stable film, with film thickness of about 100nm and containing two layers of particles inside it, can exist in this case.

Figure 11-11 shows an interferogram for the same silica system when the concentration is reduced to 10 vol%. There are only two thickness transitions to be seen in this case , the first transition at about 120nm and the second at 77nm. The film achieves its equilibrium thickness at about 40nm and contains one layer of particles inside it. The mean height of the step is about 38nm. In this case, the effective diameter of the particle including the Debye diameter is found to be about 36nm, which is closer to the mean height of the step.

The light intensity curve for the 20 vol.% silica suspensions shows the steep vertical steps as seen with the nonionic surfactant micellar solution (see Figure 11-3) which represents the case of hard sphere interactions, whereas the light intensity curve for the 10 vol% system shows gradual decrease in film thinning, indicative more of a soft particle interactions.

Upon further decrease of the particle concentration to 5 vol% we could see only one stepwise transition and, in this case, the film ruptured.

These experimental observations clearly reveal the effect of particle concentrations on the formation of the long-range ordered structure inside the stratifying film. At higher concentrations, ordered

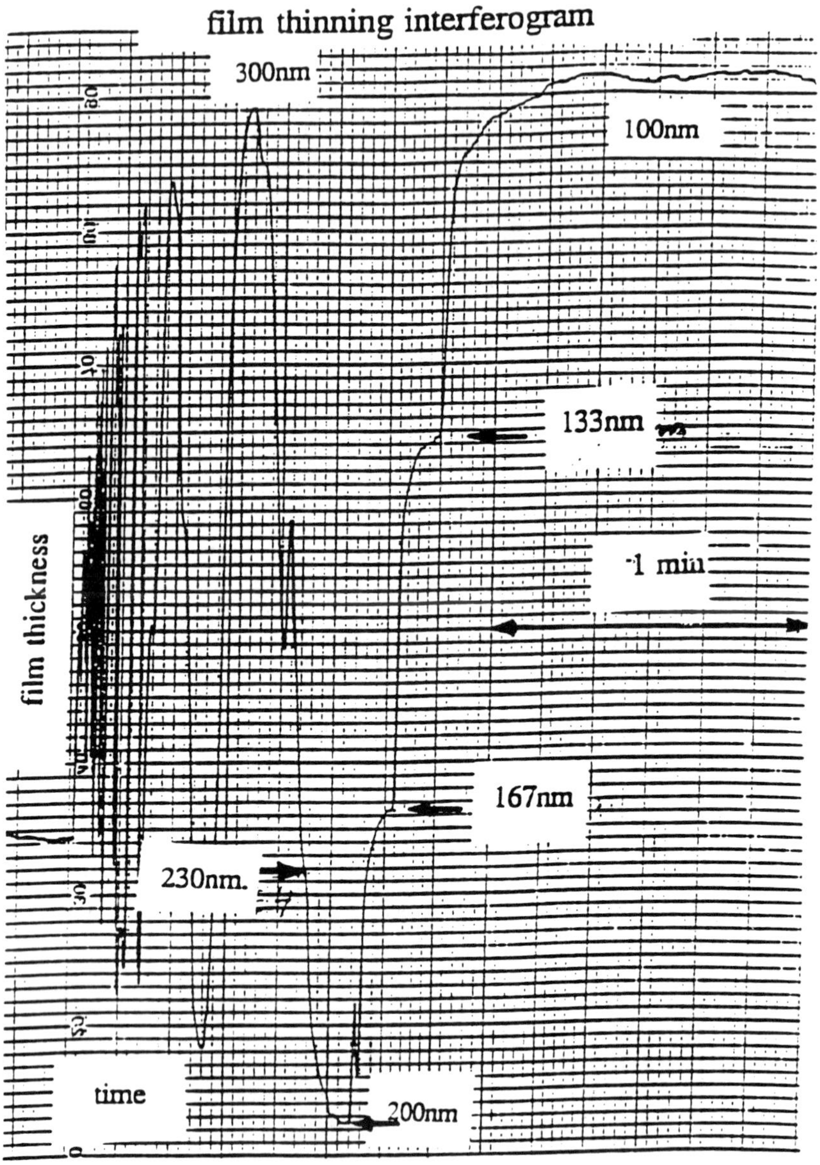

Figure 11-10. *Photocurrent vs. time interferogram of film thinning process of microscopic horizontal film (film diameter* $6x10^{-2}cm$*) stabilized by silica particles (20 V%) with particle diameter, about 19nm.*

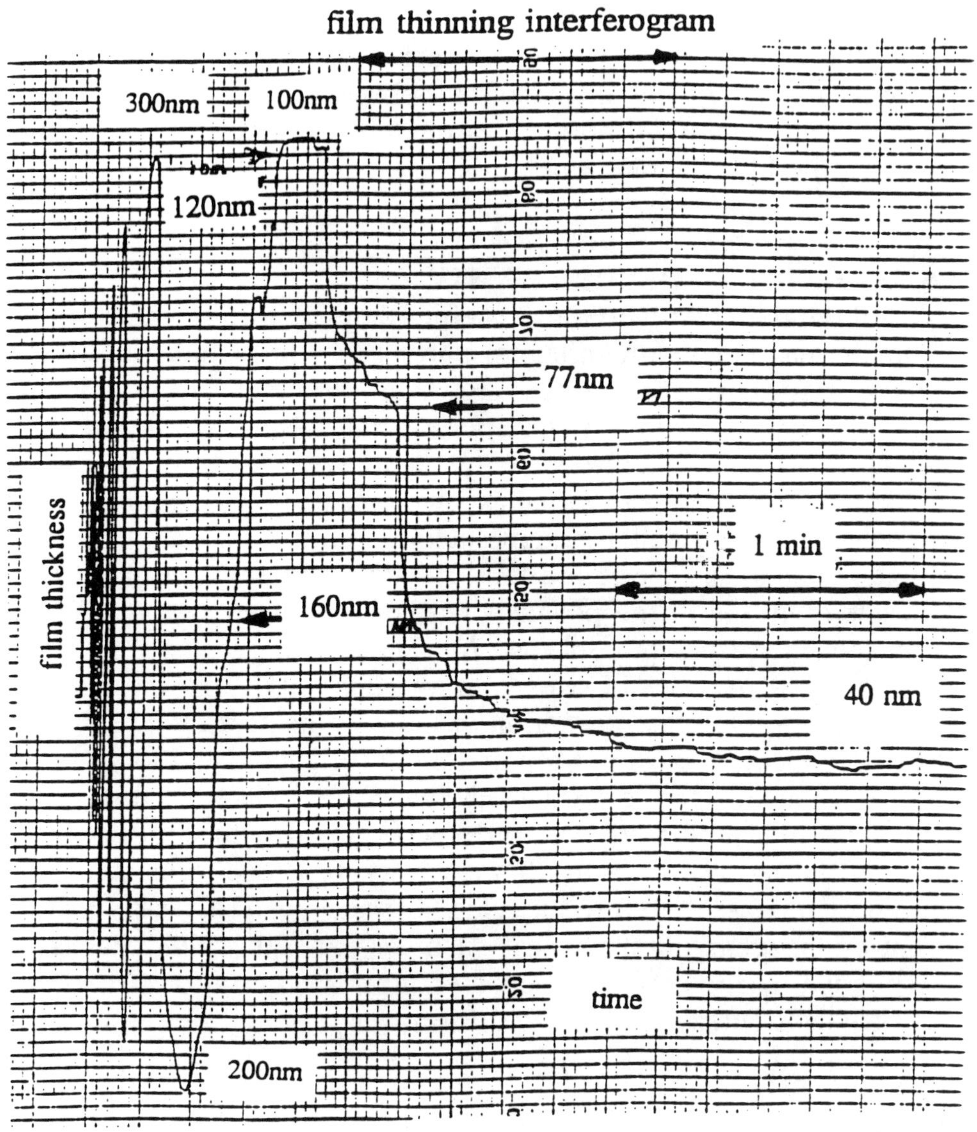

Figure 11-11. *Photocurrent vs. time interferogram of film thinning process of microscopic horizontal film (with diameter 6x10⁻²cm) stabilized with silica particles (10V%) with diameter, about 19nm.*

microstructures form at a much larger film thickness (i.e. even before the last maximum in the intensity of light, e.g., at a film thickness of about 100nm) and the film contains several layers of particles. As we decrease the particle concentration (i.e. increase interparticle spacing) we reduce the energy of interactions between the particles, thus resulting in a reduction in the number of stepwise transitions and an increase in the film thinning rate.

11.6 EFFECT OF POLYDISPERSITY

All the data obtained today on film thinning and stratification phenomena has been with spherical particles or surfactant micelles with very narrow size distribution. However, dispersions of practical interest invariably contain polydispersed particles. In order to elucidate the affect of polydispersity on long-range ordering and multilayered structuring, our initial experiments dealt with thinning of films from a mixed micellar system such as solutions of anionic and nonionic surfactants. We found that the step-wise thinning process was accelerated due to the polydispersity of the mixed system. We now present the results of our new experiments using a 20 vol%; 1:1 mixture of silica particles of sizes 7nm and 28nm. During the film thinning we observed fast formation of many spots at the same time and the stratification began at about 100nm instead of at 230nm for the monodispersed particle system. However, the spots were not very sharp, which indicated a not very ordered (rigid) structure inside the film. Furthermore, the film thinned very fast and then finally ruptured. It was therefore concluded that, even at very high concentration, charged particles such as at 20 vol% (which corresponded to about 50% effective volume concentration), the effect of polydispersity leads to the disruption of long-range order.

11.7 EFFECT OF SURFACTANT

Surfactants are commonly used to improve particle dispersion stability in industrial processes. To explore the effect of surfactants on thin film stratification process we conducted the following experiments. We have added to a 30 vol% aqueous latex suspension (with a diameter of about 91nm) surfactant micellar solutions of sodium dodecyl-sulfate. It was found that after the four thickness transitions produced by the latex particle (see Figure 11-4b), two additional transitions produced by the surfactant micelles were observed. The step-height in each of the last two transitions corresponded to the effective micellar diameter. These observations clearly showed that a mixture of micelles and latex particles leads to a more ordered microstructuring inside the film. Therefore, it can be concluded that the micellar particles, which are smaller than the latex particles, could be trapped in the dispersion and contribute to the very high stability of the dispersion. This study offers an illustration of the complex structural features one may expect to see in strongly-interacting particle dispersion systems.

11.8 VISCOSITY OF STRATIFYING VERTICAL FILM

In an earlier section in this paper, we presented the stratification phenomenon observed in a stratifying, macroscopic vertical film formed from concentrated latex particle suspensions showing a series of parallel, colored horizontal stripes of different thicknesses. We reasoned that the stratification occurs in thin films containing submicron particles with narrow size distribution and prevailing repulsive forces. A long-range ordered structure forms inside these films and the stepwise transitions are a layer-by-layer thinning of that structure. The formation of long-range ordered structures has many implications of both fundamental and practical significance. Here we will be concerned with the rheological aspect, in particular, the dynamic viscosity of such an ordered structure.

We recorded the motion of the nearly circular colored spots arising in a vertical film formed from a 44 V% aqueous latex suspension (with particle diameter of 156nm). We then tried to determine the effective viscosity of the colloid crystal-like structure inside the vertical film using a two-dimensional analogue of the known Stokes formula. The details of these calculations are given elsewhere (Basheva et al., 1990). We measured the radius, a, and the velocity, v, of a number of spots which satisfied the following conditions:

(i) The spot is far enough from the stripe boundaries so that it moves with a constant velocity; and

(i) The velocity of the spot is not perturbed by the proximity of other spots.

We equated the buoyancy force of the spot to the frictional force and found a simple equation.
where

$$-\ln Re = -\ln\left(\frac{c\eta}{\eta_0}\right) + \frac{4\eta v}{\rho g a^2} \qquad (11.6)$$

$$Re = Reynolds\ number = \frac{2\rho a v}{\eta} \qquad (11.7)$$

Here, ρ is the density, g the gravitational acceleration, C is a constant = 7.406, η = dynamic viscosity, n of the ordered structure and η_0 is the viscosity of the solvent (.01*P*). A plot of *ln Re* vs. v/a^2 should give a straight line and the dynamic or effective viscosity, η, is given by the slope of this plot.

Figure 11-12 shows our measurements. The upper line corresponds to the blue spots moving in the red stripe (film thickness about 740nm), whereas the lower line corresponds to the yellow spots moving in the blue stripe (film thickness about 555nm). From the slopes of these lines, one calculates $\eta \simeq 1.6P$. Therefore, we can conclude that value of η is at least one hundred times larger than the viscosity of the solvent. The value of η = 1.6*P* is reasonable for an ordered suspension (Basheva et al. 1990). For the purpose of this comparison, we used the experimental data (See Benzing and Russell, 1981) for the shear elastic modules.

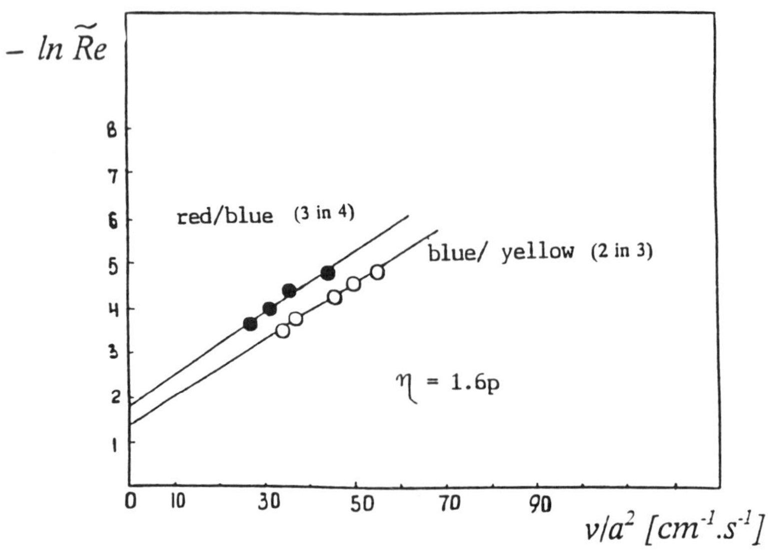

Figure 11-12. *Plot of ln Re vs. v/a^2 for blue spots in the red stripe (•) and for yellow spots in the blue stripe (o) at 25°C*

11-9 SUMMARY

Recent experiments on thinning of single liquid films formed from concentrated colloidal dispersions show that these films stratify, or become thinner, in a stepwise fashion. Stratification is observed in films formed from concentrated suspensions of polystyrene latexes and silica particles as well as from micellar surfactant solutions. This stratification, which is a universal phenomena and is due to the formation of an ordered crystal-like structure inside the film, can be observed in films containing submicron particles with narrow size distribution and prevailing repulsive forces. The stratification, and the stepwise transitions, are a layer-by-layer thinning of that structure. The dynamic process of stratification or multilayer micro-structuring in sub-micron thin liquid films is shown to be an important tool for probing the long-range structural or interaction forces in concentrated particle suspensions and colloidal dispersions. Using this technique, the effects of particle concentration, size, polydispersity, electrolyte and surfactant on the oscillatory structural stability of films can be explained. Furthermore, the dynamic viscosity of the long-range crystal-like structure inside the thinning films, formed from concentrated colloidal particle suspensions such as polystyrene latexes, is found to be much higher than that of the pure solvent.

ACKNOWLEDGEMENT

This work was supported in part by the National Science Foundation. We are grateful to Professor Ivan Ivanov for useful discussions.

REFERENCES

Basheva E.S., A.B. Nikolov, P.A. Kralchevsky, I.B. Ivanov, and D.T. Wasan, "Step-wise Drainage and Spot Formation in Liquid Films From Concentrated Micellar Solutions and Latex Suspensions," Lecture presented at the 8th International Symposium on Surfactants in Solution Gainesville, FL, U.S.A. 1990. Paper to appear in the Symposium volume (in Press).

Benzing D.W. and W.B. Russel, "The Viscoelastic Properties of Ordered Latices: A Self-Consistent Field Theory," *J. Colloid Interface Sci.* **83**, 178 (1981).

Beresford-Smith, B., C. Chan, and D.T. Mitchell, "The Electrostatic Interaction in Colloidal Systems with Low Added Electrolyte," *J. Colloid Interface Sci.* **105**, 216 (1985).

Castillo, C., R. Rajagopalan, and C. Hirtzel, "Statistical Thermodynamics of Colloidal Dispersions," *Reviews in Chemical Engineering* **2**, 237 (1984).

Hachisu, S., Y. Kobayashi, and A. Kose, "Phase Separation in Monodisperse Latexes," *J. Colloid Interface Sci.* **42**, 342 (1973).

Hirtzel, C.S. and R. Rajagopalan, "Colloidal Phenomenon-Advanced Topics," Noyes Publications, Park Ridge, N.J. (1985). Ise, N. and I. Sogami, (Eds.) "Ordering and Organization in Ionic Solutions," World Scientific, New Jersey (1988).

Israelachivili, J.N. and R.M. Pashley, "Molecular Layering of Water at Surfaces and Origin of Repulsive Hydration Forces," *Nature*, **306**, 249 (1983).

Kekicheff, P., "Direct Measurements of Interactions in Superamolecular Fluids and Liquid Crystals," Lecture presented at the 11th Scandinavian Symposium on Surface Chemistry, Bergen, Norway June 26-28 (1991).

Kralchevsky, P.A., A.D. Nikolov, D.T. Wasan, and I.B. Ivanov, "Formation and Expansion of Dark Spots in Stratifying Foam Films," *Langmuir*, **6**, 1180 (1990).

Lyklema, I., P.S. Scholten, and K. Mysels, "Flow in Thin Liquid Films," *J. Phys. Chem.*, **69**, 116 (1965).

Nikolov, A.D., P.A. Kralchevsky, I.B. Ivanov, and D.T. Wasan, "Ordered Micelle Structuring in Thin Films Formed from Anionic Surfactant Solutions. II. Model Development," *J. Colloid Interface Sci.* **133**, 13 (1989).

Nikolov, A.D. and D.T. Wasan, "Ordered Micelle Structuring in Thin Films Formed from Anionic Surfactant Solutions. I. Experimental," *J. Colloid Interface Sci.* **133**, 1 (1989).

Nikolov, A.D., D.T. Wasan, N.D. Denkov, P.A. Kralchevsky, and I.B. Ivanov, "Drainage of Foam Films in the Presence of Nonionic Micelles," *Prog. Colloid Polym. Sci.* **82**, 87 (1990).

Nikolov, A.D., D.T. Wasan, P.A. Kralchevsky and I.B. Ivanov, "Ordered Structures in Thinning Micellar Foam and Latex Films," in *Ordering and Organization in Ionic Solutions*," N. Ise and I. Sogami (Eds.), World Scientific, New Jersey (1988).

Ohtsuki, T., S. Mitku, and Okano, "Studies of Ordered Monodisperse Latexes. II Theory of Mechanical Properties *Jap. J. Appl. Phys.* **17**, 627 (1978).

Pieranski, P., "Colloidal Crystals," *Contemp. Phys.* **24**, 25 (1983).

Reiner, S. and C. Radke, "Electrostatic Interactions in Colloidal Suspensions: Tests of Pairwise Additivity," *AIChE Journal* **37**, 805 (1991).

Russel, W.B. and D.W. Benzing, "The Viscoelastic Properties of Ordered Latices: A Self-Consistent Field Theory," *J. Colloid Interface Sci.* **83**, 163 (1983).

Russel, W., D. Saville, and W. Schowalter, "Colloidal Dispersion," Cambridge (1989).

CHAPTER 12

STRUCTURE AND MECHANISMS OF ELECTRO-RHEOLOGICAL (ER) FLUIDS

H. Conrad

12.1 INTRODUCTION

Following a short literature review, this chapter presents the work by the author and his coworkers at North Carolina State University (NCSU) on electrorheological (ER) fluids. The chain-like structure of a model ER fluid is defined and described quantitatively by stereological parameters. A theoretical model for the dipole interaction force between the particles in a chain is presented and compared with the measured force between particles in a model ER fluid of glass beads in silicone oil and the measured yield stress of ER fluids containing zeolite particles. Reasonable agreement exists between the predictions of the model and the measured force and the yield strength. Based on these results an estimate of the maximum attainable yield strength of ER fluids is given. The molecular mechanism responsible for the response of zeolite/silicone oil fluids is briefly discussed. Mentioned are future areas of research and potential technological applications of ER fluids.

12.2 LITERATURE REVIEW

That electric fields can produce modest increases in the viscosity of certain fluids was first reported around the turn of the century (Duff 1896; Quinke 1897) and subsequently studied by a number of investigators (Herzog et al. 1934; Bjornstahl 1935; Dobinski 1935; Alcock 1936; Kimura 1937; Menz 1939; Dixon 1939; Andrade and Dodd 1947, 1951, 1954), who often gave contradictory results. One of the more definitive studies of this so-called **electroviscous effect** was by Andrade and Dodd, who upon application of electric fields of 1–10 kV/cm obtained moderate increases (up to ~ 100%) in the viscosity of polar liquids containing conducting impurity ions. As an explanation for the behavior, they proposed that the electric field causes the impurity ions to accumulate in the vicinity of the electrodes, around which polar molecules cluster and thereby resist shear.

In contrast to the modest increases in viscosity obtained for the conducting polar liquids, Winslow in 1949 reported orders of magnitude increase in viscosity upon applying an electric field (a.c. or d.c.) to suspensions of finely divided solids in low viscosity oils, e.g., corn starch or activated silica gel in mineral oils. He identified many of the rheological and electrical characteristics of such suspensions and attributed the large increase in viscosity to the observed field-induced fibration of the particles between the electrodes, whose rupture required an additional shear stress for flow. Subsequent studies (Klass and Martinek 1967; Uejima 1972; Arguelles et al. 1974; Arp et al. 1980; Stangroom 1983) confirmed that this novel phenomenon occurred in times of the order of milliseconds, was reversible and consumed only a small amount of electrical power. The phenomenon was initially termed the **Winslow effect**, but has more recently been called the **electrorheological (ER) effect** to distinguish it from the electroviscous effect mentioned above.

Of the early studies on ER fluids those by Klass and Martinek (1967) on suspensions of silica particles in napthenic fluid and Uejima (1972) on crystalline cellulose in chlorine insulator oil are especially noteworthy. These authors established many of the basic rheological and electrical properties of ER fluids, including the following:

- The rheological behavior approximates that of a Bingham solid (Figure 12.1) whereby

$$\tau = \tau_y^d(E) + \eta_s \dot{\gamma} \tag{12.1}$$

where τ is the shear strength, $\tau_y{}^d$ the so-called **dynamic** yield strength, which increases with the electric field E; η_s is the viscosity of the suspension and $\dot{\gamma}$ the shear rate.

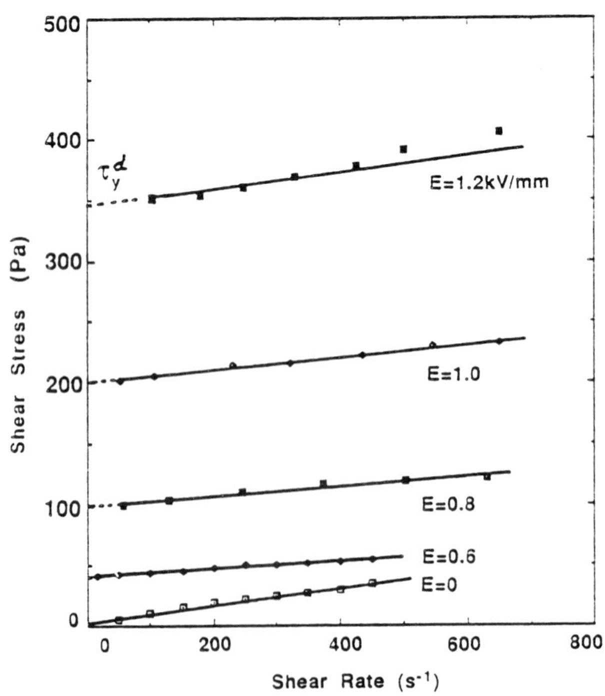

Figure 12.1 Shear stress vs shear rate as a function of electric field E for an ER fluid consisting of 5 μm particles of crystalline cellulose in chlorine insulator oil. Illustrated is a method for determining the dynamic yield stress $\tau_y{}^d$. (Uejima 1972)

- $\tau_y{}^d$ increases with the square of the electric field E, which may be d.c. or a.c.
- $\tau_y{}^d$ first begins to decrease markedly when the electric field frequency f exceeds ~1 kHz.
- $\tau_y{}^d$ increases with increase in temperature T.
- $\tau_y{}^d$ initially increases with concentration α_s of particles, but generally goes through a maximum at larger concentrations.
- Adsorbed water on the particles is needed for ER behavior in the suspensions they employed.
- The conductivity of the ER fluids is low and non-ohmic, the current increasing with voltage considerably more than ohmic.
- The important physical constants for ER response are the dielectric constants of the particles and the fluid, with the particle having a higher dielectric constant. However, factors other than a high dielectric constant of the particles must be important (e.g., double layer), since the strength of an ER fluid containing ferroelectric $BaTi0_3$ particles which have a very high dielectric constant was inferior to one containing silica particles (Klass and Martinek 1967).

Subsequent studies by Arguelles et al. (1974), Stangroom (1983) and Sprecher et al. (1987) and those presented at the two recent International Conferences on Electrorheological Fluids (Conrad et al. 1989a; Carlson et al. 1990) confirmed these generic properties of ER fluids and also the observation by Winslow (1949) that the particles in an active ER fluid align to form a fibrated or chain-like structure between the electrodes; see for example Figure. 12.2.

A summary of typical characteristics of ER fluids is here given in Table 12.1. A listing of known ER-active suspensions is found in Block and

Table 12.1 — Experimental characteristics of ER fluids (taken from Gast and Zukoski, 1989).

Suspending Fluid Properties	
Relative Dielectric Constant	$K_f = 2–15$
Low Field Strength Conductivity	$\sigma_f = 10^{-7}–10^{-13}$ mho/m
Viscosity	$\eta_f = 0.01–10$ Pa-s
Density	$\rho_f = 0.6–2$ g/cm^3
Particle Properties	
Shape	approximately cubic or spherical
Size	$a = 0.1–100$ μm
Relative Dielectric Constant	$K_p = 2–40$
Suspension Properties	
Zero Field Viscosity	$\eta_s = 0.1–10$ Pa-s
Volume Fraction of Particles	$\alpha_s = 0.05–0.5$
Conductivity	$\sigma_s = 10^{-6}–10^{-13}$ mho/m
Typical Field Strengths	$E = 0.1–5$ kV/mm

Kelley (1988). The host or carrier fluids are generally nonpolar liquids and include mineral oil, kerosene, silicone oil, halogenated hydrocarbon and even cooking oils. The suspended particles may be organic (e.g., corn starch, cellulose, ion-exchange resins and metal soaps) or inorganic (e.g., silica, zeolites, clays and other metal oxides) materials. The presence of small amounts of water adsorbed on or within these particles is generally required for a significant ER response. However, essentially anhydrous systems consisting of metal charge transfer complexes (e.g. copper phthalocyanine) and electrically conducting polymers (polyaniline and polypyrrole) have also been found to exhibit strong ER behavior (Block and Kelley 1988). Moreover, a suspension in silicone oil of metal particles covered with a thin nonconducting layer exhibits significant ER response, especially with a.c. fields (Inoue 1990).

The more interesting ER fluids exhibit yield strengths of the order of 0.1 to 1 kPa, with maximum values of ~ 15 kPa having been reported (American Cyanamid Co. Bulletin 1986; Block and Kelley 1989). Some perspective on how the maximum reported yield stress of ER fluids compares with that of a structural metal such as Al is provided in Figure 12.3. It should be noted that in making the comparison the stresses taken were those for a shear rate of 10^{-2} s^{-1}.

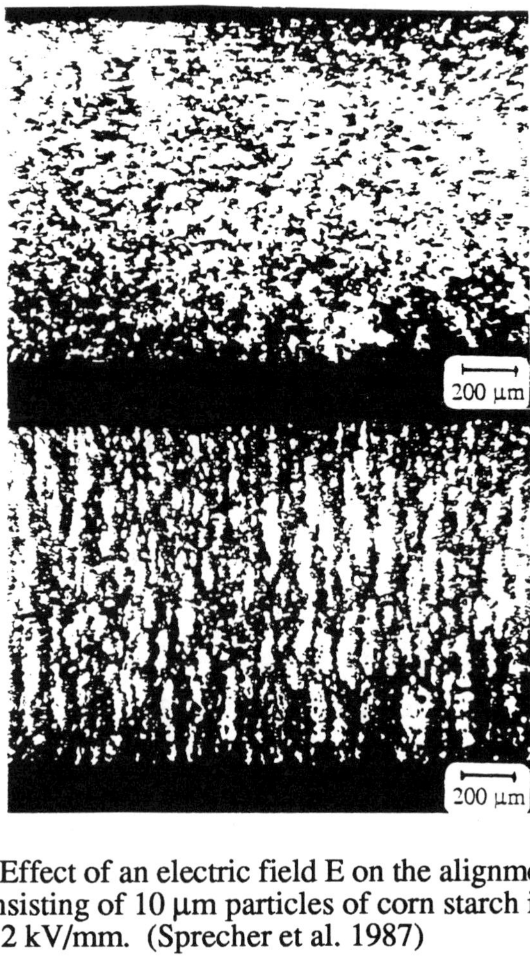

Figure 12.2 Effect of an electric field E on the alignment of the particles in an ER fluid consisting of 10 µm particles of corn starch in silicone oil: (a) E = 0 and (b) E = 2 kV/mm. (Sprecher et al. 1987)

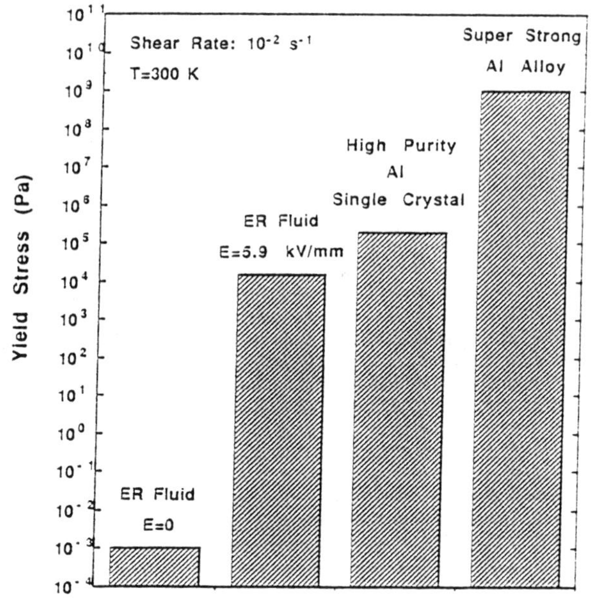

Figure 12.3 Comparison of the maximum strength of known ER fluids with that of the structural metal Al at a shear rate of $10^{-2}s^{-1}$ and T = 300K.

Two basic physical phenomena have been proposed to explain the large increase in shear strength of ER fluids, namely water bridging between particles (Figure 12.4) and polarization of particles (Figure 12.5). In the bridging model, it is assumed that water molecules associated with mobile ions reside within the pores of the particles. Upon application of an electric field the mobile ions move to one end of the particle carrying the water with them, which then forms a bridge with an adjacent particle. The interfacial tension between the host fluid and the water bridge thereby provides a source of shear resistance. Qualitative support for this model is claimed by the fact that the strength of ER fluids containing phenol-formaldehyde particles with differing mobile cations and anions varied as the hydration energy of the ions (Stangroom 1990). However, quantitative details of the model have yet to be developed. Moreover, water bridging cannot readily explain the ER response of the "anhydrous" ER fluids mentioned above. In the polarization model (which is more widely accepted) an increase in shear resistance results from the attractive forces between the electric dipoles induced on the particles by the electric field. The dipoles may develop by several mechanisms, as shown in Figure 12.6.

The force f_d between dipoles has been calculated rigorously for the case of two spherical particles separated by a distance R much greater than their radius a (Coulson and Boyd 1979)

$$f_d = \left(\frac{6}{4\pi\varepsilon_f}\right) \frac{p^2}{R^4} \tag{12.2}$$

where p is the dipole moment and ε_f the total dielectric constant of the host fluid. p is given by

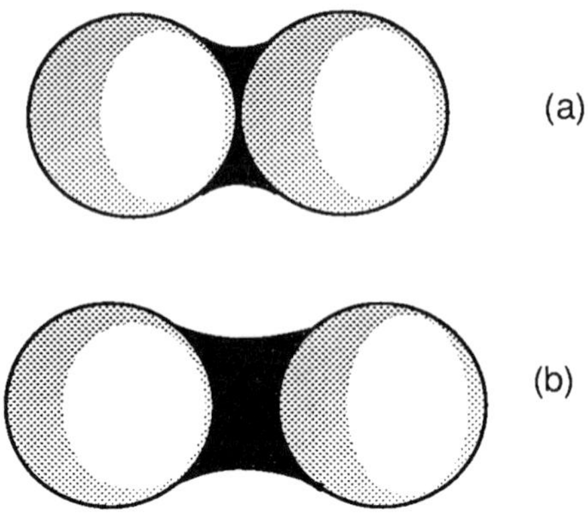

Figure 12.4 Schematic of the water bridging model for the strength of ER fluids: (a) water bridge between two particles and (b) resisting force due to interfacial tension. (Stangroom 1983)

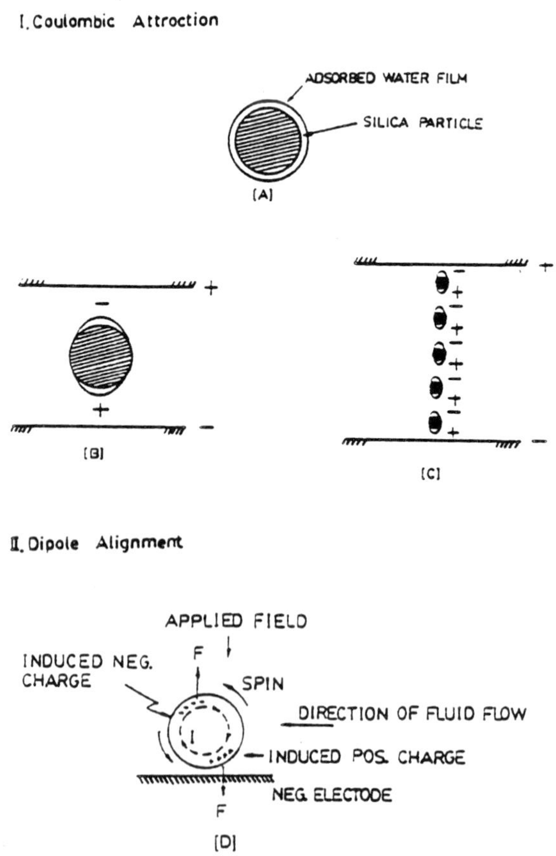

Figure 12.5 Schematic of the polarization model for the strength of ER fluids: (a) water adsorbed on surface of a silica particle, (b) dipole moment induced by the electric field, (c) attraction of dipoles to form a chain and (d) resistance of a dipole to rotation in an electric field. (Philips 1969; Stangroom 1983)

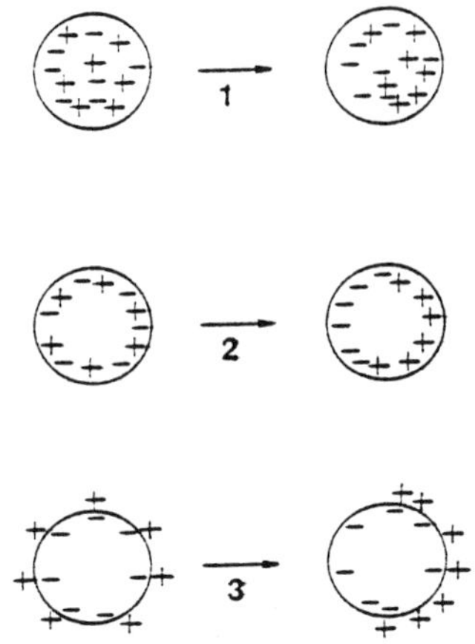

Figure 12.6 Mechanisms by which dipoles may be imparted to a particle: (1) bulk conductance, (2) surface conductance and (3) double layer distortion. (Block and Kelley 1989)

$$p = 4\pi\varepsilon_o a^3 E K_f \left[\frac{K_p - K_f}{K_p + 2K_f}\right] \tag{12.3}$$

where ε_o is the dielectric constant of free space and $K = \varepsilon/\varepsilon_o$ is the total relative permittivity, the subscripts p and f referring to the particle and host fluid respectively. Combining Equations 12.2 and 12.3 we obtain

$$f_d = \frac{24\pi a^6 \varepsilon_o K_f (\beta E)^2}{R^4} \tag{12.4}$$

where β the dielectric mismatch parameter is given by

$$\beta = \frac{(K_p - K_f)}{(K_p + 2K_f)} \tag{12.5}$$

The value of f_d given by Equation 12.4 is for particles aligned in the direction of the applied field. By considering the angle θ with respect to the field, it is found that f_d is a maximum and attractive when $\theta = 0°$, i.e. when the particles are aligned in the field and repulsive when $\theta = 90°$, i.e. when the particles are aligned perpendicular to the field. The theory is thus in qualitative accord with both the observed chain-like arrangement of the particles and the fact that the ER response is governed largely by the respective dielectric constants of the host fluid and the suspended particles.

Assuming that all of the particles are in linear chains spanning the electrode gap, Klingenberg and Zukoski (1990) obtained the following expression for the shear stress required to deform and break the chains

$$\tau_y = 18\alpha_s F_{max}\left\{1-\left(\frac{(\pi/6)^{1/2}}{\frac{\ell}{a}\tan\theta_{max}\alpha_s^{1/2}}\right)\right\} \qquad (12.6)$$

where α_s, is the volume fraction of particles, ℓ the spacing between the electrodes and F_{max} the maximum restoring force exerted between two hard dielectric spheres aligned along the applied electric field. F_{max} occurs at an angle θ_{max} or a displacement x_{max}. The magnitude of F_x given by Gast and Zukoski (1989) is

$$F_x = \frac{kT}{a}\lambda\left(\frac{2a}{R}\right)^4\left\{2(f_{||} + f_r)\sin\theta\cos^2\theta - f_\perp\sin^3\theta\right\} \qquad (12.7)$$

where $\lambda = \pi\,\varepsilon_o\,K_f\,(\beta E)^2\,a^3/kT$ and $f_{||}$, $f_\perp$ and f_r are the parallel, perpendicular and torsional force components respectively. In the point dipole limit, $f_{||}$, $f_\perp$ and f_r are all unity; in the exact calculation of Equation 12.7 they are functions of ε_p, ε_f and a/R. The agreement of Equation 12.6 with experiment is shown in Figure 12.7, which is a plot of the yield stress vs volume fraction of particles according to Equation 12.6 (solid line) and experiment (data points) for an ER fluid consisting of lithium poly(methacrylate) in chlorinated hydrocarbon oil.

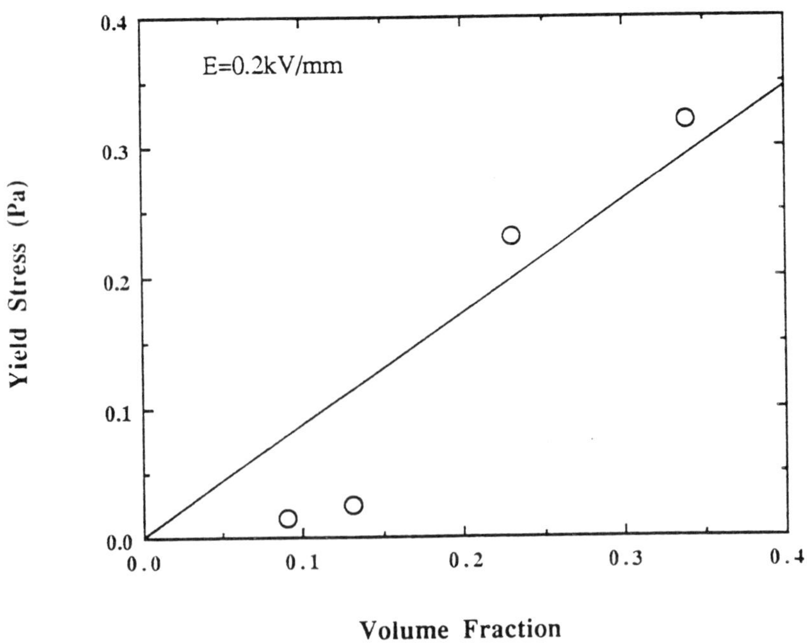

Figure 12.7 Yield stress vs volume fraction of particles according to calculation (solid line) and experiment (data points taken from Marshall et al. 1989) for lithium poly (methacrylate) in chlorinated hydrocarbon oil. (Gast and Zukoski 1989)

The forces which are active in ER fluids consist of the following (Gast and Zukoski 1989): (a) thermal, (b) London-van der Waals, (c) electrostatic, (d) polarization and (e) viscious. Of these, generally only the viscious and polarization forces are important, their ratio being given by the so-called Mason number (Marshall et al. 1989)

$$\mathrm{Mn} = \frac{\eta_s \dot{\gamma}}{2\varepsilon_o K_f (\beta E)^2} \tag{12.8}$$

named after the classic studies of Mason and his coworkers. This then leads to the result that the rheology of an ER fluid can be described by two dimensionless parameters, Mn and the concentration of particles α_s; see Figure 12.8. This behavior is in keeping with the fact that the rheology of ER fluids approximates that of a Bingham solid, which can be seen from the following. Starting with Equation 12.1 and dividing both sides by $\dot{\gamma}$, we obtain

$$\tau/\dot{\gamma} = \tau_y/\dot{\gamma} + \eta_s \tag{12.9}$$

Further, dividing both sides of Equation 12.9 by η_s and taking $\tau_y = Af_d$ (where A is a constant) gives

$$\eta_a/\eta_s = \left(\frac{A24\pi a^6 \varepsilon_o K_f (E\beta)^2}{\dot{\gamma}\eta_s R^4} \right) + 1 \tag{12.10}$$

or by dividing by η_f

$$\eta_a/\eta_f = \left(\frac{A24\pi a^6 \varepsilon_o K_f (E\beta)^2}{\dot{\gamma}\eta_f R^4} \right) + \frac{\eta_s}{\eta_f} \tag{12.11}$$

where η_f is the viscosity of the host fluid and $\eta_a = \tau/\dot{\gamma}$ is the apparent viscosity. Equations. 12.10 and 12.11 reduce to

$$\eta_a/\eta_s = (B/\mathrm{Mn}) + 1 \tag{12.10a}$$

or

$$\eta_a/\eta_f = (B/\mathrm{Mn}) + (\eta_s/\eta_f) \tag{12.11a}$$

where $B = 12A\pi a^6/R^4$ and $\mathrm{Mn} = \dot{\gamma}\eta_f/2\varepsilon_o K_f(\beta E)^2$. Since $\eta_s \approx \eta_f$, Equations 12.10a and 12.11a are essentially equivalent. A log-log plot of η_a/η_f vs Mn should thus have an initial slope of −1 and approach $\eta_a/\eta_f = 1$ at large values of Mn, which is observed as illustrated in Figure 12.8.

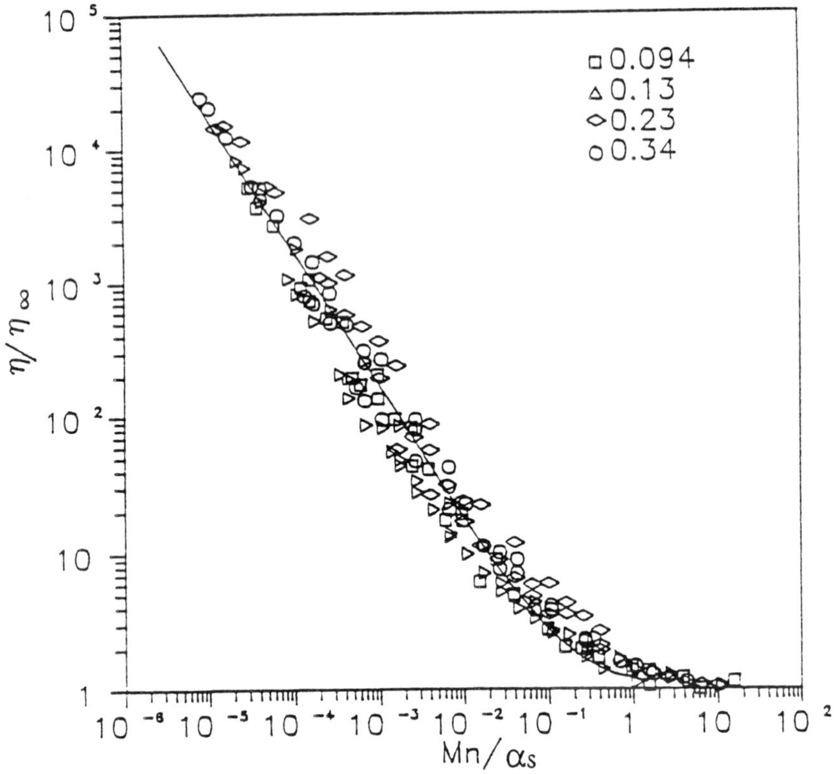

Figure 12.8 Relative viscosity η/η_∞ vs Mason number Mn divided by volume fraction α_s for an ER fluid (lithium polymethacrylate in chlorinated hydrocarbon oil) with $\alpha_s = 0.094$ to 0.34 tested at 25° to 35°C and E = 0.05 – 0.4 kV/mm. η_∞ is the viscosity of the ER fluid at large values of Mn. (Marshall et al. 1989)

12.3 RESEARCH AT NORTH CAROLINA STATE UNIVERSITY (NCSU)

12.3.1 Structure of ER Fluids

It is now well documented that upon application of an electric field the particles in an ER fluid align into a fibrous or chain-like structure roughly parallel to the field, e.g., Figure 12.2. Further, it is expected that the static strength of the fluid will be governed by the stress required to rupture these fibers. Therefore, to understand the strength of ER fluids it is important to have a **quantitative** description of the arrangement of the particles therein. As a first approach to meeting this objective, the structure of a model ER fluid consisting of humidified glass beads in 50cst. silicone oil was investigated (Conrad et al. 1990a; Fisher et al. 1991). This system has three desirable characteristics for such an investigation: (a) it exhibits ER behavior similar to other hydrous-based fluids, (b) the beads are readily visible under an optical microscope at a magnification of 50–200X and (c) the chemical composition of the glass beads is comparable to that of silica particles employed in experimental fluids being considered for practical applications (Duclos et al. 1988). The cell employed to observe the structure of the ER fluid under static conditions is shown in Figure 12.9.

The development of the chain-like structure with gradually increasing electric field in a fluid containing 0.2 vol. fraction of 27 μm mean dia. (gaussian distribution) glass beads in silicone oil is illustrated in Figure 12.10. At low fields (E = 0.15 kV/mm) the beads start to form chain segments aligned in the direction of the field. With increase in field to 0.3 kV/mm the chain segments lengthen by attachment of available free beads and by the joining of segments. At E = 0.5 kV/mm some segments have attained sufficient length to bridge the electrode gap. Upon further increase in E to 1.5 kV/mm most chains now extend from one electrode to the other, giving the typical fibrous structure. Further increases in field to 3 kV/mm produce some thickening and cross-linking of chains, but otherwise no dramatic change in the general appearance of the structure. Similar behavior was observed for other concentrations of glass beads and for smaller beads (7 μm dia.).

An example of the structure which occurred for a sudden application (within 50 ms) of E = 1.5 kV/mm to a fluid containing ~ 0.4 vol. fraction of beads is given in Figure 12.11. The main features of the structure developed faster than the eye could follow; however, adjustments and rearrangements of the beads continued for some time thereafter. The stereological features which were considered to be important pertaining to the initial dispersion at E = 0 and after application of the field are listed in Table 12.2; some of them are illustrated in the micrographs of Figure 12.11. The parameters were measured directly from photomicrographs of the structure using standard procedures (DeHoff and Rhines 1968; Underwood 1970; Russ 1986) and plotted in histograph format for which the mean, median and standard deviation were determined. Moreover, the data were analyzed statistically to ascertain whether the distribution of a given parameter was normal (gaussian) or log-normal.

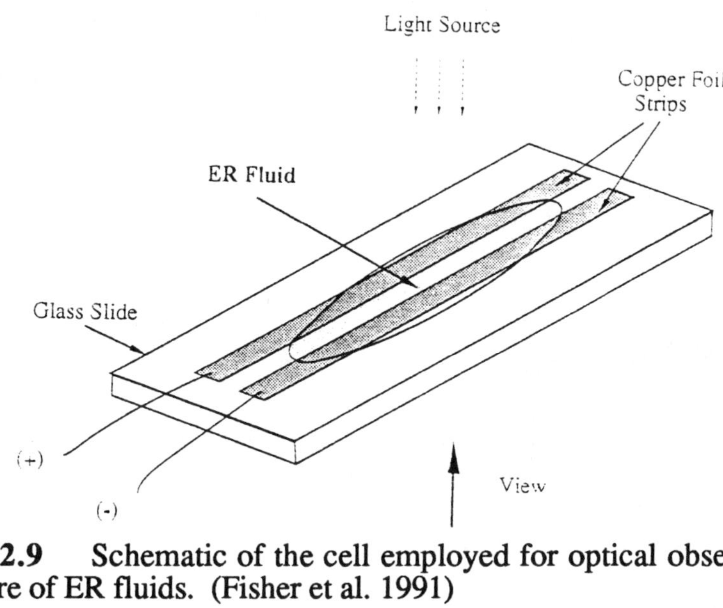

Figure 12.9 Schematic of the cell employed for optical observation of the structure of ER fluids. (Fisher et al. 1991)

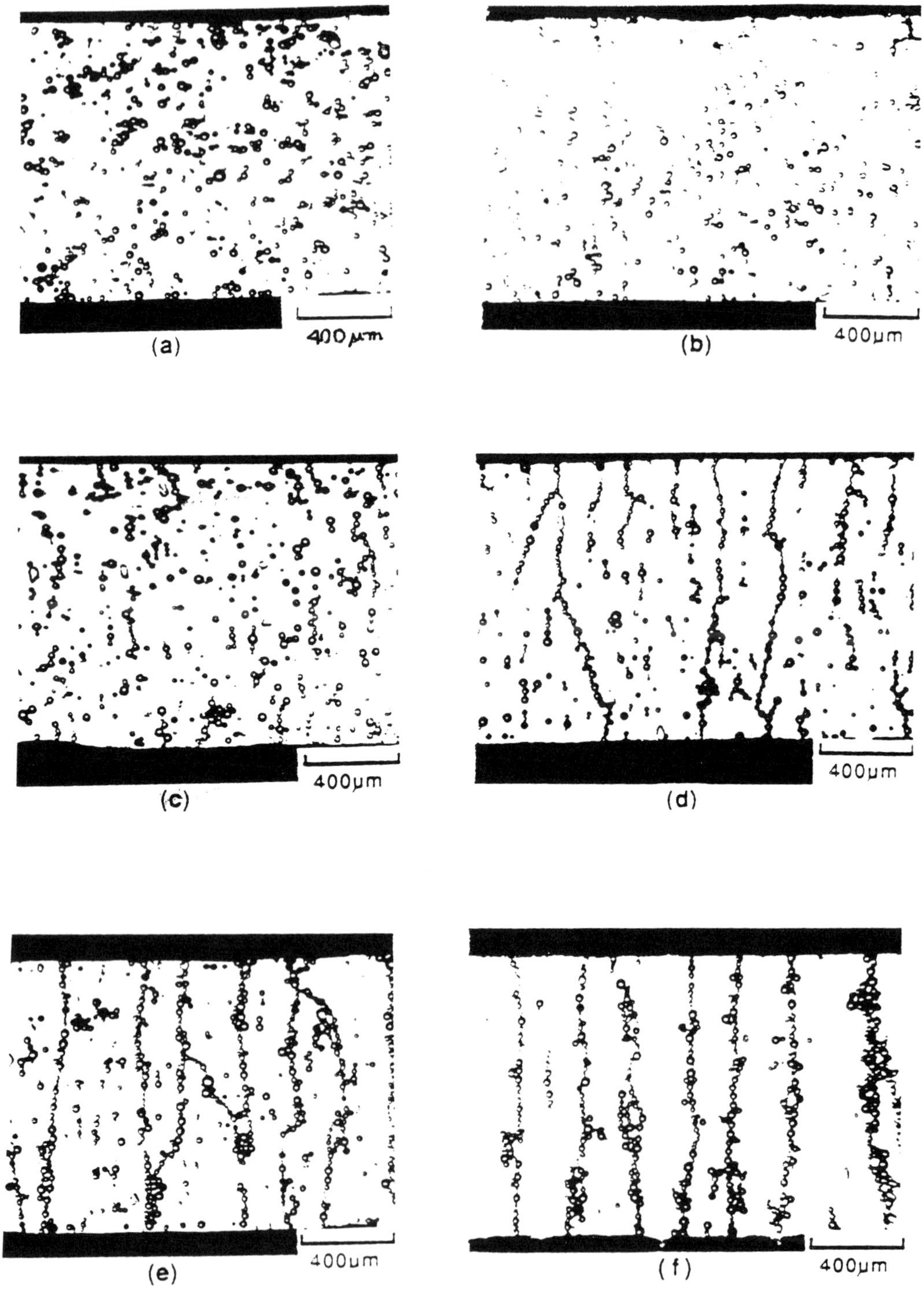

Figure 12.10 Development of the chain structure with increasing electric field E in a fluid containing 0.2 volume fraction of 27 mm glass beads: (a) E = 0, (b) E = 0.15 kV/mm, (c) E = 0.3 kV/mm, (d) E = 0.5 kV/mm, (e) E = 1.5 kV/mm and (f) E = 3.0 kV/mm. (Conrad et al. 1990a)

Table 12.2 — Stereological parameters measured on the ER fluids.

Symbol	*Definition*
A_A	The area fraction of glass beads, which is proportional to volume fraction V_V.
l_{cc}	Center-to-center spacing of the beads in the initial structure.
l_{nn}	Nearest-neighbor spacing of the beads in the initial structure.
λ	Free spacing between the beads in the initial structure.
C	Contiguity, which is the fraction of particles contacting other particles in the initial structure. $C = 2\ N_{\beta\beta}/N_A$; $N_{\beta\beta}$ is number of contacts per unit area, N_A the number of beads per unit area.
$l_c\ (\perp)$	The spacing between chain centers taken perpendicular to E.
$\lambda_c\ (\perp)$	The free spacing between chains taken perpendicular to E.
$X_c\ (\perp)$	The thickness of the chains taken perpendicular to E.
$\lambda_c\ (\|\|)$	The spacing taken parallel to E of the cells formed by cross-links between chains.
P_L	Number of points of the feature of interest per unit length, parallel $P_L\ (\|\|)$ and perpendicular $P_L\ (\perp)$ to E.
Ω_{12}	The degree of orientation or anisotropic arrangement of the particles. $\Omega_{12} = [P_L\ (\perp) - P_L\ (\|\|)]/[P_L\ (\perp) - 0.571\ P_L\ (\|\|)]$ for the two-dimensional case.

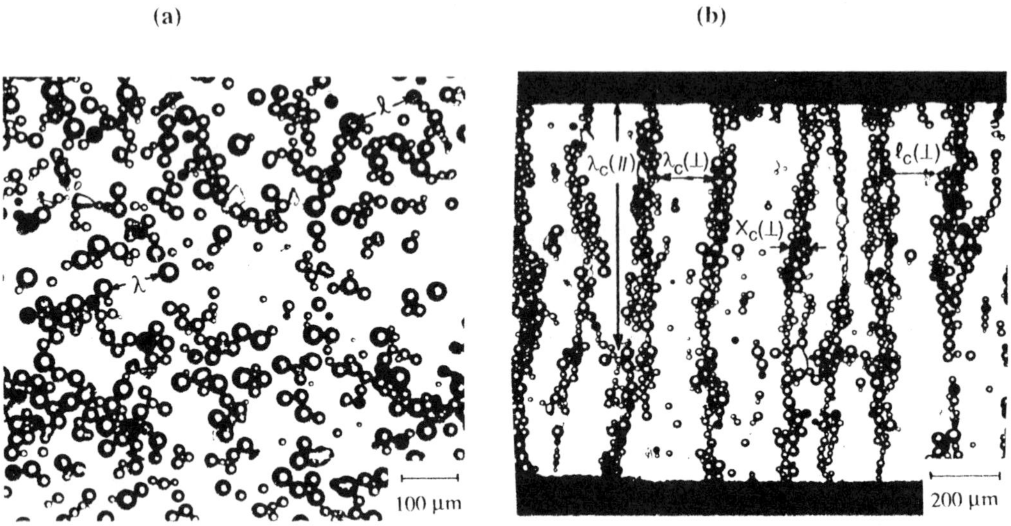

Figure 12.11 The structure prior to (a) and after (b) the sudden application of an electric field E = 1.5 kV/mm to a model ER fluid consisting of 0.4 vol. fraction of 27 μm glass beads in silicone oil. Illustrated are some of the stereological parameters that were measured. (Conrad et al. 1990a)

A structure parameter which is especially important is the orientation parameter Ω_{12}, which gives a measure of the degree of alignment of the particles in the direction of the electric field. Figure 12.12 shows that there occurs a rapid increase in Ω_{12} with E to about 0.5–1.0 kV/mm and thereafter the parameter remains relatively constant. The rapid increase in Ω_{12} with E is relatively independent of the concentration α_s of the beads (given by the area fraction A_A), but the maximum value of Ω_{12} in the relatively constant region decreases with α_s. This can be attributed to the increase in the number of cross-links which occur as α_s increases. Worthy of mention is that the saturation of Ω_{12} with respect to electric field for the glass beads occurs at a similar value of E as did the activation energy Q_{str} associated with the structure term in rheological measurements on ER fluids consisting of fine silica particles in silicone oil (Conrad et al. 1989b). Further, a saturation of the structure as determined by optical measurements (birefringence and dichroism) occurred at E = 1–2 kV/mm for ER fluids consisting of 49 and 130 nm silica particles in cyclohexane (Smith and Fuller 1989).

Meaningful values of the parameters pertaining to the chain structure such as the mean center-to-center spacing between chains $\overline{l}_c(\perp)$, the mean free spacing between chains $\overline{l}_c(\perp)$ and the mean thickness of the chains $X_c(\perp)$ were obtained once the chain structure had fully developed, i.e. for E > 0.5

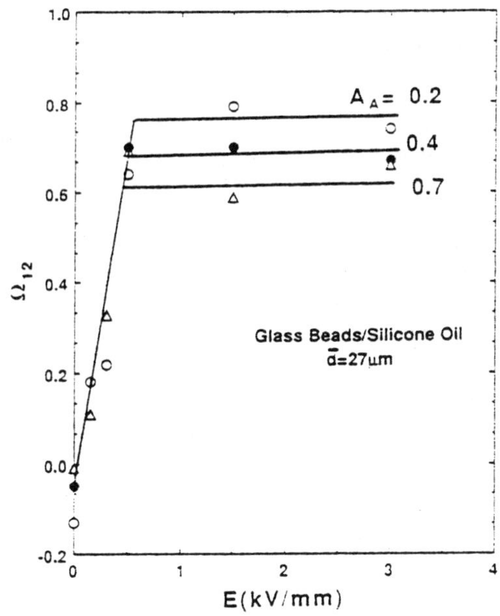

Figure 12.12 The degree of orientation Ω_{12} of the particles vs electric field E for three concentrations of 27 µm glass beads. (Conrad et al. 1990a)

kV/mm in the present case. These parameters were found to be relatively independent of field from E = 0.5 to 3.0 kV/mm, in keeping with the fact that the structure appeared to have reached a saturated condition at these field strengths. The details of the field-saturated structure did however depend on the concentration of particles; see Figure 12.13, which is a plot of $\bar{l}_c(\perp)$, $\bar{\lambda}_c(\perp)$ and $\bar{X}_c(\perp)$ vs the concentration of glass beads given by the area fraction A_A. It is here seen that the spacing between chains given by $\bar{l}_c(\perp)$ and $\bar{\lambda}_c(\perp)$ decreases significantly with A_A, whereas their thickness $\bar{X}_c(\perp)$ increases. It was also noted that the number of cross-links between the chains increased with A_A. Figure 12.14 shows that the mean free spacing between cross-links $\bar{\lambda}_c(\parallel)$ is proportional to the mean free spacing between chains $\bar{\lambda}_c(\perp)$, the proportionality constant being ~ 4. Thus, the structure for $A_A \geq 0.2$ consists of a cellular arrangement of the beads, with the cells having an aspect ratio of ~ 4. Worthy of mention is that the distribution of the values for the free spacing between chains $\lambda_c(\perp)$ was nearly normal (gaussian), whereas that for the thickness of chains $X_c(\perp)$ was nearly log-normal.

Further indication that the glass beads are not all ideally aligned in chains along the direction of the field is provided in Figure 12.15, which gives a plot of the ratio of the spacing-to-thickness of the chains vs the reciprocal of the area fraction. For the ideal case where equisized beads are all aligned in equispaced chains as shown in the insert of the figure, one obtains from the conservation of beads

$$\bar{l}_c(\perp)/\bar{X}_c(\perp) = \pi/4A_A \tag{12.12}$$

The fact that the slope of the plot of measured values is ~ 1 rather than $\pi/4$ is considered to be due mainly to the occurrence of cross-links and to a lesser degree to the existence of unattached beads and chain segments. That size distribution of the particles may also be important was indicated by the fact that a plot similar to Figure 12.15 but for 7 μm dia. beads with a wider size distribution (log-normal) had a slope of ~ 1.5 (Fisher et al. 1991).

The stereology measurements also provided some insight into the manner in which the chains formed. It was found that the mean free spacing between chains $\bar{\lambda}_c(\perp)$ was about twice the mean free spacing $\bar{\lambda}$ of the dispersed beads prior to the application of the field. This suggests that the chains form by the movement on the average of one-half of the beads towards one set of their nearest neighbors and the other half in the opposite direction to their other nearest neighbors (Conrad et al. 1990a).

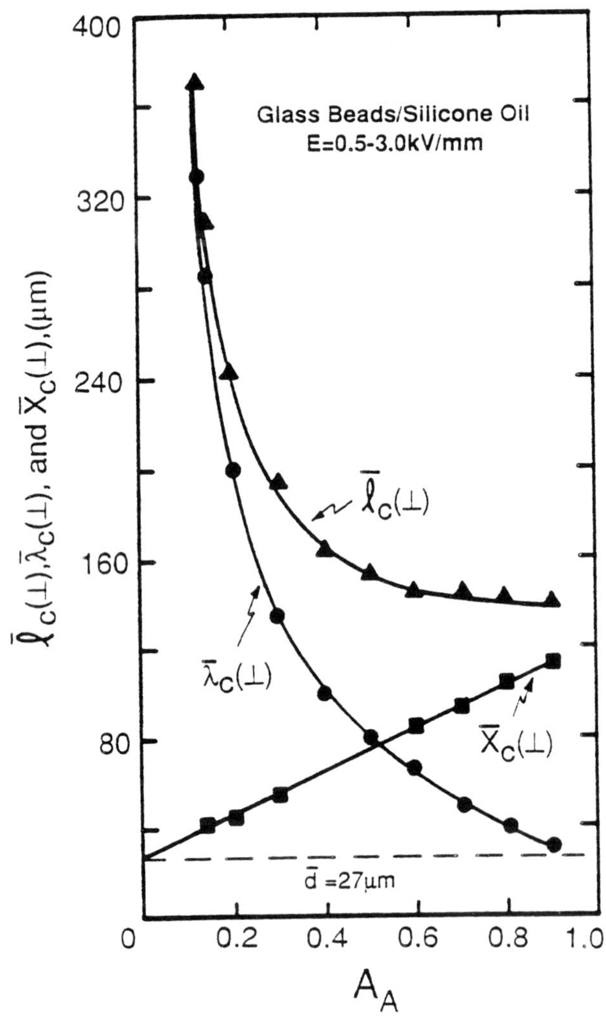

Figure 12.13 Variation of the mean center-to-center spacing $\bar{l}_c(\perp)$, mean free space $\bar{\lambda}_c(\perp)$ and mean thickness $\bar{X}_c(\perp)$ of the chains with area fraction A_A of 27 μm glass beads for electric fields in the range of 0.5–3.0 kV/mm. (Conrad et al. 1990a)

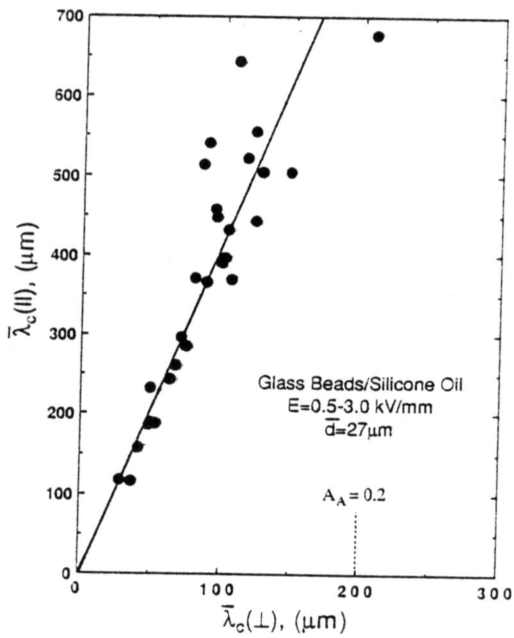

Figure 12.14 Mean free spacing between cross-links $\bar{\lambda}_c(\|)$ vs mean free spacing between chains $\bar{\lambda}_c(\perp)$ for the fluid with 27 μm glass beads. (Conrad et al. 1990a)

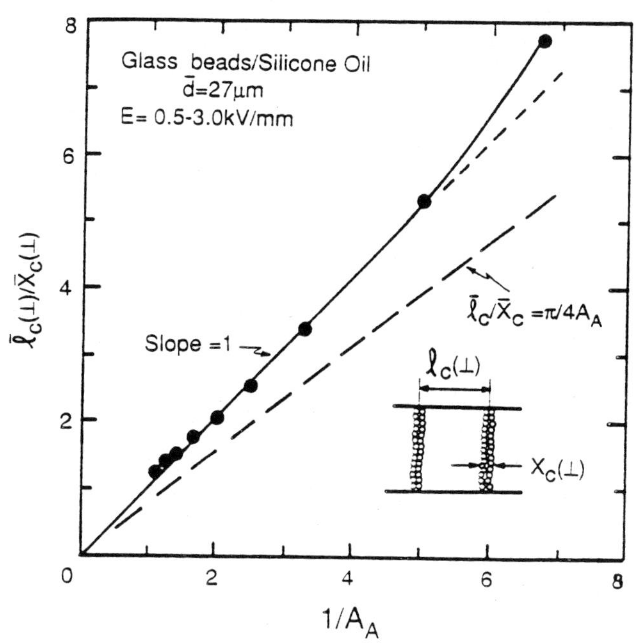

Figure 12.15 The ratio $\bar{l}_c(\perp)/\bar{X}_c(\perp)$ vs the reciprocal of the area fraction A_A for the fluid with 27 μm glass beads. The dashed line is for the ideal case of equisized spheres forming equispaced chains parallel to the electric field as shown in the insert. (Conrad et al. 1990a)

12.3.2 Theoretical Considerations

Since the particles in an ER fluid arrange themselves into a fibrous or chain-like structure, one expects that a major factor in the shear strength of ER fluids is the polarization force which exists between the particles in a chain. Initial attempts to characterize the polarization force were in terms of a point dipole approximation (Carlson 1961; Cuolson and Boyd 1979). It is now generally recognized that this is an oversimplification and leads to particle-to-particle interactions lower than observed. In an attempt to correct this, and to take into account the geometric configuration of the structure of the suspended particles, calculations have been made for a two-sphere system by solving Laplace's equation (Klingenberg and Zukoski 1990). For a many-particle system, a mean field approximation has been employed by Gast and Zukoski (1989) and Adriani and Gast (1989) and a system capacitance matrix was introduced by Bonnecaze and Brady (1990).

The approach employed in our work (Chen et al. 1991) considers an expression for the electric potential V which satisfies Laplace's equation (Reitz et al. 1973)

$$\nabla^2 V = 0 \tag{12.13}$$

The calculation assumes a single row of spherical particles with radius a, having a uniform permittivity ε_p and zero conductivity, situated in a homogeneous fluid of permittivity ε_f, dielectric breakdown strength E_b and zero conductivity. The resulting equations for the potential, electric field, polarization charge and force of interaction are in the form of an expansion which cannot be solved in closed form. However, numerical solutions were generated with a computer and distributions for the functions mapped out. Some of the more important results of the calculations are the following:

- The electric field in the vicinity of the particles E_{local} is increased above the applied field E_o, the magnitude of the increase being greatest when the particles are aligned in the direction of the field and are in contact.
- The ratio E_{local}/E_o increases with the ratio K_p/K_f; see Figure 12.16. Evident is that for particles nearly in contact (R/a = 2.05) the concentration of the electric field can attain values in excess of 20.
- The resulting interaction force between particles decreases significantly with their separation, but increases with the ratio K_p/K_f; see Figure 12.17.
- The force between particles has a maximum at $K_p/K_f = 15$; Figure 12.18. However, the magnitude of the maximum force becomes larger as the values of K_p and K_f comprising the ratio increase.
- The ratio of the force between particles calculated by the present model to that calculated by the point dipole approximation increases with increase in β^2 and with decrease in R/a; see Figure 12.19. This then gives for the force between particles in a chain

$$\frac{F}{f_d} = f(\beta^2, R/a) \tag{12.14}$$

For $R/a \geq 2.05$ the curves in Figure 12.19 are approximated by the following expression

$$f(\beta^2, R/a) = 1.56 \exp\left\{\left[14.84 - 6.16\left(\frac{R}{a}\right)\right]\beta^2\right\} \tag{12.15}$$

Substituting for f_d given by Equation 12.4, we obtain

$$F = \frac{37.44\pi a^2 \varepsilon_o K_f (E_o \beta)^2}{(R/a)^4} \exp\left\{\left[14.84 - 6.16\left(\frac{R}{a}\right)\right]\beta^2\right\} \tag{12.16}$$

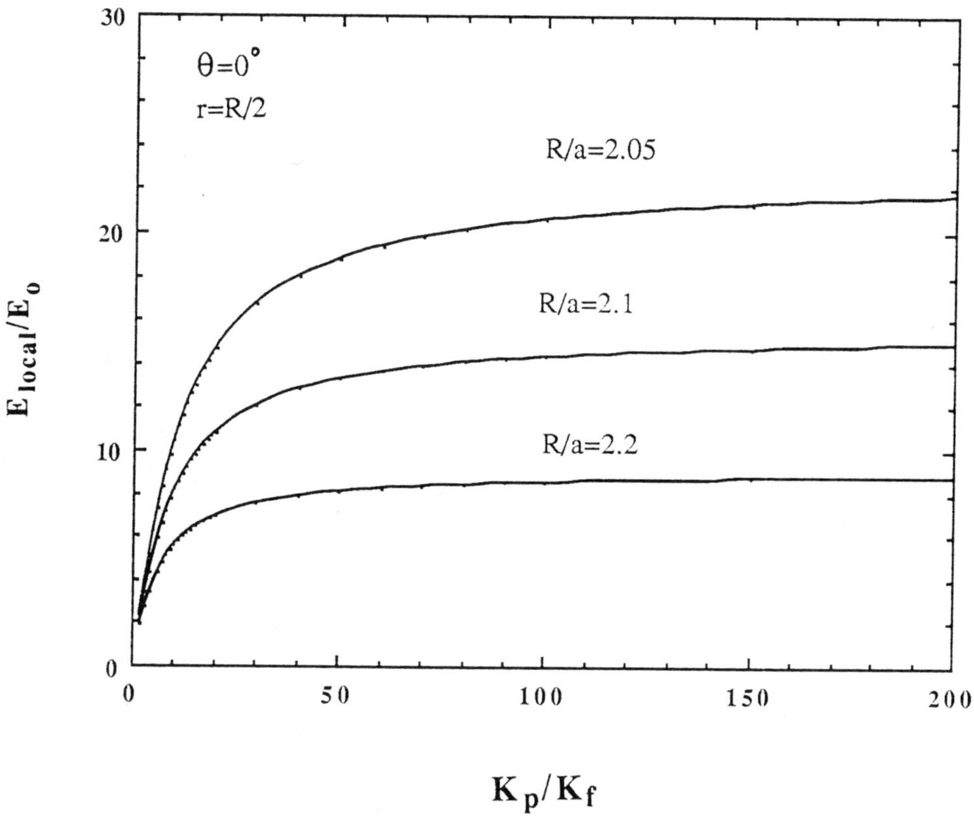

Figure 12.16 Local electric field concentration outside the particle at $\theta = 0°$ vs K_p/K_f as a function of the ratio of the center-to-center separation distance R to the particle radius a. (Chen et al. 1991)

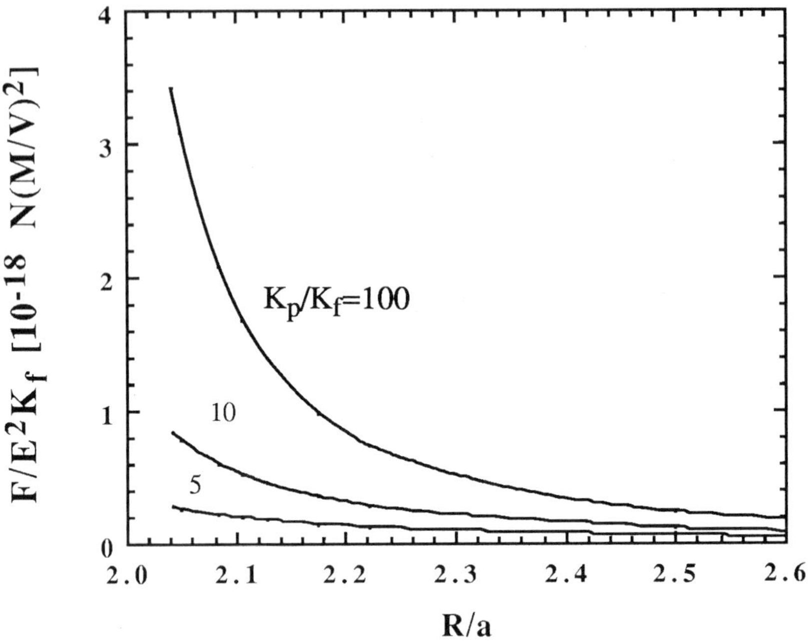

Figure 12.17 Normalized particle-particle interaction force vs separation distance R/a as a function of K_p/K_f at a constant $K_f = 2.5$. (Chen et al. 1991)

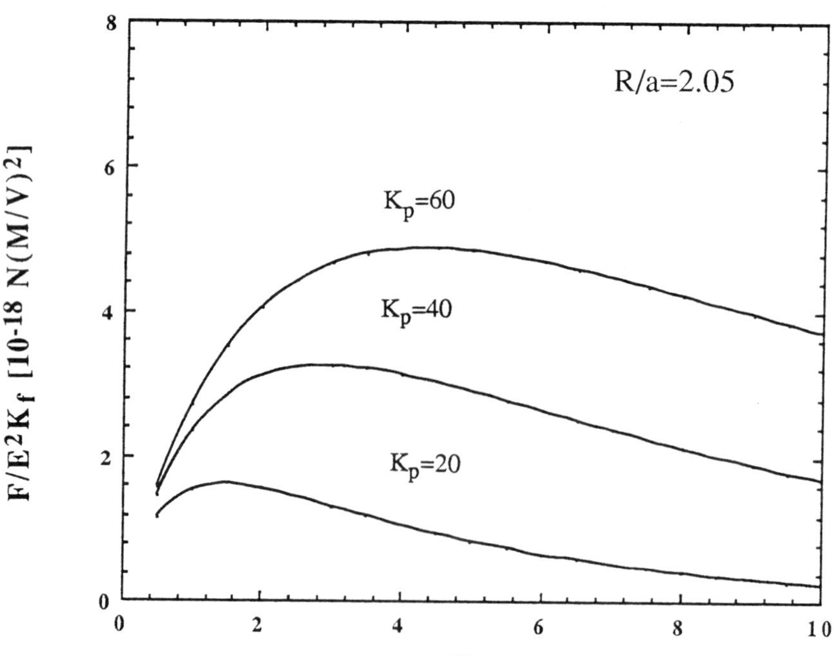

Figure 12.18 Normalized particle-particle interaction force vs K_f as a function of K_p with R/a = 2.05. (Chen et al. 1991)

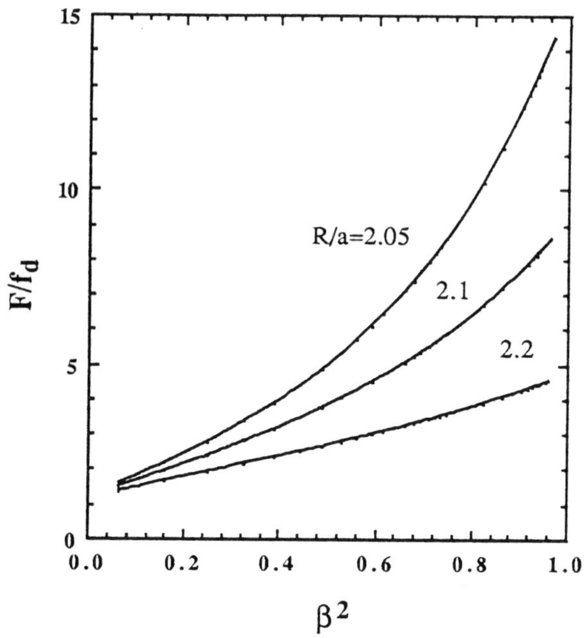

Figure 12.19 Ratio of the calculated particle interaction force F_d for a chain to that of the classical point dipole approach f_d vs the square of the dielectric mismatch parameter β. (Chen et al. 1991)

12.3.3 Measurement of the Forces Between Particles

To provide a check on the theoretical predictions of the model given above, it is desired to measure directly the forces between particles in a chain. For this purpose the apparatus shown in Figure 12.20 was designed and constructed to determine the forces between the particles in a chain of humidified 156 μm glass beads in silicone oil (Sprecher et al. 1990). Using this apparatus, the load-displacement behavior of a single chain consisting of 3 to 5 particles was measured while simultaneously viewing the chain. It was found that the force F was proportional to the shear strain γ (Figure 12.21) and increased with the square of the electric field. Moreover, rupture of the chains occurred at shear strains of 0.4–0.5, the rupture force being proportional to E^2.

A comparison of the measured shearing force with that predicted by our theoretical model and that by the point dipole approximation is presented in Figure 12.22. The agreement with our model is rather good, being an order of magnitude better than the point dipole approximation.

12.3.4 Rheology and Electrical Properties

Desired is an interpretation of the shear strength of ER fluids in terms of the theoretical considerations and fundamental measurements discussed above. For this purpose we conducted a study into the rheology of ER fluids

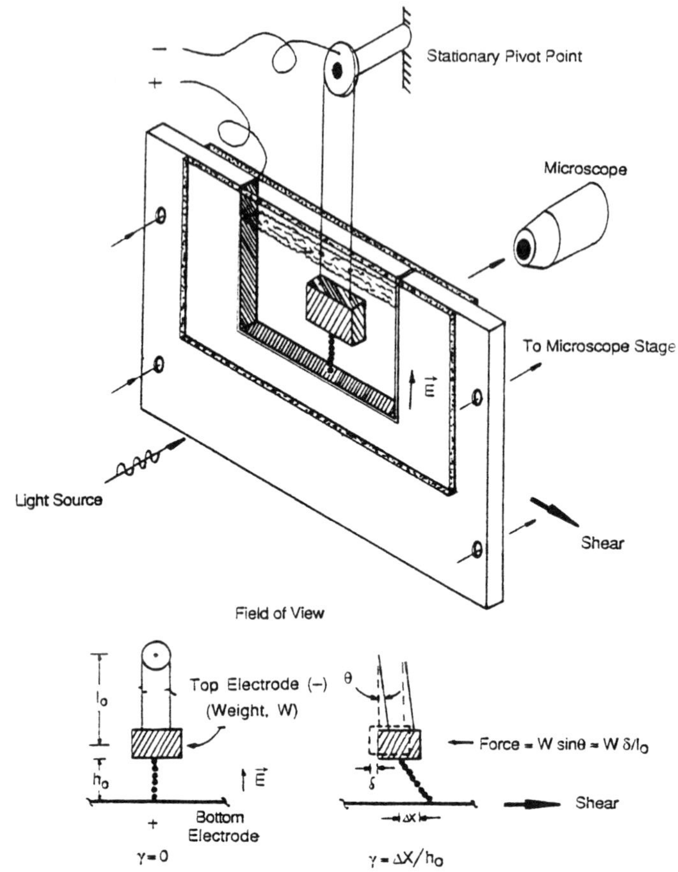

Figure 12.20 Schematic of the apparatus and related geometry used to measure the interaction force of a single chain of particles. (Sprecher et al. 1990)

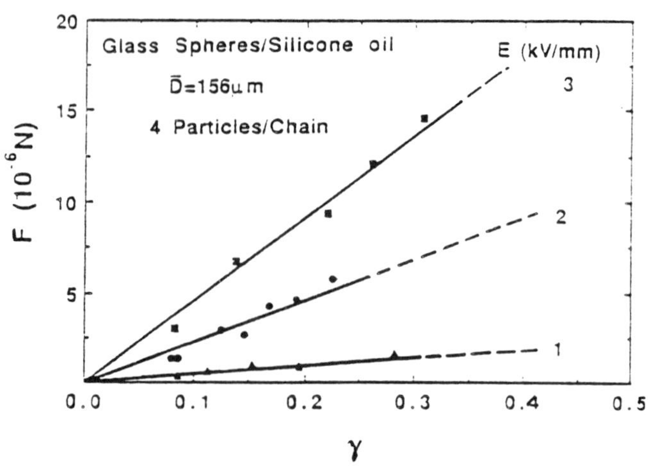

Figure 12.21 Shear force F vs shear strain γ as a function of electric field E for a 4-particle chain of 156 μm glass beads in silicone oil. (Sprecher et al. 1990)

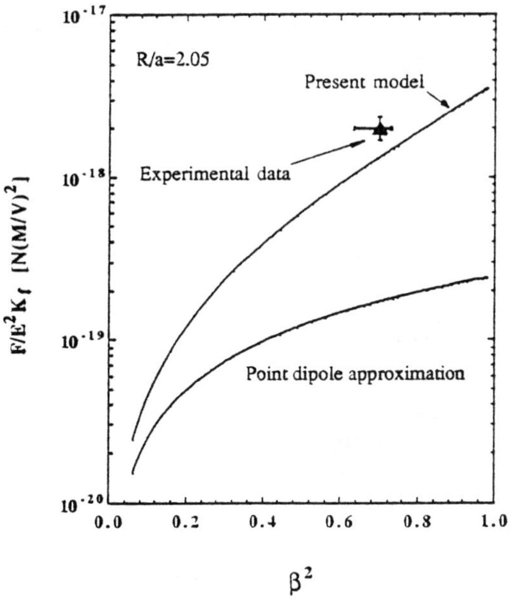

Figure 12.22 Comparison of the measured force required to rupture a single chain of 156 μm glass beads with predictions based on the model of Chen et al. (1991) and on the point dipole approximation. β is the dielectric mismatch parameter.

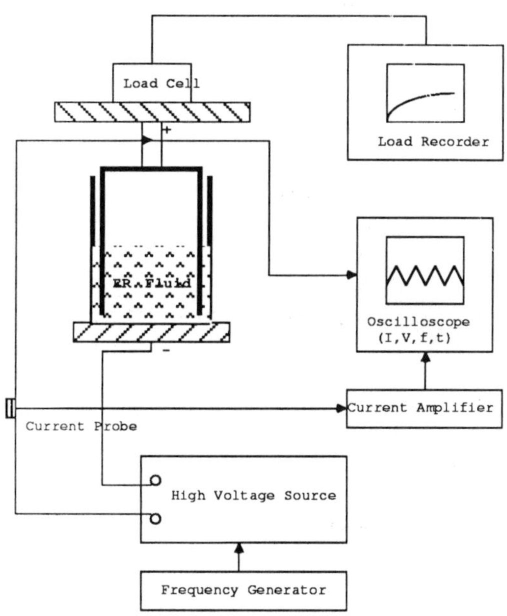

Figure 12.23 Experimental arrangement for measuring the rheology and complex permittivity of ER fluids at relatively low shear rates. (Conrad et al. 1991b)

consisting of fine (~ 2 μm) zeolite particles (Linde 4A and 13X and Sigma Z3125) in silicone oil (Conrad et al. 1990b; Conrad et al. 1991b). The rheological behavior and pertinent electrical properties were determined in the apparatus shown in Figure 12.23, which was attached to a commercial mechanical testing machine (Instron), and tests conducted at a constant shear rate. Typical shear stress τ vs shear strain γ curves obtained in such tests are illustrated in Figure 12.24. The **quasi-static** yield stress τ_y^s is defined in the manner shown. The region of the τ vs γ curve prior to attaining τ_y^s is known as the **pre-yield region**, that beyond the **post-yield region**. In Figure 12.23 the pre-yield region extends to a shear strain γ of ~ 1; however, in general it only extended to γ = 0.2–0.3, with a sharper transition to the post-yield region. It was found that the magnitude of τ_y^s depended on the amount of water which the zeolite particles contained. Therefore, the effects of other variables were determined at a fixed water content of ~ 8 wt.%, where the influence of water reached a maximum (Conrad et al. 1990b).

The effect of temperature on τ_y^s is illustrated in Figure 12.25. The yield stress increases markedly with temperature to ~ 100°C, goes through a maximum and then decreases at higher temperature. An Arrhenius-type plot of the data gave reasonably straight lines for temperatures up to ~ 100°C, the slopes of which yielded an activation energy $Q_\tau \cong 0.14$ eV. The effect of shear rate on τ_y^s is shown in Figure 12.26. It is seen that τ_y^s either decreases with shear rate in the range considered or is relatively independent of rate. This effect of $\dot{\gamma}$ on τ_y^s indicates that the effect of temperature is **not** on the kinetics of the deformation process **per se**.

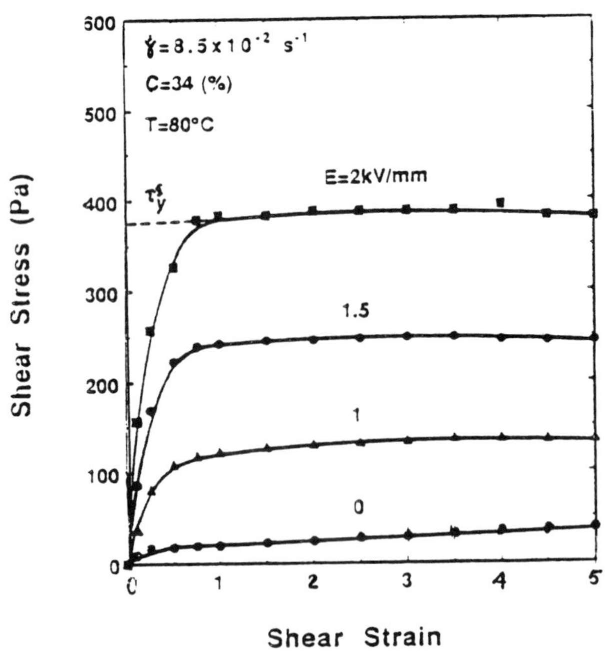

Figure 12.24 Shear stress vs shear strain curves as a function of electric field E for a zeolite ($\alpha_s = 0.2$)/silicone oil ER fluid tested at 80°C and a shear rate of $0.085 s^{-1}$. Illustrated is a method for determining the static yield stress τ_y^s. (Conrad et al. 1990b)

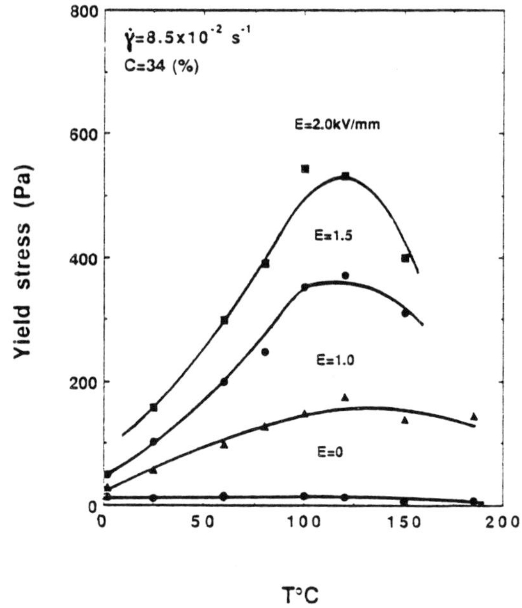

Figure 12.25 Yield stress vs temperature as a function of electric field E of a zeolite ($\alpha_s = 0.2$)/silicone oil ER fluid tested at a shear rate of $0.085s^{-1}$. (Conrad et al. 1990b)

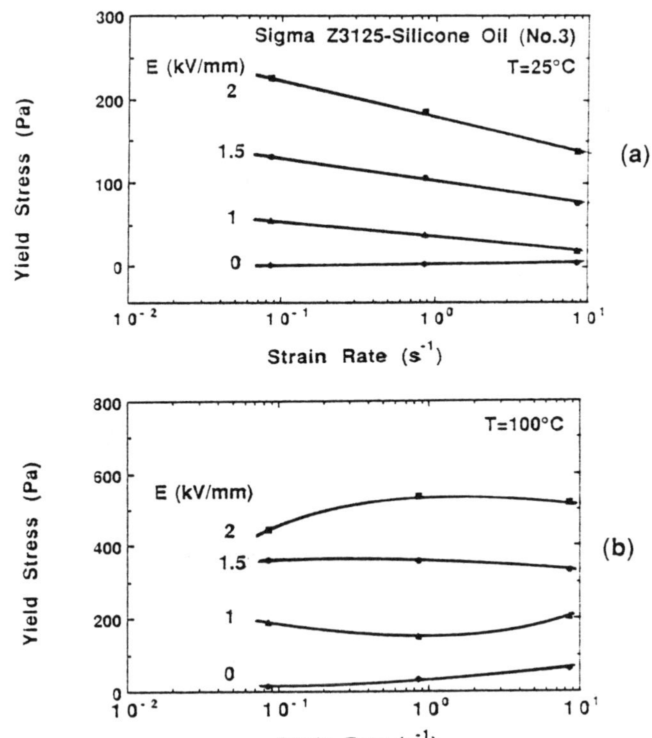

Figure 12.26 Semilog plot of the yield stress vs shear rate as a function of electric field E for a zeolite ($\alpha_s = 0.2$)/silicone oil ER fluid tested at 25° and 100°C. (Conrad et al. 1990b)

It is clear from the theoretical considerations discussed above that to compare the measured strength of an ER fluid with theoretical predictions the values of the permittivity of the host fluid and of the particles need to be known. Hence, these were determined as a function of temperature and electric field for the zeolite/silicone oil suspensions (Conrad et al. 1991b). Also determined were the current density and conductivity of the ER fluid.

The effects of temperature on the permittivity of the silicone oil K_f and on that of the zeolite particles K_p are shown in Figure 12.27. K_p was derived from measurements of the permittivity of the zeolite/silicone oil suspension, K_{ERF} employing the linear rule of mixtures (Marshall et al. 1989)

$$K_{ERF} = \alpha_s K_p + (1-\alpha_s) K_f \tag{12.17}$$

Evident in Figure 12.27 is that K_f decreases slightly with temperature, whereas K_p increases with temperature (and electric field) in a manner similar to $\tau_y{}^s$. A plot of K_p vs 1/T yields an activation energy $Q_{K_p} \cong 0.14$ eV for T < 100°C, which magnitude is the same as that for $\tau_y{}^s$. Worthy of note is that for a.c. fields of 500 Hz, K_p had a value similar to the maximum in Figure 12.27, but was relatively independent of temperature from 25° to 125°C.

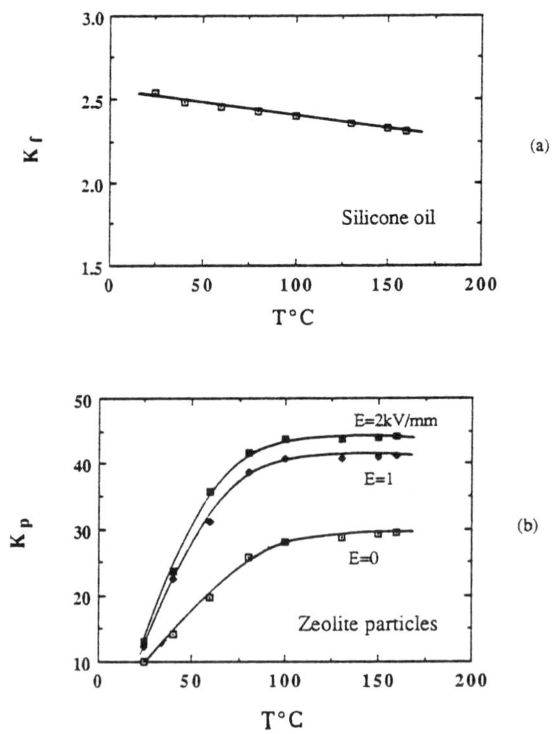

Figure 12.27 Effect of temperature on the dielectric constants of: (a) silicone oil, K_f and (b) zeolite particles in the silicone oil, K_p. (Conrad et al. 1991b)

It was found that the current density j of the zeolite/silicone oil ER fluids increased with the water content of the zeolite particles in a manner similar to the yield stress (Conrad et al. 1990b). Hence, again evaluations were made with zeolite particles having a water content of ~ 8 wt.%, where saturation of the effect of water on j appears to have occurred. The current density for this water content increased continuously with temperature up to 185°C. An Arrhenius-type plot of the effect of temperature on the current density is presented in Figure 12.28. Reasonable straight lines occur up to ~ 100°C, yielding an activation energy $Q_j = 0.73$ eV. A similar value of the activation energy (0.70 eV) was obtained in other measurements of the electrical conductivity σ (Conrad et al. 1991b). To be noted in Figure 12.28 is that although the conductivity increased continuously with temperature, the slope of the Arrhenius-type plot changed above 100°C.

Of further interest is a comparison of the measured yield stress with that predicted by our model presented above on the force of interaction between the particles aligned in a single chain. As a first approximation one expects that the static yield stress $\tau_y{}^s$ is proportional to the force F required to rupture the particles in a chain times the number of chains per unit area N_A, i.e.

$$\tau_y^s = A_s N_A F \sin\theta \qquad (12.18)$$

where A_s is a constant which is determined by the geometric arrangement of the particles (i.e., the structure) and F sin θ represents the component of the force acting in the direction of the shear as the chains are sheared through the angle θ. F is given by the curves in Figure 12.19, or by Equation 12.16. For single chains of spherical particles, it can be shown that

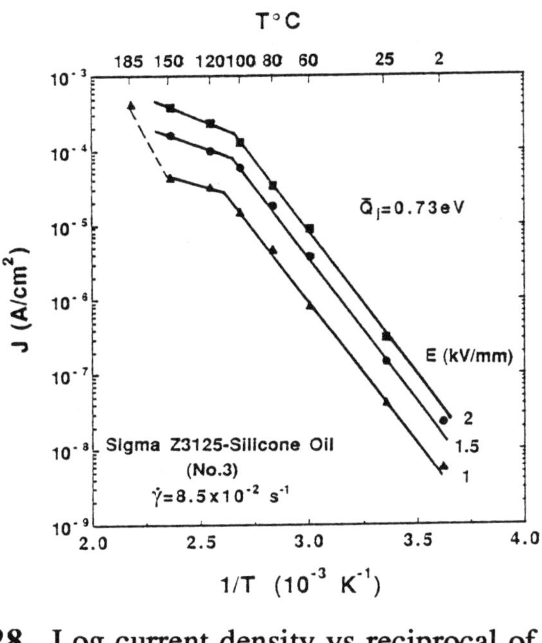

Figure 12.28 Log current density vs reciprocal of the temperature as a function of electric field for zeolite ($\alpha_s = 0.2$)/silicone oil ER fluid tested at a shear rate of $0.085 s^{-1}$. (Conrad et al. 1990b)

$$N_A = \chi 3\alpha_s/2\pi a^2 \tag{12.19}$$

where χ is the fraction of particles that are aligned in chains parallel to the electric field and α_s is the volume fraction of particles. Our studies on the structure of glass beads in silicone oil suggest that χ may have values in the range of $\pi/4$ to $\pi/6$ depending on the size distribution of particles. Substituting Equations 12.16 and 12.19 into Equation 12.18 and taking $\chi = \pi/4$, $R/a = 2.05$ ($\theta = 22°$, $\gamma = 0.4$), one obtains

$$\tau_y = 0.8\, A_s\alpha_s\varepsilon_o K_f(\beta E_o)^2 \exp(2.21\, \beta^2) \tag{12.20}$$

The measured yield stress of ER fluids containing 0.2 vol. fraction of zeolite particles in several oils (silicone, mineral, corn and caster) and tested at

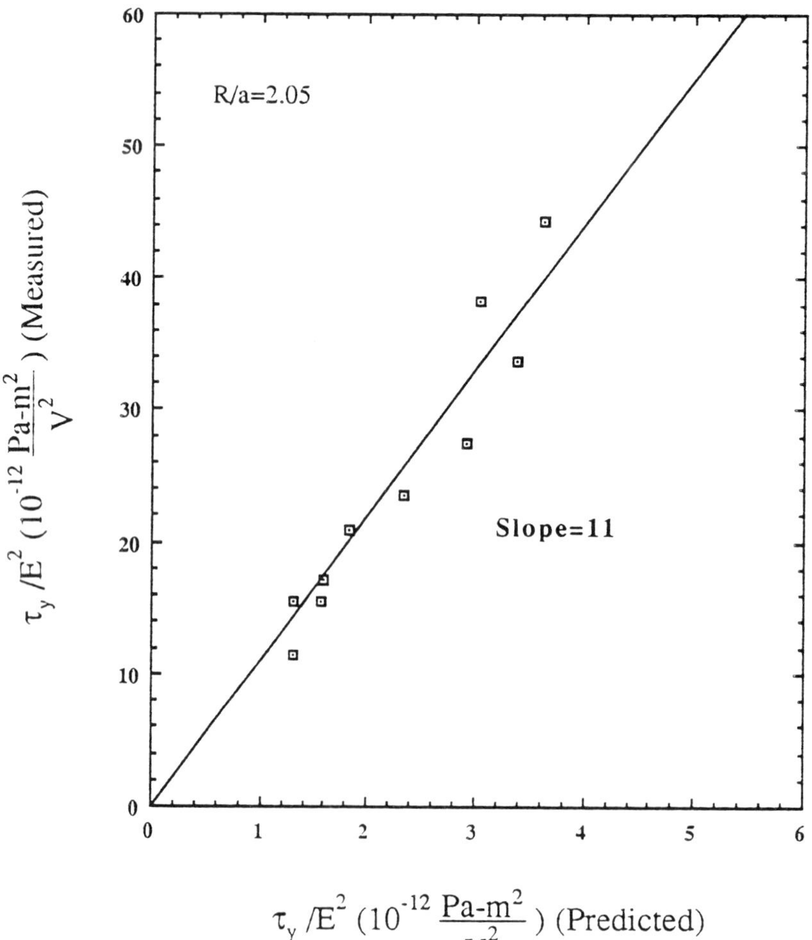

Figure 12.29 Measured vs predicted static yield stress for suspensions of zeolite particles in various oils tested at 25° to 100°C and a shear rate of 0.085s^{-1}. (Conrad et al. 1991b)

25° to 100°C is compared with that predicted by Equation 12.20 in Figure 12.29, using measured values of α_s, K_p, K_f and E. To be noted is that the measured values are about 11 times those predicted, i.e. $A_s \approx 11$. It is tentatively concluded that the increased shear stress results from the fact that the theoretical estimate of the force F is for a chain of single particle thickness, whereas in an ER fluid with a particle content $\alpha_s \geq 0.2$ the arrangement of the particles is in the form of columns, whose thickness consists of two or more particles (see for example Figures 12–2 and 12–11). In such columns, the particle-to-particle contacts with side neighbors may give an increased force compared to that for chains of single particle thickness . Also, the observed cross-linking of the chains may provide a further increase in the resistance to shearing of the columns.

12.3.5 Molecular Mechanism

The molecular mechanism by which polarization of the zeolite particles occurs is not yet clear. The fact that the strength of the zeolite/silicone oil ER fluid depends on the water content of the zeolite particles indicates that water molecules play a role. It is probably the water which resides in the zeolite crystal channels which is important rather than that adsorbed or chemisorbed at the particle surface. The activation energy Q_j of ~ 0.7 eV for the conductivity of the ER fluid is approximately equal to that for the conductivity of zeolite, which in turn is governed by the diffusion of Na^+ ions along the zeolite crystal channels (Breck 1974). This suggests that the diffusion of the Na^+ ions in the zeolite particles may be the rate controlling mechanism for conduction in the ER fluid. Further, it was found (Conrad et al. 1990b, 1991b) that the effect of temperature on the yield stress correlated with its effect on the current density (or conductivity), giving

$$\tau_y^s = Aj^{1/5} = BE^2\,[\exp - (Q_j/kT)]^{1/5} \qquad (12.21)$$

so that $Q_\tau = 1/5\ Q_j$. Since $Q_{K_p} \approx Q_\tau$, this suggests that the mechanism responsible for Q_{K_p} may also be responsible for Q_τ, and in turn both may be related to Q_j.

The fact that $\tau_y{}^s$ was either relatively independent of shear rate or decreased with shear rate indicates that Q_τ does not portray a thermally-activated deformation process. For such a process the effects of temperature and strain rate on the shear stress are inversely related through a simple Arrhenius-type rate equation of the form $\dot{\gamma} = A \exp[-\Delta H(\tau)/kT]$, where A is a constant and ΔH is the enthalpy of activation for the shearing process which decreases with τ. Rather, it appears that Q_τ reflects the polarization process, especially since $Q_\tau = Q_{K_p}$. Further, since K_p was found to be relatively independent of the frequency of the a.c. field up to 500 Hz, it appears that the kinetics of the polarization process are quite rapid. This then suggests that the activation energies Q_τ and Q_{K_p} correspond to some equilibrium concentration (or arrangement) of groups of atoms or molecules in the channels of the zeolite crystal structure which polarize. Moreover, in view of the relationship between Q_τ (and Q_{K_p}) and Q_j, it appears that those conditions which govern the polarization are also involved in the conductivity. Additional work is however needed before more can be said concerning the molecular processes which are involved in the polarization and conductivity of zeolite/silicone ER fluids. Likewise, more work is needed to determine why the yield stress goes through a maximum at ~ 100°C. This could be due to a change in either the water content of the zeolite above this temperature or to its chemical form in the zeolite channels as the temperature is increased.

12.3.6 Maximum Strength of ER Fluids

In view of the agreement between the measured values of the yield stress and those predicted by Equation. 12.20 with $A_s = 11$, it is tempting to estimate what might be the maximum yield strength attainable with ER fluids. The theoretical considerations indicate that the maximum force of interaction between the particles occurs for $K_p/K_f = 15$. Taking the reasonable maximum values $K_f = 12.5$, $K_p = 187.5$, $\alpha_s = 0.5$, $E = 5$ kV/mm and $R/a = 2.05$ one obtains using Equation 12.20, $\tau_y^s = 39$ kPa, which is 2.6 times the maximum value of 15 kPa reported to date. To realize this strength, the local dielectric breakdown strength of the fluid would need to be on the order of 65 kV/mm according to the results in Figure 12.16. Larger values of E could give significantly higher values of τ_y, providing the local dielectric breakdown strength was not exceeded.

12.4 POTENTIAL APPLICATIONS OF ER FLUIDS

The large changes in shear strength possible with ER fluids, coupled with rapid response times, offer decided advantages in a variety of mechanical systems, including clutches, high speed valves with no moving parts and active dampers for vibration control (Bullough and Stringer 1973; Bullough and Foxon 1978; Gorodkin et al. 1979; Kerr 1981; Sproston et al. 1983; Stevens et al. 1984; Scott 1984; Scott and Yamaguchi 1983, 1985). Especially noteworthy is that ER fluids offer the possibility for a rapid-response, low-power interface between electronic controls and mechanical devices, which is needed to realize the full potential of robotic systems (Stangroom 1983). Some idea of the present stage of development of mechanical devices employing ER fluids can be obtained from the papers in Section 5 of the Proceedings of the Second International Conference on ER Fluids (Carlson et al. 1990).

In addition to their possible use in mechanical devices, ER fluids offer potential for the development of "smart" structural materials, in that the mechanical response of a composite containing an ER fluid can be significantly altered upon application of an electric field (Gandhi and Thompson 1988; Coulton and Duclos 1990; Choi et al. 1990; Sprecher et al. 1991). Smart composites employing ER fluids could find application for vibration control in rotorcraft and in various aerospace and hydrospace systems.

12.5 OPPORTUNITIES FOR FUTURE WORK

Some of the subjects on which additional work is needed to further our understanding of ER fluids include the following:

- Theoretical calculations of the force between particles in a chain whose thickness consists of more than one particle, i.e., the force between particles in a column. Also, the influence of cross-links between chains on this force needs to be determined.
- Theoretical and experimental studies of the atomic or molecular mechanisms by which polarization of the particles in an ER fluid occurs. Included should be a consideration of the thermodynamics and kinetics associated with each mechanism.
- A quantitative description of the arrangement and motion of the particles during dynamic shearing, including the effects of a rapid change in the magnitude of the electric field, its frequency and in the shear rate.
- The mechanical behavior of ER fluids in the pre-yield region, giving special attention to reversibility, oscillatory straining and to the components of the shear modulus as a function of electric field, shear rate and temperature.
- A consideration of whether the alignment of particles upon application of the electric field represents a phase-change as proposed by Tao et al. (1989) and Jaggi (1990).
- An evaluation of the anisotropic nature of the mechanical properties of ER fluids.

12.6 SUMMARY

The following is a summary of the present status of our understanding of the structure and mechanisms associated with ER fluid behavior based on data available in the literature and work carried out by the author and his coworkers:

- A distinguishing characteristic of ER fluids is that upon application of an electric field the particles suspended therein form chains or fibers aligned in the direction of the field, whose thickness increases and separation decreases with increase in concentration of particles. A critical field appears to be required for the chains to attain sufficient length to span the electrode gap.

- The alignment of the particles is considered to result from polarization forces imparted to the particles by the electric field and in turn the increase in shear resistance is due to these forces.
- Reasonable agreement is obtained between the measured value of the force between spherical particles in a single chain and that predicted by calculations performed by the author and his coworkers.
- The static yield stress of ER fluids consisting of zeolite particles with 8 wt.% water in various oils was about an order of magnitude higher than that predicted based on a simple model of yielding. The difference is attributed to the fact that the arrangement of the particles in the fluids is more complex than the isolated chains of single particle thickness assumed in the model.
- No truly definitive studies are available on the detailed atomic or molecular mechanism by which polarization of the particles in known ER fluids occurs.
- Results obtained to date suggest that ER fluids may attain a yield strength of ~ 40 kPa at E = 5 kV/mm.
- Several areas for future work are proposed for obtaining a better understanding of ER fluid properties.
- ER fluids offer considerable potential for use in active damping devices, clutches, brakes, valves, activators and smart materials.

ACKNOWLEDGEMENTS

The author wishes to acknowledge that Y. Chen, Y. Choi and M. Fisher, graduate students, and Dr. A. F. Sprecher, senior research associate, performed the experiments reported here and contributed many helpful ideas throughout the course of the work. Mrs. K. Jones typed the manuscript.

This work was jointly sponsored by NSF Award #CBT–8714515, the Ford Motor Company and the NCSU University/Industry Consortium on ER fluids.

REFERENCES

Adriani, P. M., and A. P. Gast, "A Microscopic Model of Electrorheology," Phys. Fluids **31** 2757 (1989).

Alcock, E. O., "Effect of an Electric Field on the Viscosity of Liquids," Physics **7** 126 (1936).

American Cyanamid Co. Bulletin, "Electrorheological Fluids," NBO–001–86–2507–4.5M 1086 (1986).

Andrade, E. N. da C., and C. Dodd, "Effect of an Electric Field on the Viscosity of Liquids," Proc. Roy. Soc. (London) Ser. A, **187** 296 (1947); Part II **204** 449 (1951); E. N. da C. Andrade and J. Hart, Part III **225** 463 (1954).

Arguelles, J., H. R. Martin, and R. Pick, "A Theoretical Model for Steady Electroviscous Flow Between Parallel Plates," J. Mech. Engr. Sci. **16** 232 (1974).

Arp, P. A., R. T. Foister, and S. G. Mason, "Some Electrodynamic Effects in Fluid Dispersions," Adv. Colloid and Interface Sci. **12** 295 (1980).

Bjornstahl, Y., "The Influence of an Electric Field on the Viscosity of Aeolotropic Liquids," Physics **6** 257 (1935).

Block, H., and J. P. Kelley, "Electro-Rheology," J. Phys. D: Appl. Phys. (1988) p. 1661.

Block, H., and J. P. Kelley, "Materials and Mechanisms in Electro-Rheology," **Proc. 1st Int. Symp. on Electrorheological Fluids**, H. Conrad, A. F. Sprecher and J. D. Carlson, eds., NCSU Engr. Publ., Raleigh, NC (1989) p. 1.

Bonnecaze, R. T., and J. F. Brady, "Dynamic Simulation of a Suspension of Dielectric Particles Forming an Electrorheological Fluid," **Proc. 2nd Int. Conf. Electrorheological Fluids**, J. D. Carlson, A. F. Sprecher and H. Conrad, eds., Technomic Publ. Co., Lancaster, PA (1990) p. 27.

Breck, S. W., **Zeolite Molecular Sieves**, Wiley, N. Y. (1974).

Brooks, D., J. Goodwin, C. Hjelm, L. Marshall, and C. Zukoski, "Visco-Elastic Studies of an Electro-Rheological Fluid," Colloids and Surfaces **18** (1986) p. 293.

Bullough, W. A., and J. D. Stringer, "The Utilization of the Electroviscous Effect in a Fluid Power System," **3rd Int. Fluid Power Symp.** Turin, Italy (9–11 May, 1973) Paper F–3, p. 37.

Bullough, W. A., and M. B. Foxon, "A Proportionate Coulomb and Vsicously Damped Isolation System," J. Sound and Vibration **56** 35 (1978).

Carlson, Charles A, **Electricity**, Interscience, N. Y. (1961) p. 42.

Carlson, J. D., A. F. Sprecher, and H. Conrad, eds.,**Proceedings Second International Conference on Electrorheological Fluids**, Technomic Publ. Co., Lancaster, PA (1990).

Chen, Y., A. F. Sprecher, and H. Conrad, "Particle-Particle Dipole Interactions in Electrorheological Fluids", submitted J. Appl. Phys. (1991).

Choi, Y. A. F. Sprecher, and H. Conrad, "Vibration Characteristics of a Composite Beam Containing an Electrorheological Fluid, J. Intell. Mater. Syst. Struct. **1** 91 (1990).

Conrad, H., and A. F. Sprecher, "Characteristics of Electrorheological Fluids," **Proc. Adv. Materials Conf.**, J. G. Morse, ed., TMS-AIME, Warrendale, PA (1987) p. 63.

Conrad, H., A. F. Sprecher, and J. D. Carlson, eds., **Proceedings First International Symposium on Electrorheological Fluids**, NCSU Engineering Publications, Raleigh, NC (1989a).

Conrad, H., A. R. Shamala, and A. F. Sprecher, "Characteristics of Silica/Silicone Oil Electrorheological Fluids at Small Strains and Strain Rates," **Proc. 1st Int. Conf. Electrorheological Fluids**, H. Conrad, A. F. Sprecher and J. D. Carlson, eds., NCSU College of Engr. Publ. (1989b) p. 47.

Conrad, H., M. Fisher, and A. F. Sprecher, "Characterization of a Model Electrorheological Fluid Employing Stereology," **Proc. 2nd Int. Conf. on ER Fluids**, J. D. Carlson, A. F. Sprecher and H. Conrad, eds., Technomic Publ. Co., Lancaster (1990a) p. 63.

Conrad, H., Y. Chen, and A. F. Sprecher, "Electrorheology of Suspensions of Zeolite Particles in Silicone Oil," **Proc. 2nd Int. Conf. Electrorheological Fluids**, J. D. Carlson, A. F. Sprecher and H. Conrad, eds., Technomic, Lancaster, PA (1990b) p. 252.

Conrad, H., and A. F. Sprecher, "Characteristics and Mechanisms of Electrorheological Fluids", J. Statis. Phys. **64** 1073 (1991a).

Conrad, H., A. F. Sprecher, Y. Choi, and Y. Chen, "The Temperature Dependence of the Electrical Properties and Strength of Electrorheological Fluids", to be publ. J. Rheology (1991b) .

Coulson, C. A. and T. S. M. Boyd, **Electricity**, Longman, N. Y. (1979) p. 43.

Coulten, J. P., and T. G. Duclos, "Applications of Electrorheological Materials in Vibration Control," **Proc. 2nd Int. Conf. on ER Fluids**, J. D. Carlson, A. F. Sprecher and H. Conrad, eds., Technomic, Lancaster, PA (1990).

DeHoff, R. T., and F. N. Rhines, **Quantitative Microscopy**, McGraw Hill, N. Y. (1968).

Dixon, A. A., "Change of the Viscosity of Liquids Produced by an Electric Field," Phys. Rev. **56** 847 (1939).

Dobinski, S., "Uber den Einfluss eines Electrischen Feldes auf die Viskositätvon Flussigkeiten," Phys. Z. **36** 509 (1935).

Duclos, T. G., D. N. Acker, and J. D. Carlson, "Fluids That Thicken Electrically," Machine Design, Jan. 21 (1988) p. 42.

Duff, A. W., "Viscosity of Polarized Dielectrics," Phys. Rev. **4** 23 (1896).

Fisher, M. D., A. F. Sprecher, and H. Conrad, "Metallography Applied to the Structure of Electrorheological (ER) Fluids," **Metallography 75 Years Later**, to be publ. ASTM STP (1991).

Gast, A. P., and C. F. Zukoski, "Electrorheological Fluids as Colloidal Suspensions," Adv. Colloid Interface Sci. **30** 153 (1989).

Gandhi, M. V., and B. S. Thompson, "A New Generation of Revolutionary Ultra-Advanced Intelligent Composite Materials Featuring Electro-Rheological Fluids, U. S. ARO Workshop **Smart Materials, Structures and Mathematical Issues**, C. A. Rogers, ed., Virginia Polytechnic Institute, Blacksburg, VA (15–16 Sept. 1988).

Gorodkin, R. G., Ye. V. Korobko, G. M. Blokh, V. K. Gleb, G. I. Sidorova, and M. M. Ragofner, "Applications of the Electrorheological Effect in Engineering Practice," Fluid Mechanics-Soviet Res. **8** 48 (1979).

Herzog, R. O., H. Kuder, and E. Paersch, "Electrostatischer Viskositäts effect bei Flüssigkeiten," Phys. Z. **35** 446 (1934).

Inoue, A. "Study of New Electrorheological Fluids," **Proc. 2nd Int. Conf. Electrorheological Fluids**, J. D. Carlson, A. F. Sprecher and H. Conrad, eds., Technomic Publ. Co., Lancaster, PA (1990) p. 176.

Jaggi, N . K., "Possible Phase Transition in Electrorheological Fluids," **Proc. 2nd Int. Conf. ER Fluids**, J. D. Carlson, A. F. Sprecher and H. Conrad, eds., Technomic, Lancaster, PA (1990) p. 53.

Jordan, T. C., and M. T. Shaw, "Electrorheology," IEEE Trans. Electrical Insul. **24** 849 (1989).
Kerr, J., "Freeze! A Solid Chance to Jam Liquid Flow LInes," The Engineer (23 July 1981) p. 63; 67.
Kimura, O., "Etudes de effect du Electrique sur la viscosite' des Liquidses. I," Chem. Soc. Jap. **12** 147 (1937).
Klass, D. L., and T. W. Martinek, "Electroviscous Fluids. I. Rheological Properties," J. Appl. Phys. **38** 67 (1967); "Electroviscous Fluids. II. Electrical Properties", J. Appl. Phys. **38** 75 (1967).
Klingenberg, D. J., and C. F. Zukoski, "Studies on the Steady-Shear Behavior of Electrorheological Suspensions," Langmuir **6** 15 (1990).
Marshall, L., J. W. Goodwin, and C. F. Zukoski, "The Effects of Electric Fields on the Rheology of Concentrated Suspensions," J. Chem. Soc., Faraday Trans. **85** 2185 (1989).
Menz, H., "Über den Einflusz eines Electrischen Feldes auf die innere Ruibung von Flüssigkeiten," Ann. Phys. **34** 740 (1939).
Phillips, M. V., "Engineering Applications of Fluids with a Variable Yield Stress", D. Engr. Dissertation, Univ. Calif., Berkeley (1969).
Quinke, G., "Die Klebrigkeit Isolierender Flüssigkeiten im Constanten Electrischen Felde," Ann. Phys. Chem. **62** 1 (1897).
Reitz, J. R., F. J. Milford, and R. W. Christy, **Foundations of Electromagnetic Theory**, Addison-Wesley, Reading, MA (1973) p. 53.
Russ, J. C., **Practical Stereology**, Plenum Press, NY (1986).
Scott, D., "Amazing Hardening Fluid," Popular Science (Apr. 1984) p. 82.
Scott, D., and J. Yamaguchi, "Solidifying Fluid Transforms Clutches and Flow Valves," Automotive Engr. **91** 61 (1983).
Scott, D., and J. Yamaguchi, "ER Fluid Devices Near Commercial Stage," Automotive Engr. **93** 75 (1985).
Smith, K. L., and G. G. Fuller, "Kinetics of Electric Field Induced Structure in Dense Suspensions," **Proc. 1st Int. Conf. Electrorheological Fluids**, Conrad, A. F. Sprecher, and J. D. Carlson, eds., NCSU College of Engr. Publ. (1989) p. 27.
Sprecher, A. F., J. D. Carlson, and H. Conrad, "Electrorheology at Small Strains and Strain Rates of Suspensions of Silica Particles in Silicone Oil," Mat. Sci. Engr. **95** (1987) p. 187.
Sprecher, A. F., Y. Chen, and H. Conrad, "Measurement of Forces Between Particles in a Model ER Fluid," **Proc. 2nd Int. Conf. Electrorheological Fluids**, J. D. Carlson, A. F. Sprecher and H. Conrad, eds., Technomic Publ. Co., Lancaster, PA (1990) p. 82.
Sprecher, A. F., Y. Choi, and H. Conrad, "Mechanical Behavior of ER Fluid-Filled Composites in Forced Oscillation," **Proc. 1st Joint U.S./Japan Conf. on Adaptive Structures**, Nov. 13–15, 1990, Maui, Hawaii, B. Wade, ed., Technomic, Lancaster, PA (1991) p. 560.
Sproston, J. L., N. G. Stevens, and I. M. Page, "An Investigation of Torque Transmission Using Electrically Stressed Dielectric Fluids," **Inst. Phys. Conf. Ser. No. 66**, Oxford (1983) p. 53.
Stangroom, J. E., "Electrorheological Fluids," Phys. Tech. **14** 290 (1983).

Stangrom, M. V., "Basic Considerations in Flowing Electro-Rheological Fluids," presented at the symposium **Nonlinear Dynamical Structures in Simple and Complex Liquids**, NATO Workshop, Los Alamos, (June 26–29, 1990).

Stevens, N. G., J. L. Sproston, and R. Stanway, "Experimental Evaluation of a Simple Electrovsicous Damper," J. Electroelectrics **15** 275 (1984).

Tao, R., J. T. Woestman and N. K. Jaggi, "Electric Field Induced Solidification," Appl. Phys. Lett. **55** 1844 (1989).

Uejima, H., "Dielectric Mechanism and Rheological Properties of Electro-Fluids,"Jap. J. Appl. Phys. **11** 319 (1972).

Underwood,E., **Quantitative Stereology**, Addison-Wesley, Reading, MA (1970).

Winslow, W. M., "Method and Means for Translating Electric Impulses into Mechanical Force," U. S. Patent No. 2,417,850 (Mar. 25, 1947); "Induced Fibration of Suspensions", J. Appl. Phys. **20** 1137 (1949).

Zukoski, C. F., and J. W. Goodwin, "Towards the Electrical Control of Viscosity: Optimization of Electrorheological Fluids," IEEE IECON'86 (1986) p. 9.

CHAPTER 13

PERIODIC LIQUID-SOLID FLOW MICROSTRUCTURE IN A CENTRIFUGAL PUMP

M. C. Roco
T. Cader
O. Masbernat

13.1 INTRODUCTION

Periodic two–phase flows are encountered in a variety of industrial applications, such as positive displacement pumps, batch reactors, mixers and propellers. Typical situations occur in rotating fluid machinery, in which the flow periodicity is caused by the finite number of blades of the rotor and stator, as well as by the interaction between the rotor and casing flow. A centrifugal pump, which is the most used type of fluid machinery in industry, is considered here as a representative case study. Few measurements on velocity distribution are available on particulate two–phase flows in pumps (Itaya and Nishikawa, 1964; Cader et al., 1991a). Quantitative correlations revealing the microstructure are in development for much simpler two–phase flows, starting with plane Couette flow (Graham, 1980; Leighton and Rampall, 1992; Ye and Roco, 1992). Due to the flow complexity, the current predictions based on single particle dynamics in pumps are not reliable. Since it is impossible at this moment to have general conclusions on slip velocity patterns in a rotating machine, it is useful to present a set of measurements where qualitative trends can be identified. This chapter will discuss several aspects of the periodic flow structure of dilute particulate flow in a centrifugal pump casing, and its effect on the pump head and energy efficiency. The goal is to offer primary data needed for development of new theories, validation of numerical simulations, and in the long term for the prediction of pump performance.

Besides the scientific relevance of the interaction between solids and carrier liquid, the slip velocity distribution in slurry equipment is of interest for design and operation. The relative velocity determines the energy dissipation, particle–fluid segregation, particle–wall interaction and erosion wear of the wetted surfaces. Overall, it affects the performance curves of the slurry equipment. If at low slip velocities one may approximate the flow as homogeneous and use an Eulerian frame of reference when solving the governing equations, this is no longer true at large slip velocities. The equations have to be reformulated for each phase, and a Lagrangian frame of reference is generally required for solid phase. Periodical pump flow is a typical unsteady flow situation, for which currently available numerical simulations need reference data for validation.

A pioneering study on single fluid velocity distribution in pumps was reported in 1957 by Bowerman and Acosta. The instrumentation available at the time only allowed for average measurements over a full impeller rotation. More recent work on single fluid flow was directed to either point measurements with good temporal resolution and accuracy, using laser Doppler velocimetry, LDV (Durst et al., 1981 for general reference; Adler and Levy, 1979; and Hamkins and Flack, 1987 for centrifugal pumps), or field measurements that provide instantaneous field imaging, using particle image velocimetry, PIV (Liu and Adrian, 1992, chapter 2 in this book, for general reference; Paone et al., 1989; and Dong et al., 1991 for centrifugal pumps), and laser induced photochemical anemometry, LIPA (Falco and Nocera, 1992, chapter 3 in this book, for general reference).

This chapter presents results of LDV two–phase particulate flow in a centrifugal slurry pump model, including the effect of the impeller position. The results are for liquid, solids and slip velocity distributions. The point LDV measurements performed as a function of the impeller angle were reconstructed into three–dimensional field velocity distributions in space, for all impeller positions. Other studies on LDV slurry flow measurements in pumps have not been reported in literature. LDV research for particle–fluid systems has been performed mainly for pipe flow (Zisselmar and Molerus, 1979; Lee and Durst, 1982; Tsuji et al., 1984; Nouri et al., 1987; Gore and Crowe, 1989) and round jet flow (Modaress et al., 1984a and 1984b; Parthasarathy and Faeth, 1987). Barlow and Morrison (1990) applied the Doppler amplitude discrimination approach in particle laden jets. In this study, a slightly modified amplitude discrimination technique as that used by Barlow and Morrison in the jet zones of low concentration has been applied for both liquid and solids measurements (Cader et al., 1991). The velocity of the carrier fluid (naturally seeded with micron size tracers) and of solid phase (0.8 mm glass beads) have been compared at the impeller–casing flow interface and within the pump casing. Two types of sections have been selected for detailed measurements: (a) the radial casing cross–sections (windows #2 to #8 in Figure 13.1a), and (b) the cylindrical surface surrounding the impeller outlet. A two–color LDV system was employed in back–scatter mode to obtain all three velocity components by measuring along two orthogonal directions. Experimental results are shown at the nominal flow rate (Q_n) and at half of it ($Q_n/2$). The data were calibrated with results from visualization with a sheet of laser light in radial and tangential planes to the impeller circumference (Hamelin and Roco, 1988). An angular position of the impeller was attached to every validated measurement using a rotary shaft encoder. The fluid and solid–liquid slip velocity representation as a function of the impeller angle gives a synthetic view of the flow pattern during the impeller rotation and allows for the identification of specific flow structures.

The two–phase flow experiments were conducted using a dilute concentration of glass beads of 0.16% by volume. The glass beads have a narrow size distribution about 0.8 mm . The density ratio is $\rho_S/\rho_L = 2.48$. The particle Reynolds number, calculated with the beads terminal fall velocity of $v_\infty = 0.082$ m/s and water viscosity of $\nu_L = 1.01\ 10^{-6}$ m²/s², is $Re_\infty = 65$. Dilute suspension flow with relatively large particle settling velocities was chosen as a relevant limiting case for analysis. The fact that comparisons to models can be performed without the supplementary effect of particle interactions, and slip velocities can be measured with a superior accuracy as compared to dense slurry flow, makes this case relevant.

Two time scales are considered for the periodic motion:

– one corresponding to a full impeller rotation ($T_i = 60/N = 0.06$ seconds, where $N = 1000$ rpm is the rotational speed of the impeller used in the present study), and

– another for the blade–to–blade interval ($T_b = T_i/Z = 0.02$ seconds, where $Z = 3$ is the number of blades on impeller). The particle time scale in turbulent conditions, equal to $1.33\ Sd_s/(0.44\ v_\infty) = 0.072$ seconds, is much larger than T_b. Since the particle Stokes number is $0.072/T_b = 3.6$, larger than 1, the particle motion will be only slightly affected by the periodical flow of the carrier fluid.

13.2 THE PUMP MODEL

A schematic of the centrifugal pump tested in laboratory is shown in Figure 13.1 (after Roco et al., 1991). The casing and impeller are quasi–concentric. The impeller of diameter $D_{2i} = 280$ mm is shrouded, with plane shrouds and three blades of simple curvature. The width of each shroud is 25 mm, and the blade width is 35 mm. The inlet clearance between the front impeller shroud and casing is about 5 mm. The pump has an inlet diameter of $D_1 = 100$ mm, and a discharge diameter of $D_2 = 76.2$ mm. The blades have an inlet angle of 26 degrees and an outlet angle of 23 degrees. The casing flow cross–section is rectangular, with $\Delta r_{max} = 60$ mm and $B_c = 100$ mm, as shown in Figure 13.1c. The power is transmitted to the pump from a 15 kW motor with variable speed between $N = 0$ and 2000 rpm. The experiments presented here were run at $N = 1000$ rpm. The nominal flow rate is $Q_n = 63.6$ m³/h and the corresponding pump head is $H_n = 9.7$ m. The dimensionless specific speed of the pump is $NQ^{0.5}/(gH)^{0.75} = 0.46$, which is a typical value for a radial centrifugal pump. The pump Reynolds number based upon the linear velocity of the impeller outlet surface ($U_{2i} = \pi D_{2i}N/60 = 14.67$ m/s) and the impeller diameter is $Re_{pump} = 4 \cdot 10^6$.

The pump was mounted in a closed circuit, with a feeder and a separation system for solids. The water temperature was maintained at 20 ± 0.5^oC. The particle–fluid Froude number calculated with the friction velocity at the outlet diameter of the impeller (Roco et al., 1986) is about 400 (>> 1). Since the particle–fluid Froude number represents the ratio between the turbulence dispersive force and particle inertial force, its large value suggests that particles are fully in suspension within the pump channels.

The laser beams have access through eight transparent windows mounted flush into the casing wall. In order to obtain a better signal–to–noise ratio and to increase wear resistance, optical grade quartz with optimum light transmittance to the green and blue laser beams was chosen instead of plexiglass. In order to increase the amount of light back–scattered from particles, the surfaces of the 0.8 mm glass beads were made opaque by increasing their roughness using a mill with finer sand particles.

The θ–axis is tangential to the impeller circumference, the r–axis is in the impeller radial direction, and the z–axis is directed from the impeller back shroud to the front shroud (Figure 13.1). The measuring locations along the z–axis start at 5 mm away from the front wall and continue up to 75 mm from the same wall. Measurements were performed at intervals of 5 to 8 mm along five lines parallel to the axis z (see lines A,B,C,D and E in Figure 13.1c).

The axial coordinate was normalized by the axial width of the casing ($z^* = z/ B_c$) and the radial coordinate by the radial gap ($r^* = \Delta r/\Delta r_{max}$), respectively. The back shroud covers from $z^* = 0.1$ to 0.35, and the front shroud from $z^* = 0.7$ to 0.95.

The dimensionless impeller angle ϕ^*_o is measured counter–clockwise from window #2, and is normalized by the blade–to–blade interval $\Delta\phi_b = 360$ degrees / Z (number of blades) = 120 degrees:

$$\phi^*_o = (\phi - \phi_o)/\Delta\phi_b \,. \tag{13.1a}$$

The angle $\phi_o = 23.7$ degrees corresponds to the location of window #2. The normalized angle ϕ^*_o is zero when the suction side of the trailing edge of an arbitrarily chosen blade of the impeller passes by window #2, as shown in Figure 13.1a. We denote the normalized impeller angle simply by ϕ^* if the impeller angle (encoder position) is measured with reference to the casing location of a current window.

The dimensionless angle in casing, θ^*, has the same origin as ϕ. It is normalized by 360 degrees

$$\theta^* = (\theta - \phi_o)/360 \text{ degrees}, \tag{13.1b}$$

By considering the origin at window #2, the access windows have the following normalized angles:

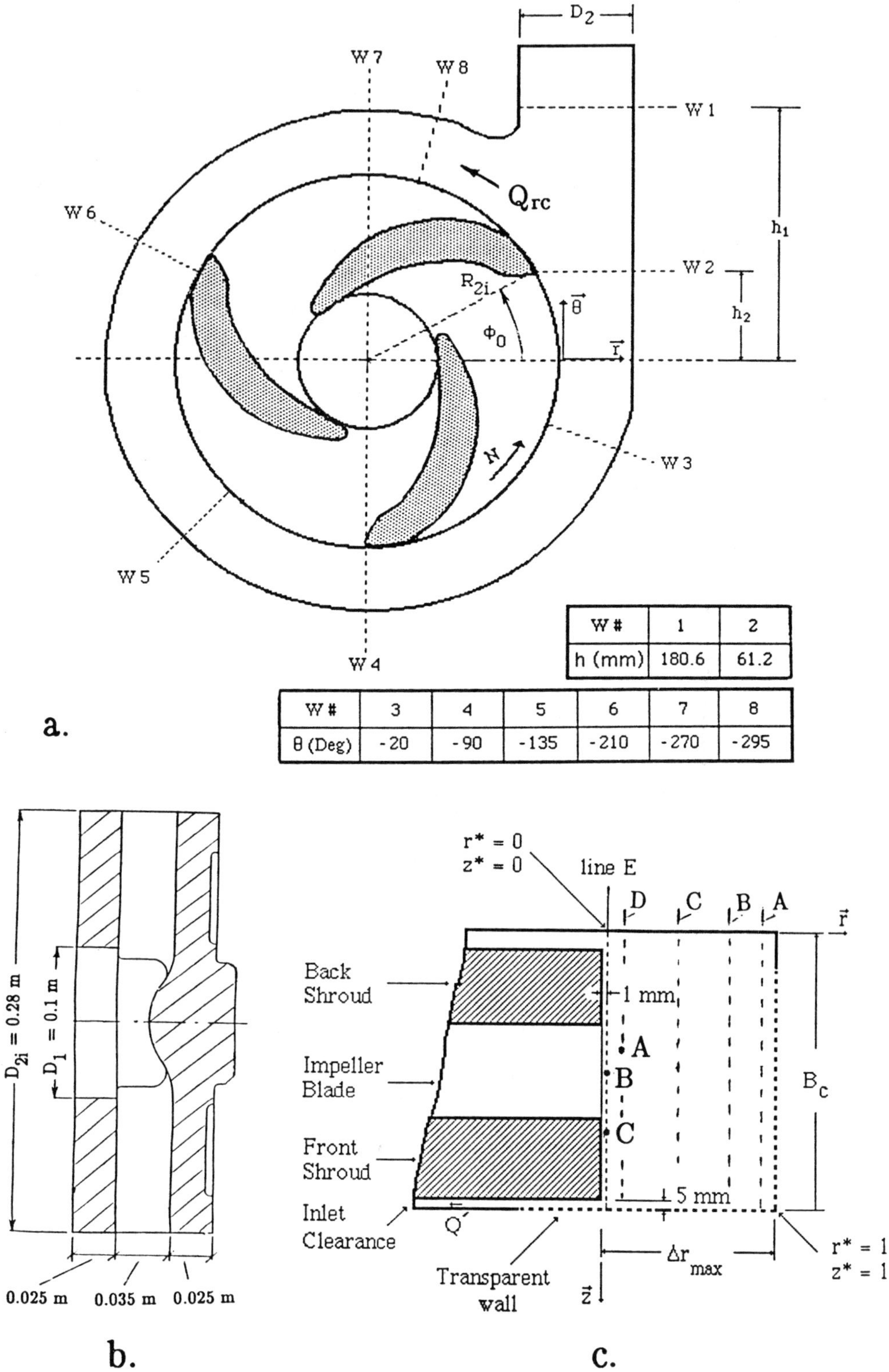

W #	1	2
h (mm)	180.6	61.2

W #	3	4	5	6	7	8
θ (Deg)	-20	-90	-135	-210	-270	-295

Fig. 13.1 Pump geometry: a. Axial cross–section; b. Impeller cross–section; c. Radial cross–section

Window #		2	8	7	6	5	4	3	2
θ^*	(–)	0.0	0.12	0.19	0.36	0.56	0.69	0.88	1.0

13.3 EFFECT OF IMPELLER POSITION: TYPICAL MEASUREMENTS

In order to determine the effect of the impeller angle on the flow, an "once per revolution" pulse was generated for every revolution of the impeller (see Figure 13.2) with a TSI low speed rotary encoder. In addition, 3600 pulses are generated between two successive "once per revolution" pulses. Each pulse interval corresponds to a 0.1 degree displacement of the impeller. Then, the impeller angular position is attached to each validated LDV measurement.

A full impeller rotation was divided into 72 angular sectors, each of $\Delta\phi = 5$ degrees, and a velocity average was calculated for each sector. The measurements were performed at eight windows. The velocity vector, turbulence intensity and Reynolds shear stresses were

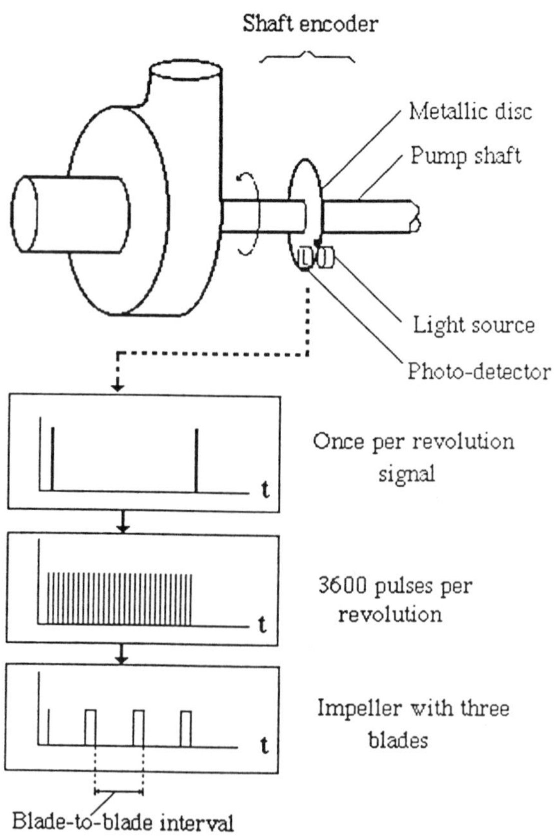

Fig. 13.2 Rotary shaft encoder

obtained at each measuring point, for each angular sector of the rotary encoder. The sample size within each angular sector is about 1400 data points (that is, 1400 X 72 intervals = 100,800 data points over 360 degrees). With these measurements, the velocity frequency distribution functions and average velocities were obtained at different points at all $\Delta\phi$ intervals .

Any **instantaneous point velocity** at a location $\mathbf{r}$, impeller angle ϕ , and time t , normalized by the peripheral impeller velocity U_{2i} , can be decomposed into three components:

$$v(\mathbf{r},\phi,t) = U(\mathbf{r}) + u_p(\mathbf{r},\phi) + u'(\mathbf{r},\phi,t) \qquad (13.2)$$

$U(\mathbf{r})$ is the point average of $v(\mathbf{r},\phi,t)$ over the blade–to–blade interval $\Delta\phi_b$ and a sufficiently long time interval. The periodic component $u_p(\mathbf{r},\phi)$ is the point average of $v(\mathbf{r},\phi,t)$ over a large number of rotations for same impeller angle ϕ . This velocity component is induced by the blade passage, with an angular periodicity $\Delta\phi_b$, and has a zero average over $\Delta\phi_b$. The measured velocities in pump at same $\mathbf{r}$ and ϕ have significant fluctuations $u'(\mathbf{r},\phi,t)$ from one rotation to another as compared to the sum of the mean and periodic velocity components, $u = U + u_p$. The velocity u defines the average of all LDV velocity measurements at a point for the same blade–to–blade angular interval from $(\phi - \Delta\phi/2)$ to $(\phi + \Delta\phi/2)$

$$u = \frac{1}{\Delta t}\sum_n (u_j \cdot \Delta t_j) \quad ; \quad \Delta t = \sum_n \Delta t_j \qquad (13.3)$$

where n = 1400 is the sample size of the LDV measurement, and u_j and Δt_j are the velocity and associated burst time of the j–th particle recorded within the same angular interval $\Delta\phi = 5$ degrees.

Particle slip velocity was obtained by subtracting liquid velocity from solid velocity. The slip velocity is normalized by the particle terminal settling velocity v_∞

$$u^*_{sL} = u_{sL}/v^*_\infty \qquad (13.4)$$

where $u_{sL} = u_s - u_L$ and $v^*_\infty = v_\infty/U_{2i} = 0.0056$.

The **frequency distribution functions** of the circumferential and radial water velocity components are illustrated in Figure 13.3 for a point A located in front of the blade width (r^* =0.12, z^* = 0.41; Fig. 13.1c) at window #2, at Q_n . The normalized impeller angle takes the following values ϕ^* = 0, 0.25, 0.50, and 0.75. The effect of the impeller angle on the mean velocity components and their standard deviations is clearly observed. In the wake of the blade, at $\phi^* = 0.25$, the functions are flatter than prior to the arrival of the next blade, at $\phi^* = 0.75$ (Figure 13.3). For both velocity components, the velocity standard deviations $u_{\theta,std}$ and $u_{r,std}$ are three times larger at

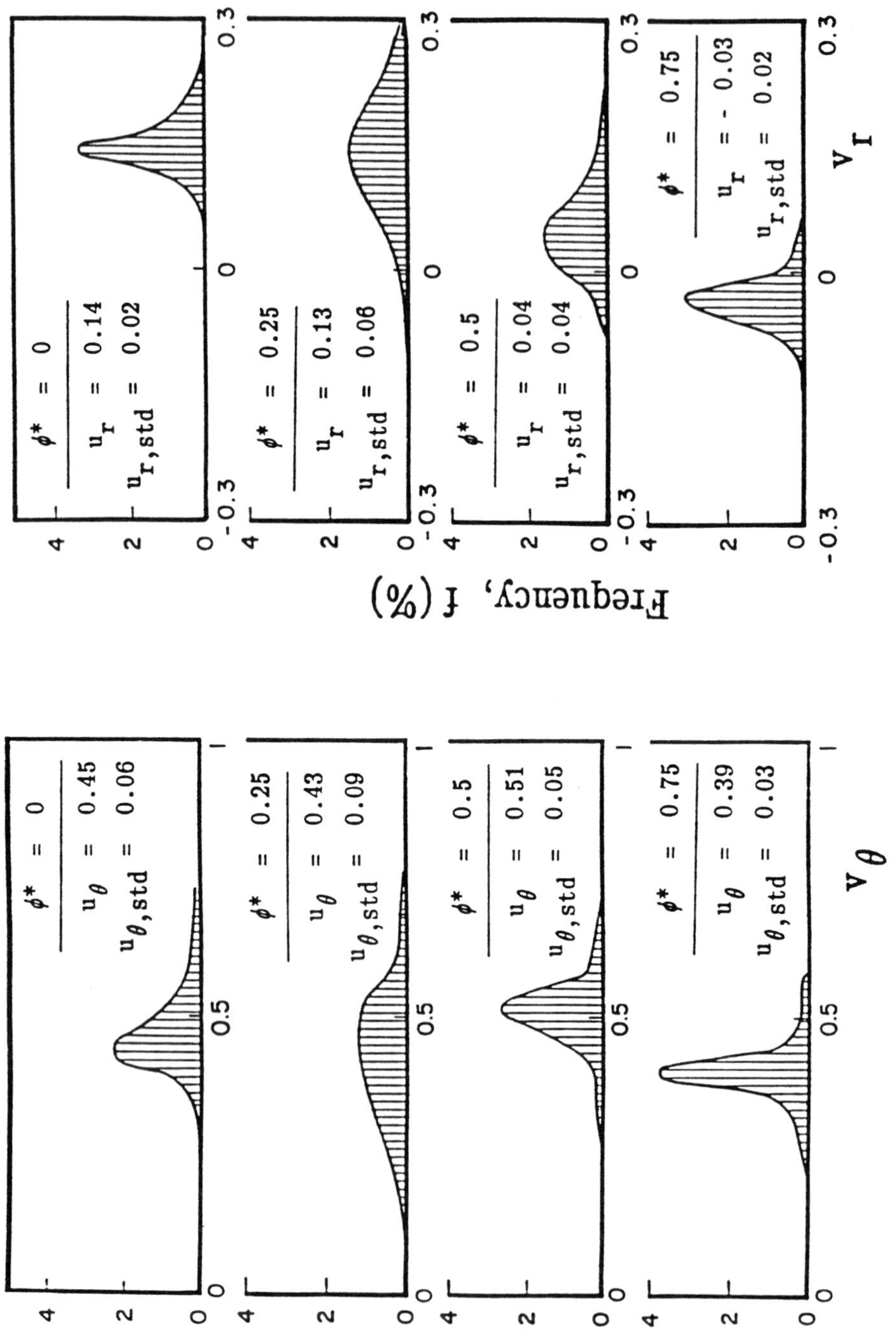

Fig. 13.3 Velocity frequency distributions at the point A, window#2, Q_{n}, for different impeller angular positions ($\phi^* = 0, 0.25, 0.5, 0.75$): a. v_θ; b. v_r

$\phi^* = 0.25$ as compared to $\phi^* = 0.75$. The velocity components increase during the displacement of the blade trailing edge surface past the measuring point ($\phi^* = 0.82$ to 1), particularly the radial component. These frequency distribution functions, as well as those for solids, are best described as log–normal. The differences between the water and solids frequency distribution functions are caused by the particle slip velocities, which are one order of magnitude smaller than the water velocities. The variation of the periodical components from one impeller position within one blade–to–blade interval to another, as a result of small differences in the blade dimensions, is not considered here.

The effect of the impeller angular position on the velocity u, its standard deviation, flow angle, and Reynolds shear stresses at the same point A within window #2 is exemplified in Figures 13.4 and 13.5. A single point on any given curve is taken from an average over thousands of revolutions, for the same ϕ^*. The position of the blade trailing edge surface at D_{2i} is marked on the ϕ^* coordinate between 0.82 and 1. The variations of u' from one rotation to another for the same blade–to–blade interval are significant, of the same order of magnitude with u_p, and their measure is given by the normalized velocity standard deviation (u_{std} in Figure 13.4 b, d and f). One can observe the fluctuations of u_θ up to 20% from U_θ for various ϕ^*, u_r up to 90% of U_r, and u_z up to 200% of U_z (Figure 13.4 c to f).

The angle $\beta_{r\theta}$ formed by the velocity with the r direction as defined in Figure 13.5a changes periodically between $\beta_{r\theta} = 66$ and 95 degrees at the chosen point A. Measurements taken at other points along the impeller shrouds show that the angle is about 90°, i.e. the flow is almost circumferential. In front of the blade width at the impeller outlet ($z^* = 0.35$ to 0.7), the flow is more radially oriented, and accordingly the angle decreases.

The Reynolds shear stresses (Figure 13.5b) are negative and generally larger in the wake region after the blade displacement. Their values change sign before the arrival of the following blade in our illustration. The flow is more chaotic, with larger energy loss in the wake (interval I_1), and becomes more ordered prior to the arrival of the next blade (interval I_2).

The frequency distribution functions for both phases vary significantly with position in the radial cross–section (r^*,z^*) and around the impeller (θ^*). Their variance is larger in front of the blade width, decrease with r^*, and is larger near the casing tongue (sector S_1; windows #7 and #8) and in the bottle neck sector (sector S_3; windows #3 and #4). For illustration, Table 1 gives the first four moments of the fluid velocity frequency distribution function for three points located at window #2: A is in the core of the flow; B ($r^* = 0.02$, $z^* = 0.51$) is also in the core region of the flow, and C ($r^* = 0.02$, $z^* = 0.73$) is located at 1 mm away from the front shroud surface at D_{2i} (Figure 13.1c). The averages over the blade–to–blade interval are

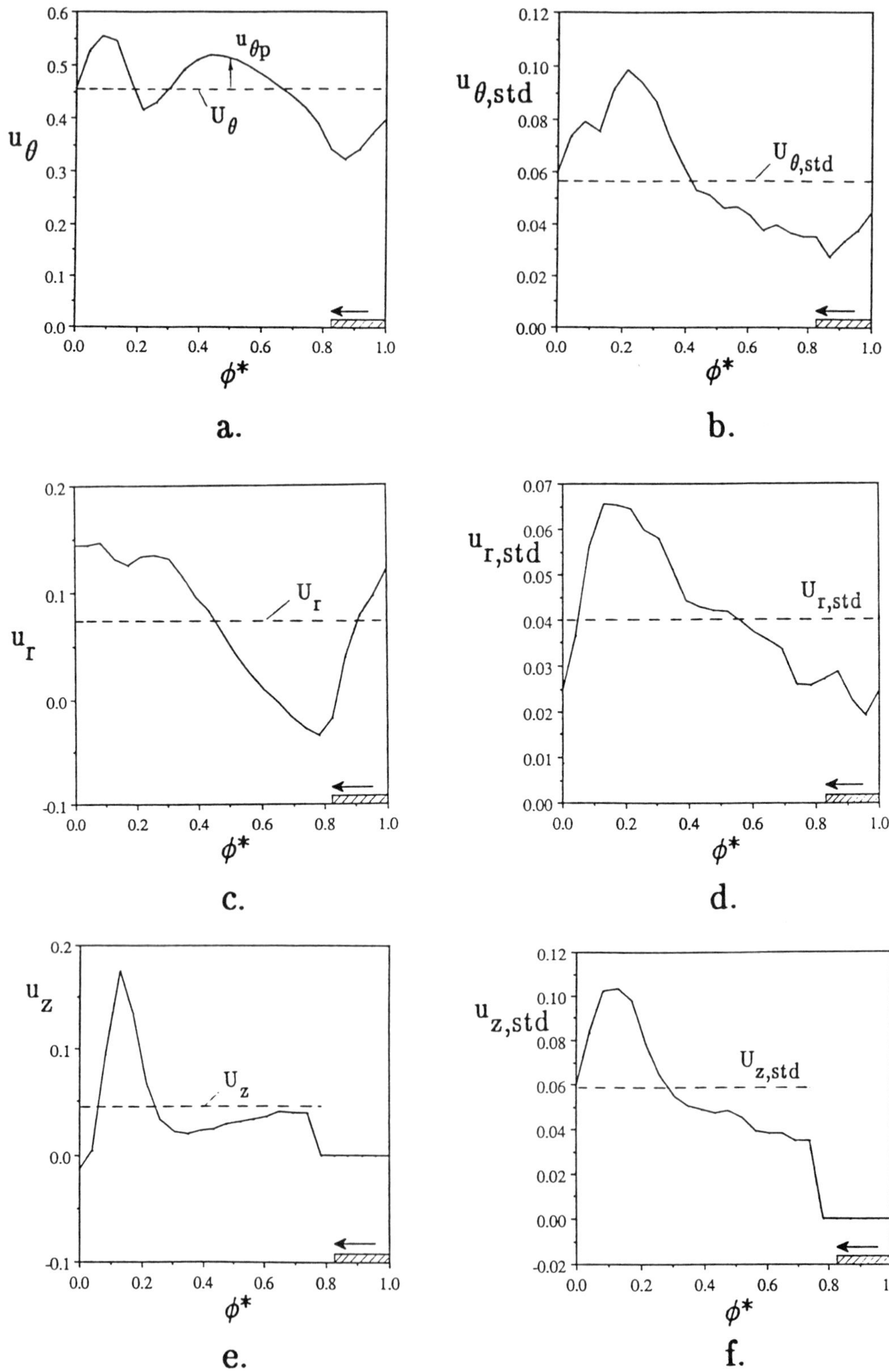

Fig. 13.4 Effect of the impeller angular position ϕ^* on the normalized velocity components (u_θ, u_r, u_z), and their standard deviations

Table 1

ϕ^* (–)	Circumferential velocity (m/s)			Circumferential velocity STD (m/s)			Skewness coefficient (–)			Flatness coefficient (–)		
Point:	A	B	C	A	B	C	A	B	C	A	B	C
0	6.7	5.3	5.3	0.9	0.9	0.6	0.0	0.0	0.2	5.1	4.1	3.9
0.25	6.3	6.1	5.6	1.4	0.7	0.8	–0.1	0.1	0.3	2.8	4.3	4.6
0.50	7.5	6.5	5.6	0.7	1.1	0.7	–0.4	0.1	0.5	6.4	3.0	9.2
0.75	6.2	6.1	5.3	0.6	0.9	0.8	–0.7	0.1	0.4	13.9	4.1	4.0
1	6.7	5.3	5.2	0.9	0.9	0.6	0.0	0.0	0.2	5.1	4.1	4.0
ave.	6.7	6.0	5.4	0.9	0.9	0.7	–0.3	0.1	0.4	7.1	3.9	5.4

given on the bottom line of that table. One notes that in the vicinity of the mid–point of a shroud (point C), the average circumferential velocity and its standard deviation are smaller, while the skewness of the frequency function is evidently larger.

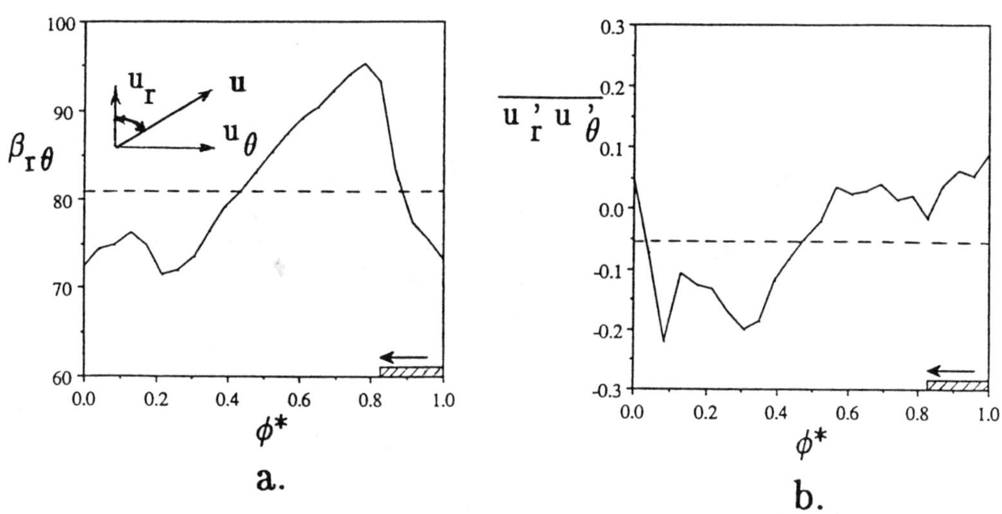

Fig. 13.5 Effect of ϕ^* on (a) the flow angle $\beta_{r\theta}$ and (b) $\overline{u_r' u_\theta'}$, at a point A within window #2 (flowrate Q_n)

From the general analysis of the blade–to–blade velocity data (section 13.5), three ϕ^* intervals have been identified:

- I_1 ($\phi^* = 0$ to 0.5), the wake interval after the displacement of a blade. This interval is marked by maxima for all velocity components and their standard deviations. The shape of the velocity profile in the θ direction clearly suggests a wake. The velocity standard deviation, $u_{\theta,std}$, increases after the blade displacement at $\phi^* = 0$ to 0.3. This increase should be related to the increase of the turbulence intensity for the same interval;
- I_2 ($\phi^* = 0.5$ to 0.82), interval before the arrival of the leading face of the next blade. In this interval the velocity and its standard deviation have a continuous decrease. This interval corresponds to the boundary layer formed on the pressure face of the blade.
- I_3 ($\phi^* = 0.82$ to 1), corresponding to the displacement of the cylindrical surface of the blade trailing edge past the measuring point.

From the analysis of the flow pattern around the impeller (section 13.6), four θ^* sectors with distinctive characteristics have been identified in the casing:

- S_1 ($\theta^* = 0.1$–0.2; windows #7 and #8), the casing tongue sector,
- S_2 ($\theta^* = 0.2$–0.7; windows #5 and #6)), flow acceleration sector,
- S_3 ($\theta^* = 0.7$–0.9; windows #3 and #4), bottle–neck sector, and
- S_4 ($\theta^* = 0.9$–1.0 and 0–0.1; about window #2), discharge nozzle sector.

The **mean velocity components** U_r and U_θ along line E at window #2 are shown in Figure 13.6 a, for Q_n and $Q_n/2$. The radial component of the mean velocity, along the blade width ($z^* = 0.35$ to 0.7), is shown to be quasi–symmetric for both flow rates. The profiles are almost flat. The normalized velocity in the radial direction, averaged over the blade width, is about $U_r = 0.09$ for Q_n and 0.045 for $Q_n/2$. The radial velocity is close to zero along line E in the proximity of the impeller shrouds. Approaching the front wall, the values decrease continuously, becoming slightly negative, then increase suddenly. This suggests that part of the flow induced by the inlet recirculation finds its way up along the front plate, creating a zone with high gradient of velocity at the wall. The circumferential velocity U_θ is asymmetric at Q_n. This asymmetry is due to the recirculation through the inlet clearance, in the gap between the impeller front shroud and the front casing wall (Q' in Figure 13.7). The flow in the inlet clearance is from the flow area with higher pressure in casing towards the suction pipe. Circumferential momentum is transported from the core of the casing cross–section towards the inlet clearance (in the z direction) by the inlet clearance recirculation stream. This

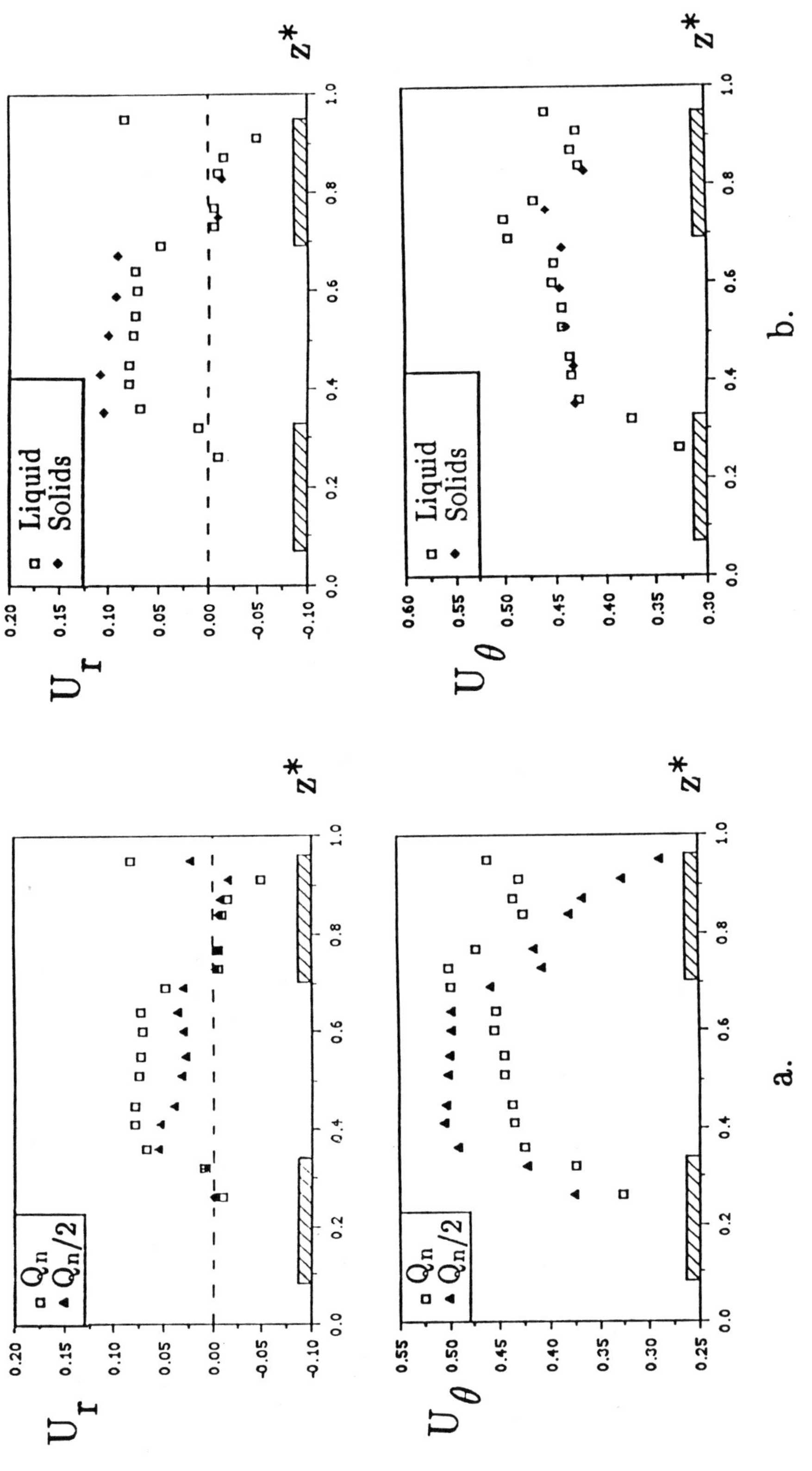

Fig. 13.6 Velocity distribution along line E at window #2:
a. U_r and U_θ for water at Q_n and $Q_n/2$; b. U_r and U_θ for water and solids at Q_n

asymmetry almost disappears at $Q_n/2$, presumably due to the reduction of the inlet clearance recirculation at window #2. The axial velocity U_z becomes negative at a location away from the mid–point between shrouds, closer to the front shroud, at about $z^* = 0.63$. This is in agreement wwith other measurements which indicate that the double helicoidal flow in casing is not symmetrical, but inclined towards the front shroud.

According to equation (13.2), a **velocity fluctuation** has two parts, one periodic (u_p) and the other random (u'). The random fluctuation of circumferential velocity averaged over the blade–to–blade interval (U'_θ) seems to be symmetric for both flow rates, at Q_n and $Q_n/2$ (Figure 13.8 a). The fluctuations U'_θ are larger near the shrouds and smaller in the core. Two local maxima are located at the extremities of the blade width near the shrouds, which is a common result for a turbulent flow in a sudden expansion (Durrett et al., 1988). Another maximum is reached close to the casing front wall. The velocity fluctuations are slightly higher for $Q_n/2$. The radial fluctuations U'_r have almost the same shape as the circumferential fluctuations, but with a smaller magnitude.

The distributions of the solid phase velocity components as compared to the corresponding liquid distributions (U_r and U_θ), at Q_n, window #2 on cylindrical surface at the impeller–casing interface are shown in Figure 13.6b. The asymmetry with respect to $z^* = 0.5$, previously observed in the liquid velocity distribution, is maintained in the solid phase velocity profile. The solids radial component U_r is generally larger than the corresponding fluid component, while the circumferential one is smaller (Figure 13.6 b).

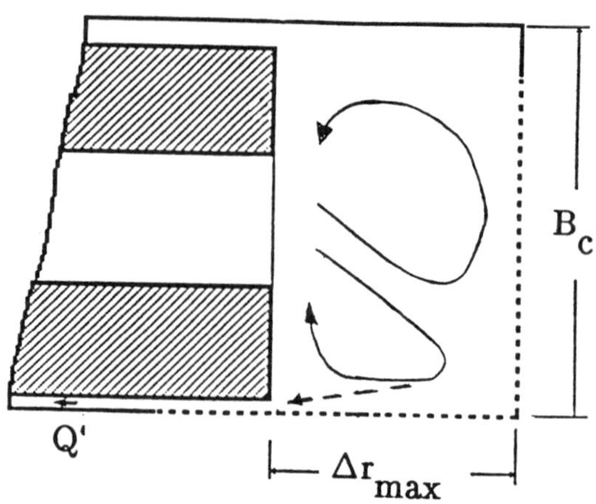

Fig. 13.7 Double helicoidal flow in a radial section

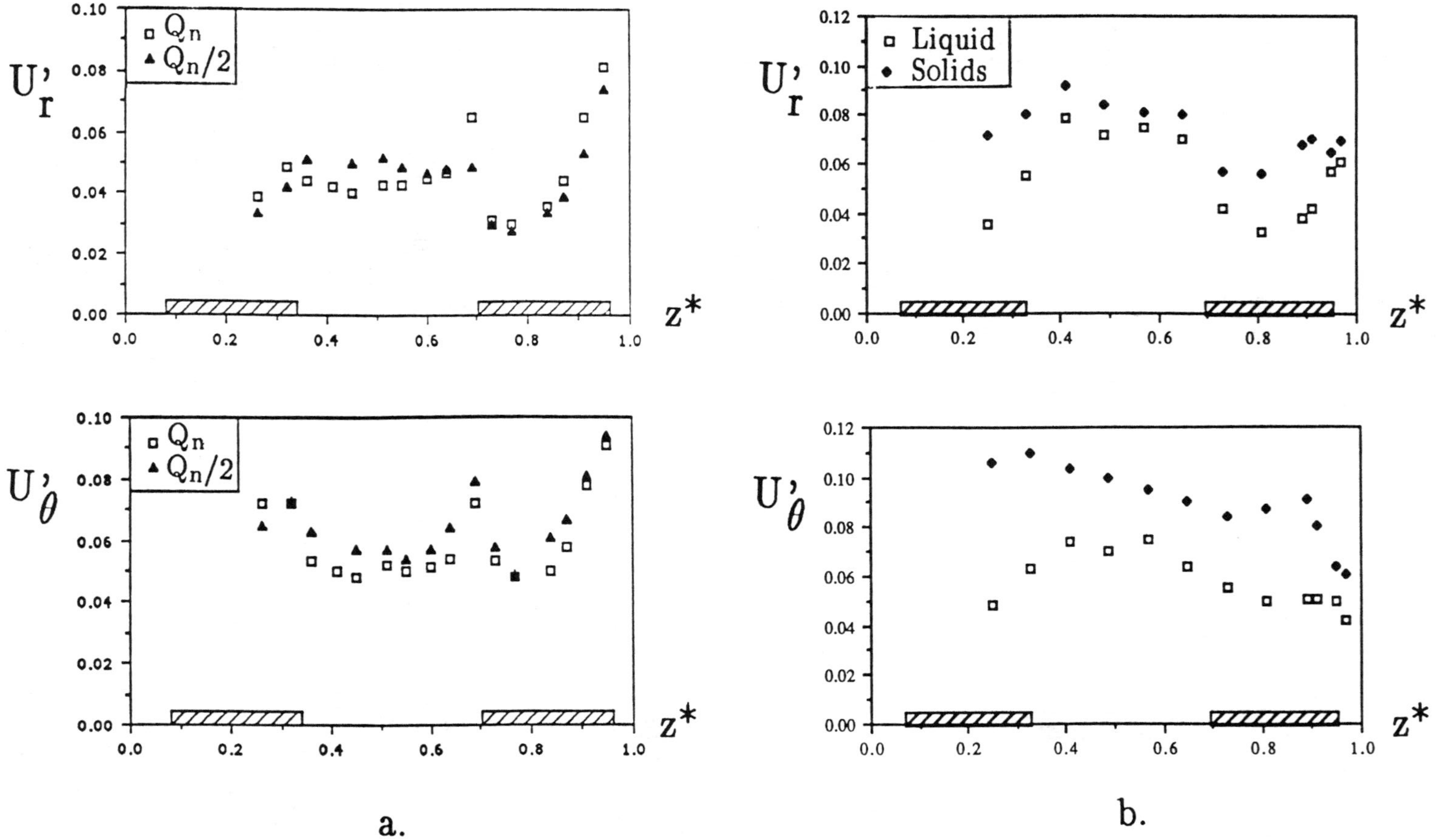

Fig. 13.8 Fluctuation velocity distribution along line E:
a. U'_r and U'_θ for water at Q_n and $Q_n/2$, at window #2;
b. U'_r and U'_θ for water and solids at Q_n , at window #7

The measurements show that the random fluctuations of the solids velocity components are larger in the bottle–neck sector (S_3) and in the tongue sector (S_1, illustrated in Figure 13.8 b) of the casing.

The measured velocity components in the **radial cross–section** of window #4 are illustrated in Figures 13.9 (for Q_n) and 13.10 (for $Q_n/2$). The results are for $\phi^* = 0$. The interpolation was performed on a two–dimensional plane (r^*, z^*). The Lawson's triangulation method and Akima's interpolation approach were used to ensure that the main feature of the velocity distribution are conserved after interpolation (Akima, 1978). In average, the spatial velocity gradient is relatively small in this pump sector (S_3). The radial velocity component exhibits two peaks near the impeller outlet (at $r^* = 0.02$). The circumferential velocity is larger at $r^* = 0.02$, and decreases to the velocity in the remaining cross–section in a relatively short distance ($r^* = 0.02$ to 0.2). The axial velocity distribution has a typical shape for the double helicoidal flow which develops in the radial cross–section. The two peaks of u_r, the slip for u_θ and the double helicoidal pattern suggested by u_z are more pronounced at $Q_n/2$ than at Q_n (see Figure 13.10 versus Figure 13.9). Figure 13.11 shows that the velocity distribution in the radial cross–section has larger spatial variance at window #2 at the beginning of the discharge nozzle (sector S_4). However, the fluctuations in time at window #2 is small as compared to window #4 (sector S_3). The flow structure in radial cross–section will be further discussed in section 13.4.

The evolution of the fluid velocity along line E, window #2, in a **blade–to–blade interval** is illustrated in Figure 13.12 (after Cader et al., 1991). The illustration is for the radial velocity u_r and circumferential velocity u_θ at Q_n. One notes that where u_r is larger u_θ is smaller, and vice–versa. The radial velocity reaches a maximum (Figure 13.12a) at about the same angular position ($\phi^* \approx 0.75$) where u_θ has its minimum in the center of the impeller outlet ($z^* \approx 0.5$). Two peaks after the passage of a blade ($\phi^* = 0$) are evident for u_θ in Figure 13.12b. The effect of the impeller angle on the velocity distribution will be discussed further in sections 13.5 and 13.6.

Figure 13.13 shows the distribution of the circumferential velocity averaged over the radial cross–section ($U_{\theta A}$) as a function of circumferential position θ, at two discharge flow rates Q_n and $Q_n/2$. The velocity $U_{\theta A}$ is larger at the flowrate $Q_n/2$ than at Q_n in most of the casing, in the section $\theta = 60$ to 360 degrees. The reason is a larger recirculation Q_{rc} between the impeller shroud and casing at flowrates smaller than Q_n. For further comments see section 13.6.

The measured **slip velocity** was of the order of the particle settling velocity, but the largest slip velocities have exceeded 10 times the settling velocity. The largest slip velocity occurs in front of the blade ($z^* = 0.35$ to 0.7) at the impeller–casing flow interface . In the

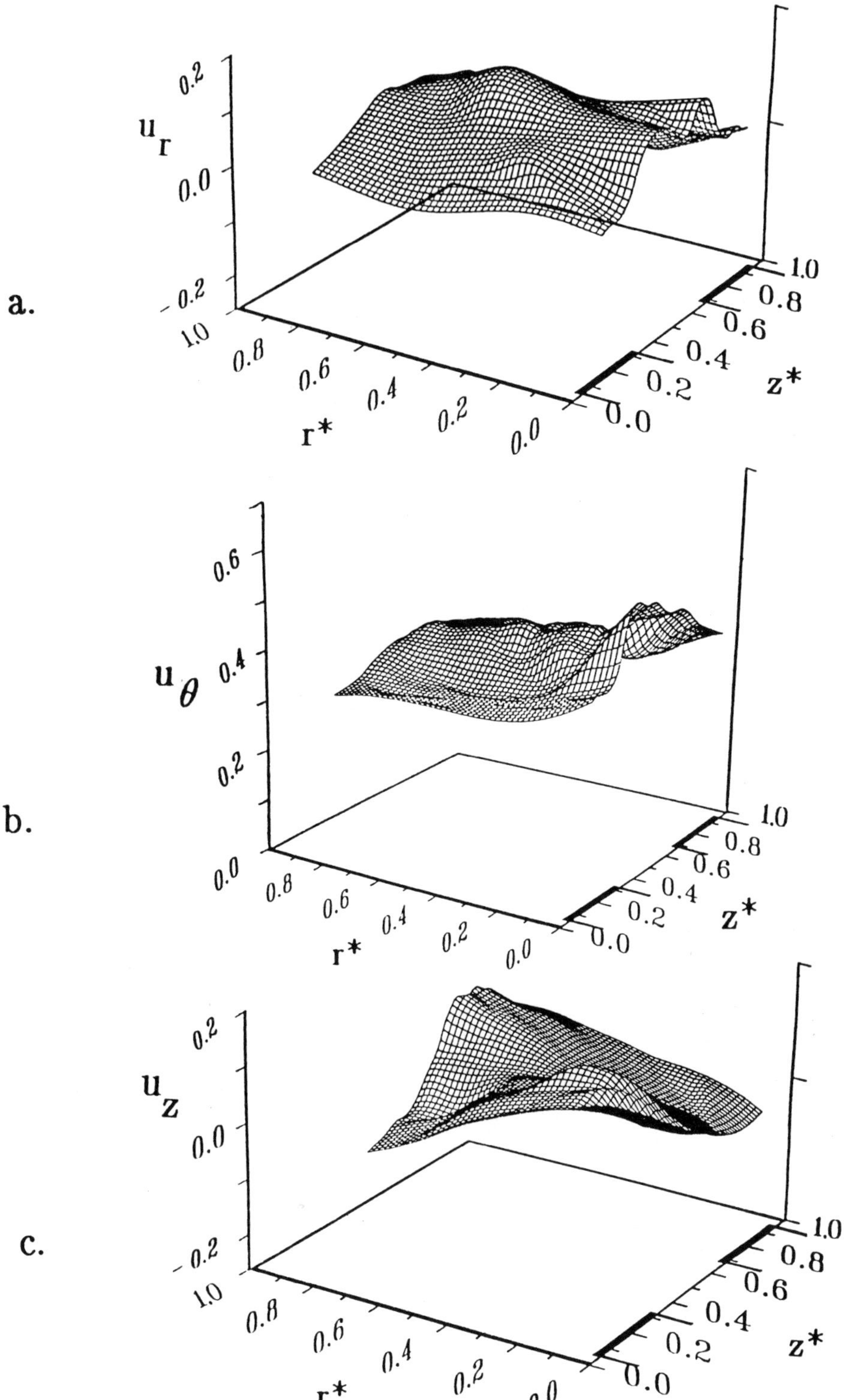

Fig. 13.9 Velocity distribution in the radial section of window #4, at Q_n and $\phi^* = 0$: a. u_r ; b. u_θ ; c. u_z

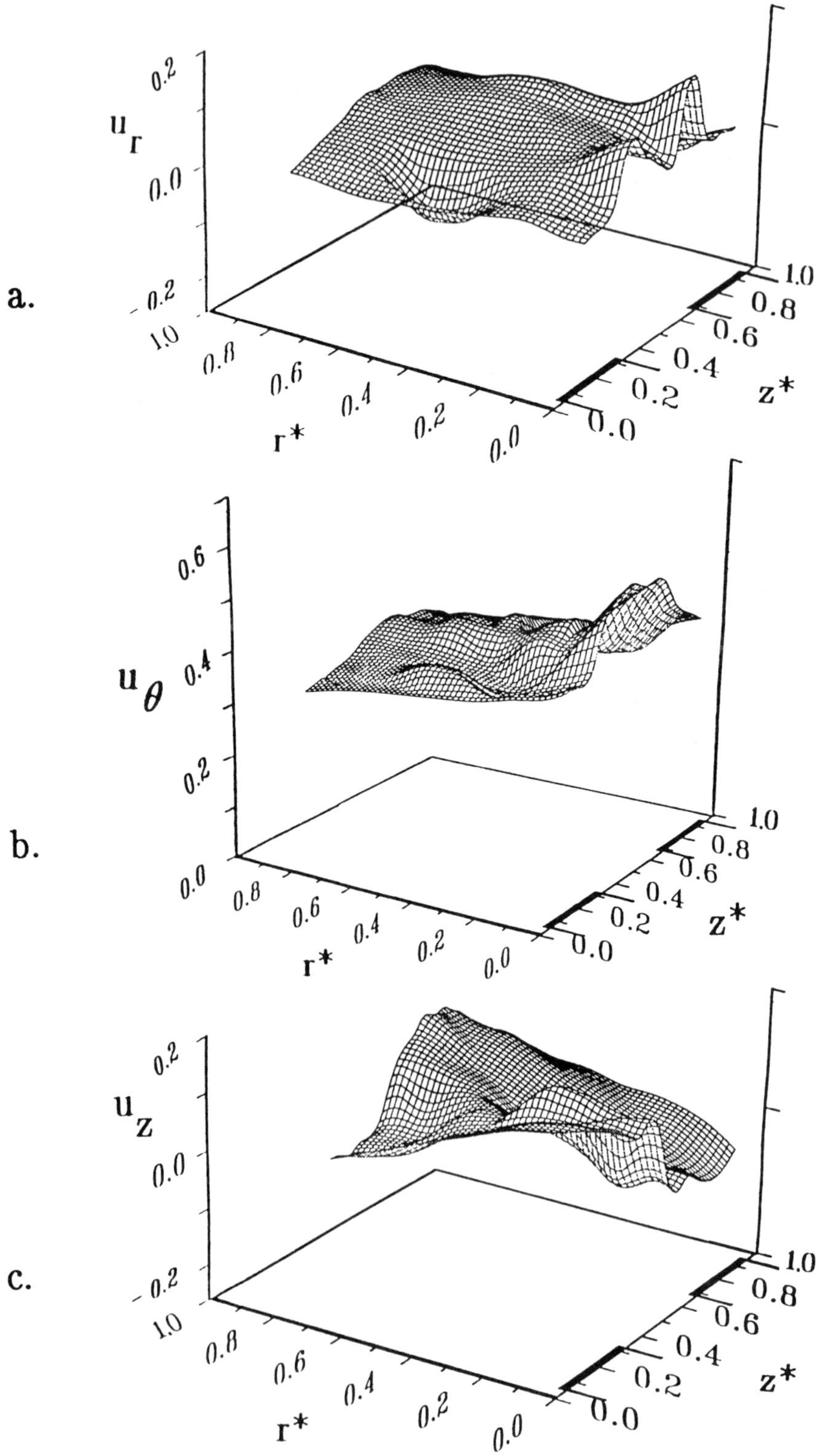

Fig. 13.10 Velocity distribution in the radial section of window #4, at $Q_n/2$ and $\phi^* = 0$: a. u_r ; b. u_θ ; c. u_z

a.

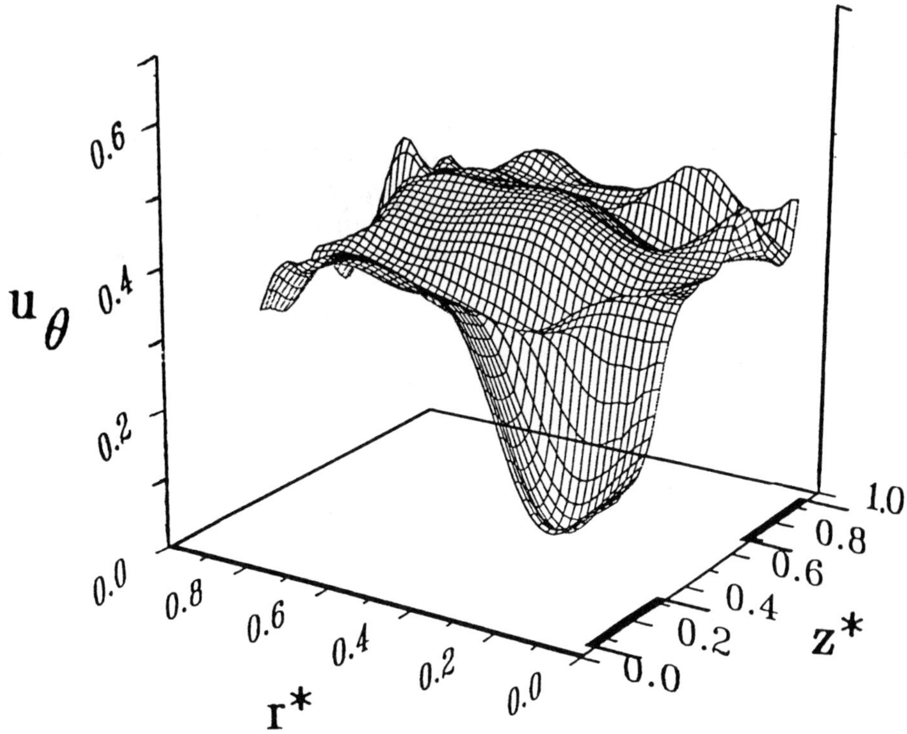

b.

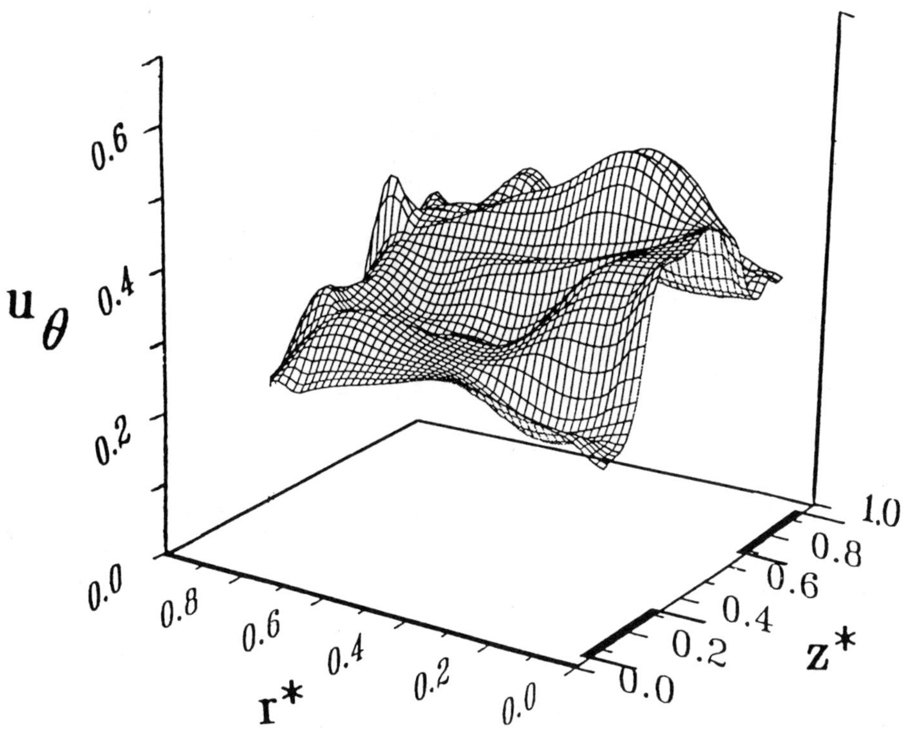

Fig. 13.11 Circumferential velocity distribution in the radial section of window #2, at $\phi^* = 0.75$: a. Q_n ; b. $Q_n/2$

a.

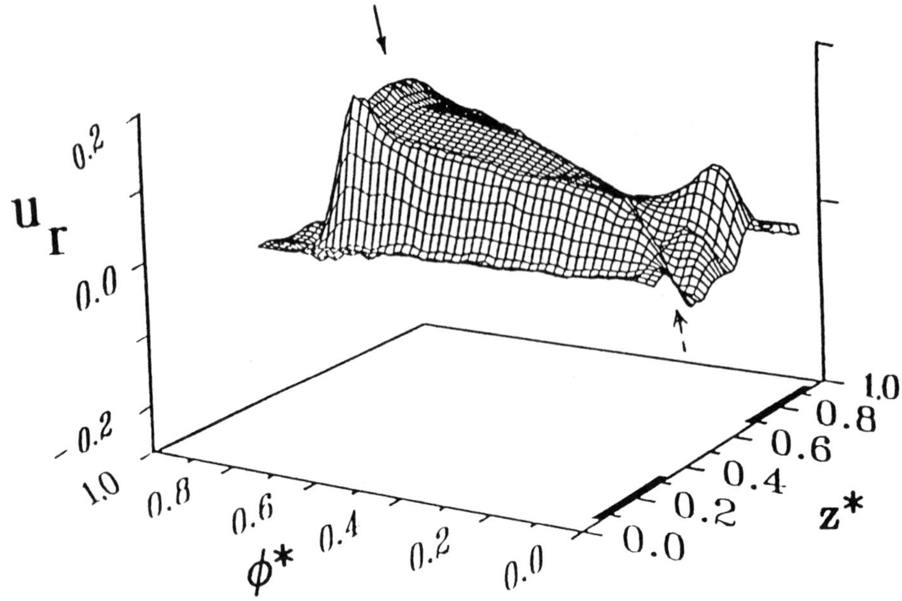

b.

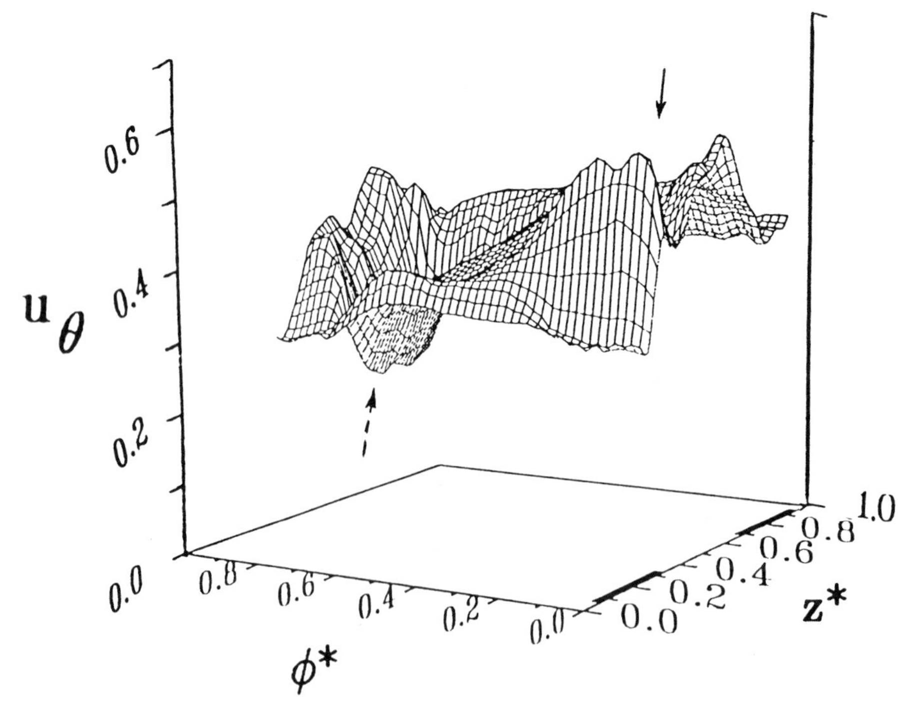

Fig. 13.12 Velocity distribution at the impeller outlet (line E) for the blade–to–blade interval ($\phi^* = 0$ to 1), at window #2 and Q_n

proximity of both shrouds the particles almost follow the mean water flow streamlines.

The radial and circumferential components of the normalized particle slip velocity, at windows #2 (sector S_4) and #5 (sector S_2), averaged over ϕ^*, are illustrated in Figure 13.14. The circumferential slip velocities averaged over $z^* = 0.3 - 0.7$ are negative and their average is about $< U^*_{sL,\theta} >_b = -1$, while the average radial slip is positive about $< U^*_{sL,r} >_b = +3$. The particle slip is due to the centrifugal and Coriolis forces, as well as the secondary currents and rapid change of the velocity direction at the impeller outlet. By varying concentration by an order of 2, the effect of solids concentration on the measured slip velocity played a minor role because the suspension is very dilute ($\alpha_s < 0.2\%$).

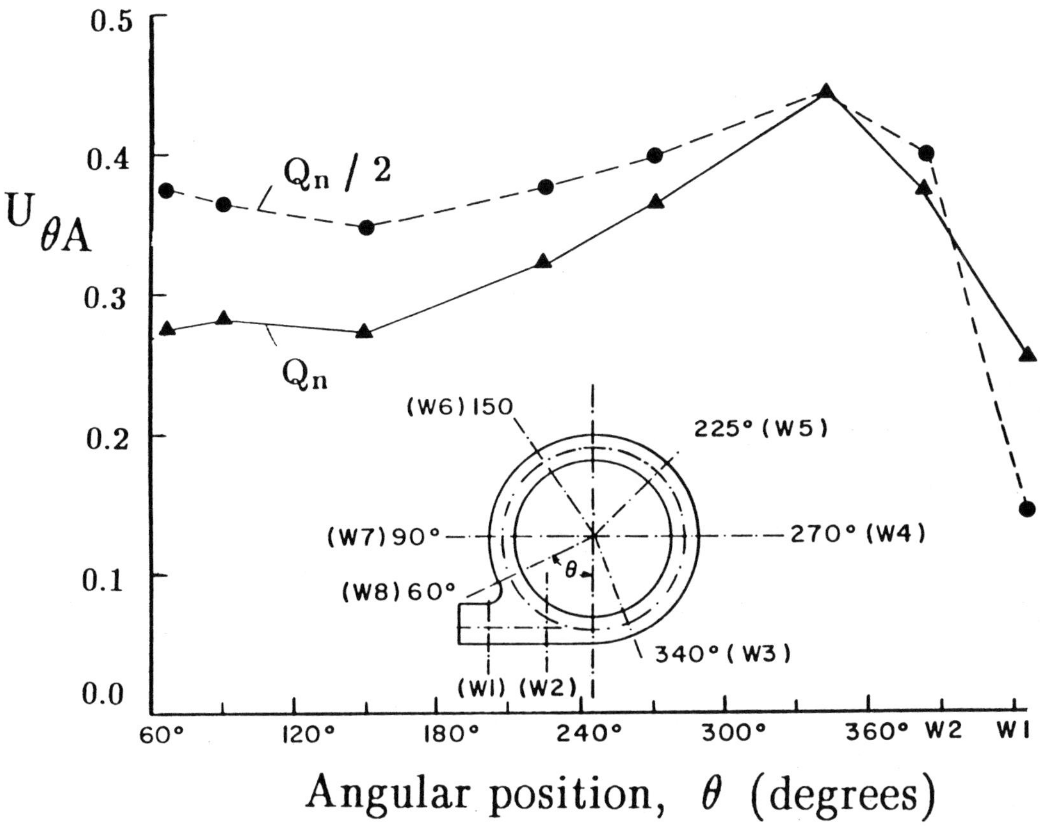

Fig. 13.13 Circumferential velocity averaged over the casing radial cross–section and a full impeller rotation

The variation of the particle slip velocities as a function of ϕ^* at three corresponding points in different casing cross–sections (S_1, window #8; S_2, window #5; and S_4, window #2), is illustrated in Fig. 13.15. The points have the same position ($r^* = 0.02$, $z^* = 0.67$) in the respective cross–sections. Overall, the slip is larger in the wake interval ($\phi^* = 0$ to 0.5), and decreases before the arrival of the following blade ($\phi^* = 0.5$ to 0.82).

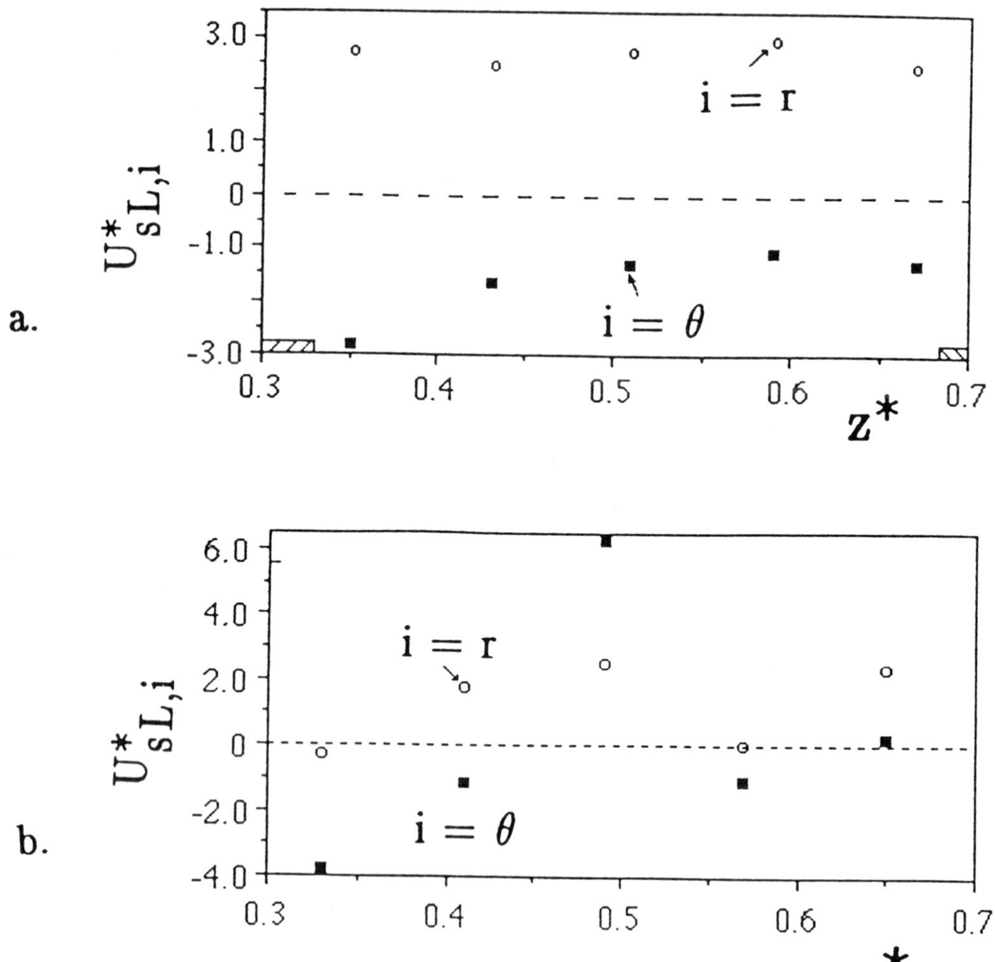

Fig. 13.14 Normalized slip velocity:
a. At window #2; b. At window #5

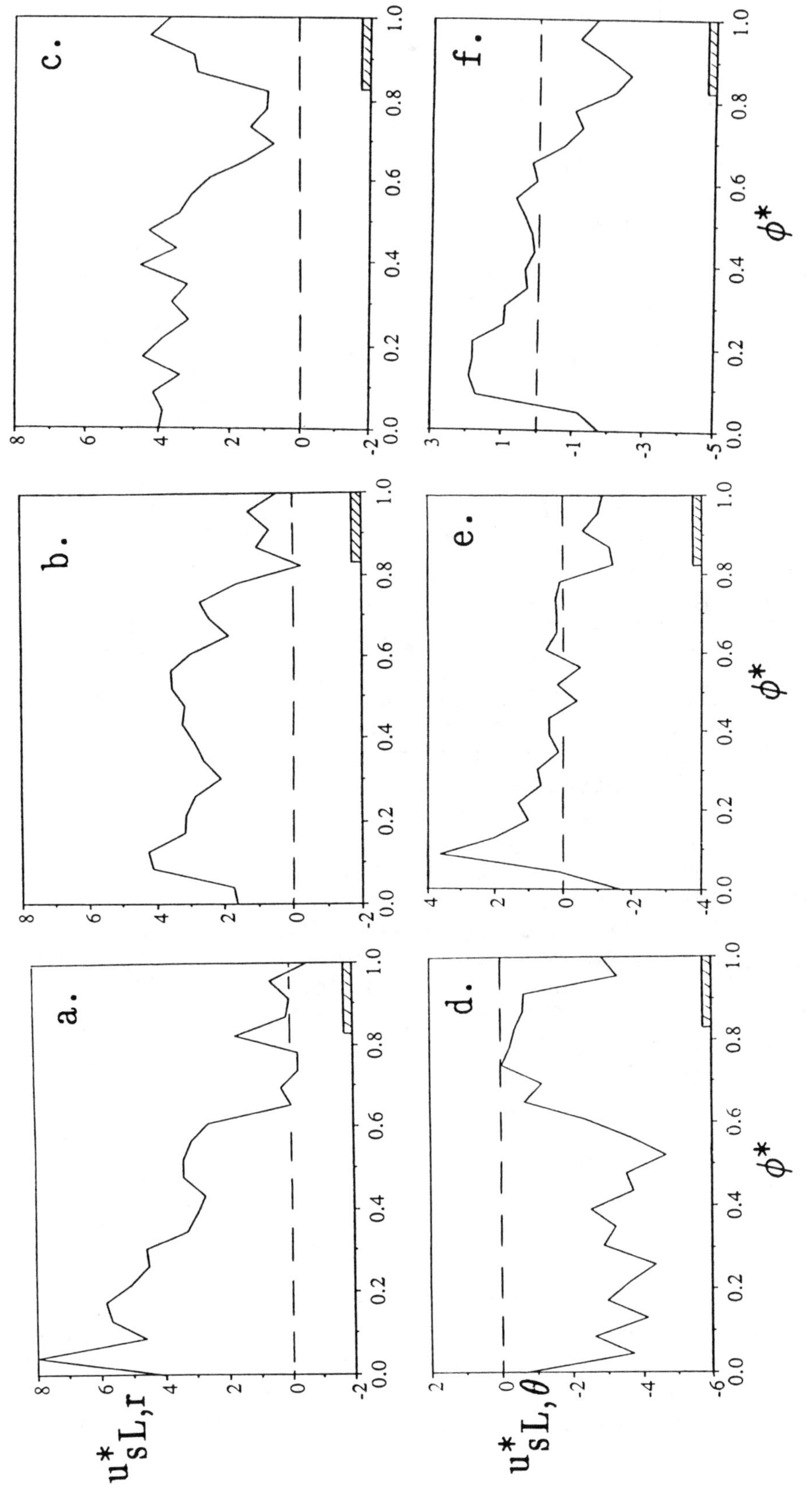

Fig. 13.15 Effect ϕ^* on the slip velocity at $z^* = 0.67$ on line E, for Q_n :
a – c. Radial slip at windows #8, #5 and #2; d – e. Circumferential slip at windows #8, #5 and #2

13.4 FLOW STRUCTURE IN A RADIAL SECTION

The flow within each radial section in casing is affected by the interaction with the impeller flow and the centrifugal effect. A double–helicoidal pattern develops (sketch in Figures 13.7), similar to curve duct flow. Evidence of this pattern is provided by the velocity distribution on line E shown in Figures 13.6 and 13.8, and by the velocity distribution in the whole cross–section shown in Figures 13.9 and 13.10. The recirculation through the inlet clearance, Q' in Figure 13.7, plays an important role in the flow balance and momentum transfer. It generates a specific asymmetry of the flow pattern in cross–section, in addition to the effect of the outflow from the impeller. The helicoidal current formed in casing, close to the front shroud, covers a smaller area than that close to the back shroud. The measured flowrate Q' around the impeller is about $0.15\ Q_n$ at the nominal flowrate (Q_n), and more than $0.2\ Q_n$ at half of the nominal flowrate ($Q_n/2$). The circumferential velocity is larger near the inlet clearance at $z^* > 0.7$ as compared to the back shroud (see Figure 13.6).

The significance of the fluid slip velocity between the impeller flow and casing flow is suggested by the circumferential velocity distribution at window #4 at Q_n and $Q_n/2$, $\phi^* = 0$, in Figure 13.9b and 13.10b. The velocity profiles $u_\theta(r^*,z^*)$ are relatively uniform and only slightly affected by the blade position if $r^* > 0.2$. The velocity distribution in impeller "adjusts" rapidly within the interval $r^* = 0 - 0.2$ to the velocity distribution in casing. This rapid change of the velocity at the impeller–casing flow interface justifies the term of "slip velocity" between the flow in impeller and that in casing. The measured fluid slip velocity is about 25% of the section average.

The periodical component u_p of the velocity, as well as the turbulence fluctuation u' and vorticity, have large fluctuations near the impeller ($r^* \approx 0$), and their fluctuations are dampened rapidly with r^*. Figure 13.16 shows a typical situation, where the periodical velocity fluctuation at window #4 and Q_n decreases significantly in the radial direction. The radial and circumferential velocity fluctuations measured at the impeller outlet, $r^* = 0.02$, are significantly dampened as they propagate to $r^* = 0.49$. The fluctuations are about 10 times smaller near the casing wall, at $r^* = 0.92$, as compared to $r^* = 0.02$. Similar results, but with relatively lower dampening, were obtained at $Q_n/2$.

The impeller angular position, ϕ^*, has a significant effect on the mean velocity $u_{\theta A}$ (Figure 13.17), vorticity $\overline{\omega}_{\theta A}$ (Figure 13.18), angular rate of deformation $\zeta_{\theta A}$ and velocity variance in a radial cross–section (after Roco et al., 1991). The notation has the following meaning. Each cross–section, of area A, was covered by a uniform mesh 48 X 48. The velocity was defined at the nodes of this mesh by interpolating the measured values. Then, the parameters of interest at

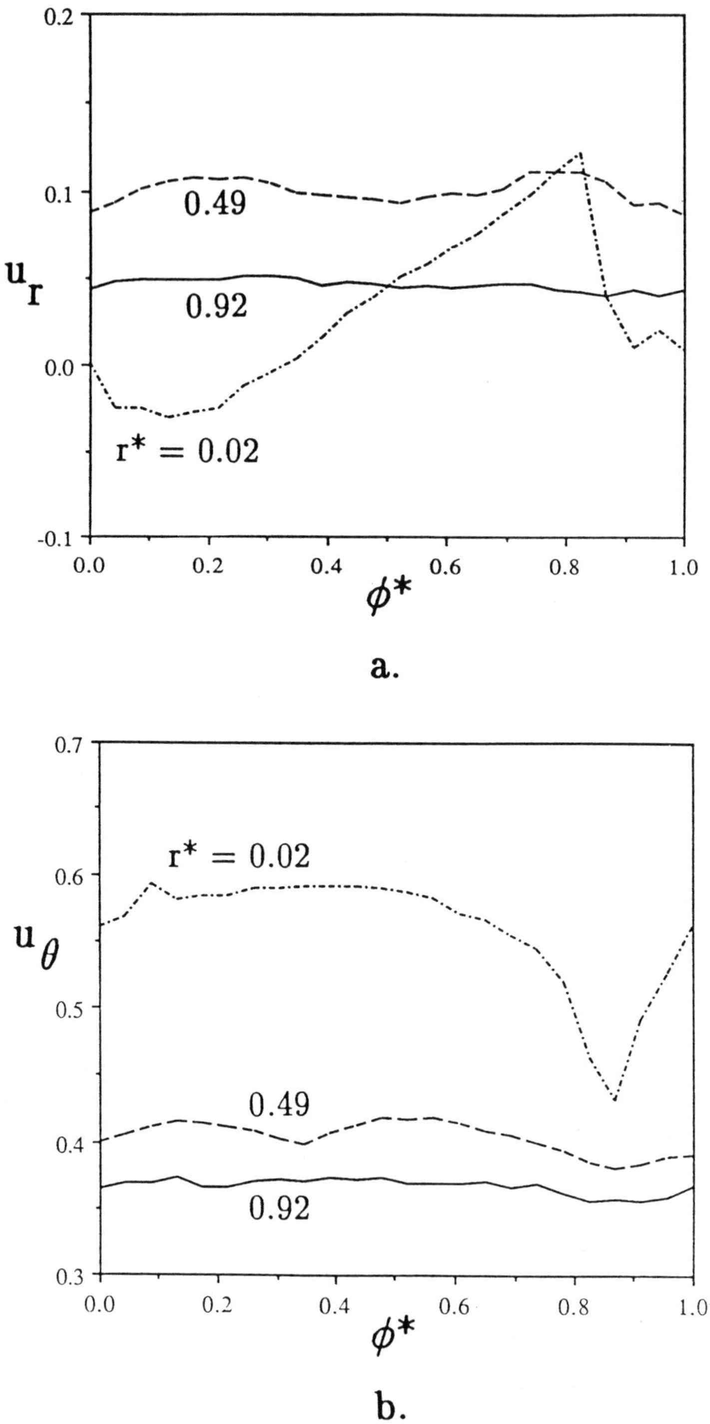

Fig. 13.16 Effect of ϕ^* on the fluid velocity at three points on a radial line ($z^* = 0.49$) at window #4, for Q_n : a. u_r ; b. u_θ

each node were calculated using a finite difference scheme. For a two–dimensional domain, $\overline{\omega}_\theta$ and ζ_θ have the well–known expressions:

$$\overline{\omega}_\theta = U_{2i} \left(\partial u_r / \partial z - \partial u_z / \partial r\right) / 2 \tag{13.5}$$

$$\zeta_\theta = U_{2i} \left(\partial u_r / \partial z + \partial u_z / \partial r\right) / 2 \tag{13.6}$$

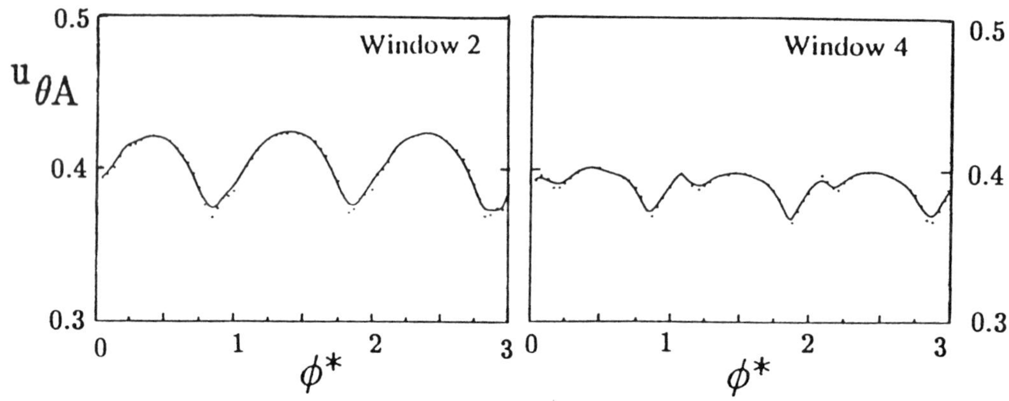

Fig. 13.17 Effect of the impeller angle on the average circumferential velocity along line E at the windows #2 and #4.

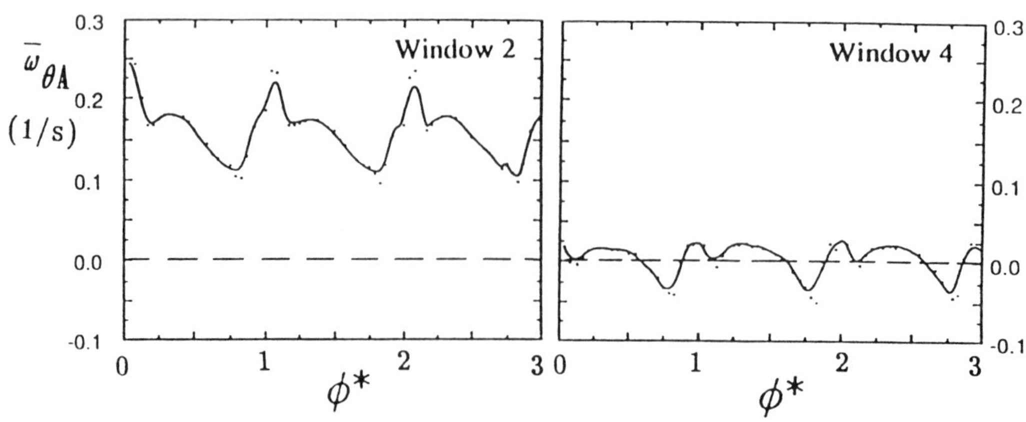

Fig. 13.18 Effect of the impeller angle on the average vorticity $\overline{\omega}_{\theta A}$ at windows #2 and #4

The averages were weighted by area:

$$u_{\theta A} = \frac{1}{A}\int_A u_\theta \, dA \ , \ \overline{\omega}_{\theta A} = \frac{1}{A}\int_A \overline{\omega}_\theta \, dA \qquad (13.7)$$

For a given impeller angle, the variance of circumferential velocity u_θ in a cross–section is:

$$\sigma_{u\theta} = U_{2i} \left[\frac{1}{n} \sum_{i=1}^{n} (u_\theta - u_{\theta A})^2\right]^{0.5} \qquad (13.8)$$

where n is the total number of nodes of the uniform mesh.

Figure 13.17 shows that the variation of $u_{\theta A}$ is about two times larger at window #2, in the proximity of the casing discharge nozzle, than at window #4. The mean vorticity is relatively large at window #2 around, around 0.2 1/s (Figure 13.18), which may suggest the beginning of a flow pattern dominated by a vortex as result of the interaction with the flow in the discharge pipe. In contrast, the averaged vorticity at window #4 for a full impeller revolution is almost zero. The angular rate of deformation averaged over the cross–section is about $\zeta_\theta = 0.13$ 1/s at window #2 as compared to $-$ 0.03 1/s at window #4. Also, the ratio of velocity variance over cross–section to mean velocity is larger at window #2, about $\sigma_{u\theta}/u_{\theta A} = 0.2$, as compared to window #4 where it is about 0.17. This should correspond to a larger rate of energy dissipation at window #2.

The blade displacement, and the wakes formed behind the blades, increases the mixing in the radial cross–sections. This mixing is beneficial in a slurry pump with large particles in suspension. The mixing effect of the wakes is stronger when blade boundary layer separation occurs.

The particularities observed in the proximity of the discharge nozzle at window #2 (sector S_4) suggests a flow interaction with the discharge pipe flow. Larger mean vorticity (Figure 13.18) could be related to the transition of the double helicoidal flow at S_3 to a flow dominated by a single helicoidal current in the discharge pipe. Larger fluctuations of the mean velocity (Figure 13.17) could be related to the fluid boundary, without a containing wall, towards the discharge pipe. Reverse flows have been measured at window #1 in the discharge nozzle, as well as for shorter intervals at window #2 towards the impeller.

At windows #3 to #8, the cross–section mean velocities $U_{\theta A}$ are up to 1.5 m/s larger at $Q_n/2$ than at Q_n (Figure 13.13). This is significant for centrifugal pumps operating at off–design flowrates, because energy losses and wear rates are functions of velocities. A common approach in design has been to oversize the pump and run it at lower flowrate (back on the pump characteristic curve) in order to maintain "lower" velocities and therefore improve wear life. The measurements given in Figure 13.13 contradict this misconception for slurry pumps.

13.5 BLADE–TO–BLADE FLOW STRUCTURE

The periodical velocity components defined in equation (13.2) determine a specific macrostructure, which propagates through a cascade process to the turbulence microstructure. The velocity distributions given in Figures 13.9 – 13.11 are only for a blade position, $\phi^* = 0$ and 0.75, respectively. These distributions change with ϕ^*. The blade–to–blade flow structure, was previously exemplified in Figure 13.12 for water velocity and in Figure 13.15 for slip velocity.

The effect of ϕ^* on the velocity distributions at the impeller–casing flow interface, averaged over the blade width and around the entire casing circumference (windows #2 to #8) is presented in Figure 13.19. The velocity distribution exemplified for window #4 in Figure 13.10 is similar to the water velocity distributions averaged over all windows (Figure 13.19 a and c). The fact that the fluid velocity periodicity is very similar in all sectors, from S_1 to S_4, shows that the impeller outflow dominates the velocity distribution at the impeller–casing flow interface. The effect of casing flow is secondary. The solids velocity profiles as a function of ϕ^* are less uniform at different windows (compare Figures 13.15 with 13.19 b and d) because of different upstream conditions and relatively large particle Stokes number.

A blade–to–blade interval may be separated into three parts, corresponding to :

– the wake ($\phi^* = 0$ to 0.5). In this interval, named I_1, the liquid radial velocity tends to increase continuously, the liquid circumferential velocity to decrease continuously, and the radial and circumferential particle slip to decrease (see Figure 13.19).

– the interval before the arrival of the next blade ($\phi^* = 0.5$ to 0.82). In this "blade–arrival" interval (I_2), all four velocity components discussed above reach either their absolute maximum or minimum. The smallest absolute values are reached for the water circumferential velocity and particle slip radial velocity. This may be caused by the boundary layer development on the pressure side of the blade.

– the "blade–displacement" interval (I_3; $\phi^* = 0.82$ to 1). In this interval the fluid radial velocity is almost zero on line E, while the other velocity components have transitory variations between the end of the interval I_2 and beginning of the interval I_1.

The influence of the flowrate is shown in Figure 13.19a. All velocity components decrease in average by decreasing the flowrate from Q_n to $Q_n/2$, with the exception of the water circumferential velocity. All velocity variations are similar at the two flowrates with the exception of the circumferential component of the particle slip velocity (Figure 13.19d).

The spatial correlations of the fluctuations, which are particularly large in a pump, are neglected in the currently applied time averaged methods. At the same time, direct numerical

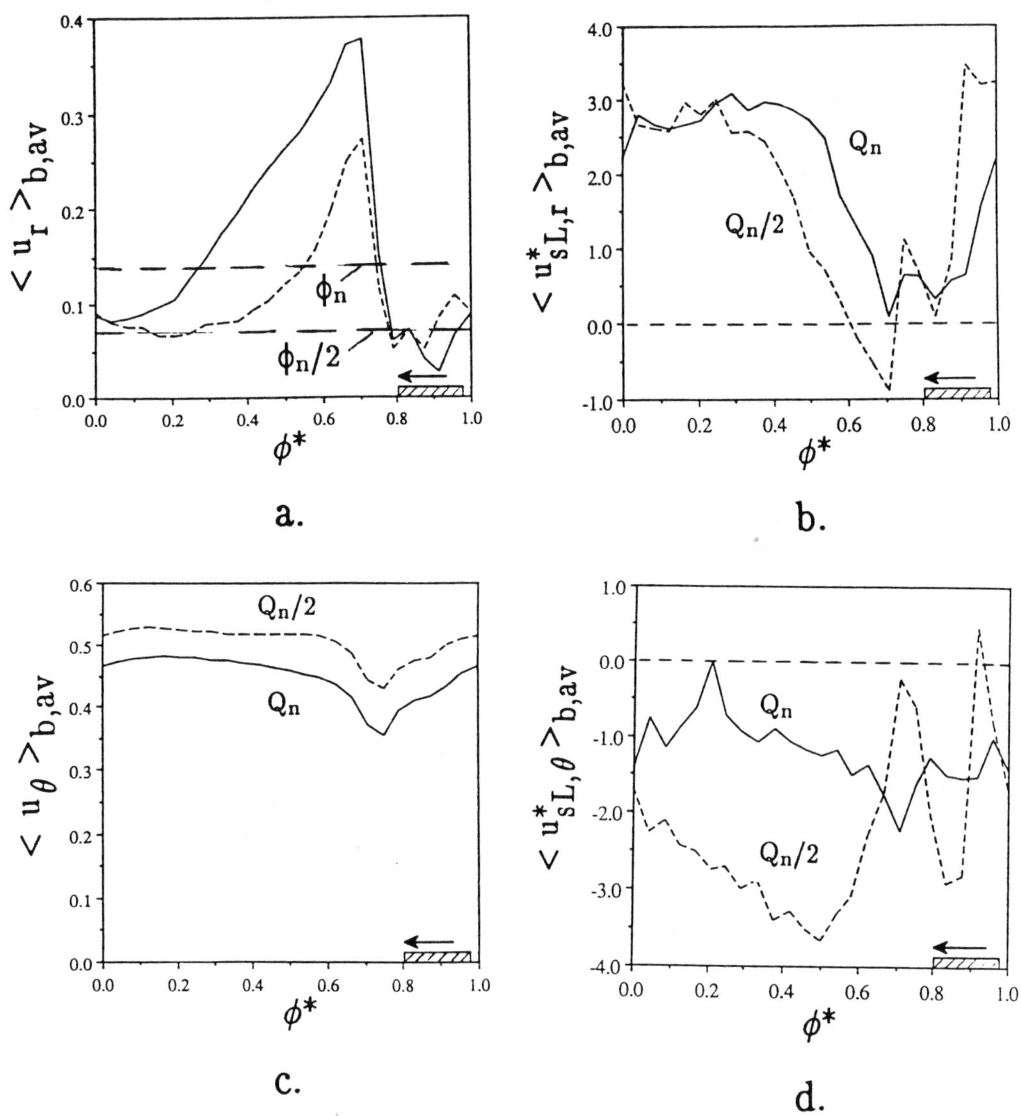

Fig. 13.19 Water and slip velocities averaged over B_{2i} and all windows around the impeller:
a. $< u_r >_{b,av}$, b. $< u^*_{sL,r} >$, c. $< u_\theta >_{b,av}$, d. $< u^*_{sL,\theta} >$

simulations are not feasible in such complex geometries. A probabilistic numerical approach, which can be used in conjunction with any other deterministic numerical method, and includes this effect was proposed by Zhang and Roco (1991).

A global indication of the effect of ϕ^* on the fluctuating velocity component can be obtained by studying the kinetic energy. Since the periodic fluctuations u_p are independent of random fluctuations u', the kinetic energy per unit mass of the total fluctuation may be separated into two parts

$$k = k_p + k_t = U_{2i}^2 \left(u_{std,p}^2 + u_{std,t}^2\right) / 2 \tag{13.9}$$

where $u_{std,p}$ denotes the standard deviation of the periodic fluctuation when the impeller angle changes, and $u_{std,t}$ is the standard deviation of the random fluctuation in time at a chosen impeller angle ϕ^*. The ratio between the turbulent kinetic energy and total kinetic energy, $k_t/(k_t + k_p)$, as a function of the impeller angle is illustrated in Figure 13.20 for averages over B_{2i} and around the impeller. The maximum and minimum envelopes for various positions z^* on line E are also shown. In the wake interval (I_1) that ratio increases up to 0.7 (in average). The difference between the envelopes also increases up to approximately 0.5. In the the blade–arrival interval (I_2) the ratio decreases to a minimum of 0.1. Overall, the periodic and random fluctuations have the same order of magnitude.

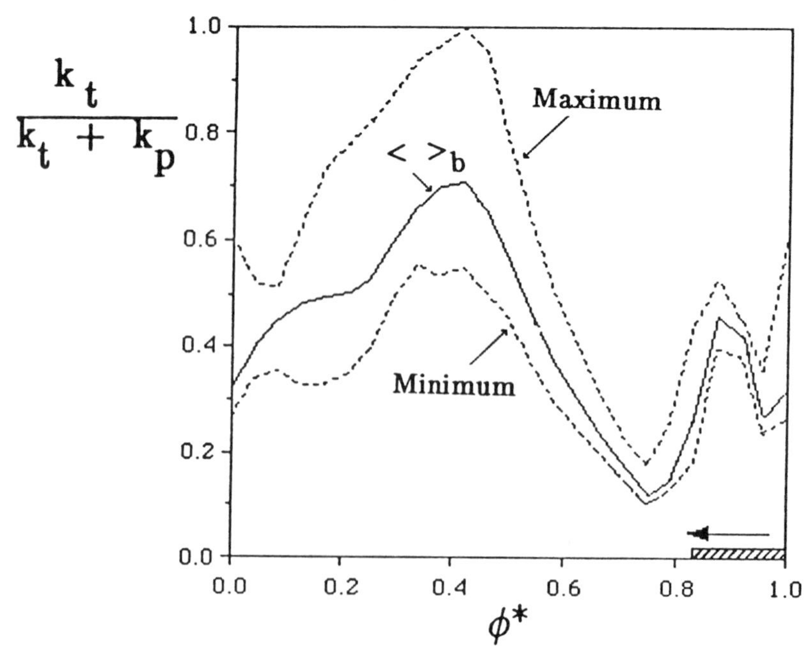

Fig. 13.20 The proportion of the turbulence kinetic energy (k_t) in the total fluctuation energy ($k_t + k_p$) as a function of ϕ^*

13.6 FLOW STRUCTURE AROUND THE IMPELLER

Now, we focus attention on the measurements on the entire cylindrical surface at the impeller–casing flow interface, as well as in the radial cross–sections around the casing.

A relatively large recirculation flowrate develops between the impeller shrouds and casing around the impeller, which can be approximately determined by subtracting the flowrate measured at window #1 from the flowrate measured at window #2 ($Q_{rc} = Q_2 - Q_1$, see Figure 13.1a). The normalized recirculation Q_{rc}/Q increases as the discharge flowrate Q decreases. The reentrained flowrate in casing from the discharge nozzle is about $Q_{rc} \approx 0.6\ Q_n$ (40 m³/h) at the nominal flowrate, and about $\approx 1.3\ Q_n$ (82 m³/h) at half of the nominal flowrate. This explains larger velocities $U_{\theta A}$ measured around the casing between $\theta = 60$ and 360 degrees at $Q_n/2$ as compared to Q_n (see Figure 13.13). By considering the inertial forces as predominant and using an inviscid flow model, the recirculation flowrate and mean velocities $U_{\theta A}$ were predicted within 7% of the experimental data at Q_n. A turbulence model had to be used for a more realistic velocity distribution in radial cross–sections (Roco and Dehaven, 1989).

Four sectors around the casing, S_1 to S_4 (section 13.3), were identified based on their flow characteristics. Particularities of those areas are illustrated in Figures 13.21 to 13.25. The water velocity and particle slip velocity distributions exhibit local maxima periodically, at certain ϕ^*, in the sectors S_1, S_3 and S_4. Illustrative circumferential velocity distributions at the impeller–casing flow interface, at Q_n and $Q_n/2$, for $\phi^* = 0.87$, are shown in Figure 13.21. Three pairs of peaks, equal to the number of blades, seem to dominate the velocity distribution. The first pair of peaks is at about $\theta^* = 0.1 - 0.2$ (sector S_1, windows #7 and #8), immediately after the change of direction at the tongue. This is the location where the outflow from impeller meets the recirculation stream coming from window #2. The peaks are pronounced in space, change periodically in time as a function of the blade displacement, and notably maintain their location. A second area with a pair of fluctuating peaks is at about $\theta^* = 0.7 - 0.8$ (sector S_3, windows #3 and #4), where the upstream casing flow arriving from the circumferential direction interacts with the outflow from impeller in a "bottle–neck" region. At Q_n, the peaks of the water velocity distribution are slightly larger and closer to each other than in the sector S_1. The circumferential velocity component reaches values over 50% higher than the average of the entire distribution. Between $\theta^* = 0.2$ and 0.7, the fluid is accelerated as a result of the continuous outflow from impeller. The measured average velocity in the casing cross–section, of constant area, increases continuously, from

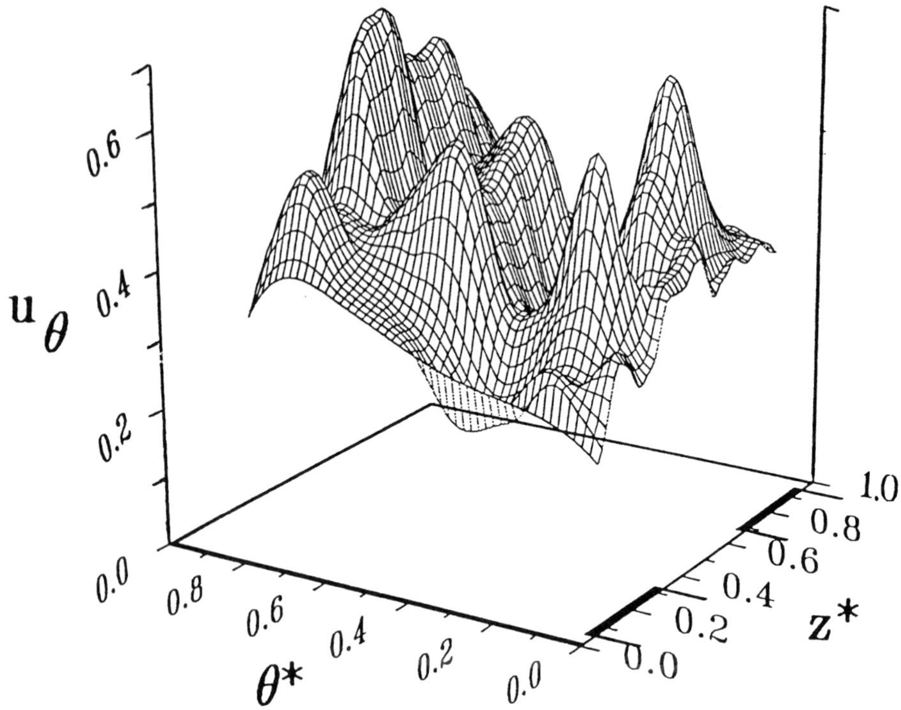

a.

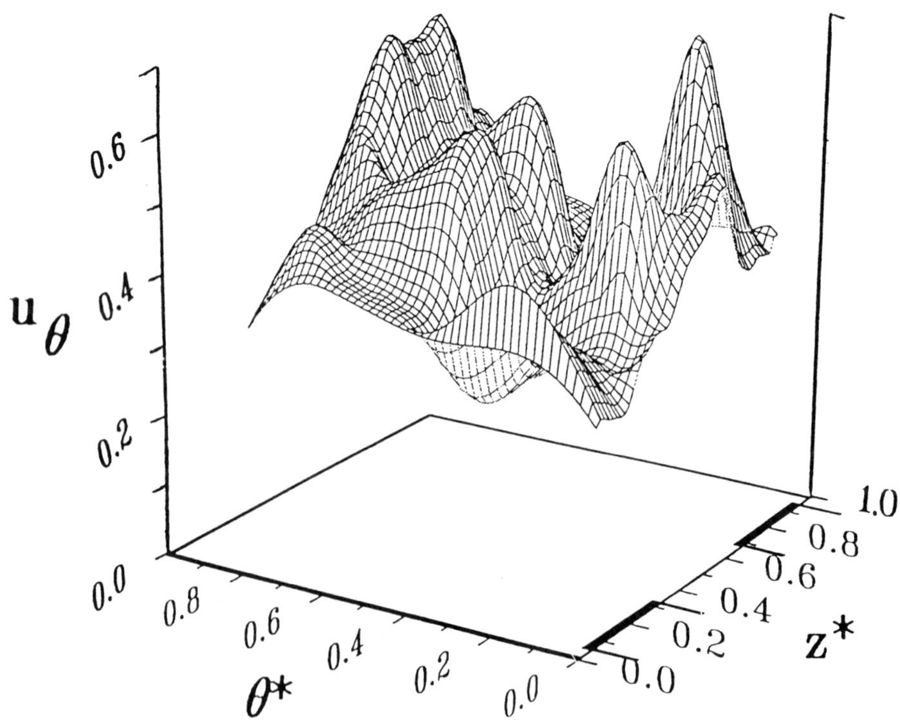

b.

Fig. 13.21 Measured circumferential velocity on line E around the impeller ($\theta^* = 0$ to 1), at $\phi^*_O = 0.87$: a. Q_n ; b. $Q_n/2$

window #8 to #3 (Figure 13.13). A less pronounced pair of peaks, which has a tendency to travel with the rotating impeller, was observed in sector S_2. An outstanding flow characteristic observed in our experiments is the presence of stationary waves with peaks in sectors S_1 and S_3, whose frequency but not location is related to the blade displacement. Their location and amplitude are related to the interaction between the impeller and casing flows. Also, the sectors S_1 and S_3 are the areas with the lowest and highest pressure around the casing, and a secondary flow through the lateral clearances between the impeller shrouds and casing develops.

The particularities of the water and particle slip velocity are well illustrated by the isovels in Figures 13.22 to 13.24. A set of isovels are shown in the plane (θ^*,z^*) in Figure 13.22, for a given position of the impeller ($\phi_0^* = 0$), at the nominal flowrate. The peaks of the circumferential and radial velocity components at $\theta^* \approx 0.2$ and ≈ 0.8 are correlated. At $\theta^* \approx 0.2$, the radial velocity reaches a minimum $u_r < -0.3$ at the same location where the circumferential component reaches its maximum $u_\theta > 0.5$. At $\theta^* \approx 0.8$, the local maximum of $u_r > 0.1$ corresponds to the absolute maximum of $u_\theta > 0.65$. At the same θ^*, the radial slip velocity has a pair of maxima of $u^*_{sL,r} > 8$, and the circumferential slip velocity also two maxima of $u^*_{sL,\theta} > 6$ at the same z^* locations. The circumferential slip is slightly asymmetrical with respect to $z^* = 0.5$.

The wakes formed behind the blades, as seen by an observer attached to the casing are illustrated in Figure 13.23 for window #5, $\theta^* = 0.5$. The wake pattern changes around the impeller. Generally, the wakes have a pair of vortices which originate from the blade edges, near their contact with the shrouds.

The water velocity triangle and corresponding particle slip vector are functions of the location in casing. The effect of ϕ^* on the water velocity and particle slip is illustrated in Figure 13.25, by representing the vector extremities at intervals of $\Delta\phi^* = 0.05$. All velocities are normalized by U_{2i} and averaged over the blade width ($<>_b$). The periodical velocity fluctuations from $\phi^* = 0$ to 0.25, 0.5, 0.75, and again 0, have a characteristic pattern. It is remarkable that that the radial velocity increases more than five times from $\phi^* = 0$ (when the trailing edge of the blade passes by window #5) to $\phi^* = 0.83$ (at the arrival of the next blade). Figure 13.25 c shows a maximum particle slip at about $\phi^* = 0.5$. While at $\phi^* = 0$ the particle slip velocity is quasi–circumferential, at $\phi^* = 0.5$ is quasi–radial.

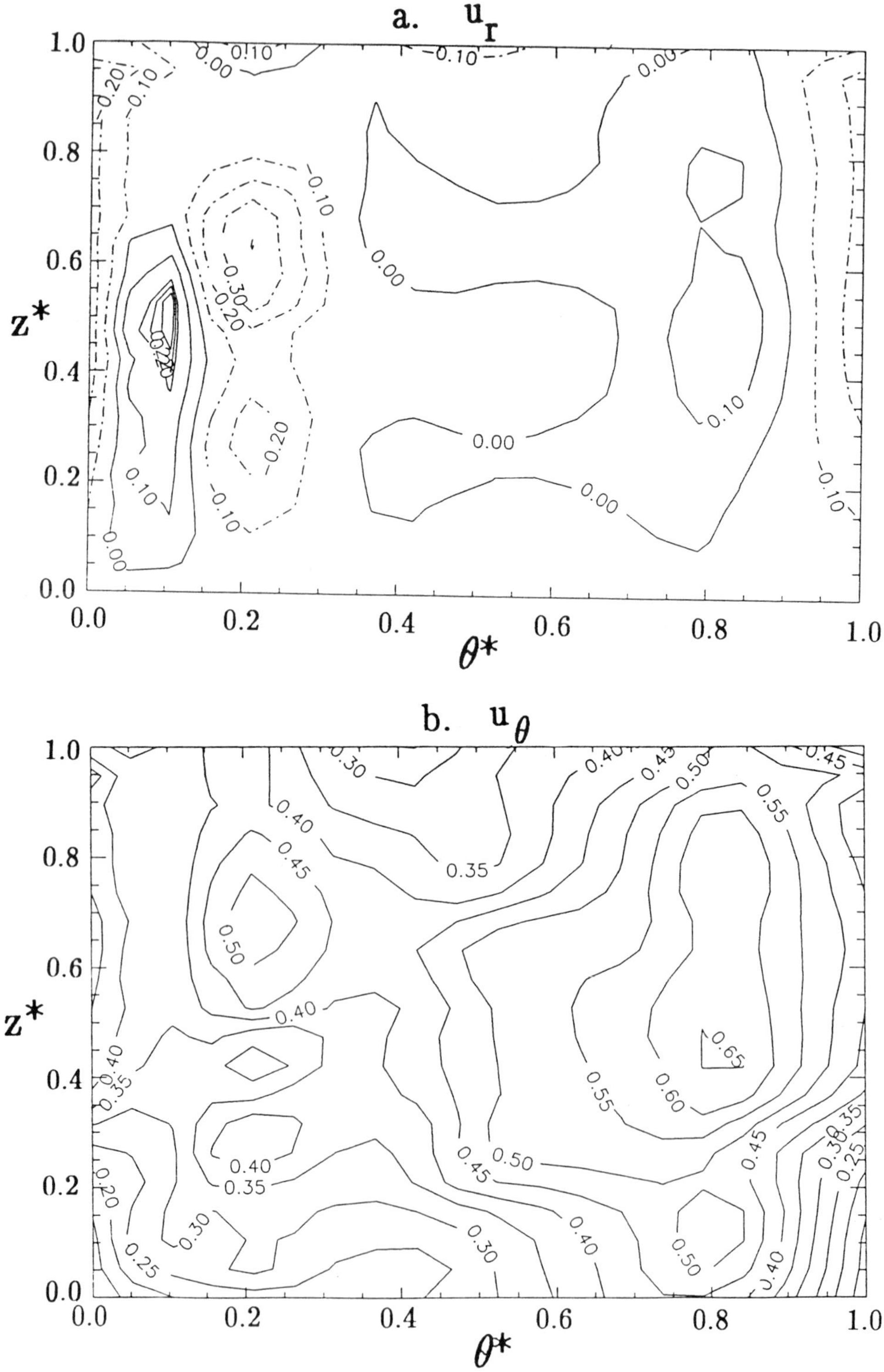

Fig. 13.22 Isovels on the cylindrical surface (θ^*,z^*) at $r^* = 0.02$, for $\phi_o^* = 0$ and Q_n : a. u_r ; b. u_θ ; c. $u^*_{sL,r}$; d. $u^*_{sL,\theta}$ → (see color Plate G) ↑

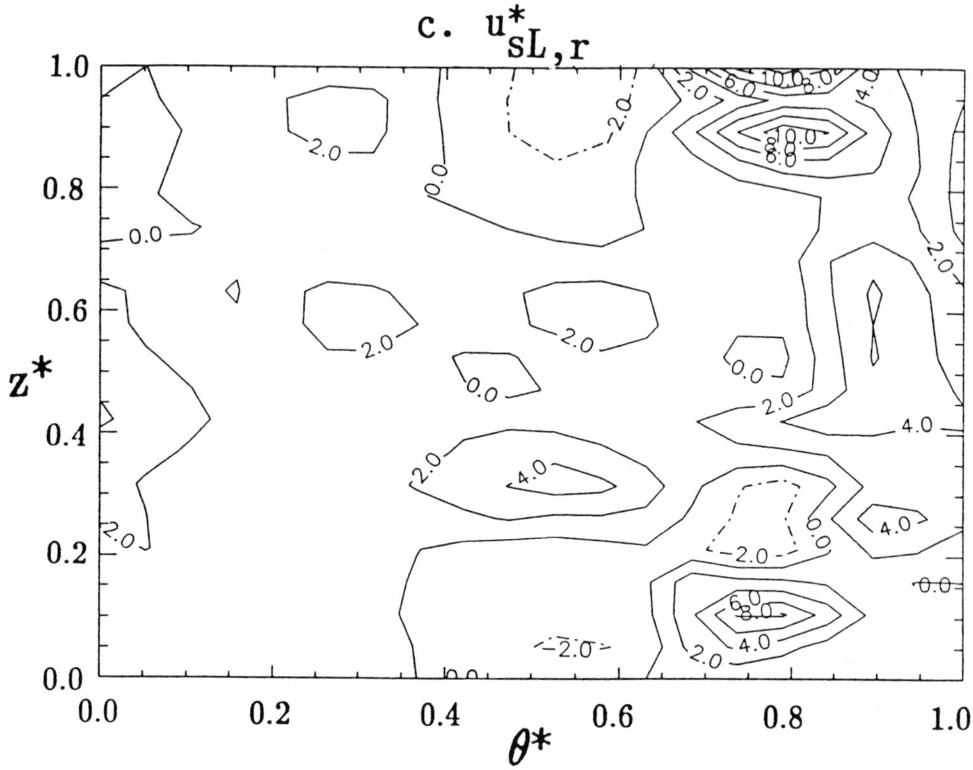
c. $u^*_{sL,r}$
z^*
θ^*
0.0
0.2
0.4
0.6
0.8
1.0

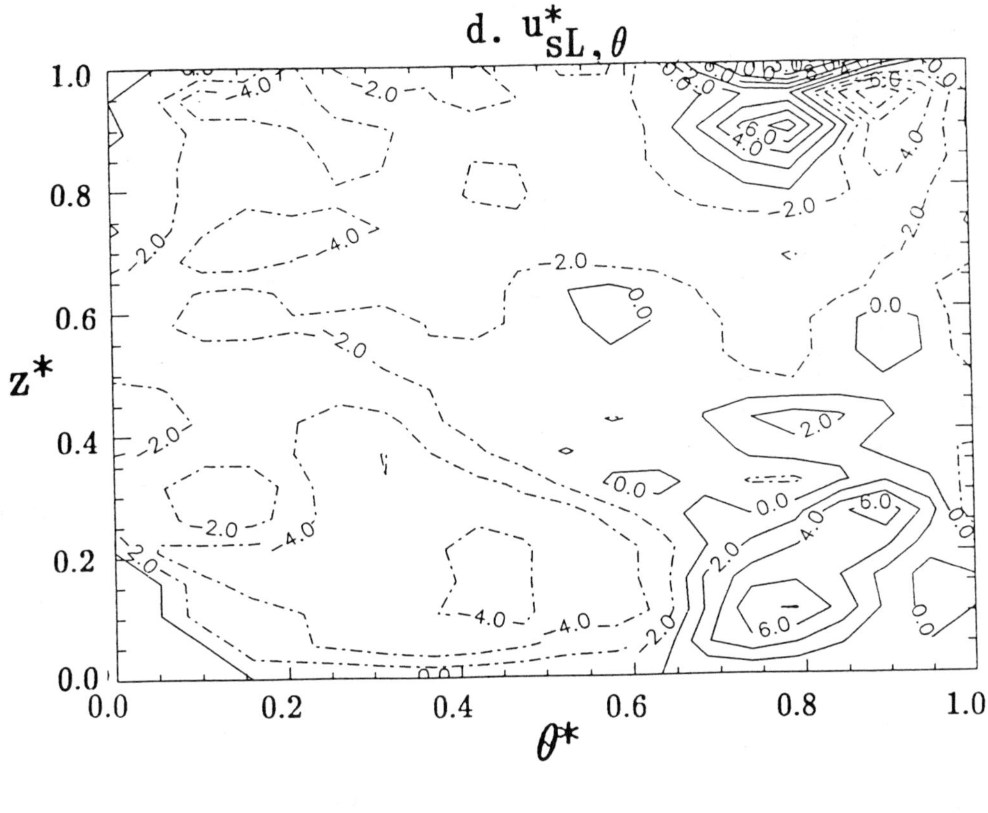
d. $u^*_{sL,\theta}$
z^*
θ^*
0.0
0.2
0.4
0.6
0.8
1.0

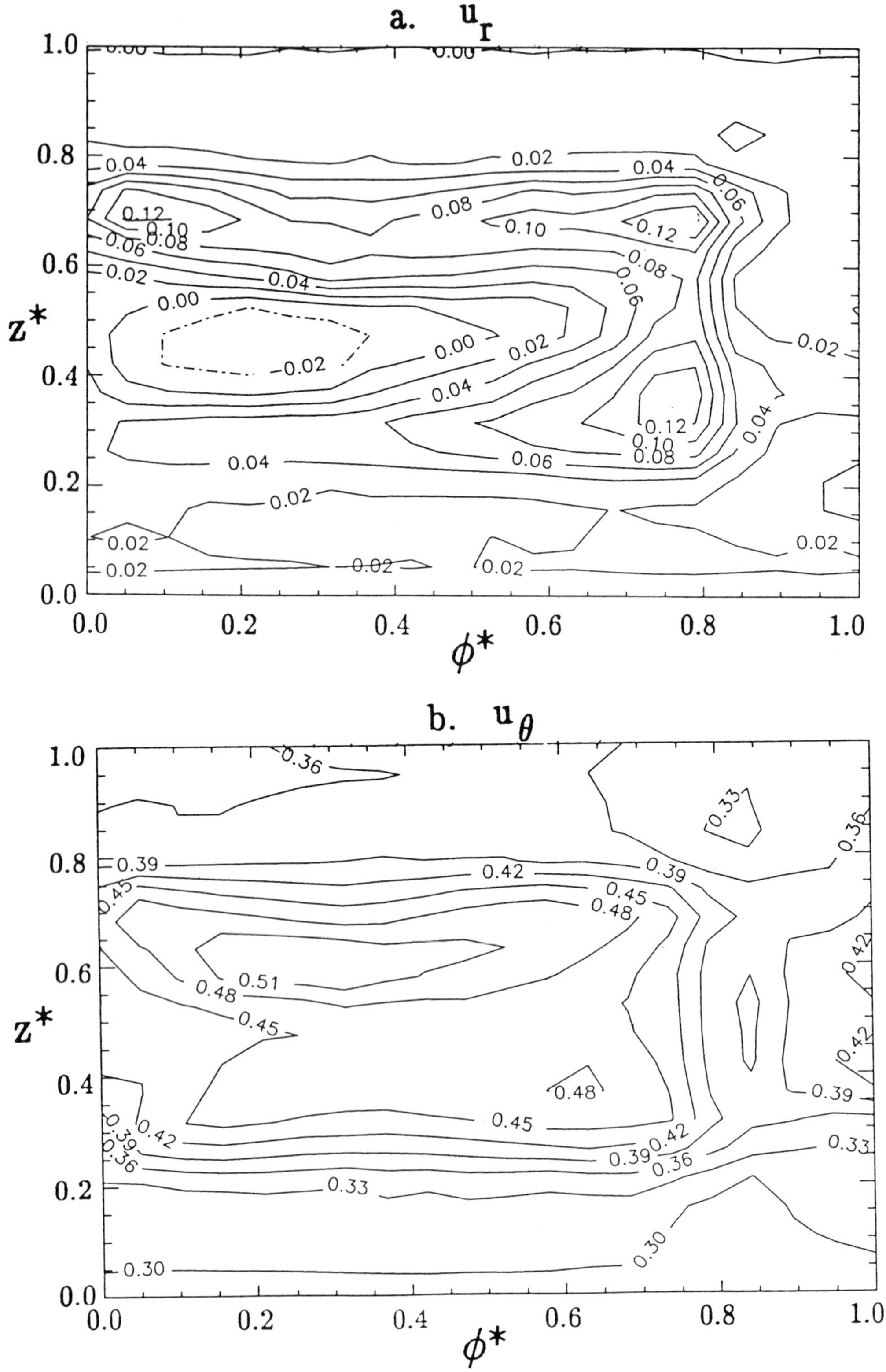

Fig. 13.23 Isovels in the plane (ϕ^*, z^*) at $r^* = 0.02$, for window #5 and Q_n : a. u_r ; b. u_θ ; c. $u^*_{sL,r}$; d. $u^*_{sL,\theta}$

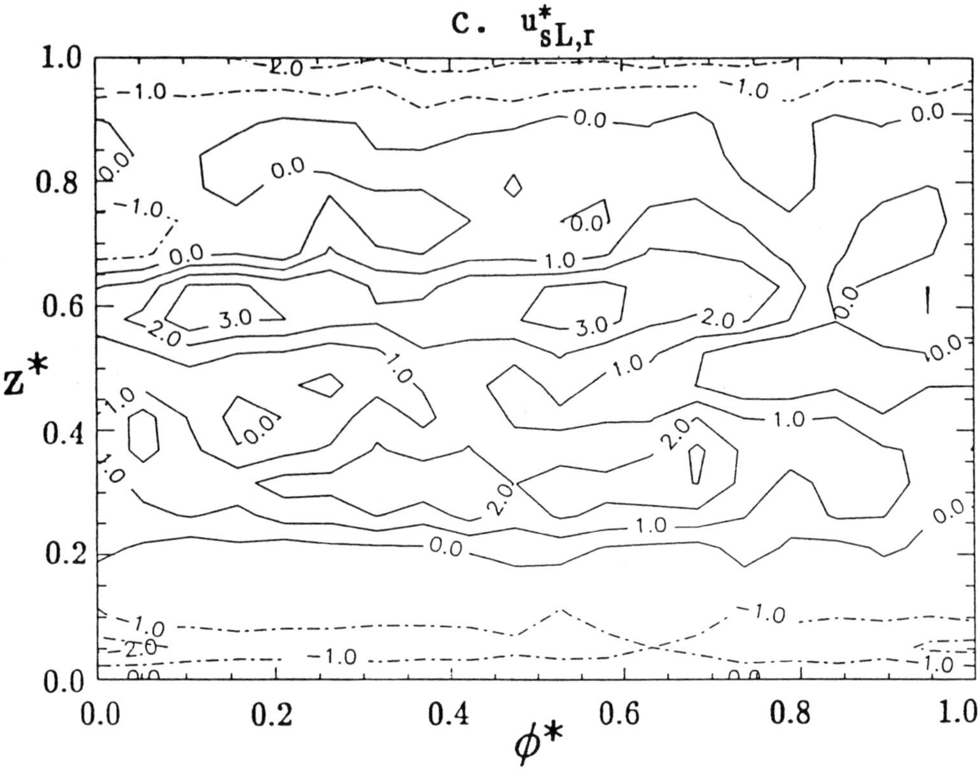
c. $u^*_{sL,r}$
z*
ϕ^*
0.0
0.2
0.4
0.6
0.8
1.0

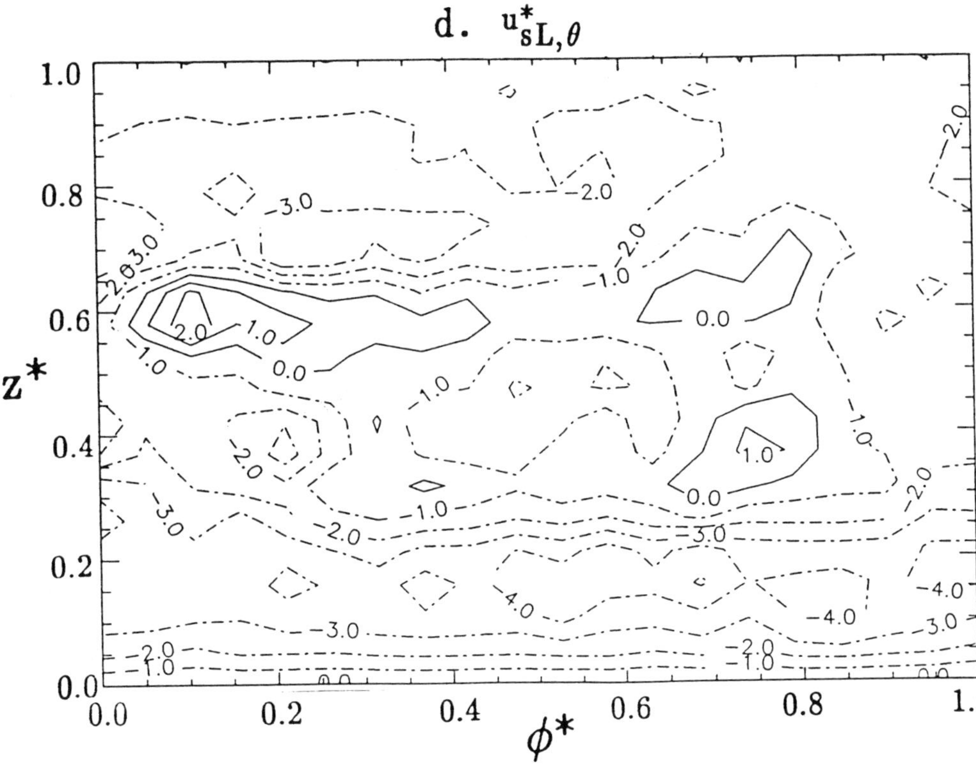
d. $u^*_{sL,\theta}$
z*
ϕ^*
0.0
0.2
0.4
0.6
0.8
1.0

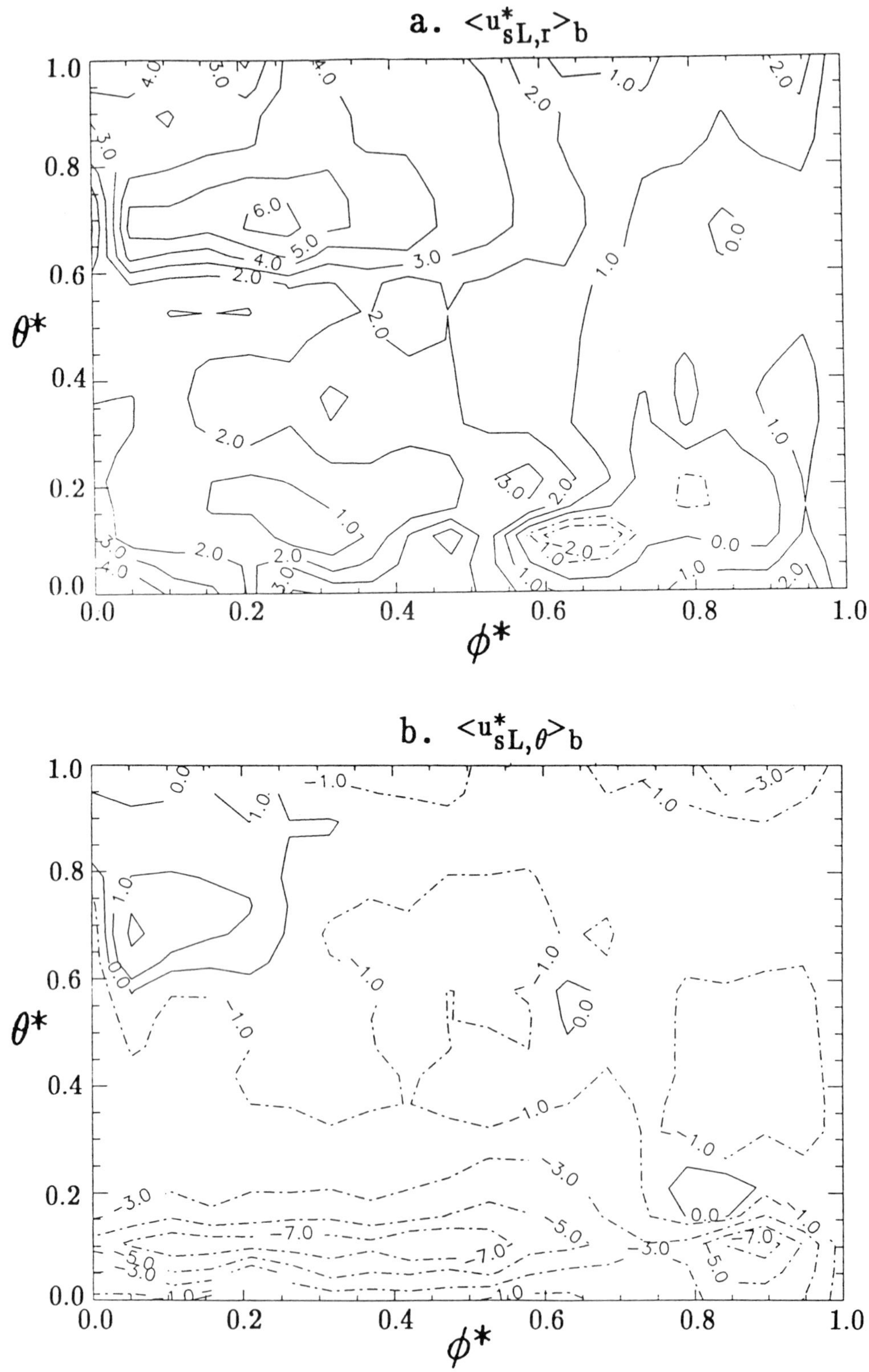

Fig. 13.24 Isovels in the plane (ϕ^*, θ^*) at $r^* = 0.02$, for Q_n, for: a. $< u^*_{sL,r} >_b$; b. $< u^*_{sL,\theta} >_b$

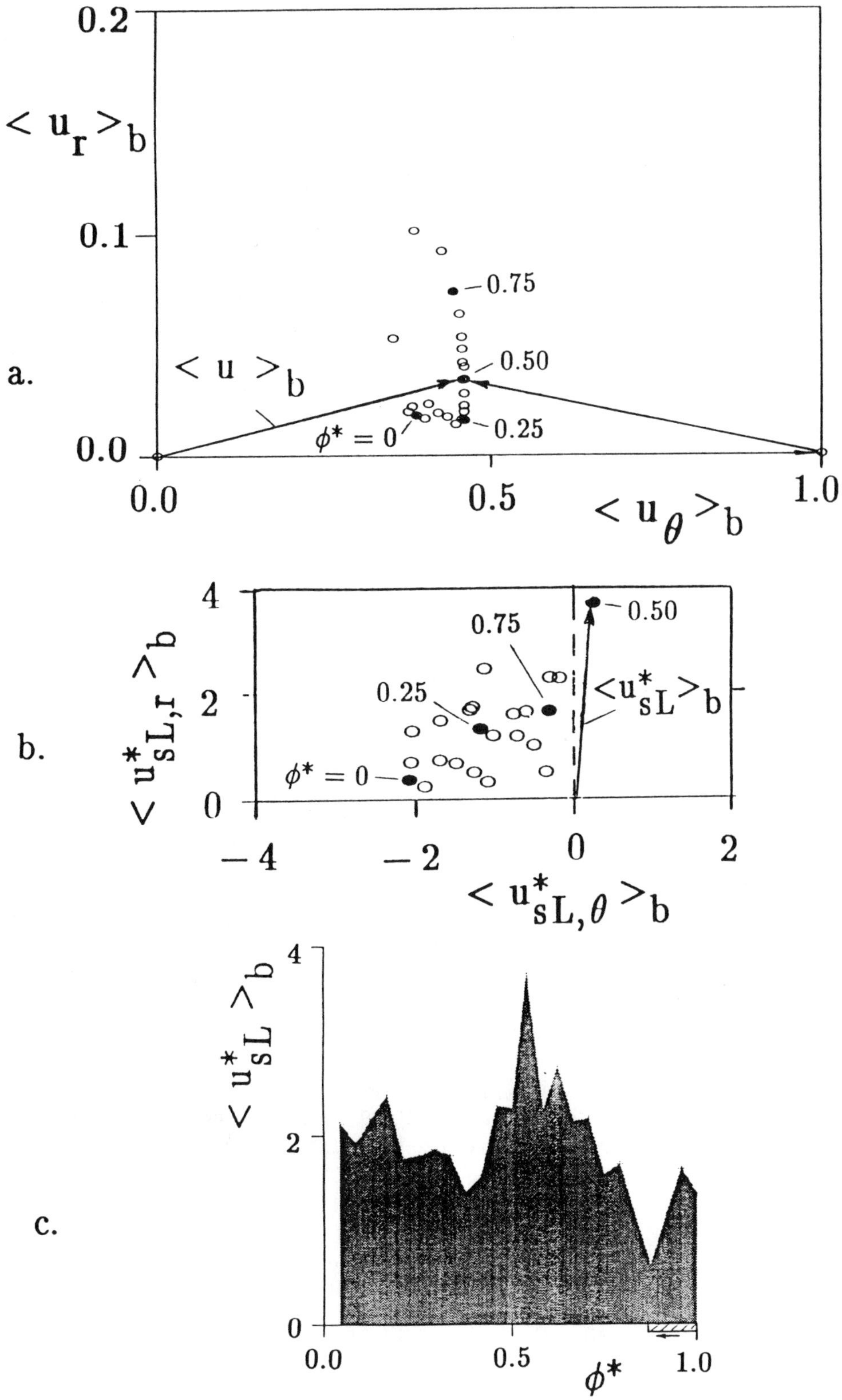

Fig. 13.25 Effect of ϕ^* on the (a) liquid velocity triangle, (b) slip velocity vector, and (c) its modulus, at window #5, for Q_n

13.7 EFFECT OF VELOCITY DISTRIBUTION ON PUMP PERFORMANCE (H, Q, η)

The two–phase flow velocity measurements presented above can be used to estimate the overall pump performance, namely the effect of solid phase on the head, flowrate and energy efficiency. The modification of the pump head and flowrate is evaluated from the velocity triangles at the impeller outlet, and the loss of efficiency from the measured slip velocity distribution at the same location.

We have found that the velocity averages over the blade width (B_{2i}) at the impeller outlet correlate well with the pump head obtained from the pressure measurements at the pump inlet and outlet. The velocity averages over the casing width (B_c) underestimate the pump head. Usually, the pump performance is given in average over a full impeller rotation, i.e averaged over ϕ^*. If the dimensionless pump flow rate and head coefficient averaged over ϕ^* are denoted, respectively, by $\phi = Q/(U_{2i}A_{2i})$ and $\Psi = gH/U_{2i}^2$, we have for dilute liquid–solid particle mixtures:

$$\phi = (1-\alpha_s) <U_r>_{b,av} + \alpha_s v_\infty^* <U_{sL,r}^*>_{b,av}, \quad \text{and}$$

$$\Psi = (1-\alpha_s)\frac{1}{S_m} <U_\theta>_{b,av} + \alpha_s \frac{S_s}{S_m} v_\infty^* <U_{sL,\theta}^*>_{b,av} \qquad (13.10)$$

where $<U_r>_{b,av}$ and $<U_\theta>_{b,av}$ are averages for the liquid velocity components in the presence of solids, and $<U_{sL,r}^*>_{b,av}$ and $<U_{sL,\theta}^*>_{b,av}$ are averages for the slip velocity components.

Figure 13.26 shows the averaged water velocity vectors at windows #2 to #8 for Q_n. By averaging the fluid and slip velocity vectors given in Figure 13.25 for window #5, one obtains a single pair of vectors corresponding to that window in Figure 13.26. At Q_n, the radial velocity at windows #4, #5 and #8 is about half of that at the remaining windows (Figure 13.26a). The circumferential velocity is larger in sector S_3 (windows #3 and #4), and smaller in sector S_1 (windows #7 and #8). At $Q_n/2$, the radial velocity is about zero at window #8, and maximum at windows #6 and #7. Again, the circumferential velocity is maximum at windows #3 and #4. The particle slip velocity vectors have negative circumferential components and positive radial components at Q_n (Figure 13.26b). At $Q_n/2$, both the norms and angular positions of the slip velocity vectors are more dispersed, with a maximum radial component at window #4, $<U_{sL,r}^*>_b > 7$. If at Q_n the maximum angle between the slip velocity vectors is 75 degrees (Figure 13.26 b), at $Q_n/2$ the corresponding angle is larger than 120 degrees. The effect of solids on the characteristic curve can be determined from the second term of ϕ and Ψ in equation (13.10).

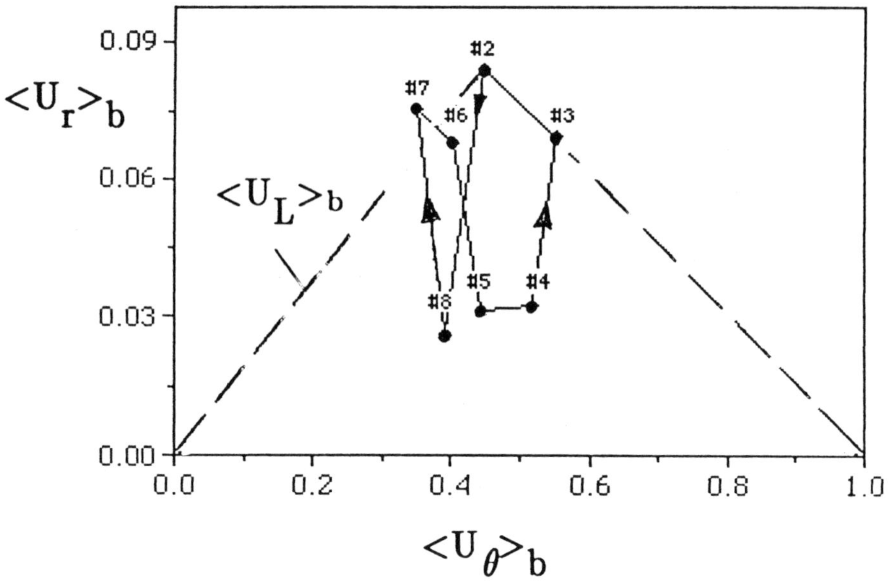

a.

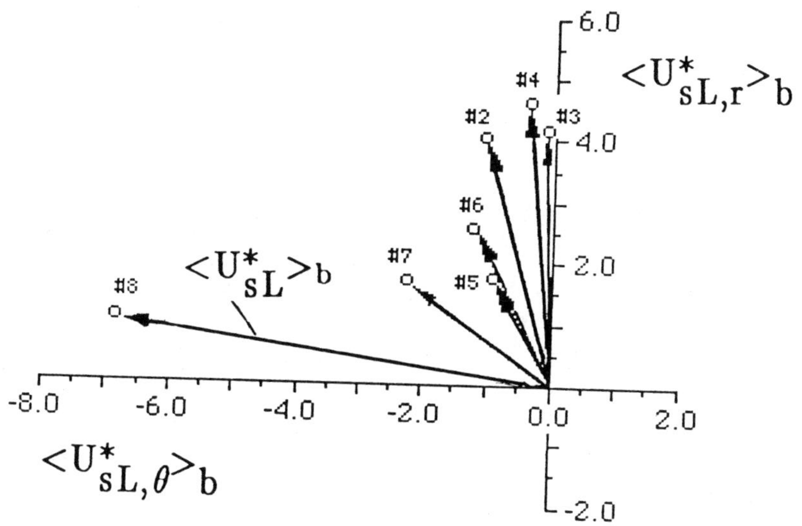

b.

Fig. 13.26 Variation of the the liquid velocity triangle and slip velocity vector, averaged over B_{2i}, at different windows (#2 to #8), at Q_n: a. $<U_L>_b$; b. $<U^*_{sL}>_b$

By averaging the data presented in Figure 13.26 for ϕ_n over the casing circumference, one obtains a point (marked by a triangle) on the pump characteristic curve. Similarly, it was obtained a point from the LDV velocity measurements at half the nominal flowrate ($\phi_n/2$). These points are compared to the data obtained from the pressure measurement at the pump discharge (solid line in Figure 13.27). By considering the velocity averaged over the impeller blade width, the agreement for Ψ is very good, with differences within 5%. The averages over the casing width predict lower head coefficients Ψ than those obtained from pressure measurements.

The pump head has a fluctuation because of the finite number of blades. The variation of the pump head with ϕ_0^* (0 to 1) is obtained by integrating the corresponding velocities over all windows at the impeller–casing flow interface. The variance of Ψ derived from the velocity distributions is about 10% at Q_n, and about 5% at $Q_n/2$.

Previous experimental results, including our work (Roco et al., 1986), were based on the pressure and flowrate measurements at the pump inlet and outlet. The present analysis shows the origin of the characteristic curve $\Psi(\phi)$, and how the internal flow structure dictate its shape.

The work performed by the solid–liquid drag forces determines a loss of pump energy efficiency, $\Delta\eta_{slip}$. This loss of efficiency caused by particle slip was evaluated assuming that the impeller–casing flow interface is a representative area in the pump. First, the energy dissipated per unit volume of fluid at the impeller outlet is obtained by multiplying the number of particles per unit volume of mixture ($6\alpha_s/(\pi d_s^3)$) with the modulus of the drag force (as a function of the particle drag coefficient) and slip velocity ($v_\infty u_{sL}^*$). The result is multiplied by the volume of mixture in the pump, and normalized by the total hydraulic power of the pump (ρgQH). For dilute slurries with negligible particle interactions, the ratio $\Delta\eta_{slip}/\alpha_s$ is independent of solids concentration. This ratio, averaged over the casing circumference, is represented as a function of the blade–to–blade angular position in Figure 13.28. The energy dissipation is larger in the blade wake (interval I_1, $\phi^* = 0 - 0.5$), and decreases before the arrival of the next blade (interval I_2, $\phi^* = 0.5$ to 0.82). The energy dissipation caused by particle slip is about three times as large at $Q_n/2$ than at Q_n. The averages over the blade–to–blade interval are $\Delta\eta_{slip}/\alpha_s =$ 0.045 for Q_n, and 0.140 for $Q_n/2$.

The periodicity due to the finite number of blades has a significant effect on the local velocity distribution. In the present experiments, the velocity fluctuations are of the same order of magnitude with the velocity mean. However, the fluctuations for Q, H, $\Delta\eta_{slip}$ are about 10% of the mean because of the averaging effect around the casing. If the impeller has a smaller number of blades, the fluctuations of the characteristic curves would increase.

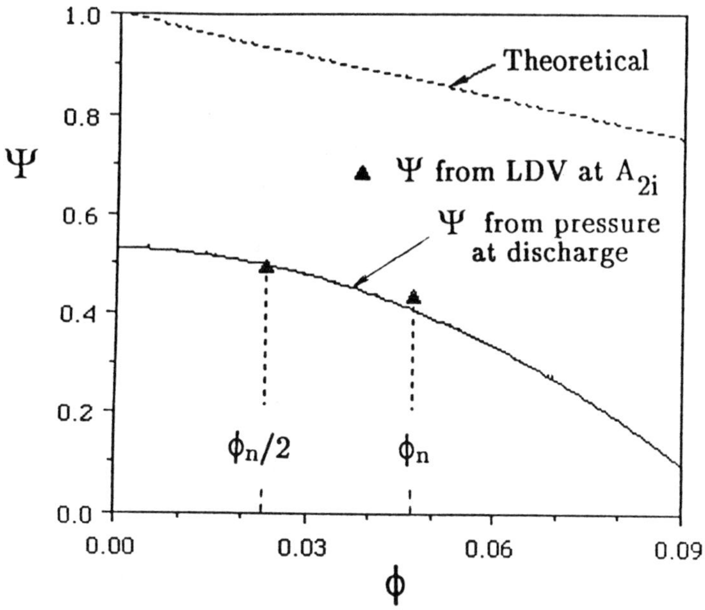

Fig. 13.27 The head coefficient – flow rate coefficient curve

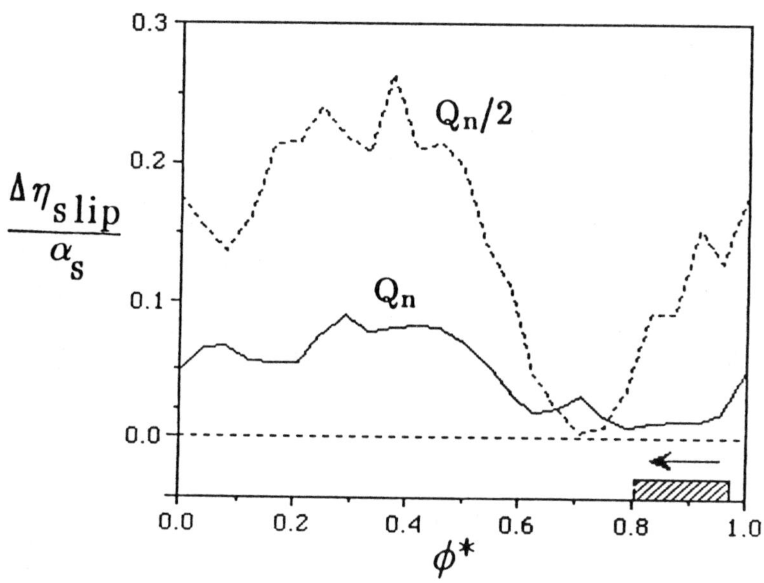

Fig. 13.28 Energy efficiency loss by particle slip

13.8 SUMMARY AND CONCLUSIONS

Back–scattering LDV has been used to investigate dilute particulate two–phase flow in the casing of a centrifugal slurry pump, which includes the impeller–casing flow interface. This interface is of practical importance because the corresponding velocity measurements can be used to determine the pump head and flowrate. The same velocity distributions can be applied as boundary conditions for impeller flow and casing flow. The liquid and solid velocities were measured using a discrimination method based on the laser Doppler signal amplitude. All three particle slip velocity components were obtained for a dilute water–glass beads mixture flow. The laboratory pump is of radial type, has a casing concentric to impeller, and has a discharge diameter of 76.2 mm. Its dimensionless specific speed is 0.46. The particles in suspension are 0.8 mm glass beads ($Re_{\infty} = 65$) and have a quasi–uniform size distribution.

Several qualitative aspects of the flow pattern have been identified in this study: (a) Large–scale periodical, two–phase flow structures develop in the entire casing, and are dominated by stationary waves; (b) The particles generally lead the fluid in the radial direction and lag in the circumferential direction; (c) The measured velocity distribution at the impeller–casing flow interface, averaged over the blade width, can be used to determine the pump head. The averaged velocity distribution averaged over the casing width determines the flowrate; (d) The periodical and turbulent velocity fluctuations have a complexity which is not fully reflected in the present flow simulations. Further comments are given below.

a. The liquid and particle slip velocity distributions show large–scale periodical structures caused by a combined effect of the impeller blade motion, impeller–casing flow interaction, and secondary currents in pump. The large scale flow structures affect the local two–phase turbulent microstructure through the cascading process. The effect of the **blade motion** was evaluated by velocity measurements at fixed locations as a function of the impeller angle. Periodic fluctuations of relatively high amplitude are superimposed on the fluid and solids velocities. As a result, also the random velocity fluctuations increase. This effect is more pronounced at Q_n than at $Q_n/2$. Overall in the pump casing, the velocity fluctuation, shear stress and energy dissipation have point fluctuations of the same order of magnitude as their mean. At the impeller–casing flow interface, the kinetic energy of the periodic motion is of the same order as the turbulent kinetic energy. The number of blades affects the velocity fluctuations of both phases and their relative velocity, with a direct influence on the energy dissipation and on erosion wear. by increasing the fluctuation level of both phases and the relative velocity of solid phase. The **impeller–casing flow interaction** determines large stationary waves at two locations. The impeller and casing are quasi–concentric, which is a typical arrangement for slurry pumps.

Four angular sectors have been identifies in casing: (S_1) tongue sector, with strong interaction between the circumferential casing flow and the outflow from impeller, particularly at lower flowrates, (S_2) acceleration sector, with weaker flow interaction, (S_3) bottle–neck sector, with strong flow interaction increasing with the flowrate, and (S_4) at the beginning of the discharge nozzle. Particle slip is a function of the angular intervals: (I_1) $\phi^* = 0$–0.5 with larger slip velocities, increasing with ϕ^*; (I_2) 0.5–0.82 with a decreasing slip velocity; and (I_3) $0.82 - 1$. The **secondary currents** identified in the present study are the inlet clearance pump recirculation, recirculation around the entire casing, blade–to–blade recirculation, and double helicoidal flow pattern in casing.

b. The particle trajectories deviate from the fluid flow streamlines. The particle response time is 3.6 times larger than the blade–to–blade time scale. In average, the particle slip velocity relative to fluid velocity in the radial direction ($U_{sL,r}/U_r \approx 0.1$; particles lead the fluid) is about 10 times larger than in the relative slip in the circumferential direction ($U_{sL,\theta}/U_\theta \approx -0.01$; particles lag the fluid).

c. The measured velocity distributions can be related to the pump performance characteristic curves. The velocity triangles over the blade width (B_{2i}) give a good estimate of the head–flowrate characteristic curve. The particle slip velocity distributions can be used to evaluate the energy loss by particle slip. By using the velocity distribution within the blade–to–blade interval, the pump head and flowrate can be determined as a function of the impeller angular position, $\Psi(\phi_0^*)$ and $\Phi(\phi_0^*)$. This group of characteristic curves identified in the present study is important when the flow fluctuations and/or vibrations in a hydraulic system are relevant.

d. The measured velocity field is much more complex than the currently applied flow simulations would suggest. The velocity histograms change as a function of the impeller angle, ϕ^*, with large transfer of energy between different zones of the energy spectrum. Anisotropy is significant. At the impeller–casing flow interface, the velocity standard deviation is 10% in the θ direction, and about 100% in the radial direction. It is expected that sudden accelerations and decelerations caused by the stationary waves cause large rates of energy dissipation. The interactions with the impeller flow and discharge pipe flow are intense and unsteady. In the case of a slurry flow, turbulence modulation by solid particles should be a key element in simulations. This complexity has ramifications on the interpretation of experimental data and on the modeling assumptions. It also has implications for pump design and wear simulation. One has to identify in any reported experiment in literature what velocity was measured (with our notations): v or u, u_p or u', U, $< >_b$ or $< >_c$, $< >_{av}$ or $< >_{b,av}$.

The periodical velocity fluctuations in sectors S_1 and S_3

dominate the wave pattern. This is the result of the interaction between the outflow from an impeller with a finite number of blades and the casing flow, in the presence of the casing recirculation (Q_{rc}) and the inlet clearance recirculation (Q'). The experimental results presented here provide a limited, qualitative source of data to better understand the flow mechanisms, calibrate numerical models, and evaluate the pump performance from the velocity distribution within the pump.

ACKNOWLEDGEMENTS

This study was partially sponsored by Dresser Pump and NSF.

MAIN NOTATION

A_{2i}	impeller outlet cross–section, $\pi D_{2i} B_{2i}$
B_{2i}	blade width at the impeller outlet
d_s	particle diameter
D_{2i}	outlet impeller diameter
H	pump head
I_1–I_3	angular intervals, defined in section 13.3
k	kinetic energy per unit mass
N	pump rotational speed
r, θ(degrees),z	coordinates
Q, Q_n	flow rate, and its nominal value
Q_{rc}	casing recirculation flowrate (Figure 13.1 a)
Q'	inlet clearance recirculation flow (Figure 13.1 c)
Re_∞	particle Reynolds number calculated with the terminal settling velocity
u	normalized velocity at a given ϕ ($u = U + u_p$)
U	average of u over $\Delta\phi_b$
U_{2i}	peripheral impeller tip velocity
v	instantaneous point velocity
v_∞	terminal settling velocity at $\alpha_s = 0$
S_s, S_m	specific density of solids and mixture, respectively
S_1–S_4	sectors in the casing, defined in section 13.3
T_i, T_b	time scales corresponding to a full impeller rotation and to blade–to–blade interval, respectively
Z	number of blades on impeller
α_s	volumetric solids concentration
$\beta_{r\theta}$	flow angle defined in Fig. 13.5a
Δr_{max}	maximum radial gap (Figure 13.1c)
$\Delta\phi$, $\Delta\phi_b$	angular interval, and the blade–to–blade angular interval, respectively

$\Delta\eta_{slip}$ loss of efficiency by particle slip
ϕ impeller angle
Φ pump flowrate coefficient
Ψ pump head coefficient
ν kinematic viscosity
θ angle in casing

Subscripts:
av averaged around the impeller (over $\theta = 0$ to 360 degrees)
A averaged over the area of a radial cross–section
b blade
i impeller
L,s,m liquid, solids and mixture, respectively
p periodic value
r,θ,z radial, circumferential and axial direction, respectively
$\langle\ \rangle_b$ average over B_{2i} ($z^* = 0.35$ to 0.7)
$\langle\ \rangle_c$ average over B_c ($z^* = 0$ to 1)

Superscripts:
r^*, θ^*, z^* normalized coordinates
u^*_{sL} normalized slip velocity, defined in (13.4)
v^*_∞ normalized v_∞
u', U' random fluctuations of u and U, respectively
ϕ^*_0, ϕ^* normalized impeller angle measured from window #2, and from a current window, respectively (defined in eq. 13.1)

REFERENCES

Adler, D. and Levy, Y., "A Laser–Doppler Investigation of the Flow Inside a Backswept, Closed, Centrifugal Impeller", *J. of* Mechanical Engineering Science, Vol. 21, No. 1, 1–6 (1979).

Akima, C.L., "A Method of Bivariate Interpolation and Smooth Surface Fitting for Irregularly Distributed Data Points", ACM Transactions on Math,. Software, Vol. 4, No. 2, 148–159 (June 1978).

Barlow, R.S. and Morrison, C.O., "Two–phase Velocity Measurements in Dense Particle Laden Jets", Experiments in Fluids, Vol. 9, 93–104 (1990).

Bowerman, R.D. and Acosta, A.J., "Effect of the Volute on Performance of a Centrifugal Pump Impeller", Trans. ASME, Vol. 79, No. 5, 1057–1069 (1957).

Cader, T., Masbernat, O. and Roco, M.C., "LDV Measurements in a Centrifugal Slurry Pump: Water and Dilute Slurry Flow", In Laser Anemometry (A. Dybbs and B. Ghorashi, eds.), Vol. 2, ASME, New York, 763–773 (1991); (also J. Fluids Eng., Vol. 114, No. 4, Dec. 1992).

Dong, R., Chu, S. and Katz, J., "Quantitative Visualization of the Flow within the Volute of a Centrifugal Pump", Joint ASME–JSME Fluids Eng. Conf., Portland, Oregon (1991).

Durrett, R.P., Stevenson, W.H., and Thompson, H.D., "Radial and Axial Turbulent Flow Measurements with an LDV in an Axisymmetric Sudden Expansion Air Flow", *J. of Fluids Engineering*, Vol. 110, 367–372 (1988).

Durst, F., Melling, A. and Whitelaw, J.H., "Principles and Practice of Laser–Doppler Anemometry", 2nd ed., Academic Press, New York (1981).

Falco, R. and Nocera, D., "Quantitative Multipoint Measurements and Visualization of Dense Liquid–solid Flows using Laser Induced Photochemical Anemometry (LIPA)", Chapter 3 in Particulate Two–phase Flow (M.C. Roco, ed.), Butterworth (1992).

Gore, R.A. and Crowe, C.T., "Effect of Particle Size on Modulating Turbulent Intensity", *Int. J. Multiphase Flow*, Vol. 15, No. 2, 279–285 (1989).

Graham, A., PhD thesis, Dept. of Chem. Eng., Univ. of Wisconsin (1980).

Hamelin, P., and Roco, M.C., "Flow Visualization in a Centrifugal Pump by Particle Tracking", Report for Dresser Pump, University of Kentucky, (1988).

Hamkins, C.P. and Flack, R.D., "Laser Velocimeter Measurements in Shrouded and Unshrouded Radial Flow Pump Impellers", J. of Turbomachinery (ASME), Vol. 109, 70–76 (1987).

Itaya, T. and Nishikawa, T., "Study on the Pump Sand: (I) On the Trajectories of Solid Particles in the Pump Impeller", Bull JSME, Vol. 7, No. 27, 577–582 (1964).

Lee, S.L. and Durst, F., "On the Motion of Particles in Turbulent Duct Flows", *Int. J. Multiphase Flow*, Vol. 8, 125–146 (1982).

Leighton, D. and Rampall, I., "Measurement of the Shear Induced Microstructure of Concentrated Suspensions of Non–Colloidal Spheres", Chapter 6 in Particulate Two–Phase Flow (M.C. Roco, ed.), Butterworth (1992).

Liu, Z.C. and Adrian, R.J., "Simultaneous Imaging of the Velocity Fields of Two–phases", Chapter 2 in Particulate Two–phase Flow (M.C. Roco, ed.), Butterworth (1992).

Modaress, D., Wuerer, J. and Elghobashi, S., "An Experimental Study of a Turbulent Round Two–Phase Jet", *Chem. Eng. Commun.*, Vol. 28, 341–354 (1984a).

Modaress, D., Tan, H. and Elghobashi, S., "Two–Component LDA Measurement in a Two–Phase Turbulent Jet", *AIAA Journal*, Vol. 22, 624–630 (1984b).

Masbernat, O., Cader, T. and Roco, M.C., "Velocity Distribution of a Dilute Slurry Around a Centrifugal Pump Impeller", Report University of Kentucky (1991).

Nouri, J.M., Whitelaw, J.H. and Yianneskis, M., "Particle Motion and Turbulence in Dense Two–Phase Flows", *Int. J. Multiphase Flow*, Vol. 13, 729–739 (1987).

Paone, N., Riethmuller, M.L. and Braembussche, R.A., "Experimental

Investigation of the Flow in the Vaneless Diffuser of a Centrifugal Pump by Particle Image Displacement Velocimetry", Experiments in Fluids, Vol. 7, 371–378 (1989).

Parthasarathy, R.N., and Faeth, G.M., "Structure of a Turbulent Particle Laden Water Jet in Still Water", *Int. J. Multiphase Flow*, Vol. 13, 699–716 (1987).

Roco, M.C. and Dehaven, S., "Marching Approach for Channels with Lateral Inflow and Secondary Currents", J. Appl. Mathematical Modelling, Vol. 13, 468–478 (1989).

Roco, M.C., Marsh, M., Addie, G.R. and Maffet, J.R., "Dredge Pump Performance Prediction", *J. of Pipelines*, Vol. 5, pp. 171–190 (1986).

Roco, M.C., Hamelin, P., Cader, T. and Davidson, G., "Animation of LDV Measurements in a Centrifugal Pump", Video Journal of Engineering Research, Wiley & Sons, Vol. 1, No. 1, 25–40 (1991).

Tsuji, Y., Morikawa, Y. and Shiomi, H., "LDV Measurements of an Air–Solid Two–Phase Flow in a Vertical Pipe", *J. Fluid Mech.*, Vol. 139, 417–434 (1984).

Ye, J. and Roco, M.C., "Particle Rotation in Couette Flow", Phys. of Fluids – A, Vol. 4, No. 2, 220–224 (1992).

Zisselmar, R. and Molerus, O., "Investigation of Solid–Liquid Pipe Flow with Regard to Turbulent Modification", *Chem. Eng. J.*, Vol. 18, 233–239 (1979).

Zhang, J. and Roco, M.C., "A Probabilistic Method for Field Solutions in Two–phase Flow", ASME Vol. FED , "Cavitation and Multiphase Flow Forum" (1991), (also Int. J. Numer. Methods in Fluids", Vol. 12, 1992).

CHAPTER 14

AEROSOL DYNAMICS

R. Flagan

14.1 INTRODUCTION

Aerosols are metastable suspensions of particles in gases, and generally lie at the dilute limit of multiphase flows, with mass fractions below 10^{-3} and particulate volume fractions below 10^{-6}. The focus of aerosol science is, therefore, generally on the behavior of the particles rather than on the influence of the particles on the gas. Questions commonly addressed include the mechanisms of particle formation and growth, the deposition of particles on surfaces, and the effects of particles on light transmission, human health, and the environment. Aerosol particles may be solid or liquid and may have simple or complex forms. In some cases all of the particles that comprise the aerosol may have the same chemical composition, but many aerosols are mixtures of particles from different sources. This is especially common in aerosols produced in chemically reacting systems.

Aerosols are ubiquitous in nature and in technology. Small atmospheric particles, known as cloud condensation nuclei, are the seeds on which cloud and fog droplets form. These particles may be derived from natural materials, or they may be the products of man's activities, either primary articles that are released directly to the atmosphere, or secondary particles formed from gaseous pollutants. Much of aerosol science has arisen from studies of the nature, origins, and fate of such environmental aerosols.

Aerosols are important technologically far beyond the obvious environmental control issues. Carbon black, which sees use in pigments and as a reinforcing agent in rubber, undoubtedly has the longest history, extending to prehistoric times. More recent entries into the domain of aerosol technology include the synthesis of TiO_2 by the chloride process, and the fabrication of the preforms from which graded refractive index optical waveguides are drawn. Both these highly profitable processes generate metal oxide particles by high temperature oxidation of metal halide vapors. This is usually carried out in a flame, but plasma reactors have also been used in both applications. Aerosol routes have also received interest for the synthesis of ceramic powders, as an intermediate in the refining of silicon for materials applications, and in the production of nanostructured materials that exhibit unique structure-scale-dependent mechanical, electrical, and optical properties.

The theory of aerosol behavior originated with studies of atmospheric particles, beginning with Aitken's work in the nineteenth century (Podzimek, 1989) and with studies of the atomistic nature of matter around the turn of the century. Much of the work has been motivated by efforts to understand the atmospheric aerosol, with emphasis on liquid particles and assumptions that the particles are, at all times, fully dense spheres. As aerosol science moves beyond the study of dust, pollutant, and cloud particles in the atmosphere, additional physical processes must be considered. The gaseous medium in which aerosol particles reside introduces a length scale that profoundly influences the aerosol dynamics. Some particles are sufficiently large that the fluid may be assumed to be a continuum medium, but most aerosols include particles for which this assumption is not valid. Particles smaller than the mean free path of the gas molecules are present in many aerosols, so the non-continuum nature of the fluid medium must be taken into account in descriptions.

Pyrogenous fumes, aerosols originating from vapor condensation in high temperature systems, have received considerable attention in recent years because of their complex microstructures. The particles are frequently agglomerates of approximately spherical *primary particles.* The structures of agglomerate particles from different sources are remarkably similar. The primary particle size is typically in the range from 10 to 100 nm diameter, and is seemingly independent of the material composition, temperature and pressure at which the particles form, and other environmental parameters. The apparent universality of the agglomerate structure has been partially explained in terms of agglomeration which produces particles that exhibit *fractal* structure, a concept that will be discussed in more detail later. The reasons for the similarity of primary particle size has not been adequately explained to this time, however.

Technological applications of aerosols place severe requirements on the structures of the aerosol particles. Tightly bonded agglomerate particles with complex structures are valuable in carbon black that is used to reinforce rubber, but undesirable for the particles to be used to fabricate ceramic structures. In ceramic applications, agglomerates lead to large, strength-robbing voids in the finished material. Powder synthesis systems must, therefore, be designed to minimize the formation of hard agglomerates. The phase of the particulate material is also important, determining mechanical, optical, and electronic properties. The theoretical description of aerosol evolution must, therefore, be extended to include agglomeration and phase transformation kinetics, and coalescence rates to facilitate process design and optimization.

Aerosols differ from commonly studied colloidal systems in a number of ways:

- Particle volume fractions in aerosols are low enough that aerosol perturbations to the gas flow can frequently be neglected.

- Ion concentrations are low enough that particle charge levels are relatively low and structures like double layers do not form. Exceptions may be found in plasmas and some high temperature systems to which low work function materials are added.

- Weak fluid-particle interactions lessen the extent of rearrangement of agglomerates in aerosols as compared to particulate systems in liquids.

- Temperatures may be high enough that solid phase diffusion significantly rearranges the particles by sintering or coalescence.

This chapter, like much of the aerosol literature will focus on the physical processes that govern particle formation and growth, *e.g.* condensation, evaporation, nucleation, and coagulation, but many of the aerosol systems of interest today can only be fully understood in the context of the chemical processes that lead to their formation. Photochemical reactions in the atmosphere produce low volatility species that ultimately condense heterogeneously causing existing aerosol particles to grow or homogeneously to form new particles. Gaseous species can also dissolve in or adsorb on aerosol particles and then undergo aerosol phase chemical reactions. One such process that has received considerable attention in the atmospheric aerosol literature is the solution phase oxidation of sulfur dioxide. Soot is formed as part of the complex sequence of combustion reactions taking place in fuel rich regions of flames. Indeed, soot particles are formed through a complex and poorly understood chemical nucleation process rather than by the physical condensation processes that have received most attention in studies of nucleation of particles from the vapor phase. A comprehensive description of the dynamics of aerosol formation and growth frequently requires coupling models of the aerosol physics with mechanisms of gas phase, surface, and solution chemistry taking place in the system of interest.

In this chapter, I will examine the physical processes that govern the dynamics of aerosols, with special emphasis on the structures of the particles that evolve.

14.2 SCALES

For the moment, let us assume that aerosol particles are dense spheres, a common assumption in much of the existing aerosol literature. The particles may range in size from a few tenths of a nanometer, mere molecular clusters, to tens or even hundreds of microns. No unique upper bound can be imposed on the sizes of aerosol particles. Rather, the range of particles included under the rubric of aerosols depends on the situation at hand. For example, particles that will be lost by sedimentation on a time scale much shorter than the experimental time would normally be excluded from the aerosol, *i.e.*, particles larger than 100 μm would not generally be counted in the atmospheric aerosol due to their high $\approx$25 cm/s sedimentation velocity.

The particle sizes must be compared with the relevant length scale of the flow under study. The length scale that characterizes the gas is the mean-free-path, λ. The aerodynamic behavior of particles much larger than λ is described using continuum fluid mechanics, while particles much smaller than λ are said to be in the free molecular regime. The particle size may, thus, be characterized by the ratio of the mean-free-path to the particle size, a quantity called the Knudsen number,

$$\mathrm{Kn} = \frac{\lambda}{r_p} = \frac{2\lambda}{d_p} \tag{14.1}$$

where the particle size has been expressed in terms of a radius or diameter and the particle has implicitly been assumed to be a sphere.

Figure 14-1 illustrates the size dependence of some aerosol properties and the instruments used to characterize aerosol particles. Diffusional interactions between aerosol particles and the surrounding gas , *i.e.*, aerodynamic drag, and heat mass transfer rates scale as r for particles in the continuum size regime, and as r^2 in the free molecular limit. Since the mean-free-path of gas molecules in air at ambient conditions is about 65 nm, the transition from one limit to the other extends over a range of particles sizes centered about this value. Similarly, the coagulation rate exhibits transitional behavior in the same size regime. Coincidentally, the regime of transition of optical behavior from geometric optics

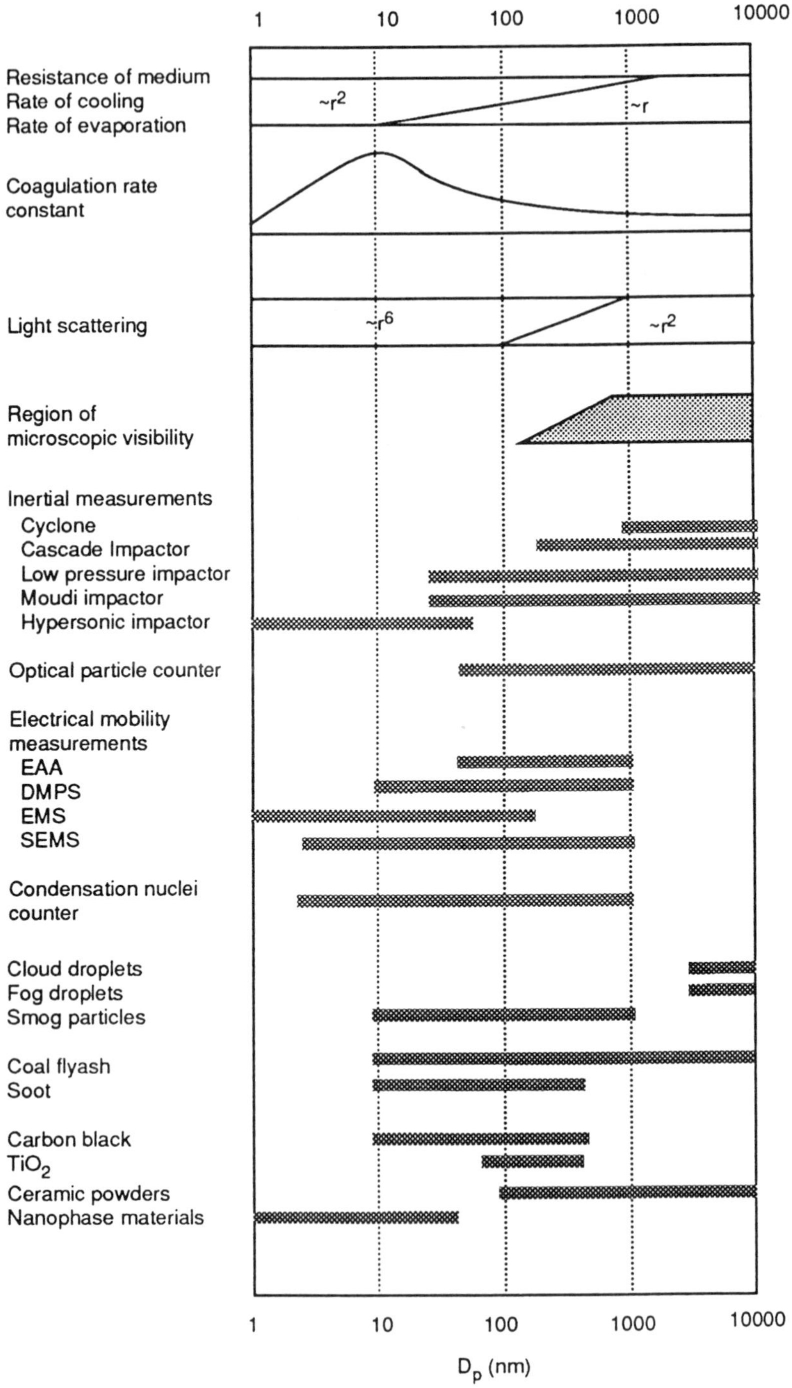

Figure 14-1: Some properties of aerosols in relation to particle size, and aerosol measurement methods. (After Fuchs, 1964)

to Rayleigh scattering by essentially point particles occurs in about the same size range due to the electromagnetic coupling between the particle and visible light with wavelengths in the 300 to 700 nm range.

Aerosol particles are measured by a number of different techniques that are applicable to different size ranges. Light scattering and inertial methods are used for particles larger than 0.1 to 1 μm. Smaller particles can be counted optically if grown by vapor condensation in a condensation nuclei counter. The size distribution of ultrafine particles can be measured in terms of their migration velocities in an electric field. Such electrical mobility measurements offer the highest resolution in both particle size and time, and have been extended to particles as small as 1 nm in diameter. No one aerosol instrument can, however, cover the entire size range of interest for either atmospheric, air pollution sources, or aerosol technologies.

Many aerosol particles are not, however, spherical. Spherical particles result when the material that comprises the particle is a relatively low viscosity liquid. Particle produced by high temperature processes, combustion, industrial processes, volcanos, etc., are frequently solids or very viscous liquid. The particles then grow as agglomerates. At the smallest scales, the particles frequently appear to be spherical, but they are characterized by a range of scales. When examined by electron microscopy, the particle size that is reported is frequently the size of the smallest units of the agglomerates. For other purposes, particularly to understand the dynamics of the aerosol, a scale that is characteristic of the entire agglomerate is more relevant. The large discrepancy between the two sizes is illustrated in Fig. 14-2 where the mean sizes of Si particles produced by silane pyrolysis are plotted as a function of the concentration of silane entering the reactor. Two sizes are presented: (*i*) the size determined by measurement of fine structure sizes from transmission electron microscope photographs; and (*ii*) the sizes determined by on-line electrical mobility analysis wherein the particles are charged and then classified according to their migration velocities in an electric field, *i.e.*, the size of a sphere with the same aerodynamic resistance to motion as the agglomerate.

The discrepancy between the two estimates of particle size indicates that the particles cannot be characterized in terms of a single length scale. The structures of agglomerate particles arising from widely different sources are, however, remarkably similar. As can be seen from the electron microscope photograph in Fig. 14-3, they consist of chain agglomerates of roughly spherical primary particles, with the primary particles ranging in size from 10 nm to 100 nm in most cases. While the

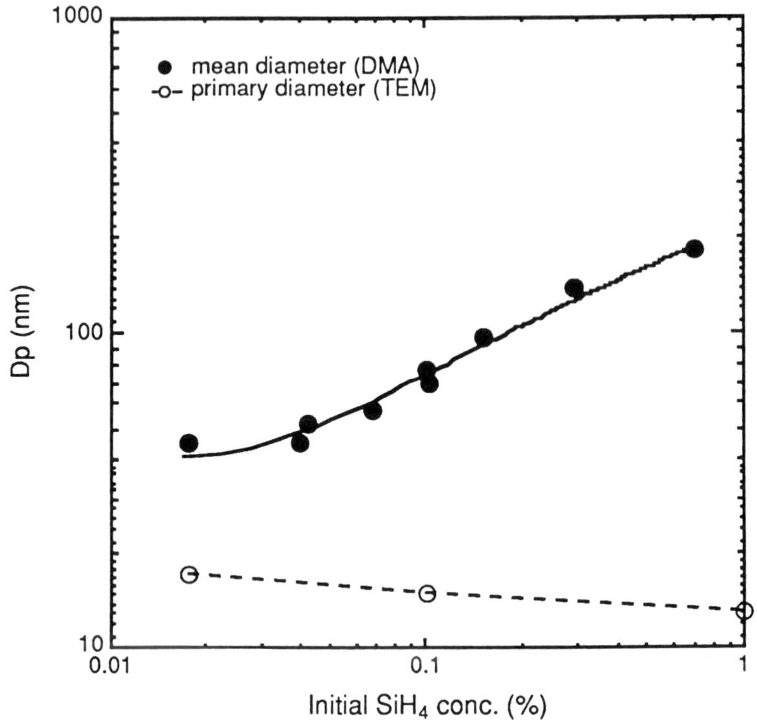

Figure 14-2: Mean sizes of particle produced by silane pyrolysis as inferred by electrical mobility analysis and by transmission electron microscopy.

detailed arrangement of the primary particles is random, the statistics of that distribution have common features. Forrest and Witten (1979) first reported that agglomerate particles appear to exhibit dilation symmetry, *i.e.*, the structures appear to be similar when viewed on different scales. This common structure has since been described using power-law models and labeled *fractal* (Mandelbrot, 1982; Meakin, 1990). The densities of pyrogenous fume agglomerates decrease with increasing size. Forrest and Witten observed that the aggregate mass varies as a power law of the size of that aggregate, *i.e.*,

$$m \propto r^D \tag{14.2}$$

where D has been called the fractal dimension which generally has a value below the Euclidean dimension of 3.

Computer simulations of aggregate formation yield similar structures and provide insights into the formation mechanisms (Meakin, 1990). Rel-

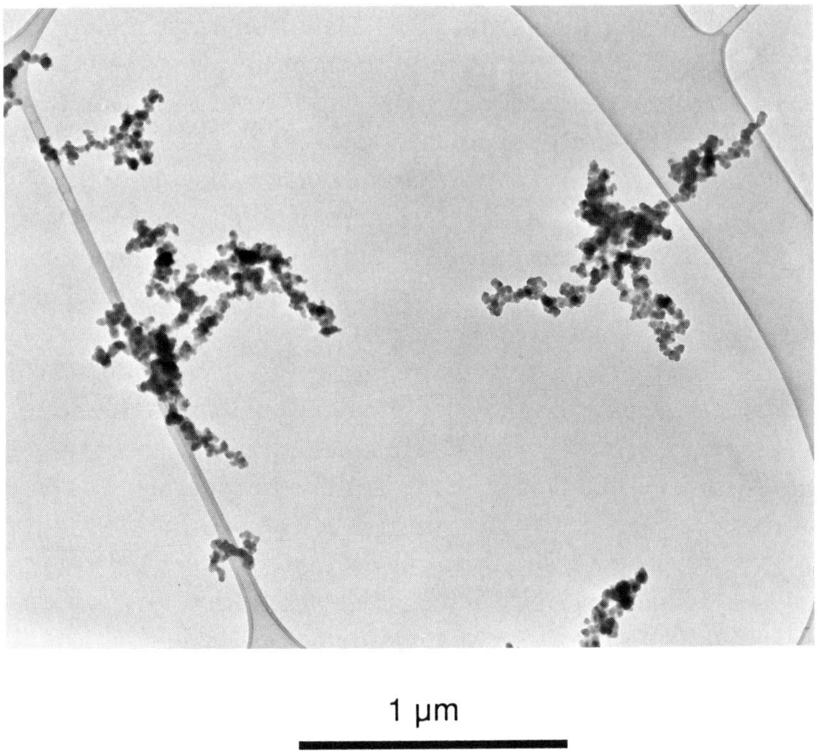

Figure 14-3: Electron microscope photographs of TiO_2 agglomerate particles.

atively dense particles ($D \sim 2.4$) are produced when individual spheres diffuse to the surface of the growing aggregate (diffusion limited aggregation). Lower density aggregates ($D \sim 1.8$) result from coagulation of like-sized aggregates (cluster-cluster aggregation). These structure calculations make one very important assumption that is not valid for many pyrogenous fumes, namely that the structure of the aggregate does change due to fusion spherules under the driving force of surface tension. This may be expected to increase the fractal dimensionality of the aggregates as they age, although this process is not yet quantitatively understood. Moreover, the range over which aerosol agglomerates exhibit scale-invariance is usually quite limited. Nonetheless, the fact that the particles are fractal-like allows one to simplify the description of complex structures, and may prove quite useful in modeling their dynamics.

More perplexing is the similarity of the fine structure scales of agglomerates produced by different systems. Most commonly, the primary particle size is in the range of 10 to 30 nm, although some systems produce larger or smaller primaries. The formation environment influences the microstructure although the fundamental mechanisms have not been elucidated. High concentrations of the vapor precursors for the particulate material that lead to particulate loadings in excess of 10% by mass lead to primary particle sizes on the order of 100 to 300 nm in TiO_2 synthesis, but precursor concentrations ranging from a few percent to parts per million do not dramatically influence the primary particle size. Agglomerate formation by evaporation of refractory materials into a low -pressure cold gas leads to microstructure scales on the order of 5 to 10 nm, essential to the synthesis of nanophase materials. All of these particles are formed by homogeneous nucleation from the vapor phase, with the ultimate primary particle size being much coarser than the critical nucleus for new particle formation. The origins of this similarity in microstructure arise from the properties of the solid materials produced and the kinetics of sintering as discussed qualitatively below.

Thus, a detailed description of particle formation, growth, and structure must include descriptions of agglomeration, coalescence or sintering, nucleation, and condensation.

14.3 PARTICLE SIZE DISTRIBUTIONS

Any physically realizable aerosol system includes a range of particle sizes. The size of a particle can be uniquely described in terms of its mass, m. The distribution of particle masses can be described in terms of the particle size distribution,

$$dN = n_m(m)dm \tag{14.3}$$

where dN is the number of particles per unit volume of gas with masses between m and $m + dm$. For very small particles, the distribution with respect to size may be described in terms of the number of molecules comprising the particle, *i.e.*, the number of particles containing g molecules (g-mers) per unit volume of gas is N_g. Ideal for the description of small molecular clusters, this approach rapidly becomes impractical as the particle size increases. Often, several decades in particle diameter must be followed to accurately describe the aerosol dynamics. The number of discrete sizes that would have to be followed rapidly exceeds the capabilities of even the largest computers. Continuous representations of the particle size distribution are a convenient way to span the broad range of sizes encountered in aerosol systems, but the computational requirements of integrating the aerosol dynamic equations over all particle sizes at each time step are severe. Gelbard (Gelbard and Seinfeld, 1979) introduced

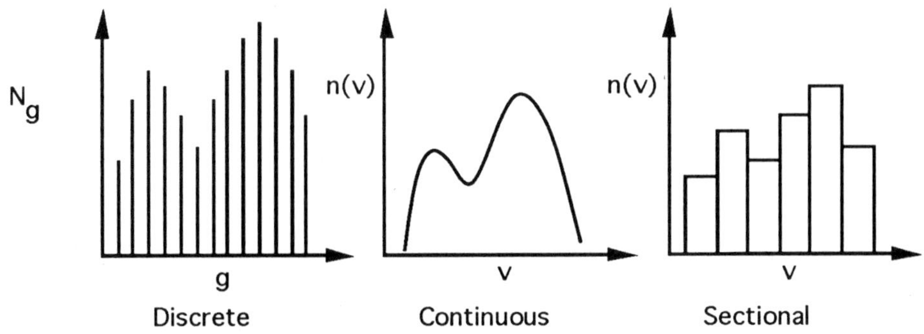

Figure 14-4: Schematic representations of the different types of particle size distribution functions.

an approximate representation known as the sectional representation to overcome this difficulty. In this approach, the size distribution is approximated by a histogram (usually of mass concentration), and the integrals are reduced to summations. Mass is conserved exactly and computation times are dramatically reduced. These three representations of the particle size distribution are illustrated schematically in Fig. 14-4. Further developments of the sectional method have led to the multicomponent sectional representation that makes it possible to follow the evolution of the composition size distribution, assuming that all particles in a section have the same chemical composition, but allowing particles in different sections to differ in composition.

An accurate description of nucleation and growth in particulate systems requires that particles ranging from molecular sizes to microscopic size be represented. None of these models can span the vast size range with uniform accuracy. Hybrid descriptions of the particle size distribution have been developed to span this range, with the smallest particle being described by the number concentration of the particular g-mer, while the largest particles are described by a continuous distribution function. This description arises naturally since the discrete sizes of successive clusters differ significantly for the smallest particles but become insignificant for larger particles as illustrated by the discrete particle size distribution shown in Fig. 14-5.

The particle size distribution only partially describes the aerosol, however. Friedlander (1970) developed general descriptions of the distribution of particles with respect to both size and chemical composition. The evolution of the composition-size distribution of the aerosol particles would be calculated using balance equations that describe the influence

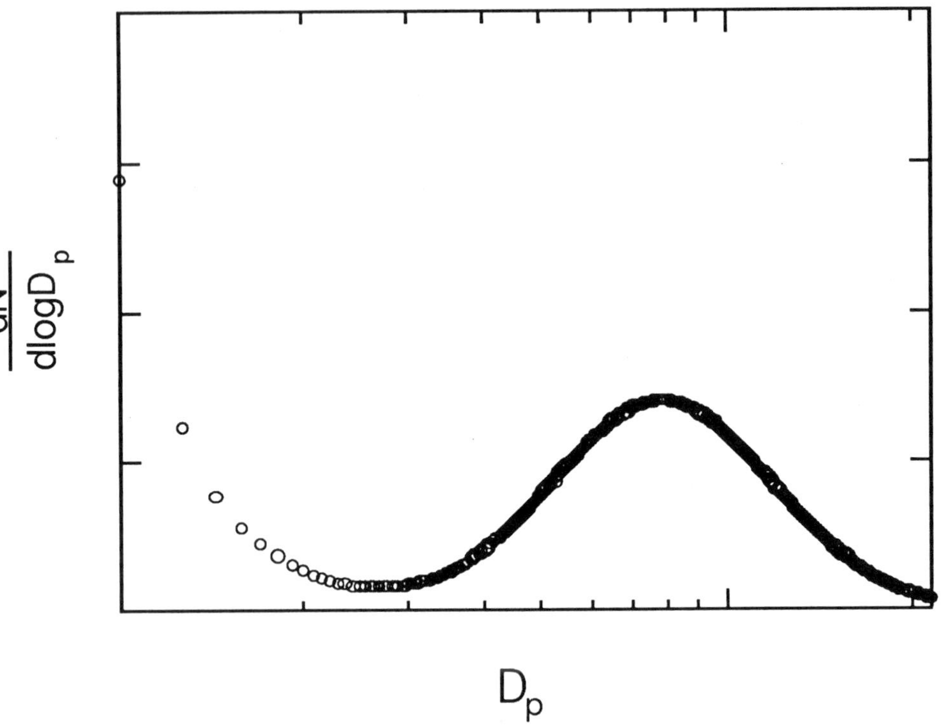

Figure 14-5: Size distribution function showing the transition from discrete sizes at the smallest cluster sizes to the continuous distribution at larger sizes.

of different physical and chemical processes on the distribution function. The distribution of particles with respect to composition and size can be described in terms of a composition distribution function, such that the number of particles of mass between m and $m + dm$ containing molar quantities per unit mass of each chemical species between η_i and $\eta_i + d\eta_i$ with $i = 1, 2, ..., k$ is

$$dN = n(m, \eta_1, \eta_2, ..., \eta_k) dm d\eta_1 d\eta_2 \cdots d\eta_k \tag{14.4}$$

or,

$$dN = n(m, \eta_i) dm d\eta_i \tag{14.5}$$

In a companion paper, Friedlander (1971) characterized the instruments that would be needed to fully characterize chemically heterogeneous aerosols. While significant improvements have been made since those papers appeared, the experimental advances needed to meet Friedlander's ideal have not yet been achieved.

General implementation of the composition–size distribution in simulations of aerosol evolution has only recently been attempted. The earliest attempt is the multicomponent aerosol dynamics model of Gelbard and Seinfeld (1978) in which all particles of a given size m are assumed to have the same composition, but the composition is allowed to vary with particle size, *i.e.*, $\eta_i = \eta_i(m)$. The number of moles of species i contained within particles with masses between m and $m + dm$ is then,

$$d\mathcal{M}_i = m\eta_i n(m) dm \tag{14.6}$$

This simpler description has been implemented using a histogram representation of the particle size distribution in the multicomponent aerosol dynamics code MAEROS. The predictions of this code are consistent in level of detail with aerosol characterization techniques that lump together particles within a narrow range of sizes for chemical analysis such as cascade impactors. More recently, Kim and Seinfeld (1991) have developed an aerosol dynamic model that not only describes the variation of particle composition with size, it also accounts for the variation of particle composition among particles of a given size.

Even the composition-size distribution does not completely describe all aerosols. Particle structure and the distribution of phases must also be considered. In a recent paper Kobata et al. (1991) made a first attempt to describe the phase distribution in a solid aerosol with a model of TiO_2 aerosol evolution. Using an approach analogous to the multicomponent sectional model of Gelbard (Gelbard and Seinfeld, 1980), the distribution of anatase and rutile in the TiO_2 aerosol was followed. Only the primary particle distribution was examined in their study, however.

A detailed description of the distribution of agglomerate particle structure and mass would account for the distribution of primary particles. A first approximation to this has been taken first by Ulrich (Ulrich et al., 1977) and more recently by Koch and Friedlander (1991) is to follow the surface area α of the particles with total mass m. Assuming all particles of a given mass have the same surface area, the total surface area of particles in size range from m to $m + dm$ is

$$dA = \alpha(m) n(m) dm \tag{14.7}$$

The surface area then changes through the action of sintering or coalescence, a process whose rate is constrained by diffusional processes in the condensed phase. This approach differs in detail but is essentially the same as that taken by Ulrich and Subramanian (1977) in an effort to explain inconsistencies in their earlier observations (Ulrich, 1971; Ulrich et al., 1976) of the rate of silica particle growth in flames. BET gas adsorption had been used to infer particle size, providing a measure of the fine structure scale (primary particle size) that suggested much less coagulation than predicted with commonly accepted coagulation models. Incomplete coalescence of the silica particles due to the very high viscosity of silica caused this discrepancy. An approximate description of the coalescence was derived by dividing the agglomerate into cells that each contain a central sphere in contact with ν other spheres. Using a characteristic coalescence time τ calculated from sintering theory, the growth in the primary particle size was calculated.

To rigorously describe the evolution of the agglomerate structure, one must know the number of primary particles and the number distribution of agglomerates. Moreover, a detailed understanding of the structure of the agglomerates is required to determine ν. Moreover, for fractal agglomerates, the agglomerate density decreases with agglomerate radius, so the coordination number may be expected to vary with position in the agglomerate. Much remains to be done in the detailed description of the evolution of agglomerate particles, particularly in chemically heterogeneous systems.

14.4 PARTICLE DYNAMICS

An aerosol evolves as the particles grow or shrink due to mass transfer between the particle and the surrounding atmosphere, collide with one another due to Brownian motion or other processes, sticking together and increasing the mean size while decreasing the total number concentration, are formed from the gas phase by homogeneous nucleation, or are lost to surfaces by a variety of deposition mechanisms.

The dynamics of aerosol systems undergoing simultaneous gas-to-particle conversion and coagulation have been studied by numerous investigators, primarily with emphasis on atmospheric and other ambient temperature aerosols. Most attention has been focussed on liquid particles that coalesce immediately upon agglomeration. Several comprehensive summaries of aerosol physics are available, including (Fuchs, 1964; Friedlander, 1977; Seinfeld, 1986; Flagan and Seinfeld, 1988).

Attention given to noncoalescing particles has been considerably more limited. Diffusion limited aggregation and cluster-cluster aggregation models have focussed on the structure of the particle, primarily through computer simulations that employed simplistic models of particle aerodynamics and collision cross sections. Direct measurement of aerodynamic parameters for aggregate particles is a prerequisite for the development of a more rigorous understanding of such aerosol systems.

Because our focus in this chapter is on the nature of the particles produced in aerosol systems, we consider first the origins of the agglomerate structure. To understand the role of particle structure on aerosol dynamics, we must first review the salient features of the classical theory of the dynamics of rapidly coalescing aerosol particles. The particle size distribution function evolves with time through the actions of particle-particle collisions (coagulation), evaporation and condensation, etc. These processes are described in the so-called general dynamic equation (GDE) for aerosols,

$$\begin{aligned}\frac{\partial n(v,t)}{\partial t} &= \frac{1}{2}\int_{v_o}^{v-v_o} K(v-\tilde{v},\tilde{v})n(v-\tilde{v},t)n(\tilde{v},t)d\tilde{v} \\ &- \int_{v_o}^{\infty} K(v,\tilde{v})n(v,t)n(\tilde{v},t)d\tilde{v} \\ &- \frac{\partial}{\partial v}\left[I_o(v)n(v,t)\right] \qquad (14.8)\end{aligned}$$

The first two terms on the right hand side of the GDE represent the formation of particles of volume v by coagulation of smaller particles and the loss of particles of volume v by coagulation with particles of any size. $K(v,\tilde{v})n(v,t)n(\tilde{v},t)$ is the collision frequency of particles of volume v with particles of volume $\tilde{v}$. The third term describes the growth of particles by condensation (or shrinkage by evaporation), surface reactions, and the like. The condensation flux is

$$I_o(v,t) = \int_0^{v_m} K(v,\tilde{v})\tilde{v}n'(\tilde{v},t)d\tilde{v} \qquad (14.9)$$

where $n'(\tilde{v},t)$ is the number distribution of small *condensing nuclei*, and $\tilde{v}_m$ is volume of the largest condensing species. This formulation treats condensation as the coagulation of molecules or, possibly, small molecular clusters with the particle. A general description of the evolution of the particle size distribution thus requires knowledge of the collision frequency function, $K(v,\tilde{v})$. Extensions that account for composition or other properties of the aerosol than number have been explored by Friedlander (1970). His recent paper with Koch (Koch and Friedlander, 1991) considers the a single moment of the microstructure, namely the total surface area of particles of a given mass. To date, however, no formulation that fully accounts for the particle microstructure has been proposed.

14.4.1 Coagulation

Coagulation is the growth of particles when two particles collide and stick to one another. Particle collisions may result from their mutual

Brownian motion, laminar or turbulent velocity gradients, differential sedimentation, or other mechanisms. For submicron fume particles, the predominant collision mechanism under most circumstances is Brownian motion. The frequency of collisions of spherical particles is determined by the diffusivities of the two particles and their collision cross sections. If we first consider continuum regime particles, the coagulation of spherical particles of diameters d_{pi} and d_{pj} is modeled as the diffusion of a point particle with diffusivity $\mathcal{D}_{ij}$ to the distance from the center of a stationary particle that corresponds to contact between the two particles, *i.e.*, to a sphere of diameter $(d_{pi} + d_{pj})$. The binary diffusion coefficient for the relative diffusion of the two particles is $\mathcal{D}_{ij} = \mathcal{D}_i + \mathcal{D}_j$. The coagulation frequency becomes

$$Z_{ij} = K(d_{pi}, d_{pj})N_i N_j \tag{14.10}$$

where

$$K(d_{pi}, d_{pj}) = 2\pi(d_{pi} + d_{pj})(\mathcal{D}_i + \mathcal{D}_j) \tag{14.11}$$

The diffusivity of small particles can be related to the aerodynamic drag forces acting on a particle moving relative to the gas in which it is immersed through the Stokes-Einstein relation,

$$\mathcal{D}_i = \frac{kT}{f_i} \tag{14.12}$$

The friction factor $f_i = B_i^{-1}$ (the inverse of the particle mobility) is the drag force per unit velocity. Spherical particles in the continuum size regime are subject to Stokesian drag,

$$f_i = 3\pi\mu d_{pi} \tag{14.13}$$

The collision frequency function in the continuum regime for particles of equal size becomes

$$K(d_p, d_p) = \frac{8kT}{3\mu} \tag{14.14}$$

which, being independent of particle size, is the natural value to use to nondimensionalize the coagulation rate coefficient.

For spherical particles with finite Knudsen numbers,

$$f = 3\pi\mu d_{pi}/C_c(\mathrm{Kn}) \tag{14.15}$$

so we may write

$$\mathcal{D}_i = \frac{kTC_c(\mathrm{Kn})}{3\pi\mu d_{pi}} \tag{14.16}$$

where $C_c(\text{Kn})$, the Cunningham slip correction factor, accounts for slip of the gas at the particle surface. For spherical particles, a commonly used expression for the slip correction is

$$C_c = 1 + \text{Kn}\left[1.257 + 0.40 \exp\left(-\frac{1.1}{\text{Kn}}\right)\right] \tag{14.17}$$

Even for particles small enough to exhibit slip, the continuum model may reasonably approximate interparticle collisions since the mean free path of the particle motion is much smaller than that of the gas molecules.

In the free molecular limit (Kn >> 1) particles undergo ballistic collisions. The collision frequency function for spherical particles with volumes v_i and v_j then becomes

$$K(v_i, v_j) = \sqrt{\frac{6kT}{\rho_p}}\left(\frac{3}{4\pi}\right)^{1/6}\left(\frac{1}{v_i} + \frac{1}{v_j}\right)^{1/2}\left(v_i^{1/3} + v_j^{1/3}\right)^2 \tag{14.18}$$

For equal size particles,

$$K(v, v) = 4\sqrt{\frac{12kT}{\rho_P}}\left(\frac{3v}{4\pi}\right)^{1/6} \tag{14.19}$$

which is a weak function of particle size.

At intermediate sizes, the particle mean free path l_D may approach the size of one or both particles. This regime is described by diffusion Knudsen numbers,

$$\text{Kn}_D = \frac{2\mathcal{D}_{ij}}{\bar{c}_{ij} r_p} \tag{14.20}$$

The flux matching method of Fuchs (1964) can be used to estimate the collision frequency. In this approximate model, noncontinuum effects are assumed to be limited to a regime Δ from the surface. Near the surface particles are assumed to behave as if in vacuum, *i.e.*, to move along ballistic trajectories, whereas outside Δ the particle motion is described by continuum diffusion. This is illustrated schematically in Fig. 14-6. Matching the fluxes at Δ leads to an approximate expression for the flux reaching the surface. Taking into account the finite sizes of the two colliding particles and detailed geometrical arguments, Fuchs (1964) used the flux-matching method to derive an expression for the collision frequency function that is commonly used to describe Brownian coagulation through the transition regime,

$$K_{12} = 2\pi(\mathcal{D}_1 + \mathcal{D}_2)(d_{p1} + d_{p2})\left[\frac{d_{p1} + d_{p2}}{d_{p1} + d_{p2} + 2g_{12}} + (d_{p1} + d_{p2})\right]^{-1} \tag{14.21}$$

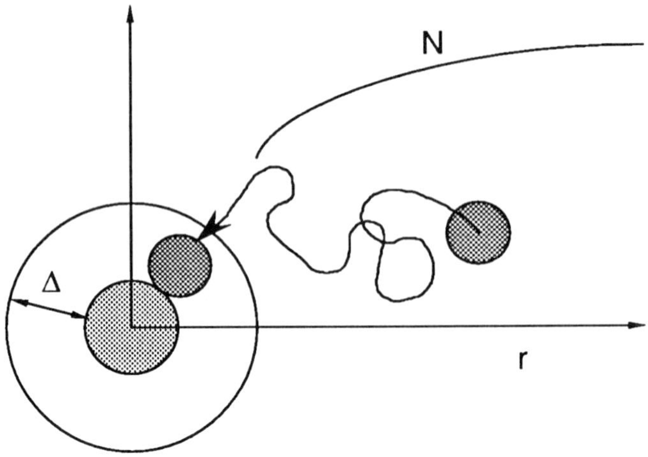

Figure 14-6: Schematic of the approach used to derive approximate collision frequency functions for transition regime particles.

where

$$\mathcal{D}_i = \frac{kT}{3\pi\mu d_{pi}}\left[\frac{5+4\mathrm{Kn}_i+6\mathrm{Kn}_i^2+18\mathrm{Kn}_i^3}{5-\mathrm{Kn}_i+(8+\pi)\mathrm{Kn}_i^2}\right] \tag{14.22}$$

$$g_{12} = \sqrt{g_1^2+g_2^2} \tag{14.23}$$

$$g_i = \frac{1}{3d_{pi}l_i}\left[(d_{pi}+l_i)^3-(d_{pi}^2+l_i^2)^{3/2}\right]-d_{pi} \tag{14.24}$$

$$l_i = \frac{8\mathcal{D}_i}{\pi\bar{c}_i} \tag{14.25}$$

$$\bar{c}_i = \sqrt{\frac{8kT}{\pi m_i}} \tag{14.26}$$

$$\mathrm{Kn}_i = \frac{2\lambda_{gas}}{d_{pi}} \tag{14.27}$$

$$\bar{c}_{12} = \sqrt{\bar{c}_1^2+\bar{c}_2^2} \tag{14.28}$$

Models of the dynamics of nonspherical particles will, necessarily, require an understanding of the mobility of the particles and the separation of the centers of mass at which the two particles come into contact.

A simpler approximation derived using the flux–matching method by Seinfeld (1986) will be useful in examining the dynamics of agglomerate particles

$$\beta = \frac{1+\mathrm{Kn}_D}{1+2\mathrm{Kn}_D(1+\mathrm{Kn}_D)} \tag{14.29}$$

14.4.2 Agglomerate Particles

Particle structure affects the collision frequency function for continuum regime particles in two ways: (*i*) through the collision size $(R_{pi} + R_{pj})$; and (*ii*) through the particle diffusivity $(\mathcal{D}_{ij})$. The influence on non-continuum effects on the coagulation can be taken into account approximately by the simplified interpolation formula given in Eq. 14.29. Thus, a description of the dynamics of agglomerate aerosols can be derived by separately considering the particle diffusivity and effective size. The diffusivity can be calculated from the Stokes-Einstein relationship (Eq. 14.12), given an understanding of the aerodynamics of the particles.

14.4.3 Aerodynamics of Aggregates

Detailed models to describe the aerodynamic drag acting on fractal agglomerates have been developed by Mountain et al. (1986), Chen et al. (1987), and Rogak and Flagan (1991a). The free molecular limit is more relevant to pyrogenous fumes, but it has received less rigorous attention. In the limit of low fractal dimensionality, the drag is proportional to the projected area of the aggregate as postulated by Mountain et al. (1986). For $D > 2$, some of the mass in the interior of the aggregate is shielded from bombardment by gas molecules, so the drag force decreases below that predicted with the simple projected area calculation. Intermediate Knudsen numbers require the development of interpolation formulae to serve the purpose of the slip correction factor. The friction factor can be expected to depend on both D and Kn, so empirical validation of any model is essential to rigorous characterization of the aerodynamics of aggregate particles. It is worth noting that even the definition of the Knudsen number becomes ambiguous for fractal agglomerates since a spectrum of length scales are required to characterize the particles. Hence, new approaches will be needed to characterize the propensity of aggregate particles to slip with respect to the gas in which they are immersed.

Schmidt-Ott (1988) made an important step towards direct measurement of the aerodynamic forces acting on aggregate particles. Fume particles were charged and then classified in terms of their migration in an electric field using a differential mobility analyzer (DMA). The migration velocity of charged aerosol particles can be expressed in terms of their electrical mobilities, the migration velocity per unit of electrical field strength,

$$B_e = \frac{u}{E} = \frac{q}{f} \tag{14.30}$$

Assuming the charge on the particle is known, *e.g.*, one elementary charge per particle (an achievable level at ambient conditions for aerosol particles smaller than about 200 nm diameter), the electrical mobility method provides a direct measure of the friction factor f. Schmidt-Ott determined the mass-equivalent size of the aggregates thus classified by passing an aerosol consisting of mono-mobility aggregates through a furnace to densify the aggregates. A second DMA was then used to determine the electrical mobilities of the resulting dense, spherical particles. Schmidt-Ott went on to argue that the two mobility diameters could be used to estimate the fractal dimensionality of the aggregates by examining the power law dependence of a plot of the aggregate mobility diameter versus the dense mobility diameter. If the fractal dimension is less than 2, Schmidt-Ott's method loses resolution, *i.e.*, straight chain ($D = 1$) and branched agglomerate ($D = 1.8$) particles containing the same number and size of primary particles have nearly the same mobility. In practice even for $D > 2$, one requires a good understanding of the mobility in order to infer structure from mobility measurements (Rogak and Flagan, 1991a). The structure-drag relationship is model dependent, so unambiguous structure determination is not possible.

More recently, we have used the differential mobility classifier to obtain a direct relationship between particle aerodynamics and particle structure. Mono-mobility particles were collected directly on electron microscope grids using a thermophoretic sampling probe in the system illustrated in Fig. 14-7. Image processing was then used to determine the fractal dimensionality from transmission electron microscope pictures of the particles. Two fume sources were employed: (*i*) a rapid mixing reactor in which TiO_2 fume particles were generated by thermal decomposition of titanium tetraisopropoxide vapor at temperatures far below the melting point of TiO_2; and (*ii*) an aerosol reactor in which silicon particles were generated by thermal decomposition of silane.

The agglomerates examined in this study were produced following relatively rapid reaction, so they grew by coagulation of comparably sized particles. This process, called cluster–cluster aggregation, leads to the formation of relatively low density agglomerates with fractal dimension of about 1.8. For these particles, the aerodynamic drag scales directly with the projected area equivalent particle size, as illustrated by comparison of the mobility equivalent and area equivalent diameters in Fig. 14-8. Rigorously valid only in the free molecular regime, this correlation is seen to be quite good throughout the transition regime as well. At low fractal dimensions, the projected area of the aggregate is close to the sum of the projected areas of the constituent primary particles. Thus, this

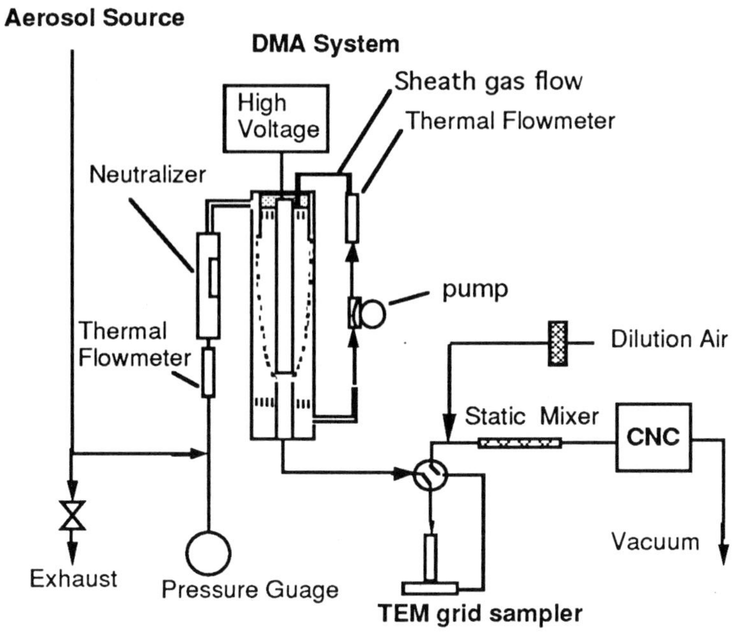

Figure 14-7: Aerosol classification system used to obtain electron microscope pictures of mono–mobility agglomerate particles. Particles were deposited thermophoretically on transmission electron microscope grids.

simple correlation between the mobility and area equivalent sizes of the particles allows one to predict important aspects of the particle dynamics with the limited information that is generally available at present.

14.4.4 Dynamics of aerosol agglomerates.

Early efforts at understanding the dynamics of aggregate aerosols and colloids employed direct simulation, computationally following an ensemble of particles as they undergo a random walk on a lattice or simulated Brownian motion (Meakin, 1990). When two particles come into contact, a bond is assumed to be formed. The structure evolves in time as additional bonds are formed. Diffusion limited aggregation describes the growth of aggregates as individual spheres attach to the growing structure. Cluster-cluster aggregation describes a system in which an ensemble of spheres forms clusters which are allowed to undergo a random walk so that aggregates grow by addition of clusters of a range of sizes. The nature of the random walk has been varied in the model calculations to probe the implications of different growth mechanisms. In an important advance over the idealized treatments that characterized early studies,

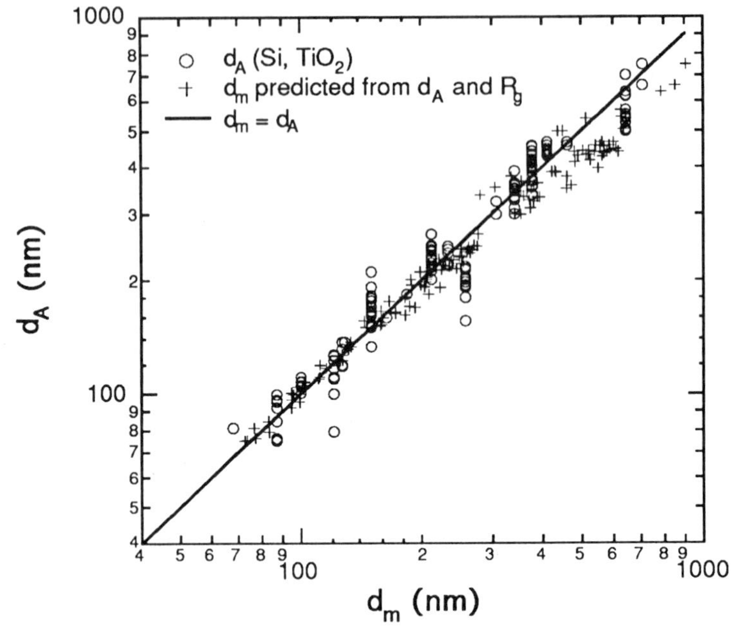

Figure 14-8: Correlation between mobility equivalent diameters and projected area equivalent diameters as determined by analysis of electron microscope images of mobility classified agglomerates.

Mountain et al. (1986) incorporated models of the aerodynamic drag on the growing aggregates that were derived from fluid mechanical models in order to explore the growth of soot and other fume particles. They also made some preliminary steps toward the development of a coagulation equation based on collision frequency expressions derived for aggregate particles. The latter approach is essential if fume aerosol evolution in practical systems is to be understood.

Rogak and Flagan (1991b) incorporated a collision frequency function derived for aggregate particles into a computationally efficient solution to the coagulation equation. This development is outlined briefly below. As a first approximation, the collision frequency function can be related to the diffusivities of aggregate particles through Eq. (7), i.e.,

$$K(D_{eff,i}, D_{eff,j}) = 2\pi(D_{eff,i} + D_{eff,j})(\mathcal{D}_i + \mathcal{D}_j)\beta(\text{Kn}_{Dij}) \qquad (14.31)$$

where effective particle sizes, $D_{eff,k}$, diffusivities, $\mathcal{D}_k$, and the non-continuum correction factor β must be defined for the particular structure under study.

The coagulation problem is usually solved by transforming the real problem into the problem of a point particle diffusing into a stationary absorbing sphere of radius R_{abs}. To find this size for nonspherical particles using direct simulation models, one could average the distance x between cluster centers of mass over many collisions. At very small Knudsen numbers, R_{abs} would equal the ensemble average of x. At high Knudsen numbers, the coagulation of spheres varies as R^2, so R_{abs}^2 is the ensemble average of x^2. In principle, one could use direct, Monte Carlo simulation to determine R_{abs} as a function of cluster size and population, shape, and diffusion mean free path of the particles, or measure coagulation rates directly in carefully controlled experiments. At this point only rough estimates of the collision radius are available.

To compare predictions of the collision frequency functions for agglomerate particles with spheres, it is necessary to develop appropriate scaling. Rogak and Flagan (1991c) define the following dimensionless groups to provide this basis for comparison:

dimensionless group	characteristic scale	
$\tilde{d}_v = \frac{d_v}{d_0}$	$d_0 = 2\lambda$:	mean free path
$\tilde{K} = \frac{K}{K_0}$	$K_0 = \frac{8kT}{3\mu}$:	continuum regime coef.
$\tilde{n} = \frac{n}{n_0}$	$n_0 = \frac{6\phi}{\pi d_0^3}$:	volume equivalent number at $d = d_0$
$\tilde{t} = \frac{t}{t_0}$	$t_0 = \frac{1}{K_0 n_0}$:	collision time
$\tilde{\rho} = \frac{\rho}{\rho_0}$	$\rho_0 = \frac{P\mathcal{M}}{\mathcal{R}T}$:	gas density

The inverse of the particle Knudsen number is the natural scaled particle diameter. In the continuum limit, the collision frequency function for particles of equal size becomes independent of particle size, providing the natural scaling for K. The number concentration follows from that leading to the same volume fraction if all particles had Knudsen number of unity. Time is scaled with the collision time of particles of the characteristic size and concentration, and the density is scaled with the gas density. With these definitions,

$$\tilde{K} = \frac{\tilde{d}_{c1} + \tilde{d}_{c2}}{4}\left(\frac{C_c(\mathrm{Kn}_1)}{\tilde{d}_{m1}} + \frac{C_c(\mathrm{Kn}_2)}{\tilde{d}_{m2}}\right)\beta(\mathrm{Kn}_0) \qquad (14.32)$$

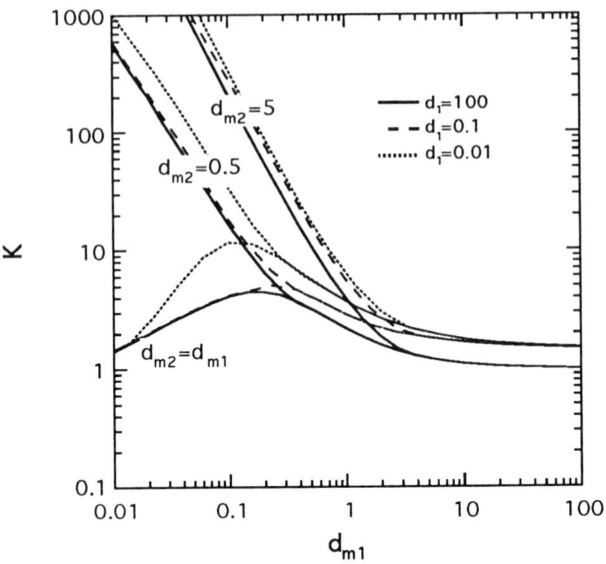

Figure 14-9: Coagulation kernel $\tilde{K}$ for agglomerates with mobility diameters $\tilde{d}_{m1}$ and $\tilde{d}_{m2}$, ($D_f = 1.8$, $\tilde{\rho} = 0.0025$). For primary particle diameter $\tilde{d}_o = 100$, the model is equivalent to the traditional model of rapidly coalescing spheres.

and

$$\text{Kn}_D = \sqrt{\tilde{\rho}} \frac{\frac{C_c(\text{Kn}_1)}{\tilde{d}_{m1}} + \frac{C_c(\text{Kn}_2)}{\tilde{d}_{m2}}}{(\tilde{d}_{c1} + \tilde{d}_{c2})\sqrt{\tilde{d}_{v1}^{-3} + \tilde{d}_{v2}^{-3}}} \tag{14.33}$$

where $C_c(\tilde{\text{Kn}}_i)$ is the slip correction.

Using these scalings, Figure 14-9 shows the variation of the $K(d_{pi}, d_{pj})$ with particle size for aggregates of fractal dimensionality $D = 1.8$, a value typical of particles grown by cluster-cluster aggregation and typical of the values determined experimentally for pyrogenous fumes. For comparison, the results for spherical particles of uniform density are shown in Fig. 14-10. Two densities are shown for the spheres, a high density typical of the materials from which the particles are comprised, and a lower density that is comparable to typical agglomerate particles. A comparison of these two figures reveals that much of the difference between the coagulation rates of agglomerates and dense particles is due to the lower density of the agglomerates and their correspondingly higher collision cross sections.

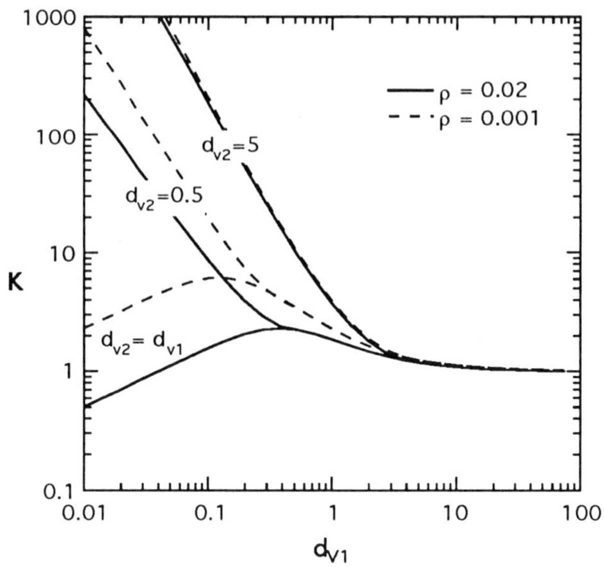

Figure 14-10: Coagulation kernel $\tilde{K}$ for spheres of diameter $\tilde{d}_{o1}$ and $\tilde{d}_{o2}$, and having material density $r\tilde{h}o$. Quantities are non-dimensionalized using the characteristic scales of the background gas.

14.4.5 Particle Coalescence

The model predictions presented above treat the coalescence of aggregate particles in a very simplistic manner. The final component in a comprehensive description of the evolution of aggregate particles will be a model of the morphological changes that occur as the particles coalesce in the hot combustion environment. This requires an understanding of the rates of coalescence or sintering of two spheres in contact with one another, a problem that has been described in the ceramics literature (Kingery et al., 1976). Coalescence of agglomerating aerosols has identified as a rate controlling process determining the fine structure size in flame synthesized silica (Ulrich and Subramanian, 1977; Ulrich and Riehl, 1982). More quantitative predictions have been made using sintering theory by Helble and Sarofim (1989) in a study of coal ash fumes. By changing the composition of the fume particles to alter the viscosity, the role of particle coalescence was unambiguously demonstrated. Koch and Friedlander (1991) have used sintering models to describe the evolution of an aerosol by following the change in the surface area of particles as they coalesce. In spite of these advances, no model presented to date fully captures the complexity of the rearrangements of agglomerate particles as they coalesce or sinter together.

Wu (1986) observed that particles retain their apparent agglomerate structures as they coalesce. Silicon particles were produced by silane pyrolysis at low temperature. The highly agglomerated particles were diluted to suppress further agglomeration and then heat treated while still entrained in the inert carrier gas. The densified particles were collected on filters and examined by electron microscopy and BET gas adsorption methods. Both of these approaches probe the microstructures of the particles. Because of the way the particles were collected and processed, no information about the agglomerate structure is available. Comparison of the agglomerates before and after heat treatment, see Fig. 14-11, reveals particles comprised of roughly spherical primary particles of fairly uniform size, although that size increased dramatically with heat treatment. Figure 14-12 shows how the primary particle size increased with heat treatment temperature. The fact that the agglomerate structure is preserved while the "primary particle" mass increases by more than an order of magnitude raises serious doubts on the use of micrographs or fractal dimension to infer the growth mechanisms of pyrogeneous fumes. Efforts to reconcile these observations with sintering theory have only recently met with qualitative success, due primarily to the limited data on the solid phase transport properties available for this material. Hence, a comprehensive picture of the structural evolution of a coagulating, sintering aerosol are not possible at this time.

The coalescence of agglomerate particles has been found to be a critical step not only in the production of dense particles in high source rate reactors, but also in the growth of seed particles by what was originally thought to be vapor deposition. Wu and Flagan (1987) showed that particles grown at essentially constant number concentration but at temperatures well below the melting point of the material (silicon in this case) can develop a microstructure that is consistent with particle growth by diffusion limited aggregation of small clusters to the growing particle. By heating the particles to the melting point, the low density agglomerates coalesced to form massive particles, e.g., a 5 μm sphere was shown.

In later experiments, Wu et al. (1987) produced dense spherical particles by silane pyrolysis in a single stage reactor. Nguyen and Flagan (1991) subsequently demonstrated that coagulation played a major role in the growth of the fine nuclei produced in that experiment, and that the dense particles were produced in the final high temperature region of the reactor. Particles produced at very low reactor temperatures were clearly agglomerates, having a mean primary particle diameter of 38 nm, as measured using transmission electron microscopy and the BET gas

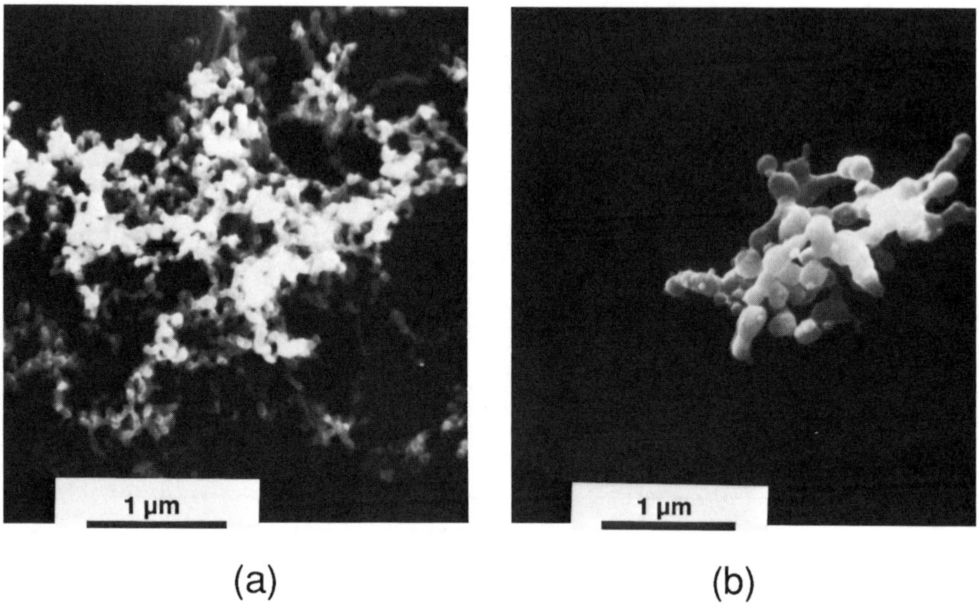

Figure 14-11: Scanning electron microscope pictures of (a) particles produced by low temperature pyrolysis of silane, and (b) those that resulted from heat treatment of the agglomerates for 4s at 1173K.

adsorption method.

Densification studies were conducted by passing the aerosol through a hot reactor stage. Within a fixed residence time at elevated temperatures, the particles underwent varying degrees of densification as evidenced by an increase in the size of the apparent primary particle size in the particles collected downstream of the heat treatment reactor. The primary particle size was measured by electron microscopy and by the BET gas adsorption method. The primary particle radius increased from 30 nm to 150 nm as the heat treatment temperature was increased from 1100K to 1400 K. We have used sintering models to examine the observed coarsening as discussed below.

Solid particles in contact with one another can interdiffuse by several different mechanisms under the driving force of the excess surface free energy in the particle system. Surface curvature variations create corresponding differences in pressure on neighboring surfaces, gradients in chemical potential, and differences in local surface tension. The total surface energy of the particles decreases as the particles change their shape by rounding sharp edges, or by developing facets with lower energies. Necks form between particles as low energy grain boundaries are

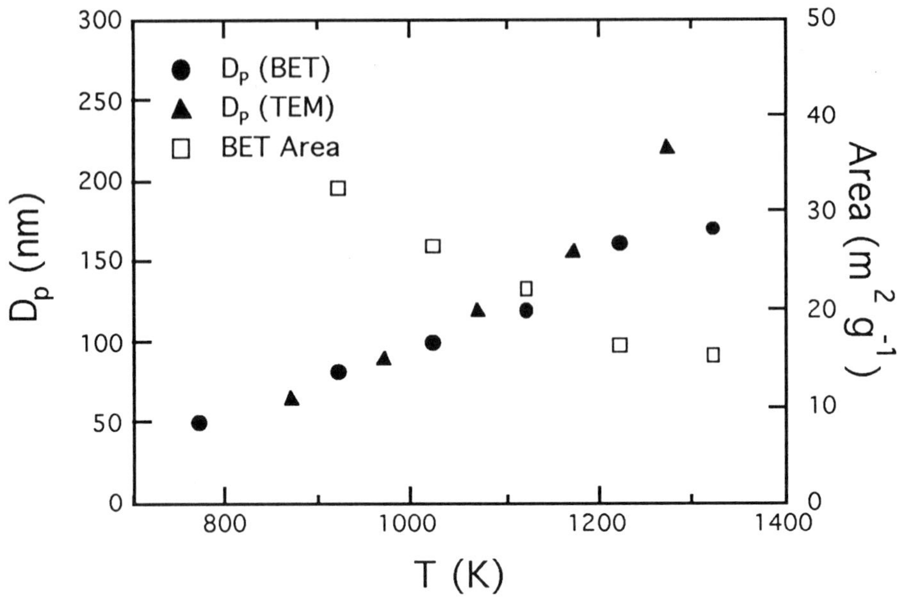

Figure 14-12: BET surface area and apparent primary particle sizes inferred from BET surface area measurements and TEM micrographs as a function of heat treatment temperature for silicon particles produced by low temperature pyrolysis of silane and heat treated at the indicated temperatures *in situ*.

substituted for surface area. Throughout the early stages of sintering, the grain boundary is pinned between the two crystallites. Eventually, however, the particles merge to the point that the free energy of the particle system is reduced by migration of the grain boundary through one of the crystallites, unpinning the grain boundary and allowing the final consolidation of the spheres (Bernholc et al., 1987). As a result, the multi-particle system will tend to grow together into a single particle. The shape of this final particle can range from a loosely connected, low density agglomerate to a dense round sphere. The shape of this final stage depends upon the sintering behavior during heating.

The mechanisms of material transport involved in this rearrangement include viscous flow, evaporation and condensation, volume diffusion, surface diffusion, and grain boundary diffusion. Theories of sintering of particulate systems have been developed by using simple geometries to describe the growth of the neck between the spheres, usually in the form of a power law:

$$\left(\frac{x}{a}\right)^n = \frac{Bt}{a^m} \tag{14.34}$$

where x is the radius of the neck, a is the original radius of the sphere, t is the time, and B is a temperature dependent constant incorporating physical and transport properties. The exponents n and m depend on the dominant sintering mechanism (Frenkel, 1945; Kingery and Berg, 1949; Kuczynski, 1949; Johnson, 1969). The precise form of these rate laws has been the subject of some debate, with many investigators making different assumptions. The power law relations between neck growth and time are constrained by the geometrical assumptions needed to obtain them and are generally not applicable to ratios of x/a greater than about 0.3.

A number of studies of the silicon, SiC, and Si_3N_4 systems have been reported (Coblenz,1990; Greskovich and Rusolowski,1976; Shaw and Heuer, 1983). Vapor phase transport was identified as the dominant mechanism in early studies, but it appears that oxide contamination may have led to the formation of SiO. (Shaw and Heuer, 1983; Coblenz, 1990) Other studies identified surface diffusion as the dominant mechanism of Si sintering, although oxide contamination clouds the interpretation of many of the experiments.

To understand the dynamics of aggregation in aerosol systems, including the growth of particles by cluster deposition at temperatures far below the melting point, a description of agglomerate sintering that rigorously describes the entire process is needed. Not only must the geometrical constraints of the early stage sintering models be relaxed, the combined effects of the various mechanisms must be taken into account. Better experiments are also needed to validate the theory, ideally this would involve the sintering of doublets of equal-sized particles. After developing and validating the theory for this idealized model system, extension to more complex aggregate particles can be undertaken with confidence. Although silicon has been an interesting proving ground, materials for which better property data are available need to be investigated. Experiments with different materials are needed to allow the dynamics of aggregate particles that sinter by different mechanisms to be explored.

14.4.6 Condensation and Evaporation

The diffusion of vapor species to the surfaces of fine aerosol particles has received considerable attention, although most attention has been given to spherical particles. For particles that are large compared to the mean-free-path of the gas molecules, λ_{AB}, vapor transport to the surface of a spherical particle is described by the diffusion equation,

$$\frac{\partial c_A}{\partial t} = \frac{\mathcal{D}_{AB}}{r^2}\frac{\partial}{\partial r}\left(r^2\frac{\partial c_A}{\partial r}\right) \tag{14.35}$$

where c_A is the vapor concentration, and $\mathcal{D}_{AB}$ is the binary diffusivity of vapor species A in background gas B. Equation (3) can readily be solved subject to vapor equilibrium at the particle surface and specified vapor concentration at r for the case of simple vaporization or condensation. From this solution, the vapor flow becomes

$$J_{Ac} = 4\pi r_p \mathcal{D}_{AB}(c_{A\infty} - c_{As}) \tag{14.36}$$

In the free molecular regime where the particle is much smaller than the mean free path, the molecular flux to the particle surface is the so-called effusion flux, $\frac{N\bar{c}_A}{4}$, where $\bar{c}_A = \sqrt{\frac{8kT}{\pi m_A}}$ is the mean molecular speed of species A. The net vapor flow then becomes

$$J_{Ak} = \pi r_p^2 \bar{c}_A(c_{A\infty} - c_{As}) \tag{14.37}$$

Approximate models have been developed to describe the mass transfer for particles in the intermediate or transition regime. It is convenient to characterize the particle size in terms of the Knudsen number, $\mathrm{Kn} = \frac{2\lambda}{d_p}$, the ratio of the mean-free-path of the gas molecules to the particle radius. The concentration of the diffusing vapor molecules and those of the background gas around a particle in the transition regime of Knudsen numbers, *i.e.*, $0.1 < \mathrm{Kn} < 10$, is rigorously described by the Boltzmann equation. Since no general solution to the Boltzmann equation is valid over the full range of Knudsen numbers for diffusing species with arbitrary masses, approximate models of the mass transfer are generally used to describe the transition regime. A commonly used interpolation formula for mass transfer in the transition regime is that derived by Fuchs and Sutugin (1971), using the so-called flux matching method, /em i.e.,

$$J_A = J_{Ac}\beta_{FS} \tag{14.38}$$

where

$$\beta_{FS} = \frac{1 + \mathrm{Kn}}{1 + 1.71\mathrm{Kn} + 1.33\mathrm{Kn}^2} \tag{14.39}$$

The derivation of this expression is based on a conceptual picture of the transport processes in which non-continuum effects are limited to a region beyond the particle surface, $d_p/2 \leq r \leq \Delta$. (Seinfeld, 1986)

These results are strictly valid only for spherical particles. To describe the condensation or adsorption of vapors on aggregate particles will require new solutions of the vapor transport problem. In the transition regime, experimental results will be required to validate any models developed. Theoretical investigations of the mass transfer to agglomerate particle in the transition and free molecular regimes have been undertaken by Hagenloch and Friedlander (1989). The effectiveness of the agglomerate surface for mass transfer was found to decrease dramatically with increasing cluster size, although it should be noted that the simulations were carried out for agglomerates of fractal dimension about 2.5. This is higher than typically observed for particles grown following a rapid burst of nucleation. Rogak et al. (1991) measured the mass transfer to mobility classified agglomerate particles with fractal dimensions of about 1.7 to 2.1. The mass transfer equivalent diameter was found to correspond closely to the mobility equivalent diameter, as might be expected for relatively transparent agglomerates.

14.5 MEASUREMENTS

One of the major differences between aerosol science and colloid science is the existence of high resolution instruments for direct measurements of particle size distributions in aerosols. Aerosol instruments have been the number of numerous reviews, e.g. (Pui and Liu, 1988) so only a few of the key features will be addressed here. The relatively low number concentrations typical of aerosol systems make single particle counting and sizing possible in aerosols where it would not be practical for many colloidal systems. For particles below about 1 μm in size, electrical mobility analysis is the method of choice, providing extremely high resolution on particle size and, with recent innovations (Wang and Flagan, 1990) time resolution below one minute. In the typical electrical mobility classifier, the aerosol is charged by exposure to bipolar gas ions, producing a steady-state charge distribution that approximates the equilibrium Boltzmann distribution. Particles are introduced near the outer wall of a coaxial electrostatic precipitator, with a particle-free sheath of clean carrier gas being introduced between the two cylinders. An applied electric field drives the particles toward the inner cylinder as they are advected along the length of the precipitator. At the downstream end of the central collection rod, a small sample flow is extracted through small holes. Particles with mobilities in a narrow range are extracted with that flow and carried to an electrometer to measure the transmitted charge, to a single particle counter (typically a condensation nuclei counter), or to another experimental systems if studies of properties of mono-sized particles are to be performed. Electrical mobility methods provide a unique determination of size for aerosols because the probability of charge transfer is very low, and below about 200 nm diameter, most particles can be made to acquire at most one charge. When an electric field is imposed on such an aerosol, the particles are subject to a known force, so the aerodynamic resistance to motion can be determined uniquely. With care, particle sizing within 10% on diameter is feasible with an instrument known as the differential mobility analyzer. Agglomerate aerosols introduce uncertainty in the interpretation of such mobility measurements, but recent experimental studies provide the necessary data base to guide interpretation of the experiments, at least for fractal dimensions below about 2.

One important feature of the mobility classifier is the ability to characterize particles of known mobility. Numerous investigators have capitalized on this feature to infer such properties as vapor pressure of the particulate material (Rader and McMurry, 1986), optical signature (discriminating between soot and nonabsorbing atmospheric particles) (Covert et al., 1990), and particle structure (Schmidt–Ott, 1988; Rogak and Flagan, 1991c).

Particles larger than several hundred nanometers in diameter tend to acquire multiple charges, making unambiguous interpretation of mobility data impossible. However, the large density difference between gases and aerosol particles make aerodynamic classification possible. Several different instruments characterize particles in terms of the Stokes number, the ratio of the aerodynamic stopping distance of a particle moving with an initial velocity U to a length characteristic of the instrument,

$$\mathrm{St} = \frac{L_{stop}}{L} = \frac{\rho_p U d_p^2}{18\mu L} C_c(\mathrm{Kn}) \tag{14.40}$$

Inertial classifiers include the cascade impactor in which particles are deposited on plates by an impinging jet flow, and the aerodynamic particle sizer which makes time of flight measurements of the ability of aerosol particles to equilibrate with a gas flow after a rapid velocity change. Most aerodynamic instruments classify particles larger than about 100 nm diameter, but smaller sizes can be resolved by operating the instrument at reduced pressure (taking advantage of particle slip) or using very small flow dimensions (the micro-orifice impactor of Marple and coworkers (Marple et al., 1981). By combining reduced pressure operation with hypersonic flow velocities, the size range can be extended to molecular levels. The size resolution of aerodynamic instruments built to date does not approach that of the differential mobility classifier, although there is no fundamental reason why a high resolution aerodynamic aerosol classifier could not be built.

Single particle optical analysis is possible, as are optical measurements based on the scattering properties of the cloud of aerosol particles. The cloud scattering measurements require assumptions about the nature of the particle size distribution to be made before the data can be interpreted, a major weakness in attempts to characterize the aerosol in detail. Optical instruments do, however, have unique abilities for *in situ* characterization of aerosol particles. In situ optical size estimates, while lacking the resolution of extractive analysis, do eliminate sampling biases that can be particularly perplexing in the study of high temperature aerosols in the presence of condensible vapors. Small angle light scattering provides estimates of the structure parameters for agglomerate particles *in situ.*

14.6 AEROSOL TECHNOLOGY

The production of commodity chemicals such as carbon black and TiO_2 using aerosol reactors is an established technology. Millions of tons of submicron particles are produced each year by these flame-based aerosol reaction technologies. Major efforts have been invested into the synthesis of more exotic materials using aerosol reactors, with one notable success. Optical waveguides have revolutionized communications systems in recent years. These waveguides are small glass fibers, typically 50 μm to 200 μm in diameter in which the refractive index is varied with radial position in a precisely controlled manner. Such fibers were first produced by a process mis-labeled as modified chemical vapor deposition (MCVD). As in chemical vapor deposition used in microelectronic device fabrication, vapor precursors are used in MCVD to grow a substrate known as a preform. Vapor deposition makes only minor contributions to preform growth, however. Instead, aerosol particles that are produced by burning precursors such as $SiCl_4$ are deposited on the preform through a process known as thermophoresis. Other applications of aerosols to the processing of advanced materials that have received considerable attention, but less commercial success to date, include the synthesis of powders for advanced ceramics. A variety of reactor types have been proposed, with plasma and laser driven reactions receiving much of the attention. Less exotic synthesis technologies have, however, demonstrated advantages for special applications. Aerosols have contributed to the synthesis of some novel materials that exhibit properties quite distinct from those of the bulk material. A particularly exciting example are the so-called *nanophase* materials. Having microstructures on the nanometer scale, these metastable materials have been synthesized by volatilization of metals into a low pressure gas. While extremely inefficient, this gas condensation method has produced materials with unusual hardness and ductility. In this section, some of the technological applications of aerosols to the processing of materials will be examined briefly.

14.6.1 High source rate aerosol reactors

A number of technologies produce fine particles from vapor precursors. Flame synthesis, as noted above, is responsible for substantial production of fine particles for pigment and structural applications. Plasma and laser driven reactions are, from a particle formation perspective, similar technologies, operating at different temperatures and, therefore, having some differences in the product characteristics. All of these systems involve rapid production of a condensible species. High number concentrations of very fine particles are generated in a brief pulse of nucleation that follows the rapid gas phase reaction. Particle growth then proceeds by coagulation and vapor deposition, with coagulation generally dominating.

While most development of high source rate aerosol reactors has been empirically based, Landgrebe et al. (1990) have recently attempted to derive general design principles for such reactors theoretically. Using a version of Wu and Flagan's (1988) discrete-sectional solution to the general dynamic equation for aerosols, they simulated particle growth under isothermal conditions assuming that the particles are dense spheres and that the precursor reactions are first order, presenting their results in the form of the nomographs shown in Fig. 14-13. Their results are presented in terms of a dimensionless reaction time

$$\theta = \frac{\text{reaction time}}{\text{collision time}} = \frac{\tau_r}{\tau_c} = \frac{\sqrt{\frac{6kTr_1}{\rho_p}}}{k_{rxn}} \qquad (14.41)$$

and a dimensionless reaction time

$$X = \frac{t}{\tau_c} \qquad (14.42)$$

where the characteristic coagulation time is defined in terms of the collision frequency of the total monomer concentration that would result if all of the precursor were reacted instantaneously, and the reaction is assumed to be first order. The mean particle size, Fig. 14-13a, increases monotonically for small θ (fast chemical reaction). Monomers are continuously produced during slower reactions (large θ), and the total number concentration remains low until large X, so the mean particle size remains small. The geometric standard deviation σ_g rapidly approaches the asymptotic value of the so-called *self-preserving particle size distribution* (Lai et al., 1972) ($\sigma_g = 1.6$) for reaction times below $\theta = 0.25$. At long residence times (X), the distribution narrows from the asymptotic value for the free–molecular regime to that expected for the continuum regime. Slower reactions lead to broader particle size distributions due to the simultaneous formation of new particles and particle growth by coagulation. Such calculations are a useful first step toward generalizing the design of aerosol reactors. The implicit assumption that the particles are dense throughout the growth process is, however, infrequently satisfied in high temperature aerosol reactors, particularly when coagulation dominates (small reaction time θ).

Dense particles can be grown by coagulation, provided it takes place at sufficiently high temperature that the particles are liquid and coalesce rapidly after coagulation. Plasma reactors or oxygen enhanced flames routinely operate at such high temperatures. More commonly in the synthesis of refractory materials, the particles do not fully coalesce, but rather form low density agglomerates. The aerosol dynamic models can still be applied in this case, however, the collision frequency function must be modified to account for the larger section and lower mobility of the agglomerate particles.

Agglomerates of some particulate materials can be densified by heating the aerosol to sufficiently high temperature. As will be shown later, heat treatment of agglomerate particles results in gradual increase in the fine structure scale (the size of the apparent primary particles), with the time required to fully densify a particle increasing dramatically with structure size. For high melting point materials, only fine structures below a threshold size can be erased by thermal treatment. Hence, there is an upper bound to the size of fully dense particles that can be produced in systems that use coagulation as the primary growth mechanism.

Dense particles of silicon carbide and silicon nitride have been produced by rapid pyrolysis of silane with ammonia or hydrocarbons at modest temperatures, but this has required a two stage reaction process. (Aoki et al., 1987) In the first stage, silicon particles are generated by nu-

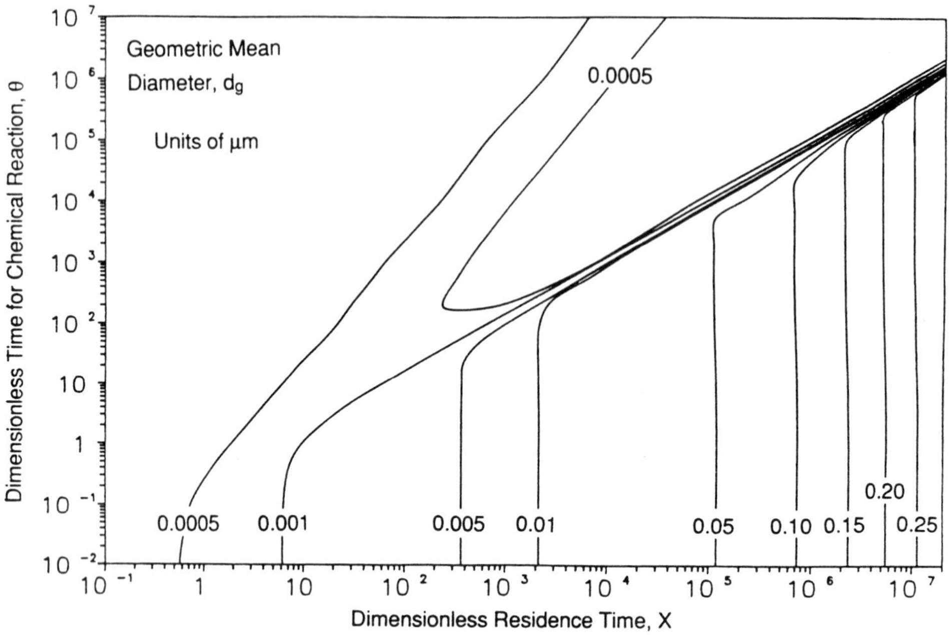

(a)

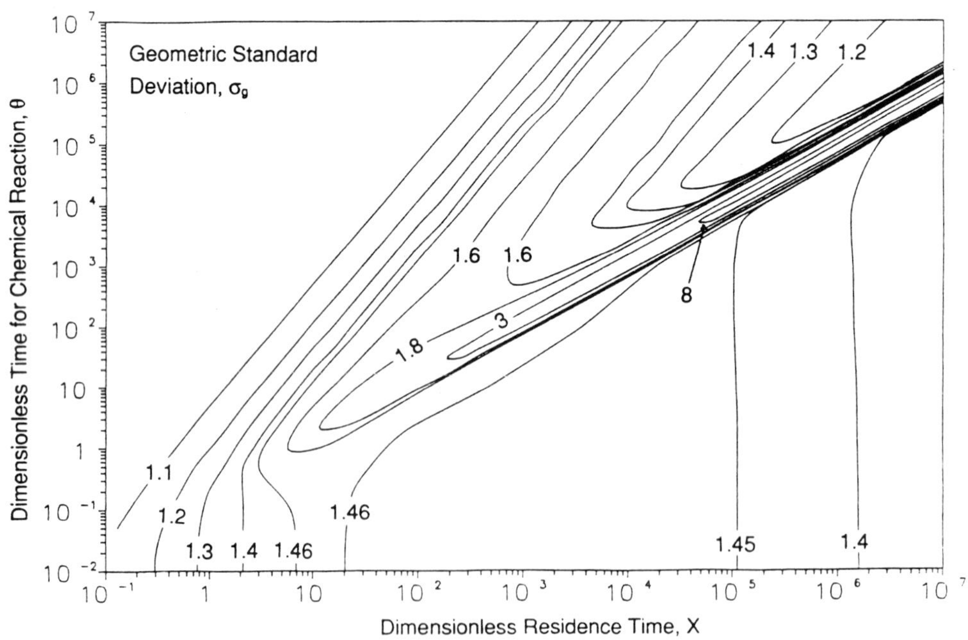

(b)

Figure 14-13: Nomographs for particle growth in isothermal aerosol reactors: (a) geometric mean diameter; (b) geometric standard deviation. Reprinted by permission from Pergamon Press.

cleation and coagulation of the products of silane pyrolysis. By delaying the exposure of the particles to the nitrogen or carbon source, advantage is taken of the low melting point of silicon. Dense particles of the desired ceramic are produced by the reactions of ammonia or hydrocarbons with the silicon particles. Attempts to produce these materials by simultaneous addition of the reactants have yielded highly agglomerated particles.

14.6.2 Particle Growth by Vapor Deposition

Agglomerate formation can be minimized by limiting the number concentration of particles in the reactor and operating under conditions that favor growth by vapor deposition rather than by coagulation. However, even very slow reactions can lead to sufficient supersaturation of nonvolatile materials that homogeneous nucleation will occur. The rate of reaction must be severely limited in such systems.

This alone is not sufficient to preclude the formation of low density agglomerates. In growth of seed particles by silane decomposition under conditions that lead to no net increase in the number of particles in the reactor, low density particles have been observed nonetheless (Wu and Flagan, 1987). That the agglomerate shown in Fig. 14-14a was generated in the aerosol reactor and was not an artifact of the sample collection was proven by densifying the particle by heating such particles above the melting point of silicon *in situ*. The resulting dense structure is shown in Fig. 14-14b. While coagulation of like sized particles was suppressed, the particle structure suggests that growth proceeded by coagulation of the growing seed with clusters that are sufficiently large that they retain their integrity after depositing. The extremely fine microstructure of the flocs was attributed to particle growth by deposition of clusters smaller than 4 nm diameter at temperatures so low that the clusters could not fuse with the substrate before being covered with additional clusters.

To grow dense particles in an environment where cluster deposition contributes significantly, sintering kinetics must be taken into account. If clusters begin to accumulate on one another instead of coalescing into the growing seed, agglomerate particles can be expected. On the other hand, if a deposited cluster coalesces into the particle on which it has deposited before another cluster strikes it, dense particles can be expected. Thus, growth of refractory particles by vapor deposition requires a fundamental understanding of the nature of the clusters produced as particles grow. Detailed aerosol dynamic models, such as the discrete continuous model of Gelbard (1979), or the discrete sectional model (Wu and

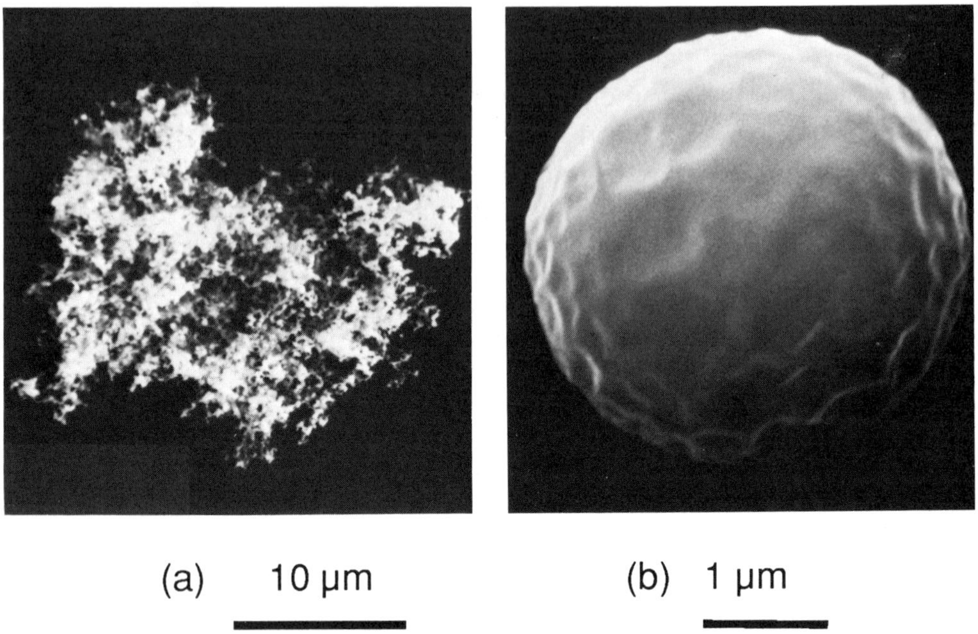

Figure 14-14: SEM picture of large agglomerate particle grown at constant number concentration in a reaction–rate–constrained aerosol reactor (a) as grown; and (b) after *in situ* densification.

Flagan, 1988) provide the necessary descriptions of the smallest clusters, although classical nucleation theory may be adequate in some cases.

An alternative approach that has the potential to grow large particles from vapor phase precursors is to employ vapor deposition as the primary growth mechanism. (Alam and Flagan, 1986) This may entail the production of a species that will condense or may proceed via surface reactions. Operating temperatures can be dramatically lower than reactors that must melt the product particles, providing a significant energy efficiency advantage. Because the diffusivities of vapor species are very much higher than those of even small particles, particles can be grown to very much larger sizes by this route than by coagulation of like-sized particles. To avoid agglomerate formation, the number concentration must be kept below that level which would lead to significant coagulation. Vapor deposition on the growing particles reduces the concentration of condensible reaction products, thereby reducing the tendency to form new particles by homogeneous nucleation.

Because the particle concentrations must be kept low to suppress coagulation, and because the condensible species produced during ceramic powder synthesis have very low equilibrium vapor pressures, the reaction

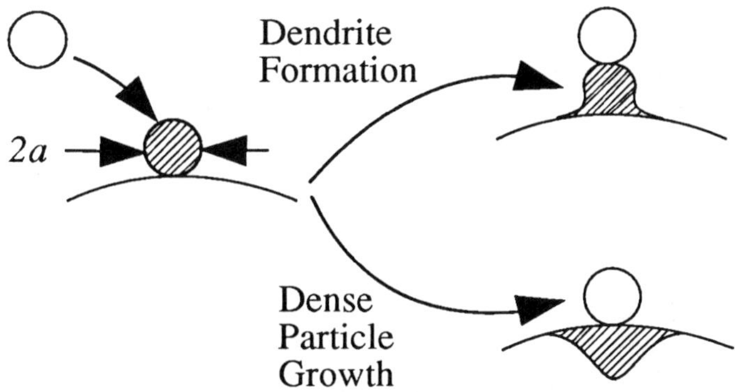

Figure 14-15: Formation of dense particles or dendrites in seed particle growth by cluster deposition.

rate must be kept low throughout the growth process. As the particles grow, however, the rate can be accelerated. Growth of seed particles to sizes well in excess of 1 μm has been achieved in a residence time of only a few seconds, and with minimal new particle formation (Alam and Flagan, 1986). The mass throughputs of laboratory reactors employed to date have been small. Still, the suitability of the powders generated for ceramic synthesis has been demonstrated in the production of silicon nitride by reaction bonding of silicon powders produced by silane pyrolysis in a single stage aerosol reactor (Gregory et al., 1987).

Subsequent studies demonstrated that the assumption that the particles grow by vapor deposition alone is not valid (Wu and Flagan, 1987). Instead, monomer deposition is augmented by the deposition of small clusters, greatly enhancing the growth rate but also raising the possibility of growing low density flocs when the clusters do not diffuse into the seed particle sufficiently rapidly. Thus, even when particles are grown by vapor deposition, the solid phase diffusion processes involved in densification must be taken into account.

Before high quality ceramic powders can routinely be synthesized in rate-constrained aerosol reactors, the fate of molecular clusters that are formed in the gas phase must be understood. Dendrite growth can be envisioned as starting when a cluster deposits on a previously deposited cluster that has not fully coalesced into the seed particle as illustrated in Fig.14-15. The mean time between arrivals of clusters into contact with a previously deposited cluster is

$$\tau_a = \frac{N_c \bar{c}_c}{4} \pi D_c^2 \tag{14.43}$$

where N_c is the cluster number concentration, $\bar{c}_c$ is the mean thermal speed of the clusters, and D_c is the cluster diameter. If $\tau_{coalescence}/\tau_a <<$ 1, one may reasonably expect that dense particles will result.

14.6.3 Nanophase Materials

The exciting properties of nanophase materials have been demonstrated using materials generated by the so-called gas–condensation method which is derived from a technique that has long been used as a source for small clusters for studies of cluster physics and chemistry. Material is vaporized into a cool low-pressure gas, and diffusion of the vapor into the cool gas leads to rapid homogeneous nucleation. Operating pressures of 0.1 – 10 Torr suppress particle growth by Brownian coagulation, making fine particle synthesis practical. The product particles can then be collected on a cool surface by thermophoresis.

Granqvist and Buhrman (1976) explored the influence on particle properties of vapor source temperature, background gas pressure, and composition. Gleiter and coworkers (Birringer, 1988; Birringer et al., 1984) adapted the gas-condensation method for production of nanophase metal and metal oxide powders. Key changes from the system of Granqvist and Buhrman that facilitated production of consolidated materials include: (*i*) changes in the geometry of the thermophoretic collector to simplify the transfer of powder to the consolidation system without exposure to atmospheric contaminants; (*ii*) *in situ* reaction to produce compounds such as oxides rather than elemental forms; and (*iii*) *in situ* consolidation. Their system is illustrated in Fig. 14-16. A small heater is used to vaporize pure metals, with oxides being formed from collected powders. The particles are collected on a liquid nitrogen cooled "cold finger". The collected particles are scraped from the cold finger after the chamber has been evacuated. They fall into a collection funnel and are transferred to an apparatus where the nanophase powder is compacted under high vacuum. *In situ* consolidation is thought to prevent adsorption of gases that might inhibit sintering, and may effect the grain boundary integrity.

Palladium particles that we have generated with a gas-condensation system reveal that the particles are sintered together into low density agglomerates. X-ray diffraction line broadening and dark field transmission electron microscopy show that the crystallite size is on the order of 6–7 nm. When nanoparticles are characterized using electron microscopy or x-ray diffraction, only the sizes of the crystallites that comprise the agglomerate particles are determined. Reported densities for nanophase

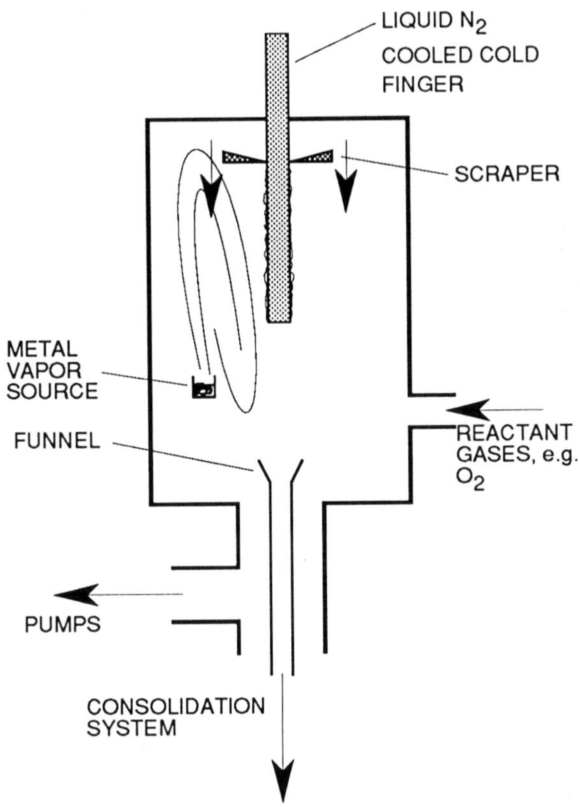

Figure 14-16: Schematic of the gas-condensation reactor of Gleiter and coworkers.

materials that have been consolidated at room temperature range from about 45% to 95%, with the smaller values for ceramics and the larger ones for metals (Siegel et al., 1988). One possible cause for the low densities of the compacted materials is the agglomerate nature of the as-synthesized nanoparticles. These agglomerates form by coagulation at sufficiently high temperatures that the particles partially sinter.

There is not yet any fundamental theoretical description of the processes that govern particle evolution in the gas condensation method. The flow fields in the gas condensation systems used to date are extremely complex. Buoyancy driven vortical flows, evident by the deposits on the cold finger by Eastman and Siegel (1990) (see Fig. 14-16), can readily be eliminated by redesigning the vapor source and particle collection systems. The ideal nanophase powder synthesis reactor would be a limited-volume reactor with clearly-defined flow fields.

The gas condensation method has proven effective for the synthesis of nanocrystalline powders in small quantities. It is, however, far from ideal for scaling up to larger production rates for several reasons, namely:

- The batch nature of the process,
- The high temperature of the vapor source,
- The inhomogeneities of the gas compositions and temperature field.

Applications of the methods of aerosol science to the design of reactors for the synthesis of nanophase materials should yield dramatic improvements in throughput, efficiency, and control of the particle properties. As with the ceramic powders, control of agglomeration may be a critical issue.

14.7 SUMMARY AND CONCLUSIONS

The physical processes that govern the evolution of aerosols have been reviewed. Aerosol dynamic theory is well established for the special case of particles that remain dense spheres throughout the growth process, but even here many approximations are required and questions remain. In many aerosol systems, particles range in size from much smaller than the mean-free-path of the gas molecules (free molecular regime) to much larger (continuum regime). While rigorous treatments of aerosol dynamics are available for the free molecular and continuum regimes, only approximate models are available for intermediate sizes. Those approximations have been validated by comparisons with experiments, but extension to nonspherical particles are questionable.

Progress has been made in the theoretical description of nonspherical particles, but much remains to be done. In many high temperature systems, particles are formed by nucleation from the vapor phase and then evolve through the combined action of coagulation, vapor deposition (condensation or surface reactions), and coalescence. When coalescence is slow, agglomerate particles result. Observations that such particles exhibit similar structures has led many investigators to probe the agglomeration process. Agglomerate particles have been shown to arise from accretion due to random collisions between particles. Much has been said about the influence of the agglomeration mechanism on the resulting structures. The particles are said to be *fractal*, and the fractal dimension is determined by the growth mechanism. Particles grown by addition of individual primary particles tend to be denser than those grown by coagulation of clusters of comparable size. Since most aerosol agglomerates arise in high temperature systems, structural rearrangements may occur through sintering. We have seen that, even after sintering has increased the primary particle size of agglomerate particles dramatically, the particles can appear to be aggregates comprised of larger primary particles than those present at the time of aggregation. The fractal dimension can be expected to increase during this densification, clouding the relationship between the fractal dimension and the agglomeration mechanism. The structural rearrangements that accompany sintering have not yet been modeled in detail, in part due to the paucity of data on the transport properties for those materials that have been studied experimentally. Still, the fractal dimension is a useful tool for describing agglomerate aerosols. Preliminary efforts have been made to derive collision frequency functions and mass transfer rates for agglomerate aerosols as a function of particle fractal dimension and size, but experimental studies of agglomeration kinetics remain to be performed.

A number of fundamental processes that are critical to particle formation and growth have not been treated in this chapter. Homogeneous nucleation of new particles directly from the vapor phase has not been addressed since that is a major topic in its own right. Similarly, interparticle forces have not been discussed. Electrostatic effects are particularly important since they offer one approach to control of the agglomeration process and may influence particle structure. The use of additives to control particle characteristics

The increasing use of aerosols in the processing of advanced materials makes a fundamental understanding of factors that govern particle structure crucial. The production of commodity chemicals by aerosol techniques clearly demonstrates that aerosol technologies can be commercially viable. The new applications that are being explored place increasingly severe constraints on the control of the properties of the product particles. Not only must size be controlled, but the formation of hard agglomerates must be limited, and the composition and phase of the particles must be controlled. Interest in materials with nanometer–scale microstructures makes it critical that the methods of aerosol science be extended to clusters approaching molecular dimensions.

ACKNOWLEDGEMENTS: This work was supported by the National Science Foundation under Grant Number CBT-8813006.

REFERENCES

M. K. Alam and R. C. Flagan. "Controlled Nucleation Aerosol Reactors: Production of Bulk Silicon." *Aerosol Sci. Technol.*, **5**, 237 (1986).

M. Aoki, J.H. Flint, and J.S. Haggerty. "Synthesis of Non-Agglomerated

Si_3N_4 Powder." *Ceramic Trans.*, **1A**, 253-260 (1988).
J. Bernholc, P. Salamon, and R. S. Barry. "Annealing of fine powders: Initial stages and grain boundary motion." In S. N. Khanna P. Jena, B. K. Rao, editor, *Physics and Chemistry of Small Clusters.* NATO ASI Series, New York (1987).
R. Birringer. "Nanocrystalline Materials." *Mater. Sci. Engr.*, **A117**, 33 (1984).
R. Birringer, H. Gleiter, H. P. Klein, and P. Marquardt. "Nanocrystalline Materials: An Approach to a Novel Structure with Gas-Like Disorder." *Phys. Lett.*, **102A**, 365 (1984).
Z. Y. Chen, P. Meakin, and J. M. Deutsch. "Comment on Hydrodynamic Behavior of Fractal Aggregates." *Phys. Rev. Lett.*, **59**, 2121-2121 (1987).
W. S. Coblenz. "The Physics and Chemistry of the Sintering of Silicon." *J. Mater. Sci.*, **25**, 2754–2764 (1990).
D. S. Covert, J. Heintzenberg, and H.-C. Hansson. "Electro-Optical Detection of External Mixtures in Aerosols." *Aerosol Sci. Technol.*, **12**, 446–456 (1990).
J. Eastman and R. W. Siegel. "Nanophase Synthesis Assembles Materials from Atomic Clusters." *Research and Development*, page 56, January (1990).
R.C. Flagan. "Aerosol Routes for Powder Synthesis." *Ceramic Transactions*, **1A**, 229–243 (1988).
S. R. Forrest and T. A. Witten. "Long-Range Correlations in Smoke Particle Aggregates." *J. Phys. A. Math. Gen.*, **12**, L109 (1979).
J. Frenkel. "Viscous Flow of Crystalline Bodies Under the Action of Surface Tension." *J. Phys. (USSR)*, **9**, 385 (1945).
S. K. Friedlander. "The Characterization of Aerosols Distributed with Respect to Size and Chemical Composition." *Aerosol Sci.*, **1**, 295–307 (1970).
S. K. Friedlander. "The Characterization of Aerosols Distributed with Respect to Size and Chemical Composition – II. Classification and Design of Aerosol Measuring Devices." *Aerosol Sci.*, **2**, 331–340 (1991).
S. K. Friedlander. *Smoke, Dust, and Haze.* Wiley, New York (1977).
N. A. Fuchs. *The Mechanics of Aerosols.* MacMillan, New York (1964).
F. Gelbard and J. H. Seinfeld. "Simulation of Multicomponent Aerosol Dynamics." *J. Colloid Interface Sci.*, **2**, 485–501 (1978).
F. Gelbard and J. H. Seinfeld. "Numerical Solution of the Dynamic Equation for Particulate Systems." *J. Comput. Phys.*, **28**, 357–375 (1979).

C. G. Granqvist and R. A. Buhrman. "Ultrafine Metal Particles." *J Appl. Phys.*, **47**, 2200–2219 (1976).

O.J. Gregory, S-B. Lee, and R.C. Flagan. "Reaction Sintering of Submicrometer Silicon Powder." *J. Am. Ceram. Soc.*, **70**, 52–55 (1987).

C. Greskovich and J. M. Rusolowski. "Sintering of Covalent Solids." *J. Amer. Ceram. Soc.*, **59**, 336 (1976).

R. Hagenloch and S. K. Friedlander. "Numerical Calculations of Collision Frequency of Molecules with DLA clusters." *J. Colloid Interface Sci.*, **133**, 185–191 (1989).

J. J. Helble and A. F. Sarofim. "Factors Determining the Primary Particle Size of Flame-Generated Inorganic Aerosols." *J. Colloid Interface Sci.*, **128**, 348–362 (1989).

D. L. Johnson. "New Method of Obtaining Volume, Grain-Boundary, and Surface Diffusion Coefficients from Sintering Data." *J. Appl. Phys.*, **40**, 192 (1969).

W. D. Kingery and M Berg. "Study of the Initial Stages of Sintering Solids by Viscous Flow, Evaporation-Condensation, and Self-Diffusion." *J. Appl. Phys.*, **26**, 1205 (1949).

W. D. Kingery, H. K. Bowen, and D. R. Uhlmann. *Introduction to Ceramics.* John Wiley & Sons, New York (1976).

A. Kobata, K. Kusakabe, and S. Morooka. "Growth and Transformation of TiO_2 Crystallites in Aerosol Reactor." *AIChE J.*, **37**, 347–359 (1991).

W. Koch and S. K. Friedlander. "Particle Growth by Coalescence and Agglomeration." *Part. Part. Syst. Charact.*, **8**, 86–89 (1991).

G. C. Kuczynski. "Self Diffusion in Sintering of Metallic Particles." *Trans. AIME*, **185**, 169 (1949).

F.S. Lai, S.K. Friedlander, J. Pich, and G.M. Hidy. "The Self-Preserving Size Distribution for Brownian Coagulation in the Free-Molecular Regime." *J. Colloid and Interface Science*, **39**, 395-405 (1972).

J. D. Landgrebe, S. E. Pratsinis, and S. V. R. Mastrangelo. "Nomographs for Vapor Synthesis of Ceramic Powders." *Chem. Eng. Sci.*, **45**, 2931–2941 (1990).

B. Mandelbrot. *The Fractal Geometry of Nature.* W. H. Freeman, New York (1982).

V. A. Marple, B. Y. H. Liu, and G. A. Kuhlmey. "A Uniform Deposit Impactor." *J. Aerosol Sci.*, **12**, 333 (1981).

P. Meakin. "Fractal Structures." *Prog. Solid State Chem.*, **20**, 35–233 (1990).

R. D. Mountain, G. W. Mulholland, and H. Baum. "Simulation of Aerosol Agglomeration in the Free Molecular and Continuum Flow

Regimes." *J. Colloid Interface Sci.*, **114**, 67–81 (1986).

H. V. Nguyen and R. C. Flagan. "Particle Formation and Growth in Single-Stage Aerosol Reactors." *Langmuir*, **7**, 1807 (1991).

J. Podzimek. "John Aitken's Contribution to Atmospheric and Aerosol Sciences – One Hundred Years of Condensation Nuclei Counting." *Bull. Amer. Meterol. Soc.*, **70**, 1538–1545 (1989).

D. Y. H. Pui and B. Y. H. Liu. "Advances in Instrumentation for Atmospheric Aerosol Measurement." *Physica Scripta*, **37**, 252–269 (1988).

D. J. Rader and P. H. McMurry. "Application of the Tandem Differential Mobility Analyzer to Studies of Droplet Growth or Evaporation." *J. Aerosol Sci.*, **17**, 771–787 (1986).

S. N. Rogak, U. Baltensperger, and R. C. Flagan. "Measurement of Mass Transfer to Agglomerate Aerosols." *Aerosol Sci. Technol.*, **14**, 447–458 (1991).

S. N. Rogak and R. C. Flagan. "The Mobility and Structure of Aerosol Aglomerates." *submitted to Aerosol Sci. Technol.* (1991a).

S. N. Rogak and R. C. Flagan. "Coagulation of Aerosol Agglomerates in the Transition Regime." *J. Colloid Interface Sci., in press* (1991b).

S. N. Rogak and R. C. Flagan. "Characterization of the structure of agglomerate particles." *Particle Characterization and Particle Systems, in press* (1991c).

A. Schmidt-Ott. "*In Situ* Measurement of the Fractal Dimensionality of Ultrafine Aerosol Particles." *Appl. Phys. Lett.*, **52**, 954–956 (1988).

J. H. Seinfeld. *Atmospheric Chemistry and Physics of Air Polllution.* Wiley-Interscience, New York (1986).

N. J. Shaw and A. M. Heuer. "On Particle Coarsening During Sintering of Silicon." *Acta. Metall.*, **31**, 55 (1983).

R. W. Siegel, S. Ramasamy, H. Hahn, Z. Q. Li, and T. Lu. "Synthesis, Characterization and Properties of Nanophase Ti0−2." *J. Mater. Res.*, **3**, 1367-1372 (1988).

G. D. Ulrich. "Theory of Particle Formation and Growth in Oxide Synthesis Flames." *Combust. Sci. Technol.*, **4**, 47 (1971).

G. D. Ulrich, B. A. Milnes, and N. S. Subramanian. "Particle Growth in Flames. II: Experimental Results for Silica Particles." *Combust. Sci. Technol.*, **14**, 243 (1976).

G. D. Ulrich and J. W. Riehl. "Aggregation and Growth of Submicron Oxide Particles in Flames." *J. Colloid Interface Sci.*, **87**, 257 (1982).

G. D. Ulrich and N. S. Subramanian. "Particle Growth in Flames. III: Coalescence as a Rate-Controlling Process." *Combust. Sci. Technol.*, **17**, 119 (1977).

S. C. Wang and R. C. Flagan. "Scanning Electrical Mobility Spectrometer." *Aerosol Sci. Technol*, **13**, 230 (1990).

J. J. Wu. *Powder Synthesis in Aerosol Reactors.* PhD thesis, California Institute of Technology (1986).

J. J. Wu and R. C. Flagan. "The Onset of Runaway Nucleation in Aerosol Reactors." *J. Appl. Phys*, **61**, 1365–1371 (1987).

J. J. Wu and R.C. Flagan. "A Discrete-Sectional Solution to the Aerosol Dynamic Equation." *J.Colloid Interface Sci.*, **123**, 339 (1988).

J. J. Wu, H.V. Nguyen, and R.C. Flagan. "A Method for Synthesis of Submicron Particles." *Langmuir*, **3**, 266–271 (1987).

CHAPTER 15

FORMATION OF UNIFORM PRECIPITATES: ADVANCES AND UNRESOLVED QUESTIONS

C. Zukoski

15.1 INTRODUCTION

Processing of fine particles lies at the core of many technologies including photographic, paint, ceramic, paper coatings, magnetic recording, and pigments industries. The quality of the final products resulting from these technologies is often significantly affected by the size and shape distribution of the particles used in material formulation. As a result, a great deal of effort is expended on new and better particle preparation techniques. Recently, driven by the need for chemical purity and lower sintering temperatures, synthetic as opposed to mechanical routes to preparing uniform submicron particles have seen increased attention (see for example reviews by LaMer 1952; Sugimoto 1987; Matijevic 1981, 1986; Gunn 1976; and Brinker and Scherer 1990). In addition, particles in the "nano" (i.e., sub-100 nm) size range have unique optical and mechanical properties (Steigerwald and Brus 1989; Henglein 1989). Two commonly considered routes to particle synthesis are gas phase aerosol and precipitation processes. This review will focus on particle formation from precipitation reactions.

While many recipes are available for precipitating particles for a host of chemistries (see for example the work of E. Matijevic 1981, and 1986 for a wonderful array of particle sizes and shapes and fig. 1), these routes are often limited to a small scale where the yield in terms of solid mass is quite low. Attempts to scale up these reactions have had limited success. In addition, when faced with a new chemistry where uniform precipitates are sought, few rules of thumb are available for discovering reaction conditions resulting in the desired particles. As a result, tedious experiments are often required where controlling parameters such as concentration, pH, and temperature are varied over wide ranges

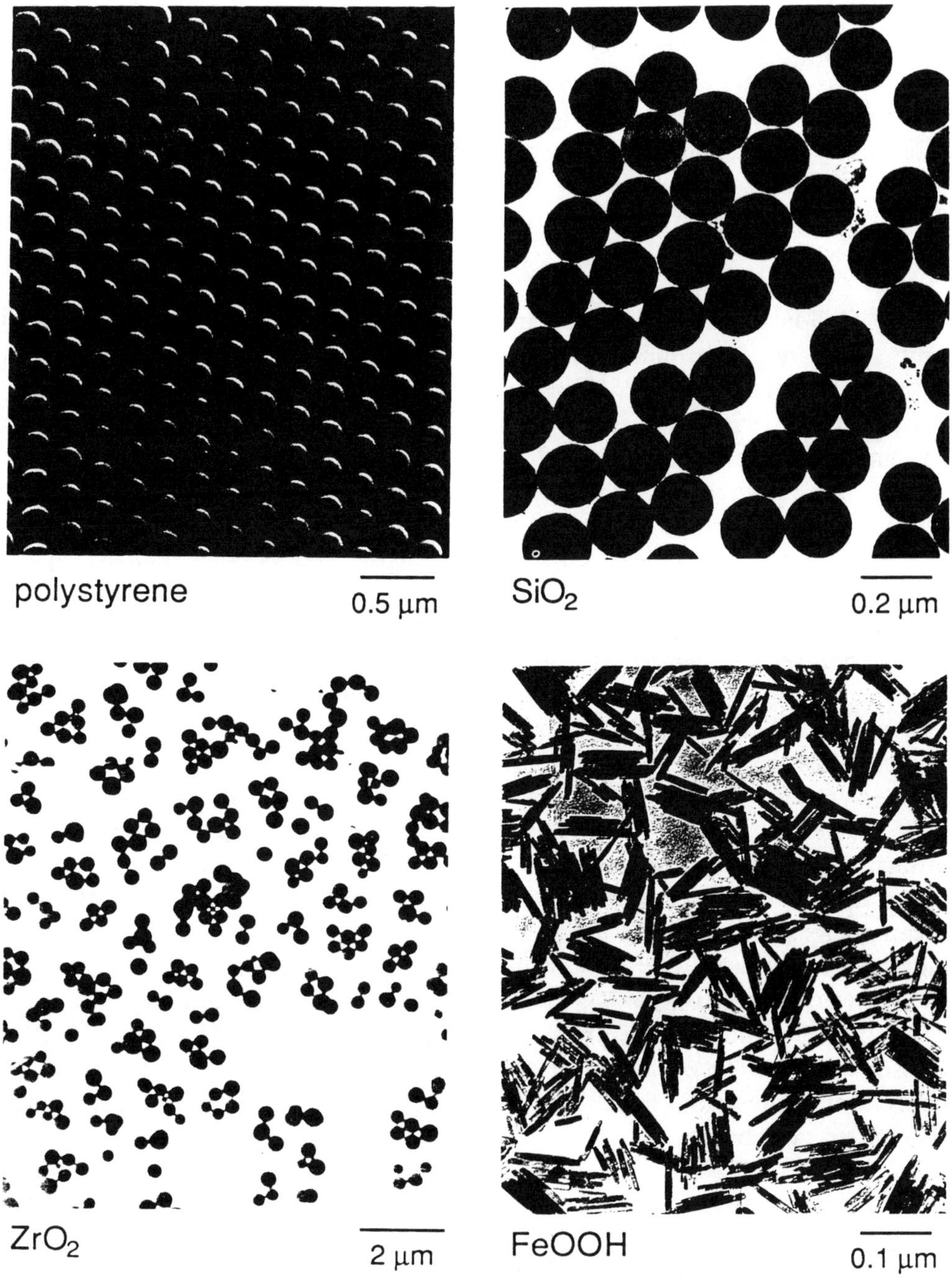

Fig. 1 Examples of submicron particles formed through precipitation reactions. The polystyrene particles are formed from emulsifier free emulsion polymerization while the silica and zirconia particles are formed from the hydrolysis of metalalkoxides. These three particles are amorphous. The iron hydroxide rods are crystalline and are formed through the hydrolysis of iron chloride in acidic aqueous solutions.

in small steps. While the reaction chemistry giving rise to precipitates is quite complex and development of a-priori general synthetic routes is unlikely, if progress is to be made in scaling up known recipes, a better understanding is required of the chemical and physical processes controlling precipitate morphology. Studies leading to such a fundamental knowledge level are limited to a small number of chemistries but technological opportunities suggest the need to expand this knowledge base such that precipitation reactions can be developed on a rational basis.

Colloidal interactions during the precipitation and in post synthetic processing are often found to control the usefulness of many reaction products. While colloidal stability has been extensively studied (for example see the text of Russel et al., 1990), studies of changes in interaction potential during particle formation processes have been limited. There is, however, little doubt that aggregation plays an important role in controlling precipitate morphology. Due to the pervasive nature of attractive van der Waals forces, the formation of discrete precipitate particles requires that repulsive interactions which are typically developed through steric or electrostatic mechanisms. Many unresolved questions concerning colloidal stability in precipitating suspensions remain unanswered. These include mechanims of polymer adsorption to growing particles and the dynamics of the electrostatic double layer in reacting interfaces are poorly understood. In addition, the role of solvation or hydration forces in the nucleation and stabilization of sub - 10 nm particles has only recently attracted attention.

The colloidal properties of the uniform particles after synthesis are also attracting increasing attention. Many studies of model colloidal suspensions have relied on polymer lattices due to particle uniformity and ease of synthesis. A recent increase of interest in inorganic colloids can be traced to their technological significance as well as to their unique surface properties (see for example, Steigerwald and Brus (1989), Henglein (1989), Marshall and Zukoski (1990), deKruif et al (1986), O'Brien et al. (1990), and references in Matijevic (1985) and Russel et al. (1990)). However these investigations are often limited by low concentrations resulting from many synthetic routes and difficulties in controlling precipitate morphology.

Here a general overview of recent advances in understanding of the physical basis for precipitate morphology will be presented and key issues identified. A series of specific chemistries is used to illustrate current understanding and point to examples where further research is needed. The paper does not delve deeply into the details of bond breaking and making but instead focusses attention on underlying physical conditions that must be met if particles of uniform size and shape are to be formed.

15.2 PRECIPITATION PHENOMENA

Here precipitation will be defined as the formation of a new solid phase from one or more fluid phases. The primary requirement for a precipitation to occur and the product to be stable is that the formation of the new phase must lower the free energy of the system. For many precipitation chemistries however, equilibrium arguments are not sufficient to uniquely define the state of the solid phase. Precipitate shape, size, and degree of crystallinity are primarily determined by the relative rates of many steps which occur during the solid's formation and growth. As a result, a wide of variety of solids can be formed which are not truly thermodynamically stable, but which are trapped by kinetic barriers in useful, metastable states which have long lifetimes on a human scale.

Precipitation can be broken into three general stages (LaMer 1952; Nielson 1964). These are; i) the formation of a supersaturated solution (i.e., the creation of a thermodynamic driving force for the formation of a new phase), ii) nucleation to relieve the supersaturation, and iii) growth of the new solid phase. These stages and their influence on precipitation of uniform particles are discussed below.

15.3 SUPERSATURATION AND NUCLEATION

In the initial stage of a precipitation reaction material is introduced to the reactor. As the solubility of the material exceeds its equilibrium value, there is a driving force for the formation of a new phase. Major thermodynamic variables used to alter solubility include temperature, concentration, and the manipulation of solution chemistry variables such as ionic strength and pH. Variations on this are to create new chemical species through chemical reactions occurring in the solution (such as the forced hydrolysis of metal ions, or the creation of large molecular weight species as in free radical polymerization reactions) which become insoluble and precipitate. The rate of generation of a supersaturated solution greatly influences nucleation and subsequent solid growth. As a result, the route to achieving supersaturation is often considered as a key to formation of uniform precipitates.

Once the equilibrium solubility is exceeded, there is a thermodynamic driving force for the creation of a new phase. By this it is meant that the free energy of the system can be reduced if a new phase is formed from the molecules in the host liquid. Rates of formation of the new phase have been extensively studied from both theoretical and experimental perspectives. The classic theory of Becker and Doring, Volmer, and Zeldovich (as reviewed extensively by Nielson, 1964 and Abraham; 1979) describes the rate of nucleation from an initially homogeneous fluid. As many precipitation reactions leading to uniform submicron sized particles occur under homogeneous nucleation conditions, a discussion will not be presented of heterogeneous nucleation where the new phase grows on extraneous surfaces already present in the host fluid or of spinodal decomposition where nucleation takes place from a thermodynamically unstable state (Abraham, 1979).

Homogeneous nucleation here will refer to the case where the new phase is produced from a metastable solution. This implies that the free energy of the system can be lowered by the formation of a solid phase. In the metastable region however, small concentration fluctuations tend to decay. For the new phase to be formed, clusters (or concentration fluctuations) with a large composition difference from the host phase must occur. Due to the low probability that such concentration excursions can occur, the clusters are invariably small. This leads to the concept of a small particle or nucleus being the first physical manifestation of the new phase in the supersaturated fluid. In nucleation from an unstable phase, as in spinodal decomposition, concentration fluctuations of any magnitude grow and thus the first stages of nucleation are associated with small concentration fluctuations which have a large extent.

The classical theory of homogeneous nucleation considers a solution to consist of a host solvent and the species which is supersaturated (Nielson, 1964; Abraham, 1979). Clusters of supersaturated molecules are continually forming and degrading. The free energy change associated with the formation of a cluster containing n molecules is given as the sum of a negative term associated with driving force for phase transition which scales as n and a positive term associated with the formation of an interface between the cluster and the host phase. This second term is the free energy of formation of the surface and is given as the product of the surface tension and the surface area of the cluster. The surface energy term scales as $n^{2/3}$. For small values of n the positive surface energy term grows faster with n than does the negative volumetric term. As a result, small clusters have a higher free energy than the separate molecules. However at a critical value of n, say n^*, the change of free energy in putting n molecules together into a cluster goes through a maximum and decreases with increasing n. Clusters smaller than n^* tend to degrade while clusters of size greater than n^* tend to grow. In the classical formulation of nucleation kinetics, a cluster of size n^* is considered a nucleus and thus represents the first presence of the new phase.

The rate of formation of clusters of size n is determined by calculating a cluster size distribution where clusters are formed by binary reactions of single molecules with a cluster of size n-1. A conservation equation for clusters of size n is solved using the assumption that the cluster size distribution during steady state nucleation is not much different than it would be at equilibrium. The resulting expression for the steady state nucleation rate can then be written as

$$J = J_o \exp\left[-\Delta G^*/kT\right] \tag{3.1}$$

where $\Delta G^* = 16\pi\gamma^3 v/\{3[\ln C/C_{eq}]^2\}$, γ is the surface tension of the clusters (assumed to be independent of cluster size), C/C_{eq} is the supersaturation ratio (the ratio of the solute concentration to its equilibrium value), v is solute molar volume, and kT is the product of Boltzmann's constant and the absolute temperature. The pre-exponential factor, J_o, depends on the kinetics of cluster formation.

The critical assumptions of classical nucleation theory are that it relies on the solid/liquid interfacial tension to describe the energy barrier to clustering and that it assumes micro-reversibility in cluster formation kinetics. The result of these arguments is a Boltzmann distribution of equilibrium cluster sizes (i.e., that distribution which would exist if there was no net flux of molecules into the precipitated state). While the thermodynamic derivation of the changes in free energies of cluster formation has been questioned, the classical formulation has been successful in providing insight into the trends that control the rates of formation of precipitated particles from supersaturated solutions (Abraham, 1979).

The basic success of classical nucleation theory lies in its ability to predict the extreme sensitivity of J to γ and C/C_{eq}. Tested most extensively for vapor/liquid transitions, applications to precipitation reactions typically rely on using measurements of the critical supersaturation concentration to estimate surface tensions. The nucleation rate is found to increase sharply with C/C_{eq} such that if a minimum level of detectability of say 1 particle per mm^3 is chosen, and the supersaturation giving rise to nucleation of particles at this concentration measured, γ can be estimated from eqn 3.1. Surface tensions for common inorganic materials in aqueous solutions are found to lie in the range of ten to several hundred dynes/cm^2 (Nielson, 1964). Due to the sharp dependence of J on C/C_{eq}, γ is relatively insensitive to choice of the minimum level of particle dectectability. On the other hand, as a result of the extreme sensitivity of J to γ, small uncertainties give rise to large variations in nucleation rate.

Measurement of the critical cluster size ultimately relies on determination of the size where particles spontaneously grow rather than decay. In the past, the critical supersaturations were determined by detecting particle formation primarily by optical techniques. However, as aggregation is available as a growth pathway for clusters larger than the critical size, estimates of nucleus size based on final particle size and number density are uncertain unless carried out under extremely dilute conditions. As a result of these difficulties, detailed confirmation of classical nucleation theory in precipitation reactions is difficult.

There is growing evidence that classical nucleation theory fails to capture nucleation events in many precipitation reactions. One example of this can be seen by considering the work of Reiss and coworkers (1983, 1984) on vapor phase free radical polymerization reactions carried out in the presence of slightly supersaturated monomer vapor ($1.5 < C/C_{eq} < 5$). Here oligimers initiated by exposure to UV radiation grow to a size where the polymers act as a seed for nucleation of monomer droplets. As the supersaturation of the vapor is increased, the size of the polymer capable of nucleating a droplet is decreased. Reiss and Chowdhury (1983, 1984) have developed a modified Flory-Huggins (Flory, 1953) model for such a nucleation process which demonstrates that as a result of the continued growth of the polymer within a condensing droplet, all polymers growing to a critical size will nucleate and the rate of nucleation is a measure of the rate of polymer propagation. In this process nucleation occurs only because monomer is being consumed in a irreversible reaction. The clustering giving rise to the polymer lowers the system's free energy and if monomer condensation on the polymer did not occur, high molecular weight polymers would grow in the vapor phase. While the nucleation event, i.e., the condensation of monomer on the growing polymer, can be treated from the classical perspective (and is not that far removed from heterogeneous nucleation) the growth process producing the polymer (a nonreversible cluster) does not face the size barrier seen in classical nucleation.

Emulsion and dispersion polymerization of vinyl monomers offer other examples of nonclassical nucleation (Lichti et al. 1983; Feeney et al. 1984, 1985). During the initial stages of the reaction, polymerization proceeds in the continuous phase. Due to propagation reactions a critical molecular weight (typically thought to be near a degree of polymerization of the order of 5-10 for an aqueous continuous phase and hydrophobic monomers such as styrene) is exceeded and oligimers undergo a type of Flory-Huggins sol/gel phase transition to produce polymer rich and polymer poor phases (Flory 1952). If the oligimers are dilute, the nucleation event will initially involve only one oligimer. (If two reacting oligimers combine there is a termination reaction and the resulting particle stops growing.)

The rate of production of polymer latex nuclei in emulsion polymerization reactions does not follow the basic assumptions of the classical nucleation theory. First the reactions giving rise to a "cluster" of a critical size are essentially irreversible. Secondly, there is no equilibrium "cluster" size (oligomer molecular weight) distribution. Finally, the concept of a supersaturated solution is of limited utility as a single oligomer will precipitate if it grows to sufficient size and the nucleus will be large enough to be recognized as a separate phase. Thus nucleation does not depend on cooperative gathering of many oligomers of a particular molecular weight to produce a cluster of a size large enough to overcome the surface tension generated barrier. Instead nucleation depends on rates of molecular addition reactions and the nucleation event is associated with a conformational change of the polymer. As a result, if the number of free radicals formed at the initial stages of the emulsion polymerization and the rates of free radical addition are known, a prediction could be made of the number and formation rates of nuclei. Indeed, estimates following these lines of argument have proven successful in the case of emulsion polymerization of styrene (Feeney et al., 1984).

A third example of non-classical nucleation can be found for silicon alkoxides which Klemperer and Ramamurthi (1988, 1990) have demonstrated to condense according to Flory-Stockmayer theory (Flory, 1953). In this model, the most probable molecular weight distributions are determined for a given state of reaction under the assumptions that all monomer sites have the same reactivity and that no cyclic species are formed. While these assumptions fail at late times in the alkoxide hydrolysis reactions, Klemperer and Ramamurthi demonstrate that the model is followed at short times, both in terms of the molecular weight distribution and the speciation within isomers. Under basic reaction conditions the alkoxides condense as if they have 3-4 reactive sites per monomer. The resulting molecular weight distribution is peaked at the monomer and has a very long tail extending to high molecular weights. The solid phase produced in these reactions has been linked with a sol/gel transition of the large molecular weight species which are thought to form 2-10 nm nuclei that aggregate to form gels or dense particles depending on the reaction conditions (Bogush and Zukoski, 1991a and b). As in the cases of emulsion and dispersion polymerization, solid phase formation is associated with a "polymer" growing in a poor solvent where, if a critical molecular weight is exceeded, a phase separation will lower the system's free energy. The concept that a critical cluster size determined by a balance of surface and volumetric energy terms determines the rate of solid phase nucleation appears to have limited utility.

These three examples and many more which can be found in the inorganic and organic precipitation literature demonstrate that initiation of phase changes through fluctuations in local density to form clusters larger of a critical size may not be sufficient to characterize precipitation reactions where clusters are formed from interactions of a nonphysical origin (i.e., those formed by forces other than van der Waals or hydrogen bonding interactions). Research in this area has been limited but offers opportunities for considerable advancement.

15.4 CONTROL OF CRYSTAL PHASE

Many materials are known to precipitate in a number of different forms. For example, de Bruyn and coworkers (van Stratten et al., 1984; Vermeulen et al., 1975; Stol et al., 1976) have studied precipitation from supersaturated aluminate solutions and found that the predominant species precipitated tends to follow Ostwald's rule of stages which states that the phase which precipitates is not the most thermodynamically stable phase but rather a metastable phase. Thus, for example, van Stratten et al. (1984) found that aluminate solutions precipitated in the sequence, amorphous, psuedoboehemite, bayerite, gibbsite as the supersaturation was reduced and/or the precipitation temperature was raised.

Eqn. 1 argues a particular phase will nucleate faster as its supersaturation is raised. The equilibrium solubility of a species decreases as the phase becomes more thermodynamically stable. Thus, there is always a greater driving force for the formation of the most thermodynamically stable phase. Two factors offset the formation of the most stable phase. The first is the relative rates of nucleation of two phases are modulated by both the pre-exponential factor governing rates of reaction during cluster formation and the surface tension of the clusters. Gutzow and Toschev (1968) argue that isotropic nuclei will grow faster than nuclei having a large degree of asymmetry due to both changes in surface tension and rates of molecular incorporation into growing clusters. Thus, for example, an amorphous phase will nucleate more easily than a crystalline phase. The second factor controlling the phase that precipitates is a phase's growth rate after nucleation has occurred. Feenstra and de Bruyn (1981) suggest that the decision about which phase will predominate is made at some later time after critical nucleus formation has occurred. This argument is based on the very small size of nuclei formed in highly supersaturated solution. These authors argue that, due to their small size, the crystallography of the nucleus is poorly determined. As a consequence, kinetic factors and thermodynamic conditions of the bulk must control which phase dominates at the end of the precipitation.

Control of the phase that precipitates is clearly important to any number of particle production technologies. While the general trends predicted by Ostwald's rule of stages are followed for many systems, there are many counter examples. As a result, a comprehensive set of guiding principles has not been developed despite the obvious need.

Major obstacles to developing such rules lie in the complex chemistry occurring during nucleation of new phases. These systems are far from equilibrium where the composition of the mother liquor is difficult to characterize and thus poorly understood. Important events occur on small length scales and on short time scales making the gathering of meaningful kinetic data difficult. While the classical theory of homogeneous nucleation provides a good semiquantitative description, its verification is based on critical supersaturation measurements and other indirect data (Nielson, 1964 and Abraham, 1979). As described above there are classes of nucleation phenomena that are not readily encompassed into this formalism. Even the rates and mechanisms of nuclei growth are poorly characterized. As a result, it is not surprising that there has been little progress in developing a consistent set of rules for determining which phase will precipitate when a series is possible.

15.5 PARTICLE GROWTH

Extensive research has gone into investigating precipitate growth mechanisms. As point out by Nielson (1964), who provides an analysis of several growth mechanisms, no single mechanism is expected to hold if the particles grow indefinitely. The particles grow at the rate of the slowest reaction step. Thus, for example, particles growing by a surface reaction limited mechanism will show a crossover to diffusion limited growth when the particles are of such a size that there are mass transfer limitations in getting species from the bulk to the surface.

Studies of particle growth mechanisms ask questions similar to those investigated during studies of nucleation including:

What controls the rate of growth?
What is the subunit which attaches to the precipitate surface?, and
How can the rates be manipulated?

Many of these questions can be answered by studying deposition on large surfaces but, while important, answers to these questions do not provide insight into to growth mechanisms controlling particle size distribution.

15.6 CONTROL OF PRECIPITATE MORPHOLOGY

Size, shape and porosity of fine particles determine their usefulness. Often controlled particle size distributions are desired for ease in handling, formulation of flowable dense suspensions, and uniformity of particle compacts. Development of large scale precipitation reactions capable of producing useful quantities of submicron particles of controlled size distribution remains a major stumbling block to the advancement of many technologies. For example, methods of preparing uniform metal oxide particles hinders development of advanced ceramics where lower sintering temperatures, finer microstructures and higher quality optical, electronic and mechanical properties have been demonstrated as resulting from the use of precursor particles of narrow size distributions. Uniform particles of controlled porosity are desired in the development of catalyst supports. Nanophase materials of narrow size distribution are sought for their optical and structural properties. Despite the great technological demands for methods of producing large quantities of narrow size distribution particles, little is understood about parameters which determine size and shape of particles produced from in precipitation reactions.

In the inorganic particle literature the most enduring model used to describe precipitation reactions is that of LaMer (1952) who postulated that if the particles were nucleated in a single short burst, the final precipitate would have a narrow size distribution. Recognizing that the period of time at which the supersaturation would be above the level capable of producing a finite rate of nucleation, the LaMer model suggests that uniformity will result if the particles grow by a self-sharpening growth mechanism. Here small particles increase in size at a faster rate than larger particles and the particle diameter distribution narrows as the supersaturation is reduced. Reactions which are limited by the diffusion of reactive species to the particle surface and those which are limited by surface reactions meet the self-sharpening growth requirement. As a consequence, for particles growing by commonly observed molecular addition mechanisms, the LaMer model suggests that narrow size distributions result from a short nucleation period where each nucleus survives at the end of the reaction as a single particle.

The conceptual model of LaMer has endured since it was postulated and is commonly evoked when developing a framework on which to build ideas of how to control particle size distributions during precipitation reactions. The short nucleation model is appealing because of its simplicity and because its postulates, if followed, would undoubtedly result in particles of narrow size distribution. However, the LaMer model has seen little verification. Indeed in the few cases where sufficient data has been taken, the model has been shown to fail. Even in

As inorganic materials can "polymerize" by wide range of mechanisms resulting in an equally wide range of isomers, such detailed studies are more difficult. However, the multiplicity of particle sizes and shapes which can be made with inorganic materials testifies to the richness of this field of work and the rewards of developing an understanding of the governing mechanisms. Recent interest in alkoxide derived inorganic materials demonstrates that many of the general features encountered in the organic precipitation reactions extend to these inorganic materials.

As mentioned above, silicon alkoxides polymerize following the rules developed by Flory and Stockmayer for organic polymers (Klemperer and Ramamurtui, 1988, 1990). In basic media particles of narrow size distribution are formed (Stober et al., 1968; Bogush et al., 1988). This reaction can be viewed as an inorganic analogy to dispersion polymerization in that the monomer is soluble in the continuous phase but the polymer is not. Clearly, even for the nucleation period is short, if the resulting primary particles are not colloidally stable, the final solid will suffer the same fate as the organic polymer in a dispersion polymerization carried out in the absence of stearic stabilizers (i.e., the final particles will have a broad, size distribution). Recently, a series of experiments have suggested that the nucleation period in the hydrolysis and condensation of silicon alkoxides is not short and may last for up to 2/3^{rds} of the over all reaction (Bogush and Zukoski 1991a and b). In addition, particle size is found to be strongly influenced by the addition of electrolyte while reaction rates are not. These studies indicate that colloidal stability plays a key role in determining final particle size distributions.

A model has been developed (Bogush and Zukoski 1991b) for alkoxide derived particle growth kinetics based on the assumption that supersaturation is relieved solely by the production of primary particles and final size distribution is determined by aggregation kinetics. Uniformity is the result of size dependent aggregation rates. Using electrostatic, van der Waals and solvation interaction potentials, the model is able to capture reagent concentration dependencies of final particle diameters with primary particle size being the sole adjustable parameter. Primary particle sizes in the range of 2-10 nm are found to give the observed final particle diameters. These physically reasonable primary particle sizes are in line with those predicted from classical nucleation theories using critical nucleation supersaturation data to estimate the primary particle/continuous phase surface tension (Look et al., 1990). While capable of predicting final particle diameters, the model consistently under predicts the breadth of the final size distribution indicating that further work is required before a comprehensive model is complete.

the case of the sulfur sols for which the model was originally developed, later work has demonstrated that over the course of the precipitation reaction, particle number density passes through a maximum; an indication that each particle produced as a nucleus does not survive to the end of the reaction (Rowell et al., 1968; Levit and Rowell, 1975). Three examples will be drawn from the literature to demonstrate that final particle size distributions are not determined by the rate of nucleation.

Perhaps the most widely studied class of precipitation reactions are those involving emulsion and dispersion polymerization of vinyl monomers. In emulsion polymerization, a water insoluble monomer is mixed with an aqueous electrolyte solution and a free radical initiator added. Extensive studies have shown that the initial stages of the reaction occur in the aqueous phase where, as mentioned above, oligimers grow to a size where they become insoluble and precipitate (see for example, Lichti et al., 1983; Feeney et al., 1984, 1985). Due to their low charge, the particles aggregate until they reach a size where they are stable to further aggregation with particles of similar size. Growth of the stable particles is determined by particle swelling, the rates of propagation, termination, chain transfer, and entry of free radicals into the particle. The final particle number density, which determines the average particle size, is controlled by the colloidal properties of the primary particles.

In dispersion polymerization a continuous phase is chosen in which the monomer is soluble but the polymer is not. Again oligimers grow to a size where they undergo a sol/gel transition and collapse to form primary particles. Final particle size distributions are largely determined by the colloidal properties of the growing polymer rich phase where if stearic stabilizers are not added the final solid consists of large agglomerated particles of poorly defined size and shape. In the presence of stearic stabilizers, particles of narrow size distribution can be formed (Croucher and Winnik, 1990; Barrett, 1975).

Studies of the basic rate processes in organic free radical precipitation reactions are aided by the limited number or pathways that can be followed. Methods of determining rates of free radical initiation, propagation, termination and free radical entry have been developed and models for particle growth and size distributions tested. The success of these models can be seen in large commercial scale emulsion polymerization reactors that routinely turn out concentrated dispersions of uniform latex particles.

Not surprisingly, a poorer understanding exists of the kinetics and mechanisms at work in precipitation reactions that have seen less intense scrutiny.

The third example where a short nucleation period does not control precipitate morphology which again emphasizes the importance of particle interaction potentials in the preparation of uniform particles can be found in alkoxide derived titania (Look et al. 1990; Zukoski et al. 1990; Look 1991). In the absence of attention to details of parameters controlling colloidal stability, typical precipitates are composed of large open agglomerates of relatively uniform particles which are held together by well formed necks. Addition of HCl or NaCl in up to milimolar concentrations is found to increase colloidal stability of the growing particles and give rise to narrow size distribution precipitates. In addition, growing particles are very sensitive to shear. In the absence of shear particles grow monotonically to a final size. However, under the action of small shear rates (15 s^{-1}) particles grow to a diameter near 0.6 microns and then agglomerate and fuse.

These observations have been interpreted in terms of size dependent particle interaction potentials where small particles are stabilized by solvation interactions (Look et al., 1990; Zukoski et al., 1990). As the particles grow, a secondary minimum develops as the result of summing van der Waals, electrostatic and short range repulsive interactions. In the absence of shear, Brownian aggregation into the secondary minimum is limited. However with increasing shear rate, viscous forces drive particles into the secondary minimum where they are captured. In the presence of continued precipitation, the agglomerate fuses. These studies suggest that short range repulsions play an important role in controlling colloidal interaction of metal oxide precipitates. Unfortunately, the origin and magnitude of these short range interactions are poorly understood.

These detailed studies of latex formation and the kinetics of growth of alkoxide derived silica and titania particles indicate that the formation of uniform precipitates is governed by more diverse phenomena than the length of the nucleation period. While the short nucleation time model remains conceptually attractive, little evidence has accumulated that chemical or physical systems can be found where the length of the nucleation period is the sole parameter on which attention must be focussed if one wishes to form precipitates of narrow size distribution.

15.7 CONTROLLING PARTICLE SIZE DISTRIBUTION: BEYOND LAMER

For a wide range of chemistries and reaction conditions giving rise to uniform precipitates, two common themes emerge. Primarily, what is meant by nucleation and the physical state of the nucleus is questioned. Secondly, the importance of establishing a population of colloidally stable particles early in the precipitation reaction is clear.

These themes become intermingled in the study of formation of nanophase materials which may be considered large molecules or small particles. Here the goal is to produce solids with sizes in the 1-10 nm range for use in studies of their physical properties. The resulting solids are in the size range of nuclei inferred from kinetic studies from many precipitation chemistries. As a result, investigation of the growth mechanisms and physical chemistry of these systems is likely to lead to a greater understanding of the growth processes at work in a diverse range of precipitation reactions (Steigerwald and Brus, 1989; Henglien, 1989).

In developing synthetic routes for preparing semiconducting particles where the band gap becomes a function of particle size, stabilizing the 1-10 nm particles is recognized as a primary concern. Successful synthetic routes can be rationalized in the context of stearic stabilization where, due to the small particle size, a stearic layer of small thickness (< 1 nm) is required. For example, in the formation of large clusters containing 10-100 metal atoms, ligands can be considered as preventing aggregation. The uniformity of these truly nanophase colloids is evident in that they are routinely crystallized for X-ray determination of internal structure (Henglein, 1989).

Preparation and characterization of large molecules (small particles) opens up questions about what are meant by a soluble species and nucleation and the role solvation interactions play in controlling not only particle stability but also particle size and shape in precipitation reactions. The continuum of particle (molecular) sizes which have been prepared with metal sulfides and metals suggests concepts commonly employed in the colloids literature can be successfully extended to the synthesis of large clusters. For example, unless displaced, ligands (stearic stabilizers) render the metal clusters thermodynamically stable and a nucleation event may not be relevant in cluster formation (i.e., no insoluble species are produced).

On the other hand, these stable particles can be made to crystallize under conditions indicating the existence of attractive particle interactions. As pair potentials change qualitatively with particle size, small but growing stable particles may reach a size where attractive interactions produce clustering. In the presence of reactive species, particles within the clusters may fuse and the clusters will then become recognizable as large particles. Such a growth pathway is observed for a range of chemistries (Klemperer and Ramamurthi, 1988 and 1990; Van der Woude and deBruyne, 1984; Dousma and deBruyne, 1976 and 1978; Murphy, Posner and Quirk, 1976a-d; Blier and Cannon, 1986; Ugeda et al., 1973; Chow, 1991; Sugimoto and Matijevic, 1980; and Edelstein et al., 1991).

In most metal oxide precipitations, the hydrolysis of metal ions results in large molecular weight species. Often these are referred to as polymers. However it is incorrect to consider these species as random coils of a linear molecule. The degree of cross linking varies from system to system. deBruyn and coworkers report on precipitation reactions resulting in iron (III) and aluminum(III) hydrous oxides (van Stratten et al., 1984; Vermeulen et al., 1975; van der Woude and deBruyn, 1984; Dousma and deBruyn, 1978), where, as base is added to an electrolyte solution containing unhydrolyzed metal species, initially rapid, reversible reactions occur that are attributed to the formation of low molecular weight complexes. At longer times and with the addition of more base, reactions take place that are not easily reversible but which do not immediately lead to the formation of a new solid phase. It is not until much later or with the addition of larger amounts of base that precipitation is observed. The reactions that appear irreversible are associated with the formation of large molecular weight "polymeric" hydroxy metal species. These species apparently grow into "nuclei" once precipitation begins and the aggregation of such nuclei can result in the formation of narrow size distribution rods in the case of goethite (Dousma and deBruyne, 1978; Murphy, Posner and Quirk 1976a-d).

In the hydrolysis of silicon alkoxides there are two generally recognized reaction pathways (Brinker and Scherer, 1990; Klemperer and Ramamurthi, 1988 and 1990). If the reaction is acid catalyzed, hydrolysis and condensation results in the formation of a polymeric network or gel. There does not appear to be a phase transition associated with the formation of this gel, i.e., the polymers forming the gel remain soluble in the host phase up to the gel point. If the hydrolysis and condensation of the silicon alkoxides is carried out under basic conditions, either gels or discrete particles are formed depending on the base and reaction conditions used. The gels produced under basic conditions are, however, composed of open agglomerates of 4-10 nm diameter silica particles which have phase separated (precipitated) from the host phase. A similar growth pathway is implicated in the formation of titania particles from the hydrolysis and condensation of titanium alkoxides (Look et al., 1990; Zukoski et al., 1990; Look, 1991). Other examples can be found in ZrO_2 (Blier and Cannon, 1986), metallic gold (Uyeda et al., 1973; Chow, 1991), magnitite (Sugimoto and Matijevic, 1980), and aerosol derived MgO (Edelstein et al., 1991). In all of these cases agglomeration of slowly growing 2-10 nm particles results in colloidally stable units of narrow size distribution.

15.8 CONCLUSIONS

The richness of the field of small particle formation chemistry and physics is exemplified by the diverse range of sizes and shapes that can be formed and the technological uses of these materials. While many recipes are available for synthesizing particles in suspensions of low weight fraction, understanding the growth processes and methods of scaling up these reactions to commercial scales remain limited. Recent advances in analytical capability which allow for rapid, routine characterization of particle size and stoichiometry are demonstrating that models developed in the past are insufficient to explain the observed particle growth kinetics and mechanisms. The failure of previous models to explain these observations may be at the heart of the problems encountered in attempts to develop large scale precipitation reactions.

While the chemistries used to prepare uniform particles remain of fundamental importance, understanding physical limitations on growth processes resulting in narrow size distributions has been largely overlooked. Understanding the nature of interactions between sub- 100 nm particles and the role of solvation interactions in modulating aggregation and floc rearrangement will play a key role in expanding our ability to synthesize particles of tailored size and chemistry.

ACKNOWLEDGEMENTS

The author would like to thank G. H. Bogush, J-L. Look, and M. K. Chow for their interest and stimulating discussions and the U.S. Department of Energy which supported the research through the Materials Research Laboratory at the University of Illinois with grant DOE AC0276ER01198. Development of this review was partially supported by the International Fine Particle Research Institute.

REFERENCES

Abraham, F. F., Homogeneous Nucleation Theory: The Pretransition Theory of Vapor Condensation, Academic Press, NY (1979).

Barrett, K. I. J., Dispersion Polymerization in Organic Media, Wiley, NY (1975).

Blier, A., and Cannon, R. M., "Nucleation and Growth of Uniform m-ZrO_2", *Mat. Res. Soc. Symp. Proc.* 73 71-78 (1986).

Bogush, G. H., Tracy, M. A., and Zukoski, C. F., "Preparation of Monodisperse Silica Particles - Control of Size and Mass Fraction", *J. Non-Crystalline Solids* 104 95-106 (1988).

Bogush, G. H., and Zukoski, C. F., "Studies of the Kinetics of the Precipitation of Uniform Silica Particles through the Hydrolysis and Condensation of Silicon Alkoxides", *J. Colloid Interface Sci.* 142 1-18 (1991a).

Bogush, G. H., and Zukoski, C. F., "Uniform Silica Particle-Precipitation - An Aggregative Growth", *J. Colloid Interface Sci.* 142 19-34 (1991b).

Brinker, C. J., and Scherer, G. W., Sol-Gel Science, Academic Press, NY (1990).

Chow, M. K., M.S. Thesis, Department of Chemical Engineering, University of Illinois, 1991.

Chowdhury, M. and Reiss, H., "Nucleation in Supersaturated Monomer Vapors Due to Polymers", *J. Phys. Chem.* 87 4599-609 (1983).

Croucher, M. D., and Winnik, M. A., in An Introduction to Polymer Colloids, Kluwer Academic Publishers, Dordreeth, 1990.

deKruif, C. G. van Iersel, E. M. F., Vrij, A., and Russel, W. B., "Hard Sphere Colloidal Dispersions: Viscosity as a Function of Shear Rate and Volume Fraction", *J. Chem. Phys.* 83 4717-4725 (1986).

Dousma, J., and de Bruyn, P. L., "Hydrolysis-Precipitation Studies of Iron Solution. II. Aging Studies and the Model for Precipitation from Fe(111) Nitrate Solutions", *J. Colloid Interface Sci.* 64 154-170 (1978).

Dousma, J., and de Bruyn, P. L., "Hydrolysis-Precipitation Studies of Iron Solutions. I. Model for Hydrolysis and Precipitation from Fe(111) Nitrate Solutions", *J. Colloid Interface Sci.* 56 527-539 (1976).

Edelstein, A. S., Chow, G. M., Altman, E. I., Colton, R. J., and Hwang, D. M., "Self-Arrangement of Molybolenum Particles into Cubes", *Science* 251 1590-92 (1991).

Feenstra, T. P., and de Bruyn, P. L., "The Ostwald Rule of Stages in Precipitation from Highly Supersaturated Solutions: A Model and its Application to the formation of the NonStoichiometric Amorphous Calcium Phosphate Precursor Phase", *J. Colloid Interface Sci.* 84 66-72 (1981).

Feeny, P. J., Napper, D. H., and Gilbert, R. G., "Coagulative Nucleation and Particle Size Distribution in Emulsion Polymerization", *Macromolecules* 17 2520-29 (1984).

Feeny, P. J., Gilbert, R. G., and Napper, D. H., "Periodic Nucleation Processes in Emulsion Polymerization Systems", *J. Colloid Interface Sci.* 107 159-173 (1985).

Flory, P. J., Principles of Polymer Chemistry, Cornell University Press, Ithaca, 1953.

Gunn, D. J., "Mechanism for the Formation and Growth of Ionic Precipitates from Aqueous Solution", *Faraday Disc.* 61 133-140 (1976).

Gutzow, I., and Toschev, S., "Non-Steady State Nulceation in the Formation of Isotropic and Anisotropic Phases, *Krist and Tech.* 3 485-497 (1968).

Henglein, A., "Small-Particle Research-Physiochemical Properties of Extremely Small Colloidal Metal and Semiconductor Particles", *Chem. Rev.* 89 1861-1873 (1989).

Klemperer, W. G., and Ramamurthi, S. D., "Molecular Growth Pathways in Silica Sol-gel Polymerization", *Mat. Res. Soc. Symp.* 121 1-13 (1988).

Klemperer, W. G., and Ramamurthi, S. D., "A Flory-Stockmayer Analysis of Silica Sol-gel Polymerization", *J. Non-crystalline Solids* 121 16-20 (1990).

LaMer, V. K., "Nucleation in Phase Transitions", *Ind. and Engr. Chem.* 44 1270-1277 (1952).

Levit, A. B., and Rowell, R. L., "Time Dependence of the Size Distribution, Number Concentration, and Surface Area in LaMer Sulfur Sols", *J. Colloid Interface Sci.* 50 162-9 (1975).

Lichti, G., Gilbert, R. G., and Napper, D. H., "Mechanisms of Latex Particle Formation and Growth in the Emulsion Polymerization of Styrene Using Surfactant Sodium Dodecyl Sulfate", J. Polymer Sci. Polym. Chem. Edn., 21 269-291 (1983).

Look, J-L., Bogush, G. H., and Zukoski, C. F., "Colloidal Interactions during the Precipitation of Uniform Submicrometre Particles", *Faraday Disc.* 90 345-357 (1990).

Look, J-L., Ph.D. Dissertation, Department of Chemical Engineering, University of Illinois, 1991.

Marshall, L. and Zukoski, C. F., "Experimental Studies on the Rheology of Hard-Sphere Suspensions Near the Glass Transition", *J. Phys. Chem.* 94 1164 (1990).

Matijevic, E., "Monodispersed Metal (Hydious) Oxides - A Fascinating Field of colloid Science", *Acc. Chem. Res.* 14 22-29 (1981).

Matijevic, E., "Production of Monodispersed Colloidal Particles", *Ann. Rev. Mat. Sci.* 15 483-516 (1985).

Morris, J. W., in Chemical Vapor Deposition, 2nd International Conference, J. M. Blocker ed., *Electrochem. Soc.*, NY (1970).

Murphy, P. J., Posner, A. M., and Quirk, J. P., "Characterization of Partially Neutralized Ferric Perchlorate Solutions", *J. Colloid Interface Sci.*, 56 298-311 (1976a).

Murphy, P. J., Posner, A. M., and Quirk, J. P., "Characterization of Partially Neutralized Ferric Nitrate Solutions", *J. Colloid Interface Sci.* 56 270-283 (1976b).

Murphy, P. J., Posner, A. M., and Quirk, J. P., "Characterization of Hydrolyzed Ferric Ion Solutions", *J. Colloid Interface Sci.*, 56 312-319 (1976c).

Murphy, P. J., Posner, A. M., and Quirk, J. P., "Characterization of Partially Neutralized Ferric Chloride Solutions", *J. Colloid Interface Sci.* 56 284-297 (1976d).

Nielson, A. E., Kinetics of Precipitation, Pergammon Press, New York, 1964.

O'Brien, W., Midmore, B. R., Lamb, A. and Hunter, R. J., "Electroacoustic Studies of Moderetly Concetrated Colloidal Suspensions", *Faraday Discussions* 90 261-270 (1990).

Reiss, H. and Chowdhury, M. A., Squire, D. R. and Stannett, V., "Measurement of the Kinetics of Chain Polymerization of Single Poly(vinyl acetate) Molecules in a Cloud Chamber", *Macromolecules* 17 1436-47 (1984).

Rowell, R. L., Kratohvil, J. P., and Kerker, M. J., "Time Dependance of Particle-Size Distribution and Number Concentration in LaMer Sulfur Sols", *J. Colloid Interface Sci.* 27 501-6 (1968).

Russel, W. B., Saville, D. A., and Schowalter, W. R., Colloidal Dispersions, Cambridge University Press, Cambridge, 1989.

Steigerwald, M. L., and Brus, L. E., "Synthesis, Stabilization, and Electronic-Structure of Quantum Semiconductor Nanoclusters", *Ann. Rev. Mater. Sci.* 19 471-495 (1989).

Stober, W., Fink, A., and Bohn, E., "Controlled Growth of Monodisperse Silica Spheres in the Micron Size Range", *J. Colloid Interface Sci.* 26 62-9 (1968).

Stol, R. J., van Helden, A. K., and de Bruyn, P. L., "Hydrolysis-Precipitation Studies of Aluminum (III) Solutions. 2. A Kinetic Study and Model", *J. Colloid Interface Sci.* 57 115-31 (1976).

Sugimoto, T., and Matijevic, E., "Formation of Uniform Spherical Magnetite Particles by Crystallization from Ferrons Hydroxide Gels", *J. Colloid Interface Sci.* 74 227-43 (1980).

Sugimoto, T., "Preparation of Monodispersed Colloidal Particles", *Adv. Colloid Interface Sci.* 28 65-108 (1987).

Uyeda, N., Nishino, M., and Suito, E., "Nucleus Interaction and Fine Structures of Colloidal Gold Particles", *J. Colloid Interface Sci.* 43 264-76 (1973).

van der Woude, and de Bruyn, P. L., "Formation of Colloidal Dispersions From Supersaturated Iron (III) Nitrate Solutions. IV. Analysis of Slow Flocculation of Goethite", *Colloids and Surfaces* 11 391-399 (1984).

Van Stratten, H. A., Holtkamp, B. T. W., and de Bruyn, P. L., "Precipitation from Supersaturated Aluminate Solutions. I. Nucleation and Growth of Solid Phases at Room Temperature", *J. Colloid Interface Sci.* 98 342-62 (1984).

Vermeulen, A. C., Geus, J. W., Stol, R. J., and de Bruyn, P. L., "Hydrolysis-Precipitation Studies of Aluminum (III) Solutions I. Titration of Acidified Aluminum Nitrate Solutions", *J. Interface Sci.* 51 449-58 (1975).

Zukoski, C. F., Chow, M. K., Bogush, G. H., and Look, J-L., "Precipitation of Uniform Particles: The Role of Aggregation", *Mat. Res. Symp.* 180 131 (1990).

PART II

MODELING AND NUMERICAL SIMULATIONS

CHAPTER 16

ANALYTICAL MODELING OF MULTIPHASE FLOW

D. A. Drew
R. T. Lahey, Jr.

16.1 INTRODUCTION

The occurrence of multiphase flows is very widespread. Indeed, vapor/liquid flows are common in power production and utilization technologies and in chemical processing. Likewise, solid/fluid slurry flows, fluidized beds and various combustion processes are other common examples of important multiphase flows.

It is vital for industrial applications that multiphase flows be thoroughly understood. For example, gas turbines blades may be eroded by particulate matter suspended in the gas stream which impacts on turbine blades. The trajectories of individual particles moving through the gas turbine are very complicated, and depend on where and when the particles enter the device. Fortunately, such predictions are not normally required. A prediction of interest to the designer is the *expected* value of the particle flux near components susceptible to erosion in the gas turbine. Thus, perdictions of the average, or *expected* values, of the concentration and velocities of the particles are needed. The local concentration of particles is proportional to the probability that particles will be at the various points in the device at various times. The particle velocity field will be the mean velocity that the particle will have if they are at that position in the device. With this information, the design engineer will be able to assess the places where erosion due to particle impact may occur. Note that there may be no times that there will be many particles in some representative control volume. Thus, averaging schemes that depend on the concept of many representative particles will fail. Ensemble averaging, on the other hand, is appropriate. In this case,

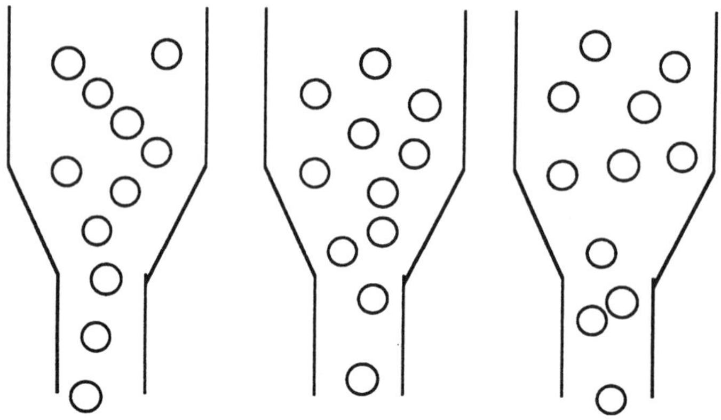

Figure 16-1 Three "Equivalent" Flows

the ensemble is the set of motions of a single particle through the device, given that it started at a random point at the inlet (with some distribution of position associated with the dynamics of the particle moving through the inlet flow) at a random time during the transient flow through the device. It is clear that the solution for the average concentration and average velocity gives little information about the behavior of a single particle in the device; however, the information is quite appropriate for assessing the probability of damage to the device.

Often in the literature, averages such as time averages (Ishii, 1975), space averages (Delhaye, 1981), and combinations of such averages (Lahey & Drew, 1989) are used. Such averages sometimes have utility in certain specific situations but have shortcomings.

A prime characteristic of multiphase flows is that there is uncertainty in the exact locations of particular constituents at any given time. This means that, for equivalent gross flows, there will be uncertainty in the locations of particular constituents for *all* times. For instance, consider a suspension of small particles in a liquid. The exact distribution of the locations of the particles is immaterial as long as they are reasonably "spread out." See Figure 16.1. We would not allow the particles to be lumped in some way that is not consistent with the initial conditions appropriate for the flow.

Consider the set of all experiments with the same boundary conditions, and initial conditions with some (undefined) properties that we would like to associate with the mean and distribution of the particles and their velocities. We call this set an *ensemble*. Such ensembles are reasonable sets over which to perform averages because variations in the details of the flows are assured in all situations, while at the same time variations in the gross flows cannot occur.

We take the approach here that we wish to predict averages over ensembles of flows. If, indeed, an average is justified, it does not matter that we allow the particles, bubbles or droplets to be rearranged in space within reason.

We start by deriving the balance equations by applying an averaging operation to the equations of motion for N continua separated by an interface across which the densities, velocities, *etc.* may jump. We then define appropriate averaged variables, and write the averaged equations in terms of them. As a special example, we shall consider these equations for two materials, specifically, we shall consider a particle/fluid mixture. We will then consider constitutive equations for the flow of a dilute mixture of solid spheres.

16.2 EXACT EQUATIONS

16.2.1 Local Balance Equations

The exact equations of motion, valid inside each material are

Mass Balance

$$\frac{\partial \rho}{\partial t} + \nabla \cdot \rho \mathbf{v} = 0, \tag{16.1}$$

Momentum Balance

$$\frac{\partial \rho \mathbf{v}}{\partial t} + \nabla \cdot \rho \mathbf{v}\mathbf{v} = \nabla \cdot \mathbf{T} + \rho \mathbf{g}, \tag{16.2}$$

Energy Balance

$$\frac{\partial}{\partial t}\rho(u + \frac{1}{2}v^2) + \nabla \cdot \rho \mathbf{v}(u + \frac{1}{2}v^2) = \nabla \cdot (\mathbf{T} \cdot \mathbf{v} - \mathbf{q}'') + \rho(r + \mathbf{g} \cdot \mathbf{v}), \tag{16.3}$$

Entropy Balance

$$\frac{\partial \rho s}{\partial t} + \nabla \cdot \rho \mathbf{v} s \geq \nabla \cdot \frac{\mathbf{q}''}{\theta} + \frac{r}{\theta}, \tag{16.4}$$

where ρ is the density, $\mathbf{v}$ is the velocity, $\mathbf{T}$ is the stress tensor

$$\mathbf{T} = -p\mathbf{I} + \boldsymbol{\tau} \tag{16.5}$$

Here p is the pressure and $\boldsymbol{\tau}$ is the shear stress; u is the internal energy; $\mathbf{q}''$ is the heat flux; r is the distributed (per unit mass) heating source; s is the entropy density; θ is the temperature; $\mathbf{q}''/\theta$is the entropy flux; and r/θ is the entropy supply.

Balance of angular momentum implies

$$\mathbf{T} = \mathbf{T}^t. \tag{16.6}$$

An alternate form of the energy equation can be obtained by eliminating the kinetic energy. This leads to the so-called internal energy equation

Internal Energy Balance

$$\frac{\partial \rho u}{\partial t} + \nabla \cdot \rho u \mathbf{v} = -\nabla \cdot \mathbf{q}'' + \rho r + \mathbf{T} : \nabla \mathbf{v}\,. \tag{16.7}$$

These equations are assumed to hold in the interiors of each material involved in the flow.

The canonical form of the equations of motion for the flow of a pure substance, or for the flow within a region consisting entirely of one phase in a multiphase mixture, can be written conveniently as

$$\frac{\partial \rho \Psi}{\partial t} + \nabla \cdot \rho \Psi \mathbf{v} = \nabla \cdot \mathbf{J} + \rho f\,, \tag{16.8}$$

where Ψ is the quantity conserved, $\mathbf{J}$ is its molecular flux, and f is its source density. The so-called molecular or diffusive fluxes $\mathbf{J}$ involve transport properties. For example, the heat-flux vector involves thermal conductivity; the stress, or momentum flux vector, involves the viscosity.

16.2.2 Jump Conditions

At an interface between phases, properties are discontinuous, although mass, momentum and energy must be conserved. Neglecting storage terms on the interface, the jump conditions valid across the interface are

Mass Jump

$$[\rho(\mathbf{v} - \mathbf{v}_i) \cdot \mathbf{n}]_{kl} = 0\,, \tag{16.9}$$

Momentum Jump

$$[(\rho \mathbf{v}(\mathbf{v} - \mathbf{v}_i) + \mathbf{T}) \cdot \mathbf{n}]_{kl} = \mathbf{m}_i^{\sigma}\,, \tag{16.10}$$

Energy Jump

$$[(\rho \left(u + \frac{1}{2}v^2\right)(\mathbf{v} - \mathbf{v}_i) + (\mathbf{T} \cdot \mathbf{v} - \mathbf{q}'')) \cdot \mathbf{n}]_{kl} = \epsilon_i^{\sigma}\,, \tag{16.11}$$

Entropy Jump

$$[(\rho s(\mathbf{v} - \mathbf{v}_i) + \frac{\mathbf{q}''}{\theta}) \cdot \mathbf{n}]_{kl} = \frac{\epsilon_i^\sigma}{\theta_i}, \tag{16.12}$$

where $\mathbf{v}_i$ is the velocity of the interface, $\mathbf{n}$ is the unit normal, and $[\ ]_{kl}$ denotes the jump across the interface between material k and material l. Here $\mathbf{m}_i^\sigma$ is the traction associated with surface tension; ϵ_i^σ is the surface energy associated with the interface, and θ_i is the temperature of the interface. Implicit in eq. (16.12) is the condition of thermal equilibrium at the interface, which forces the temperature to be continuous at the interface.

The surface traction, which has the dimensions of stress, is defined as (Aris, 1962)

$$\mathbf{m}_i^\sigma = (\mathbf{t}_\alpha a^{\alpha\beta}\sigma)_{,\beta} \tag{16.13}$$

where $\mathbf{t}_\alpha$ is the hybrid tensor; $a^{\alpha\beta}$ is the metric tensor in the surface; σ is the surface tension; and $(\ )_{,\beta}$ denotes the covariant derivative. The surface energy source term is given by

$$\epsilon_i^\sigma = (\mathbf{t}_\alpha a^{\alpha\beta}\sigma \cdot \mathbf{v}_i)_{,\beta}\,. \tag{16.14}$$

The generic balance principle for an interface is expressed by the following jump condition:

$$[(\rho\Psi(\mathbf{v} - \mathbf{v}_i) + \mathbf{J}) \cdot \mathbf{n}]_{kl} = M_i \tag{16.15}$$

where $\mathbf{v}_i$ is the velocity of the interface, $\mathbf{n}$ is the unit normal, and M_i is the interfacial source of Ψ.

16.2.3 Summary of the Exact Equations of Motion

The generic form of the equations of motion for the exact motions of the materials involved have been shown to be

$$\frac{\partial \rho\Psi}{\partial t} + \nabla \cdot \rho\Psi\mathbf{v} = \nabla \cdot \mathbf{J} + \rho f, \tag{16.16}$$

with corresponding jump conditions

$$[(\rho\Psi(\mathbf{v} - \mathbf{v}_i) + \mathbf{J}) \cdot \mathbf{n}]_{kl} = M_i\,. \tag{16.17}$$

The usual values for Ψ, $\mathbf{J}$, f and m are given in Table 16.1.

Balance Principle	Ψ	$\mathbf{J}$	f	M_i
Mass	1	0	0	0
Momentum	$\mathbf{v}$	$\mathbf{T}$	$\mathbf{g}$	$\mathbf{m}_i^\sigma$
Energy	$u + \frac{1}{2}v^2$	$\mathbf{T} \cdot \mathbf{v} - \mathbf{q}''$	$\mathbf{g} \cdot \mathbf{v} + r$	ϵ_i^σ
Entropy	s	$\mathbf{q}''/\theta$	r/θ	$\epsilon_i^\sigma/\theta_i$

TABLE 16.1. Variables in Generic Balance and Jump Equations

16.3 ENSEMBLE AVERAGING

A very elementary concept of averaging involves simply adding the observed values and dividing by the number of observations. Ensemble average is a generalization of adding the values of the variable for each realization, and dividing by the number of observations. We shall refer to a "process" as the set of possible flows that could occur, given that the initial and boundary conditions are those appropriate to the physical situation that we wish to describe. We refer to a "realization" of the flow as a possible motion that could have happened. Generally, we expect an infinite number of realizations of the flow, consisting of variations of position, attitude, and velocities of the discrete units and the fluid between them.

If f is some field (*i.e.*, a function of position $\mathbf{x}$ and time t) for some particular realization μ of the process, then the average of f is

$$\overline{f}(\mathbf{x}, t) = \int_{\mathcal{E}} f(\mathbf{x}, t; \mu) dm(\mu) \tag{16.18}$$

where $dm(\mu)$ is the measure (*i.e.*, probability) of observing process μ and $\mathcal{E}$ is the set of all processes of interest. We refer to $\mathcal{E}$ as the ensemble. The ensemble average is an average that allows the interpretation of the phenomena in terms of the repeatability of experiments. Any one exact experiment or realization will not be repeatable; however, any repetition of the experiment will lead to another member of the ensemble.

16.3.1 Other Averages

The literature in fluid mechanics contains many different types of averages. These are often motivated by the type of application. In this section we shall discuss some of these averaging processes.

The time average is defined by taking one realization, μ^*, integrating over the time interval from $t - T$ to t, and normalizing:

$$\overline{f}^t \equiv \frac{1}{T} \int_{t-T}^{t} f(\mathbf{x}, t - \tau; \mu^*) d\tau. \tag{16.19}$$

The use of this average relies on the ergodic hypothesis, and has the advantages that we need not sample the ensemble repeatedly but need make only one detailed observation to infer averaged values.

A similar interpretation can be given to the volume average. Thus,

$$f(\mathbf{x}, t; \mu_{\boldsymbol{\eta}}) = f(\mathbf{x} + \boldsymbol{\eta}, t; \mu^*) \,. \tag{16.20}$$

The ensemble average is the more fundamental averaging process. Both the time and volume average can be viewed as approximations to the ensemble average, which can be justified for steady flow or homogeneous flow, respectively.

16.3.2 Averaging Procedures

In order to apply the ensemble averaging procedure to the equations of motion, we shall need some results about the averaging procedure.

Gauss and Liebniz Rules

In order to average to the exact equations, we need expressions for $\overline{\partial f / \partial t}$ and $\overline{\nabla f}$. If f is "well behaved," then it is clear from the definition of the ensemble average that

$$\overline{\frac{\partial f}{\partial t}} = \frac{\partial \overline{f}}{\partial t} \tag{16.21}$$

and

$$\overline{\nabla f} = \nabla \overline{f} \tag{16.22}$$

Functions are generally discontinuous at the interface in most multiphase flows. They are well behaved within each phase, however. Thus, consider $\overline{X_k \nabla f}$, where X_k is the phase indicator function for phase k:

$$X_k = \begin{cases} 1, & \text{if } \mathbf{x} \in k; \\ 0, & \text{otherwise.} \end{cases} \tag{16.23}$$

Then

$$\begin{aligned} \overline{X_k \nabla f} &= \overline{\nabla X_k f} - \overline{f \nabla X_k} \\ &= \nabla \overline{X_k f} - \overline{f \nabla X_k}. \end{aligned} \tag{16.24}$$

and

$$\begin{aligned} \overline{X_k \frac{\partial f}{\partial t}} &= \overline{\frac{\partial X_k f}{\partial t}} - \overline{f \frac{\partial X_k}{\partial t}} \\ &= \frac{\partial \overline{X_k f}}{\partial t} - \overline{f \frac{\partial X_k}{\partial t}}. \end{aligned} \tag{16.25}$$

The second term on the right hand side in both of these equations is related to the surface average of f, evaluated on the phase k side, over the interface.

The Topological Equation

In the averaging process we use require the result

$$\frac{D_i X_k}{Dt} = \frac{\partial X_k}{\partial t} + \mathbf{v}_i \cdot \nabla X_k = 0\,. \tag{16.26}$$

This relation (16.26) has a reasonable physical explanation. Note that it is the "material" derivative of X_k following the interface. If we look at a point that is not on the interface, then either $X_k = 1$ or $X_k = 0$. In either case, the partial derivatives are both zero, and hence the expression (16.26) is zero. If we consider a point on the interface, and move with the interface, we see the function X_k as a constant jump. Thus, its material derivative is zero.

In this Chapter, we shall assume that all materials are inert, such as forsolid particles in a transporting fluid. Thus, we have no phase change or chemical reaction occuring at the interface. This implies that

$$\mathbf{v} \cdot \mathbf{n} = \mathbf{v}_i \cdot \mathbf{n} \tag{16.27}$$

at the interface. This in turn gives

$$\frac{\partial X_k}{\partial t} + \mathbf{v} \cdot \nabla X_k = 0\,. \tag{16.28}$$

16.3.3 Averaged Equations

The process of deriving the averaged equations involves multiplying the exact balance equation, eq. (16.8) by X_k and ensemble averaging. The result of this procedure can be seen by manipulating the phase function X_k. We have:

$$\frac{\partial \overline{X_k \rho \Psi}}{\partial t} + \nabla \cdot \overline{X_k \rho \Psi \mathbf{v}} - \nabla \cdot \overline{X_k \mathbf{J}} - \overline{X_k \rho f} = \overline{\rho \Psi \left(\frac{\partial X_k}{\partial t} + \mathbf{v} \cdot \nabla X_k \right) - \mathbf{J} \cdot \nabla X_k}\,. \tag{16.29}$$

Subtracting the average of $\rho\Psi$ times the result in (16.26) reduces the right hand side of (16.29) to

$$-\overline{\mathbf{J} \cdot \nabla X_k}\,. \tag{16.30}$$

This is the interfacial source of Ψ, and is due to the molecular flux $\mathbf{J}$.

Using $\Psi = 1$, $\mathbf{J} = 0$, and $f = 0$ results in the equation of balance of mass; using $\Psi = \mathbf{v}$, $\mathbf{J} = \mathbf{T}$, and $f = \mathbf{g}$ results in the equation of balance of mass; and using $\Psi = e = u + \frac{1}{2}v^2$; $\mathbf{J} = \mathbf{T}\cdot\mathbf{v} - \mathbf{q}''$; and $f = \mathbf{g}\cdot\mathbf{v} + r$ gives the equations of balance of mass, momentum and energy for each component. The entropy inequality for component k can be obtained by taking $\Psi = s$, $\mathbf{J} = -\mathbf{q}''/T$ and $f = r/T$ and by changing the equality in eq. (16.29) to a positive semidefinite (≥ 0) relationship.

The resultant ensemble averaged equations are

Mass Balance

$$\frac{\partial \overline{X_k \rho}}{\partial t} + \nabla \cdot \overline{X_k \rho \mathbf{v}} = 0 \tag{16.31}$$

Momentum Balance

$$\frac{\partial \overline{X_k \rho \mathbf{v}}}{\partial t} + \nabla \cdot \overline{X_k \rho \mathbf{v}\mathbf{v}} = \nabla \cdot \overline{X_k \mathbf{T}} + \overline{X_k \rho \mathbf{g}} - \overline{\mathbf{T}) \cdot \nabla X_k}. \tag{16.32}$$

Energy Balance

$$\begin{aligned}\frac{\partial \overline{X_k \rho (u + \frac{1}{2}v^2)}}{\partial t} \quad + \quad & \nabla \cdot \overline{X_k \rho (u + \frac{1}{2}v^2)\mathbf{v}} = \nabla \cdot \overline{X_k (\mathbf{T}\cdot\mathbf{v} - \mathbf{q}'')} \\ & + \overline{X_k \rho (\mathbf{g}\cdot\mathbf{v} + r)} - \overline{(\mathbf{T}\cdot\mathbf{v} - \mathbf{q}'') \cdot \nabla X_k}\,.\end{aligned} \tag{16.33}$$

Entropy

$$\frac{\partial \overline{X_k \rho s}}{\partial t} + \nabla \cdot \overline{X_k \rho s \mathbf{v}} + \nabla \cdot \overline{X_k \frac{\mathbf{q}''}{T}} - \overline{X_k \rho \frac{r}{T}} - \overline{\left(\frac{\mathbf{q}''}{T}\right) \cdot \nabla X_k} \geq 0\,. \tag{16.34}$$

Instead of using a balance equation for the total energy, it is equally valid to use one for the internal energy. This gives

Internal Energy Balance

$$\begin{aligned}\frac{\partial \overline{X_k \rho u}}{\partial t} + \nabla \cdot \overline{X_k \rho u \mathbf{v}} \quad = \quad & -\nabla \cdot \overline{X_k \mathbf{q}''} + \overline{X_k \mathbf{T} : \nabla \mathbf{v}} \\ & + \overline{X_k \rho r} + \overline{\mathbf{q}'' \cdot \nabla X_k}\,.\end{aligned} \tag{16.35}$$

16.3.4 Definition of Average Variables

In this section, we define the appropriate average variables describing multiphase flow.

First, the geometry of the exact, or microscopic situation is defined in terms of the phase indicator function X_k. The average of X_k is the average fraction of the occurrences of phase k at point $\mathbf{x}$ at time t.

$$\alpha_k = \overline{X_k} \tag{16.36}$$

It has become customary to call this variable the volume fraction of phase k, even though technically the volume fraction is the ratio of the volume of phase k in a small region, divided by the total volume. We note that this concept is intimately tied to volume averaging, and so is correct only for spatially homogeneous situations.

Another important geometric variable is the interfacial area density. The interfacial area density of phase k is defined by

$$A_k''' = -\overline{\partial X_k/\partial n_k} \tag{16.37}$$

where $\mathbf{n}_k$ is the unit external normal to material k. The area of contact between material k and material l is defined to be

$$A_{kl}''' = -\overline{X_l \partial X_k/\partial n_k}\,. \tag{16.38}$$

Thus,

$$A_k''' = \sum_{l\neq k} A_{kl}'''\,. \tag{16.39}$$

All the remaining variables are defined in terms of weighted averages. The main variables are either weighted with the component function X_k or mass-weighted (or Favré) averaged (weighted by $X_k\rho$). Other variables are weighted with the interface variable ∇X_k.

The "conserved" variables are the appropriately weighted density,

$$\overline{\rho}_k^x = \overline{X_k\rho}/\alpha_k, \, , \tag{16.40}$$

velocity,

$$\overline{\mathbf{v}}_k^{x\rho} = \overline{X_k\rho\mathbf{v}}/\alpha_k\overline{\rho}_k^x\,, \tag{16.41}$$

internal energy,

$$\overline{u}_k^{x\rho} = \overline{X_k\rho u}/\alpha_k\overline{\rho}_k^x\,, \tag{16.42}$$

and entropy,

$$\overline{s}_k^{x\rho} = \overline{X_k\rho s}/\alpha_k\overline{\rho}_k^x\,. \tag{16.43}$$

The variables representing the averaged effects of the molecular fluxes are stress,

$$\overline{\mathbf{T}}_k^x = \overline{X_k\mathbf{T}}/\alpha_k\,, \tag{16.44}$$

heat flux,

$$\overline{\mathbf{q}}''^{x}_{k} = \overline{X_k \mathbf{q}}/\alpha_k \,, \tag{16.45}$$

and entropy flux,

$$\overline{\boldsymbol{\phi}}^{x}_{k} = \overline{X_k(\mathbf{q}''/\theta)}/\alpha_k \,. \tag{16.46}$$

The average body sources are body force,

$$\overline{\mathbf{g}}^{x\rho}_{k} = \overline{X_k \rho \mathbf{g}}/\alpha_k \overline{\rho}^{x}_{k} \,, \tag{16.47}$$

energy source,

$$\overline{r}^{x\rho}_{k} = \overline{X_k \rho r}/\alpha_k \overline{\rho}^{x}_{k} \,, \tag{16.48}$$

and entropy source,

$$\Sigma^{x\rho}_{k} = \overline{(X_k \rho r/\theta)}/\alpha_k \overline{\rho}^{x}_{k} \,. \tag{16.49}$$

The molecular fluxes to the interface act as sources of mass, momentum, energy or entropy to the interface. These terms are important in the theory of multicomponent flows since they represent the interactions between the materials. The interfacial momentum source is defined by

$$\mathbf{M}_k = -\overline{\mathbf{T} \cdot \nabla X_k} \,, \tag{16.50}$$

the interfacial heat source is defined by

$$E_k = \overline{\mathbf{q}'' \cdot \nabla X_k} \,, \tag{16.51}$$

the interfacial work is defined by

$$W_k = -\overline{\mathbf{T} \cdot \mathbf{v} \cdot \nabla X_k} \,, \tag{16.52}$$

and the interfacial entropy source is defined by

$$S_k = \overline{(\mathbf{q}''/\theta) \cdot \nabla X_k} \,. \tag{16.53}$$

It is also useful to have expressions for the interactions between phases k and l. We have

$$\mathbf{M}_{kl} = -\overline{X_l \mathbf{T} \cdot \nabla X_k} \,, \tag{16.54}$$

the interfacial heat source is defined by

$$E_{kl} = \overline{X_l \mathbf{q}'' \cdot \nabla X_k} \,, \tag{16.55}$$

the interfacial work is defined by

$$W_{kl} = -\overline{X_l \mathbf{T} \cdot \mathbf{v} \cdot \nabla X_k} \,, \tag{16.56}$$

and the interfacial entropy source is defined by

$$S_{kl} = \overline{(X_l \mathbf{q}''/\theta) \cdot \nabla X_k}\,, \tag{16.57}$$

where

$$\mathbf{M}_k = \sum_{l \neq k} \mathbf{M}_{kl}\,, \tag{16.58}$$

$$E_k = \sum_{l \neq k} E_{kl}\,, \tag{16.59}$$

$$W_k = \sum_{l \neq k} W_{kl}\,, \tag{16.60}$$

$$S_k = \sum_{l \neq k} S_{kl}\,. \tag{16.61}$$

The relative motion of the various phases give rise to velocities that are not "laminar" in general. The effect of the fluctuations of a variable from its mean value is accounted for by introducing its fluctuating field (denoted by the prime superscript). The variables defined in this way are the so-called Reynoids stress

$$\mathbf{T}_k^{Re} = -\overline{X_k \rho \mathbf{v}_k' \mathbf{v}_k'}/\alpha_k, \tag{16.62}$$

and the fluctuation (*i.e.*, Reynolds) energy flux

$$\mathbf{q}_k^{Re} = \hat{\mathbf{q}}_k^{Re} + \mathbf{q}_k^{p} + \mathbf{q}_k^{\tau} + \mathbf{q}_k^{K}\,. \tag{16.63}$$

The pieces of the fluctuation energy flux are the fluctuation (*i.e.*, Reynolds) kinetic energy flux

$$\mathbf{q}_k^{K} = \overline{X_k \rho \mathbf{v}_k' \frac{1}{2} (v_k')^2}/\alpha_k\,, \tag{16.64}$$

the fluctuation (*i.e.*, Reynolds) pressure-velocity correlation

$$\mathbf{q}_k^{p} = \overline{X_k p \mathbf{v}_k'}/\alpha_k\,, \tag{16.65}$$

and the fluctuation (*i.e.*, Reynolds) shear working

$$\mathbf{q}_k^{\tau} = -\overline{X_k \boldsymbol{\tau} \cdot \mathbf{v}_k'}/\alpha_k\,, \tag{16.66}$$

where

$$\mathbf{v}_k' = \mathbf{v} - \overline{\mathbf{v}}_k^{x\rho}\,, \tag{16.67}$$

$$p_k' = p - \overline{p}_k^{x}\,. \tag{16.68}$$

We can combine the terms in eqs. (16.65) and (16.66) as the fluctuation (*i.e.*, Reynolds) shear work

$$\mathbf{q}_k^{T} = -\overline{X_k \mathbf{T} \cdot \mathbf{v}_k'}/\alpha_k\,. \tag{16.69}$$

There is an analogous term in the entropy inequality, the fluctuation (*i.e.*, Reynolds) entropy flux

$$\overline{\boldsymbol{\phi}}_k^{Re} = -\overline{X_k \rho s \mathbf{v}'_k}/\alpha_k\,. \tag{16.70}$$

In the energy equation, it is convenient to combine other terms with the turbulent energy flux. We denote these other effective turbulent heat flux terms with carets. They are the fluctuation (*i.e.*, Reynolds) internal energy flux

$$\hat{\mathbf{q}}_k^{Re} = \overline{X_k \rho \mathbf{v}'_k u'_k}/\alpha_k\,, \tag{16.71}$$

and the fluctuation (*i.e.*, Reynolds) enthalpy flux

$$\hat{\hat{\mathbf{q}}}_k^{Re} = \overline{X_k \rho \mathbf{v}'_k h'_k}/\alpha_k\,. \tag{16.72}$$

Thus we may write

$$\mathbf{q}_k^{Re} = \hat{\hat{\mathbf{q}}}_k^{Re} + \mathbf{q}_k^{T} + \mathbf{q}_k^{K}\,. \tag{16.73}$$

Some other terms involving fluctuations that appear are the fluctuation (*i.e.*, Reynolds) kinetic energy

$$u_k^{Re} = \frac{1}{2}\overline{X_k \rho v'^2_k}/\alpha_k \overline{\overline{\rho}}_k^{x} \tag{16.74}$$

and the dissipation

$$D_k = \overline{X_k \boldsymbol{\tau} : \nabla \mathbf{v}'_k}/\alpha_k. \tag{16.75}$$

Note that

$$\mathrm{tr}\ \mathbf{T}_k^{Re} = -\overline{X_k \rho_k \mathbf{v}'_k \cdot \mathbf{v}'_k}/\alpha_k = -2u_k^{Re}\,. \tag{16.76}$$

The averaged interfacial pressure p_{ki} and shear stress $\boldsymbol{\tau}_{ki}$ are introduced to separate mean field effects from local effects in the interfacial force. The interfacial pressure is defined by

$$p_{ki} = \overline{p \partial X_k/\partial n_k}/A_i''' \tag{16.77}$$

and the interfacial shear stress is

$$\boldsymbol{\tau}_{ki} = \overline{\boldsymbol{\tau}_k \partial X_k/\partial n_k}/A_i''' \tag{16.78}$$

Thus,

$$\begin{aligned}
\mathbf{M}_k &= -\overline{\mathbf{T}\cdot\nabla \mathbf{X_k}} \\
&= \overline{p\nabla X_k} - \overline{\boldsymbol{\tau}\cdot\nabla X_k} \\
&= p_{ki}\overline{\nabla X_k} - \boldsymbol{\tau}_{ki}\cdot\overline{\nabla X_k} - \overline{\mathbf{T}'_{ki}\cdot\nabla X_k} \\
&= p_{ki}\nabla\alpha_k - \boldsymbol{\tau}_{ki}\nabla\alpha_k + \mathbf{M}'_k\,,
\end{aligned} \tag{16.79}$$

where we have defined an interfacial momentum source as

$$\mathbf{M}'_k = \mathbf{M}_k - p_{ki}\nabla\alpha_k + \boldsymbol{\tau}_{ki} \cdot \nabla\alpha_k \tag{16.80}$$

and have introduced

$$\mathbf{T}'_{ki} = -p'_{ki}\mathbf{I} + \boldsymbol{\tau}'_{ki} = -(p - p_{ki})\mathbf{I} + (\boldsymbol{\tau} - \boldsymbol{\tau}_{ki})\,. \tag{16.81}$$

Similarly, the interfacial work term is rearranged to read

$$\begin{aligned} W_k &= -\overline{\mathbf{T} \cdot \mathbf{v} \cdot \nabla X_k} \\ &= -\overline{\mathbf{T}\overline{\mathbf{v}}_k^{xp} \cdot \nabla X_k} - \overline{\mathbf{T} \cdot \mathbf{v}'_k \cdot \nabla X_k} \\ &= W'_k + \mathbf{M}_k \cdot \overline{\mathbf{v}}_k^{xp}\,, \end{aligned} \tag{16.82}$$

where the so-called interfacial extra work is defined by

$$W'_k = -\overline{\mathbf{T} \cdot \mathbf{v}'_k \cdot \nabla X_k}\,. \tag{16.83}$$

One term appears in the equation for the averaged internal energy that has no counterpart in the other equations. We choose to split it into two terms, the internal dissipation

$$D_k = \frac{\overline{X_k \boldsymbol{\tau} : \nabla \mathbf{v}'_k}}{\alpha_k}\,, \tag{16.84}$$

and a "p dV" type work term

$$P_k = \frac{\overline{X_k p \nabla \cdot \mathbf{v}'_k}}{\alpha_k}\,. \tag{16.85}$$

16.3.5 Averaged Equations

We now present the averaged equations. The averaged equations governing each phase are

Mass Balance

$$\frac{\partial \alpha_k \overline{\rho}_k^x}{\partial t} + \nabla \cdot \alpha_k \overline{\rho}_k^x \overline{\mathbf{v}}_k^{xp} = 0 \tag{16.86}$$

Momentum Balance

$$\frac{\partial \alpha_k \overline{\rho}_k^x \overline{\mathbf{v}}_k^{xp}}{\partial t} + \nabla \cdot \alpha_k \overline{\rho}_k^x \overline{\mathbf{v}}_k^{xp} \overline{\mathbf{v}}_k^{xp} = \nabla \cdot \alpha_k \left(\overline{\mathbf{T}}_k^x + \mathbf{T}_k^{Re}\right) + \alpha_k \overline{\rho}_k^x \mathbf{g} + \mathbf{M}_k\,, \tag{16.87}$$

Energy Balance

$$\frac{\partial}{\partial t}\alpha_k\overline{\rho}_k^x\left(\overline{u}_k^{xp}+\frac{1}{2}(\overline{v}_k^{xp})^2+u_k^{Re}\right)+\nabla\cdot\alpha_k\overline{\rho}_k^x\overline{\mathbf{v}}_k^{xp}\left(\overline{u}_k^{xp}+\frac{1}{2}(\overline{v}_k^{xp})^2+u_k^{Re}\right)=$$
$$\nabla\cdot\alpha_k\left((\overline{\mathbf{T}}_k^x+\mathbf{T}_k^{Re})\cdot\overline{\mathbf{v}}_k^{xp}-\overline{\mathbf{q}''}_k^x-\mathbf{q}_k^{Re}\right)$$
$$+\alpha_k\overline{\rho}_k^x\left(r_k+\mathbf{g}\cdot\overline{\mathbf{v}}_k^{xp}\right)+E_k+W_k\,. \tag{16.88}$$

The equation for the internal energy is
Internal Energy Balance

$$\frac{\partial\alpha_k\overline{\rho}_k^x\overline{u}_k^{xp}}{\partial t}+\nabla\cdot\alpha_k\overline{\rho}_k^x\overline{\mathbf{v}}_k^{xp}\overline{u}_k^{xp} = \alpha_k\overline{\mathbf{T}}_k^x:\nabla\overline{\mathbf{v}}_k^{xp}-\nabla\cdot\alpha_k(\overline{\mathbf{q}''}_k^x+\hat{\mathbf{q}}_k^{Re})$$
$$+\alpha_k\overline{\rho}_k^x r_k+E_k+\alpha_k D_k-\alpha_k P_k\,. \tag{16.89}$$

The entropy inequality becomes
Entropy Balance

$$\frac{\partial\alpha_k\overline{\rho}_k^x\overline{s}_k^{xp}}{\partial t}+\nabla\cdot\alpha_k\overline{\rho}_k^x\overline{s}_k^{xp}\overline{\mathbf{v}}_k^{xp}\geq\nabla\cdot\alpha_k\left(\overline{\boldsymbol{\phi}}_k^x+\boldsymbol{\phi}_k^{Re}\right)+\alpha_k\overline{\rho}_k^x\Sigma_k+S_k\,. \tag{16.90}$$

The equation of balance of mass for phase k eq. (16.86) can be used in the momentum equation eq. (16.87) to yield the Lagrangian form of the momentum equation:

$$\alpha_k\overline{\rho}_k^x\frac{D_k\overline{\mathbf{v}}_k^{xp}}{Dt} = \alpha_k\overline{\rho}_k^x\left(\frac{\partial\overline{\mathbf{v}}_k^{xp}}{\partial t}+\overline{\mathbf{v}}_k^{xp}\cdot\nabla\overline{\mathbf{v}}_k^{xp}\right)$$
$$= \nabla\cdot\alpha_k(\overline{\mathbf{T}}_k^x+\mathbf{T}_k^{Re})+\alpha_k\overline{\rho}_k^x\mathbf{g}+\mathbf{M}_k\,. \tag{16.91}$$

Similarly, the equation of balance of mass applied to the energy equation yields

$$\alpha_k\overline{\rho}_k^x\frac{D_k}{Dt}\left(\overline{u}_k^{xp}+\frac{1}{2}(\overline{v}_k^{xp})^2+u_k^{Re}\right) = \nabla\cdot\alpha_k\left(\overline{\mathbf{T}}_k^x+\mathbf{T}_k^{Re})\cdot\overline{\mathbf{v}}_k^{xp}-\overline{\mathbf{q}''}_k^x-\mathbf{q}_k^{Re}\right)$$
$$+\alpha_k\overline{\rho}_k^x(r_k+\mathbf{g}\cdot\overline{\mathbf{v}}_k^{xp})+E_k+W_k\,. \tag{16.92}$$

With the aid of the momentum equation, eq. (16.87), and the identity

$$\frac{D_k}{Dt}\left(\frac{1}{2}(\overline{v}_k^{xp})^2\right)=\overline{\mathbf{v}}_k^{xp}\cdot\frac{D_k\overline{\mathbf{v}}_k^{xp}}{Dt}\,, \tag{16.93}$$

the mechanical energy equation can be derived by taking the dot product of the averaged velocity with the momentum equation. Then eq. (16.91) yields

$$\alpha_k \overline{\rho}_k^x \frac{D_k \frac{1}{2}(\overline{v}_k^{xp})^2}{Dt} = \overline{\mathbf{v}}_k^{xp} \cdot \nabla \alpha_k (\overline{\mathbf{T}}_k^x + \mathbf{T}_k^{Re}) + \alpha_k \overline{\rho}_k^x \overline{\mathbf{v}}_k^{xp} \cdot \mathbf{g} + \overline{\mathbf{v}}_k^{xp} \cdot \mathbf{M}_k \,. \quad (16.94)$$

Using this, the following form of the energy equation can be obtained:

$$\alpha_k \overline{\rho}_k^x \frac{D_k}{Dt} (\overline{u}_k^{xp} + u_k^{Re}) = \alpha_k (\overline{\mathbf{T}}_k^x + \mathbf{T}_k^{Re}) : \nabla \overline{\mathbf{v}}_k^{xp} - \nabla \cdot \alpha_k (\overline{\mathbf{q}}''^x_k + \mathbf{q}_k^{Re}) + \alpha_k \overline{\rho}_k^x r_k + E_k + W_k' \,. \quad (16.95)$$

Subtracting the equation for the averaged internal energy gives the equation for the average fluctuation kinetic energy

$$\frac{\partial \alpha_k \overline{\rho}_k^x u_k^{Re}}{\partial t} + \nabla \cdot \alpha_k \overline{\rho}_k^x u_k^{Re} \overline{\mathbf{v}}_k^{xp} = \alpha_k \mathbf{T}_k^{Re} : \nabla \overline{\mathbf{v}}_k^{xp} - \nabla \cdot \alpha_k (\mathbf{q}_k^K + \mathbf{q}_k^T) + W_k' - \alpha_k D_k + \alpha_k P_k \,. \quad (16.96)$$

This equation is of special interest in multiphase flows since it is the motivation for the structure of the theory with an entropy, or "disorder," at the scale of the molecular vibrations, and a separate independent entropy, or "disorder," at the scale of the phase motion fluctuations. We note that there is an adequate definition of temperature at the molecular level, but that there is no counterpart of "temperature" in the derivation of the averaged equations of motion.

16.3.6 Jump Conditions

The jump conditions are derived by multiplying the exact jump condition by $\mathbf{n}_1 \cdot \nabla X_1$ and averaging. This process yields the following conditions:

Momentum Jump

$$\sum_{k=1}^{N} \mathbf{M}_k = \overline{\mathbf{m}_i^\sigma} \,, \quad (16.97)$$

Energy Jump

$$\sum_{k=1}^{N} (E_k + W_k) = \overline{\epsilon_i^\sigma} \,, \quad (16.98)$$

Entropy Jump

$$\sum_{k=1}^{N} S_k \geq 0 \,, \quad (16.99)$$

where $\overline{\mathbf{m}_i^\sigma}$ is the surface tension source, and $\overline{e_i^\sigma}$ is the interfacial energy source. For solid/fluid flows, these interfacial source terms are zero.

The ensemble averaging process has generated many new terms which must be constituted in order to achieve closure. Thus, let us now turn our attention to the closure laws for two-fluid models of two-phase flows.

16.4. CLOSURE CONDITIONS

16.4.1 Completeness of the Formulation

The three-dimensional, unsteady multiphase model given by (16.86), (16.87), (16.88), and (16.90), must be supplemented with state equations, constitutive equations such as those discussed in Section 16.5 and boundary and initial conditions. Thus let us discuss a general approach by which closure can be achieved.

Initial conditions specify how the experiment or the multiphase flow starts. The boundary conditions specify how the flow interacts with its environment, specifically with the inlet and outlet flow devices and with the walls of conduits.

The constitutive equations generally specify how the materials (*i.e.*, phases) interact with themselves and with each other. State equations specify the thermodynamic state of the material as functions of those (state) variables that determine it. For example, in a single phase fluid, the internal energy is determined by the density and the entropy. Boundary and initial conditions are important, in the sense that if one is specified incorrectly, the predicted flow will be in error. However, the key to the accurate modeling of multiphase flows is to specify correctly the phase interaction terms ($\mathbf{M}_k$, E_k, ...), and the self-interaction terms ($\overline{\mathbf{T}}_k^x$, $\mathbf{T}_k^{Re}$, $\overline{\mathbf{q}''}_k^x$, $\mathbf{q}_k^{Re}$, ...) in terms of the state variables of the (α_k, $\overline{\mathbf{v}}_k^{xp}$, ...) .

16.4.2 Constitutive Equations

The variables still to be specified involve the phasic interaction terms, and the self-interaction terms, which, in turn, depend on transport properties and on turbulence effects. In view of the interfacial jump conditions, not all of the phase interaction and phase change terms are independent. The first step in completing the formulation, then, requires the choice of a set of dependent variables and a set of independent variables whose values determine them. Once this functional dependence is established, we must find the functions which describe the constitution of the materials. The following principles may be applied to guide the development of these constitutive relations.

Guiding Principles

The principles of constitutive equations (Truesdell & Toupin, 1960) provide a rational means for obtaining descriptions of classes of materials, without inadvertently neglecting an important dependence. We shall consider here the principles of equipresence, well-posedness, phase separation and frame indifference. These are the most fundamental and restrictive of the guiding principles provided by continuum mechanics.

The principle of equipresence states that all variables should be included in each constitutive equation, unless another principle shows that a particular dependence cannot occur.

The principle of well-posedness states that the description of the motion should be such that a solution to the initial boundary value problem exists and depends continuously on the initial and boundary conditions. Moreover, it implies that all eigenvalues of the system of partial differential equations be real.

The principle of frame indifference asserts that the constitutive equations cannot depend on the reference frame. The motivation is that reference frames are inventions of the observer and are not intrinsic to the materials being described. Thus, the functional dependence of the constituted variable on the variables describing the mechanical/thermal state must be independent of frame (*i.e.*, frame indifferent), and the variables used to describe the mechanical/thermal state must also be objective, and thus independent of coordinate system. We shall discuss this principle in some detail next.

Frame Indifference

Consider two reference frames, so that $\mathbf{x}$ and $\mathbf{x}'$, are different representations of the same vector. See Figure 16.2. The two representations are related by

$$\mathbf{x}' = \mathbf{Q} \cdot \mathbf{x} + \mathbf{b}, \tag{16.100}$$

where $\mathbf{Q}(\mathbf{t})$ is an orthonormal transformation, and $\mathbf{b}$ is a translation. A scalar is objective if its value is invariant to a change of coordinate system, that is, if $s' = s$. A vector is objective if it changes coordinates according to $\mathbf{V}' = \mathbf{Q} \cdot \mathbf{V}$. Tensors are objective if they transform each phase correctly. Specifically, a second order tensor is objective if $\mathbf{T}' = \mathbf{Q} \cdot \mathbf{T} \cdot \mathbf{Q}^T$.

Velocity is objective. If we differentiate eq. (16.100) following material

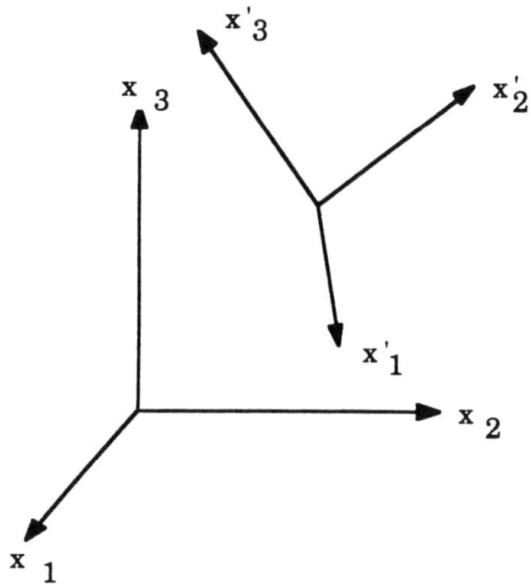

Figure 16-2 Two Different Coordinate Frames

k we see that

$$\mathbf{v}'_k = \mathbf{Q} \cdot \mathbf{v}_k + (\dot{\mathbf{Q}} \cdot \mathbf{x} + \dot{\mathbf{b}}) . \tag{16.101}$$

If eq. (16.101) is evaluated for two different values of k and the results subtracted, the velocity difference becomes

$$\mathbf{v}'_k - \mathbf{v}'_l = \mathbf{Q} \cdot (\mathbf{v}_k - \mathbf{v}_l) . \tag{16.102}$$

Thus, velocity differences are objective.

In multiphase flows there are reasons to believe that the interactions can depend on accelerations. However, in general, accelerations are not objective. Following phase l and differentiating eq. (16.101) leads to

$$\frac{D_l \mathbf{v}'_k}{Dt} = \mathbf{Q} \cdot \frac{D_l \mathbf{v}_k}{Dt} + \dot{\mathbf{Q}} \cdot (\mathbf{v}_k + \mathbf{v}_l) + (\ddot{\mathbf{Q}} \cdot \mathbf{x} + \ddot{\mathbf{b}}) . \tag{16.103}$$

Again subtracting, we have

$$\frac{D_k \mathbf{v}'_l}{Dt} - \frac{D_l \mathbf{v}'_k}{Dt} = \mathbf{Q} \cdot \left(\frac{D_k \mathbf{v}_l}{Dt} - \frac{D_l \mathbf{v}_k}{Dt} \right) . \tag{16.104}$$

Thus, particular acceleration differences are objective.

In exactly the same manner as in classical continuum mechanics, since

$$\nabla' \mathbf{v}'_k = \mathbf{Q} \cdot (\nabla \mathbf{v}_k) \cdot \mathbf{Q}^T + \dot{\mathbf{Q}} \cdot \mathbf{Q}^T , \tag{16.105}$$

we have

$$\mathbf{D}_k = \frac{1}{2} \left(\nabla \mathbf{v}_k + (\nabla \mathbf{v}_k)^T \right) \tag{16.106}$$

is objective, whereas

$$\begin{aligned}\mathbf{W}'_k &= \mathbf{Q}\cdot\mathbf{W}_k\cdot\mathbf{Q}^T + \frac{1}{2}\left(\dot{\mathbf{Q}}\cdot\mathbf{Q}^T - \mathbf{Q}\cdot\dot{\mathbf{Q}}^T\right)\\ &\neq \mathbf{Q}\cdot\mathbf{W}_k\cdot\mathbf{Q}^T. \qquad (16.107)\end{aligned}$$

Thus, the rotation tensor is not objective. However, there are other objective tensors in multiphase continuum mechanics. If we add eq. (16.105) for different values of k, we have

$$\nabla'\mathbf{v}'_k + (\nabla'\mathbf{v}'_l)^T = \mathbf{Q}\cdot[\nabla\mathbf{v}_k + (\nabla\mathbf{v}_l)^T]\cdot\mathbf{Q}^T, \qquad (16.108)$$

where we have used the identity $\mathbf{Q}\cdot\mathbf{Q}^T = \mathbf{I}$ to obtain $\dot{\mathbf{Q}}\cdot\mathbf{Q}^T + \mathbf{Q}\cdot\dot{\mathbf{Q}}^T = 0$. Hence

$$\mathbf{D}_{kl} = \frac{1}{2}[\nabla\mathbf{v}_k + (\nabla\mathbf{v}_l)^T] \qquad (16.109)$$

is objective for any k and l. It is also clear from eq. (16.107) that

$$\mathbf{W}_{kl} = \mathbf{W}_k - \mathbf{W}_l \qquad (16.110)$$

is objective.

We also note that

$$\begin{aligned}&\frac{D_k\mathbf{v}'_k}{Dt} - \frac{D_l\mathbf{v}'_l}{Dt} - 2\mathbf{W}'_m\cdot(\mathbf{v}'_k - \mathbf{v}'_l) =\\ &\mathbf{Q}\cdot\left(\frac{D_k\mathbf{v}_k}{Dt} - \frac{D_l\mathbf{v}_l}{Dt} - 2\mathbf{W}_m\cdot(\mathbf{v}_k - \mathbf{v}_l)\right), \qquad (16.111)\end{aligned}$$

so that the quantity

$$\frac{D_k\mathbf{v}_k}{Dt} - \frac{D_l\mathbf{v}_l}{Dt} - 2\mathbf{W}_m\cdot(\mathbf{v}_k - \mathbf{v}_l) \qquad (16.112)$$

is objective.

One implication of the principle of frame indifference is that constituted variables can depend on only objective variables. The second implication is that the function giving the dependence of the constituted variable on the objective variables must be independent of the coordinate system, or frame indifferent. The complicated forms which are assumed for the interfacial interaction terms and the average molecular and Reynolds fluxes can be made objective by assuming that they depend on only objective quantities. This will ensure that the total is objective. It is the total quantity (interfacial force, for example) which should be compared with physical observations.

16.5 MICROSCOPIC CONSIDERATIONS

Determination of constitutive equations for particle-fluid flows is difficult. In this Section, we study the exact flow of an inviscid fluid around a single sphere, in order to motivate the choices of constitutive equations given in the following Section. The calculations should give valid values for the constituted quantities for dilute flows.

16.5.1 Inviscid Flow Around a Sphere

Consider a sphere of radius a located at a point $\mathbf{z}$ in a flow field, moving with velocity $\mathbf{v}_p(\mathbf{z}, t)$. The velocity of the flow field is assumed to be such that it would be $\mathbf{v}_0 + \mathbf{e}_f \cdot \mathbf{x}$ if the sphere were not there. The velocity potential for the flow of fluid around the sphere is given by Voinov (1973) as

$$\begin{aligned}\phi &= \mathbf{v}_0(t) \cdot \mathbf{x} + \frac{1}{2}\mathbf{x} \cdot \mathbf{e}_f \cdot \mathbf{x} \\ &\quad + \frac{1}{2}\left(\mathbf{v}_p(\mathbf{z}, t) - \mathbf{v}_0(t) - \mathbf{z} \cdot \mathbf{e}_f\right) \cdot (\mathbf{x} - \mathbf{z}) \left(\frac{a^3}{r^3}\right) \\ &\quad + \frac{1}{3}(\mathbf{x} - \mathbf{z}) \cdot \mathbf{e}_f \cdot (\mathbf{x} - \mathbf{z}) \left(\frac{a^5}{r^5}\right) . \end{aligned} \tag{16.113}$$

Averaged Velocities

The ensemble average for the flow of a dilute suspension of spheres can be approximated by the "cell" average (Arnold, *et al.*, 1989), obtained by averaging the value of the variable to be averaged, f, calculated with a sphere at location $\mathbf{z}$. The averaging is performed by integrating the $\mathbf{z}$ variable over all positions in a sphere of raduis R surrounding the field point $\mathbf{x}$. See Figure 16.3. The radius of this "cell" is chosen so that

$$\alpha_p = \frac{\frac{4}{3}\pi a^3}{\frac{4}{3}\pi R^3} . \tag{16.114}$$

Thus, the average velocity of the fluid phase is given by

$$\overline{\mathbf{v}}_f^{xp} = \frac{1}{\frac{4}{3}\pi(R^3 - a^3)} \int_a^R \int\int_{\Omega(r)} \mathbf{v}(\mathbf{x}, t|\mathbf{z})\, d\Omega\, dr , \tag{16.115}$$

where the integration is over the $\mathbf{z}$ variable.

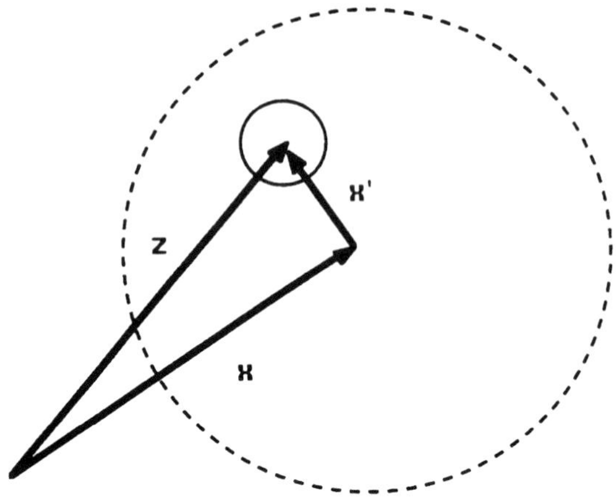

Figure 16-3 A "Cell".

In order to evaluate the integrals appearing in eq. (16.115), we must express the $\mathbf{z}$ dependence of the velocities in terms of $\mathbf{x}$ and $\mathbf{x}' = \mathbf{x} - \mathbf{z}$. We have

$$\mathbf{v}_f(\mathbf{z}) = \mathbf{v}_f(\mathbf{x}) - \mathbf{x}' \cdot \mathbf{e}_f \tag{16.116}$$

and

$$\mathbf{v}_p(\mathbf{z}) = \mathbf{v}_p(\mathbf{x}) - \mathbf{x}' \cdot \mathbf{e}_p \,, \tag{16.117}$$

where $\mathbf{e}_p$ is the velocity gradient tensor for the motion of the sphere. We shall assume that this tensor is constant and symmetric.

Using these results in eq. (16.115) gives

$$\overline{\mathbf{v}}_f^{xp}(\mathbf{x}, t) = \mathbf{v}_f(\mathbf{x}, t) \,. \tag{16.118}$$

The interfacial averaged velocity of the fluid is given by

$$\overline{\mathbf{v}}_{fi}(\mathbf{x}, t) = \frac{1}{4\pi a^2} \int_{\Omega(a)} \mathbf{v}(\mathbf{x}, t|\mathbf{z}) \, d\Omega \,. \tag{16.119}$$

Substituting and performing the integrations lead to the result

$$\overline{\mathbf{v}}_{fi}(\mathbf{x}, t) = \mathbf{v}_f(\mathbf{x}, t). \tag{16.120}$$

Averaged Pressures

Now let us compute averaged pressures using this formalism. The exact pressure can be computed by Bernoulli's equation

$$p = p_0 - \rho_f \left(\frac{1}{2} |\nabla \phi|^2 - \frac{\partial \phi}{\partial t} \right) . \tag{16.121}$$

Calculating the average pressures is tedious,but results in (Arnold, 1988)

$$\overline{p}_f^x = p_0 - \rho_f \frac{\partial \mathbf{v}_f}{\partial t} \cdot \mathbf{x} - \frac{1}{2} \mathbf{v}_f(\mathbf{x}) \cdot \mathbf{v}_f(\mathbf{x}) \tag{16.122}$$

$$\overline{p}_{fi} = \overline{p}_f^x - \frac{1}{4} \rho_f |\mathbf{v}_f - \mathbf{v}_p|^2 . \tag{16.123}$$

where we have ignored terms of order a/R in addition to those ignored previously.

Interfacial Force

The interfacial force density $\mathbf{M}_p$ is given by $\mathbf{M}_p = \overline{p \nabla X_p}$. Thus,

$$\mathbf{M}_p = -\frac{1}{\frac{4}{3}\pi R^3} \int_{\Omega(a)} \mathbf{n} p(\mathbf{x}, \mathbf{z}, t) \, d\Omega . \tag{16.124}$$

Note that the integration is again over the $\mathbf{z}$ variable. this corresponds to letting the sphere be tangent to the point $\mathbf{x}$, so that $\mathbf{z}$ is integrated over the spherical shell of radius a centered at $\mathbf{x}$. If we substitute Bernoulli's equation for the pressure, and recognize that $\mathbf{n} = \mathbf{x}'/a$, the result is that

$$\begin{aligned} \mathbf{M}_p &= \alpha_p \rho_f \left(\frac{1}{2} \left[\frac{\partial \mathbf{v}_0}{\partial t} - \frac{\partial \mathbf{v}_p}{\partial t} + \mathbf{v}_f \cdot \mathbf{e}_f - \mathbf{v}_p \cdot \mathbf{e}_p \right] \right. \\ &\quad \left. - \frac{7}{20} (\mathbf{v}_f - \mathbf{v}_p) \cdot (\mathbf{e}_f - \mathbf{e}_p) \right) . \end{aligned} \tag{16.125}$$

Dispersed Component Elastic Stresses

If a distribution of stresses is applied to the surface of an elastic body, there results a distribution of stresses inside the body. These induced stresses can be important in computing constitutive equations for solid-fluid mixtures. The average stress inside the sphere is given by

$$\overline{\mathbf{T}}_p^x(\mathbf{x}, t) = \frac{1}{\frac{4}{3}\pi a^3} \int_0^a \int \int_{\Omega(r)} \mathbf{T}(\mathbf{x}, t|\mathbf{z}) d\Omega \tag{16.126}$$

where $\mathbf{T}(\mathbf{x}, t|\mathbf{z})$ is the stress at point $\mathbf{x}$ inside a sphere having its center at $\mathbf{z}$ at time t. We shall assume that the dispersed spheres are linearly elastic solids, but we shall assume that the deformation is sufficiently small that the fluid motion is unaffected by the deformation of the spheres. Then the stress-strain relation is given by

$$\mathbf{T} = \mu_s [\nabla \mathbf{u} + (\nabla \mathbf{u})^T] + \lambda_s \nabla \cdot \mathbf{u} \mathbf{I} \tag{16.127}$$

where $\mathbf{u}$ is the local displacement within the solid sphere. The stress can be written as

$$\mathbf{T} = \boldsymbol{\sigma} + \Theta \mathbf{I} \tag{16.128}$$

where

$$\boldsymbol{\sigma} = \mu_s[\nabla \mathbf{u} + (\nabla \mathbf{u})^T] - \frac{2}{3}\nabla \cdot \mathbf{u}\mathbf{I} \tag{16.129}$$

The spherical part of the stress satisfies (Love, 1932)

$$\nabla^2 \Theta(\mathbf{x}, t|\mathbf{z}) = 0 \tag{16.130}$$

$$\Theta(\mathbf{x}, t|\mathbf{z}) = -p(\mathbf{x}, t|\mathbf{z}) \quad \text{on} \quad |\mathbf{x} - \mathbf{z}| = a\,. \tag{16.131}$$

Solving and performing the integration yields

$$\overline{\Theta_p^x}(\mathbf{x}, t) = -\overline{p}_{fi}(\mathbf{x}, t)\,. \tag{16.132}$$

The solution for $\boldsymbol{\sigma}$ in the solid phase is also given in Love (Love, 1932) and can be averaged to give

$$\overline{\boldsymbol{\sigma}_p^x}(\mathbf{x}, t) = \rho_f[-\frac{9}{20}(\mathbf{v}_f - \mathbf{v}_p)(\mathbf{v}_f - \mathbf{v}_p) + \frac{8}{20}|\mathbf{v}_f - \mathbf{v}_p|^2\mathbf{I}]\,. \tag{16.133}$$

Thus, the average particle phase stress is given by

$$\begin{aligned}
\overline{\mathbf{T}_p^x} &= \overline{\boldsymbol{\sigma}_d^x} + \overline{\Theta_p^x}\mathbf{I} \\
&= -\overline{p}_{fi}\mathbf{I} + \rho_f[-\frac{9}{20}(\mathbf{v}_f - \mathbf{v}_p)(\mathbf{v}_f - \mathbf{v}_p) + \frac{8}{20}|\mathbf{v}_f - \mathbf{v}_p|^2\mathbf{I}]
\end{aligned} \tag{16.134}$$

Reynolds Stresses

We next turn to computation of the Reynolds stress in the fluid because of the velocity fluctuations induced due to the inviscid flow around a sphere. Using the expression for the velocity potential, eq. (16.113), we see that

$$\begin{aligned}
\mathbf{v}_f'(\mathbf{x}, t|\mathbf{z}) &= \frac{1}{2}(\mathbf{v}_f(\mathbf{z}) - \mathbf{v}_p(\mathbf{z}))\left(\frac{a^3}{r^3}\right) \\
&\quad - \frac{3}{2}(\mathbf{v}_f(\mathbf{z}) - \mathbf{v}_p(\mathbf{z})) \cdot (\mathbf{x} - \mathbf{z})\left(\frac{a^3}{r^5}\right)(\mathbf{x} - \mathbf{z}) \\
&\quad + \frac{2}{3}(\mathbf{x} - \mathbf{z}) \cdot \mathbf{e}_f\left(\frac{a^5}{r^5}\right) \\
&\quad - \frac{5}{3}(\mathbf{x} - \mathbf{z}) \cdot \mathbf{e}_f \cdot (\mathbf{x} - \mathbf{z})\left(\frac{a^5}{r^7}\right)(\mathbf{x} - \mathbf{z})\,.
\end{aligned} \tag{16.135}$$

Therefore $\mathbf{T}_f^{Re} = -\overline{X_f \rho \mathbf{v}'\mathbf{v}'}$ can be computed, and the result is

$$\mathbf{T}_f^{Re} = -\frac{1}{20}\alpha_p \rho_f [(\mathbf{v}_f - \mathbf{v}_p)(\mathbf{v}_f - \mathbf{v}_p) + 3(\mathbf{v}_f - \mathbf{v}_p)\cdot(\mathbf{v}_f - \mathbf{v}_p)\mathbf{I}] \,. \quad (16.136)$$

The fluid fluctuation kinetic energy is $u_f^{Re} = \frac{1}{2}\overline{\mathbf{v}' \cdot \mathbf{v}'}^x$, and can be computed by taking the trace of eq. (16.136) for $\mathbf{T}_f^{Re}$. The result is

$$u_f^{Re} = \frac{1}{4}\alpha_p(\mathbf{v}_f - \mathbf{v}_p)\cdot(\mathbf{v}_f - \mathbf{v}_p)\,. \quad (16.137)$$

Interfacial Work

The interfacial work $\mathbf{W}_f = \overline{p\mathbf{v}' \cdot \nabla X_f}$ can be computed in a similar way; the result is

$$\mathbf{W}_f = \frac{1}{2}\alpha_p \rho_f(\mathbf{v}_f - \mathbf{v}_p)\cdot\left(\frac{\partial \overline{\mathbf{v}}_f}{\partial t} + \overline{\mathbf{v}}_f \cdot \nabla \overline{\mathbf{v}}_f - \frac{\partial \overline{\mathbf{v}}_p}{\partial t}\right). \quad (16.138)$$

All of the terms derived in this section are comparable in magnitude to the inertial terms; consequently a model capable of describing effects that depend on inertia would be incomplete without closure terms of this type.

16.5.2 Effect of Concentration Gradients

In this section, we assume that the distribution of positions within a cell depends on the gradient of the volume fraction of the particles. This accounts for non-uniform distributions of spherical particles in a natural way, in determining the density function. Thus, we assume that

$$\frac{dV}{\frac{4}{3}\pi R^3}\left[1 - \mathbf{x}' \cdot \frac{\nabla \alpha_p(\mathbf{x},t)}{\alpha_p(\mathbf{x},t)}\right] \quad (16.139)$$

is the probability of finding the sphere in a volume dV surrounding the point $\mathbf{z}$, where $\mathbf{x}' = \mathbf{x} - \mathbf{z}$.

The result of including the effects of volume fraction gradients is to add terms to the quantities computed in the previous section. For convenience of notation, we shall denote the average taken without allowing particle velocity fluctuations by $\overline{g}_0^x(\mathbf{x}, t; \mathbf{v}_p)$. Thus, for example, the average interfacial force density, as calculated before, evaluated without accounting for a gradient in the volume fraction, will be denoted by

$$\mathbf{M}_{p0}(\mathbf{x},t) = \alpha_p \rho_f \left(\frac{1}{2} \left[\frac{\partial \overline{\mathbf{v}}_f^{xp}}{\partial t} - \frac{\partial \overline{\mathbf{v}}_p^{xp}}{\partial t} + \overline{\mathbf{v}}_f^{xp} \cdot \nabla \overline{\mathbf{v}}_f^{xp} - \overline{\mathbf{v}}_p^{xp} \cdot \nabla \overline{\mathbf{v}}_p^{xp} \right] - \frac{7}{20} (\overline{\mathbf{v}}_f^{xp} - \overline{\mathbf{v}}_p^{xp}) \cdot (\nabla \overline{\mathbf{v}}_f^{xp} - \nabla \overline{\mathbf{v}}_p^{xp}) \right) . \tag{16.140}$$

Using the probability distribution given by eq. (16.139), the resulting integrals can be computed. Since we use only the linear approximation to the velocity fields, and the volume fraction gradient, we shall neglect the products of $\mathbf{e}_f$, $\mathbf{e}_p$ and $\nabla \alpha_p$. The results are that all averaged quantities are unchanged except the interfacial force density, and it is given by

$$\mathbf{M}_p(\mathbf{x},t) = \mathbf{M}_{p0}(\mathbf{x},t) + \overline{p}_{fi0} \nabla \alpha_p . \tag{16.141}$$

16.5.3 Effect of Particle Velocity Fluctuations

Models that treat assemblages of solid particles have been proposed by Jenkins and Savage (1983). These models result in the particles behaving like a pseudo-gas, with a pressure due to the fluctuations in the velocities that is attributed to collisional motions of the individual particles. The model in Section 16 for inertial effects does not include this effect. In this section we derive terms representing the effects of the particle velocity fluctuations when the particle motions are assumed to be at a sufficiently high Reynolds number that the fluid motion is inviscid, but viscous effects such as boundary layer separation are neglected. Other models of this type have also been proposed (Lundgren, 1984).

Let the sphere located at a point $\mathbf{z}$ move with velocity $\mathbf{v}_p$. We now introduce the averaging process. The first aspect of the ensemble average when there are particle velocity fluctuations present is to average over the sphere velocities, with a distribution function $f^{(1)}(\mathbf{v}_p, \mathbf{x}, t)$. The second aspect is to allow the sphere to lie anywhere in the "cell" discussed in Section 16.5.1. Thus, the average of a quantity $g(\mathbf{x}, t; \mathbf{z}, \mathbf{v}_p)$ is performed in two parts: first, we perform a conditional average of g for given sphere position $\mathbf{z}$, integrating over the velocity space $\mathbf{v}_p$, followed by the average over the spatial positions the sphere can have. Thus, for the average over the fluid of a quantity g, we have

$$\overline{g}(\mathbf{x}, t|\mathbf{z}) = \int_{R^3} g(\mathbf{x}, t; \mathbf{z}, \mathbf{v}_p)\, f^{(1)}(\mathbf{v}_p, \mathbf{z}, t)\, dV_{\mathbf{v}_p} . \tag{16.142}$$

Here the notation $\overline{g}(\mathbf{x}, t|\mathbf{z})$ is intended to suggest the conditional average assuming the sphere is located at $\mathbf{z}$. The average of g over the fluid phase is then given by

$$\overline{g}_f^x(\mathbf{x},t) = \frac{1}{\frac{4}{3}\pi(R^3 - a^3)} \int_a^R \int\int_{\Omega(r)} \overline{g}(\mathbf{x},t|\mathbf{z})\, d\Omega\, dr. \qquad (16.143)$$

It will be convenient to introduce the average particle velocity and the fluctuation particle velocity as

$$\overline{\mathbf{v}}_p(\mathbf{x},t) = \int_{R^3} \mathbf{v}_p f^{(1)}(\mathbf{v}_p,\mathbf{x},t)\, dV_{\mathbf{v}_p}\,, \qquad (16.144)$$

$$\mathbf{v}_p'(\mathbf{x},t) = \mathbf{v}_p - \overline{\mathbf{v}}_p(\mathbf{x},t)\,, \qquad (16.145)$$

where we note that

$$\overline{\mathbf{v}_p'} = 0\,. \qquad (16.146)$$

The particle kinetic energy per unit particle mass is

$$u_p^{Re}(\mathbf{x},t) = \frac{1}{2}\int_{R^3} |\mathbf{v}_p'|^2 f^{(1)}(\mathbf{v}_p,\mathbf{x},t)\, dV_{\mathbf{v}_p}\,, \qquad (16.147)$$

and the Reynolds stress for the particles is defined by

$$\mathbf{T}_p^{Re}(\mathbf{x},t) = -\rho_p \int_{R^3} \mathbf{v}_p'\mathbf{v}_p' f^{(1)}(\mathbf{v}_p,\mathbf{x},t)\, dV_{\mathbf{v}_p}\,. \qquad (16.148)$$

The result of including the effects of particle velocity fluctuations is to add terms to the quantities computed in the previous section. For convenience of notation, we shall denote the average taken without allowing particle velocity fluctuations by $\overline{g}^x(\mathbf{x},t;\mathbf{v}_p)$. Thus, for example, the average interfacial pressure, as calculated before, evaluated at the average particle velocity, will be denoted by

$$\overline{p}_{fi}^x(\mathbf{x},t,\overline{\mathbf{v}}_p^{x\rho}) = \overline{p}_f^x(\mathbf{x},t,\overline{\mathbf{v}}_p^{x\rho}) - \frac{1}{4}\rho_f|\overline{\mathbf{v}}_f^{x\rho} - \overline{\mathbf{v}}_p^{x\rho}|^2\,. \qquad (16.149)$$

We have

$$\begin{aligned}\mathbf{v}(\mathbf{x},t;\mathbf{z},\mathbf{v}_p) &= \nabla\phi(\mathbf{x},t;\mathbf{z},\mathbf{v}_p)\\ &= \mathbf{v}_f(\mathbf{x}) + \frac{1}{2}(\mathbf{v}_f(\mathbf{x}) - \overline{\mathbf{v}}_p(\mathbf{x}) - \mathbf{v}_p' - \mathbf{x}'\cdot(\mathbf{e}_f - \mathbf{e}_p))\left(\frac{a^3}{r^3}\right)\\ &\quad - \frac{3}{2}(\mathbf{v}_f(\mathbf{x}) - \overline{\mathbf{v}}_p(\mathbf{x}) - \mathbf{v}_p' - \mathbf{x}'\cdot(\mathbf{e}_f - \mathbf{e}_p))\cdot\mathbf{x}'\left(\frac{a^3}{r^5}\right)\mathbf{x}'\\ &\quad + \frac{2}{3}\mathbf{x}'\cdot\mathbf{e}_f\left(\frac{a^5}{r^5}\right) - \frac{5}{3}\mathbf{x}'\cdot\mathbf{e}_f\cdot\mathbf{x}'\left(\frac{a^5}{r^7}\right)\mathbf{x}'\,.\end{aligned} \qquad (16.150)$$

In order to obtain the average velocities from eq. (16.150), we note that the average over the velocity fluctuations gives no contribution. Therefore,

$$\overline{\mathbf{v}}_f^{xp}(\mathbf{x}, t) = \mathbf{v}_f(\mathbf{x}, t) \tag{16.151}$$

and

$$\overline{\mathbf{v}}_{fi}(\mathbf{x}, t) = \mathbf{v}_f(\mathbf{x}, t). \tag{16.152}$$

The pressure is again given by Bernoulli's equation. Substituting $\mathbf{v}_p = \overline{\mathbf{v}}_p(\mathbf{z}, t) + \mathbf{v}_p'$ and averaging over the velocity fluctuations gives

$$\begin{aligned}\overline{p}(\mathbf{x}, t|\mathbf{z}) &= p(\mathbf{x}, t|\mathbf{z}, \overline{\mathbf{v}}_p) \\ &\quad - \rho_f \left\{ u_p^{Re}(\mathbf{z}, t) \left(\frac{a^3}{r^3}\right) - \left[\frac{3}{2} \frac{\mathbf{x}' \cdot \mathbf{T}_p^{Re}(\mathbf{z}, t) \cdot \mathbf{x}'}{\rho_p}\right] \left(\frac{a^3}{r^5}\right) \right. \\ &\quad \left. + \frac{1}{4} u_p^{Re}(\mathbf{z}, t) \left(\frac{a^6}{r^6}\right) - \frac{9}{8} \frac{\mathbf{x}' \cdot \mathbf{T}_p^{Re}(\mathbf{z}, t) \cdot \mathbf{x}'}{\rho_p} \left(\frac{a^6}{r^{10}}\right) \right\}.\end{aligned} \tag{16.153}$$

The spatial integration is now quite similar to that done previously, and results in

$$\overline{p}_f^x = p_0 - \rho_f \frac{\partial \mathbf{v}_f}{\partial t} \cdot \mathbf{x} - \frac{1}{2} \mathbf{v}_f(\mathbf{x}) \cdot \mathbf{v}_f(\mathbf{x}) = \overline{p}_f^x(\mathbf{x}, t, \overline{\mathbf{v}}_p). \tag{16.154}$$

Thus, there is no contribution to the average fluid pressure due to the particle velocity fluctuations. We also obtain

$$\begin{aligned}\overline{p}_{fi}(\mathbf{x}, t) &= \overline{p}_f^x - \frac{1}{4} \rho_f |\overline{\mathbf{v}}_f(\mathbf{x}, t) - \overline{\mathbf{v}}_p(\mathbf{x}, t)|^2 + \frac{1}{2} \rho_f u_p^{Re} \\ &= \overline{p}_{fi}(\mathbf{x}, t, |\overline{\mathbf{v}}_p) + \frac{1}{2} \rho_f u_p^{Re}.\end{aligned} \tag{16.155}$$

For the interfacial force density $\mathbf{M}_p$, we must also expand the terms $u_p^{Re}(\mathbf{z}, t) = u_p^{Re}(\mathbf{x}, t) - \mathbf{x}' \cdot \nabla u_p^{Re}(\mathbf{x}, t)$ and $\mathbf{T}_p^{Re}(\mathbf{z}, t) = \mathbf{T}_p^{Re}(\mathbf{x}, t) - \mathbf{x}' \cdot \nabla \mathbf{T}_p^{Re}(\mathbf{x}, t)$. The result is that

$$\mathbf{M}_p(\mathbf{x}, t) = \mathbf{M}_p(\mathbf{x}, t, \overline{\mathbf{v}}_p) - \nabla \left(\frac{4}{5} \alpha_p \rho_f u_p^{Re}\right) - \nabla \cdot \left(\alpha_p \frac{9}{20} \frac{\rho_f}{\rho_p} \mathbf{T}_p^{Re}\right) \tag{16.156}$$

The average stress inside the solid sphere is given by

$$\overline{\mathbf{T}}_p^x(\mathbf{x}, t) = \overline{\mathbf{T}}_p^x(\mathbf{x}, t, \overline{\mathbf{v}}_p) - \frac{9}{20} \frac{\rho_f}{\rho_p} \mathbf{T}_p^{Re} + \frac{4}{5} \rho_f u_p^{Re} \mathbf{I}. \tag{16.157}$$

The Reynolds stress can be computed using the velocity fluctuations derived from eqs. (16.150) and (16.151). We have

$$\begin{aligned}
\mathbf{T}_f^{Re}(\mathbf{x},t|\mathbf{z}) &= -\rho_f\overline{\mathbf{v}'_f(\mathbf{x},t|\mathbf{z})\mathbf{v}'_f(\mathbf{x},t|\mathbf{z})} \\
&\quad + \rho_f\left[\frac{1}{4}\left(\frac{a^6}{r^6}\right)\frac{\mathbf{T}_p^{Re}}{\rho_p} + \frac{9}{4}\left(\frac{a^6}{r^{10}}\right)\mathbf{x}'\mathbf{x}'(\mathbf{x}'\cdot\frac{\mathbf{T}_p^{Re}}{\rho_p}\cdot\mathbf{x}')\right. \\
&\quad \left. - \frac{3}{4}\left(\frac{a^6}{r^8}\right)[\mathbf{x}'(\mathbf{x}'\cdot\frac{\mathbf{T}_p^{Re}}{\rho_p}) + (\mathbf{x}'\cdot\frac{\mathbf{T}_p^{Re}}{\rho_p})\mathbf{x}']\right] \quad (16.158)
\end{aligned}$$

Hence

$$\mathbf{T}_f^{Re}(\mathbf{x},t) = \mathbf{T}_f^{Re}(\mathbf{x},t,\overline{\mathbf{v}}_p) + \frac{1}{20}\alpha_p\frac{\rho_f}{\rho_p}\mathbf{T}_p^{Re} + \frac{3}{10}\alpha_p\rho_f u_p^{Re}\,. \qquad (16.159)$$

The fluid fluctuation kinetic energy can be computed by taking the trace of eq. (16.159). The result is

$$u_f^{Re} = \frac{1}{4}\alpha_p|\overline{\mathbf{v}}_f^{xp} - \overline{\mathbf{v}}_p^{xp}|^2 + \frac{1}{2}\alpha_p u_p^{Re}\,. \qquad (16.160)$$

While these closure relations are only valid for a dilute suspension of solid spheres and are based on some restrictive assumptions (*e.g.*, inviscid fluid), they are representative of the type of constitutive equations needed for closure.

16.6 CONSTITUTIVE ASSUMPTIONS

Some general properties of constitutive equations were discussed in the previous section. Additional guidance in formulating realistic constitutive equations can be obtained by assuming that the quantities computed from these "exact" solutions can be taken as the proper forms for constitutive equations. For example, the dynamics of the irrotational flow of an inviscid fluid about a sphere (see Section 16.5) can be used to compute the average force on a sphere and the average interfacial pressure. Constitutive equations arrived at in this way should reduce to the appropriate limit for dilute flows having high particle Reynolds number.

As is evident from the state of our abilities to calculate the flow around assemblages of bodies of arbitrary shape, we cannot hope to obtain exact solutions that can be averaged to give appropriate constitutive equations for general motions. The ultimate solution to the problem of obtaining constitutive equations rests in empiricism. The calculated forms from "exact" solutions provide a guide to the appropriate forms for empirical testing. In order to obtain a system of equations that will predict the flow of a dispersed multiphase material, it is necessary to find experiments that isolate appropriate phenomena and give (relatively) direct measurements of unknown coefficients involved therein. For best results, the physical experiment should be simple, and should correspond to a simple solution of the equations that depends mostly on the particular constitutive form being evaluated. These flows are sometimes called "viscometric," or "separate effects" experiments. It is hopefully clear that experiments which isolate effects, and the corresponding exact solutions are highly desirable.

16.6.1 Stress

The stress in a multiphase mixture represents the force per unit area on phase k across an element of area having normal $\mathbf{n}$. When both materials "flow," fluid models seem appropriate, but particle stresses are due to a different mechanism than fluid stresses. We first discuss the averaged stress.

In order for the mixture to flow, one of the materials present must be a fluid. We shall refer to this material as material c, for "continuous" phase. In particulate flows, the other phase is a solid, while in gas/liquid flows, both phases are fluids. In both of these cases, we refer to this phase as phase d, for dispersed.

We shall assume that the microscale stress can be written as a pressure plus a viscous stress

$$\mathbf{T}_k = -p_k\mathbf{I} + \boldsymbol{\tau}_k \,. \tag{16.161}$$

Similarly, the Reynolds stress is written as

$$\mathbf{T}_k^{Re} = -p_k^{Re}\mathbf{I} + \boldsymbol{\tau}_k^{Re} \,. \tag{16.162}$$

We shall first discuss constitutive equations for the pressure in both a fluid material and a particulate phase. We then discuss the shear stresses in a viscous fluid phase. Finally, we discuss the extra stress in the particle phase of a particle/fluid flow.

In addition to the bulk stresses in both materials it is convenient to have expressions for the interfacial averaged values of the stresses, broken up into the interfacial pressure and the interfacial shear stress. Thus, we write

$$\mathbf{T}_{ki} = -p_{ki}\mathbf{I} + \boldsymbol{\tau}_{ki} \,. \tag{16.163}$$

The utility of this relation will become more evident when we discuss the interfacial force later in this section.

16.6.2 Pressure

The pressure in a compressible fluid is related to the temperature and density of the fluid through a constitutive equation called the *equation of state*. When a phase is incompressible, then extra assumptions are required to give closure. We treat these extra assumptions as constitutive equations. If one phase is finely dispersed throughout the other, the pressure in the dispersed phase cannot differ much from its value at the interface. Thus, for dispersed flows, one often assumes

$$p_{di} = p_d \,. \tag{16.164}$$

With constant surface tension and spherical bubbles or drops, one normally uses the classical Laplace equation:

$$p_{ci} = p_{di} - H\sigma \,, \tag{16.165}$$

where H is the mean curvature of the interface. If the dispersed phase is spherical, $H = 2/r_d$, where r_d is the radius of the dispersed phase elements (bubbles or drops), and σ is the surface tension.

Bernoulli's equation for the variation of pressure in a flowing (inviscid) fluid suggests that the interfacial pressure should be given by

$$p_{ci} = p_c - \xi \rho_c |\mathbf{v}_c - \mathbf{v}_d|^2 \,. \tag{16.166}$$

The calculations for the averaged fields for the flow of an inviscid fluid around a single sphere given in Section 16 suggest that for dilute flows, we have $\xi = 1/4$ when the boundary layer remains attached to the spherical particle. For low Reynolds number flows, Givler (1987) calculates the value $\xi = -9/32$. We speculate that since eq. (16.166) holds in both limits of the inviscid and very viscous flows, then we might expect a similar form to hold in somewhat more generality. Thus we use

$$\xi = \xi(\alpha_k, Re_k) \,. \tag{16.167}$$

It is more difficult to interpret the "pressure" of a solid phase. In the flow of a particle/fluid mixture, the solid particles have a variety of roles in transmitting momentum. First, the particles may act as "stress transmitters," allowing the pressure and the shear stress in the fluid to be transmitted and concentrated. This is sometimes called the "particle presence stress" (Hwang, *et al.*, 1989). Another means of transporting particle momentum is through particle/particle contacts (Jenkins and Savage, 1983, Lun, *et al.*, 1984). If the particles touch, they behave collectively as a solid, in that they can transmit stresses elastically. The particles can also exchange momentum with the conduit walls during collisions (Jenkins 1987). Finally, they can transport momentum by undergoing random fluctuations about the mean motion. In this mode, collisions are usually important, and the model will resemble the models used in the kinetic theory of gases, where the "particles" act like the molecules. In some sense, contact stresses and collisional stresses are not different, although quite different models are often used for them (Lun, *et al.*, 1984).

As noted previously, the microscale pressure is normally assumed to be

$$p_{di} = p_d \,. \tag{16.168}$$

If the particles are so concentrated that they form a coherent solid material with spaces or pores where the fluid is, then the behavior may be more appropriately modeled using the concepts of elastic and plastic deformation, as done in the soil mechanics community. A simple model of this type ignores the plastic behavior, and assumes that if the particles are at sufficiently high concentration, then the pressure is related to the volume fraction (Lambe & Whitman, 1969). That is, we assume

$$p_d = \begin{cases} 0, & \text{for } \alpha_d < \alpha_{cont} \\ p_{cont}(\alpha_d) & \text{for } \alpha_d \geq \alpha_{cont} \end{cases} \tag{16.169}$$

where $p_{cont}(\alpha_d)$ can be determined experimentally by compacting the particles statically and measuring the force necessary to change the volume of the sample.

The most complicated case is when collisions occur between the particles. In this case, stress is transmitted through the particles, and the fluctuation motions of the particles themselves transport momentum. We shall discuss only the model for the Reynolds stress, which is due to the velocity fluctuations.

The pressure is the spherical part of the momentum tensor, and, in analogy with kinetic theory (Chapman & Cowling, 1970; Jenkins & Savage, 1983), it is given in terms of the fluctuation kinetic energy as

$$p_d^{Re} = \alpha_d \rho_d \mathcal{R}_d u_d^{Re} \tag{16.170}$$

where $\mathcal{R}_d = \frac{2}{3}$ for frictionless collisions.

16.6.3 Shear Stress

Viscous Fluid Stress

It is a common observation that a particle/fluid mixture has a different viscosity from the suspending fluid. Mud is a common example, with the viscosity of the suspension being many times greater than that of water. Viscous forces can give a significant contribution to flow resistance, often through a (two-phase) turbulent boundary layer having a single-phase laminar sublayer. The effects are also important in the analysis of phase distribution for bubbly flows (Drew & Lahey, 1982) which, in turn, is important in determining profile parameters. Profile parameters are key elements in one-dimensional modeling.

In dilute flows, the stress in the dispersed phase can differ little from its value on the interface. Hence,

$$\boldsymbol{\tau}_{di} = \boldsymbol{\tau}_d . \tag{16.171}$$

The assumption of no Marangoni effects, such as those caused by temperature or concentration-driven surface tension gradients, results in

$$\boldsymbol{\tau}_{ci} = \boldsymbol{\tau}_{di} . \tag{16.172}$$

A fairly general assumption for the form of stress tensor for the continuous phase is

$$\begin{aligned}\boldsymbol{\tau}_c = &\sum_j \mu_{cj}\mathbf{D}_j + \sum_j\sum_k\sum_l \gamma_{cjkl}\left[\nabla\alpha_j(\mathbf{v}_k - \mathbf{v}_l) + (\mathbf{v}_k - \mathbf{v}_l)\nabla\alpha_j\right] \\ &+ \sum_j\sum_k\sum_l\sum_m a_{cjklm}\left[(\mathbf{v}_j - \mathbf{v}_k)(\mathbf{v}_l - \mathbf{v}_m) + (\mathbf{v}_l - \mathbf{v}_m)(\mathbf{v}_j - \mathbf{v}_k)\right] ,\end{aligned} \tag{16.173}$$

where $\mu_{cc}(\alpha_d)$ is the effective fluid viscosity of the fluid, and $\mu_{cd}(\alpha_d)$ is the fluid cross-viscosity. Note that we obtain Lundgren's (1972) expression for the effective viscosity if we take $\mu_{cc} = \mu_f$, $\mu_{cd} = \alpha_d/(1 - \alpha_d)\mu_f$, and $\gamma_{cddc} = \mu_f$.

Solid Phase Shear Stress

In the case of a flow of a dilute suspension of solid particles in a fluid, the particles act as transmitters for the fluid stress. Einstein (1906) was the first to show that the fluid (viscous) shear stress is "concentrated" by the presence of the particles, resulting in the "effective" viscosity of a suspension being significantly larger than that of a clear fluid. A simple model for this that may be appropriate is that the interfacial averaged fluid stress is a factor β times the averaged fluid stress. That is,

$$\boldsymbol{\tau}_{ci} = \beta\boldsymbol{\tau}_c . \tag{16.174}$$

The shear stress in a compacted bed of particles with fluid in the pores is often modeled by an elastic stress-strain law for small shears, and by a plastic model for stresses reaching some yield stress for the particles (Lambe & Whitman, 1969).

In a situation where the particle momentum transport is dominated by collisions, the stress can be related to the fluctuation kinetic energy u_d^{Re}. Kinetic theory results (Jenkins & Savage, 1983) suggest that

$$\boldsymbol{\tau}_d = \gamma_d \rho_d a (u_d^{Re})^{1/2} (\nabla \mathbf{v}_d + (\nabla \mathbf{v}_d)^T) \tag{16.175}$$

where a is the equivalent radius of the particles.

Abou-Arab and Roco (1990) have proposed a model for the turbulent, or Reynolds stresses in a particle/fluid mixture. It may be written in the form

$$\begin{aligned}\boldsymbol{\tau}_k^{Re} &= \sum_j \mu_{kj}^{Re} \mathbf{D}_j + \sum_j \sum_l \sum_m \gamma_{kjlm}^{Re} \left[\nabla \alpha_j (\mathbf{v}_l - \mathbf{v}_m) + (\mathbf{v}_l - \mathbf{v}_m) \nabla \alpha_j\right] \\ &+ \sum_j \sum_l \sum_m \sum_n a_{kjlmn}^{Re} \left[(\mathbf{v}_j - \mathbf{v}_l)(\mathbf{v}_m - \mathbf{v}_n) + (\mathbf{v}_m - \mathbf{v}_n)(\mathbf{v}_j - \mathbf{v}_l)\right] ,\end{aligned} \tag{16.176}$$

where $\mu_{kl}^{Re}(\alpha_d, u_k^{Re})$ is an effective Reynolds viscosity.

16.6.4 Interfacial Force

The interfacial force, $\mathbf{M}_k$, arises from stresses acting on the interface. It is the crucial term in modeling two-phase flows, since it is the presence of the interface which makes the discipline of multiphase fluid mechanics so challenging and interesting.

It has become customary (Ishii and Mishima, 1981) to write the interfacial force in the form given by eq. (16.80). That is,

$$\mathbf{M}_k = p_{ki} \nabla \alpha_k - \boldsymbol{\tau}_{ki} \cdot \nabla \alpha_k + \mathbf{M}_k', \tag{16.177}$$

where p_{ki} and $\boldsymbol{\tau}_{ki}$ are the interfacially averaged pressure and the shear stress of phase k. We shall assume

$$\mathbf{M}_k' = \mathbf{M}_k'\left(\alpha_k, \nabla \alpha_k, \mathbf{v}_d, \mathbf{v}_c, \frac{\partial \mathbf{v}_d}{\partial t}, \frac{\partial \mathbf{v}_c}{\partial t}, \nabla \mathbf{v}_d, \nabla \mathbf{v}_c, \ldots\right), \tag{16.178}$$

where ... includes scalar properties, such as the viscosities and the densities of the phases, as well as intrinsic parameters, such as distorted particle radius and interfacial area density.

We shall represent $\mathbf{M}'_d$ as a sum of the forces associated with drag, virtual mass, lift, and unsteady viscous effects,

$$\mathbf{M}'_d = \sum \mathbf{F}_n, \tag{16.179}$$

Drag

The drag force on a single particle, drop or bubble is the force felt by that object as it moves steadily through the surrounding fluid. The concept is clouded when an array of particles moves through a fluid, but the drag force is still attributed to steady effects. The drag force is usually given in terms of a dimensionless drag coefficient C_D, where the drag force is defined as

$$\mathbf{F}_D = \frac{3}{8}\frac{\alpha_d \rho_c C_D}{r_d}|\mathbf{v}_c - \mathbf{v}_d|\,(\mathbf{v}_c - \mathbf{v}_d)\,. \tag{16.180}$$

Here, the drag coefficient C_D is assumed to be a function of α_d and

$$Re_d = \frac{r_d|\mathbf{v}_d - \mathbf{v}_c|}{\nu_c},$$

where ν_c is the fluid viscosity. The dependence of C_D on α_d and on Re_d for bubbly, droplet and particle flows has been studied by Ishii and Zuber (1979), among others.

Virtual Mass and Lift

The virtual mass force is the added force exerted on a moving object when it accelerates through a fluid. If an object is immersed in fluid and accelerated, it must accelerate some of the surrounding fluid. This results in an interaction force on the object. This force was calculated explicitly for a single sphere in Section 16. We shall take this force to be

$$\mathbf{F}_{vm} = C_{vm}\alpha_d\rho_c\left[\left(\frac{\partial \mathbf{v}_c}{\partial t} + \mathbf{v}_c\cdot\nabla\mathbf{v}_c\right) - \left(\frac{\partial \mathbf{v}_d}{\partial t} + \mathbf{v}_d\cdot\nabla\mathbf{v}_d\right)\right]. \tag{16.181}$$

The parameter C_{vm} is called the virtual volume coefficient For a spherical particle, it is

$$C_{vm} = \frac{1}{2}\,. \tag{16.182}$$

This is valid for dilute suspensions of spheres in a fluid.

In addition, if a particle moves through a fluid that is in shearing motion then the particle experiences a force transverse to the direction of motion.

Such a force is often called a "lift force." If the particle is spherical, the force is given by Drew and & Lahey (1990) as

$$\mathbf{F}_L = -C_L \alpha_d \rho_c (\mathbf{v}_c - \mathbf{v}_d) \times (\nabla \times \mathbf{v}_c). \tag{16.183}$$

The coefficient C_L is called the lift coefficient. It has the value

$$C_L = C_{vm} = \frac{1}{2} \tag{16.184}$$

for a dilute suspension of spheres in an inviscid continuous phase. Note that the virtual mass and lift forces as proposed above are not frame-indifferent separately. However, if we take the calculated values $C_{vm} = C_L = \frac{1}{2}$, the sum $\mathbf{F}_{vm} + \mathbf{F}_L$ *is* frame-indifferent and agrees with calculations for the force on a single sphere in an inviscid fluid undergoing strain. Thus, the principle of objectivity requires that

$$C_L = C_{vm}. \tag{16.185}$$

The virtual volume coefficient C_{vm} is believed to be a function of at least α_d and Re_d. For the high Reynolds number (nearly inviscid) flow of a dilute suspension of spheres, Zuber (1964) proposed

$$C_{vm} \approx \frac{1}{2} + \frac{3}{2}\alpha_d. \tag{16.186}$$

The last term in eq. (16.186) accounts for interaction effects between the dispersed spheres.

Other Forces

Other forces can be calculated from the flow of an inviscid fluid around a single sphere. These forces are not usually included in multiphase flow models. We include them here for completeness, and because some of them are needed to satisfy the second law of thermodynamics (Arnold, *et al.*, 1989). We write

$$\begin{aligned} F_O \; = \; & C_r \rho_c \left(\frac{\partial \alpha_d}{\partial t} + \mathbf{v}_d \cdot \nabla \alpha_d \right) (\mathbf{v}_d - \mathbf{v}_c) \\ & + C_c \alpha_d \rho_c (\nabla \cdot (\mathbf{v}_d - \mathbf{v}_c))(\mathbf{v}_d - \mathbf{v}_c) \\ & + C_e \alpha_d \rho_c (\mathbf{v}_d - \mathbf{v}_c) \cdot \nabla (\mathbf{v}_d - \mathbf{v}_c) \\ & + C_g \alpha_d \rho_c (\mathbf{v}_d - \mathbf{v}_c) \cdot \left(\nabla (\mathbf{v}_d - \mathbf{v}_c) + (\nabla (\mathbf{v}_d - \mathbf{v}_c))^T \right) \\ & + C_j \rho_c |\mathbf{v}_d - \mathbf{v}_c|^2 \nabla \alpha_d \\ & + C_n \rho_c (\mathbf{v}_d - \mathbf{v}_c) \cdot \nabla \alpha_d (\mathbf{v}_d - \mathbf{v}_c). \end{aligned} \tag{16.187}$$

Arnold, *et al.*, (1989) give the values $C_r = \frac{1}{2}$, $C_c = \frac{1}{20}$, $C_e = \frac{5}{4}$, $C_g = -\frac{9}{20}$, $C_j = \frac{2}{5}$, and $C_n = -\frac{9}{20}$.

Other forces which are sometimes included in the interfacial force are the so-called Faxén and Basset forces. Both forces arise from viscous effects in the continuous phase. The Faxén force is taken to be (Brenner, 1964)

$$\mathbf{F}_F = C_F \alpha_d \mu_c \nabla^2 \mathbf{v}_c. \tag{16.188}$$

Single-sphere calculations indicate $C_F = 3/4$, this value may be expected to be valid for low concentrations of the dispersed phase.

The Basset force is an unsteady force associated with a viscous wake and is due to boundary layer formation. If a flat plate is impulsively started from rest moving parallel to itself, it experiences a drag force which relaxes proportional to $t^{-1/2}$ as the boundary layer grows. Basset (1861) calculated the analogous force on a single sphere. This force can be expressed as

$$\mathbf{F}_B = \alpha_d \frac{9}{2r_d}\left(\frac{\rho_c \mu_c}{\pi}\right)^{1/2} \int_{-\infty}^{t} \frac{\mathbf{a}(\mathbf{x},\tau)d\tau}{\sqrt{(t-\tau)}}, \tag{16.189}$$

where the appropriate acceleration $\mathbf{a}$ is taken here to be the frame-indifferent quantity

$$\begin{aligned}\mathbf{a} &= \left(\frac{\partial \mathbf{v}_c}{\partial t} + \mathbf{v}_c \cdot \nabla \mathbf{v}_c\right) - \left(\frac{\partial \mathbf{v}_d}{\partial t} + \mathbf{v}_d \cdot \nabla \mathbf{v}_d\right) \\ &\quad - (\mathbf{v}_c - \mathbf{v}_d) \times (\nabla \times \mathbf{v}_c).\end{aligned} \tag{16.190}$$

16.6.5 Heat Flux

Conductive Heat Flux

Averaging the heat flux $\mathbf{q}'' = -k_k \nabla \theta$ yields

$$\mathbf{q}_k'' = -k_k \nabla \theta_k^m + k_k \frac{\gamma_k(\alpha_k)}{2\alpha_k} \nabla \alpha_k (\theta_i - \theta_k^m), \tag{16.191}$$

where $\gamma_k(\alpha_k)$ is the "thermal mobility" of phase k, and θ_i is the interfacial average temperature.

Turbulent Heat Flux

The turbulent heat flux can be modeled in analogy with the Reynolds stresses:

$$\hat{\mathbf{q}}_c^{Re} = d_c \alpha_d \rho_c (\mathbf{v}_c - \mathbf{v}_d)(\theta_c - \theta_d) + k_c^{Re} \nabla \theta_c^M \tag{16.192}$$

$$\hat{\mathbf{q}}_d^{Re} = d_d \alpha_d \rho_d (\mathbf{v}_c - \mathbf{v}_d)(\theta_c - \theta_d) + k_d^{Re} \nabla \theta_d^M \tag{16.193}$$

where d_k are is a parameter, k_k^{Re} is the eddy conductivity, c_k is the specific heat of phase k, at constant volume.

Turbulent Kinetic Energy Flux

The turbulent kinetic energy flux is assumed to be

$$\mathbf{q}_c^K = f_c \alpha_d \rho_c |\mathbf{v}_c - \mathbf{v}_d|^2 (\mathbf{v}_c - \mathbf{v}_d) \tag{16.194}$$

$$\mathbf{q}_d^K = f_d \alpha_d \rho_d |\mathbf{v}_c - \mathbf{v}_d|^2 (\mathbf{v}_c - \mathbf{v}_d) \tag{16.195}$$

A calculation similar to that of Section 16 gives $f_c = 1/40$. No models are known for f_d; consequently, it is often neglected.

16.6.6 Interfacial Energy Source

We write

$$E_k = \mathbf{q}_{ki}'' \cdot \nabla \alpha_k + E_k', \tag{16.196}$$

where $\mathbf{q}_{ki}''$ is the average interfacial energy flux. It is often ignored for solid/fluid flows. We propose

$$\mathbf{q}_{di}'' = \mathbf{q}_d'' \tag{16.197}$$

$$\mathbf{q}_{ci}'' = \mathbf{q}_{di}'' . \tag{16.198}$$

The term E_k' is usually taken to be

$$E_k' = A_i''' H_k (\theta_i - \theta_k^m), \tag{16.199}$$

where θ_i is the interfacial temperature, and H_k is the appropriate heat transfer coefficient for phase k.

16.6.7 Interfacial Work

The interfacial extra work is given by

$$\begin{aligned} W_k' = {} & C_w \alpha_d \rho_c \mathbf{a} \cdot (\mathbf{v}_d - \mathbf{v}_c) + (W_p p_c + K_w \rho_c |\mathbf{v}_d - \mathbf{v}_c|^2) \left(\frac{\partial \alpha_d}{\partial t} + \mathbf{v}_d \cdot \nabla \alpha_d \right) \\ & + H_w p_c (\mathbf{v}_d - \mathbf{v}_c) \cdot \nabla \alpha_d + J_w \rho_c |\mathbf{v}_d - \mathbf{v}_c|^2 (\mathbf{v}_d - \mathbf{v}_c) \cdot \nabla \alpha_d \\ & + B_w \alpha_d \rho_c (\mathbf{v}_d - \mathbf{v}_c)(\mathbf{v}_d - \mathbf{v}_c) : D_c + Q_w \alpha_d \rho_c (\mathbf{v}_d - \mathbf{v}_c)(\mathbf{v}_d - \mathbf{v}_c) : D_d \\ & + R_w \alpha_d \rho_c |\mathbf{v}_d - \mathbf{v}_c|^2 (\nabla \cdot (\mathbf{v}_d - \mathbf{v}_c)) + P_w \alpha_d \rho_c |\mathbf{v}_d - \mathbf{v}_c|^2 (\nabla \cdot \mathbf{v}_c) . \end{aligned} \tag{16.200}$$

The coefficients in (16.200) have been calculated by Arnold (1988) and are given by $C_w = 1/2$, $W_p = 1$, $H_w = 1$, $J_w = 1/5$, $K_w = 1/4$, $B_w = -9/20$, $Q_w = -1/10$, $R_w = 1/5$, $P_w = 3/20$.

16.7. SOLUTIONS TO THE EQUATIONS

So far in this Chapter, we have derived a model for particle/fluid flows based on obtaining the basic balance equations by applying an averaging process to the exact equations of motion, and deriving constitutive equations by applying the ensemble average to the solution for the flow around a single sphere. This approach to modeling multicomponent particle/fluid flows new and is almost completely untested. In order to predict particle/fluid flows accurately, we must be assured that the formulation reproduces the essential physics of the problem.

16.7.1 The Force on a Sphere due to Inertia

The force on a sphere at $\mathbf{z}$ is computed by integrating over $\mathbf{x}$. This gives

$$\mathbf{F}_p(\mathbf{x}, t) = \int_{\Omega(a)} \mathbf{n} \left(p_0 - \rho_f \left[\frac{1}{2} |\nabla \phi|^2 + \frac{\partial \phi}{\partial t} \right] \right) d\Omega \,, \tag{16.201}$$

where the integration is over the variable $\mathbf{x}'$, with the sphere centered at $\mathbf{x}$, and $|\mathbf{x}'| = a$. This results in

$$\mathbf{F}_p(\mathbf{z}) = \frac{4}{3} \pi a^3 \rho_p \left(\frac{\partial \mathbf{v}_f}{\partial t} + \mathbf{v}_f \cdot \mathbf{e}_f + \frac{1}{2} \left[\frac{\partial \mathbf{v}_f}{\partial t} - \frac{\partial \mathbf{v}_p}{\partial t} + \mathbf{v}_f \cdot \mathbf{e}_f \right] \right) . \tag{16.202}$$

Note that this force agrees with Taylor's (1928) calculation of the force necessary to hold a sphere at rest in an accelerating stream.

Using the relations between the quantities $\mathbf{v}_f$, $\mathbf{e}_f$, $\mathbf{v}_p$ and $\mathbf{e}_p$ and the average velocities $\overline{\mathbf{v}}_f^{xp}$ and $\overline{\mathbf{v}}_p^{xp}$ and their gradients, in the constitutive equations in this Section, the momentum equation for the dispersed particle phase becomes

$$\begin{aligned}
\alpha_p \rho_p \left(\frac{\partial \overline{\mathbf{v}}_p^{xp}}{\partial t} + \overline{\mathbf{v}}_p^{xp} \cdot \nabla \overline{\mathbf{v}}_f^{xp} \right) &= -\alpha_p \nabla (\overline{p}_f^{x} - \tfrac{1}{4} \rho_f |\overline{\mathbf{v}}_f^{xp} - \overline{\mathbf{v}}_p^{xp}|^2) \\
&\quad + \nabla \cdot \alpha_p \Big\{ \rho_f \Big[-\tfrac{9}{20} (\overline{\mathbf{v}}_f^{xp} - \overline{\mathbf{v}}_p^{xp})(\overline{\mathbf{v}}_f^{xp} - \overline{\mathbf{v}}_p^{xp}) \\
&\qquad + \tfrac{3}{20} |\overline{\mathbf{v}}_f^{xp} - \overline{\mathbf{v}}_p^{xp}|^2 \mathbf{I} \Big] \Big\} \\
&\quad + \alpha_p \rho_f \Bigg(\frac{1}{2} \left[\frac{\partial \overline{\mathbf{v}}_f^{xp}}{\partial t} - \frac{\partial \overline{\mathbf{v}}_p^{xp}}{\partial t} + \overline{\mathbf{v}}_f^{xp} \cdot \nabla \overline{\mathbf{v}}_f^{xp} - \overline{\mathbf{v}}_p^{xp} \cdot \nabla \overline{\mathbf{v}}_p^{xp} \right] \\
&\qquad - \tfrac{7}{20} \left[\overline{\mathbf{v}}_f^{xp} - \overline{\mathbf{v}}_p^{xp} \right] \cdot \left[\nabla \overline{\mathbf{v}}_f^{xp} - \nabla \overline{\mathbf{v}}_p^{xp} \right] \Bigg)
\end{aligned} \tag{16.203}$$

It is evident that by setting $\overline{\mathbf{v}}_p^{xp} = 0$, assuming α_p is small and the flow is steady, and using the continuity equation for the fluid in the form $\nabla \cdot \overline{\mathbf{v}}_f^{xp} = 0$ that eq. (16.203) correctly reduces to eq. (16.202) in the limit as $\alpha_p \to 0$. Thus, the closure conditionss used in the momentum equations yield the correct limit.

16.7.2 Waves

The one-dimensional model equations for incompressible flow can be rearranged into the matrix form

$$\mathbf{A}\frac{\partial \mathbf{\Psi}}{\partial t} + \mathbf{B}\frac{\partial \mathbf{\Psi}}{\partial z} = \mathbf{C} \qquad (16.204)$$

where

$$\mathbf{\Psi} = [\alpha, v_p, v_f, p_f]^T \;, \qquad (16.205)$$

and $\mathbf{A}(\mathbf{\Psi})$ and $\mathbf{B}(\mathbf{\Psi})$ are 4×4 square matrices; and $\mathbf{C}(\mathbf{\Psi})$ is a column vector, none of which contains derivatives of $\mathbf{\Psi}$.

Characteristics

The characteristics of this one dimensional system are given by

$$det\,(\mathbf{A}(\mathbf{\Psi})\lambda - \mathbf{B}(\mathbf{\Psi})) = 0 \qquad (16.206)$$

in which we have introduced the eigenvalues λ_j, which determine the characteristic curves given by

$$\frac{dz_j}{dt} = \lambda_j. \qquad (16.207)$$

If the values of λ that satisfy (16.206) are complex, the initial value problem for (16.204) is ill-posed, and the predictions from such a model are meaningless (Garabedian, 1964).

The one-dimensional model that includes the inertia effects derived in this Chapter can be written in the form

$$\frac{\partial \alpha}{\partial t} + \frac{\partial \alpha v_p}{\partial z} = 0 \qquad (16.208)$$

$$\frac{\partial (1-\alpha)}{\partial t} + \frac{\partial (1-\alpha) v_f}{\partial z} = 0 \qquad (16.209)$$

$$\alpha\rho_p\left(\frac{\partial v_p}{\partial t}+v_p\frac{\partial v_p}{\partial z}\right)=-\alpha\frac{\partial p}{\partial z}+2\alpha\xi\rho_f(v_p-v_f)\frac{\partial}{\partial z}(v_p-v_f)$$
$$+\alpha C_{vm}\rho_f\left[\left(\frac{\partial v_f}{\partial t}+v_f\frac{\partial v_f}{\partial z}\right)-\left(\frac{\partial v_p}{\partial t}+v_p\frac{\partial v_p}{\partial z}\right)\right]$$
$$+\alpha C_{gd}\rho_f(v_p-v_f)\frac{\partial}{\partial z}(v_p-v_f)+C_{ad}\rho_f(v_p-v_f)^2\frac{\partial \alpha}{\partial z}+F_p \tag{16.210}$$

$$(1-\alpha)\rho_f\left(\frac{\partial v_f}{\partial t}+v_f\frac{\partial v_f}{\partial z}\right)=-(1-\alpha)\frac{\partial p}{\partial z}+\xi\rho_f(v_p-v_f)^2\frac{\partial \alpha}{\partial z}$$
$$+\alpha C_{vm}\rho_f\left[\left(\frac{\partial v_p}{\partial t}+v_p\frac{\partial v_p}{\partial z}\right)-\left(\frac{\partial v_f}{\partial t}+v_f\frac{\partial v_f}{\partial z}\right)\right]$$
$$-\frac{\partial}{\partial z}\left(\alpha(a_f+b_f)\rho_f(v_p-v_f)^2\right)$$
$$-\alpha C_{gc}\rho_f(v_p-v_f)\frac{\partial}{\partial z}(v_p-v_f)-C_{ac}\rho_f(v_p-v_f)^2\frac{\partial \alpha}{\partial z}-F_p \tag{16.211}$$

where, based on inviscid calculation, $C_{vm}=\frac{1}{2}$, $\xi=\frac{1}{4}$, $a_f+b_f=\frac{1}{5}$, $C_{ad}=\frac{11}{20}$, $C_{gc}=-\frac{19}{20}$, $C_{ac}=\frac{17}{20}$ and $C_{gc}=\frac{7}{20}$. Here, F_p is the drag force, and only algebraic terms.

The "simplest" inviscid two-fluid model is obtained for $C_{vm}=0$, $a_f+b_f=0$, $\xi=0$, $C_{gd}=0$, $C_{ad}=0$, $C_{gc}=0$, and $C_{ac}=0$; its characteristics are complex, unless $v_p=v_f$. If $C_{vm}>0$, but the remaining coefficients are set to zero, the characteristics are also complex, unless $v_p=v_f$ again. If $C_{vm}>0$, $a_f+b_f=0$ and $\rho_p/\rho_f<<1$ (as in bubbly flow), the characteristics are real for ξ sufficiently large. For $\xi=\frac{1}{4}$, $C_{vm}=\frac{1}{2}$ and $a_f+b_f=0$, the characteristics are real for $0<\alpha<0.26$. For $C_{vm}=\frac{1}{2}$, $\xi=\frac{1}{4}$ and $a_f+b_f=\frac{1}{5}$, the characteristics are real for $0<\alpha<0.27$. For $C_{vm}=\frac{1}{2}$, $\xi=\frac{1}{4}$ $a_f+b_f=\frac{1}{5}$, $C_{ad}=\frac{11}{20}$, $C_{gc}=-\frac{19}{20}$, $C_{ac}=\frac{17}{20}$ and $C_{gc}=\frac{7}{20}$, the characteristics are real for $0<\alpha<0.446$.

For the 4×4 system given by eqs.(16.208-16.211), we shall use the integrated form of the equation of conservation of total volume with the volumetric flow j_0 assumed to be constant. Thus,

$$(1-\alpha)v_f+\alpha v_p=j_0\ . \tag{16.212}$$

If the pressure is eliminated, the system can be written as

$$\frac{\partial \mathbf{\Psi}}{\partial t}+\mathbf{B}\frac{\partial \mathbf{\Psi}}{\partial z}=\mathbf{C}\,, \tag{16.213}$$

where

$$\mathbf{\Psi} = \begin{bmatrix} 1-\alpha \\ v_f \end{bmatrix}, \tag{16.214}$$

$$\mathbf{B} = \begin{bmatrix} v_f & 1-\alpha \\ (v_p - v_f)^2 K_1(\alpha) & v_f + (v_p - v_f) K_2(\alpha) \end{bmatrix}, \tag{16.215}$$

$$\mathbf{C} = \frac{\alpha}{\alpha(1-\alpha) + C_{vm}} \begin{bmatrix} 0 \\ -g\alpha(1-\alpha) + \alpha S(\alpha)(v_p - v_f) \end{bmatrix},$$

$$K_1 = \frac{-C_{vm} + (2-\alpha)\xi + \alpha k}{\alpha(1-\alpha) + C_{vm}}, \tag{16.216}$$

$$K_2 = \frac{2(1-\alpha)(C_{vm} - \xi) - \alpha k}{\alpha(1-\alpha) + C_{vm}}, \tag{16.217}$$

$$(v_p - v_f) = \frac{j_0 - v_f}{\alpha}. \tag{16.218}$$

This system has characteristics given by

$$\det(\mathbf{B} - \lambda \mathbf{I}) = 0. \tag{16.219}$$

If we nondimensionalize the characteristics with

$$\lambda^* = \frac{\lambda - v_f}{v_p - v_f}, \tag{16.220}$$

then $\lambda^* = \lambda^*(\alpha)$ turns out to be a function of α only. The expression for λ^* is

$$\lambda^* = \frac{1}{2}\left(K_2(\alpha) \pm \sqrt{K_2^2(\alpha) + 4(1-\alpha)K_1(\alpha)}\right). \tag{16.221}$$

The graph of λ^* *versus* α is shown in Figure 16.4. Note that for each value of α, there are two values of λ^*, each corresponding to a curve in x-t space given by $dx/dt = \lambda$.

We note here that the right eigenvectors can be expressed as

$$\mathbf{R} = \begin{bmatrix} 1-\alpha \\ v_f - \lambda \end{bmatrix}. \tag{16.222}$$

Inlet Step Response

We wish to study the behavior of the system for $z > 0$, $t > 0$, subject to the initial condition $\mathbf{\Psi}(z,0) = \mathbf{\Psi}^+$ for $z > 0$, and $\mathbf{\Psi}(0,t) = \mathbf{\Psi}^-$ for $t > 0$.

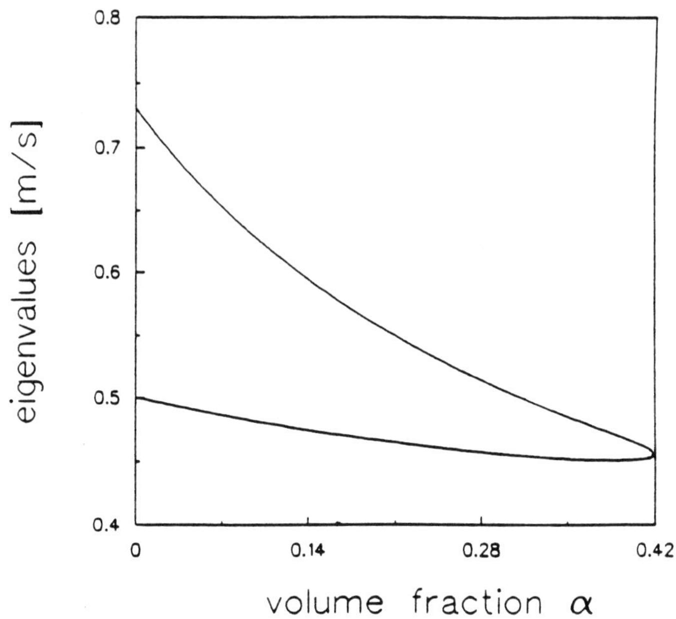

Figure 16-4 Characteristics

This corresponds to allowing the system to reach a steady state $\mathbf{\Psi} = \mathbf{\Psi}^+$ and then, at $t = 0$, changing the conditions at the inlet to $\mathbf{\Psi} = \mathbf{\Psi}^-$. In general, we expect to see several waves eminate from the z-t origin.

The theory is relatively well-developed for homogeneous, or unforced systems, with $\mathbf{C} = 0$, which are in conservation form. A system is said to be in conservation form if it is in the form

$$\frac{\partial \mathbf{\Upsilon}}{\partial t} + \frac{\partial \mathbf{Q}(\mathbf{\Upsilon})}{\partial z} = 0 \tag{16.223}$$

The system (16.219) is not in this form; however, a transformation of dependent variables yields an equivalent system that is in conservation form.

We note that the equation of conservation of mass for the continuous phase is in conservation form; consequently,

$$\Upsilon_1 = 1 - \alpha\,, \tag{16.224}$$

$$Q_1 = (1 - \alpha)v_f\,. \tag{16.225}$$

A second conserved quantity Υ_2 can be found by assuming that the conserved quantity is of the form

$$\Upsilon_2 = f(\alpha)(j_0 - v_f)\,, \tag{16.226}$$

and that the flux Q_2 is given by

$$Q_2 = f(\alpha)\left(\frac{1}{2}v_f^2 + \frac{K_2(j_0 v_f - \frac{1}{2}v_f^2)}{\alpha}\right) - (1-\alpha)(j_0 v_f - \frac{1}{2}v_f^2)f'(\alpha) + g(\alpha)\,. \tag{16.227}$$

Then $f(\alpha)$ is any solution of

$$\frac{\partial}{\partial\alpha}\left[(1-\alpha)f'(\alpha) - \frac{f(\alpha)K_2}{\alpha}\right] - 2f'(\alpha) - \frac{2K_1 f(\alpha)}{\alpha^2} = 0\,, \tag{16.228}$$

and

$$g'(\alpha) = \frac{f(\alpha)K_1 j_0^2}{\alpha^2}\,. \tag{16.229}$$

The second conserved quantity is proportional to the relative velocity $v_p - v_f$. The "density" of the conserved quantity is a complicated function of α.

According to the theory for conservation laws for systems of equations (Lax, 1960), the wave consists of a total of n shocks and rarefactions. For $n = 2$, there will be a shock, followed by a rarefaction, or a rarefaction, followed by a shock. To determine the structure of the wave, consider the characteristics emanating from behind the discontinuity at $t = 0$ in the $x - t$ plane, with conditions given by $\boldsymbol{\Upsilon}^-$. Note that there are two sets of these, since there are two characteristics $\lambda*$ for each α. Similarly, there are two sets of characteristics emanating from just ahead of the discontinuity. According to the theory, one of these sets will come together, and one set will diverge. By shock stability arguments, the set of characteristics that come together lead to the propagation of a shock, while the set that diverge from each other lead to a rarefaction. The intermediate state $\boldsymbol{\Upsilon}^i$ must be compatible with both the shock and the rarefaction. We illustrate the wave for the situation where the shock is "fast," and the rarefaction is "slow."

Shock Wave If the speed of the shock is C_s, conservation of $\boldsymbol{\Upsilon}$ across the shock requires that

$$-C_s[\boldsymbol{\Upsilon}] + [\mathbf{Q}] = 0\,. \tag{16.230}$$

where [ ] denotes the jump across the shock. Thus,

$$C_s = \frac{[(1-\alpha)v_f]}{[(1-\alpha)]} = \frac{[\Upsilon_2]}{[Q_2]}\,. \tag{16.231}$$

For given α^- and v_f^-, eq. (16.231) gives v_f^i as a function of α^i. Indeed, in the system described above, the equation for v_f^i is quadratic, where the coefficients involve the function $f(\alpha)$. If we treat α^i as a parameter, the

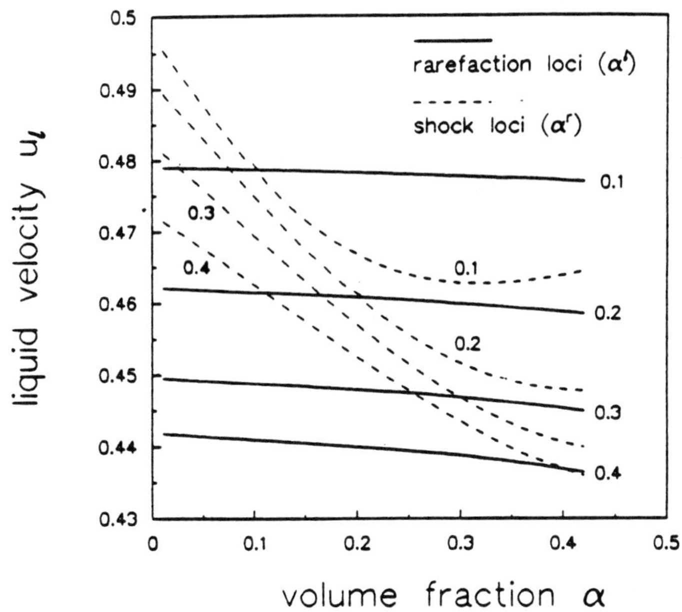

Figure 16-5 Shock and Rarefaction Loci

curve $v_f^i(\alpha^i)$ describes the possible intermediate states that can be reached from (α^-, v_f^-). Note that the point (α^-, v_f^-) lies on this curve. Several of these curves, for different (α^-, v_f^-), are shown in Fig. 16.5.

Rarefaction Wave Since the "fan" emanates from the origin in the $z - t$ plane, $\mathbf{\Upsilon} = \hat{\mathbf{\Upsilon}}(z/t)$, where $z/t = \sigma$ is the slope of the characteristic through the origin. Thus, on any characteristic having eigenvalue λ, we have $\sigma = \lambda$. If we differentiate $\hat{\mathbf{\Upsilon}}(z/t)$ with respect to t and z, and substitute into the system of equations, we have

$$\frac{\partial \mathbf{\Upsilon}}{\partial \sigma}\left(-\frac{z}{t^2}\right) + \frac{\partial \mathbf{Q}}{\partial \mathbf{\Upsilon}}\frac{\partial \mathbf{\Upsilon}}{\partial \sigma}\left(\frac{1}{t}\right) = 0\,. \tag{16.232}$$

Thus, $\partial \mathbf{\Upsilon}/\partial \sigma$ is a right eigenvector of the matrix $\partial \mathbf{Q}/\partial \mathbf{\Upsilon}$, and, as suggested above, $\sigma = z/t = \lambda$ is an eigenvalue.

If we write the system as

$$\frac{\partial \mathbf{\Upsilon}}{\partial t} + \frac{\partial \mathbf{Q}}{\partial \mathbf{\Upsilon}}\frac{\partial \mathbf{\Upsilon}}{\partial z} = \frac{\partial \mathbf{\Upsilon}}{\partial \mathbf{\Psi}}\frac{\partial \mathbf{\Psi}}{\partial t} + \frac{\partial \mathbf{Q}}{\partial \mathbf{\Upsilon}}\frac{\partial \mathbf{\Upsilon}}{\partial \mathbf{\Psi}}\frac{\partial \mathbf{\Psi}}{\partial z}\,. \tag{16.233}$$

Thus,

$$\mathbf{B} = \frac{\partial \mathbf{\Psi}}{\partial \mathbf{\Upsilon}}\frac{\partial \mathbf{Q}}{\partial \mathbf{\Upsilon}}\frac{\partial \mathbf{\Upsilon}}{\partial \mathbf{\Psi}}\,, \tag{16.234}$$

so that the eigenvalues of $\partial \mathbf{Q}/\partial \boldsymbol{\Upsilon}$ are the eigenvalues of $\mathbf{B}$, and if $\mathbf{R}$ is a right eigenvector of $\partial \mathbf{Q}/\partial \boldsymbol{\Upsilon}$, then $(\partial \boldsymbol{\Psi}/\partial \boldsymbol{\Upsilon}) \cdot \mathbf{R}$ is an eigenvector of $\mathbf{B}$. We note that since $\partial \boldsymbol{\Upsilon}/\partial \sigma$ is a right eigenvector of $\partial \mathbf{Q}/\partial \boldsymbol{\Upsilon}$, then

$$\frac{\partial \boldsymbol{\Psi}}{\partial \sigma} = \frac{\partial \boldsymbol{\Psi}}{\partial \boldsymbol{\Upsilon}} \frac{\partial \boldsymbol{\Upsilon}}{\partial \sigma} \tag{16.235}$$

is a right eigenvector of $\mathbf{B}$.

Since one right eigenvector is given by eq. (16.222), we have

$$\frac{d(1-\alpha)}{d\sigma} = 1 - \alpha\,, \tag{16.236}$$

$$\frac{dv_f}{d\sigma} = v_f - \lambda = -\frac{\lambda^*(\alpha)}{\alpha}(j_0 - v_f)\,. \tag{16.237}$$

Dividing the second equation by the first, we have

$$\frac{dv_f}{d\alpha} = \frac{\lambda^*(\alpha)}{\alpha(1-\alpha)}(j_0 - v_f)\,. \tag{16.238}$$

The solution satisfying $v_f = v_f^+$ at $\alpha = \alpha^+$ is

$$v_f(\alpha) = j_0 + (v_f^+ - j_0)\exp\left(-\int_{\alpha^+}^{\alpha} \frac{\lambda^*(\alpha')}{\alpha'(1-\alpha')}d\alpha'\right)\,. \tag{16.239}$$

The possible intermediate states starting from (α^+, v_f^+) then lie on the integral curves given by eq. (16.239). Several integral curves are shown in Fig. 16.5. The intermediate state separating the shock and the rarefaction is then given by the intersection of the shock curves and the rarefaction curves.

Nonlinear Volumetric Waves

The inlet step response solution in the previous subsection ignores the effect of the right-hand-side of the system. If the initial conditions on the volume fraction and the velocity are sufficiently sharp that they may be treated as a discontinuity, then the solution will have large derivatives, at least for a short time after the initial instant. The shock and the rarefaction travel at different speeds, and the rarefaction weakens, so that a steep gradient shortly after the initial instant will become less and less steep. Eventually, the terms on the right-hand-side will become important. In this subsection, we examine the situation that should be valid after a very long time. Specifically, we assume a solution of the form $\boldsymbol{\Psi} = \boldsymbol{\Psi}(\zeta)$, where

$\zeta = z - C_\alpha t$. We seek value(s) of the speed C_α to give nontrivial solutions satisfying

$$\frac{\partial \Psi}{\partial z} \to 0 \text{ as } z \to \pm\infty . \tag{16.240}$$

Such solutions are variously called "solitons," or, more generally, travelling waves.

From the first of eqs. (16.213), we see that

$$(1-\alpha)(v_f - C_\alpha) = k_0 , \tag{16.241}$$

where k_0 is a constant. Thus,

$$v_f = v_f(\alpha) = C_\alpha + \frac{k_0}{1-\alpha} . \tag{16.242}$$

From eq. (16.212), we see that

$$v_p = v_p(\alpha) = v_f(\alpha) + \frac{j_0 - v_f(\alpha)}{\alpha} . \tag{16.243}$$

The second of eqs. (16.213) yields

$$H(\alpha)\frac{d\alpha}{d\zeta} = C_2(\alpha) , \tag{16.244}$$

where

$$H(\alpha) = K_1(\alpha)[v_p(\alpha) - v_f(\alpha)]^2 + K_2(\alpha)[v_p(\alpha) - v_f(\alpha)]\frac{dv_f}{d\alpha} - C_\alpha , \tag{16.245}$$

and $C_2(\alpha)$ is the second component of the vector right-hand-side of eq. (16.213), with v_f and v_p replaced by the functions of α given by eqs. (16.242) and (16.243).

The solution of eq. (16.244) is given by a quadrature, with

$$\int_{\alpha_0}^{\alpha} \frac{H(\alpha')}{C_2(\alpha')} d\alpha' = \zeta , \tag{16.246}$$

where the arbitrary location of the place where $\zeta = 0$ has value α_0. Note that eq. (16.240) requires that, if $\lim_{\zeta\to\infty} \alpha = \alpha^+$, and $\lim_{\zeta\to-\infty} \alpha = \alpha^-$, we must have $C_2(\alpha^\pm) = 0$. If we examine C_2, we see that $C_2 = 0$ when

$$S(\alpha)[v_p(\alpha) - v_f(\alpha)] = (1-\alpha)\rho_f g . \tag{16.247}$$

If we take S =constant, and substitute (16.242) and (16.243), we see that the roots of C_2 are given by

$$\hat{S}[(j_0 - C_\alpha)\alpha - k_0] = \alpha(1-\alpha)^2 , \tag{16.248}$$

where $\hat{S} = S/\rho_f g$. Thus, there are two, one, or zero roots between $\alpha = 0$ and $\alpha = 1$, depending on the values of j_0, k_0, and C_α. We also note that $d\alpha/d\zeta$ should not change sign in the interval in α. Thus, $H(\alpha)$ must have one sign. We further note that $H(\alpha) = 0$ when α and $v_f(\alpha)$ give a characteristic. We further note that whether α^+ is greater than, or less than, α^- must be consistent with the sign of the derivative $d\alpha/d\zeta = H(\alpha)/C_2(\alpha)$.

The wave analyses done in the previous two subsections fit together in the following way. The solution to the conservation form of the equations (*i. e.*, zero right-hand-side) is correct for short times, starting from a sudden change in conditions. The initial jump evolves into a shock and a rarefaction, which propagate at different speeds. The shock remains a shock "forever," while the rarefaction weakens until the right-hand-side terms become important. The rarefaction evolves into a traveling wave which propagates at the speed C_α.

16.7.3 Solutions for Shearing Flows

When a nondilute particle-fluid flow occurs in a viscometric testing device, the flow that is produced is not a viscometric flow, that is, the velocity gradient is not necessarily constant, and the concentration is not uniform. Instead, relatively thin layers of fluid with relatively few suspended particles accumulate near the boundaries. These layers exhibit high shear rates. The suspended particles accumulate at some distance from the boundaries, giving a high local viscosity and a low shear rate. These phenomena are so robust that they appear to be exhibited by a large number of theories for suspensions. Here, we consider two properly invariant theories of multicomponent fluids. Passman and Drew (1991) have investigated steady flow between parallel plates with one plate stationary and the other plate moving parallel to it at constant speed. These solutions exhibit the phenomenon of a core of concentrated suspension surrounded by a layer of clear fluid.

Viscous Stresses It is possible to use averaging techniques to motivate the existence of cross-viscosities. One such model, that of Lundgren (1972), is of particular interest. His model for the fluid stress reduces to

$$\begin{aligned}(1-\alpha)\boldsymbol{\tau}_f^m &= \mu_{fs}\mathbf{D}_s + \mu_{ff}\mathbf{D}_f \\ &\quad -\mu_{f\nabla}\left[\nabla\alpha(\mathbf{v}_f - \mathbf{v}_s) + (\mathbf{v}_f - \mathbf{v}_s)\nabla\alpha\right] \quad (16.249)\end{aligned}$$

where

$$\mu_{ff}(\alpha) = (1-\alpha)\mu_f \,, \quad (16.250)$$

$$\mu_{fs}(\alpha) = \alpha\mu_f \,, \quad (16.251)$$

$$\mu_{f\nabla}(\alpha) = -\mu_f \,, \quad (16.252)$$

where μ_f is the viscosity of the "pure" fluid component.

In analogy with the fluid stress, we write the particle stress as

$$\begin{aligned}\alpha \boldsymbol{\tau}_s^m &= \mu_{ss}\mathbf{D}_s + \mu_{sf}\mathbf{D}_f \\ &\quad -\mu_{s\nabla}\left[\nabla\alpha(\mathbf{v}_f - \mathbf{v}_s) + (\mathbf{v}_f - \mathbf{v}_s)\nabla\alpha\right] . \qquad (16.253)\end{aligned}$$

Nothing is known about the coefficients introduced in this model. In order to estimate these values, assume that the particles move with the fluid, so that $\mathbf{D}_s = \mathbf{D}_f$. Adding Equations (16.249) and 16.253 gives

$$\alpha\boldsymbol{\tau}_s^m + (1-\alpha)\boldsymbol{\tau}_f^m = \left[\mu_{ss}(\alpha) + \mu_{sf}(\alpha) + \mu_{fs}(\alpha) + \mu_{ff}(\alpha)\right]\mathbf{D}_f \ . \qquad (16.254)$$

We assume that there is a function $\beta(\alpha)$ with the properties

$$\mu_{ss} = \alpha\beta\mu_{fs} \ , \qquad (16.255)$$

$$\mu_{sf} = \alpha\beta\mu_{ff} \ . \qquad (16.256)$$

The function β is called an "amplification factor" for the stress. Motivation from averaging theory indicates that it arises because of the rigidity of the individual particles. Then Equation (16.254) should reduce to an "effective viscosity" for a suspension. If α is small, then it is plausible that the mixture stress should reduce to the formula of Einstein, (1906) which is valid for small concentrations of spherical particles. That is, we should have

$$\mu_f(1+\alpha\beta)\mathbf{D}_f = \left(1 + \frac{5}{2}\alpha\right)\mathbf{D}_f \qquad (16.257)$$

This is true if $\beta = \frac{5}{2}$. Furthermore, there is a widely accepted formula of Frankel and Acrivos (1967), valid for large concentrations. A form that agrees with Einstein's for small α and with that of Frankel and Acrivos for large concentrations is given by Graham (1981) as

$$\beta = \frac{5}{2}\alpha + \frac{9}{4}\left[1 + \left(\frac{h}{2a}\right)\right]^{-1}\left[\frac{1}{(h/a)} - \frac{1}{[1+(h/a)]} - \frac{1}{[1+(h/a)]^2}\right] , \qquad (16.258)$$

where a is the radius of the spherical particles, and h is the interparticle spacing. For a simple cubic packing,

$$\frac{h}{a} = 2\frac{1-(\alpha/\alpha_m)^{1/3}}{(\alpha/\alpha_m)^{1/3}} , \qquad (16.259)$$

where α_m is the solid volume fraction for maximum packing of spheres. It is to be determined experimentally.

While this set of assumptions has some pleasing features, we need not restrict matters this much at this stage. Rather, we shall retain the generality contained in arbitrary cross viscosities.

Reynolds Stresses The single sphere calculations in Section 16.5 indicate that the "Reynolds" stresses should depend on deformation rates and forces of interaction. Thus the form

$$\begin{aligned}\alpha \mathbf{T}_s^{Re} &= -\alpha p_s^{Re}\mathbf{I} + \mu_{ss}^{Re}(\alpha)\mathbf{D}_s + \mu_{sf}^{Re}(\alpha)\mathbf{D}_f \\ &\quad + \alpha\rho_s\left[a_s|\mathbf{v}_f - \mathbf{v}_s|^2\mathbf{I} + b_s(\mathbf{v}_f - \mathbf{v}_s)(\mathbf{v}_f - \mathbf{v}_s)\right] ,\end{aligned} \tag{16.260}$$

$$\begin{aligned}(1-\alpha)\mathbf{T}_f^{Re} &= -(1-\alpha)p_f^{Re}\mathbf{I} + \mu_{fs}^{Re}(\alpha)\mathbf{D}_s + \mu_{ff}^{Re}(\alpha)\mathbf{D}_f \\ &\quad + \alpha\rho_f\left[a_f|\mathbf{v}_f - \mathbf{v}_s|^2\mathbf{I} + b_f(\mathbf{v}_f - \mathbf{v}_s)(\mathbf{v}_f - \mathbf{v}_s)\right] ,\end{aligned} \tag{16.261}$$

is plausible.

In our constitutive equations for the stresses, the two types of pressures play intrinsically different roles. The pressures p_f^m and p_s^m are reactions to the constraint of incompressibility of the fluid and the solid, and are determined as part of the solution to a boundary-value problem. The pressures $p_s^{Re}(\alpha)$ and $p_f^{Re}(\alpha)$, on the other hand, can arise due to the fluctuation motions of the particles, and the fluid turbulence, and are given by constitutive equations. Generally it would be necessary to specify such an equation precisely to solve boundary-value problems. However, in the particular structure presented here, it suffices that p_s^{Re} depend on α only.

Momentum Exchange We consider forces of interaction for drag, buoyancy, the Faxén force, virtual mass, and lift. Thus

$$\begin{aligned}\mathbf{M}_s &= p_{si}\nabla\alpha + \alpha S(\mathbf{v}_f - \mathbf{v}_s) + \alpha k\mu_f\nabla^2\mathbf{v}_f - \nabla\alpha\cdot\tau_{si} \\ &\quad + \alpha\rho_f C_{vm}\left[\frac{\partial\mathbf{v}_f}{\partial t} + \mathbf{v}_f\cdot\nabla\mathbf{v}_f - \frac{\partial\mathbf{v}_s}{\partial t} - \mathbf{v}_s\cdot\nabla\mathbf{v}_s + (\mathbf{v}_s - \mathbf{v}_f)\times(\nabla\times\mathbf{v}_f)\right] .\end{aligned} \tag{16.262}$$

In this section, we assume no body forces, so that $\mathbf{g} = 0$.

We define a *rectilinear shearing flow* to have the form

$$\begin{aligned}\mathbf{v}_k &= \hat{v}_k(y)\mathbf{e}_x , \\ \alpha &= \hat{\alpha}(y) ,\end{aligned} \tag{16.263}$$

where, for example, $\mathbf{v}_k$ means both of the velocities $\mathbf{v}_s$ and $\mathbf{v}_f$, $\mathbf{e}_x$ is the unit vector in the positive x-direction. Thus

$$\nabla\alpha = \frac{d\hat{\alpha}}{dy}\mathbf{e}_y , \tag{16.264}$$

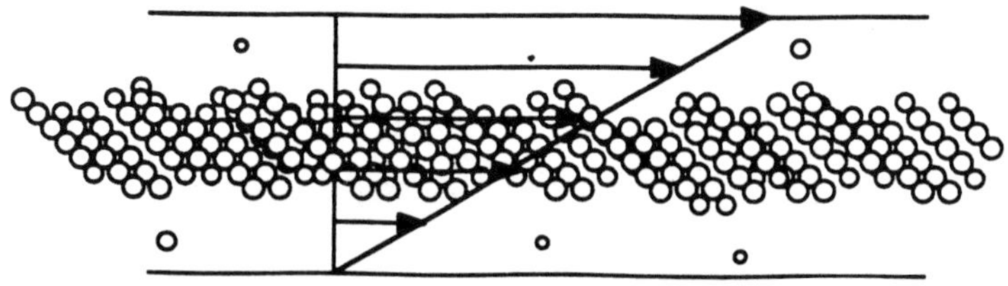

Figure 16-6 Simple Shearing Flow

and

$$\nabla \mathbf{v}_k = \mathbf{e}_y \mathbf{e}_x \frac{d\hat{v}_k}{dy} \ . \tag{16.265}$$

Then

$$2\mathbf{D}_k = \frac{1}{2}\frac{d\hat{v}_k}{dy}(\mathbf{e}_x \mathbf{e}_y + \mathbf{e}_y \mathbf{e}_x) \ . \tag{16.266}$$

The boundary conditions appropriate for rectilinear shearing flow are dependent on the material of which the walls are constructed, whether it is smooth, or bumpy, or sticky. In at least some situations, observations indicate that α vanishes or nearly does so at boundaries. Moreover, the average value of α across a cross-section defines the average concentration of solids in the material, and may be specified at the outset. Thus we have the conditions

$$\begin{aligned} &(1) \quad \hat{\alpha}(0) = 0 \ , \\ &(2) \quad \hat{\alpha}(2L) = 0 \ , \\ &(3) \quad \textstyle\int_0^{2L} \hat{\alpha} \, dy = 2L\langle \alpha \rangle \ , \end{aligned} \tag{16.267}$$

where $\langle \alpha \rangle$ is a given positive constant representing the average value of α across the cross-section of the flow.

We require an adequate number of boundary conditions for the velocities. For the fluid and the solid, we assume the classical boundary condition, that is, the velocity at a solid boundary matches the velocity of the boundary. We assume sufficient symmetry so that conditions at the centerline replace conditions at the top boundary $y = 2L$. These symmetry conditions are that the velocity of the fluid *and* of the solid at the centerline both are half of the velocity of the top boundary. Thus

$$\begin{array}{ll} (1) & \hat{v}_f(0) = 0 \, , \\ (2) & \hat{v}_s(0) = 0 \, , \\ (3) & \hat{v}_f(L) = V \, , \\ (4) & \hat{v}_s(L) = V \, . \end{array} \tag{16.268}$$

The equations are obtained by making the following choices for pressures,

$$\begin{aligned} \alpha p_s^{Re} &= \mathcal{P}_s(\alpha) \, , \\ (1-\alpha) p_f^{Re} &= 0 \, , \\ \alpha p_s^m &= 0 \, , \\ (1-\alpha) p_f^m &= \mathcal{P}_f \, , \end{aligned} \tag{16.269}$$

for the Reynolds stresses,

$$\begin{aligned} \boldsymbol{\tau}_p^{Re} &= 0 \, , \\ \boldsymbol{\tau}_f^{Re} &= 0 \, , \end{aligned} \tag{16.270}$$

for coefficients in the force of interaction,

$$\begin{aligned} \alpha S &= \lambda_1(\alpha) \, , \\ p_{si} &= \lambda_2(\alpha) \, , \\ p_{fi} &= \lambda_2(\alpha) \, , \\ C_{vm} &= 0 \, , \\ k &= 0 \, , \end{aligned} \tag{16.271}$$

for interfacial shear stresses,

$$\tau_{di} = \tau_{ci} = 0 \, , \tag{16.272}$$

and for the microscale viscosities, we assume that $\mu_{ss}(\alpha)$, $\mu_{sf}(\alpha)$, $\mu_{fs}(\alpha)$, and $\mu_{ff}(\alpha)$ are obtained from observations. In particular, for this model, the virtual mass, lift, and Faxén forces are assumed to vanish. In fact, in the particular flows studied here, the virtual mass force vanishes identically, and the Faxén force has no qualitative effect on the flow, so the only significant force that is ignored is the lift force.

It is convenient to use the notation

$$\begin{aligned} \mu_S &= \mu_{sf} + \mu_{ss} \, , \\ \mu_F &= \mu_{ff} + \mu_{fs} \, . \end{aligned} \tag{16.273}$$

We substitute assumptions (16.263) into the field equations, and assume that $\mathcal{P}_f$ is a function of y only. The equations of conservation of mass are satisfied identically. The remaining field equations come from substituting the constitutive Equations (16.261) for stress, and the constitutive

Equations (16.262) for momentum interaction into the balance of momentum equations. The resulting equations are most easily dealt with by using their sum and one of the individual equations. Here, we choose to do that, using the sum and the equation for the momentum of the fluid. The resulting system is

$$\frac{d}{dy}[\mu_F \frac{d\hat{v}_f}{dy} + \mu_S \frac{d\hat{v}_s}{dy}] = 0 , \tag{16.274}$$

$$\frac{d}{dy}[\mathcal{P}_f + \mathcal{P}_s] = 0 , \tag{16.275}$$

and

$$\frac{d}{dy}[\mu_{ff} \frac{d\hat{v}_f}{dy} + \mu_{fs} \frac{d\hat{v}_s}{dy}] + \lambda_1(\hat{v}_s - \hat{v}_f) = 0 , \tag{16.276}$$

$$\frac{d\mathcal{P}_f}{dy} + \lambda_2 \frac{d\alpha}{dy} = 0 . \tag{16.277}$$

To begin solving this system, let us consider Equations (16.275) and (16.277). Notice that this part of the system is decoupled from the rest in the sense that it involves no dependence on the velocities. Its solution involves solution of the ordinary differential equations and application of the appropriate boundary conditions. Equation (16.277) becomes

$$\left[\frac{d\mathcal{P}_s}{d\hat{\alpha}} + \lambda_2(\hat{\alpha})\right] \frac{d\hat{\alpha}}{dy} = 0 , \tag{16.278}$$

so that $d\hat{\alpha}/dy = 0$. If we assume that the function $\hat{\alpha}(y)$ is continuous, there is no solution that agrees with the boundary condition (16.267). Thus, we assume that the solution $\hat{\alpha}(y)$ is piecewise continuous. Then there is a solution of the form

$$\hat{\alpha} = \begin{cases} 0, & 0 \leq y < h \\ \alpha_0, & h < y < 2L - h \\ 0, & 2L - h < y < 2L \end{cases} \tag{16.279}$$

for some $h < L$, and with $\hat{\alpha}$ undefined at $y = h$ and $y = 2L - h$. This makes the constitutive functions λ_1, λ_2, μ_{ff}, μ_{fs}, μ_{sf}, and μ_{ss} constant also. Here, α_0 and h are quantities to be determined. Not all of the parameters α_0, $\langle\alpha\rangle$, and h are independent. Suppose for example that the steady shearing flow evolves at long time from a homogeneous flow with uniform volume fraction $\langle\alpha\rangle$. Then the total fraction of the solid constituent in the cross section is given by

$$\langle\alpha\rangle L = \alpha_0(L - h) . \tag{16.280}$$

The velocities can be found by solving the differential equations, and are given by

$$\hat{v}_f = \hat{A}y + \hat{B} \;, \tag{16.281}$$

for $0 \leq y < h$, and

$$\begin{aligned} \hat{v}_f &= \frac{1}{\mathrm{T}\mu_S}(Ay + B) \;+\; (C\cosh\beta y + D\sinh\beta y) \;, \\ \hat{v}_s &= \frac{1}{\mathrm{T}\mu_S}(Ay + B) \;-\; \frac{\mu_F}{\mu_S}(C\cosh\beta y + D\sinh\beta y) \;, \end{aligned} \tag{16.282}$$

for $h < y < L$. Here

$$\bar{\mu}_f = \frac{\mu_{ff}(\mu_{sf} + \mu_{ss} - \mu_{fs}) - \mu_{fs}^2}{\mu_{sf} + \mu_{ss}} \;, \tag{16.283}$$

and

$$\mathrm{T} = \frac{\mu_{ff} + \mu_{fs} + \mu_{ss} + \mu_{sf}}{\mu_{sf} + \mu_{ss}} \;. \tag{16.284}$$

The quantity $\hat{A}$ is the shear rate in the clear fluid. To complete the formal solution, we require an adequate number of boundary conditions to specify A, B, C, D, $\hat{A}$, and $\hat{B}$. For the clear fluid, we assume the classical boundary condition, that is, the velocity at a solid boundary matches the velocity of the boundary. In this particular boundary-value problem, there is no solid boundary for the mixed material regime. Thus only interface conditions, not boundary conditions, are needed. Our assumption is that the clear fluid at the interface affects only the fluid in the mixture at the interface, and the solid at the interface acts as if it were at a traction-free boundary. We assume sufficient symmetry so that conditions at the center line replace conditions at the interface $y = 2L - h$ and the top boundary $y = 2L$. These symmetry conditions are that the velocity of the fluid *and* of the solid at the interface both are half of the velocity of the top boundary. Thus

$$\begin{aligned} &(1)\quad \hat{v}_f(0) = 0 \;, \\ &(2)\quad \lim_{y\to h^+} \hat{v}_f(y) = \lim_{y\to h^-} \hat{v}_f(y) \;, \\ &(3)\quad \lim_{y\to h^+} \left(\mu_{ff}\frac{\hat{v}_f(y)}{dy} + \mu_{fs}\frac{\hat{v}_s(y)}{dy} \right) = \lim_{y\to h^-} \mu_f \frac{\hat{v}_f(y)}{dy} \;, \\ &(4)\quad \lim_{y\to h^+} \left(\mu_{sf}\frac{d\hat{v}_f(y)}{dy} + \mu_{ss}\frac{\hat{v}_s(y)}{dy} \right) = 0 \;, \\ &(5)\quad \hat{v}_f(L) = V \;, \\ &(6)\quad \hat{v}_s(L) = V \;, \end{aligned} \tag{16.285}$$

where μ_f is the viscosity of the clear fluid. These boundary conditions, when applied to the Equations (16.282) gives a nonsingular set of six linear algebraic equations in the six unknown constants A, B, C, D, $\hat{A}$, and $\hat{B}$.

Applying boundary conditions (1), (5) and (6) gives

$$\hat{v}_f = \hat{A}y \tag{16.286}$$

for $0 \leq y < h$, and

$$\begin{aligned} \hat{v}_f &= V + A(y-L) + \quad C\left(\frac{\sinh\beta(y-L)}{\sinh\beta L}\right) , \\ \hat{v}_s &= V + A(y-L) - \quad C\frac{\mu_F}{\mu_S}\left(\frac{\sinh\beta(y-L)}{\sinh\beta L}\right) , \end{aligned} \tag{16.287}$$

for $h < y \leq L$. The remaining three constants, $\hat{A}$, A and C are obtained from the interface conditions (2), (3) and (4), and are somewhat complicated to write out. Let

$$\begin{aligned} \Delta \quad = \quad & \beta[(L-h)\mu_f(\mu_{ss}\mu_S - \mu_{fs}\mu_F) \\ & +h(\mu_S + \mu_F)(\mu_{ss}\mu_{ff} - \mu_{fs}\mu_{sf})]\cosh\beta(L-h) \\ & +\mu_f\mu_F\mu_S\sinh\beta(L-h) \ . \end{aligned} \tag{16.288}$$

These expressions are the determinants resulting from solving three linear algebraic equations in three unknowns. For $\hat{A}$, A and C then

$$\hat{A} = \frac{\beta(\mu_S + \mu_F)(\mu_{ss}\mu_{ff} - \mu_{fs}\mu_{sf})\cosh\,\beta(L-h)}{\Delta}V , \tag{16.289}$$

while

$$A = \frac{\beta\mu_f({\mu_{ss}}^2 + \mu_{sf}\mu_{ss} - \mu_{fs}\mu_{sf} - \mu_{ff}\mu_{sf})\cosh\,\beta(L-h)}{\Delta}V , \tag{16.290}$$

and

$$C = -\frac{\mu_f(\mu_{fs} + \mu_{ff})(\mu_{ss} + \mu_{sf})\sinh\,\beta L}{\Delta}V . \tag{16.291}$$

The quantity $\hat{A}$ is the shear rate in the clear fluid. The situation is slightly more complicated in the mixed material because the velocity profiles are nonlinear. Nonetheless, the quantity A is essentially the shear rate in the mixed material.

16.8 SUMMARY

This Chapter has presented a rigorous framework for the multifluid modeling of multiphase flows. Two-fluid models of this type have been applied successfully to vapor/liquid flows (Ishii & Mishima, 1981; Lopez, *et al.*,

1990). Similar models are starting to be developed for the multidimensional analysis of solid/fluid flows.

The model is based on balance equations for both the particles and fluid. The balance equations are derived by applying an ensemble average to the exact equations of motion. Constitutive equations are presented, and some forms for them are motivated by considering potential flow around a single sphere. Finally, we present some results for these equations, including calculations of the characteristics, the speeds of kinematic waves and profiles of velocity and volume fraction in shear flows.

The success of these models depends on the appropriateness of the closure laws proposed. It is hoped that the material presented in this Chapter will help motivate researchers in the field to contribute to the development of the needed closure laws so that future multifluid predictions of solid/fluid flows can be made with precision. The ability of the model to predict a wide variety of motions seems promising.

REFERENCES

Abou-Arab, T. W. and M. C. Roco, "Solid Phase Contribution in the Two-phase Turbulence Kinetic Energy Equations," *J. Fluids Engineering*, Vol. 112, 1990, pp.351-361.

Aris, R., *Vectors, Tensors and the Basic Equations of Fluid Mechanics*, Prentice Hall, Englewood Cliffs, New Jersey, 1962.

Arnold, G. S., D. A. Drew and R. T.Lahey, Jr., "The Assessment of Multiphase Flow Models Using the Second Law of Thermodynamics," *Int. J. Multiphase Flow*, Vol. 16, 1990, pp. 481–494.

Arnold, G. S., D. A. Drew and R. T. Lahey, Jr., "Derivation of Constitutive Equations for Interfacial Force and Reynolds Stress for a Suspension of Spheres Using Ensemble Cell Averaging," *Chemical Engineering Communications*, Vol. 86, 1989, pp. 43–54.

Basset, A. B., *Treatise on Hydrodynamics*, Dieghton Bell (1888).

Brenner, H., "The Stokes Resistance of an Arbitrary Particle IV Arbitrary Fields of Flow," *Chem. Eng. Sci.*, Vol. 19, 1964, pp.703-727.

Chapman, S. and T. G. Cowling, *The Mathematical Theory of Non-Uniform Gases*, Third Edition, Cambridge University Press, 1970.

Delhaye, J. M., "Basic Equations for Two-Phase Flow Modeling," *Two Phase Flow and Heat Transfer in the Power and Process Industries*, A. E. Bergles, et al., Eds. Hemisphere Publishing, Washington, 1981.

Drew, D. A., and R. T. Lahey, Jr., "Phase Distribution Mechanisms in Turbulent Two-Phase Flow in Channels of Arbitrary Cross Section," *Journal of Fluids Engineering*, Vol. 103, 1981, pp. 583–589.

Drew, D. A., and R. T. Lahey, Jr., "Some Supplemental Analysis Concerning the Virtual Mass and Lift Force on a Sphere in a Rotating and Straining Flow," *Int. J. Multiphase Flow*, Vol. 16, 1990, pp. 1127-1130.

Einstein, A., "Eine Neue Bestimmung der Molekúdimensionen." *Annln Phys.* Vol. 19, 1906, p.289.

Frankel, N. A., and A. Acrivos, "On the Viscosity of a Concentrated Suspension of Solid Spheres," *Chem. Eng. Sci.*, Vol. 22, 1967, pp. 847–853.

Garabedian, P., *Partial Differential Equations*, Wiley, New York, 1964.

Givler, R. "An Interpretation for the Solid Phase Pressure in Slow, Fluid-Particle Flows," *International Journal of Multiphase Flow*, Vol. 13, 1987, pp. 717–722.

Graham, A. L., "On the Viscosity of Suspensions of Solid Spheres," *Appl. Sci. Res.*, Vol. 37, 1981, pp. 275-286.

Hwang, G. J. and H. H. Shen, "Modeling the Solid Phase Stress in a Fluid-Solid Mixture," *Int. J. Multiphase Flow*, Vol. 15, 1989, pp. 257-268.

Ishii, M., *Thermo-Fluid Dynamic Theory of Two-Phase Flow*, Eyrolles, Paris, 1975.

Ishii, M. and K. Mishima, "Two-Fluid Model and Hydrodynamic Constitutive Equations," *Nuclear Engineering and Design*, Vol. 82, 1981, pp. 107–126.

Ishii, M. and N. Zuber, "Relative Motion and Interfacial Drag Coefficient in Dispersed Two-Phase Flow of Bubbles, Drops and Particles," *A. I. Ch. E. Journal*, Vol. 25, 1979.

Jenkins, J. T. and S. B. Savage, "A Theory for the Rapid Flow of Identical, Smooth, Nearly Elastic, Spherical Particles," *J. Fluid Mech.* Vol. 130, 1983, pp. 187–202.

Jenkins, J. T., "Rapid Flows of Granular Materials," *Non-Classical Continuum Mechanics: Abstract Techniques and Applications* (Eds. Knops & Lacey) Cambridge University Press, 1987, pp. 213-225.

Lambe, T. W. and R. V. Whitman, *Soil Mechanics*, Wiley, New York 1969.

Lax, P. D., *Hyperbolic Systems of Conservation Laws and the Mathematical Theory of Shock Waves.* Regional Conference Series in Applied Mathematics, Vol. 11, Society for Industrial and Applied Mathematics, Philadelphia, 1973.

Lopez de Bertodano, M., S-J. Lee, R. T. Lahey, Jr., and D. A. Drew, "The Prediction of Two-Phase Turbulence and Phase Distribution Phenomena Using a Reynolds Stress Model," *J. Fluids Engineering*, Vol. 12,

1990, pp. 107-113.

Love, A. E. H., *A Treatise on the Mathematical Theory of Elasticity*, Dover, New York, 1944.

Lun, C. K. K., S. B. Savage, D. J. Jeffreys, and N. Chepurney, "Kinetic Theories of Granular Flows: Inelastic Particles in Couette Flow and Slightly Inelastic Particles in a General Flow Field," *J. Fluid Mech.*, Vol. 140, 1984, pp. 223-256.

Lundgren, T. S., "Slow Flow Through Stationary Random Beds and Suspensions of Spheres," *J. Fluid Mech.*, Vol. 51, 1972, pp. 273–299.

Passman, S. L. and D. A. Drew, "An Exact Solution for Shearing Flow of Multicomponent Mixtures," *Chem. Eng. Sci.*, Vol. 46, 1991.

Taylor, G. I., "The Forces on a Body Placed in a Curved or Converging Stream of Fluid," *Proc. Roy. Soc.*, A Vol. 120, 1928, pp. 260-283.

Truesdell C. and R. Toupin, "The Classical Field Theories." *Handbuch der Physik*, Vol. III, Pt. 1, Springer-Verlag, Berlin 1960.

Truesdell C. and W. Noll, "The Non-Linear Field Theroies of Mechanics." *Handbuch der Physik*, Vol. III, Pt. 3, Springer-Verlag, Berlin 1966.

Voinov, O. V., "Force Acting on a Sphere in a Inhomogeneous Flow of an Ideal Incompressible Fluid," *Journal of Applied Mechanics and Technical Physics*, Vol. 14, 1973, pp. 592–594.

Whitham, G. B., *Linear and Nonlinear Waves*, Wiley, New York, 1973.

Zuber, N., "On the Dispersed Two-Phase in the Laminar Flow Regime," *Chem Eng. Sci.*, Vol. 19, 1964, pp. 897-903.

CHAPTER 17

DYNAMICS OF CONCENTRATED COLLOIDAL DISPERSIONS: STATISTICAL MECHANICAL APPROACHES

W. B. Russel

17.1 INTRODUCTION

Colloidal dispersions consist of small particles, generally 5 nm to 10 μm, dispersed in a liquid, perhaps water or an organic solvent. Brownian motion is just observable for the largest particles but drives rapid fluctuations at the lower end of the range. Even at the nanometer level, though, the viscous solvent strongly damps thermal motion; one normally detects the resulting random displacements rather than the velocities characteristic of the kinetic energy.

At finite concentrations the particles interact through a variety of forces. Van der Waals, or dispersion, attractions arise from a dielectric mismatch between particles and solvent. Generally the magnitude becomes comparable to kT at separations roughly a tenth of the particle size. When possible, though, matching the refractive indices in the visible effectively eliminates the effect. In polar solvents acquisition of charge is inevitable, causing electrostatic repulsions between like particles. The range, characterized by the Debye length, is governed independently by the ionic strength and extends to microns for carefully deionized dispersions; hence, strong interactions at low volume fractions are possible. In both polar and nonpolar solvents introduction of soluble polymers profoundly affects particle interactions. Expulsion of nonadsorbing polymer from the gap between particles generates an osmotic attraction. For adsorbing polymer the interparticle potential depends on the surface coverage and the time scale of the interaction, with strong attractions or repulsions possible. Irreversible attachment of chains yields repulsion in good solvents and generally attraction in poor solvents. So the total interaction potential can be purely repulsive, either hard or soft, or can exhibit an attractive minimum of depth from less than 1 kT to more than 100 kT (c.f. Hunter 1987; Russel, et al. 1989).

With this variety of interactions, concentrated colloidal dispersions exhibit phase behavior and rheology as rich as those for molecular or polymeric systems (e.g. Goodwin and Ottewill 1991). Even the simplest of systems, monodisperse spheres, undergoes disorder-order (liquid-crystal) transitions and separates into coexisting dilute fluid (gas) and dense solid (crystal) phases with attractions of 2-3 kT. Furthermore, rapid consolidation across the liquid-crystal phase boundary produces a persistent disordered phase (glass), while rapid quenching across the gas-crystal boundary results in a volume filling solid (gel).

The rheology of a dispersion closely reflect the nature of the structure at rest (e.g. Russel, et al. 1989; Goodwin and Hughes 1990). The equilibrium fluid possesses a Newtonian low shear viscosity; however, as the rate of convection by shear exceeds the rate of relaxation via Brownian motion the microstructure is noticably displaced from equilibrium and the viscosity decreases. A second Newtonian plateau is often attained, or expected. In some cases, further increases in shear rate cause the viscosity to rise, smoothly or discontinuously, above the low shear plateau. Increasing the strength of interactions, whether attractive or repulsive, enhances the low shear viscosity but affects much less that at high shear rates. Crossing a phase boundary, into a glassy/crystalline or gel region, dramatically increases the viscosity at low shear and extends the relaxation times. Because of the latter the exact nature of the behavior of such phases, which are either metastable or highly dislocated, remains poorly defined.

Despite a significant and growing body of experimental data, conclusive links between the rheology and the characteristics of the dispersion are limited to the simplest of systems. Likewise, connections between the rheology and light scattering measurements, of the dynamics of the equilibrium structure and the static non-equilibrium structure, are only beginning to emerge. Much remains ahead in defining the phenomenology and then relating the dynamics to the interparticle potential. The latter is essential for the formulation of materials with desired properties.

Theoretical treatments of the dynamics of molecular, polymeric, and colloidal systems have a common basis but, in the end, emphasize different features. The Fokker-Planck equation governs, in general, the evolution of a system in configuration and momentum space. For molecular fluids analyses focus on collisional dynamics on the inertial time scale and the origin of algebraic tails in the density and velocity autocorrelation functions (c.f. Boon and Yip 1980). For polymeric liquids, molecular details on the inertial time scale are generally ignored, in favor of diffusive motions on the macromolecular scales. For dilute solutions the dynamics resemble those for colloidal systems but with additional internal degrees of freedom. However, in concentrated solutions and melts unique reptational modes emerge, i.e. translation along the backbone to escape constraints imposed by neighboring chains (c.f. Doi and Edwards 1986; Bird, et al. 1987). Colloidal particles lack the internal degrees of freedom of macromolecules and manifest inertial effects only at time scales too short to affect macroscopic behavior. So a description on the diffusion time scale, recognizing only the positions and not the momenta of the particles, is normally presumed to suffice.

Several derivations or arguments establish the Smoluchoski equation as the proper starting point. Deutch and Oppenheim (1971) begin with the Liouville equation and find that for particles (a) much larger than the solvent molecules and (b) with much slower relaxations the latter may be replaced by an inert solvent governed by the classical equations of fluid mechanics. Murphy and Aguirre (1972) conclude that further reduction to the Smoluchoski equation is appropriate when the viscous relaxation time of the fluid, $\rho a^2/\mu$, greatly exceeds the time required for significant change in particle configurations due to Brownian motion, $a^2/D_o = 6\pi\mu a^3/kT$. Under these conditions Hess and Klein (1983), with followup by Cichocki and Hess (1986, 1987), derived identical forms for the density-density autocorrelation function from the Fokker-Planck and Smoluchoski equations. Thus one can confidently analyze the dynamics of colloidal dispersions on the basis of the Smoluchoski equation.

The different backgrounds and interests of the individuals drawn to the subject have produced several independent approaches to the theory. Those interested in fluctuations about equilibrium typically emphasize thermodynamic interactions, drawing heavily on the statistical mechanics of dense fluids in their analyses (c.f. Hess and Klein 1983; Pusey and Tough 1985). Those interested in transport processes, particularly rheology, emphasize hydrodynamic interactions, capitalizing on developments in low Reynolds number fluid mechanics (c.f. Happel and Brenner 1983; Kim and Karilla 1990). Here Batchelor (1970, 1972, 1977) led the way, blending statistical treatments of configurations and proper renormalization of interactions with rigorous hydrodynamics (c.f. Russel, et al. 1989). A third group relies on computer simulations on the diffusive time scale, known as Brownian dynamics (Ermak and McCammon 1978). At present the statistical mechanical approaches, most suitable for weak flows, are beginning to converge and ultimately should complement simulations, which fare better for strong flows (e.g. Brady 1992).

Nonetheless predictions of the dynamics of concentrated dispersions are scarce, leaving little to review. So we seek instead to identify and, to a limited extent, pursue promising paths for the future. Section 2 lays out the basic Smoluchoski formalism and defines the phenomena and properties of interest. Section 3 then briefly notes the status of equilibrium statistical mechanics, as reviewed by Hansen and McDonald (1986) and Evans (1989, 1991). Section 4 addresses fluctuations about equilibrium through both the Smoluchoski and Fokker-Planck approaches. For the former we extend the dilute analysis of Rallison and Hinch (1986) to finite concentrations via new closures proposed by Scherwinski (1990). For the latter we adopt for comparison the mode-mode coupling approximation as described by Hess and Klein (1983). The long-time self-diffusion coefficient provides a common focus relevant to rheological phenomena. Section 5 considers the steady perturbations from equilibrium induced by shear flows, with the analysis restricted to the Smoluchoski equation because the linear response theory is limited (e.g. Machta, et al. 1980). Closures employed in the literature are compared within Scherwinski's framework, for both real and Fourier space formulations in the absence of hydrodynamic interactions. Section 6 summarizes predictions of rheology from the simplest of models, dilute dispersions with only far-field hydrodynamics, to suggest what might be learned by robust approximations for concentrated systems. Section 7 returns to the hydrodynamics, proposing a self-consistent field closure for predicting purely hydrodynamic transport coefficients such as the high frequency limiting viscosity and the short-time self-diffusion coefficient. This extension of Hinch's (1977) preaveraged equations offers rigor and possibly accuracy.

So the agenda is a speculative one, offering promising but as yet untried paths for the future.

17.2 FORMULATION OF THE SMOLUCHOSKI THEORY

The statement of conservation for N particles in configuration space $(\boldsymbol{x}_1,\ldots,\boldsymbol{x}_N)$

$$\frac{\partial P_N}{\partial t} + \sum_{i=1}^{N} \nabla_i \cdot \boldsymbol{j}_i = 0 \tag{2.1}$$

governs $P_N d\boldsymbol{x}_1 \ldots d\boldsymbol{x}_N$, the probability of the particles within the region

$$\left(\boldsymbol{x}_1 \pm d\boldsymbol{x}_1,\ldots,\boldsymbol{x}_N \pm d\boldsymbol{x}_N\right)$$

The fluxes $\boldsymbol{j}_i$ follow from the Brownian and interparticle forces and the rate of strain $\boldsymbol{E}$ and vorticity Ω of the bulk fluid motion according to

$$\boldsymbol{j}_i = P_N\left((\boldsymbol{E}+\boldsymbol{\Omega})\cdot \boldsymbol{x}_i + \sum_{j=1}^{N}\left[\boldsymbol{\omega}_{ij}\cdot\left(-\nabla_j kT ln P_N + \boldsymbol{F}_j\right) - \boldsymbol{C}_{ij}\cdot\boldsymbol{E}\cdot\left(\boldsymbol{x}_i - \boldsymbol{x}_j\right)\right]\right) \tag{2.2}$$

with

$$\boldsymbol{F}_j = -\nabla_j \Phi = -\sum_{i<j} \nabla_j \Phi_{ij}$$

denoting the total interparticle force, approximated as pairwise additive. The mobilities ω_{ij}, relating the velocity induced in particle i to the force on j, depend on the fluid viscosity and particle size and, together with the tensors $\boldsymbol{C}_{ij}$, vary with the configuration of particles i and j and their neighbors due to hydrodynamic interactions. As noted above, the Smoluchoski equation, with suitable boundary conditions, suffices for all phenomena on the diffusion time scale, i.e. $t > \rho a^2/\mu$.

Conversion into a tractable and useful theory, however, necessitates reducing the dimensionality of the problem, normally by integrating over all but two particles within a closed system as

$$\int\ldots\int\left(\frac{\partial P_N}{\partial t} + \sum_{i=1}^{N}\nabla_i\cdot\boldsymbol{j}_i\right)d\boldsymbol{x}_3\ldots d\boldsymbol{x}_N = 0$$

with

$$P_n = \frac{1}{(N-n)!}\int P_N d\boldsymbol{x}_{n+1}\ldots d\boldsymbol{x}_N$$

Since normal fluxes vanish on the boundaries, the two-particle Smoluchoski equation follows as

$$\frac{\partial P_2}{\partial t} + \sum_{i=1}^{2} \nabla_i \cdot \boldsymbol{J}_i = 0 \tag{2.3}$$

with $(i,j=1,2;i\neq j)$

$$\begin{aligned}\boldsymbol{J}_i = {} & P_2\Big((\boldsymbol{E}+\boldsymbol{\Omega})\cdot \boldsymbol{x}_i - 2\langle \boldsymbol{C}_{ij}\rangle_{N-2}\cdot \boldsymbol{E}\cdot(\boldsymbol{x}_i-\boldsymbol{x}_j)\Big) \\ & -P_2\Big(\langle \boldsymbol{\omega}_{ii}\rangle_{N-2}\cdot \nabla_i + \langle \boldsymbol{\omega}_{ij}\rangle_{N-2}\cdot \nabla_j\Big)(\Phi + kT\ln P_2) \\ & -\int\Big(P_3\Big(\langle \boldsymbol{\omega}_{ii}\rangle_{N-3}\cdot \nabla_i \Phi_{i3} + \langle \boldsymbol{\omega}_{ij}\rangle_{N-3}\cdot \nabla_j \Phi_{j3}\Big) + kT\langle \boldsymbol{\omega}_{i3}\rangle_{N-3}\cdot \nabla_3 P_3\Big) d\boldsymbol{x}_3 \\ & -\int P_4 \langle \boldsymbol{\omega}_{j3}\rangle_{N-4}\cdot \nabla_3 \Phi_{34}\, d\boldsymbol{x}_3 d\boldsymbol{x}_4\end{aligned} \tag{2.4}$$

The brackets indicate

$$P_n <\ldots>_{N-n} = \frac{1}{(N-n)!}\int \ldots P_N d\boldsymbol{x}_{n+1}\ldots d\boldsymbol{x}_N$$

i.e. an ensemble average over the positions of N-n particles. Subsequently, we set

$$\langle \boldsymbol{\omega}_{ij}\rangle_{N-3} \cong \langle \boldsymbol{\omega}_{ij}\rangle_{N-2}$$

$$\langle \boldsymbol{\omega}_{i3}\rangle_{N-4} \cong \langle \boldsymbol{\omega}_{i3}\rangle_{N-3}$$

which is consistent with the equilibrium limit and probably represents the best hydrodynamic approximation tractable. Furthermore, symmetry considerations permit a decomposition of the hydrodynamic tensors into components along and perpendicular to line of centers as

$$\begin{aligned}\langle \omega_{ij}\rangle_{N-2} &= \frac{1}{6\pi\mu a}\left\{\langle A_{ij}\rangle_{N-2}(r_{21})\frac{\boldsymbol{r}_{21}\boldsymbol{r}_{21}}{r_{21}^2} + \langle B_{ij}\rangle_{N-2}(r_{21})\left(\delta - \frac{\boldsymbol{r}_{21}\boldsymbol{r}_{21}}{r_{21}^2}\right)\right\} \\ \langle \boldsymbol{C}_{ij}\rangle_{N-2} &= \langle A\rangle_{N-2}(r_{21})\frac{\boldsymbol{r}_{21}\boldsymbol{r}_{21}}{r_{21}^2} + \langle B\rangle_{N-2}(r_{21})\left(\delta - \frac{\boldsymbol{r}_{21}\boldsymbol{r}_{21}}{r_{21}^2}\right)\end{aligned} \tag{2.5}$$

The boundary conditions,

$$\begin{aligned} & P_2 \to n^2 \quad \text{as} \quad r^2 = \boldsymbol{r}\cdot\boldsymbol{r} \to \infty \\ & \frac{\boldsymbol{r}}{r}\cdot(\boldsymbol{J}_2 - \boldsymbol{J}_1) = 0 \quad \text{for} \quad r = 2a\end{aligned} \tag{2.6}$$

with $\boldsymbol{r}=\boldsymbol{x}_2-\boldsymbol{x}_1$, correspond to a disordered structure in the bulk, i.e. a liquid, and no interpenetration of the hard cores.

The form of the Smoluchoski equation [2.3,4] immediately signals the difficulties in devising a theory for concentrated dispersions. The integrals dictate closures, inevitably approximate, relating the three- and four-particle densities to the pair density. Likewise, configuration-dependent hydrodynamic functions must be constructed to capture the many-body interactions. In both cases, reduction to the pair level can introduce a nonlinear dependence on the pair density, so numerical solutions are often necessary. The task here is to formulate the full problem, consider possible closures, and discuss the results available for particular limiting cases.

The measurable macroscopic properties of primary interest for concentrated dispersions are the static structure factor, S_0, and equation of state, Π/nkT, at equilibrium; the dynamic structure factors, F and F_S, characterizing relaxation of thermal fluctuations about equilbrium; and the nonequilibrium structure factor, S, and the stresses, $<\sigma>$, generated by shear flows.

17.2.1 Equilibrium: Static Properties

Without the external fields, the individual fluxes vanish identically, leaving

$$0 = -P_2^o\langle\omega_{1i}\rangle_{N-2}\cdot\nabla_i\left(\Phi + kT\ln P_2^o\right) - \int P_3^o\langle\omega_{1i}\rangle_{N-3}\cdot\nabla_i\Phi_{i3}dx_3$$

and (2.7)

$$0 = -\int\left(kT\langle\omega_{i3}\rangle_{N-3}\cdot\nabla_3 P_3^o + \int P_4^o\langle\omega_{i3}\rangle_{N-4}\cdot\nabla_3\Phi_{34}dx_4\right)dx_3$$

Since

$$nP_2^o\left\langle\omega_{ij}\right\rangle_{N-2} = \int P_3^o\left\langle\omega_{ij}\right\rangle_{N-3}dx_3$$

$$nP_3^o\langle\omega_{i3}\rangle_{N-3} = \int P_4^o\langle\omega_{i3}\rangle_{N-4}dx_4$$

these obtain if the equilibrium probability densities satisfy

$$kT\nabla_i P_2^o = -P_2^o\nabla_i\Phi_{ij} - \int P_3^o\nabla_i\Phi_{i3}dx_3 \tag{2.8}$$

known variously as the YBG equation or the second equation in the BBGKY heirarchy (e.g. Hansen and McDonald 1986). With a suitable closure on the triplet correlation this determines the equilibrium pair distribution or, equivalently, the radial distribution function $g(r)$ defined by

$$P_2^o = n^2 g(r)$$

At finite concentrations the appropriate closure depends on the nature of the interparticle potential as described in Section 3.

The Fourier transform of the radial distribution function comprises the equilibrium static structure factor

$$S_o(q) = 1 + 4\pi n \int_0^\infty r^2 (g(r) - 1) \frac{\sin qr}{qr} dr \qquad (2.9)$$

which has the limiting values

$$S_o(0) = 1 + 4\pi n \int_0^\infty r^2 (g(r) - 1) dr$$

$$S_o(\infty) = 1$$

$S_o(q)$ characterizes, for the equilibrium dispersion, the magnitude of density fluctuations with wavenumber q.

The equation of state follows from the radial distribution function via either the pressure (virial) equation

$$\frac{\Pi}{nkT} = 1 - \tfrac{2\pi}{3} n \int_0^\infty r^3 g(r) \frac{d\Phi_{12}}{dr} dr \qquad (2.10)$$

or the compressibility equation

$$kT \frac{dn}{d\Pi} = 1 + 4\pi n \int_0^\infty (g(r) - 1) r^2 dr \qquad (2.11)$$

Note that the zero wavenumber limit of the static structure factor coincides with the osmotic compressibility. Since approximate closures do not yield exact values for $g(r)$ the pressures obtained from the two equations will differ, providing one test for optimizing the closure.

17.2.2 Equilibrium: Fluctuations

At equilibrium Brownian motion causes the local number density to fluctuate about the mean. Representing the instantaneous density as

$$n(\boldsymbol{x}, t) = \sum_{i=1}^{N} \delta(\boldsymbol{x} - \boldsymbol{x}_i(t))$$

leads to Fourier components of the fluctuations of the form

$$\hat{n}(\boldsymbol{q}, t) = \int e^{i\boldsymbol{q}\cdot\boldsymbol{x}} (n(\boldsymbol{x}, t) - n) d\boldsymbol{x}$$

$$= \sum_{i=1}^{N} e^{i\boldsymbol{q}\cdot\boldsymbol{x}_i(t)} - \delta(\boldsymbol{q})$$

Defining the density-density autocorrelation function as the ensemble average of the product of two Fourier components, shifted by time t, yields the dynamic structure factor as (Hess and Klein 1983; Rallison and Hinch 1986)

$$\begin{aligned} F(\boldsymbol{q},t) &= \tfrac{1}{N}\left\langle \hat{n}(\boldsymbol{q},t)\hat{n}(-\boldsymbol{q},0)\right\rangle \\ &= \tfrac{1}{NN!}\int\!\!\int \sum_{i,j=1}^{N} e^{i\boldsymbol{q}\cdot\left(\boldsymbol{x}_i(t)-\boldsymbol{x}_j(0)\right)} P_N\left(\boldsymbol{X}(t)|\boldsymbol{X}(0)\right)d\boldsymbol{X}(t)d\boldsymbol{X}(0) \\ &= \tfrac{1}{N!}\int\!\!\int\left(e^{i\boldsymbol{q}\cdot(\boldsymbol{x}_1(t)-\boldsymbol{x}_1(0))} + (N-1)e^{i\boldsymbol{q}\cdot(\boldsymbol{x}_2(t)-\boldsymbol{x}_1(0))}\right)P_N d\boldsymbol{X}(t)d\boldsymbol{X}(0) \\ &\equiv F_s(\boldsymbol{q},t) + \tfrac{N-1}{N!}\int e^{i\boldsymbol{q}\cdot(\boldsymbol{x}_2(t)-\boldsymbol{x}_1(0))}P_N d\boldsymbol{X}(t)d\boldsymbol{X}(0) \end{aligned} \tag{2.12}$$

with

$$\boldsymbol{X}(t) = \left(\boldsymbol{x}_1(t),\ldots,\boldsymbol{x}_N(t)\right)$$

Here P_N describes the time-dependent decay of the probability density from an initial state with

$$P_N\left(\boldsymbol{X}(0)|\boldsymbol{X}(0)\right) = P_N^o \prod_{i=1}^{N} \delta\left(\boldsymbol{x} - \boldsymbol{x}_i(0)\right)$$

Note that F_S characterizes the fluctuation of tracers, i.e. self-diffusion, while F monitors a wave-number dependent mutual diffusion process. With zero time delay these autocorrelation functions reduce to the static limits

$$F_s(\boldsymbol{q},0) = \tfrac{1}{N!}\int P_N^o d\boldsymbol{X} = 1$$

$$\begin{aligned} F(\boldsymbol{q},0) &= 1 + \tfrac{N-1}{N!}\int e^{i\boldsymbol{q}\cdot(\boldsymbol{x}_2-\boldsymbol{x}_1)}P_N^o d\boldsymbol{X} \\ &= 1 + \frac{4\pi n}{q}\int_0^\infty r(g(r)-1)\sin qr\, dr = S_o(q) \end{aligned} \tag{2.13}$$

The equilibrium static structure factor, $S_O(q)$, is measurable by static light scattering, while photon correlation spectroscopy is capable of detecting both dynamic structure factors. Then the wavenumber varies with the scattering angle θ and the wavelength λ according to $q=(4\pi/\lambda)\sin\theta/2$. However, this interpretation of both the static and dynamic measurements suffices only for dispersions that scatter weakly, i.e. that do not interact electromagnetically and generate multiple scattering. In the limit of extensive multiple scattering, an alternative approach, known as diffusive wave spectroscopy, becomes feasible, but requires a quite different interpretation.

17.2.3 Nonequilibrium States: Stresses

When a shear flow perturbs the structure from equilibrium all the forces acting on the particles - viscous, Brownian, and interparticle - contribute to the stress. Then dispersions differ fundamentally from molecular liquids, which possess only kinetic energy and interparticle forces. Derivations of the stress by Batchelor (1970, 1977) and Felderhof (1987) differ in approach but concur on the final form, specialized by both to the dilute limit. Here we detail the formal expressions for arbitrary volume fractions.

The mechanical portion of the stress, arising from the viscous nature of the continuous fluid, comprises the ensemble average of a volume average over fluid plus particles, expressed as

$$\begin{aligned} \langle \sigma \rangle &= \tfrac{1}{V}\left\langle \int \sigma dx \right\rangle \\ &= 2\mu \boldsymbol{E} + n\langle \boldsymbol{S} \rangle \end{aligned} \tag{2.14}$$

The stress dipole, $\boldsymbol{S}$, the source of the extra contribution from the particles, can be converted to an integral of the fluid stress over the surfaces of the particles, i.e.

$$\begin{aligned} \boldsymbol{S} &= \int_{|x-x_1|\le a} (\sigma - 2\mu e) dx \\ &= \int_{|x-x_1|=a} \left(\sigma \cdot \boldsymbol{n}(x - x_1) - \tfrac{1}{3}\boldsymbol{\delta}(x - x_1)\cdot \sigma \cdot \boldsymbol{n}\right) dx \\ &= \boldsymbol{S}_o + \sum_{i=1}^{N} \left(\left\langle \boldsymbol{S}'_{1i} \right\rangle_{N-2} - (x_i - x_1)\left\langle C_{1i} \right\rangle_{N-2} \cdot \left(-\nabla_1 kT \ln P_N + F_1\right) \right) \end{aligned} \tag{2.15}$$

The first term is the dipole for an isolated sphere

$$\boldsymbol{S}_o = \tfrac{20\pi}{3} \mu a^3 \boldsymbol{E}$$

while the second accounts for the additional response induced by interactions and attains the limit

$$\boldsymbol{S}'_{ij} \approx \tfrac{20\pi}{3} \mu a^3 \left(1 + \tfrac{a^2}{10}\nabla^2\right) \boldsymbol{e}'_{ij}$$

for $r_{ij} >> 2a$. The perturbation rate of strain, $\boldsymbol{e}'_{ij}$, generated at the position of sphere j by an isolated sphere i, decays as $(r_{ij})^{-3}$. The third term captures the mechanical stress engendered by the relative translation due to interparticle forces.

Batchelor (1977) derived the thermodynamic stress from the increase in the Helmholtz free energy accompanying a deformation ε of the equilibrium configuration,

$$\begin{aligned} dA &= d\left(kT\ln P_N + \Phi\right) \\ &= \sum_{i=1}^{N} \boldsymbol{x}_i \cdot \boldsymbol{\varepsilon} \cdot \nabla_i\left(kT\ln P_N + \Phi\right) \\ &\equiv V\boldsymbol{\varepsilon} : \boldsymbol{\tau} \end{aligned} \tag{2.16}$$

Thus the thermodynamic stress also assumes the dipole form,

$$\langle \boldsymbol{\tau} \rangle = -\tfrac{1}{V}\left\langle \sum_{i=1}^{N} \boldsymbol{x}_i\left(-kT\nabla_i P_N + \boldsymbol{F}_i\right)\right\rangle \tag{2.17}$$

with $\boldsymbol{F}_i$ from (2.2)

The treatment of interactions, even at the pair level, introduces nonconvergent integrals, due to the decay as $(r_{ij})^{-3}$ of the dipole $\boldsymbol{S}'_{ij}$ and $-kT\boldsymbol{x}_i\nabla_i\ln P_2$. Fortunately, proper renormalizations leading to unique, finite results follow from any of several different approaches, each recognizing the presence of a bulk suspension with properties reflecting cumulative many-body interactions (Batchelor and Green 1972; Batchelor 1977; Hinch 1977; O'Brien 1979). With these results the stresses above may be regrouped into contributions from the individual forces and expressed as

$$\begin{aligned} \langle \boldsymbol{\sigma} \rangle^H &= 2\mu \boldsymbol{E}\left(1 + \tfrac{5}{2}\phi + \tfrac{5}{2}\phi^2\right) + \int\left(\left\langle \boldsymbol{S}'_{12}\right\rangle_{N-2} P_2 - \tfrac{20\pi}{3}a^3 n^2 \mu \langle \boldsymbol{e}' \rangle_{N-2}\right) d\boldsymbol{r} \\ \langle \boldsymbol{\sigma} \rangle^I &= \tfrac{1}{2}\int\left(1 - \langle A \rangle_{N-2}\right) r \frac{d\Phi_{12}}{dr}\left(\frac{\boldsymbol{rr}}{r^2} - \frac{1}{3}\boldsymbol{\delta}\right) P_2 d\boldsymbol{r} \\ \langle \boldsymbol{\sigma} \rangle^B &= -\tfrac{1}{2}\int \langle W \rangle_{N-2}\left(\frac{\boldsymbol{rr}}{r^2} - \frac{1}{3}\boldsymbol{\delta}\right) P_2 d\boldsymbol{r} \end{aligned} \tag{2.18}$$

For the isotropic equilibrium structure the latter two integrate to zero, since the interparticle force then contributes to the osmotic pressure, not the deviatoric stresses considered here. The first, however, is non-zero and linear in the imposed rate of strain.

One can derive an alternative form for the interparticle stress (Wagner and Russel 1989)

$$\langle\sigma\rangle^I = \tfrac{1}{2}\int\left\{r_{12}\left(\delta + 2\langle C_{12}\rangle_{N-2}\right)\cdot P_2\nabla\Phi_{12} + \int r_{12}\left(\delta + 2\langle C_{12}\rangle_{N-3}\right)\cdot P_3\nabla\Phi_{23}dx_3\right\}dr$$

$$\cong \tfrac{1}{2}\int r_{12}\left(\delta + 2\langle C_{12}\rangle_{N-2}\right)\cdot\left\{P_2\nabla\Phi_{12} + \int P_3\nabla\Phi_{23}dx_3\right\}dr$$

simply by averaging in a different order. The stress still vanishes at equilibrium, but comparison of the two forms now constrains the nonequilibrium closure relating P_3 to P_2.

The effect of hydrodynamic interactions on the interparticle and Brownian stresses is noteworthy. Since $\langle A\rangle_{N-2}=1$ at $r_{12}=2a$ for spheres with no-slip boundary conditions, the lubrication stresses eliminate any direct contribution from the hard sphere repulsion. On the contrary, $\langle W\rangle_{N-2}=0$ in the absence of hydrodynamic interactions, so the same effects are entirely responsible for the Brownian stresses in dispersions of spheres. Thus, hydrodynamic interactions should have profound effects for short range potentials or high concentrations. However, for long range repulsions and dilute concentrations hydrodynamic interactions become negligible, driving the Brownian stress to zero and making the interparticle stress dominant.

17.2.4 Common Limiting Cases

While the full problem has yet to be addressed, two limiting cases have received considerable attention: negligible hydrodynamic but strong thermodynamic interactions and dilute concentrations.

Dispersions of charged particles at low ionic strengths interact strongly at very low volume fractions, comprising an interesting model system with only far-field hydrodynamic interactions. Setting $\omega_{ij}=\omega_o\delta\delta_{ij}$ and $\boldsymbol{C}_{ij}=\boldsymbol{0}$ leaves only the three-body coupling through the interparticle potential and

$$J_i = P_2\left((E+\Omega)\cdot x_i - \omega_o\nabla_i\left(kT\ln P_2 + \Phi_{ij}\right)\right) - \omega_o\int P_3\nabla_i\Phi_{i3}dx_3 \quad (2.19)$$

Nonequilibrium analogues to the closures developed for (2.8) will be discussed in Section 4 in the context of fluctuations about equilibrium and in Section 5 for perturbations caused by shear flows.

This limit also permits analyses in Fourier space, by transforming the pair distribution function into the nonequilibrium static structure factor as

$$S(q,t) = 1 + n\int h(r)e^{iq\cdot r}dr \qquad \text{with} \qquad h(r) = P_2/n^2 - 1$$

and applying the same treatment to the Smoluchoski equation to generate

$$\frac{\partial S}{\partial t} + (E+\Omega):q\nabla_q S = 2D_o\left\{-q^2(S-1) + iq\cdot\hat{j}\right\} \quad (2.20)$$

with $D_o = \omega_o kT$. The flux $\hat{j}$ driven by the total interparticle force,

$$2D_o n\hat{j} = \omega_o \int P_2(F_2 - F_1)e^{iq\cdot x}dx \qquad (2.21)$$

contains the higher order couplings that necessitate a closure approximation. In Section 5 we note the relationship between one such approximation and those developed for (2.19).

Rigorous theory is possible at dilute concentrations and suggests the interesting predictions that might result from well formulated treatments of higher concentrations. Omission of the higher order couplings from (2.7) leaves the Smoluchoski equation in differential form with

$$J_i = P_2\Big((E + \Omega)\cdot x_i - 2C_{ij}\cdot E\cdot(x_i - x_j) - (\omega_{ii}\cdot\nabla_i + \omega_{ij}\cdot\nabla_j)(kT\ln P_2 + \Phi_{ij})\Big) \qquad (2.22)$$

with i,j=1,2. The variation of the hydrodynamic functions with separation is known exactly for spheres (c.f. Kim and Karilla 1990). Without hydrodynamic interactions (2.22) simplifies to (2.19) without the three-body term.

In Section 4 this proves a useful limit for comparing closure approximations in the treatment of fluctuations about equilibrium. In Section 6 we demonstrate that analyses of weak and strong flows in this limit reveal considerable about the effect of the interparticle potential on the rheology.

With this general formulation of the problem we are now prepared to address specific situations for which some results are available and begin to outline means by which more complex situations might be addressed.

17.3 EQUILIBRIUM STRUCTURE AND PROPERTIES

In principle, (2.8) determines the equilibrium radial distribution function, but solutions can be extracted at finite concentrations only with a closure for the triplet distribution. The standard route to developing these actually bypasses the YBG equation, focussing instead on the direct correlation function c_{12} defined by the Ornstein-Zernicke equation

$$h_{12} = c_{12} + n\int c_{13}h_{32}dx_3 \qquad (3.1)$$

This decomposes the total correlation between two particles, represented by $h_{12}=g_{12}-1$, into direct correlations with range comparable to that of the pair potential and indirect correlations propagated through other particles over larger separations. Solving or expanding (3.1) as

$$h_{12} = c_{12} + n\int c_{13}c_{23}dx_3 + n^2\int c_{13}c_{34}c_{42}dx_3dx_4 + \ldots \qquad (3.2)$$

demonstrates clearly the hierarchy of clusters contributing to the indirect couplings.

One way of developing the closure approximations (Hansen and McDonald, 1986) is to construct exact infinite series in powers of the density for c_{12} and h_{12} to prove that

$$h_{12} = c_{12} + \frac{\Phi_{12} - \Phi_{mf}}{kT} - b_{12} \tag{3.3}$$

where the potential of mean force is defined by

$$\frac{\Phi_{mf}}{kT} = -\ln g_{12}$$

and b_{12}, an infinite series with leading term of $O(n^2)$ that cannot be expressed in closed form, is known as a bridge function.

The *hypernetted chain approximation* then follows from setting b_{12}=0, leaving the direct correlation function as

$$c_{12} = h_{12} - \frac{\Phi_{12}}{kT} - \ln(h_{12} + 1) \tag{3 4}$$

This closes the Ornstein-Zernicke equation as a nonlinear integral equation

$$\ln g_{12} + \frac{\Phi_{12}}{kT} = n\int (g_{23} - 1)\left(g_{13} - 1 - \frac{\Phi_{13}}{kT} - \ln g_{13} \right) d\boldsymbol{x}_3 \tag{3.5}$$

that must be solved numerically. Comparison with the integrated form of the YBG equation identifies the corresponding approximation for the triplet correlation function as

$$\int P_3^o \nabla_1 \Phi_{13} d\boldsymbol{x}_3 = -n^3 kT g_{12} \int h_{23} \nabla_1 c_{13} d\boldsymbol{x}_3$$

Still neglecting the bridge function but rewriting the closure as

$$g_{12} = e^{h_{12} - c_{12} - \Phi_{12}/kT}$$

and expanding for h_{12}-c_{12}<<1 yields the *Percus-Yevick approximation,*

$$c_{12} = \left(1 - e^{-\Phi_{12}/kT}\right) g_{12} \tag{3.6}$$

This reduces the Ornstein-Zernicke equation to a form

$$g_{12} = e^{-\Phi_{12}/kT}\left(1 + n\int h_{32} g_{13}\left(1 - e^{\Phi_{13}/kT}\right) d\boldsymbol{x}_3\right) \tag{3.7}$$

that can be solved analytically for singular potentials.

For the Percus-Yevick approximation the direct correlation function is identically zero when the interparticle potential vanishes but remains finite when the potential diverges, e.g. within the hard core. Comparisons with simulations indicate greatest accuracy for relatively hard repulsions, particularly hard spheres (Figure 1), suggesting a fortuitious cancellation of errors. The hypernetted chain approximation proves more successful for soft, longrange repulsions, as in charged systems.

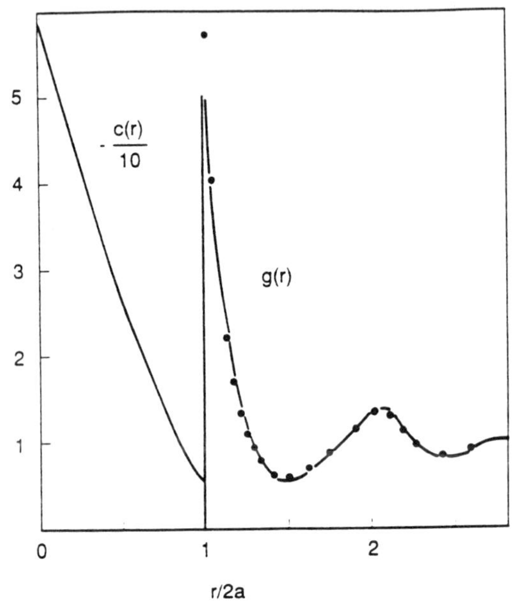

Figure 1 Equilibrium radial distribution function and direct correlation function for hard spheres at ϕ=0.49: (—) Percus-Yevick approximation, (•) Monte Carlo simulations (adapted from Hansen and McDonald 1986).

For large separations where $h_{12} << 1$ both closures reduce to the Debye-Huckel approximation

$$c_{12} \cong -\frac{\Phi_{12}}{kT}$$

This asymptotic form is the basis for a third, quite useful closure, known as the *mean spherical approximation*, which sets

$$\begin{aligned} g_{12} &= 0 \qquad && r_{12} < 2a \\ c_{12} &= -\frac{\Phi_{12}}{kT} \qquad && r_{12} > 2a \end{aligned} \tag{3.8}$$

This corresponds to the Percus-Yevick approximation exactly for $r_{12}<2a$ but requires expanding the exponential and setting $g_{12}=1$ for $r_{12}>2a$. The linearization of the integral equation permits analytical solutions for smooth potentials.

Implementation with the Yukawa potential

$$\frac{\Phi_{12}}{kT} = \alpha \frac{e^{-\kappa r_{12}}}{\kappa r_{12}}, \tag{3.9}$$

which corresponds to the superposition approximation for electrostatic interactions, accurately reproduces experimental data at moderate volume fractions, but fails for very dilute charged spheres at low ionic strengths. The *rescaled mean spherical approximation* devised by Hansen and Hayter (1982) removes this difficulty and has been widely applied.

Earlier we noted that substituting radial distribution functions from approximate closures into the virial and compressibility equations generates two numerically different functions for the pressure. For purely repulsive potentials the PY and HNC approximations bracket the "exact" g_{12} from simulations, suggesting a simple interpolation. The *Rogers-Young* (1980) *approximation*, advocated by Klein and coworkers (e.g. D'Aguanno, et al. 1990), recognizes the advantages of the PY form at small separations and the HNC in the far-field through the closure

$$g_{12} = e^{-\Phi_{12}/kT}\left(1+\frac{e^{f_{12}(h_{12}-c_{12})}-1}{f_{12}}\right) \tag{3.10}$$
$$f_{12} = 1-e^{-\gamma r_{12}}$$

with γ adjusted to obtain thermodynamic consistency at each set of conditions. Comparison with simulations also suggests improved predictions for the radial distribution function or, equivalently, the static structure factor. The procedure fails for attractions, however, since both closures err in the same direction (Evans 1989).

Extensive results, derived with one or more of these closures, are now available for several model potentials. *Hard spheres,* with

$$e^{-\Phi_{12}/kT} = H(r_{12}-2a)$$

and $H(x)$ the Heaviside step function, are the most widely studied and exemplify the success of the Percus-Yevick closure (Figure 1). *Adhesive hard spheres,* with (Baxter 1967)

$$e^{-\Phi_{12}/kT} = H(r_{12}-2a)+\tfrac{1}{6\tau}\delta(r_{12}-2a)$$

and $\delta(x)$ the Dirac delta function, provide the simplest model for a short range attraction and have also been solved in the Percus-Yevick approximation. The Yukawa, generally solved through the mean-spherical, rescaled mean spherical, or Rogers-Young approximations, serves for soft repulsions.

More faithful representations of specific interparticle potentials can be constructed through perturbation theories. These expand the free energy and the radial distribution function about a reference state, generally taken as the hard sphere. Soft repulsions are accommodated by proper choice of the effective hard sphere diameter d and attractions through a regular perturbation. For example, the original version due to Barker and Henderson (1967 a,b) resolves a colloidal interparticle potential into the hard core (r<2), a soft repulsion ($2a<r<\sigma$), and an attraction ($\sigma<r$). Then

$$d = 2a + \int_{2\alpha}^{\sigma} \left(1 - e^{-\Phi_{12}/kT}\right) dr_{12}$$

and (3.13)

$$g_{12} = g_{12}^{hs} e^{-\Phi_{12}'/kT} + n \int_{\sigma}^{\infty} h_{23} \frac{\Phi_{13}'}{kT} g_{13}^{hs} dx_3 \qquad r_{12} > d$$

with the perturbation potential defined by

$$\frac{\Phi_{12}'}{kT} = \begin{cases} 0 & d<r_{12}<\sigma \\ \dfrac{\Phi_{12}}{kT}(r_{12}) & \sigma<r_{12} \end{cases}$$

Actually the Boltzmann factor in the first term of g_{12} should be linearized but this poses the risk of negative values for modest attractions. The approach has been used extensively for detailed treatments of electrostatic repulsions and van der Waals or depletion attractions (van Megen and Snook 1984).

Despite the elegance and popularity of these closures, the predictions inevitably deviate from those produced by computer simulations as the many-body interactions become $O(1)$. Applications, to interprete scattering experiments, to predict phase transitions via perturbation expansions or density functional theories, or to calculate time correlation functions or glass transitions, often demand more accurate representations. Thus considerable effort has been expended either parameterizing the simulation results or correcting the integral equation theories.

For hard spheres, the Verlet-Weiss (1972) correction for the radial distribution function, based on results from computer simulations, maintains the proper value at contact and the positions of the peaks up to the freezing transition. The correction can be extended through the metastable fluid region, preserving the contact value but not adjusting the behavior at larger separations.

A more versatile approach constructs approximations for the bridge functions neglected in the conventional closures. A variety of proposals are reviewed by Hansen and McDonald (1986) and by Evans (1989, 1991). The bridge functions contribute an additional interparticle repulsion of quite short range. Several versions succeed in accurately reproducing thermodynamic properties at liquid densities. Though still empirical, these represent a viable route for obtaining accurate predictions. More recent density functional approaches offer a systematic interative process for constructing very accurate and self-consistent approximations, but involve substantial computation.

17.4 THE DYNAMICS OF QUIESCENT DISPERSIONS

Many studies of interacting Brownian particles focus on fluctuations in the particle number density, as characterized by the dynamic structure factor. Although equilibrium phenomena, their spatial and temporal variations define wavenumber dependent diffusion coefficients $D(q)$ and $D_s(q)$. Most analyses have addressed charged particles at low ionic strengths and volume fractions, for which hydrodynamic interactions are nonexistent or secondary relative to interparticle forces.

Ackerson (1976, 1978) and Dieterich and Peschel (1979) initiated the Smoluchoski treatment of fluctuations in colloidal systems by constructing short time expansions of the dynamic structure factor. While others adopted Langevin approaches (Pusey and Tough 1985), Rallison and Hinch (1986) have extended the Smoluchoski analysis to long times, but only for dilute concentrations. Here we reconstruct this development within the framework of Section 2 and then examine the alternative approach derived from the Fokker-Planck equation, outlined by Hess and Klein (1983).

17.4.1 Smoluchoski Analysis

Extracting the relaxation of fluctuations from the Smoluchoski equation means solving an initial value problem with

$$P_n = P_n^o \prod_{i=1}^{n} \delta(x - x_i)$$

at t=0. Transforming the distribution function according to

$$\tilde{P}_n = \int P_n e^{iq\cdot x_1(0)} dx_1(0)...dx_n(0)$$

simplifies the form of the dynamic structure factors to

$$\begin{aligned} F_s(q,t) &= \frac{1}{N(N-1)} \int \tilde{P}_2 e^{iq\cdot x_1} dx_1 dx_2 \\ F(q,t) &= F_s(q,t) + \frac{1}{N} \int \tilde{P}_2 e^{iq\cdot x_2} dx_1 dx_2 \end{aligned} \tag{4.1}$$

with the transformed distribution function still satisfying (2.3) and

$$\tilde{P}_2 = P_2^o e^{iq\cdot x_1}$$

at t=0.

The cumulants or moments about zero time are constructed by differentiating (4.1) with respect to time, substituting the transformed pair conservation equation, and integrating by parts for a closed system with no-flux boundary conditions. Thus the first derivatives

$$\frac{dF_s}{dt} = -\tfrac{1}{N(N-1)} i\boldsymbol{q}\cdot\int e^{i\boldsymbol{q}\cdot\boldsymbol{x}_1} \boldsymbol{J}_1 d\boldsymbol{x}_1 d\boldsymbol{x}_2$$

$$\frac{dF}{dt} = \frac{dF_s}{dt} - \tfrac{1}{N} i\boldsymbol{q}\cdot\int e^{i\boldsymbol{q}\cdot\boldsymbol{x}_2} \boldsymbol{J}_2 d\boldsymbol{x}_1 d\boldsymbol{x}_2 \tag{4.2}$$

have initial values

$$\frac{dF_s}{dt}(q,0) = -\tfrac{1}{N(N-1)}\int \boldsymbol{q}\cdot\langle\omega_{11}\rangle_{N-2}\cdot \boldsymbol{q}P_2^o d\boldsymbol{x}_1 d\boldsymbol{x}_2$$

$$\equiv -q^2 D_s^o F_s(q,0) \tag{4.3}$$

$$\frac{dF}{dt}(q,0) = -\tfrac{1}{N(N-1)}\int \boldsymbol{q}\cdot\left(\langle\omega_{11}\rangle_{N-2} + (N-1)e^{i\boldsymbol{q}\cdot(\boldsymbol{x}_2-\boldsymbol{x}_1)}\langle\omega_{12}\rangle_{N-2}\right)\cdot \boldsymbol{q}P_2^o d\boldsymbol{x}_1 d\boldsymbol{x}_2$$

$$\equiv -q^2 D(q) F(q,0)$$

that define the short-time self-diffusion coefficient $D_s{}^o$ and the wavevector dependent mutual diffusion coefficient $D(q)$.

This limit reflects the equilibrium structure, since the spheres have yet to move from the initial configuration. At short times self diffusion merely senses the average mobility of a single particle amidst force-free neighbors, while

$$D(q) = D_o \frac{K(q)}{S_o(q)}$$

includes hydrodynamic effects through the generalized translational mobility $K(q)$ and thermodynamic interactions in the equilibrium static structure factor. In the limit of small wavenumbers the former reduces to the mobility characterizing uniform settling and the latter to the osmotic compressibility, demonstrating that $D(0)$ is the normal mutual diffusion coefficient. For very large wavenumbers $S_o(q)$->1 and $D_o K(q)$->$D_s{}^o$ and the two diffusion coefficients become equivalent (Rallison and Hinch 1986).

For slightly longer times the relative motion of particles generates a nonequilibrium microstructure and additional forces that slow the relaxation of the density fluctuation. To describe these at finite concentrations requires evaluating the three- and four-body terms in the Smoluchoski equation. To address these we first rewrite the flux using (2.2) as

$$\boldsymbol{J}_i = -\sum_{j=1}^{2} \boldsymbol{\omega}_{ij} \cdot \left(kT\nabla_j P_2 + P_2 \boldsymbol{F}_j\right) - \int \boldsymbol{\omega}_{i3} \cdot \left(kT\nabla_3 P_3 + \int P_4 \nabla_3 \Phi_{34} d\boldsymbol{x}_4\right) d\boldsymbol{x}_3 \quad (4.4)$$

with the two mobilities now understood to represent ensemble averages over N-2 and N-3 spheres, respectively, and

$$P_2 \boldsymbol{F}_i = -P_2 \nabla_i \Phi_{ij} + \int P_3 \nabla_i \Phi_{i3} d\boldsymbol{x}_3 \quad (4.5)$$

comprising the total interparticle force acting on sphere i.

17.4.2 Closures for Higher Order Terms

Following earlier work by Kremp et al. (1983), Scherwinski (1990) has derived two approximate closures for $\boldsymbol{F}_i$ by the procedure sketched above for the equilibrium case. He defines the direct correlation force, $\boldsymbol{C}_{ij}$, as the analogue to the gradient of the direct correlation function. The derivation leads to coupled integral equations for $\boldsymbol{C}_{ij}$ and $\boldsymbol{F}_i$ with alternative forms corresponding to the *hypernetted chain* and *Percus-Yevick approximations*,

$$P_2 \boldsymbol{F}_i + P_2 \nabla_i \Phi_{ij} = nP_2 \int \boldsymbol{C}_{i3} h_{j3} d\boldsymbol{x}_3 \qquad \text{(HNC)}$$

$$= n^3 e^{-\Phi_{ij}/kT} \int \boldsymbol{C}_{i3} h_{j3} d\boldsymbol{x}_3 \qquad \text{(PY)} \quad (4.6)$$

$$n^2 \boldsymbol{C}_{ij} + P_2 \nabla_i \Phi_{ij} = n\left(P_2 - n^2\right) \int \boldsymbol{C}_{i3} h_{j3} d\boldsymbol{x}_3 \qquad \text{(HNC)}$$

$$= n^3 \left(e^{-\Phi_{ij}/kT} - 1\right) \int \boldsymbol{C}_{i3} h_{j3} d\boldsymbol{x}_3 \qquad \text{(PY)}$$

with

$$h_{ij} = \frac{1}{n^2} P_2\left(\boldsymbol{r}_{ij}\right) - 1$$

Note that the triplet distribution function disappears from the expression and, hence, can be determined in terms of the direct correlation force by equating (4.5) and (4.6). At equilibrium

$$\boldsymbol{C}_{ij} = kT\nabla_i c_{ij}$$

and setting (4.7)

$$P_2^o \boldsymbol{F}_i - kT\nabla_i P_2^o = \boldsymbol{0}$$

yields the algebraic forms for the direct correlation function noted in Section 3. Away from equilibrium the integral equation must be solved for $\boldsymbol{C}_{ij}$, since the simple force balance no longer holds. Scherwinski notes that these forms are not unique approximations but do satisfy the available constraints.

For the analysis of density fluctuations the pair distribution function and the direct correlation force take the forms

$$P_2 = n^2 e^{-iq\cdot X_{ij}} p_{ij}$$

$$C_{ij} = e^{-iq\cdot X_{ij}} c_{ij}$$

with $\quad X_{ij} = \frac{1}{2}\left(x_i + x_j\right)$

denoting the center-of-mass of the pair. Substituting into (4.6) , for example, gives the Percus-Yevick approximation

$$\begin{aligned} c_{ij} &= -p_{ij}\nabla_i \Phi_{ij} + n\left(e^{-\Phi_{ij}/kT} - 1\right)\int c_{i3} e^{-\frac{1}{2}iq\cdot r_{3j}}\left(p_{j3} - 1\right)dx_3 \\ P_2 F_i &= n^2 e^{-iq\cdot X_{ij}}\left\{-p_{ij}\nabla_i \Phi_{ij} + n e^{-\Phi_{ij}/kT}\int c_{i3} e^{-\frac{1}{2}iq\cdot r_{3j}}\left(p_{j3} - 1\right)dx_3\right\} \end{aligned} \tag{4.8}$$

For the hydrodynamic couplings in the pair conservation equation, lower order approximations along the lines above seem appropriate. For example, by analogy to (4.6) we can construct

$$-\int P_4 \nabla_3 \Phi_{34} dx_4 \cong -n^4 e^{-iq\cdot X_{12}} e^{-\Phi_{12}/kT}\int p_{34} e^{-iq\cdot\frac{1}{2}\left(r_{31} + r_{42}\right)}\nabla_3 \Phi_{34}\left(p_{14}p_{24} - 1\right)dx_4$$

and (4.9)

$$P_3 = n^3 e^{-\Phi_{12}/kT} e^{-iq\cdot X_{12}} p_{13} e^{-iq\cdot\frac{1}{2}r_{32}}\left(p_{23} - 1\right)$$

representing the Percus-Yevick form with first-order approximations for the direct correlation force. Together with the conservation equation (2.8) these closures provide a closed problem for the pair distribution function, once the hydrodynamic functions are specified.

17.4.3 Long Time, Small Wavenumber Limit

Clearly a general solution of the full problem, including time-dependence, hydrodynamic interactions, and the higher order couplings, is a daunting task. So, for the present, we will neglect hydrodynamic interactions and focus on conditions most relevant to rheological considerations. Deformation in shear flows with length scales large relative to the particles corresponds to fluctuations with very long wavelengths, $aq \ll 1$. In this limit a regular perturbation expansion readily demonstrates that $F(q,t)$ decays exponentially for all time, unaffected by relaxations, but $F_s(q,t)$ develops nonequilibrium effects quite similar to those responsible for interparticle and Brownian contributions to the bulk stresses.

Without hydrodynamic interactions the flux reduces to

$$\boldsymbol{J}_i = -D_o\left(\nabla_i P_2 - \tfrac{1}{kT} P_2 \boldsymbol{F}_i\right)$$

and with (4.8) the full conservation equation becomes

$$\frac{\partial p_{12}}{\partial t} = -\tfrac{1}{2} q^2 D_o p_{12} + 2D_o \left\{ \nabla^2 p_{12} + \nabla \cdot p_{12} \nabla \frac{\Phi_{12}}{kT} \right.$$
$$\left. - n\nabla \cdot e^{-\Phi_{12}/kT} \int \frac{\boldsymbol{c}_{23}}{kT} e^{-\frac{1}{2} i\boldsymbol{q}\cdot\boldsymbol{r}_{32}} \left(p_{13} - 1\right) d\boldsymbol{x}_3 \right\} \tag{4.10}$$

The dynamic structure factor of interest can be expressed in terms of p_{12} by substituting the flux into (4.8) and carrying out the appropriate integrations with

$$\nabla e^{-\frac{1}{2} i\boldsymbol{q}\cdot\boldsymbol{r}_{21}} p_{12} = \left(-\tfrac{1}{2} i\boldsymbol{q} p_{12} + \nabla p_{12}\right) e^{-\frac{1}{2} i\boldsymbol{q}\cdot\boldsymbol{r}_{21}}$$

$$\int \nabla e^{-\frac{1}{2} i\boldsymbol{q}\cdot\boldsymbol{r}_{21}} p_{12} d\boldsymbol{r}_{21} = 0$$

$$n \int e^{-\frac{1}{2} i\boldsymbol{q}\cdot\boldsymbol{r}_{21}} p_{12} d\boldsymbol{r}_{21} = F_s$$

The result,

$$\frac{dF_s}{dt} = -q^2 D_o F_s + nD_o i\boldsymbol{q} \cdot \int e^{-\frac{1}{2} i\boldsymbol{q}\cdot\boldsymbol{r}_{21}} \left\{ p_{12} \nabla \frac{\Phi_{12}}{kT} \right.$$
$$\left. + ne^{-\Phi_{12}/kT} \int \frac{\boldsymbol{c}_{13}}{kT} e^{-\frac{1}{2} i\boldsymbol{q}\cdot\boldsymbol{r}_{32}} \left(p_{23} - 1\right) d\boldsymbol{x}_3 \right\} d\boldsymbol{r}_{21} \tag{4.11}$$

cleanly distinguishes the initial, nominally exponential, decay controlled by the equilibrium structure from the relaxation effects arising from the nonequilibrium structure.

The analysis for $aq \ll 1$ is most convenient with the expansions

$$F_s \approx 1$$

$$e^{-\frac{1}{2} i\boldsymbol{q}\cdot\boldsymbol{r}_{ij}} \approx 1 - \tfrac{1}{2} i\boldsymbol{q}\cdot\boldsymbol{r}_{ij}$$

$$p_{ij} \approx g_{ij}\left(1 - \tfrac{1}{2} i\boldsymbol{q}\cdot\boldsymbol{r}_{ij} + i\boldsymbol{q}\cdot\boldsymbol{r}_{ij} f_{ij}\right)$$

$$p_{ij} - 1 \approx g_{ij}\left(1 + i\boldsymbol{q}\cdot\boldsymbol{r}_{ij} f_{ij}\right) - 1$$

$$\boldsymbol{c}_{ij} \approx kT\left(\frac{\boldsymbol{r}_{ij}}{r_{ij}} \frac{dc_{ij}}{dr_{ij}}\left(1 - \tfrac{1}{2} i\boldsymbol{q}\cdot\boldsymbol{r}_{ij}\right) + i\boldsymbol{q} c_{ij}^{\perp} + i\boldsymbol{q} \cdot \frac{\boldsymbol{r}_{ij}\boldsymbol{r}_{ij}}{r_{ij}^2}\left(c_{ij}^{\parallel} - c_{ij}^{\perp}\right)\right)$$

The penultimate expansion acknowledges a modest inconsistency in the previous notation since the factor omitted represents a variation in the single particle number density that does not belong in (p_{12}-1). Substitution into (4.11) and integration over the various angular variables with

$$\boldsymbol{r}_{13} = a\left(z+\tfrac{1}{2}r\right)\frac{\boldsymbol{r}}{r} + a\rho\boldsymbol{n} \qquad \boldsymbol{r}_{23} = a\left(z-\tfrac{1}{2}r\right)\frac{\boldsymbol{r}}{r} + a\rho\boldsymbol{n} \qquad a\boldsymbol{r} = \boldsymbol{r}_{21}$$

$$\int d\boldsymbol{n} = 2\pi \qquad \int \boldsymbol{n}\boldsymbol{n}d\boldsymbol{n} = \pi\left(\boldsymbol{\delta} - \frac{\boldsymbol{r}\boldsymbol{r}}{r^2}\right)$$

and the usual orthogonality relations for $\boldsymbol{r}/r$ yields

$$\frac{dF_s}{dt} = -q^2 D_o \left\{ 1 - \phi \int r^3 gf \frac{d}{dr}\frac{\Phi}{kT} dr + \tfrac{3}{2}\phi^2 \int e^{-\Phi_{12}/kT} r^2 \right.$$
$$\left. \int \left(\frac{a^2\left(z^2 - \frac{1}{4}r^2 + \rho^2\right)}{r_{13}r_{23}} r_{23} g_{23} f_{23} \frac{dc_{13}}{dr_{13}} + \left(2c_{13}^{\perp} + c_{13}^{\parallel}\right)(g_{23} - 1) \right) \rho d\rho dz dr \right\} \tag{4.12}$$

The two components for the direct correlation force satisfy the coupled equations

$$c_{ij}^{\perp} = \tfrac{3}{2}\phi\left(e^{-\Phi_{12}/kT} - 1\right) \int \left\{ \left(c_{i3}^{\perp} + \frac{1}{2}\frac{a^2\rho^2}{r_{i3}^2}\left(c_{i3}^{\parallel} - c_{i3}^{\perp}\right)\right)(g_{3j} - 1) \right.$$
$$\left. - \frac{1}{2}\frac{a^2\rho^2}{r_{i3}r_{j3}} g_{3j} f_{3j} r_{3j} \frac{dc_{i3}}{dr_{i3}} \right\} \rho d\rho dz$$
$$\tag{4.13}$$
$$c_{ij}^{\parallel} = g_{ij} f_{ij} r_{ij} \frac{d}{dr_{ij}}\frac{\Phi_{12}}{kT} + \tfrac{3}{2}\phi\left(e^{-\Phi_{12}/kT} - 1\right) \int \left\{ \left(c_{i3}^{\parallel} - 2\frac{a^2\rho^2}{r_{i3}^2}\left(c_{i3}^{\parallel} - c_{i3}^{\perp}\right)\right)(g_{3j} - 1) \right.$$
$$\left. - \frac{a^2\left(z^2 - \frac{1}{2}r^2\right)}{r_{i3}r_{j3}} g_{3j} f_{3j} r_{3j} \frac{dc_{i3}}{dr_{i3}} \right\} \rho d\rho dz$$

For times long relative to the microstructural relaxations but short relative to the full decay of the density fluctuation, $a^2/D_o << t << 1/q^2 D_o$, the flux can be written as

$$\boldsymbol{J}_r = -2D_o i\boldsymbol{q} \cdot \left(\boldsymbol{\delta}\left(-\tfrac{1}{2}g_{12} + j^{\perp}\right) + \frac{\boldsymbol{r}\boldsymbol{r}}{r^2}\left(j^{\parallel} - j^{\perp}\right) \right) \tag{4.14}$$

with

$$j^{\perp} = g_{12}f_{12} - \tfrac{3}{2}\phi e^{-\Phi_{12}/kT}\int\left(-\frac{a^2\rho^2}{2r_{13}r_{23}}r_{13}g_{13}f_{13}\frac{dc_{23}}{dr_{23}}\right.$$
$$\left.+\left(c_{23}^{\perp} + \frac{a^2\rho^2}{2r_{23}^2}\left(c_{23}^{\parallel} - c_{23}^{\perp}\right)\right)(g_{13}-1)\right)\rho d\rho dz$$

$$j^{\parallel} - j^{\perp} = r\frac{df_{12}g_{12}}{dr} + rf_{12}g_{12}\frac{d}{dr}\frac{\Phi_{12}}{kT} - \tfrac{3}{2}\phi e^{-\Phi_{12}/kT}\int\left(-\frac{a^2\left(z^2-\frac{1}{2}r^2\right)}{r_{13}r_{23}}r_{13}g_{13}f_{13}\frac{dc_{23}}{dr_{23}}\right.$$
$$\left.+\frac{a^2\left(\left(z-\frac{1}{2}r\right)^2-\frac{1}{2}\rho^2\right)}{r_{23}^2}\left(c_{23}^{\parallel} - c_{23}^{\perp}\right)c_{23}'(g_{13}-1)\right)\rho d\rho dz$$

and the perturbed microstructure attains a steady form governed by

$$\nabla \cdot \boldsymbol{J}_r = -2D_o i\boldsymbol{q}\cdot\frac{\boldsymbol{r}}{r}\left(-\frac{1}{2}\frac{dg_{12}}{dr} - \frac{2}{r}j^{\perp} + \frac{1}{r^2}\frac{d}{dr}r^2 j^{\parallel}\right) = 0 \qquad (4.15)$$

Solution and substitution into (4.12) yields

$$F_s = 1 - q^2 D_s^{\infty} t \qquad (4.16)$$

with

$$D_s^{\infty} = D_o\left\{1 - \phi\int r^3 gf\frac{d}{dr}\frac{\Phi}{kT}dr + \tfrac{3}{2}\phi^2\int e^{-\Phi_{12}/kT}r^2\right.$$
$$\left.\int\left(\frac{a^2\left(z^2-\frac{1}{4}r^2+\rho^2\right)}{r_{13}r_{23}}r_{23}g_{23}f_{23}\frac{dc_{13}}{dr_{13}} + \left(2c_{13}^{\perp} + c_{13}^{\parallel}\right)(g_{23}-1)\right)\rho d\rho dz dr\right\} \qquad (4.17)$$

Evaluating this long-time self-diffusion coefficient is clearly not a trivial process; the best approach may be eliminate the integral terms in the conservation equation, through the closures, before differentiating to obtain the explicit form for the integrodifferential equation governing f_{12}.

17.4.4 Dilute and Weakly Coupled Limits

At this point we note only two limits. At dilute concentrations only pair interactions need be considered so that $g_{12}=exp(-\Phi_{12}/kT)$ and the equation governing f_{12} reduces to

$$\frac{d^2 f_{12}}{dr^2} + \frac{4}{r}\frac{df_{12}}{dr} + \frac{d\ln g_{12}}{dr}\left(\frac{df_{12}}{dr} + \frac{f_{12}}{r}\right) = \frac{1}{2r}\frac{d\ln g_{12}}{dr} \qquad (4.18)$$

with

$$f_{12} \to 0 \qquad \text{as } r \to \infty$$

$$r^2 g\left(-\frac{1}{2} + f_{12} + r\frac{df_{12}}{dr}\right) \to 0 \qquad \text{as } r \to 0$$

The expression for the diffusion coefficient reduces to

$$D_s^\infty = D_o\left\{1 + \phi \int_0^\infty r^3 f_{12} \frac{dg_{12}}{dr} dr + O\left(\phi^2\right)\right\} \tag{4.19}$$

For hard spheres straightforward solution yields f_{12}=-2/r^3 and

$$D_s^\infty = D_o\left(1 - 2\phi + O\left(\phi^2\right)\right) \tag{4.20}$$

Note that the exact result (Batchelor 1983), with full hydrodynamic interactions, is $D_o\{1\text{-}2.1\phi+O(\phi^2)\}$, indicating that the hard sphere repulsion compensates for the missing viscous forces. Others have derived the dilute limit for hard spheres and other interparticle potentials (e.g. Ohtsuki 1982; Sonnenburg, et al. 1990; Cichocki and Felderhof 1991a,b) but this list is far from comprehensive.

In the *weak coupling limit*, i.e. Φ_{12}/kT<<1 everywhere, the higher order interactions also vanish so that g_{12}~1-Φ_{12}/kT and

$$f_{12} = -\frac{1}{2r^3}\int_0^r r^2 \frac{\Phi_{12}}{kT} dr$$

so that (4.21)

$$D_s^\infty = D_o\left\{1 - \tfrac{1}{2}\phi \int (\Phi/kT)^2 r^2 dr\right\}$$

17.4.5 Fokker-Planck Analysis

Hess and Klein (1983) derive from the Fokker-Planck equation formal relationships among the correlation functions and the memory functions of linear response theory. Rigorous expansions in the short time limit serve as the benchmarks for closures capable of estimating the response over the full range of wavenumbers and times. The approach thereby bypasses the nonequilibrium structure responsible for these relaxation processes, yet predicts at least qualitatively some very interesting phenomena including the density fluctuations in charged latices at low ionic strengths.

Motivated by the success for molecular systems Klein and coworkers (Nagële, et al. 1988; D'Aguanno, et al. 1990; Genz and Klein 1991) have harnessed the approach for predicting the dynamics of colloidal systems with weak or negligible hydrodynamic interactions. Here we limit our attention to the dynamics in the long time limit, for comparison with the Smoluchoski treatment described above.

The generalized hydrodynamic or linear response theory incorporates the time dependence of the dynamic structure into the memory functions $\Delta D(q,t)$ and $\Delta D_S(q,t)$ as

$$\frac{\partial F}{\partial t}(q,t) = -\frac{q^2 D_o}{S(q)} F(q,t) + q^2 \int_0^t \Delta D(q,t-t') F(q,t') dt'$$

$$\frac{\partial F_s}{\partial t}(q,t) = -q^2 D_o F_s(q,t) + q^2 \int_0^t \Delta D_s(q,t-t') F_s(q,t') dt' \qquad (4.22)$$

Supplementary equations relate the memory functions to generalized friction coefficients

$$\Delta D(q,t) = \frac{D_o}{S(q)} \frac{\Delta\zeta(q,t)}{\zeta_o} - \int_0^t \Delta D(q,t-t') \frac{\Delta\zeta(q,t')}{\zeta_o} dt'$$

$$\Delta D_s(q,t) = D_o \frac{\Delta\zeta_s(q,t)}{\zeta_o} - \int_0^t \Delta D_s(q,t-t') \frac{\Delta\zeta_s(q,t')}{\zeta_o} dt' \qquad (4.23)$$

Thus the first term in (4.22) indicates a purely exponential decay at short times, before memory effects accumulate in the integral to alter the behavior. Note that the equations for the memory functions resemble those for the direct correlation force in the Smoluchoski theory.

These can be expressed more conveniently in terms of Laplace transforms, since the integrals comprise convolutions. Then

$$\Delta\tilde{D}_s = D_o \frac{\Delta\tilde{\zeta}_s/\zeta_o}{1+\Delta\tilde{\zeta}_s/\zeta_o}$$

$$\tilde{F}_s = \frac{1}{s + q^2 D_o\left(1 - \Delta\tilde{D}_s/D_o\right)} = \frac{1}{s + \dfrac{q^2 D_o}{1+\Delta\tilde{\zeta}_s/\zeta_o}} \qquad (4.24)$$

Thus the long-time self-diffusion coefficient is simply

$$D_s^\infty = \frac{D_o}{1+\Delta\tilde{\zeta}_s(0,0)/\zeta_o} \qquad (4.25)$$

with $\Delta\tilde{\zeta}_s(0,0) = \int_0^\infty \Delta\zeta_s(0,t)dt$.

17.4.6 Mode-Mode Coupling Approximation

Describing the nonexponential decay requires a closure approximation relating the generalized friction coefficients to the equilibrium structure factor $S_o(q)$ and the dynamic structure factors. Phenomenological closures motivated by short time expansions exist, e.g. the linear viscoelastic approximation discussed by Hess and Klein (Arauz-Lara and Medina-Noyola 1983, 1986), while the weak-coupling approximation connects the generalized friction coefficients directly to the interparticle potential. Hess and Klein use the latter to motivate the more popular mode-mode coupling approximation (MMCA), that predicts several interesting features of simple liquids, including the hard sphere glass transition.

Providing either a physical or mathematical explanation of the mode-mode coupling approximation lies beyond this review; nonetheless, one senses a similarity with other closures discussed here, accounting for many-body hydrodynamic interactions and three-body thermodynamic couplings in the Smoluchoski equation. Like these the MMCA factors higher order correlations into pair correlations and imposes self-consistency, leading to an expression for the memory function in terms of the dynamic structure factor itself. In the overdamped limit appropriate for Brownian particles, and without hydrodynamic interactions, the isolated friction and diffusion coefficients reduce to the Stokes drag coefficient and the Stokes-Einstein diffusion coefficients, respectively,

$$\zeta_o = 6\pi\mu a$$

$$D_o = \frac{kT}{\zeta_o}$$

and the additional friction due to relaxation effects takes the form

$$\Delta\zeta(\boldsymbol{q},t) = -\frac{2kT}{(2\pi)^3 n}\int G(\boldsymbol{q},\boldsymbol{q}')G(-\boldsymbol{q},-\boldsymbol{q}')F(\boldsymbol{q}'_1,t)F(\boldsymbol{q}'_2,t)d\boldsymbol{q}' \quad (4.26)$$

with $\boldsymbol{q}'_{1,2} = \frac{1}{2}\boldsymbol{q} \mp \boldsymbol{q}'$ and

$$G(\boldsymbol{q},\boldsymbol{q}') = \frac{ikT}{2S_o(q'_1)S_o(q'_2)}\left(\boldsymbol{q}'_1\hat{h}(q'_1) + \boldsymbol{q}'_2\hat{h}(q'_2)\right)\cdot\frac{\boldsymbol{q}}{q}$$

and (4.27)

$$\Delta\zeta_s(\boldsymbol{q},t) = -\frac{kT}{(2\pi)^3 n}\int\left(\frac{\boldsymbol{q}'\cdot\boldsymbol{q}\hat{h}(q')}{qS_o(q')}\right)^2 F_s(\boldsymbol{q}-\boldsymbol{q}',t)F(\boldsymbol{q}',t)d\boldsymbol{q}'$$

depending on the static structure factor through

$$\hat{h}(q) = \frac{S_o(q) - 1}{S_o(q)}$$

Thus one must specify $S_O(q)$, calculate $G(\boldsymbol{q},\boldsymbol{q}')$, and then solve the coupled integro-differential equations for the memory function and the dynamic structure factor.

The potential for these approximations is suggested by the calculations reported by Hess and Klein (1983) for charged particles interacting through a Yukawa potential. A measure of the nonexponential decay of F

$$\Delta(q) = \frac{\Delta\tilde{D}(q,0)}{D_o / S_o(q)}$$

is plotted in Figure 2 as calculated through the mode-mode coupling approximation and measured via photon correlation spectroscopy.

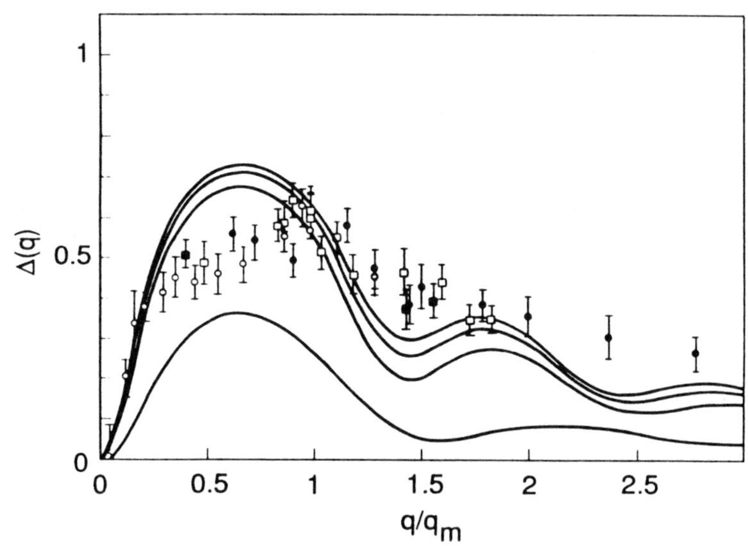

Figure 2 Nonexponential decay of dynamic structure factor for charged spheres with ϕ=1-5x10^{-3} (Hess and Klein 1983): (——) mode-mode coupling approximation, (● ○ ■ □) data from photon correlation spectroscopy.

The prediction of glass transitions (Bengtzelius, et al. 1984; Leutheusser 1984; Bengtzelius 1986; Bengtzelius and Sjögren 1986; Götze and Sjögren 1986a,b, 1988; Kim and Mazenko 1990) involves solving the nonlinear equations for the dynamic structure factor to determine conditions for which the autocorrelation function does not decay to zero. Recently van Megen, et al. (1991) have compared the predictions for hard spheres with measurements via photon correlation spectroscopy with PMMA spheres stabilized by poly-

(hydroxy stearic acid) layers and dispersed in index matching mixtures of cis- and trans-decalin (Figure 3).

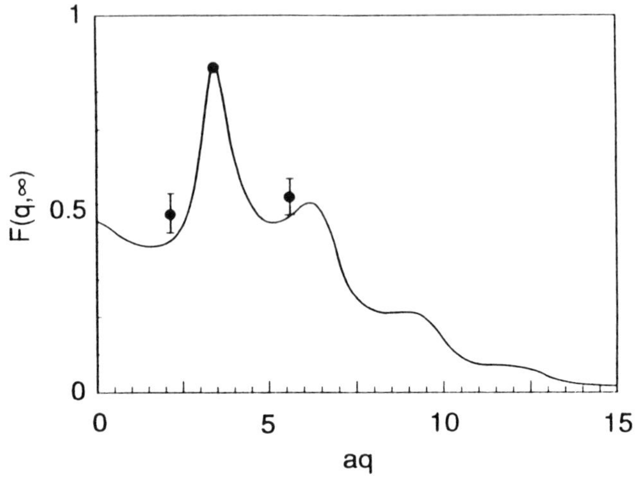

Figure 3 Amplitude, $F(q,\infty)$, of the density fluctuations frozen in at the hard sphere glass transition (van Megen, et al. 1991): (——) mode-mode coupling approximation, (•) data from photon correlation spectroscopy.

Here we focus on the small wavenumber limit, q->0, in which the generalized friction coefficient for the mutual diffusion process vanishes, as in the Smoluchoski approach, leaving

$$F(q,t) = S_o(q)exp\left(-\frac{q^2 D_o t}{S_o(q)}\right)$$

For the self-diffusion process

$$F_s \approx \exp\left(-q^2 D_s^\infty(q)t\right)$$

so that

$$\Delta\zeta_s(0,t) = -\frac{kT}{(2\pi)^3 n}\frac{\boldsymbol{qq}}{q^2} : \int \boldsymbol{q'q'}\frac{(S_o(q')-1)^2}{S_o(q')} exp\left(-q'^2 t\left(\frac{D_o}{S_o(q')} + D_s^\infty(q')\right)\right) dq' \tag{4.28}$$

and

$$\Delta D_s(0,t) \approx D_o \frac{\Delta\zeta_s(0,t)}{\zeta_o}$$

for $t\Delta\zeta_s(0,0)/\zeta_o << 1$. Substitution into (4.25) and integration over the angular variables gives

$$D_s^\infty = D_o \left\{ 1 + \frac{1}{6\pi^2 n} \int \frac{\left(S_o(q') - 1\right)^2}{1 + S_o(q') D_s^\infty(q')/D_o} q'^2 \, dq' \right\}^{-1} \tag{4.29}$$

Hess and Klein (1983) suggest setting $D_S{}^\infty = D_O$ within the integral to estimate the long time limit. Alternatively, Medina-Noyola (1988) takes $D_S{}^\infty << D_O$ and substitutes $D_O = D_S{}^O$ to account approximately for hydrodynamic interactions and obtains rather nice agreement with data for hard spheres (Figure 4).

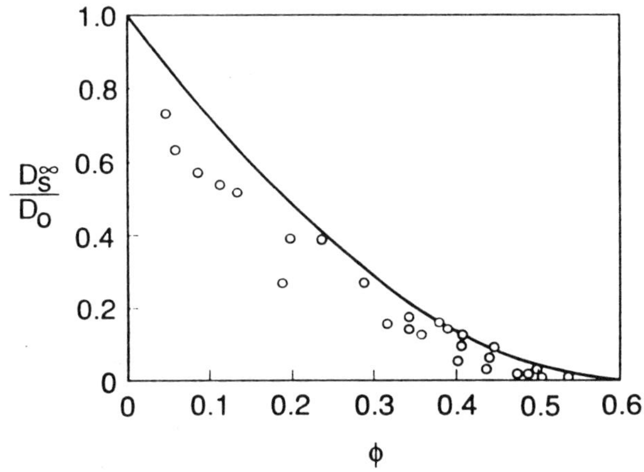

Figure 4 Long-time self-diffusion coefficient for hard spheres (Medina-Noyola 1988): (——) mode-mode coupling approximation for $D_S{}^\infty / D_S{}^O$, (○) data from photon correlation spectroscopy.

Comparison with the complex expression (4.17) from the Smoluchoski approach clearly demonstrates the appeal of the mode-mode coupling approach. The only current point of comparison lies in the dilute limit. For example, with hard spheres $g(r) = H(r - 2a)$ and

$$S_o(q) - 1 = \frac{24\phi}{(2aq)^2} \left(\cos 2aq - \frac{\sin 2aq}{2aq} \right) \tag{4.30}$$

so that

$$D_s^\infty = D_o \left(1 - \tfrac{4}{3}\phi\right) \tag{4.31}$$

which differs from the exact result (4.20) but not drastically. With $\Phi/kT << 1$, linearizing the Boltzmann factor yields

$$D_s^\infty = D_o \left\{ 1 - \tfrac{1}{2}\phi \int (\Phi/kT)^2 r^2 dr \right\} \tag{4.32}$$

i.e. the correct weak-coupling approximation. So the mode-mode coupling approximation becomes exact for weak interactions, erring quantitatively, but not dramatically, in the dilute limit for hard spheres and, presumably, other strong repulsions or attractions.

Thus the mode-mode coupling appoximation and the Smoluchoski approach describe the same the relaxation processes, but through different closure approximations. Whether the comparison provides clues to better approximations remains to be seen.

17.4.7 Combined Approaches

Several efforts have sought to join the Fokker-Planck and Smoluchoski approaches, to derive better approximations for the memory functions of the former and to generalize or simplify the latter. Hess and Klein (1983) formally expressed the Smoluchoski results in terms of memory functions, but derived somewhat different forms for the diffusion coefficients than from the over-damped limit of the Fokker-Planck equation. About the same time Felderhof and Jones (1983a,b, 1987) extracted explicit memory functions from solutions to the Smoluchoski equation, but failed to recover the generalized Stokes-Einstein diffusion coefficient in the q=0 limit. Subsequently, Cichocki and Hess (1986, 1987) resolved the paradox, uncovering the flaw in the original derivation and proving conclusively the validity of the Smoluchoski equation for particles dispersed in a viscous liquid. On this basis Cichocki (1988a,b) derived explicit forms for the memory functions with the proper behavior in both the dilute and small wavenumber limits.

Cichocki (1988b) also suggested invoking an Enskog approximation, "in which two particle dynamical process are taken into account exactly and more particle processes only via static correlations". Within the context of the the present development this means rewriting (4.11) with the HNC closure as

$$\frac{dF_s}{dt} = -q^2 D_o F_s + nD_o i\boldsymbol{q} \cdot \int e^{-\frac{1}{2}i\boldsymbol{q}\cdot\boldsymbol{r}_{21}} p_{12} \left\{ \nabla_1 \frac{\Phi_{12}}{kT} - n\int (p_{23} - 1) \frac{C_{13}}{kT} d\boldsymbol{x}_3 \right\} d\boldsymbol{r}_{21} \tag{4.33}$$

and setting terms within the three-body integral to their equilibrium values,

$$p_{23} = g_{23}$$
$$C_{13} = kT\nabla_1 c_{13}$$

From the equilibrium force balance

$$\nabla_1 \ln g_{12} = -\nabla_1 \frac{\Phi_{12}}{kT} + n\int (g_{23} - 1) \nabla_1 c_{13} d\boldsymbol{x}_3$$

so that

$$\frac{dF_s}{dt} = -q^2 D_o F_s + nD_o i\mathbf{q} \cdot \int e^{-\frac{1}{2}i\mathbf{q}\cdot\mathbf{r}_{21}} p_{12} \nabla_1 \ln g_{12} d\mathbf{r}_{21} \tag{4.34}$$

Following through the remainder of the procedure leads to

$$D_s^\infty = D_o \left\{ 1 + \phi \int_0^\infty r^3 f_{12} \frac{dg_{12}}{dr} dr \right\} \tag{4.35}$$

rather than (4.17).

Similar approximations in (4.10) governing the nonequilibrium pair distribution function yield

$$\frac{\partial p_{12}}{\partial t} = -\tfrac{1}{2} q^2 D_o p_{12} + 2D_o \left\{ \nabla^2 p_{12} + \nabla \cdot p_{12} \left(\nabla \frac{\Phi_{12}}{kT} - n \int \nabla c_{23} (g_{13} - 1) d\mathbf{x}_3 \right) \right\} \tag{4.36}$$

and ultimately (4.18) for f_{12}. However, g_{12}, now evaluated at finite volume fraction, has longer range than the pair potential.

The results for hard spheres prove interesting. Extending the approximation to neglect many-body effects in the calculation of the nonequilibrium structure leaves f_{12}=$-2/r^3$ as in the dilute limit. Then, with g_{12}=$g(r)H(r$-$2)$, the diffusion coefficient becomes

$$\begin{aligned} D_s^\infty &= D_o \left\{ 1 - 2\phi \int_0^\infty \left(g(2)\delta(r-2) + \frac{dg}{dr} H(r-2) \right) dr \right\} \\ &= D_o (1 - 2\phi) \end{aligned} \tag{4.37}$$

which is identical to the dilute limit. So extracting the nonequilibrium structure from (4.36) with the full g_{12}, is essential to retaining higher order effects in the diffusion coefficient. One should also note that a slightly more careful derivation would yield (Cichocki 1988b)

$$D_s^\infty = D_o \left\{ 1 - \phi \int_0^\infty r^3 f_{12} \frac{dg_{12}}{dr} dr \right\}^{-1} \tag{4.38}$$

17.5 STRUCTURE AND STRESSES IN SHEAR FLOW

Shear flows convect particles at a velocity that depends on the spatial position, and thereby perturb the microstructure from equilibrium. The perturbation scales on the strength of the flow relative to Brownian motion, as measured by the Peclet number,

$$Pe = \frac{3\pi\mu a^3 \gamma}{kT}$$

As $Pe \rightarrow \infty$ hydrodynamic forces dominate, producing a highly nonequilibrium structure and rendering negligible the Brownian and interparticle contributions to the stress. As $Pe \rightarrow 0$ the perturbation becomes small but remains important in determining the $O(1)$ nonhydrodynamic contributions to the stress. In this section, we first discuss the calculation of the nonequilibrium pair distribution function and stresses, focussing on weak flows without near-field hydrodynamic interactions to highlight the closures. Then we examine briefly the Fourier space alternative.

17.5.1 Weak Flow Expansion

For a weak steady or oscillatory flow with rate of strain $\boldsymbol{E}\exp(i\omega t)$ expansion of the pair distribution function as

$$P_2(\boldsymbol{r}) = n^2 g(r)\left\{1 - Pe\frac{\boldsymbol{r}\cdot\boldsymbol{E}\cdot\boldsymbol{r}}{r^2} f(r) e^{i\omega t}\right\} \tag{5.1}$$

motivates a similar expansion for the direct correlation force,

$$\boldsymbol{C} = kT\nabla_1 c + Pe\frac{kT}{r}e^{i\omega t}\left\{\frac{\boldsymbol{E}\cdot\boldsymbol{r}}{r}c^{\perp} + \frac{\boldsymbol{rr}\cdot\boldsymbol{E}\cdot\boldsymbol{r}}{r^3}\left(c^{\parallel} - c^{\perp}\right)\right\} \tag{5.2}$$

Note that $\boldsymbol{r}$ and $\boldsymbol{E}$ are now scaled such that r=2 for two spheres in contact and $\boldsymbol{E}:\boldsymbol{E}$=1. Substitution of both into the Percus-Yevick approximation determines the longitudinal and transverse components of the latter as

$$c_{12}^{\perp} = \tfrac{3}{2}\phi\left(e^{-\Phi_{12}/kT} - 1\right) r \int\left\{\left(\frac{a^2\left(z+\frac{1}{2}r\right)}{r_{13}^2}c_{13}^{\perp} + \frac{a^4\left(z+\frac{1}{2}r\right)\rho^2}{r_{13}^4}\left(c_{13}^{\parallel} - c_{13}^{\perp}\right)\right)\left(g_{23} - 1\right)\right.$$

$$\left. - \frac{a^4\rho^2\left(z-\frac{1}{2}r\right)}{r_{13}r_{23}^2} g_{23} f_{23}\frac{dc_{13}}{dr_{13}}\right\}\rho\, d\rho\, dz \tag{5.3}$$

$$c_{12}^{\parallel} - c_{12}^{\perp} = r f_{12} g_{12}\frac{d}{dr}\frac{\Phi_{12}}{kT} + \tfrac{3}{2}\phi\left(e^{-\Phi_{12}/kT} - 1\right) r \int\left\{\frac{a^4\left(z+\frac{1}{2}r\right)\left(\left(z+\frac{1}{2}r\right)^2 - \frac{3}{2}\rho^2\right)}{r_{13}^4}\left(c_{13}^{\parallel} - c_{13}^{\perp}\right)\right.$$

$$\left. \times\left(g_{23} - 1\right) - \frac{a^4\left(\left(z+\frac{1}{2}r\right)\left(z-\frac{1}{2}r\right)^2 - \frac{3}{2}\rho^2\left(z-\frac{1}{6}r\right)\right)}{r_{13}r_{23}^2} g_{23} f_{23}\frac{dc_{13}}{dr_{13}}\right\}\rho\, d\rho\, dz$$

with

$$c_{13} = g_{13}\left(1 - e^{\Phi_{13}/kT}\right)$$

and $\boldsymbol{r}_{13}$ and $\boldsymbol{r}_{23}$ defined above (4.12) so that the other position variables are scaled on the particle radius.

With these the $O(Pe)$ flux can be written as

$$J = 2D_o n^2 Pe\, e^{i\omega t} \frac{\boldsymbol{r}\cdot\boldsymbol{E}}{r} \cdot \left\{ \frac{\boldsymbol{rr}}{r^2} G\left(g\frac{df}{dr} - \frac{\phi\bar{c}^{\parallel}}{r} \right) + \left(\delta - \frac{\boldsymbol{rr}}{r^2} \right) H \left(\frac{2}{r} fg - \frac{\phi\bar{c}^{\perp}}{r} \right) \right\}$$

$$+ n^2 g\{(\boldsymbol{E}+\boldsymbol{\Omega})\cdot\boldsymbol{r} - \boldsymbol{C}\cdot\boldsymbol{E}\cdot\boldsymbol{r}\} \qquad (5.4)$$

with

$$\phi\bar{c}^{\perp} = c^{\perp} \Big/ \left(1 - e^{\Phi/kT}\right)$$

$$\phi\bar{c}^{\parallel} = \left(c^{\parallel} - rf\frac{dc}{dr} - 2rfg\frac{d}{dr}\left(\frac{\Phi}{kT}\right)\right) \Big/ \left(1 - e^{\Phi/kT}\right)$$

and $G=A_{11}-A_{12}$ and $H=B_{11}-B_{12}$.. At this point the hydrodynamic couplings to the third and fourth particles, the last two terms in (2.4), have been omitted. The $O(Pe)$ perturbation to the Smoluchoski equation follows from substitution into (2.3) as

$$-\frac{i\omega a^2}{2D_o} f + \frac{1}{gr^2}\frac{d}{dr}\left\{ r^2 G\left(g\frac{df}{dr} - \frac{\phi\bar{c}^{\parallel}}{r} \right)\right\} - \frac{3H}{r^2}\left(2f - \phi\bar{c}^{\perp}\right) = -W - r(1-A)\frac{d\ln g}{dr} \qquad (5.5)$$

with the usual boundary conditions,

$$G\left(g\frac{df}{dr} - \frac{\phi\bar{c}^{\parallel}}{r} \right) + r(1-A)g = 0 \qquad \text{at} \quad r=2$$

$$f \to \infty \qquad \text{as} \quad r \to \infty$$

The governing equation is now inhomogeneous with the divergence of the relative velocity, $W=3B-r^{-2}d(r^3A)/dr$, and convection of the nonuniform equilibrium distribution both driving the relaxation processes on the lefthand side.

Certain aspects of the perturbed microstructure are determined by the form of the governing equation and boundary conditions. For $a^2\omega/D_o \ll 1$ the transient term is negligible, leaving the diffusional relaxations to balance the convective forcing. Hence, f is real and $O(1)$ with the asymptotic form

$$f \approx \frac{b_1}{r^3} \text{ as } r \to \infty$$

As noted by Dhont (1987) an interesting singularity arises in the far-field, since convection, $O(r^{-4})$, eventually dominates diffusion, $O(r^{-5})$, producing an outer region for $r=O(Pe^{-1})$ in which the perturbation decays with more complex symmetry.

For $a^2\omega/D_o \gg 1$, however, the transient term dominates, requiring that f become imaginary and small to balance convection according to

$$f = -i\frac{2D_o}{\omega a^2}\left(W + r(1-A)\frac{d\ln g}{dr}\right) \tag{5.6}$$

providing, of course, the boundary conditions are satisfied. A simple affine deformation of the structure by the imposed strain would produce

$$f = -i\frac{2D_o}{\omega a^2} r\frac{d\ln g}{dr} \tag{5.7}$$

Thus the deformation is in phase with and of similar magnitude to the strain, but can be affected substantially by hydrodynamic interactions.

Substitution of the weak flow expansions into the stresses (2.18) provides expressions that are linear in the rate of strain and independent of the vorticity:

$$\begin{aligned}
\langle\sigma\rangle^H &= 2\mu E e^{i\omega t}\left(1 + \tfrac{5}{2}\phi + \tfrac{5}{2}\phi^2 + \tfrac{15}{2}\phi^2\int_2^\infty r^2 J(r)g(r)dr\right) \\
\langle\sigma\rangle^I &= -\tfrac{9}{20}\mu E e^{i\omega t}\phi^2\int_2^\infty r^3(1-A)\frac{d}{dr}\left(\frac{\Phi}{kT}\right)g(r)f(r)dr \\
\langle\sigma\rangle^B &= \tfrac{9}{20}\mu E e^{i\omega t}\phi^2\int_2^\infty r^2 W g(r)f(r)dr
\end{aligned} \tag{5.8}$$

The scalar function J represents the angular integral of the corresponding term in (2.18).

In the low frequency limit, a dispersion with the rest state of an equilibrium fluid will behave as a Newtonian fluid with all three stresses contributing to the low shear viscosity as

$$\begin{aligned}
\frac{\eta_o}{\mu} &= 1 + \tfrac{5}{2}\phi + \tfrac{5}{2}\phi^2 + \tfrac{15}{2}\phi^2\int_2^\infty r^2 J(r)g(r)dr \\
&\quad + \tfrac{9}{40}\phi^2\int_2^\infty r^2\left[W - r(1-A)\frac{d}{dr}\left(\frac{\Phi}{kT}\right)\right]g(r)f(r)dr
\end{aligned} \tag{5.9}$$

In the high frequency limit, the phase shift in the nonequilibrium structure generates an elastic response captured in the shear modulus, part of the complex viscosity defined by

$$\langle\sigma\rangle = \left(\eta'_\infty - i\frac{G'_\infty}{\omega}\right)E e^{i\omega t} \tag{5.10}$$

From (5.6), the high frequency dynamic viscosity equals the hydrodynamic component of the low shear viscosity, i.e.

$$\frac{\eta'_\infty}{\mu} = 1 + \tfrac{5}{2}\phi + \tfrac{5}{2}\phi^2 + \tfrac{15}{2}\phi^2 \int_2^\infty r^2 J(r) g(r) dr \tag{5.11}$$

while the shear modulus derives entirely from the interparticle and Brownian forces according to

$$\frac{a^3 G'_\infty}{kT} = \tfrac{3}{40\pi}\phi^2 \int_2^\infty r^2 \left[W - r(1-A)\frac{d}{dr}\left(\frac{\Phi}{kT}\right)\right]\left[W + r(1-A)\frac{d\ln g}{dr}\right] g(r) dr \tag{5.12}$$

In the absence of hydrodynamic interactions this coincides with the result of Zwanzig and Mountain (1965)

$$\frac{a^3 G'_\infty}{kT} = \tfrac{3}{40\pi}\phi^2 \int_0^\infty g \frac{d}{dr}\left[r^4 \frac{d}{dr}\left(\frac{\Phi}{kT}\right)\right] dr \tag{5.13}$$

derived from linear response theory. Note, however, that the integral diverges for hard repulsions.

Although the shear modulus provides the most direct link between a rheological measurement and the interparticle potential, two factors complicate the interpretation: the limited range of frequencies accessible experimentally and the presence of hydrodynamic interactions. For a dispersion with a fluid rest state, the high frequency plateau must be attained and the range of the interaction potential must significantly exceed the particle radius for the simple result (5.13) to be valid. Otherwise, the hydrodynamics intervene. For a dispersion with a glassy rest state, a static deformation is unaffected by hydrodynamics and may involve few local relaxations. Then the modulus would approximate the form of (5.13).

17.5.2 Truncated Closures

Solution of the full problem requires the input of suitable equilibrium structure and hydrodynamic functions and numerical integration of the three coupled integro-differential equations, (5.3) and (5.5), for the nonequilibrium structure and the two components of the direct correlation force.

Several approximations proposed in the literature for the thermodynamic closures can be compared in the context of the formal development above. The Enskog approximation, as defined by Cichocki (1988), corresponds to setting

$$c_{12}^{\perp} = 0$$

$$c_{12}^{\parallel} = g_{12} f_{12} r_{12} \frac{d\Phi_{12}}{dr_{12}}$$

such that

$$\phi\bar{c}^{\perp} = 0$$

$$\phi\bar{c}^{\parallel} = -rf\left(\frac{dg}{dr} + g\frac{d\Phi}{dr}\right)$$

This eliminates the three-body integrals and leaves a differential conservation equation. The approximation of Russel and Gast (1986) goes further, taking the dilute limit of the Enskog closure, i.e. setting g=exp(-Φ/kT) such that

$$\phi\bar{c}^{\perp} = \phi\bar{c}^{\parallel} = 0.$$

The next higher level of approximation is to evaluate the integrals in (5.3) with the Enskog forms for the components of the direct correlation force, obtaining

$$\phi\bar{c}^{\perp} = \tfrac{3}{2}\phi e^{-\Phi/kT} r\int\left\{\frac{a^4\left(z+\frac{1}{2}r\right)\rho^2}{r_{13}^3} f_{13}g_{13}\frac{d}{dr_{13}}\left(\frac{\Phi_{13}}{kT}\right)(g_{23}-1) - \frac{a^4\rho^2\left(z-\frac{1}{2}r\right)}{r_{13}r_{23}^2} g_{23}f_{23}\frac{dc_{13}}{dr_{13}}\right\}$$
$$\times\rho d\rho dz \qquad (5.14)$$

$$\phi\bar{c}^{\parallel} = -rf\left(\frac{dg}{dr} + g\frac{d}{dr}\left(\frac{\Phi}{kT}\right)\right) + \tfrac{3}{2}\phi e^{-\Phi/kT} r\int\left\{\frac{a^4\left(z+\frac{1}{2}r\right)\left(\left(z+\frac{1}{2}r\right)^2 - \frac{3}{2}\rho^2\right)}{r_{13}^3} f_{13}g_{13}\right.$$
$$\left.\times\frac{d}{dr_{13}}\left(\frac{\Phi_{13}}{kT}\right)(g_{23}-1) - \frac{a^4\left(\left(z+\frac{1}{2}r\right)\left(z-\frac{1}{2}r\right)^2 - \frac{3}{2}\rho^2\left(z-\frac{1}{6}r\right)\right)}{r_{13}r_{23}^2} g_{23}f_{23}\frac{dc_{13}}{dr_{13}}\right\}\rho d\rho dz$$

This leads to a conservation equation of integro-differential form, but still explicit in f. The closures proposed by Wagner and Russel (1989) truncate the above by setting

$$c^{\perp} = c^{\parallel} = 0$$

on the righthand sides of (5.14) and deleting the first term in the second equation to obtain

$$\phi\bar{c}^{\perp} = -\tfrac{3}{2}\phi r e^{-\Phi/kT}\int\frac{a^4\rho^2\left(z-\frac{1}{2}r\right)}{r_{13}r_{23}^2} g_{23}f_{23}\frac{dc_{13}}{dr_{13}}\rho d\rho dz$$

$$\phi\bar{c}^{\parallel} = -\tfrac{3}{2}\phi r e^{-\Phi/kT}\int\frac{a^4\left(\left(z+\frac{1}{2}r\right)\left(z-\frac{1}{2}r\right)^2 - \frac{3}{2}\rho^2\left(z-\frac{1}{6}r\right)\right)}{r_{13}r_{23}^2} g_{23}f_{23}\frac{dc_{13}}{dr_{13}}\rho d\rho dz \qquad (5.15)$$

Clearly this misses a portion of the nonequilibrium force. The superposition approximation, implemented by Ohtsuki (1981, 1983) and Wagner and Russel (1989), also leaves an integro-differential conservation equation for f but does not fit into this framework.

17.5.3 Hydrodynamic Approximations

If one neglects many-body hydrodynamic interactions, the various hydrodynamic functions in the conservation equation, boundary conditions, and the stresses reduce to the pair functions, which are known exactly for spheres and approximately for a few other geometries. This amounts to pairwise addition of the velocities induced in a single particle by all others. Such treatments typically neglect the hydrodynamic couplings to a third and fourth particle, the last two terms of (2.4). The Oseen approximation for interactions comprises a far-field form of pairwise additivity.

Simply neglecting hydrodynamic interactions yields

$$-\frac{i\omega a^2}{2D_o}f+\frac{1}{r^2 g}\frac{d}{dr}\left[r^2\left(g\frac{df}{dr}-\phi\frac{\bar{c}^{\parallel}}{r}\right)\right]-\frac{3}{r^2}\left(2f-\phi\bar{c}^{\perp}\right)=-r\frac{d\ln g}{dr}$$

with

$$f\to 0 \quad \text{as} \quad r\to\infty \tag{5.16}$$

$$g\frac{df}{dr}-\frac{\bar{c}^{\parallel}}{r}+rg=0 \quad \text{at} \quad r=2$$

Now only the convection of the nonuniform equilibrium microstructure by the imposed flow and the no-flux boundary condition remain to drive the relaxation processes.

17.5.4 Results

Predictions of the nonequilibrium microstructure at finite concentrations exist only for charged spheres without hydrodynamic interactions, using the superposition closure (Ohtsuki, 1981), and for hard spheres with pairwise additive hydrodynamics, using superposition and the truncated closures noted above (Russel and Gast 1986, Wagner and Russel 1989). For hard spheres, the mode of closure profoundly affects the form of $f(r)$ at high volume fractions, with the value at contact and the shape and amplitude of the oscillations differing significantly. The nonequilibrium structure factor has the proper form but deviates substantially in magnitude from measurements (Figure 5) (Ackerson, et al. 1989; Wagner and Russel, 1990). Whether the systematic closures described above rectify this remains to be seen.

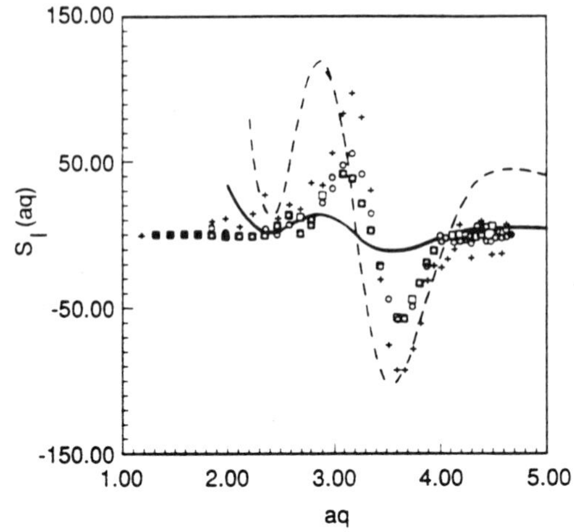

Figure 5 The *O(Pe)* component of the nonequilibrium static structure factor defined as

$$S(\boldsymbol{q}) = S_o(q)\left(1 + \phi Pe\frac{\boldsymbol{q}\cdot\boldsymbol{E}\cdot\boldsymbol{q}}{q^2}S_1(q)\right)$$

for a dispersion of hard spheres at ϕ=0.40 (Wagner and Russel 1990): (○) static light scattering measurements at Pe=0.01 (+), 0.025 (○), 0.03 (□), (——) Smoluchoski theory x3 (——) and x27 (---).

The results for viscosities and moduli have also been disappointing. For hard spheres, for which lubrication stresses eliminate the interparticle stress according to (5.8), both the low shear viscosity and the high frequency modulus fall well below measured values at volume fractions beyond the dilute regime (e.g. Wagner and Russel, 1989). The fault must lie with the assumption of pairwise additive hydrodynamics, which underestimates the effects of interactions on the nonequilibrium structure as well as the contributions to the hydrodynamic and Brownian stresses through $J(r)$ and $W(r)$.

Comparison of data for the high frequency and low shear viscosities (de Kruif, et al. 1986; van der Werff, et al. 1989) reveals a marked divergence in the latter, but not the former, about random close packing. This is manifest in Figure 6 by the data for η_o reaching the limiting curve but that for $\eta\text{э}_\infty$ falling well short. Thus the Brownian stress, solely responsible for the difference between them, apparently drives the divergence at low shear rates. Clearly, acceptable predictions for hard spheres require accurate many-body hydrodynamics for all the functions determining the structure and stresses.

Calculations for charged particles at low ionic strengths and low volume fractions avoid these difficulties, highlighting the role of the thermodynamic closures. To date only Ohtsuki (1981) has pursued the Smoluchoski approach, but with the less robust superposition closure. At the two volume fractions considered the low shear viscosity and the high frequency shear modulus were predicted to increase markedly with decreasing ionic strength, in accord at least qualitatively with the corresponding data. Calculations at intermediate frequencies revealed a relaxation for $a^2\omega/2D_0$ of $O(1)$ as expected.

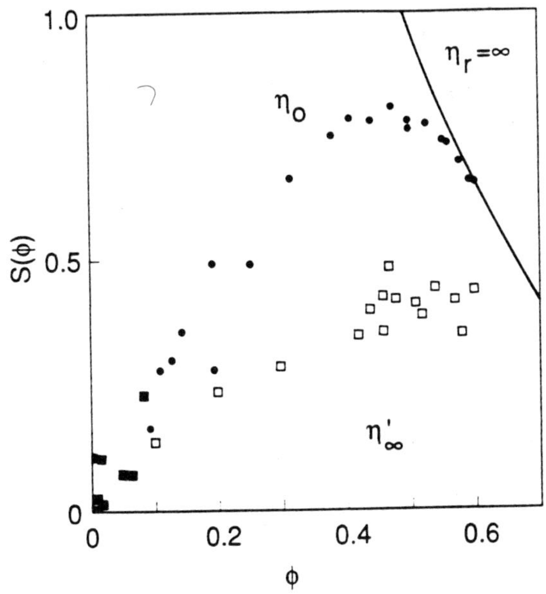

Figure 6 Viscosities of hard sphere dispersions as

$$\frac{\eta/\mu - 1}{\eta/\mu + \frac{3}{2}} = \phi\left[1 + S(\phi)\right]$$

versus volume fraction for (•) low shear and (□) high frequency limits (van der Werff, et al. 1989). Solid curve corresponds to $\eta/\mu \rightarrow \infty$ limit.

17.5.5 Fourier Space Analyses

Alternative analyses in Fourier space for charged particles at low volume fractions, based on (2.20), also bear mention (Ronis 1984a, 1986; Dhont 1987; Dhont, et al. 1989). Adopting the HNC closure (4.6), appropriate for soft repulsions, and carrying out the requisite transforms yields, with

$$\hat{h} = S(q) - 1$$

the direct correlation force and the flux defined by (2.21) as

$$\hat{C} = iq\hat{\Phi} + \frac{1}{(2\pi)^3}\int \hat{h}(q-q')\left(iq'\hat{\Phi}(q') + n\hat{C}(q')\hat{h}(q')\right)dq'$$
$$j = nS\hat{C} \tag{5.17}$$

At equilibrium (5.17) must balance the diffusion flux as specified by (2.8), so that

$$j = -iqn(S_o - 1) \tag{5.18}$$

Neglecting relaxation by evaluating the direct correlation force at equilibrium then yields an Enskog approximation for the flux as

$$j \cong iqnS(1/S_o - 1)$$

Now the governing equation takes the simple form,

$$\frac{\partial S}{\partial t} - (\boldsymbol{E} + \boldsymbol{\Omega}):\boldsymbol{q}\nabla_q S = -q^2 \frac{D_o}{S_o}(S - S_o) \tag{5.19}$$

Ronis (1984a, 1986) proposed this equation on phenomenological grounds and extracted an exact analytical solution for a steady simple shear flow, with several interesting features. First, the zero shear limit

$$S - S_o = \frac{S_o}{2D_o}\frac{\boldsymbol{q}\cdot\boldsymbol{E}\cdot\boldsymbol{q}}{q^3}\frac{dS_o}{dq} \tag{5.20}$$

holds except in the singular region about q=0. Second, the nonequilibrium structure factor predicted at finite shear rates resembles at least qualitatively light scattering measurements on latices at low ionic strengths (Ronis 1984b). The calculated shear viscosities remain modest, i.e. $O(\mu)$, but increase dramatically with increasing intensity of the interactions, as expected. Some issues linger, however, since the predicted shear thinning may lie beyond the validity of the approximation and the form of the stress raises questions about material objectivity.

Subsequently, Dhont (1987) revisited the problem, treating explicitly both the electrostatic fields within an ionic solution containing charged particles and the singular nature of the equation. His solution for pair interactions in a steady simple shear treats quite elegantly the electrostatic interactions at low ionic strengths, capturing the primary electroviscous effect due to perturbation of the ion densities and the secondary electroviscous effect arising from the interparticle repulsion.

The solution also accounts for the singularity about q=0, due to the dominance of convection at large separations. In the end, though, he finds the simple approximation (5.20) to capture the non-equilibrium structure over much of wavenumber space. Nonetheless the full treatment unifies nicely the nonequilibrium aspects of both the ion and particle distributions, with the potential for treating both in the many-body regime.

17.6 RHEOLOGY: DILUTE CONCENTRATIONS WITH FAR-FIELD HYDRODYNAMICS

At sufficiently dilute concentrations that three-body and higher order interactions are negligible, the questions of appropriate closures are moot and exact descriptions of the hydrodynamic and thermodynamic interactions are possible. The rigorous treatment of this limit comprises the basis for addressing concentrated dispersions as outlined above (Russel 1980; Russel, et al. 1989). The success in predicting $O(\phi^2)$ coefficients quantitatively and anticipating qualitatively effects at higher concentrations promotes optimism for reasonable predictions at much higher concentrations with approximate treatments of many-body interactions. Here we review the dilute limit in its simplest form, i.e. with only long range hydrodynamic interactions and hard interparticle potentials such that $A=B=G=H=W=J=0$ for all r and $\Phi=0$ for $r>2$ to illustrate the physics and, through comparison with exact results, to suggest when near-field hydrodynamics are most important.

17.6.1 Microstructure

For these conditions the pair conservation equation (2.3) in dimensionless form reduces to

$$\frac{\partial P_2}{\partial t} + \nabla \cdot P_2(\boldsymbol{E} + \boldsymbol{\Omega}) \cdot \boldsymbol{r} = Pe^{-1}\nabla^2 P_2 \tag{6.1}$$

with the no-flux boundary condition

$$\frac{\boldsymbol{r}}{r} \cdot \left\{ \boldsymbol{E} \cdot \boldsymbol{r} P_2 - Pe^{-1} \nabla P_2 \right\} = 0 \tag{6.2}$$

at $r=2$ and

$$P_2 \rightarrow n^2$$

as $r\text{->}\infty$. Since the relative velocity and mobility do not vanish at $r=2$, the former is not satisfied trivially as with full hydrodynamics; imposing zero flux recognizes the existence of a hard sphere repulsion.

For weak flows with $Pe<<1$ (5.16) without the three-body terms governs f, with the boundary conditions

$$f \rightarrow 0 \qquad \text{as} \quad r \rightarrow \infty$$

$$r + \frac{df}{dr} = 0 \quad \text{at} \quad r = 2$$

Thus diffusion must balance convection at contact.

For strong flows with $Pe \gg 1$ only the convection term persists as

$$\nabla \cdot P_2(\boldsymbol{E}+\boldsymbol{\Omega})\cdot \boldsymbol{r} = 0 \tag{6.3}$$

with

$$P_2 \to n^2 \quad \text{as} \quad r \to \infty$$

At r=2 the non-zero radial velocity generates a population of doublets P_d governed by

$$2P_2 \frac{\boldsymbol{r}\cdot\boldsymbol{E}\cdot\boldsymbol{r}}{r^2} + \nabla_s \cdot P_d\left[(\boldsymbol{E}+\boldsymbol{\Omega})\cdot\boldsymbol{r} - \frac{\boldsymbol{r}}{r}\frac{\boldsymbol{r}\cdot\boldsymbol{E}\cdot\boldsymbol{r}}{r}\right] = 0 \tag{6.4}$$

17.6.2 Stresses

The expressions (2.18) for the stresses also simplify considerably. With J=W=0 the Brownian stress disappears and the hydrodynamic contribution reduces to

$$\Sigma^H = 2\mu E\left(1+\tfrac{5}{2}\phi+\tfrac{5}{2}\phi^2\right) \tag{6.5}$$

Setting A=B=0 leaves that from the interparticle force as

$$\Sigma^I = \tfrac{1}{2}\int r\frac{d\Phi}{dr}\left(\frac{\boldsymbol{rr}}{r^2}-\frac{1}{3}\delta\right)P_2 d^3\boldsymbol{r} \tag{6.6}$$

Note that the hard sphere repulsion now can contribute but only through the nonequilibrium structure, since an isotropic structure yields no deviatoric stress, only the second virial term in the osmotic pressure.

For weak flows at low frequencies both hydrodynamic and interparticle forces affect the low shear viscosity as

$$\frac{\eta_o}{\mu} = 1+\tfrac{5}{2}\phi+\tfrac{5}{2}\phi^2+\tfrac{9}{40}\phi^2\int_2^\infty r^3\frac{dg}{dr}f dr \tag{6.7}$$

In the high frequency limit

$$\frac{\eta'_\infty}{\mu} = 1+\tfrac{5}{2}\phi+\tfrac{5}{2}\phi^2+\tfrac{9}{40}\phi^2\int_2^\infty r^3\frac{dg}{dr}Re f dr$$

$$\frac{a^3G'_\infty}{kT} = \frac{3}{40\pi}\phi^2\frac{a^2\omega}{D_o}\int_2^\infty r^3\frac{dg}{dr}Im\, f dr \tag{6.8}$$

with the split of the interparticle stress between the viscous and elastic components depending on the phase of the nonequilibrium structure relative to the rate of strain.

For strong flows the convection and the hard sphere excluded volume generate an anisotropic limiting microstructure

$$P_2 \rightarrow P_2^{\infty}$$

independent of the rate of strain. The stress then reaches an asymptote

$$\Sigma = 2\mu E\left(1+\tfrac{5}{2}\phi+\tfrac{5}{2}\phi^2\right)+\tfrac{1}{2}\int r\frac{d\Phi}{dr}\left(\frac{rr}{r^2}-\frac{1}{3}\delta\right)P_2^{\infty}d^3r \qquad (6.9)$$

with the interparticle force contributing a constant term, i.e. a residual or Bingham stress. Though independent of the magnitude of the rate of strain the value may well depend on the principal axes of strain and vorticity.

17.6.3 Hard Spheres

With only the hard sphere excluded volume and no attraction or longer range repulsion $g(r) = H(r\text{-}2a)$. For $Pe \ll 1$ and $a^2\omega/D_o \ll 1$ (5.16) then yields $f=32/3r^3$ and

$$\frac{\eta_o}{\mu} = 1+\tfrac{5}{2}\phi+\tfrac{49}{10}\phi^2 \qquad (6.10)$$

Comparison with the exact $O(\phi^2)$ coefficient of 6.0 (Batchelor 1977) indicates that the interparticle stress partially compensates for the hydrodynamic stress lost with the neglect of near-field interactions.

For $Pe \ll 1$ but $a^2\omega/D_o \gg 1$ a diffusion flux is still needed to negate the convective flux at contact; hence a boundary layer of thickness $(D_o/a^2\omega)^{1/2} \ll 1$ develops leading to (Russel 1989; Felderhof and Cichocki 1990)

$$f = 2(1-i)\left(D_o/a^2\omega\right)^{1/2}\exp\left(-(1+i)(r-2)\left(a^2\omega/D_o\right)^{1/2}\right) \qquad (6.11)$$

The real part contributes an extra term to the viscosity as

$$\frac{\eta'_\infty}{\mu} = 1+\tfrac{5}{2}\phi+\tfrac{5}{2}\phi^2+\tfrac{36}{5}\left(\frac{D_o}{a^2\omega}\right)^{1/2}\phi^2 \qquad (6.12)$$

and the imaginary part determines the shear modulus as

$$\frac{a^3 G'_\infty}{kT} = \frac{6}{5\pi}\left(\frac{a^2\omega}{D_o}\right)^{1/2}\phi^2 \tag{6.13}$$

Thus the modulus diverges at infinite frequency, since rigid spheres in contact cannot translate affinely with the fluid.

In this case, near-field hydrodynamic interactions suppress the convective flux at contact, hiding the hard sphere repulsion and altering the moduli profoundly to

$$\frac{\eta'_\infty}{\mu} = 1 + \tfrac{5}{2}\phi + 5.0\phi^2$$

and (6.14)

$$\frac{a^3 G'_\infty}{kT} = 0.202\phi^2$$

Interestingly, recent measurements for hard spheres at high volume fractions (van der Werff, et al. 1989) reveal an $\omega^{1/2}$-dependence similar to (6.13) (Figure 7).

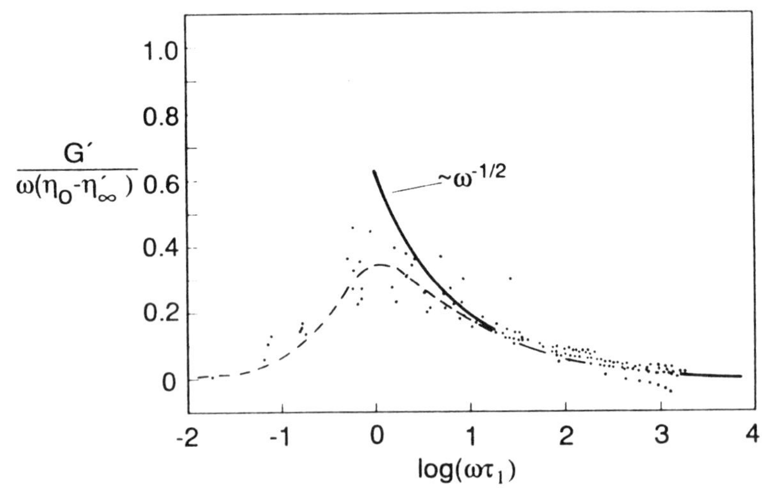

Figure 7 Correlation of shear moduli for dispersions of hard spheres (van der Werff, et al. 1989): (•) measurements with high frequency resonators, (----) fit to Rouse spectrum, (—) high frequency asymptote $G' \approx 0.63(\omega\tau_1)^{1/2}(\eta_o - \eta'_\infty)/\tau_1$.

At high shear rates, i.e. *Pe*>>1, the flow convects particles into contact, forming a doublet that rotates with the local vorticity and then separates once the velocity along lines of centers becomes positive. The pair density in a simple shear flow with $\boldsymbol{u}=(\gamma y,0,0)$ develops a highly asymmetric microstructure with a concentrated layer of doublets at contact and a shadow zone depleted of pairs downstream. At contact the hard sphere repulsion must balance the viscous force driving particles together as

$$-\frac{d\Phi}{dr} = -2Pe\frac{\boldsymbol{r}\cdot\boldsymbol{E}\cdot\boldsymbol{r}}{r^2}$$

when $\boldsymbol{r}\cdot\boldsymbol{E}\cdot\boldsymbol{r}<0$. Surprisingly, integration of the interparticle stress and addition to the hydrodynamic component yields

$$\frac{\eta_\infty}{\mu} = 1 + \tfrac{5}{2}\phi + \tfrac{49}{10}\phi^2 \tag{6.15}$$

Thus, pair interactions between hard spheres cause anisotropic microstructure but no shear thinning, at least without near-field hydrodynamics.

The exact result is not available for simple shear, since closed trajectories render the pair distribution function indeterminate (Batchelor and Green 1972). Data show significant differences between the high and low shear limiting viscosities only for $\phi>0.25$ (Krieger 1972, de Kruif, et al. 1986), presumably where three-body interactions become $O(1)$.

17.6.4 Charged Spheres

The long range repulsion between charged spheres in electrolyte solutions decays exponentially with separation, with the asymptotic form

$$\frac{\Phi}{kT} = \frac{\alpha}{a\kappa r}e^{-a\kappa r} \tag{6.16}$$

valid for $a\kappa(r\text{-}2)>1$. When $\alpha>>1$

$$g = \exp\left(-\frac{\alpha}{a\kappa r}\exp(-a\kappa r)\right) \approx H(r - s_o)$$

with the effective hard sphere diameter s_o determined by

$$1 \cong \frac{\alpha}{a\kappa s_o}\exp(-a\kappa s_o)$$

The second virial coefficient in the osmotic pressure then follows as

$$A_2 = \tfrac{1}{2}s_o^3 \tag{6.17}$$

For steady shear with $Pe<<1$ the previous results for hard spheres lead to (Russel 1978)

$$\frac{\eta_o}{\mu} = 1 + \tfrac{5}{2}\phi + \left(\tfrac{5}{2} + \tfrac{3}{40}s_o^5\right)\phi^2 \tag{6.18}$$

Since $s_o>2$ the $O(\phi^2)$ coefficient can be quite large, with values up to 10^3 measured for latices at low ionic strengths (Stone-Masui and Watillon 1968).

For the high frequency limit the hard sphere model is insufficient. However, for $\alpha >>1$ $g(2)<<1$, so an affine deformation of the equilibrium structure still approximates no-flux at contact. Evaluation of the stresses then gives

$$\frac{\eta'_\infty}{\mu} = 1 + \tfrac{5}{2}\phi + \tfrac{5}{2}\phi^2 \tag{6.19}$$

and (5.13) for the shear modulus. The low dynamic viscosity accords with some measurements. Goodwin and coworkers (Buscall, et al. 1982; Goodwin and Hughes 1990) predict, or at least correlate, the high frequency modulus for many charged dispersions at low ionic strengths with (5.13).

With increasing shear rate viscous forces overcome the Brownian forces, thereby reducing the characteristic separation below s_O. When $Pe>>1$ Brownian motion is negligible, leaving the electrostatic repulsion to balance the viscous forces and determine the characteristic separation as (Blachford, et al. 1969)

$$\frac{Pe}{\alpha(a\kappa)^2} = \frac{1 + a\kappa s}{(a\kappa s)^3}\exp(-a\kappa s) \tag{6.20}$$

and the shear viscosity from (6.18) with s replacing s_O. Thus the viscosity decreases from η_O for $Pe<<1$ to η_∞ for hard spheres (6.15) when s->2 for stresses greater than

$$\sigma_c = -\frac{1}{3\pi a^3}\frac{d\Phi}{dr}(2)$$

Ample data demonstrate shear thinning in qualitative, but not yet quantitative, accord.

17.6.5 Adhesive Hard Spheres

A square well of depth Φ_m and range Δ comprises the simplest attraction, reducing to the sticky or adhesive sphere model of Baxter (1968) with

$$\frac{1}{6\tau} = \frac{\Delta}{a}\left(\exp(\Phi_m/kT) - 1\right)$$

in the limit of Δ/a->0. The radial distribution function and second virial coefficient have the forms

$$g(s) = H(r-2) + \frac{1}{6\tau}(r-2)$$

and (6.21)

$$A_2 = 4 - 1/\tau$$

The singularity in $g(r)$ at contact necessitates careful treatment of the non-equilibrium problem (Cichocki and Felderhof, 1990). The most direct approach is to solve separately for $P_2^*(r)$ within $2<r<2+\Delta/a$ and $P_2(r)$ for $2+\Delta/a<r$, apply the boundary conditions at $r=2$ to P_2^* and for $r\to\infty$ to P_2, and then require continuity of fluxes and

$$P_2 e^{\Phi_m/kT} = P_2^*$$

at $r=2+\Delta/a$.

For steady shear with $Pe<<1$ the non-equilibrium structure with far-field hydrodynamics yields the low shear viscosity

$$\frac{\eta_o}{\mu} = 1 + \frac{5}{2}\phi + \left(\frac{49}{10} + \frac{1}{\tau}\frac{\frac{1}{4\tau}-1}{\frac{1}{6\tau}+1}\right)\phi^2 \tag{6.22}$$

Exact hydrodynamics do not suppress the singular attraction at contact but alter the viscosity to

$$\frac{\eta_o}{\mu} = 1 + \frac{5}{2}\phi + \left(6.0 + \frac{1.9}{\tau}\right)\phi^2 \tag{6.23}$$

Not surprisingly, the near-field hydrodynamics affect the result profoundly. According to (6.22) the attraction increases the viscosity only for $A_2 = 4\text{-}1/\tau < 0$, while (6.23) indicates a monotonic increase in viscosity as A_2 decreases from the hard sphere value (Figure 8).

For $Pe>>1$ the pair conservation function conforms to the hard sphere limit and the interparticle stress follows from (6.9). For simple shear the stress assumes the form

$$\sigma_{12} = \mu\gamma\left(1 + \frac{5}{2}\phi + \frac{49}{10}\phi^2\right) + \frac{6}{\pi^2}\frac{\Phi_m}{a^3}\phi^2 \tag{6.24}$$

This result offers the intriguing possibility of extracting the strength of the attraction from viscosity measurements. Rarely, however, do measurements conform fully to this limit, since the stress must be sufficient to separate all pairs and attain the hard sphere viscosity. Partridge and Goodwin (Goodwin 1990) did achieve this limit, with polystyrene latices bearing a surfactant layer and flocculated by dispersion forces, and observed a Bingham stress linear in ϕ^2. However, the prefactor exceeds than predicted by an order of magnitude, perhaps another effect of near-field hydrodynamic interactions.

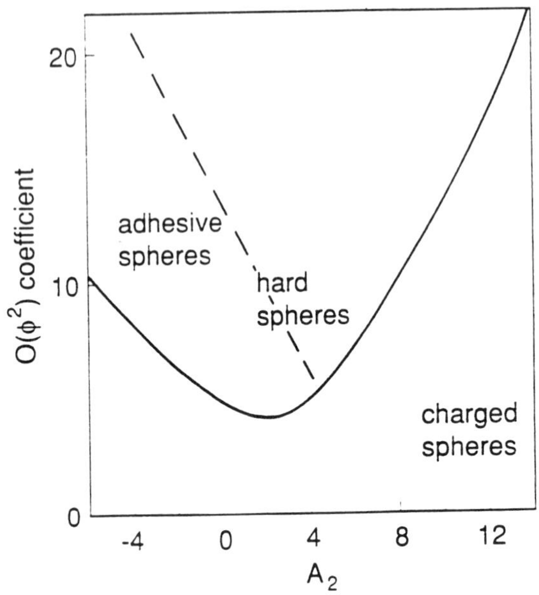

Figure 8 $O(\phi^2)$ coefficient in the low shear viscosity as function of the second virial coefficient for charged spheres (A_2>4) and adhesive hard spheres (A_2<4): (——) with far-field hydrodynamic interactions, (----) with exact hydrodynamic interactions.

17.7 SELF-CONSISTENT FIELD THEORY

As noted several times, a realistic non-equilibrium theory for concentrated dispersions requires an accurate and tractable description of many-body hydrodynamics. Two promising approaches for dealing with these problems have emerged over the past decade. Mazur and coworkers (Mazur 1987) address the many-body problem directly, cleverly summing multiple reflections to obtain the far-field interactions among many particles. Their approach is appealing for its elegance and success in predicting some macroscopic properties at moderate volume fractions. However, the complexity, the need for higher order distribution functions, and the uncertain accuracy at high volume fractions present problems. Alternatively, Brady (1992) adds in a pairwise fashion the forces induced by particle motion, calculated via two-particle friction coefficients that capture lubrication forces near contact, and correct the far-field behavior through a multipole expansion. The demonstrations of accuracy up to close packing are convincing, but the approach requires knowledge of the full N-particle configuration obtainable only from simulations.

Fortunately, a third possibility more suitable for statistical mechanical theories exists, the older and generally less respectable self-consistent field theories. Here we adopt this approach with the objective of predicting the dependence on volume fraction and interparticle potential of those dynamical properties sensitive only to the equilibrium distribution function, e.g. the high frequency limiting viscosity and the short-time self-diffusion coefficient. These will provide a basis for calculating the ensemble averaged pair hydrodynamics defined in Section 2.

In this section we derive the self-consistent field approximation from the governing equations through the preaveraging formalism of Hinch (1977), identifying spheres with slip boundary conditions as the appropriate model for interactions in concentrated dispersions. This description will be limited to the high frequency limiting viscosity, but the derivation of the short-time self-diffusion coefficient would follow similar lines.

17.7.1 Preaveraged Equations

The derivation begins with the momentum equation, averaged over the positions of N-1 spheres with one fixed at the origin (Hinch 1977),

$$-\nabla p(\boldsymbol{x})+\mu\nabla^2\boldsymbol{u}(\boldsymbol{x}) = \int_{r\geq a} P_2(r) \oint_{|x'-r|=a} \boldsymbol{\sigma}(\boldsymbol{x}'|\boldsymbol{r})\cdot\boldsymbol{n}'\,\delta(\boldsymbol{x}'-\boldsymbol{x})d\boldsymbol{x}'\,d\boldsymbol{r} \tag{7.1}$$

The lefthand side is the Stokes operator for the solvent and the righthand side represents the extra forces due to the other particles. Boundary conditions on the velocities or stresses at the sphere surface and at infinity complete the formulation.

To develop the self-consistent field approximation we follow Hinch's second renormalization, but with constant effective viscosity outside the sphere at the origin such that

$$-\nabla p(\boldsymbol{x})+\mu\left(1+\left(\frac{\eta_{eff}}{\mu}-1\right)\right)\nabla^2\boldsymbol{u}(\boldsymbol{x}) = \int_{r\geq a}\{P_2(r) \oint_{|x'-r|=a} \boldsymbol{\sigma}(\boldsymbol{x}'|\boldsymbol{r})\cdot\boldsymbol{n}'\,\boldsymbol{\delta}(\boldsymbol{x}'-\boldsymbol{x})d\boldsymbol{x}'$$

$$-2\left(\eta_{eff}-\mu\right)\boldsymbol{e}(\boldsymbol{r})\cdot\nabla\boldsymbol{\delta}(\boldsymbol{x}-\boldsymbol{r})\}d\boldsymbol{r} \tag{7.2}$$

Now the Stokes operator corresponds to a uniform effective medium surrounding the sphere and the righthand side corrects for the difference between the extra stress in the effective medium and that generated by the other spheres. The equation remains exact, but requires that the distribution of stress at $\boldsymbol{x}$ on the surface of a particle centered at $\boldsymbol{r}$ and $P_2(r)$ be known.

17.7.2 Closure Approximations

Self-consistent field theories assume the righthand side to be negligible and proceed to solve the boundary value problem and calculate stresses, ultimately determining the medium viscosity from the bulk stress. However, this approximation for the effective medium, invoked by Brinkman (1947), fails at high volume fractions because the righthand side remains too large to neglect. To correct for this we introduce a slip boundary condition at $\boldsymbol{x}\cdot\boldsymbol{x}=a^2$

$$\frac{\mu}{\ell}(\boldsymbol{u}-\boldsymbol{n}\boldsymbol{u}\cdot\boldsymbol{n})=2\left(\eta_{eff}-\mu\right)(\boldsymbol{e}\cdot\boldsymbol{n}-\boldsymbol{n}\boldsymbol{n}\cdot\boldsymbol{e}\cdot\boldsymbol{n}) \tag{7.3}$$

and later will adjust the slip length ℓ to minimize the residual on the righthand side of (7.2). So the task is to characterize the effective medium by determining η_{eff} and ℓ.

The first step is to solve the boundary value problem without the residual, calculate the hydrodynamic contribution to the bulk stress, and then add the contribution from the righthand side of (7.2) to obtain

$$\boldsymbol{\Sigma}^H=-p\boldsymbol{\delta}+2\mu\boldsymbol{E}+5\eta_{eff}S_o(\lambda)\phi\boldsymbol{E}+n\int_{r\geq a}\left\{ng(r)S_{12}'(r)-2\left(\eta_{eff}-\mu\right)e'(r)\right\}d\boldsymbol{r} \tag{7.4}$$

Note the distinct terms from the pure solvent, the extra stress in the effective medium with

$$\lambda=5\ell\left(\eta_{eff}-\mu\right)/a\mu$$

and

$$S_o(\lambda)=\frac{1+\frac{2}{5}\lambda}{1+\lambda}$$

accounting for slip, and the residual with $S_2(\boldsymbol{r})$ denoting the excess dipole induced in the sphere at $\boldsymbol{r}$ derived from the stress distribution $\boldsymbol{\sigma}(\boldsymbol{x}|\boldsymbol{r})$. Since both this dipole and the rate of strain $\boldsymbol{e}'(\boldsymbol{r})$ decay as r^{-3}, the last term converges if and only if

$$\lim_{r\to\infty}\left[nS_2(r)-2\left(\eta_{eff}-\mu\right)e'(r)\right]=\left[5\phi\eta_{eff}S_o\left(\frac{(\eta_{eff}-\mu)\ell}{\mu a}\right)-2\left(\eta_{eff}-\mu\right)\right]e'(r)$$
$$=0$$

dictating

$$\eta_{eff}-\mu=\frac{5}{2}\phi\eta_{eff}S_o\left(\frac{(\eta_{eff}-\mu)\ell}{\mu a}\right) \tag{7.5}$$

Defining

$$nS'_{12} = 5\phi\eta_{eff}S_o(e'+s')$$

with

$$\lim_{\phi\to 0} nS'_{12} = \tfrac{20\pi}{3}a^3\mu\left(e'_o + s'_o\right)$$

permits us to write

$$\Sigma^H = -p\delta + 2\mu E + 5\phi\eta_{eff}S_o\{E(1+k\phi)+\delta E\} \tag{7.6}$$

with

$$k\phi E = n\int\left(g_o s'_o + (g_o - 1)e'_o\right)dr$$

$$\delta E = n\int\left((g-g_o)s' - g_o\left(s'_o - s'\right) + (g-1)e' - (g_o-1)e'_o\right)dr$$

The $O(\phi^2)$ term can be evaluated explicitly for spheres with the exact s'_O and g_O from the pair values.

Next we propose to choose the slip length by eliminating the residual δE. The key lies in suitably approximating s', for example, by considering two spheres with slip boundary conditions interacting through the effective medium. In this case the appropriate Faxen's law (Kim and Karrila 1990)

$$\lim_{r\to\infty} s' = \frac{1}{10(1+2\lambda)}\nabla^2 e'$$

provides the far-field form but integrates to zero in (7.6). So at least the next term must be calculated, in order to determine the slip coefficient as a function of volume fraction with

$$\lim_{\phi\to 0} \ell = 0$$

Early work by Felderhof (1977) may prove helpful.

With the properties of the effective medium thus determined by (7.5) and $\delta E = 0$ the bulk stress yields the high frequency limiting viscosity as

$$\frac{\eta'_\infty}{\mu} = 1 + \tfrac{5}{2}\phi(1+k\phi)\eta_{eff}S_o(\lambda) \tag{7.7}$$

Note that the virial expansion, for example, for hard spheres

$$\frac{\eta'_\infty}{\mu} = 1 + \tfrac{5}{2}\phi\left(1 - \tfrac{1}{2}\phi\right)\left(1 + \tfrac{5}{2}\phi\right) = 1 + \tfrac{5}{2}\phi + 5\phi^2$$

will conform to the exact result. The behavior with increasing volume fraction depends on the increase in the slip coefficient. Comparison of the predictions with the data of van der Werff, et al. (1989) and the predictions of Phillips, et al. (1989) will test the approximation.

17.7.3 Pair Interactions

The final step is to derive the scalar functions governing the interactions between two spheres in the effective medium. Here we illustrate the approach with the function $G(r)$ for motion along the line of centers due to equal but opposite forces on the two spheres

$$(\boldsymbol{U}_2 - \boldsymbol{U}_1)\cdot\frac{\boldsymbol{r}}{r} = \frac{1}{3\pi\mu a}G(r)\boldsymbol{F}\cdot\frac{\boldsymbol{r}}{r} \tag{7.8}$$

Far-field interactions will be governed by generalized forms of the well-known Faxen's laws for rigid spheres with no slip. Indeed one can easily show that spheres with slip behave as viscous droplets with the ratio of external to internal viscosity replaced by λ. Thus a sphere held fixed in fluid with undisturbed velocity $\boldsymbol{u}(\boldsymbol{x})$ will experience a force (Kim and Karilla 1990)

$$\boldsymbol{F} = 3\pi\eta_{eff}a\frac{1+\frac{2}{3}\lambda}{1+\lambda}\left(1+\frac{a^2}{6\left(1+\frac{2}{3}\lambda\right)}\nabla^2\right)\boldsymbol{u}(\boldsymbol{x}) \tag{7.9}$$

and acquire a induced dipole indicated by Faxen's law. Together with the corresponding velocity fields generated by a force-free sphere subject to the force and dipole these lead to

$$\begin{aligned} G(r) &\approx \frac{\mu}{\eta_{eff}}\left\{\frac{1+\lambda}{1+\frac{2}{3}\lambda} - \frac{3}{2}\frac{a}{r} + \frac{1}{1+\frac{2}{3}\lambda}\left(\frac{a}{r}\right)^3 - \frac{15}{4}\frac{(1+2\lambda)}{(1+\lambda)}\left(\frac{a}{r}\right)^4\right\} \\ &\equiv \frac{D_s^o(\phi)}{D_o}G_\infty(r) \end{aligned} \tag{7.10}$$

Note that slip does weaken the interaction, though not dramatically.

Near contact lubrication analyses yield

$$\lim_{r\to 2} G(r) = 2g_o(\lambda)(r-2) \equiv G_o(r)$$

with $g_o(\lambda)$ yet to be determined.

These limiting forms resemble those from computer simulations (Figure 9), i.e. conforming to the pair function near contact and with similar asymptotic form but reduced prefactor in the far-field. Note, however, that the functional form of $G(r)$ at intermediate separations does not suggest a simple interpolation. So construction of the complete function will take further thought. Nonetheless the general approach seems a reasonable one.

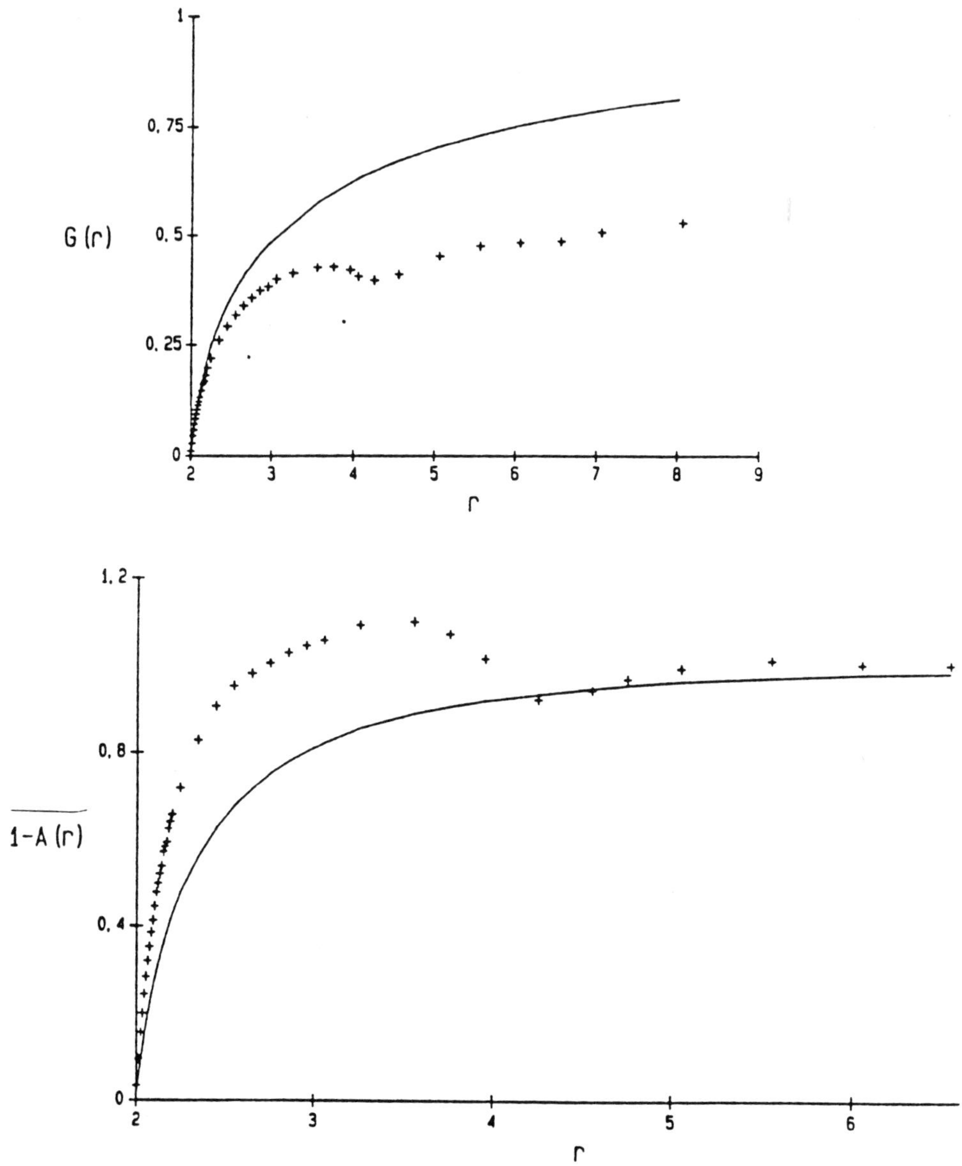

Figure 9 Mobilities for motion of two spheres along lines of center due to interparticle force (G) and shear flow (1-A) (Brady, et al. 1988): (——) pair functions, (+) Stokesian dynamics simulations at ϕ_a=0.45.

17.8 SUMMARY

This review addresses the dynamics of colloidal dispersions broadly. The primary accomplishments are twofold:

(1) combining the classical Smoluchoski formulation, integrated to the pair level, with recently derived closures for the three-body couplings to provide a framework completely analogous to integral equation theories for equilibrium structure and properties and

(2) applying Hinch's method for averaging the Stokes equations to identify a self-consistent field closure for calculating the necessary hydrodynamic functions, for two particles interacting amongst many others.

Thus a basis now exists for (a) comparing approximate treatments in the literature and (b) constructing more accurate theories in the future. Some comparisons and examples are offered.

ACKNOWLEDGEMENTS

The author thanks members of the Colloid Science Group at Bristol University for their hospitality during his sabbatical; Unilever, the National Science Foundation, and the International Fine Particles Research Institute for financial support; Robert Evans for a stimulating discussion at the outset of this effort; and several colleagues, particularly, Rudolph Klein and Ubo Felderhof, for a continuing stream of preprints on these topics.

REFERENCES

Ackerson, B.J., Correlations for interacting Brownian particles. *J. Chem. Phys.* (1976) **64** 242-6.

Ackerson, B.J., Correlations for interacting Brownian particles II. *J. Chem. Phys.* (1978) **69** 684-90.

Ackerson, B.J., C.G. de Kruif, N.J. Wagner, and W.B. Russel, Comparison of small shear flow-small wave vector static structure factor data with theory. *J. Chem. Phys.* (1989) **90** 3250-3.

Arauz-Lara, J.L., and M. Medina-Noyola, Sum rules for S_{ab} for two diffusing species. *Physica* (1983) **122A** 547-62.

Arauz-Lara, J.L., and M. Medina-Noyola, Theory of self-diffusion of highly charged Brownian particles. *J. Phys. A* (1986) **19** L117-21.

Barker, J.A., and D. Henderson, Perturbation theory and equation of state for fluids: I. The square well potential. *J. Chem. Phys.* (1967a) **47** 2856-61; II. a successful theory of liquids. *J. Chem. Phys.* (1967b) **47** 4714-21.

Batchelor, G.K., The stress system in a suspension of force-free particles. *J. Fluid Mech.* (1970) **41** 545-70.

Batchelor, G.K., The effect of Brownian motion on the bulk stress in a suspension of spherical particles. *J. Fluid Mech.* (1977) **83** 97- 117.

Batchelor, G.K., Diffusion in a dilute polydisperse system of interacting spheres. *J. Fluid Mech.* (1983) **131** 155-75.

Batchelor, G.K., and J.T. Green, The determination of the bulk stress in a suspension of spherical particles to order c^2. *J. Fluid Mech.* (1972) **56** 401-72.

Baxter, R.J., Percus-Yevick equation for hard spheres with surface adhesion. *J. Chem. Phys.* (1968) **49** 2770-4.

Bengtzelius, U., Dynamics of a Lennard-Jones system close to the glass transition. *Phys. Rev. A* (1986) **34** 5059-69.

Bengtzelius, U., W. Götze, and A. Sjölander, Dynamics of supercooled liquids and the glass transition. *J. Phys. C: Solid State* (1984) **17** 5915-34.

Bengtzelius, U., and L. Sjögren, Changes of thermodynamic quantities at the glass transition. *J. Chem. Phys.* (1986) **84** 1744-51.

Bird, R.B., H.C. Curtiss, R.C. Armstrong, and O. Hassager, *Dynamics of Polymeric Liquids Vol. II Kinetic Theory* (1987) Wiley.

Blachford, J., F.S. Chan, and D.A.I. Goring, Secondary electroviscous effect. Paths of approach of two charged spheres in a viscous medium. *J. Phys. Chem.* (1969) **73** 1062-65.

Boon, Y.P., and S. Yip, *Molecular Hydrodynamics* (1980) McGraw-Hill.

Brady, J.F., Stokesian dynamics simulation of particulate flows. *Particulate Two-Phase Flow* (1992) (ed. M.C. Roco) Chap. 26. Butterworth.

Brady, J.F., G. Bossis, and C. Mathis, Shear-induced structure in colloidal suspensions I. Numerical simulations. *J. Colloid Interface Sci.* (1988) **126** 1-15.

Brinkman, H.C., A calculation of the viscous force exerted by a flowing fluid on a dense swarm of particles. *Appl. Sci. Res.* (1947) **A1** 27-34.

Buscall, R., J.W. Goodwin, M.W. Hawkins, and R.H. Ottewill, Visco-elastic properties of concentrated latices I. Methods of examination; II. Theoretical analysis. *J. Chem. Soc. Far. Trans.* I (1982) **78** 2873-87; 2889-99.

Cichocki, B., Linear kinetic theory of a suspension of interacting Brownian particles. I. Enskog renormalization, II. Density-density autocorrelation function. *Physica* (1988) **148A** 165-90, 191-207.

Cichocki, B., and B.U. Felderhof, Diffusion coefficients and effective viscosity of suspensions of sticky hard spheres with hydrodynamic interactions. *J. Chem. Phys.* (1990) **93** 4427-32.

Cichocki, B., and B.U. Felderhof, Self-diffusion of Brownian particles with hydrodynamic interaction and square step or well potential. *J. Chem. Phys.* (1991a) **94** 563-68.

Cichocki, B., and B.U. Felderhof, Diffusion of Brownian particles with hydrodynamic interaction and hard core repulsion. *J. Chem. Phys.* (1991b) **94** 556-62.

Cichocki, B., and W. Hess, Comment: On the memory function for a colloidal liquid. *J. Chem. Phys.* (1986) **85** 1705-6.

Cichocki, B., and W. Hess, On the memory function for the dynamic structure factor of interacting Brownian particles. *Physica* (1987) **141A** 475-88.
D'Aguanno, B., R. Klein, and N.J. Wagner, Charge and size polydispersity effects on the scattering properties and the high-frequency elasticity of colloids. *J. Chem Soc. Far. Trans.* (1990) **87** 379-90.
de Kruif, C.G., E.M.F. van Iersel, A. Vrij, and W.B. Russel, Hard sphere colloidal dispersions: viscosity as a function of shear rate and volume fraction. *J. Chem. Phys.* (1986) **83** 4717-25.
Deutch, J.M., and I. Oppenheim, Molecular theory of Brownian motion of several particles. *J. Chem. Phys.* (1971) **54** 3547-55.
Dhont, J.K.G., Steady state viscosity coefficients of charged colloidal sys tems. *Physica* (1987) **146A** 541-81.
Dhont, J.G.K., J.C. van der Werff, and C.G. de Kruif, The shear thinning behavior of colloidal dispersions I. Some theoretical considerations. *Physica A* (1989) **160** 195-204.
Dieterich, W., and I. Peschel, Memory function approach to the dynamics of interacting Brownian particles. *Physica* (1979) **95A** 208-24.
Doi, M., and S.F. Edwards, *The Theory of Polymer Dynamics* (1986) Oxford University Press.
Ermak, D.L., and J.A. McCammon, Brownian dynamics with hydrodynamic interactions. *J. Chem. Phys.* (1978) **69** 1352-60.
Evans, R., Microscopic theories of simple fluids and their interfaces. *Liquids at Interfaces*, (1989) Les Houches, Session XLVIII, 1988 (ed. J. Charvolin, J.F. Joanny, and J. Zinn-Justin) 1-98.
Evans, R., Density functionals in the theory of non-uniform fluids. *Inhomogeneous Fluids* (1991) (ed. D. Henderson) Dekker (to appear).
Felderhof, B.U., Hydrodynamic interactions between spheres. *Physica* (1977) **89A** 373-84.
Felderhof, B.U., Brownian motion and creeping flow on the Smoluchoski time scale. *Physica* (1987) **147A** 203-18.
Felderhof, B.U., and R.B. Jones, Cluster expansion of the diffusion kernal of a suspension of interacting Brownian particles. *Physica* (1983a) **121A** 329-44; Diffusion in hard sphere suspensions. *Physica* (1983b) **122A** 89-104.
Felderhof, B.U., and R.B. Jones, Linear response theory of the viscosity of suspensions of spherical Brownian particles. *Physica* (1987) **146A** 417-32.
Genz, U., and R. Klein, Collective diffusion of charged spheres in the presence of hydrodynamic interaction. *Physica* (1991) **171A** 26-42.
Goodwin, J.W., Rheology of ceramic materials. *Ceramic Bulletin* (1990) **69** 1694-8.
Goodwin, J.W., and R.W. Hughes, The flow properties of concentrated colloidal suspensions. *Mat. Res. Soc. Symp. Proc.* (1990) **177** 187-98.
Goodwin, J.W., and Ottewill, R.H., Properties of concentrated colloidal dispersions. *J. Chem. Soc. Far. Trans.* (1991) **87** 357-69.

Götze, W., and L Sjögren, a-relaxation near the liquid-glass transition. *J. Phys. C* (1987a) **20** 879-94.
Götze, W., and L Sjögren, The glass transition singularity. *Z. Phys. B* (1987b) **65** 415-27.
Götze, W., and L Sjögren, Scaling properties in supercooled liquids near the glass transition. *J. Phys. C* (1988) **21** 3407-21.
Hansen, J.-P., and J.B. Hayter, A rescaled MSA structure for dilute charged colloidal dispersions. *Mol. Phys.* (1982) **46** 651-6.
Hansen,J.-P., and I.R. McDonald, *Theory of Simple Liquids.* (1986) 2nd Edition, Academic Press.
Happel, J., and H. Brenner, *Low Reynolds Hydrodynamics* (1983) Martinus Nijhoff, Boston.
Hess, W., and R. Klein, Generalized hydrodynamics of systems of Brownian particles. *Adv. Phys.* (1983) **32** 173-283.
Hinch, E.J., An averaged-equation approach to particle interactions in a fluid suspension. *J. Fluid Mech.* (1977) **83** 695-720.
Hunter, R.J., *Foundations of Colloid Science, Vol. I,* (1987) Oxford University Press.
Kim, B., and G.F. Mazenko, Fluctuating nonlinear hydrodynamics, dense fluids and the glass transition. *Adv. Chem. Phys.* (1990) **78**, 129-70.
Kim, S., and S.J. Karrila, *Microhydrodynamics: Principles and Selected Applications.* (1990) Butterworths.
Kremp, D., W. Ebeling, H. Krienke, and R. Sändig, HNC-type approximation for transport processes in electrolyte solutions. *J. Stat. Phys.* (1983) **33** 99-106.
Krieger, I., Rheology of monodisperse latices. *Adv. Colloid Interface Sci.* (1972) **3** 111-36.
Leutheusser, E., Dynamical model of the liquid-glass transition. *Phys. Rev. A* (1984) **29** 2765-73.
Machta, J., I. Oppenheim, and I. Procaccia, Statistical mechanics of stationary states V. Fluctuations in systems with shear flow. *Phys. Rev. A* (1980) **22** 2809-17.
Mazur, P., Many-sphere hydrodynamic interactions. *Far. Disc. Chem. Soc.* (1987) **83** 33-46.
Medina-Noyola, M., Long-time self diffusion in concentrated colloidal dis persions. *Phys. Rev. Lett.* (1988) **60** 2705-8.
Murphy, T.J., and J.L. Aguirre, Brownian motion of N-interacting particles I. Extension of the Einstein diffusion relation to the N-particle case. *J. Chem. Phys.* (1972) **57** 2098-2104.
Nägele, G., M. Medina-Noyola, R. Klein, and J.L. Arauz-Lara, Time dependent self diffusion in model suspensions of highly charged Brownian particles. *Physica* (1988) **149**A, 123-63.
O'Brien, R.W., A method for the calculation of the effective transport properties of suspensions of interacting particles. *J. Fluid Mech.* (1979) **91** 17-39.
Ohtsuki, T., Dynamical properties of strongly interacting Brownian particles I. Dynamic shear viscosity. *Physica* (1981) **108A** 441-58.
Ohtsuki, T., Dynamical properties of strongly interacting Brownian particles II. Self-diffusion. *Physica* (1982) **110A** 606-16.

Ohtsuki, T., Dynamical properties of strongly interacting Brownian particles III. Binary mixtures. *Physica* (1983) **122A** 212-30.
Phillips, R.J., J.F. Brady, and G. Bossis, Hydrodynamic transport properties of hard sphere dispersions I. Suspensions of freely mobile particles. *Phys. Fluids* (1988) **31** 3462- 77.
Pusey, P.N., and R.J. Tough, Particle interactions. *Dynamic Light Scattering: Applications of Photon Correlation Spectroscopy* (1985) (ed. R. Pecora) Plenum Press, 85-179.
Rallison, J.M., and E.J. Hinch, The effect of particle interactions on dynamic light scattering from a dilute suspension, *J. Fluid Mech.* (1986) **167** 131-68.
Rogers, F.J., and D.A. Young, New, thermodynamically consistent, integral equation for simple fluids. *Phys. Rev. A* (1984) **30** 999-1007.
Ronis, D., Theory of fluctuations in colloidal suspensions undergoing steady shear flow. *Phys. Rev. A* (1984a) **29** 1453-60.
Ronis, D., Debye-Scherrer light scattering patterns for dilute colloidal suspensions undergoing shear flow. *Phys. Rev. Lett.* (1984b) **52** 473-5.
Ronis, D., Configurational viscosity of dilute colloidal suspensions. *Phys. Rev. A* (1986) **34** 1472-80.
Russel, W.B., The rheology of suspensions of charged rigid spheres. *J. Fluid Mech.* (1978) **85** 209-32.
Russel, W.B., A review of the role of colloidal forces in the rheology of suspensions. *J. Rheology* (1980) **24** 287-317.
Russel, W.B., High frequency limiting elastic modulus of dispersions of hard spheres. *63rd Colloid and Surface Science Symposium*, (1989) Seattle.
Russel, W.B., and A.P. Gast, Nonequilibrium statistical mechanics of concentrated colloidal dispersions: Hard spheres in weak flows. *J. Chem. Phys.* (1986) **84** 1815-.26.
Russel, W.B., D.A. Saville, and W.R. Schowalter, *Colloidal Dispersions* (1989) Cambridge University Press.
Scherwinski, K., Two approximations for the classical triplet distribution function. *Mol. Phys.* (1990) **70** 797-809.
Sonnenburg, J., D. Kremp, and R. Sändig, On the density dependence of the self-diffusion coefficient of interacting particles. *Mol. Phys.* (1991) **72** 873-92.
Stone-Masui, J., and A. Watillon, Electroviscous effects in dispersions of monodisperse polystyrene latices. *J. Colloid Interface Sci.* (1968) **28** 187-202.
van der Werff, J., C.G. de Kruif, C. Blom, and J. Mellema, Linear viscoelastic behavior of dense hard-sphere dispersions. *Phys. Rev. A* (1989) **39** 795-807.
van Megen, W., and I. Snook, Equilibrium properties of dispersions. *Adv. Colloid Interface Sci.* (1984) **21** 119-94.
van Megen, W., S.M. Underwood, and P.N. Pusey, Dynamics of hard spherical colloids from the fluid to the glass. *J. Chem. Soc. Far. Trans.* (1991) **87** 395-402.
Verlet, L., and J.J. Weis, Equilibrium theory of simple liquids. *Phys. Rev. A* (1972) **5** 939-52.

Wagner, N.J., and W.B. Russel, Nonequilibrium statistical mechanics of concentrated colloidal dispersions: Hard spheres in weak flows with many-body thermodynamic interactions. *Physica* (1989) **155A** 475-518.

Wagner, N.J., and W.B. Russel, Light scattering measurements of a hard sphere suspension under shear. *Phys. Fluids A* (1990) **2** 491-502.

Zwanzig, R., and R.D. Mountain, High frequency elastic moduli of simple fluids. *J. Chem. Phys.* (1965) **43** 4464-71.

CHAPTER 18

PARTICLE DISPERSION BY ORGANIZED TURBULENT STRUCTURES

C. Crowe
J. N. Chung
T. R. Troutt

18.1 INTRODUCTION

Particle dispersion in turbulent flows play a key role in a wide variety of industrial and energy conversion processes. Thermal power plants utilizing pulverized coal operate by blowing the coal particles into the combustion chamber. The degree of dispersion of the particles through the turbulence produced by the feed jets is important to maintaining high combustion efficiency and controlling soot formation. The efficiency of liquid fueled combustion devices often depends on the degree of dispersion of the fuel droplets. The spread of forest fires is a result of firebrands being dispersed in large scale turbulent structures created by the atmospheric winds. The efficient operation of a wire- and-plate electrostatic precipitator is dependent on the degree of dispersion of the particulates in the turbulence created by the supporting plate structure. Obviously the study of particle dispersion and mixing in organized, large turbulent structures represents an very important area to support the improved design of many industrial processes.

The use of the words, mixing, dispersion and diffusion, are often used in terchangeably in the literature which can lead to confusion. In this paper, the following definitions will be used. The word *diffusion* will be restricted to processes taking place on the molecular scale such as the mixing of two gaseous species. *Dispersion*, on the other hand, will refer to processes associated with the droplets or particles (dispersed phase). This is a general term which actually describes three identifiable processes; mixing, spreading and transport as shown schematically in Figure 18-1. *Mixing* is the interpenetration of particle path lines towards the production of a homogeneous mixture. For example, if the incoming particle stream had two types of initially separated particles, A and B, mixing would ultimately yield a homogeneous mixture of A and B. *Spreading*, on the other hand, is the movement of the particles into regions which previously were devoid of particles. In this case, the number density of the particles will decrease as the particles spread. Spreading also includes the situation when particles move inward (negative spreading). *Transport* will refer to the bulk movement of the particles from one spatial location to the other. In principle, this movement could take place with neither spreading nor mixing. If the particles emerged from a jet and the field expanded with no crossing of the pathlines, spreading and transport would be occurring simultaneously. In a turbulent, particle-laden jet, spreading ,transport and mixing typically take place at the same time. Particles flowing in the convergent section of a nozzle tend to concentrate on the centerline of the nozzle. This focusing phenomenon is a combination of transport and negative spreading.

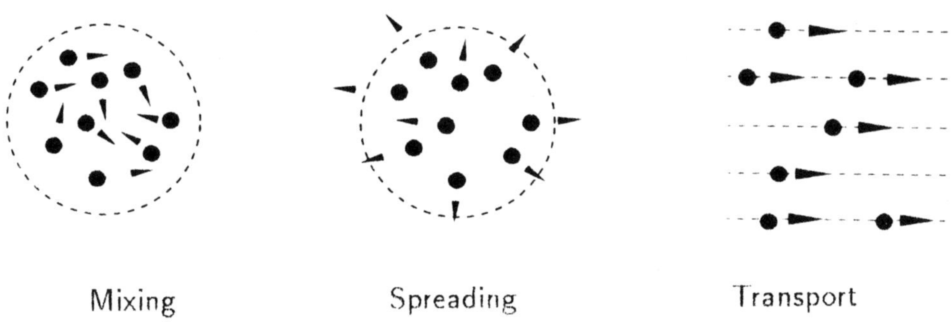

FIGURE 18-1 *The Mixing, Spreading and Transport Processes*

Turbulence is the key mechanism responsible for the mixing and spreading of particles in a fluid flow. The classic approach to the treatment of turbulence is to regard the turbulent flow as a mean velocity with a superimposed fluctuation component. The statistical description of the fluctuation velocities define the intensity and the isotropic nature of the field while the spatial variation of these quantities relates to the homogeneity. Numerous studies have been reported on predictions and measurements of particle spread in homogeneous and near-homogeneous turbulence. Analytic and numerical studies (Wang 1990) utilize representations of the turbulence energy spectrum to predict particle motion and dispersion. Also a primary goal in the study of "classical" turbulence is to develop more universally viable models for the Reynolds shear stress which can be used with confidence in computational fluid dynamics.

A more recent area of turbulence research involves the study of organized structures which are presumably produced by nonlinear dynamic flow processes. Brown and Rosko (1974) were the first researchers to clearly demonstrate the existence of organized structures produced in the flow at the downstream edge of a splitter plate. These structures first appear due to a Kelvin-Helmholtz instability and later merge in pairs to produce even larger structures as the flow proceeds downstream. This phenomenon has been referred to as "vortex pairing" by Winant and Brownand (1974). Similar large scale, organized structures occur at the edge of a jet issuing into a quiescent medium. In this case the vortex structures are circumferential rings around the centerline of the jet. Wakes also produce large scale structures by a mechanism often referred to as "vortex shedding". A classic case is the vortices shed behind a cylinder. In this well known and much studied case, the vortices produced by each side of the cylinder have opposite rotations and there is no strong vortex pairing mechanism. In this field of turbulence, researchers are more concerned with understanding the flow structure and the dynamic interactive mechanisms and less concerned with the statistical description of the turbulence field. Excellent reviews are available in the literature, such as Laufer (1975), on the early works associated with organized structures evolving from free shear layers.

Even though there is a clear division between the efforts of those researchers involved with organized, dynamic structures and those with the small scale, statistical approach, there is the commonality that, even at the small scale, the turbulent flow may also consist of small recognizable vortices which interact, grow and dissipate.

The traditional approach to treating particle dispersion in turbulence is to regard the process as a Fickian diffusion process. The problem centers on finding a suitable dispersion coefficient to use in the continuity equation. Many studies have been reported in the literature on identifying parameters which control the dispersion coefficient. Early works suggest the importance of the ratio of the particle response time to a time associated with the turbulent motion of the fluid. Sophisticated analytical studies (Reeks 1977) using models for the turbulent energy spectrum in homogeneous, isotropic and stationary turbulence show that, after sufficient time, the diffusion of "heavy" particle can exceed that of a fluid particle. Yudine (1959) was the first person to identify the "crossing trajectory" effect ; that is, the reduced diffusion when the particles move with a velocity different that the mean flow velocity. This crossing trajectory effect has been confirmed by experimentation. Still, one cannot select, with confidence, a dispersion coefficient which adequately models particle dispersion in an arbitrary turbulent flow field.

A popular method for including particle dispersion in a numerical model is the Monte Carlo approach (Faeth 1987). In this approach a local velocity fluctuation is selected randomly for each particle. The flow velocity experienced by the particle is the sum of the mean velocity and the fluctuation velocity. The fluctuation is assumed to persist until either the eddy decays or the particle passes from the eddy (crossing trajectory effect). This approach, popular because of its easy implementation, does not correctly model the turbulent field and changes in empirical coefficients often have to be made to yield good correlations with experimental data.

The approach for treating particle dispersion by organized large scale structures is quite different. A conceptual model for the dispersion of particles is the entrainment of particles by the vortex structure and ensuing centrifuging effect as the particles rotate with the organized structures. A pictorial representation is shown in Figure 18-2. Small particles will closely follow the the flow path lines and the particles will spread and mix as the fluid. Larger particles will be entrapped by the vortices and centrifuged beyond the vortex effecting a spreading rate exceeding that of a fluid particle. Even larger particles will pass through the vortex structures with little change in their trajectory. Thus, the potential exists to have a situation where the particles spread more than the carrier fluid.

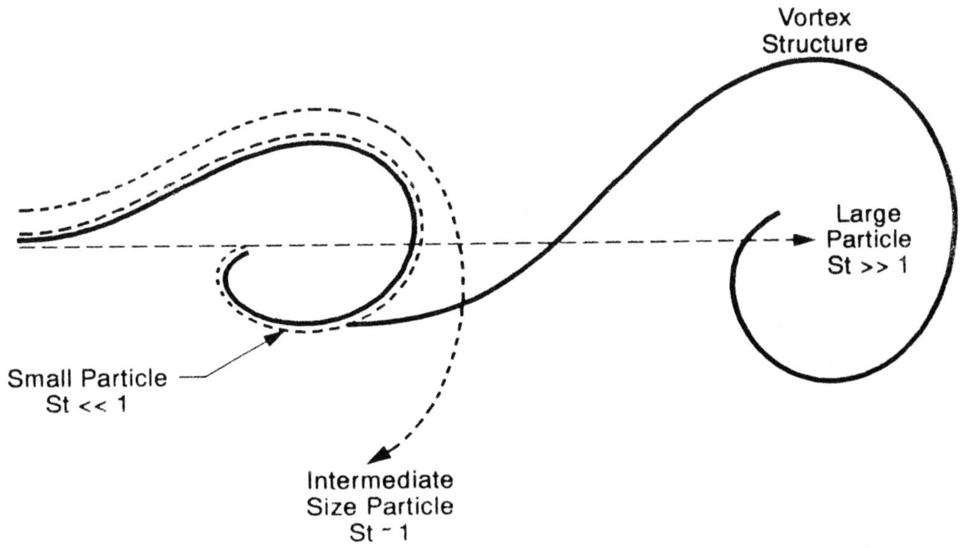

FIGURE 18-2 *Conceptual Model for Stokes Number Effect on Particle Spreading in Organized Structures*

In an effort to establish those parameters which affect the particle dispersion in large scale, organized structures, Crowe *et al.* (1985) proposed a Stokes number defined as

$$St = \frac{\tau_s}{\tau_F} \tag{18.1}$$

where τ_s is the aerodynamic response time of a particle and τ_F is the time associated with the motion of the large scale structure. The aerodynamic response time is scaled as

$$\tau_s = \frac{\rho_p d^2}{18\mu} \tag{18.2}$$

where ρ_p is the material density of the particle, d is the partic le diameter and μ is the dynamic viscosity of the carrier fluid. The flow time is modeled as

$$\tau_F = \frac{\delta}{U} \tag{18.3}$$

where δ is the size of the structure and ΔU is the velocity difference across the shear layer. Thus the Stokes number becomes

$$St = \frac{\rho_p d^2 \Delta U}{18\mu\delta} \tag{18.4}$$

If the $St << 1$ then the particle response time is much less that the flow time and the particles will faithfully follow the fluid path lines. If $St \sim 1$, the particles will tend to be centrifuged by the structures giving rise to an augmented spread rate. For $St >> 1$, the particles have insufficient time to respond to changes in fluid velocity and will continue in near rectilinear trajectories. Both numerical and experimental studies have verified the appropriateness of Stokes number as the scaling factor for particle dispersion in large scale structures.

The early experimental studies on the dispersion of particles in organized structures have been presented in a previous review (Crowe and Chung and Troutt 1988) and will only be mentioned here for purposes of continuity. Kamalu *et al.* (1988) and Wen (1990) report on flow visualization and point measurements of particle released in a plane shear layer. The size of the solid particles corresponded to a Stokes number near unity. The flow visualization (using smoke lines for the carrier phase) revealed that the particles are centrifuged to the peripheries of the vortices leaving a near particle void at the vortex center, as postulated by the conceptual model. Laser Doppler measurements of the cross stream particle velocities showed that the velocities were outward indicative of the centrifuging effect of the rotating structures.

A similar experiment has been reported by Kobayashi *et al.* (1987) who used solid particles from 30 to 50 microns in diameter and fine water drops for visualization of the gaseous phase. Laser Doppler measurements indicated that the particles moved away from the center while the water droplets moved towards the center which agreed with the previous data (Kamalu *et al.* 1987) and further corroborated the conceptual model. Other experiments by Lazaro and Lasheras (1989) using light scattering from a laser sheet also show that the vortical structures control the particle dispersion process.

The first numerical simulation of particle dispersion in an organized structure was carried out by Gore *et al.* (1985) using a vortex street model proposed by Stuart (1967). A subsequent model (Gore *et al.* 1989) used a pseudo-spectral analysis of the flow produced by a shear layer. Both these models showed the tendency of the particles to achieve maximum spread for a Stokes number near unity.

The majority of numerical models for particle dispersion in a large scale structure have been based on the discrete vortex method. Chien and Chung (1988a) used this method for simulation of a vortex pair which simulates the flow field development in a coordinate system which moves with the average velocity of the two velocity streams. They integrated the equations for particle motion in this flow field to predict particle trajectories. They found that the particles dispersed more than a fluid particle for a range of Stokes number between 0.5 and 5. Chien and Chung also applied the discrete vortex method to calculating particle trajectories in a developing shear layer and found that the particles dispersed more than the carrier fluid at a Stokes number of unity. They also observed the tendency of the particles to accumulate at the peripheries of the vortices which was also demonstrated in flow visualization studies. Chung and Troutt (1989) applied the same analysis to axisymmetric flows and found similar trends.

Laitone (1983) applied the discrete vortex method to modeling the wake of a circular cylinder with particles suspended in the flow. The results also showed the tendency of the particles to be centrifuged by the large scale vortices but were not correlated with Stokes number. Davis *et al.* (1983) modeled gas particle flow around a square cylinder in a two-dimensional channel using a finite-difference scheme. They showed the trend for particles to be captured by the large scale structures when the Stokes number was of order 0.1.

It is instructive to point out that the dispersion coefficient has little significance in modeling the dispersion of particles in organized turbulent structures. The use of dispersion coefficient implies the dispersion of particles is due to a concentration gradient. However, particle dispersion in organized turbulent structures is primarily due to transport and spreading so the concept of a dispersion coefficient is inapplicable. For example, if particles were introduced into a flow in the base region of a wake, analyses based on the average flow properties and dispersion coefficient would predict a monotonically varying particle concentration distribution across the flow with a maximum concentration near the centerline. A transient flow analysis with a Stokes number near unity will predict locally high concentrations of particles near the peripheries of the periodically shed vortices. The latter case represents the true physical phenomenon.

Only a few studies have appeared in the literature on the motion of bubbles in large scale turbulent structures. Thomas *et al.* (1983) used the discrete vortex method for bubble transport in a two-dimensional mixing layer. The simulation did not yield the observed pairing phenomena of the structures but the results did show the entrainment of the bubble by the structures.

The analysis of the motion of bubbles or particles in a developing shear layer is performed by integrating the equations of motion of the dispersed phase in the flow field produced by the organized structures. The equation of motion for a bubble or a solid is quite different leading to dissimilar dispersion characteristics in the flow structures.

The equation of motion for a particle equates the product of the mass and acceleration to the forces acting on the particle. The equation of motion for small particles (Reynolds numbers the order of unity) is expressed as(Maxey and Riley 1983),

$$m_d \frac{d\vec{u}_d}{dt} = -V\nabla p + 3\pi\mu d(\vec{u}_f - \vec{u}_d) + \rho_f \frac{V}{2}\left(\frac{d\vec{u}_f}{dt} - \frac{d\vec{u}_d}{dt}\right) + \frac{3}{2}d^2\sqrt{\pi\rho_f\mu}\int_0^t \frac{\left(\frac{d\vec{u}_f}{dt} - \frac{d\vec{u}_d}{dt}\right)}{\sqrt{t-t'}}dt' + m_d\vec{g} \tag{18.5}$$

where m_d and V are the mass and volume of the particle (dispersed phase) and the subscripts d and f refer to the particle and fluid respectively. The first term on the right is the buoyancy term due to the pressure gradient in the fluid. The second term is the drag force assuming Stokes drag law is valid. The third term is the virtual mass term which represents the additional force required to accelerate the fluid. The fourth term is the Basset term which accrues from the unsteady viscous flow around the sphere. The fifth term is the body force term which may be due to gravity or some other body force. One term not included is the Faxen force which arises because of the local curvature of the fluid velocity profile.

The pressure gradient in the fluid can be approximated by Euler's equation,

$$\nabla p = -\rho_f \frac{d\vec{u}_f}{dt} + \rho_f \vec{g} \tag{18.6}$$

Substituting this expression for pressure gradient into the equation for particle motion and dividing by the particle mass gives

$$\left(1 + \frac{\rho_f}{2\rho_d}\right)\frac{d\vec{u}_d}{dt} = \frac{3\rho_f}{2\rho_d}\frac{d\vec{u}_f}{dt} + \frac{1}{\tau_s}(\vec{u}_f - \vec{u}_d) + \frac{3}{\sqrt{2\pi}}\frac{1}{\tau_s}\sqrt{\frac{\rho_f}{\rho_d}}\int_0^t \frac{\left(\frac{d\vec{u}_f}{dt} - \frac{d\vec{u}_d}{dt}\right)}{\sqrt{t-t'}}dt' + \vec{g}\left(1 - \frac{\rho_f}{\rho_d}\right) \tag{18.7}$$

where τ_s is the aerodynamic response time of the particle.

For the motion of particles in a gas, the ratio of the fluid density to the material density of the particle is of the order of 10^{-3}. In this case, the only term s which are important are the drag force and the body force so the equation of motion reduces to

$$\frac{d\vec{u}_p}{dt} = \frac{f}{\tau_s}(\vec{u}_f - \vec{u}_p) + \vec{g} \tag{18.8}$$

where f is a factor to correct for the drag coefficient beyond the Stokes regime. Particles which become entrained in a rotating turbulent structure will be moved outward from the vortex core due to the centrifugal acceleration and the motion will be resisted by the aerodynamic drag. For very small particles, the aerodynamic response time becomes very small and the particle velocity approaches the fluid velocity, $\vec{u}_d \sim \vec{u}_f$.

The equation of motion for a bubble in a transient flow field represents the situation where the density of the surrounding fluid is much larger than the density of the gas in the bubble. In the case of bubble motion the Basset term is not significant so the forces in the equation of motion reduce to the virtual mass force, a viscous force and force due to pressure gradient in the fluid. The viscous force is same as that for a sphere at low Reynolds numbers (Wallis, 1969). The density for the dispersed phase in Eqn. (18.7) is the gas density inside the bubble so $\rho_d/\rho_f << 1$. In this case the equation of motion becomes

$$\frac{d\vec{u}_d}{dt} = 3\frac{d\vec{u}_f}{dt} + \frac{2}{\tau_s}(\vec{u}_f - \vec{u}_d) - 2\vec{g} \tag{18.9}$$

where now the response time τ_s is based on the density of the carrier phase (liquid) and not the material density of the gas in the bubble. It is interesting to note that the acceleration of a bubble can approach three times that of the carrier fluid.

The motion of the bubble is strongly influenced by the local acceleration of the fluid and this term becomes predominant as a bubble approaches the core of a vortex. The acceleration vector is towards the core of the rotating vortex, thus the bubble will move toward the core and will not be centrifuged outward as the solid particles.

The study of particle dispersion in organized, turbulent structures requires a coordinated experimental, numerical and analytic study. The processes controlling dispersion are interactive and transient which complicates both the experimental and numerical studies. The status of both numerical and experimental studies are reviewed in this paper.

18.2 INDUSTRIAL RELEVANCE

The improvement in technology to be derived by the study of particle dispersion in organized turbulent structures is significant. The essential element of a gas phase mixing process is the merging of two streams of unequal velocity. This may be a feed jet for furnace burning pulverized coal, a two-fluid atomizer, an eductor for pumping multiphase mixtures or a nozzle for a spray casting process. Dispersion in large scale structures is also important in spread of atmospheric pollution. Other applications can be found in biotechnology. Wake flows are also used industrially for mixing such as a flame holder for a combustion system. A dump combustor is another example. Ribs for structural support in electrostatic precipitators also create large scale turbulent structures which have significant effect on the collection efficiency.

The treatment of particle dispersion in these systems must be treated as a transient process in which dispersion is due primarily to transport and spreading. Improved predictions of the local concentration of droplets (and stochiometry) will lead to improved designs of combustion systems. Control of the large scale structures by acoustic, or other, means may lead to high combustion efficiency over a wide turn-down ratio. Understanding the mechanisms responsible for spatial droplet size distributions in the vicinity of a two-fluid atomizer may lead to configurations of atomizers to meet specific design requirements. Understanding particle dispersion in turbulent structures in electrostatic precipitators will enhance optimization of configurations for improved collection efficiency.

18.3 EXPERIMENTAL STUDIES

Several recent experimental efforts have been carried out to examine the importance of vortex structures in the turbulent dispersion of particles. These experiments have involved various shear flow situations such as plane mixing layers (Kamal u 1989; Kamalu *et al.* 1989; Lazaro & Lasheras 1989; Wen 1990), plane wakes (Tang *et al.* 1990), axisymmetric jets, (Longmire & Eaton 1990; Hardalupas *et al.* 1989) and recirculation regions, (Hardalupas *et al.* 1991)

Numerous qualitative and quantitative experimental techniques have been used to characterize the particle dispersion process in these investigations. These techniques have employed visualization efforts, using diffuse and laser light sheet illumination (Kamalu 1989; Kamalu *et al.* 1989; Lazaro & Lasheras 1989; Longmire & Eaton 1990; Wen *et al.* 1990), laser Doppler and laser phase Doppler anemometry techniques, (Kamalu *et al.* 1989; Longmire & Eaton 1990; Wen 1990), and laser attenuation and diffraction techniques, (Lazaro & Lasheras 1989).

These experimental investigations have focussed on situations involving solid particles or droplets in air flows such that the material density of the particulates is large compared to the fluid flow. Under this condition, as discussed previously in this chapter, the drag term dominates the particle force relation. This situation then simplifies the analysis and interpretation of the experimental results. At the present time there are very few reported results of either experimental or numerical investigations conce rning particle dispersion where the particle/fluid material density ratio is allowed to vary over a broad range.

The results of these experimental investigations appear to agree on several important issues relating to the dispersion of high density ($\rho_d/\rho_f >> 1$) particles by organized vortex structures.

1. The large scale organized vortex structures appear to be the important controlling mechanism in the dispersion process.
2. The particle dispersion patterns produced by the vortices may be highly non- homogeneous and strongly dependent on particle size or response time.
3. Intermediate or relatively large scale particles are found to be the most likely to be concentrated in specific regions near the outer edges of the vortex structures.

The importance of organized structures in the dispersion process is illustrated by instantaneous photographic results shown in Figure 18-3 for a bluff body wake flow containing solid particles with very small and intermediate Stokes number (Yang 1991). These photographs clearly illustrate the high degree of non-homogeneity associated with the organized vortex structures in a bluff body wake flow for particles with Stokes number near unity. These intermediate size particles appear to concentrate near the outer edges of the vortex structures in agreement with recent numerical predictions, whereas the small Stokes number particles tend to concentrate in the vortex cores. This tendency for the intermediate-size particles to migrate away from the center of the vortex structures in free shear layers has also been supported by time average laser velocity measurements of particle concentration profiles using laser interdata validation (Wen *et al.* 1990), laser attenuation measurements (Lazaro & Lasheras 1989), and digital particle counting techniques, (Longmire & Eaton 1990).

Time average particle concentration profiles obtained from laser velocimeter interdata times from experiments in a plane mixing layer are shown in Figure 18.4 from Wen *et al.* (1990). All distances are normalized by the momentum thickness (θ) at the end of the plate. The concentration profiles are highly non Gaussian indicating the complex movement by the vortex structures. In addition the strong effect on the concentration profiles by subharmonic forcing is evidence of not only the importance of vortex structures in the dispersion process but, in addition, their controlability.

Experimental information concerning relative lateral dispersion rates of high density particles in the same flow has been obtained over a range of Stokes numbers by several investigators (Kamalu *et al.* 1989; Hardalupas *et al.* 1991). The experimental results, in general, indicate that there can be a substantial increase in lateral particle dispersion as particle Stokes number increases from a very small value to order of unity for a given shear flow environment. However the experimental results at very high Stokes numbers are difficult to obtain because gravitational effects tend to become increasingly important for large particles or, conversely, high speed flows make experiments considerably more difficult . Thus it is difficult at this time to determine from the experimental results the exact shape of the particle dispersion function over a broad range comparable to the numerical predictions. Novel experimental techniques are needed to further explore this area such that an extended Stokes number range can be experimentally explored.

Another general area in which more experimental emphasis is needed is the precise identification of the important physical mechanisms involved in the particle dispersion process. These experiment s will need to examine both two-dimensional and three-dimensional particle-flow processes. This will probably require extensive three- dimensional and time dependent global visualization techniques. These types of experimental techniques are challenging and hold considerable promise for revealing new phenomena not predictable by current two-dimensional simulations. One further area in which experimental efforts are badly needed involves two-way coupling effects between the particles or droplets and the flow. These coupling effects involve mass, momentum and energy transfer making experimental investigations rather complex.

18.4 NUMERICAL STUDIES

Numerical simulations of particle dynamics in large-scale structure dominated turbulent flows have been advanced rapidly during the past

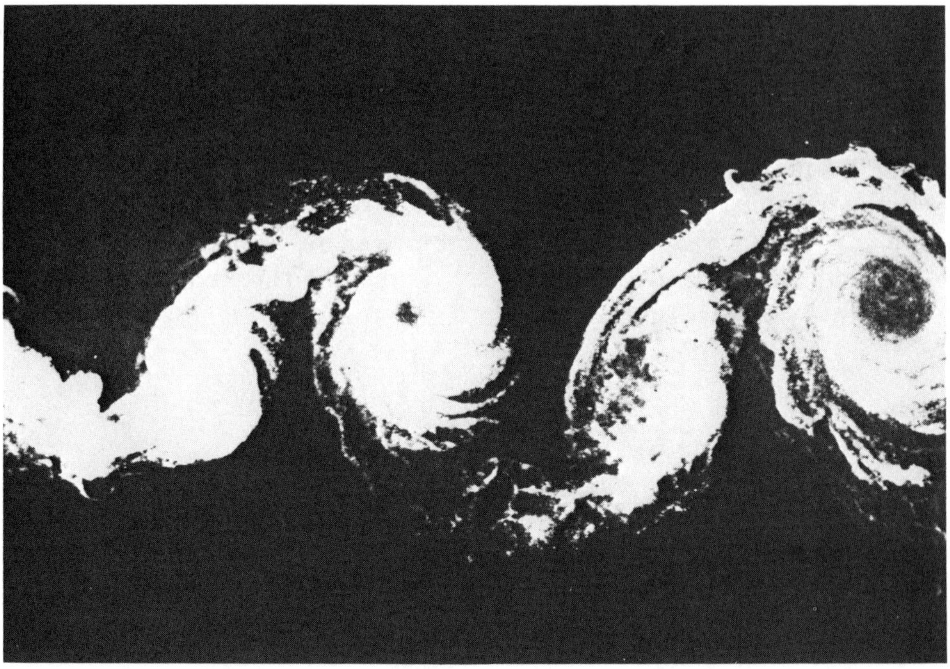

a)

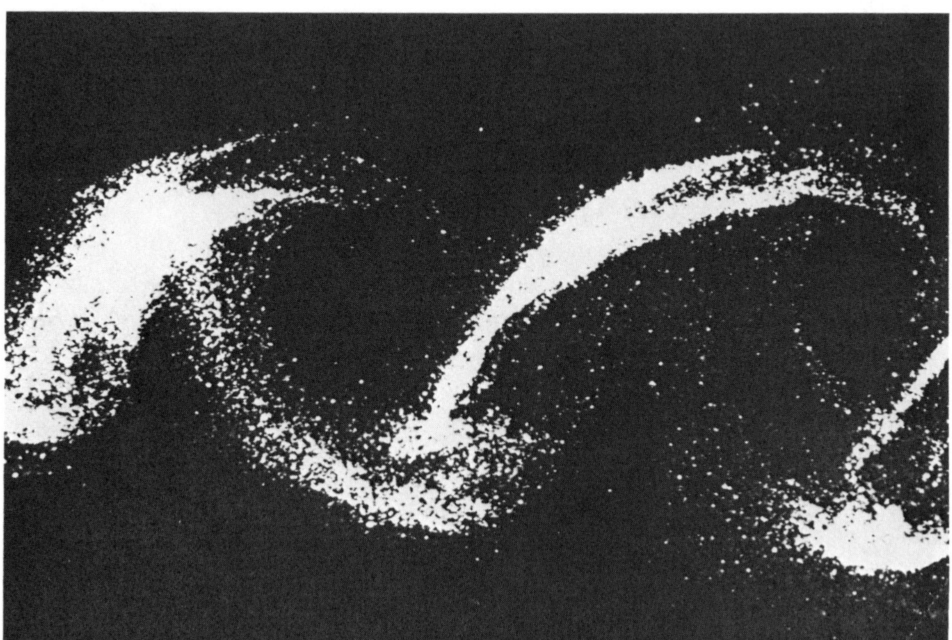

b)

FIGURE 18-3 *Instantaneous Photographs of Flow Field in a Wake; a) Particles with Small Stokes Number (smoke) and b) Particles with Intermediate Stokes Number.*

decade mainly due to the improvements in computer speed and storage capacity. Direct numerical simulations with the discrete vortex method and spectral methods have surpassed the traditional modeling approach which require closure schemes for the turbulence equations. The direct numerical simulations enable direct tracking of the particles through the flow field without the uncertainties associated with the stochastic approach. The previous work as been reviewed in an earlier publication (Crowe and Chung and Troutt 1988), and only the achievements over the past few years will be covered in this paper.

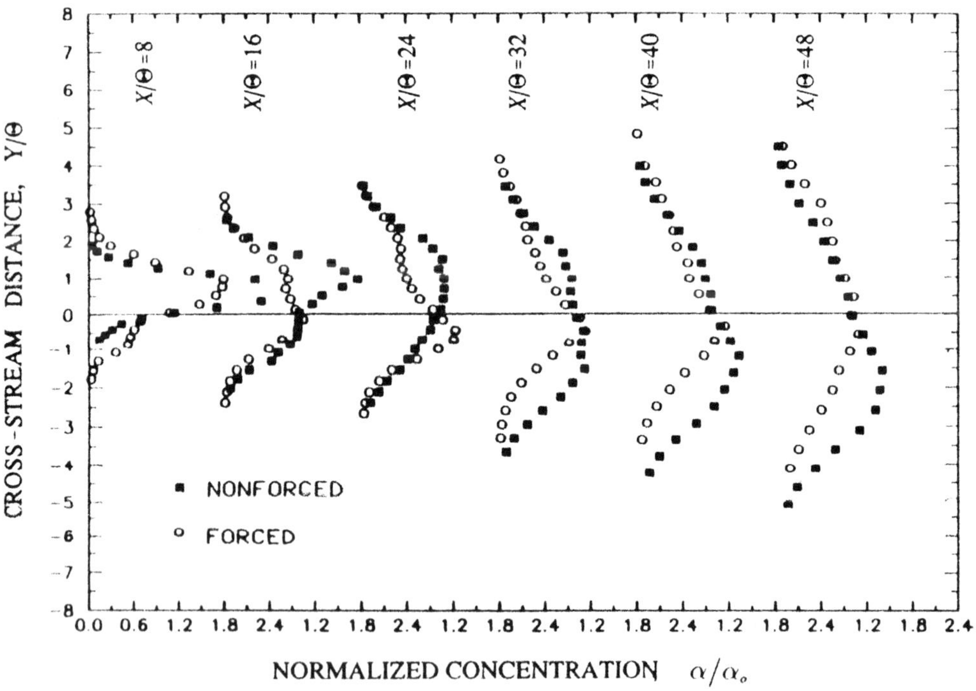

FIGURE 18-4 *Particle Concentration Profiles in a Natural Free Shear Layer and a Shear Layer Forced at a Subharmonic*

18.4.1 Flow Field Simulation

The discrete vortex method and spectral method are two numerical schemes that have been used to simulate the time dependent nature of the flow field. The flow field generated by these methods can then be used to track the particle motion.

Discrete Vortex Method

The discrete vortex method has been applied to numerous flow fields with impressive success. This method was found to be most effective for incompressible inviscid flows with strong vorticity elements such as shear layers, jets, and wakes. The discrete vortex method is not plagued by numerical diffusion and appears to work well for high Reynolds numbers. The most recent comprehensive review on the theoretical background and application of the discrete vortex method was given by Sarpkaya (1988). The application of this method to combustion research was provided by Ghoniem (1991) who, with his co-workers, has systematically advanced the vortex simulation technique for reacting shear flows. Two recent improvements are worth noting. The first concerns the confined flow field where the presence of the solid boundary walls should be included. In the past, almost all the simulations were carried out for unconfined flows where the presence of walls was neglected. Ghonien and Ng (1987) used the conformal mapping method to treat the irrotational potential flow field confined by the boundary. This scheme is error-free and relatively computationally inexpensive. The conformal mapping method produces exact solutions for the potential flow by defining an image system of vortex elements. Moreover, the scheme is grid independent and compatible with the discrete vortex method. It can provide higher resolution close to the boundaries and at points of separation where it is essential to determine accurately the flux of vorticity into the interior of the domain.

The second improvement involves a modified algorithm that makes the discrete vortex simulation more representative of the continuous vortex elements in the actual flow. Rangel and Sirignano (1989) demonstrated that for line vortices, redistribution of vorticity after each time step should be employed to control the numerical stability. They suggested that a new discrete vortex element be introduced in the array whenever the separation distance between any two adjacent vortices exceeds the initial separation by a given factor (5 percent in their calculation) while the total circulation must be conserved according to the Kelvin's theorem. For vortex blobs, they suggested merging or adding of vortices be implemented as two elemental vortices get too close or too far from each other, respectively.

Spectral Method

In certain areas of computational fluid dynamics, spectral methods have become the prevailing numerical tool for large-scale three-dimensional transient fluid mechanics problems. This is certainly the case for such three-dimensional applications as direct simulation of homogeneous turbulence and computation of transition in shear flows. The more rapid convergence and higher spatial resolution make spectral methods more attractive than the finite difference and finite element methods. Numerical spectral methods for partial differential equations were originally developed by meteorologists. Though this approach was proposed by Blinova (1944) and Haurwitz and Craig (1952), the first numerical computations were conducted by Siberman (1954). The expense of computing nonlinear terms remained a severe drawback until Orszag (1969) and Eliasen *et al.* (1970) developed efficient transform methods that still form the basis of many large-scale spectral computations. These methods and others used in fluid mechanics prior to 1970 are now termed spectral Galerkin methods. The fundamental unknowns in these methods are the expansion coefficients. The governing equations for these coefficients are derived by the techniques used in classical analysis. As the power of the digital computer increased, Kreiss and Oliger (1971) and Orszag (1972) suggested an alternative discretization, termed the spectral collocation technique, in which the fundamental unknowns are the solution values at selected collocation points, and the series expansion is used solely for the purpose of approximating the derivatives.

Many useful versions of spectral methods have been developed since 1971 and especially during the 1980s. The monograph by Gottlieb and Orszag (1977) describes the theory and applications developed prior to 1977. The following five years are covered by Voigt *et al.* (1984). Fluid-dynamic applications, especially multigrid techniques, are discussed by Zang and Hussaini (1985). The most recent review is given by Zang and Hussaini (1987). The books by Voigt *et al.* (1984) and by Canuto ıt et al. (1988) contain a detailed description of many spectral algorithms and present an exhaustive discussion of the theoretical aspects of these numerical methods.

To-date the spectral method has found limited use in modeling particle dispersion in organized turbulent structures. Gore *et al.* (1989) used the pseudo- spectral method developed by Riley and Metcalf (1980) to generate a flow field simulating vortex pairing. They found the same tendency for "heavy" particles to spread more than the fluid particles for Stokes number near unity. Squires and Eaton (1990) also applied the pseudo-spectral technique to predict the effects of particles on turbulence in homogeneous turbulence. Obviously spectral methods provide a much more accurate representation of the flow field than the discrete vortex method and will be especially useful to include three dimensional effects. Their use is dependent on the availability of new generation computers.

Another numerical scheme which holds considerable promise for modeling particle dispersion in turbulence is that developed by Gerz *et al.* (1989) and applied to gas-particle flows by Truesdell and Elghobashi (1991). The Navier-Stokes equations are expressed as second order finite difference equations and integrated in time using the Adams-Bashforth scheme. Pressure is solved for implicitly using a fast Poisson solver. This scheme will also be more useful with the availability of high speed computer systems.

18.4.2 Particle Dispersion in a Plane Shear Layer

As stated in the introduction the earlier numerical simulations for the plane mixing layer have all provided strong confirmations for the concept proposed by Crowe *et. al.* (1985). A unique dependence of the particle dispersion pattern on the Stokes number, the central element in the model, has been reinforced further by the recent work at Washington State University (Wen *et. al.* 1991; Wen 1990). A conformal mapping technique was adopted to simulate the walls of the test facility. Using conformal transformations, the flow region between the solid surfaces was mapped onto a semi-infinite plane and the reflection principle was applied to locate vortices to ensure that the walls are streamlines. Uniform size particles were released at a constant rate into the simulated flow stream from cross-stream positions at X=0. The instantaneous particle distribution patterns for Stokes numbers of 0, 0.1, 1.0 and 10 are shown in Figure 18-5. All distance variables are nondimensionalized by the width of the simulation field. Fluid tracer particles are represented by the zero Stokes number. The fluid tracer pattern is very similar to their smoke visualization pictures (Figure 18-3). Larger particles with a Stokes number of 0(0.1) tend to move towards the edges of the large scale structures. Particles with a Stokes number of 0(1) are clearly centrifuged to the boundaries of the vortex structures and yield the largest spread. Further increase in Stokes number results in a sharp decrease in the spread rate, as shown in the case of a Stokes number of 0(10). The limit of large Stokes number is pure transport of the particles. The particle dispersion patterns predicted by the numerical simulations are generally in good qualitative agreement with experimental visualizations (Figure 18-3). This indicates that the Stokes number, St, is indeed a controlling parameter for the dispersion process of particles in vortical flows dominated by large scale structures.

The relative particle flux distribution at several downstream locations for a Stokes number of 0(1) were also predicted by Wen *et. al.*

(1991) as shown in Figure 18-6. The simulated profiles show much of the same character as the experimental profiles; a double peak developed near the downstream position followed by the movement of a single peak gradually below the centerline.

These results suggest that time-averaged methods are inadequate for treating the particle dynamics and dispersion in flow fields dominated by large-scale vortex structures. The prediction from a time-averaged model would tend to be a monotonic variation in velocity across the shear layer. Assuming a Fickian diffusion process for the discrete phase would also lead to a monotonic change in particle concentration distribution. Obviously for Stokes numbers unity, the monotonic variation does not exist and time-averaged models do not yield the correct details of the mixing process.

An important discovery concerning the dispersion mechanism was also reported by Wen *et. al.* (1991) in which it was suggested that the dispersion sequence could be separated into two mechanisms, *stretching* and *folding* as shown in Figure 18-7. The first is the stretching associated with the high velocity gradient regions near the high and low speed boundaries of the adjacent vortex structures. This result is similar to that obtained by Squires and Eaton (1990) who found, using a pseudo spectral technique to model a homogeneous turbulent field, that the particles tend to concentrate in areas of high shear. The rotational nature of the large vortex structures tends to rapidly separate neighboring particles in these regions. Sheets of particles, which ride in the upstream side of the large scale vortices or are centrifuged into the high speed stream as shown in Figure 18-7, as a direct result of the stretching mechanism. The second mechanism appears to be associated with the pairing interactions of adjacent large scale vortices. These pairing interactions tend to fold the particle sheets down onto themselves towards the low speed side of the mixing layer as demonstrated in Figure 18-8. Further proof of the stretching and folding mechanism is given in Figure 18-9. The tendency for particles to concentrate in sheets located in the outer boundaries of the vortex structures and separated in a regular arrangement is shown in Figure 18-9a. Then, as shown in Figure 18-9b, the particle sheets are folded into the center of the mixing layer due to the vortex pairing process. These two figures thus directly illustrate the connection between the instantaneous vortex development and the stretching and folding mixing process. The stretching and folding mechanism discovered here for the two-phase mixing layer is qualitatively the same as the stretching and folding phenomena associated with chaotic dynamic systems in phase space. Moreover, Ottino (1990) has recently pointed out that the mixing of fluid interfaces may be closely connected to chaotic stretching and folding processes. The discovery by Wen (1991), however, represents the first time this mixing mechanism has been directly associated with particle dispersion in free shear flows.

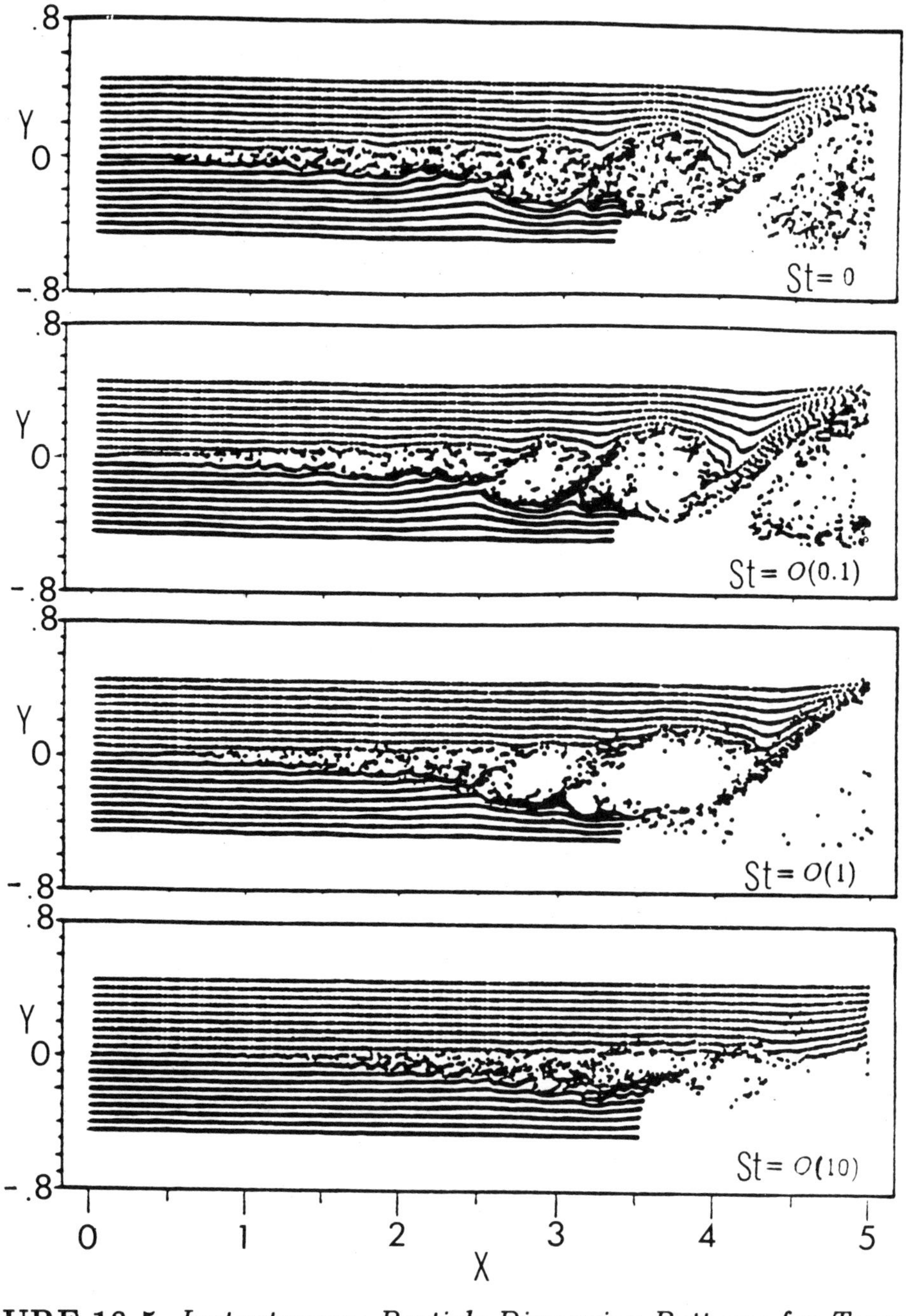

FIGURE 18-5 *Instantaneous Particle Dispersion Patterns for Tracers and Particles with Stokes Numbers of 0.1, 1 and 10 (from top to bottom)*

18.4.3 Particle Dispersion in the Wake

The wake flow is dominated by large scale vortex structures but differs from a shear layer in that vortex pairing is absent. A schematic representation of the wake vortex pattern is shown in Figure 18-10. Wake flows are characterized by the alternate shedding of opposite sign vortices from both sides of a body. Unlike the mixing layer, the large scale vortex structures do not pair and merge but rather move downstream in as eparate, but organized, pattern. Particle dispersion in the wake of a vertical or an inclined plate (Chein and Chung 1988b) and behind a cylinder (Laitone 1981) has been reviewed previously (Crowe *et al.* 1988).

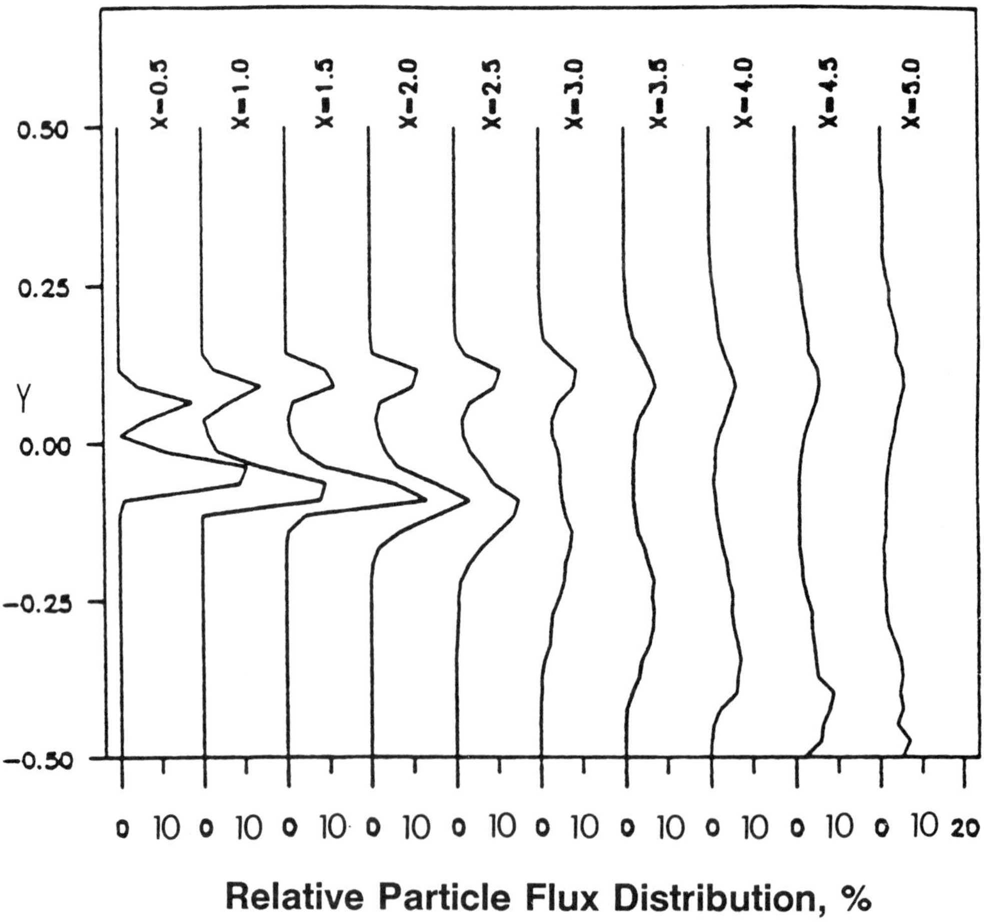

FIGURE 18-6 *Relative Particle Flux Distribution at Various Downstream Locations from Numerical Simulations with Stokes Number of Unity*

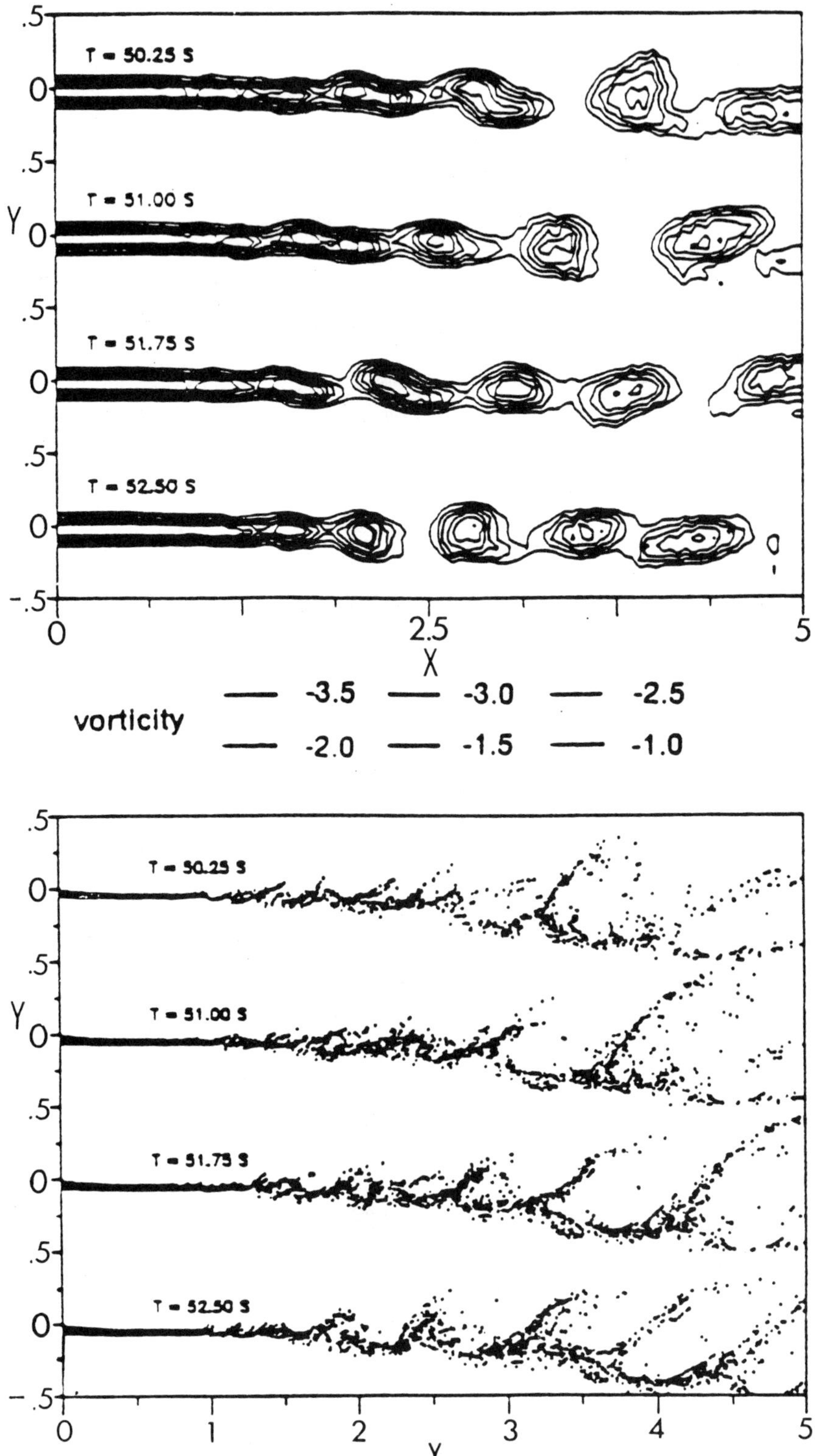

FIGURE 18-7a *Instantaneous Vorticity Contours and Particle Streaklines with Stokes Numbers of O(1) at Successive Times illustrating Stretching*

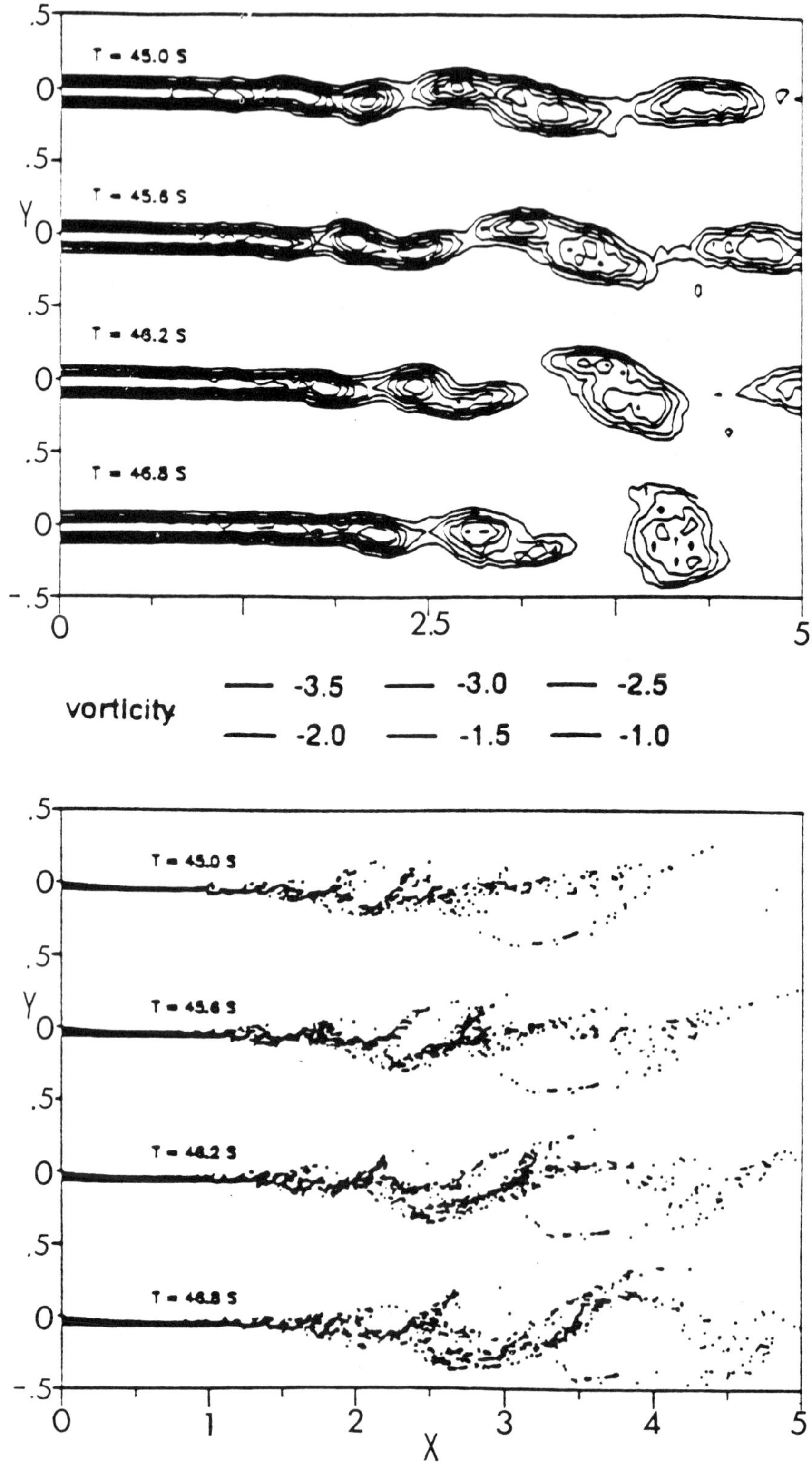

FIGURE 18-7b *Instantaneous Vorticity Contours and Particle Streaklines with Stokes Numbers of O(1) at Successive Times illustrating Folding*

Yang *et. al.* (1991) recently reported numerical simulation results for particle dispersion in a plane wake downstream of a thick plate. A numerical simulation using the discrete vortex method was carried out with the geometry of the body mapped onto a semi-infinite plane according to the Schwartz-Christoffel transformation. The vortex motion in the transformed plane is accompanied by reflected vortex motion in the other semi-infinite plane to ensure the boundary is a streamline. The flow field in the wake resulting from the discrete vortex simulation is given in Figure 18-11a. One notes the large scale vortex structures which compare well with the smoke visualization picture (Figure 18-11b) from Tang *et. al.* (1991).

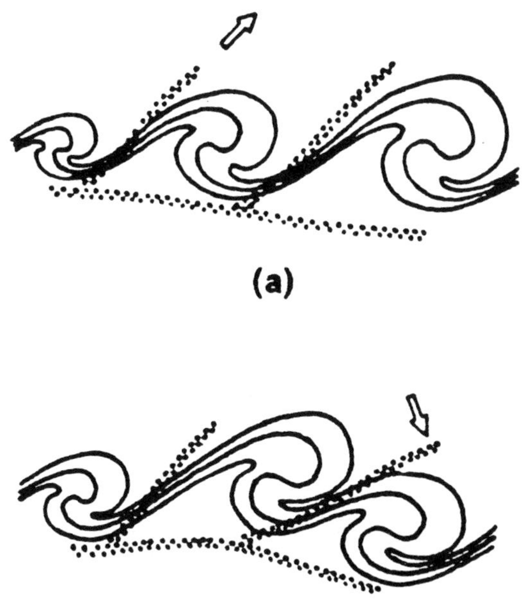

FIGURE 18-8 *Schematic diagram illustrating particle dispersion mechanism: a) stretching of particle streaklines and b) folding of particle streaklines*

Instantaneous particle patterns for particles released among the downstream edge of the bluff body for several Stokes numbers are shown in Figure 18-12. As expected, particle dispersion patterns in the wake are closely correlated with the Stokes number. At a Stokes number of 0.01 the particles follow the flow field showing the vortex structures which form and are convected downstream. At a Stokes number of unity, the particles are focused into thin regions at the edges of the vortex structures - a rather dramatic phenomenon, which was termed "focusing" by Tang *et. al.* (1991). As the Stokes number is further increased to 10, the entrainment of the particles in the vortices is delayed and they become more distributed through the wake region. Finally, at a Stokes number of 100, the particle trajectories demonstrate some waviness but little spreading or mixing as they move downstream. The focusing of the particles at a Stokes number near unity emphasizes that conventional wisdom concerning the utility of the wake flow as a mixing process is, under certain conditions, incorrect. In fact the large scale structures in the wake can lead to a very organized particle concentration pattern.

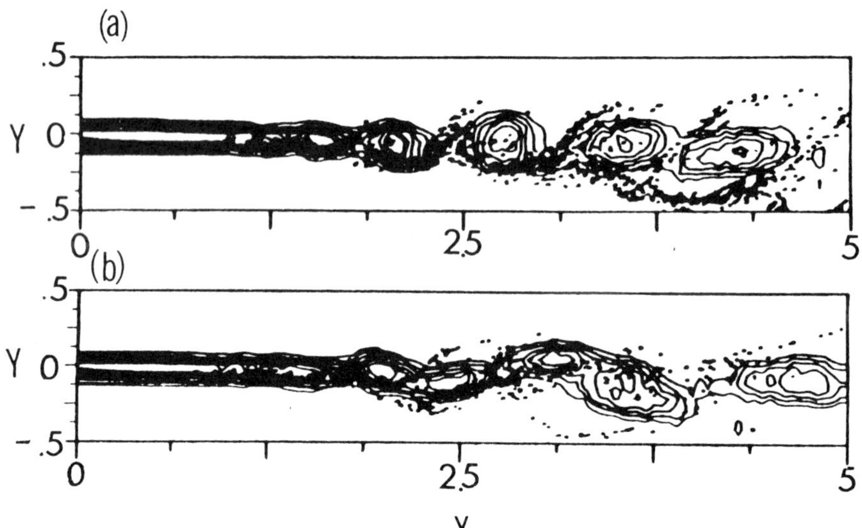

FIGURE 18-9 *Instantaneous superimposed vorticity contours and particle patterns: a) stretching process and b) folding process*

FIGURE 18-10 *Schematic Diagram of Plane Wake Flow*

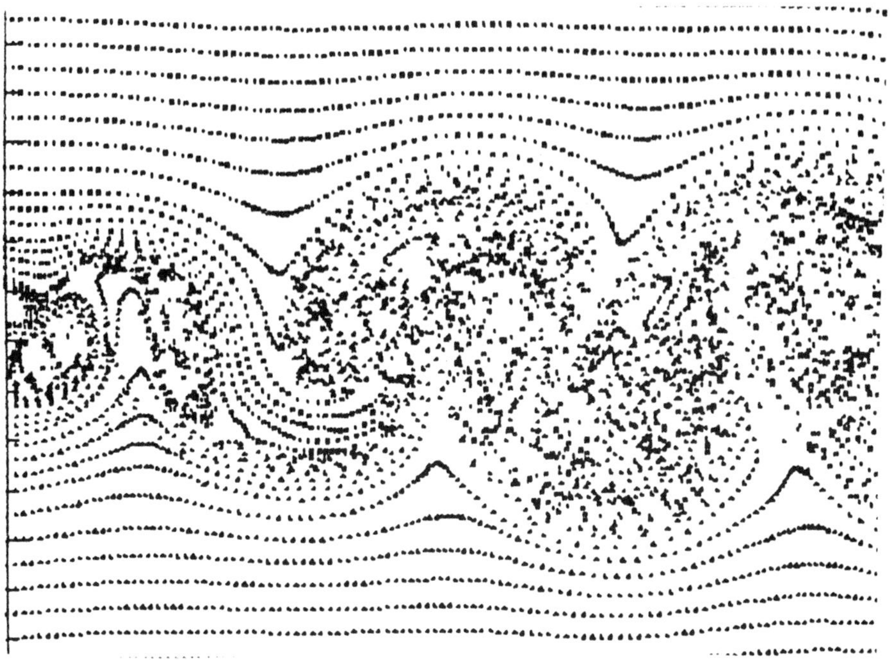

a)

b)

FIGURE 18-11 *Experimental and Numerical Flow Streakline Visualizations; a) Numerical Result and b) Experimental Result*

18.4.4 Correlation Dimension of the Dispersed Phase

The correlation dimension introduced by Grassburger and Procaccia (1983) was applied to the particle dispersion process in a mixing layer and a bluff body wake by Tang *et al.* (1991). One of the difficulties with the analysis of particle dispersion in nonstationary flows is the classification of the degree of particle dispersion or "focusing." The correlation dimension provides a quantitative measure of the degree of concentration of particles in the flow. It was introduced initially for the analysis of chaotic systems and defined as

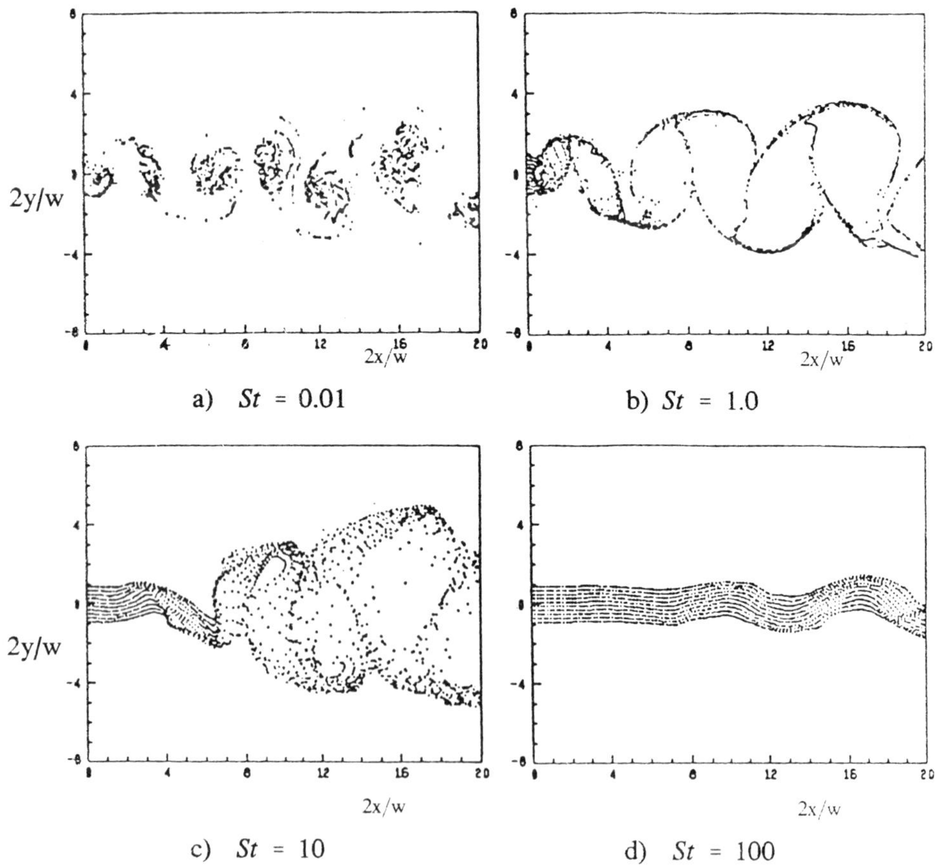

FIGURE 18-12 *Instantaneous Particle Dispersion Patterns from Numerical Simulation of Plane Wake Flow*

$$D = \lim_{l \to 0} \left(\frac{1}{\log l} \log \sum_i p_i^2 \right) \tag{18.10}$$

where l is a variable length scale and p_i is the probability that a particle is within a spatial cell of area l^2. The correlation dimension is obtain ed by graphing the sum $\sum_i p_i^2$, the correlation integral, as a function of length scale l on log-log coordinates and taking the slope of the resulting curve. For example, a correlation dimension of unity corresponds to all the particles lying on lines. Tang *et. al.* (1991) provided some typical examples for the correlation integral as a function of length scale for the plane mixing layer and the wake (Figures 18-13a and 18-13b) at a Stokes number of unity. The length l_0 is the reference length which, for these calculations, was the distance between successive vortices entering the flow field. The correlation dimension as a function of the Stokes number for both mixing layers and wakes was a lso given in Tang *et al.* (1991) and is shown here as Figure 18-14. Similar trends are shown for both shear flows. The minimum correlation dimensions correspond to Stokes numbers of the order of unity. As a matter of fact, the minimum correlation dimension falls right at the Stokes number of unity, the focusing point. The results imply that the particle dispersion patterns become increasingly organized as the particle time scales become closely matched to those of the large scale flow motions. For particle time scales above and below the flow time scale, the correlation dimension increases indicating a defocusing or more disorderly dispersion patterns.

18.4.5 Hydrodynamic Stability of a Two-Phase Flow

Stability analyses for single-phase shear flows such as mixing layers, wakes and jet have been under continuing and intensive development for the past several decades, Ho and Huerre (1984). The stability of two-phase flows has not received much at tention. The stability characteristics of a two-phase flow is vital to the design and efficient operation of the entire system. Pioneering contributions on two-phase stability have been made by Acrivos and his co-workers, Herbolzheimer (1983); Shaqfeh and Acrivos (1987); and Borahn and Acrivos (1988), for sedimentation flows inside an inclined settler. They developed linear theory models for a wide range of system parameters to predict the effects of particles on the stability characteristics during sedimentation. Recently, Yang *et. al.* (1990) reported some analytical and numerical results on the effects of particles on the spatially stability of two-phase mixing layers. A modified Rayleigh equation to allow the coupling between the two phases through drag interaction was developed. The mathematical model is based on the assumption that the mean flow profile can be approximated as that of the particle- free single-phase mixing layer and the two phases are in dynamic equilibrium at the start of the perturbation. The resulting eigenvalue problem was solved by numerical means. The major finding from the analysis is that the presence of the particles enhances the stability of the two-phase flow and decreases the amplification rate of the perturbation in the flow . The spatial amplification rate versus the angular frequency for various particle loadings is shown in Figure 18-15. The particle loading is represented by the parameter, A, defined as

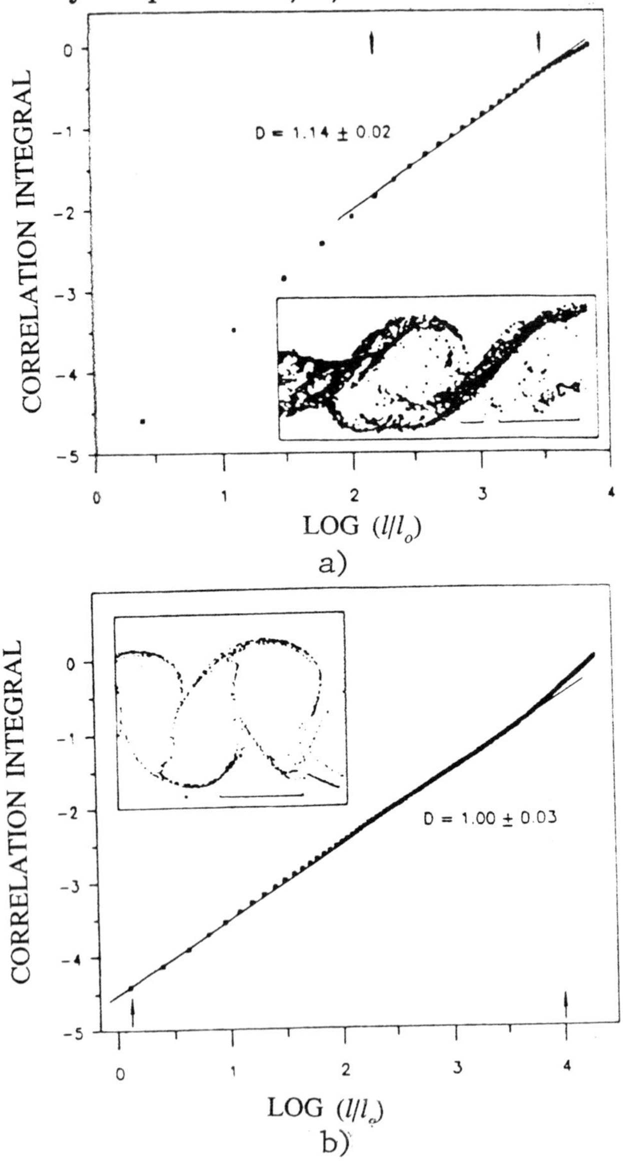

FIGURE 18-13 *Typical Logarithmic Plots for Computing Correlation Dimension; a) Plane Mixing Layer and b) Plane Wake*

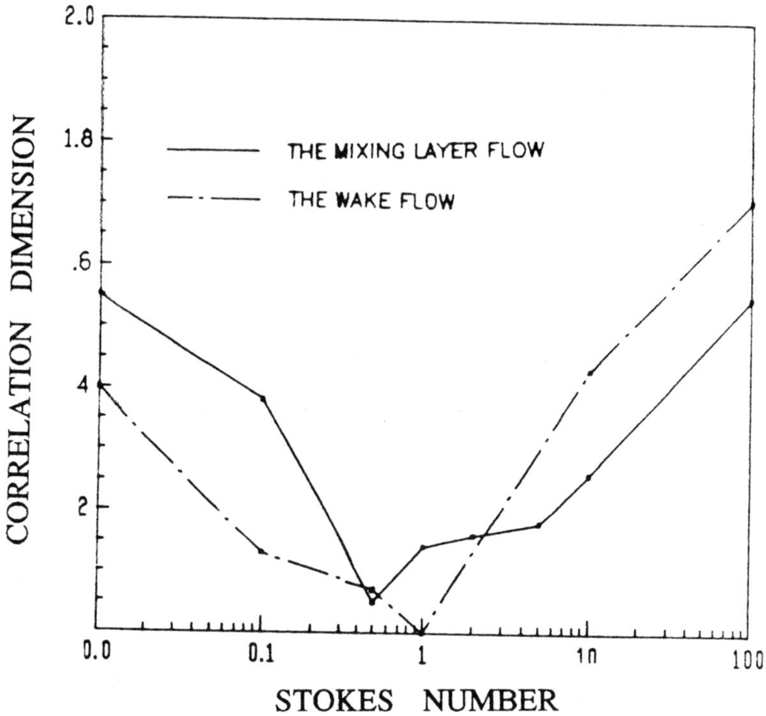

FIGURE 18-14 *Fractal Correlation Dimension for Particle Dispersion Patterns as a Function of Particle Time Scale Ratio*

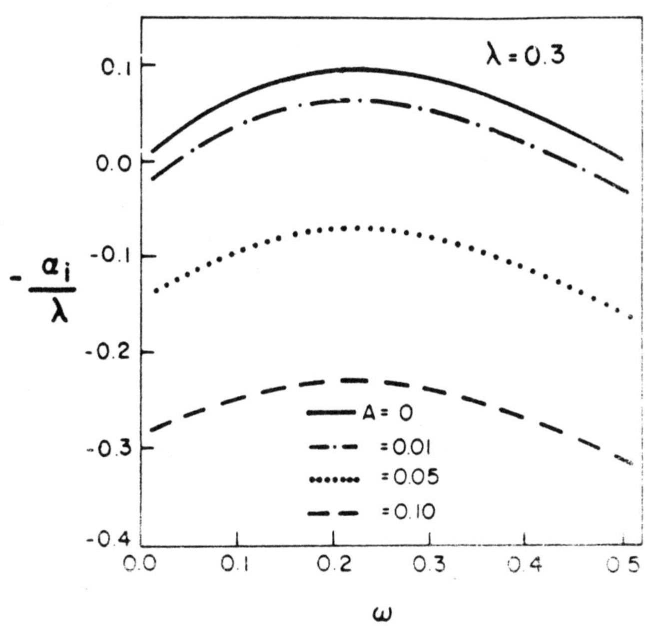

FIGURE 18-15 *The Spatial Amplification Rate, $-\alpha_i$, normalized by λ versus the Angular Frequency, ω for different A's*

$$A = \frac{2\lambda \rho'_p f}{\rho_f \; St}$$

where λ is equal to $\Delta U / 2\bar{U}$, ΔU is the free stream velocity difference and $\bar{U}$ is the average velocity of the two streams, ρ'_p is the particle bulk density, ρ_f is the material density of the fluid and f is the ratio of the actual drag on the particles to the Stokes drag. The Stokes number, St, is the ratio of the particle aerodynamic response time to the flow characteristic time. As indicated in Figure 18-16, the maximum values of $-\alpha_i/\lambda$ (the most amplified rate) occur at frequencies w, ranging between 0.21 and 0.22 for values of λ at 0.3, 0.5, 0.7 and 1.0. The most significant implication is that the presence of particles enhances the stability of the two-phase flow. For unstable flows, the presence of the particles attenuates the amplification rate, but affects angular frequency only slightly. The most amplified growth rates occur near the same angular frequency as that of the particle-free flow. The effectiveness of enhancement by the particles on the flow stability increases with decreasing λ, as shown in Figure 18-17, for a constant A value of 0.05 and various λ of 0.3 0.5, 0.7, and 1.0. In other words, the presence of the particles can mitigate the instability caused by the velocity shear (free-stream velocity difference) in the mixing layer. Figure 18-18 shows explicitly the effects of particles on the most amplified growth rate for $\lambda = 0.3, 0.5, 0.7$, and 1.0. For all these cases, the relationship is near linear for A ranging from 0 to 0.1. The effect of particles on the perturbation phase velocity is most significant for small values of the angular frequency. However, the presence of particle does not change the phase velocity at the angular frequency corresponding to the most amplified rate of the single-phase flow. The phase velocity of the perturbation is slower compared to particle-free flows for modes with frequencies below the maximum amplified frequency, while it is the opposite for modes at frequencies above the maximum amplified frequency. The profiles of streamwise and cross-stream velocity fluctuations are also modified by the existence of particles. The magnitude of effects is related closely to the particle loading. Also for two-phase mixing layers, the temporal growth rate decreases linearly with increasing A, the particle loading. The temporal and spatial results are consistent except for the phase velocity at frequencies apart from that of the most amplified mode.

The amplification of the spatial disturbances in the mixing layer stability analysis of Yang *et. al.* (1990) were assumed to vary sinusoidally in time and to be damped or amplified exponentially as they were convected downstream. Huerre and Monkewitz (1985) suggested that the mixing layer is convectively unstable if the free stream velocity ratio, λ, is less than 1.315. For convectively unstable flows, the spatial stability theory can be applied to predict the stability characteristics of the flow according to Yang *et. al.* (1990). For a wake flow, Monkewitz (1988) pointed out that both spatial and temporal modes are important, therefore the convective and absolute stability characteristics of flow should be addressed.

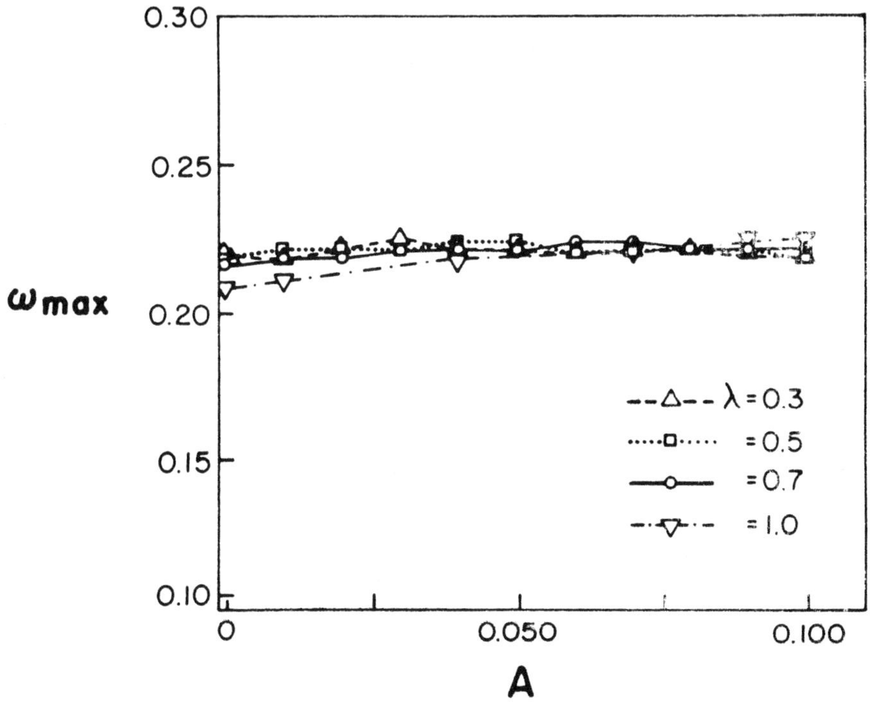

FIGURE 18-16 *The Most Amplified Frequency, ω_{max}, versus A for Different Velocity Ratios, λ.*

Yang *et. al.* (1991) recently attempted to analyze of two-phase stability for a wake flow. They applied the same assumptions as those used in their mixing layer analysis except for the mean flow profile and stability modes. They adopted the hyperbolic secant mean flow profile suggested by Mattingly and Criminale (1972) for a plane wake. Both the wave number and the frequency are complex numbers in their analysis. They solved the modified Rayleigh equation by the shooting method using a standard fourth-order Runge-Kutta scheme with a constant step size. Yang *et. al.* (1991) first investigated the case for uniform particle size distribution. The effects of particles at various loadings of A $=$ 0, 0.05, and 0.1 are shown in Figures 18-20 and 18-21. These maps show the wave number, α, versus the angular frequency, ω, at the downstream location where the centerline velocity is 0.0012 of the free stream velocity. As α_r (the real pare of α) increases along

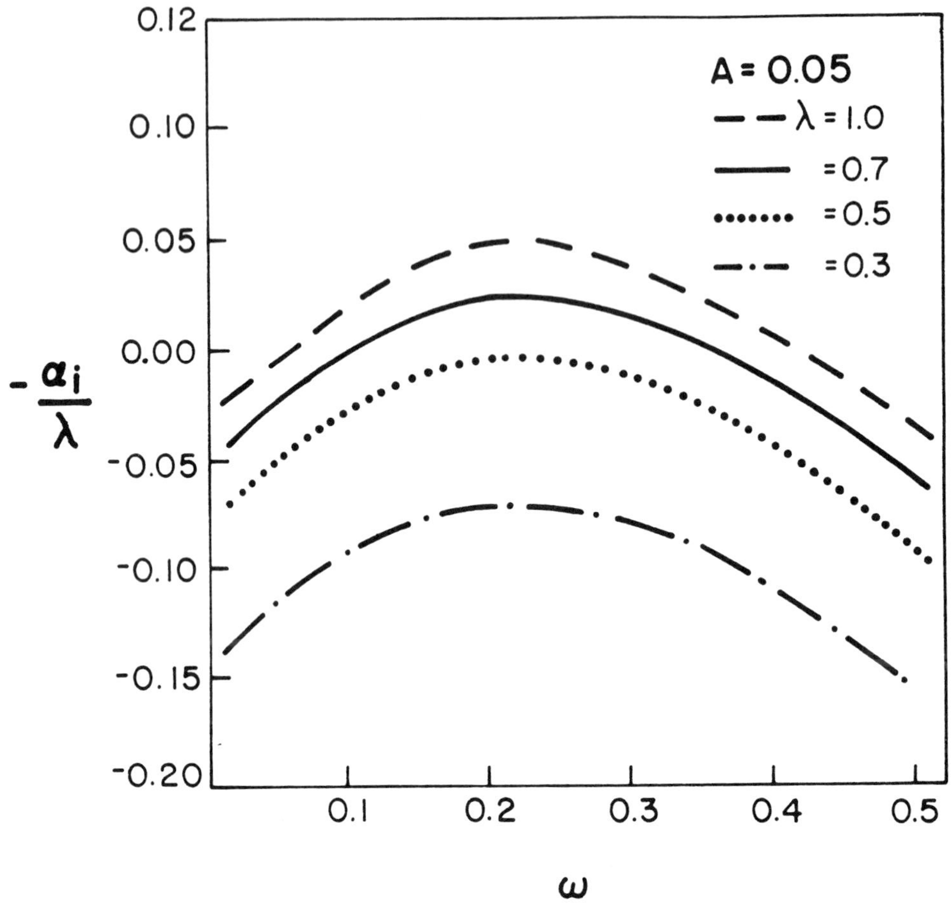

FIGURE 18-17 *The Spatial Amplification Rate, $-\alpha_i$ at A=0.05 normalized by λ versus the Angular Frequency.*

the α_i (the imaginary part of α) lines, curved-clockwise contours are traced about the point ω_o, the branch point. As A is increased, the imaginary part of the branch point, $Im(\omega_o)$, decreases from 0.0454786 ($A = 0$), to -0. 0045215 ($A = 0.05$) and then to -0.0545217 ($A = 0.1$). It is noted that the particle free flow ($A = 0$) is absolutely unstable and the two-phase flows ($A = 0.05$ *and* 0.1) are convectively unstable. The presence of the particles effectively improves the stability characteristic of the flow. For the disturbances in the form of an impulse, the amplified disturbances will eventually be convected downstream and out of the system if the flow is convectively unstable. The amplified disturbances will propagate throughout the entire domain of the absolutely unstable flow. For particle-free flows, the stability characteristic will switch from absolute instability to convective instability at some downstream point. The presence of particles will hasten the switch-over. For $A = 0.05$, the switch over point is almost at the trailing edge of the bluff body. Yang *et. al.* (1991) also investigated the effectiveness of the particles on the flow stability as a function of the ir position in the flow field. It was found that particles are more effective the closer they are to the centerline.

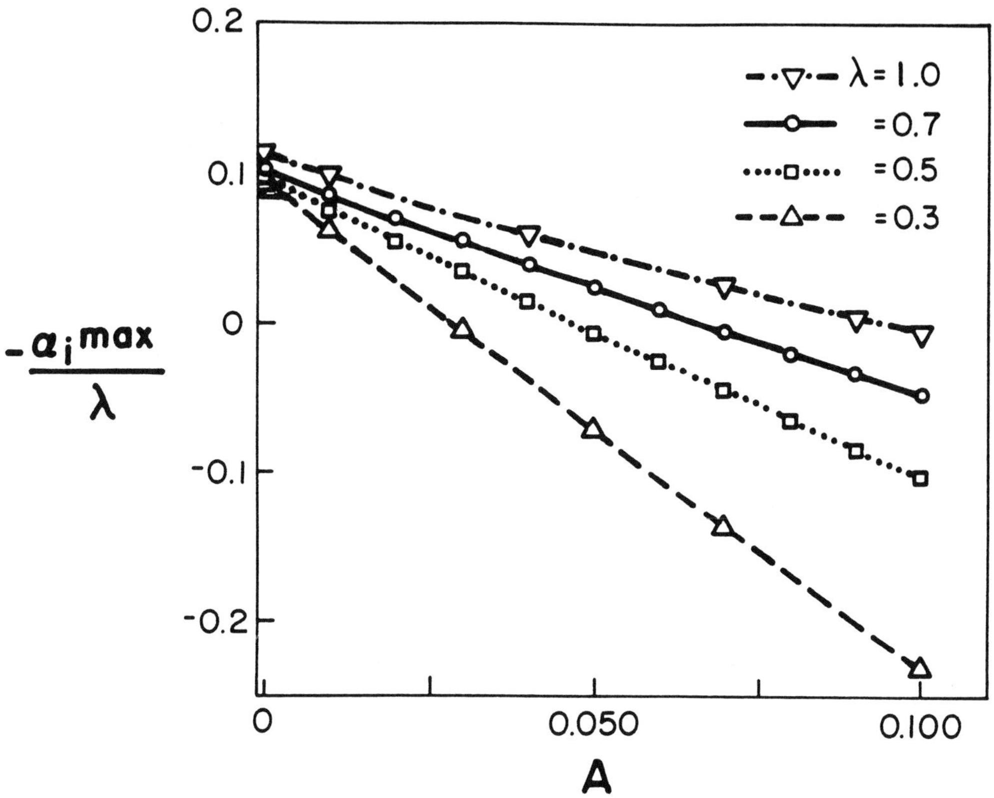

FIGURE 18-18 *The Maximum Spatial Amplification Rate, $-\alpha_i^{max}$, normalized by λ versus A for Different Velocity Ratios, λ*

18.4.6 Coupling Effects

Besides affecting the unstable nature of a shear layer, the presence of particles or droplets can also affect the growth and vorticity distribution in the developing layer. The effect may be regarded as change in the strength (circulation) of the vortices as they travel through the field. The interaction between the particles and the fluid is commonly referred to as coupling. One-way coupling is the case when the carrier fluid is responsible for the particle or droplet motion while the dispersed phase has no effect on the properties of the carrier fluid. One-way coupling is a valid assumption for low dispersed-phase concentrations. However, with increased concentrations, the dispersed phase can affect the fluid motion and two-way coupling is important.

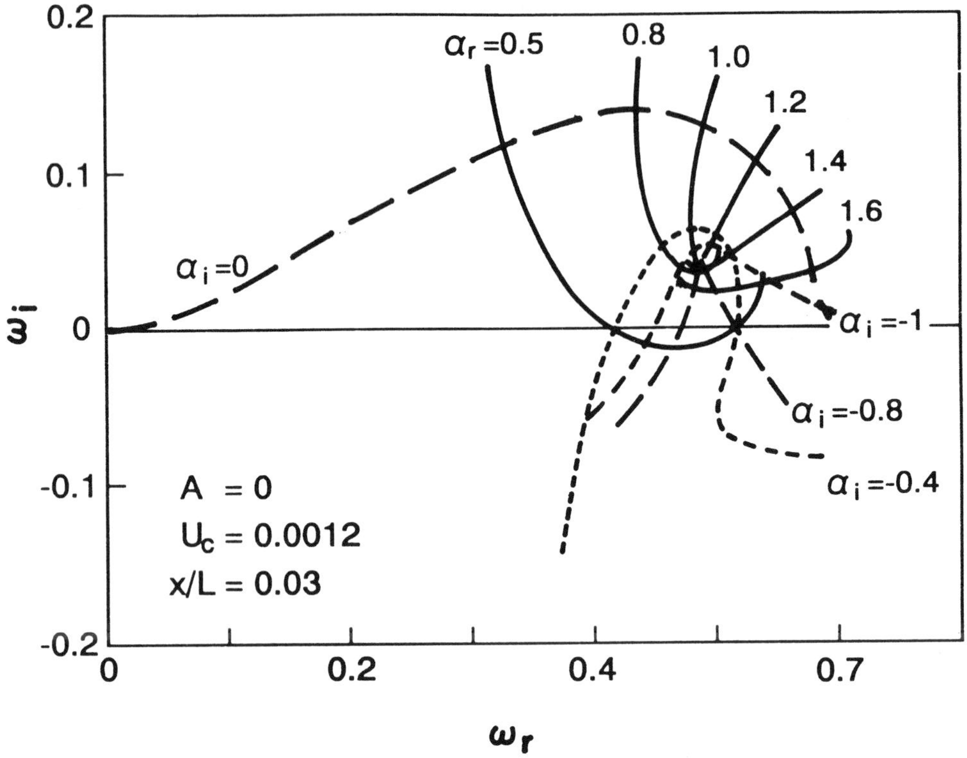

FIGURE 18-19 *Map of α on the ω plane, for A=0, U_c=0.0012 and Uniform Particle Distribution.*

An attempt has been made to incorporate two-way coupling into the discret e vortex models for particle dispersion (Tang 1990). Euler's equation for a fluid with a suspended disperse phase can be expressed as

$$\frac{d\vec{u}_f}{dt} = -\frac{1}{\rho_f}\nabla p + \frac{1}{\rho_f}\vec{f}_D + \frac{S_m}{\rho_f}(u_d - u_f) \tag{18.11}$$

where $\vec{f}_D$ is the force on the fluid per unit volume due to the drag on the dispersed phase and S_m is the mass released into the fluid due to mass transfer (evaporation or condensation) of the dispersed phase. The equation does not account for the change in fluid density due to the addition of the vapor phase. Of course, the viscous effects are confined only to the immediate neighborhood of the particles.

By taking the curl of this equation one can show that the rate of change of the vortex strength is given by

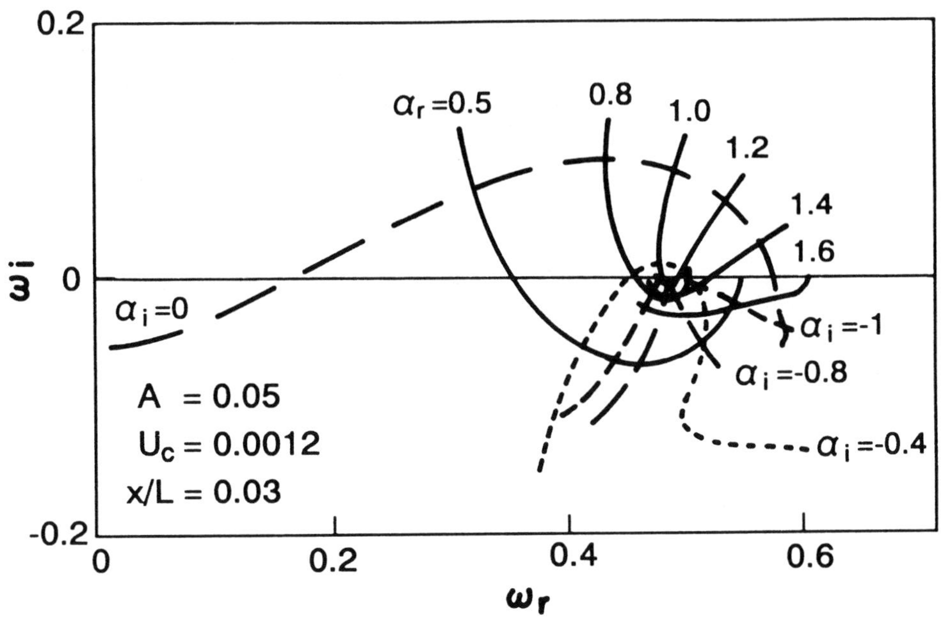

FIGURE 18-20 *Map of α on the ω Plane for A=0.05, U_c=0.0012 and Uniform Particle Distribution.*

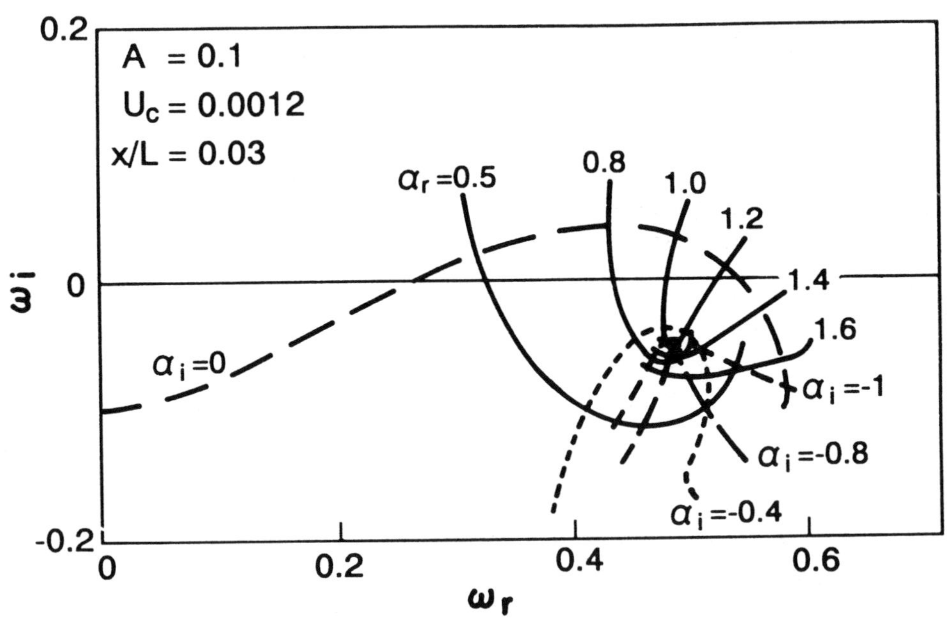

FIGURE 18-21 *Map of α versus ω for A=0.1, U_c=0.0012 and Uniform Particle Distribution.*

$$\frac{d\Gamma}{dt} = \frac{A}{\rho_f}[\nabla \times \vec{f}_D + \nabla \times S_m(\vec{u}_d - \vec{u}_f)] \tag{18.12}$$

where A is the area of the vortex element. The first term on the right is the effect of the drag of the particles on the vortex strength and the second term yields the effect of mass addition (or depletion).

This scheme has been incorporated into the discrete vortex method by Tang (1990). As the particles are tracked in the flow field, the drag force and mass transfer are ascribed to computational cells and distributed to the vortices in the cells. The vortex strength is then updated and the new strengths are used to establish the velocity field in keeping with the discrete vortex method.

The effect of momentum coupling on the development of eight vortices in a cell has been reported by Tang *et al.*, (1989). These calculations were carried out assuming a particle/fluid mass concentration ratio of 10 and a Stokes number of unity (based on the wave length of the initial disturbance). The initial ratio of particle velocity to fluid velocity was 0.5.

The interaction of the vortices, subsequent pairing and growth into a single vortex is shown in Figure 18-22. The length scale, L, is the wave length of the initial sinsusoidal distribution. The nondimensional time is the wave length divided by the half the freest ream velocity difference across the mixing layer. It is noted that the particles, which were initially distributed uniformly throughout the field, are centrifuged from the the vortex core to the edge of the vortex. The presence of the particles does not influence the basic features of the merging and pairing process but does slow the rate of flow development.

A measure of the time rate of development of the vortex is provided by the variation of the momentum thickness, as defined by Browand and Weidman (1974), with time. The momentum thickness is defined as

$$\theta = \int (\frac{1}{4} - (\frac{\bar{u}}{\Delta U})^2)dy \tag{18.13}$$

where $\bar{u}$ is the mean velocity along a $y = const$ line and ΔU is the velocity difference across the layer and the integration is performed over the computational field. The variation in momentum thickness with time for a flow with different particle concentrations is shown in Figure 18-23. The concentration is defined as the initial ratio of the particle bulk density to the fluid density. Also the initial particle/fluid velocity ratio is 0.5. One notes that increasing the particle concentration significantly slows the development of the organized vortex structures. It was found that the effect is considerably reduced as the initial velocity ratio approaches unity. Further numerical studies with evaporating droplets in a shear layer have shown that mass coupling has a significant effect on the vortex shape, tending to elongate the structures and impede the pairing phenomena (Tang *et al.* 1990).

The analysis of coupling effects on large scale organized structures is only in its infancy but represents a fruitful area for model development because of its importance in combustion and reacting systems.

18.4.7 Bubble Motion

The analysis of bubble motion in organized vortex structures is difficult because of the sensitivity of the bubble acceleration to the acceleration of the fluid. From Eqn. (18.9) it was noted that the bubble acceleration can approach three times that of the local fluid. The acceleration of the fluid in a developing and rotating vortex flow is toward the center of the vortex so the bubble will travel quickly in this direction. The calculation is made difficult by the very small time steps required to integrate the bubble motion equation.

A preliminary calculation has been done to predict the temporal change in the bubble distribution caused by a Stuart's vortex (Stuart 1967). The bubbles were initially distributed uniformly throughout the field and then released in the flow field. The path line of a single bubble released at the edge of the Stuart vortex is shown in Figure 18-24a. The Stokes number of the bubble is unity. The length scale is the distance between adjacent vortices and the time scale is the length scale divided by the velocity across the shear layer. One notes that the bubble spirals toward the center of the vortex. The distribution of the bubbles after 50 nondimensional time steps is shown in Figure 18-24b. An interesting "S" pattern is observed. Contrast this trend with that for heavy particles which move away from the vortex core as shown in Figure 18-22. Considerable effort is needed to develop numerical schemes which can be u sed to track bubbles through vortices. The computation is made difficult by not only the high acceleration rates but also the need to include the effect of pressure and Reynolds number on bubble size and shape.

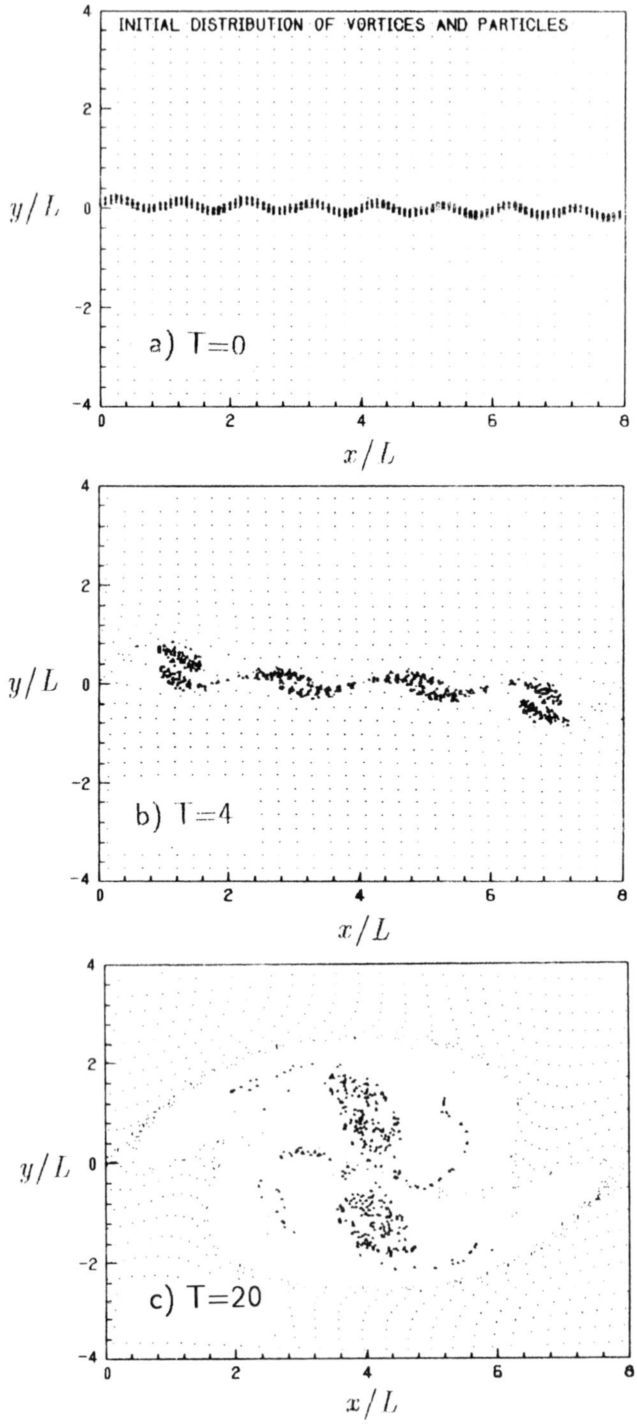

FIGURE 18-22 *Development of Vortex Structure and Particle Dispersion Pattern for an Initial Sinusoidal Distribution.*

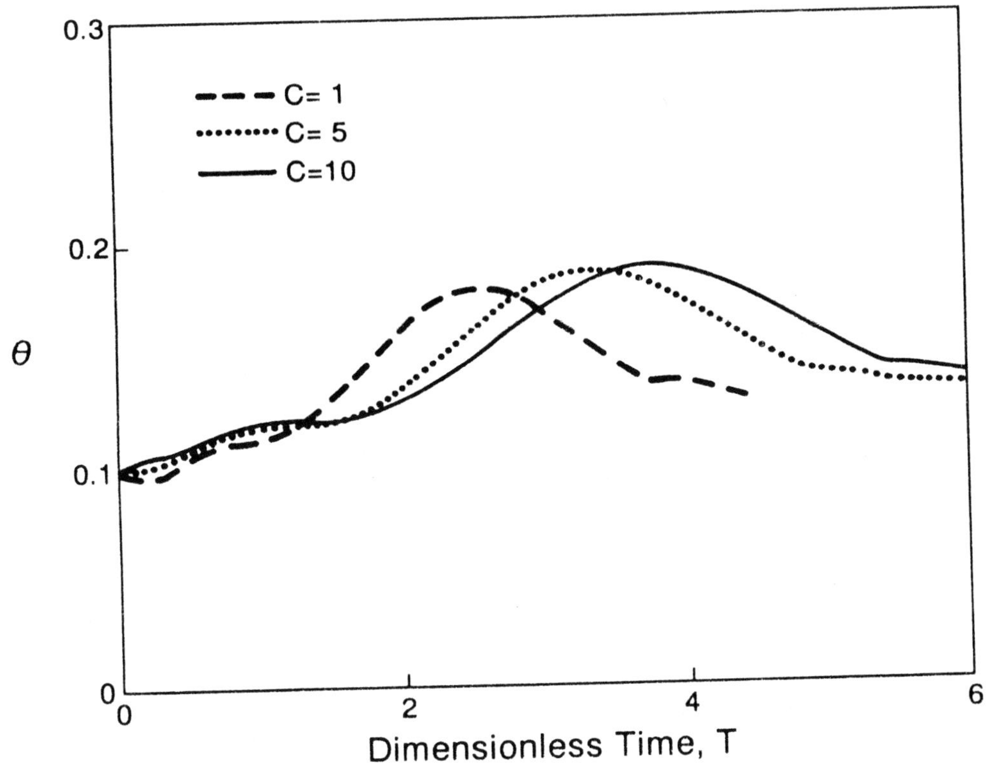

FIGURE 18-23 *Variation of Momentum Thickness with Time for Developing Vortex Pattern for Different Initial Particle Concentrations*

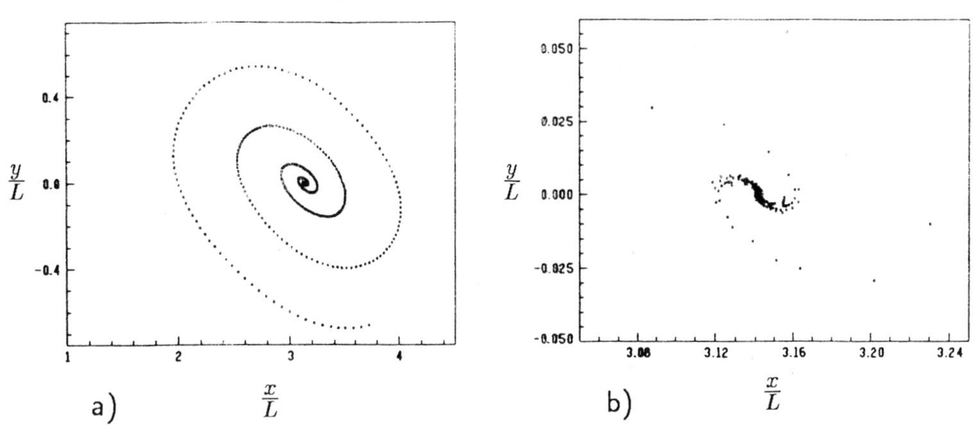

FIGURE 18-24 *Bubble Motion and Distribution in a Stuart Vortex Field; a) Pathline of an Isolated Bubble and b) Bubble Distribution after 50 Nondimensional Times.*

18.5 FUTURE DIRECTIONS

Significant advances have been made in identifying the basic mechanisms responsible for particle dispersion in large scale organized structures. However considerable more effort is needed to achieve an improved understanding and quantification of particle dispersion mechanisms. The specific studies include;

- Extension of numerical models to three-dimensional flows to enable predictions for industrially relevant problems.
- Measurements of instantaneous particle and droplet velocity and concentration profiles in three dimensional fields.
- Inclusion of coupling effects for reactive flows with application to heterogeneous combustion.
- Measurement and analysis of bubble motion in organized structures.

18.6 CONCLUSION

Particle dispersion in organized structures is important because of the many industrial processes that depend on the mixing, spreading and transport of particles by turbulence. Experimental and numerical studies show that heavy particles entrained in large, scale organized structures are centrifuged from the vortex cores and tend to concentrate on the peripheries of the vortices. This is especially apparent in wakes where a "focusing" effect is predicted. The degree of dispersion is scaled with the Stokes number. It is quite apparent that, at Stokes numbers near unity, the turbulent structures tend to concentrate the particles into localized regions thus contradicting the general belief that turbulence will tend to produce a homogeneous mixture.

ACKNOWLEDGEMENTS

The authors acknowledge the support of DOE Grant DE-FG06-86ER13567 under the direction of Dr. Oscar Manley and Dr. Daniel Frederick and equipment grant support from National Science Foundation, grants No. CBT-8506618 and CTB-8806525. The contributions of Dr. L. Tang, Dr. F. Wen, Dr. N. Kamalu, Dr. R. Chien, Dr. R. Gore, Y. Yang, and R. Hanle are recognized.

REFERENCES

Blinova, E.N. 1944. Hydrodynamic theory of pressure and temperature waves and center of action of the atmosphere. *Transl. No. 113. Reg. Control Off., Second Weather Reg.*, Patterson Field, Dayton Ohio

Borham, A. and A. Acrivos, 1988, The Sedimentation of Non-dilute Suspensions in Inclined Settlers,*Phys. Fluids*, **31**, 3488-3501.

Browand, F.K. and P.D.Weidman, P.D., 1976, Large Scales in the Developing Mixing Layer, *Jnl. Fluid Mech.*,**97**, 127-144.

Brown, G.L. and A. Roshko, 1974, On the Density Effects and Large Structure in Turbulent Mixing Layers, *Jnl. Fluids Mech.*,**64**, 775-816.

Canuto, C., M.Y. Hussiaini, A. Quarteroni and T.A. Zang, 1988. *Spectral Methods in Fluid Dynamics.* Berlin: Springer-Verlag.

Chien, R. and J.N. Chung, 1988a, Particle Dynamics in a Gas-Particle Flow over Normal and Inclined Plates, *Chem. Engr. Sci.*, **43**, 1621-1636.

Chein, R. and J.N. Chung, 1988b, Effect of Vortex Pairing on Particle Dispersion in Turbulent Shear Flow, *AIChE Jnl.*,**34**, 785-802.

Chung, J.N. and T.R. Troutt, 1988, Simulation of Particle Dispersion in an Axisymmetric Jet, *Jnl. of Fluid Mech.*, **186**, 199-222.

Crowe, C.T., R.A. Gore, and T.R. Troutt, 1985, Particle Dispersion by Coherent Structures in Free Shear Flows, *Particle Science and Tech.*,**3**, 149.

Crowe, C.T., J.N. Chung and T.R. Troutt, 1988, Particle Mixing in Free Shear Layers, *Prog. Energy Combust. Sci.*, **14**, 171-194.

Davis, R.W., E.F. More and C.T. Crowe, 1983, *Proc. of Third Intl. Congress on Numerical Methods in Laminar and Turbulent Flow*, Pineridge Press, 1037-1046.

Eliasen, E., E. Machenauer and E. Rasmussen, 1970, On a numerical method for integration of the hydrodynamical equations with a spectral representation of the horizontal fields. *Rep. No.2*, Dep. Meteorol., Copenhagen Univ., Denmark.

Faeth, G.M., 1987, Mixing, Transport and Combustion in Sprays, *Prog. Energy Comb. Sci.*, **13**, 293-346.

Gerz, T., U. Schuman and S.E. Elghobashi, 1989, Direct Numerical Simulation of Stratified Homogeneous Turbulent Shear Flows, *Jnl. Fluid Mech.,* **200**, 563-594.

Ghaddar, N.K., M. Magen, B.B. Mikic, and A.T. Patera, 1986. Numerical investigation of incompressible flow in grooved channels. Part 2.

Resonance and oscillator y heat-transfer enhancement. *Jnl. Fluid Mech.*, **168**, 541-567.

Ghoniem, A.F. and K.K. Ng, 1987, Numerical Study of the Dynamics of a Forced Shear Layer, *Phys. Fluids*, **30**, 706-721.

Ghoniem, A.F., 1991, Vortex Simulation of Reacting Shear Flow, to appear as Chapter in *Numerical Approaches in Combustion* AIAA Publication Series, edit ed by E. Oran and J. Boris.

Gore, R.A., C.T. Crowe, N. Kamalu. T.R. Troutt and J.J. Riley, 1989, Particle Dispersion by Large Scale Vortex Structures, *Particle Sci. and Tech.*, **7**, 201-207.

Gottlieb, D., S.A. Orszag, 1977. *Numerical Analysis of Spectral Methods: Theory and Applications, CBMS-NSF Reg. Conf. Ser. in Appl. Math., Vol. 26.* Philadelphia: SIAM.

Grassberger, P. and L. Procaccia, 1983,Measuring the Strangeness of Strange Attractors, *Physica*, **9D**, 189-196.

Hardalupas, Y., A.M.K.P. Taylor, and J.H. Whitelaw, 1989, Velocity and Particle Flux Characteristics of Turbulent Particle-laden Jets, *Proc. Roy. Soc., London*, **A426**, 31-78.

Hardalupas, Y., A.M.K.P. Taylor, and J.H. Whitelaw, 1990, Particle Dispersion in a Vertical Round Sudden Expansion Flow, submitted for publication to the *Jnl of Royal Soc.*

Haurwitz, B. and R.A. Craig, 1952. Atmospheric flow patterns and their representation by spherical surface harmonics. AFCRL Geophys. Res. Pap. No. 14, Hanscom AFB Mass.

Herbolzheimer, E., 1983, Stability of the Flow during Sedimentation in Inclined Channels, *Phys. Fluids*, 2043-2054.

Huerre, P. and P.A. Monkewitz, 1985, Absolute and Convective Instabilities in Free Shear Layers, *Jnl. Fluid Mech.*, **159**, 151-168.

Ho, C.M. and P. Huerre, 1984, Pertrubed Free Shear Layers, *Ann. Rev. Fluid Mech.*, **16**, 365.

Kamalu, N., F. Wen, T.R. Troutt and C.T. Crowe, 1988, Particle Dispersion by Ordered Motion in Turbulent Mixing Layers, *ASME Cavitation and Multiphase Flow Forum, FED*, **64**, 150-154.

Kamalu, N., 1989, Visualization Techniques for Understanding Particle Dispersion in Free Shear Flows, PhD Dissertation, Washington State University.

Kamalu, N., L. Tang, T.R. Troutt, J.N., Chung, and C.T. Crowe, C.T., 1989, Particle Dispersion in Developing Shear Layers, *Proc. Intl. Conf. on Mech. of Two-Phase Flows*, National Taiwan Univ., Taiwan, 199-202.

Kobayashi, H., S.M. Masutani, S. Azuhata, N. Arashi and Y. Hishinuma, 1988, Dispersed Phase Transport in a Plane Mixing Layer, Transport

Phenomena in Turbulent Flows - Theory, Experimemnt and Simulation, Hemisphere Pub., 433-456.

Kreiss, H.O. and J. Oliger, 1971, Comparison of accurate methods for the integration of hyperbolic equations.*Rep. No. 36*, Dep. Comput. Sci., Uppsala Univ., Sweden.

Laufer, J., 1975, New Trends in Experimental Turbulence Research, *Ann. Rev. Fluid Mech.*,**7**, 307-326.

Lazaro, B.J. and J.C. Lasheras, 1989, Particle Dispersion in a Turbulent, Plane Shear Layer, *Phys. Fluids*, **A.1**(6) 1035-1044.

Laitone, J.A., 1981, A Numerical Simulation for Gas-Particle Flow at High Reynolds Numbers, *Jnl. Appl. Mech.*, **48**, 465-471.

Leonard, A. 1981. Divergence-free vector expansions for 3-D flow simulations. *Bull. Amer. Phys. Soc.* **26**, 1247

Leonard, A. and A. Wray, 1982. A new numerical method for the simulation of three-dimensional flow in a pipe. *Int. Conf. on Num. Meth. in Fluid Dyn.*, Aachen, W. Germany, June 1982.

Longmire, E.K. and J.K. Eaton, 1990, Structure and Control of a Particle-laden Jet, Thermosciences Division, Dept. of Mech. Engr., Stanford University, Rep. No. MD-58.

Mattingly, G.E. and W.O. Criminale, 1972, The Stability of an Incompressible Two-Dimensional Wake, *Jnl. Fluid Mech.*, **51**, 233-272.

Maxey, M.R. and J.J. Riley, 1983, Equation of Motion for a Small Rigid Sphere in a Uniform Flow, *Phys. Fluids*, **26**(4), 883-889.

Monkewitz, P.A., 1988, The Absolute and Convective Nature of Instability in Two-Dimensional Wakes at Low Reynolds Numbers, *Phys. Fluids*, **51**, 233-272.

Moser, R.D., P. Moin, P. and A. Leonard, 1983, A spectral numerical method for the Navier-Stokes equations with applications to Taylor-Couette flow. *Jnl. Comp. Phys.*, **52**, 524-544

Orszag, S.A. 1969. Numerical methods for the simulation of turbulence. *Phys. Fluids*, **12**, 250-257 (Suppl. II)

Orszag, S.A. 1972. Numerical simulation of incompressible flows within simple boundaries: I. Galerkin (spectral) representations. *Stud. Appl. Math.*, **50**, 293-327.

Orszag, S.A., Kells, L.C. 1980. Transition to turbulence in plane Poiseuille and plane Coutte flow. *Jnl. Fluid Mech.*, **96**, 159-205.

Ottino, J.M., 1990, Chaotic Advection and Turbulence, *Ann. Rev. Fluid Mech.*, **22**, 207-253.

Rangel, R. and W.A. Sirignano, 1989, The Dynamics of Vortex Pairing and Merging, AIAA Paper 89-0128, 27th Aerospace Sciences Meeting, Reno, NV.

Reeks, M.W., 1977, On the Dispersion of Small Particles Suspended in an Isotropic Turbulent Field, *Jnl. Fluids Mech.*, **83**, part 3, 529-546.

Sarpkaya, T., 1988, Computational Methods with Vortices, The 1988 Freeman Scholar Lecture presented at the 1988 ASME Winter Annual Meeting in Chicago, Nov., 1988.

Shaqfeh, E.S.G. and A. Acrivos, 1987, The Effects of Inertia on the Stability of the Convective Flow in Inclined Particle Settlers, *Phys. Fluids*, **26**, 2043-2054.

Siberman, I. 1954. Planetary waves in the atmosphere. *J. Meteorol.* **11**, 27-34.

Spalart, P. R. 1986. Numerical simulation of boundary layers: Part 1. Weak formulation and numerical method. *NASA TM* 88222

Squires, K.D. and J.K. Eaton, 1990, Particle Response and Turbulence Modification in Isotropic Turbulence, *Phys. Fluids A*, **2**(7), 1191-1203

Stuart, J.T., 1967, On Finite Amplitude Oscillations in Laminar Mixing Layers, *Jnl. Fluid Mech.*, **3**, 347-370.

Tang, L., C.T. Crowe, J.N. Chung, and T.R. Troutt, 1989, Effect of Momentum Coupling on the Developmemt of Shear Layers in Gas-particle Flows, *Proc. Intl. Conf. on Mech, of Two-Phase Flows*, National Taiwan University, 387-391.

Tang, L., 1990, Simulation of Multiphase Free Shear Layers, PhD Disseration, Washington State University.

Tang, L., C.T. Crowe, J.N. Chung and T.R. Troutt, 1990, A Numerical Model for Droplets Dispersing in a Developing Plane Shear Layer Including Coupling Effects, Numerical Methods for Multiphase Flows, ASME FED, Vol. 91, 27-33.

Tang, L., F. Wen, Y. Yang, C.T. Crowe, J.N. Chung and T.R. Troutt, 1991, Self-organized Particle Dispersion Mechanisms in Free Shear Flows, submitted to *Phys. Fluids.*

Thomas, N.H., T.R. Anton, K. Sene and J.C.R. Hunt, 1983, *Proc. Intl. Conf. Physical Modeling of Multiphase Flow*, Coventry, Eng., **169**, 176-183.

Truesdell, G.C. and S.E. Elghobashi, 1991, Direct Numerical Simulation of a Particle Laden Homogeneous Turbulent Flow, *Gas-Solid Flows -1991*, ASME-FED **121**, 11-17.

Voigt, R.G., D. Gottlieb and M.Y. Hussaini, eds. 1984. *Spectral Methods for Partial Differential Equations.* Philadelphia: SIAM-CBMS.

Wallis, G.B., 1969, *One Dimensional Two-Phase Flow*, McGraw-Hill

Wang, Lian-Ping, 1990, On the Dispersion of Heavy Particles by Turbulent Motion, PhD Dissertation, Washington State University, Pullman, WA.

Wen, F., 1990, A Study on Particle Dispersion in a Two-Dimensional Mixing Layer, PhD Dissertation, Dept. of Mech. and Matls Engr., Washington State University.

Wen, F., N. Kamalu, R. Hanle, J.N. Chung, C.T. Crowe and T.R. Troutt, 1991, Particle Dispersion by Vortex Structures in Plane Mixing Layers, submitted to *Int. Jnl. Multiphase Flows.*

Wen, F., N. Kamalu, R. Hanle, J.N. Chung and C.T. Crowe, 1990, Visualization of Particle Dispersion in a Plane Mixing Layer, submitted to the *Intl. Jnl. of Multiphase Flow.*

Winant, C.D. and F.K. Browand, 1974, Vortex Pairing; Mechanisms of Mixing Layer Growth at Moderate Reynolds Number, *Jnl. Fluid Mech.*,**63**, 237-255.

Yang, Y., J.N. Chung, T.R. Troutt and C.T. Crowe, 1990, The Influence of Particcles on the Spatial Stability of Two-Phase Mixing Layers, *Phys. Fluids A*, **2**, 1839-1845.

Yang, Y., J.N. Chung, T.R. Troutt and C.T. Crowe, 1991, The Absolute and Convective Instabilities of a Two-Phase Wake, to be submitted to *Phys. Fluids.*

Yang, Y., Particle Dispersion by Organized Vortices in Plane Wakes, PhD Dissertation (in progress).

Yudine, M.I., 1959, Physical Considerations on Heavy Particle Diffusion, Atmospheric Difussion and Air Pollution, *Adv. Geophysics*, **6**, 185-191.

Zang, T.A. and M.Y. Hussiaini, 1985. Recent Applications of Spectral Methods in Fluid Dynamics. *Lect. Appl. Math.* **22**, 379-409.

Zang, T.A and M.Y. Hussiaini, 1987. Spectral Methods in Fluid Dynamics. *Ann. Rev. Fluid Mech. 1987* **19**, 339-367.

CHAPTER 19

USE OF LAGRANGIAN STATISTICS TO DESCRIBE TURBULENT DISPERSED FLOWS

T. J. Hanratty
J. L. Binder

19.1 INTRODUCTION

A critical problem in understanding the behavior of turbulent flows containing droplets or solid particles is the prediction of the distribution of the dispersed phase and its rate of impaction on boundaries. The most common method of analysis is to use a diffusion model. This is implemented within an Eulerian framework whereby the diffusion flux is set equal to the product of the local concentration gradient and a turbulent diffusivity, and a differential equation is developed from a species balance equation.

It is well known that this approach is not fundamentally correct, because the large scale characterizing turbulent mixing suggests that the turbulent flux does not depend just on the local behavior of the turbulence. The only theory which takes account of the dependency of turbulent diffusion on previous history in an exact way is the analysis of diffusion from a point source by Taylor (1921). The turbulence is assumed to be homogeneous and the time-dependency of turbulent diffusion is described by relating the mean-square of the x-component of the displacement of fluid particles, $\overline{X^2}$, to the mean-square of the turbulent velocity fluctuations and the Lagrangian correlation.

This paper explores the use of this Lagrangian approach to describe the behavior of a dispersed phase. The main feature is that the concentration field is pictured as resulting from a distribution of sources of particles. The chief theoretical problem, then, involves the description of the behavior of one of these sources.

The haphazard motion of the particles associated with diffusion results from the turbulent motion of the fluid. The ability of the particles to follow the fluid turbulence is measured by a reciprocal time constant, β, which gives the stopping distance of a particle moving with velocity, v_P, in a stationary fluid as $v_P\beta$. From the equation of motion of the particle the following relation is obtained:

$$\beta = \frac{3C_D \rho}{4d_p \rho_p} | U_R | \tag{19.1}$$

where C_D is the drag coefficient, ρ, the fluid density, ρ_P, the particle density, d_P, the particle diameter and $| U_R |$, the absolute value of the relative velocity between the fluid and the particle. For the Stokes flow range,

$$\beta = \frac{18\mu}{\rho_p d_p^2} \tag{19.2}$$

where μ is the fluid viscosity.

The inertial characteristics of the particle are represented by two dimensionless groups: The ratio of the time scale of the fluid turbulence to the time scale of the particle is given by $\beta\tau_{LF}$, where τ_{LF} is the Lagrangian time scale of the fluid turbulence. For small $\beta\tau_{LF}$ (large particle diameters or small fluid velocities) the particles do not follow the turbulence, so the turbulence properties of the particles will be different from the turbulence characteristics of the fluid. For large $\beta\tau_{LF}$ the fluid and particle turbulence are the same.

The other dimensionless group is

$$\tau_p^+ = \frac{u_* \tau_p}{\upsilon / u_*} \tag{19.3}$$

where $\tau_P = 1/\beta$, υ is the kinematic viscosity of the fluid, and u_* is the friction velocity. It is a measure of the ratio of the stopping distance of a particle in a still fluid to the thickness of the viscous wall region. For $\tau_P^+ > 20$ (see McCoy & Hanratty, 1975) the stopping distance is larger than the thickness of the viscous wall region.

Aerosol particles are characterized by small values of τ_P^+. As a consequence, deposition is strongly influenced by non-homogeities in the fluid turbulence close to the wall. The dispersed flow problems considered in this paper, however, are characterized by large enough τ_P^+ that particles move in free-flight through the viscous wall region. Fluid turbulence in the viscous wall region is not important. Consequently, the influence of turbulence non-homogeneities can, to a first approximation, be ignored. The simplifying assumptions are, therefore, made that the turbulence is homogeneous and that the mean velocity profile may be approximated as a plug flow. Furthermore, it is assumed that the concentration of particles is small enough that they are not affecting the fluid turbulence or interacting with one another.

The methods outlined in this chapter could have an impact on analyses where a knowledge of particle distribution and deposition is needed: An important environmental problem is the determination of the amount of sediment carried by flowing water and this, in turn, depends on understanding the influence of settling and turbulence on particle dispersion.

Vertical gas/liquid annular flows occur in the process industries and in the cooling systems associated with nuclear reactors. Here part of the liquid flows along the wall as a liquid film and part as droplets in the gas. The critical problem in analyzing these flows is the prediction of the fraction of the liquid entrained in the gas flow. An understanding of the rate of deposition of drops is an important ingredient because the entrainment may be visualized as a balance between the rate of deposition and the rate of atomization of the liquid film.

Horizontal gas/liquid annular flows occur in gas transfer lines and in heat transfer operations involving boiling and condensation. Additional complications arise because gravitational effects cause asymmetric film and droplet distributions. An understanding of the variation of the thickness of liquid film around the pipe circumference is of primary importance. This, in turn, depends on the circumferential variation of the deposition rate. There are strong similarities in the distribution of droplets in horizontal gas/liquid flow and the distribution of solid particles in sediment transport, so the methodologies for analyzing these two problems are the same.

The principal motivation for using the Lagrangian approach in such problems is that the physics emerges in a more natural way, wherein an understanding of the time-dependency of turbulent diffusion is substituted for the determination of the spatial variation of a diffusion coefficient. However, the avoidance of a number of difficulties associated with the use of Eulerian equations provides additional incentives: The definition of boundary conditions in the Eulerian formulation is not clear so that they are often empirically derived (Jobson & Sayre, 1970; Hjelmfelt & Lenau, 1970; Parker, 1978; VanRijn, 1986; Celik & Rodi, 1988; Hunt, 1954, 1969). In gas-liquid annular flows two types of droplet transport have been identified, whereby very massive drops move in straight line trajectories from one wall to another and small drops assume a haphazard motion characteristic of turbulent diffusion (Andreussi & Azzopardi, 1983). These two mechanisms are addressed, in terms of Taylor's formulation of point source diffusion, as limiting behaviors for small and large diffusion times. The gravitational settling in horizontal flows, which causes asymmetries in the concentration profile, is usually taken into account by assuming the settling velocities of the particles are equal to their terminal values. This can lead to error because the particles might not have been in the field long enough to attain free-fall. The use of an equation for the particle trajectory in the description of diffusion from a small wall source takes this into account.

19.2 POINT SOURCE DIFFUSION OF FLUID PARTICLES

Taylor's (1921) analysis of point source diffusion in a homogeneous turbulent field considers a number of experiments in which the path of an element, which is at the origin at time zero, is followed. The x-coordinate of this element sometime later is given as X(t). The average value of this coordinate for a large number of elements, $\overline{X}(t)$, is zero because it is just as likely that it will have a plus or minus value. However, $\overline{X^2}(t)$ is not zero and its variation with time is a measure of the rate of mixing.

For the case of molecular dispersion in a stagnant fluid, the element whose path is followed is a molecule. Einstein showed that

$$\frac{1}{2}\frac{d\overline{X^2}}{dt} = D \tag{19.4}$$

where D is the molecular diffusion coefficient defined in the diffusion equation,

$$\frac{\partial C}{\partial t} = D\,\Delta^2\,C \tag{19.5}$$

Taylor considered dispersion of elements which are fluid particles. The x-component of the velocity of a particle whose path is followed is given as v(t). Again, the average for a large number of fluid particles, $\overline{v}(t)$, is zero. However, the average of the square of the velocity, $\overline{v^2}$, is a finite value which is a measure of the intensity of the turbulence. It is independent of time in a homogeneous stationary field.

Taylor derived the following relation for the mean-squared displacement of fluid particles:

$$\varepsilon(t) = \frac{1}{2}\frac{d\overline{X^2}}{dt} = \overline{v^2}\int_o^t R_L(t)dt \tag{19.6}$$

Here $R_L(t)$ is the Lagrangian correlation coefficient defined as

$$R_L(t) = \frac{\overline{v(o)v(t)}}{\overline{v^2}} \tag{19.7}$$

The numerator is the average product of the velocities of a fluid particle at times zero and t. For small times, $v(o) \cong v(t)$ so that $\overline{v(o)v(t)} = \overline{v^2}$ and $R_L(t) \rightarrow 1$. For large times, v(t) is not related to v(o) so that v(o) v(t) can be plus as often as it is minus. Consequently $\overline{v(o)v(t)} = 0$ and $R_L = 0$ for $t \rightarrow \infty$. The integral for $t \rightarrow \infty$ is a constant which is defined as the Lagrangian time scale of the fluid,

$$\tau_{LF} = \int_0^\infty R_L(t)\, dt \tag{19.8}$$

From (19.6) it can be seen for $t \rightarrow 0$ that

$$\varepsilon(t) = \overline{v^2}\, t \tag{19.9}$$

and for $t \rightarrow \infty$ that

$$\varepsilon = \overline{v^2}\, \tau_{LF} \tag{19.10}$$

The diffusivity is time-dependent. It increases linearly with time for small time and is constant for large time.

Batchelor (1949) has shown that if the distribution function describing the dispersion of fluid particles is Gaussian for all times then the concentration field resulting from a source in a homogeneous stationary field is given as a solution to the following differential equation:

$$\frac{\partial C}{\partial t} = \frac{1}{2}\frac{d\overline{X^2}}{dt}\nabla^2 C + Q\delta(\vec{x}\,/\,\vec{x}')\delta\ (t/t') \tag{19.11}$$

where Q is the source strength. Equation (19.11) is a convenient way to analyze the behavior of a differential source with a complicated geometry.

19.3 APPLICATION OF TAYLOR'S THEORY TO HEAT TRANSFER PROBLEMS

The application of Taylor's analysis to more general problems than point source diffusion has been considered in a number of papers from this laboratory (Hanratty, 1956, 1958; Hanratty & Flint, 1958; Eckelman & Hanratty, 1972). Figure 19.1 depicts heat transfer through a flowing fluid from a hot wall to a cold wall. In the fully-developed condition that exists far downstream there is no change of temperature in the flow direction and the heat flux, q, in the fluid is a constant.

For a laminar flow the temperature profile is a straight line given by the equation

$$\frac{dT}{dx} = -\frac{q}{k} \tag{19.12}$$

where k is the molecular conductivity. For a turbulent flow, measurements by Sage and his coworkers (1952) give the type temperature profile sketched in Figure 19.1. This can be represented with an Eulerian approach in which the turbulent diffusion coefficient is a function of x:

$$-q = \left[\frac{k}{\rho c} + E(x)\right]\frac{d(\rho c T)}{dx} \tag{19.13}$$

with q being a constant independent of x. Very close to the walls $E(x) \cong 0$. Consequently transport is controlled by molecular conduction and dT/dx is large. In the central region of the channel E is much larger than $(k/\rho c)$ and dT/dx assumes much smaller values than at the wall. In regions where $E \gg (k/\rho c)$ the increase in the slope of the temperature profile can be associated with decreases in E(x) caused by changes in the turbulence properties close to the wall.

If a Lagrangian approach is used the hot and cold walls are represented by series of line sources and sinks of heat. The behavior of one of these source-sink pairs is described; the fully developed temperature

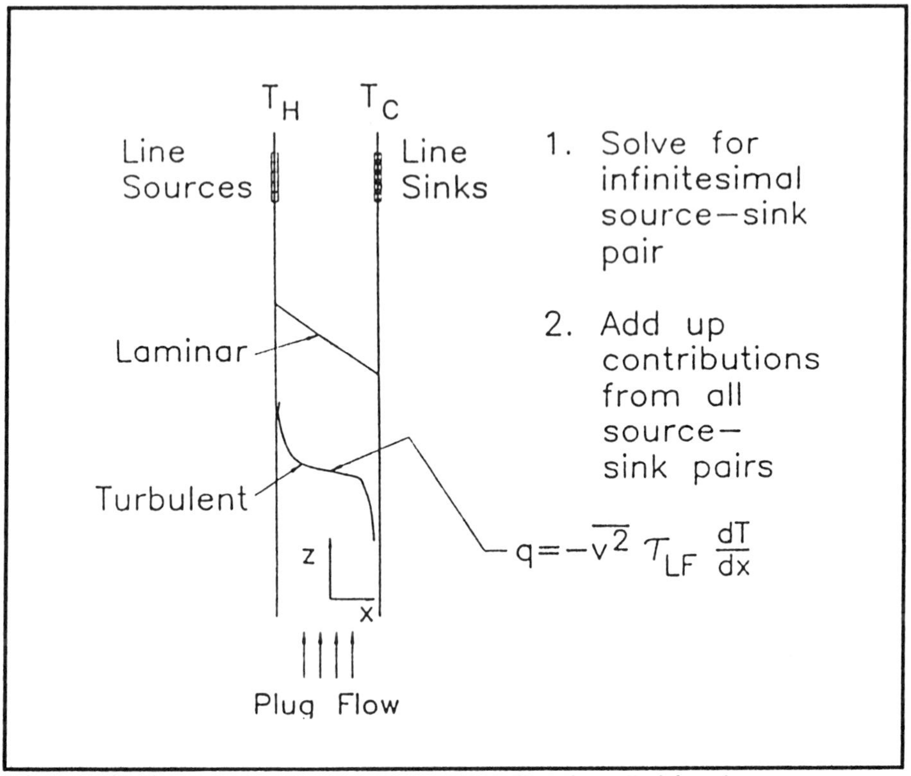

Figure 19.1 Heat transfer through a flowing fluid - the application of Taylor's theory on point source diffusion

profile is then calculated as resulting from contributions of an infinite number of these source-sink pairs upstream of the location being considered. In this context, the temperature field can be interpreted in the following way: The temperature profile in the center of the channel is influenced by hot or cold fluid particles that have been in the field for a long time. As a consequence, the Eulerian turbulent diffusivity, E, may be represented by the long time value of the Lagrangian eddy diffusivity, $E = \overline{v^2}\,\tau_{LF}$. Close to the wall the temperature field is influenced by hot or cold fluid particles that have been in the field for smaller times, so the effective diffusivity is smaller. The Lagrangian approach, therefore, argues that the increase in the slope of the temperature profile could be influenced greatly by the time-dependency of turbulent diffusion.

Hanratty (1956) carried out such a calculation by assuming homogeneous turbulence and a plug flow for which

$$dt = dz/U \tag{19.14}$$

where U is the fluid velocity and z is the coordinate in the direction of mean flow. Equation (19.11), with $C = \rho cT$, was solved to obtain the behavior of a source-sink pair that is introducing hot and cold particles for the interval between t' and $t' + dt'$. The boundary conditions were formulated by specifying the source strength to be $\pm\, q\, dt'$ during the interval dt' and to be zero for $t > t' + dt'$. The latter condition was satisfied by using the method of images.

The Lagrangian correlation coefficient was assumed to be of exponential form,

$$R_L = \exp\left(-(t-t')/\tau_{LF}\right) \qquad (19.15)$$

so that, from (19.6), the relation for $\overline{X^2}$ is

$$\overline{X^2} = \overline{v^2}\,\tau_{LF}\left[(t-t') - \tau_{LF}\left(1-e^{-(t-t')/\tau_{LF}}\right)\right] + \overline{X_0^2} \qquad (19.16)$$

The solution of (19.11) for a single source-sink pair gives $T = T_{SS}\, qdt'$, where T_{SS} is a function of x and of $\overline{X^2}$ [or of $(t-t')$]. The temperature profile at some value of $t = z/U$ is then obtained by summing the contributions of all of the source-sink pairs

$$T = \int_0^t T_{SS}\, q\, d t' \qquad (19.17)$$

If $\overline{X_0^2} = 0$ the integral represented by (19.17) diverges because for $(t-t')\to 0$, $\overline{X^2}\sim(t-t')^2$ and $T_{SS}\sim(t-t')^{-1}$. The inclusion of $\overline{X_0^2} \neq 0$, which is the same as assuming a source with a finite x-dimension avoids the divergence. Another way to accomplish this is to include the contribution of molecular diffusion so that $\overline{X_0^2}$ in (19.16) is replaced by $2(t-t')(k/\rho c)$. In this case $\overline{X^2} \sim (t-t')$ for $(t - t') \to 0$ and $T_{SS} \sim (t-t')^{-1/2}$.

Results of the calculations by Hanratty (1956) are shown in Figure 19.2. The lower part of the figure is a plot of the turbulent diffusion coefficient, made dimensionless with the half width of the channel, a, and the friction velocity, u_*, against the distance from the wall, made dimensionless with a. The points were obtained from (19.13) and temperature measurements by Sage (1952) for an air flow. The top part

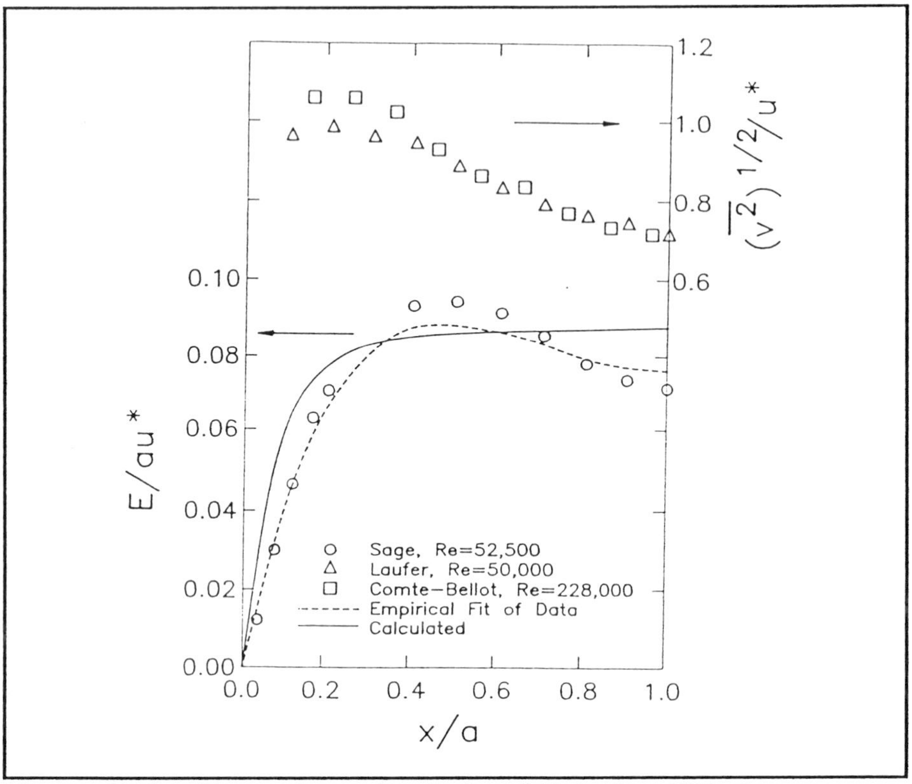

Figure 19.2 Comparison of calculated Eulerian diffusion coefficients with measurements by Sage

of the figure is a plot of measurements by Laufer (1954) and by Comte-Bellot (1965) of the root-mean-square of the x-component of the turbulent velocity fluctuations. It is noted that for x/a > 0.4 the decrease in E/a u_* is consistent with the decrease in $(\overline{v^2})^{1/2}/u_*$. However, in the region x/a<0.2 the sharp increase in turbulent diffusion coefficient with increasing x/a cannot be explained by a change in the magnitude of the turbulent velocity fluctuations.

The solid curve in the lower part of Figure 19.2 was obtained from (19.13) and the temperature profile was calculated by the method outlined by Hanratty for plug flow of homogeneous turbulence (Eckelman & Hanratty, 1972). The values of $\overline{v^2}\,\tau_{LF}$ and $\overline{v^2}$ used in these calculations were selected so as to agree with the measurements at x/a = 0.7. Equation (19.16) with $\overline{X_0^2} = 0$ was used to represent $\overline{X^2}$. Thus temperatures of ± ∞ were obtained at the two walls because of the singularities at x = 0 and x = 2a. However, this is of no consequence since Figure 19.2 is only considering measurements outside the viscous wall region. Of particular interest is the large variation of E for x/a <0.2. The conclusion from this calculation is that the large increase in E with increasing x/a observed for measurements in the region x/a <0.2 can be partially associated with the time-dependency of turbulent diffusion, defined by Taylor.

Eckelman & Hanratty (1972) considered in more detail the differences between the calculations of E and the measurements. They concluded that errors associated within the plug flow assumption, t = z/U, occur for x/a < 0.2. They suggested a modification in the model for line-source diffusion in a homogeneous turbulence which could take account of the variations of the mean velocity that exist in this region. The differences in the calculations and measurements for x/a >0.2 are not serious and are associated with non-homogeneities in the turbulence.

The value of these studies using Taylor's theory to calculate temperature fields in heat transfer problems is that they provide a basis for obtaining physical understanding of empirically derived variations of turbulent diffusivities. However (at present), because of the difficulties in taking into account spatial variations of the turbulence and of the mean velocity, Taylor's theory is not useful in practical calculations.

This is not the case for dispersed flows. Here, the inertia of the particles tends to smear out influences of spatial variations of the velocity field, as discussed in Section 19.1.

19.4 TURBULENCE PROPERTIES OF THE DISPERSED PHASE

A first step in using Lagrangian methods to analyze the behavior of dispersed phases is to define their turbulence characteristics. Laboratory studies of dispersion from a point source of solid particles in a liquid [Young & Hanratty (1991a) (1991b)] and of droplets in a gas [Lee, Adrian & Hanratty (1989a) (1989b)], and theoretical studies by Reeks (1977), Pismen & Nir (1978) and Nir & Pismen (1979) provide considerable help in this direction. Two dimensionless groups are of importance, $\beta\tau_{LF}$, and $v_{s\infty}/u_*$, where $v_{s\infty}$ is the terminal velocity in a still fluid. The group $v_{s\infty}/u_*$ is proportional to the ratio of the slip velocity and the magnitude of fluid turbulent velocity fluctuations. It is a measure of the influence of the crossing of fluid trajectories defined by Yudine (1959).

The measurements of Lee et al (1989a) of particle turbulence give the following correlation:

$$\frac{\overline{v_P^2}}{\overline{v^2}} = \frac{\beta\tau_{LF}}{0.7 + \beta\tau_{LF}} , \qquad (19.18)$$

where $\overline{v_P^2}$ and $\overline{v^2}$ are turbulence intensities for the particle and for the fluid. This shows that if $\beta\tau_{LF} < 1$ the inertia of the particle is large enough that it does not follow the turbulence.

For $v_{s\infty}/u_* < 1$ it is found that $\overline{v_P^2}$ decreases and that the Langrangian length scale, $L_P = (\overline{v_P^2})^{1/2}\ \tau_{TL}$, increases with decreasing $\beta\tau_{LF}$. However the product remains constant so that $\varepsilon_P = \mathcal{L}_P(\overline{v_P^2})^{1/2}$ is approximately equal to the long time turbulent diffusion coefficient for the fluid.

For $v_{s\infty}/u_* > 1$ it is found that $\overline{v_P^2}\ \tau_{LP}$ is less than $\overline{v^2}\tau_{LF}$ because of the crossing of trajectories effect.

19.5 DISPERSED FLOW IN A VERTICAL PIPE

Binder & Hanratty (1991) explored the application of a Lagrangian analysis to a vertical dispersed flow. The system considered was upward gas/liquid annular flow. In this configuration part of the liquid flows along the wall as a liquid layer and part as entrained droplets. There is an exchange of liquid between the annular liquid layer and the core. Under a fully developed condition the rate of atomization of the liquid layer, R_A, equals the rate of deposition of droplets from the core, R_D. Of particular interest is the relation between R_D and the concentration of liquid in the core, $\alpha\rho_l$. Binder & Hanratty considered this problem for dilute enough droplet concentrations that a diffusion model could be applicable and the rate of atomization is linearly dependent on the droplet concentration; i.e.,

$$R_D = k_D\ \alpha\rho_l \tag{19.19}$$

Measurements of dropsize in vertical annular flow have been presented by Azzopardi (1985). If the Sauter-mean diameter is accepted as the characteristic drop size, it is found that $\beta\tau_{LF}$ is typically in the range of 0.01 to 0.2, that $\tau_P^+ > 20$, and that $v_{s\infty}/u_* < 1$. This means that droplets are not following the turbulence but that $\varepsilon_P \cong \varepsilon$. Vames & Hanratty (1988) reviewed fluid turbulence measurements in a pipe and suggested that

$$\frac{\tau_{LF}\ u_*}{2a} = 0.046 \tag{19.20}$$

and

$$\overline{v^2} \cong 0.9\ u_*\ \text{ at } r \cong a \tag{19.21}$$

where a is the pipe radius. Equations (19.18) and (19.21) can be used to calculate $\overline{v_P^2}$. Equation (19.20) and

$$\frac{\tau_{LP}}{\tau_{LF}} = \frac{\overline{v^2}}{\overline{v_P^2}} \tag{19.22}$$

give τ_{LP}.

The film flowing along the wall is a source of droplets. The droplet field can be viewed as resulting from a distribution of ring sources, at the wall, of differential length. The solution of (19.11) for a ring source in a cylindrical polar coordinate system has been discussed by Hanratty & Flint (1956) and by Binder & Hanratty (1991). The source strength was specified as $R_A dt'$. The mean flow of the droplets was assumed to be uniform, so that dz = dt/U, and the turbulence properties of the droplets were assumed to be homogeneous.

The details on how the droplets enter the field are not firmly established. Therefore, Binder & Hanratty assumed that they are fully entrained in the turbulence at a distance from the wall which is at least one droplet diameter. Equation (19.16) with $\overline{v_P^2}$ and τ_{LP} substituted for $\overline{v^2}$ and τ_{LF}, was used to calculate $\overline{X_P^2}$. Term $\overline{X_0^2}$ has to be at least of the order of d_P^2. Fortunately it turns out that the results are not terribly sensitive to the choice of $\overline{X_0^2}$.

The boundary condition at r = a imposed on the ring source solution requires special consideration. The film in annular flow is usually assumed to be a perfect absorber. For a process involving only molecular diffusion the assumption of perfect absorption implies a zero concentration at the wall. However, this is not the case for particle deposition, for which the length scale of particle motion can be large compared to the scale characterizing the droplet concentration. This can be seen by using a radiation boundary condition at the wall

$$-\varepsilon_P \left.\frac{\partial C}{\partial r}\right|_{r=a} = VC(a) \quad , \tag{19.23}$$

where V is the velocity with which particles move to the wall. If ε_P is represented by the product of V and a characteristic length L then (19.23) can be rewritten as

$$- VL \frac{\partial C}{\partial r} \Big|_{r=a} \cong VC\ (a) \tag{19.24}$$

If L is small (as is the case for molecular diffusion) and $(\partial C/\partial r)_a$ is finite, then C(a) = 0. However, if L is of the order of the length scale characterizing $\partial C/\partial r$, then C(a) needs to be a finite number.

Lee, et al (1989) formulated this boundary condition by specifying the mean number of particles per second per unit area that move toward the wall at r = a as

$$R_D = C(a) \langle v_r \mid v_r > 0 > f \rangle \quad , \tag{19.25}$$

where $\langle v_r \mid v_r > 0 \rangle$ is the conditionally averaged radial velocity, given that the radial velocity is positive, and f is the probability that the radial velocity is positive. It is assumed that the velocity with which the particles move toward the wall is proportional to the r.m.s. turbulent radial velocity component, $\left(\overline{v_r^2}\right)^{1/2}$:

$$\langle v_r / v_r > 0 \rangle = A\ (\overline{v_r^2})^{1/2} \tag{19.26}$$

If the probability distribution of v_r is a joint normal with zero mean then $A = \sqrt{2/\pi}$. Boundary condition (19.23) is

$$- \varepsilon_P \frac{\partial C}{\partial r} = \sqrt{\frac{2}{\pi}} \left(\overline{v_P^2}\right)^{1/2} Cf \tag{19.27}$$

at r = a.

Binder & Hanratty assumed

$$f\ (t) = \frac{1}{2} \left[1 - \exp\left(- \frac{t}{\tau_{LP}} \right) \right] \tag{19.28}$$

Initially all of the particles are moving away from the wall so f(0) = 0 . The function f(t) approaches 1/2 asymptotically at large times with the

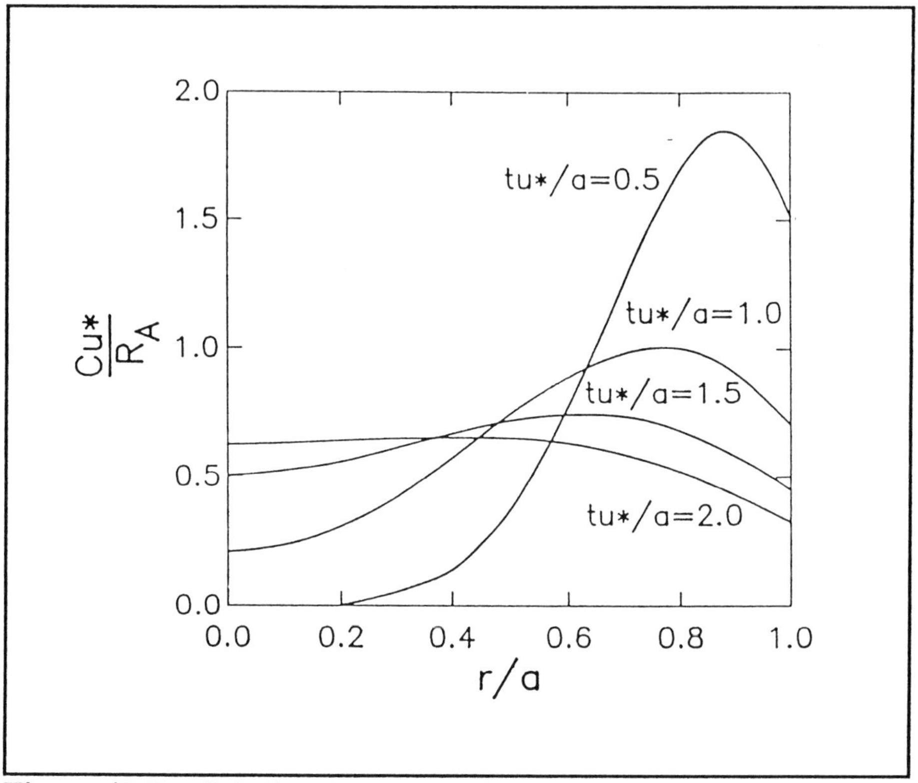

Figure 19.3 Calculated concentration fields for a ring source in vertical annular flow; $\beta\tau_{LF} = 1.0$

same time dependence as the diffusivity. The choice of this function was made for mathematical convenience, although it is physically reasonable since it would not be expected that f(t) equals 1/2 until the dispersing droplets have completely obtained the haphazard motion characteristic of diffusion at large times.

Calculations for a single ring source, with $C=\alpha\rho_l$, are shown in Figures 19.3 and 19.4, where $\alpha\rho_l\, u_*/R_A$ is plotted against r/a. The parameter for the different curves is the dimensionless time. Figure 19.3 is for $\beta\tau_{LF} = 1.0$. At times close to zero f = 0, so that there is no deposition at the wall. A maximum occurs at the wall and there is a diffusion of droplets away from the wall. As f takes on values different from zero, deposition occurs and there is diffusion both to the wall and away from the wall region. As a consequence, a peak in the droplet concentration profiles occurs close to the wall. As time increases the concentration profile becomes more diffuse. At long enough times the maximum is at the pipe center and diffusion occurs only toward the wall. Eventually, all the droplets redeposit and the concentration is zero over the whole pipe cross section.

Figure 19.4 represents very sluggish droplets for which $\beta\tau_{LF} = 0.01$. The profiles are calculated at the same values of dimensionless time as in Figure 19.3. The same qualitative behavior is found for $\beta\tau_{LF} = 0.01$ and for $\beta\tau_{LF} = 1.0$. However, the time development is much slower for $\beta\tau_{LF} = 0.01$, as would be expected for a particle with a larger time scale or, from (19.18), a smaller $(\overline{v_P^2})^{1/2}$.

Fully developed concentration profiles at large t can be calculated by summing the contributions from sources at different t′:

$$C = \int_0^t C_{s.s.}\,(t-t')\,dt' \tag{19.29}$$

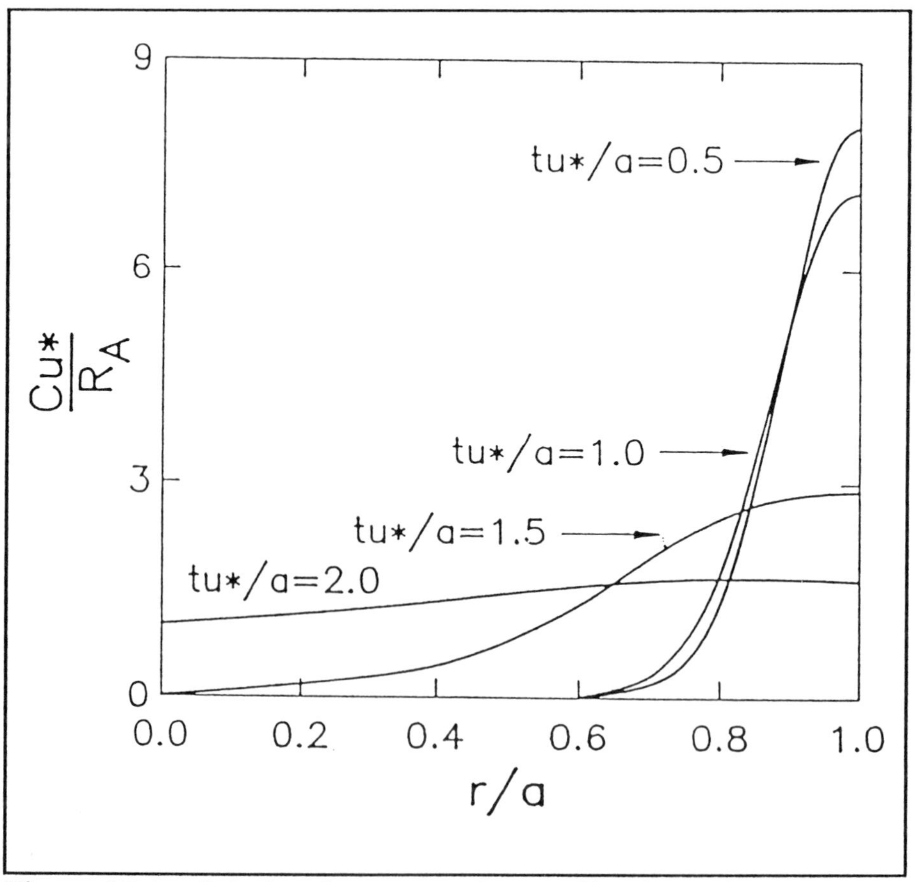

Figure 19.4 Calculated concentration profiles for a ring source of droplets in vertical annular flow; $\beta\tau_{LF} = 0.01$

Figure 19.5 shows results for values of $\beta\tau_{LF}$ calculated from droplet size measurements of Azzopardi (1985) to correspond to experiments of Gill, et al (1965). The calculations and the measured profiles are seen to agree reasonably well. A relatively flat profile with a maximum near the wall is obtained. A simple model of the type used to describe molecular diffusion would use a constant diffusion coefficient and a concentration of zero at the wall; a flat profile would not be predicted. Consideration of the time dependency of the turbulent diffusion process and the use of boundary condition (19.27) yields the correct interpretation of concentration profiles in vertical annular flow.

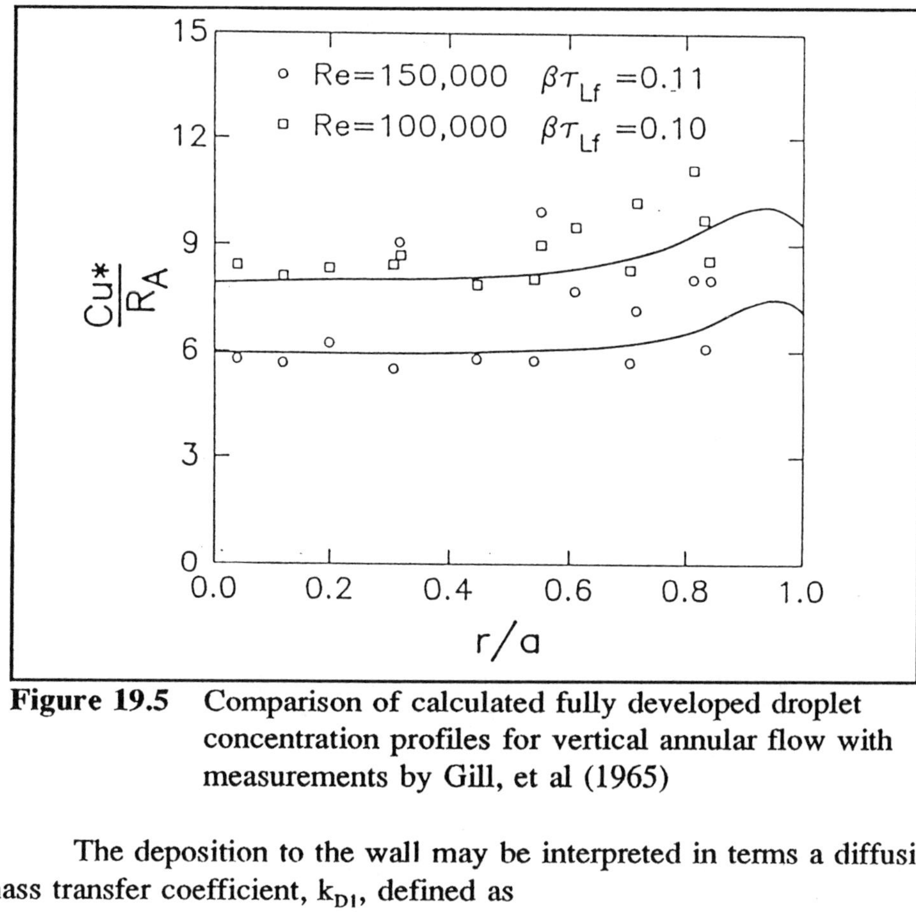

Figure 19.5 Comparison of calculated fully developed droplet concentration profiles for vertical annular flow with measurements by Gill, et al (1965)

The deposition to the wall may be interpreted in terms a diffusive mass transfer coefficient, k_{D1}, defined as

$$R_D = k_{D1}\left[\alpha_b\rho_l - \alpha\rho_l(r/a = 1)\right] \tag{19.30}$$

and a free flight mass transfer coefficient, k_{D2}, defined as

$$R_D = k_{D2}\,\alpha\rho_l(r = a) \quad , \tag{19.31}$$

where $\alpha_b\,\rho_l$ is the bulk concentration of liquid and $\alpha\rho$ is the local concentration of liquid. From (19.30), (19.31) and (19.19)

$$\frac{1}{k_D} = \frac{1}{k_{D1}} + \frac{1}{k_{D2}} \tag{19.32}$$

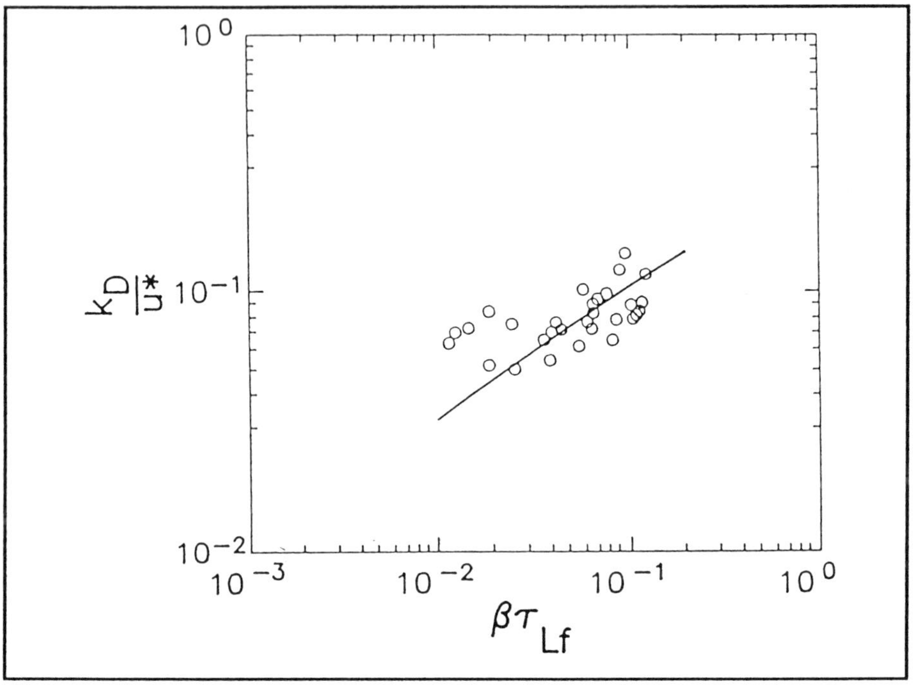

Figure 19.6 Comparison of calculated deposition constants for vertical annular flow with measurements by Schadel, et al (1990)

The concentration profile demonstrates that for a fully developed flow the concentration at the wall is just slightly larger than the bulk concentration. Therefore, the deposition is controlled by free flight to the wall. Thus $k_D \approx k_{D2}$ and, according to (19.26), k_D scales with $(\overline{v_P^2})^{1/2}$. From (19.18) it follows that k_D/u_* should be a function of $\beta\tau_{LF}$. Figure 19.6 compares calculated k_D with measurements by Schadel, et al (1990) for dilute droplet loadings. Values of $\beta\tau_{LF}$ were calculated for each of the experiments by (19.1) (19.2) and (19.20), using the drop size correlation of Azzopardi (1985) and the friction factor correlation of Asali, et al (1985).

19.6 DISPERSED FLOW IN A HORIZONTAL ANNULAR FLOW

For horizontal annular flows the gravitational field causes a stratification of the droplets and an asymmetric distribution of the liquid flowing along the wall. Binder (1991) extended the analysis presented in the previous section to the case of annular flow in a rectangular channel. Of particular interest are the droplet concentration profiles and the deposition rates at the top and bottom walls.

Figure 19.7 illustrates the system being considered. A differential line source on the bottom wall has a strength R_{AB} dt′, where R_{AB} is the rate of atomization from the bottom wall. Droplets originating from a source on the bottom wall can deposit either on the bottom wall or on the top wall. The fraction depositing on the top wall, F_{BT}, is expected to be less

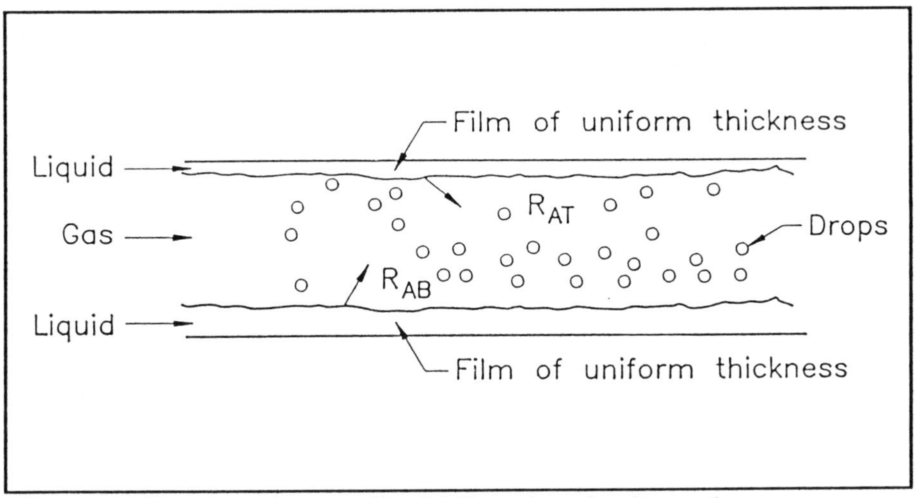

Figure 19.7 Annular gas-liquid flow in a horizontal two-dimensional channel

than the fraction depositing on the bottom wall, 1-F_{BT}.

Similar sources, of strength R_{AT} dt′, exist on the top wall. Under fully developed conditions, conservation of mass shows that

$$\frac{R_{AT}}{R_{AB}} = \frac{F_{BT}}{F_{TB}} \ . \tag{19.33}$$

It is expected that, because of gravitational settling, that $F_{BT} < F_{TB}$. As a consequence, the rate of atomization and the film thickness for the bottom wall are larger than for the top wall.

Gravity is expected to act independently of the turbulence. Therefore an average particle trajectory can be calculated from the following equation:

$$m_P \frac{dV_P}{dt} = -\frac{1}{2} \rho \, C_D \, A_p \, V_P^2 - g m_P \left(1 - \frac{\rho}{\rho_P} \right) \tag{19.34}$$

Here V_P is the time-averaged velocity in the x-direction, m_P, the particle mass, C_D, the drag coefficient, A_p, the projected area of the drop and g, the acceleration of gravity. At t = t′ the particles are assumed to be fully entrained in the field so that

$$V = V_0 = \pm \left(\overline{v_p^2} \right)^{1/2} . \tag{19.35}$$

The plus sign is for a source on the bottom wall and the minus sign is for a source on the top wall. At $t = \infty$, $dV_P/dt = 0$ and the particle reaches its terminal velocity. Since the particles will not be in the field for infinite time it is expected that $V < V_{s\infty}$.

The concentrations of particles originating from a wall source are calculated from the following modification of (19.11):

$$\frac{\partial C}{\partial t} + V_P \, (t-t') \, \frac{\partial C}{\partial x} = \varepsilon_P \, (t-t') \, \frac{\partial^2 C}{\partial x^2} + Q \, \delta \, (x/x') \, \delta \, (t/t') \, , \tag{19.36}$$

This is solved using boundary condition (19.28) to obtain the behavior of a line source on the bottom or top wall.

As for vertical annular flow, droplet sizes have a range so that $0.01 < \beta\tau_{LF} < 0.2$, $v_{s\infty} / u_* < 1$ and $\tau_p^+ > 20$. Since $\rho/\rho_p \ll 1$, two dimensionless groups characterize the system, $\beta\tau_{LF}$ and Fr $\beta\tau_{LF}$, where the Froude number is defined as $Fr = u_*^2/gH$. For Stokesian particles,

$$Fr \, \beta \, \tau_{LF} = 0.046 \, u_*/v_{s\infty} \tag{19.37}$$

Therefore, the reciprocal of this group may be looked upon as a measure of the ratio the settling velocity to the magnitude of turbulent velocity fluctuations.

Figures 19.8ab show concentration profiles downstream of a source. The calculations in 19.8a are for $Fr\beta\tau_{LF} = 0.1$ and represent the case of a large $v_{s\infty}/u_*$. It is noted that the influence of gravity is so strong that droplets do not reach the top wall, at x/H = 1. Deposition on the bottom wall occurs almost entirely by gravitational settling.

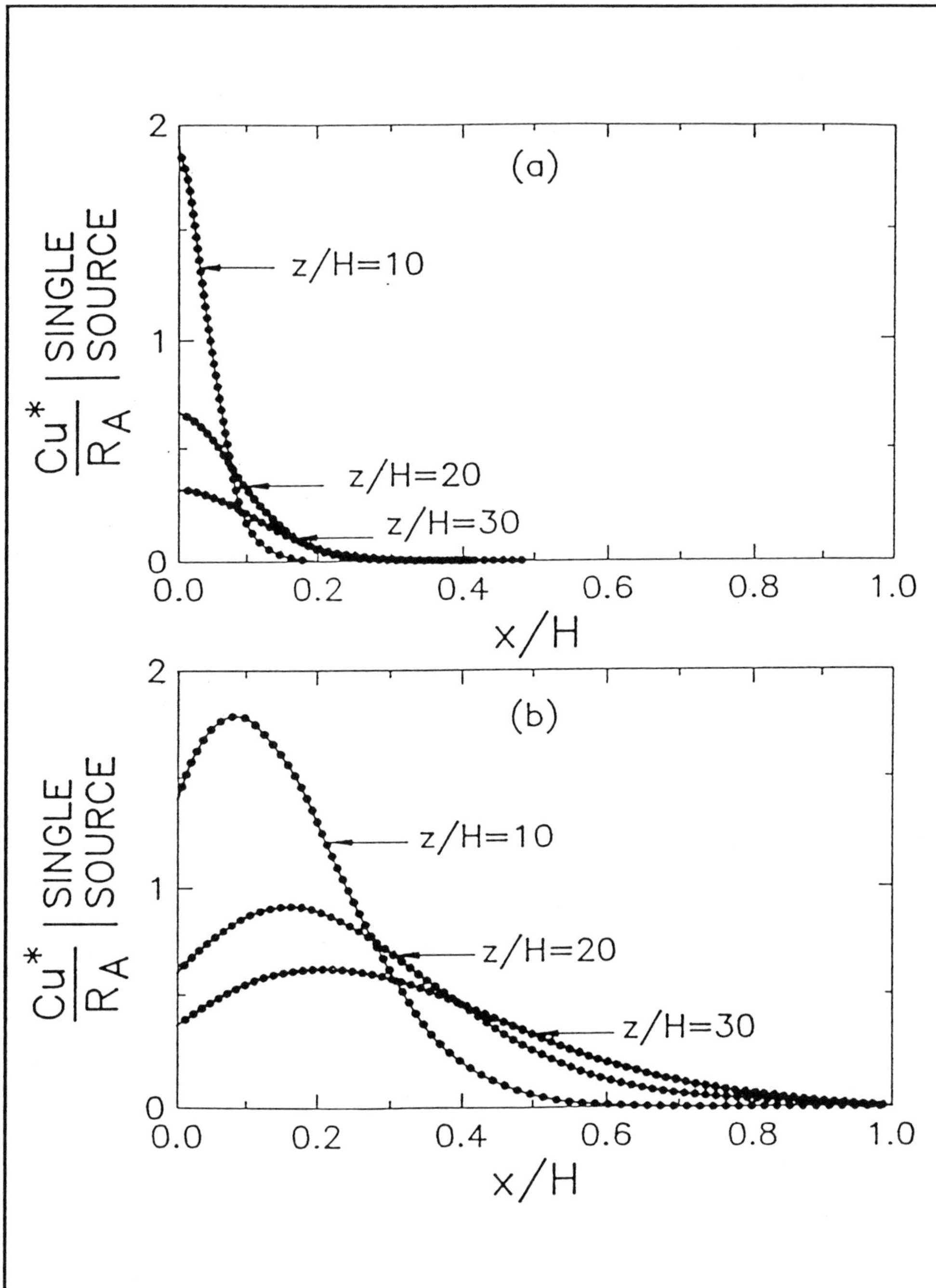

Figure 19.8a and 19.8b Concentration profile of an instantaneous source on the bottom wall. a) Fr = 10, $\beta\tau_{LF}$ = 0.01, Re = 100,000; b) Fr = 10, $\beta\tau_{LF}$ = 1.0, Re = 100,000

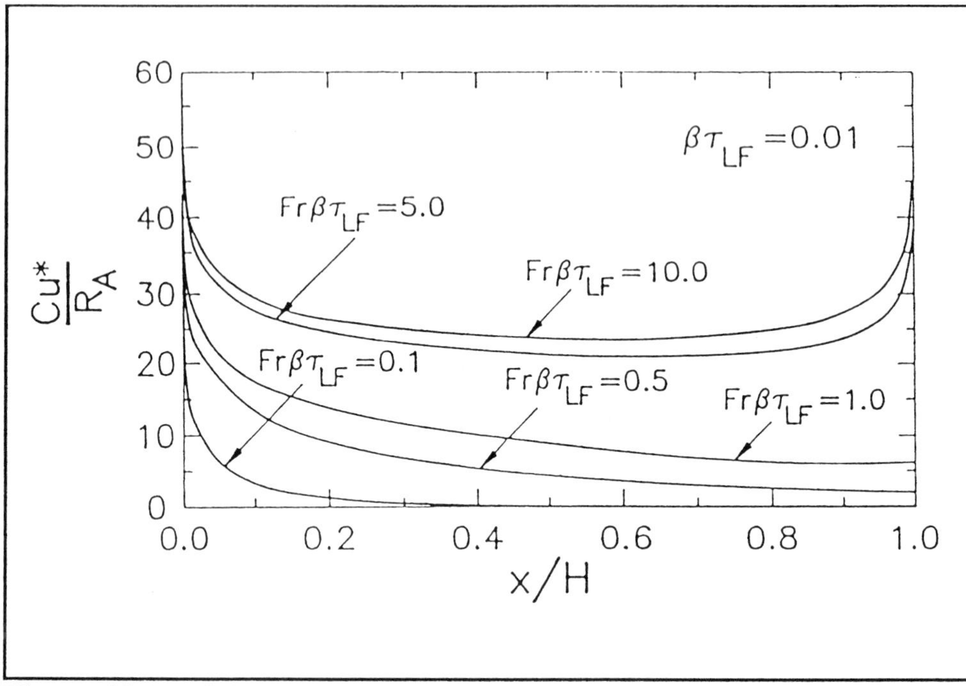

Figure 19.9 Calculated fully-developed droplet concentration profiled for gas-liquid annular flow in a horizontal channel

The calculations in Figure 19.8b are for a large $Fr\beta\tau_{LF}$ or for a small settling velocity. In this case turbulent mixing is strong compared to settling so that droplets reach the top wall to form a liquid layer. Deposition by turbulence is much stronger than by gravitational settling, as evidenced by the strong concentration gradients at the bottom wall.

Fully developed concentrations profiles, calculated by summing the contributions of a number of sources, are shown in Figure 19.9 for $\beta\tau_{LF} = 0.01$ and for different Fr. It is noted that with increasing $Fr\beta\tau_{LF}$ the flow pattern changes from one in which the droplets are located in a region close to the liquid at the bottom of the channel to one in which the droplets are uniformly dispersed.

Figure 19.10 shows deposition constants based on the rate of deposition on the bottom wall (k_{DB}) and on the rate of deposition on the top wall (k_{DT}). For $Fr\beta\tau_{LF} > 10$ a fully dispersed pattern similar to what is observed in vertical flows is obtained. Deposition is caused by turbulence and the rates to the bottom and top walls are the same. They increase with increasing $\beta\tau_{LF}$ because of the increased ability of the particles to follow the fluid turbulence.

For $Fr\beta\tau_{LF} < 0.2$ to 1 (depending on $\beta\tau_{LF}$) the drops are located in the bottom part of the channel and all of the deposition occurs on the bottom wall. The deposition rate is controlled by gravitational settling and the rate of settling is roughly proportional to $(Fr\beta\tau_{LF})^{-1}$ or to the settling velocity. This means that, in contrast to the case of large $Fr\beta\tau_{LF}$, the rate of settling increases with decreasing $\beta\tau_{LF}$, or increasing droplet size.

For intermediate $Fr\beta\tau_{LF}$ a stratified dispersed flow exists for which deposition occurs on both walls. Both turbulence and gravitational settling are important in determining the deposition rate.

Of interest are the opposite effects of $\beta\tau_{LF}$ (or droplet size) at high and low $Fr\beta\tau_{LF}$. A consequence of this is that for a fixed Fr (or a fixed

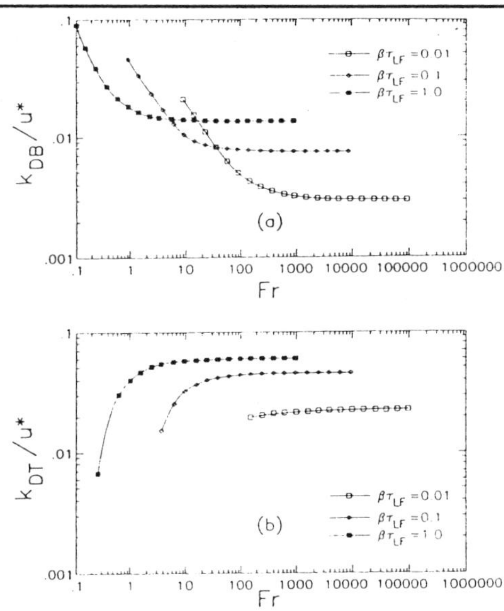

Figure 19.10 Deposition constants for sources on the bottom wall. a) Bottom wall b) Top wall

fluid flow rate) there is a drop size for which the deposition rate is a minimum. These drops will have the largest lifetime and it is to be expected that the dropsize distribution of the drops in the flow will be different from that characterizing the drops entering the flow.

19.7 SEDIMENT TRANSPORT

The approach outlined in the previous section has also been used by Binder (1991) to analyze sediment transport in a liquid.

In this case, conditions are such that $1 < \beta\tau_{LF} < 100$ and $v_{s\infty}/u_* < 1$. As a consequence, the turbulence characteristics of the solid particles are the same as the turbulence characteristics of the fluid: $\overline{v_P^2} \cong \overline{v^2}$ and $\varepsilon_P \cong \varepsilon$. One can expect that the behavior will be different from droplets in horizontal gas-liquid annular flow in that it will be most strongly affected by the dimensionless group, $\beta\tau_{LF}$ Fr.

The method of analysis is pictured in Figure 19.11. The bottom of the channel through which the liquid is flowing is a source of solid particles. As shown in Figure 19.11, particles entering the field at $t = t'$ could redeposit on the bottom boundary. However, some particles could reach the top wall at $t = t_S$. Unlike the case of annular flow, these

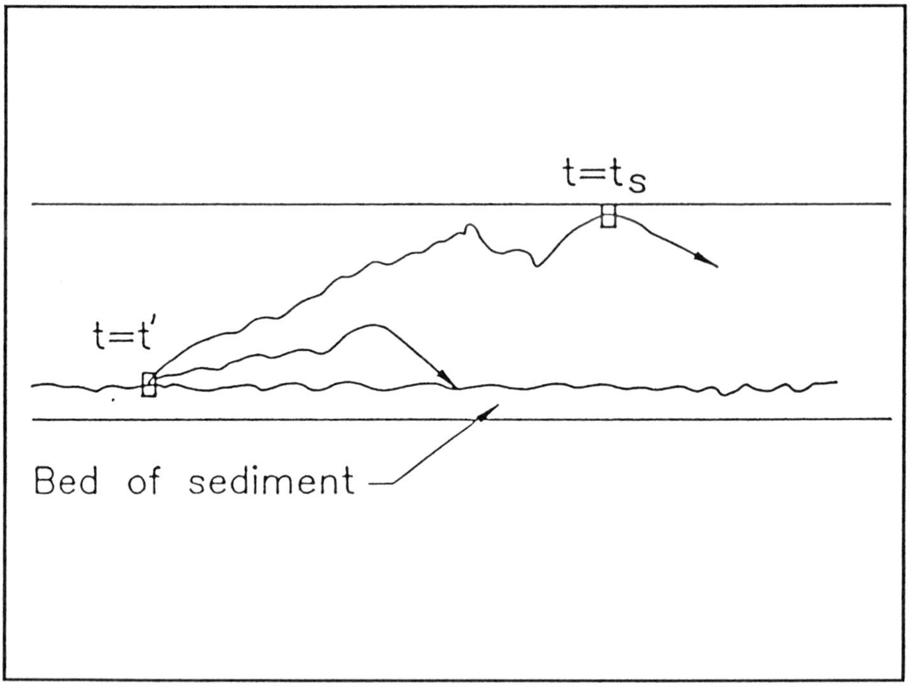

Figure 19.11 Sediment transport in a horizontal channel; typical trajectories from a wall source

particles do not lose their identity but are carried away from the wall by gravity and turbulence.

Thus, entrained solid particles from the bottom boundary enter the field at $t = t'$ with a velocity

$$V_O = \left(\overline{v_P^2}\right)^{1/2} \tag{19.38}$$

The deterministic velocity of the particles, $V(t-t')$ is calculated with (19.34) with an initial condition given by (19.38). Particles that reach the top wall are pictured to form new sources which immediately enter the field with velocity

$$V_o = V(t_S - t') - \left(\overline{v_P^2}\right)^{1/2} \tag{19.39}$$

The deterministic velocity of these particles, $V(t-t_S)$, is calculated from (19.34) using initial condition (19.39) at $t = t_S$.

The behavior of a single source (including the contributions from sources it creates at the top boundary) of solid particles is calculated from (19.38) using a source strength of $R_{AB}dt'$ and boundary conditions given by (19.27) at both boundaries, $y = 0$ and $y = H$. These particles deposit at the top and bottom boundaries at rates given by

$$R_{DT}\ (t-t') = -\ \varepsilon_P \left.\frac{\partial C}{\partial x}\right|_{y=H} \tag{19.40}$$

$$R_{DB}\ (t-t') = -\ \varepsilon_P \left.\frac{\partial C}{\partial x}\right|_{y=0}, \tag{19.41}$$

where $C\ \alpha\rho_s$. The strength of sources created at top boundary by the bottom source at $t = t'$ are given as

$$R_{AT}\ dt' = R_{DT}\ (t_S - t')\ dt' \tag{19.42}$$

The behaviors of these secondary sources for $t>t_S$ are calculated from (19.36), again using boundary condition (19.27).

The calculated behavior of single of solid particles sources are similar to those in Figure 19.8. The fully developed concentration profiles shown in Figure 19.12 were calculated by summing the contributions of different sources. It is noted that these become more stratified as $Fr\beta\tau_{LF}$ decreases, or as $v_{s\infty}/u_*$ increases.

For a fully developed flow the rate of deposition is equal to the rate of entrainment of particles under these conditions. A deposition constant can be calculated as

$$R_{AB} = k_D\,\alpha_b\rho_s \tag{19.43}$$

where $\alpha_b\rho_s$ is the bulk mass concentration of solid particles. For the range of parameters being considered, the calculations show that

$$\frac{k_D}{u_*} = f\left(\beta\tau_{LF}\,Fr\right) \quad , \tag{19.44}$$

where the right side is a monotinically increasing function of $(\beta\tau_{LF}\,Fr)^{-1}$. Therefore, a dimensionless bed load is calculated as

$$\frac{\alpha_b\,\rho_s\,u_*}{R_{AB}} = f^{-1}\left(\beta\tau_{LF}\,Fr\right) \tag{19.45}$$

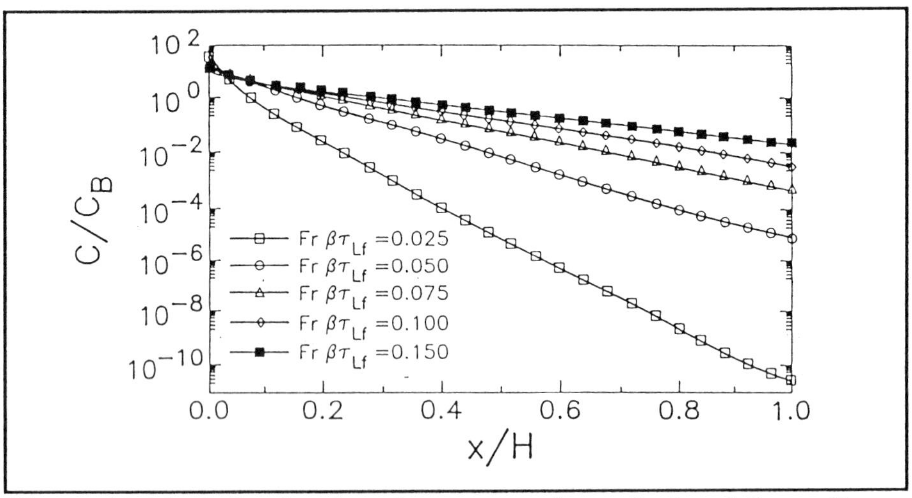

Figure 19.12 Fully developed concentration calculated for sediment transport in a horizontal channel

19.8 CONCLUSIONS

The main theoretical work done on dispersed flows has focussed on the transport of dilute suspensions of sediment. Consequently, it is of interest to conclude this article by comparing the Lagrangian analysis with approaches currently followed in the literature.

The classical theoretical treatment (O'Brien, 1933; Rouse, 1977; Yalin, 1977) is to assume that the settling is balanced by turbulent diffusion so that

$$E_p(x)\,\frac{dC}{dx} + v_{s\infty} = 0 \ , \qquad (19.46)$$

with $C = \alpha\, \rho_p$. Far enough away from the boundary the diffusivity may be considered to be approximately constant so that

$$\frac{d\,\ell n\,C}{d\,x} = -\,\frac{v_{s\infty}}{E_p} \qquad (19.47)$$

Comparison of (19.47) with measurements (Karabelas, 1977; Hunt, 1954, 1969) often gives values of E_p which are larger than the diffusion coefficient for fluid particles. From the slopes of the concentration profiles

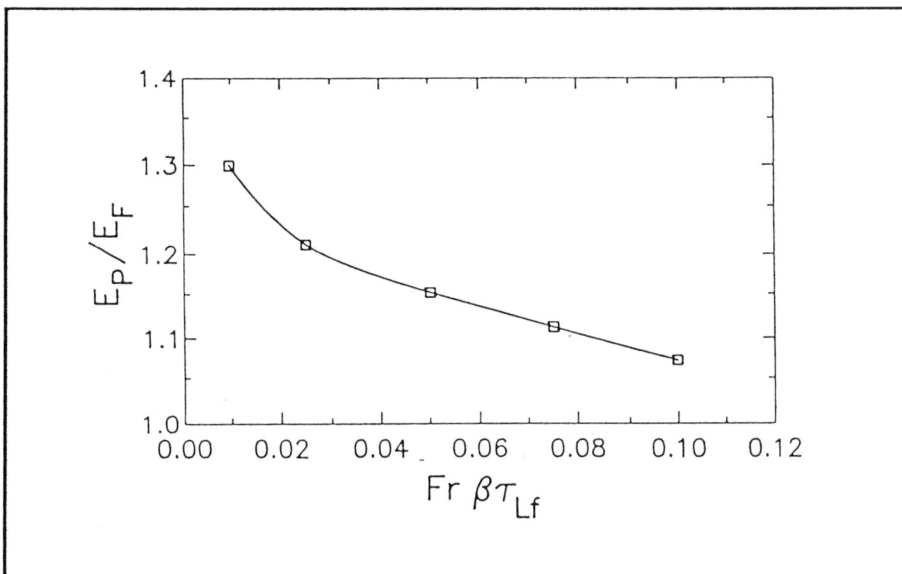

Figure 19.13 Ratios of the Eulerian turbulent diffusion coefficients calculated for the profiles in Figure 19.12 for 0.2<y/H,0.6 by using equation 19.46

in Figure 19.12 in the region from $y/H = 0.2$ to $y/H = 0.6$ values of E_p can be calculated from (19.47). These are shown in Figure 19.13. They are consistent with laboratory measurements in that they are larger than the long time fluid diffusivity. For the range of parameters explored in this paper it is expected that $\varepsilon_p \cong \varepsilon_f$. Consequently, the calculated $(E_p/E_F) > 1$, shown in Figure 19.13, could arise because (19.46) should use some average V which is less than $v_{s\infty}$, since the solid particles redeposit before they reach their terminal velocity. The rate of deposition and, therefore, the discrepancy between V and $v_{s\infty}$ increases with decreasing $Fr\beta\tau_{LF}$.

The integration of (19.46) provides a concentration profile which is singular at y = 0. This difficulty is usually avoided by specifying empirically a boundary condition at some finite x.

The extension of (19.46) so as to describe developing concentration profiles (Jobson & Sayre, 1970; Celik & Rodi 1988; Hjelmfelt & Lenau, 1970; Parker, 1978; Van Rijn, 1986) is carried out by solving the diffusion equation,

$$U \frac{\partial C}{\partial z} + V \frac{\partial C}{\partial x} = \frac{\partial}{\partial x}\left(E_P \frac{\partial C}{\partial x}\right), \tag{19.48}$$

where $C = \alpha\,\rho_p$. For a fully developed flow U is only a function of x and V is constant, usually taken as $v_{s\infty}$. The boundary condition at the top boundary is specified by assuming at y = H, that

$$V\,C - E_P \frac{\partial C}{\partial x} = 0 \tag{19.49}$$

Another boundary condition at a finite distance from the lower boundary needs to be specified empirically.

The analysis presented in this paper differs from the above approach in a number of ways:

The difficulties of specifying a boundary condition is sidestepped by specifying the strength of a source. The present analysis focusses attention on the process by which particles enter the flow. What is the rate of entrainment? How do the particles enter the field? In particular, what is their velocity and how rapidly do they become entrained in the turbulence?

Equation (19.49) does not appear to be an appropriate boundary condition since it can be derived directly from (19.48) if it is assumed that $\partial C/\partial z = 0$. This type assumption seems inconsistent if the flow is developing. The present analysis uses boundary condition (19.27), which does not contain the deterministic velocity, V. A rigorous justification for this choice can be obtained by considering the integral of (19.36) over the entire spatial domain (Binder, 1991).

The Eulerian analysis, represented by (19.47), provides no way to specify how V is related to the time the particles have been in the field.

The consideration of diffusion from a small source, along with the determinate equation for V, introduces this effect in a natural way.

The turbulence properties of solid particles in a liquid flow are usually the same as the fluid turbulence. However, this is not the case for solid particles or droplets in a gas flow. As has been discussed, most often $\varepsilon_P \cong \varepsilon$ but $\overline{v_P^2} < \overline{v^2}$. Equation (19.48) and the boundary conditions usually employed in an Eulerian approach account for $E_P \neq E$, but do not account for $\overline{v_P^2} \neq \overline{v^2}$.

In the Eulerian approach the turbulent diffusion coefficient is allowed to vary with distance from the boundary. In the Lagrangian method, outlined in this paper, the turbulence properties are assumed to be uniform. The time-dependency of diffusion from a small source gives rise to an effective spatial variation of the turbulent diffusion coefficient defined from an Eulerian framework. The assumption of a homogeneous turbulence and a plug flow, needed to implement the Lagrangian analysis in a convenient way, is a limitation. However, the ability to account for the time-dependency of turbulent diffusion in an exact way gives additional physical insights. For example, in vertical annular flow the two methods of droplet mixing (straight line trajectories versus haphazard trajectories) are taken into account as the short time and long time behavior of turbulent diffusion from a small source.

The analysis presented in this chapter must be looked upon as a first step because of the simplifying assumptions of a dilute particle field, a plug velocity profile, uniform turbulence characteristics and zero slip in the x-direction. The effect of slip can be taken into account by solving an equation similar to (19.34) for the particle velocity component in the flow direction, but the initial velocity of the particles must be specified. The effects of flow non-homogeneities and of particle-particle interactions seem like longer range problems which will require a careful examination of experimental data in the context of the Lagrangian analysis.

To the authors, the most interesting aspect of the analysis, from the viewpoint of future research, is that it focuses attention on the need to understand how particles are entrained. What physical processes control the rate of entrainment? How and where do the particles enter the field?

ACKNOWLEDGEMENT

This work has been supported by the National Science Foundation under Grant NSF CTS 89-19843 and by the Department of Energy under Grant DOE DEF G02-86ER 13556.

REFERENCES

Andreussi, P. and B. J. Azzopardi, "Droplet Deposition and Interchange in Annular Flow", *Int. J. Multiphase Flow*, Vol. 9, 1983, pp. 681-695.

Asali, J. C., T. J. Hanratty, and P. Andreussi, "Interfacial Drag and Film Height for Vertical Annular Flow", *AIChE Journal*, Vol. 31, 1985, pp. 895-902.

Azzopardi, B. J., "Drop Sizes in Annular Two-Phase Flow", *Experiments In Fluids*, Vol. 3, 1985, pp. 53-59.

Batchelor, G. K., "Diffusion in a Field of Homogeneous Turbulence", *Australian Journal of Scientific Research*, Vol. 2, 1949, p. 437.

Binder, J. L., *Use of Lagrangian Methods to Describe Particle Deposition and Distribution in Dispersed Flows*, Ph. D. thesis, University of Illinois, Urbana, Illinois 1991.

Binder, J. L. and T. J. Hanratty, "A Diffusion Model for Droplet Deposition in Gas/Liquid Annular Flow, *Int. J. Multiphase Flow*, Vol. 17, 1991, pp. 1-11.

Celik, I. and W. Rodi, "Modeling Suspended Sediment Transport in Nonequilibrium Situations", *Journal of Hydraulic Engineering*, Vol. 114, 1988, pp. 1157-1191.

Comte-Bellot, G., "Ecoulement Turbulent Entre Deux Parois Paralleles", *Publications Scientifiques et Techniques De Ministère De L'Air*, No. 419, 1965, Paris.

Gill, L. E., G. F. Hewitt, and P. M. C. Lacey, "Sampling Probe Studies of the Gas Core in Annular Two Phase Flow: Part 2 Studies of the Effect of Two-Phase Rates on Phase and Velocity Distribution", *Chemical Engineering Science*, Vol. 19, 1965, pp. 665-682.

Hanratty, T. J., "Heat Transfer Through a Homogeneous Isotropic Turbulent Field", *AIChE Journal*, Vol. 2, 1956, pp. 42-45.

Hanratty, T. J., "Note on the Analogy Between Momentum Transfer and Heat or Mass Transfer for a Homogeneous Isotropic Turbulent Field, *AIChE Journal*, Vol. 4, 1958, pp. 495-496.

Hanratty, T. J. and D. L. Flint, "Velocity Profile for Fully Developed Turbulent Flow in a Pipe", *AIChE Journal*, Vol. 4, 1958, pp. 132-136.

Hjelmfelt, A. T. and C. W. Lenau, "Nonequilibrium Transport of Suspended Sediment", *Journal of Hydraulic Engineering*, Vol. 96, 1970, pp. 1567-1586.

Hunt, J. N., "The Turbulent Transport of Suspended Sediment in Open Channels", *Proceedings of the Royal Society of London*, Vol. A 224, 1954, pp. 332-335.

Hunt, J. N., "On the Transport of a Heterogeneous Sediment", *Quarterly Journal of Mechanics and Applied Mathematics*, Vol. 22, Pt. 2, 1969, pp. 236-246.

Jobson, H. E. and W. W. Sayre, "Vertical Transfer in Open Channel Flow", *Journal of Hydraulic Engineering*, Vol. 96, 1970, pp. 703-724.

Karabelas, A. J., "Vertical Distribution of Dilute Suspensions in Turbulent Pipe Flow", *AIChE Journal*, Vol. 23, 1977, pp. 426-434.

Laufer, J., "The Structure of Turbulence in Fully Developed Pipe Flow", *NACA Technical Report*, 1174, 1954.

Lee, M. M., T. J. Hanratty, and R. J. Adrian, "An Axial Viewing Photographic Technique to Study Turbulence Characteristics of Particles", *Int. J. Multiphase Flow*, Vol. 15, 1989a, pp. 787-802.

Lee, M. M., T. J. Hanratty, and R. J. Adrian, "The Interpretation of Droplet Diffusion Measurements with a Diffusion Model", *Int. J. Multiphase Flow*, Vol. 15, 1989b, pp. 459-469.

McCoy, D. D. and T. J. Hanratty, "Rate of Deposition of Droplets in Annular Dispersed Flow", *Int. J. Multiphase Flow*, Vol. 3, 1975, pp. 319-331.

Nir, A. and L. M. Pismen, "The Effect of a Steady Drift on the Dispersion of a Particle in a Turbulent Fluid", *J. Fluid Mech.*, Vol. 94, 1979, pp. 369-381.

O'Brien, M. P., "A Review of the Theory of Turbulent Flow and Its Relation to Sediment Transport", *Transactions of the American Geophysical Union*, Vol. 14, 1933, pp. 487-491.

Parker, G., "Self-Formed Straight Rivers with Equilibrium Banks and Mobile Bed. Part I. The Sand-Silt River", *J. Fluid Mech.*, Vol. 89, 1978, pp. 109-125.

Pismen, L. M. and A. Nir, "On the Motion of Suspended Particles in Stationary Turbulence", *J. Fluid Mech.*, Vol. 84, 1978, pp. 193-206.

Reeks, M. W., "On the Dispersion of Small Particles Suspended in an Isotropic Turbulent Fluid", *J. Fluid Mech.*, Vol. 83, 1977, pp. 529-546.

Rouse, H., "Modern Conceptions of the Mechanics of Fluid Turbulence", *Transactions of the American Society of Civil Engineering*, Vol. 102, 1937, pp. 463-505.

Sage, B. H., F. Page, W. H. Corcoran, and W. G. Schlinger, "Temperature and Velocity Distributions in Uniform Flow Between Parallel Plates", *Industrial Engineering Chemistry*, Vol. 44, 1952, p. 419.

Schadel, S. A., G. W. Leman, J. L. Binder, and T. J. Hanratty, "Rates of Atomization and Deposition in Vertical Annular Flow", *Int. J. Multiphase Flow*, Vol. 16, 1990, pp. 363-374.

Taylor, G. I., "Diffusion by Continuous Movements", Proceedings of the London Mathematical Society, Vol. 151, 1926, pp. 196-211.

Vames, J. S. and T. J. Hanratty, "Turbulent Dispersion of Droplets for Air Flow in a Pipe", *Experiments in Fluids*, Vol. 6, 1988, pp. 94-104.

Van Rijn, L. C., "Mathematical Modelling of Suspended Sediment in Non-Uniform Flows", *Journal of Hydraulic Engineering*, Vol. 112, 1986, pp. 433-455.

Yalin, M. S., "Mechanics of Sediment Transport", Pergamon Press, Toronto, 1977.

Young, J. B. and T. J. Hanratty, "Optical Studies on the Turbulent Motion of Solid Particles in a Pipe Flow", *Journal of Fluid Mechanics*, Vol. 231, 1991, pp. 665-688.

Young, J. B. and T. J. Hanratty, "Trapping of Solid Particles at the Wall in a Turbulent Flow", *AIChE Journal*, October, 1991.

Yudine, M. I., "Physical Considerations on Heavy-Particle Diffusion in Atomospheric Diffusion and Air Polluting", *Advances In Geophysics*, Academic Press, New York, Vol. 6, 1959, pp. 185-191.

CHAPTER 20

GAS-SOLID FLOW IN PIPES

R. Jackson

20.1 INTRODUCTION

When a single phase fluid flows in a pipe the walls of the pipe influence the flow in two ways. First, they *confine* the flow; in other words they demand that the normal component of the velocity should vanish at their surface. Second, they exert a tangential force on the fluid in contact with them, usually large enough to ensure that there is negligible slip between this fluid and the solid surface. A treatment of this situation at the most naive level, neglecting viscous stresses in the fluid and recognizing only the confining function of the pipe walls, leads to virtually no results that are even qualitatively correct. Thus for example, for fully developed flow it predicts that the pressure gradient in the axial direction is independent of the flow rate and is simply hydrostatic. In contrast, a treatment of the flow of a suspension of particles in a gas at a comparably simple level, allowing only for forces exerted between the two phases and neglecting effects due to the walls, other than their function of confining the flow, generates a qualitative pattern of behavior that is complicated and, in many important respects, realistic. Such a treatment also indicates that the nature of the suspension flow depends on the characteristics of the device used to drive the gas through the pipe, in a way that has no analog in the flow of a single phase fluid. This coupling is so close that it often has little meaning to discuss the mechanics of two phase flow in the pipe alone, without reference to the devices used to feed the fluid and the particles. Much of the confusion to be found in the literature of empirical observations on these systems can probably be attributed to failure to recognize this.

However, it would be wrong to conclude, from these comments, that suspension flows can be analyzed successfully without regard to interactions between the flowing material and the walls. On the contrary, *quantitative* predictions of this sort, such as pressure gradients and particle phase concentrations, are usually enormously in error, and of course it is impossible to predict details of velocity and concentration profiles over the cross section, which are sometimes of great technical importance.

In this chapter we will first illustrate these remarks by a simple, and generally familiar example, then we will present a case in which a very complicated overall pattern of behavior can be elucidated adopting only the simplest picture of flow within the pipe, provided that careful account is taken of the interaction between the pipe and its feed device. We will then examine factors which influence the cross-sectional distributions of particle concentration and the velocities of the two phases, and will show how a model including some of these factors generates a variety of behavior quite inaccessible to simpler treatments. Finally, we will comment on entrance effects, the local stability of the flow field and the likelihood that other prominent features of the flow are related to instabilities of this sort.

No attempt will be made to survey the considerable literature describing empirical correlations for the prediction of the pressure drop accompanying gas-particle flow in particular situations (see, for example, Zenz and Othmer, 1960 or Kunii and Levenspiel, 1969). This is not with any intent of minimizing the importance of such results; indeed, at present they are virtually the only reliable resource for designers of practical devices. However, the present purpose is to look forward at the possibility of replacing, or at least supplementing them with something more closely related to the physics of the system, which will lead to an understanding of the mechanisms behind the rather complicated phenomena we observe.

20.2 ONE DIMENSIONAL THEORY OF PIPE FLOW

The simplest model of fully developed gas-particle flow in a pipe neglects all details of velocity and concentration profiles and is formulated entirely in terms of cross-sectional average values of the particle concentration and the velocities of the two phases. The walls of the pipe serve to contain the flow but additional, axially directed forces exerted between the walls and the adjacent flowing material are accounted for, if at all, by empirical contributions to the overall pressure gradient formulated in terms of "friction factors". There is, of course, an analogous treatment of the flow of a single fluid phase, with results that are summarized in the familiar dimensionless relation between friction factor and Reynolds number. However, there is an important difference between the two cases. For a single phase fluid, if the drag force between the fluid and the pipe wall is neglected the results reduce to the trivial prediction that the pressure gradient in the pipe is simply hydrostatic. For the two phase flow, on the other hand, even without the wall drag it is possible to predict a complicated and interesting pattern of behavior that is, in many respects, realistic. We shall start, therefore, at this extreme level of simplification.

Let $\bar{V}_g$ denote the volume flow rate of gas, divided by the cross sectional area of the pipe, while $\bar{V}_s$ similarly denotes the volume flow rate of solid particulate material, per unit cross sectional area. Then if α_s denotes the average fraction of the cross section occupied by solid material, the drag force per unit total volume of the suspension exerted by the gas on the particles will be assumed to be given by

$$f = \beta(\alpha_s)\left[\frac{\bar{V}_g}{1 - \alpha_s} - \frac{\bar{V}_s}{\alpha_s}\right] \tag{1}$$

Here the term $\beta(\alpha_s)$ represents a drag coefficient that depends on the average concentration of the particles, while the term in square brackets would be the relative velocity of the two phases if conditions were uniform over the cross section. (This algebraic form is appropriate only if the Reynolds number, determined by the particle size and the relative velocity, is sufficiently small: for large values of the Reynolds number the second power of the relative velocity would appear. However, for the purpose of the present qualitative discussion it serves no purpose to complicate the algebra to accommodate all Reynolds numbers.) The form of $\beta(\alpha_s)$ will depend on the way in which the velocities and concentration are distributed over the cross section. We will initially make the simplest assumption that they are uniform, in which case the well known Richardson-Zaki (1954) correlation gives

$$\beta(\alpha_s) = \frac{\rho_s \alpha_s g}{v_t(1 - \alpha_s)^n} \tag{2}$$

where v_t denotes the terminal velocity of fall of a single particle through a large volume of gas at rest, n is an exponent whose value depends on the Reynolds number for the particle at velocity v_t, and ρ_s denotes the intrinsic density of the solid material.

Neglecting drag forces exerted on either phase by the wall of the pipe, a force balance on the particle phase then gives

$$\beta(\alpha_s)\left[\frac{\bar{V}_g}{1 - \alpha_s} - \frac{\bar{V}_s}{\alpha_s}\right] = \rho_s \alpha_s g$$

or, using (2)

$$\frac{\bar{V}_g/v_t}{1 - \alpha_s} - \frac{\bar{V}_s/v_t}{\alpha_s} = (1 - \alpha_s)^n \tag{3}$$

Since the drag exerts an equal and opposite reaction on the gas, a force balance for the gas phase gives

$$\beta(\alpha_s)\left[\frac{\bar{V}_s}{\alpha_s} - \frac{\bar{V}_g}{1-\alpha_s}\right] = \frac{dp}{dz} \tag{4}$$

or, using the force balance on the particles

$$\frac{dp}{dz} = -\rho_s\alpha_s g \tag{5}$$

where z is a coordinate measured upward along the axis of the pipe and p is the pressure of the gas.

Clearly α_s can be eliminated between equations (3) and (5) to give a relation between the three variables that describe the macroscopic mechanical behavior of the pipe, namely $\bar{V}_g$, $\bar{V}_s$ and dp/dz. The form of this relation can be seen most easily by noting that, for each value of α_s, equation (3) defines a straight line of slope $(1 - \alpha_s)/\alpha_s$ in the $(\bar{V}_s, \bar{V}_g)$-plane. These lines are sketched in Figure 1: they form a one-parameter set bounded by the ordinate axis for $\alpha_s = 0$, and a line of smallest slope for $\alpha_s = \alpha_{max}$, where α_{max} is the largest possible value of the volume fraction of solids. This depends on the geometry of the particles and cannot exceed a value corresponding to random close packing. Since the pressure gradient is related to α_s by equation (5) each line corresponds to a fixed value of dp/dz, so the diagram represents the relation between $\bar{V}_g$, $\bar{V}_s$ and dp/dz in the form of contours of constant dp/dz (or, equivalently α_s) in the $(\bar{V}_s,\bar{V}_g)$-plane. The segments of the lines lying in the first, second, and third quadrants represent cocurrent upflow, countercurrent flow, and cocurrent downflow, respectively. In the second quadrant the lines are confined to a region adjacent to the origin, bounded by their envelope. Physically this corresponds to the fact that countercurrent flow is not possible if the magnitude of the flow rate of either phase is too large. The envelope therefore represents the conditions referred to as "flooding" in countercurrent flow.

Though the above exhibits the behavior in a simple and compact form, for the purpose of describing the mechanics of pipe flow it is easier to work with sections of this diagram at constant values of $\bar{V}_s$; that is, curves relating α_s (or equivalently $-(dp/dz)/\rho_s g$) to $\bar{V}_g/v_t$, for various fixed values of $\bar{V}_s/v_t$. Such a plot is sketched in Figure 2. The accentuated curve corresponds to zero solids flow. It meets the axis $\alpha_s = 0$ at a point B where $\bar{V}_g = v_t$, which corresponds to the point also labeled B in Figure 1. Similarly it meets the line $\alpha_s = \alpha_{max}$ at a point A where $\bar{V}_g = v_t(1 - \alpha_{max})^{n+1}$, which corresponds to the point also labeled A in Figure 1. The region between the accentuated curve and the ordinate axis corresponds to countercurrent flow, the region to the right of this curve represents cocurrent upflow, while that to the left of the ordinate axis represents cocurrent downflow. All the curves for $\bar{V}_s > 0$ approach the positive axis $\alpha_s = 0$ asymptotically for large positive values of $\bar{V}_g$, while the curves for $\bar{V}_s < 0$ approach the negative axis $\alpha_s = 0$ for large negative values of $\bar{V}_g$.

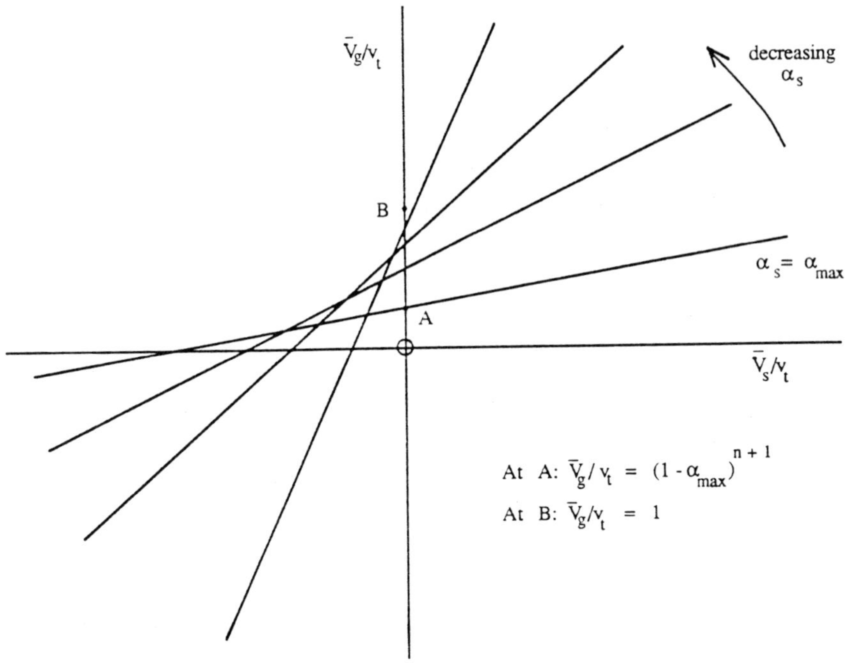

Figure 1. Contours of constant pressure gradient in the $(\bar{V}_s, \bar{V}_g)$ - plane.

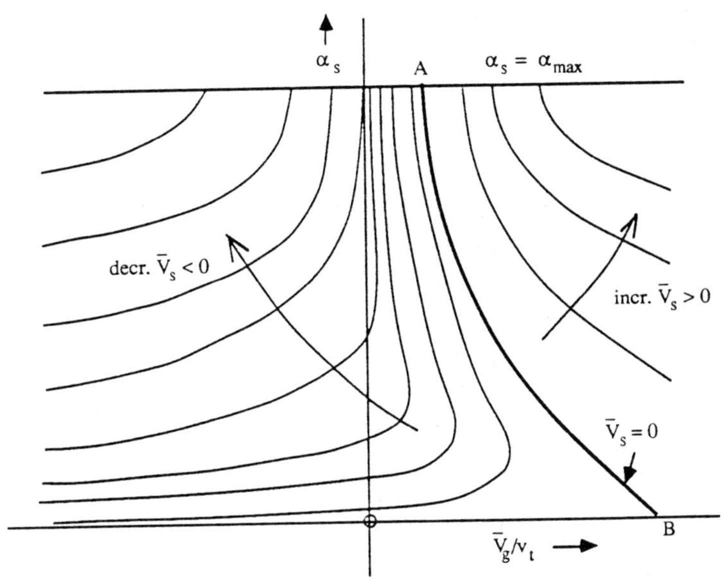

Figure 2. Contours of constant $\bar{V}_s$ in the $(\bar{V}_g/v_t, \alpha_s)$ - plane.

At A: $\bar{V}_g = v_t(1-\alpha_{max})^{n+1}$

At B: $\bar{V}_g = v_t$

As an aside, it is interesting to note that the whole of the information embodied in Figures 2 and 3, or equivalently equations (3) and (5), can be derived from the special case for which $\bar{V}_s = 0$ by a Galilean transformation. Thus, in a frame where $\bar{V}_s = 0$, equation (3) reduces to $\bar{V}_{go}/v_t = (1 - \alpha_s)^{n+1}$, but if we change to a second frame whose origin has velocity -c relative to this, we have

$$V_g = V_{go} + c, \qquad V_s = c$$

where V_g and V_s are the velocities of the gas and particles, seen from the second frame, while α_s remains unchanged under the change of frame, so

$$\bar{V}_g = (1 - \alpha_s)V_g = (1 - \alpha_s)(V_{go} + c) = \bar{V}_{go} + (1 - \alpha_s)c \ ,$$

and

$$\bar{V}_s = \alpha_s V_s = \alpha_s c$$

Eliminating c between these gives

$$\bar{V}_g = \bar{V}_{go} + \left[\frac{1 - \alpha_s}{\alpha_s}\right]\bar{V}_s$$

and introducing the expression for $\bar{V}_{go}/v_t$ given earlier then reduces this to equation (3), thus proving our contention.

If Figure 2 is to be used to predict the fully developed flow in a given pipe it is necessary to know something of the behavior of the devices used to supply the gas and the particles to the pipe. This is also true for single phase flow, but here the need is much more compelling, since the qualitative pattern of observed behavior depends on the nature of these devices. To illustrate this simply let us first confine attention to cocurrent upflow, and assume that the device feeding the particles controls their flow rate $\bar{V}_s$ directly. The gas, on the other hand can be supplied by various devices such as a positive displacement compressor, a rotary blower, or a pressurized reservoir of large volume, and control valves may also be connected in series or parallel with the device to provide regulation of the flow rate. In general, therefore, the gas pressure rise over the feed system will depend on the flow rate, and the form of this dependence constitutes the *characteristic* of the gas feed system. Figure 3 shows sketches of idealized characteristics for four common types of delivery system. The characteristics of real systems of each type differ somewhat from these idealizations; for example, those of a real reciprocating compressor have a negative slope because of clearance in the cylinder, but the forms sketched are near enough to the truth to illustrate our point. We therefore consider a vertical pipe, fed with particles and gas at its lower end, which is closed. If the pipe is long the pressure drop in the entry length that precedes the establishment of fully developed flow can be neglected, and the overall pressure drop in the pipe can then be related to the gas flow using that curve from Figure 2 whose parameter value matches the specified particle feed rate. The ordinate is scaled to give the pressure gradient, then multiplied by the pipe length to give the pressure drop as a function of the gas flow rate. The actual flow rate in the pipe is then given by the intersection of this curve with the pressure rise characteristic of the gas feed device.

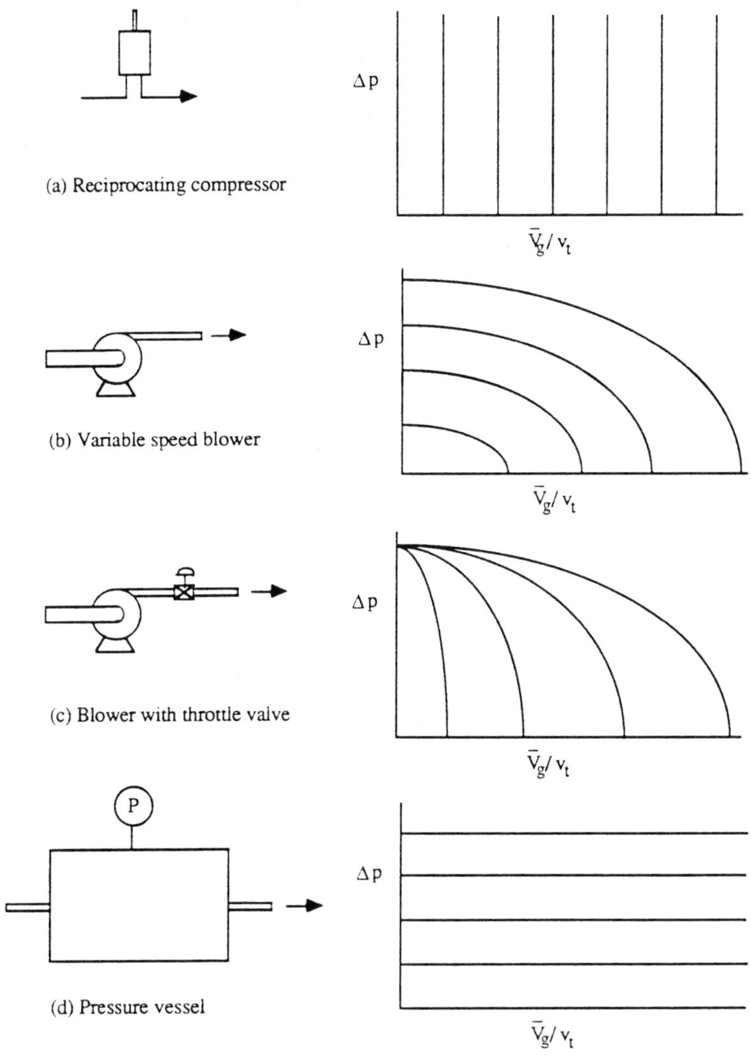

Figure 3. Examples of gas delivery devices and their characteristics.

For characteristics of the sort shown in Figures 3a and 3d the pressure drop curve and the characteristic clearly intersect in a single point, representing unique operating conditions for the pipe, but for the characteristics of Figures 3b and 3c the number of intersections may be two or zero. This is illustrated by Figure 4, in which a set of characteristics of type 3b are superimposed on a single pressure drop curve for the pipe, derived from Figure 2. For higher values of the blower speed the curves intersect in two points, such as a,a' and b,b'. As the speed is reduced the intersection points move closer together until a critical condition is reached at which the curves touch at a single point, labeled c in the diagram. For lower values of the blower speed there is no intersection. This means that there is then no solution for which the flow is steady and fully developed and the particles are entirely supported by the forces exerted on them by the gas. We would, therefore, expect a discontinuous change in the nature of the flow when the blower speed is reduced through this value, though the present simple theory does not reveal the nature of the flow at lower blower speeds.

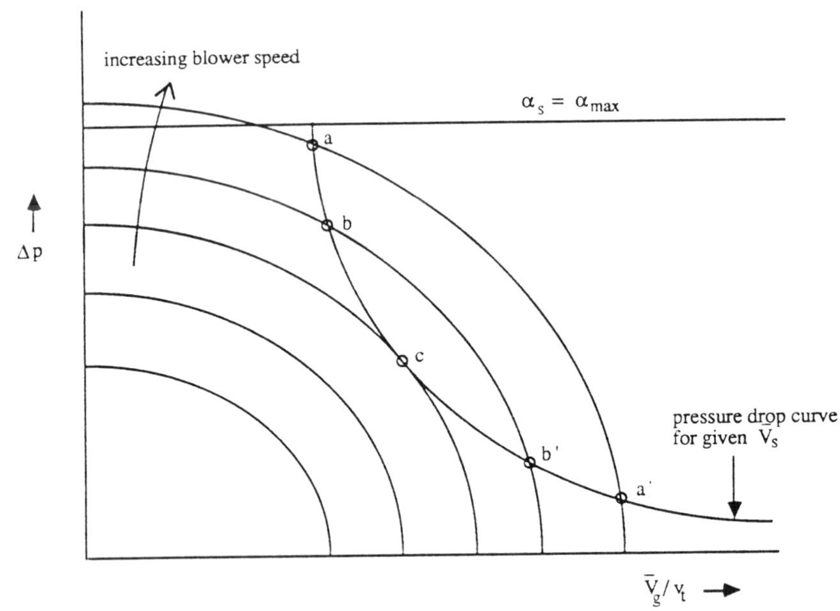

Figure 4. Operating conditions from pressure drop curve and characteristic of gas feed device

It is important also to consider the stability of the flows represented by intersections of the curves. Intuitively, if the pressure rise in the gas feed device becomes larger than the pressure drop in the pipe when the gas flow is increased slightly above the value corresponding to an intersection point, we would expect this intersection to represent an unstable state. Conversely, if the pressure rise in the gas feed device becomes smaller than the pressure drop in the pipe when the gas flow is increased, we would expect the intersection to represent a stable flow. These conclusions can easily be verified by a more formal stability analysis (see e.g. Matsumoto, 1986). It is seen that this criterion divides the pressure drop curve of Figure 4 into two arcs, separated by the point c. Intersections lying on the arc to the right of c correspond to stable operation, while those on the arc to the left of c are unstable. in particular, a and b represent unstable states, while the states corresponding to a' and b' are stable. It must be emphasized that this classification refers to the pipe in interaction with this particular gas feed device. If the gas were fed by an ideal reciprocating compressor (Figure 3a) all points on the pressure drop curves would represent stable states, since the characteristics of the compressor have large negative slopes. On the other hand, if the gas feed device were an ideal pressure tank (Figure 3d) all points would represent unstable states, since the characteristics of the tank have small negative slopes.

The above example, though simple and well known, underlines the vital importance of taking due account of the influence of the feed devices on the observed behavior when interpreting experimental observations of gas-particle flows. The experiments characterize, not the pipe itself, but the pipe in interaction with whatever feed device is used. For many years accounts of cocurrent upflow of gas and particles in vertical pipes have distinguished between "choking" and "non-choking" systems. In the former, as the gas flow is progressively reduced, a point is reached where there is a large and discontinuous increase in the mean density of the suspension in the pipe and, at the same time, the steady motion is replaced by a flow with marked, time-dependent fluctuations in bulk density and pressure drop. In the latter case, the mean density of the suspension increases continuously as the gas flow is reduced, eventually leading to a dense, rather inhomogeneous distribution of particles called a "fast fluidized bed". This is just the sort of distinction predicted above between the behavior to be expected when the gas is fed from a variable speed blower, and when it is fed from a variable speed reciprocating compressor. From this point of view, whether choking or non-choking behavior will be observed depends on the shape and steepness of the characteristics of the gas feed device, in relation to the pressure drop curve for the pipe. In fact, this is an oversimplification. The mechanics of the choking phenomenon, and the nature of the flow patterns at high bulk densities, certainly involve other factors in addition to those considered in the above simple analysis. Nevertheless, it is clear that the influence of the feed device cannot be neglected.

So far forces exerted between the pipe walls and the flowing material have been neglected. However, for a fixed value of the particle flow rate we would expect these to increase monotonically with the gas flow rate, diverging quadratically as the gas flow tends to infinity. Indeed, in the absence of particles a relation of just this sort can be predicted from the familiar curve giving the friction factor in terms of the Reynolds number. Without going any further into quantitative detail we can, therefore, guess how Figure 2 should be modified to take account of wall friction. The modified curves are sketched in Figure 5, with attention confined to those corresponding to cocurrent upflow. Note that most of the *qualitative* results regarding numbers of solutions and their stability, previously obtained without considering wall friction, are not changed greatly by extending the theory in this way, though the number of intersections of each of these curves with the characteristics for a pressure tank gas feed device is increased from one to two. Note also that the curve $\bar{V}_S = 0$ now has a branch that starts from the origin, with the pressure gradient rising monotonically as the gas flow increases. This represents gas flow in the absence of any particulate material (in Figure 2 the corresponding curve was also present, but it coincided with the horizontal coordinate axis). The other branch of $\bar{V}_s = 0$, in contrast, represents fluidized suspensions at rest in the pipe. If we begin to probe more deeply into the origins of wall stresses in cross-sectional velocity profiles, the existence of such a branch becomes problematic, and depends on a variety of physical assumptions.

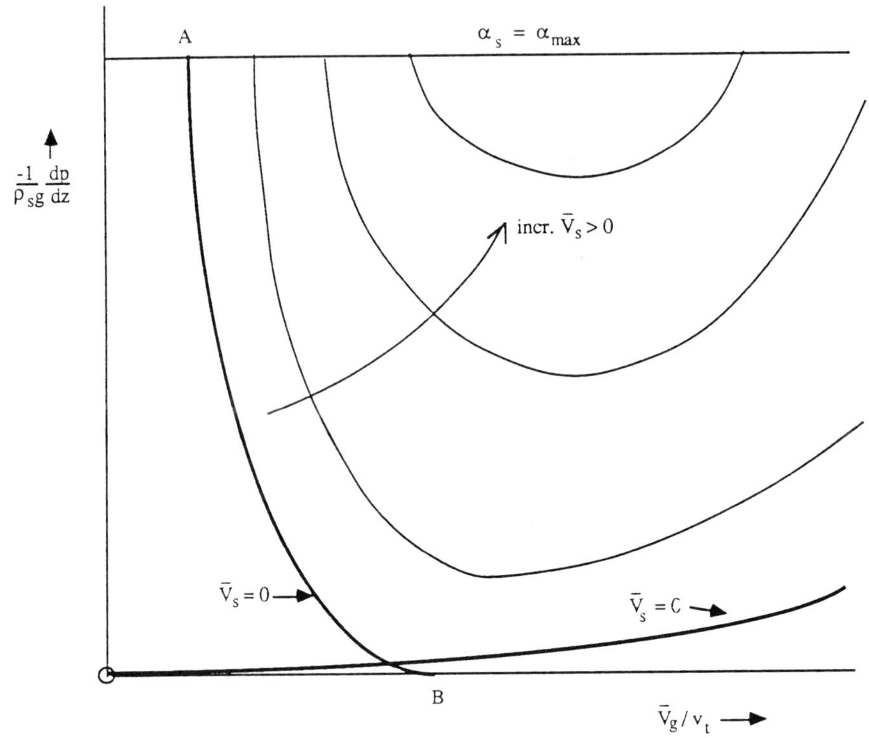

Figure 5. Contours of constant $\bar{V}_s$ in the plane of gas flow and pressure gradient, including the effects of wall friction.

The information summarized by Figure 5 can also be presented in a form analogous to Figure 1, and is so sketched in Figure 6, where we have again shown only the first quadrant, corresponding to cocurrent upflow. The straight lines, which represented the contours of constant pressure gradient in Figure 1, are now replaced by curves which spring from the ordinate axis between points A and B, and are tangent to the corresponding lines from Figure 1 where they leave this axis. However, the contours of constant pressure gradient now curve upward and eventually turn back to intersect the ordinate axis again, at a much higher value of the gas flow. The part of the axis $\bar{V}_s = 0$ lying above the point B represents flow of gas in the absence of any particles, while the intersections of the pressure gradient contours with the part of the axis between points A and B represents stationary fluidized beds. There are also solutions describing flow of gas alone which are represented by points on the ordinate axis below B. These are simply the continuation of the solutions represented by the points above B on this axis. As a consequence of this, each point on the axis between A and B represents two different solutions. The situation is best visualized by comparing Figures 5 and 6.

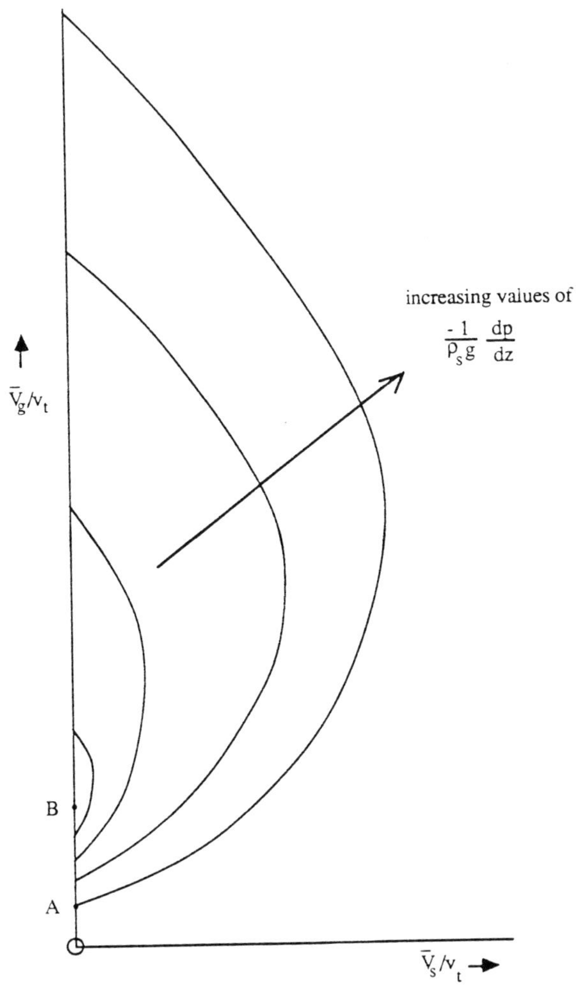

Figure 6. Contours of constant pressure gradient in the $(\bar{V}_s, \bar{V}_g)$ - plane (co-current upflow only) taking account of wall friction.

At A $\bar{V}_g/v_t = (1 - \alpha_{max})^{n+1}$, at B $\bar{V}_g/v_t = 1$

When the particles are at rest, forming a stationary fluidized bed, our procedure for combining the pressure drop curve for the pipe with the characteristic of the gas feed device is slightly modified. When the particles are flowing they necessarily fill the pipe completely, whatever its length, but when $\bar{V}_s = 0$ the length of pipe occupied by the suspension is determined by the gas flow and the total mass of particulate material present. To see this note that equation (3) reduces to $\bar{V}_g/v_t = (1 - \alpha_s)^{n+1}$ when $\bar{V}_s = 0$, and this relates α_s to the gas flow rate. The length L of the pipe occupied by the suspension is then related to the total mass of solid material charged, M, and the cross sectional area, A, by $L = M/\rho_S\alpha_S A$, and equation (5) shows that the total pressure drop across the section of pipe occupied by the suspension is given by $\Delta p = \rho_S\alpha_S gL = Mg/A$. This is equal to the weight of the suspended particles and is independent of the gas flow rate, so it will have just one intersection with any characteristic curve for a gas feed device, provided only that this curve has negative slope everywhere. This is illustrated by Figure 7. Since the pressure rise in the feed device is expected to decrease with increasing delivery flow, whatever the nature of this device, we can be confident that there is no possibility of multiple operating states or instabilities of the kind we have discussed in the case of fluidized beds.

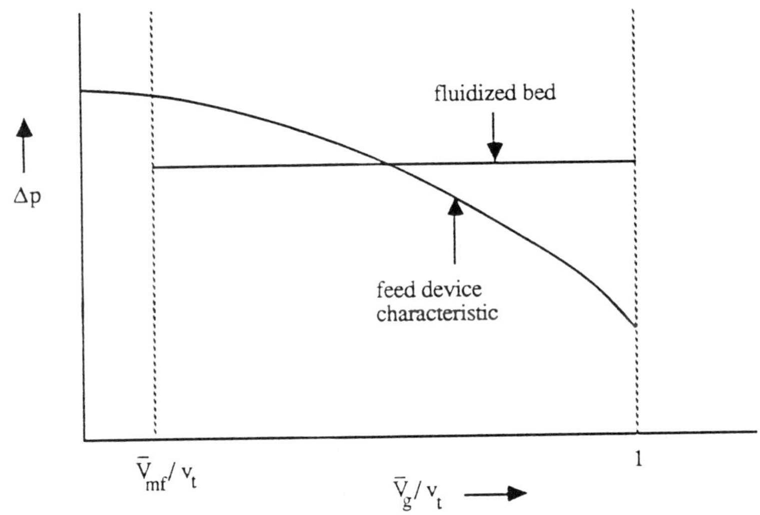

Figure 7. Determinationof the operating condtiions for a fluidized bed.

The example discussed above is simple and well-known, but it serves to illustrate important differences between single phase flow and gas-particle flow which can be recognized without explicit consideration of the details of the boundary value problems whose solution yields the full structure of these flows. Reasoning of the same sort can be applied to more complicated systems, and it leads to qualitative conclusions about the structure of their behavior that appear to be robust (in the sense that they are not changed, in essentials, by further elaboration of the theory) and correct. This will now be illustrated by the remarkable, and technically important example of standpipe flow.

20.3 STANDPIPE FLOW

In the context of gas-particle flow a standpipe is a device in which particles are made to flow from a region of lower gas pressure to a region of higher pressure, with the aid of gravity. The device in question is simply a length of pipe, ideally vertical, though practical constraints often force standpipes to depart from this orientation. Early investigations revealed the nature of flow patterns within the pipe (see the review article of Jones and Leung (1985) which summarizes most of this work), but this is an excellent example of a case in which the interaction between the pipe and its feed system is all important in determining the overall pattern of behavior. Standpipes have been considered from this point of view by Ginestra et al. (1980), Rangachari and Jackson (1982), Chen et al. (1984), and Mountziaris and Jackson (1991). The first three of these papers worked out the complete pattern of behavior for a simple, vertical pipe fed from a conical hopper, and revealed the importance of taking account of the compressibility of the gas. The last explored the complex, but technically important matter of the influence of secondary aeration injected at an intermediate level into the pipe.

Some of the characteristic features of standpipe flow can be introduced by an account of the observed behavior of a conical hopper fitted with a vertical tailpipe, of the same diameter as the discharge hole at the bottom of the hopper. If the time of emptying of the hopper alone is compared with the time required for the same mass of granular material to be discharged with the pipe attached to the hopper, it is found that the latter is much shorter; typically a pipe of length little more than one meter will halve the time of discharge for a small hopper. This phenomenon depends on the existence of a good gas seal at the junction of the hopper and the tailpipe. If this seal is broken, the discharge rate immediately falls to a value somewhat less than that for the hopper alone, so it is clear that the flow of the interstitial air is implicated in this effect. Indeed, the physics involved is simple. The suspension of particles falling down the tailpipe entrains a downward flow of air. The downward velocity of the entrained air is less than that of the entraining particles, so the drag forces acting between the two phases cause the air pressure to decrease on moving up the pipe. Since the pressure is atmospheric below the end of the pipe, this means that there is a vacuum at the top of the pipe, which coincides with the exit from the hopper. The atmospheric pressure above the fill of particles in the hopper then drives a flow of air downward through these slow moving particles, and consequently a downward drag force is exerted on them. This has an effect similar to that of an increase in the specific gravitational force; namely, it increases the discharge rate. Thus, the mechanism for flow enhancement is a coupling between the pipe and its feed device (the hopper) through the medium of the interstitial air: the observed effect cannot be explained by studying the behavior of the hopper alone or the characteristics of gas-particle flows in the pipe alone.

The simple experiment just described can be elaborated by permitting the pipe to discharge into a region where the air pressure exceeds that above the hopper by an amount Δp, and studying the dependence of the observed discharge rate on the value of Δp. Figure 8 shows the results of such an experiment with fine sand of 154μm mean diameter, discharging from a conical, mass flow hopper of half-angle 9.5°, through a vertical pipe of internal diameter 0.0254m and length 3.27m. (The values of $\overline{V}_s$ and Δp in this diagram correspond to dimensionless forms of these variables, as defined below.) The result for $\Delta p = 0$ corresponds to the experiment previously described, and the discharge rate of the same hopper without the attached pipe is also indicated on the ordinate axis. It is seen that there is an amplification of the flow rate by a factor of about eight in this case, so the effect of the pipe is very large. As Δp is increased from zero the flow of particulate material at first decreases gradually, but a critical value of Δp is reached at which there is a sudden decrease in the discharge rate by about a decimal order of magnitude. Beyond this point there is, once more, a continuous decrease in the discharge rate as Δp is increased. If the value of Δp is now decreased progressively the curve giving the discharge rate is at first retraced in detail, but the decrease in Δp can now be continued well beyond the previous critical value without any discontinuous jump in the discharge rate. Only at a much lower value of Δp is there a second discontinuous change, this time a sudden increase back to the values found in the initial part of the experiment, before the first critical point. The system therefore exhibits a very large hysteresis phenomenon in the relation between discharge rate and pressure rise. Unlike the flow amplification at Δp = 0, discussed above, it is not easy to find a qualitative explanation for this hysteresis. Rather surprisingly, we shall see later that it is a consequence of the compressibility of the gas.

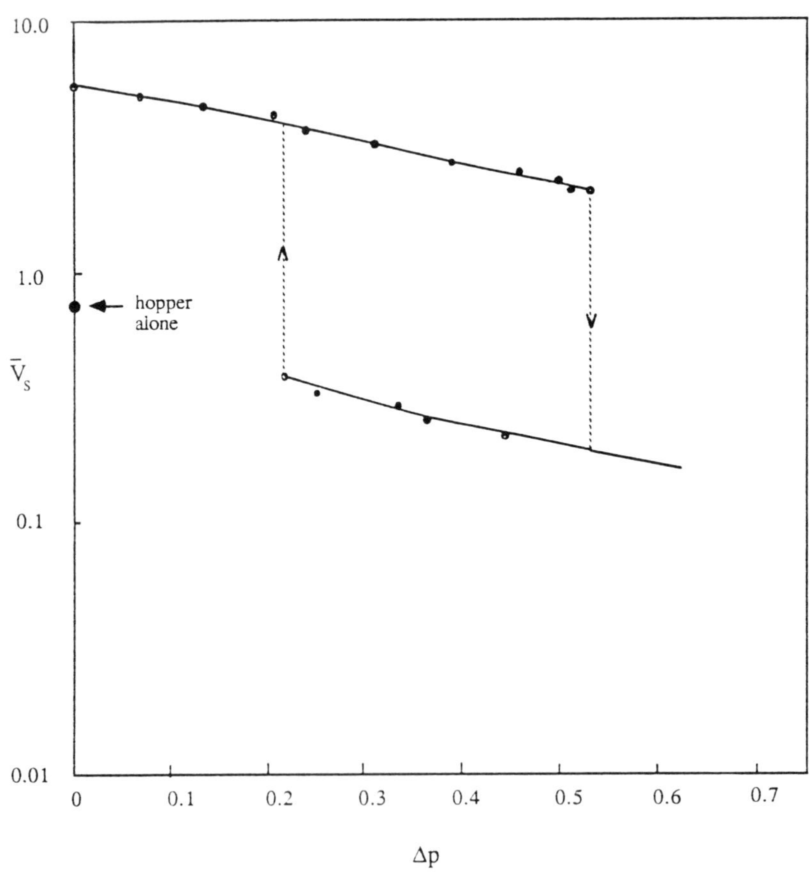

Figure 8. Discharge rate versus pressure rise for a hopper with a tailpipe.

For the purpose of modeling we will consider a generalization of the arrangement discussed above, in which a vertical pipe of radius r_p and length H is fed with particles from a conical, steep-walled hopper at its upper end, as shown in Figure 9. In addition, the lower end may be restricted by a circular orifice of radius r_v, coaxial with the pipe, and gas may be injected at one intermediate level to supplement the air entrained by the particles entering from the hopper. A coordinate y is measured downward from the top of the pipe, and the aeration point is at $y = y_a$. The pressure of gas above the fill of particles in the hopper is p_a and that below the orifice at the bottom of the pipe is p_b, so the pressure rise which the particles must overcome is given by $\Delta p = p_b - p_a$.

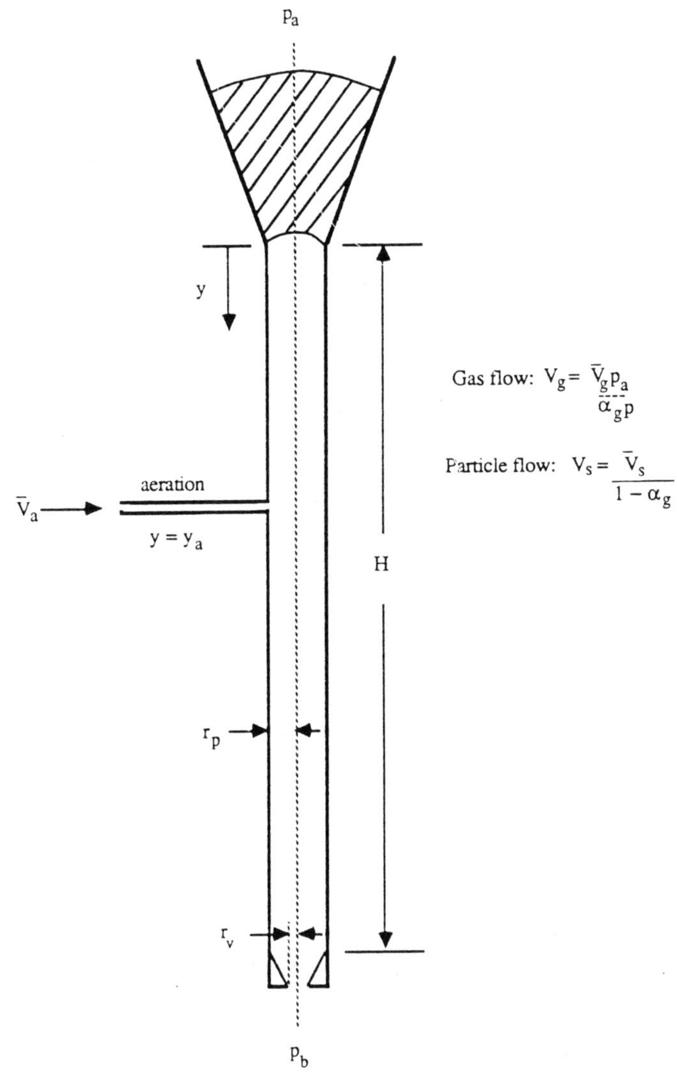

Figure 9. Simple aerated standpipe system.

In practice the detailed structure of the interdispersed gas and particles is very complex and can take a number of different forms. However, for the present purpose it will be sufficient to distinguish two simplified flow patterns, to one or other of which the dispersion belongs everywhere in the system.

(a) *Moving bed flow* - a dense flow with the particles retaining contact with their neighbors and transmitting forces through the points of contact. The void fraction is assumed to be essentially constant, taking a value $1 - \alpha_{max}$, and the picture is one of a moving packed bed of particles. Just as the particles exert forces on each other, they also press against the walls of the pipe and hopper, and the associated axial forces due to wall friction must be taken into account. The particles are therefore supported against gravity by a combination of normal stresses transmitted through their points of contact, tangential stresses at their contacts with the pipe walls, and drag forces exerted on them by the gas.

(b) *Suspension flow* - a less dense flow in which the particles are suspended in the gas. Forces are still exerted between pairs of particles, and between particles and the wall, during collisions but these are secondary to the drag forces exerted between the gas and the particles and will be neglected entirely in this simple treatment. The void fraction α_g is now variable, but is constrained by $1 - \alpha_{max} < \alpha_g < 1$.

In reality, of course, there is no such clear cut distinction. As the concentration of the particles increases collisions between them contribute more and more to the mechanics, and longer term contacts become increasingly common, eventually leading to the typical moving bed flow. Nevertheless, this simple classification is adequate to model many important features of standpipe behavior, and it will further be assumed that both phases within the pipe are in plug flow, with flat velocity profiles. Because of the relatively low density of the gas, gravitational and inertial effects will be neglected for this phase and viscous drag forces between the gas and the pipe walls will also be neglected. For suspension flow the model is, therefore, comparable in simplicity to that used in the previous section when discussing pipe flow. Only steady flow will be treated though, in contrast to the earlier discussion of pipe flow, it will not be possible to restrict attention to fully developed flows. The drag force between the gas and the particles can be represented by the Richardson-Zaki equation, as in equations (1) and (2) above, or by some modification of this which is more applicable at low void fractions. As we shall see, the choice is not critical in view of other approximations that are made.

For the rest of this section it will be convenient to use dimensionless variables throughout. Lengths are all scaled by the length H of the standpipe, densities by ρ_s the intrinsic density of the solid material, velocities by $(gr_p)^{1/2}$, where r_p is the radius of the pipe, and pressures and stresses by the hydrostatic head corresponding to a particle bed of void fraction $1 - \alpha_{max}$ filling the pipe, namely $\rho_s gH\alpha_{max}$. The same symbols will be used for the dimensionless quantities as for the original dimensional quantities.

Within the pipe the equations of continuity for the two phases can be integrated, subject to the above simplifying assumptions, to give

$$V_g = \frac{\bar{V}_g p_a}{\alpha_g p} \qquad V_s = \frac{\bar{V}_s}{\alpha_s} \tag{6}$$

where V_g and V_s are the linear velocities of gas and particles, respectively, in the y-direction, and α_g is the void fraction. $\bar{V}_s$ is the volume flow rate of particulate material and $\bar{V}_g$ the corresponding flow of gas, measured at the pressure p_a that exists above the feed hopper. p is the local value of the gas pressure, and the equation for V_g is correct if the gas flow may be assumed to be isothermal. This is a good approximation when the thermal capacity of the particulate material greatly exceeds that of the gas, and the particles are small enough that thermal equilibration of the two phases is rapid compared with the time scale over which conditions vary as the material passes through the system.

The form taken by the momentum equations depends on whether the particles move as a suspension or a moving bed in the section of the pipe in question. For a moving bed there is no acceleration in steady flow, so the momentum balances reduce to force balances and take the form

$$\frac{dp}{dy} = -B(1-\alpha_{max})\left[\frac{\bar{V}_g p_a}{(1-\alpha_{max})p} - \frac{\bar{V}_s}{\alpha_{max}}\right] \tag{7}$$

for the gas, and

$$\frac{d\sigma}{dy} + \frac{2J\sigma}{r_p} = 1 + B(1-\alpha_{max})\left[\frac{\bar{V}_g p_a}{(1-\alpha_{max})p} - \frac{\bar{V}_s}{\alpha_{max}}\right] \tag{8}$$

for the particles. Here $B(\alpha_g)$ denotes the drag coefficient, given by the Richardson-Zaki correlation or any variant of this appropriate to the circumstances, and reduced to dimensionless form in the manner specified above. σ is the cross-sectional average of the yy-component of normal stress exerted within the bed of particles, and J is the Janssen coefficient, which relates the tangential stress exerted between the particles and the wall to σ. These two equations can be integrated from initial conditions to determine the variation of gas pressure and particle phase stress on moving down the pipe. When the pipe is occupied by a suspension, on the other hand, under the simplifying assumptions set out above σ vanishes. However, it is replaced as a variable by α_g and the momentum balances now take the form

$$\frac{\alpha_{max}}{1-\alpha_g}\frac{dp}{dy} = -B(\alpha_g)\left[\frac{\bar{V}_g}{\alpha_g p} - \frac{\bar{V}_s}{1-\alpha_g}\right] \tag{9}$$

and

$$\left[\frac{r_p \bar{V}_s^2}{(1-\alpha_g)^3}\right]\frac{d\alpha_g}{dy} = 1 + B(\alpha_g)\left[\frac{\bar{V}_g p_a}{\alpha_g p} - \frac{\bar{V}_s}{1-\alpha_g}\right] \tag{10}$$

The left hand side of equation (10) represents the rate of change of momentum of the particle phase, after replacing the velocity by the void fraction using the second of equations (6). These equations can be integrated from initial conditions to determine the variation of gas pressure and void fraction with distance down a section of the pipe occupied by a suspension.

Everywhere within the feed hopper the particles form a moving bed, and continuity and momentum balances analogous to equations (6) through (8) are applicable, though these should be written in polar coordinates, with origin at the vertex of the cone formed by the hopper walls. With this geometry components of the stress other than σ (representing σ_{rr} in this case) enter the equations, so it is necessary to introduce some constitutive assumptions for stresses in the moving bed. Also, a satisfactory solution of the complete boundary value problem for hopper flow has not yet been found, so it is necessary to introduce a number of mathematical simplifying assumptions. These technicalities are discussed in more detail by Chen et al. (1984) and by Mountziaris and Jackson (1991). The approximations used are essentially those of Nguyen et al. (1979), modified to take account of drag forces exerted as a result of the relative motion of the particles and the interstitial gas. The final result is an algorithm for calculating the gas pressure p and the particle phase stress σ (= σ_{rr}) at the hopper exit in terms of the flow rates of the two phases. If the depth of the fill of particles in the hopper is large compared with the standpipe radius, the value of this depth has negligible influence on these relations.

The modeling of flow through the constricting orifice at the foot of the standpipe presents a more difficult problem. Here the flowing material must pass through a converging channel formed by shoulders of stationary material accumulated above the annular shelf that surrounds the orifice. (This is indicated in Figure 9) Though the shape of this channel should be determined as part of the problem, for the present purpose it will be assumed to be conical, so that the shoulders of stationary material behave like the walls of a rather short, rough-walled hopper. The treatment can then follow the same lines as that used for the feed hopper, except that the simplifications arising from the large depth are no longer applicable. The size of the exit orifice is specified by the ratio $w_v = r_v/r_p$, and it is assumed to be spanned by a "free fall surface" in the form of a spherical cap, on which the particle phase stress vanishes. This, together with the requirement that the gas pressure should match the specified value p_b, provides the exit boundary conditions for the problem. Details of the treatment of the discharge region are given by Chen et al. (1984) and by Mountziaris and Jackson (1991).

With separate models available for the three parts of the standpipe system it remains only to link them together, and this is done by imposing matching conditions across the surfaces of junction between the feed hopper and the standpipe, and between the standpipe and the conical discharge region. It is, of course, necessary that the total flow rate of each phase approaching and leaving each of these surfaces should be the same, and we also require the jump in momentum flux of the particle phase across each surface to match the corresponding jump in the particle phase normal stress. The momentum flux for the gas is negligible, because of its low density, so we require the gas pressure to be continuous across each surface. Similar matching conditions must be imposed across any interface separating sections of the standpipe occupied by a moving bed and a suspension, respectively. The primary object of the computations is to relate the particle discharge rate $\bar{V}_s$ and the gas flow rate $\bar{V}_g$ to the geometry of the system and the operating conditions, namely the pressure rise Δp and the orifice size w_v. However, in order to do this it is necessary to integrate the momentum equations down the standpipe, find whether it is occupied entirely by a suspension, entirely by a moving bed, or partly by each and, in the last case, to locate the interfaces separating those sections of the pipe occupied by the different flow patterns. Thus, each computation also determines a flow regime for the system and the gas pressure profile within the pipe.

Figure 10, from Mountziaris and Jackson (1991), shows the results of computations of this sort for the system of Figure 9, presented in a form called a "performance diagram" by Chen et al. (1984). In this the volume flow rate of solids, expressed as a dimensionless superficial velocity, is plotted against the overall dimensionless pressure rise, for each of a sequence of values of the constricting orifice diameter, represented by w_v. There is no flow of aeration gas, so the curve for $w_v = 1$ can be compared with the experimental results of Figure 8. The theoretical curve has three branches, the second of which actually corresponds to unstable states (Rangachari and Jackson, 1982). The upper branch agrees well with the observations before the first critical point, the hysteresis phenomenon, and even the two critical values of Δp are predicted well, and the lower branch is qualitatively similar to the observations. Thus, the theory generally gives a good account of the experimental observations. However, if the gas compressibility is neglected, replacing the terms $\bar{V}_g p_a/\alpha_g p$ in equations (6) through (10) by $\bar{V}_g/\alpha_g$, the same computations predict no hysteresis phenomenon. Instead the flow rate is found to be a smooth, monotone decreasing function of the overall pressure rise. Thus, the gas compressibility is a vital factor in predicting even the correct *qualitative* pattern of behavior. As the orifice size is decreased (w_v <1), the flow rate decreases for all values of Δp, of course, and the hysteresis loops decrease in size and shift to the left until, for $w_v \leq 0.8$, the upper branches disappear altogether. These predicted changes in form of the performance curves have also been verified experimentally (Chen et al., 1884).

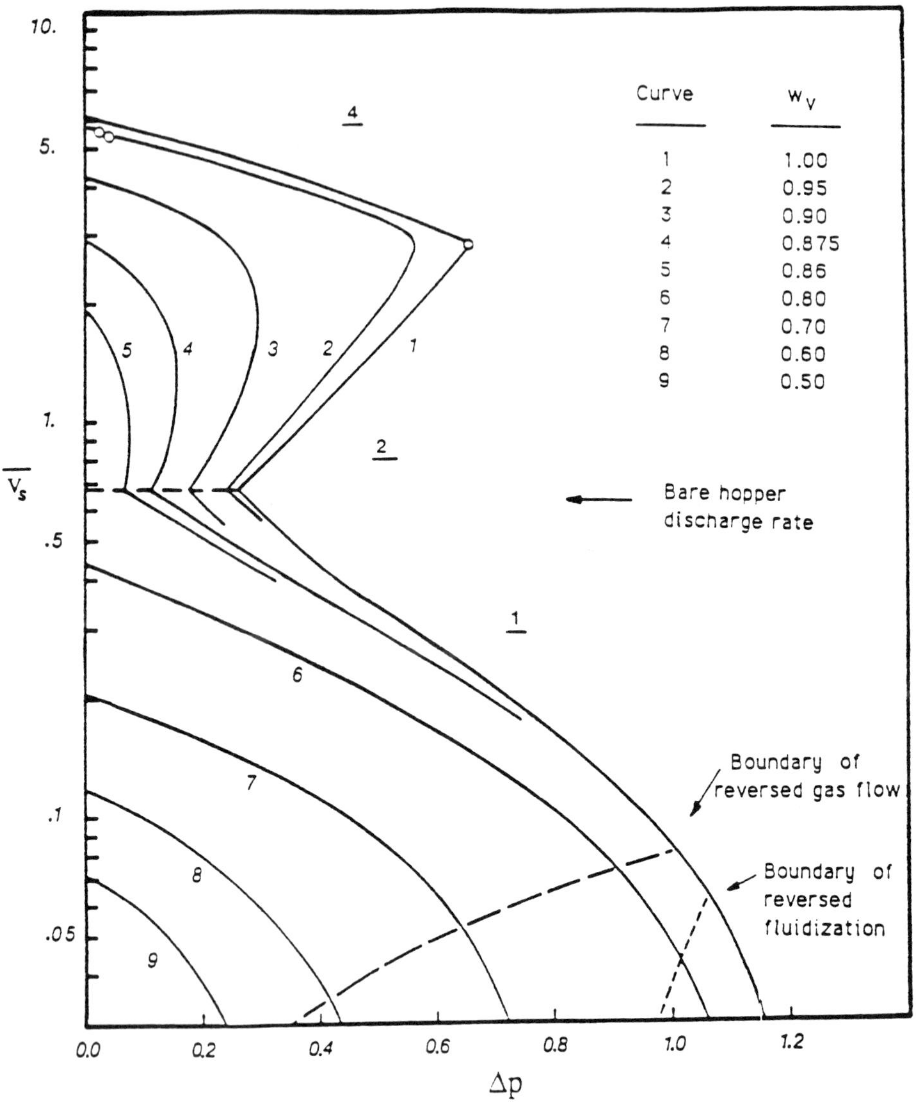

Figure 10. Theoretical performance diagram for unaerated standpipe.

The effects of aeration could be represented by plotting a sequence of performance diagrams like Figure 10, one for each value of the aeration gas flow, but it is more instructive to present the results in a different way, as "aeration diagrams" (Mountziaris and Jackson, 1991), in which the discharge rate is plotted against the aeration gas flow, represented as a dimensionless superficial velocity $\overline{V}_a$. Each curve then refers to fixed values of orifice size w_v and pressure rise Δp. Figure 11, from Mountziaris and Jackson, shows four such diagrams for the system of Figure 10. In each case the aeration is injected just below the mid point of the pipe ($y_a = 0.54$), while $w_v = 0.8$ and the curves belong to successively increasing values of Δp, as specified on the diagram. Each curve is divided into segments labeled by a pair of digits.

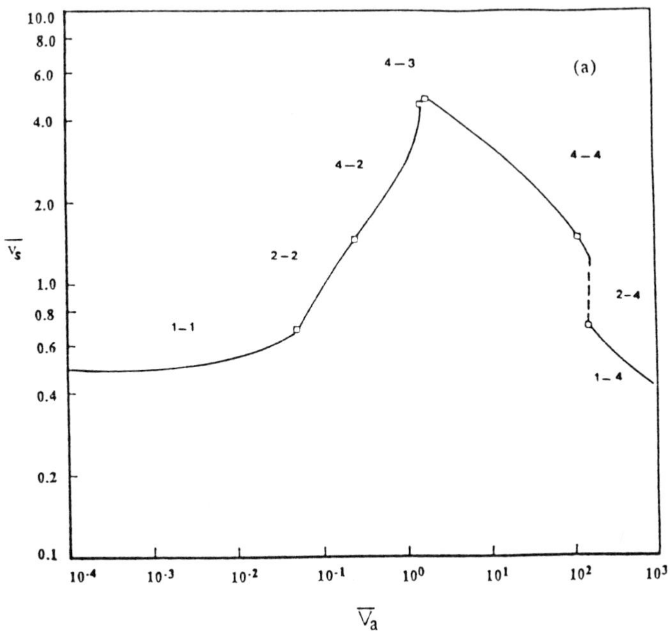

(a)
4 – 3
4 – 2
4 – 4
2 – 2
1 – 1
2 – 4
1 – 4
10.0
8.0
6.0
4.0
2.0
1.0
0.8
0.6
0.4
0.2
0.1
$\overline{v_s}$
10^{-4} 10^{-3} 10^{-2} 10^{-1} 10^{0} 10^{1} 10^{2} 10^{3}
$\overline{V_a}$

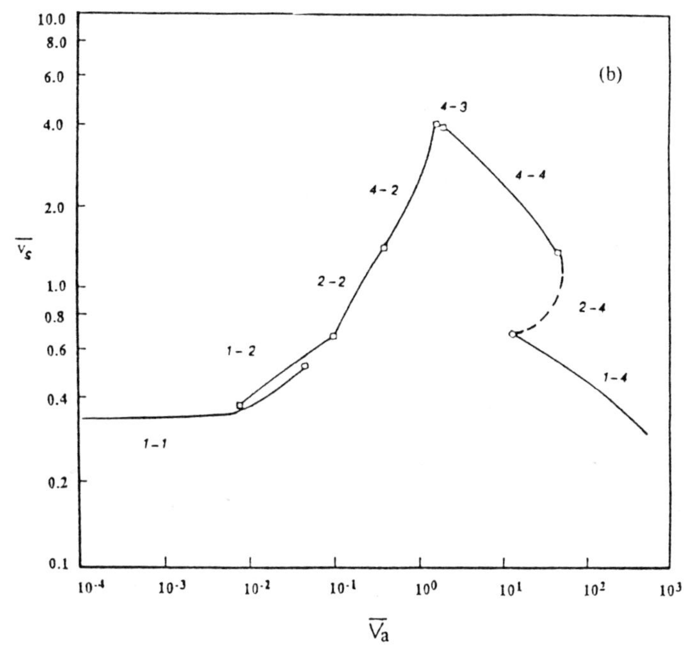

(b)
4 - 3
4 - 2
4 - 4
2 - 2
2 - 4
1 - 2
1 - 4
1 - 1
10.0
8.0
6.0
4.0
2.0
1.0
0.8
0.6
0.4
0.2
0.1
$\overline{v_s}$
10^{-4} 10^{-3} 10^{-2} 10^{-1} 10^{0} 10^{1} 10^{2} 10^{3}
$\overline{V_a}$

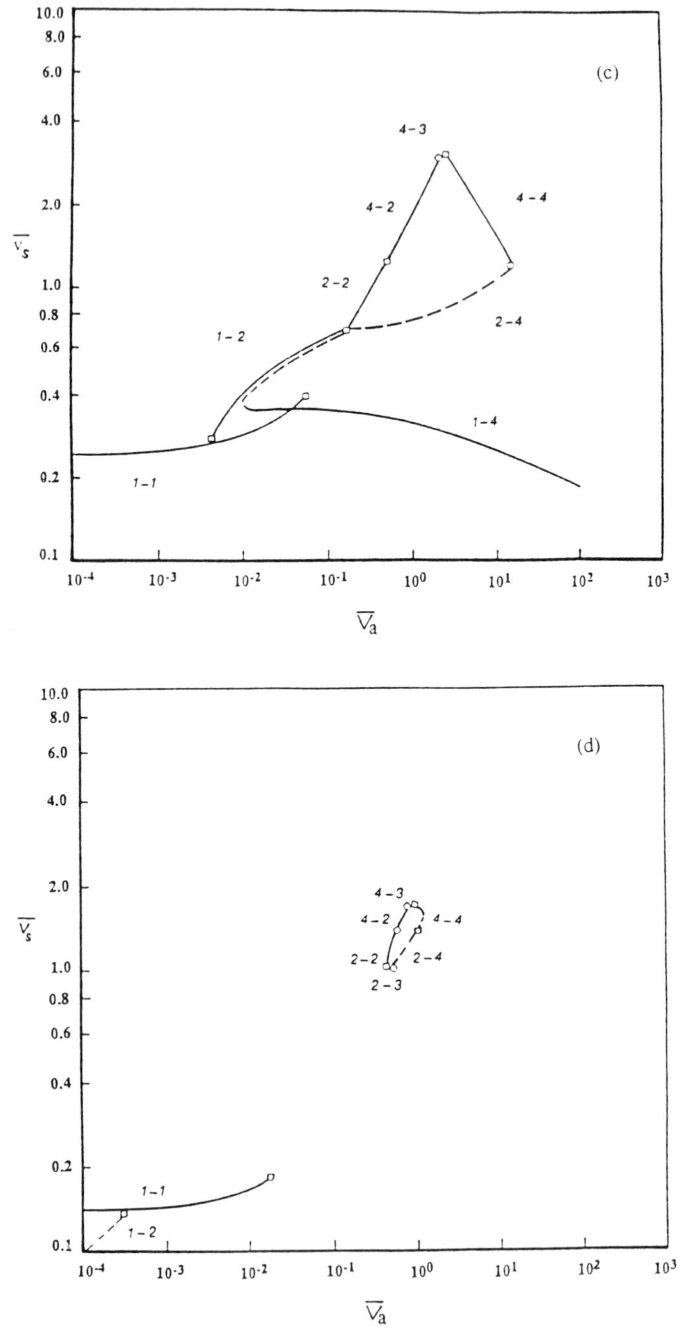

Figure 11. Theoretical aeration diagrams for $y_a = 0.54$ and $w_v = 0.8$;
(a) $\Delta p = 0$;
(b) $\Delta p = 0.2$;
(c) $\Delta p = 0.4$;
(d) $\Delta p = 0.7$.

Squares denote transitions between different flow regimes.

These serve to identify the predicted flow pattern in the pipe from those sketched in Figure 12, which represents all possible patterns for a single aeration point, with those parts of the system occupied by moving beds distinguished by line shading and those occupied by suspension flow distinguished by dot shading. For this value of w_v the unaerated system is filled with a moving bed of particles and the corresponding curve in Figure 10 has no upper branch.

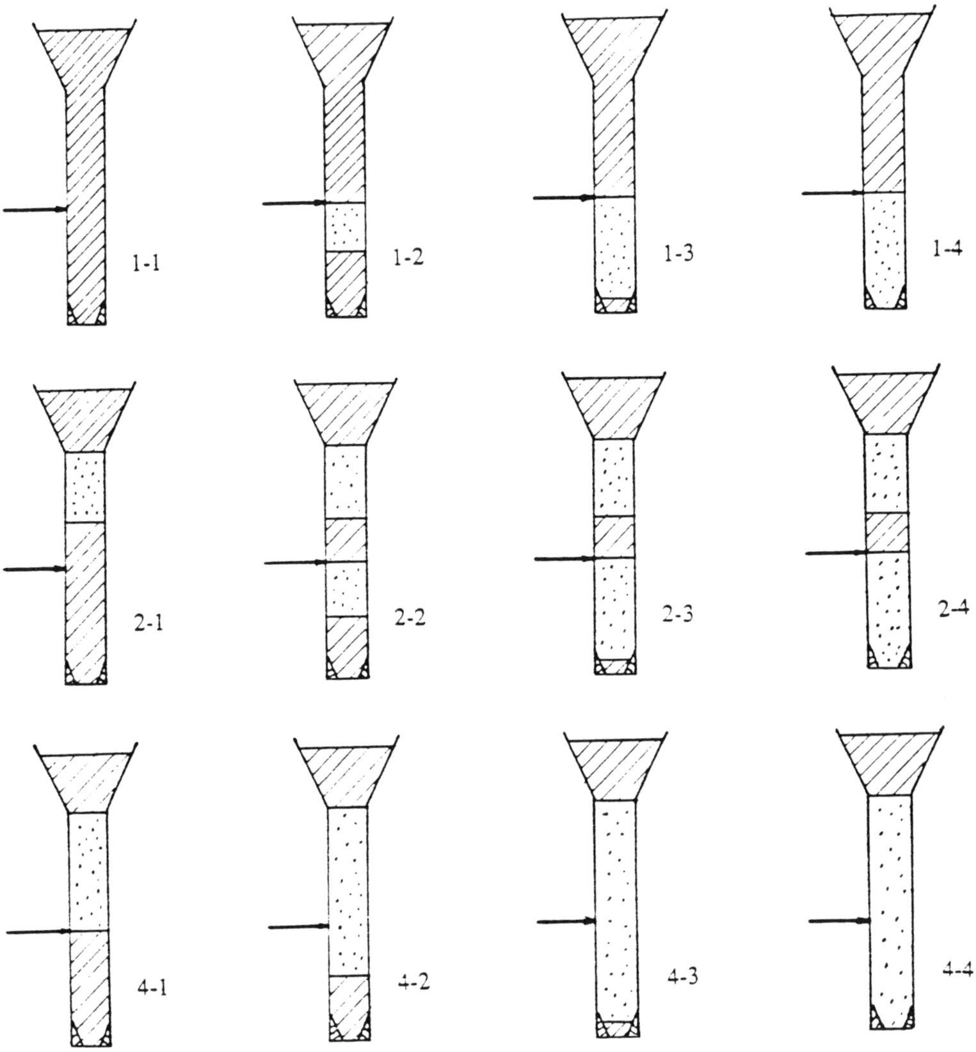

Figure 12. Flow regimes for a standpipe with a single aeration point.

For zero pressure rise (Figure 11a) it is seen that the discharge rate first increases with increasing aeration flow, with the moving particle bed being progressively replaced within the pipe by a suspension. At the point where the last remnant of moving bed vanishes from the pipe the discharge rate reaches a maximum, which is approximately ten times the rate attainable without aeration, and for larger aeration flows the discharge rate decreases quite rapidly as the suspension becomes diluted by the injected air. Ultimately there is a discontinuous drop in the discharge rate at a value of the aeration flow where a moving bed reappears in the section of the pipe above the aeration point, and with further increase in the aeration flow the discharge rate continues to drop quite steeply, though continuously. In the presence of a small, but non-zero pressure rise the aeration diagram is modified to the form shown in Figure 11b. The main features are the same but now there is a small hysteresis loop at low values of the aeration flow, while the hysteresis at high flows, already present in Figure 11a, becomes much broader. At a larger value of the pressure rise (Figure 11c) there has been a further broadening of the two hysteresis loops until they now overlap, leading to a quite complicated sequence of jumps if the aeration flow is first increased, then decreased. Finally, at a still larger value of the pressure rise, Figure 11d shows that the curve splits into completely disjoint parts, with a closed loop or "isola" at high values of the aeration flow. States on this loop correspond to quite high values of the discharge rate, but they cannot be reached by increasing the aeration flow from an initially low value. Instead the pressure rise must be decreased until the separate parts of the diagram are joined. The aeration flow must then be adjusted to maximize the discharge rate, after which the pressure rise can be increased again, thus "stranding" the system at a point on the isola.

These predicted patterns of behavior can be compared with the results of experiments on the same system. Figure 13, from Mountziaris and Jackson (1991), shows two experimentally determined aeration diagrams, whose similarity to the predicted diagrams of Figure 11 is quite striking. There is an upper bound on the aeration flows that can be used in the experiments because of the jet effect of the injected air at high flow rates, so the parts of the theoretical diagrams corresponding to the highest aeration flows are not seen in the experiments. Nevertheless, the existence of an optimum aeration rate, at which the discharge rate peaks, is very clear, and even the predicted small hysteresis loop at small aeration rates is observed experimentally. Furthermore, the progression in the form of the diagram between Figures 13a and b, as Δp is increased, matches the progression in the theoretical diagrams of Figure 11. Mountziaris and Jackson also studied the influence of the location of the aeration point on the performance of the system. There was found to be a continuous increase in the optimum discharge rate as the aeration point was moved down the pipe, terminating in a catastrophic, discontinuous decrease in the rate if the aeration point was moved below some critical position. This was associated with the sudden appearance of a moving bed occupying a major section of the pipe. The optimum aeration strategy is, therefore, to delay the addition of supplementary air as long as possible without introducing moving bed flow into parts of the pipe.

The standpipe system provides an excellent example of the power of rather simple models of flow in ducts to catch the general features of the behavior of a gas-particle flow system, even when these are quite complicated, and it underlines the importance of considering the whole system, including the devices which control the input and removal of each phase, rather than the

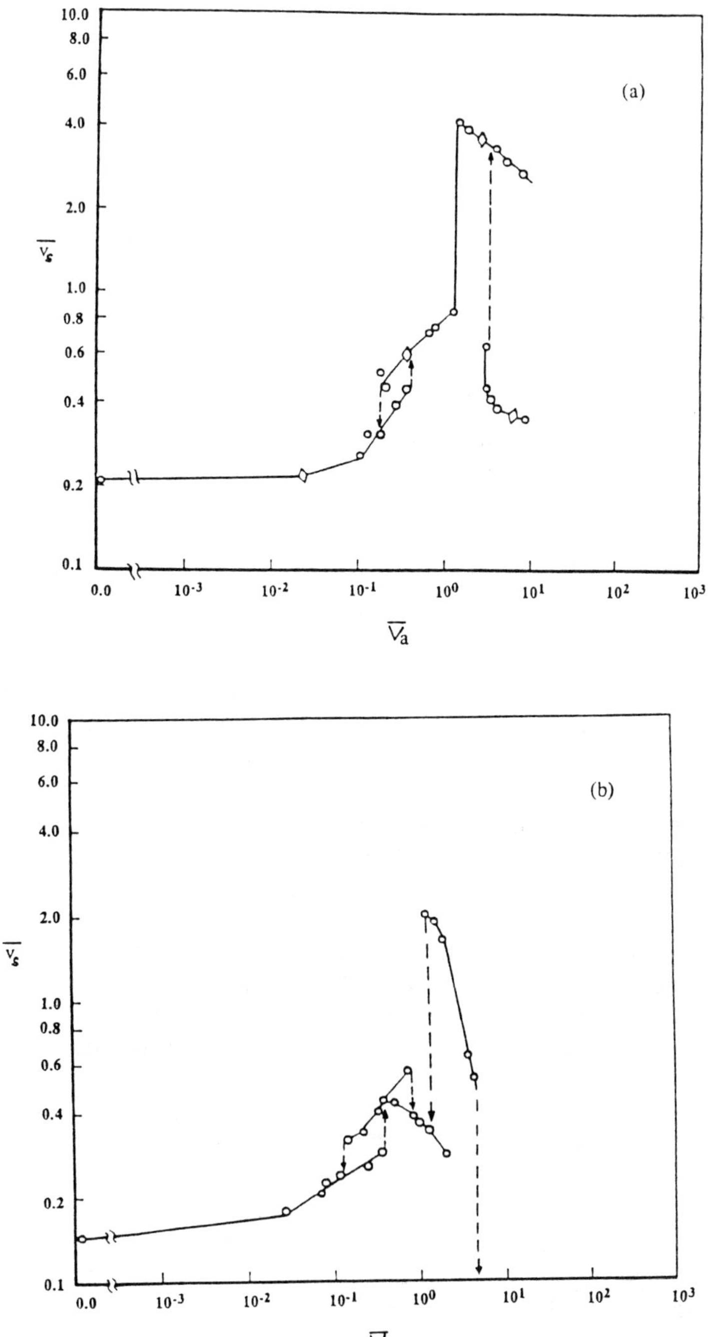

Figure 13. Experimental aeration diagrams for $y_a = 0.54$ and $w_v = 0.92$:

(a) $\Delta p = 0$
(b) $\Delta p = 0.08$.

duct in isolation. Nevertheless, closer examination of the relation between predictions and observations also reveals the weakness of these simple models. As an example of this we point out that the fair quantitative agreement between the performance curve for $w_V = 1$, in Figure 10, and the experimental results of Figure 8, depends on the adjustment of a parameter in the computations to a value quite inconsistent with the physical picture adopted for the flow in the pipe. In the model the particles are viewed as a uniform suspension in plug flow, for which the drag coefficient $B(\alpha_g)$ can be estimated with reasonable accuracy. However, to secure the agreement between Figure 10 and the observed discharge rates, it was necessary to multiply this estimated drag coefficient by a number much smaller than one. In other words, for given values of the cross-sectional average velocities of the two phases, the cross-sectional average drag force is much smaller than would be expected on the basis of a uniform suspension and plug flow of each phase. Visual examination of the flow reveals the reason for this: the particles are by no means uniformly distributed over the section of the pipe. Very often they appear to be concentrated in dense streamers near the walls, while the gas bypasses these streamers near the center of the pipe, where the concentration of particles is much smaller. Other patterns are also seen, but the common feature is a marked segregation between the distributions of the two phases, leading to a much reduced value for the average drag force. Furthermore, the degree of segregation varies with the system considered and with the operating conditions for a given system, and there is no simple way of predicting it in advance. Thus, despite its qualitative success, the simple standpipe theory presented above is essentially valueless for *a priori* quantitative predictions of standpipe behavior (and hence for design purposes), and not much better for extrapolating from one set of operating conditions to another. The segregation effects just described for standpipes are, in fact, ubiquitous for gas-particle flows in ducts of any kind. Consequently, significant progress in quantitative prediction of their behavior, beyond the level of empirical correlations with severely restricted ranges of applicability, depends entirely on our ability to understand the causes of segregation and to model it reliably. The next section is devoted to an account of some early efforts in this direction.

20.4 SEGREGATION AND ITS EFFECTS

When particles and gas flow in a duct inclined to the vertical there is expected to be some segregation due to gravitational sedimentation, with the concentration of particles highest near the lower surface of the duct. However, as noted above, there is also observed to be quite marked segregation even in vertical ducts. For co-current upflow in ducts of large size this can be quite spectacular, with a dense curtain of particulate material adjacent to the wall of the duct recirculating counter to the gas flow. The upward gas velocity is large at the axis of the duct but the descending particles near the wall slow the gas, and may even drag it downward, causing recirculation of the gas phase. A number of quantitative studies of these effects have been published (Bartholomew and Casagrande, 1957; Saxton and Worley, 1970; Yerushalmi et al., 1978; Youchou and Kwauk, 1980; Weinstein et al., 1984; Bader et al., 1989) and there have been quite extensive unpublished studies by industrial laboratories; in particular, those in the petroleum refining industry, where the catalytic cracking of gas oil is carried out on a very large scale in vertical tubular reactors with cocurrent upflow of the reaction mixture and the catalyst. In this case, in addition to needing the relation between the pressure gradient and the flow rates of the phases, it is very important to be able to predict the detailed flow patterns of both phases, since these affect the conversion and selectivity for the chemical reactions taking place.

The dominant physical mechanism responsible for segregation may be different for particles of different sizes and densities. Fluid cracking catalysts used in the oil industry typically have particles with mean diameter of the order of 70μm, particle density near 1 g/cm^3 and a fairly broad size distribution. For this material, under typical conditions used for cracking, the Reynolds number for the gas flow, based on the duct diameter, is very large and there seems to have been a general feeling in the industry that the strong segregation of the catalyst toward the wall is a result of turbulence in the gas flow. This idea has been formalized by Berker and Tulig (1986), using a two-equation model for the gas turbulence, but it is difficult to assess the influence of the burden of particulate material on this turbulence. More recently Sinclair and Jackson (1989) and Louge, Mastorakos and Jenkins (1990) have studied rather different models, with the common feature that direct collisions between the solid particles play a major role in determining the distribution of particle concentration.

Sinclair and Jackson noted that the shearing of the particle assembly due to the drag force exerted by the gas, which varies over the cross section, would be expected to cause collisions between particles so that a random component of velocity for individual particles would be superimposed on the local convective velocity of the particle assembly. This concept has recently been exploited extensively in the mechanics of rapidly flowing granular materials, where the kinetic energy of the random motion is thought of as a "granular temperature". The assembly of particles then has transport properties analogous to those of a molecular fluid, and can be treated using the methods of kinetic theory. A vector representing the flux of the pseudo-thermal energy of random motion can be defined and a balance equation can be written, whose solution should determine the distribution of granular temperature. The momentum transferred by translation of the particles as a result of their random motion, and by collisions between the particles, can be formulated as an effective stress tensor for the particle phase, and this in turn can be related to spatial derivatives of the components of the convective velocity. The relation involves a "viscosity" for the particle phase which depends on the granular temperature, so a force balance equation for the particles must be coupled with the balance of pseudo-thermal energy, solving both together.

Since velocity profiles are now to be taken into account viscous stresses must also be included in the force balance for the gas phase, with an appropriate modification of the viscosity to take account of the presence of suspended particles. As in our earlier discussions, the gas phase and particle phase force balances are also coupled through the drag force that results from the relative motion of the two phases.

Sinclair and Jackson (1989) explored in some detail the consequences of the simplest possible model of this sort. Only fully developed flow in a vertical pipe was considered and turbulence was neglected completely, so their solutions might be expected to bear a similar relation to reality as the Poiseuille solution for single phase flow. In addition, the only aspect of gas-particle interactions taken into account was a drag force proportional to the difference between the local values of the convective velocities of the two phases. The damping effect of lubrication forces on collisions between particles was neglected (which is justifiable only for large, dense particles), and no account was taken of any other inter-particle forces exerted through the medium of the interstitial gas. With these simplifications the equations of motion can be formulated as follows in cylindrical coordinates centered on the axis of the pipe. An axial force balance for the particle phase gives

$$-\frac{\partial \sigma_{rz}}{\partial r} - \frac{\sigma_{rz}}{r} + \beta(\alpha_s)(v_g - v_s) - \rho_s \alpha_s g = 0 \tag{11}$$

and the corresponding balance in the radial direction is

$$\frac{\partial \sigma_{rr}}{\partial r} + \frac{\sigma_{rr}}{r} - \frac{\sigma_{\theta\theta}}{r} = 0 \tag{12}$$

The axial force balance for the gas phase gives

$$\mu_{eg}(\alpha_s)\left(\frac{\partial^2 v_g}{\partial r^2} + \frac{1}{r}\frac{\partial v_g}{\partial r}\right) - \beta(\alpha_s)(v_g - v_s) - \frac{\partial p}{\partial z} = 0 \tag{13}$$

and the radial force balance simply demands that the gas pressure be independent of r. Finally, a balance for the pseudo-thermal energy of random particle motion takes the form

$$\frac{\partial q_{PT}}{\partial r} + \frac{q_{PT}}{r} + \sigma_{rz}\frac{\partial v}{\partial r} + \gamma = 0 \tag{14}$$

In these equations the symbols represent dimensional quantities and have the following meanings.

v_g, v_s - axial velocities of gas and particles, respectively
p - gas pressure

σ_{ij} - components of the particle phase stress tensor
α_s - volume fraction of space occupied by the particles
μ_{eg} - effective viscosity of particle-laden gas
q_{PT} - radial component of the flux of pseudo-thermal energy
γ - rate of dissipation of energy, per unit total volume, due to inelastic collisions between particles

The form of the first three equations is self explanatory. (The gravitational pressure gradient in the gas is neglected because of the small value of the density.) In equation (14) the third term represents the rate of generation of pseudo-thermal energy of random motion by working of the particle phase stresses, while the fourth term represents its rate of dissipation due to the inelasticity of collisions between particles. It is this last term which distinguishes the present equations from those for transport in a monatomic gas.

To close the set of equations expressions are needed for $\boldsymbol{\sigma}$, γ, μ_{eg}, q_{PT} and $\beta(\alpha_s)$. For β the Richardson-Zaki equation (2) above can be used, while $\boldsymbol{\sigma}$, γ and q_{PT} are given by kinetic theory in the form

$$\boldsymbol{\sigma} = p_s(\alpha_s, T)\,\mathbf{I} + \mu_s(\alpha_s, T)\,\mathbf{D} \tag{15}$$

$$\gamma = f(\alpha_s, T) \tag{16}$$

$$q_{PT} = -\,k(\alpha_s, T)\,\nabla T \tag{17}$$

where p_s, μ_s, f and k are functions whose forms are determined by kinetic theory (see e.g. Lun et al., 1984, for details), T denotes the granular temperature, **I** is the unit tensor, and **D** denotes the rate of deformation tensor associated with the particle phase velocity field ($\mathbf{D} = 1/2[\nabla\mathbf{v} + \nabla\mathbf{v}^T]$). A possible form for μ_{eg} is proposed by Sinclair and Jackson (1989), but is less certain than those above. The inelasticity of collisions between particles is taken into account by the kinetic theory through a coefficient of restitution e, representing the ratio of the speed of separation of a pair of particles, along their line of centers, to their speed of approach. The expressions used by Sinclair and Jackson are valid only when $1 - e \ll 1$, that is, for almost elastic collisions.

Equations (11) through (14) must be supplemented by appropriate boundary conditions at the wall of the pipe. A momentum boundary condition for the particle phase can be obtained by equating the transfer rate of axial momentum in collisions between the particles and the wall to the limit of the tangential stress in the particle phase on approaching the wall, with the result

$$\frac{\mathbf{v}_{sl}}{|\mathbf{v}_{sl}|}\cdot(\boldsymbol{\sigma}\,.\,\mathbf{n}) + \frac{\sqrt{3}\,\pi\alpha_s\rho_s\,T^{1/2}|\mathbf{v}_{sl}|}{6\alpha_{max}[1 - (\alpha_s/\alpha_{max})^{1/3}]} = 0 \tag{18}$$

where **n** is the unit normal to the wall, directed into the pipe, $\mathbf{v}_{sl}$ is the slip velocity between the particles and the wall, α_{max} is the maximum possible value of α_s, and Φ is a "specularity factor", representing the mean fraction of the tangential momentum of a particle transferred to the wall during a collision. Similarly, balancing the flux of pseudo-thermal energy to the wall with the difference between the rates of generation and dissipation at the wall gives

$$- \mathbf{n}.\mathbf{q}_{PT} = \gamma_w + \mathbf{v}_{sl}.(\sigma \cdot \mathbf{n}) \tag{19}$$

where γ_w represents the rate of dissipation of pseudo-thermal energy, per unit wall area, due to inelastic collisions between particles and the wall. This is given by

$$\gamma_w = \frac{\sqrt{3}\,\pi\alpha_s\rho_s T^{3/2}(1 - e_w^2)}{4\alpha_{max}[1 - (\alpha_s/\alpha_{max})^{1/3}]} \tag{20}$$

where e_w is a coefficient of restitution for collisions between particles and the pipe walls. A boundary condition for the gas velocity at the wall is also needed, and this is harder to formulate. Of course, the point value of the gas velocity vanishes everywhere on the wall, and correspondingly its ensemble average must also vanish. On the other hand, the spatial average of the gas velocity, taken over a small, three dimensional region, will not vanish in general when the center of that region lies on the wall. The choice of the no-slip condition precludes the possibility of solutions representing classical fluidized beds, where the particles are uniformly distributed and at rest, supported by the drag forces exerted on them by the gas. For this reason Sinclair and Jackson gave a heuristic argument in support of a more complicated condition (equation (12) of their paper), which imposes no constraint on the gas velocity at the wall when the concentration of particles is high and reduces to the no-slip condition when their concentration is low. Such a boundary condition admits fluidized beds as solutions, but the question of whether or not solutions of this sort should exist is far from settled.

Numerical solution of the complete boundary value problem defined by the above equations and boundary conditions gives the relation between the flow rates of the two phases and the gas pressure gradient and, for each set of values of these quantities, it provides profiles of the velocities of the separate phases, the granular temperature, and the concentration of the particles, as functions of r. In presenting the results it is convenient to use dimensionless velocities, scaled by the terminal velocity v_t for an isolated particle. The following dimensionless forms are also introduced for the pressure gradient and the flow rates

$$\left(\frac{\partial p}{\partial z}\right)^* = \frac{1}{\rho_s g}\frac{\partial p}{\partial z}; \quad Q^*_s = \frac{Q_s}{a^2 v_t}; \quad Q^*_g = \frac{Q_g}{a^2 v_t}$$

where a is the radius of the pipe, while Q_s and Q_g are the volume flow rates of solid and gas, respectively.

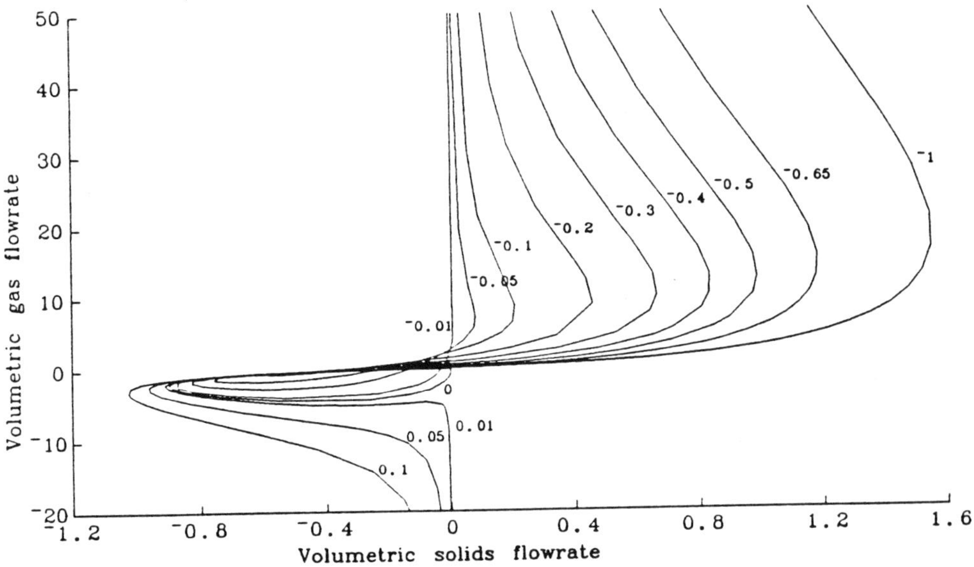

Figure 14. Computed contours of $(dp/dz)^*$ in the (Q^*_s, Q^*_g)-plane.

Figure 14 then gives an overview of the relation between these three quantities, represented by contours of constant pressure gradient in the plane of the two flow rates, in a manner analogous to Figures 1 and 6. Indeed, within the first quadrant the computed contours are similar to those sketched in Figure 6, but even on this small scale some of the contours show interesting loops in the third quadrant, and an enlargement of the region near the origin would show a great deal of complicated structure not visible on the scale of Figure 14. Plots on expanded scales are presented by Sinclair and Jackson, but for the present purpose we illustrate the nature of the results with a sketch (Figure 15), which distorts the contours enough to show all aspects of their qualitative behavior on a single scale without changing the topology of the diagram in any way. From this we see that each contour crosses from the first to the second quadrant at a point corresponding to zero net particle flow rate, but after moving down for some distance it reverses direction and returns to the ordinate axis at a point not far above that where it crossed from the first quadrant. Thus there is a second point of zero net particle flow, and here the contour reverses direction once more and moves down through the second quadrant, eventually to return, yet again, to the ordinate axis , with or without traversing a loop, depending on the value of the pressure gradient. The region of multiplicity of steady states generated by the zigzag of a contour in the second quadrant was called the "crossback region" by Sinclair (1989). However, there are clearly other, quite complicated possibilities for multiple steady states when the characteristics of a gas supply system are combined with those of the pipe, as discussed earlier. It is seen that even this, grossly simplified physical model of the system has generated an amazingly rich variety of behavior.

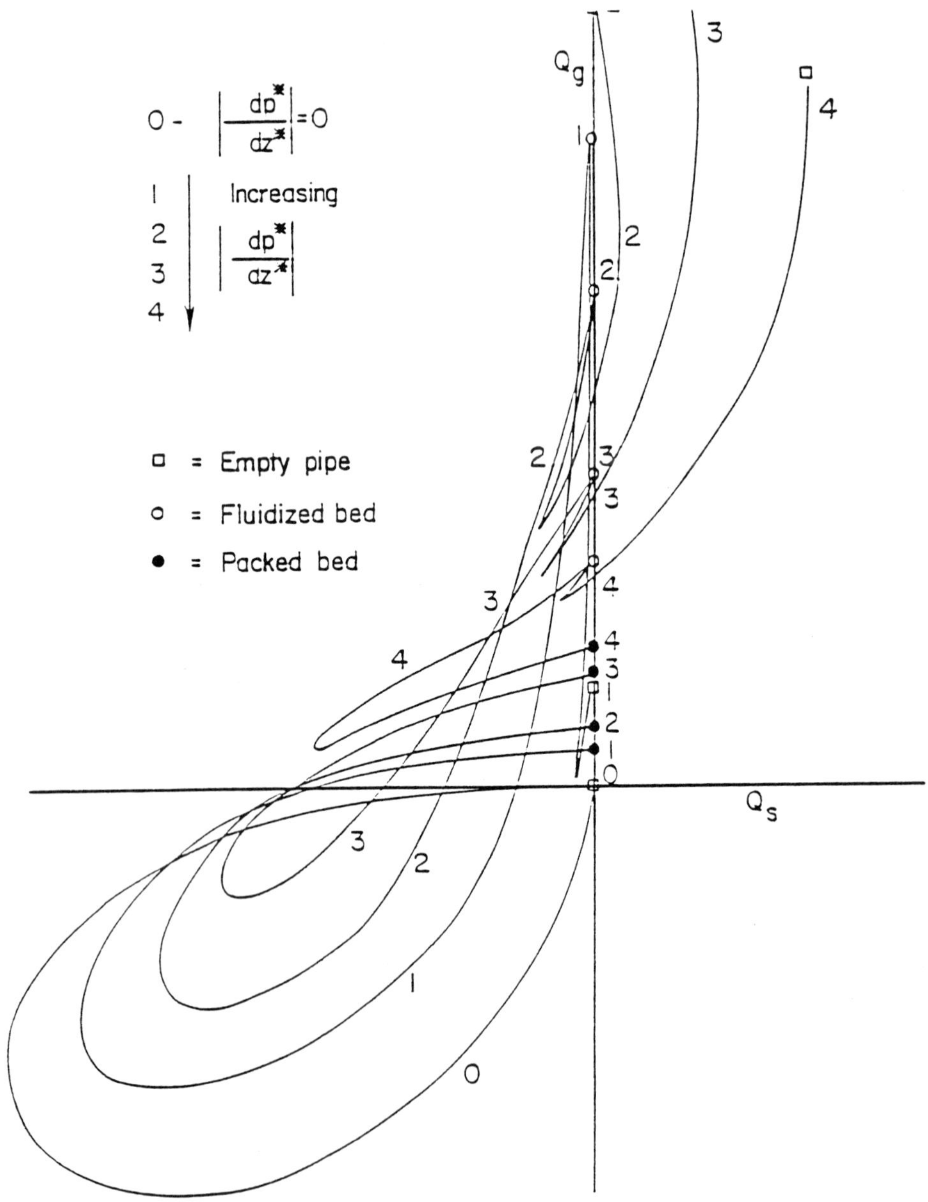

Figure 15. Sketch of the (Q^*_s, Q^*_g)-plane to illustrate the geometry of the $(dp/dz)^*$ contours.

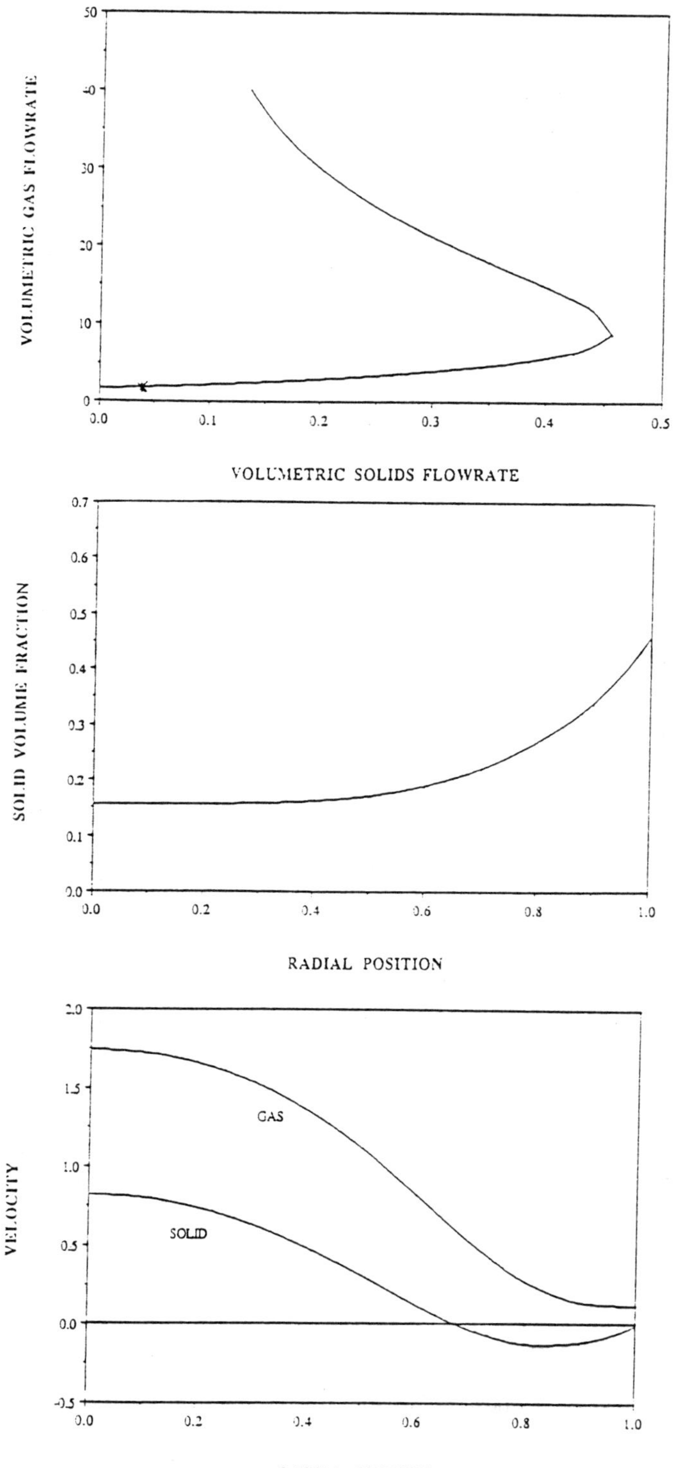

Figure 16. Velocity and particle concentration profiles for $(dp/dz)^* = -0.2$, $Q^*_s = 0.04$ and $Q^*_g = 1.7$. Mean solids volume fraction is 0.25.

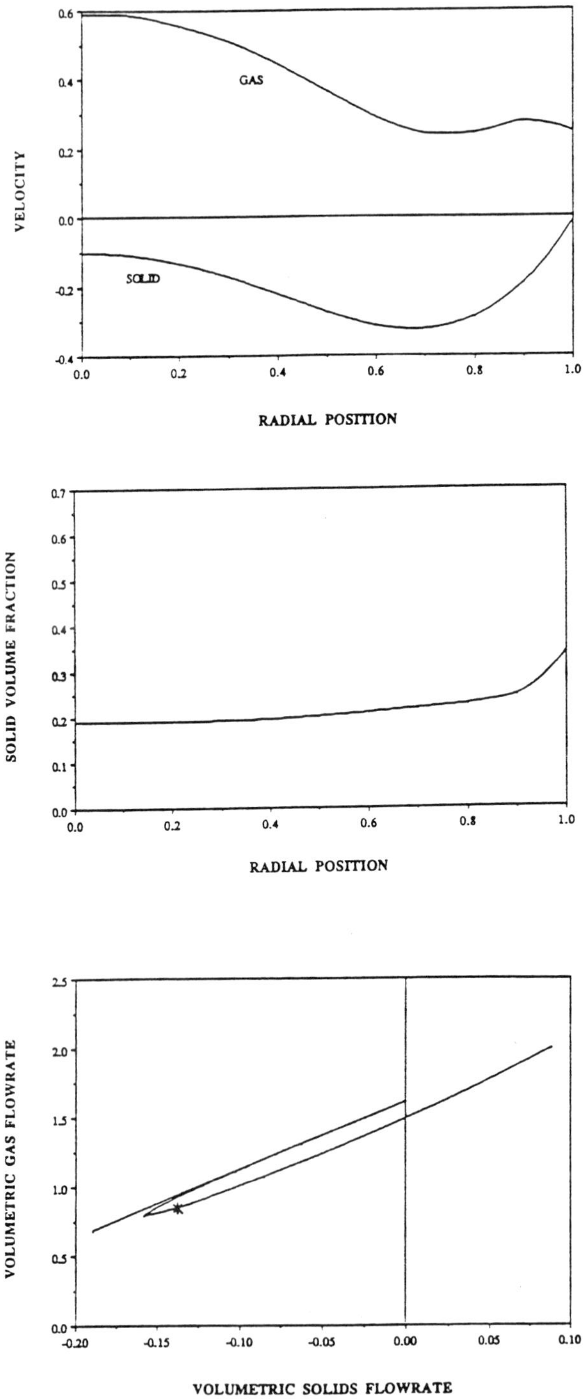

Figure 17. Velocity and particle concentration profiles for $(dp/dz)^* = -0.2$, $Q^*_s = -0.16$ and $Q^*_g = 0.83$. Mean solids volume fraction is 0.22.

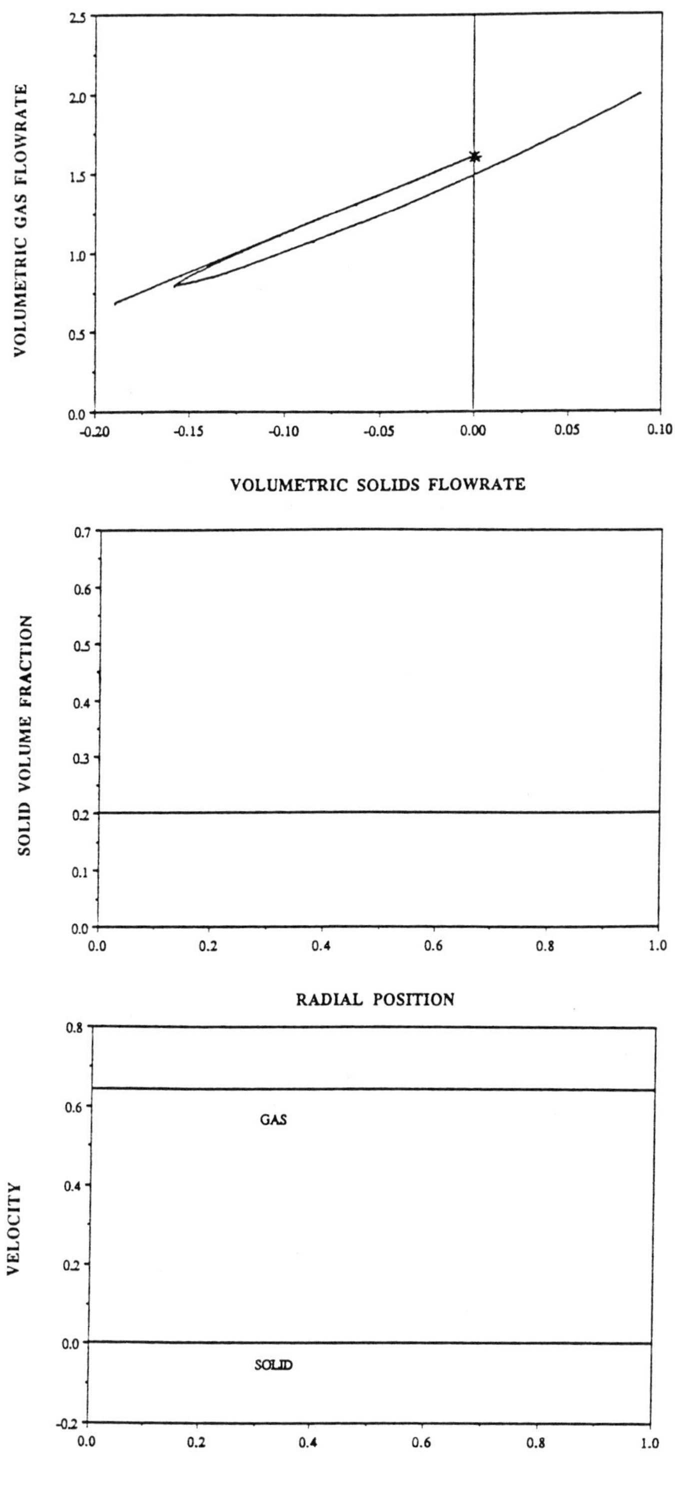

Figure 18. Velocity and particle concentration profiles for $(dp/dz)^* = -0.2$, $Q^*_s = 0$ and $Q^*_g = 1.61$. Mean solids volume fraction is 0.20.

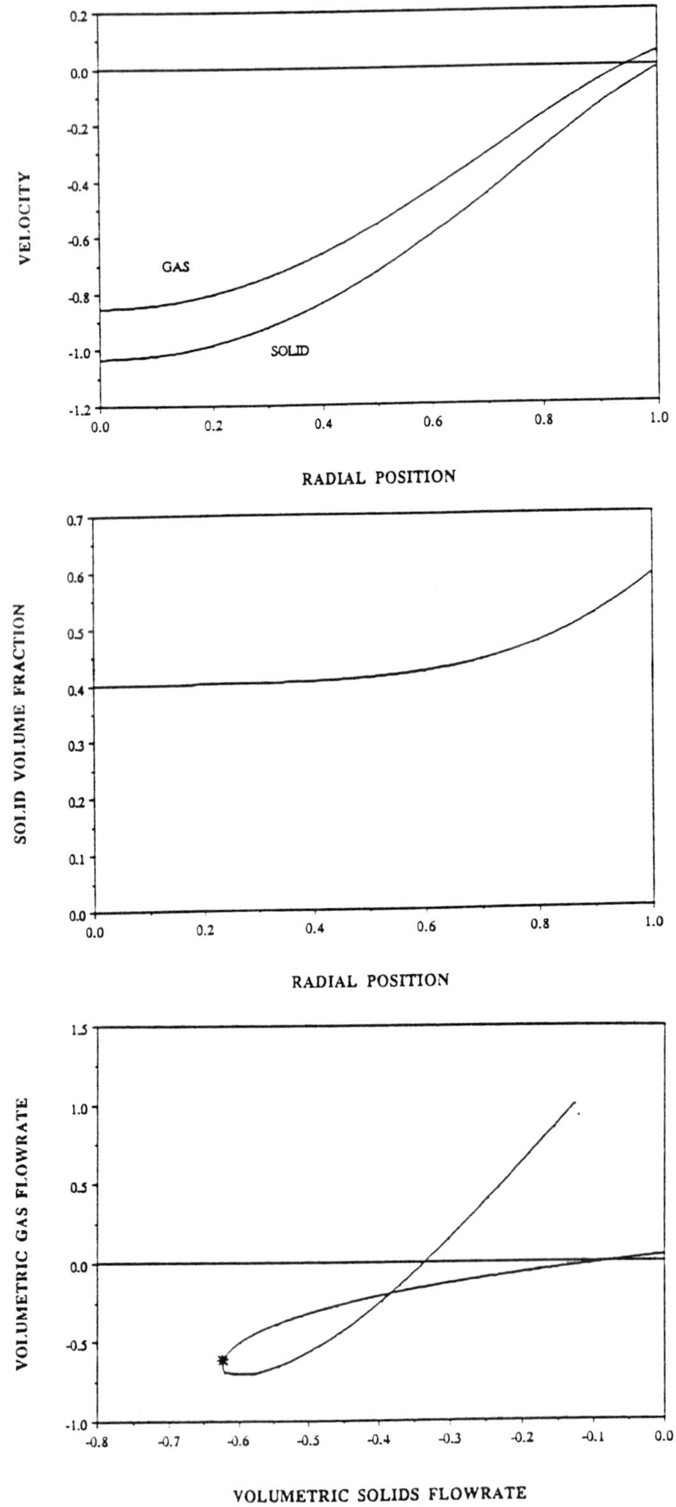

Figure 19. Velocity and particle concentration profiles for $(dp/dz)^* = -0.2$, $Q^*_s = -0.62$ and $Q^*_g = -0.61$. Mean solids volume fraction is 0.46.

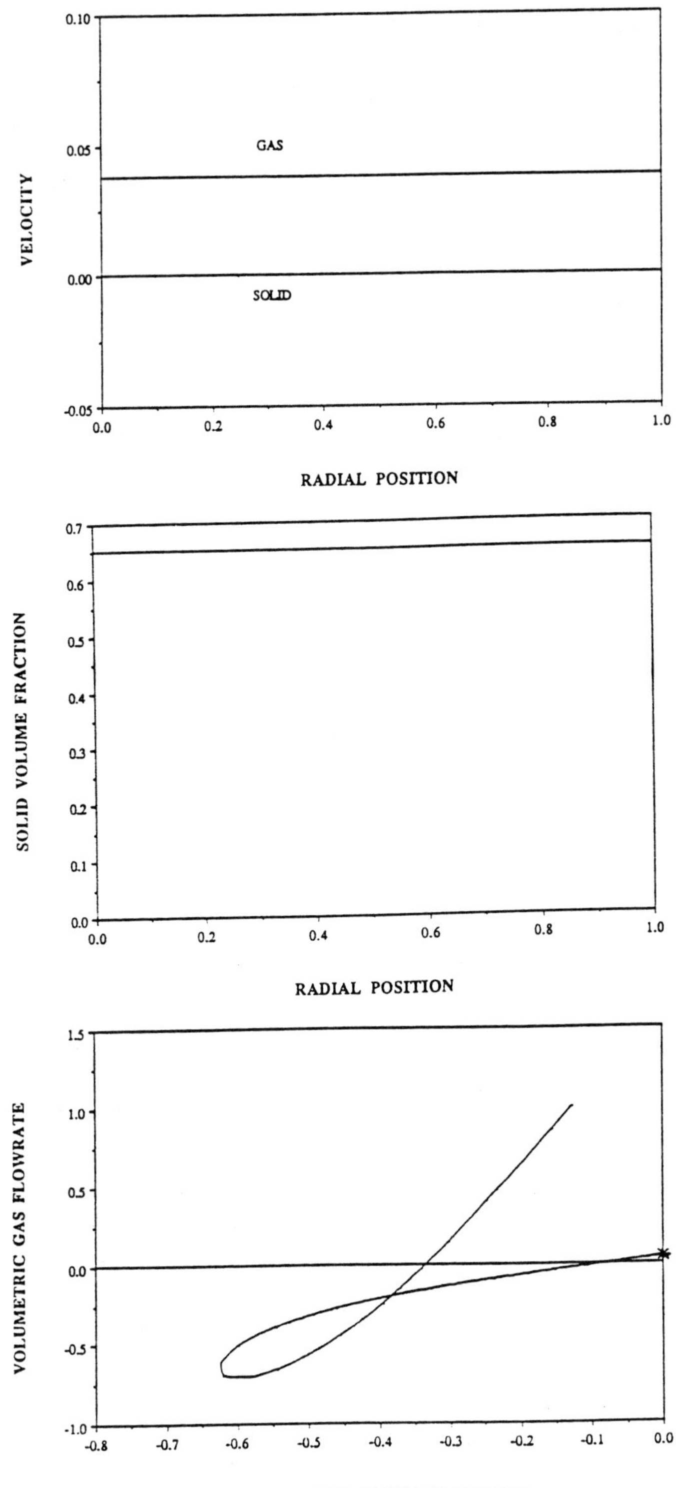

Figure 20. Velocity and particle concentration profiles for $(dp/dz)^* = -0.2$, $Q^*_s = 0$ and $Q^*_g = 0.04$. Mean solids volume fraction is 0.65.

The physical behavior responsible for the complex shape of a typical contour can be understood by plotting velocity and particle concentration profiles for a sequence of points along a selected contour, and this is done for the contour $(\partial p/\partial z)^* = -0.2$ in Figures 16 through 20. Figure 16 corresponds to a point (identified by a star) near the crossing from the first to the second quadrant, and it is seen that there is strong segregation of particles and gas, with the particles concentrated near the wall, where gravity is sufficient to make them move downward despite the upwardly directed drag force. At the axis, on the other hand, they move upward, giving a net upflow which is the difference between a large upflow near the axis and a somewhat smaller downflow at the periphery. Where the contour crosses into the second quadrant it is now clear that the net particle flow vanishes because of balancing upflow and downflow in different parts of the cross section. On moving into the crossback region the velocity and concentration profiles become progressively flatter, as shown by Figure 17, and at the apex of this region, where the contour again meets the ordinate axis, we have a classical fluidized bed, as seen in Figure 18. Beyond the crossback region non-uniform profiles of concentration and velocity develop once more as shown in Figure 19, which corresponds to a point on the loop of the contour in the third quadrant. Though the net flow of both phases is now downward there is an upflow of gas in a small annulus adjacent to the wall. Finally, when the contour returns to the ordinate axis for the last time (Figure 20), there is a uniform concentration $\alpha_s = \alpha_{max}$ representing a stationary packed bed of particles supported by the gas flowing through it.

Figures 14 through 20 all correspond to computations for perfectly elastic particles ($e = 1$), and inelastic particle-wall collisions ($e_w < 1$), and in all cases segregation takes the form of an increased concentration of particles near the wall of the pipe. However, the results are very sensitive to the values assumed for these parameters. For example, if e is reduced to a value slightly below unity, though still larger than e_w, the concentration profile is found to have maxima both on the axis and at the wall, with a ring of lower concentration in between.

Though the results of Sinclair and Jackson demonstrate the potential for interactions between the particles to generate complicated behavior of the general sort observed in practice, in most technical applications the Reynolds number, based on the gas flow alone, lies well within the range for turbulent flow. Of course, the presence of the suspended particles may damp the turbulence, but it is unlikely that its influence can be neglected entirely in most cases. Furthermore, the great sensitivity of the results to the values assumed for the coefficients of restitution is a source of concern, and this would almost certainly be blunted by the effects of turbulence. Recently Louge, Mastorakos and Jenkins (1990) have made a start in this direction. The mutual influence of the particles and the turbulent fluctuations of the gas velocity will depend on the relation between the relaxation time for changes in velocity of a particle in the gas, and the time scale of the turbulent fluctuations. In the limit of short relaxation times the particles and gas will move in synchronism, but at the opposite limit, when the particles are large and heavy, the particle motion will be essentially uninfluenced by the gas phase turbulence, though the presence of the suspended particles will certainly influence the turbulence. This second limit is the one considered by Louge et al., who compute complete solutions for fully developed flow in this case.

When the axis of the pipe is inclined to the vertical the situation is further complicated by the effect of gravitational sedimentation, which leads to an increase of concentration near the downward facing part of the wall. For horizontal pipes this effect may be large, with a moving bed of particles in mutual contact, or even a stationary packed bed, occupying the lower part of the pipe. In this region the particle concentration may be high enough for stress to be transmitted predominantly through long term sliding and rolling contacts, rather than collisions, and attempts to model this case theoretically should, therefore, allow for this possibility. Recently Ocone et al. (1990) have reported some results obtained for ducts of arbitrary inclination by extending the work of Sinclair and Jackson in this direction.

20.5 OTHER PROBLEMS

With the exception of the case of suspension flow in a standpipe, the work described above is all directed to modeling fully developed flow in straight ducts of essentially unbounded length. However, there remains a question whether real flows will ever approach these solutions and, if so, whether this will happen in a distance which is short compared with the total length of the pipe. There are also questions of what happens at bends, or obstructions in the pipe, or downstream of points where extra gas or particles are injected into the flow. Visual observation suggests that entry effects in these systems can be both complex and persistent, and in many instances they are precursors of flow patterns which are time-dependent, rather than steady as assumed in the simple models. For example, the flow pattern ultimately achieved may be quasi-periodic, as in slug flow where regions of high and low particle concentration alternate and move along the pipe. In the important example of catalytic cracking, referred to above, it appears that a large fraction of the reactor is occupied by a developing flow pattern in which a major part of the chemical reactions of interest take place.

Developing flow has been treated by Arastoopour and Gidaspow (1979), using a one-dimensional model of the sort discussed in the second section above, but allowing for acceleration of the material. Of course, this cannot show how the segregation pattern of gas and particles emerges on moving along the pipe, so quantitative agreement with observations of pressure drop will depend on adjustment of the effective drag force or, what amounts to the same thing, the effective slip velocity.

The whole question of the stability of the fully developed flows also remains open, and is likely to be very complicated. Unlike the case of a single phase fluid, where we are concerned only with shear instabilities that are presumably related to turbulence, the two phase mixture may be unstable even when there is no shearing and the particles are uniformly distributed, as we know from the stability theory of fluidized beds (see, for example, Jackson, 1985). This essentially two-phase instability is thought to be associated with the phenomenon of bubbling, and it seems possible that either a turbulent or a bubbling type of instability may predominate, depending on the circumstances. Modeling then becomes very difficult, but direct simulation by numerical solution of the full, time-dependent equations of motion is becoming a possibility, as shown by recent work on the simulation of fluidized beds (Gidaspow, 1986). To what extent it will be possible to bypass this, and devise steady state models that can predict the time average behavior, remains to be seen.

20.6 SUMMARY AND CONCLUSIONS

Two phase gas-particle mixtures are intrinsically more complicated than single phase fluids but, rather paradoxically, we have seen that useful results can be obtained from very simple, one dimensional models of flow which would yield only trivialities for a single phase fluid. The reason for this is that the interaction force between the two phases, rather than forces exerted at the confining walls of the duct, are primarily responsible for much physical behavior of interest, and these have no analog in a single phase fluid.

In analyzing any system in which particles and gas flow through a duct, we have also seen that it is vital to take into account the physical characteristics of the devices which feed the two phases to the duct, treating the duct itself and its feed devices as a single system. Without this, much of the complexity of the system's behavior is missed, as is very clearly shown by the example of flow in a vertical standpipe.

Perhaps the most striking qualitative feature of gas-particle flows is the heterogeneity of the suspension. This manifests itself as time-dependent effects, such as the gas bubbles which are typically observed in dense suspensions or the moving "dunes" seen in certain circumstances in horizontal ducts, but time averaged values of the particle concentration also shows large spatial variations, even when there is no obvious physical force, such as gravity, to generate them. Because of these heterogeneities there is a second layer of structure, in addition to that resulting from overall force balances and the interaction of different parts of the system, which cannot be accounted for by simple, one-dimensional models that treat the suspension as uniform at each section of the duct. These intrinsic segregation effects defeat efforts to make *a priori* quantitative predictions of behavior, or even to scale from one size of duct to another, using theories which do not contain the physical mechanisms responsible for the segregation. Thus, the present challenge is to understand these physical mechanisms, and to incorporate this understanding, in quantitative form, in mathematical models.

REFERENCES

Arastoopour, H. & Gidaspow, D., "Vertical pneumatic conveying using four hydrodynamic models," *I&EC Fund.*, **18**, 123-130 (1979).

Bader, R., J. Findlay, & T.M.Knowlton, "Gas-solid flow patterns in a 30.5cm diameter circulating fluidized bed,"*2nd International Circulating Fluidized Bed Conference*, Compiegne, France, 1989.

Bartholomew, R. & R. Casagrande, "Measuring solids concentration in fluidized systems by gamma-ray absorption," *Ind. Eng. Chem.*, **49**, 428-431 (1957).

Berker, A. & T. Tulig, "Hydrodynamics of gas-solids flow in a catalytic cracker riser: implications for reactor sensitivity performance," *Chem. Eng. Sci.,* **41**, 821-827 (1986).

Chen, Y-M., S. Rangachari, & R. Jackson, "A theoretical and experimental investigation of flow in a vertical standpipe," *I&EC Fund.,* **23**, 354-370 (1984).

Gidaspow, D., "Hydrodynamics of fluidization and heat transfer: supercomputer modeling," *Appl. Mech. Rev.,* **39**, no.1, 1-11 (1986).

Ginestra, J. C., S. Rangachari, & R. Jackson, "A one-dimensional theory of flow in vertical standpipes," *Powder Tech.,* **27**, 69-84 (1980).

Jackson, R., "Hydrodynamic theory of fluid-particle systems," in *Fluidization* (J.F. Davidson, R. Clift and D. Harrison, eds.) 2nd ed., Chap. 2., pp. 47-72, Academic Press, London (1985).

Jones, P. J. & L. S. Leung , "Downflow of solids through pipes and valves," in *Fluidization* (J.F. Davidson, R. Clift and D. Harrison, eds.) 2nd ed., Chap. 8., pp. 293-329, Academic press, London (1985).

Kunii, D., & O. Levenspiel, *Fluidization Engineering*, John Wiley & Sons Inc., New York (1969).

Louge, M., M. Mastorakis, & J.T. Jenkins , "The role of particle collisions in pneumatic transport," Private communication (1990).

Lun, C., S. B. Savage, D. Jeffrey, & N. Chepurniy, "Kinetic theories for granular flow: inelastic particles in Couette flow and slightly inelastic particles in a general flow field," *J. Fluid Mech.,* **140**, 223-256 (1984).

Matsumoto, S., "Stability analysis of choking in pneumatic conveying," in *Encyclopedia of Fluid Mechanics, vol. 4, Solids and Gas-Solids Flows* (N.P. Cheremisinoff, ed.), Chap. 16, pp. 485-496, Gulf Publishing Company, Houston (1986).

Mountziaris, T. J. & R. Jackson, " The effects of aeration on the gravity flow of particles and gas in vertical standpipes," *Chem. Eng. Sci.,* **46**, 381-407 (1991).

Nguyen, T. V., C. Brennen, & R. H. Sabersky , "Gravity flow of granular materials in conical hoppers," *J. Appl. Mech.,* **46**, 529-535 (1979).

Ocone, R., S. Sundaresan, & R. Jackson, "The structure of fully developed particle-gas flows in vertical and inclined ducts," in *NSF-DOE Workshop on Flow of Particulates and Fluids*, pp. 327-351, Gaithersburg (1990).

Rangachari, S. & R. Jackson, "The stability of steady states in a one-dimensional model of standpipe flow," *Powder Tech.,* **31**, 185-196 (1982).

Richardson, J. F. & W. Zaki, "Sedimentation and fluidization: Part I," *Trans. Instn. Chem. Engrs.* **32**, 35-53 (1954).

Saxton, A. & A. Worley, "Modern catalytic cracking design," *Oil and Gas J.*, **68**, No. 20, 82-89 (1970)

Sinclair, J. L., "Vertical transport of gas and solids with radial solid density variations," *PhD Diss., Princeton Univ.* (1989).

Sinclair, J. L. & R. Jackson, "Gas-particle flow in a vertical pipe with particle-particle interactions," *AIChE Jl.*, **35**, 1473-1486 (1989).

Tsuo, Y. P. & D. Gidaspow, "Computation of flow patterns in circulating fluidized beds," *AICHE Jl.*, **36**, 885-896 (1990).

Weinstein, H., M. Shao, & L. Wasserzug, "Radial solid density variation in a fast fluidized bed," *AIChE Symp. Ser.*, **80**, 117-121 (1984).

Yerushalmi, J. M., D. Cankurt, D. Geldart, & B. Liss, "Flow regimes in vertical gas-solid contact systems," *AIChE Symp. Ser.*, **74**, 1-13 (1978).

Youchou, L. & M. Kwauk, "The dynamics of fast fluidization," in *Fluidization* (J. Grace and J. Matsen, eds.), pp. 537-544, Plenum Press, New York (1980).

Zenz, F.A. & D.F. Othmer, *Fluidization and Fluid Particle Systems*, Reinhold Publ. Corp., New York (1960).

CHAPTER 21

EXPERT SYSTEMS IN SOLIDS HANDLING

G. Klinzing
S. V. Dhodapkar

21.1 INTRODUCTION

Computers, since their conception, have been instrumental in the refinement and verification of ideas. Early machines were primarily used for performing repetitive calculations with more speed and accuracy. The emphasis on data manipulation persisted until recent past when it was realized that a certain class of problems cannot be solved by faster computers. These problems involved elaborate search trees where a combinatorial explosion could bring even the fastest computer to its knees. This realization that we need to make our computers smarter and not just faster has motivated the research in Artificial Intelligence.

The efforts in Artificial Intelligence do not conflict or undermine the great strides made in hardware and conventional computer science. It merely provides us a tool for data interpretation (in contrast to data manipulation) using the explicit and implicit relationships between the elements. One of the objectives of AI is to put a hundred years of human thought process into every decision that is made. Expert systems are an important outgrowth of AI with the potential for improving the consistency and quality of the task. Since some systems do not live up to their expectation of emulation of expert behavior, the term "knowledge-based systems" appears to be more universally accepted.

Expert systems are highly specialized computer systems capable of simulating that element of a human specialist's knowledge and reasoning which can be formulated into knowledge chunks characterized by a set of facts and heuristics (Bowerman and Glower 1988). Heuristics are rules of thumb generally accumulated by a human expert through intensive problem solving in the domain of a particular task. The central idea in building expert systems is to develop methods and techniques for such special problem expertise. The expertise consists of understanding of domain problems and the skill required to solve them.

Human thought process is often guided by previous experiences and intuition. The decision making process is directed by strategies which blends the knowledge with experience. Such expertise is very difficult to document and it is often loss through job transfer, transfer, retirement and lay-offs. The importance of expertise can never be understated in the area of solids handling. Despite continuing effort in fundamental sciences, the gap between theory and practice still exists. The growth of complexity of solids processing technology has further compounded this problem. It has been found that solids processing technology is the bottleneck for improvement in process performance (Knowlton 1987; Marrow 1984). Most of the aspects of solids handling are not amenable to exact mathematical modeling. Because of the complexity of interactions, much of the reasoning has to be done qualitatively. Decisions based on detailed reasoning at the conception of the project results in large savings during operations. Important activities such as selection, troubleshooting and configuration can be automated to improve the consistency and quality of results. It must be kept in mind that knowledge-based systems do not generate knowledge, but merely aid in the use of knowledge more effectively.

From our experience in the area of computer applications in solids handling (and particularly in pneumatic transport), it was observed that improvement in performance can be achieved without resorting to bench scale tests or pilot plant studies. Computer simulations have often found use in predicting the effect of operating variables on the performance of a system or to determine the optimal combination of operating conditions. This results in significant savings of time and money for the user concerned. Oftentimes it is found that the source of problems and inefficiency is solid handling systems lies not in the choice of operating parameters, but in the correct selection and specification of components.

Reduction in down-time for a process can be achieved if the problem which has resulted in shut-down can be identified, located and corrected. Such a task requires qualitative reasoning assisted by quantitative estimations. Some of these problems can be avoided by monitoring the data on the system performance from a multitude of sources and making a reliable prognosis. To solve these problems, one needs to establish the relationships between the variables.

Artificial intelligence, in particular expert systems, have changed the way we use computers. Those problems which did not require number crunching are now included in the domain of computer applications (Stephanopoulas 1987; Barnwell and Ertl 1987). Although the term "AI" was coined by McCarthy in 1955 during the Dartmouth conference, it was not until the 1970's that various complex applications began to emerge. They were instrumental in developing new knowledge representation schemes and reasoning mechanisms. Unlike conventional programs, which have the knowledge and inference strategies implicitly coded in the program, these knowledge-based systems had the domain-specific knowledge as a separate entity from the control structure. This provided much greater flexibility in developing intelligent systems. Some of the successful systems are given in Table 21.1. A comprehensive classification of knowledge engineering applications is given by Hayes-Roth et al. (Hayes-Roth et al. 1983) (Table 21.2).

TABLE 21.1 Some Existing Expert Systems

System	*Area of Application*
DENDRAL	Molecular Structure from Mass Spec.
EL	Electric
MYCIN	Diagnosing Blood Diseases
MACSYMA	Mathematical Formula Manipulation
MOLGEN	Planning DNA Experiments
R1/XCON	Computer Configuration

With the advent of personal computers in the early 1980's and subsequently improved machines in the mid-1980's, PC-based expert system building tools have flooded the market. The race for better user interface and more powerful tools has almost kept up with the improvement in the hardware. These days, one no longer needs to work with specialized hardware or use languages to develop expert systems.

TABLE 21.2 Classification of Knowledge Engineering Applications (Hayes-Roth, et al. 1983)

Category	*Problem Addressed*
Interpretation	Inferring situation descriptions from sensor data
Prediction	Inferring likely consequences of given situations
Selection	Selection of most suitable match
Diagnosis	Inferring system malfunctions from observables
Design	Configuring objects under constraint
Planning	Designing actions
Monitoring	Comparing observations to plan vulnerables
Debugging	Prescribing remedies for malfunctions
Repair	Executing plan to administer a remedy
Instruction	Diagnosis, debugging, and repairing student behavior
Control	Interpretation, prediction and monitoring system behavior

21.2 BUILDING EXPERT SYSTEMS

21.2.1 Choosing an Application

One of the most common mistakes that results in failure of an expert system application is the wrong choice of the problem. It is erroneous to select a problem area and then look for a specific problem for AI application. One must instead choose a specific problem which needs to be solved and then explore further whether it is a good problem or not (Waterman 1986).

Problem identification is the first step in developing an expert system as shown in Figure 21.1. Although almost all the texts on building expert systems (Hayes-Roth, et al. 1983; Waterman 1986; Pedersen 1989) deal with this problem extensively, we will point out the most salient considerations.

1. The problem should be real.

2. It should be sufficiently complex, but of manageable size.

3. The problem should be doable (at least by the experts) in finite time.

4. Experts should be available and ready to work with the knowledge engineer. It is often beneficial to involve the end user because such involvement eliminates the apprehension of the user about any black-box technology.

5. Information should not continually change.

6. The cost of development should be justifiable by the usage or the quality of decisions made by the system.

7. Information from various sources should be mostly consistent.

8. Ask yourself, "Does the solution to the problem rely too much on common sense?". If the expert relies too heavily on common sense, intuition, or hunches, which cannot be easily formulated into a set of rules and relationships, even using uncertainty, then that application is unsuitable for expert system development (Casey 1989; Keyes 1989; Smith 1989).

In essence, remember to keep the application simple but significant.

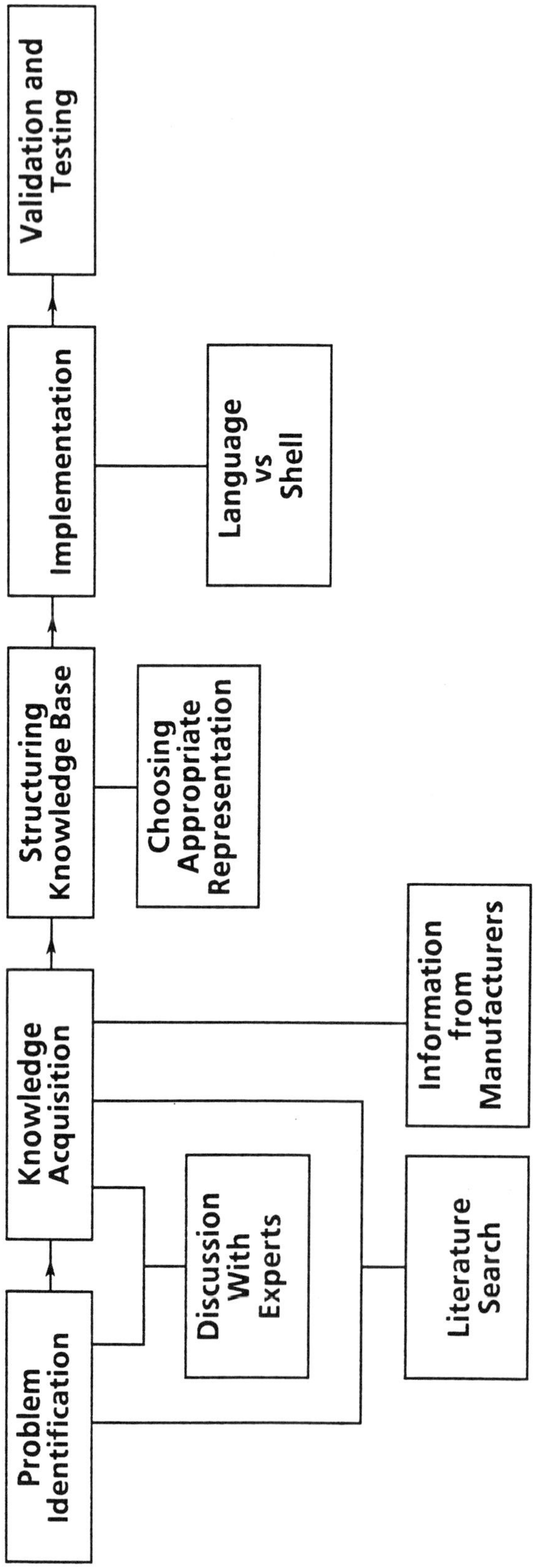

Figure 21.1. Development cycle of expert system.

21.2.2 Knowledge Acquisition

As mentioned earlier, knowledge based systems operate on knowledge obtained from a variety of sources. To code the knowledge, the knowledge engineer should be familiar with the domain and its terminology. Our familiarity with the field of bulk solids handling greatly helped in understanding the complex decision-making tasks performed by the experts. Some excellent strategies are available in literature (Pedersen 1989) which help us to elicit and structure the cognitive processes of an expert.

21.2.3 Choosing the Right Representation and Right Tool

Once a good problem has been selected and a reasonable amount of knowledge in that domain is acquired, a suitable Expert System Building Tool (ESBT) should be chosen. It is unwise to select a problem that a particular tool can solve or to bend the tool and/or the problem to fit each other. Unlike database spreadsheet or word processing packages which are fairly standard across a company, no standard ESBT can be chosen. Choice of an expert system building tool is also guided by the knowledge representation strategy that suits the knowledge base.

Other important considerations are:

- Hardware Requirements
- Maximum Size of Rule Base Permissible by the Shell
- High Level Language Interface
- Graphics Interface
- Interface With Other Data Base and Spreadsheets
- Cost and Portability
- Documentation and Technical Support

The issue of choosing ESBT is irrelevant if a language (such as LISP, PROLOG, SMALLTALK, etc.) is used for development (Degroff 1987). It must be noted that conventional programs can be written in AI languages while AI programs can be written in conventional languages, too.

Therefore, it is the content of the program that classifies it as AI/non-AI and not the language.

Language could be chosen as a development tool due to many reasons. If more flexibility in representation is desired or the code needs to be compacted or greater customization is desired or new strategies have to be incorporated later, then language is the way to go. Although they are difficult to maintain, uneconomical for small to mid-size applications and require too much effort for prototyping, they might be the only way to go for certain a class of applications. Shells, on the other hand, are easy to maintain and easy to prototype (Gevarter 1987). Their limited capabilities, however, surpass most of our needs. New products with very sophisticated strategies are flooding the market so fast that it seems futile to develop one's own shell. So it is recommended that an exhaustive market survey should be conducted before any decision to use a language is made (Freedman 1987; PCPlus Manuals 1986).

21.2.4 Implementation

Once the knowledge representation strategy is decided, the effort required in coding is not much. Most of the shell provides a very user-friendly interface to ease the coding of the knowledge base.

21.2.5 Testing and Validation

Testing or verification means that the system does not have technical errors or "conceptual bugs". Validation means that the program performs the functions it is intended for and conforms to the user's requirements. Validation also means that the answers provided by the program are acceptable to the end user (Geissman and Schultz 1988).

21.3 APPLICATIONS

The field of bulk solids handling is very amenable to expert system applications. As stated earlier, the generic categories of applications of knowledge engineering encompasses a wide range of activities performed by experts in bulk solids handling. Solids handling operations mainly involve storage, transport (or conveying) and processing of bulk solids,

Figure 21.2. Other common activities, such as separation and material classification, can also be included. In this section, we have pointed out some possible applications.

21.3.1 Storage Related Problems

Problems that can be best handled by the expert system approach can be further classified according to the generic classification. See Table 21.2.

21.3.1.1 Interpretation

Interpretation of shear cell results or interpretation of wall shear stress sensor to understand flow patterns in bins/silos.

21.3.1.2 Prediction

To assess the need to do predictive maintenance.

21.3.1.3 Selection

Selection of an appropriate flow aid device is a complicated task. One must take into account the material properties, bin configuration and desired performance to make a selection among many mechanical, electrical and non-mechanical flow aid devices. Proper choice of such devices as retrofits can also result in significant improvement in performance.

Feeders and hoppers should be treated as an integral unit. Selection of feeders depends on the type of hopper, material properties, downstream conditions and performance expectation. We will discuss it in greater detail in the following section.

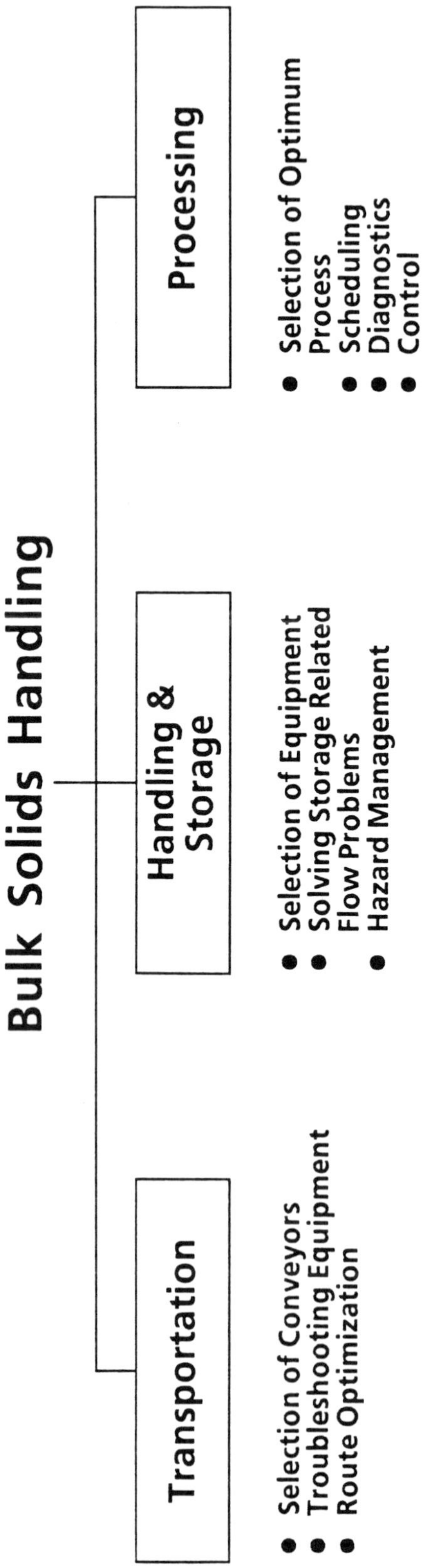

Figure 21.2. Areas of application for expert systems in solids handling.

21.3.1.4 Diagnosis and Debugging

It is not unusual to encounter a bin/silo with poor flow characteristics. Solution to these problems is not obvious because of the numerous interactions of material properties and silo geometry. In many occasions, more than one solution is feasible. The complexity of solution space and the need to optimize the solution due to its capital intensive nature, makes it an interesting problem for expert system.

21.3.1.5 Design and Configuration

The success of solids feeding systems depends on the right choice of the combination of bins, hoppers and feeders. Qualitative reasoning required to suggest the proper combination can be emulated in expert systems.

21.3.1.6 Repair

In high risk applications, where large capital expenditure is involved or hazardous materials are handled and the success must be ensured, knowledge based systems can be useful.

21.3.1.4 Instruction

An intelligent tutoring system can be developed to train a new generation of experts. These systems can be developed in any of the above-mentioned areas. One of the outstanding features of expert system is its ability to explain WHY and HOW a decision has been made.

21.3.2 Transport Related Problems

21.3.2.1 Interpretation

Interpretation of data from sensors to determine conveying conditions at various points in a complex network of pneumatic conveying lines is vital. We are exploring this aspect in our laboratory at the present time.

21.3.2.2 Prediction

From the behavior of the conveying system, one can project what components are likely to fail and hence, need replacement. Erosion and wear are the prime reasons for limited life of equipment. Knowledge based systems can make reasonable judgements to prevent mechanical failures.

21.3.2.3 Selection

Selection is generally the first capital intensive activity involved in any project. Proper selection of conveying systems (mechanical or pneumatic) will alleviate future problems. As a sub-task, one must select appropriate components for the system. Selection process is generally accompanied by specification of the components. We will discuss various issues involved in the next section.

21.3.2.4 Diagnosis and Debugging

Diagnostics or troubleshooting is one of the main reasons for which the advice of an expert is sought. Troubleshooting is both an art and science. It is like trying to find the identity of an animal from its footprints on wet sand. Many observations that are required to diagnose the problems have to be made by human operators. An ideal diagnostic system would be one which replaces this human intermediate with automatic sensors to "look and listen".

21.3.2.5 Design or Configuration

In many instances, it is required to configure a system for required duty using a combination of hopper, feeder, pipes and collector from a large inventory. Using concepts similar to the classical expert system XCON from DEC, which configures a VAX computer according to customer needs using the large inventory of modules, conveying systems can be quickly customized.

21.3.2.6 Monitoring

Monitoring conveying conditions at various points in a conveying loop is possible by monitoring the pressures and fluctuations.

21.3.2.7 Instruction

Training systems can be developed on all the above aspects.

21.3.3 Processing Related

Solids processing includes a wide range of activities, such as separation, communition, drying, crystallization, etc. Each process must be evaluated individually under the classifications suggested in Table 21.2. Most of the applications will be similar to that discussed earlier and we leave it to the reader to enumerate them for relevant processes.

21.4 CASE STUDIES

From the previous section, it is evident that numerous problems in solids handling can be addressed by this new technology. In conformity with our interests in solids transport, we have developed several systems with differing degrees of complexity. These can serve as guiding examples for those interested in developing similar applications.

21.4.1 Nu-Select

Pneumatic conveying systems can be categorized into various classes based on the density of suspension, system pressure, type of air supply, type of feed system, etc. Not all types of systems meet the performance criteria for a given material. Selection of correct type of conveying system will prevent costly modifications.

Information on suitability of conveying systems is available in the literature. However, most concise representation of this information is given by Caldwell, 1976. According to Caldwell, four main considerations are:

A. Process or Physical Requirements

B. Distance to be Conveyed

C. Material Properties (Friability, Abrasiveness, Density and Size)

D. Type of Operation (Unloading, Reclaiming, Simple Transfer, etc.)

One of the drawbacks of this chart is the inability to distinguish between various conclusions on the basis of their relative merit. For example, both Fan Pressure Systems and Stream Flow Pressure Systems might be feasible for a given set of conditions. The current representation does not elucidate the difference in degrees of confidence for each answer. Further information was therefore acquired from an expert by providing a more elaborate table with special slots for certainty factors. Rules were then formulated to transform the knowledge into a rule base. A generic form of the rule is as follows:

IF:

PROCESS REQUIREMENTS = . . . AND/OR . . . AND/OR NONE

AND OPERATION TYPE = . . . AND/OR . . . AND/OR NONE

AND DISTANCE CONVEYED = ...

AND MATERIAL CHARACTERISTICS = . . . AND/OR . . . AND/OR . . . AND/OR

THEN:

CONVEYING SYSTEM = SYSTEM1 WITH CERTAINTY FACTOR CF1

CONVEYING SYSTEM = SYSTEM2 WITH CERTAINTY FACTOR CF2

AND SO ON.

This system was implemented on PC+/PC Easy environment (Caldwell 1976). PC Plus is a LISP-based environment with an easy to use user interface. NU-SELECT attempts to determine the value of the goal parameter (type of conveying system) during each consultation. The rules that contain the goal parameter in the consequent part of the rule are marked for tracing. Further, backward chaining is initiated if some of the parameters in the antecedent part can be evaluated from some other rules. Backward chaining is a goal driven process in which a goal is hypothesized and effort is made to validate or invalidate the goal. Suitable prompts are generated in case it is not possible to evaluate some of the parameters from implicit information. Goal properties can be changed to change the nature of consultation (from all possible solutions to best first solution).

21.4.2 EXCON

EXCON (Dhodapkar and Klinzing 1989) is an expert system for mechanical conveyor selection. The structure and formulation of this problem is very similar to NU-SELECT. Although the structure remains the same, the knowledge base has been replaced by that for mechanical conveyors. This shows the versatility of expert systems to adapt to newer but similar problems without making fundamental changes in the logic and strategy. The goal of the consultation in EXCON is to suggest a suitable type of mechanical conveying equipment for a given duty.

21.4.3 Feeder

Feeders are an integral part of solids processing plants. Apart from being an essential component of pneumatic conveying systems, they are used in many solid handling operations to feed, meter and regulate the flow of solids from bins and hoppers. An extensive amount of information is available on individual units and numerous claims are made by the manufacturers regarding their applicability. Selection of an appropriate feeder is a challenging problem because of the sparseness of the information and lack of consensus on validity of available information (Dhodapkar and Klinzing 1990).

Lacking a systematic taxonomy of feeders, we decided to classify the feeders into two major categories -- feeders for pneumatic conveying applications and feeders for solids handling applications, Figure 21.3. Pneumatic conveying feeders were further sub-classified as low pressure (less than 1 bar differential pressure) and high pressure (greater than 1 bar differential pressure). Feeders for general solids handling can be classified either as volumetric or gravimetric. Gravimetric feeders are also volumetric feeders but with weighing capability. It can be seen some

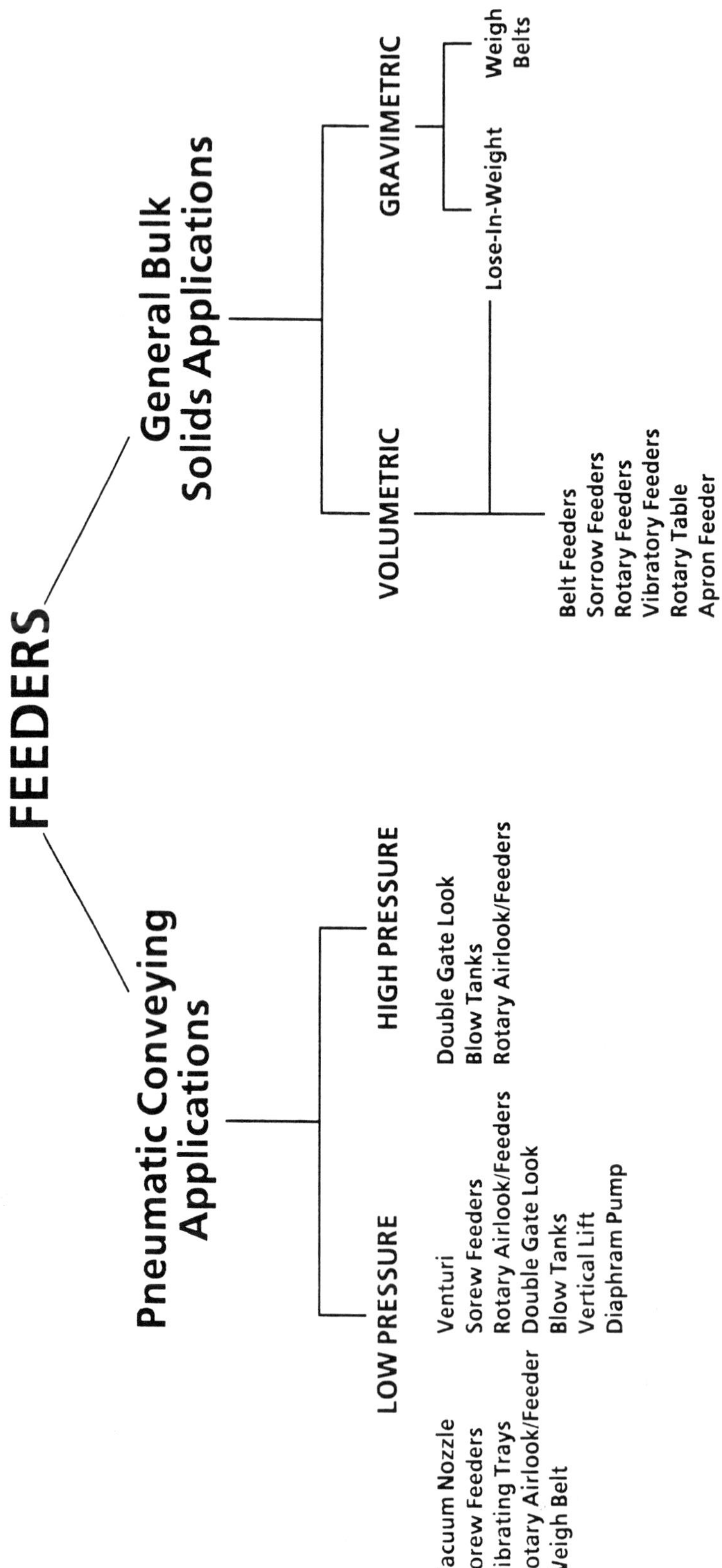

Figure 21.3. Classification of feeders.

feeders belong to more than one branch of the tree, thus depicting their broad range of application. Such a classification is most essential to structure the information which is available in chunks.

Using the frame-based knowledge representation strategy of PC Plus, each feeder can therefore be separated from the general information. The overall strategy that appears to work very well is given here:

1. Based on application type or other evidence, either pre-select the frames to instantiate or explore them one by one, based on their ubiquity in actual practice.
2. Once enough positive evidence is gathered, look for negative pointers for that selection.
3. Select the feeder if no evidence against that selection is found or if the user decides in favor of it.
4. Instantiate that feeder's sub-frame for further specifications and special recommendations. Call external programs for design and specification if necessary.
5. Do 2 until the list is exhausted or the user terminates the session.

The HOW and WHY capabilities of this expert system help the user to understand the knowledge and decision making process.

As a test case, an appropriate feeder for feeding plastic pellets in a pneumatic conveying system was sought. Table 21.3 shows a typical consultation with the expert system.

TABLE 21.3
CASE STUDY

Suggest a feeder for a pneumatic conveying system to feed plastic pellets at 500 lb/h.

The following text does not recreate the actual prompts or screens. It is merely indicative of the problem solving strategy employed by the expert system for a given problem.

COMPUTER --> TYPE OF FEEDER APPLICATION?

USER ** PNEUMATIC CONVEYING

> TYPE OF SYSTEM?
** POSITIVE PRESSURE

User selectable or use NU-SELECT (Dhodapkar and Klinzing 1989)

> PRESSURE AT FEED POINT?
** 10 psi

Use UPSD (Dhodapkar and Klinzing 1988) or OPSD (Plasynski 1989) if necessary

> OPENING LOW PRESSURE FEEDER FRAME

> CHOOSE AREA OF APPLICATION
** PLASTIC PELLETS

reduce the search space using explicit knowledge

> ROTARY AIRLOCK/FEEDER IS RECOMMENDED

selection complete; specification begins

> ABRASIVENESS?
** NON-ABRASIVE

> FRIABILITY?
** NON-FRIABLE

TABLE 21.3
CASE STUDY
(Cont.)

> MATERIAL DESCRIPTION (GRANULES, PELLETS, FLAKES, POWDER, FIBRES)?
PELLETS - O.K.?
** YES

> IS LEAKAGE A MAJOR CONCERN?
** YES

> FLOWABILITY?
** FREE-FLOWING

get help for explanation of terms

> OTHER PROPERTIES - (E.G. STICKY, GREASY, DAMP, HYGROSCOPIC..)
** NONE

> OUTLET SIZE OF HOPPER (feet)?
** 1.5

> FEED RATE (lb/hour)?
** 500

> BULK DENSITY (lb/ft^3)?
** 40

> DOES THE MATERIAL SMEAR?
** YES

> TYPE OF FEED CONTROL?
** APPROXIMATE METERING

> SPECIFICATION FOR THE ROTARY FEEDER

Rotor End Suggest:
Shrouded Rotor (closed pocket)

TABLE 21.3
CASE STUDY
(Cont.)

Rotor Volume:
    Standard Full Volume Rotor
Number of Blades:
    Ten Blades recommended
Type of Blade:
    Relieved tip or back beveled rotor
Housing Type:
    Side Entry Configuration
Venting:
    Top Venting
R.P.M. (operation):
    Operate between 15-30 RPM
Special Considerations:
    For temperatures less than 150°F, no special recommendations

Expert System continues to explore further for feasibility of other feeders unless the session is terminated by user.

21.4.4 Panacea

Troubleshooting complex chemical processes and networks using a computer based approach is not new. Much work has been done in the area of fault tree diagnosis since the 1960s. A gradual shift toward knowledge based systems for diagnosis and troubleshooting has been observed since the last 1970s. It has been found that comparable capabilities of fault detection can be achieved much easier if a knowledge based approach is taken. Essential elements of troubleshooting are looking, feeling, listening and measuring. These activities are qualitative in nature and are not amenable to a regular number crunching approach. Searching for a remedy to a troubleshooting problem is like looking for a needle in a haystack. The economics of search, both in time and equipment costs, is so prohibitive that it is imperative to resolve the situation without too many iterations. We have therefore made a systematic effort to elucidate and structure the information on troubleshooting pneumatic conveying systems. Most importantly, we can now transform these ideas into computer programs usable on portable computers. It is much cheaper to send a computer disk across a continent than to dispatch a team of experts. The economical advantage is tremendous, especially when timely actions can save a lot of money. Such computer programs should be made an integral part of any training program where a systematic approach to troubleshooting can be taught more like a basic skill.

Knowledge structuring indicates the limits of interpolation, extrapolation and analogy of concepts in the domain. It must be born in mind that **there is no one way of representing knowledge**. As long as we are efficient, concise and complete in our representation, half the battle is won. Essential to every knowledge domain is the underlying structure. Troubleshooting information specific to systems that worked before and systems that have never worked can be categorized separately. In each of these categories, we further distinguish between positive and negative pressure systems as shown in Figure 21.4 We also like to distinguish between troubleshooting and optimization. Troubleshooting refers to remedying a situation when no solids are being transported from source to destination. Optimization refers to fine tuning a system which is not meeting the desired performance goals.

The problems can be classified into two major categories:

I. For systems that worked before:

a. Problems related to equipment wear and failure
b. Problems related to material bulk properties
c. Problems related to operating procedures

Special Case: System that has been shut down temporarily and does not work while start-up. Mostly the problems are due to improper start-up procedures.

II. For systems that never worked before:

a. Problems related to bad operating procedures
b. Problems related to wrong selection of equipment
c. Problems related to line configuration

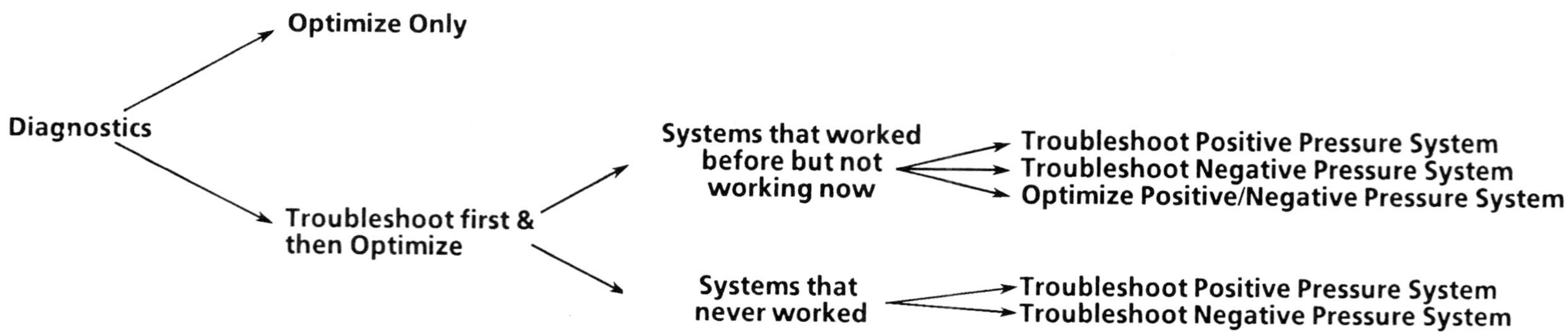

Figure 21.4. Knowledge structure for pneumatic conveying troubleshooting domain.

In practice, no two pneumatic conveying systems are exactly the same. However, we can broadly classify various pneumatic conveying systems and then assign certain components. We are interested in generating a hierarchial tree to direct our search. For example, if a pneumatic conveying system is blow-tank arrangement, then feeder related air leakage problems are not expected. Any pneumatic conveying system can be further subdivided into the following sections:

1. Material Feeder
2. Air Supply
3. Air Material Mix
4. Conveying Line
5. Air Material Separator
6. Control System

Engineering systems are generally well defined and the knowledge about the system can be organized at various levels of abstraction. For example, a complete system can be further subdivided into its components, namely; feeder, air supply, air material mixer, conveying line (includes bends, splitters, etc.), air material separator and control unit. At a lower level of abstraction, we can identify various components of each unit. At further lower level of abstraction, we can concern ourselves with mechanical details of each unit. It is obvious that lower the lower the level of abstraction, the more non-hierarchial knowledge becomes. For example, nuts and bolts may be found in the feeder, air supply or air material separator. Therefore, if a broken bolt is the cause of a malfunction, it is not enough to specify only the cause, but its location is required. The amount of information necessary to pinpoint a cause is dependent on the level of abstraction.

Any indication of the system behavior that suggests the existence of a problem is called a symptom. Before we brand a particular system behavior as a "symptom", we must be clear about our definition of a "problem". Is the troubleshooter trying to enhance the performance of a system which is operating as desired, is he trying to start a system which never worked, or is he trying to do preventive maintenance? For all the above reasons, one might seek the advice of an expert.

With years of experience, troubleshooters may gather a lot of handy rules. These rules associate obvious symptoms with non-obvious symptoms or causes. Implicit knowledge can always be deduced from explicit knowledge when the complexity of reasoning and time required to reach a conclusion is inconsequential. One must strike a balance between operating with explicit information versus generating deductions from first principles.

It might take unnecessarily complex reasoning to find the non-obvious causes. Certain actions are generated by the troubleshooters very quickly. For example, given "line plug in the pipe", one must look for "air leakage at the coupling or flange or a broken pipe". From a theoretical viewpoint, if a good representation of the system is available, it is possible to derive every symptom. But such a process may be slow and circuitous, and therefore, error-prone. Problems, causes and remedies are related in a tree-like structure, Figure 21.5. One can appreciate the possibility of a combinatorial explosion in such cases.

If the system is operating at part with its original installed capacity, and you would like to further fine tune its performance, you have the following choices:

1. Improve horsepower requirements
2. Improve capacity
3. Improve flow smoothness
4. Reduce system wear
5. Reduce particle degradation

Follow are some of the fine tuning suggestions:

1. Better operating parameters
2. Equipment changes or modifications
3. Line size/configuration changes

Complexity and ubiquity of a malfunctioning system justifies the effort to develop PANACEA. As can be seen, PANACEA requires a lot of input from a human operator, often making suggestions to look, listen and feel the symptoms. Therefore, the human operator becomes an interface between the conveying system and PANACEA. This has been one of the limitations for PANACEA. The information that the operator provides (such as flow patterns or sounds that the system makes) is vital to assess the conveying conditions and problems. We have made significant progress in providing this "hearing and visual" capability by developing techniques of flow pattern identification using pressure fluctuations. This will be of tremendous help in complete automation of the diagnostic system, Figure 21.6.

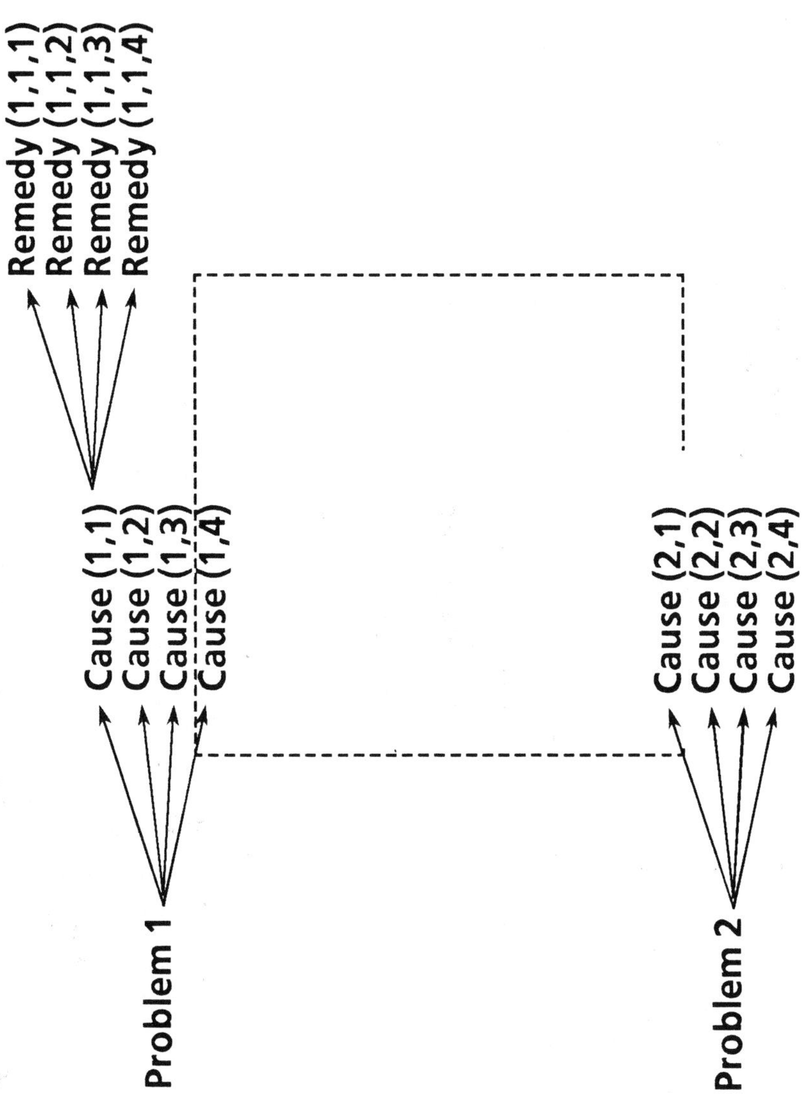

Figure 21.5. Problem-cause-remedy tree structure

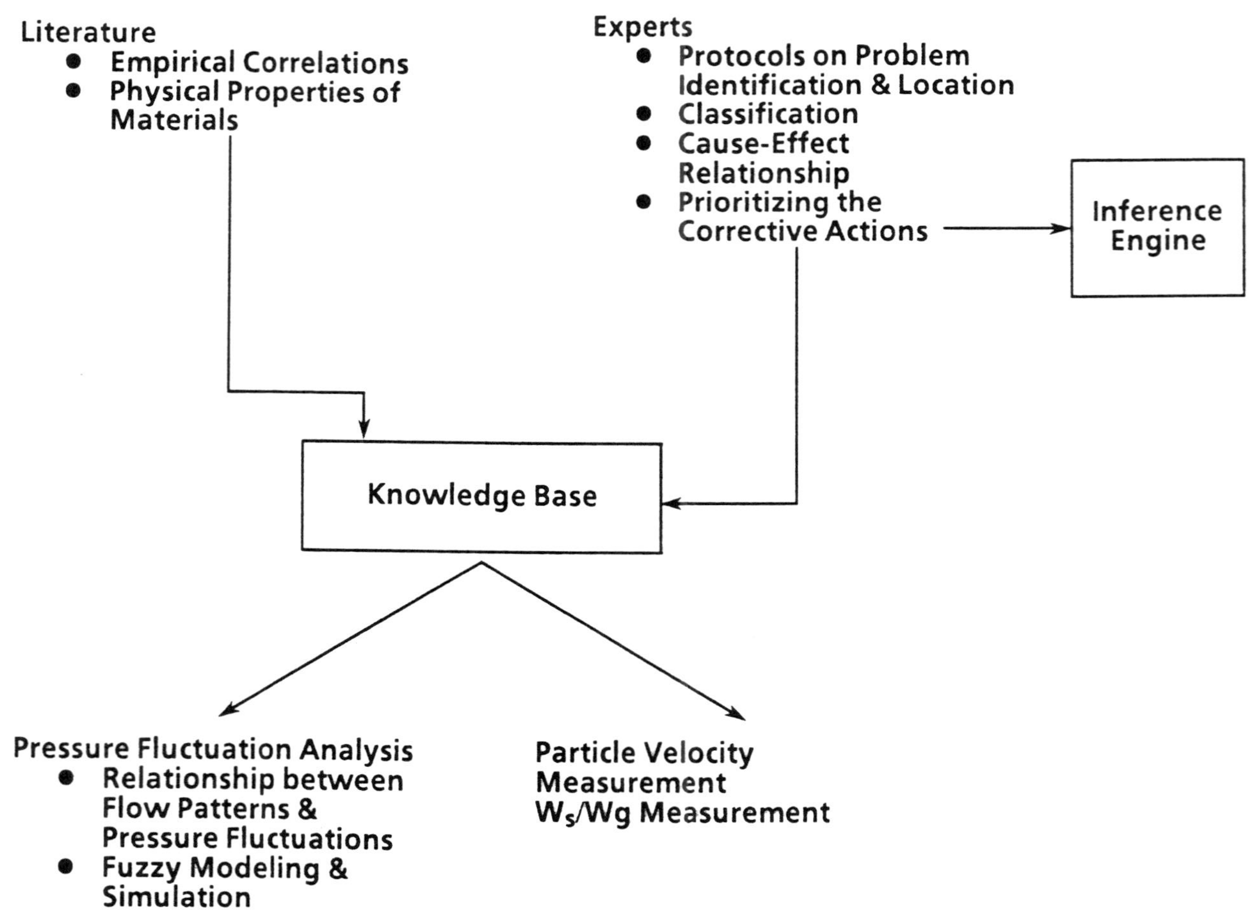

Figure 21.6. Structure of a diagnostic system for pneumatic conveying systems

21.5 FLOW PATTERN IDENTIFICATION

This section discusses a fuzzy logic based approach for flow pattern identification in pneumatic conveying systems.

On-line determination of flow patterns gains significance when a network of conveying lines is involved. Measurement of total gas flow rate and solids flow rate at the inlet does not ensure smooth flow in the conveying line. For example, the conveying air may leak out of a diverter valve or a rotary air lock valve. This will cause flow instabilities in the conveying line even though the measurement of gas and solids mass flow rates suggest otherwise. The determination of flow patterns in the conveying lines becomes very significant when non-equal distribution of gas and solids occurs due to availability of parallel flow channels (such as in injection systems of combustors or flow splitters).

One of the most obvious methods of flow pattern identification is visual observations. However, it is rarely possible to have transparent conveying lines or viewing sections at different locations. The physical location of these conveying lines may also make the visual observations difficult. Measurement of capacitance or particle velocity is a viable alternative, but cost of such set-ups and complexity of instrumentation strongly weighs against them. Using pressure fluctuations as a means of flow pattern identification has been used as a viable alternative in gas-liquid systems (Tutu 1982). Their use to detect and identify the flow patterns in gas-liquid systems has been limited and needs more work. The aim of this work is to study the effects and interactions of various parameters on pressure fluctuations. It is also necessary to outline a procedure to analyze these signals. Fuzzy algorithms have been found useful for such analysis. These algorithms will be discussed later in this section.

Apart from the paucity of available data, the analysis of pressure fluctuation data and determination of various data acquisition parameters is challenging. Published information on analysis of pressure fluctuation in gas-liquid systems gives a good starting point. A summary of different types of flow regime detection techniques has been given by Rouhani and Sohal (1983). Various flow regime indicators have been explored so far. The flow regime indicators may be derived either from frequency domain analysis or amplitude domain analysis. Frequency domain analysis includes power spectral density function, cross spectral density function and coherence analysis. Amplitude domain analysis includes probability density or distribution analysis, various moments of the mean and other derived quantities. However, it can be concluded from the available information on gas-liquid systems that there is no single effective indicator of flow regime which is universal.

At present, most of the techniques employed for flow regime identification using pressure fluctuations are empirical or statistical in nature. Some qualitative work has been done by Tsuji and Morikawa (1982) and Matusmoto and Morikawa (1987), but no workable methodology or algorithm has been proposed. Pressure fluctuations in pneumatic conveying systems are a result of some complex interactions of two phases (Satija et al. 1985). and the configurations of the system. Oftentimes, it is simply unwise to model each of these interactions using an oversimplified approach. In such systems, which are beyond the complexity threshold of the classical model, no realistic modeling can be done. It is of interest to note that Summer et al. (1989), in their study on slurry pipeline flow, measured the concentration fluctuations, finding that the low frequency spectra showed significant concentration and particle diameter effects. For fine particles, they found that at high concentrations, the amplitudes increase with increasing concentration. These effects could be employed in flow regime identification for slurry flow.

An alternative approach is to use fuzzy modeling. Fuzzy set theory is an extension of classical set theory. A major distinction from the classical theory is the acceptance of partial belongingness of an element to a set. Fuzzy models are a collection of fuzzy implication of the form "IF A THEN B". Fuzzy models described Dohnal (1982, 1983a, 1985, 1983b) are useful in such situations. New applications have been proposed recently by Dohnal in the field of chemical engineering as an alternative modeling approach. Fuzzy simulation of ill-defined problems, especially in engineering when there is little data available, is a practical alternative to modeling.

It is necessary to enumerate the possible sources of pressure fluctuations before such an experimental set up can be conceived. Pressure fluctuations (both static and differential) can be generated in horizontal systems by:

1. Fluctuating solids flow rate

    - causing voidage variations
    - solids distribution across the cross section

    Relative contribution of the above to the total fluctuations depends on the difference between operating velocity and saltation.

2. Due to air supply and air line feed section.

3. Turbulence due to particle collision at the wall and interparticle collisions.

4. Solids deposition at the feed point.

5. Collector characteristics.

6. Other elements in the system, especially bends and tees.

The pressure signals were acquired using A/D board and IBM PC/AT. The record length was 4096 points and the frequency of sampling depended on the flow conditions in the pipe (25 Hz to 1000 Hz). The power spectral estimation and probability distribution functions were evaluated using standard methods (Bendat and Piersol 1986).

21.5.1 Flow Pattern Maps

Flow pattern maps are flux-flux diagrams demarcating the regions of similar flow patterns. These flow pattern maps have to be experimentally determined and the regions of similar flow patterns change with particle size, orientation of the pipe and particle density. Figure 21.7 is a typical flow pattern map for coarse material. The solids flow rate, W_s, is plotted against the gas velocity, U_g. These maps can be useful if they are generalized enough to account for the effects of particle size and density. They can also be used to predict the possible flow pattern when the solid and gas flux are known. However, there are two major drawbacks with their use. Firstly, an extensive amount of experimental data is required to statistically demarcate the boundaries between the regions. Secondly, no generalized coordinates have yet been proposed to generate such maps. The effects of system components in pneumatic conveying systems on flow patterns makes the choice more difficult. It also requires the knowledge of local gas flux and solids flux which is not easy. Therefore, the flow pattern maps are useful only for first approximation of flow patterns.

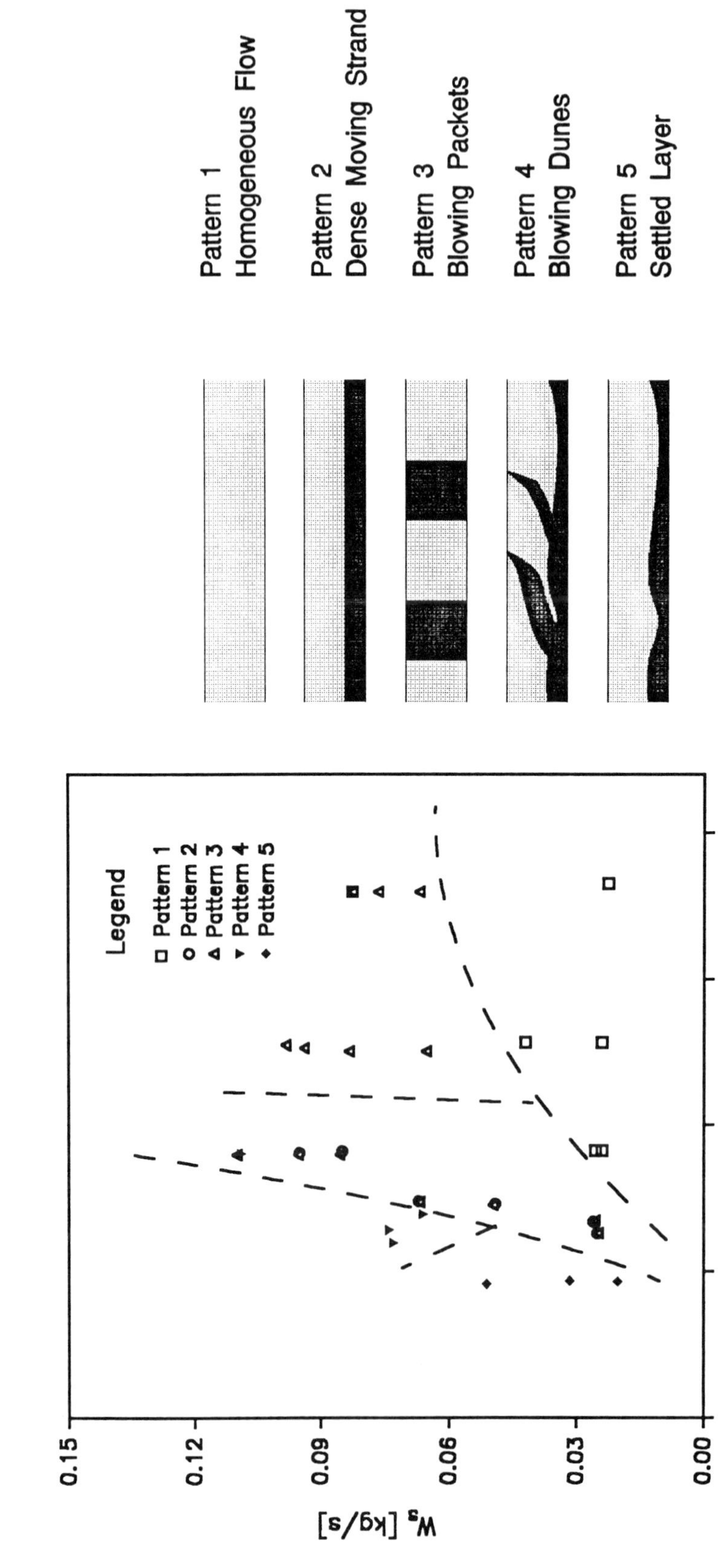

Figure 21.7. Flow pattern map for 450 μm glass beads in 0.025 m pipe in horizontal orientation.

21.5.2 Application of Fuzzy Mathematics

Failure to use standard parameters to predict the possible flow patterns in conveying lines suggests that a new approach should be taken. Classical modeling approaches are futile because oversimplification of the interactions (necessary to make it viable for on-line analysis) reduces the accuracy and reliability. Deductions from on-line measurements of pressure signals is more realistic. In this case, what we measure is what we have. Time series signal of pressure can be analyzed in amplitude domain and frequency domain. These analyses have to be correlated with observed flow patterns. Flow pattern observations are subjective. There is also uncertainty involved in analyzing the time series signals. It is quite obvious that fuzzy simulation is a good alternative to classical modeling in this case.

The fuzzy algorithm used in this analysis was proposed by Dubois and Prade (1980) and uses various implications. Yamashita, et al. (1987) have proposed a similar model for flow pattern classification of gas-liquid systems. However, their analysis is based on flow pattern maps which were found to be non-universal in gas-solid systems.

Figure 21.8 represents the power spectrum (P_a^2) of the pressure fluctuations with frequency (Hertz) for a test case whose flow pattern needs to be determined. It can be seen (from visual comparison) that this spectrum is quite similar to the spectrum obtained from an experimental run with similar operating conditions. Exact duplication of spectrums for almost identical operating conditions is not possible.

After the first scan through the rule base, not much success can be reported. The grade of membership function (GMF) is still zero, which implies that no match was found. To proceed further, fuzzification of input data is required. Fuzzification broadens the scope of input data by widening the base of GMF. Various functions are available to achieve this. An additional factor based on the results of the pressure fluctuations has been added. The addition of this weight factor suggests that the variations in low frequency fluctuations is largest. Feeder related instabilities account for most of these fluctuations. Therefore, by using a higher fuzzification factor for low frequency components is more appropriate. The result of the fuzzy simulation is shown in Figure 21.9. In this plot, GMF is the grade of membership function plotted against the pattern numbers shown in Figure 21.7. A good match is found in less than three fuzzifications.

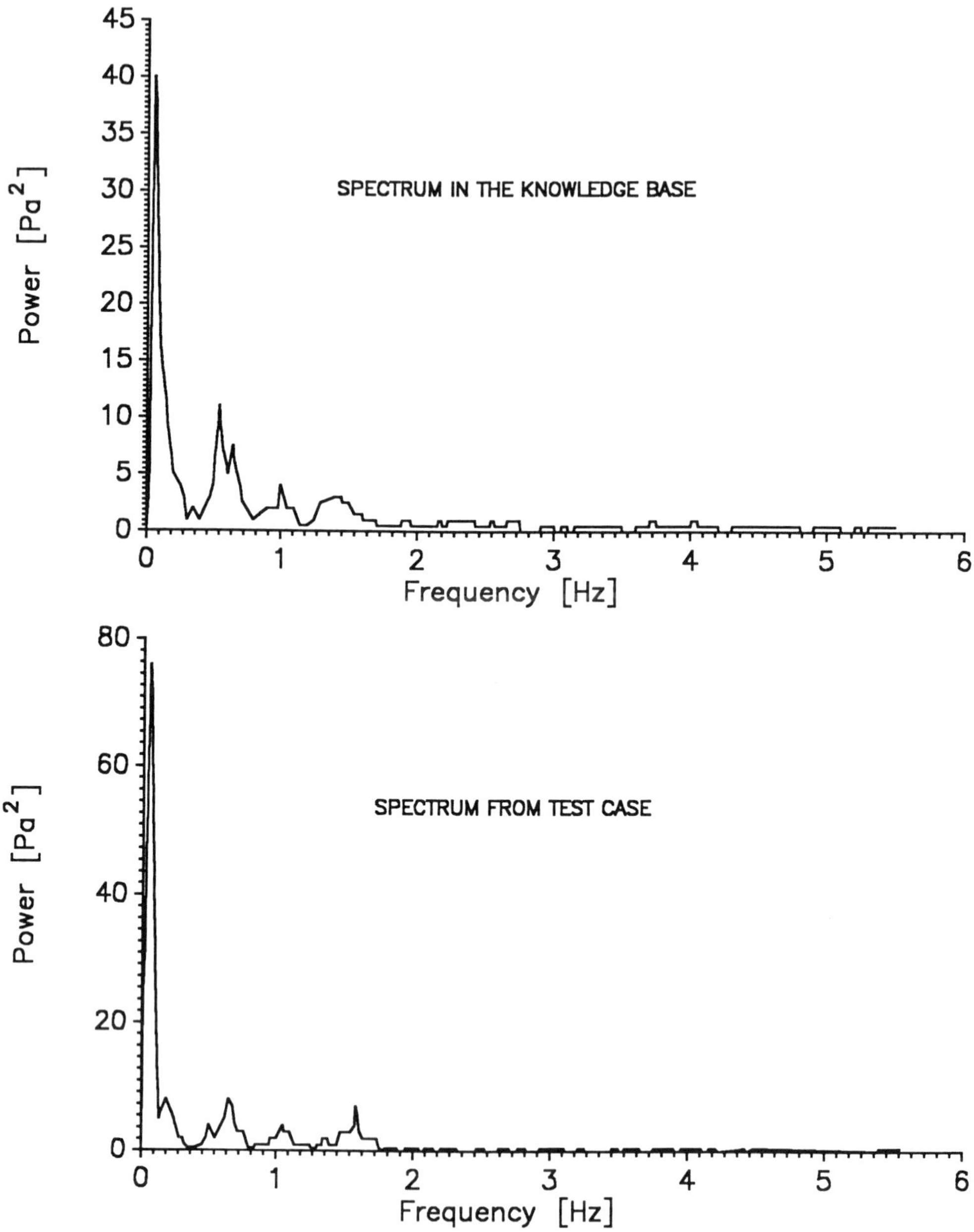

Figure 21.8. Power spectral density function of static pressure fluctuations for two runs with similar operating conditions.

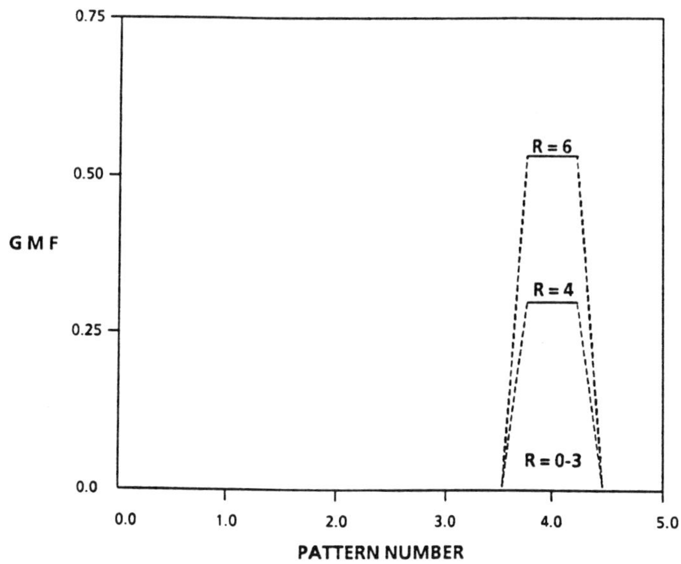

Figure 21.9. Fuzzy simulation for flow pattern identification; Patern 4 corresponds to blowing dunes (R is the number of fuzzifications).

21.6 SUMMARY

Knowledge based systems are being commonly used, either as stand alone systems or as embedded systems. A vast amount of knowledge, both fundamental and applied, has been accumulated in the area of solids handling during the past century. Even if a systematic effort is made to compile this information, it will largely remain unusable unless an **intelligent tool** is devised to manipulate and analyze it. To alleviate the problems facing us, we need to apply some of the techniques that have evolved as a part of AI. Apart from discovering the inherent structure of the knowledge and the inter-relationships of various parameters, this approach also helps to identify the research areas which need more attention. We have made an effort to project the future areas of application based on our experience; however, this list is by no means exhaustive.

Thus, the potential gain from applying expert systems to solids handling is great and the potential value is immeasurable. Advances will be made using this approach such that we will be able to consider solids handling more routinely in our plant design, with increased reliability of the overall plant operation. One must also note that while approaching solids handling problems with AI, basic studies are still essential to fully understand and predict the behavior of these bulk solids. Approaching the problem from the two aspects of basics and AI will bring, in the long run, a union of the two approaches.

REFERENCES

Barnwell, J. and B. Ertl, "Expert System and the Chemical Engineer," The Chemical Engineer, September 1987, p. 41.

Bendat, J.S. and A.G. Piersol, "Random Data: Analysis and Measurement Procedures," John Wiley & Sons, New York, 1986.

Bowerman, , R.G. and D.E. Glover, "Putting Expert Systems into Practice," Van Nostrand Reinhold Company, New York, 1988.

Caldwell, L.G., "A Pneumatic Conveying Primer," Chem. Engr. Progress, March 1976, p. 63.

Casey, J., "Picking the Right Expert System Application," AI Expert, September 1989, p. 44.

Dhodapkar, S. and G.E. Klinzing, "EXCON - An Expert System to Select Mechanical Conveyors," Research Report, University of Pittsburgh, 1989.

Dhodapkar, S. and G.E. Klinzing, "Expert System for Selection of Feeders for Pneumatic Conveying," Bulk Solids Handling, Vol. 10, No. 2, May 1990, pp. 149-154.

Degroff, L., "Conventional Languages & Expert Systems," AI Expert, April 1987, p. 32.

Dohnal, M., "Fuzzy Models of Unit Operations," Chem. Eng. Commun., Vol. 19, 1982, pp. 129-139.

Dohnal, M., "Linguistics and Fuzzy Models," Computers in Industry, Vol. 4, 1983a, pp. 341-345.

Dohnal, M., "Fuzzy Simulation of Industrial Problems," Computers in Industry, Vol. 4, 1983b, pp. 347-352.

Dohnal, M., "Application of a Universal Expert System in Industry," Computers in Industry, Vol. 6, 1985, pp. 115-121.

Dubois, D. and H. Prade, "Fuzzy Sets and Systems," Academic Press, London, 1980, pp. 1-300.

Freedman, R., "Evaluating Shells: 27 Products Wrap-up," AI Expert, September 1987, p. 69.

Geissman, J.R. and R.D. Schultz, "Verification & Validation of Expert Systems," AI Expert, February 1988, p. 26.

Gevarter, W.B., "The Nature and Evaluation of Commercial Expert System Building Tools," Computer, May 1987, p. 24.

Hayes-Roth, F., D.A. Waterman, and D. Lenat, "Building Expert Systems," Addison-Wesley Publishing Company, Inc., Reading, MA, 1983.

Keyes, J., "Why Expert Systems Fail," AI Expert, November 1989, p. 50.

Knowlton, T.M., "Solids Transport Research - Why it is Important and Research Needs," presented at Solids Transport Contract Review Meeting, Pittsburgh, PA, September 17, 1987.

Marrow, E.W., "A Quantitative Assessment of R&D Requirements for Solids Processing Technology," DOE report, WD-2339-DOE/PSSP, September 1984.

Matsumoto, S. and H. Harakawa, "Statistical Analysis of Transition of Flow Patterns in Vertical Pneumatic Conveying," Int. J. Multiphase Flow, Vol. 13, 1987, pp. 123-129.

PCPLUS Manuals, documentation on PCPlus by Texas Instruments, 1986.

Pedersen, K., "Expert System Programming: Practical Techniques for Rule-based Systems," John Wiley and Sons, In., New York, 1989.

Rouhani, S.Z. and M.S. Sohal, "Two Phase Flow Patterns: A Review of Research Results," Progress in Nuclear Energy, Vol. 11, No. 3, 1983, pp. 219-259.

Satija, S, J.B. Young, and L.S. Fan, "Pressure Fluctuations and Choking Criterion for Vertical Pneumatic Conveying," Powder Technology, Vol. 43, 1985, pp. 257-271.

Smith, D.A., "Implementing Real World Expert System," AI Expert, December 1988, p. 51.

Stephanopoulos, G., "The Future of Expert Systems in Chemical Engineering," Chemical Engineering Progress, September 1987, p. 44.

Sumner, R.S., C.A. Shook, and M.C. Roco, "Concentration Fluctuation Spectra in Turbulent Slurry Pipeline Flow," J. of Fluid Engr., Vol. 111, 1989, pp. 317-323.

Tsuji, Y. and Y. Morikawa, "Flow Pattern and Pressure Fluctuation in Air/Solid Two Phase Flow in Pipe at Low Air Velocities," Int. J. Multiphase Flow, Vol. 8, 1982, pp. 329-341.

Tutu, N.K., "Pressure Fluctuation and Flow Pattern Recognition in Vertical Two Phase Gas-Liquid Flow," Int. J. Multiphase Flow, Vol. 8, 1982, pp. 443-447.

Waterman, D.A., "A Guide to Expert Systems," Addison-Wesley Publishing Company, Reading, MA, 1986.

Yamashita, Y., S. Matsumoto, and M. Suzuki, "Classification of Flow Patterns in Two-Phase Flow by Fuzzy Simulator," Chemical Engineering Communication, Vol. 59, 1987, pp. 325-331.

CHAPTER 22

HYDRODYNAMIC MODELING OF CIRCULATING AND BUBBLING FLUIDIZED BEDS

D. Gidaspow

22.1 INTRODUCTION

Quantitative understanding of the hydrodynamics of fluidization is needed for the design and scale-up of efficient new reactors in the petroleum, chemical and in the electric power industries (Avidan et al., 1989; Squires et al., 1985). For example, although bubbling fluidized bed combustors have reached a commercial status, an understanding of solids motion and its effect on erosion of the heat exchanger tubes immersed in fluidized beds remains inadequate. Hydrodynamic models have been developed (Bouillard et al., 1989a, b) that have the potential of predicting erosion patterns in fluidized beds. These two fluid models are a generalization of the celebrated Navier-Stokes equations used in single-phase fluid mechanics. They were originally developed by Jackson (1963), Soo (1967), Garg and Prichett (1975), and others. Gidaspow (1986) has given a complete review and has shown that these models, with zero gas and solids viscosities, were able to predict a great deal of the behavior of bubbling beds because the dominant mechanism of energy dissipation is the drag between the particles and the fluid.

Previously, we (Lyczkowski, et al. 1982) had reviewed the hydrodynamic multiphase models in a 150 page chapter. Recently, Richner, et al. (1990) used a computer code developed by Systems, Science and Software to compute the large bubbles measured in the Westinghouse 3 meter rig of Yang et al. (1983). Inviscid models were used to predict the formation, the growth, and the bursting of bubbles (Gidaspow et al., 1983a, b) in qualitative agreement with experiments. Computed wall to bed heat transfer coefficients (Syamlal and Gidaspow, 1985) and velocity profiles of jets (Ettehadieh et al., 1984) agreed with measurements. Time-averaged porosity distributions (Gidaspow et al., 1983b; Bouillard et al., 1989b) agreed with measurements done using gamma ray densitometers without the use of adjustable parameters and without prescribing the porosity at any boundaries in the computation. However, analogous to classical fluid dynamics, inviscid models do not predict the forces acting on the tubes immersed in fluidized beds. Therefore, to overcome this deficiency a solids viscosity was added to the model, similar to the treatment of Jackson (1985), to compute erosion rates around the tubes. Unfortunately, this solids viscosity and also the solids stress were not known and had to be estimated based on very limited data.

While in the case of bubbling beds the solids viscosity plays the role of a correction, modeling of a circulating fluidized bed (CFB) without a viscosity is not possible.

Estimates of solids viscosity were used in the two fluid hydrodynamic model, to predict radial solids distributions and solids velocities which match the experimental distributions (Gidaspow, et al. 1989). Most important, the model predicted cluster formation and transient internal circulation which is responsible for the favorable characteristics of CFBs, such as good wall-to-bed heat transfer. Video tape movies of computations compared favorably with high speed movies of the experiments.

In the dense phase regime of the CFBs, corresponding to volume fractions of about ten percent of solids, a core-annular type of flow regime, with solids descending down at the wall, was computed. (Tsuo and Gidaspow, 1990). The computed solids velocity profiles and radial solids concentrations agreed with measurements done at IGT (Bader, et al., 1988).

22.2 VISCOUS HYDRODYNAMIC MODELS

The physical principles used are the conservation laws of mass, momentum and energy for each phase, the fluid phase and the particulate phases. This approach is similar to that of Soo (1967) for multiphase flow and of Jackson (1985) for fluidization. A Newtonian or power law constitutive equation for the surface stress of phase "k" is an appropriate first approximation, since the surface stress of phase "k" will depend at least on its symmetrical gradient of velocity. The emerging kinetic theory of granular flow provides a physical motivation for such an approach. Hence the general balance laws of mass and momentum for each phase, with phase change, are given by Equations (22.1) and (22.2) and the constitutive equation for the stress is given by Equation (22.3).

CONTINUITY EQUATION FORT PHASE k

$$\frac{\partial}{\partial t}(\alpha_k \rho_k) + \nabla \cdot (\alpha_k \rho_k v_k) = \dot{m}_k \tag{22.1}$$

MOMENTUM EQUATION FOR PHASE k

$$\frac{\partial}{\partial t}(\alpha_k \rho_k v_k) + \nabla \cdot (\alpha_k \rho_k v_k v_k) = \alpha_k \rho_k g + \nabla \cdot [\tau_k] + \sum_1 \beta(v_1 - v_k) + \dot{m}_k v_k \tag{22.2}$$

acceleration of phase k = gravity+stress+drag force +phase change momentum

CONSTITUTIVE EQUATION FOR STRESS

$$[\tau_k] = [-p_k + \xi_k \nabla \cdot v_k][I] + 2\mu_k[S_k]$$

$$[S_k] = \frac{1}{2}[\nabla v_k + (\nabla v_k)^T] - \frac{1}{3}\nabla \cdot v_k[I] \qquad (22.3)$$

These equations are similar to Bowen's (1976) balance laws for multicomponent mixtures. The principle difference is the appearance of the volume fraction of phase "k" denoted by α_k. In the case of phases not all the space is occupied at the same time by all the phases, as it is by components due to the negligible size of molecules. As in the case of the mixture equations for components, the mixture equations for phases show that the sum of the phase change productions in Equation (22.1) zero and the sum of the drag forces in Equation (22.2) is zero. In convective form the phase change momentum in Equation (22.2) is zero, insuring invariance under a change of frame of reference for translation. Equation (22.3) is the usual Newtonian expression for the stress which arises from the assumption that the stress is a function of its own symmetrical gradient of velocity. For the fluid p_k is the fluid pressure. When this form is substituted into the momentum equation, the result is not the usual momentum balance presented by Gidaspow (1986) and widely used in gas-liquid two phase flow. It is a slightly modified version of the momentum balance, called model B by Bouillard et al. (1989). This model is unconditionally well-posed. It does not require the presence of $p_k(\alpha_k)$ for stability and well-posedness in the inviscid limit.

The corresponding particulate momentum equations with zero particulate viscosities are the trajectory equations for particulate flow. Bouillard, et al. (1989) have shown that this set of equations produces essentially the same numerical answers for fluidization as did the earlier conditionally stable model which has the fluid pressure in the gas and in the solid phases. In this model the force exerted by gravity has to be altered to satisfy Archimidis' buouyancy principle. Also the interphase friction coefficient for model A shown in this chapter has to be divided by the volume fraction of the fluid phase to obtain the β for model B, since in model A the pressure gradient of the fluid or continuous phase is multiplied by the fluid volume fraction, while in model B the pressure gradient of the continuous phase appears totally in the continuous phase. With this modification the equations are well posed for gas-liquid dispersed flow, as well as for neutrally buoyant particles previously not allowed in the earlier model B version of Bouillard, et al. (1989).

In this model the characteristics are real and distinct for one dimensional transient flow. (Tsuo and Gidaspow, 1990.) The problem is well-posed as an initial value problem. For the particular phase, p_k consists of the static normal stress and the dynamic stress, called solids pressure, which arises due to the collision of the particles. Model B was used for computing the behavior of the circulating fluidized bed in the pipe, while our earlier model A was used to compute the bubbling bed behavior, as well as the circulating fluidized bed with both fluid turbulence and with our new kinetic theory constitutive equations.

22.3 KINETIC THEORY MODEL

Savage (1983, 1988) and Jenkins and Savage (1983) have shown how to obtain constitutive equations and conservation laws for granular flow. Ding and Gidaspow (1990) extended their approach to gas-particle flow. Since Ding and Gidaspow (1990) used a Maxwellian distribution their approach was restricted to dense flow. Recently Gidaspow extended this approach to cover dilute and dense particle-fluid flow. The mathematics is very similar to that described in detail in the classical text on kinetic theory by Chapman and Cowling (1961). A summary of the techniques and the results is presented below.

We begin with the Boltzmann integral-differential equation for the frequency of velocities distribution, f.

$$\frac{\partial f}{\partial t} + \vec{c} \cdot \frac{\partial f}{\partial \vec{r}} + \vec{F} \frac{\partial f}{\partial \vec{c}} = (\frac{\partial f}{\partial t})coll \tag{22.4}$$

where $\vec{c}$ is the random particle velocity and $\vec{F}$ is the force per unit mass. The collision integral for binary collision of rigid spheres is

$$(\frac{\partial f}{\partial t})_{\text{collbinary}} = \int \int (f^{(2),} \vec{c}^{\,1}_{12} \cdot \vec{k} - f^{(2)} \vec{c}_{(12)} \cdot \vec{k}) d_{12}^2 d\vec{k} d\vec{c}_1 \tag{22.5}$$

$$before\ collision - after\ collision$$

where $f^{(2)}$ is the collisional pair distribution function, d_{12} is the average particle diameter of the two colliding spheres and $\vec{k}$ is a unit vector.

Multiplication of the Boltzmann equation, (22.4), by a quantity ψ, such as mass, momentum and energy, integration over the random translational velocity, $\vec{c}$ and expansion of the collisional integral term in a Taylor series gives the dense phase transport theorem shown below, where n is average particle number density $n = \int f dc; < \psi >$, is average quantity defined as, $< \psi > = \frac{1}{n} \int \psi f dc$

DENSE PHASE TRANSPORT THEOREM

$$\frac{\partial}{\partial t}(n < \psi >) + \nabla \cdot (n < \vec{c}\psi > + P_c) = < n\vec{F}_s \frac{\partial \psi}{\partial \vec{c}} > + N_c \tag{22.6}$$

where $\vec{F}_s$ is the force acting on the system and,

$$P_c = -\frac{1}{2}g_o(\alpha_s)d_p^3 \int\int\int_{\vec{c}12\cdot\vec{k}>0} (\psi_1 - \psi_1')f_1 f_2 \vec{k}(\vec{c}_{12}\cdot\vec{k})d\vec{k}d\vec{c}_1 d\vec{c}_2 \tag{22.7}$$

$$-\frac{1}{4}g_o(\alpha_s)d_p^4 \int\int\int_{\vec{c}12\cdot\vec{k}>0} (\psi_1 - \psi_1')f_1 f_2 \nabla ln(\frac{f_2}{f_1})$$

$$\vec{k}(\vec{c}_{12}\cdot\vec{k})d\vec{k}d\vec{c}_1 d\vec{c}_2$$

and

$$N_c = \frac{d_p^2}{2}\int\int\int_{\vec{c}12\cdot\vec{k}>0} (\psi_2' + \psi_1' - \psi_2 - \psi_1)(\vec{c}_{12}\cdot\vec{k})$$

$$f^{(2)}(\vec{r} - d_p\vec{k}, \vec{c}_1; \vec{r}, \vec{c}_2)d\vec{k}d\vec{c}_1 d\vec{c}_2 \tag{22.8}$$

The collisional pair distribution function was expressed as a product of the single distribution functions f_1 and f_2 and an empirical equilibrium transfer function,

$$f^{(2)}(\vec{r}, \vec{c}_1; \vec{r}+d_{12}\vec{k}, \vec{c}_2) = x(\vec{r} - \frac{1}{2}d_{12}\vec{k})\cdot f_1(\vec{r}, \vec{c}_1)\cdot f_2(\vec{r}+d_{12}\vec{k}, \vec{c}_2) \tag{22.9}$$

and Chapman and Cowling's x

$$\mathrm{x}(r - \frac{1}{2}d_{12}\vec{k}) = g_o(\epsilon_s) = [1 - (\frac{\alpha_s}{\alpha_{smax}})^{\frac{1}{3}}]^{-1} \tag{22.10}$$

The particle conservation of mass, Equation (22.1) for no phase change is obtained by setting $\psi = m$ in Equation (22.1). Note that $nm = \alpha_s\rho_s$.

The particle conservation of momentum equation is obtained by settling $\psi = m\vec{c}$ and performing standard algebraic manipulations (Chapman and Cowling, 1961). This equation reduces itself to,

$$\rho_s\frac{Dv_s}{Dt} + \nabla(P_k + P_c) = \rho_s F_s \tag{22.11}$$

where P_k is the kinetic pressure and P_c is the collisional contribution. F_s consists of the body forces and the drag.

The collisional contribution consists of two integrals shown in Equation (22.7) Following Chapman and Cowling (1961) these integrals yield the result

$$\bar{\bar{P}}_c = \bar{\bar{P}}c_1 + \bar{\bar{P}}c2 \tag{22.12}$$

$$\bar{\bar{P}}c_1 = \frac{2(1+e)}{5}g_0\rho_p\alpha_s^2(2 < \vec{C}\vec{C} > + C^2 I) \tag{22.13}$$

$$\overset{=}{P}_{c2} = 2\alpha_s \mu_{sc} \overset{o\ s}{\nabla}\ \vec{v}_s + \alpha_s \xi_s \nabla \cdot \vec{v} I \tag{22.14}$$

where $\overset{o\ s}{\nabla}$ is the rate of shear tensor, [S] and μ_{sc} is the coefficient of viscosity, simply called viscosity by Ding and Gidaspow (1990). For a Maxwellian distribution the first integral does not contribute to the viscosity since the non-diagonal terms of $< CC >$ vanish. The expressions for the viscosity and the granular conductivity given here have been generalized to the first perturbation for the frequency distribution. They are thus valid for dilute and dense flow. Equation (T1.5C) gives this shear viscosity. The standard dilute phase viscosity is given by Equation (T1.5d). Note that the kinematic viscosity is simply a product of the mean free path times a random velocity, as expressed below

$$\frac{\mu_s}{\rho_s \alpha_s} = \text{constant} \cdot (\frac{d_P}{\alpha_s}) \cdot \theta^{\frac{1}{2}} \tag{22.15}$$

$$\text{Kinematic viscosity} = \text{mean free path x random velocity} \tag{22.15a}$$

A conservation equation for this random velocity can be obtained by substituting $\psi = \frac{1}{2} m \vec{c}^2$ into Equation (22.6). The result is the fluctuating energy equation for the granular temperature θ given by Equation (T1.6). The conductivity given by Equation (T1.6c) was obtained from equation (22.7) and the collisional dissipation γ due to the non-elasticity of the particles was obtained from Equation (22.8). The restitution coefficient "e" is the normal restitution coefficient. There also exists a tangential restitution coefficient associated with particle rotation that was not considered in the theory developed so far.

22.4 BUBBLING FLUIDIZATION USING GRANULAR TEMPERATURE

As a test of the theory a two dimensional fluidized bed with a jet roughly approximating the operation of a gasifier or fluidized bed combustor was constructed, instrumented and tested (Gidaspow, et al., 1983).

Figure 22-1 shows a schematic of our two dimensional bed that was used to obtain time averaged void fraction distributions using a gamma ray densitometer. In the past plastic and steel beds were used. Figure 22-2 shows such data for 520 micron glass beads at a jet velocity of 5.78 m/sec. On the right side of this figure we see similar computer generated results using our new granular temperature model. The figure on the bottom shows that the agreement between the new theory and the experiment is good. It is similar to the computation with the inviscid model of Bouillard, et al.

Figure 22-3 shows the experimental porosity oscillations determined using our gamma ray densitometer at the location in the bed with an

TABLE 1

GOVERNING EQUATIONS FOR PARTICULATE MULTIPHASE FLOW

---

1. CONTINUITY EQUATION FOR PHASE k(= g,s)

$$\frac{\partial}{\partial t}(\alpha_k \rho_k) + \nabla \cdot (\alpha_k \rho_k v_k) = \dot{m}_k \qquad (22-T1.1)$$

$$\sum_k \alpha_k = 1 \qquad (22-T1.1a)$$

2. MOMENTUM EQUATION FOR PHASE k(= g,s; l= g,s)

$$\frac{\partial}{\partial t}(\alpha_k \rho_k v_k) + \nabla \cdot (\alpha_k \rho_k v_k v_k) = \alpha_k \rho_k g + \nabla \cdot \overline{\overline{\tau}}_k + \beta(v_l - v_k) + \dot{m}_k v_k \qquad (22-T1.2)$$

acceleration of phase k = gravity + stress + drag force + phase change momentum

3. CONSTITUTIVE EQUATION FOR STRESS

$$\overline{\overline{\tau}}_k = [-P_k + \xi_k \nabla \cdot v_k]\,\overline{\overline{I}} + 2\mu_s \overline{\overline{S}}_k \qquad (22-T1.3)$$

where

$$\overline{\overline{S}}_k = \frac{1}{2}\left[\nabla v_k + (\nabla v_k)^T\right] - \frac{1}{3}\nabla \cdot v_k \overline{\overline{I}} \qquad (22-T1.3a)$$

4. Gas Phase Stress

$$\overline{\overline{\tau}}_g = 2\epsilon_g \mu_{ge} \overline{\overline{S}}_g \qquad (22-T1.4)$$

Turbulent Viscosity

$$\mu_{ge} = \mu_t + \mu_g = \rho_g (c_t \Delta)^2 (2\overline{\overline{S}}_g \cdot \overline{\overline{S}}_g)^{\frac{1}{2}} + \mu_g \qquad (22-T1.4a)$$

with $c_t = 0.1$ and $\Delta = \sqrt{(\Delta x \Delta y)}$

Table 1 continued

5. Solid Phase Stress

Kinetic Theory Model (Gidaspow's 1991 extension of Savage,1983 and Ding and Gidaspow's, 1990 expressions to dilute and dense flow)

Solids Phase Pressure

$$P_s = \rho_p \alpha_s \Theta \left[1 + 2(1+e) g_o \alpha_s\right] \qquad (22-T1.5a)$$

Solids Phase Bulk Viscosity

$$\xi_s = \frac{4}{3}\alpha_s^2 \rho_p d_p g_o (1+e)(\frac{\Theta}{\pi})^{\frac{1}{2}} \qquad (22-T1.5b)$$

Solids Phase Shear Viscosity

$$\mu_s = \frac{2\mu_{s_{dil}}}{(1+e)g_o}\left[1+\frac{4}{5}(1+e)g_o\alpha_s\right]^2 + \frac{4}{5}\alpha_s^2\rho_p d_p g_o(1+e)(\frac{\Theta}{\pi})^{\frac{1}{2}} \qquad (22-T1.5c)$$

Solids Phase Dilute Viscosity

$$\mu_{s_{dil}} = \frac{5\sqrt{\pi}}{96}\rho_p d_p \Theta^{\frac{1}{2}} \qquad (22-T1.5d)$$

Radial Distribution Function

$$g_o = \frac{3}{5}\left[1-(\frac{\alpha_s}{\alpha_{s_{max}}})^{\frac{1}{3}}\right]^{-1} \qquad (22-T1.5e)$$

6. FLUCTUATING ENERGY ($\frac{3}{2}\Theta = \frac{1}{2} < C^2 >$) EQUATION

$$\frac{3}{2}\left[\frac{\partial}{\partial t}(\alpha_s\rho_s\Theta) + \nabla\cdot(\alpha_s\rho_s v_s\Theta)\right] = \bar{\bar{\tau}}_s : \nabla v_s - \nabla\cdot q - \gamma \quad (22-T1.6)$$

Collisional Energy Dissipation γ

$$\gamma = 3(1-e^2)\alpha_s^2\rho_s g_o\Theta\left(\frac{4}{d_p}(\frac{\Theta}{\pi})^{\frac{1}{2}} - \nabla\cdot v_s\right) \qquad (22-T1.6a)$$

Table 1 continued

Flux of Fluctuating Energy q

$$q = -\kappa \nabla \Theta \qquad (22 - T1.6b)$$

Conductivity of Fluctuating Energy

$$\kappa = \frac{2}{(1+e)g_o}\left[1 + \frac{6}{5}(1+e)g_o\alpha_s\right]^2 \kappa_{dil} + 2\alpha_s^2\rho_p d_p g_o(1+e)(\frac{\Theta}{\pi})^{\frac{1}{2}} \qquad (22 - T1.6c)$$

Dilute Phase ("Eddy Type") Granular Conductivity

$$\kappa_{dil} = \frac{75}{384}\sqrt{\pi}\rho_p d_p \Theta^{\frac{1}{2}} \qquad (22 - T1.6d)$$

7. GAS - SOLID DRAG COEFFICIENTS

For $\alpha < 0.8$ (based on Ergun equation)

$$\beta = 150\frac{\alpha_s^2\mu_g}{\alpha d_p^2} + 1.75\frac{\rho_g\alpha_s|v_g - v_s|}{d_p} \qquad (22 - T1.7a)$$

For $\alpha > 0.8$ (based on empirical correlation)

$$\beta = \frac{3}{4}C_d\frac{\alpha\alpha_s\rho_g|v_g - v_s|}{d_p}\alpha^{-2.65} \qquad (22 - T1.7b)$$

where

$$C_d = \frac{24}{Re_p}\left[1 + 0.15Re_p^{0.687}\right], \text{for} \quad Re_p < 1000 \qquad (22 - T1.7c)$$

$$C_d = 0.44, \text{for} \quad Re_p > 1000 \qquad (22 - T1.7d)$$

$$Re_p = \frac{\alpha\rho_g|v_g - v_s|d_p}{\mu_g} \qquad (22 - T1.7e)$$

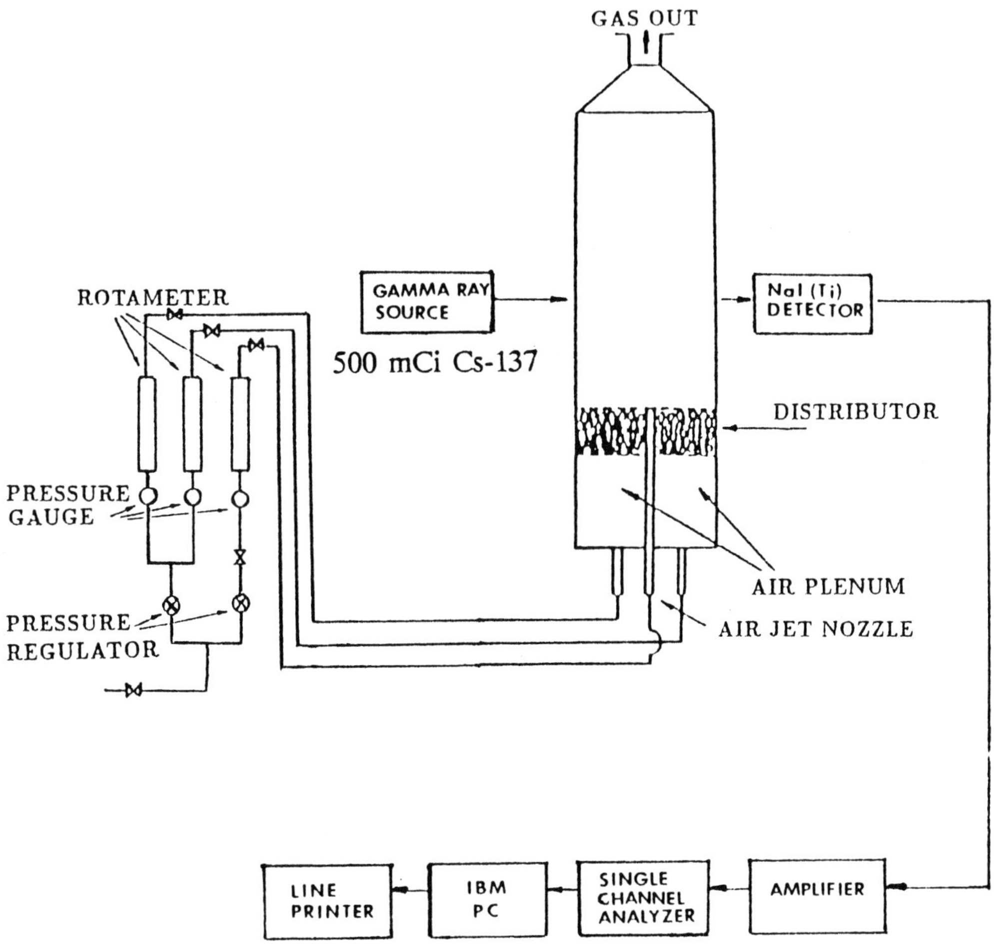

Figure 22-1. Apparatus for Measuring Void Fraction in IIT Laboratory

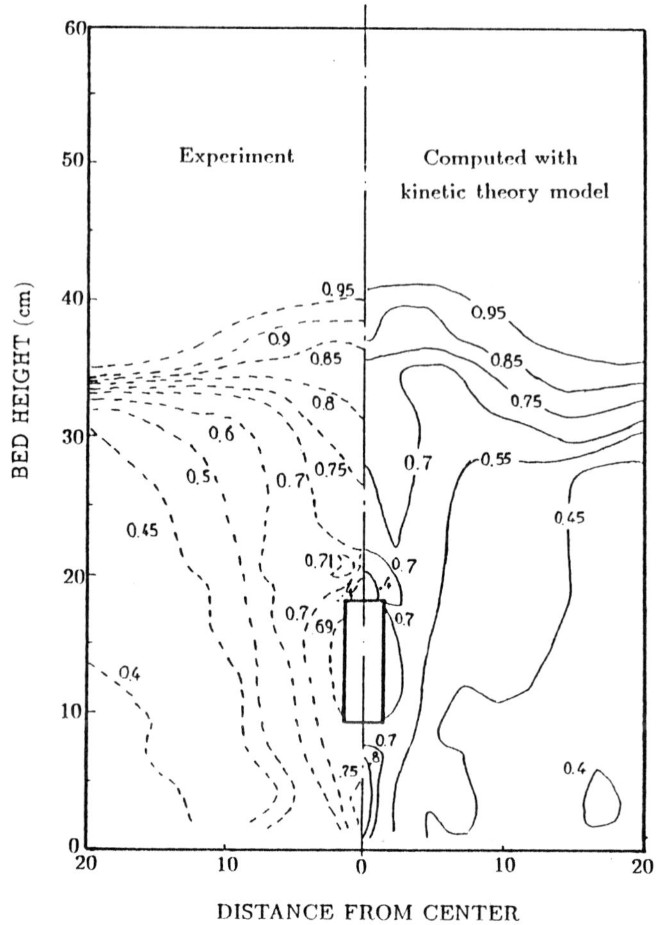

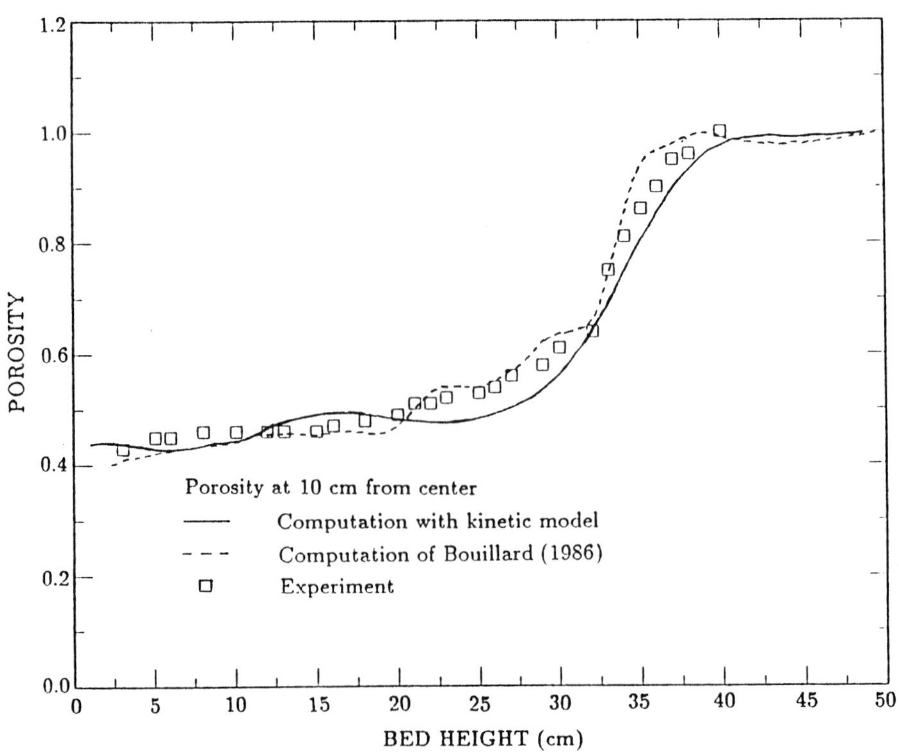

Figure 22-2. Experimental and Computed Time-averaged porosity

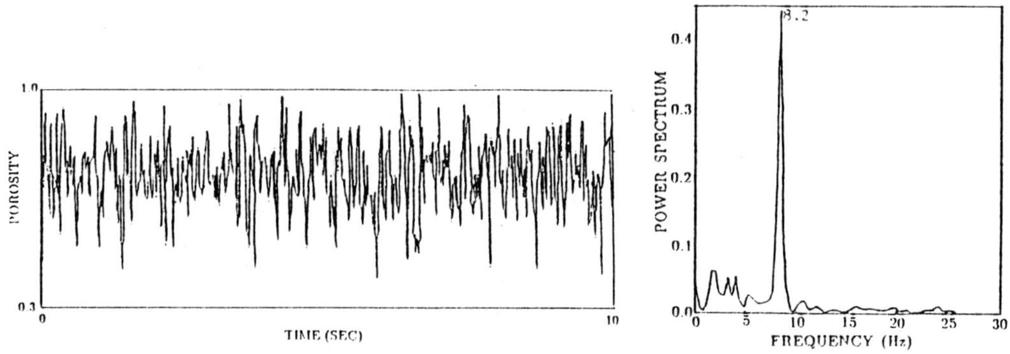

Figure 22-3. Experimental Porosity Oscillation at x = 0.3, y = 1.0 cm

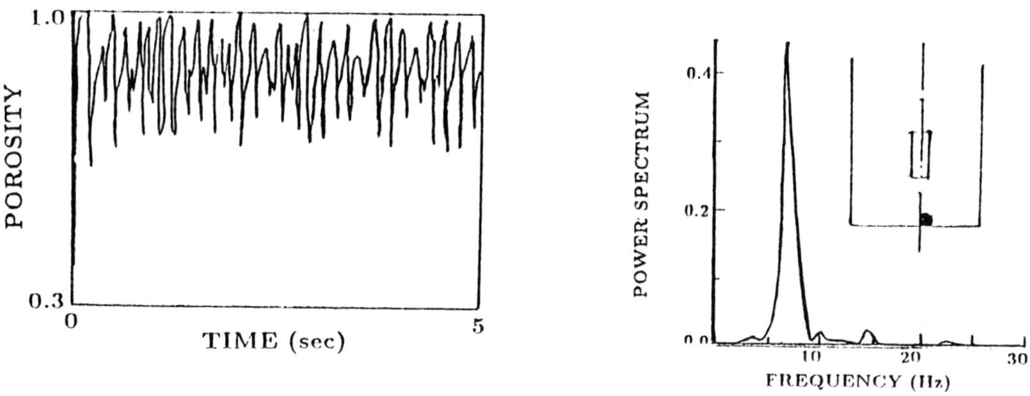

Figure 22-4. Computed Porosity Oscillation at x = 0.3, y = 1 cm (Ding and Gidaspow, 1990)

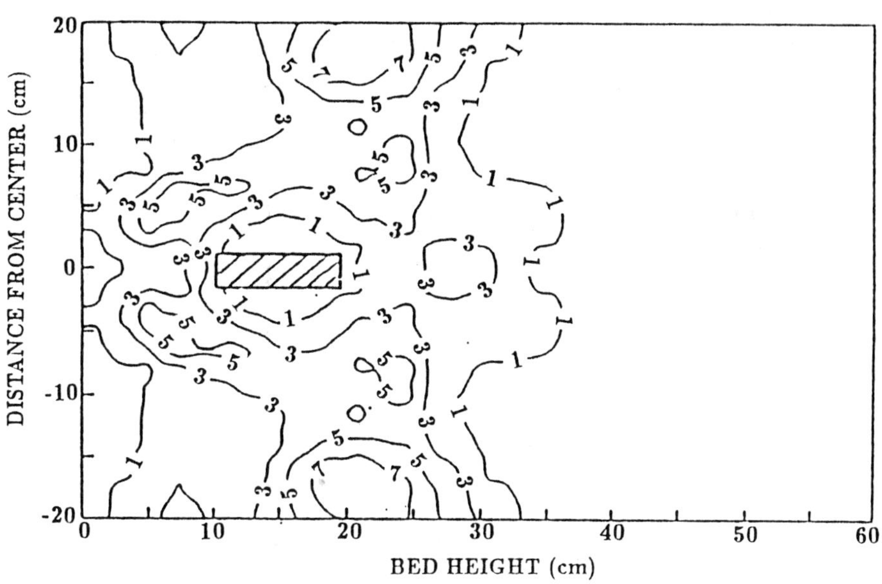

Figure 22-5. Computed Time-averaged Solid Viscosity (poise) in the Bed (Ding and Gidaspow, 1990)

obstacle near the jet inlet indicated in Figure 22-4. The experimental frequency determined from a power spectrum was 8 Hertz. As shown in Figure 22-4 the granular flow model produces similar results.

The computations also agree with previous computations that use estimated solids viscosities and granular stresses (Bouillard, et al. 1989b). Such agreement is due to the fact that bubbling fluidization with a jet can be described by means of inviscid equations as a first approximation. The computed viscosity for the particulate phase is shown in Figure 22-5. The computed values are close to the effective constant viscosity used in simulations done at the Argonne National Laboratory. (Bouillard et al. 1989b). Figure 22-6 shows the computed granular temperature. Near the obstacle immersed in the bed there is a lot of action. Hence the granular temperature is high.

Bubbling Bed with Uniform Grid Flow

As discussed in the introduction section of this paper, most of our modeling work dealt with fluidization with jets, where the bubbles form near the jet. The jet defines the flow pattern in the system. However, with a gas entering a fluidized bed through a uniform grid, flow patterns vary with flow rates and geometry. Figure 22-7 shows upward flow near the center of the bed. This flow pattern occurs because bubbles had moved to the center of the bed, as shown in Figure 22- 8. However, Figure 22-9 shows that consistent with photographic observation and generally accepted state of the art, bubbles can form between the center of the bed and the walls.

22.5 CIRCULATING FLUIDIZED BED FLOW REGIMES

Circulating fluidized beds (CFB) are common in the petroleum industry in which the older bubbling fluidized bed reactors are being replaced by the circulating fluidized bed risers due to the recently developed highly active catalysts. The second major emerging application is to circulating fluidized bed combustors which are being built to burn U.S. Eastern high sulfur coals. These combustors offer the potential of burning coal and incinerating wastes with low SO_2 and NO_x emissions. However, to minimize limestone utilization, to maximize heat transfer and to properly control NO_x and other potential pollutants, a mathematical model of the combustor is needed. This is recognized by boiler manufacturers since today the design of CFB combustors is more of an art rather than a science. The first step in understanding CFB operation is to comprehend and model its hydrodynamics.

In our first attempt to model circulating fluidized beds, we have used guessed solids viscosities. With such assumed viscosities we (Tsuo and Gidaspow, 1990) have shown that there exist two flow regimes in a CFB. In the dilute flow regime particles move down near the wall in the form of clusters. See Figure 22- 10. In the dense regime we obtain the more typical core-annular flow. See Figure 22-11. The

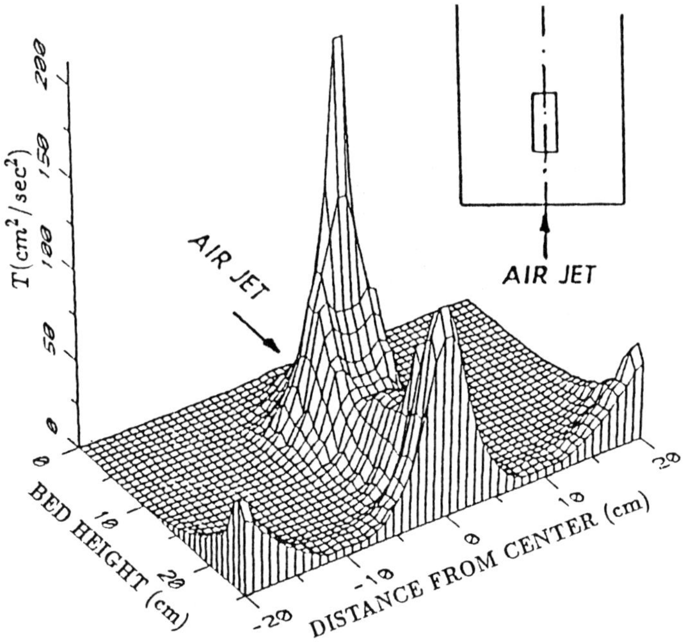

Figure 22-6. Computed Time-averaged Granular Temperature, Θ, Around the Obstacle Where 3Θ = < C^2> (Ding and Gidaspow, 1990)

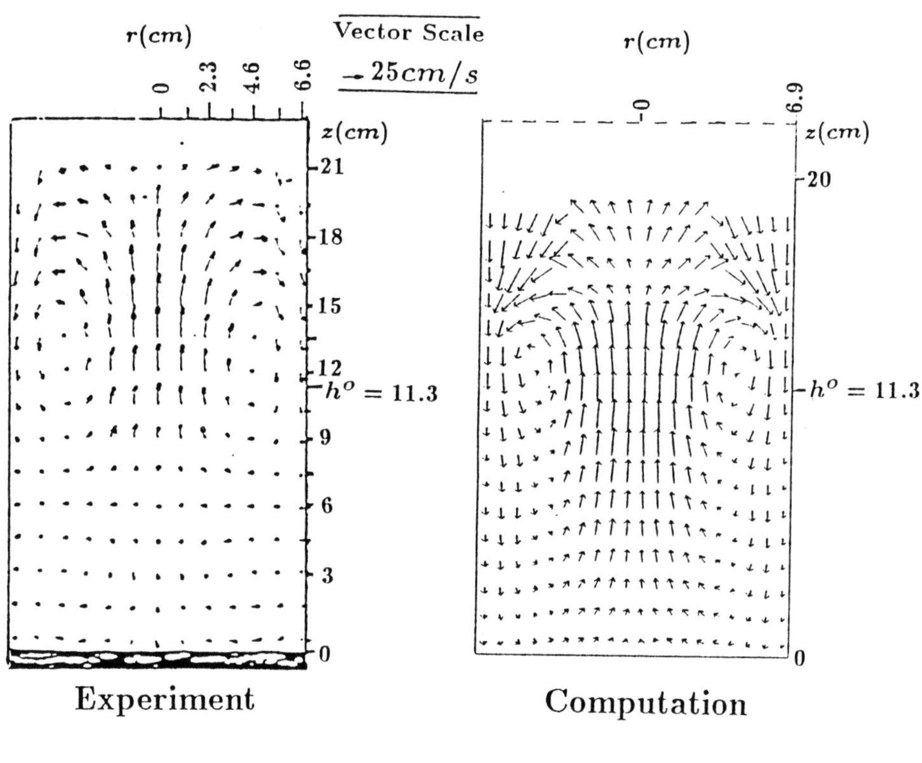

Figure 22-7. Experimental and Computed Solid Velocities in the Cylindrical Bed with Inlet Air Velocity = 64.1 cm/s. Experiment: Lin et al. (1985)

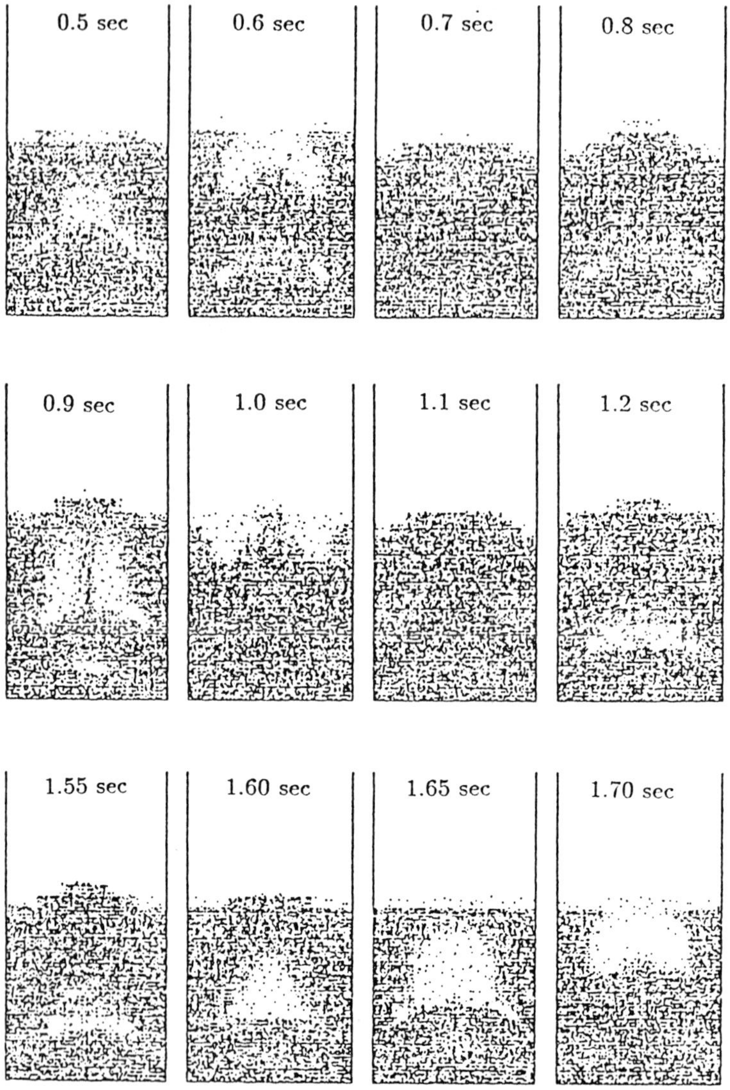

Figure 22-8. Computed Bubbles in the Cylindrical Bed at Several Times with Inlet Air Velocity = 64.1 cm/sec. (Ding and Gidaspow, 1990)

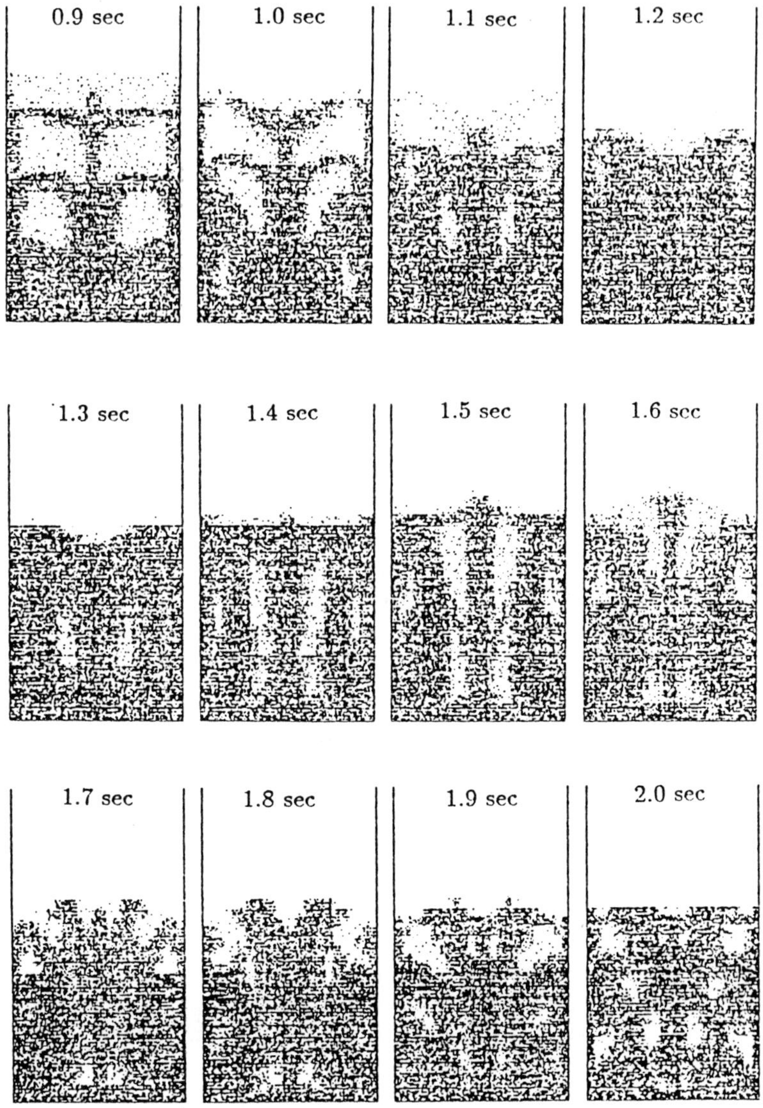

Figure 22-9. Computed Particle Distributions at Several Times in the Rectangular Two-Dimensional Bed with Uniform Inlet Air Velocity of 100cm/sec. (Ding and Gidaspow, 1990).

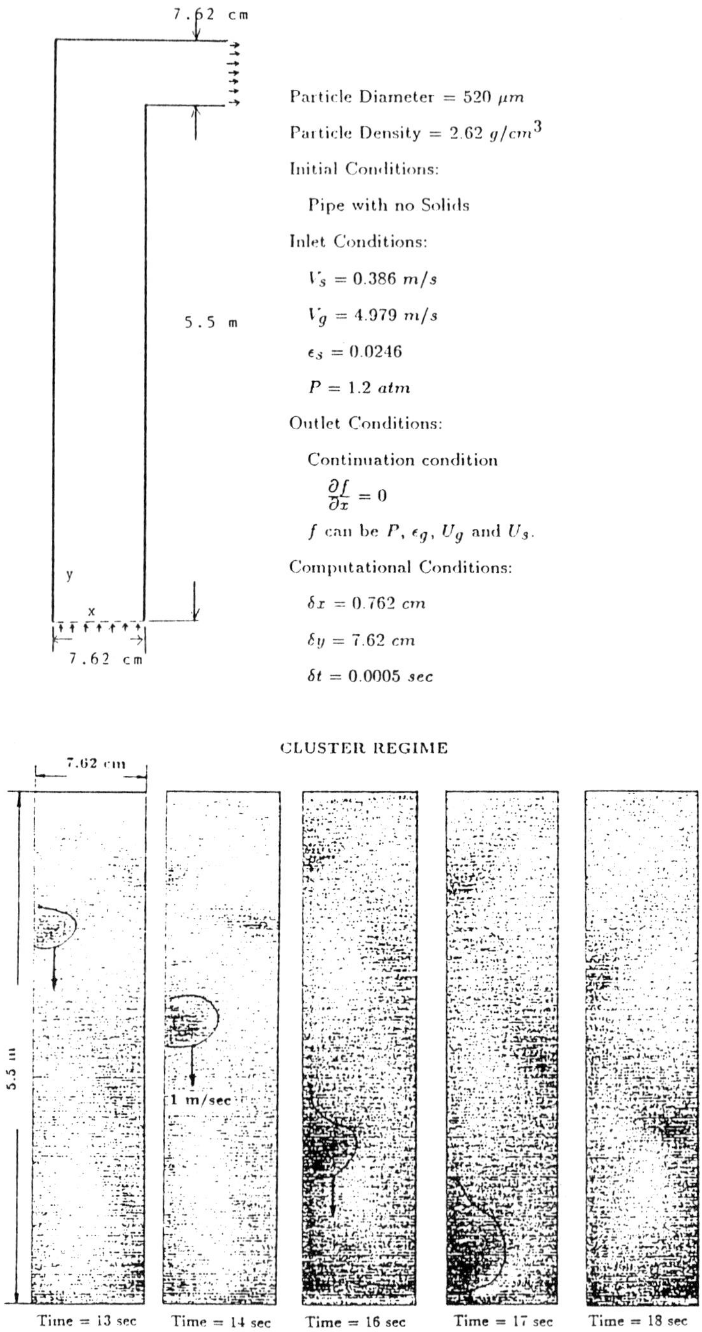

Figure 22-10. Simulation of Observed (Gidaspow, et al. 1989) Cluster Regime in Dilute Vertical Transport of Particles (Tsuo and Gidaspow, 1990).

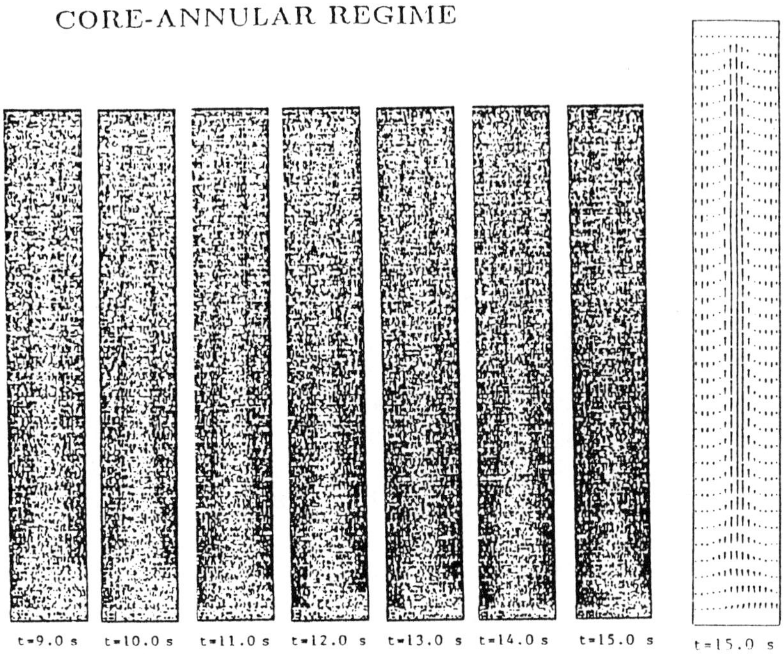

Figure 22-11. Core-annular Flow Regime in the IGT (Bader et al. 1988) 30.5 cm Diameter Riser Fed With 76 μm FCC Particles

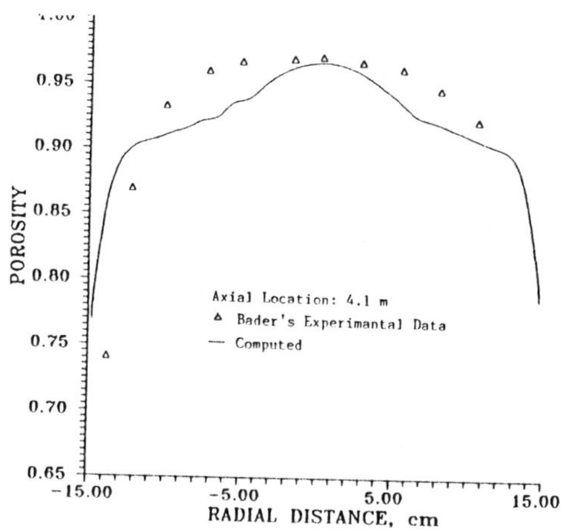

Figure 22-12. Radial Variation of Computed and Measured Void Fraction in the Riser at 4.1m. Averaged from 12 to 18 s for Computed Results.

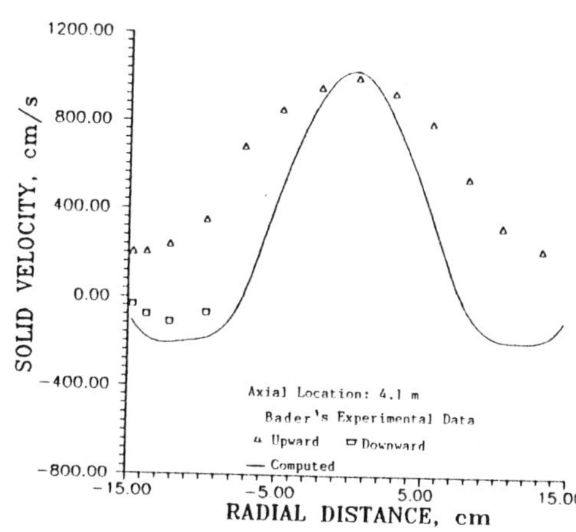

Figure 22-13. Radial Variation of Computed and Measured Axial Solids Velocity at 4.1 m. (Tsuo and Gidaspow, 1990).

particles are transported up the pipe near the center. They run down near the walls. Our computations agree with the experimental radial profiles of Bader, et al. (1988) and with that of Weinstein, et al. (1986).

Bader, et al. (1988) measured fluxes and velocity profiles using an extraction probe and a pilot tube, respectively in a 30.5 cm diameter riser with 76 μm FCC catalyst particles. From these measurements they estimated radial variations of porosity. Figure 22-12 shows a comparison of our computation to their data at one axial location. We predict the existence of the dilute core and the dense wall region. Figure 22-13 shows a comparison of our computation to Bader's solids velocity. We predict the higher then the inlet velocity at the center and the downward velocity at the wall. Weinstein, et al. (1986) used a non-intrusive X-ray densitometer to measure the radial porosity distribution in a riser at several fluxes. Figure 22-14 shows a comparison of our computation (Gidaspow, et al., 1991) to their experiment.

22.6 FULL SCALE CFB SIMULATION

A pyroflow CFB system built for Goodrich Co. in Henry, Illinois (Johnk and Wietske, 1989) was used to illustrate the capability of our modeling (Gidaspow, et al. 1990). In this simulation we used a primitive turbulence model in the dilute phase and dense phase kinetic theory with a Maxwellian distribution, as described fully by Ding and Gidaspow (1990) for bubbling beds.

To simulate the real process of PYROFLOW CFB, a number of simplifications have to be made. The dimensions of the CFB are estimated from a man shown in the Pyroflow system (Johnk and Wietske, 1989) at the left bottom of Figure 22-15. The bed is assumed to be isothermal and running at room temperature. The diameter and density of the solid particles are 150 μ and 1714 kg/m^3, respectively. The terminal velocity of this particle is about 1.1 m/s. The top of the standpipe is covered by a screen to prevent the particles from leaving the system. The system is open to the atmosphere. At the bottom of the standpipe, gas is fed in at a minimum fluidizing velocity of 2.5 cm/s. An air inlet velocity at the bottom of the riser is set to be V_{gin} = 2 to 5 m/s. Initially, the system is static. The particles are packed in the riser and the standpipe with a porosity $\alpha = 0.42$ and a height $h_o = 4.5$ m. Non-uniform finite difference grids are used in the computation.

Figure 22-16 shows typical computed solid flow and gas flow patterns as a function of time, for an inlet air velocity of $V_{gin} = 200$ cm/s. Due to initial effects, the solids in the standpipe are blown up at 1 second of real time. Then the solids begin to discharge from the standpipe to the riser. It takes about 5 seconds for most particles to reach the top of the riser. Since the outlet of the riser is at the right side, the solids near the left wall of the riser move down to the bottom and are then carried up by the inlet gas. We can see a downflow near the right wall of the riser. The radial profiles of time averaged axial gas and solid velocities at a height of 10 m in the riser are shown

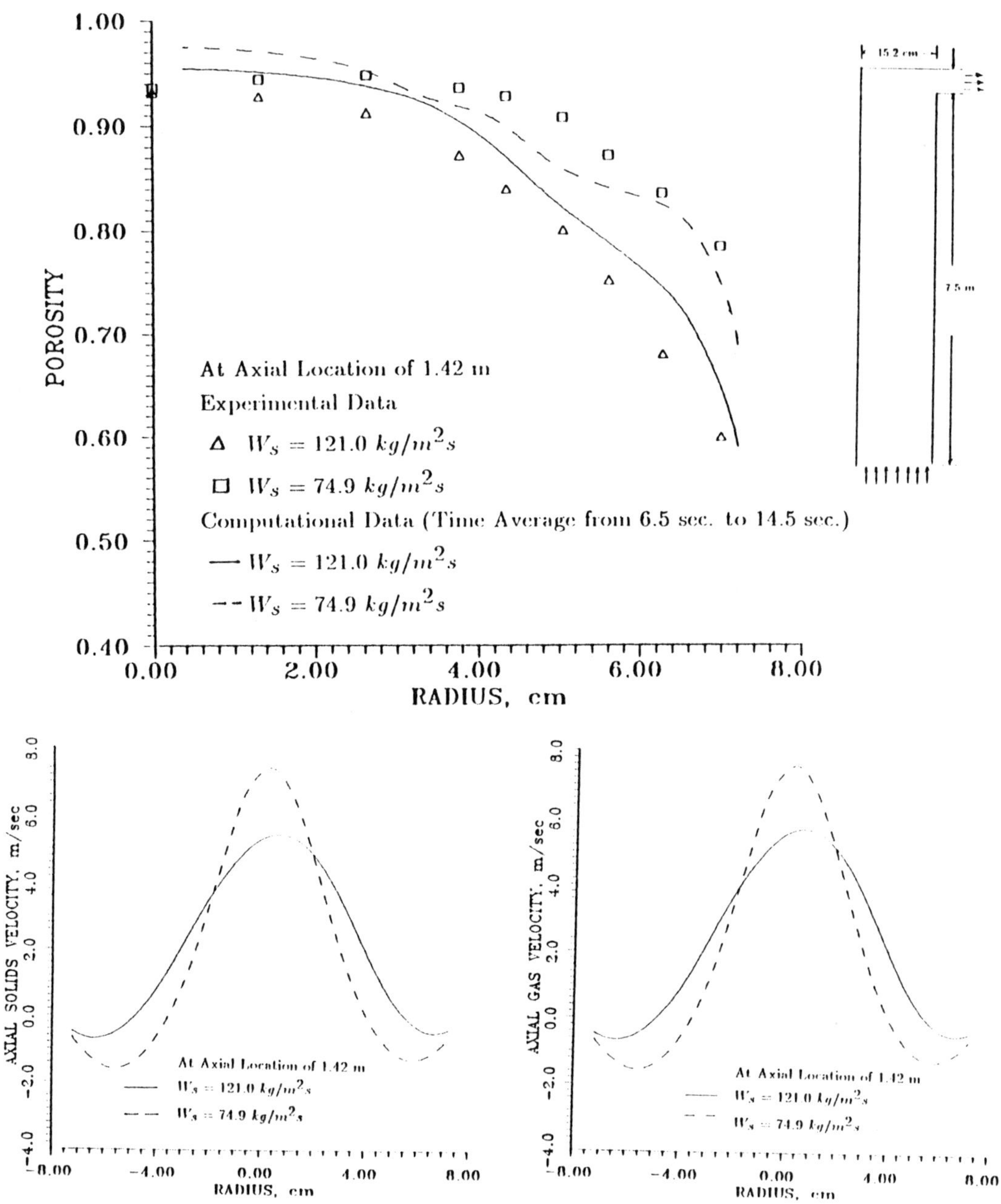

Figure 22-14. Simulation of the CCNY Fast Bed (Weinstein, et al. 1986). HF2-33 Cracking Catalyst, 59 μm. for 121 Kg/m²s, the Inlet Conditions Chosen Were: $V_g = 3.1$ m/s; $V_s = 43.62$ cm/s; $\alpha_s = 0.19$; and P = 1.2015 atm. (Gidaspow, et al. 1989).

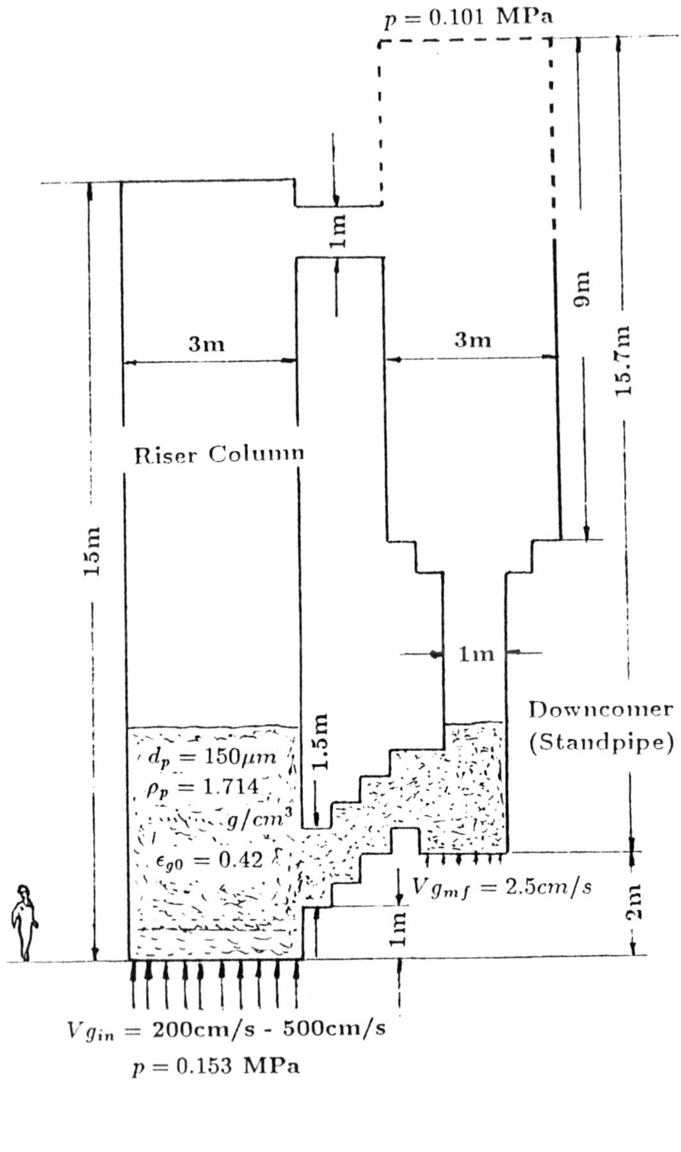

Figure 22-15. The Simplified Full-Scale PYROFLOW CFB with Dimensions and Flow Conditions (Gidaspow, Ding and Jayaswal, 1990).

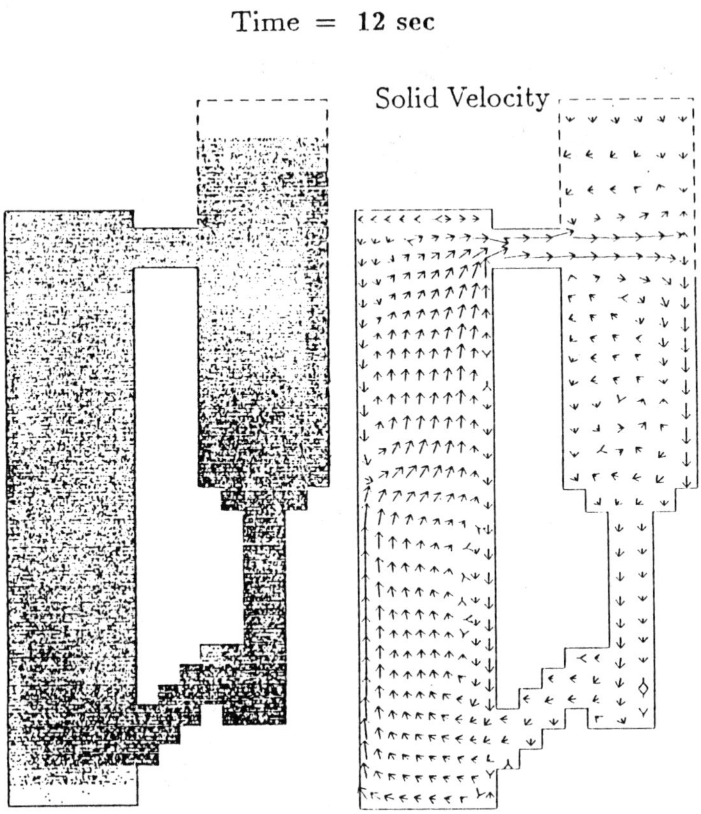

Figure 22-16. Particle Distribution and Solid Velocity at Time = 12 sec. with V_{gin} = 200 cm/s.

in Figure 22-17. The slip velocities are generally not large. In the lower part of the riser, there is downflow near the right wall, whereas, towards the top of the riser, the downflow shifts to the left wall. The asymmetric behavior is due to the asymmetric inlet and outlet locations. Figure 22-18 shows the radial profiles of solid volume fraction in the riser at heights of 5 m and 10 m. We see again the asymmetric distribution of particles in the radial direction. Figure 22-19 shows the radial profiles in the riser of solid mass flux at heights of 5 m and 10 m. The maximum solid mass flux is obtained near the center.

Figure 22-20 shows the solid discharge velocities from the standpipe to the riser as a function of time for inlet gas velocities of $V_{gin} = 2$ m/s, 3 m/s and 5 m/s. The negative discharge velocity is due to initial effects. After 5 seconds, the discharge rates are of the order of 1 m/s, which is consistent with our measured data of solids discharge from hoppers.

The inlet solid mass flux into the riser for $V_{gin} = 200$ cm/s is shown in Figure 22-21. After 7 seconds, the inlet mass flow rates are approximately equal to the outlet mass flow rates. In Figure 22-21 the dashed line shows the maximum flux obtained from a pressure balance derived by Gidaspow, et al. (1983).

The cross-sectional area averaged solid volume fractions and the net flux of the solid mass are shown in Figures 22-22 and 22-23. The figures show that the flows are close to fully developed between heights of 5 m and 12 m. The pressure distribution in the CFB for $V_{gin} = 500$ cm/s is shown in Figure 22-24. The high pressure rises due to the conversion of the kinetic energy of the solid discharged into the riser into a static head.

We see that the simulation predicted a reasonable behavior. Unfortunately, no data for a comparison to the theory are available in the open literature at this time.

22.7 CFB SIMULATION USING NEW KINETIC THEORY

The state of the art of turbulence modeling with large number of particles in the fluid is such that the observed flow regimes cannot be computed with these models. Although fluid turbulence will clearly play a role as the particle concentration is reduced to, say, one hundreds volume fraction of particles or smaller, it is fruitful to explore the potential of the new kinetic theory without fluid turbulence. Gidaspow and Bezbaruah modified our previous computer model developed primarily by Ding and have compared the computations discussed in the previous section to those of using the kinetic theory as described in this chapter (Gidaspow, et al. 1991). The system behavior of the PYROFLOW CFB is very similar for both models. For example, Figure 22-25 shows a comparison of the new kinetic theory model results to Ding's model described in the previous paragraph for a gas velocity of 2 m/sec. The biggest surprise was that to obtain reasonable values of solids viscosity, we had to use a restitutive coefficient near one. Figure 22-26 shows the effect of the restitution coefficient on the computed viscosity for the PYROFLOW CFB with

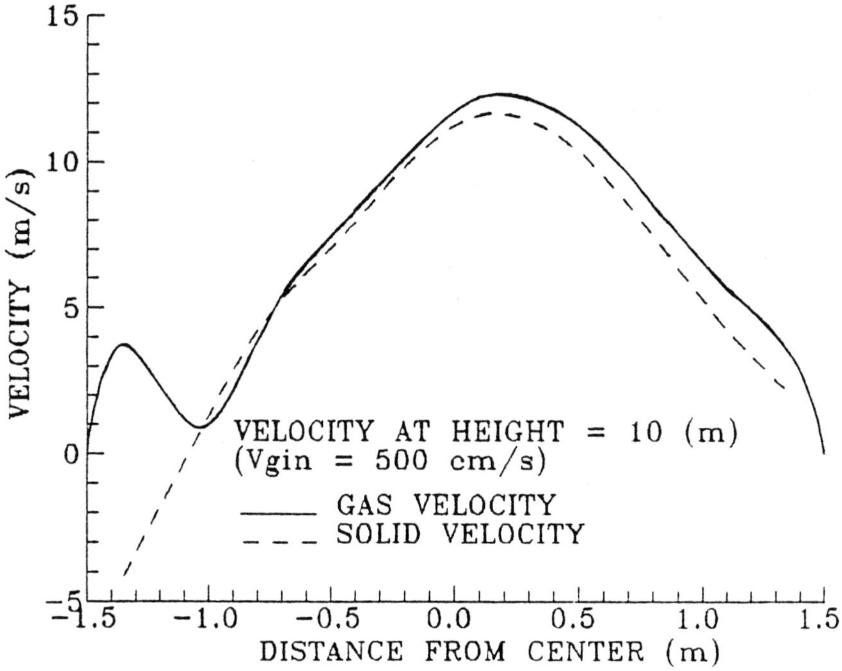

Figure 22-17. Radial Profiles of Axial Velocities at Height = 10 m with V_{gin} = 500 cm/s

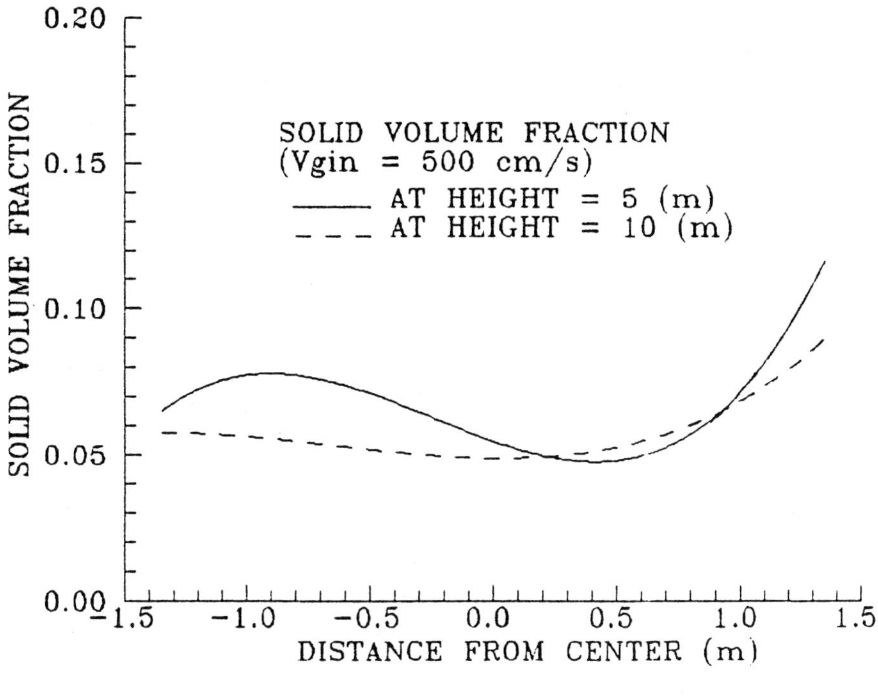

Figure 22-18. Radial Profiles of Solid Volume Fraction with V_{gin} = 500 cm/s

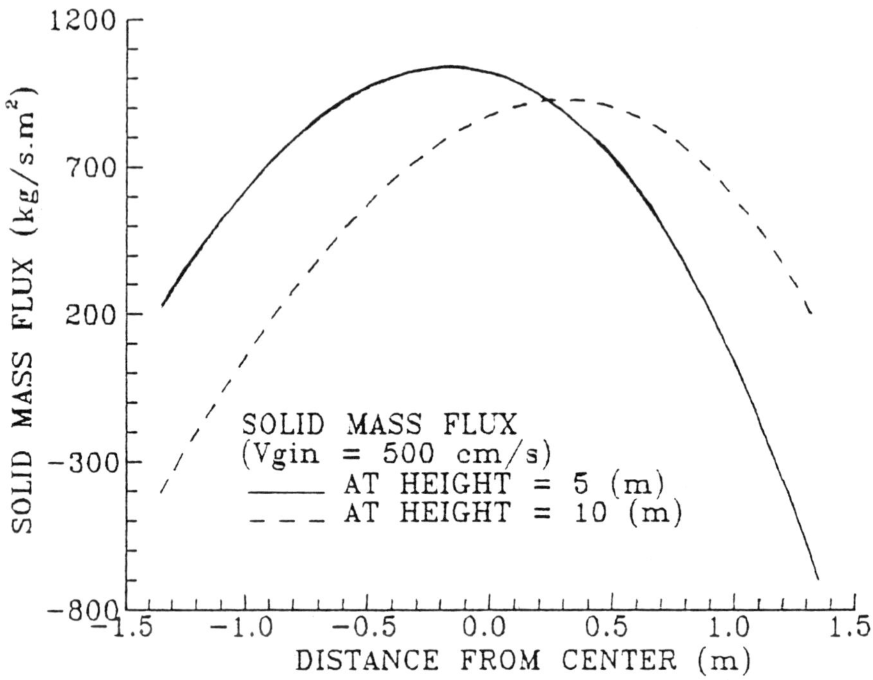

Figure 22-19. Radial Profiles of Solid Mass Flux with V_{gin} = 500 cm/s

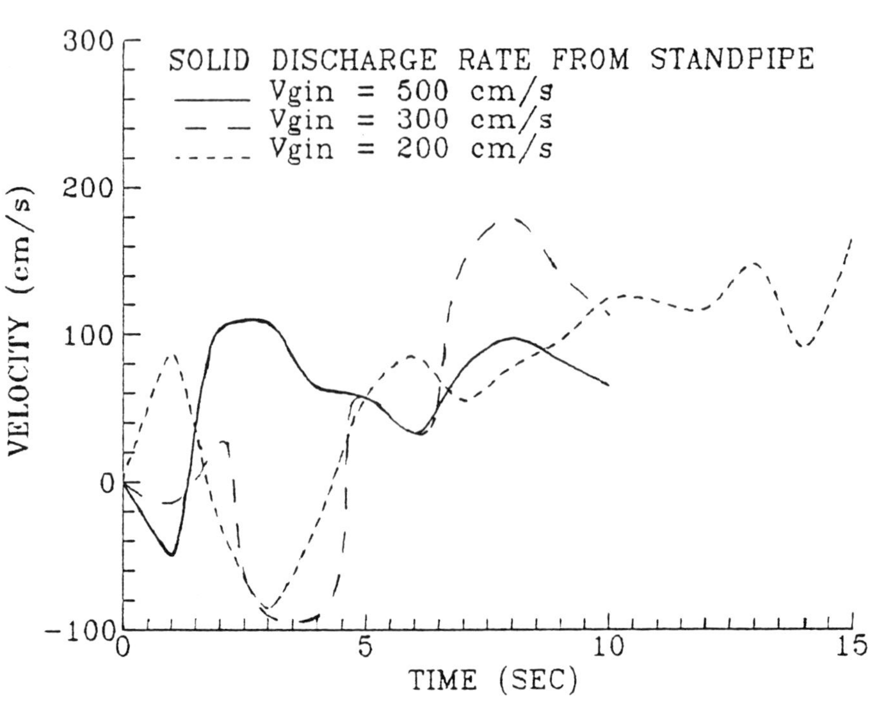

Figure 22-20. Solid Discharge Rate From the Standpipe to the Riser as a Function of Time

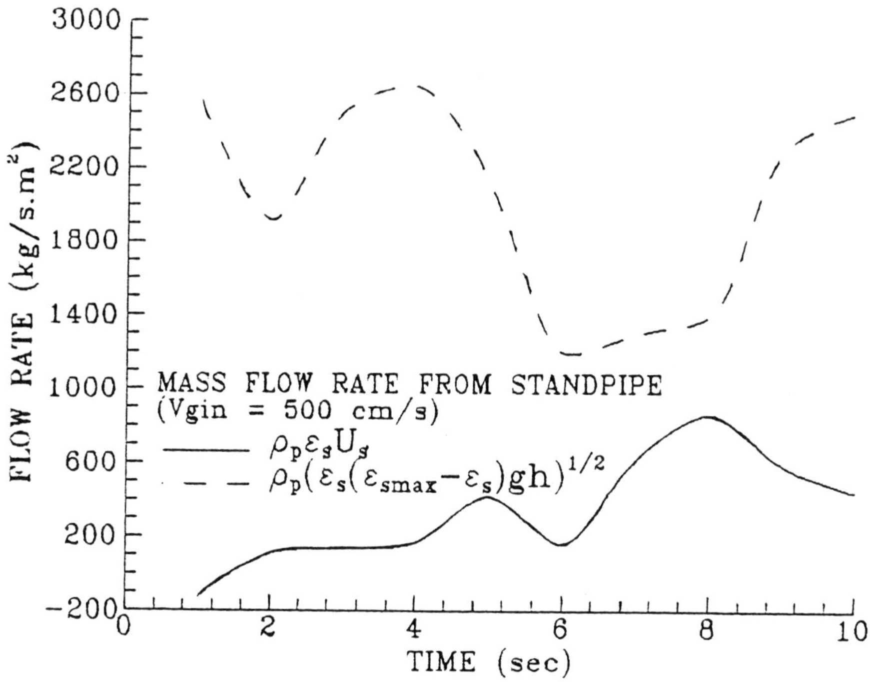

Figure 22-21. Inlet Solid Mass Flux versus Time (V_{gin} = 500 cm/s)

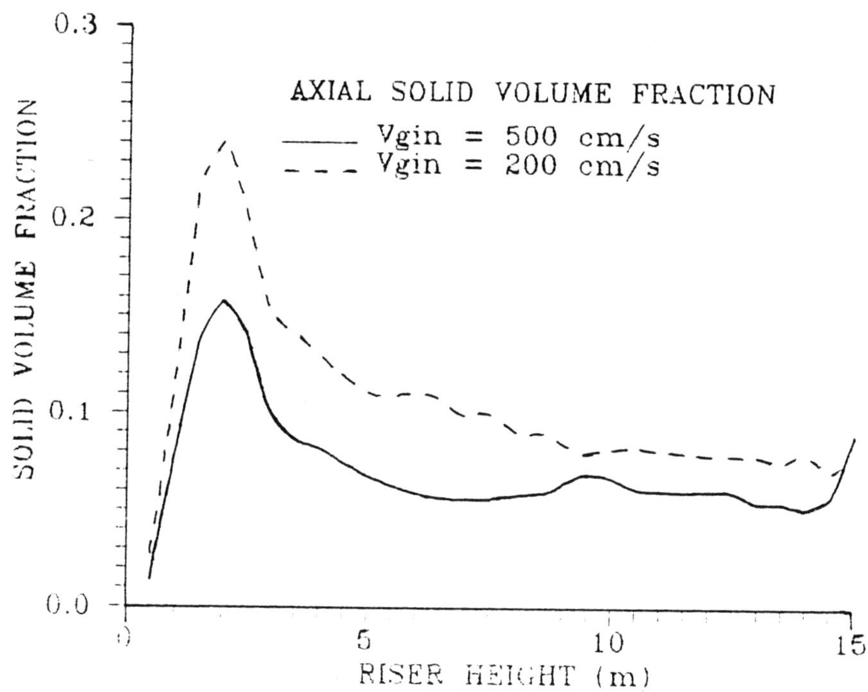

Figure 22-22. Area Averaged Solid Volume Fraction versus Riser Height

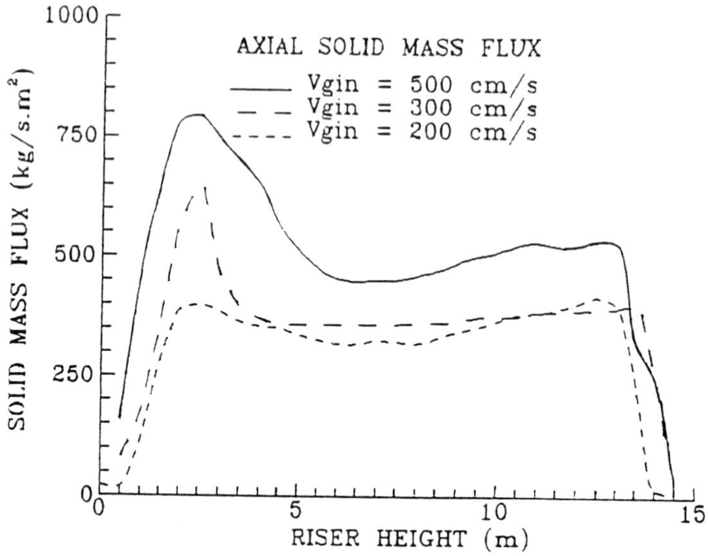

Figure 22-23. Area Averaged Solid Mass Flux versus Riser Height

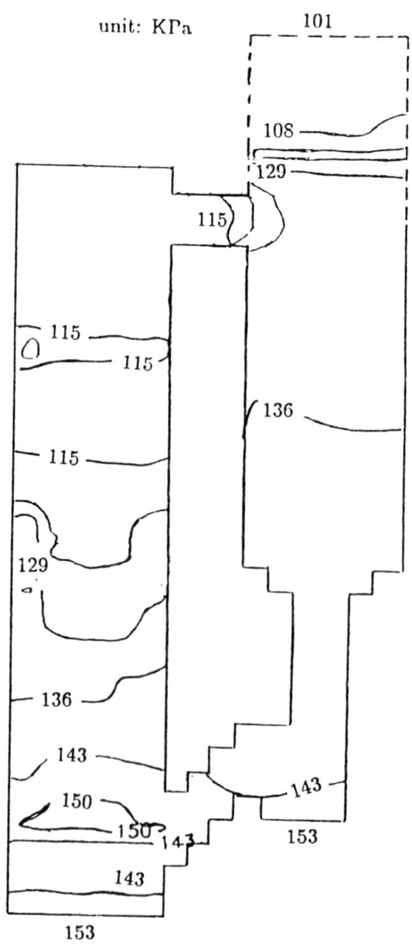

Figure 22-24. Time Averaged Pressure Distribution in the CFB with V_{gin} = 500 cm/s

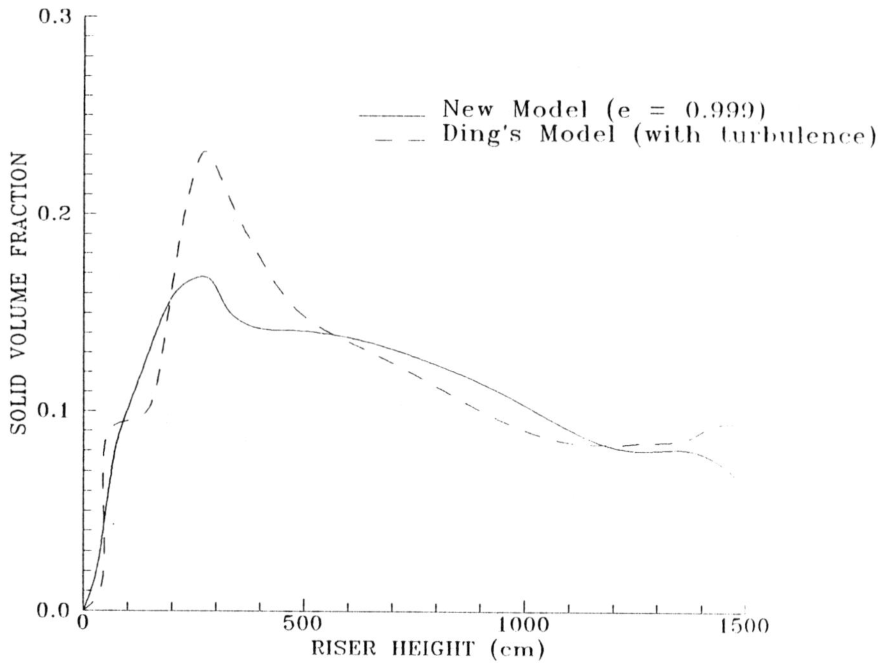

Figure 22-25. Area Averaged Axial Solids Volume Fraction along the Riser Height

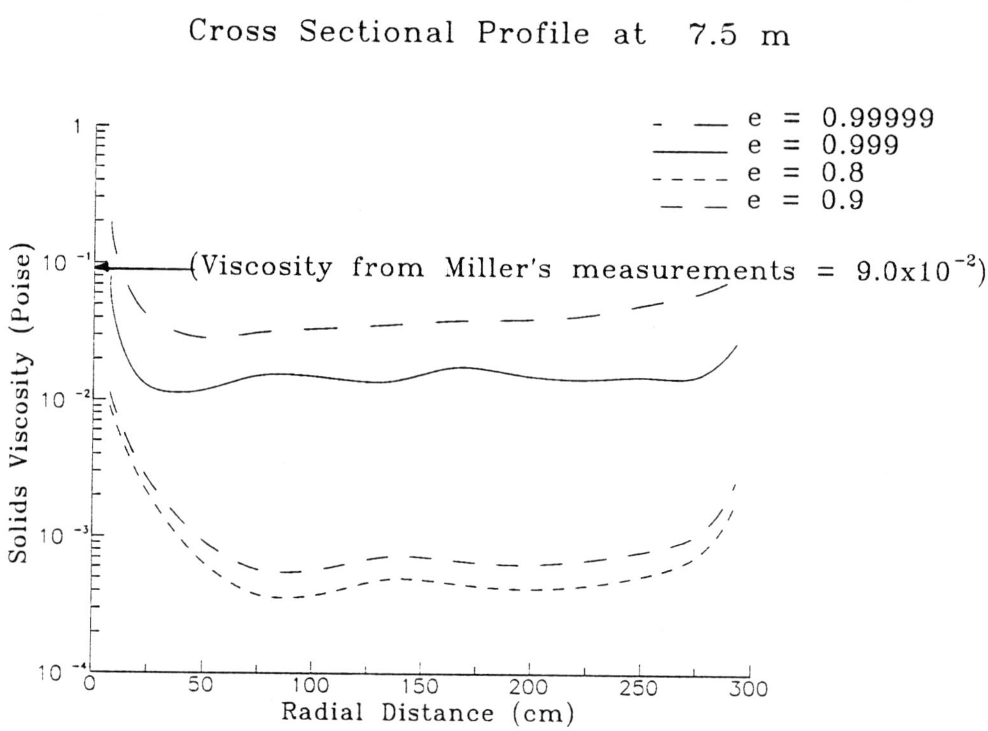

Figure 22-26. Effect of Restitution Coefficient on Solids Viscosity

an inlet gas velocity of 2 m/sec. With a restitution coefficient of 0.8 or 0.9 used by Savage and others the computed solids viscosities are much too low. In a Ph.D. investigation being completed, Miller (1991) measured solids viscosities in a riser built at Illinois Institute of Technology. His viscosities range between two and ten centipoises for 75 μm FCC particles in the concentration range of one to four percent particles by volume. With a restitution coefficient of 0.999 we obtain values of viscosity that are in the range of measurement. The computed viscosities using this value of the restitution coefficient at 10 second after start-up are shown in Figure 22-27. In the dilute regime on the left side of the riser wall and at the top of the CFB in the dilute region, the viscosities are high. Here the granular temperature, that is random particle oscillations, are also high. The kinetic theory of Lun, et al. (1984) and direct numerical simulation of particle motion also yield a viscosity that increases as the particle concentration decreases in the dilute regime. Clearly, the kinetic theory approach must be extended to include effects such as rotation, that have not been incorporated into the theory before we can conclude that we have a predictive theory.

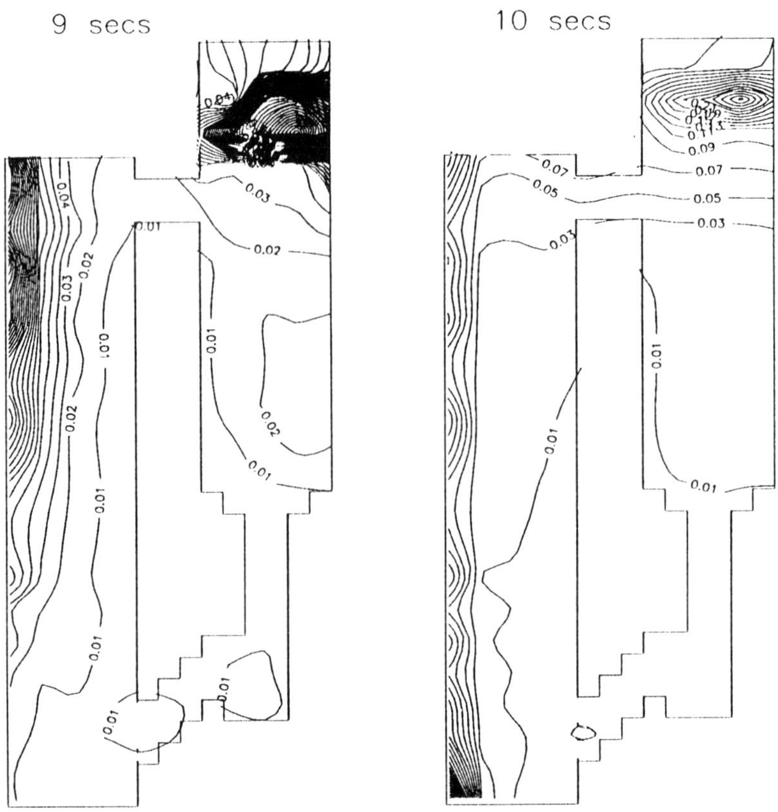

Figure 22-27. Solids Viscosity (Poise) at 9 and 10 Seconds

22.8 CONCLUSIONS

A potentially predictive theory for dense particulate multiphase flow based on the conservation laws of mass, momentum and energy for each phase has been developed. With a minimum amount of empirical input, such as with estimated effective solids viscosities, this theory can predict bubble growth and flow patterns in bubbling fluidized bed systems. It can also predict the observed flow regimes in dense vertical gas-solid transport pipes. An extension of kinetic theory of granular flow pioneered by Professor Savage and collaborators to multiphase flow offers the great potential of predicting the constitutive laws from a knowledge of particle properties such as from normal and the tangential restitution coefficients. Such a complete theory, however, is the subject of future research.

Despite some incomplete features of the theory, the wide spread use of powerful work stations allows the application of this theory to practical engineering systems. The author believes that such simulations should lead to a better understanding of plants processing solids that are known to be very unreliable (Merrow, 1985), and hence to more efficient and reliable designs.

ACKNOWLEDGEMENTS

This study was supported by the National Science Foundation and by a U.S. Department of Energy Grant, DE-FG22-89PC89769.

NOTATION

c random particle velocity
C particle fluctuating velocity relative to mean velocity
C_d drag coefficient
d_p particle diameter
e coefficient of restitution
f velocity distribution function
F force
G solids stress modulus
g gravitational acceleration
g_o radial distribution function
k unit vector along the line from center of particle one to two
m mass of particle
$\dot{m}$ rate of phase production per unit volume
n number density of particles
p pressure
$\overline{\overline{P}}$ pressure tensor
q fluctuating energy flux vector
$\vec{r}$ position vector

Re_p particle Reynolds number
[S] deformation rate tensor
t time
v velocity

Greek Letters

α porosity
α_k volume fraction of phase k
β drag coefficient between phases
γ collisional energy dissipation
τ stress
μ shear viscosity
κ conductivity of fluctuating energy
ξ bulk viscosity
ρ density
ϕ_s particle sphericity
θ granular temperature $= \frac{1}{3} < C^2 >$

REFERENCES

Avidan, A.A., M. Edwards and H. Owen, "Fluid Catalytic Cracking - Past and Future Challenges," Sixth Int. Conf. Fluidization, Banff, Alberta, Canada (1989).

Bader, R.J., Findlay and T.M. Knowlton, "Gas/Solid Flow Patterns in a 30.5-Cm-Diameter Circulating Fluidized Bed," in "Circulating Fluidized Bed Technology II," edited by Basu, P. and J.F. Large, Pergamon Press, 123-137 (1988).

Bouillard, J.X., Lyczkowski, R.W. and Gidaspow, D., "Porosity Distributions in a Fluidized Bed with an Immersed Obstacle," AIChE Journal, 35(6), 908 (1989a).

Bouillard, J.X., R.W. Lyczkowski, S. Folga, D. Gidaspow and G.F. Berry, "Hydrodynamics of Erosion of Heat Exchanger Tubes in Fluidized Bed Combustors," Canad. J. Chem. Eng., 67, 218 (1989b).

Bowen, R.M., Theory of Mixtures in Continuum Physics, editor, A.C. Eringen, Vol. III, pp. 1-117, Academic Press, New York (1976).

Chapman, S. and T.G. Cowling, "The Mathematical Theory of Non-Uniform Gases," 2nd Ed., Cambridge Univ. Press (1961).

Ding, J. and D. Gidaspow, "A Bubbling Fluidization Model Using Kinetic Theory of Granular Flow," AIChE Journal, 36, 523-538 (1990).

Ettehadieh, B., Gidaspow, D. and Lyczkowski, R.W., "Hydrodynamics of Fluidization in a Semi-circular Bed with a Jet," AIChE Journal, **30**, 529-536 (1984).

Garg, S.K., and Pritchett, J.W. (1975) Dynamics of Gas-fluidized Beds, J. Appl. Phys., **46**, 4493-4500 (1975).

Gidaspow, D., Lin, C. and Seo, Y.C., "Fluidization in Two Dimensional Beds with a Jet. Part I. Experimental Porosity Distributions," I&EC Fundamentals, 22, 187-193 May (1983a).

Gidaspow, D., and B. Ettehadieh, "ibid, Part II. Hydrodynamic Modeling," I&EC Fundamentals, 22, 193-201 May (1983b).

Gidaspow, D., "Hydrodynamics of Fluidization and Heat Transfer: Supercomputer Modeling," Appl. Mech. Rev., 39, No. 1, 1-23, (1986).

Gidaspow, D., M. Syamlal and Y. Seo, "Hydrodynamics of Fluidization of Single and Binary Size Particles: Supercomputer Modeling," Fluidization V." Ostergaard, K. and Sorensen, A., editors, Engineering Foundation, 1-8 (1986).

Gidaspow, D. and Y.P. Tsuo, and K.M. Luo, "Computed and Experimental Cluster Formation and Velocity Profiles in Circulating Fluidized Beds," "Fluidization VI." Grace, J.R., Shemilt, L.W. and Bergougnou, M.A. editors, Engineering Foundation, 81-88 (1989).

Gidaspow, D., Y.P. Tsuo and Ding, "Hydrodynamics of Circulating and Bubbling Fluidized Beds," Presented at the "Materials Issues in Circulating Fluidized Bed Combustors," EPRI-Argonne National Lab Symposium, June 1989 and preprinted for the U.S.-Korea Joint Seminar on Fluids Engineering and Science, September 1989. Published by EPRI (1990).

D. Gidaspow, J. Ding and U.K. Jayaswal, "Multiphase Navier-Stokes Equation Solver," in "Numerical Methods for Multiphase Flows," FED-Vol. 91, 47-46, American Society of Mechanical Engineers (1990).

D. Gidaspow, R. Bezbaruah and J. Ding, "Hydrodynamics of Circulating Fluidized Beds," submitted to Seventh International Conference on Fluidization (1991).

Jackson, R. "The Mechanics of Fluidized Beds," Trans. Inst. Chem. Eng., **41**, 13-28 (1963).

Jackson, R., "Hydrodynamic Stability of Fluid-Particle Systems," in Fluidization, edited by J.F. Davidson, R. Clift and D. Harrison, Academic Press, 47-72, (1985).

Jenkins, J.T. and Savage, S.B. (1983). "A Theory of the Rapid Flow of Identical, Smooth, Nearly Elastic, Spherical Particles." J. Fluid Mech., 130, 187-202 (1983).

Johnk, C. and Wietzke, D., "Ahlstrom PYROFLOW Circulating Fluidized Bed Boiler Tube Erosion Experience," pp. 15-1 to 15-25, in Proceedings: Workshop on Materials Issues in Circulating Fluidized Bed Combustors," EPRI-GS-6747, Feb. 1990 (1989).

Lin, J.S., M.M. Chen, and B.T. Chao, "Novel Radioactive Particle Tracking Facility for Measurement of Solids Motion in Gas-Fluidized Beds," AIChE J., 31 (3), 465 (1985).

Lun, C.K.K., Savage, S.B., Jeffrey, D.J. and Chepurniy, N. (1984). Kinetic Theories for Granular Flow: Inelastic Particles in Couette Flow and Singly Inelastic Particles in a General Flow Field, J. Fluid Mech., **140**, 223.

Lyczkowski, R.W., D. Gidaspow and C.W. Solbrig, "Multiphase Flow-Models for Nuclear, Fossil and Biomass Energy Production, Chapter in Vol. II of Advances in Transport Processes, A.S. Mujumdar and R.A. Mashelkar, Editors, Wiley-Eastern Publisher, pp. 198-351, (1982).

Merrow, E.W., "Linking R&D Problems Experienced in Solids Processing," CEP 81, 14-22 (1985).

Miller, Aubrey, Ph.D. Thesis, Illinois Institute of Technology, December 1991.

Richner, D.W., Minaura, T., J.W., Pritchett and T.R. Blake, "Computer Simulation of Isothermal Fluidization in Large-Scale Laboratory Rigs," AIChE Journal 36, 361-369 (1990).

Savage, S.B., "Granular Flows at High Shear Rates." In "Theory of Dispersed Multiphase Flow," ed. R.E. Meyer, Academic Press. pp. 339-358 (1983).

Savage, S.B. "Streaming Motions in a Bed of Vibrationally Fluidized Dry Granular Material." J. Fluid Mech., 194, 457-478 (1988).

Soo, S.L., "Fluid Dynamics of Multiphase Systems," Blaisdell Publishing Corporation, Waltham, MA (1967).

Squires, A.M., M. Kwauk, and A.A. Avidan, "Fluid Beds: At Last, Challenging Two Entrenched Practices," Science, 230 4732, 1329-1337 (1985).

Syamlal, M. and Gidaspow, D., "Hydrodynamics of Fluidization: Prediction of Wall to Bed Head Transfer Coefficients," AIChE J., **31**, 127-135 (1985).

Tsuo, Y.P. and Gidaspow, D., "Computation of Flow Patterns in Circulating Fluidized Beds." Presented at the 26th National Heat Transfer Conference, Philadelphia, PA, August 9 (1989). AIChE Journal, 36, 885-896 (1990).

Weinstein, H., M. Shao, M. Schnitzlein, and R.A. Graff, "Radial Variation in Void Fraction in a Fast Fluidized Bed," "Fluidization V," edited by Ostergaard, K. and A. Sorensen, Engineering Foundation, 329-336 (1986).

Yang, W.-C. and D. Revay, R.G. Anderson, E.J. Chelen, D.L. Keairns and D.C. Cicero, "Fluidization Phenomena in a Large-Scale, Cold-Flow Model," Fluidization, pp. 77-84, D. Kunii and R. Toei, editors, Engineering Foundation (1983).

CHAPTER 23

BUBBLE FLOW IN LIQUID-SOLID SUSPENSIONS

L.-S. Fan
K. Tsuchiya

23.1 INTRODUCTION

A variety of physical and chemical processes, such as those for fermentation, biological wastewater treatment, flue gas desulfurization and resids hydrotreating (Fan, 1989), utilize gas-liquid-solid contacting devices whose operations involve complex interactions among phases. The most prominent interaction occurs between the rising gas bubbles and the surrounding liquid-solid medium. For instance, in reactive systems, gas bubbles are usually a source of reactant species whose transport phenomena often depend on the fluid flow around the bubble; in both reactive and non-reactive systems, gas bubbles induce intimate liquid/solids mixing. It has been specifically recognized that the bubble wake located immediately underneath the bubble base is the dominating factor contributing to these interactions. It is, thus, of primary importance to thoroughly understand the fluid dynamic behavior of the rising bubbles and their wakes so that a sound, fundamental framework can be established for modeling, simulation and design of such multiphase reactor systems.

There exists a symbiotic relationship between a rising bubble and its wake. In general, consideration of the flow associated with the bubble wake near the bubble rear, whether laminar or turbulent, is essential in determining the complete behavior of the rising bubble including its motion. Conversely, examining the shape, rise velocity and motion of the bubble can provide an indirect understanding of the dynamics of the liquid-solid dispersion around the bubble, or more specifically the dynamics of the bubble wake. In the ensuing section, the behavior of rising bubbles in liquid-solid suspensions is discussed along with that in liquids. The discussion focuses on a freely rising single bubble since the rise characteristics of an individual bubble are primarily controlled by the flow immediately around it.

The rise of gas bubbles through an otherwise "homogeneous" liquid-solid flow medium establishes regions of "heterogeneity" in the flow field. The most marked effects are flow instability induced in the downstream of the bubbles, and the resulting large-scale vortical motion and local solids concentration gradient. The fundamental characteristics associated with the flow instability under discussion include the wake instability and shedding frequency, vortical wake structure, dynamic variations in wake sizes, the solids concentration in the wake, and pressure distributions around the bubble.

Interactions between neighboring bubbles are usually recognized in terms of bubble coalescence and/or breakup. Direct causes for these phenomena, however, lie in the interactions between the bubbles and the surrounding flow: bubble coalescence is preceded by the acceleration of trailing bubbles in the wake of leading bubbles; bubble breakup occurs due to inherent instability induced along the bubble frontal surface or due to continuous external disturbances such as shear stresses present in the liquid flow and penetration through the bubble of large, heavy particles in liquid-solid fluidized media. The mechanisms of bubble coalescence and breakup are discussed in light of the interactions between bubbles and wakes and/or between particles and bubble-wakes.

Updated information is presented in this chapter beyond that given in the recent book of the authors (Fan and Tsuchiya, 1990). Specifically, the effects of surface properties, i.e., non-wettability of the solid particles on the fluid mechanic behavior of the bubble and its wake, the effect of the presence of solid particles on the pressure distributions around the bubble, and the characteristics of the liquid-solid flow induced by the movement of multiple chains of rising bubbles are included. The ultimate design of multiphase systems for industrial applications can be strengthened with comprehensive knowledge of the above-mentioned fluid dynamic/transport phenomena. While these systems generally consist of bubbles rising in swarms, and exhibit complex system characteristics, they can be approached by considering the behavior of a single bubble and its wake together with any effects related to its interaction with other bubbles and/or other bubble wakes.

23.2 BUBBLE RISE CHARACTERISTICS

The characteristics of a rising bubble can be described in terms of the shape, rise velocity and motion of the bubble. These rise characteristics are closely associated with the flow and physical properties (mainly viscosity or presence/absence of solid particles) of the surrounding medium as well as the interfacial properties (i.e., presence/absence of surfactants) of the bubble surface.

The interaction between a rising gas bubble and the surrounding liquid-solid medium determines the shape of the bubble and the extent of the disturbance in the surrounding flow field. It is probably most convenient to describe the behavior of a bubble and its wake in light of its shape.

23.2.1 Bubble Shape

Bubbles in motion are generally classified by shape as spherical, oblate ellipsoidal, and spherical/ellipsoidal cap; the observed bubble shape is a result of an intricate balance among forces acting on the bubble, viz., surface tension, viscous and inertial (or buoyancy) forces. Figure 23.1 shows sketches of bubble shapes thus classified in addition to other diversified shapes. As the dominant forces change with increasing bubble size, the bubble shape goes through a transition from spherical to ellipsoidal to spherical-cap shape.

When the bubble size is small (for example, d_e less than 1 mm in water), surface tension forces predominate and the bubble shape is approximately spherical. The flow around the bubble is controlled by viscous forces and can be described, provided Re_e $(= \rho_l d_e v_{b\infty}/\mu_l) \ll 1$, by the creeping flow approximation, i.e., by the Hadamard-Rybczynski theory for purified systems or by Stokes' law for contaminated systems (Clift et al., 1978). For small spherical bubbles at very low Reynolds numbers, therefore, the flow does not separate from the bubble surface and the wake does not exist.

For bubbles of intermediate size, the effects of both surface tension and the inertia of the medium flowing around the bubble are important. In addition, the bubble dynamics are profoundly influenced by

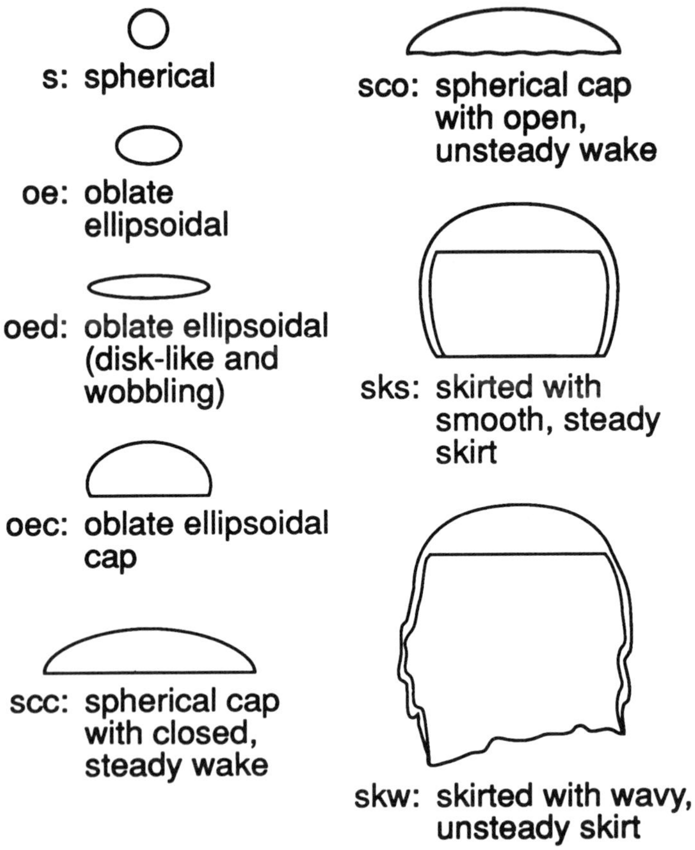

Figure 23.1 *Sketches of various bubble shapes observed in infinite Newtonian liquids (after Bhaga and Weber, 1981).*

liquid viscosity and the presence of surface-active contaminants. As a result, intermediate-size bubbles exhibit very complex shape and motion characteristics. Though called ellipsoidal bubbles, they often lack in fore-and-aft symmetry and, in extreme circumstances, cannot be described by any simple regular geometry due to significant shape fluctuations. Understandably, the behavior of wakes behind intermediate-size bubbles is also complicated.

Large bubbles, whose volumes are greater than 3 cm^3 (i.e., $d_e > 18$ mm) in general (Clift et al., 1978), are dominated by inertial or buoyancy forces with negligible effects of surface tension, viscosity and purity of the liquid media. The bubble shape is approximately a spherical cap with an included angle of about 100°, provided $\mu_l < 50$ mPa·s, and a relatively flat or sometimes indented base. These spherical-cap bubbles generate a relatively large-volume wake.

In systems other than gas-liquid systems, such as gas-slurry bubble columns (Dayan and Zalmanovich, 1982) and gas-liquid-solid fluidized beds of small, light particles (Stewart and Davidson, 1964; Ostergaard, 1973), bubble behavior has been often observed to resemble that in viscous liquids, e.g., large bubbles assume the spherical-cap shape mentioned above. This similarity is attributed to the premise that the liquid-solid mixture in such systems can be regarded as a pseudo-homogeneous medium of apparently higher viscosity compared to the suspending liquid. In three-phase fluidized beds of large and/or heavy particles, on the other hand, gas bubbles behave in different ways. Stewart and Davidson (1964) observed, in a bed of dense particles such as lead shot or iron shot, that a liquid bubble could be stabilized by either small gas bubbles or a thin layer of gas lining its roof.

23.2.1.1 Effect of Physical Properties on the Bubble Shape

Although the flattening of the bubble with increasing bubble size can be explained qualitatively (Fan and Tsuchiya, 1990), quantitative predictions are not possible, especially when a variety of physical properties of the surrounding medium are involved. In this regard, the state-of-the-art knowledge about bubble shapes in various liquids is established strictly from experimental observations.

The most widely recognized information, although still somewhat qualitative, is presented graphically in terms of three dimensionless groups: the Reynolds number (Re_e); the Eötvös number ($Eo = g\rho_l d_e^2/\sigma$); and the Morton number ($Mo = g\mu_l^4/\rho_l\sigma^3$) (Grace, 1973; Bhaga and Weber, 1981). Figure 23.2 shows such a diagram presented by Bhaga and Weber (1981) for single bubbles rising freely in unbounded Newtonian liquids; for each representative bubble shape (denoted by its acronym), refer to the corresponding sketch presented in Fig. 23.1. With the bubble size and the physical properties of the surrounding liquid specified, this diagram indicates the expected bubble shape. In addition, Fig. 23.2 may be used to estimate terminal velocities of bubble rise since only the Re_e contains the velocity term.

23.2.1.2 Correlations for the Bubble Shape

At present complete theoretical prediction of the bubble shape or the geometric parameter of the bubble is not available over the entire range of bubble sizes or Reynolds numbers. The prediction also becomes difficult in the presence of solid particles. The bubble shape parameters are thus, in general, empirically correlated in terms of bubble size, rise velocity and the surrounding medium properties.

For bubbles in low-Mo liquids ($Mo < 10^{-3}$), Fan and Tsuchiya (1990) extended Vakhrushev-Efremov's (1970) formula, which is given by

$$\frac{h}{b} = \begin{cases} 1 & Ta < Ta_1 \\ \{c_1 + c_2 \tanh\left[c_3\left(c_4 - \log_{10} Ta\right)\right]\}^m & Ta_1 \le Ta < Ta_2 \\ \left(\frac{h}{b}\right)_{cap} & Ta_2 \le Ta \end{cases} \tag{23.1}$$

to predict the bubble aspect ratio for a variety of systems including purified and contaminated and two- and three-dimensional systems. Here Ta ($= Re_e Mo^{0.23}$) is the Tadaki number, and c_1–c_4 are empirical parameters. Table 23.1 summarizes the parameters in Eq. (23.1) determined based on the experimental data available in the literature.

Due to experimental difficulties involved in the measurement of bubble shape in three-dimensional liquid-solid suspensions, especially when a high concentration of nontransparent solid particles are present in the system, very limited data are available (Miyahara et al., 1988). Figure 23.3 compares the bubble aspect ratio in water predicted by Vakhrushev-Efremov's (1970) correlation with Miyahara et al.'s data in a tap water-fluidized bed containing various types of particles. As can be seen in the figure, the bubble shape in a liquid-solid fluidized bed with low (≤ 0.1)

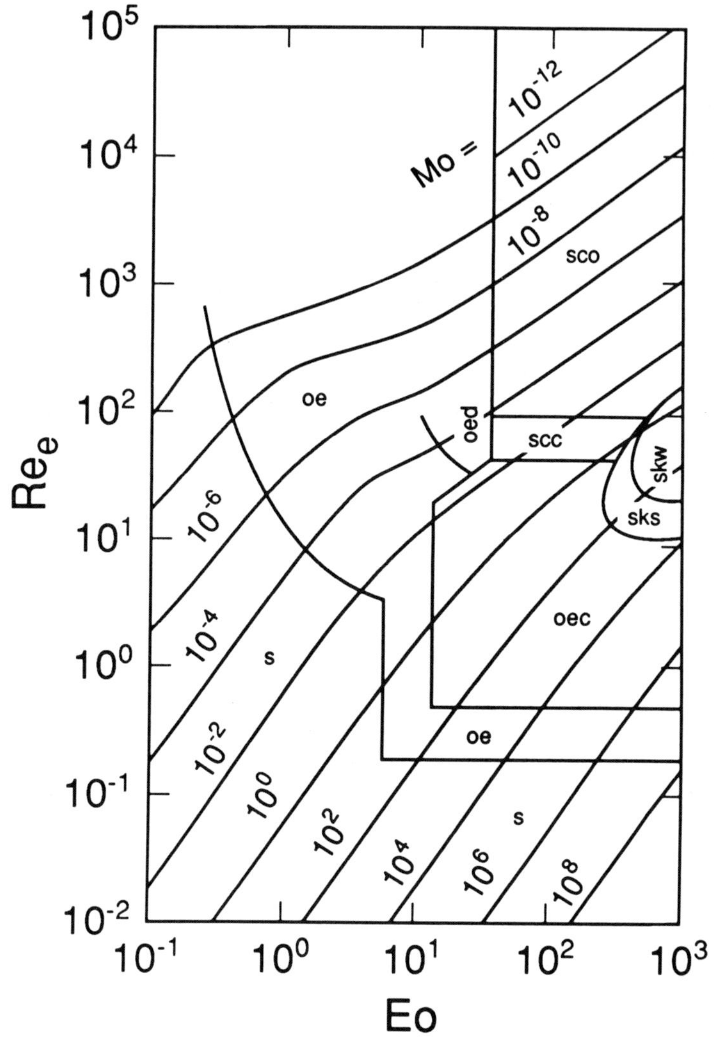

Figure 23.2 *Shape-regime map for single bubbles rising in infinite Newtonian liquids. For acronyms representing bubble shapes, refer to sketches given in Fig. 23.1 (from Bhaga and Weber, 1981).*

Table 23.1 Numerical Values for Parameters in Eq. (23.1) for Various Systems

System	m	$(h/b)_{cap}$	Ta_1	Ta_2	c_1	c_2	c_3	c_4
3D impure[a]	3	0.24	1.0	40	0.81	0.20	2.0	0.80
3D pure	3	0.24	0.3	20	0.81	0.20	1.8	0.40
2D impure	2	0.30	1.0	40	0.77	0.24	2.1	0.80
2D pure	2	0.30	0.3	20	0.77	0.24	1.9	0.40

[a] Identical to Vakhrushev and Efremov (1970).

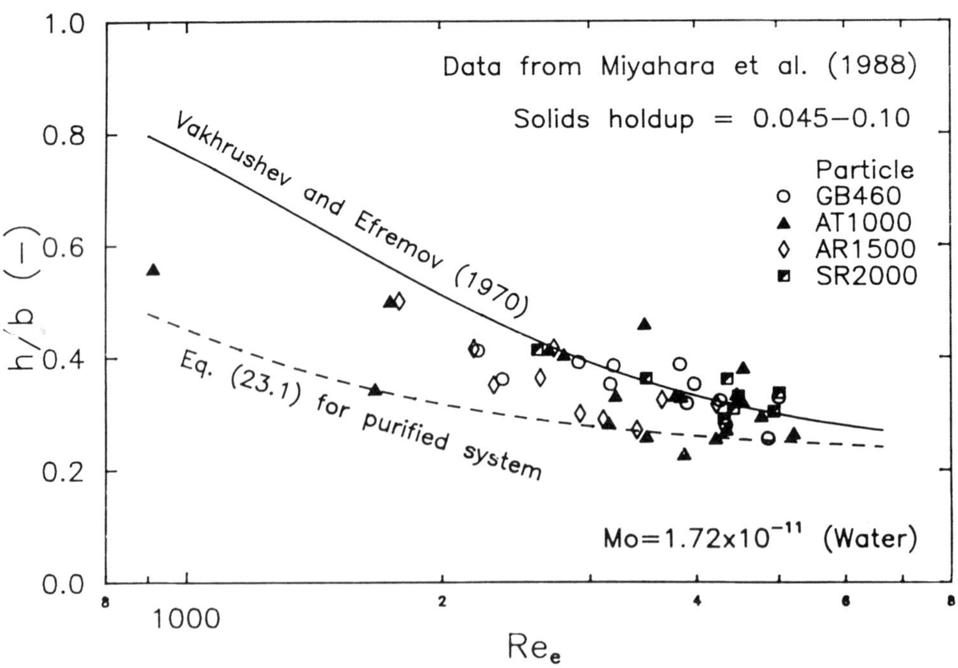

Figure 23.3 *Bubble aspect ratio in a three-dimensional water-solid fluidized bed (after Miyahara et al., 1988).*

solids holdup can be reasonably represented by the Vakhrushev-Efremov correlation, provided Re_e is moderately high (> 2000). At lower Re_e (< 2000), bubbles in the liquid-solid fluidized bed tend to flatten to a greater extent and apparently behave like those in somewhat purified water; note the agreement between the data and Eq. (23.1) with parameters evaluated for pure systems given in Table 23.1. Under the experimental conditions Miyahara et al. (1988) examined, no systematic variation with respect to the particle properties was apparent.

To circumvent the difficulties involved in three-dimensional systems, two- dimensional systems are often employed (e.g., Henriksen and Ostergaard, 1974; Tsuchiya and Fan, 1986; Song, 1989; Tsuchiya et al., 1990). Among the investigators, Song (1989) and Tsuchiya et al. (1990) systematically studied the effects of solids properties and holdup on the bubble shape over a wide range of bubble Reynolds number. Figure 23.4 shows the results of Tsuchiya et al. for glass beads of four different sizes, each at three different bed voidages ($\alpha_l = 1 - \alpha_s$), in a two-dimensional (tap) water-solid fluidized bed. Each set of data is compared with Eq. (23.1) with the appropriate parameters given in Table 23.1.

Figure 23.4 indicates that, for a given particle size, the variation in bed voidage has a minimal influence on the bubble aspect ratio over the studied range $0.5 < \alpha_l \leq 1$. The effect of particle size is also not significant for most particles Tsuchiya et al. (1990) examined; however, note that at higher Re_e (> 3000) the bubbles flatten less extensively for the largest particles (1 mm glass beads with terminal velocity of 0.17 m/s). At lower Re_e (< 2000), the same trend as in the three-dimensional system is seen; the bubble shape in the liquid-solid fluidized bed tends to be flatter than that in liquid alone, represented by the solid curve, at the same Re_e. The effect of wettability of the particle was recently investigated by Tsutsumi et al. (1990) using 460 and 774 μm glass beads (wettable) and Teflon-coated 460 and 774 μm glass beads (non-wettable). They found no appreciable difference in the bubble aspect ratio and the same qualitative dependence on bed voidage and particle size between these two types of particle systems.

23.2.2 Bubble Rise Velocity

Estimating the bubble rise velocity is extremely important; however, the theoretical development of a universal prediction equation over a wide

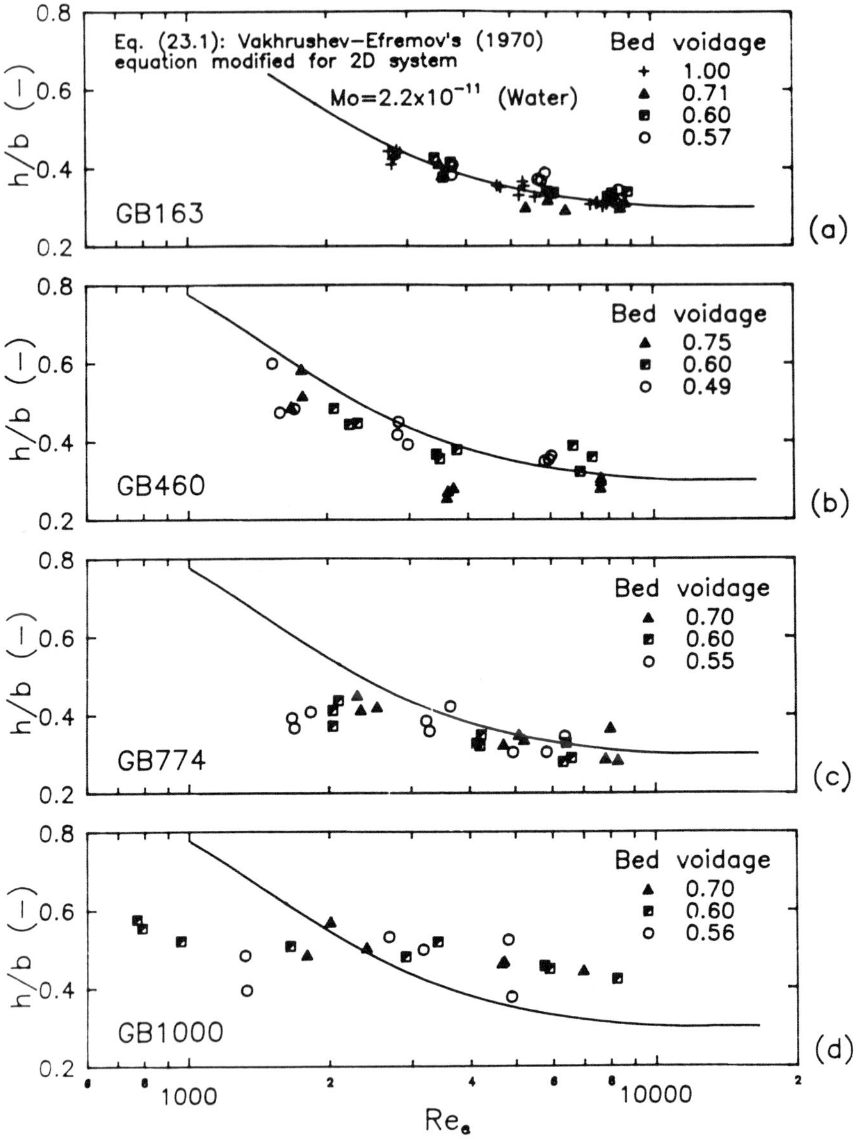

Figure 23.4 *Effects of solids properties and holdup on bubble aspect ratio in a two-dimensional water-solid fluidized bed (from Tsuchiya et al., 1990).*

range of bubble sizes or Reynolds numbers has been hampered due to the diversified effect of the wake flow. At low and high Re_e, the rise velocity can be determined without considering the wake effect from the fore-and-aft symmetry of the flow around spherical bubbles and based on the assumption of potential flow around the nose of spherical-cap bubbles. At intermediate Re_e, the presence of the bubble wake increases the form drag and has a significant effect on determining the rise velocity. Consequently, it is practical to develop empirical correlations which feature the dominant force(s) acting on the bubble over wide ranges of bubble sizes and physical properties of the surrounding medium. Among the correlations in the literature that have been shown to be applicable to gas-liquid systems (e.g., Grace et al., 1976), the one proposed by Fan and Tsuchiya (1990) possesses this feature. The correlation is given by

$$v_{b\infty} = \left(v_{b1}^{-n} + v_{b2}^{-n}\right)^{-1/n} \tag{23.2}$$

where

$$v_{b1} = \frac{\rho_l g d_e^2}{K_b \mu_l} \tag{23.2a}$$

$$v_{b2} = \sqrt{\frac{2c\sigma}{\rho_l d_e} + \frac{g d_e}{2}} \tag{23.2b}$$

Three parameters n, K_b and c depend on physical properties of the liquid and/or contamination level of the liquid.

As mentioned earlier, a liquid-solid mixture of small, light particles is often regarded as a pseudo-homogeneous medium. Whether a rising bubble behaves as if it were in a "homogeneous" or "heterogeneous" medium depends primarily on the ratio of the particle diameter to bubble diameter (d_s/d_e). Figure 23.5 shows the terminal velocity of single bubbles in water-fluidized beds of (a) 500 and (b) 1000 μm sand particles at various bed voidages. The lines are predictions based on Eq. (23.2) assuming that the bubble rises in a homogeneous Newtonian medium of higher density and viscosity than those of the fluidizing liquid alone. When using Eq. (23.2), the surface tension of water is used since the particles are wettable. The apparent bed density (ρ_m) is defined as

$$\rho_m = \rho_l \alpha_l + \rho_s \alpha_s \tag{23.3}$$

and the apparent bed viscosity (μ_m) is evaluated by

$$\mu_m = \mu_l \exp\left(36.15\alpha_s^{2.5}\right) \qquad (\alpha_s > 0.2) \tag{23.4}$$

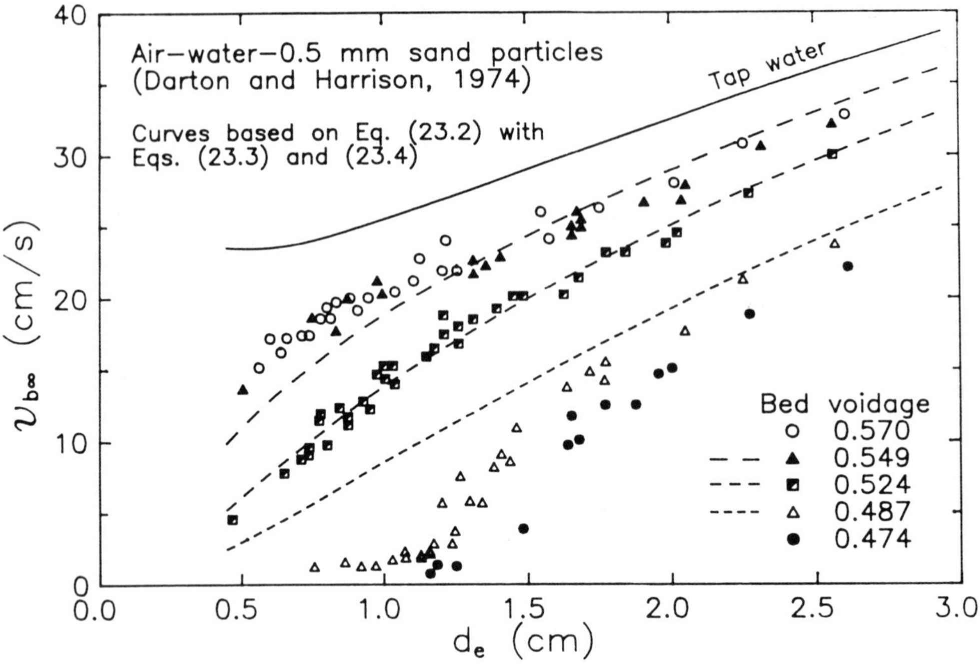

Figure 23.5(a) *Terminal velocity of air bubbles in a water-fluidized bed of 500 μm sand particles at various bed voidages (data from Darton and Harrison, 1974).*

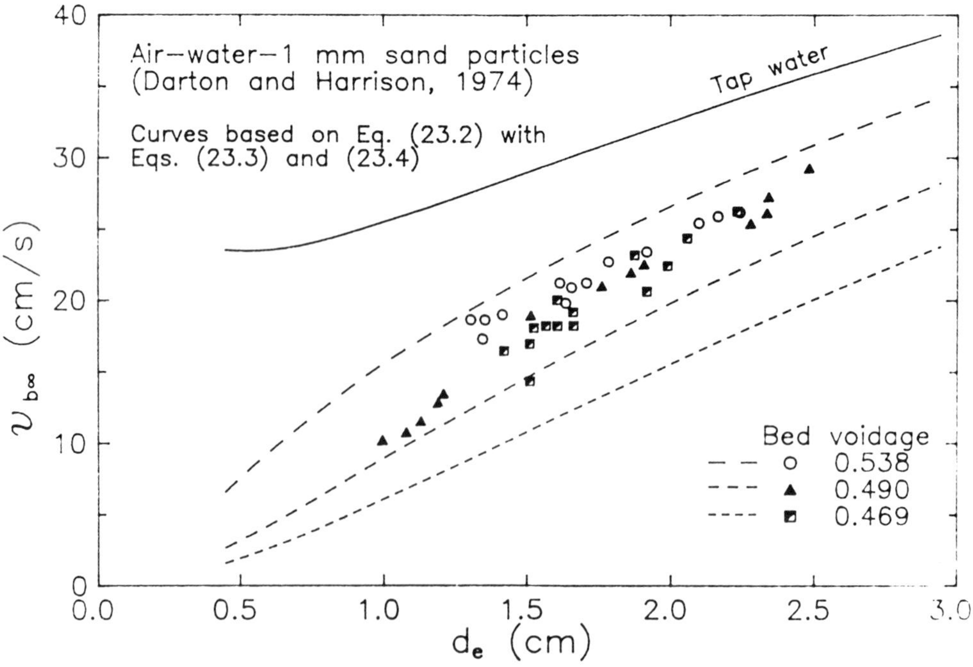

Figure 23.5(b) *Terminal velocity of air bubbles in a water-fluidized bed of 1000 μm sand particles at various bed voidages (data from Darton and Harrison, 1974).*

Equation (23.4) is an empirical correlation proposed by Darton (1985) and is applicable for particles of moderate densities ($\rho_s < 3$ g/cm^3) and sizes ranging from 50 to 1000 μm.

As seen in Fig. 23.5, the assumption of bed homogeneity applies only under limited conditions. For 500 μm sand particles [Fig. 23.5(a)], the prediction deviates from the data when the bubble size is small (d_e less than approximately 15 mm or $d_e/d_s < 30$) except when the bed voidage is 0.524. At higher bed voidages, a small bubble tends to find its way through interstices between the particles. This leads to a heterogeneous effect of the solids on the rise of bubbles; the bubbles rise faster than they would rise through a "fictitious" homogeneous medium [$d_e < 10$ mm at $\alpha_l > 0.524$ in Fig. 23.5(a)]. At lower bed voidages, on the other hand, small bubbles rise much slower than they would through an equivalent Newtonian medium. Darton (1985) attributed this effect to non-Newtonian behavior of the fluidized beds near the point of incipient fluidization.

For 1000 μm sand particles [Fig. 23.5(b)], the prediction deviates from the data significantly while Eq. (23.4) is still supposed to be applicable; the bed voidage effect is not appropriately taken into account. Apparently in this case, the particle size plays a more significant role in determining $v_{b\infty}$ than through its effects on the bed voidage/solids holdup alone. Since the apparent bed properties are often considered independent of the particle size at given phase holdups [see Eqs. (23.3) and (23.4)], the pseudo-homogeneous concept may not be utilized for providing the general correlation for $v_{b\infty}$ in liquid-solid suspensions over the entire range of solids properties.

An alternative description of the particle effect on $v_{b\infty}$ is that the liquid-solid medium is regarded as heterogeneous and the bubble rise velocity is strongly influenced by the impact of solid particles upon the bubble roof. Jean and Fan (1990) developed a theoretical model based on this concept and showed that their model could predict $v_{b\infty}$ successfully for small particles ($d_s < 500$ μm) and large spherical-cap bubbles ($d_e >$ 15 mm). Due to the limitations of bubble size and shape, however, Jean and Fan's model cannot be directly extended to a wide range of bubble sizes.

A systematic measurement of the bubble rise velocity over wide ranges of bubble and particle sizes was recently conducted by Jang (1989) in a water-glass bead fluidized bed. His experimental ranges include $d_e =$ 2–

20 mm, $d_s = 0.163$, 0.460, 0.774, 1.0 and 2.0 mm, $\rho_s = 2.50$–2.79 g/cm^3, and $\alpha_l = 0.48$, 0.53 and 0.58. Based on his results, Jang proposed an empirical correlation for $v_{b\infty}$ which had the same form as Eq. (23.2); the parameters in the equation were evaluated as functions of particle diameter and bed voidage. The final correlation is given by

$$v_{b\infty} = \left[\left(K_{bs}d_e^2\right)^{-n} + \left(\frac{2\sigma}{\rho_l d_e} + \frac{gd_e}{2}\right)^{-n/2}\right]^{-1/n} \tag{23.5}$$

where

$$K_{bs} = \left\{(604\alpha_l - 235)\exp\left[\left(1.08\times10^4 - 2.35\times10^4\alpha_l\right)d_s\right] + (430\alpha_l - 211)\right\}$$
$$\bullet\left\{190 - \frac{120}{\pi}\tan^{-1}\left[1.23\times10^6\left(d_s - 3.11\times10^{-4}\right)\right]\right\} \tag{23.5a}$$

The parameter n is correlated graphically with d_s and α_l (Jang, 1989). The variables in the above equations are in the following units: $v_{b\infty}$ (m/s), d_e (m), d_s (m), σ (N/m), ρ_l (kg/m^3) and g (m/s^2). The prediction given by Eq. (23.5) was found to be in reasonable agreement with that by Eq. (23.2) for tap (contaminated) water, the model of Jean and Fan (1990) for spherical-cap bubbles ($d_e > 15$ mm) and smaller particles ($d_s < 460$ μm), and the correlation of El-Temtamy and Epstein (1980) for spherical-cap bubbles and larger particles ($d_s > 774$ μm).

Figure 23.6 compares the experimental results and the predictions obtained by Jang (1989) for various particle sizes at bed voidages of (a) 0.48 and (b) 0.58. At the lower bed voidage [Fig. 23.6(a)], the reduction of $v_{b\infty}$ for small bubbles ($d_e < 5$ mm) due to the presence of particles is drastic for all the particles examined ($d_s \geq 460$ μm) regardless of the particle size, while for larger bubbles $v_{b\infty}$ approaches that in water as d_s decreases even at this high solids holdup ($\alpha_s = 0.52$). At the higher bed voidage [Fig. 23.6(b)], the particle-size effect appears to be somewhat different. For small bubbles, the extent of $v_{b\infty}$ reduction depends on the particle size; it decreases with decreasing d_s and, for 163 μm glass beads, the difference in $v_{b\infty}$ between water and the water-solid medium is dramatically reduced. For larger bubbles ($d_e > 7$ mm), the presence of solid particles does not significantly reduce $v_{b\infty}$, and the extent of reduction is a rather weak function of the particle size. The observed variations of $v_{b\infty}$ with d_e for both bed voidages are satisfactorily predicted by Eq. (23.5) as shown in Fig. 23.6. Although the prediction may be in

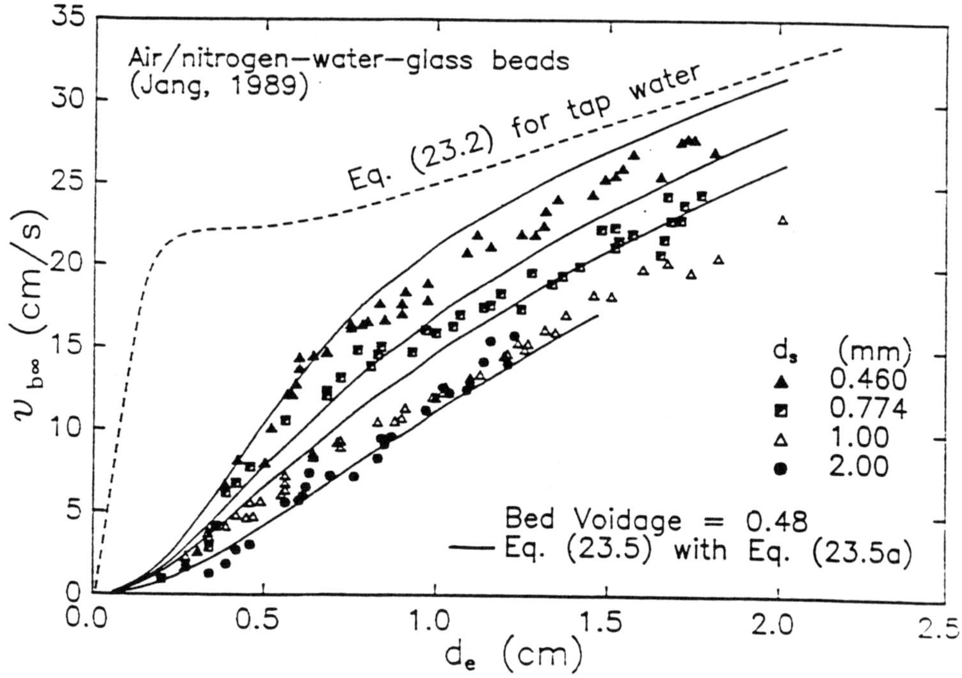

Figure 23.6(a) *Terminal velocity of gas bubbles in a water-glass bead fluidized bed at a bed voidage of 0.48 (data from Jang, 1989).*

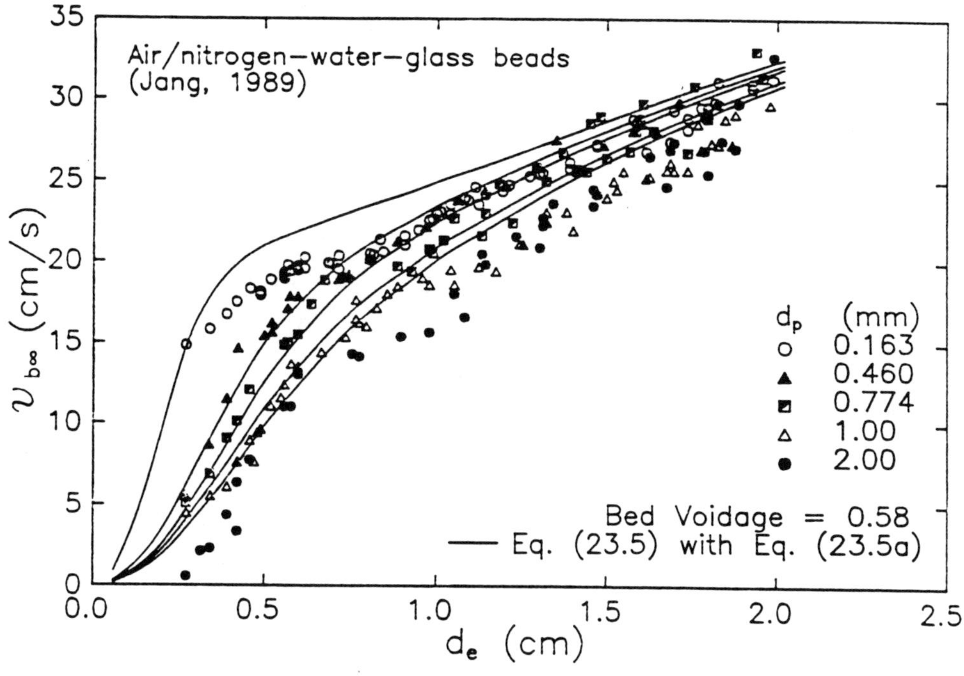

Figure 23.6(b) *Terminal velocity of gas bubbles in a water-glass bead fluidized bed at a bed voidage of 0.58 (data from Jang, 1989).*

error by as much as 30%, Eq. (23.5) can be used as a general correlation for $v_{b\infty}$ in water-glass bead media. A more general correlation requires more comprehensive investigations on the particle effects over wide ranges of solids properties.

The effect of particle wettability on the bubble rise velocity was examined in a two-dimensional (8.4 mm thick) system by Tsutsumi et al. (1991). Non-wettable particles have a larger contact angle compared to wettable particles. Thus, non-wettable particles favor the contact between bubbles and solids. Tsutsumi et al. observed that in small bubble systems, bubbles were usually attached by non-wettable particles forming a monolayer on the bubble surface. Consequently, the bubble rise velocity is reduced compared to that for a wettable particle. On the other hand, little effect of particle wettability was observed on the rise velocity of bubbles of $b > 15$ mm; for large bubbles of a circular-cap shape, the attachment of particles occurred only at the bubble base, and it was not observed on the bubble roof due to fluid shear effects caused by fast rising bubbles.

23.2.3 Bubble Motion

The rise path (defined as the trajectory of the bubble centroid) and change in orientation (defined as the angle between the bubble major axis and the vertical axis of the system) of the bubble are known to be strongly related to the bubble shape. The motion of spherical bubbles is usually rectilinear. Once the bubble becomes deformed into an oblate spheroid, instability sets in and results in a spiral or zigzag trajectory; at the same time the bubble orientation changes in such a way that the bubble major axis tends to be perpendicular to the direction of instantaneous motion (Miyagi, 1925; Haberman and Morton, 1953), although some phase shift in the orientation angle is observed for some bubble sizes (Hibino, 1969; Tsuge and Hibino, 1971). As the bubble changes from an ellipsoidal to spherical-cap shape, the radius of the spiral or the amplitude of the zigzag gradually decreases and the motion becomes rectilinear, but with rocking. The spherical-cap bubble, as its size increases, eventually rises rectilinearly.

It is generally accepted that the above oscillating motion is closely associated with vortex shedding behind the bubble. The onset and frequency of oscillation in the bubble motion can be reasonably estimated

(Fan and Tsuchiya, 1990); however, the detailed understanding of the oscillation modes has been hindered and is often inconclusive, since they strongly depend on the presence/absence of surface-active contaminants as well as the initial bubble motion. The effect of solid particles on these bubble oscillation properties is also not well understood, thus requiring more extensive investigations.

23.3 BUBBLE-INDUCED LIQUID-SOLID FLOW

A major factor contributing to the complexity of the liquid-solid flow near rising bubbles is the instability induced in the wakes behind the bubbles. Through this instability, the wake flow becomes unsteady and interactive with the external flow. The phenomenon is often characterized by a cycle of vortex formation and shedding and, for free-rising bubbles, is intimately related to the bubble oscillating movement. Thus, it is necessary to discuss the periodic/dynamic nature of the wake.

23.3.1 Wake Instability

Within a brief time after bubble injection, the bubble wake consists of a closed laminar region of a toroidal vortex ring (main portion) followed by a narrow tail. As the bubble accelerates and the wake grows in size by continuously accumulating material from outside the wake, the symmetry of the wake flow will be perturbed and eventually some wake material will be discharged. Such wake instability sets in as the bubble Reynolds number exceeds a certain critical value: for large bubbles in viscous media, the critical Re_e for wake shedding is approximately 100; for small bubbles in low-viscosity media, the critical Re_e depends on the medium purity and, for contaminated systems, the critical Re_e is approximately 200.

Besides its inherent nature, wake instability can be induced by external disturbances. Thus, wake shedding is expected to take place in a shorter time and /or at lower Reynolds numbers, and more frequently in flow systems such as gas-liquid-solid fluidized beds than in stationary systems.

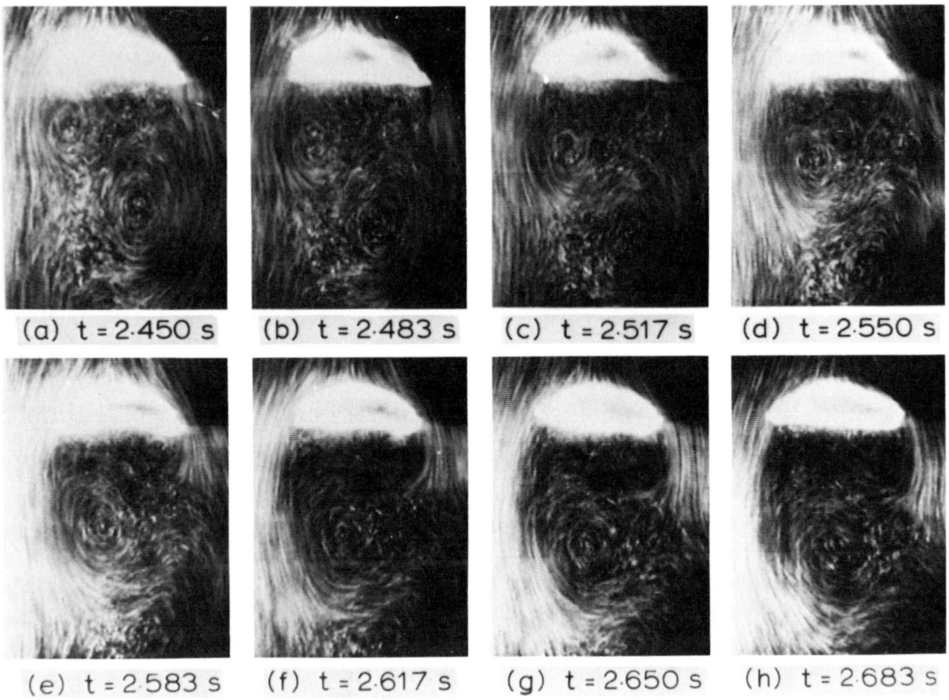

Figure 23.7 *Photographs of a circular-cap bubble and its wake in stationary water over approximately a one-half cycle of the vortex formation-shedding process. Instantaneous streamline patterns are visualized through hydrogen-bubble liquid tracer (from Fan and Tsuchiya, 1990; reproduced with permission).*

23.3.1.1 Mechanisms of Wake Formation and Shedding

The dynamics of vortical motion play a key role in wake formation and shedding. In two-dimensional systems, periodic wake shedding can take place either asymmetrically or symmetrically, though the asymmetric mode appears to be more prevalent in the real flow. The asymmetric vortex formation-shedding mechanism consists of the roll-up of a vortex sheet, the growth of a circular vortex and the cut-off of the vorticity supply to the growing vortex. Figure 23.7 shows a sequence of photographs representing instantaneous streamline patterns over approximately one-half the cycle of the vortex formation-shedding process. Hydrogen bubbles were used to trace the liquid motion around a large two-dimensional bubble in water. Note in the figure the ordered process of the formation, growth and shedding of the vortices. On the other hand, the vortex sheet instability (Crowe at al., 1991) occurring independently on each side of a large bubble is responsible for the symmetric shedding.

In three-dimensional systems, the formation-shedding mechanism consists of a cycle of buildup and release of vorticity. Plausibly, this cycle can be realized through emergence of a daughter vortex filament out of a mother vortex ring and reformation of the mother ring with the aid of a series of topological transformations such as pinching-off, disconnection and reconnection of vortex filaments.

23.3.1.2 Wake Shedding Frequency

The frequency of wake shedding behind rising bubbles is often estimated from that of bubble rocking or gyration due to the synchronization, i.e., the coincidence of frequencies between the wake shedding and bubble motion. For single bubbles in two- and three-dimensional systems at low solids holdups (< 0.4), the following trends prevail (Fan and Tsuchiya, 1990): (1) the shedding frequency is not significantly affected by the presence of solid particles; (2) at low Reynolds numbers the shedding frequency is uniquely defined due to the occurrence of only asymmetric shedding; and (3) at high Reynolds numbers two distinct shedding frequencies can be obtained for a given bubble due probably to the coexistence of asymmetric and symmetric shedding modes.

The particle effect, however, becomes appreciable for high solids holdups, large inertia particles and small bubbles. Figure 23.8 shows the relationship between the Strouhal number ($Sr_b = f_v b / v_{b\infty}$, the reduced shedding frequency) and the Reynolds number ($Re_b = \rho_l b v_{b\infty} / \mu_l$) for glass beads of four different sizes fluidized by tap water. It can be seen in the figure that, although the effect of solids holdup is small, the solids properties (particle size in this case) influence the relationship systematically; Sr_b generally increases with increasing particle size at a given Re_b. In the figure, the dashed lines represent the correlation for Sr_b-Re_b relationship in the absence of the particle effect (Tsuchiya and Fan, 1986).

Considering the particle effect, Fan and Tsuchiya (1990) extended the correlation to the following form:

$$\frac{1}{Sr_b} = \frac{s_1}{Re_b} + \frac{1}{Sr_{b0}} \tag{23.6}$$

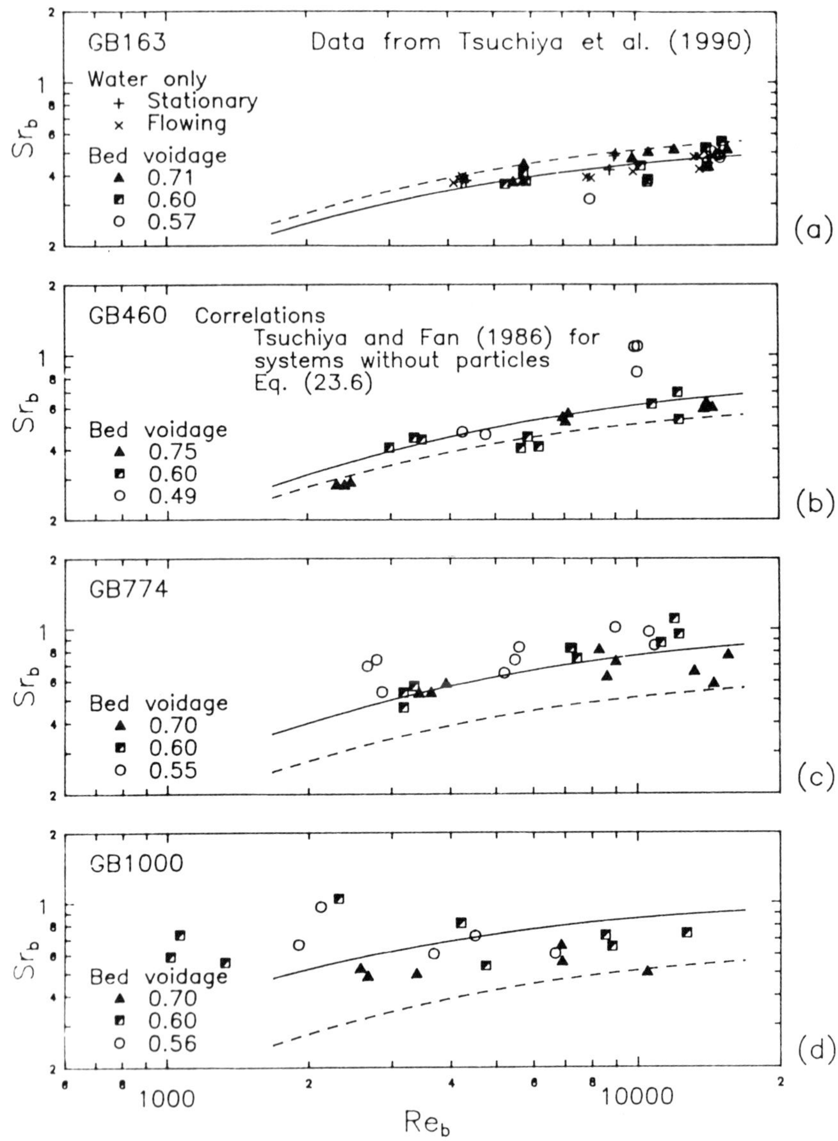

Figure 23.8 *Effects of solids properties and holdup on Sr_b–Re_b relationship for vortex shedding from single bubbles in a two-dimensional water-solid fluidized bed (from Fan and Tsuchiya, 1990).*

From the linear relation between $1/Sr_b$ and $1/Re_b$ for each type of particle, the parameters in Eq. (23.6), i.e., s_1 and Sr_{b0}, can be determined and correlated with the particle terminal velocity as

$$s_1(v_{s\infty}) = 6500 - \frac{26700 v_{s\infty}}{\tanh(v_{s\infty}/0.075)} \tag{23.6a}$$

and

$$Sr_{b0}(v_{s\infty}) = \{1.45 + 0.47 \tanh[29(0.05 - v_{s\infty})]\}^{-1} \tag{23.6b}$$

In these correlations, $v_{s\infty}$ is in m/s. Equations (23.6a,b) can be used for particles of $v_{s\infty} = 0$–0.17 m/s fluidized by a low-viscosity liquid in the bed voidage range above 0.5. Figure 23.8 includes the calculated values of Sr_b using Eq. (23.6) for the four cases, which can reasonably account for the particle effect and the variation with Re_b except at low Re_b (< 2000). A marked reduction in $v_{b\infty}$ for large $v_{s\infty}$ particles (e.g., 1 mm glass beads) and high α_s (> 0.4) is the major contribution to the observed increase in Sr_b, especially at low Re_b.

An additional effect of the particle properties is due to particle wettability. Tsutsumi et al. (1990) found that the wake-shedding frequency for glass beads coated with Teflon was smaller than that for glass beads without Teflon coating under the same fluidization conditions. The reduction of the shedding frequency is probably due to the attachment of non-wettable particles at the bubble base. Two effects may result from this particle adhesion: (1) reduction in the bubble mobility by virtue of higher bubble inertia, thus in the frequency of bubble oscillation or vortex shedding; (2) stabilization (or weakening) of the lateral outward flow of liquid along the bubble base. As the bubble size, thus, Re_b, increases, the bubble-base wobbling action can impede particle attachment (Tsutsumi et al., 1991).

23.3.2 Wake Structure

In general, the observed diversity of wake structures (Wasowski and Blass, 1988) can be described based on the dual-wake-structure concept, that is, the wake consists of a primary wake moving in close association with the bubble and a secondary wake extending far downstream. This concept is established based on the general findings that the bubble wake flow is usually characterized by the cyclic process of vortex formation

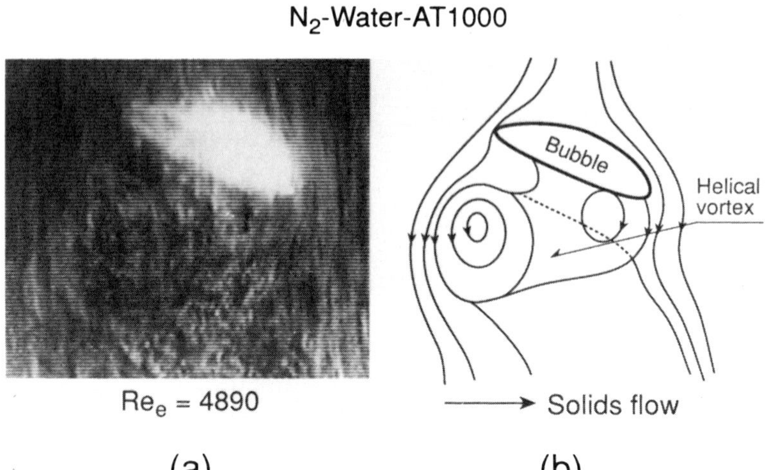

Figure 23.9 *(a) Photograph of a spherical-cap bubble and its wake (helical vortex wake) in a water-1mm acetate particle fluidized bed. (b) Schematic interpretation of the wake flow (form Miyahara et al., 1988; reproduced with permission).*

and shedding described above. A typical dynamic structure for a two-dimensional bubble is shown in Fig. 23.7.

The specific details of the wake structures, especially in three dimensions, however, are not easily described due to varying degrees of flow stability occurring over a wide range of Reynolds numbers. Under this condition, the wake structure is commonly represented according to five categories depending on the Reynolds number. These categories are: (1) a steady wake with a negligible (re)circulation region; (2) a steady wake with a well-developed circulation region followed by a laminar streamwise tail; (3) an unsteady wake with large-scale vortical structures; (4) an unsteady wake with a high degree of turbulence; and (5) a highly turbulent wake. For the wake behind gas bubbles, most observations are limited to categories (1), (2) and (3) due to instability of large bubbles. For categories (1) and (2), once vortices are formed in the primary wake, they do not shed downstream. Category (3) can be represented by several configurations depending on the bubble rise characteristics.

For ellipsoidal and spherical-cap bubbles rising with gyration motion, the wake configuration, at least in the near wake, consists of a helical vortex. Figure 23.9 shows a photograph of gyrating bubble ($Re_e < 5000$) in a water-1 mm acetate particle fluidized bed; also presented is a schematic interpretation of the flow field surrounding the bubble, including the wake. As can be seen in the figure, the structure of the wake can be represented by an asymmetric wake about the vertical axis of the bubble movement. Bubbles in this regime usually experience significant shape oscillations. The shape asymmetry and rather periodic variation in the bubble orientation may yield non-uniform flow separation along the bubble edge. This "biased" flow separation then induces locally strong vorticity generation, which eventually forms a vortex filament constantly discharged from the portion of the bubble edge with the "strongest" flow separation. Observation of dynamic motion of the separation region revealed that this strong separation region actually rotates around the bubble edge. The frequency of this rotation was found to match the bubble-gyration frequency (Miyahara et al., 1988).

For large spherical-cap bubbles rising rectilinearly, the wake configuration can be characterized as either a discharge of horseshoe-type vortex loops or an axisymmetric discharge of vorticity. In the lower range of Reynolds numbers for category (3), i.e., at Reynolds numbers slightly higher than that for the onset of wake oscillations or shedding, the horseshoe vortex loops occur most plausibly. At high Reynolds numbers ($Re_e > 5000$; see Fig. 23.10), the structure of the wake can be represented by a symmetric wake about the vertical axis of the bubble movement. As Re_e increases beyond 5000, the bubble tends to discharge vorticity symmetrically in the form of either a set of apparently discontinuous toroidal vortices (or vortex rings) or two continuous helical vortices (or bispiral vortices). Research is indeed needed for the understanding of the definitive configurations of the large bubble wake.

23.3.3 Wake Sizes

While the frequency of vortex shedding reflects that of the interaction or exchange between the wake phase and the surrounding phase, it does not reflect the extent of the exchange. One of the wake properties pertaining to the extent of this exchange is wake size.

23.3.3.1 Dynamic Variation in Primary-Wake Size

Due to its close association with a rising bubble, the primary wake

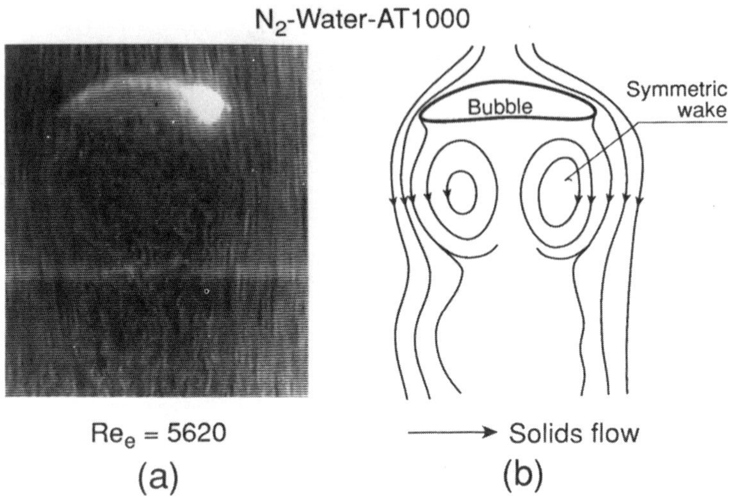

Figure 23.10 *(a) Photograph of a spherical-cap bubble and its wake (symmetric wake) in a water-1 mm acetate particle fluidized bed. (b) Schematic interpretation of the wake flow (from Miyahara et al., 1988; reproduced with permission).*

often dictates the transport phenomena in multiphase flow systems. In many practical operations of such systems, the wake exhibits unsteady variations in the internal structure. Its effective size, usually represented by the primary-wake size, must be estimated based on the mechanism of vortex formation and shedding, taking into account the temporal variation caused by wake shedding.

The primary wake generally consists of the confined turbulent wake (CTW) region and the shedding vortical wake (SVW) region. Figure 23.11 shows the relative locations of the two regions within the primary wake. The CTW region includes the stable liquid layer immediately beneath the bubble base and the chaotic turbulent region about the wake central axis. It collects some particles expelled from the SVW region due to centrifugal force and gradually discharges some particles downward along the central axis. The SVW region consists of growing vortice(s) or vortical blob(s) downstream of the bubble edge/rim and the spouting region at the center. It constantly sheds a certain volume of solids, usually asymmetrically, into the secondary wake and thus, eventually into the liquid-solid flow region further downstream.

Tsuchiya and Fan (1986) measured the dynamic variation in the

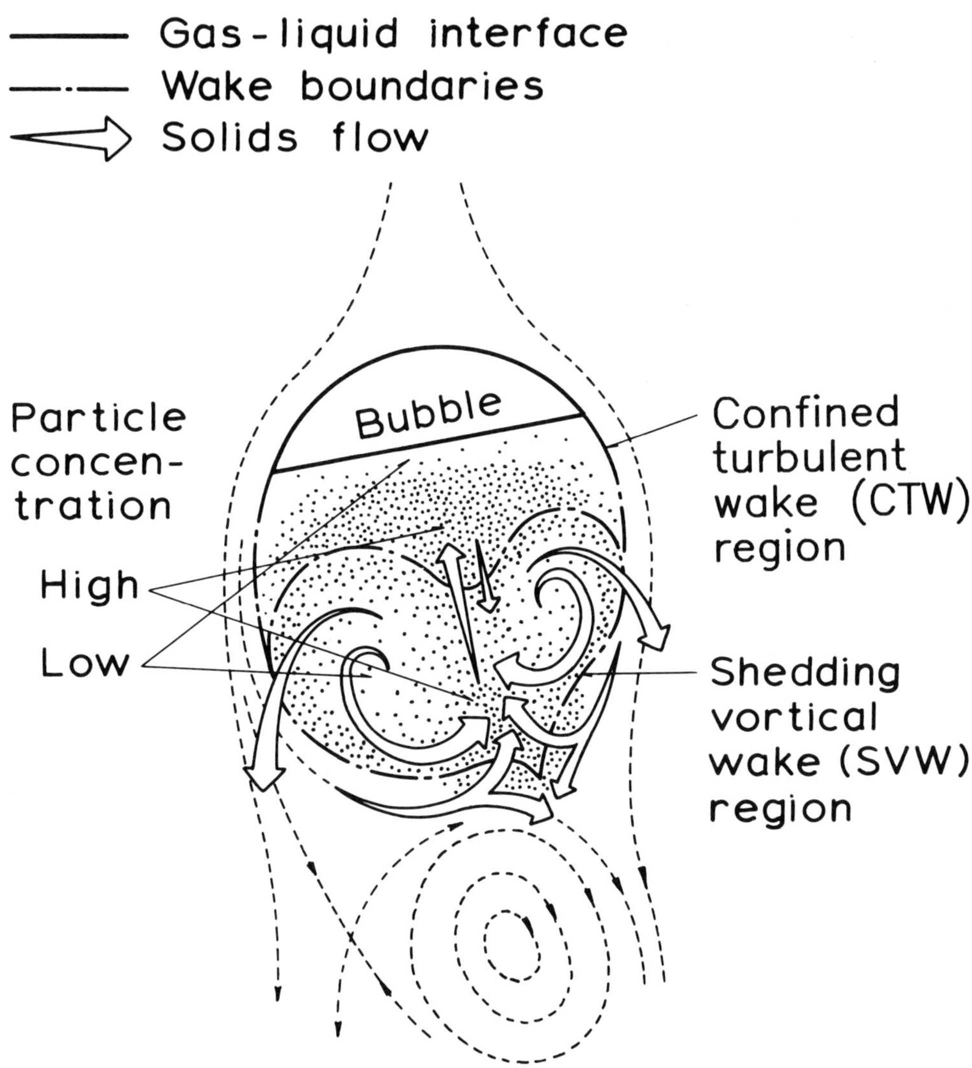

Figure 23.11 *Schematic representation of postulated primary-wake structure (from Tsuchiya and Fan, 1986).*

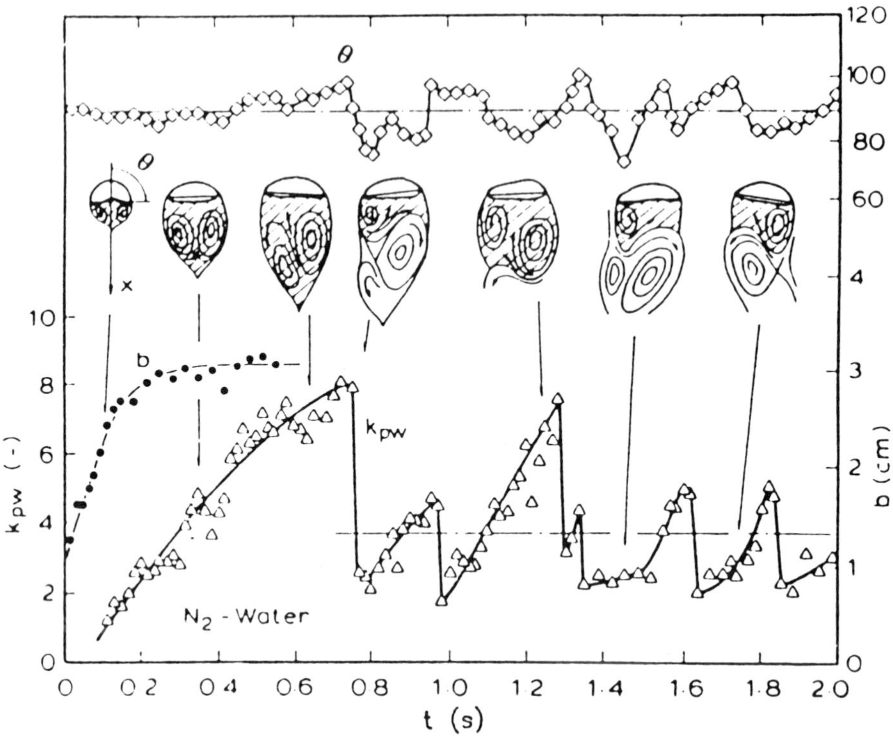

Figure 23.12 *Dynamic variations in primary-wake area and angle of attack in stationary water (from Tsuchiya and Fan, 1986).*

primary-wake size in a two-dimensional system (8 mm nominal gap thickness). They monitored a rising bubble and its wake with a video camera which moved vertically at the same speed as the bubble. Representative results are discussed in the following.

Figure 23.12 depicts the variations over the 2.0 s period right after the injection of the bubble into stationary water. Although the bubble reaches its steady-state geometry quickly (in less than 0.3 s), the bubble wake steadily grows in size (to about 8 times the bubble area) for about 0.75 s before it sheds any wake material—the steady growth period. Note that during this period, the angle of attack θ (see the figure for definition) remains almost constant at around 90°, which characterizes the symmetry of the wake flow. When θ starts deviating from 90°, the wake is considered to have initiated its shedding. Once this asymmetry sets in (see Section 23.3.1), the periodic vortex shedding and the periodic variation in the primary-wake size follow—the steady shedding period. The figure includes 5 cycles of such a variation. The size and

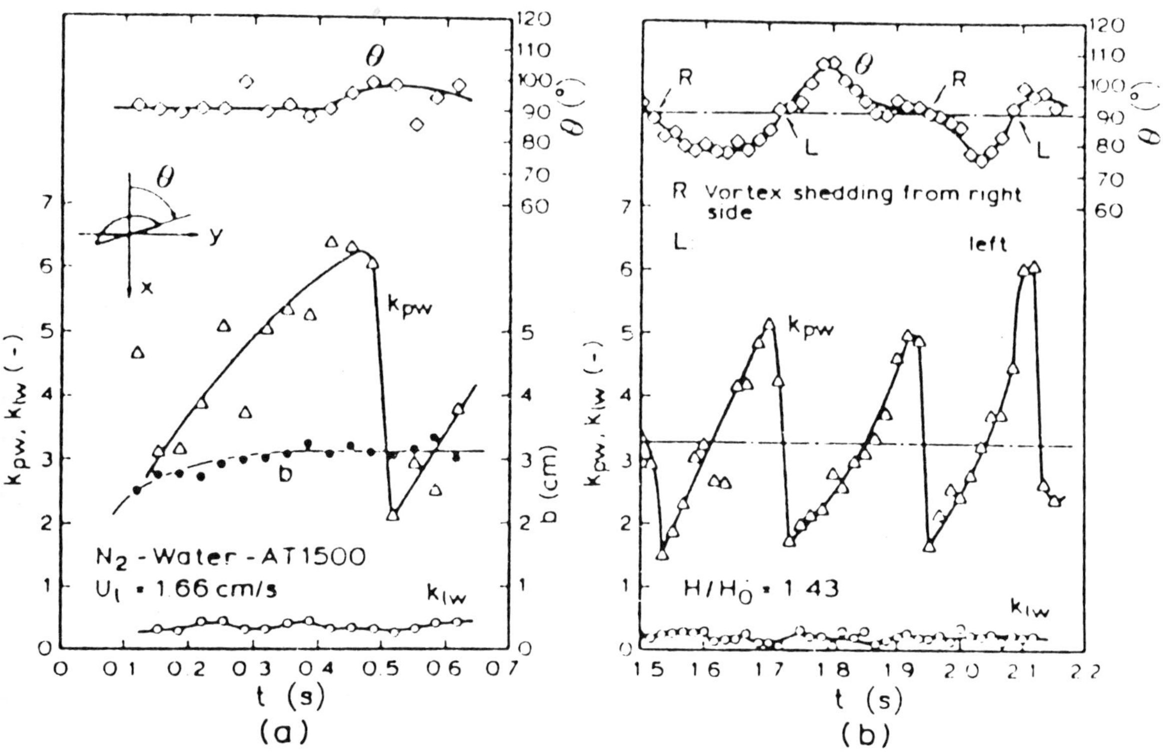

Figure 23.13 *Dynamic variations in primary-wake area, liquid-wake area and angle of attack during (a) steady growth period and (b) steady shedding period in a water-1.5 mm acetate particle fluidized bed (from Tsuchiya and Fan, 1986).*

duration of each peak are not uniform in this case since shedding does not exactly occur alternately; however, the cycle can be characterized by a saw-tooth wave function. The primary-wake area varies from about 2.3 to 5.4 times the bubble area with the mean value around 3.7 and at an average frequency of 6.2 s^{-1}. The angle of attack varies correspondingly, although the frequencies of the size and angle variations do not match exactly.

Figure 23.13 shows the increase in primary-wake area in the steady growth period for a bubble rising in a water-1.5 mm acetate particle fluidized bed. Results are similar to those in the previous case: the primary wake grows to about 6 times the bubble area, then drops in size to about twice the bubble area. Dynamic variations in the steady shedding period are shown in Fig. 23.13(b). The figure includes three cycles of purely asymmetric vortex shedding with a frequency of about 5 s^{-1}. The primary-wake area again periodically varies in the form of a saw-tooth wave function from about 1.5 to 5.5 times the bubble area. Also clearly demonstrated in this example is that the shedding frequency can be estimated by the corresponding bubble-rocking frequency if vortex shedding takes place in a clearly alternating sequence. Notice a one-to-one correspondence between the frequency in the variation in k_{pw} and that in θ as shown in the figure; a decrease in θ from 90° corresponds to vortex shedding from the right side of the bubble and an increase in θ indicates shedding from the left side.

23.3.3.2 Confined Turbulent Wake Size

Unlike the primary-wake size as a whole, the size of the stable liquid wake may not exhibit cyclic variation. Tsuchiya and Fan (1986) noted no systematic variation in k_{lw}, the size ratio of liquid wake to bubble, under the given conditions, as shown in Fig. 23.13. Some slight fluctuations shown in the figure stem from random scatter in experimental measurement and do not correspond to the variation due to vortex shedding. Although not conclusive, the size of the confined turbulent wake appears to be also constant. Thus, the dynamic variation in the size of the confined turbulent wake or the stable liquid layer can be neglected. The height of the liquid-wake layer, which is particularly important in three-phase fluidization, can be estimated based on steady wake flow patterns, more specifically by the difference in the streamlines between liquid and solids flows. The model prediction (Darton and Harrison, 1976) has been found to be satisfactory for large cap-shaped bubbles. For small bubbles, more theoretical and experimental work is required to provide a quantitative account of the role of the liquid-wake layer in three-phase fluidization.

23.3.4 Wake Solids Concentration

In bubbling systems containing solid particles, a close relationship exists between the local flow patterns and the local solids holdup in the bubble wake. For example, the local maxima and minima in the solids holdup profiles behind rising bubbles are strong indications of the specific flow characteristics prevailing in the bubble wake. The average solids concentration in the bubble wake obtained from the solids holdup distributions reflects quantitatively the extent of solids transport between the liquid-solid dispersion region and the bubble wake region.

When a bubble rises through a liquid-solid suspension, the dispersed solid particles are redistributed under the strong influence of the local flow field in the bubble wake. Immediately beneath the bubble base, the solids holdup can be higher than that in the undisturbed liquid-solid region if the sum of the upward inertial force and the form drag acting upward on the particles is stronger than the downward net gravity, and vice versa. Further beneath the bubble, the solids holdup tends to be higher due to the overall pressure defect. Locally, e.g., near the vortex center, however, the solids holdup can be lower due to centrifugal forces. Within the free shear layer, the solids holdup tends to be lower due to particle migration induced by the liquid velocity gradient across the shear layer.

23.3.4.1 Solids Holdup Distributions in Single-Bubble Systems

The local solids concentration around bubbles can be measured with an optical fiber probe (Yamazaki et al., 1978; Kitano and Fan, 1988; Kreischer et al., 1988; Tsutsumi et al., 1988) or through an imaging technique (Song, 1989; Song et al., 1991). The vertical solids holdup variation in the bubble wake obtained with an optical fiber probe roughly follows a sinusoidal function, with its amplitude attenuating, wavelength increasing, and the wave form becoming more irregular as the vertical distance from the bubble base increases. Such a variation reflects the existence of an unsteady viscous-decaying vortex configuration in the wake.

Kitano and Fan (1988), after analyzing all their measurements in a two-dimensional system, proposed the general characteristics of the near wake in terms of solids holdup as depicted in Fig. 23.14. That is, the near wake can be subdivided in terms of solids concentration into four regions: (1) a stable liquid wake region; (2) a stable solids wake region; (3) a vortex sheet region; and (4) a fluctuating solids wake region.

General similarities exist between two- and three-dimensional systems in the local solids holdup distributions around a large bubble. Figure 23.15 presents three data sets obtained in a three-dimensional fluidized bed, and the following conditions are identical among them: the gas-injection volume, Re_e of 7200, and a mean solids holdup ($\overline{\alpha}_s$) of 0.34

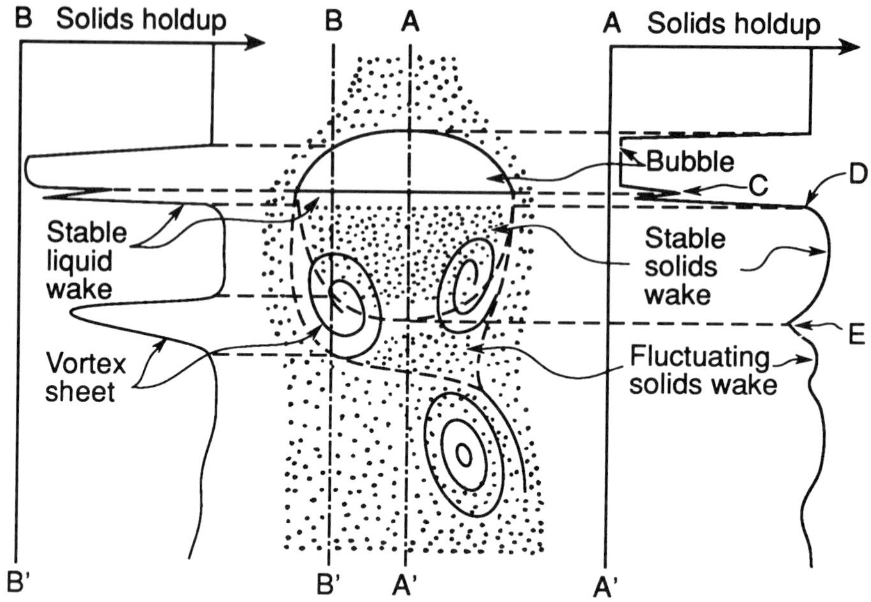

Figure 23.14 *Relationship between wake solids holdup distribution and internal structure of the primary wake (from Kitano and Fan, 1988).*

(Kreischer et al., 1988). The difference between the individual data sets lies in the position along the bubble roof (or base) where the probe intersects. Different positions result in different values of the bubble chord length (ξ) and thus, different solids holdup profiles along longitudinal axes of different lateral locations within the wake. For bubbles of a given volume (and shape), a larger chord length indicates that the probe has passed through the bubble at a location closer to the central axis.

The similarity in the near-wake solids holdup distribution shown in Figs. 23.15(a) and (b), compared to Fig. 23.15(c), suggests a relatively large, near-uniform solids concentration region about the wake central axis. Based on the measured ξ and an assumed spherical-cap geometry ($\theta_w = 100°$) of the bubble, this uniform concentration region may extend to a lateral position of $0.3b$, or 60% of the bubble basal radius, from the central axis. The near-wake solids concentration decreases, however, once the lateral position exceeds $0.4b$ as shown in Fig. 23.15(c). The behavior shown in Fig. 23.15 gives a general trend that the lateral variations in solids concentration within the stable wake regions around the wake central axis are relatively small, while the variations in the vortex sheet

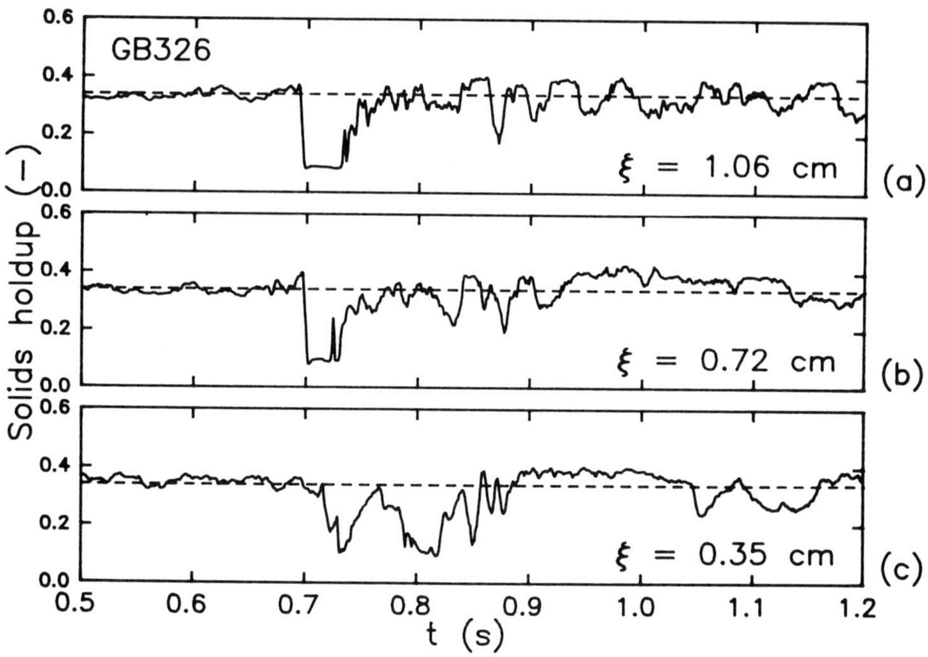

Figure 23.15 *Solids holdup distributions around single bubbles injected under identical conditions into a three-dimensional water-fluidized bed of 326 μm glass beads (from Kreischer et al., 1988).*

region and the fluctuating solids wake region are significant, in both lateral and axial directions.

23.3.4.2 Solids Holdup Distributions in Multi-Bubble Systems

The solids holdup behavior around a single bubble (especially in its immediate vicinity) provides a fundamental framework for the understanding of that in a freely bubbling system. Kreischer (1989) extended the study of single-bubble systems to multi-bubble systems of low gas holdups ($\alpha_g < 0.025$) for 326 μm glass beads. He compared typical solids holdup profiles in the following three systems (see Fig. 23.16): (a) a liquid-solid fluidized bed with 0.43 mean solids holdup; (b) a single bubble injected into an identical liquid-solid fluidized bed as (a); and (c) a free-bubbling three-phase fluidized bed with $\overline{\alpha}_s = 0.39$. In each system, parts of the signal corresponding to large bubbles are clearly evident; note, however, that in the multi-bubble system the peaks for small bubbles may not be distinguishable from other types of peaks in the distribution, such as those due to large solids concentration variations

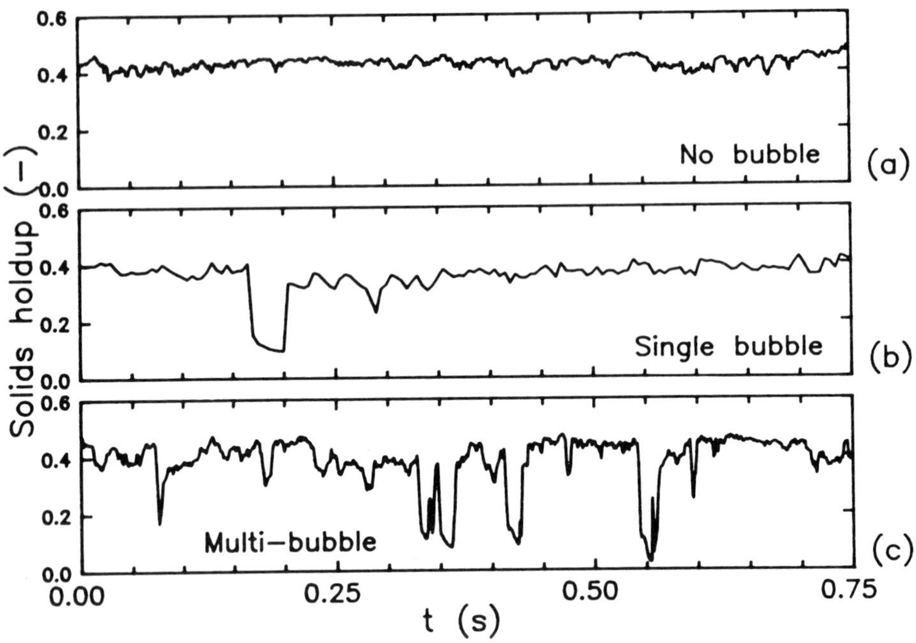

Figure 23.16 *Typical solids holdup distributions in (a) a water-326 μm glass bead fluidized bed, (b) a water-glass bead fluidized bed with a single bubble, and (c) a free-bubbling nitrogen-water-glass bead fluidized bed (from Kreischer, 1989).*

caused by vortical liquid motion. More data on solids holdups in the bubble wake are indeed needed for multi-bubble systems to clarify this point, especially for cases where bubble interactions become important.

23.3.5 Pressure Distributions in the Wake

The changes in the bubble wake flow imply variations in the pressure distribution in the bubble wake. The pressure in the bubble base region is directly involved in the force balance of the bubble, thus influencing the bubble rise velocity and shape deformation. The pressure in the primary wake indirectly affects such wake properties as wake sizes and wake solids concentration.

The existing experimental results for the local pressure around bubbles are limited to large single bubbles rising rectilinearly in two-dimensional systems (Lazarek and Littman, 1974; Bessler, 1984; Bessler and Littman, 1987; Raghunathan et al., 1990) and in three-dimensional systems (Jin and Kim, 1990). For bubbles in liquids, the pressure measurement was conducted by Lazarek and Littman (1974) for a circular-cap air bubble in stationary water contained in a 12.7 mm thick column and was extended to liquids of various viscosities by Bessler (1984) and Bessler and Littman (1987). For bubbles in liquid-solid media, Raghunathan et al. (1990) measured the pressure distribution in the wake and related the pressure distribution to the vortical structure in the primary wake.

In Lazarek and Littman's (1974) measurements, two different types of pressure minima in the near wake were detected: (1) pressure minima occurring symmetrically with respect to the bubble centerline; and (2) a sharp minimum pressure layer with quick, localized recovery immediately beneath the bubble base. The former result provides indirect evidence that circulating flow patterns, i.e., a pair of standing vortices, exist in the near wake; no vortex shedding was observed in their experiments due probably to the extreme symmetric flow field established through their special bubble injection scheme along with the dominant wall effects stabilizing the wake flow, thus delaying the onset of vortex shedding. The latter result, along with the observations of a wavy bubble base and a nearly constant drag coefficient (Bessler and Littman, 1987), indicates that the bubble base region is occupied with a liquid layer of turbulent flow. This turbulent liquid layer, characterized by a small lateral variation in pressure gradient, extends laterally as far as the bubble edges.

Figure 23.17 shows the effect of liquid viscosity on the axial pressure distributions along the wake central axis examined by Bessler and Littman (1987). The pressure is represented by the pressure coefficient defined as

$$C_p = \frac{p_T - p_\infty}{\frac{1}{2}\rho_l v_{b\infty}^2} \tag{23.7}$$

where p_∞ is the static pressure due to the liquid head above the measurement point. As seen in the figure, the pressure distributions are similar when the liquid kinematic viscosity, ν, is less than 10 mm^2/s. The bubble rising in such a medium is characterized by a circular-cap shape with a rippled base. When ν exceeds 20 mm^2/s, the bubble base no longer wobbles (Bessler and Littman, 1987) and the pressure distribution loses the characteristic of the thin turbulent liquid layer.

Measuring the pressure distributions in both water and water-solid systems, Raghunathan et al. (1990) found the distributions to be qualitatively similar between the two systems. It is noted that the wake in both systems of Raghunathan et al.'s (1990) study has the open wake

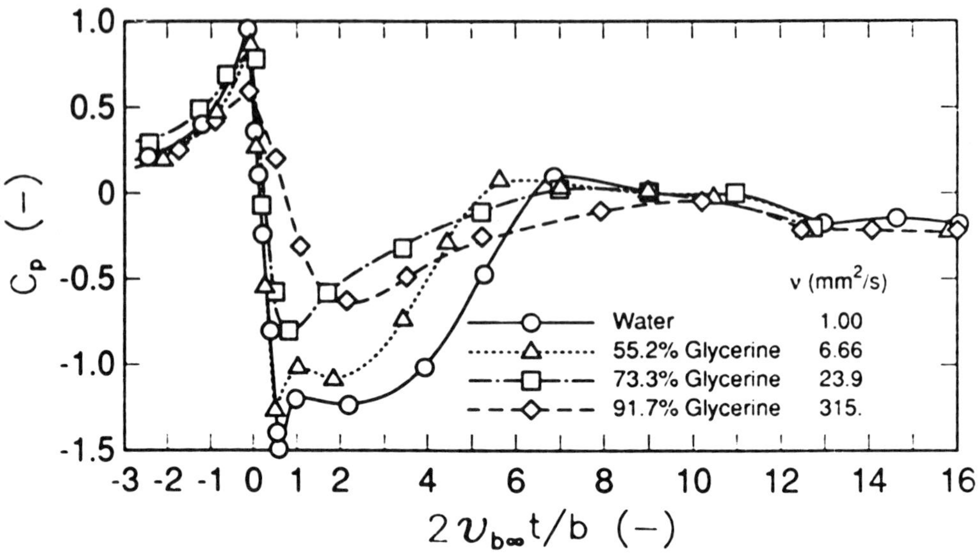

Figure 23.17 *Axial pressure distributions along the bubble central axis for bubbles in water and glycerine solutions (from Bessler and Littman, 1987).*

structures with shedding vortices in contrast to the closed wake structures of Lazarek and Littman's (1974) study. Figure 23.18 shows the pressure distributions for water-solid fluidized beds containing 163 and 326 μm glass beads. It is seen that both signals in the figure have a local minimum immediately after the global minimum. The pressure recovery is nearly complete for both particle systems, indicating a smaller secondary wake effect compared to that of the open wake for particle-free systems.

Bessler and Littman (1987) reported that for a viscous liquid the value of C_p at the bubble nose is less than unity (see Fig. 23.17) when the pressure coefficient C_p is defined by an inviscid flow condition. Raghunathan et al. (1990) found that the same behavior exists for the pressure signals for 163 μm glass beads [see Fig. 23.18(a)] when the apparent density of the liquid-solid mixture (ρ_m) [see Eq. (23.3)] instead of ρ_l is used in Eq. (23.7). The results suggest that the liquid-solid medium may be treated as a pseudo-homogeneous mixture for these particles. Raghunathan et al., however, noted that for 326 μm glass beads the value of C_p at the nose of the bubble is consistently greater than unity implying that for these particles the liquid-solid medium may not be treated as a pseudo-homogeneous mixture. This heterogeneous effects due to particles tend to

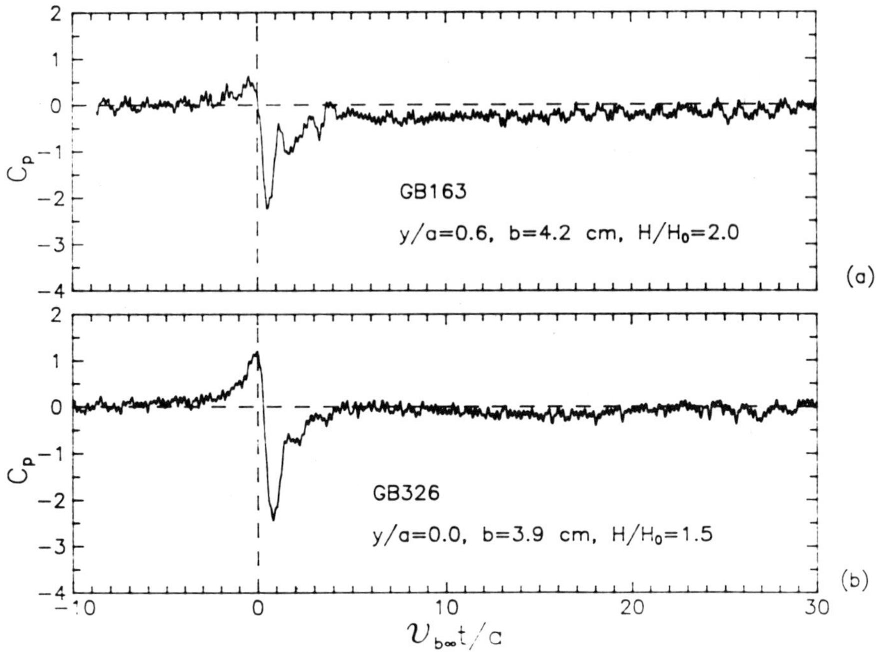

Figure 23.18 ***Axial pressure distributions around the bubble for water-solid suspensions of 163 and 326 μm glass beads (from Raghunathan et al., 1990).***

disrupt the organized motion, such as recirculation of the liquid medium.

23.3.6 Macroscopic Flow Characteristics

The behavior of bubbles, generated in the form of multiple chains, in liquid-solid media, has been studied recently by Fan et al. (1992) in a two-dimensional column (12.7 mm gap thickness). Utilizing high-speed cinematography and a computerized image processing system, they presented the dynamic characteristics of macroscopic flow structures.

At high gas velocities, they identified four flow regions based on local liquid velocity profiles. They are the central bubbly upward flow region, bubble coalescence region, vortical flow region and descending flow region. The central bubbly upward flow region is characterized by the presence of non-coalescing, dispersed bubbles arriving from the area around the bed center. In the bubble coalescence region, located at the outer boundaries of the central bubbly region, the lateral migration of bubbles near the wall towards the inner bed and intense bubble-bubble interactions (see the ensuing section) take place resulting in coalescence with minimum breakup. Fan et al. observed, with the aid of liquid-tracing and colored solids-tracing particles, that no direct exchange of the liquid and the solids across this region from either the central bubbly flow region or the vortical flow region was evident.

The vortical flow region is represented by the presence of vortices which are steadily formed near both corners of the upper free surface of the column. At low gas velocities, the vortices steadily descend along the wall; at high gas velocities, the vortices become unstable due to considerable small bubbles trapping in the vortical flow field and to the disturbance caused by ascending large bubbles traveling adjacent to the vortex core.

23.4 BUBBLE COALESCENCE AND BREAKUP

Several possible mechanisms of bubble coalescence and/or breakup have been proposed for gas-liquid-solid systems (Fan and Tsuchiya, 1990). These include: coalescence enhancement due to an increase in apparent density and viscosity of the liquid-solid mixture compared to suspending/fluidizing liquid alone in a bed of small, light particles (Massimilla et al., 1961; Ostergaard, 1966; Rigby et al., 1970); extensive bubble disintegration due to penetration of large, heavy particles through the bubble roof (e.g., Lee, 1965; Chen and Fan, 1989a); breakup due to Rayleigh-Taylor instability along the bubble roof (Henriksen and Ostergaard, 1974); and coalescence/breakup under the influence of the bubble wake flow (Otake et al., 1977; Tsuchiya et al., 1988; Miyahara et al., 1991). These mechanisms reflect the important interactive roles of bubbles, wakes and particles.

23.4.1 Bubble and Wake Interactions

Tsuchiya et al. (1988) employed a two-dimensional system to monitor the location, shape deformation, coalescence and breakup of a pair of bubbles in a stream of bubbles after the trailing bubble (denoted as bubble 2) moved to the near wake of the leading bubble (denoted as bubble 1). Figures 23.19 and 23.20 show, respectively, a series of photographs illustrating bubble coalescence and breakup phenomena in a water-778 μm activated carbon particle fluidized bed. In the photographs, the bubbles appear as bright blobs whose outline is sometimes blurred due to

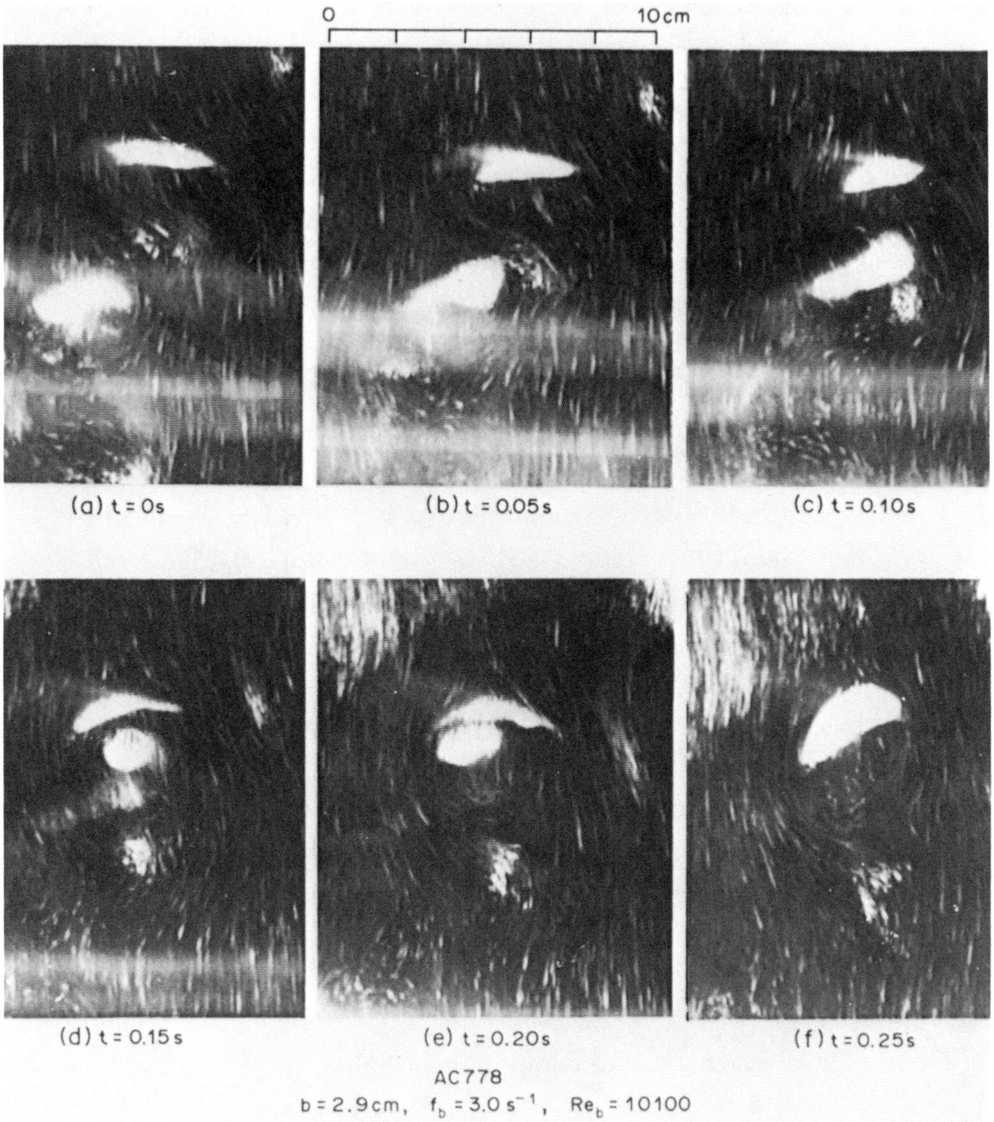

Figure 23.19 *Sequence of bubble coalescence in a two-dimensional water-778 μm activated carbon particle fluidized bed (from Tsuchiya et al., 1988; reproduced with permission).*

halation, or persistence of vision (ghost), from the previous image in the video system. The particle flow around the bubbles is represented by streaks of colored particles with the same properties.

The first three photographs in Fig. 23.19, spanning a period of 0.1 s, depict a period of rapid acceleration of bubble 2 toward bubble 1. In Fig. 23.19(a) bubble 2 is still outside the primary wake of bubble 1. Note an almost enclosed region (marked by the flow of the tracer particles) existing immediately beneath bubble 1 with a vortex (appearing less dark due to lower solids concentration) at the lower bound of this region. As bubble 2 crosses the primary-wake boundary [Fig. 23.19(b)], it deforms in the direction of its movement and accelerates rapidly. Immediately after bubble 2 moves to the primary wake of bubble 1 [Fig. 23.19(c)], the speed of bubble 2 reaches a maximum, as evidenced by an apparent tail or "ghost" behind it. The last three photographs in Fig. 23.19 show the final stage of the bubble collision-coalescence process. Bubble 2 first invades the central region of the primary wake of bubble 1 [Fig. 23.19(d)]. As the thin liquid film between the two bubbles [Fig. 23.19(e)] drains, they collide and coalesce [Fig. 23.19(f)]. Note an enclosed region in Figs. 23.19(e) and (f) whose size (relative to bubble size) is about the same (extending about one bubble breadth downstream of the bubble base) as that in Fig. 23.19(a), indicating that the relative primary-wake size is almost invariant throughout the bubble coalescence process. The time-averaged primary-wake area was reported to be 3.3 ± 1.2 times the bubble area and insensitive to Re_b over the range $1500 < Re_b < 8150$ (Tsuchiya and Fan, 1986).

Figure 23.20 shows the phenomenon of splitting of bubble 2 by the shear flow created by bubble 1. In this case, initially bubble 2 has not reached the primary wake of bubble 1 nor has it overlapped with bubble 1 in the vertical direction [Fig. 23.20(a)]. Splitting of bubble 2 is triggered by lateral stretching of a part of bubble 2 around the left edge [Figs. 23.20(a) and (b)], followed by flattening [Fig. 23.20(b)], indentation [Fig. 23.20(c)], necking [Fig. 23.20(d)] and eventual breakage [Figs. 23.20(e) and (f)]. Very similar observations of the bubble coalescence and breakup processes through the wake effects were reported by Miyahara et al. (1991) in a three-dimensional system.

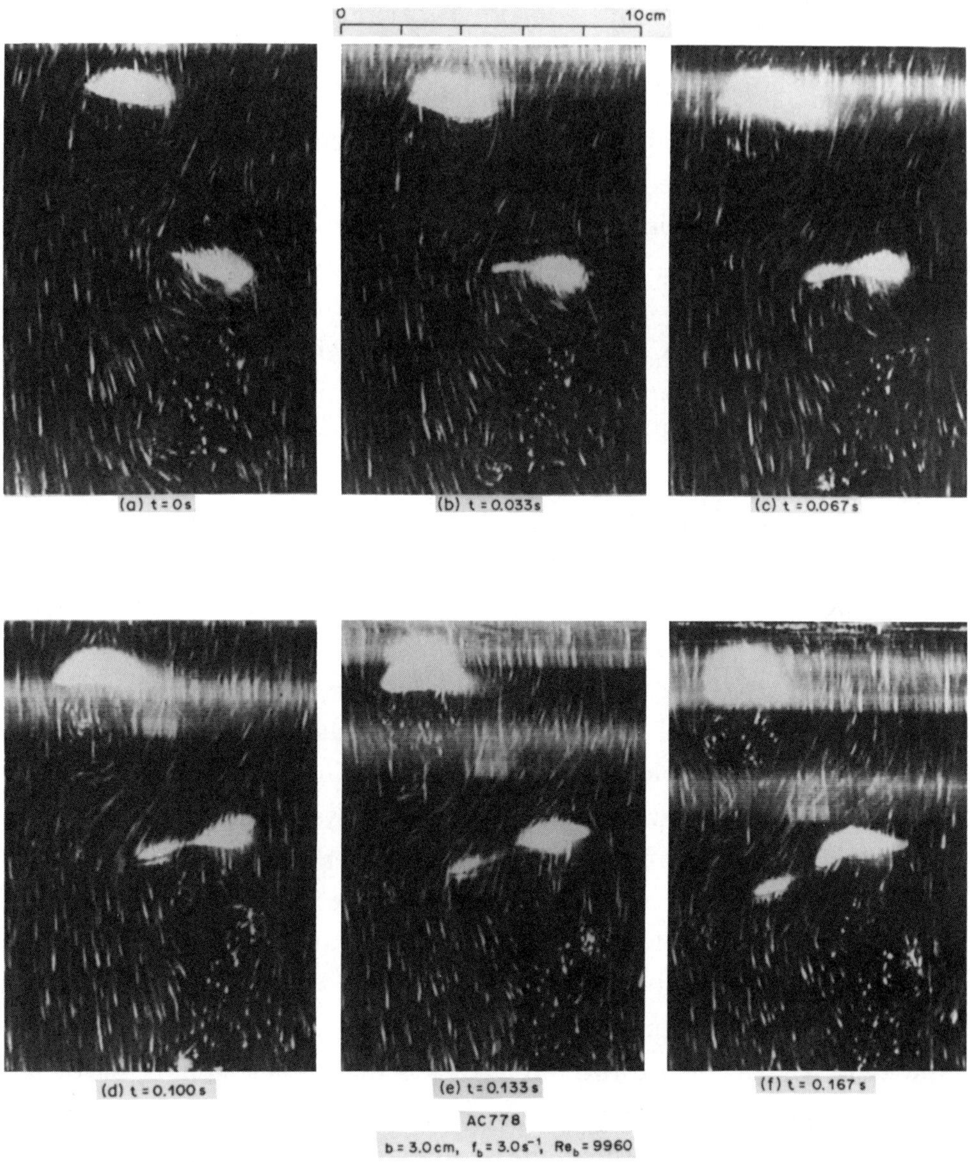

Figure 23.20 *Sequence of bubble breakup in a two-dimensional water-778 μm activated carbon particle fluidized bed (from Tsuchiya et al., 1988; reproduced with permission).*

23.4.2 Particle and Bubble-Wake Interactions

The basic mechanism of bubble breakup due to a particle-bubble collision can be most readily described by considering a spherical particle colliding with a spherical-cap bubble (Chen and Fan, 1989a, 1989b). Particle surface properties, e.g., wettable or non-wettable, affect bubble breakage. Chen and Fan analyzed the collision between the particle and the bubble in terms of the virtual masses of the both objects. For a gas bubble moving in a liquid medium, the real mass of the bubble is negligible; its virtual mass includes the mass of the bubble-wake and the added mass of the liquid induced by the bubble and wake motion. Based on the force balances on the particle and the bubble during collision, Chen and Fan (1989a) obtained a set of theoretical criteria for particle penetration. The criteria stem from the following physical arguments. If a particle does not penetrate a bubble, the depth of particle indentation will increase with time, reach a maximum and then decrease. Therefore, in order for a particle not to penetrate a bubble, the point of maximum indentation depth, which corresponds to zero relative velocity, should be located within the bubble. The criteria imply three conditions for particle penetration. They are (1) the particle always accelerates downward with respect to the bubble and the point of zero relative velocity does not exist; (2) when the particle inertia is sufficient, it penetrates deep enough into the bubble that the liquid head behind the particle forces it through the bubble; and (3) the maximum penetration of the particle is greater than the bubble height. If any one of these conditions is attained, the particle will penetrate the bubble. It is important to note, however, that particle penetration through the bubble is considered as a necessary but not sufficient condition for bubble breakage. Chen and Fan (1989a) claimed that the bubble will break only if the penetrating particle has a diameter greater than the height of the bubble being penetrated, which can be assumed to deform into a doughnut shape.

A photographic representation of bubble and particle interaction is shown in Figs. 23.21 and 23.22. Figures 23.21(a)–(d) show the collision between the bubble and a 3 mm glass sphere, which is a medium size of particle. In Figs. 23.21(b) and (c), the penetration of the particle is clearly depicted, showing a funnel-shaped hole in the bubble. In Fig. 23.21(d), after the particle penetration, the funnel-shaped hole in the bubble starts to merge with the bubble base and the penetrated bubble

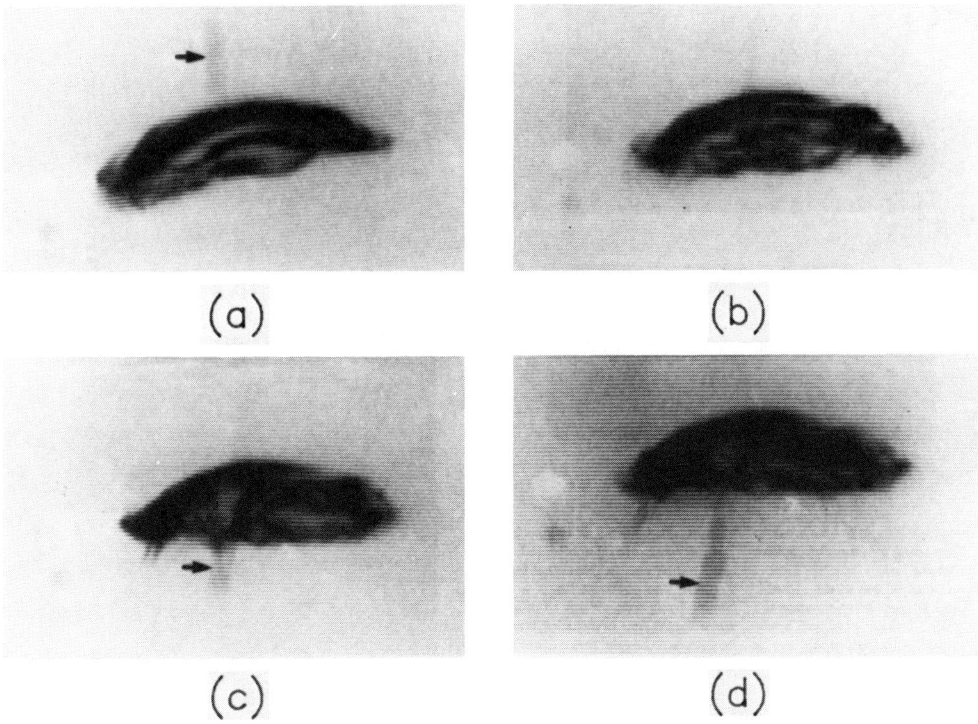

Figure 23.21 *Photographs of collision sequence of a 3-mm glass sphere and a spherical-cap bubble (d_e = 22.2 mm); the particle, indicated by the arrow, penetrates but does not break the bubble (from Chen and Fan, 1989a; reproduced with permission).*

eventually recovers to its original shape. The particle is able to penetrate but is unable to break the bubble. Note that the particle pathline in Fig. 23.21(b) is deflected only very slightly by the bubble. Similar results of particle penetration without bubble breakup were also observed when using a 4 mm glass sphere and a 7 mm alumina sphere.

Figures 23.22(a)–(e) show the collision between the bubble and an 11.5 mm iron oxide particle, a large and heavy particle. In Fig. 23.22(c), the bubble deformation into a doughnut shape after particle penetration is clearly depicted. It also shows that a small portion of the doughnut-shaped bubble is torn off by the particle because the collision is not perfectly centered. The portion torn off immediately breaks into several small bubbles below the mother bubble, as depicted in Fig. 23.22(d). This occurrence of a small portion of a bubble tearing off due to an off-center collision is possible for all size particles; however, it should not be considered as a major bubble breakup event since the size of the mother

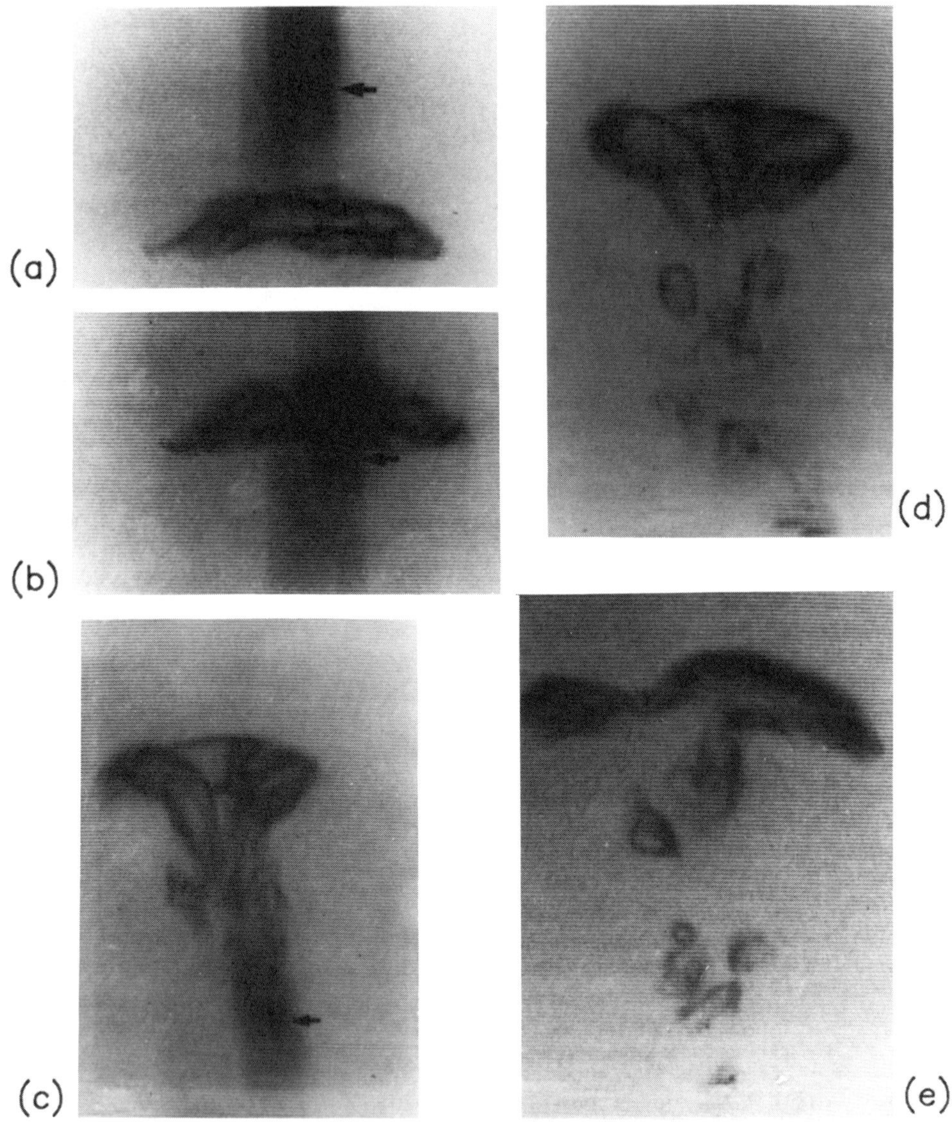

Figure 23.22 *Photographs of collision sequence of a 11.5-mm iron oxide particle and a spherical-cap bubble (d_e = 22.4mm); the particle, indicated by the arrow, penetrates and breaks the bubble (from Chen and Fan, 1989a; reproduced with permission).*

bubble remains almost unchanged. The major bubble breakup is through the mechanism of necking of the doughnut-shaped bubble, as depicted in Figs. 23.22(d) and (e). In this case, as shown, the major portion of the doughnut-shaped bubble splits into two bubbles side by side. The particle penetration or bubble breakup shown in Figs. 23.21 and 23.22 was accurately predicted by Chen and Fan's (1989a) theory discussed above.

23.5 CONCLUSIONS

The success in modeling, simulation and design of a gas-liquid-solid flow system requires a fundamental understanding of interactive behavior between the gas bubbles and the surrounding liquid-solid suspension. Though the three-phase system is often being treated as a pseudo-two-phase system consisting of a gas phase and a liquid-solid phase, the latter phase may exhibit homogeneous or heterogeneous characteristics depending on the relative magnitudes of solid density to liquid density and of solid particle size to bubble size. Solid properties including wettability of particle surface and particle shape play an important role in the interaction among phases.

The rise characteristics of bubbles in a liquid-solid medium such as bubble shape depends more on the size of an individual rising bubble and the surrounding liquid properties, and less on the surrounding solid properties. The bubble aspect ratio, which represents the extent of bubble flattening, can be correlated with the bubble Reynolds number for a given liquid. Some particle effects, however, need to be taken into account: bubbles at lower Reynolds numbers tend to flatten to a greater extent in the presence of solid particles; at higher Reynolds numbers the bubble flattening is less extensive for large particles; both the solids holdup and the particle wettability appear to have a minimal influence on the bubble aspect ratio.

The effect of solid particles on the bubble rise velocity can be accounted for by considering the homogeneous or heterogeneous behavior. The homogeneous behavior prevails when the ratio of the particle diameter to bubble diameter is small; the bubble rise velocity can be reasonably predicted from the correlation for Newtonian liquids with apparent density and viscosity of the liquid-solid medium. At high solids holdups, the medium exhibits non-Newtonian properties. The heterogeneous behavior prevails when the ratio of the particle diameter to bubble diameter is large. In this situation, bubbles of smaller sizes move through the interstices of surrounding fluidized particles and the impact of particles upon the bubble roof becomes frequent. Quantitative demarcation between the homogeneous and heterogeneous behavior, however, is yet to be clarified. The effect of particle wettability exists on the rise of small bubbles. The bubble rise velocity can be significantly reduced due to the attachment of non-wettable particles on the bubble surface.

The flow of a liquid-solid mixture induced by a rising bubble exhibits a variety of flow characteristics especially behind the bubble. The formation and shedding of the wake behind the bubble results from instability induced in the liquid-solid flow near the bubble surface. While the frequency of the wake formation-shedding cycle depends mainly on the bubble Reynolds number, it is affected to some extent by the particle inertia and particle wettability (though to less extent by the solids holdup). The structure and the effective size of the bubble wake are strongly influenced by the wake shedding phenomenon; the direct influence of solid particles on these wake properties appears to be secondary.

Other important flow characteristics include the local gradients in solids holdup and pressure in the wake. When a bubble rises through a liquid-solid suspension, the dispersed particles are redistributed and the static pressure profile is disturbed under a strong influence of the local flow field in the bubble wake. The resulting axial solids holdup distribution behind the bubble can be characterized approximately by a sinusoidal function, with its amplitude attenuating, wavelength increasing, and the wave form becoming more irregular with increasing distance below the bubble base. The observed pressure distribution is characterized by pressure minima located either immediately beneath the bubble base or around the vortex center. Increasing liquid viscosity reduces the extent of these pressure minima. The solid particles renders the pressure recovery occurring closer to the bubble base, thus reducing the secondary wake effect.

Among various mechanisms of bubble coalescence and/or breakup, bubble-bubble or bubble-wake interactions signify the symbiotic relationship between the rising bubbles and their wakes. Bubble coalescence/collision is essentially the process of entrapment of the trailing bubble into the effective/primary wake of the leading bubble. The process is controlled by the strength of the bubble-induced liquid-solid (mainly liquid) flow and thus, by the extent of pressure defect in the bubble wake.

Bubble breakup can be caused by the shear flow generated in the liquid-solid medium flowing near the leading bubble surface. Bubble breakage can also be caused by the interaction between particles and bubbles. The phenomenon of bubble disintegration due to penetration of large and/or heavy particles can be described using Newton's second law. The breakup criterion can be expressed in terms of virtual masses of the particle and bubble, size ratio of the particle to bubble and particle wettability.

23.6 FUTURE WORK

Besides the research needs indicated in some of the sections given above (Sections 23.2.3, 23.3.2, 23.3.3.2 and 23.5), further investigations needed include:

(1) incorporating the fluid dynamic behavior associated with the local interactions between bubbles and the surrounding medium discussed in this chapter into the mechanistic description of the overall behavior of multi-bubble systems,

(2) elucidating the effect of the surrounding medium properties including the viscosity and non-Newtonian nature of the liquid phase on the bubble wake characteristics as well as the solid properties, and

(3) illustrating the bubble-wake behavior when other interaction effects such as the wall proximity effect prevails.

ACKNOWLEDGEMENT

The assistance of Mr. J.-W. Tzeng in the preparation of this chapter is gratefully acknowledged. This work was supported by the Pittsburgh Energy Technology Center of the U.S. Department of Energy through Argonne National Laboratory under Contract # 83052401.

NOMENCLATURE

a	half bubble width (or breadth)
b	bubble breadth
C_p	pressure coefficient defined by Eq. (23.7)

c	empirical parameter introduced in accounting for the deviation of the surface tension term in Eq. (23.2b) from the observed bubble rise velocity: $c = 1.2$ for monocomponent liquids or 1.4 for multicomponent liquids
d_e	equivalent bubble diameter: diameter of a sphere having the same volume as the bubble in three dimensions (3D) or diameter of a circle having the same area as the bubble in two dimensions (2D)
d_s	particle diameter
Eo	Eötvös number (or Bond number) $= g\rho_l d_e^2/\sigma$
f_b	bubble-formation frequency
f_v	shedding frequency of vortex pairs from a bubble
g	gravitational acceleration
H	effective height of bed expansion
H_0	static bed height
h	bubble height
$(\frac{h}{b})_{cap}$	aspect ratio of a spherical-/circular-cap bubble
K_b	parameter accounting for surface conditions (interfacial mobility and/or boundary-layer formation) of a rising spherical bubble defined in Eq. (23.2a): $K_b = 12$, 18 and 36 for Hadamard-Rybczynski, Stokes and Levich theories, respectively
K_{bs}	empirical parameter defined in Eq. (23.5)
k_{lw}	ratio of stable liquid wake size to bubble size (size: volume in 3D or area in 2D)
k_{pw}	ratio of primary-wake size to bubble size
Mo	Morton number (or property number) $= g\mu_l^4/\rho_l\sigma^3$
m	parameter accounting for system dimensions: $m = 3$ in 3D or 2 in 2D
n	empirical parameter accounting for system impurity [see Eq. (23.2)]: $n = 1.6$ for purified system or 0.8 for contaminated system ($0.8 < n < 1.6$ for partially contaminated system); empirical parameter defined in Eq. (23.5)
p_T	total pressure
p_∞	static pressure due to liquid head
Re_b	bubble Reynolds number based on the bubble breadth $= bv_{b\infty}/\nu$
Re_e	bubble Reynolds number based on the equivalent bubble diameter $= d_e v_{b\infty}/\nu$
Sr_b	Strouhal number for vortex-pair shedding based on the bubble breadth $= f_v b/v_{b\infty}$

Sr_{b0}	asymptotic value of Sr_b as Re_b approaches infinity
Ta	Tadaki number $= Re_e Mo^{0.23}$
Ta_1	Tadaki number demarcating spherical/circular and ellipsoidal/elliptical regimes
Ta_2	Tadaki number demarcating ellipsoidal/elliptical and spherical-/circular-cap regimes
t	time elapsed from the moment of bubble injection; or time elapsed from arbitrary moment
U_l	superficial liquid velocity
$v_{b\infty}$	bubble terminal velocity: bubble rise velocity relative to the liquid phase
v_{b1}	predicted $v_{b\infty}$ in the spherical regime [Eq. (23.2a)]
v_{b2}	predicted $v_{b\infty}$ in the Mendelson regime [Eq. (23.2b)]
$v_{s\infty}$	particle terminal velocity
x	vertical downward distance from the bubble base
y	horizontal right-hand distance from the center of the bubble base

Greek Letters

α_g	gas holdup
α_l	liquid holdup
α_s	solids holdup
$\overline{\alpha}_s$	mean solids holdup
θ	angle of attack
θ_w	bubble included angle
μ_l	liquid viscosity
μ_m	apparent viscosity of a liquid-solid mixture
ν	liquid kinematic viscosity $= \mu_l/\rho_l$
ξ	bubble chord length
ρ_l	liquid density
ρ_m	apparent density of a liquid-solid mixture defined by Eq. (23.3)
ρ_s	solid density
σ	interfacial tension between the gas and liquid phases

REFERENCES

Bessler, W. F. and H. Littman, "Experimental Studies of Wakes behind Circularly Capped Bubbles," *J. Fluid Mech.* **185**, 137–151 (1987).

Bhaga, D. and M. E. Weber, "Bubbles in Viscous Liquids: Shapes, Wakes and Velocities," *J. Fluid Mech.* **105**, 61–85 (1981).

Chen, Y.-M. and L.-S. Fan, "Bubble Breakage Mechanisms due to Collision with a Particle in a Liquid Medium," *Chem. Eng. Sci.* **44**, 117–132 (1989a).

Chen, Y.-M. and L.-S. Fan, "Bubble Breakage due to Particle Collision in a Liquid Medium: Particle Wettability Effects," *Chem. Eng. Sci.* **44**, 2762–2767 (1989b).

Clift, R., J. R. Grace, and M. E. Weber, *Bubbles, Drops, and Particles*, Academic Press, New York (1978).

Crowe, C. T., J. N. Chung, and T. R. Troutt, "Particle Dispersion by Organized Turbulent Structures," in *Particulate Two-Phase Flow* (M. C. Roco, ed.), Chap. 18, Butterworth-Heinemann, Stoneham, MA (1991).

Darton, R. C., "The Physical Behaviour of Three-Phase Fluidized Beds," in *Fluidization*, 2nd ed. (J. F. Davidson, R. Clift, and D. Harrison, eds.), Chap. 15, pp. 495–528, Academic Press, London (1985).

Darton, R. C. and D. Harrison, "Bubble Wake Structure in Three-Phase Fluidization," in *Fluidization Technology* (D. L. Keairns ed.), pp. 399–403, Hemisphere (1976).

Dayan, A. and S. Zalmanovich, "Axial Dispersion and Entrainment of Particles in Wakes of Bubbles," *Chem. Eng. Sci.* **37**, 1253–1257 (1982).

El-Temtamy, S. A. and N. Epstein, "Rise Velocities of Large Single Two-Dimensional and Three-Dimensional Gas Bubbles in Liquids and in Liquid Fluidized Beds," *Chem. Eng. J.* **19**, 153–156 (1980).

Fan, L.-S., *Gas-Liquid-Solid Fluidization Engineering,* Butterworth Publishers, MA (1989).

Fan, L.-S. and K. Tsuchiya, *Bubble Wake Dynamics in Liquids and Liquid-Solid Suspensions,* Butterworth-Heinemann, Stoneham, MA (1990).

Fan, L.-S., J.-W. Tzeng, and H. T. Bi, "Macroscopic Flow Characteristics in Three-Phase Fluidization," to be presented at the Seventh International Conference on Fluidization, Broadbeach, Australia, May 3–8 (1992).

Grace, J. R., "Shapes and Velocities of Bubbles Rising in Infinite Liquids," *Trans. Inst. Chem. Engrs* **51**, 116–120 (1973).

Grace, J. R., T. Wairegi, and T. H. Nguyen, "Shapes and Velocities of Single Drops and Bubbles Moving Freely through Immiscible Liquids," *Trans. Inst. Chem. Engrs.* **54**, 167–173 (1976).

Haberman, W. L. and R. K. Morton, "An Experimental Investigation of the Drag and Shape of Air Bubbles Rising in Various Liquids," *David W. Taylor Model Basin* Report 802, Navy Dept., Washington, DC (1953).

Henriksen, H. K. and K. Ostergaard, "Characteristics of Large Two-Dimensional Air Bubbles in Liquids and in Three-Phase Fluidized Beds," *Chem. Eng. J.* **7**, 141–146 (1974).

Hibino, S., "The Behavior of Single Bubbles," in *Advances in Chemical Engineering 3: Dynamics of Bubbles and Drops* (Soc. Chem. Eng. Japan, eds.), Chap. 2, pp. 3–26, Nikkan Kogyo Shinbun, Tokyo, Japan (1969).

Jang, C.-S., "Hydrodynamics of Liquid-Solid Fluidization," M.S. Thesis, Ohio State Univ., Columbus, OH (1989).

Jean, R.-H. and L.-S. Fan, "Rise Velocity and Gas-Liquid Mass Transfer of a Single Large Bubble in Liquids and Liquid-Solid Fluidized Beds," *Chem. Eng. Sci.* **45**, 1057–1070 (1990).

Jin, G. T. and S. D. Kim, "Bubble Properties and Pressure Fluctuations of a Single Bubble in Water," *J. Chem. Eng. Japan*, **23**, 523–530 (1990).

Kitano, K. and L.-S. Fan, "Near-Wake Structure of a Single Gas Bubble in a Two-Dimensional Liquid-Solid Fluidized Bed: Solids Holdup," *Chem. Eng. Sci.* **43**, 1355–1361 (1988).

Kreischer, B. E., "Bubble and Bubble Wake Characteristics in a Gas-Liquid-Solid Fluidized Bed," Ph.D. Dissertation, Ohio State Univ., Columbus, OH (1989).

Kreischer, B. E., H. Moritomi, and L.-S. Fan, "Wake Solids Holdup Characteristics behind a Single Bubble in a Three-Dimensional Liquid-Solid Fluidized Bed," *Int. J. Multiphase Flow* **16**, 187–200 (1990); Paper (22o) presented at AIChE Ann. Mtg, Washington, DC, Nov. 27–Dec. 2 (1988).

Lazarek, G. M. and H. Littman, "The Pressure Field due to a Large Circular Capped Air Bubble Rising in Water," *J. Fluid Mech.* **66**, 673–687 (1974).

Lee, J. C., "Discussion in the Paper of Adlington and Thompson," in *Proc. 3rd European Symp. Chem. React. Eng.* p. 211, Pergamon Press, Oxford (1965).

Massimilla, L., A. Solomando, and E. Squillace, "Gas Dispersion in Solid-Liquid Fluidized Beds," *Brit. Chem. Eng.* **6**, 232-239 (1961).

Miyagi, O., "The Motion of an Air Bubble Rising in Water," *Phil. Mag.* (Ser. 6) **50**, 112–140 (1925).

Miyahara, T., K. Tsuchiya, and L.-S. Fan, "Wake Properties of a Single Gas Bubble in a Three-Dimensional Liquid-Solid Fluidized Bed," *Int. J. Multiphase Flow* **14**, 749–763 (1988).

Miyahara, T., K. Tsuchiya, and L.-S. Fan, "Effect of Turbulent Wake on Bubble-Bubble Interaction in a Gas-Liquid-Solid Fluidized Bed," *Chem. Eng. Sci.* **46**, 2368-2373 (1991).

Ostergaard, K., "Flow Phenomena of Three-Phase (Gas-Liquid-Solid) Fluidized Beds," *AIChE Symp. Ser. No. 128* **69**, 28–29 (1973).

Ostergaard, K., "On the Growth of Air Bubbles Formed at a Single Orifice in a Water Fluidized Bed," *Chem. Eng. Sci.* **21**, 470–472 (1966).

Otake, T., S. Tone, K. Nakao, and Y. Mitsuhashi, "Coalescence and Breakup of Bubbles in Liquids," *Chem. Eng. Sci.* **32**, 377–383 (1977).

Raghunathan, K., S. Kumar, and L.-S. Fan, "Pressure Distribution and Vortical Structure in the Wake behind Gas Bubbles in Liquid and Liquid-Solid Systems," *Int. J. Multiphase Flow* (in press, 1991); paper (309i) presented at AIChE Ann. Mtg, Chicago, IL, Nov. 11-16 (1990).

Rigby, G. R., G. P. Van Block Land, W. H. Park, and C. E. Capes, "Properties of Bubbles in Three PHase Fluidized Beds as Measured by an Electroresistivity Probe," *Chem. Eng. Sci.* **25**, 1729–1741 (1970).

Song, G.-H., "Hydrodynamics and Interfacial Gas-Liquid Mass Transfer of Gas-Liquid-Solid Fluidized Beds," Ph.D. Dissertation, Ohio State Univ., Columbus, OH (1989).

Song, G.-H., K. Tsuchiya, and L.-S. Fan, "Image Processing Technique for Measurement of Solids Holdup in Near Wake behind a Single Bubble in a Liquid-Solid Fluidized Bed," *Chem. Eng. Sci.* (in press, 1991).

Stewart, P. S. B. and J. F. Davidson, "Three-Phase Fluidization: Water, Particles and Air," *Chem. Eng. Sci.* **19**, 319–322 (1964).

Tsuchiya, K. and L.-S. Fan, "Near-Wake Structure of a Single Gas Bubble in a Two-Dimensional Liquid-Solid Fluidized Bed: Vortex Shedding and Wake Size Variation," *Chem. Eng. Sci.* **43**, 1167–1181 (1988); Paper (10b) presented at AIChE Ann. Mtg, Miami Beach, FL, Nov. 2–7 (1986).

Tsuchiya, K., T. Miyahara, and L.-S. Fan, "Visualization of Bubble-Wake Interactions for a Stream of Bubbles in a Two-Dimensional Liquid-Solid Fluidized Bed," *Int. J. Multiphase Flow* **15**, 35–49 (1989); in *Fluidization'88 Science and Technology* (M. Kwauk and D. Kunii, eds.), pp. 204–215, 3rd China-Japan Symp., Beijing, China, Sep. 12–15, Science Press (1988).

Tsuchiya, K., G.-H. Song, and L.-S. Fan, "Effects of Particle Properties on Bubble Rise and Wake in a Two-Dimensional Liquid-Solid Fluidized Bed," *Chem. Eng. Sci.* **45**, 1429–1434 (1990).

Tsuge, H. and S. Hibino, "The Motion of Single Gas Bubbles Rising in Various Liquids," *Kagaku Kogaku* **35**, 65–71 (1971).

Tsutsumi, A., K. Uebo, and K. Yoshida, "Wake Behavior in Three-Phase Reactors," Paper presented at 4th Pacific Area Chem. Eng. Cong. (PACHEC), Acapulco, Mexico (1988).

Tsutsumi, A., J.-Y. Nieh, and L.-S. Fan, "Particle Wettability Effects on Bubble Wake Dynamics in Gas-Liquid-Solid Fluidization," *Chem. Eng. Sci.* **46**, 2381-2384 (1991); Paper (278d) presented at AIChE Ann. Mtg, Chicago, IL, Nov. 11–16 (1990).

Tustsumi, A., A. G. Dastidar, and L.-S. Fan, "Some Remarks on Hydrodynamics of Gas-Liquid-Solid Fluidization with Non-wettable Particles," AIChE J., **37**, 951-952 (1991).

Vakhrushev, I. A. and G. I. Efremov, *Chem. Technol. Fuels Oils (USSR)* **5/6**, 376–379 (1970); Cited by Clift et al. (1978).

Wasowski, T. and E. Blass, "Wake Phenomena behind Solid and Liquid Particles," *Int. Chem. Eng.*, **28**, 593-607 (1988).

Yamazaki, M., T. Sugino, and M. Jinbo, "On Voidage Distribution in Dense Phase in a Three-Phase Fluidized Bed," in *Proc. Soc. Chem. Eng. Japan 43rd Ann. Mtg*, pp. A207–A208 (1978).

CHAPTER 24

LIQUEFACTION AND SOLIDIFICATION

H. Kytomaa

24.1 INTRODUCTION

Over the past few decades, all industrial processes have continually been pushed by energy limitations and increasingly demanding applications to handle particulate materials at higher solids concentrations, be it in transporting mineral ores in a pipeline, or in injection molding of highly loaded composites. As a result, these processes have become faced with the oddities associated with concentrated solid-liquid mixtures. In working with such materials, engineers have observed that they are highly sensitive to their handling history. A mixture may flow readily, and after a brief pause in the process at hand, it may take on the character of a solid fully plugging the pipe containing it. Conversely, the porous medium that makes-up chromatographic columns has been observed to lose its structural integrity and collapse to a fraction of its original volume, wreaking havoc with the on-going separation process. In common to these observations is their poor repeatability and predictability, which force the engineer to operate processes far from the conditions that have been to empirically noted to cause this behavior.

This chapter presents an overview of the current understanding of the solidification of concentrated suspensions and the liquefaction of cohesionless granular media, and attempts to describe the physical processes that are common to these phenomena in terms of recent advances in fluid and solid mechanics. While this article does not pertain to be complete in acknowledging all the work in this broad area, it draws from the recognized mechanisms of mixture deformation to distinguish flowing suspensions from "solid" granular media. Further, this article identifies the areas of greatest uncertainty and ones of future research in this multidisciplinary subject.

Important developments in the description of deforming and flowing saturated granular masses have been made in the quest for better models for soils, and these are shown to provide a framework of understanding for other disciplines where particle interactions play an important role, as in pneumatic and slurry transport, bed-load sediment transport in oceans and rivers only to mention the most obvious. The ability of suspensions of high solids content to sustain shear stresses has been long recognized and is best exemplified by the asymptotic behavior of the mixture viscosity as the solids concentration tends to a maximum packing value (e.g. Thomas 1965, Rudgers 1962 a,b, Jeffrey and Acrivos 1976, and many others since) above which the mixture can no longer sustain steady shearing. In the presence of gravity there is a natural tendency for mixtures of dissimilar solid and liquid densities to sediment and for the settled matrix to assume stress bearing capacity. Upon mechanical excitation of sufficient magnitude, which is highly sensitive to the history of the particle assembly, the matrix can lose this property and undergo transition to a liquid like state. This phenomenon is called liquefaction. In describing the transition between solid and liquid states, the behavior specifically associated with flowing concentrated suspensions is discussed in Section 2. In Section 3, attention is turned to the micromechanics of transition from solid to liquefied states from a quasi-steady view point, identifying the limitations of viscometric studies with stress bearing particle matrices. Finally, dynamic effects are discussed in Section 4 in the context of stability of flow distributions and the evolution of liquefied regions.

24.2 ON THE LIMITS OF FLUIDITY OF CONCENTRATED SUSPENSIONS

In rheological studies of the effective viscosity of suspensions, such as those conducted for industrial process design, it is well agreed that mixtures of non-Brownian particles in Newtonian fluids below a certain threshold concentration typically behave in a Newtonian fashion exhibiting little viscosity variation with shear strain rate. However, near a critical solids volume fraction or above it, the traditional cone-and-plate or bob-and-cup Couette flow experiments become more erratic exhibiting significant unsteadiness and a generally poor repeatability. For example, Cheng and Richmond (1978) observed fluctuations and step like changes in the measured stresses at high solids loading, which they attributed to locking, jamming, and sudden changes in the arrangement of the particles. This apparent chaotic behavior is also accompanied by hysteretic behavior in the effective viscosity during a cycle of normal loading and unloading (Savage and McKeown 1983). The difficult interpretation of such results is further complicated by the presence of wall effects and slip. In all cases, however, the onset of poor repeatability coincides with the development of normal stresses proportional to the shear stress, which were first documented by Bagnold (1954). This results from the kinematic constraint imposed on the mixture by the configuration of the viscometer, that prevents particle assemblies from sustaining shear strains at concentrations above a threshold value. As it is sheared, an unconstrained mixture is accompanied by an expansion that results in a lower solids fraction that **can** sustain shearing. Difficulties, intermittence, and associated stress state indeterminacy arise when the sample in the viscometer is prevented from expanding, thus generating normal stresses in the particle matrix, and a drop in the interstitial fluid pressure. The phenomenon of matrix dilation during shearing was first observed by Reynolds (1885) who named it *dilatancy*, which is not to be confused with shear thickening, nor Bagnold dilatancy associated with the inertial particle flow regime.

24.2.1 Rigidity threshold

In simple experiments (Onoda and Liniger 1990), the threshold concentration at which monodisperse suspensions begin to transmit shear stresses was determined in two manners; first by measuring the maximum concentration at which mixtures will sustain constant straining and secondly by measuring the minimum packing concentration of the same spheres under gravity, in the limit of matched densities. This threshold concentration of 0.55 ± 0.005, named the **rigity-percolation threshold** or the Random Loose Packing concentration, is also the concentration above which the liquid will develop negative static pressures upon stirring, which demonstrates that the solids begin to transmit forces at this concentration. This condition is therefore also associated with a geometrical constraint on the mixture, such that particles are only just able to move relative to one another without forming stress transmitting particle chains. It follows from the definition of a solid as a shear stress transmitting material that a monodisperse suspension can only solidify at solids concentrations greater than 0.555 ± 0.005 .

At concentrations near or above this threshold, suspensions exhibit shear thickening behavior, in which the mixture effective viscosity increases with the strain rate. At high strain rates, this has been associated with a transition from a viscous regime to one dominated by particle inertia (Bagnold 1954). It is interesting to note that colloidal suspensions also exhibit shear thickening, although this has been attributed to a transition in the packing geometry from an ordered "two-dimensional layered arrangement" to a disordered "random three-dimensional form" (Hoffman 1972). Enlightening reviews have been written of this effect that support the claim that not only colloidal suspensions can demonstrate shear thickening at low strain rates (Barnes 1989, Boersma 1990). In non-Brownian mixtures, it is often speculated that shear thickening at low strain rates, may also be associated with Reynolds dilatancy and the onset of a stress bearing capacity in the solids matrix (Barnes 1989), although this has not been proven. In light of the importance of the individual constituent behavior, the lack of a unifying theory for shear thickening and dilatancy may be due to the difficulty in detecting the solids stress state with traditional viscometers, which only provide a combined measure of the normal stress for the mixture (solid and liquid).

As in the experiments of Hoffman which showed dramatic changes in the structure of colloidal suspensions upon shear thickening, it has long been speculated that non-Brownian viscous suspensions develop a sustained microstructure during steady shearing. Strong evidence for the existence of such microstructure in non-Brownian mixtures was first presented by Gadala-Maria and Acrivos (1980) in their work on shear reversal in concentrated suspensions. This study, conducted at solids fractions up to 55%, also identified the time scales of its formation. Their results are discussed below.

24.2.2 History effects and transient structure formation

The Gadala-Maria and Acrivos (1980) experiments tested concentrated neutrally buoyant suspensions up to 55% solids fraction by volume in a bob-and-cup viscometer with reversible shearing direction. The samples were sheared in the viscometer at a steady rate in one direction, stopped, sheared in the same direction, stopped, then sheared in the opposite direction. The resulting stress required to sustain this motion was recorded and is shown in Figure 6-1. The results conclusively show that upon reversal, and at no other time, the effective viscosity is found to be lower for a finite period, during which the sample offers less resistance to shearing. However as the shearing progresses, the higher steady value of effective viscosity is reached.

The results of numerous experiments at 50% solids fraction are assembled in dimensionless form in Figure 24-1. The shear stress, normalized to one at its maximum value, is presented against the strain, measured from shear reversal. For this solids concentration, the results for a broad range of strain rates fall together with distinct repeatability. This effect was documented for solids concentration as low as 30%, and was generally observed to result in greater initial viscosity reductions and more rapid resumption of the steady value at higher concentrations. The duration of the transient was found to be best quantified in terms of strain or the product of the strain rate and time. In

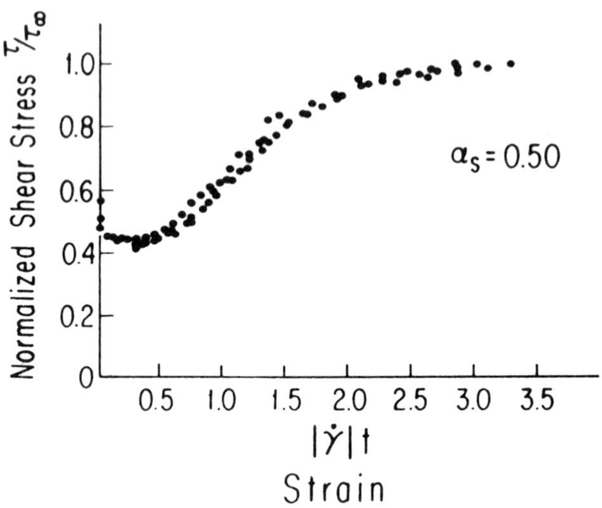

Figure 24-1 *Compiled shear reversal data for a suspension of 50% solids fraction for a range of* $\dot{\gamma}$ *of 0.15-2.4* sec^{-1} *from Gadala-Maria and Acrivos (1980). The duration of the transient corresponds to a net strain of approximately 2.*

these terms steady conditions were reached after a strain of the order of one, with a greater duration at lesser solids loadings. As the spacing between particles is greater at lower concentrations, this finding is consistent with formation of anisotropy and its dependence on the direction of straining.

While Bagnold's ground breaking work has predominantly been noted for identifying the inertial granular flow regime, he also indicated that in the viscous regime, straining a mixture at sufficiently high solids content will result in normal stress generation, as has been later speculated by Jenkins and Mc Tigue (1990) by means of scaling arguments. This has received little attention although the related effect which pertains to the particle flow that results from these stresses has been analyzed more rigorously (Leighton and Acrivos 1986, 1987). On the basis of the results of Onoda and Liniger, one may speculate that the shear transients described by Gadala-Maria and Acrivos may also be accompanied by transients in the normal stress state of the particulate matrix. Unfortunately, the normal stress state of the suspension is significantly more difficult to measure in conventional viscometers that cannot differentiate between the contributions to the global stress from the two constituents, the solids and the liquid. A few apparati exist to the author's knowledge that have the capability of independently imposing and monitoring the liquid and solids normal stresses (Savage 1983, Hanes and Inman 1985, Poutiatine 1990). Only such facilities can be expected to ultimately provide the necessary critical experimental information to address this issue.

The apparent low viscosity transient upon reversal is consistent with the behavior of stress bearing soils under shear in their common relative ease of initial deformation immediately following a change in the direction of shearing. The viscometer tests were limited to concentrations of 55% solids, and therefore a gap in knowledge remains for the range of solids concentrations that are too high to sustain large amplitude shearing but lower than the typical solids concentrations encountered in soil mechanics. Understanding the behavior of mixtures in this intermediate concentration range of 50-60% solids fraction may well be facilitated with an overview of the current understanding of liquefaction in soils.

At the very high solids concentrations pertinent to the study of soil mechanics, much has been done to describe and understand the interplay between the states of the particulate assembly and that of the saturating fluid in the context of liquefaction, in which a soil looses its shear strength. Since a liquefied soil is an assembly of particles in suspension, there is an intimate relation between the fields of suspension rheology and soil mechanics. Below, the mechanics of deformation of granular materials is discussed and ultimately related to the aforementioned developments in suspension rheology.

24.3 DEFORMATION OF COHESIONLESS STRESS BEARING GRANULAR MEDIA

24.3.1 Dry granular materials

As an introduction to the description of solid-liquid mixture behavior, an overview is first presented on the deformation of dry granular materials in the absence of a saturating fluid. This brief review does not pertain to be complete but rather, presents the qualitative elements that will be useful in later discussions of the liquid saturated cases.

The shearing characteristics of drained (dry) particulate material, which is central to understanding the process of liquefaction, has long been recognized to be sensitive to its packing (Taylor 1948, Terzhagi and Peck 1948). Loosely packed sands subjected to normal loading first undergo densification when sheared, followed by an expansion, which was first described by Reynolds (1885), who coined the term *dilatancy*. In contrast to this behavior, densely packed beds exhibit very limited densification if any at all, as shown in Figure 24-2 and will dilate at strains as low as 1% compared to *circa* 10% for loosely packed particles.

The simplest approach used to describe the continuum behavior of bulk solids relies on the Mohr-Coulomb theory (Jenike and Shield 1959) which has been extensively used to delimit failure lines separating flowing and static regions in bulk granular hopper flows (Brennen and Pearce 1978, Nguyen *et al.* 1979). However the Mohr-Coulomb theory is unable to describe volume changes that arise with densification and dilatancy. These difficulties were first overcome by Rowe (1962) in accounting for granular interactions during bulk deformation of geometrically arranged spheres under triaxial loading. In this classical work, the sliding process between adjacent particles was modeled in terms of a local Coulomb friction description using a representative orientation of the plane of contact between spheres, which in this study was constant for the prescribed geometrical arrangement of the spheres. By varying this orientation, presented in terms of an angle of dilatancy, Rowe was able to explain densification and dilatancy. In actual soil samples in which the grains are randomly arranged, the active contact angle may assume all possible values, although the force transmitted by each contact may vary significantly as a function of the angle. This observation has led to the embellishment of Rowe's model to include the random nature of the active contact angle. Elements of this theory are described below.

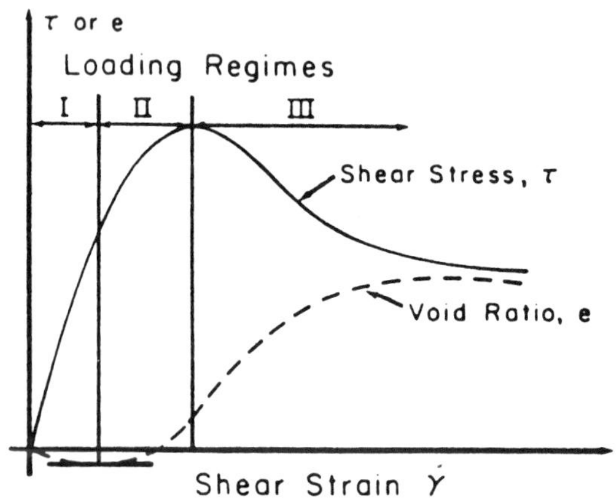

Figure 24-2 *Schematic diagram of the three regimes of deformation of dry granular masses. In I, shearing results in densification, in II, dilatancy begins, and, failure accompanied by continued dilatancy and a fall in shear stress demarks III (Nemat-Nasser and Tobita 1982).*

24.3.2 The contact angle distribution

Rowe's approach was generalized by applying it at a **microscopic** level to active grain pairs which have reached a critical stress state for slip according to the Mohr-Coulomb criterion (Nemat-Nasser 1980, Nemat-Nasser and Tobita 1982). It should be noted that the definition of an active contact contains information regarding its stress state and therefore the sought active contact angle distribution is a variable representative of both the geometrical and stress states of the mixture at a microscopic level. The active contact population can be quantified according to the fraction of all contacts, $p(\nu)d\nu$, that fall in a specific range $(\nu, \nu + d\nu)$ where $0 < \nu < 2\pi$. This defines the density distribution function $p(\nu)$, which is also called the fabric scalar in this two-dimensional case. Geometry dictates that the grains will slide relative to one another along the line specified by their contact angle (angle of dilatancy) as shown in Figure 24-3. For a sliding contact the normal and tangential forces are related by the Coulomb friction law: $T^* = N^* \tan \phi_\mu$, where ϕ_μ is the true, material specific friction angle. For a uniformly deforming body, it is further argued that each active pair will deform at the bulk strain rate. These arguments result in the following dilatancy equation which is valid for plane strain deformations of cylindrical particles:

$$\frac{1}{V}\frac{\dot{V}}{\dot{\gamma}} = \int_{\nu^-}^{\nu^+} p(\nu)\frac{\cos(\phi_\mu + \nu)\sin\nu}{\cos\phi_\mu}d\nu, \tag{24.1}$$

where the left hand side is the ratio of volume to shear strains. For a general description of two-dimensional behavior of a granular material, knowledge of the angular distribution of the dilatancy angle, $p(\nu)$, is needed. In practice this function has a periodicity of π rather than 2π as the surface normals of two grains in contact are always π apart. It follows that the range of integration $(\nu^+ - \nu^-)$ must equal π to avoid counting the same contacts twice. The above

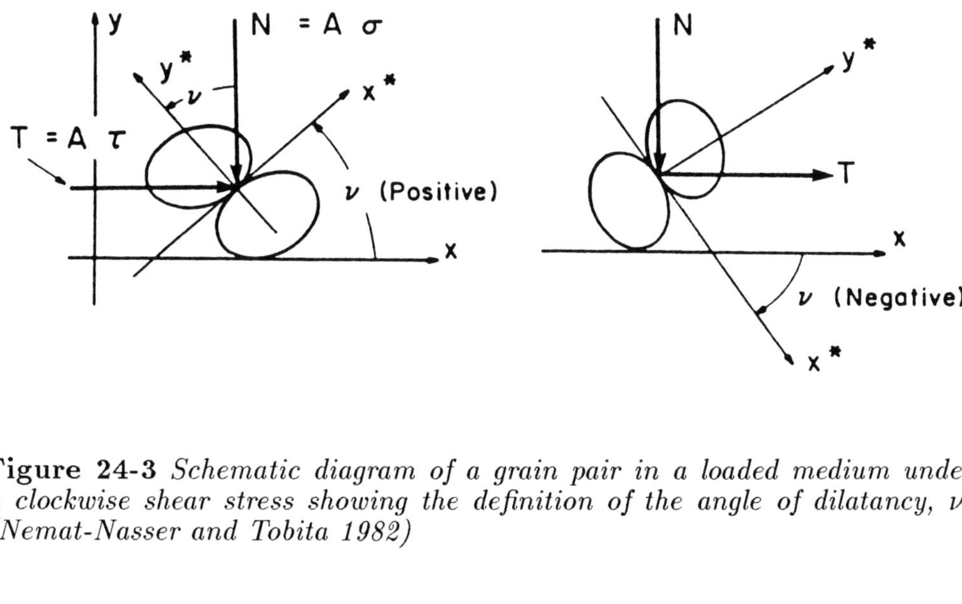

Figure 24-3 *Schematic diagram of a grain pair in a loaded medium under a clockwise shear stress showing the definition of the angle of dilatancy, ν, (Nemat-Nasser and Tobita 1982)*

equation can then be used to describe the volume behavior upon shearing. It is immediately evident that the change in volume is very sensitive to this distribution, and that both positive and negative values can result, which is consistent with the initial densification typically observed with loosely packed particles as opposed to the immediate dilatancy of denser packings. Experiments using photoelasticity have been conducted to quantify the dilatancy angle distribution function (Oda and Konishi 1974, Konishi *et al.* 1983, Matsuoka 1983, Matsuoka *et al.* 1988, Konishi and Naruse 1988, Mehrabadi *et al.* 1988) in two-dimensional pure shearing. These experiments have predominantly focussed on geometrical microstructure in terms of contact angle or void anisotropies. A good example of the variation of the fabric during cyclic straining is given by Matsuoka *et al.* (1988) and is shown in Figure 24-4.

These contact angle distributions demonstrate the tendency for this geometrical anisotropy to align with the principal stresses rather than the strain state of the granular medium.

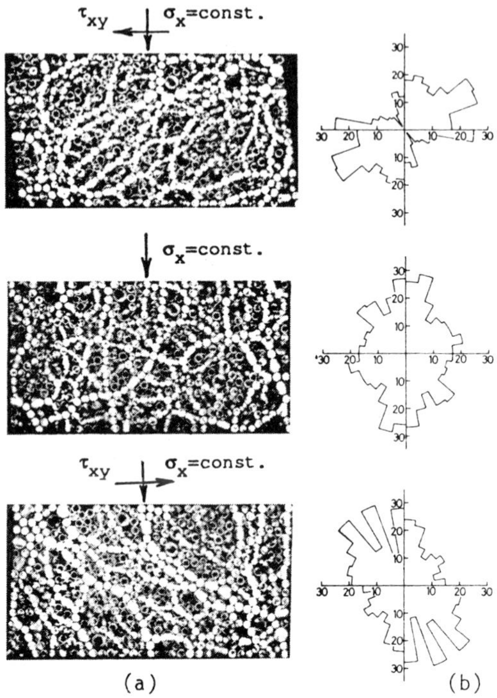

Figure 24-4 *Photoelastograph of circular rods undergoing cyclic simple shear shown with the corresponding contact angle distribution. Anisotropy is evident in the contact distribution as a result of the anisotropic stress state (Matsuoka et al. 1988).*

In introducing $p(\nu)$ as a new state variable, an additional equation is necessary to describe its variation with strain during which certain contacts fail and no longer are "critical", and others become active. The development of an evolution equation for the structure and anisotropies associated with the stress and geometrical states of the mixture remains a subject of ongoing research, and contributions are beginning to be made to this end. One such model has been proposed by Matsuoka *et al.* (1988) in which the infinitesimal principal stress rotation has been quantified for an infinitesimal change in the global strain state.

24.3.3 Quasi-steady liquefaction of uniformly sheared sediments

In the case of a loosely packed liquid-saturated assembly of particles, the total sample volume cannot readily change upon shearing due to the incompressible nature of the pore fluid. Consequently, as this mixture is sheared, the volume occupied by the solids will initially decrease according to Figure 24-2, while the total mixture volume does not change appreciably, and thus the stress transmitting particle "bridges" between the shearing surfaces are lost. This results in the loss of the sample's ability to bear a shear stress (Finn *et al.* 1970). As this occurs, the applied load is taken up by the fluid, and the pore pressure increases up to the magnitude of the applied normal stress, which corresponds to the state where the solids no longer support any of the normal stress. In this extreme situation, the grains or grain clusters have come apart and are separated by pore fluid, and the liquefied mixture will readily flow at least infinitesimally. As the flow evolves, the solid matrix will ultimately begin to expand (as in Figure 24-2), and reach a sufficient volume to regain its shear and normal strength. When this occurs, the weight is taken off the fluid and the pore pressure will decrease. This qualitative description is elaborated upon below in terms of quantitative measurements of shear strain, stress and the pore fluid pressure.

24.3.4 Phenomenological observations during shear reversal under saturated conditions

Numerous experiments have been conducted to describe the behavior of fluid saturated granular media subjected to shear reversal (Nemat-Nasser and Tobita 1982, Ishihara and Towata 1983). Such studies of simple two-dimensional shear under normal loading have been instrumental in demonstrating the phenomenon of liquefaction. For purposes of graphical simplicity in representing the solids stress state, an effective mean principal stress is defined as the imposed stress tensor minus the measured isotropic pore pressure. Representative experiments by Ishihara and Towata (1983) of the behavior of this variable and of the mixture stress-strain history during shearing of a loaded saturated medium are presented in Figure 24-5a and b. These experiments were performed in a triaxial torsion shear apparatus, and similar results have been obtained by others (e.g. Nemat-Nasser and Tobita 1982) in different testing configurations. The data shown are for a quasi-steady oscillatory shear stress

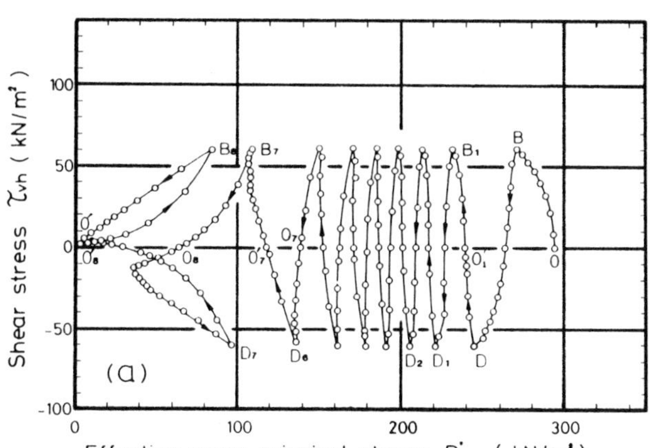

Figure 24-5a *Effective stress of the solid matrix in a saturated torsion shear test (Ishihara and Towata 1983).*

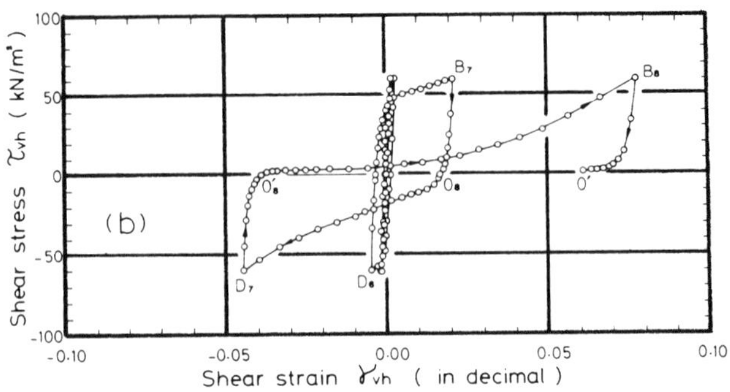

Figure 24-5b *Measured stress-strain behavior in the same saturated test (Ishihara and Towata 1983).*

of 60 kN/m^2 in amplitude imposed upon a saturated particulate sample. Initially, such a stress history results in a gradual monotonic decrease in the effective solids stress as the pore pressure builds up. Over the first few cycles, the resulting strains of $\pm$ 0.005 remain approximately linear with respect to the stress with only limited hysteresis. As the mixture is progressively cycled, the decrease in solids stress persists for the first seven full cycles, the end of which is labeled O_7'. In the first quadrant of the eighth cycle, the same trend is evident, but in addition, the sample is seen to fail just prior to reversal. Following reversal, the principal solids stress decreases dramatically and shear stress decreases sharply drops to zero. This is the first indication of incipient liquefaction. In the third quadrant of this cycle, the solid matrix regains its strength. This does not occur at the zero strain position but slightly beyond the zero stress condition. This is evident from the re-establishment of the finite slope in the stress-strain curve that coincides with a rather sudden increase of the effective solids principal stress (note the kink in Figure 24-5a) due to a sharp redistribution of the normal stress from the liquid back to the solid. This transition from a liquefied state back to stress bearing one is associated with sufficient Reynolds dilatancy in the solid matrix (see region II in Figure 24-2) to reform the stress bearing structure between the shearing boundaries. In the second reversal of this same cycle, the solids principal stress plunges to zero near O_8', and the mixture is fully liquefied as the normal stress is fully born by the fluid. This condition again is short lived as the liquid pore pressure drops and the solids regains a share of the normal load. In subsequent cycles, the shows the same response with two transfers of normal load between solid and the pore liquid per full cycle.

After the first occurrence of liquefaction, the mixture typically exhibits low resistance to shearing upon reversal and liquefies due to a strong bias toward negative dilatancy angles. As straining persists the mixture resolidifies over a finite strain of the order of 0.05 in this example, as the active contacts rearrange to favor dilatancy. This is evident in the increase in slope in Figure 24-5b between reversals, which reflects a resumption of stiffness.

24.3.5 Effects of elasticity and compressibility

In the examples discussed, there has been no mention of grain deformation or pore fluid compressibility. Consider the situation where a mixture, confined to a constant volume is subjected to simple shear under a normal stress. As the particle matrix assumes a finite stress state, the particles begin to deform at the contacts, which results in a change in actual solids volume. An indeterminacy in the isotropic stress state of both the liquid and the solid arises. If the particle contacts are sufficiently compliant or if the pore fluid is compressible, the resulting magnitude of the pore pressure drops will be smaller. To quantify this effect, the elastic properties the pore fluid and the mechanics of grain contact interactions must be addressed. For brevity, these important effects are only mentioned here. The reader should find the following works useful in studying these effects.

The elastic deformation of assemblies of contacting spheres has been addressed in the context of two- and three-dimensional particle assemblies using frictionless Hertz and frictional Hertz-Mindlin contact theories (Jenkins 1988, Jenkins *et al.* 1989) although these remain to be generalized to include fluid compressibility effects which are essential to quantify the pore pressure. Other developments have also been made by Ko and Scott (1968) and Goddard (1990) in describing the effects of increased contact population under load, and asperities at contacts on the global stress-strain behavior of dry particle masses. These effects have not been incorporated into liquefaction models although they are believed to be of importance in geophysical applications.

24.3.6 Large amplitude effects

The preceding body of works on packed particle assemblies demonstrates the mechanism by which they may lose their stress bearing capacity, and hence become liquefied. The key element that has been emphasized in these studies is the governing role taken by the microstructure of the mixture during deformation processes. It is this character that is the cause of the apparent phase change. Specifically, in the event of dominant negative dilatancy angles, the solids will tend to densify and connectivity between the confining plates will be lost, and they will undergo unresisted relative motion. Existing experimental studies of mixture deformation have been limited to small strains which correspond to regions I and II in Figure 24-2, and thus little information is available regarding the failure process from small to very large deformations spanning regions I, II and III. However, much information is available on sustained steady shearing of concentrated suspensions, which is an extreme case of large deformations.

To describe the behavior of concentrated suspensions undergoing steady shear, a state variable analogous to the distribution of the angle of active contacts has been introduced, namely the **pair distribution function**, which describes the statistical distribution of particles in continuous shearing (Husband and Gadala-Maria 1987, Parsi and Gadala-Maria 1987). Measurements of this quantity have shown suspensions to assume anisotropy which is consistent with computations (e.g. Brady and Bossis 1985) and observations of hysteresis in the bulk properties of concentrated suspensions, that have been interpreted as a global reflection of the dynamics of structure evolution. Although the pair distribution function has been quantified for steady flows, no theory exists for its evolution in unsteady circumstances, as is the case for the evolution of the distribution of contact angles in stress bearing matrices. A chapter in this volume has been devoted to the current state of understanding of the steady nature of anisotropy in sheared viscous suspensions.

24.4 LIQUEFACTION IN UNSTEADY SHEARING

24.4.1 Stability considerations

When a liquid saturated assembly of stress transmitting particles is subjected to a deviatoric stress, failure is known to occur in a localized fashion (e.g. Rowe 1962). This type of failure is the result of the inherent instability and growth of small local deformations (e.g. Vardoulakis 1985, Jenkins 1990). While the location of failure is difficult to predict in uniformly loaded samples, it can be expected to occur at imperfections within the sample or at its boundaries, which tend to introduce local ordered structure. The particles at the boundary thus perceive the boundary as being different from the particles to the interior. This phenomenon is often blamed for the non-uniform strain distributions referred to as "slip" observed in viscometric studies with suspensions of high concentrations. These exhibit high shear regions near boundaries, with limited shear deformations in the core of the flow. If localized shear zones are the result of an instability in flowing suspensions, which can only be described in terms of their dynamic character, it follows that an understanding of these phenomena will not be found on the sole basis of steady, or time invariant arguments. Rather, we should seek to describe and isolate the favored solutions as well as the decay of the unstable solutions. While the analysis is currently limited by uncertainty in constitutive behavior of suspensions, the stability of various flow fields in time could be observable in carefully designed flow transient experiments in which the time behavior of each "mode" or eigenfunction can be followed.

Research efforts on the shear deformation of concentrated mixtures have historically emphasized steady or quasi-steady motions while dynamic effects in unsteady shearing has received little attention in comparison. Unsteady effects have mostly been studied in the context of theoretical stability analyses of various flow induced structures, and their wave propagation character. For example, most recently, Savage (1991) provided analytical confirmation of numerical calculations by Hopkins and Louge (1989) that indicated the presence and propagation of low wavenumber structure in inertially dominated granular shear flows. Despite such significant contributions, experiments on unsteady flows are severely lacking. In an ongoing effort to improve the current understanding, finite height particle beds have been studied under harmonic excitation from the lower containing boundary (Kytömaa 1991). These experiments, the scales of which are consistent with ocean sediment and coarse slurry transport applications, confirm the large influence of boundaries on the deformation of the mixture.

24.4.2 Harmonic excitation of an unconstrained saturated particle bed

In an experiment, a 3" submerged bed of 1 mm glass beads was constrained only by a horizontally mobile, rough lower wall as shown in Figure 24-6. At the top, the bed forms an unconstrained interface with water. The wall was artificially roughened with cylindrical indentations in its surface,

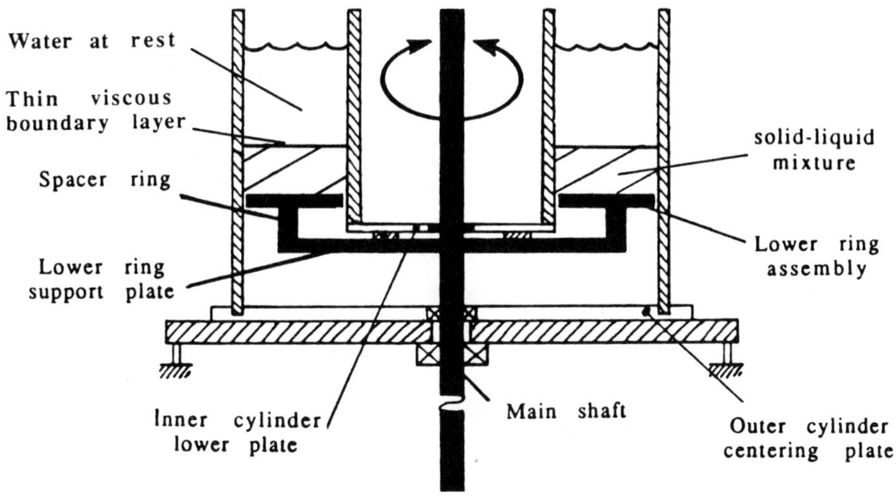

Figure 24-6 *Schematic of the Dynamic Shear Cell used for oscillatory tests of unconstrained saturated particle beds (Poutiatine 1990).*

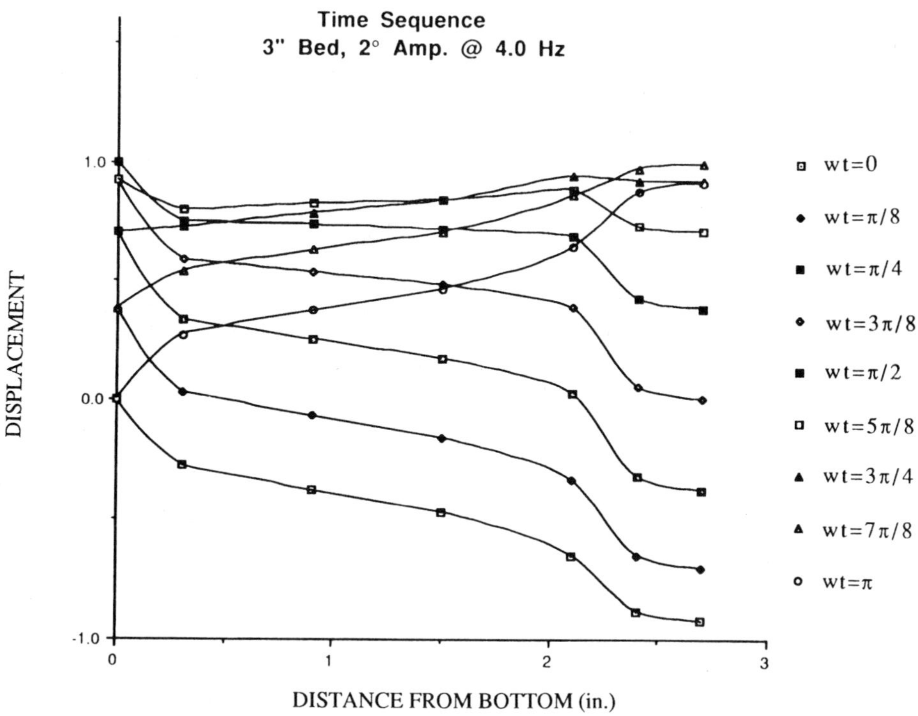

Figure 24-7 *Displacement record for 3" submerged bed of 1 mm glass spheres undergoing 4 Hz oscillations of 6 mm peak-to-peak amplitude (Kytömaa 1991). This record shows the distinct effects of the wall and the submerged free surface.*

of 4 mm diameter, 1.8 mm deep holes that occupy 55% of the area of the steel plate (Poutiatine 1990). In the results described below, the bed was subjected to horizontal oscillatory motions of the bottom boundary, and the resulting displacement of the mixture was recorded in real time at various internal locations using a displacement probe (Kytömaa 1991) designed to match the inertial properties of the particulate material. This simple experiment was devised to study dynamic effects in the presence of gravity (as is the case in most engineering applications), and is presented here to illustrate the complex nature of the displacement field.

At each height, the phase and amplitude of the displacement were recorded and are shown in Figure 24-7. Despite its poor resolution which was only limited by choice, this graph shows three distinct regions of motion: At the top and bottom boundaries, large strains are present while the central region of the bed remains relatively undeformed. Central bed motion not only shows a smaller amplitude of total displacement (about 80% of wall displacement 1/4" above bottom) but it also lags the wall by about 0.35 radians at this location. The central region invariably shows a near linear strain field, in contrast with the sharp displacement gradients at either boundary which remain unresolved due to the coarse spacing of the data.

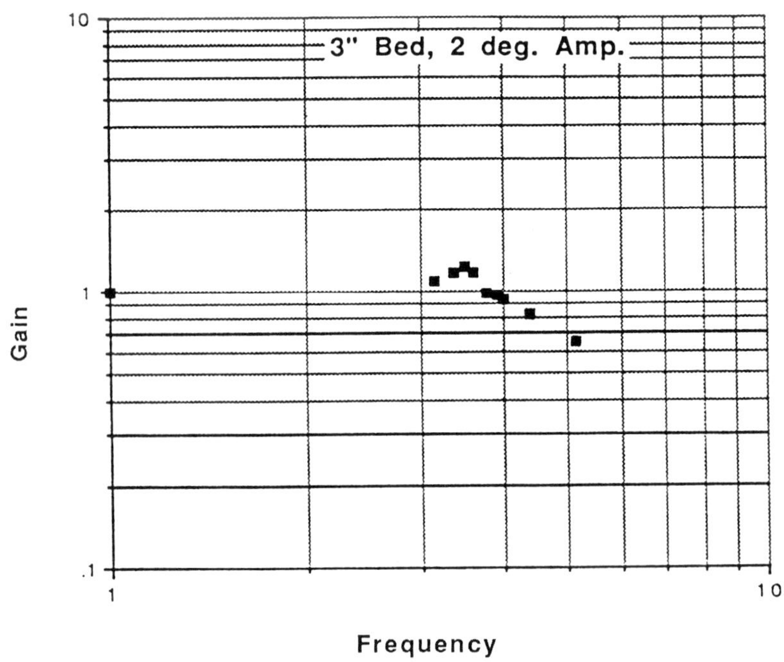

Figure 24-8 *Normalized displacement of the submerged free surface versus frequency for a 3" bed of 1 mm glass spheres at 6 mm peak-to-peak amplitude motion of the lower wall (Kytömaa 1991). The motion of the bed exhibits a resonance at a frequency of 3.4 Hertz.*

While the upper layer in turn lags the core region, the bed upper surface amplitude has been consistently measured to exceed that of the wall under certain conditions that reflect resonant characteristics, as shown in Figure 24-8. At the natural frequency of 3.4 Hz, the surface oscillates at a 25 % larger amplitude than the wall. In addition to the resonant peak it has been noted that the fall-off in amplitude consistently follows a w^{-2} dependence. This character suggests that effects of mixture shear strength, or stiffness are present in this system, which together with inertia are able to exhibit behavior similar to that of a distributed mass-spring resonator. The presence and importance of stiffness in the motion of particle beds such as slurry flows, and ocean sediment transport applications is often discarded. In steady situations it may be a reasonable approximation, but in unsteady regimes it clearly plays an important role. In an attempt to identify the mode of vibration that corresponds to the observed resonance, it is speculated that the surface "boundary layer" is the principal contributor, rather than the bottom slip layer, because the resonant frequency was found to be nearly independent of the particle bed height. In addition to shedding some light on the interfacial dynamics of dense suspended load transport, these ongoing experiments will ultimately provide direct experimental information on the phenomenon of unsteady slip at a wall, which continues to plague most experiments at high concentrations. The mode shapes of beds executing steady oscillatory motion will be documented to provide a data base for comparisons and validations to be made with theoretical developments. It is studies of this unsteady modal behavior of beds and confined mixtures in Couette devices that will lead to an understanding of the preferred deformation and velocity profiles.

In an unsteady situation where the containing boundary accelerates relative to the mixture, as in the above example or when the supporting formation below a liquefiable soil stratum moves during an earthquake, the initial localized failure at the lower boundary constitutes the initiation of a liquefaction event (the wall slip region in Figure 24-7) which can result in subsequent pore fluid flows. The large localized pore pressures that result from liquefaction act as a driving potential for these flows. In anisotropic formations, the pore fluid will flow along cracks and other formation features and in some instances can reach the surface, as is manifested by sand boils which are deposits of fines that are brought up with pore fluid. Below the homogeneous, isotropic situation is discussed in terms of diffusion of pore fluid.

24.4.3 Diffusive build-up of pore fluid

In an isotropic and homogeneous material made up of a linearly elastic porous medium, the one-dimensional rearrangement of the constituents can be modeled in terms of the interplay between Darcy drag and the elastic stress in the particle assembly. Consider an infinitesimal element of a saturated stress bearing particle assembly of constant stiffness E, permeability k and solids fraction $\alpha_s(= 1 - \alpha_l)$, where α_l is the porosity. Unsteady variations of pore pressure in this linear elastic medium are governed by the equation

$$\frac{\partial p}{\partial t} = \frac{k\alpha_s}{\mu E}\frac{\partial^2 p}{\partial y^2} \qquad \text{for} \quad y > 0. \tag{24.2}$$

This equation was first derived by Terzaghi (Terzaghi and Peck 1948) to describe the consolidation of soil under its own weight and the accompanying upward pore fluid flow. For a uniform semi-infinite region of saturated cohesionless particles that experiences a sudden rise in pressure to the lithostatic value at y = 0 (the lithostatic pore pressure is the fluid pressure when it supports the entire overlying formation), the solution to the above diffusion equation takes the form

$$\eta = \frac{p - p_{hydrostatic}}{p_{lithostatic} - p_{hydrostatic}} = [1 - \mathrm{erf}(\xi)] \,, \tag{24.3}$$

where

$$\xi = \frac{y}{2\sqrt{\nu_H t}}, \tag{24.4}$$

and

$$\nu_H = \frac{k\alpha_s}{\mu E}. \tag{24.5}$$

The variable ν_H is called the **hydraulic diffusivity** or the **coefficient of consolidation** of the porous medium. A simple interpretation of this solution can be given in terms of a characteristic time, τ associated with a change in pressure of $\eta = 0.5$ at a location y above the position of initial high pore pressure:

$$\tau = \frac{y^2}{0.91\nu_H}. \tag{24.6}$$

This characteristic time serves as a measure of the duration of penetration of pressure information to a prescribed depth y. Sands (grain size ≥ 0.075 mm) have a ν_H range of $10^2 - 10^5$ cm^2/sec, while silts (grain size < 0.075 mm) in which pressure diffusion is appreciably slower, typically have values below 1 cm^2/sec (Lambe and Whitman 1969). In recent field measurements of liquefaction during the Superstition Hill earthquake (Holzer *et al.* 1988, 1989) the diffusive nature of pore pressure evolution was confirmed *in situ*. In a fine sand with $\nu_H = 50$ cm^2/sec, the time τ for pressure information to penetrate 3 m of soil is of the order of 20 seconds. For comparison, the rise time during the above earthquake was between 20 and 30 seconds for a sediment stratum of the same thickness. This is consistent with the instrumented silty sand composition of the liquefiable stratum. The quantitative analysis of these data was plagued by a speculated low diffusivity region at the pressure transducers formed during installation, that prevented a more conclusive study of the stratum itself.

Of the contributing properties to the hydraulic diffusivity, the matrix modulus E and the permeability k stand out as the most sensitive to loading conditions and pore size respectively. While the linear theory has served to describe the phenomena of pore pressure diffusion and consolidation which are central to liquefaction and resolidification of saturated particle masses, the resulting large changes in the solids stress state and geometrical microstructure can cause significant deviations of the material properties about their assumed constant values. The modulus of a stress bearing matrix of spherical particles has been measured to scale with the 1/2 power of the confining pressure, which has independently been theoretically confirmed by Ko and Scott 1968 and Goddard 1990, who reconciled the inconsistency with the classical Hertzian contact theory result which predicts a 1/3 power dependence. Unfortunately, there exists very limited information of the granular medium modulus at the low stress states appropriate to ocean sediment and slurry transport processes, although it is known that the stiffness grows in a non-linear fashion with loading at low stresses (Lambe and Whitman 1969, Adorjan 1975, Bascur 1989, Kytömaa 1990, 1991), and as described in Section 3, that the properties of such media are also sensitive to their preparation and loading histories.

The permeability of soils, on the other hand, is mostly dependent on their geometrical configuration. It is proportional to the square of the pore pressure dimension. This reflects a strong dependence on particle size which is particularly difficult to predict for all naturally occurring size distributed soils. The presence of fines can sway permeability by orders of magnitude. These factors explain the wide range of the magnitude of hydraulic diffusivity and thus the associated uncertainty in predicting it.

24.4.4 Sedimentation waves and consolidation

In experiments on the behavior of fully liquefied soil samples in a centrifuge by Whitman and Lambe (see National Research Council 1985), the

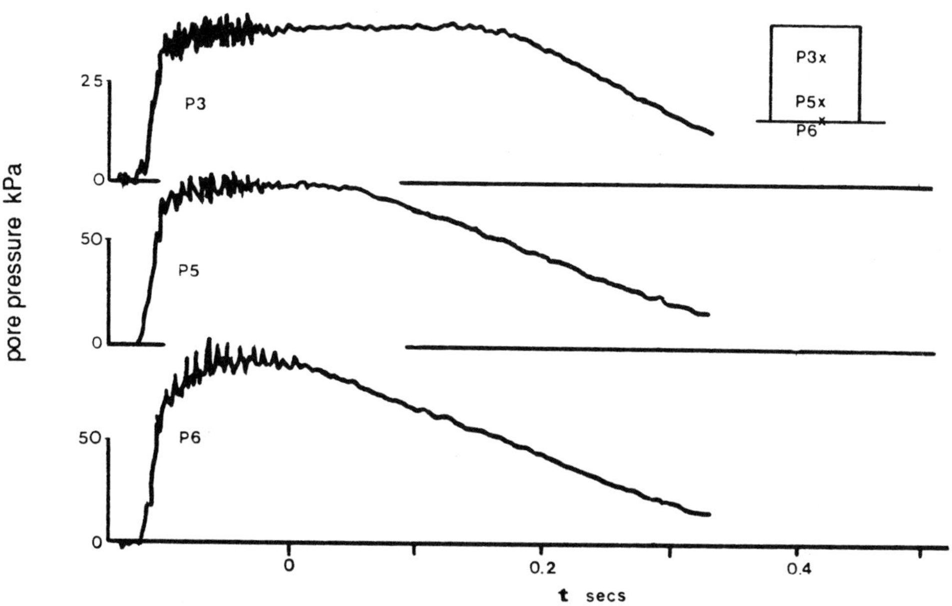

Figure 24-9 *Post-liquefaction pore pressure behavior in centrifuge test. There is a marked time delay in the initial decrease in pore pressure associated with vertical location (Lambe 1981).*

time trace of the pore pressure during the settling or consolidation event revealed a distinct character that did not resemble that of the simple theory described above. The pore pressure at the two vertically spaced locations of the container, shown in Figure 24-9, are distinctly different. There is a noticeable time delay preceding the fall of pore pressure at the upper location.

According to the linear consolidation model, both traces should exhibit an immediate, monotonic shape. While quantitative deviations could easily be accounted for in terms of non-linearities in elasticity or permeability and hence in the hydraulic diffusivity, such a qualitative difference with an associated time delay could not.

This inconsistency was first explained by Scott (1986a, 1986b) in terms of a sedimentation argument to explain the wave-like nature of the solidification front. The liquefied suspension which has lost all its stress bearing capacity simply sediments under gravity to form a growing solidified layer on the bottom of the container. As the solidified layer grows, it consolidates under its own weight according to the Terzhagi consolidation theory, thus explaining the gradual, concave pore pressure behavior after initial solidification. Although this phenomenon has not been identified *in situ*, it is interesting to note that in the Superstition Hills earthquake (Holzer *et al* 1988, 1989), the Wildlife Liquefaction Array recorded information that supports this theory. Figure 24-10 shows the vertical pore pressure distribution in time, in a manner different from Figure 24-9 although the two contain similar information. The pressure profile measured 19 minutes after initial liquefaction shows a clear decrease in pore pressure from the lithostatic value at a depth of 6.5 m, while the transducers at higher elevations remain at their maximum respective values. This initial decrease in pore pressure, and hence solidification, at lower locations suggests that the solidification wave has not reached the upper measuring stations, and is therefore consistent with the wave-like character of solidification, although this claim would be more conclusive with finer time discretization in the data.

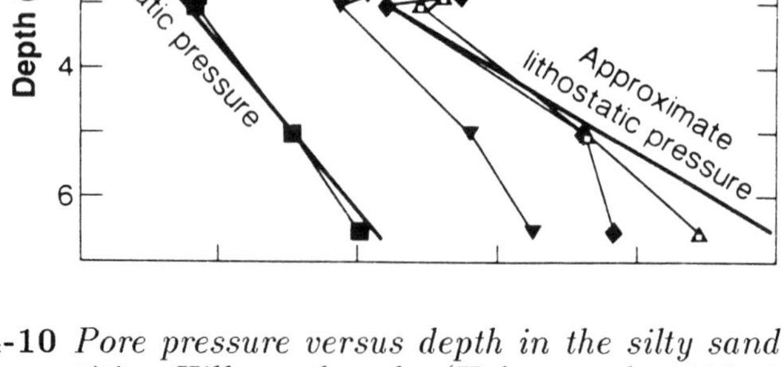

Figure 24-10 *Pore pressure versus depth in the silty sand layer during and after the Superstition Hills earthquake (Holzer* et al. *1988, 1989). At 26.5 seconds, the pore pressure has not fully reached lithostatic, and after 19 minutes, only the lower transducer has begun to see a fall in pore pressure.*

24.5 SUMMARY AND CONCLUSION

To date the field of rheology continues to describe concentrated suspensions in terms of effective single phase continuum properties by invoking recognized non-Newtonian characteristics. In contrast, in soil mechanics it has long been recognized that saturated soils truly behave as two interacting media, the particle matrix and the fluid, which can assume relative motion and different stress states. As a consequence of this realization, significant contributions have been made to the understanding of the phenomenon of liquefaction. It follows therefore, that suspensions of higher concentrations, at which the solid matrix begins to bear shear and normal stresses, will behave as a mixture of interacting media, and any attempt to describe their behavior as single continua will have severe limitations. While developments in this approach are expected to further our modeling ability of dense mixtures, there remains a strong need to develop standardized instrumentation that is able to accommodate and quantify this behavior, which is believed to be the source of experimental difficulty in conventional Couette viscometry.

Experiments with soils have mostly been limited to small strains, while shearing of flowing suspensions remain limited to concentration near 55%, above which uncertainty, poor repeatability and unsteadiness are often observed. There is an evident interplay between the strain magnitude that a suspension can sustain and its maximum solids hold-up. Future viscometers should therefore be designed to independently monitor the stress and strain states of the two constituents in prescribed oscillatory shearing with the option of operating in both stress or strain control modes. This builds upon recent innovations in Couette viscometry which provides this option for the bulk mixture alone.

As is described in this book, developments continue to be made in describing the microstructure in flowing particle assemblies. The pair distribution function has indeed indirectly been demonstrated to affect the bulk behavior of suspensions. In parallel to these findings, contact angle distributions and fabric are becoming widely used to describe the anisotropy in loaded soils, and these have proven to be reliable predictive tools for phase change. To bridge the two fields, efforts to adopt a universal means of representing internal structure that can be used for both liquefied and solidified states would be of great benefit. It would create the ability to analytically and numerically model mixtures as they undergo phase change within the same mathematical framework. However, before such capabilities can exist, the evolution of microstructure in an unsteady sense must first be understood. Evolution equations for the fabric tensor of stress bearing particle assemblies is a subject of on-going research, and the approach to this work could greatly contribute to the field of suspension rheology, which currently lacks this added sophistication.

Finally, it can be safely said that dynamic effects remain effectively unexplored in suspension rheology, although for good reason. Actual engineering applications where they are important are understandably rare. However, in the context of flow profile stability which addresses the question put forward by observations of non-uniform velocity profiles (e.g. plug flow) much can be learned from dynamic considerations which may ultimately explain the transient development of these flow profiles. Along with such flow profiles, the concentration profiles may also be expected to be non-uniform. This is the speculated reason for the wall-lubricated slurry flow regime, for example. Therefore, it is reasonable to expect redistribution of the constituents to be an integral part of the necessary dynamic modeling of mixtures, an issue that is always ignored.

REFERENCES

Adorjan, L.A., "A Theory of Sediment Compression," *Proc. of 11th Mineral Processing Conference*, Cagliari, 1975, pp. 297-318.

Bagnold, R.A., "Experiments on a Gravity-Free Dispersion of Large Solid Spheres in a Newtonian Fluid Under Shear," *Proc. R. Soc.*, Vol. 225, 1954, pp. 49-63.

Barnes, H.A., "Shear-Thickening ("Dilatancy") in Suspensions of Nonaggregating Solid Particles Dispersed in Newtonian Liquids," *J. of Rheology*, Vol. 33, 1989, pp. 329-369.

Bascur, O.A., "A Unified Solid/Liquid Separation Framework," First Regional Meeting of the American Filtration Society, Houston, Texas, 1989.

Boersma, W.H., "Shear Thickening of Concentrated Dispersions," Doctoral Dissertation, Technische Universiteit Eindhoven, 1990.

Brady, J.F., and G. Bossis, "The Rheology of Concentrated Suspensions of Spheres in Simple Shear Flow by Numerical Simulation," *J. Fluid Mech*, Vol. 155, 1985, pp. 105-129.

Brennen, C. and J.C. Pearce, "Granular Material Flow in Two- Dimensional Hoppers," *J. of Applied Mechanics*, Vol. 45, 1978, pp. 43-50.

Cheng, D.C.-H. and R.A. Richmond, "Some Observations on the Reological Behavior of Dense Suspensions," *Rheol. Acta*, Vol. 17, 1978, pp. 446-453.

Finn, W.D.L., P.L. Bransby and D.J. Pickering, " Effect of Strain History on Liquefaction of Sand," *J. Soil Mech. Fdns. Div., ASCE*, Vol. 96, No. SM6, 1970, p. 1917.

Frankel, N.A. and A. Acrivos, "On the Viscosity of a Concentrated Suspension of Solid Spheres," *Chem. Eng. Sci.*, Vol. 22, 1967, pp. 847-853. Gadala-Maria, F. and A. Acrivos, "Shear-Induced Structure in a Concentrated Suspension of Solid Spheres," *J. of Rheology*, Vol. 24, 1980, pp. 799-814.

Goddard, J.D., "Nonlinear Elasticity and Pressure-Dependent Wave Speeds in Granular Media," *Proc. R. Soc. Lond. A*, Vol. 430, 1990, pp. 105-131.

Hanes, D.M. and D.L. Inman, "Observations of Rapidly Flowing Granular-Fluid Materials," *J. Fluid Mech.*, Vol. 150, 1985, pp. 357-380.

Hoffman, R.L. " Discontinuous and Dilatant Viscosity Behavior in Concentrated Suspensions. I. Observation of a Flow Instability," *Trans. Soc. Rheol.*, Vol. 16, No. 1, 1972, pp. 155-173.

Holzer, T.L., T.L. Youd, and T.C. Hanks, "Dynamics of Liquefaction During the 1987 Superstition Hills, California, Earthquake," *Science*, Vol. 244, 1989, pp. 56-59.

Holzer, T.L., T.L. Youd, and M.J. Bennett, "In Situ Measurement of Pore Pressure Build-up During Liquefaction," *Proc. of the 20th Joint Meeting of the U.S.-Japan Cooperative Program in Natural Resources Panel on Wind and Seismic Effects*, NBS Gaithersburg, 1988, pp. 118-130.

Hopkins, M.A. and M.Y. Louge, "Inelastic Microstructure in Rapid Granular Flows of Granular Materials," *Phys. Fluids*, Vol. A 3, 1991, pp. 47-57.

Husband, D.M. and F. Gadala-Maria, "Anisotropic Particle Distribution in Dilute Suspensions of Solid Spheres in Cylindrical Couette Flow," *J. Rheology*, Vol. 31, No. 1, 1987, pp. 95-110.

Ishihara, K. and I. Towata, "Cyclic Behavior of Sand during Rotation of Principal Stress Axes," *Mechanics of Granular Materials: New Models and Constitutive Relations*, eds. J.T. Jenkins and M. Satake, Elsevier, Amsterdam, 1983, pp. 53-73.

Jeffrey, D.J. and A. Acrivos, "The Rheological Properties of Suspensions of Rigid Particles," *AIChE Journal*, Vol. 22, No. 3, 1976, p. 417.

Jenike, A.W. and R.T. Shield, "On the Plastic Flow of Coulomb Solids Beyond Original Failure," *J. of Applied Mechanics*, 1959, pp. 599-602.

Jenkins, J.T., "Volume Change in Small Strain Axisymmetric Deformations of a Granular Material," *Micromechanics of Granular Materials*, Elsevier Science Publishers B.V., Amsterdam, 1988, pp. 243-252.

Jenkins, J.T., P.A. Cundall and I. Ishibashi, "Micromechanical Modeling of Granular Materials with the Assistance of Experiments and Numerical Simulations," *Powders and Grains*, Proc. of the International Conference on Micromechanics of Granular Media, Sept. 1989, pp. 257-264.

Jenkins, J.T. and D.F. Mc Tigue, "Transport in Concentrated Suspensions: The Role of Particle Fluctuations," *Two Phase Flows and Waves*, Joseph, D.D and D.G. Schaeffer, eds., Springer-Verlag, 1991.

Jenkins, J.T., "Localization in Granular Materials," *Appl. Mech. Rev.*, Vol. 43, No. 5, Part 2, 1990, pp. 194-195.

Ko, H-Y, and R.F. Scott,"Deformation of Sand in Hydrostatic Compression," *J. Soil Mech. Fdns. Div., ASCE*, No. SM3, 1967, pp. 137-156.

Konishi, J., M. Oda and S. Nemat-Nasser, "Induced Anisotropy in Assemblies of Oval Cross-Sectional Rods in Biaxial Compression," *Mechanics of Granular Materials: New Models and Constitutive Relations*, eds. J.T. Jenkins and M. Satake, Elsevier, Amsterdam, 1983, pp. 31-39.

Konishi, J. and F. Naruse, "A Note on Fabric in Terms of Voids," *Micromechanics of Granular Materials* , eds. M. Satake and J.T. Jenkins, Elsevier, Amsterdam, 1988, pp. 39-46.

Kytömaa, H.K., "Propagation and Structure of Solidification Waves in Concentrated Suspensions," *Mechanics of Materials*, Vol. 9, 1990, pp. 205-215.

Kytömaa, H.K., "Effects of Internal Reordering on Sedimentation Waves in Concentrated Incompressible Suspensions," *Fluid/Particle Separation Journal*, Vol. 4, No. 1, 1991, pp. 37-46.

Kytömaa, H.K., "Response of Submerged Particle Beds to Harmonic Excitation" (in preparation).

Lambe, T.W. and R.V. Whitman, *Soil Mechanics*, John Wiley and Sons, 1969.

Lambe, P.C., "Dynamic Centrifuge Modeling of a Horizontal Sand Stratum," Sc.D. Thesis, Department of Civil Engineering, MIT, Cambridge, MA, USA, 1981.

Leighton, D. and A. Acrivos, "The Shear-Induced Migration of Particles in Concentrated Suspensions," *Fluid Mech.*, Vol. 181, 1987, pp. 415-439.

Leighton, D. and A. Acrivos, "Viscous Resuspension," *Chem. Eng. Sci.*, Vol. 41, 1986, pp. 415-439.

Matsuoka, H., "A Stress-Strain Model for Granular Materials Considering the Mechanism of Fabric Change," *Mechanics of Granular Materials: New Models and Constitutive Relations*, eds. J.T. Jenkins and M. Satake, Elsevier, Amsterdam, 1983, pp. 99-115.

Matsuoka, H., K. Sakabira and Y. Suzuki, "A Constitutive Model for Granular Materials Evaluating Principal Stress Rotation," *Micromechanics of Granular Materials*, eds. M. Satake and J.T. Jenkins, Elsevier, Amsterdam, 1988, pp. 287-296.

Mehrabadi, M.M., S. Nemat-Nasser, H.M. Shodja and G. Subhash, "Some Basic Theoretical and Experimental Results on Micromechanics of Granular Flow," *Micromechanics of Granular Materials*, eds M. Satake and J.T. Jenkins, Elsevier, Amsterdam, 1988, pp. 253-262.

National Research Council, *Liquefaction of Soils during Earthquakes*, prepared by the Committee on Earthquake Engineering, Commission on Engineering and Technical Systems, National Academy Press, 1985.

Nemat-Nasser, S. "On Behavior of Granular Materials in Simple Shear," *Soils and Foundations*, Vol. 20, No. 3, 1980, p. 59.

Nemat-Nasser, S. and Y. Tobita, "Influence of Fabric on Liquefaction and Densification Potential of Cohensionless Sand," *Mechanics of Materials*, Vol. 1, 1982, pp. 43-62.

Nguyen, T.V., C. Brennen and R.H. Sabersky, "Gravity Flow of Granular Materials in Conical Hoppers," *J. of Applied Mechanics*, Vol. 46, No. 3, 1979, pp. 529-535.

Oda, M. and J. Konishi, "Microscopic Deformation Mechanism of Granular Material in Simple Shear," *Soils and Foundations*, Vol. 14, No. 4, 1974, p. 25.

Onoda, G.Y. and E.G. Liniger, "Random Loose Packings of Uniform Spheres and the Dilatancy Onset," *Physical Review Letters*, Vol. 64, 1990, pp. 2727-2730.

Parsi, F. and F. Gadala-Maria, "For-and-Aft Asymmetry in a Concentrated Suspension of Solid Spheres," *J. Rheology*, Vol. 31, 1987.

Poutiatine, A., "Dynamic Shear Cell for Transient Liquefaction and Solidification of Highly Concentrated Two-Phase Flows," Master of Science Thesis, Massachusetts Institute of Technology, 1990.

Reynolds, O., "On the Dilatancy of Media Composed of Rigid Particles in Contact," *Phil Mag. and J. of Science*, Vol. 20, 1885, pp. 469-481.

Rowe, P.W., "The Stress-Dilatancy Relation for Static Equilibrium of an Assembly of Particles in Contact," *Proc. R. Soc. of London, Series A*, Vol. 269, 1962, p. 500.

Rutgers, R., "Relative Viscosity of Suspensions of Rigid Spheres in Newtonian Liquids," *Rheol. Acta*, Vol. 2, 1962a, pp. 202-210.

Rutgers, R., "Relative Viscosity and Concentration," ibid., 1962b, pp. 305-348.
Savage, S.B. and S. McKeown, "Shear Stresses Developed During Rapid Shear of Concentrated Suspensions of Large Spherical Particles Between Concentric Cylinders." *J. Fluid Mech.*, Vol. 127, 1983, pp. 453-472.
Savage, S.B. and M. Sayed, "Stresses Developed by Dry Cohesionless Granular Materials Sheared in an Annular Shear Cell," *J. Fluid Mech.*, Vol. 142, 1984, pp. 391-430.
Savage, S.B., "The Mechanics of Rapid Granular Flows," *Advances in Applied Mechanics*, Vol. 24, 1984, pp. 289-367.
Savage, S.B., "Instability of an 'Unbounded' Uniform Granular Shear Flow" (submitted for publication).
Scott, R.F., "Soil Properties from Centrifuge Liquefaction Tests," *Mechanics of Materials*, Vol. 5, 1986, pp. 199-205.
Scott, R.F., "Solidification and Consolidation of a Liquefied Sand Column," *Soils and Foundations, Japanese Society of Soil Mechanics and Foundation Engineering*, Vol. 26, No. 5, 1986, pp. 23-31.
Seed, H.B., "Design Problems in Soil Liquefaction," *J. of Geotechnical Engineering*, Vol. 113, No. 8, 1987, pp. 827-843.
Taylor D.W., *Fundamentals of Soil Mechanics*, J. Wiley and Sons, New York, 1948.
Terzhagi, K, and R.B. Peck, "Soil Mechanics in Engineering Practice," J. Wiley and Sons, New York, 1948.
Thomas, D.G., "Transport Characteristics of Suspensions: VIII. A note on the Viscosity of Uniform Spherical Particles," *J. Colloid Sci.*, Vol. 20, 1965, pp. 207-277.
Vardoulakis, I., "Stability and Bifurcation of Undrained, Plane Rectilinear Deformations on Water-Saturated Granular Soils," *Int'l J. for Numerical and Analytical Methods in Geomechanics*, Vol. 9, 1985, pp. 399-414.

CHAPTER 25

NUMERICAL SIMULTION OF INELASTIC, FRICTIONAL PARTICLE-PARTICLE INTERACTIONS

O. R. Walton

25.1 INTRODUCTION

Understanding the flow behavior of particulate suspensions as well as pneumatic and dry granular flows is important for such diverse applications as in-plant and long-distance transport, manufacturing of ceramics, casting of solid-fuel rocket propellant, preparation of pharmaceuticals and a wide variety of manufacturing, food, chemical and mineral processing operations. Current lack of understanding makes scaling from laboratory bench-top prototype operations to large-scale commercial plants more of a "cut-and-try" art, than a rational design process. New approaches that specifically take into account the micro structural nature of these materials have the potential to provide us with the ability to predict their flow behavior.

The focus of this volume is suspensions and multiphase flows, emphasizing how suspended solid particles affect the hydrodynamic stresses in the fluid. In many situations, however, direct contacts between the particles themselves strongly affect the stresses in the fluid-solid mixture and the interactions with boundaries. Hydrodynamic *lubrication* forces theoretically will prevent smooth spheres in suspension from coming in physical contact with each other, since such forces diverge as surfaces approach each other. However, depending on the surface roughness and the inertia of the particles, the density and viscosity of the fluid media, and on the deformation rate, actual physical contact between real particles is not only possible, but may be highly likely during dynamic flow conditions. In gas fluidized systems most particles exceeding a few microns in diameter are likely to experience particle-particle or particle-surface contacts. Similarly, in aqueous (or other similar liquid-) solid systems, particles exceeding a few 10's of microns in diameter are likely to experience some particle-particle contacts if the deformation rate is significant. Real particles come into contact with each other because their surface roughness exceeds the theoretical gap width that would produce a large enough hydrodynamic lubrication force to prevent such a contact. Leighton (1992) discusses this point further in Chapter 6 of this volume. At high deformation rates particle-particle collisions may even dominate the behavior of many multiphase systems. Bagnold (1954) termed the regime of flows where particle collisions dominate the behavior the *grain-initeria* regime. Ahn and Brennen (1992) discuss Bagnold's criteria in Chapter 7 of this volume. After a brief review of methods used to calculate the behavior of fluid-solid systems this chapter discusses in detail various numerical and theoretical models of particle-particle interactions and the resulting behavior of assemblies of particles undergoing such interactions. Subsequent chapters (*i.e.*, Brady, 1992; Kim and Fuentes, 1992; Prosperetti and Sangami, 1992) discuss various approaches being taken to numerically simulate the behavior of suspensions, specifically including the hydrodynamic interactions carried by the interstitial fluid.

25.1.1 Stress Tensor in Fluids

The momentum flux density tensor in an *ideal* fluid is given by

$$\bar{\Pi} = p\bar{\mathbf{I}} + \rho\mathbf{u}\mathbf{u} \tag{25.1}$$

where $\rho\mathbf{u}$ is the momentum of a unit volume and p is the equilibrium pressure. The term $\rho\mathbf{u}\mathbf{u}$ represents the momentum flux due to the bulk motion of the fluid. For a *viscous* fluid a term, $-\mathbf{T}'$, which gives the irreversible "viscous" transfer of momentum in the fluid (from regions of high momentum to regions of low momentum), must be added to the total momentum flux density. The stress tensor, $\mathbf{T}$, which is that part of the momentum flux tensor that is not due to the bulk fluid motion (*i.e.*, the negative of the pressure tensor) then is given by (Landau and Lifshitz, 1959)

$$\bar{\mathbf{T}} = -p\bar{\mathbf{I}} + \bar{\mathbf{T}}' . \tag{25.2}$$

For a Newtonian fluid the stress tensor has the form (Hoover, 1991)

$$\bar{\mathbf{T}} = (-p + \lambda\, \nabla\cdot\mathbf{u})\bar{\mathbf{I}} + \eta(\nabla\mathbf{u} + \nabla\mathbf{u}^t) \tag{25.3}$$

where the equilibrium pressure p and the two viscosity coefficients η and λ all depend on the local thermodynamic state, and the superscript t indicates the transpose of the velocity gradient tensor $\nabla\mathbf{u}$. The "bulk" viscosity, which gives the excess pressure in compression and excess tension in expansion is

$$\eta_V = \lambda + (2/3)\eta . \tag{25.4}$$

If the fluid flow is incompressible, then $\nabla\cdot\mathbf{u} = 0$, so the bulk viscosity does not influence the flow, and the stress tensor simplifies to

$$\bar{\mathbf{T}} = -p\bar{\mathbf{I}} + \eta(\nabla\mathbf{u} + \nabla\mathbf{u}^t). \tag{25.5}$$

25.1.2 Stress Tensor in Particulate Systems

For a system of particles with no interstitial fluid, interacting via pairwise additive forces $\mathbf{f}_{ij}$, the instantaneous stress tensor averaged over the volume V, is given by (Evans, 1979; Irving and Kirkwood, 1950)

$$\bar{\mathbf{T}}_p = \frac{1}{V}\sum_i m_i(\mathbf{v}_i - \mathbf{u}_i)(\mathbf{v}_i - \mathbf{u}_i) + \frac{1}{V}\sum_{i>j} \mathbf{f}_{ij}\,\mathbf{r}_{ij} \tag{25.6}$$

where the first term represents the momentum carried by the particles themselves in their fluctuations about the mean velocity field $\mathbf{u}$. This term is often called the *kinetic* contribution to the stress. The subscript i on $\mathbf{u}$ indicates the value of the mean field at the instantaneous location of the ith particle. The velocity and mass of the ith particle are given by $\mathbf{v}_i$ and m_i, respectively. The second term, called the *potential* contribution, represents the rate of momentum transfer from one particle to another due to the interaction force $\mathbf{f}_{ij}$, where $\mathbf{r}_{ij}$ is a vector from the centroid of the *i*th particle to the centroid of the *j*th particle. The sum is over all interacting near neighbors. The first term is symmetric but the second term can include non-symmetric components if the interaction force is not central (Evans, 1979). If the interactions between particles are approximated by instantaneous collisions, then the last term is replaced by an average of momentum transfers due to collisions between particles (Chapman and Cowling, 1952).

25.1.3 Approximate Stress Tensor in Suspensions

In suspensions the stress distribution around each suspended particle is influenced by all other particles in the system. The stress tensor, $\mathbf{T}$, of a suspension of rigid particles in a Newtonian medium is given by Batchelor (1970),

$$\bar{\mathbf{T}} = -p\,\bar{\mathbf{I}} + \eta(\nabla\mathbf{u} + \nabla\mathbf{u}^t) + \frac{1}{V}\sum_i \int_{s_i} \mathbf{T}\cdot\mathbf{n}(\mathbf{r}-\mathbf{r}_i)d^2\mathbf{r} \tag{25.7}$$

where $\mathbf{T}$ is the local stress tensor. The summation is performed over all particles i contained in the reference volume V. The surface of the ith particle is denoted by s_i. The vector $\mathbf{r}_i$ indicates the center of the ith particle and $\mathbf{n}$ denotes the normal vector.

To be able to calculate the above integral, the stress distribution in the fluid surrounding the particles must be known. In a very dilute suspension the flow field and the stress distribution around a spherical particle can be calculated exactly. In the case of a concentrated suspension some approximations usually need to be made. The most simplifying approximation, applied in the case of concentrated suspensions where the particles nearly touch each other, is that the stress in the interstitial holes between particles is negligible compared to the stress generated in the narrow gaps separating particles (*i.e.*, only the nearly pairwise additive *lubrication* forces need be considered). The particle contribution to the stress tensor $\mathbf{T}_p$ thus becomes (Goddard, 1977; van den Brule and Jongschaap, 1991)

$$\bar{\mathbf{T}}_p = \frac{1}{V}\sum_{i>j} \mathbf{f}_{ij}\,\mathbf{R}_{ij} \tag{25.8}$$

where $\mathbf{f}_{ij}$ is the lubrication force acting between the *i*th and *j*th particles and the summation is carried out over all near pairs.

For specific configurations of particles, estimates of the effective viscosity of a suspension have been made assuming that only the normal direction portion of the lubrication force is important (Frankel and Acrivos, 1967) or including both the normal and shear components (van der Brule and Jongschaap, 1991). The latter compared favorably to an exact calculation of the viscosity of a periodic cubic array of spheres, taking into account all hydrodynamic interactions (Nunan and Keller, 1984) indicating that, at least in a cubic configuration, the lubrication forces do dominate the behavior at high concentrations.

25.1.4 Numerical Simulation of Complete Hydrodynamic Interactions

25.1.4.1 Quasi static Deformations -- Stokesian Dynamics

For particles suspended in a viscous fluid at zero Reynolds number (*i.e.*, Stokes) flow the linearized hydrodynamic equations can be solved throughout the space occupied by the fluid (and the particles). A set of "induced" forces can be applied to the fluid at the "surfaces" of the suspended particles to reproduce stick boundary conditions. The crux of the technique known as Stokesian dynamics is the computation of the induced forces that will reproduce the desired fluid velocity field that matches the surface motion of the translating, rotating particles. For spherical particles this is accomplished numerically by multipole moment expansions of particle-surface force distributions. In theory, any desired accuracy is obtainable by simply using higher moment approximations. In practice, the technique becomes computationally unwieldy as forces diverge. Significant efficiency is gained for high concentrations by calculating the short range lubrication forces between nearest-neighbor particles as separate pairwise forces; effectively utilizing the multipole expansion solutions primarily in the fluid outside the small "gap" regions. The method does not include inertial effects, and thus, is able to successfully avoid contacts or overlapping of spherical particles through the inclusion of diverging hydrodynamic lubrication forces. Brady (1992) discuses this method in detail in the next chapter; also, see Ladd (1990).

The Stokesian dynamics method generally has limited flexibility and is thus, difficult to extend to arbitrary particle shapes or to complex flow geometries. Also, the numerical algorithms usually scale as the square or even the cube of the system size, making simulations of large systems intractable. An alternative approach, with the potential to handle larger systems and more complex flow situations, is to utilize a lattice-gas cellular-automata model for the fluid motion combined with a moving-boundary algorithms for the suspended particles.

25.1.4.2 Low Reynolds Number Flows -- Lattice Gas Models

In a lattice-gas fluid space is filled with a fixed lattice of "sites" of which ~50% are "occupied" by lattice-gas (fluid) "particles". These particles move at fixed speed to adjacent sites and "collide" with other lattice-gas particles via micro-rules that conserve momentum and energy. Averages of velocities and concentration over several lattice sites reproduce Navier-Stokes fluid behavior (Frisch, *et al.*, 1986).

In suspension models using this approach a moving solid body (*i.e.*, a suspended particle) interacts with the lattice-gas via additional micro-rules which represent the collisions of the lattice-gas particles with the surface of the solid body (Ladd and Frenkel, 1990). On average these rules force the fluid next to the solid particle to move with the local velocity of the particle surface, modeling a hydrodynamic stick boundary condition. As a consequence, the lattice-gas particles exert forces and torques on the suspended particles which are then used to update the particle velocities and angular velocities, according to the masses and moments of inertia of the solid particles.

These models include inertial effects and appropriately handle lubrication force divergence for approaches up to a small fraction of a lattice cell spacing.

For dynamic situations and very close approaches, however, the approximate lubrication forces of the lattice-gas model are insufficient to prevent occasional overlapping of the particles. In such situations Ladd and Frenkel (1990) employ elastic hard-sphere (molecular dynamics) collision operators to prevent such overlaps. The method has the flexibility to include more complex interaction models for particle-particle collisions, such as the models described later in this chapter. Full hydrodynamic coupling between the fluid and the suspended particles is achieved and good agreement has been obtained with existing data on dissipative and fluctuating hydrodynamic interactions (Ladd, 1991). The method is computationally limited to flows with relatively high effective viscosities for the fluid, and thus, can only be applied to relatively small Reynolds' number flows.

25.1.4.3 Dynamic Flows -- Lattice Boltzman Models

Another method currently under development for suspensions (Ladd, 1992) is to utilize a lattice-Boltzman equation for the fluid (McNamara and Zanetti, 1988; Higuera, *et al.*, 1989; McNamara and Alder, 1992), with coupling between the phases handled analogous to the lattice-gas models. In this method the computational domain is divided into zones by a regular three-dimensional mesh; at each node a density of "fluid-like" particles is defined moving in a set of 18 lattice directions, corresponding to the nearest and second nearest neighbors of a simple cubic lattice. The population densities are a discretization in velocity space of the one-particle Boltzman distribution function. The hydrodynamic fields, mass density, momentum density and stress, are obtained directly as moments of this discrete distribution function.

The method was originally derived from lattice-gas models; however, it has several advantages over them. There are many fewer constraints on the choice of collision operators, allowing high Reynolds' number flows to be modeled. Recent lattice Boltzman simulations of a Kelvin-Helmholtz instability, at a Reynolds' number of 10000 were in near perfect agreement with a state-of-the-art Navier-Stokes solver (McNamara and Alder, 1992). The large statistical fluctuations present in a lattice-gas are absent in the lattice-Boltzman equation; thus, spatial or temporal averaging of the fluid dynamics is no longer necessary. The divergence of the lubrication forces is expected to be valid for gaps down to a fraction of a lattice grid spacing as it is for lattice-gas models. For closer approaches, or highly dynamic collisions, some additional particle-particle collision operator will need to be included.

The remainder of this chapter will discuss various approaches for modeling particle-particle interactions and/or collisions. To date macroscopic particle interaction models have been used primarily in simulations or theories of dry granular flows; however, they could also be added as supplementary collision operators for lattice-gas or lattice-Boltzman simulations of multi-phase flows. In the limit that the viscous effects of the fluid are negligible (*i.e.*, either Bagnold's grain-inertia regime, or extremely slow deformation where the gradients in velocity are small) the particle contribution to the stress tensor can dominate the stress and the deformation behavior. In truly quasi static situations with gravity and boundaries, particles will eventually settle to the point of physical contact with each other and the container walls. In such situations the continuous-contact soft-particle models discussed below could be used to prevent particles from "overlapping".

25.2 SIMULATION OF PARTICULATE SYSTEMS

Direct numerical simulation of particulate motions has been a viable research tool for the study of the macroscopic behavior of assemblies of *molecular-scale* particles for over 3 decades. The molecular-dynamics literature is extensive, with literally thousands of papers. Several recent books on the subject cover not only the history (Ciccotti and Hoover, 1986; Ciccotti et al., 1987) but also the methods (Hockney and Eastwood, 1981; Allen and Tildesley, 1987). Allen and Tildesley provide a thorough discussion of the details of the numerical techniques and "tricks of the trade" and include over 600 references. Also, Hoover (1991) discusses both classical statistical mechanics and modern computational approaches providing information on basic algorithms and advanced non-equilibrium and constrained-dynamics techniques.

Before any numerical simulation of particulate motion can be undertaken, the nature of the particle-particle interactions must be specified. In their simplest form particle-dynamics simulations consist of a set of "hard-spheres" following piece wise-linear trajectories between isolated momentum-conserving collisions. More complex simulations numerically integrate Newton's equations of motion to determine the trajectories as particles interact via short or long range forces. In general, the hard-sphere models are considerably faster than continuous force models. At high densities, however, hard-sphere collision frequencies increase dramatically and, when inelastic particles are simulated, care must be exercised to avoid situations where clusters "collapse" to a zero vibrational energy state wherein collision frequencies diverge as relative velocities vanish.

On a molecular scale a wide variety of energy-conserving force models have been utilized, ranging from repulsive power-law potentials to empirical functions approximating measured or *ab initio* calculated potential functions. The major difference between these molecular-scale interaction models and models used for macroscopic interactions is that, in a macroscopic collision, a portion of the incident kinetic energy is converted into either plastic work, deforming the particles' surfaces, or into molecular-scale thermal energy, during the collision process. As a consequence, inelastic (energy dissipating) collision models are employed to simulate macroscopic interactions. These can be rigid-body models or deformable 'soft' particles. For continuous-force 'soft'-particle models the energy absorption can take the form of a viscous damping or a hysteretic force-displacement path, in either the normal or tangential direction, or in both directions.

25.3 MACRO-SCALE ELASTIC AND INELASTIC CONTACTS

25.3.1 Hertzian Contacts and Impacts of Elastic Spheres

Hertz solved the linear elasticity equations for elastic bodies in contact (Hertz, 1881; Timoshenko and Goodier, 1970; Johnson, 1985). The resulting quasi static force displacement relation for displacements along the line of centers between two contacting, identical elastic spheres is given by

$$F = \frac{1}{3}\,\frac{E}{1-\nu^2}\sqrt{2R}\;\alpha^{3/2} \tag{25.9}$$

where E is Young's modulus, ν is Poisson's ratio, R is the spheres' radius, and α is the relative displacement after initial contact, (*i.e.*, the "virtual" overlap of the two original spherical surfaces as the elastic spheres deform at the point of contact).

25.3.2 Elastic Spheres: Theory, Experiments and Finite Element Calculations of Impacts

It has been experimentally, theoretically and calculationally verified that, for low velocity normal-direction collisions between two elastic spheres, very little energy goes into acoustic or other vibrational internal energy inside the spheres; almost all of the initial kinetic energy is recovered in the post-collisional velocities. Walton and Hagen (1984) calculated the response of an elastic sphere impacting a rigid wall using the DYNA2D finite element code (Halquist, 1978). Figure 25-1 shows representative zoning for those calculations, which are equivalent to a head-on collision between two identical spheres. They found that if the yield strength is not exceeded in the contact region of the spheres, then 99% or more of the initial kinetic energy is recovered. For spheres of nearly equal size, the duration of the contact is usually any times the sound transit time across one of the spheres, so that the assumptions of Hertz's quasi static analysis are satisfied. The same finite element calculations confirmed that, for collisions between elastic spheres, the contact time varies nearly as the inverse 1/5 power of the relative impact velocity -- as would be predicted by the Hertz contact theory if Equation 25.9 represents the force-displacement behavior of a nonlinear spring acting between two rigid spheres (Timoshenko and Goodier, 1970),

$$\tau = 2.94 \left(\frac{5\sqrt{2}\pi\rho}{4}\, \frac{1-\nu^2}{E}\right)^{2/5} \frac{R}{v_n^{\,1/5}} \,. \qquad (25.10)$$

where τ is the contact time, ρ is the density of the spheres and v_n is the incident relative velocity in the normal direction.

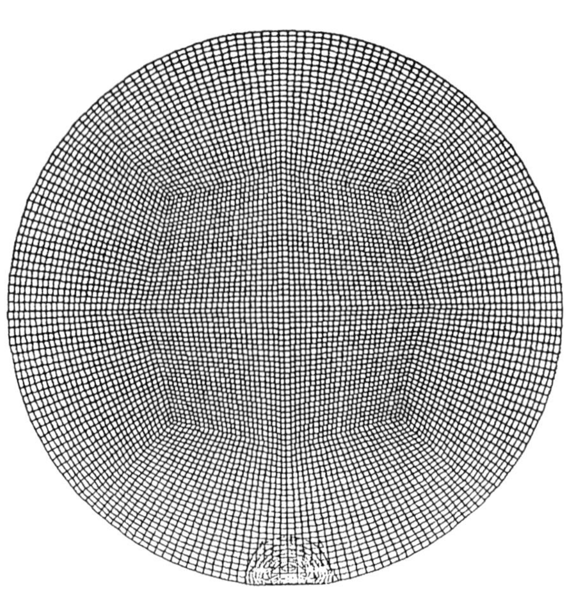

FIGURE 25-1 *Representative zoning used in dynamic finite element calculations of elastic and elastic-plastic spheres impacting a rigid wall (Walton and Hagen, 1984).*

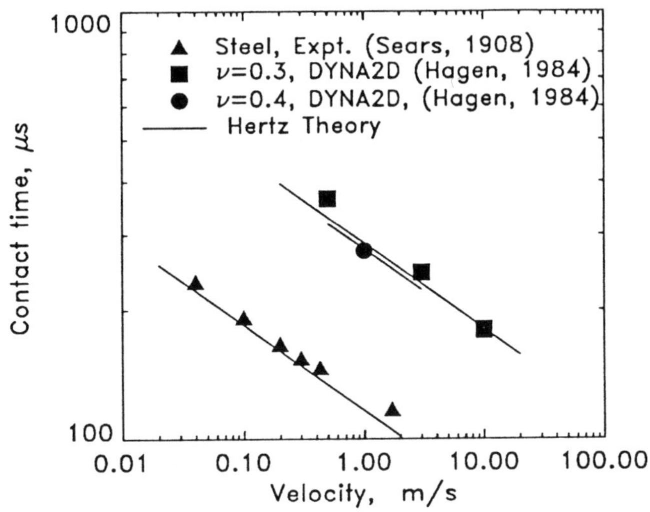

FIGURE 25-2 *Calculated contact time for impacts of elastic spheres, (Walton and Hagen, 1984), and measured contact times, for impacts of rods with spherical ends (Sears, 1908), compared with Hertz theory predicted inverse 1/5 power dependence on incident velocity, solid lines.*

Figure 25-2 shows the variation of contact time with impact velocity obtained in elastic finite element model calculations (Hagen and Walton, 1984), in experimental impacts of rods with spherical ends (Sears, 1908, as quoted in Goldsmith, 1960), and Hertz' contact theory (e.g. Equation 25.10). As can be seen, both the experiments and the finite element calculations are reasonably close to Hertz' contact theory, indicating that the quasi static assumptions of that theory are nearly satisfied in these elastic collisions.

25.3.3 Inelastic Spheres: Calculation of Elastic-Plastic Contact Forces and Impacts

There is ample experimental, theoretical and calculational evidence that for any substantial impact velocity between two colliding spheres, yield strengths will be exceeded in the contact area and significant energy will go into irreversible work deforming the surfaces in the region of contact (Goldsmith, 1960). Walton and Hagen (1984) repeated their elastic sphere finite element calculations using an elastic-perfectly-plastic constitutive model. The yield strength for these generic calculations was arbitrarily set to match the maximum calculated deviatoric stress in the contact region during a previously calculated elastic impact with an incident velocity of 0.3 m/s. For impacts at speeds less than this 'plastic yield threshold' value, the elastic-plastic results were essentially identical with the previous perfectly-elastic calculations, with less than 1 percent of the initial kinetic energy being lost upon recoil, either to numerical error or elastic waves inside the rebounding spheres. For velocities exceeding the threshold where plastic flow developed in the contact region, the effective coefficient of restitution, *e*, (defined as the negative of the ratio of recoil to incident speeds) falls approximately as the inverse 1/4 power of the velocity; a relation proposed for fully plastic collisions (Johnson, 1986).

The calculations were repeated with a slightly higher yield strength, set to allow plastic flow to initiate at an impact velocity of 0.5 m/s. These calculations produced similar results. Figure 25-3 shows the coefficient of restitution obtained in these calculations and representative experimental values obtained for collisions between ductile metal spheres (Goldsmith, 1960). Also shown on this Figure is the behavior of the empirical 'partially latching spring' model of Walton and Braun (1986).

Most discrete particle simulation calculations to date have used a constant coefficient of restitution, independent of incident velocity. The primary reason for using a fixed value for *e* is that most kinetic theory models of granular flow also employ a constant value for *e,* and there are very few experimental results available that include sufficient measurements of collision properties to adequately characterize the coefficient of restitution, except by a single average value, valid in the range of velocities observed in the experiments.

The same elastic and elastic-perfectly-plastic constitutive models were used in quasi static finite-element calculations to determine the effective normal-force displacement behavior for this idealized configuration (Walton

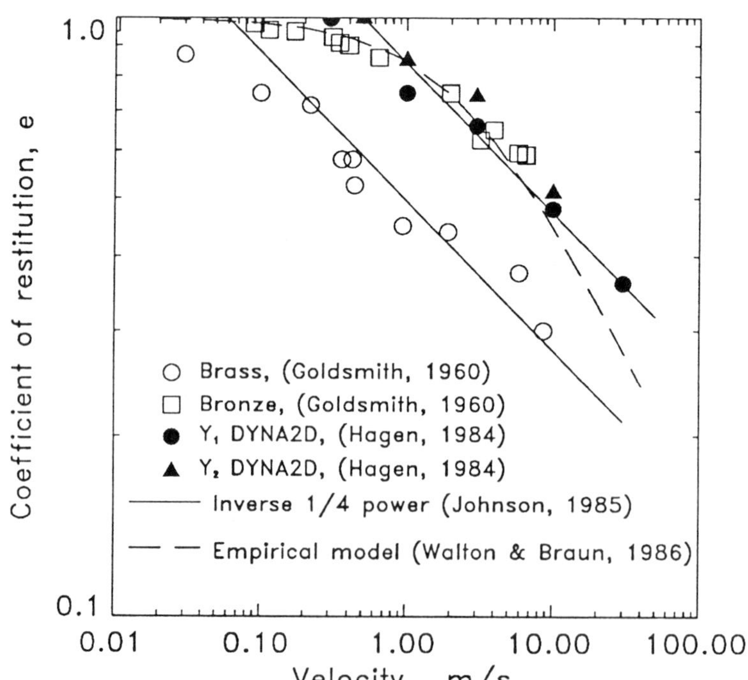

FIGURE 25-3 *Variation of coefficient of restitution with impact velocity. Experimental results (open symbols) are for metal spheres (Goldsmith, 1960). Finite element calculations for elastic perfectly-plastic material model are shown as filled circles (yield strength set so plastic deformation starts at impact velocities exceeding 0.3m/s) and pluses (yield set to allow plastic deformation only above 0.5m/s). Dashed line is representative curve for empirical formula of Walton and Braun (1986). Solid lines are inverse 1/4 power of velocity, as predicted by fully plastic theory (Johnson, 1986).*

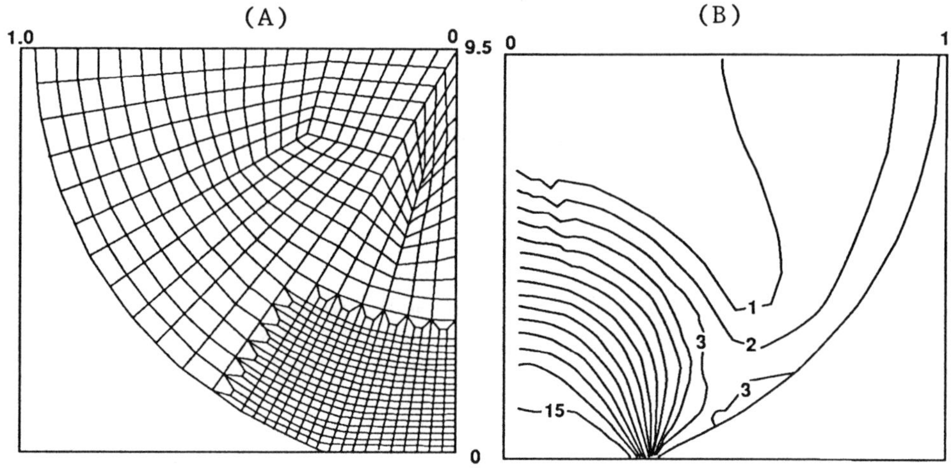

FIGURE 25-4 ***a)*** *Representative zoning used in finite element calculations of a hemisphere impinging on a rigid wall (Walton and Brandeis, 1984), and* ***b)*** *Calculated equal pressure contours using elastic-perfectly-plastic constitutive model.*

and Brandeis, 1984). Figure 25-4a shows representative zoning used in these axisymmetric, two-dimensional calculations utilizing the NIKE2D finite element model (Halquist, 1979). Coarser and finer meshes were also used. Since most of the deformation was localized in the contact region only the contacting half of the sphere was modeled. Using a perfectly-elastic constitutive model in the calculations, Walton and Brandeis verified that, to within one or two percent, the finite element force-displacement curve followed the predicted Hertzian 3/2 power relationship given by Equation 25.9.

Using an elastic-perfectly-plastic constitutive model in the NIKE2D code, the hemisphere was moved, in successive steps, toward the wall, withdrawn, and then back toward the wall again, to produce loading, unloading and reloading force-displacement curves. Figure 25-4b shows contours of constant pressure near the maximum indentation experienced by the elastic-plastic hemisphere. Figure 25-5 shows the resulting quasi-static force displacement behavior for the *elastic-plastic* body compared to the theoretical *elastic* Hertzian 3/2 power force-displacement curve. The wavy character of the elastic-plastic loading curve appears to be the result of the coarseness of the finite element mesh. When coarser or finer elements were used, the general nearly-linear loading curve was reproduced; but, the magnitude of the "wiggles" on the curve scaled with the zone size. It appears that as each new annular zone comes into contact with the rigid planer surface it initially responds elastically (*i.e.*, nearly like the theoretical Hertzian curve) but, as the elastic limit is reached and plastic flow occurs in the newly loaded zone, the force *vs.* displacement curve exhibits a lower slope until the next annular zone starts bearing a significant fraction of the load, etc. The dashed line on Figure 25-5 is a straight line drawn approximately through the 'average' calculated force-displacement loading curve.

Nearly linear loading behavior for plastically deforming contacts of spherical surfaces has also been observed experimentally. Goldsmith (1960) reports similar behavior measured during impacts of a steel sphere on an aluminum rod with a spherical end. Mullier, *et al.* (1991) measured normal-direction force *vs* micro-displacement for 6mm diameter cellulose acetate spheres in contact. While they report fitting a 3/2 power curve to the data, subsequent analysis done by Drake and Walton (1992) has inducted at least as good a fit to the observed data with a linear loading curve. Also, scanning electron microscope (SEM) examination of the contact surface region of these cellulose acetate spheres showed evidence of plastic deformation.

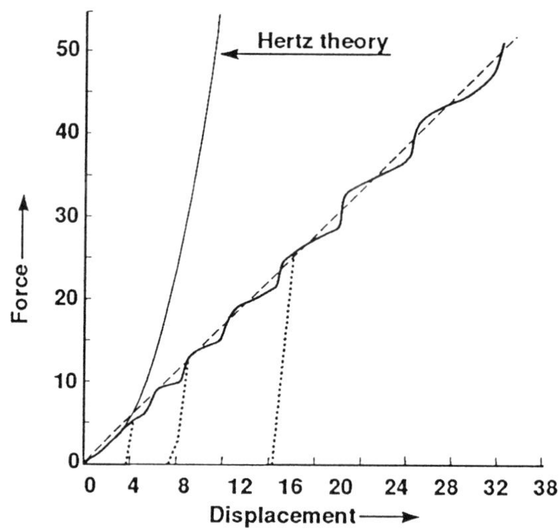

FIGURE 25-5. *Loading and unloading force-displacement behavior for elastic-plastic spheres during quasi static normal displacement as calculated using NIKE2D finite element model (Walton and Brandeis, 1984)*

25.3.4 Hysteretic Normal Force Models

Walton and Braun (1986) use a simple empirical normal force model that approximates the behavior observed in the above experiments and finite element calculations. In their model the normal force has a linear loading curve, with a slope K_1, and a steeper linear unloading slope, K_2. The normal force, N, is given by,

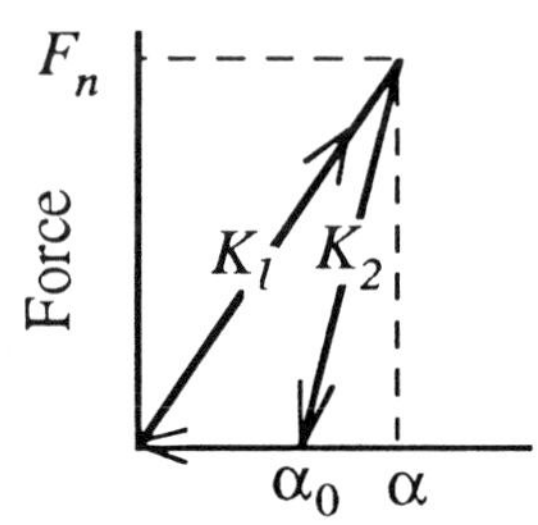

FIGURE 25-6

$$N = \begin{cases} K_1 \alpha & \text{for loading, and} \\ K_2(\alpha - \alpha_0) & \text{for unloading.} \end{cases} \tag{25.11}$$

Unloading follows this steeper slope until the normal force is zero (with a finite "overlap," α_0, remaining). Upon further unloading the force remains at zero, or, upon reloading, it increases with slope K_2 until the original loading curve (K_1) is reached, (see Fig. 25-6). The slope of the unloading/reloading curve, K_2, can be set to increase linearly with the maximum force reached during a contact, or it can be set to a constant value. The area between the loading and unloading paths represents the energy lost to plastic deformation. If the unloading slope K_2 is a fixed value (independent of the past load history) then a constant coefficient of restitution given by $e = \sqrt{K_1/K_2}$ results for this model. If the unloading slope, K_2, is allowed to increase linearly with the magnitude of the maximum force ever experienced by the contact (e.g., $K_2 = K_o + S\ F_{max}$) where the slope factor, S, is a fixed model parameter, empirically determined, then the coefficient of restitution behavior is as shown in Figure 25-3. Similar models can be easily constructed using non-linear (e.g. Hertzian-like) loading and unloading paths, by simply changing the power of the displacement terms in Equation 25.11. If the displacement terms are raised to the 3/2 power in Equation 25.11, then the initial loading is the same form as Hertz' elastic solution and the coefficient of restitution is given by $e = \sqrt[3]{K_1/K_2}$.

25.3.5 Visco-elastic Normal Force Models

An alternative to a position dependent contact model is to utilize a viscous damping term to dissipate energy. The damping can be added to an otherwise elastic normal force model. A linear damped harmonic oscillator is the simplest such approach, letting the normal force, N, be given by

$$N = K\alpha + D\dot{\alpha} \tag{25.12}$$

where K and D are constants., and α and $\dot{\alpha}$ are the relative normal displacement and velocity at the contact point. Such a model produces a constant coefficient of restitution, independent of impact velocity. As usually implemented, however, the force is purely repulsive and the viscous term is set to zero whenever there is no overlap (see, Cundall, 1974 and 1979). Such an approach leads to a discontinuity in the force as initial contact (*i.e.*, overlap) is established since $\dot{\alpha}$ is a maximum when $\alpha = 0$. The discontinuity can be eliminated by "weighting" the viscous term by the displacement. A general power law relation

$$N = K\alpha^n + D\ \alpha^m\dot{\alpha} \tag{25.13}$$

has been proposed to model real inelastic contacts (Hunt and Crossley 1975, Oden and Martins 1984). Such a power law relationship can be Hookeian (*i.e.*, with n =1) or Hertzian (n = 3/2) and can have a discontinuity on initial contact (m = 0) or it can have the damping "weighted" by the elastic force (m = n). For implementation, the values of all four parameters K, n, D and m need to be determined empirically. Oden and Martins (1984) discuss the energy losses obtained with such a model.

25.4 FRICTIONAL CONTACTS

25.4.1 Analysis of Frictional, Elastic Contacts

The problem of arbitrary oblique contacts between two frictional elastic spheres has never been completely solved. Despite the fact that more than 100 years have passed since Hertz solved the linear elastic contact equations for the distribution on normal-direction stress and strain in the region where two elastic spheres contact each other, there is still no complete solution to the stress and strain distribution if Coulombic friction is assumed to apply on the contact surface, and arbitrary rotation and tangential sliding are combined with changing normal direction contact between elastic spheres. Mindlin (1949) derived expressions for tangential compliance *vs* displacement for spheres with elastic Hertzian normal stress distributions in their contact regions. The resulting force *vs* displacement relationship is of the form

$$F = \mu N \left[1 - \left(1 - \frac{16Ga}{3(2-\nu)\mu N} \delta s \right)^{3/2} \right] \qquad (25.14)$$

up to the full sliding limit at $\delta s = \frac{3(2-\nu)}{16Ga} \mu N$ when the entire contact experiences slip and the tangential force is just equal to the Coulomb friction limiting value, $F = \mu N$; where G is the shear modulus, ν is Poisson ratio, N is normal load, μ is the coefficient of friction, δs is tangential displacement and a is the Hertzian contact radius given by

$$a = (E^* N R)^{1/3}$$

where $E^* = \frac{3}{2}\left(\frac{1-\nu^2}{E}\right)$, and E is young's modulus.

If contacting spheres simultaneously experience rotation, either about their contact normal, or "rolling," coupled with tangential sliding, then the tangential strain distribution is modified and Mindlin's theory does not adequately describe the tangential force or compliance. Various approximate models, that are relatively straight forward to calculate, have been proposed for frictional contacts such as the "elastic bed" normal force model combined with a "wire brush" tangential friction model (see Johnson, 1990) and an "incrementally slipping" model approximating Mindlin's tangential functional form for initial tangential loading (Walton and Braun, 1986; Walton, 1992). It is beyond the scope of this chapter to thoroughly review various "soft" particle friction models; however we will examine the nature of the rotational coupling that occurs for binary collisions between frictional spheres -- the feature most relevant for numerical models of granular flow. The reader is referred to Johnson (1985), and Oden and Martins (1984) for further discussion on frictional contact models. Also, Thornton (1988) and Patrakas *et al* (1990) have implemented numerical algorithms that approximate Mindlin and Deresiewicz (1953) force-displacement behavior for simultaneous tangential and normal direction displacements. (These models still do not include the confounding effects of rotations on the stress distribution in the contact region).

25.4.2 Experiments and Numerical Analysis of Frictional Collisions

Experiments involving carefully controlled impacts of an air table "puck" (made from a section of a sphere) with a planar wall, demonstrate that release of tangential elastic strain energy stored during a collision can, indeed, reverse the direction of the original rotation and the relative surface velocity at the contact point (Maw *et al.*, 1976, 1981). This is a phenomenon familiar to anyone who has played with a "superball" (e.g. see Garwin,1969), and corresponds to positive values of Lun and Savage's (1987) parameter, β, discussed below.

Maw *et al* numerically integrated the tangential stress over a Hertzian contact area during collisions that involved both sliding and rotation. Their numerical solution to the frictional contact problem was essentially equivalent to Mindlin's analysis of oblique contacts between elastic spheres (Mindlin, 1949; Mindlin and Deresiewicz, 1953). They assumed that the tangential surface displacement due to either rotation or sliding had identical effects on the distribution of tangential stress in the circular contact area (equivalent to assuming the displacement is all effectively "sliding"). This approach ignores "rolling" resistance that occurs even when no relative surface velocity exists, but both spheres are counter rotating so that their surfaces "roll" on each other and also ignores any effects due to rotation about the contact normal. Nonetheless, Maw *et al's.* analysis was able to capture the most important physical phenomena for impacts restricted to a plane. Their careful numerical integration of the contact forces acting between frictional, elastic spheres, demonstrated that the degree of tangential coupling depends strongly on the ratio of the effective tangential-direction stiffness to the stiffness in the normal-direction (Maw *et al.* 1981).

The continuously varying contact force models used in most discrete particle simulations such as Cundall (1974), Cundall and Strack(1979), Walton and Braun (1986a and Walton (1992) automatically allow reversal of the tangential surface velocity as tangential strain energy is recovered during rebound. No explicit assumptions are made in such models about rotational restitution coefficients; however, depending on the ratio of tangential to normal stiffnesses assumed, behavior very similar to that predicted by Maw *et al* can be obtained in these simulated collisions (see Walton & Braun, 1986a).

25.5 HARD SPHERE COLLISIONS WITH FRICTION

In the interest of computational efficiency, it is often preferable to specify an instantaneous collision operator for rigid particles instead of evaluating a continuously varying force-displacement relationships 40 or 50 times during each collision in the simulation. Such hard-sphere collision operators are also utilized in kinetic theory models of granular flow. Various models have been proposed to approximate collisions between inelastic, frictional spheres.

On a molecular scale perfectly rough, perfectly elastic, spheres have been a standard, energy conserving, model for molecules that include rotational degrees of freedom for decades (Chapman and Cowling 1952). More recently, various partially rough collision models have been proposed (e.g., Berne, 1977; Veseley, 1980); however, since these models conserve energy, they are not appropriate for modeling interactions between macroscopic sized particles that collide inelastically and with sliding friction. Goldsmith (1960) describes the usual model assumed for *frictional* impacts (*i.e.*, that friction forces can only slow or stop the relative tangential velocity between colliding particles). Two-dimensional hard-disk simulation models such as Hawkins(1983) and theoretical models such as Nakagawa (1987) utilize collision operators that are equivalent to Goldsmith's treatment, combined with a non-unity coefficient of restitution in the normal direction. Others, such as Campbell and Brennen (1985), have assumed that all collisions result in zero relative surface velocity upon disengagement (*i.e.*, all contacts are rolling contacts). Brach (1988) utilizes a coefficient, μ, to characterize the tangential momentum transferred in oblique impacts. His model can allow the direction of rotation to reverse, but his single parameter is not directly related to contact friction and is difficult to calibrate against measured properties.

Lun and Savage (1987) describe a rough, inelastic sphere collision operator incorporating a rotational restitution coefficient, β, defined analogous to the normal direction coefficient of restitution, e. According to their rough inelastic sphere model, two particles of diameter σ, having translational velocities $\mathbf{v}_1$ and $\mathbf{v}_2$, angular velocities $\vec{\omega}_1$ and $\vec{\omega}_2$, have a total relative velocity, $\mathbf{g}_{12}$, at their contact point just prior to collision given by,

$$\mathbf{g}_{12} = \mathbf{v}_{21} - \frac{\sigma}{2}(\hat{\mathbf{r}}_{12} \times \vec{\omega}_{12}) \tag{25.15}$$

where $\hat{\mathbf{r}}_{12}$ is a unit vector from particle 1 to 2, $\vec{\omega}_{12} = \vec{\omega}_1 + \vec{\omega}_2$, and $\mathbf{v}_{21} = \mathbf{v}_1 - \mathbf{v}_2$. During a collision the components of velocity and angular velocity are changed such that,

$$\hat{\mathbf{r}}_{12} \cdot \mathbf{v}'_{21} = -e\ (\hat{\mathbf{r}}_{12} \cdot \mathbf{v}_{21}) \tag{25.16}$$

and

$$\hat{\mathbf{r}}_{12} \times \mathbf{g}'_{12} = -\beta\ (\hat{\mathbf{r}}_{12} \times \mathbf{g}_{12}). \tag{25.17}$$

where the prime indicates the post-collision values and they assume the coefficients e and β are fixed constants in the ranges, $0 \leq e \leq 1$ and $-1 \leq \beta \leq 1$, with $\beta = -1$ being perfectly smooth, $\beta = 1$ perfectly rough spheres (Chapman and Cowling, 1952), and $\beta = 0$ corresponding to Campbell and Brennen's (1985) rolling contact model. Fixed values for e and β were used by Lun and Savage to represent average values over a spectrum of collisions and served to allow sensitivity of stresses to inelasticity and roughness to be studied, but there was no direct correspondence between rotational restitution coefficient, β, and a contact friction coefficient.

25.5.1 Frictional Hard-Sphere Collision Operator

A more realistic inelastic, frictional collision operator can be described using three parameters:

1) the coefficient of restitution, e, in the normal direction;
2) a coefficient of rotational restitution, β_o, for those contacts that are not continuously sliding during the entire collision, and,
3) a coefficient of sliding friction, μ, acting during sliding or grazing collisions.

The model is essentially an extension of the disk collision operators of Hawkins(1983) or Nakagawa (1987) or Hopkins (1988) to three dimensions with the additional feature that, during a collision, tangential velocity differences can not only decrease, but can also reverse direction. This three parameter model provides a reasonably accurate description of collisions between real macroscopic spheres, and it can easily be extended (by making β into a function of the angle of incidence) to be nearly an exact representation of collisions between frictional, elastic spheres interacting with Hertz-Mindlin contacts (see Mindlin, 1949; Maw *et al.* 1976, 1981; Walton and Braun, 1986a; or Walton, 1992).

Consider a collision between two frictional, inelastic spheres with mass m_a and m_b, centers located at $\mathbf{r}_a$ and $\mathbf{r}_b$, traveling with velocities $\mathbf{v}_a$ and $\mathbf{v}_b$ before impact and having rotational velocities $\vec{\omega}_a$ and $\vec{\omega}_b$. Let,

$$\hat{\mathbf{r}}_{ab} = \frac{\mathbf{r}_b - \mathbf{r}_a}{|\mathbf{r}_b - \mathbf{r}_a|} = \text{unit vector from } a \text{ to } b,$$

$$\mathbf{v}_{ab} = \mathbf{v}_b - \mathbf{v}_a = \text{relative velocity},$$

$$v_n = \mathbf{v}_{ab} \cdot \hat{\mathbf{r}}_{ab} = \text{normal component of relative velocity},$$

$$\mathbf{v}_n = v_n \hat{\mathbf{r}}_{ab} = \text{normal direction relative velocity},$$

$$\mathbf{v}_t = \hat{\mathbf{r}}_{ab} \times (\mathbf{v}_{ab} \times \hat{\mathbf{r}}_{ab}) = \mathbf{v}_{ab} - \mathbf{v}_n = \text{tangential direction relative velocity},$$

$$\mathbf{v}_s = \mathbf{v}_t + \frac{\sigma_a}{2}(\hat{\mathbf{r}}_{ab} \times \vec{\omega}_a) + \frac{\sigma_b}{2}(\hat{\mathbf{r}}_{ab} \times \vec{\omega}_b) = \begin{array}{l}\text{relative surface velocity}\\ \text{(tangential direction)}\end{array}$$

$$\hat{\mathbf{k}}_s = \frac{\mathbf{v}_s}{|\mathbf{v}_s|} = \text{unit vector in direction of incident surface velocity}$$

$$v_s = \mathbf{v}_s \cdot \hat{\mathbf{k}}_s = \text{tangential component of relative surface velocity.}$$

Conserving translational momentum and utilizing the usual definition of the coefficient of restitution, $e \equiv -v'_n/v_n$, and defining the rotational restitution coefficient as, $\beta \equiv -v'_s/v_s$, where the prime denotes post collision values, we obtain the changes in normal direction velocities,

$$\Delta \mathbf{v}_{na} = \frac{m_b(1+e)}{m_a + m_b} \mathbf{v}_n \tag{25.18}$$

$$\Delta \mathbf{v}_{nb} = \frac{-m_a(1+e)}{m_a + m_b} \mathbf{v}_n \tag{25.19}$$

Conserving angular momentum about the contact point we obtain three vector equations relating the four vector quantities giving the changes in rotational velocities $\vec{\omega}_a$ and $\vec{\omega}_b$ and the tangential direction translational velocities $\mathbf{v}_{ta}$ and $\mathbf{v}_{tb}$,

$$\Delta \vec{\omega}_a = \frac{2}{K\sigma_a} \hat{\mathbf{r}}_{ab} \times \Delta \mathbf{v}_{ta} \tag{25.20}$$

$$\Delta \vec{\omega}_b = \frac{2}{K\sigma_b} \hat{\mathbf{r}}_{ab} \times \Delta \mathbf{v}_{tb} \tag{25.21}$$

$$\Delta \mathbf{v}_{tb} = -\frac{m_a}{m_b} \Delta \mathbf{v}_{ta} \tag{25.22}$$

where $K = 4I_o/m\sigma^2 = 2/5$ or uniform density spheres and I_o is the moment of inertia. In order to solve for these four vector quantities an additional vector relation is needed. We use two forms for the fourth relation depending on the conditions of the impact -- one for *sliding* contacts and one for *rolling* contacts.

During frictional impacts the tangential force is always equal to or less than the product of the coefficient of friction, μ, and the normal force acting at the contact. The *sliding* solution assumes that during the entire contact the tangential force is always at the friction limit so that ,

$$\text{(tangential impulse)} = \mu \text{ (normal impulse)}.$$

However, the assumption of fully sliding contacts is not always valid since sliding may dissipate sufficient energy for the two spheres to lose all relative tangential velocity (*i.e.*, to be rolling together). In real collisions relative tangential velocities can not only go to zero, but the stored tangential strain energy in the contact region can often cause the direction of the relative surface velocity to actually reverse before separation, leading to positive values of β (e.g., see Maw *et al.* 1976, 1981; and Drake, 1988).

The solution procedure consists of assuming a *sliding* contact, obtaining the changes in tangential and rotational velocities, then checking to see if the rotational restitution coefficient resulting from the sliding solution, β^*, given by

$$\beta^* = -1 + \mu(1+e)\left(1+\frac{1}{K}\right)\frac{v_n}{v_s} \quad \text{(sliding solution)}, \tag{25.23}$$

exceeds a predetermined (constant) maximum value, β_o, where $1 \geq \beta_o \geq 0$. If $\beta^* > \beta_o$, then a *rolling* solution equivalent to Lun and Savage's collision operator with a fixed $\beta = \beta_o$ is employed to determine the post collision velocities.

25.5.2 Numerical Implementation of Collision Operator

A numerical algorithm implementing this model for equal size, equal mass, inelastic, frictional spheres is as follows:

$$\hat{\mathbf{r}}_{ab} = \frac{1}{\sigma}(\mathbf{r}_b - \mathbf{r}_a) \quad \text{(at contact)}$$

$$\mathbf{v}_{ab} = \mathbf{v}_b - \mathbf{v}_a$$

$$v_n = \mathbf{v}_{ab} \cdot \hat{\mathbf{r}}_{ab}$$

$$\mathbf{v}_t = \mathbf{v}_{ab} - v_n \hat{\mathbf{r}}_{ab}$$

$$\mathbf{v}_s = \mathbf{v}_t + \frac{\sigma}{2}\hat{\mathbf{r}}_{ab} \times \left(\vec{\omega}_a + \vec{\omega}_b\right)$$

if $(\mathbf{v}_s \cdot \mathbf{v}_s) = 0$ then

$$\beta = -1$$

else

$$\left(1+\beta^*\right)^2 = \mu^2\,(1+e)^2\left(1+\frac{1}{K}\right)^2 \frac{v_n^2}{\mathbf{v}_s\cdot\mathbf{v}_s}$$

if $\left(1+\beta^*\right)^2 > \left(1+\beta_o\right)^2$ then

$$\beta = \beta_o \qquad \text{(rolling solution)} \qquad (25.24)$$

else

$$\beta = \beta^* \qquad \left[\text{sliding solution} = -1+\sqrt{(1+\beta^*)^2}\,\right]$$

endif

endif

where the sliding solution evaluation of β^* is the only step that requires a square root call. The resulting changes in normal, tangential and rotational velocity components for each sphere are given by,

$$\Delta\mathbf{v}_{na} = -\,\Delta\mathbf{v}_{nb} = \frac{1}{2}(1+e)\,v_n\hat{\mathbf{r}}_{ab} \qquad (25.25)$$

$$\Delta\mathbf{v}_{ta} = -\Delta\mathbf{v}_{tb} = \frac{K(1+\beta)}{2(K+1)}\,\mathbf{v}_s \qquad (25.26)$$

$$\Delta\vec{\omega}_a = \Delta\vec{\omega}_b = \frac{(1+\beta)}{\sigma(K+1)}\,\hat{\mathbf{r}}_{ab}\times\mathbf{v}_s \qquad (25.27)$$

The effect of this frictional, inelastic hard-sphere collision operator can be plotted as two straight lines giving the effective rotational restitution coefficient, β, as a function of the ratio of incident normal to tangential velocities, v_n/v_s, as shown on Figure 25-7 (e.g., Equations 25.23. and 25.24).

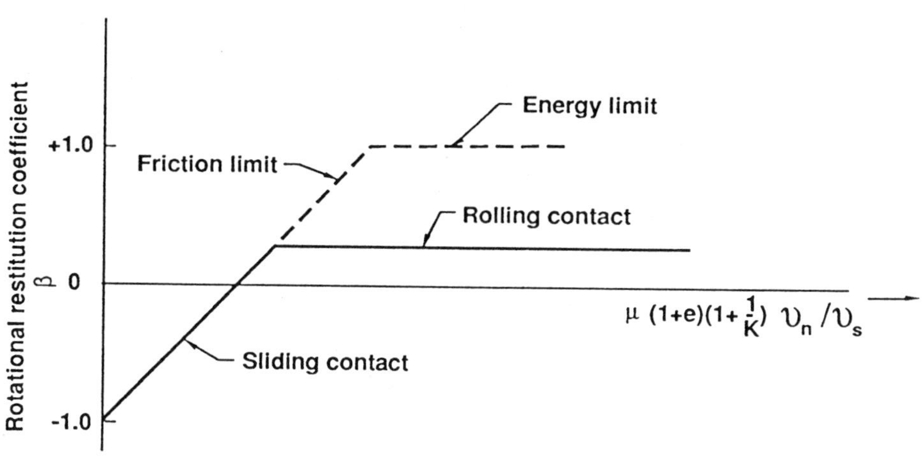

FIGURE 25-7 *Representative curve for rotational restitution coefficient, β, as given by the inelastic frictional collision operator (i.e., Equations 25.23 and 25.24).*

Application of the operator also produces two straight lines if plotted as the tangent of the effective recoil angle, $-v_s'/v_n'$, *vs* the tangent of the effective incident angle, v_s/v_n, as shown in Figure 25-8, which is equivalent to the ψ_2 *vs.* ψ_1 plots of Maw *et al.* (1976,1981). The rolling solution line in Figure. 25-8 is determined by the equation,

$$\frac{-v'_s}{v'_n} = -\frac{\beta_o}{e}\frac{v_s}{v_n}. \tag{25.28}$$

and the equation for the sliding solution line on Figure. 25-8 is,

$$\frac{-v'_s}{v'_n} = -\mu\left(1+\frac{1}{e}\right)\left(1+\frac{1}{K}\right)+\frac{1}{e}\frac{v_s}{v_n} \tag{25.29}$$

The slightly deformable particle models of Walton and Braun (1986), the experimental measurements and analysis of sphere-wall collisions of Maw *et al.* (1976, 1981), and the measurements of Drake (1988), all show a similar character for collisions of frictional spheres, with the transition from sliding to rolling occurring less abruptly for real spheres than for the collision operator model. Drake's data for plastic sphere impacts indicate that a value of β_o near 0.35 is reasonable for the "rolling" branch.

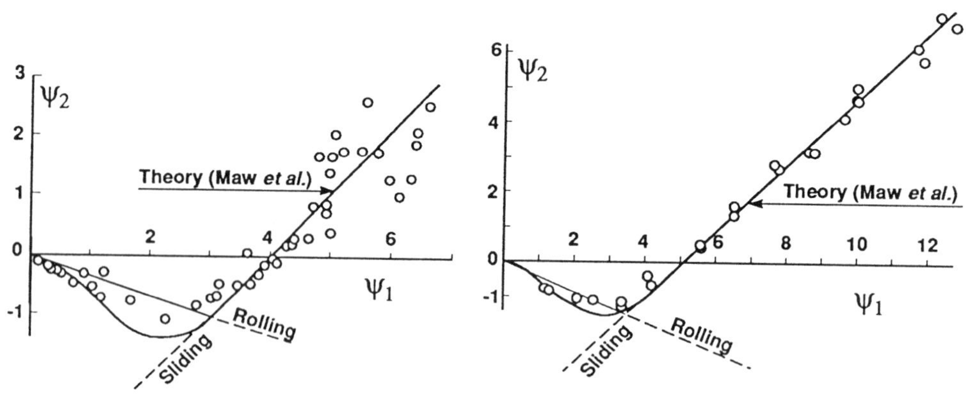

FIGURE 25.8 *Representative curves for tangent of effective recoil angle,* $-v'_s/v'_n$, *(i.e.,* ψ_2*) vs. tangent of effective incident angle,* v_s/v_n, *(i.e.,* ψ_1*) as determined by inelastic collision operator (equations 25.28 and 25.29), straight lines, by numerical analysis of Maw et al., solid curves, and in experiments of Maw et al. with rubber and steel (1976, 1981) symbols.*

25.6 STRESSES IN STEADY SHEARING FLOWS

Numerous numerical simulation studies of rapidly flowing dry granular solids have been performed over the past decade (e.g. see Campbell, 1990). Among other parameters examined, the stress tenser, given by a time average of the expression in Equation 25.6, has perhaps been the most thoroughly examined. For systems of *inelastic* particles that are at a packing density below that which produces continuous contacts (around 0.5) the *equilibrium* stress is zero since, for dissipative systems, the equilibrium *state* has no kinetic energy. Granular material in steady shearing flow, on the other hand, is not at equilibrium and numerical simulations have confirmed Bagnold's analysis that in the *grain inertia* regime all stress tensor components vary with the square of the strain rate and particle size and directly with the density of the solid material. Most steady shearing flow results are non-dimensionalized by dividing the stress by $\sigma^2 \rho_s \dot{\varepsilon}^2$ where σ is the particle diameter, ρ_s is the material density and $\dot{\varepsilon}$ is the strain rate.

Numerical simulations have corroborated kinetic theory predictions that show strong sensitivity to packing fraction and coefficient of restitution (see for example Walton and Braun, 1986a, 1986b; Walton, 1990). Initially unexpected, but later included in improved kinetic theories, were the large first normal stress differences found in steady shearing flows at very low solids fractions (Walton, Kim and Rosato, 1990). Other effects such as sensitivity to friction and recently observed clustering phenomena at low solids packing are currently the subjects of ongoing research efforts.

We have utilized the rigid sphere model to examine the effects of coefficient of friction, μ, and the maximum rotational restitution coefficient, β_o on stresses in rapid shearing flows of frictional spheres. Figure 25.9 shows how stresses in simulations at two different solids packings, $\nu = 0.4$ and 0.5, depend on coefficient of friction, μ, in the range, $0 \leq \mu \leq 1$, (stresses shown on this figure have been non-dimensionalized by dividing by $\sigma^2 \rho_s \dot{\varepsilon}^2$). The results of three different rotational coupling assumptions are shown on this figure. The curves for solid packing of 0.4 and the dashed curve at the higher density are from simulations with 125 slightly deformable, equal-sized, inelastic spheres, using an incrementally-slipping friction model (Walton and Braun, 1986a, Walton, 1992). The coefficient of restitution for all interactions in this calculational series was 0.8. The stiffness of the slightly-deformable spheres (*i.e.*, K_I in Equation 25.11) was selected so that maximum "overlaps" were a fraction of one percent of a sphere radius. This stiffness was varied by one order of magnitude to demonstrate that the results were insensitive to the specific value selected. The stresses calculated using spheres with the highest stiffness were generally within 2 percent of the values obtained with the hard sphere model (without friction).

The calculated non-dimensional normal stress, P^*_{yy}, was found to decrease monotonically with increasing friction at a packing fraction of $\nu =$ 0.4. This is apparently because friction adds another energy loss mechanism, reducing the vibrational kinetic energy in the system, and thus, the stresses. However, the additional shear coupling between colliding particles compensates somewhat for this effect in the shear direction so that the shear stress is almost unaffected by the value of friction coefficient at a solids packing of $\nu = 0.4$ Although not shown on this figure, representative calculations were done at $\mu = 0.4$ with the hard-sphere model as well. They were generally within 2% of the values shown for the slightly deformable (*i.e.*, soft) sphere model.

At the higher density the results are more interesting, with the slightly-deformable model producing normal stresses that first decrease, then increase, as the coefficient of friction increases, and shear stresses that increase nearly monotonically with the coefficient of friction. The hard-sphere model with $\beta_o = 0.3$ shows the same qualitative behavior, but the increase in stresses is much more rapid as the coefficient of friction increases

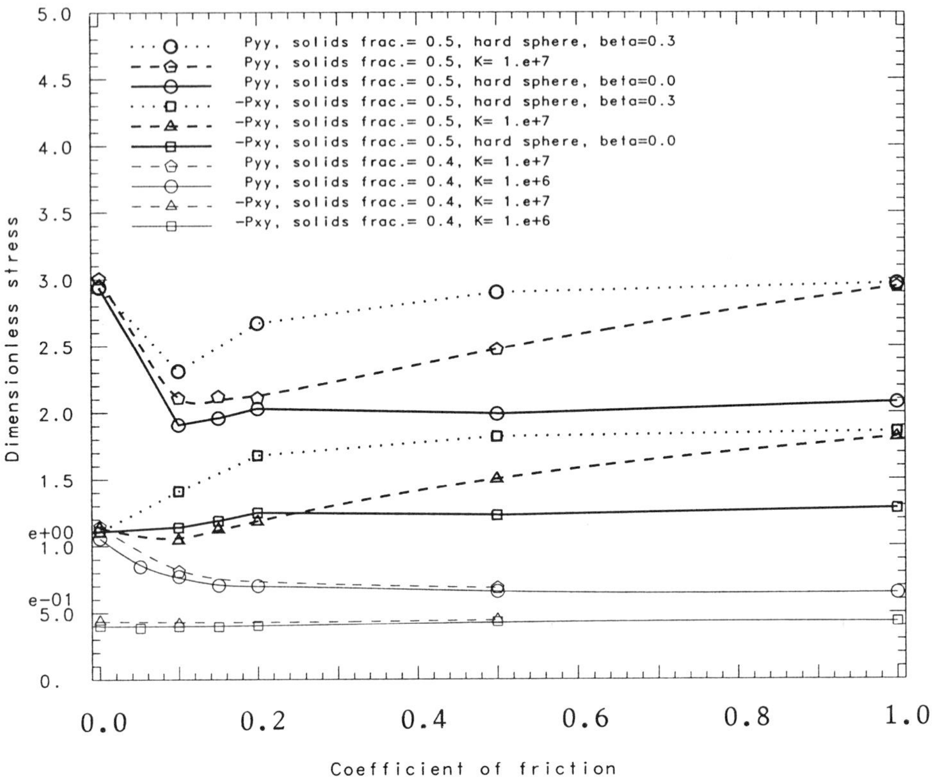

FIGURE 25-9 *Effect of coefficient of friction on non-dimensional shear and normal stresses in steady shearing flow of uniform sized, inelastic, frictional spheres as calculated for 108 hard spheres with two choices of the maximum rotational restitution coefficient, β_o, and for 125 slightly deformable spheres interacting with an incrementally-slipping friction model (Walton, 1992) similar in effect to that of Mindlin and Deresiewicz [1953].*

from 0.1 to 0.5. The hard-sphere model with $\beta_o = 0$ (corresponding to Nakagawa's model) does not show a significant increase in either stress component with increasing friction coefficient. Instead, the curves are qualitatively similar to those at the lower density. In almost all cases the stresses calculated by the incrementally-slipping deformable-sphere model lie between the two hard-sphere cases of $\beta_o = 0$ and $\beta_o = 0.3$, as we might expect since the calculated β values for "rolling branch" collisions with the soft particle model are usually between the two β_o values used in the hard sphere simulations (see Walton and Braun, 1986a. The large differences in calculated hard-sphere stresses ($\approx$50%) for the two choices of β (*i.e.*, 0 and 0.3) implies that stresses at high solids packings are more sensitive than previously thought to the details of the tangential coupling during collisions. Thus, at high packings it is important to correctly model the effects of the true interaction forces. The algorithm implementing the hard-sphere model collision operator is designed so that it can easily incorporate a non-constant β for the rolling solution; one, for example, which depends on the effective angle of incidence. This dependence could be an analytic expression or an interpolation table based on more exact collision calculations. Such an addition would only marginally decrease the computational efficiency of the hard-sphere model.

25.7 MICROSTRUCTURE AND STRESS DIFFERENCES

Recent simulations of rapid shearing flows of highly inelastic spheres have exhibited both large first normal stress differences and the formation of clusters of particles (Hopkins and Louge, 1989; Walton et al, 1991) The formation of such clusters is evidently an inherent feature of assemblies of inelastic particles even though no attractive potential is included in the interaction models (Goldhirsch, 1991; Savage, 1992; Babic, 1991). Recent simulations of Zanetti and Goldhirsch indicate that any isolated system of initially uniformly distributed, inelastic particles, with a uniform distribution of velocities will eventually exhibit large variations in the spatial distribution of mass (*i.e.*, large clusters will form as the system 'cools' via collisional energy losses).

Figure 25.10 shows the particles in two identical 'slices' from a calculation of steady shearing flow of inelastic, smooth spheres, with periodic boundaries on all sided. On average, a uniform shear is imposed by moving the image particles in the cell above the primary calculational cell to the right, and those in the image cell below the primary one to the left at constant speed. Each frame shown represents 1/5 of the total calculation space, which contains 4096 particles. The coefficient of restitution in these simulations was 0.01. The particles were initially nearly uniformly distributed throughout the

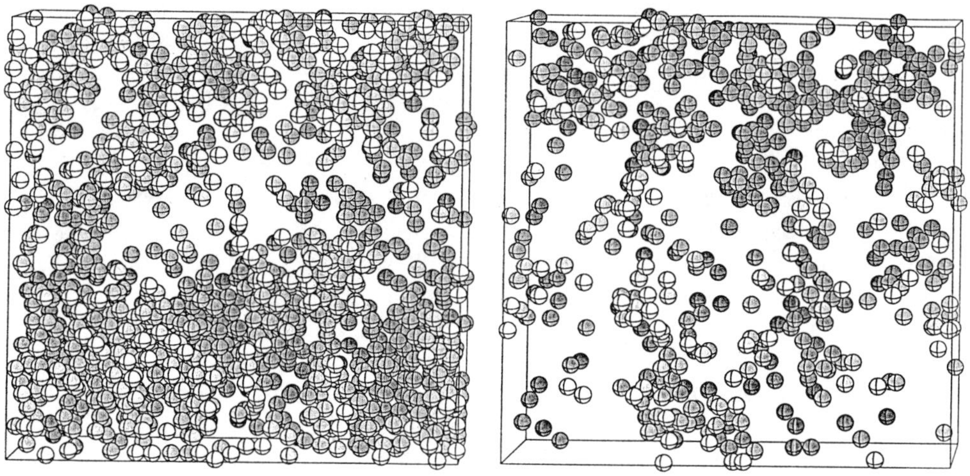

FIGURE 25-10 *Simulation of steady shearing flow of 4096 highly inelastic, e = 0.01, spheres.Two identical size slices, each showing 1/5 of the total volume of the primary calculational cell. after a shear strain of 90. Periodic boundaries on all sides. Note large variation in the distribution of particles.*

cell. The configuration shown is after a total shear deformation of 90 (*i.e.*, distance of 90 cell widths at the time of the image shown). Clearly there are large variations in the density distribution in the calculational cell. The 'cluster' size in this case is on the same order as the calculational cell size -- indicating that the simulation will be of little assistance in determining if there is a 'natural' maximum cluster size that forms in these shearing flows.

Larger sample size simulations have been used in studying cluster formation and growth (Goldhirsch, 1991) and further numerical study of such phenomena will require systems with on the order of 10^6 particles, like the models of Hoover et al. (1990).

25.8 SUMMARY

Discrete particle simulations are providing insight into mechanics that affect the flow behavior of granular solids. Good quantitative agreement has been achieved with chute flow measurements and with theories when the same assumptions are made in the models and the theories. Similar good agreement has been obtained with annular shear cell measurements of shear stress (Walton, 1990)

New phenomena, such as the clusters that were recently "discovered" in low-density simulations (Goldhirsch, 1991, Hopkins and Louge, 1991 and Walton *et al*, 1991) are now the focus of new theoretical studies (Babic, 1991, Savage, 1992, Goldhirsch (& Zanetti) , 1991). Other phenomena, such as large first normal stress differences in steady shearing flows are now generally accepted, and have been incorporated into kinetic theory models for granular materials (Richman and Chou, 1989).

In simulations of systems of interacting particles the physics that distinguishes one material from another is contained in the interaction parameters assumed. Thus, it is important to include a sufficiently realistic model of the "true" interactions if we want the simulated bulk behavior to follow the "true" bulk behavior of the material. Eliminating such details as interparticle friction (and thus all particle rotation) may produce models that are simple and straight forward to evaluate; however, leaving out such important effects may produce a "model material" that bears little resemblance to any real granular material.

A particularly useful feature of discrete particle numerical simulations like the ones described here is that sensitivity studies can be performed in a straight forward manner to determine how individual features of interparticle interactions affect the bulk behavior of the assembly. For example, to see if friction and particle rotations, are important to include, one need merely simulate the flow configuration of interest with, and without, friction in the interparticle interaction.

Based on our present knowledge, it is not possible to make a blanket statement about what effect friction has on the stresses, even in flows as simple as steady rectilinear shear flow. The addition of friction effects can either decrease or increase stresses, depending on the solids concentration and the magnitude of the friction coefficient. This is but one example showing that even dry granular flows are more complex than originally envisioned and that microstructural analyses can provide insight into their behavior.

Dry granular flows remain an important subset of two-phase particulate flow behavior. It is important that we understand the behavior of the particulate phase as well as the fluid phase. Discrete particle simulations are aiding in that understanding.

25.9 FUTURE WORK

With the advent of massively parallel computers and improved software, we are seeing significant improvements in our ability to simulate the complex phenomena in particulate flows. For example, Hoover *et al*, (1990) describe million-atom molecular dynamics simulations. Such algorithms and parallel processors could also be applied to macroscopic particle simulations. Kim and Fuentes (1992) describe new boundary integral algorithms applicable to suspensions and Ladd and Frenkel (1990) are developing new lattice-Boltzman methods for suspensions. As massively parallel computers become more widely available we can expect to see even more rapid proliferation of new algorithms to take advantage of them, and a corresponding significant improvement in our ability to simulate these complex systems. Direct simulation of flows of engineering interest will soon be possible and future design calculations may well utilize discrete particle flow simulations as routinely as today's fluid flow designers utilize finite element flow codes.

ACKNOWLEDGMENTS

This work was partially supported through the US Department of Energy Office of Fossil Energy, Advanced Research and Technology Development, Solids Transport Program and is funded through the Pittsburgh Energy Technology Center. The DYNA2D calculations and elastic/plastic collision analysis is from unpublished work of Deborah Hagen, conducted during a summer internship at Lawrence Livermore National Laboratory in 1983. The quasistatic calculations and analyses are from previously unpublished work of Julius Brandies in 1984. The author appreciates being able to utilize their work. The original hard-sphere molecular-dynamics code, used with the collision operator described here, was supplied by Tony Ladd. Carlos Caronera provided programming assistance and much of the soft sphere work was done in collaboration with Robert Braun. Work performed under the auspices of the U.S. Department of Energy by Lawrence Livermore National Laboratory under Contract W-7405-Eng-48.

REFERENCES

Ahn, H. and C.E. Brennen, "Channel flows of granular materials and their rheological implications," in *Particulate Two-Phase Flow* (M.C. Roco, ed.) Chapter 7, Butterworth-Heinemann, Boston, (1992).

Alder, B.J. and T. Wainwright, "Studies in Molecular Dynamics, I. General Method," *J. Chem. Phys* **31**, 459 (1959).

Allen, M.P. and D.J. Tildesley, *Computer Simulation of Liquids*, Clarendon Press, Oxford, (1987).

Babic, M. "Particle clustering: an instability of rapid granular flows," in Proceedings U.S.-Japan Seminar, *Micromechanics of Granular Materials*, Potsdam, NY, Aug 5-9, (1991)

Bagnold, R.A., "Experiments on a gravity-free dispersion of large solid particles in a Newtonian fluid under shear," *Proc. R. Soc. Lond.*,**A225**, 49-63 (1954).

Batchelor, G.K., *J. Fluid Mech.*, **41**,545-570 (1970).

Berne, B.J. "Molecular dynamic of the rough sphere fluid II, Kinetic models of partially sticky sphere, structured spheres, and rough screwballs," *J. Chem. Phys.*, **66**, 2821-30 (1977); also *J. Chem Phys.*, **66**(7) 1745-1754 (1977).

Brach, R.M. "Rigid Body Collisions," *J. Appl. Mech.* (1988).

Brady, J.F. "Stokesian dynamics simulation of particulate flows," in *Particulate Two-Phase Flow* (M.C. Roco, ed.) Chapter 26, Butterworth-Heinemann, Boston, (1992).

Campbell, C.S., "Rapid Granular Flows," *Ann. Rev. Fluid Mech.*, **22**, 57-92,(1990).

Campbell, C.S. and C.E. Brennen "Computer Simulation of Granular Shear Flows," *J. Fluid Mech.*, **151**, 167-188 (1985).

Chapman, S. and T.G. Cowling, *The Mathematical Theory of Non-Uniform Gases,* 2nd ed., Cambridge Univ. Press.(1952, 1970).

Ciccotti, G. and W.G. Hoover, *Molecular Dymnamics Simulations of Statistical-Mechanical Systems*, North-Holland, Amsterdam (1987).

Ciccotti, G., P. Frenkel, and I.R. McDonald, *Simulation of Liquids and Solids*, North-Holland, Amsterdam (1987).

Cundall, P.A., "Rational Design of Tunnel Supports: A Computer Model for Rock Mass Behavior Using Interactive Graphics for the Input and Output of "geometrical Data," Tech. Rept. MRD US Army Corps of Eng., Univ of Minnesota, Rept. Sept. (1974).

Cundall, P.A., and O.D.L Strack, "A discrete numerical model for granular assemblies," *Geotechnique*, **29**, 47-65 (1979); also: *The distinct element method as a tool for research in granular media, Part I.*, Univ. Minn. Rpt. to Nat. Sci. Found. Grant ENG 76-20711 (1979).

Drake, T.G. and R. Shreve "High Speed Motion Pictures of Nearly Steady Uniform, Two-dimensional, Inertial Flows of Granular Material," *J. Rheology*, **30**(5) 981-993 (1986).

Drake, T.G. *Experimental flows of granular materials*, PhD dissertation, UCLA (1988).

Drake, T.G., and O.R. Walton "Comparison of Experimental and Simulated Grain Flows" (in preparation, 1992)

Evans, D.J., "Nonsymmetric Pressure Tensor in Polyatomic Fluids," *J. Stat. Phys.* **20**(5), p547 (1979) .

Frankel, N.A. and A. Acrivos, *Chem. Eng. Sci.*, **22**, 847-853 (1967).

Frisch, U., B. Hasslacher, and Y. Pomeau, *Phys. Rev. Lett.*, **56**, 1505 (1986).

Garwin, R.L., "Kinematics of an Ultraelastic Rough Ball," *Am. J. Phys.*, **37**(1) 88-92 (1969).

Goldhirsch, I., "Clustering Instabilities in Granular Gases," in proceedings *DOE/NSF Workshop on Flow of Particulates and Fluids*, Oct 22-24 1991. S.I. Planski, W.C Peters and M.C. Roco, eds., Nat. Tech., Info. Serv., Springfield, VA, (1991).

Goldsmith, W., *Impact*, E. Arnold Pub., London. (1960).

Goddard, J.D., *J. Non-Newtonian Fluid Mech.*, **142**, 269-287 (1984).

Halquist,J., "DYNA2D--An Explicit Finite Element and Finite Difference Code for Axisymmetric and Plane Strain Calculations," Lawrence Livermore Nat. Lab. Rept. UCRL-52429 (1978).

Halquist,J., "NIKE2D: An Implicit Finite Deformation, finite Element Code for Analyzing the Static and Dynamic Response of Two-Dimensional Solids," Lawrence Livermore Nat. Lab. Rept. UCRL-52678 (1979).

Hawkins, G.W., "Simulation of Granular Flow," in *Mechanics of Granular Materials, New Models and Constitutive Relations*, J.T. Jenkins and M. Satake, eds., Elsevier, 305-312 (1983).

Hertz, H., *J. Math (Crelles's J.)*, **92**, (1881); see also; Timoshenko, S.P. and J.N. Goodier, *Theorgy of Elasticity*, 3rd ed., McGraw-Hill, NY, (1970).

Higuera, F.J. S. Succi and R. Benzi, *Europhys. Lett.*, **9**, 345 (1989).

Hockney, R.W. and J.W. Eastwood, *Computer simulation using particles*, McGraw Hill, NY (1981).

Hoover, Wm. G., *Computational Statistical Mechanics*, Elsevier.Sci. (1991).

Hoover, Wm. G., A.J. DeGroot, C.G. Hoover, I.F. Stowers, T. Kawai, and B.L. Holian, "Large-Scale Elastic Plastic Indentation Simulations via Nonequilibrium Molecular Dynamics," *Phys. Rev. Lett.*,**42** (10) 5844 (1990)

Hopkins, M.A., and H.H. Shen, "A Monte Carlo Simulation of Simple Shear Flow of Granular Materials," in *Micromechanics of Granular Materials*, M. Satake and J.T. Jenkins, eds., Elsevier Sci. Pub. (1988)

Hopkins, M.A., and M.Y. Louge, *Phys. Fluids, A*, **3**(1) p47-57 (1991).

Hunt, K.H. and F.R.E. Crossley, "Coefficients of Restitution Interpreted as Damping in Vibroimpact," *J.Appl. Mech.*,pp440-445, June (1975).

Irving, J.H. and J.G. Kirkwood, *J. Chem. Phys.* **18**(6), 817 (1950).

Johnson, K.L., "Classical vs. Numerical Methods of Elastic Contact Stress Analysis," in *Contact Stress Analysis*, A.D. Roberts and J.E. Muttershead, eds., Institute of Physics, UK Short Mtg Series No. 25, p1-14 (1990).

Johnson, K.L., *Contact Mechanics*, Cambridge Univ. Press, (1985).

Kim, S. and Y.O. Fuentes, "Numerical simulation of suspension flow on high performance computers," in *Particulate Two-Phase Flow* (M.C. Roco, ed.) Chapter 27, Butterworth-Heinemann, Boston, (1992).

Ladd, A.J.C., "Lattice-Boltzman Simulations of Solid-Fluid Processes," to be published, (1992).

Ladd, A.J.C., "Hydrodynamic interactions in a suspension of spherical particles," *J. Chem. Phys.*, **88**, 5051 (1988).

Ladd, A.J.C., "Hydrodynamic transport coefficients of random dispersions of hard spheres," *J. Chem. Phys.*. **93**, 3484 (1990).

Ladd, A.J.C., "Dissipative and fluctuating hydrodynamic interactions between suspended solid particles," in Proc. NATO Adv. Study Institute, *Computer Simulations in Material Science*, M. Meyer and V. Pontikis, eds. (Kluwer Publishers, Dordrecht) (1991).

Ladd, A.J.C. and D. Frenkel, "Dissipative hydrodynamic interactions via lattice-gas cellular automata," *Phys. Fluids*, **A2**, 1921 (1990).

Landau, L.D. and E.M. Lifshitz, *Fluid Mechanics*, Pergamon Press, London (1959).

Leighton, D. and I. Rampall, "Measurement of the shear induced microstructure of concentrated suspensions of non-colloidal spheres," in *Particulate Two-Phase Flow* (M.C. Roco, ed.) Chapter 6, Butterworth-Heinemann, Boston, (1992).

Lun, C.K.K. and S.B. Savage, "A Simple Kinetic Theory for Granular Flow of Rough, Inelastic, Spherical Particles," *J. Appl. Mech.* (1987).

Maw, N., J.R. Barber and J.N. Fawcett "The Oblique Impact of Elastic Spheres," *Wear*, **38** (1) 101-114 (1976).

Maw, N., J.R. Barber and J.N. Fawcett (1981) "The role of Elastic Tangential Compliance in Oblique Impacts,"*Trans ASME, J. Lub. Tech.*, **20**, 327-344.

McNamara, G. and G. Zanetti, *Phys. Rev. Lett.*, **61** 2332 (1988).

McNamara, G. and B.J. Alder, to be published (1992).

Mindlin, R.D. and H. Deresiewicz, *J. Appli. Mech (Trans. ASME)* **20**,327 (1953).

Mindlin, R.D., *J. Appl. Mech (Trans. ASME)* **16**, 259 (1949).

Mullier, M., U. Tuzun and O.R. Walton, "A single-particle friction cell for measuring contact frictional properties of granular materials," *Powder Technology,* **65**, 61-74, 1991.

Nakagawa, N. "Kinetic Theoretical Approach for Rapidly Deforming Disk Assembly," in *Micromechanics of Granular Materials*, M. Satake and J.T. Jenkins, eds., Elsevier Sci. Pub. (1988); see also: N. Nakagawa, PhD. Thesis, Cornell University (1987).

Nunan, K.C. and J.B. Keller, "Effective Viscosity of a Periodic Suspension," *J. Fluid Mech.*, **142**, 269-287 (1984).

Oden,J.T. and J.A.C. Martins, "Models and Computational Methods for Dynamic Friction Phenomena," Proceedings: Fenomech III in: *Computer Methods in Applied Mechanics and Engineering*, North Holland, Amsterdam (1984).

Patrakis, E., R. Dobry, T.T. Ng and L. Liu, *A Study of the Behavior and Micromechanical Modeling of Granular Solids*, Civil Eng. Rept., Rensselaer Polytech. Inst., Troy, NY, USAF Contract No. AFOSR-89-0350.(1991).

Prosperetti, A. and A.S. Sangani, "Numerical simulation of the motion of particles at large Reynolds numbers," in *Particulate Two-Phase Flow* (M.C. Roco, ed.) Chapter 28, Butterworth-Heinemann, Boston, (1992).

Richman, M.W., and C.S. Chou, *J.Rheology,* **33** (8), 1292 (1989).

Savage, S.B. "Instability of an unbounded uniform granular shear flow," *J. Fluid Mech.* (in press 1992).

Sears, J.E., "On the Longitudinal Impact of Metal Rods with Rounded Ends," *Trans. Camb. Phil. Soc.*, **21**, 49, (1908).

Timoshenko, S.P. and J.N. Goodier, *Theory of Elasticity*, 3rd ed. McGraw-Hill, NY (1970).

Thornton, C., and C.W. Randall, "Application of Theoretical Contact Mechanics to Solid Particle System Simulations," in *Micromechanics of Granular Materials*, M. Satake and J.T. Jenkins, eds., Elsevier Sci. Pub. (1988).

van den Brule, B.H.A.A., and R.J.J. Jongschaap, "Modeling of Concentrated Suspensions," *J. Stat. Phys.*, **62**(5/6) 1225-1237 (1991).

Veseley, F.S. *Mol. Phys.*, **41**, 5, 959-1015 (1980).

Walton, O.R. and R.L. Braun, "Viscosity and Temperature Calculations for Assemblies of Inelastic Frictional Disks," *J. Rheology,* **30**(5), 949-980 (1986a).

Walton, O.R., R.L. Braun, R.G. Mallon and D.M. Cervelli, "Particle-Dynamics Calculations of Gravity Flow of Inelastic, Frictional Spheres," in *Micromechanics of Granular Materials*, M. Satake and J.T. Jenkins, eds., Elsevier Sci. Pub. (1988)

Walton, O.R., R.L. Braun, "Stress Calculations for Assemblies of Inelastic Spheres in Uniform Shear," *Acta Mechanica*, **63**, 73-86 (1986b)

Walton, O.R., K.A. Hagen, and J.M. Cooper, "Interparticulate Force Models for Computational Simulation of Granular Solids Flow," presented at 16th International Congress of Theoretical and Applied Mechanics, Lyngby, Denmark 19-25, Aug (1984a).

Walton, O.R., J. Brandeis, and J.M. Cooper, "Modeling of Inelastic, Frictional, Contact Forces in Flwoing Granular Assemblies," *Abstracts*, 21st Annual Mtg. Soc of Eng. Sci., Blacksburg, VA, p257, (1984).

Walton, O.R., "Computer Simulation of Rapidly Flowing Granular Solids," *Science on Form,* S. Ishizaka, ed., KTK Sci. Pub., Tokyo, p175-189 (1990).

Walton, O.R., "Numerical Simulation of Inclined Chute Flows of Monodisperse, Inelastic Frictional Spheres," *Mech. of Mat.* (in press, 1992).

Walton, O.R., H. Kim, and A.D. Rosato, "Microstructure and Stress Differences in Shearing Flows," *Mechanics Computing in 1990's and Beyond*, H. Adeli and R.L. Sierakowski, eds., ASCE, NY, Vol 2, p1249-1253 (1991).

CHAPTER 26

STOKESIAN DYNAMICS SIMULATION OF PARTICULATE FLOWS

J. Brady

26.1 INTRODUCTION

Understanding the behavior of suspensions of particles immersed in a fluid medium is essential for the characterization of a variety of natural and industrial processes. Specific examples include the flow of coal slurries, polymer processing and the manufacture of composite materials and ceramics. The aim is the prediction of macroscopic transport properties (e.g. sedimentation or aggregation rates, self-diffusion coefficients, rheological behavior, etc.) from the microstructural mechanics—the hydrodynamic, colloidal, Brownian or externally imposed forces acting on the particles, as well as their distribution in space and time.

This chapter describes how the molecular-dynamic-like simulation method known as *Stokesian Dynamics* can be used to study suspension behavior. In Stokesian dynamics particles in suspension interact through both hydrodynamic and nonhydrodynamic forces, where the latter may be any type of Brownian, colloidal, interparticle or external force. Stokesian dynamics is capable of predicting both static and dynamic microstructural properties, as well as macroscopic properties in either dilute or concentrated systems. The applications of Stokesian dynamics are many, and we shall discuss several in this chapter ranging from the rheology of Brownian dispersions to particle migration in pressure driven channel flow to yield stress phenomena in electrorheological (ER) fluids. Stokesian dynamics is designed to provide the same theoretical and computational basis for multiphase, dispersed systems as does molecular dynamics for statistical descriptions of matter.

The problems associated with the dynamical behavior of suspensions are similar to those encountered in statistical mechanics when studying the equilibrium and transport properties of condensed matter, and it is

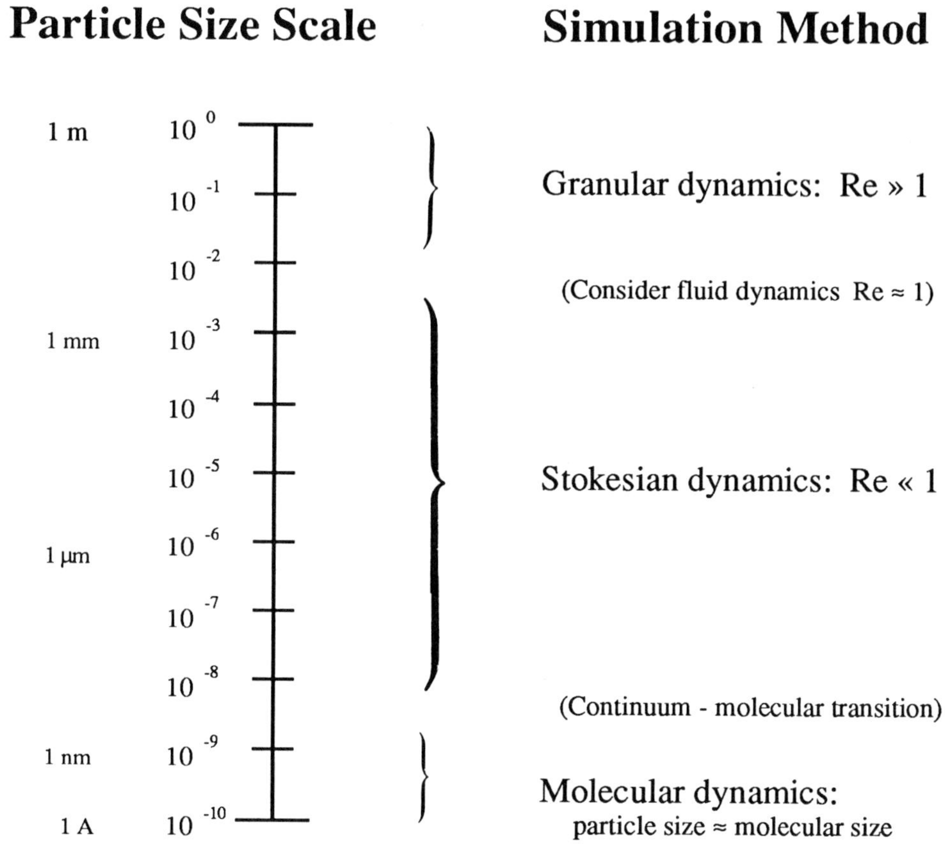

Figure 26.1: *Schematic diagram of particle size versus simulation method*

natural to ask where Stokesian dynamics fits into the broader area of dynamic simulation in general. Dynamic simulation of matter can be divided into three main areas distinguished by the length scale on which the phenomena are represented as illustrated in Figure 26.1.

On the molecular scale, where the "particles" of interest are the molecules themselves, Newton's laws of motion are simulated with particles in vacuum interacting through intermolecular potentials. Both equilibrium and nonequilibrium simulations are possible and a wide variety of phenomena have been investigated (Allen and Tildesley 1989). As the particles of interest increase in size and become larger than the solvent molecules, for example, proteins, colloids, fibers, glass beads, etc. dispersed in a continuum solvent of small molecules, the physics changes as the "macroparticles" interact via continuum-scale interparticle forces, such as London-van der Waals, screened electrostatic, etc., and through hydrodynamic forces transmitted via the continuum fluid. These macroparticles may also be subject to fluctuating thermal forces giving rise to Brownian motion. The central elements at this level are the hydrodynamic forces, and when the particle Reynolds number is small, these are governed by the Stokes equations—hence the name Stokesian dynamics.

One could imagine trying to carry out conventional molecular dynamics at this scale by simultaneously simulating the macroparticles and all the surrounding solvent particles. However, the number of solvent particles per macroparticle scales as $(a_p/a_s)^3$, where a_p is the size of a macroparticle and a_s that of a solvent molecule. Further, the time scale for the motion of the solvent is much faster than that for the macroparticles, with the ratio scaling as $(a_p/a_s)^2$. Hence, in conventional molecular dynamics one would need to follow all the solvent molecules for a very long time before any motion of the macroparticles was observed. Not only is one not interested in this rapid motion of the solvent, but the resulting excess computational demand scales (at least) as $(a_p/a_s)^5$. Clearly, when the particles of interest are larger than the solvent molecules, one needs a different approach. One in which the solvent is modeled as a continuum and only the dynamics of the macroparticles are followed. This is precisely what Stokesian dynamics accomplishes. As Figure 26.1 indicates Stokesian dynamics is applicable over a very wide range of particle sizes, from roughly 10^{-8} to 10^{-2}m; the precise conditions being that $a_p \gg a_s$ and the particle Reynolds number $Re \ll 1$.

Moving up in particle scale, when the interest is in the motion of large particles such as gravel, rocks or billiard balls, one enters the domain of "granular dynamics." While one is still on the continuum scale, it is presumed that the fluid (usually air) separating the grains plays no role in the (high-Reynolds-number) dynamics. Grain-grain collisions dominate and, along with gravity or other external forces, determine the behavior. The structure of the dynamical equations is very similar to that of conventional molecular dynamics, except care is needed to include the elasticity of the particles and friction, both macroscopic concepts (Campbell 1990).

In this rapid assessment of simulation techniques I have skipped two areas. The first is the molecular-continuum transition where the particles are only slightly larger than molecules and the time scales of interest are comparable. Here there is no clear separation of scales and both the solvent and the particles must be considered together as in conventional molecular dynamics. This poses only a computational, not a conceptual, problem. The second area is on the continuum scale when the behavior of the fluid cannot be ignored (as in granular dynamics) and when the Reynolds number is not small (as in Stokesian dynamics), so that the full Navier-Stokes equations must be addressed. There is no simulation method available, and none conceived, for this most difficult of scales.

With this perspective I would now like to describe Stokesian dynamics and illustrate some of its applications.

26.2 SIMULATION METHOD

A detailed derivation of the simulation method can be found elsewhere (Durlofsky et al. 1987; Bossis and Brady 1987; Brady and Bossis 1988; Brady et al. 1988; Bossis and Brady 1989); here, I shall proceed rather quickly. For N rigid particles suspended in an incompressible Newtonian fluid of viscosity η and density ρ, the motion of the fluid is governed by the Navier-Stokes equations, while the particle motion is described by the coupled N-body Langevin equation:

$$\mathbf{m} \cdot \frac{d\mathbf{U}}{dt} = \mathbf{F}^H + \mathbf{F}^P + \mathbf{F}^B, \tag{26.1}$$

which simply states: mass × acceleration equals the sum of the forces. In (26.1) $\mathbf{U}$ is the particle translational/rotational velocity vector of dimension $6N$, and the $6N$ force/torque vectors $\mathbf{F}$ represent: 1) the hydrodynamic forces $\mathbf{F}^H$ exerted on the particles due to their motion relative to the fluid, 2) the deterministic non-hydrodynamic forces $\mathbf{F}^P$, which may be either interparticle or external, and 3) the stochastic forces $\mathbf{F}^B$ that give rise to Brownian motion.

When the motion is such that the particle Reynolds number is small ($Re = \rho a^2 \dot{\gamma}/\eta \ll 1$ for shear flow for example), the hydrodynamic force exerted on the particles in a suspension undergoing a bulk linear flow is

$$\mathbf{F}^H = -\mathbf{R}_{FU} \cdot (\mathbf{U} - \mathbf{U}^\infty) + \mathbf{R}_{FE} : \mathbf{E}^\infty. \tag{26.2}$$

Here, $\mathbf{U}^\infty$ is the imposed flow at infinity evaluated at the particle center, $\mathbf{E}^\infty(t)$ is the symmetric part of the velocity gradient tensor and is constant in space, though it may be an arbitrary function of time. $\mathbf{R}_{FU}(\mathbf{x})$ and $\mathbf{R}_{FE}(\mathbf{x})$ are the configuration dependent resistance tensors that give the hydrodynamic force/torque on the particles due to their motion relative to the fluid and due to an imposed shear flow, respectively. The vector $\mathbf{x}$ represents the generalized configuration vector specifying the location *and* orientation of all N particles.

The deterministic, non-hydrodynamic force $\mathbf{F}^P$ is arbitrary and may be most any form of interparticle or external force. The stochastic or Brownian force $\mathbf{F}^B$ arises from the thermal fluctuations in the fluid and is characterized by

$$\langle \mathbf{F}^B \rangle = 0 \qquad \text{and} \qquad \langle \mathbf{F}^B(0)\mathbf{F}^B(t) \rangle = 2kT\mathbf{R}_{FU}\delta(t). \tag{26.3}$$

In (26.3) the angle brackets denote an ensemble average, k is Boltzmann's constant, T is the absolute temperature and $\delta(t)$ is the delta function. The amplitude of the correlation between the Brownian forces at time 0 and at time t results from the fluctuation-dissipation theorem for the N-body system.

The evolution equation for the particles is obtained by integrating (26.1) over a time step Δt that is large compared with τ, the Brownian relaxation time ($\tau = m/6\pi\eta a$), but small compared with the time over which the configuration changes. A second integration in time produces the evolution equation for the particle positions (both translational and orientational) with error of $O(\Delta t^2)$:

$$\Delta \mathbf{x} = Pe\{\mathbf{U}^{\infty} + \mathbf{R}_{FU}^{-1}\cdot[\mathbf{R}_{FE} : \mathbf{E}^{\infty} + \dot{\gamma}^{*-1}\mathbf{F}^{P}]\}\Delta t + \nabla\cdot\mathbf{R}_{FU}^{-1}\Delta t + \mathbf{X}(\Delta t),$$

$$\langle\mathbf{X}\rangle = 0 \qquad \text{and} \qquad \langle\mathbf{X}(\Delta t)\mathbf{X}(\Delta t)\rangle = 2\mathbf{R}_{FU}^{-1}\Delta t. \qquad (26.4)$$

Here, $\Delta\mathbf{x}$ is the change in particle position during the time step Δt, and $\mathbf{X}(\Delta t)$ is a random displacement due to Brownian motion that has zero mean and covariance given by the inverse of the resistance tensor. In (26.4) $\mathbf{x}$ has been nondimensionalized by the characteristic particle size a; the time by the diffusive time scale a^2/D_0, where $D_0(= kT/6\pi\eta a)$ is the diffusion coefficient of a single isolated particle; the shear forces by $6\pi\eta a^2\dot{\gamma}$, where $\dot{\gamma} = |\mathbf{E}^{\infty}|$ is the magnitude of the shear rate; and the interparticle forces by their magnitude $|\mathbf{F}^P|$. $Pe = \dot{\gamma}a^2/D_0 = 6\pi\eta a^3\dot{\gamma}/kT$ is the Péclet number, which measures the relative importance of the shear and Brownian forces, and $\dot{\gamma}^* = 6\pi\eta a^2\dot{\gamma}/|\mathbf{F}^P|$ is a nondimensional shear rate giving the relative importance of shear and interparticle or externally imposed forces.

Equation (26.4) shows clearly that the suspension's behavior depends on the dimensionless parameters: Pe, $\dot{\gamma}^*$ and α_s, the volume fraction of particles. No restriction has been made to particles of identical size or shape; they need not be spherical, and if not, other dimensionless parameters characterizing their shape would be present.

The macroscopic properties are found from appropriate definitions and averages over particles and over time in a dynamic simulation. Here we shall be primarily interested in diffusion and rheology. Several "particle diffusivities" may be defined. The short-time self-diffusivity $\mathbf{D}_0^s$, which measures the average instantaneous mobility of a particle, is given by an average over all configurations: $\mathbf{D}_0^s = \langle\mathbf{D}_{ii}\rangle$, where the subscript ii (no sum on i) indicates that only the diagonal or self terms are included in the sum. The N-particle diffusion tensor $\mathbf{D}$ is given by the Stokes-Einstein relation:

$$\mathbf{D} \equiv kT\mathbf{R}_{FU}^{-1}. \qquad (26.5)$$

The long-time self-diffusivity $\mathbf{D}_\infty^s$, which measures the ability of a particle to wander far from its starting point, is defined as the limit as time approaches infinity of one half of the time rate of change of the mean-square position of a particle:

$$\mathbf{D}_\infty^s = \lim_{t\to\infty} \frac{1}{2}\frac{d}{dt}\langle(\mathbf{x} - \langle\mathbf{x}\rangle)^2\rangle. \tag{26.6}$$

Both diffusivities are accessible by light scattering in tracer-diffusion experiments, since they are the short- and long-wavelength scattering limits, respectively (Rallison and Hinch 1986; van Megen et al. 1986). Intermediate-time definitions are not possible, as the motion of a particle is not, in general, diffusive except at short and long times.

For rheology, the bulk stress $\langle\Sigma\rangle$ is needed. This is defined as an average over the volume V containing the N particles and is given by

$$\langle\Sigma\rangle = \mathrm{IT} + 2\eta\mathbf{E}^\infty + \frac{N}{V}\{\langle\mathbf{S}^H\rangle + \langle\mathbf{S}^P\rangle + \langle\mathbf{S}^B\rangle\}. \tag{26.7}$$

Here IT stands for an isotropic term of no interest for incompressible suspensions. The particles make three contributions to the bulk stress; (*a*) a mechanical or contract stress transmitted by the fluid due to the shear flow, $\langle\mathbf{S}^H\rangle$; (*b*) a stress due to the interparticle forces, $\langle\mathbf{S}^P\rangle$; and (*c*) a direct contribution from Brownian motion, $\langle\mathbf{S}^B\rangle$. Note that the contributions to the bulk stress parallel the forces in the Langevin equation (26.1) and are given by

$$\langle\mathbf{S}^H\rangle = -\langle\mathbf{R}_{SU}\cdot\mathbf{R}_{FU}^{-1}\cdot\mathbf{R}_{FE} - \mathbf{R}_{SE}\rangle : \mathbf{E}^\infty, \tag{26.8a}$$
$$\langle\mathbf{S}^P\rangle = -\langle\mathbf{R}_{SU}\cdot\mathbf{R}_{FU}^{-1} + \mathbf{x}\mathbf{I}\rangle\cdot\mathbf{F}^P, \tag{26.8b}$$
$$\langle\mathbf{S}^B\rangle = -kT\langle\nabla\cdot(\mathbf{R}_{SU}\cdot\mathbf{R}_{FU}^{-1})\rangle. \tag{26.8c}$$

$\mathbf{R}_{SU}(\mathbf{x})$ and $\mathbf{R}_{SE}(\mathbf{x})$ are configuration-dependent resistance tensors, similar to $\mathbf{R}_{FU}$ and $\mathbf{R}_{FE}$, relating the particle "stresslet" $\mathbf{S}$ to the particle velocities and to the imposed rate of strain, respectively. The stresslet is the symmetric and traceless first moment of the force distribution integrated over the particle surface.

The evolution equation (26.4) and the macroscopic properties (26.6)–(26.8c) are the heart of the dynamic simulation. They are an exact description for N particles of arbitrary size and shape suspended in a volume V interacting through hydrodynamic, interparticle or external and Brownian forces. Given an initial configuration, (26.4) is simply integrated in time to follow the dynamic evolution of the suspension microstructure, and (26.6)–(26.8c) evaluated to determine the diffusive and rheological behavior. In addition to these macroscopic properties, a large number of particle distribution functions (*e.g.* $g(\mathbf{r})$, cluster sizes, etc.) can be determined.

In order to make use of the evolution equation and calculate macroscopic properties, the hydrodynamic resistance tensors $\mathbf{R}_{FU}$, $\mathbf{R}_{FE}$, etc. must be determined. A method that accurately and efficiently accounts for the near-field lubrication effects and the dominant many-body interactions has been developed (Durlofsky et al. 1987; Brady et al. 1988). Lubrication forces, as the name implies, result from the thin layer of viscous fluid separating particle surfaces and result in, among other effects, the relative motion of two particles approaching zero as their surfaces touch.

The resistance tensors $\mathbf{R}_{FU}$, $\mathbf{R}_{FE}$, $\mathbf{R}_{SU}$ and $\mathbf{R}_{SE}$ can be written as part of a "grand resistance" tensor $\mathcal{R}$, which relates the force/torque ($\mathbf{F}$) and stresslet ($\mathbf{S}$) exerted by the fluid on the particles to the particle velocities and the rate of strain. The corresponding inverse or "grand mobility" tensor $\mathcal{M}$ is $\mathcal{M} = \mathcal{R}^{-1}$. The method proceeds by constructing an approximation to $\mathcal{M}$ by combining the Faxén laws for particle velocities with a truncated mulitpole moments representation of the particles. The grand mobility tensor, denoted $\mathcal{M}^{\infty}$, is then inverted to yield a far-field approximation to the grand resistance tensor. This many-body approximation to the resistance tensor lacks, however, lubrication. Lubrication would only be reproduced upon inversion of the mobility tensor if all multipole moments were included. Because of their short-range nature, lubrication forces are two-body interactions and are introduced in a pairwise additive fashion in the resistance tensor. Thus, the approximate grand resistance tensor that includes near-field lubrication and far-field many-body interactions is

$$\mathcal{R} = (\mathcal{M}^{\infty})^{-1} + \mathcal{R}_{lub}, \tag{26.9}$$

where $\mathcal{R}_{lub}$ stands for the near-field lubrication interactions. The grand resistance tensor is then partitioned and used in the evolution equation (26.4) and the macroscopic properties (26.6)–(26.8c). This procedure captures both the near- and far-field physics and has given excellent results for all situations in which a comparison has been possible (Durlofsky et al. 1987; Brady et al. 1988; Phillips et al. 1988).

The long-range $(1/r)$ nature of the hydrodynamic interactions requires care in simulating infinite suspensions, i.e. letting $N \to \infty$, $V \to \infty$, keeping $n = N/V$ fixed. A simple summation of interactions results in badly divergent expressions. While there are several alternate ways to overcome this convergence problem, the method presented by

O'Brien (1979) can conveniently be used in dynamic simulation. Used with periodic boundary conditions this method "renormalizes" all divergent and conditionally convergent hydrodynamic interactions and accelerates the convergence of the interactions using the Ewald summation technique. (This amounts to removing the $\mathbf{k} = 0$ terms from the reciprocal lattice sums.)

I shall now illustrate the utility of Stokesian dynamics for modeling particulate flows by discussing four examples: 1) the behavior of sheared Brownian suspensions of spherical particles, 2) the determination of the hydrodynamic transport properties of suspensions of rod-like particles, 3) pressure driven flow of a suspension between parallel walls, and 4) yield stress phenomena in electrorheological fluids. Where possible comparison with experiment will be made.

26.3 BROWNIAN DISPERSIONS

The first studies that served to calibrated the Stokesian dynamics method with regard to the basic formulation and the treatment of the hydrodynamic interactions were for equilibrium hard-sphere microstructures. In the absence of a shear flow, a suspension of Brownian particles is distributed according to the equilibrium Boltzmann distribution: $P_N(\mathbf{x}) \sim \exp(-V(\mathbf{x})/kT)$, where $P_N(\mathbf{x})$ is the N-particle probability density and $V(\mathbf{x})$ is the N-particle interparticle potential. For the hard-sphere system the potential is zero when particles do not touch and infinite if the particles were to overlap. For this equilibrium distribution one can determine the short-time self diffusivity and the high-frequency dynamic viscosity as a function of the volume fraction. This was done in Phillips et al. (1988), and the results are in excellent agreement with experiment (Ottewill and Williams 1987; Pusey and van Megen 1983; Beenakker and Mazur 1984; Batchelor 1976).

The simulations I would like to discuss here are for the simple shear flow of a suspension of Brownian hard spheres at a volume fraction $\alpha_s = 0.45$. The only parameter to be varied is the Péclet number, measuring the relative importance of hydrodynamic to Brownian forces. The number of particles in the unit cell varies from 27 to 123, and the runs are for at least 50,000 times steps with a step size of 10^{-3}.

The first results presented are for the long-time self diffusivity defined in (26.6). Figure 26.2 is a log-log plot of the simulation results for the yy component of the long-time self-diffusion coefficient $\mathbf{D}^s_\infty$, nondimensionalized by D_0, as a function of Pe. (The velocity gradient of the shear flow is in the y−direction.) At small values of Pe, the system is essentially

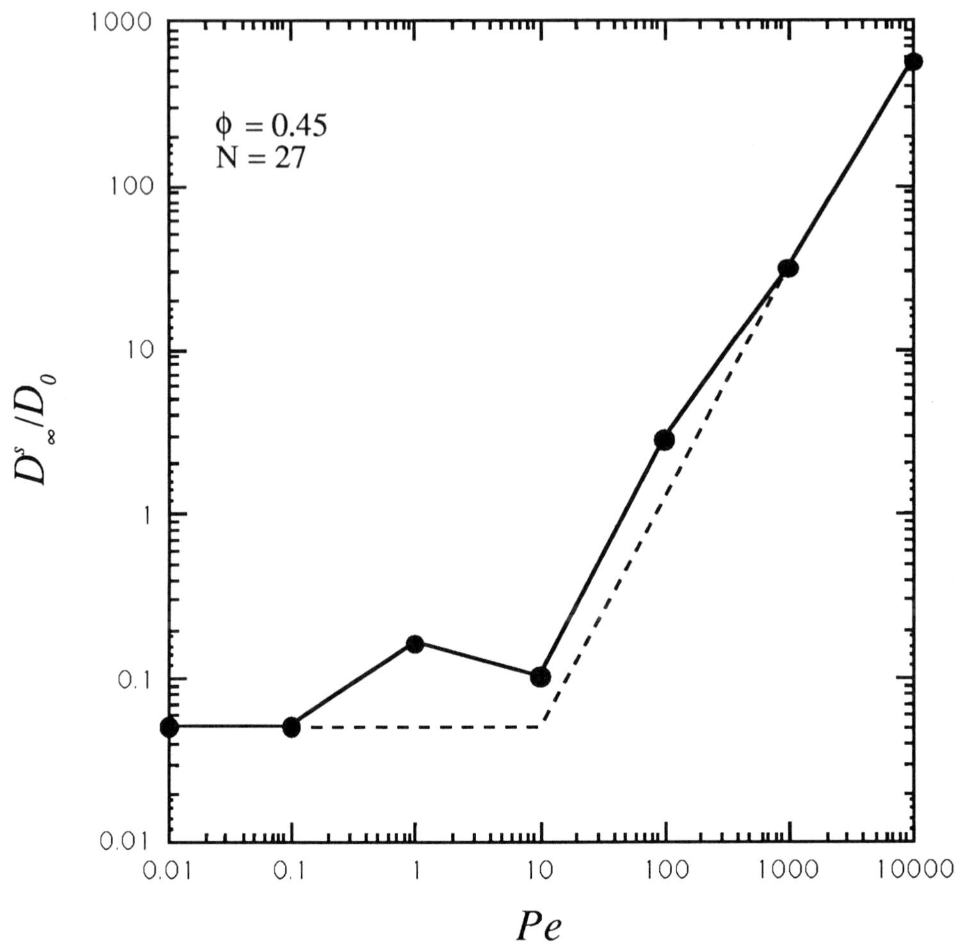

Figure 26.2: *The long-time self diffusivity normalized by the infinite dilution diffusivity* D_0 *is plotted as a function of the Péclet number for a hard-sphere suspension at* $\alpha_s = 0.45$. *The limiting asymptotes are:* $D^s_\infty \sim 0.048$ *as* $Pe \rightarrow 0$ *and* $D^s_\infty \sim 0.055Pe$ *as* $Pe \rightarrow \infty$. *Note, as* $Pe \rightarrow \infty$, *the dimensional long-time self diffusivity scales as* $\dot{\gamma}a^2$

Brownian, and the dimensionless diffusion coefficient is less than unity as expected. At higher Péclet number, however, the long-time self diffusivity behaves quite differently. On purely dimensional arguments, as $Pe \rightarrow \infty$, the *dimensional* $\mathbf{D}^s_\infty$ would be expected to scale as $\dot{\gamma}a^2$; so $\mathbf{D}^s_\infty$, nondimensionalized by D_0, should scale with Pe as $Pe \rightarrow \infty$. Reference to Figure 26.2 shows that this is indeed the case. The transitional Péclet number, where the behavior changes from a Brownian motion dominated regime to a hydrodynamically dominated regime, occurs at $Pe \approx 10$.

Recent experiments by Eckstein et al. (1977) and by Leighton and

Acrivos (1987a) have demonstrated and quantified hydrodynamic dispersion of non-Brownian particles in shear flow. This hydrodynamically induced diffusive motion comes about from the deterministic chaos displayed by the nonlinear evolution equation for the particle positions (Equation (26.4) as $Pe \rightarrow \infty$). Although there seems to be no question regarding the presence of hydrodynamic dispersion, the experimental measurements are not in complete agreement. For $\alpha_s < 0.2$ both sets of experiments agree, but for higher α_s the results of Leighton and Acrivos grow rapidly with α_s, while on the other hand Eckstein et al. find the diffusivity to saturate at a value of approximately 0.03 for $\alpha_s > 0.2$. Recently Leighton (1991) measured anew the long-time self diffusivity and found it to saturate at a value of 0.09 for $\alpha_s > 0.3$. Our result, 0.055, (*cf.* Figure 26.2) lies between the results of these two groups. Further, we found little change in this limiting value for a simulation at $\alpha_s = 0.316$, and a slight decrease at $\alpha_s = 0.49$. It should be noted, however, that we have not extrapolated our simulation results to infinite system size, which resulted in an increase in the short-time self diffusivities, and that there is a fair degree of statistical uncertainty in the simulation determination of the long-time self diffusivities owing to the very long runs required. Never the less, the agreement between simulation and experiment is quite reasonable.

The viscosity of a suspension is defined with the aid of (26.7)–(26.8c) for the bulk stress. Specifically, for a shear flow in the xy plane the relative viscosity η_r, the viscosity of the suspension divided by the fluid viscosity, is the relation between the xy component of the bulk stress and the xy component of the bulk rate of strain. From (26.7) in the absence of interparticle forces

$$\eta_r = 1 + \eta_r^H + \eta_r^B, \tag{26.10}$$

where η_r^H and η_r^B correspond to the contributions from the terms in (26.7). In addition to the sum over particles, an average over time in a dynamic simulation is also performed. The nondimensionalization is such that for an isolated spherical particle $\mathbf{S}^H = \frac{20}{3}\pi\eta a^3\mathbf{E}$, and $\eta_r^H = \frac{5}{2}\alpha_s$, which gives Einstein's viscosity correction $\eta_r = 1 + \frac{5}{2}\alpha_s$. Although the full stress is calculated in the simulation, here only the viscosity will be discussed.

In Figure 26.3 the evolution of the hydrodynamic and Brownian contributions to the viscosity with the Péclet number is shown. The total relative viscosity resembles quite strikingly experimental curves: shear thinning at low Péclet numbers ($Pe < 1$), a region where the viscosity does not vary appreciably, and finally a shear-thickening region at large Péclet number ($Pe > 10^2$). Although there is a plateau region in

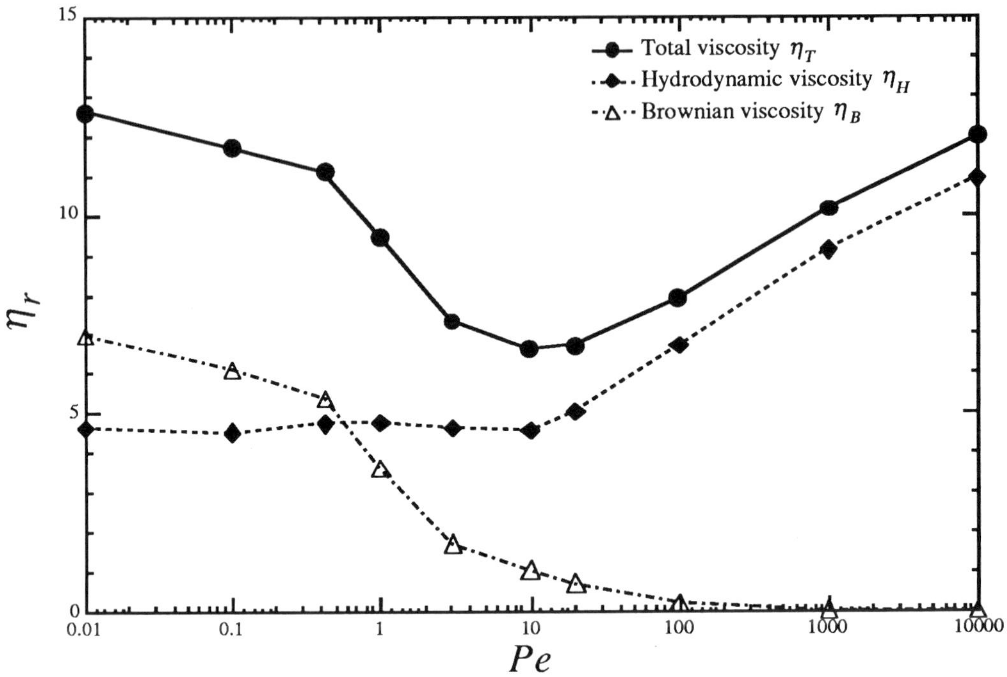

Figure 26.3: *The relative viscosity η_r of a suspension of hard spheres at volume fraction $\alpha_s = 0.45$ obtained by Stokesian dynamics as a function of the Péclet number: (–●–) total relative viscosity; (–△–) Brownian contribution η_r^B, and (–◆–) hydrodynamic contribution η_r^H.*

Pe where the total viscosity is roughly constant ($1 < Pe < 10^2$), both the hydrodynamic and Brownian components vary significantly in this region, but in such a way so as to compensate each other. For $Pe > 10$, the Brownian viscosity has essentially decayed away to zero. At low Péclet numbers, $\eta_r^B \sim \eta_r^B(0) - APe^2$, as $Pe \to 0$, where the Pe^2 scaling comes from considerations of reversing the direction of shear in a simple shear flow. Although theoretically the perturbation to the viscosity should be quadratic in Pe, our simulation results are not sufficiently accurate to check precisely this scaling. At the lowest value simulated, $Pe = 0.01$, the deformation of the equilibrium structure is so small that there is a large degree of statistical uncertainty in the Brownian contribution to the stress.

The hydrodynamic contribution to the viscosity shows a trend completely opposite that of the Brownian contribution. η_r^H remains constant for $0 < Pe < 1$, being equal to the high frequency dynamic viscosity mentioned earlier, and then continuously increases with increasing Pe. The increase in η_r^H for $Pe > 1$ is due to the formation of large clusters,

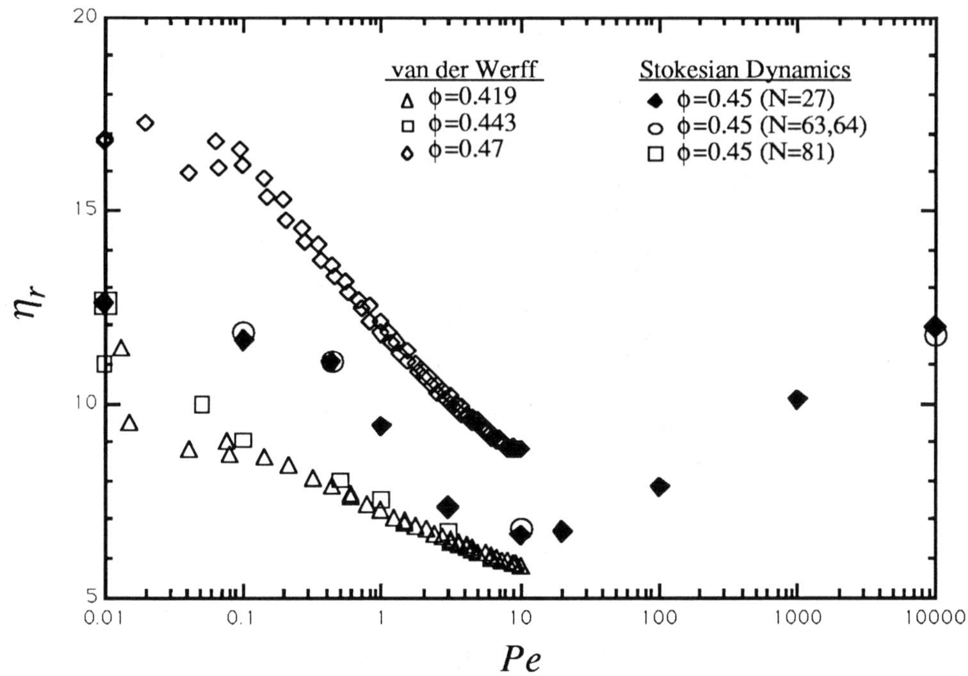

Figure 26.4: *Comparison of the simulation results for the steady shear viscosity as a function of the Peclét number at a volume fraction* $\alpha_s = 0.45$ *(*$\blacklozenge, N = 27; \bigcirc, N = 64; \square, N = 81$*) with the experiments of van der Werff and de Kruif (1989):* $\diamondsuit, \alpha_s = 0.47; \square, \alpha_s = 0.443; \triangle, \alpha_s = 0.419$.

whose size grows with the Péclet number.

Shown in Figure 26.4 is a comparison of the steady shear viscosities of Figure 26.3 with the experimental results of van der Werff and de Kruif (1989) for silica hard-spheres. The volume fractions in the experiments are $\alpha_s = 0.47 \pm 0.01, 0.443 \pm 0.01$ and 0.419 ± 0.01. The Péclet number range studied was $10^{-2} \leq Pe \leq 10$. The solid diamonds are Stokesian dynamics simulations at $\alpha_s = 0.45$ for 27 particles in the periodic cell, the open circles are for 64 particles, and the open square is for 81 particles in the unit cell. The simulation results for the different number of particles in the unit cell show little change from 27 to 81 particles. As can be seen, both the qualitative and quantitative agreement with the experiments is excellent. The major source of uncertainty in the experiments is in the determination of the volume fraction, which leads to a large uncertainty in the viscosity at high volume fraction due to the extreme sensitivity of the viscosity on volume fraction.

Both the experiments and the simulations show the suspension shear thins due to the fact that the Brownian contribution to the stress decays

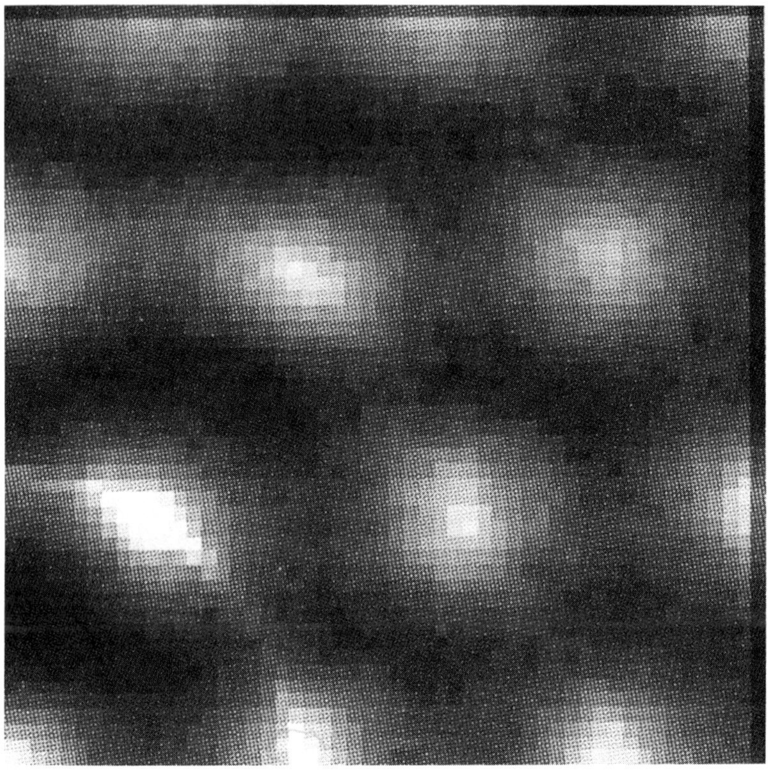

Figure 26.5: *A plot of probability density for finding a particle at a position in the $y-z$ plane given that there is a particle at the origin, $g(y,z)$, at $Pe = 10$. There are 123 particles in the unit cell. Regions of light represent high probability and dark low.*

as the shear rate increases as discussed above. The simulations show a further shear thickening region for $Pe \geq 100$. This region was not seen in the experiments of van der Werff and de Kruif because experimental limitations did not allow them to go to high enough shear rates. Shear thickening has been observed, however, in many other systems (Hoffman 1972; Laun 1988; Boersma et al. 1990). Shear thickening in Brownian suspensions is due to the formation of large clusters as the Péclet number increases (Bossis and Brady 1989). This behavior is generic, as can be appreciated by noting that the viscosities measured for spherical particles at infinite Péclet number (Jeffrey and Acrivos 1976; Pätzold 1980) are larger than the minimum in the shear thinning viscosity of the Brownian suspensions shown above in Figure 26.4; thus, the suspension *must* shear thicken.

Finally, I present some microstructural information from a simulation at $Pe = 10$ for 123 particles in the unit cell. The most revealing is a plot of the probability density for finding a second particle at a point in the $y - z$ plane given that there is a particle at the origin. Recall that the flow direction is x, the velocity gradient y and the vorticity z. Such a density plot is shown in Figure 26.5. The light areas are high probability and the dark low. This plot shows that the particles are lined up in "strings" in the flow direction, *i.e.* along the x-axis. We see quite clearly the hexagonal packing of the particles. This packing is such that it is easy for the particles to slide relative to one another following the imposed shear flow. (Note, one is looking back down the flow direction, so that particles coming out at you in the x-direction are situated directly above those shown in the $y - z$ plane.) This string-like formation has not been observed for the steady shear of hard spheres at this solids fraction. In oscillatory shear flow, however, this order has been seen experimentally (Ackerson 1989, 1991) and it may be that the periodic boundary conditions, with particles running into their images, is much more analogous to large amplitude oscillatory flow. String formation has also been observed in nonequilibrium molecular dynamics simulations of atomic liquids (Heyes 1988; Weider et al. 1991), and the tendency for the particles to arrange themselves in this fashion, in which it is easiest to flow, has been used to explain the shear thinning phenomenon observed in these computer simulations. Although our Brownian suspensions appear to form this string-like order, the shear thinning we observe is *not* due to this structural arrangement; rather, the shear thinning is a direct consequence of the disappearance of the Brownian contribution to the stress. Indeed, the hydrodynamic contribution to the stress is virtually unchanged from its zero Péclet number value even though the microstructure is very different.

26.4 ROD-LIKE SUSPENSIONS

Stokesian dynamics is not limited to suspensions of spherical particles and in this section I illustrate its application to suspensions of prolate spheroids. The equations introduced in Section 26.2 are valid for particles of any size or shape; all that is needed are expressions for the resistance tensors. This has recently been accomplished (Claeys and Brady 1991a,b,c; Claeys 1991) by using a nonlocal multipolar and Faxén law representation for spheroidal particles combined with near-field lubrication interactions for particles of arbitrary shape (Claeys and Brady 1989). Conceptually the procedure is straightforward, but its implementation is not trivial. This simulation method is in its early stages, so here we must be content with results for averages over static configurations, specifically, regular

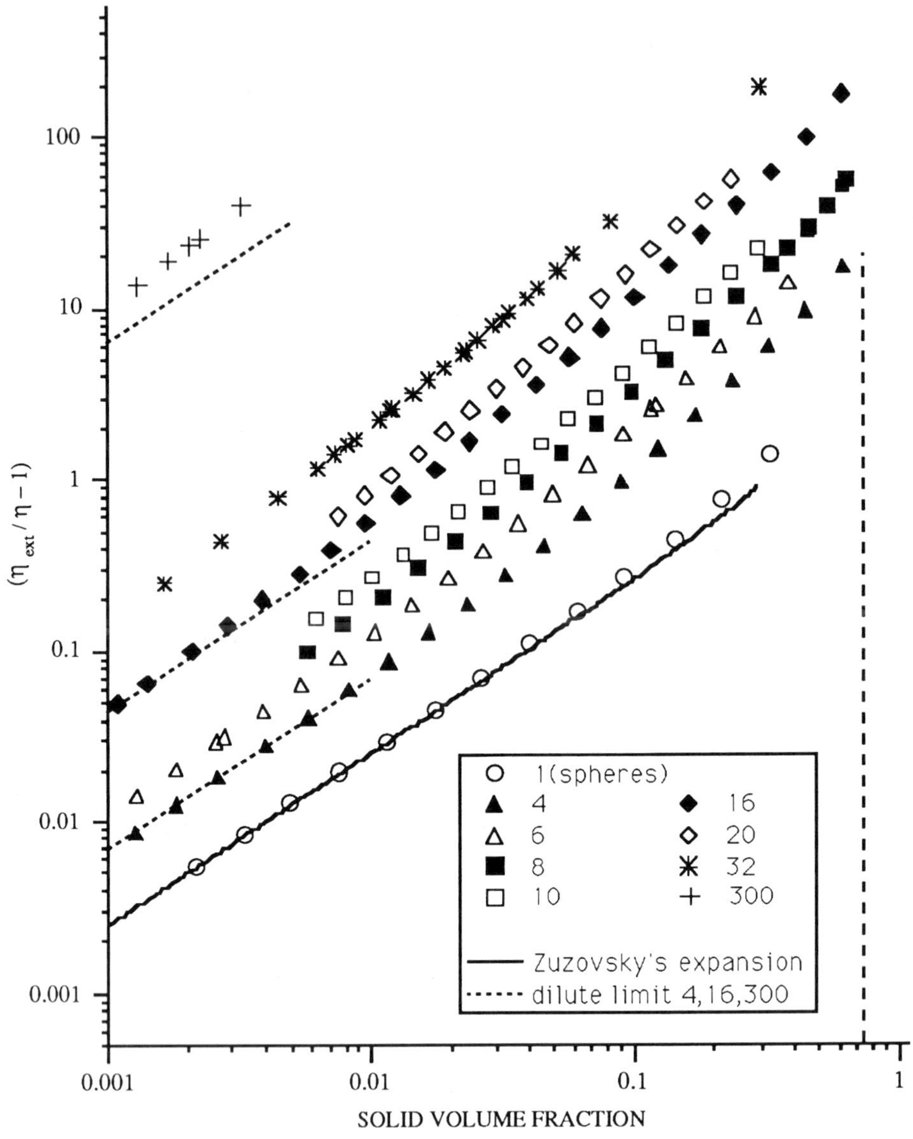

Figure 26.6: *The extensional viscosity of* efc-*lattices of prolate spheroids. The aspect ratio of the particles is given in the legend. The solid line represents the expansion for spheres of Zuzovsky et al. (1983). The dilute limit expansions for* $r_p = 4, 16$ *and* 300 *are shown as dotted lines.*

and random hard-rods dispersions.

In Figure 26.6 the extensional viscosity of an *efc* lattice is shown. An *efc* lattice is an "expanded" face centered cubic lattice of identical aligned spheroids, expanded in such a way that the maximum solids packing for prolate spheriods of arbitrary aspect ratio is the same as for spheres,

$\alpha_s^{max} = 0.7405$. If one used a conventional *fcc* lattice then the maximum packing fraction would depend on aspect ratio, r_p, scaling as r_p^{-2}. The aspect ratio is defined as the ratio of the lengths of the major and minor axes. Liquid crystals at very high solids content may order into *efc*-like lattices. The extensional viscosities in Figure 26.6 correspond to straining the lattice along the axes of the aligned particles. One sees the well-known (Batchelor, 1971; Mewis and Metzner 1974; Powell 1991) dramatic increase in the extensional viscosity with the addition of just as a small amount of solids for high aspect ratio rods.

Next, Figure 26.7 shows a comparison of the shear viscosity of random hard-rod dispersions of various aspect ratios with the experimental results of Bibbo et al. (1985) for the viscosity measured upon inception of steady shear. Under steady shear a suspension of rod-like particles will not, in general, remain in the random hard-rod microstructure, and in order to compare with the simulations ideally we need measurements at high frequency, which capture only the hydrodynamic contribution to the stress and do not distort the equilibrium microstructure. We have not been able to find experimental data at high-frequency, but the instantaneous viscosity upon the start of steady shear should be the same. We see in Figure 26.7 that there is fair agreement between simulation and experiment over a range of concentrations. Here, n is the number density of rods and a their half length, which is the relevant measure of solids content for viscosity (the viscosity scales as a^3 at infinite dilution). The open symbols for the experiments account for the presumed cylindrical shape of the particles.

A hard-rod dispersion will spontaneously go from a disordered to an ordered state, with the particles orientationally ordered but translationally disordered—the nematic state—, upon increasing the solids content. Figure 26.8 shows the effect of this transition on the extensional viscosity of the dispersion (with the axis defined as along the axis in the nematic state). At an aspect ratio of 6 the disordered-nematic transition occurs at a solids volume fraction of roughly 0.27 and is evident in the figure by the rapid increase in the viscosity. The disordered-nematic transition occurs when $\alpha_s r_p^2$ exceeds an $O(1)$ number, and thus as $r_p \to \infty$ the solids fraction approaches zero. Correspondingly, since the extensional viscosity scales as $\alpha_s r_p^2$ the transition will become increasingly sharp with a pronounced jump for large r_p. While larger than in the disordered state, the extensional viscosities in the nematic state are still quite small compared to those of an *efc*-lattice and compared to those of a perfectly aligned nematic (the asterisks), showing the relative importance of orientational and translational order.

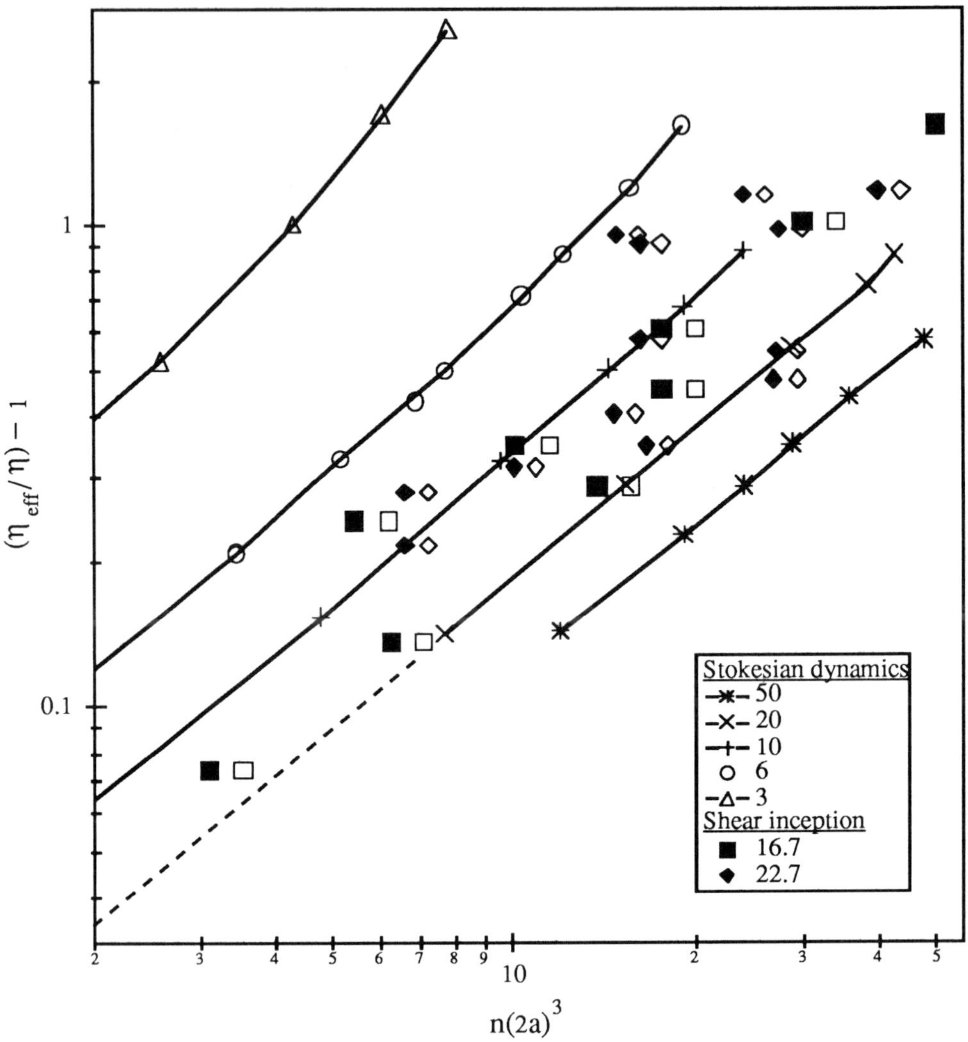

Figure 26.7: *Double logarithmic plot of the viscosity of isotropic dispersions of spheroids* vs. *concentration for various aspect ratios. Also shown are the measurements of the viscosity at inception of shear by Bibo et al. (1985). The filled symbols are the data from their Figure 2; the open symbols account for the presumed cylindrical shape of the rods.*

26.5 PRESSURE-DRIVEN CHANNEL FLOW

The above two examples have shown some of the results that can be obtained with Stokesian dynamics for unbounded infinite suspensions. I would now like to turn to a different problem, namely, the pressure driven flow of nonBrownian particles through a channel. The interest in

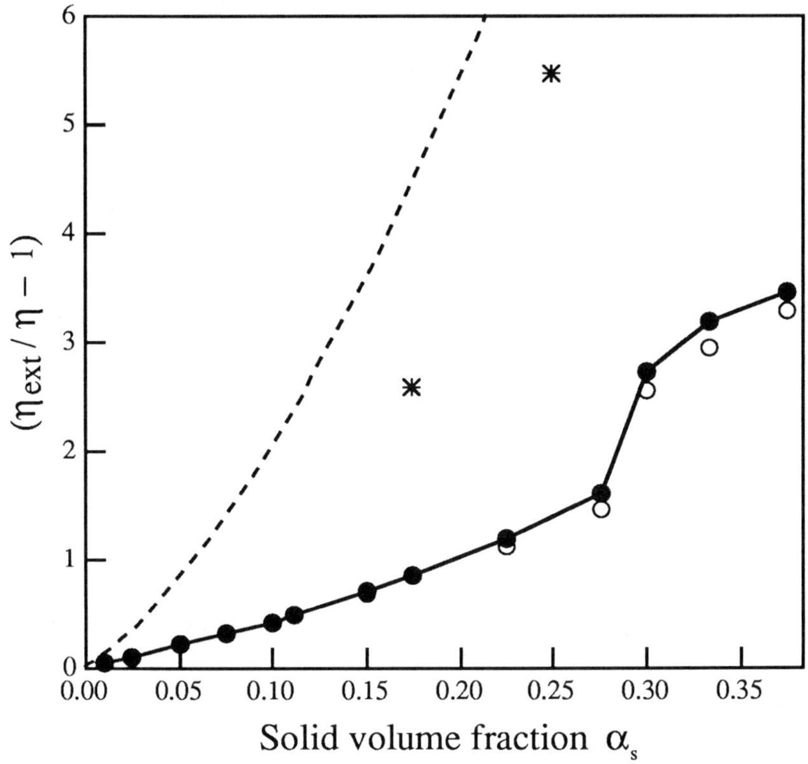

Figure 26.8: *Concentration dependence of the extensional viscosity of equilibrium dispersions of spheroids of aspect ratio 6. In the nematic region the principal axis of strain lies along the director of the liquid crystal. The asterisks are the results for random microstructures of parallel ellipsoids. The dotted line shows the data for* efc-*lattices. Lubrication interactions have been suppressed for the open symbols.*

channel flow is two fold. First, in many practical applications of suspension processing one is interested in pressure- as opposed to shear-driven flows, and so simulations of microstructure, phase slip and pressure drop are of intrinsic interest. Second, and more importantly from a fundamental perspective, in a pressure-driven flow the suspension is subject to inherently inhomogeneous stress and shear fields; the stress and shear rate vary from zero at the centerline to a maximum at the walls. The question is: Do inhomogeneous flows give rise to new physics? That is, because of the finite size of the particles and the fact that their motion is hydrodynamically driven do particles migrate and produce their own intrinsic concentration inhomogeneities?

Leighton and Acrivos (1987b) have postulated that particle migration would occur from regions of high to low shear rate or (shear stress) owing to the increased number (and severity) of particle "collisions" from the high shear rate (or stress) side. One of the motivations for this study is to examine this hypothesis. Furthermore, from a macroscopic modeling perspective suspension behavior in inhomogeneous flow leads to qualitatively new effects in constitutive relations. In a "simple" fluid the deviatoric stress and particle flux may be modeled as:

$$\boldsymbol{\tau} = \eta(\alpha_s)(\nabla\mathbf{u} + \nabla\mathbf{u}^\dagger), \tag{26.11}$$

$$\mathbf{j} = \mathbf{u}\alpha_s - D\nabla\alpha_s, \tag{26.12}$$

where $\eta(\alpha_s)$ is the concentration-dependent viscosity, $\mathbf{u}$ is the mean suspension velocity and D is the shear-induced diffusivity, for example as reported in Section 26.3. If a suspension were "simple," then (26.11) and (26.12) along with the conservation laws for momentum and particles would be sufficient to predict behavior in any type of flow, homogeneous or inhomogeneous (presuming boundary conditions are known, of course). However, particle migration due to flow inhomogeneities suggests that to (26.12) a term

$$\mathbf{j} \sim \nabla\nabla\mathbf{u} \qquad \text{(particle flux} \sim \text{velocity gradients)}$$

should be added. And in a similar manner, to the stress should be added

$$\boldsymbol{\tau} \sim \nabla\nabla\alpha_s \qquad \text{(stress} \sim \text{concentration gradients).}$$

Equations (26.11) and (26.12) must be fundamentally modified in order to capture this phenomenon.

In channel flow this migration mechanism implies that particles should move away from the walls (regions of high shear) leading to a depletion there and an increased concentration near the centerline. Correspondingly, the particle velocity profile should become blunted—plug flow in the middle and rapid shear flow near the walls. Experiments by Karnis et al. (1966), Koh (1991) and Koh et al. (1991) have recorded this blunting of the particle velocity profile in tube and channel flow. Koh also measured concentration profiles, noting a maximum at the centerline and a depletion near the walls. Karnis et al. on the other hand did not measure any concentration variations; the profiles were reported to be uniform across the tube. A possible explanation for this apparent discrepancy between these two sets of measurements is the time needed to reach steady state.

Since the profile blunting and concentration build up at the centerline are supposed to come about from particle migration as proposed by Leighton and Acrivos, a simple estimate of the time to reach steady state is as follows: The time it takes a particle to diffuse a quarter of the way across the channel is $t = H^2/64D$, where the shear-induced diffusivity is given by $D = \dot{\gamma}a^2 f(\alpha_s)$. $f(\alpha_s)$ is a nondimensional function of the solids content only; from the results of Section 26.3 $f(\alpha_s) = 0.05$ at $\alpha_s = 0.45$. The shear rate varies across the channel and its average value is $\dot{\gamma} = 3\langle u\rangle/H$, where $\langle u\rangle$ is the mean suspension velocity. Hence, the time to reach steady state is

$$t_{SS} = \left(\frac{H}{a}\right)^3 \frac{1}{192 f(\alpha_s)} \left(\frac{a}{\langle u\rangle}\right). \tag{26.13}$$

For the conditions here ($\alpha_s \geq 0.3$), $192 f(\alpha_s) \approx 10$. Thus we see that a time of roughly $\frac{1}{10}(H/a)^3$ is needed on average for a particle to diffuse one quarter of the way across the channel and consequently for steady state to be reached. This strong $(H/a)^3$ scaling has important implications for achieving steady state in both experiment and simulation. Indeed, using this time to determine the channel length required to reach steady state we have

$$\left(\frac{L}{H}\right)_{SS} \sim \frac{1}{10}\left(\frac{H}{a}\right)^2. \tag{26.14}$$

Using this criterion, none of the experiments of Karnis et al. are at steady state, while only half of those of Koh are. This may well be the explanation of the differing behaviors reported by these two groups (see below).

The Stokesian dynamics methodology described in Section 26.2 needs modification to treat the problem of pressure driven flow because of the presence of the walls, but fortunately this modification is easy to carry out. There are at least two different ways to account for rigid walls. The first, which was used by Durlofsky and Brady (1989), is to discretize the plane walls in the integral formulation of Stokes equations and include near-field sphere-wall lubrication interactions just as was done for the sphere-sphere interactions. The second, which is computationally more efficient, is simply to designate M of the N particles in the unit cell to form a plane and to move at the same prescribed velocity, $\mathbf{U}^W$, as illustrated in Figure 26.9. These rigidly connected moving particles will then generate a flow, and if we require that the net volumetric flux of material across a cross-section be zero, there will be a pressure gradient set up that pumps the fluid and the other $N - M$ suspended particles in

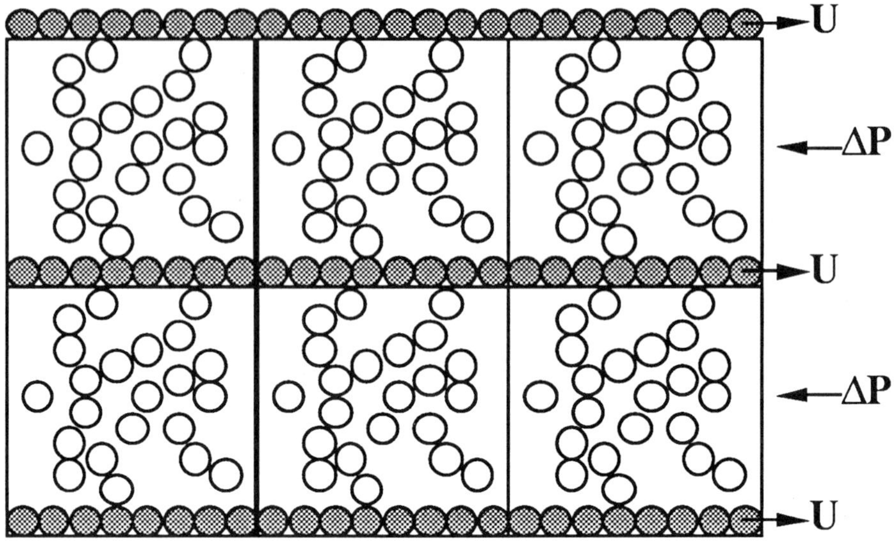

Figure 26.9: *A schematic diagram of the periodic cells used to simulate pressure driven flow through a rough-walled channel. The walls move at a specified velocity and conservation of volume requires that there be a pressure gradient that pumps the fluid and freely suspended particles in the opposite direction so that the net volume flux of material is zero.*

the direction opposite to the moving walls. It turns out that the evolution equation (26.4) is simply modified to read

$$\Delta \mathbf{x} \;=\; -\,(\mathbf{R}_{FU}^{II})^{-1}\cdot\mathbf{R}_{FU}^{IW}\cdot\mathbf{U}^{W}\Delta t, \tag{26.15}$$

where we have suppressed Brownian motion and interparticle forces. (These could, of course, be included if one so desired.) In (26.15) $\mathbf{R}_{FU}^{II}$ stands for the hydrodynamic resistance tensor for the interactions among the interior particles, while $\mathbf{R}_{FU}^{IW}$ describes the interior particle-wall interactions. $\mathbf{U}^{W}$ is the prescribed velocity of the channel walls and we have required that the volumetric averaged suspension velocity be zero. Equation (26.15) is of the same form as (26.4) if we note that in place of the imposed shear flow $(\mathbf{U}^{\infty}, \mathbf{R}_{FE}:\mathbf{E}^{\infty})$ we have the motion of the walls $(\mathbf{R}_{FU}^{IW}\cdot\mathbf{U}^{W})$. To determine the force required to move the walls, and thus by an overall momentum balance the pressure gradient required to pump the suspension in the opposite direction, we have

$$\mathbf{F}^{W} \;=\; (\mathbf{R}_{FU}^{WW}\cdot(\mathbf{R}_{FU}^{II})^{-1}\cdot\mathbf{R}_{FU}^{IW})\cdot\mathbf{U}^{W}, \tag{26.16}$$

where $\mathbf{R}_{FU}^{WW}$ is the resistance tensor describing the interactions among the wall particles. For the simulations discussed here, we have left the wall particles as true particles of the same size as the interior particles; thus, we have a rough-walled channel (*cf.* Figure 26.9). To make smooth-walled channels, we need only add in the near-field interactions appropriate for a wall, rather than those for spheres, as was done by Durlofsky and Brady.

In the simulations to be discussed, for computational convenience the suspension (both wall and interior particles) has been periodically replicated in the z or third direction as a close-packed lattice of one particle diameter spacing, forming a "two-dimensional" simulation. Thus, the problem reduces to one where there is only motion in the x and y directions and the relevant parameter is the areal fraction α_A. For this periodic in z arrangement the volume fraction is precisely 2/3 the areal fraction, *i.e.* $\alpha_s = \frac{2}{3}\alpha_A$, so only the areal fractions will be quoted. All simulations are for monodisperse nonBrownian hard spheres at an areal fraction $\alpha_A = 0.4$ in the absence of external or interparticle forces. The only forces acting are hydrodynamic. The simulation cell is square in the xy plane, the number of particles making up a wall is $N_W = 9$, and the number of interior particles is $N_I = 42$, giving a gap spacing to particle radius $H/a = 18.3$. Thus, from (26.13) a time of approximately 600 is needed to reach steady state.

Figure 26.10 shows the time evolution of the force, expressed as the equivalent viscosity of a Newtonian fluid, required to move the walls at a constant velocity. The force or pressure drop reaches a steady state in a time of roughly 75. Here time has been nondimensionalized by the particle radius divided by the wall velocity, so that in a dimensionless time of 18.3 one wall particle, and thus an average interior particle, will have been convected through the simulation cell. The large fluctuations in the instantaneous force required to maintain the constant wall velocity is a direct result of the formation and breakup of clusters that span the gap between the walls. When a cluster of particles spans, a large force is required to continue the motion. Spanning clusters and associated fluctuations in wall forces were also seen in simulations of shearing motion between parallel walls (Durlofsky and Brady 1989; Durlofsky 1986) and are suggestive that a percolation-like phenomena is operating in concentrated suspensions.

Both the force in Figure 26.10 and the particle velocity profile shown in Figure 26.11 reached steady state rapidly, long before the estimate given by (26.13). By a time of 100 or so these two properties were at steady state, but the concentration profile, shown in Figure 26.12 had hardly developed. Only a slight depletion near the walls was evident. It was not until a time very close to 600 that the particle concentration profile fully developed. The pressure drop is given by the shear stress

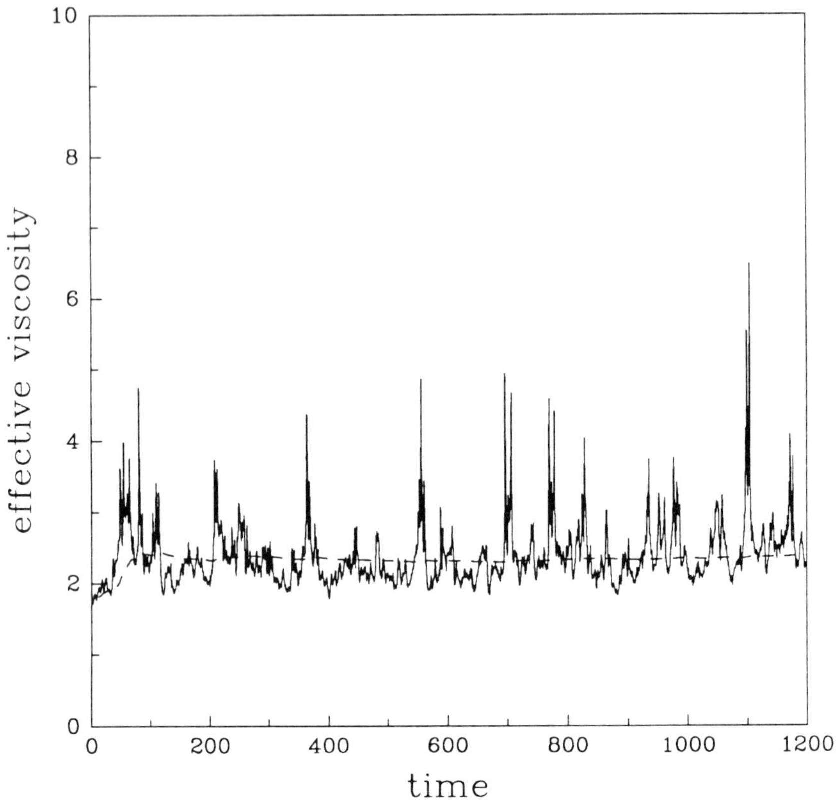

Figure 26.10: *The tangential force, expressed as the equivalent viscosity of a Newtonian fluid, required to move the walls is plotted as a function of time for a suspension at* $\alpha_A = 0.4$.

at the wall and is thus only sensitive (apart from fluctuations) to the local concentration. The fact that the pressure drop and the velocity profiles develop long before the concentration profile (and remember that the estimate (26.13) is for the concentration profile) may explain why Karnis et al. saw no concentration variations but did see velocity profile blunting—they were never at steady state.

The Stokesian dynamics simulation results compare qualitatively very well with the velocity and concentration profiles observed experimentally by Koh (1991). Quantitatively, however, there are significant differences. Most importantly, Koh sees a much higher concentration on the centerline, as high as close packing with $\alpha_s \approx 0.67$, even for a bulk solids concentration of $\alpha_s = 0.1$. Note, the monolayer simulations at $\alpha_A = 0.4$ correspond to a three-dimensional concentration of 0.27. Further, he observed a significant "phase slip" with the particle flux being only 60% of what one would expect for the flow of a simple Newtonain-like material.

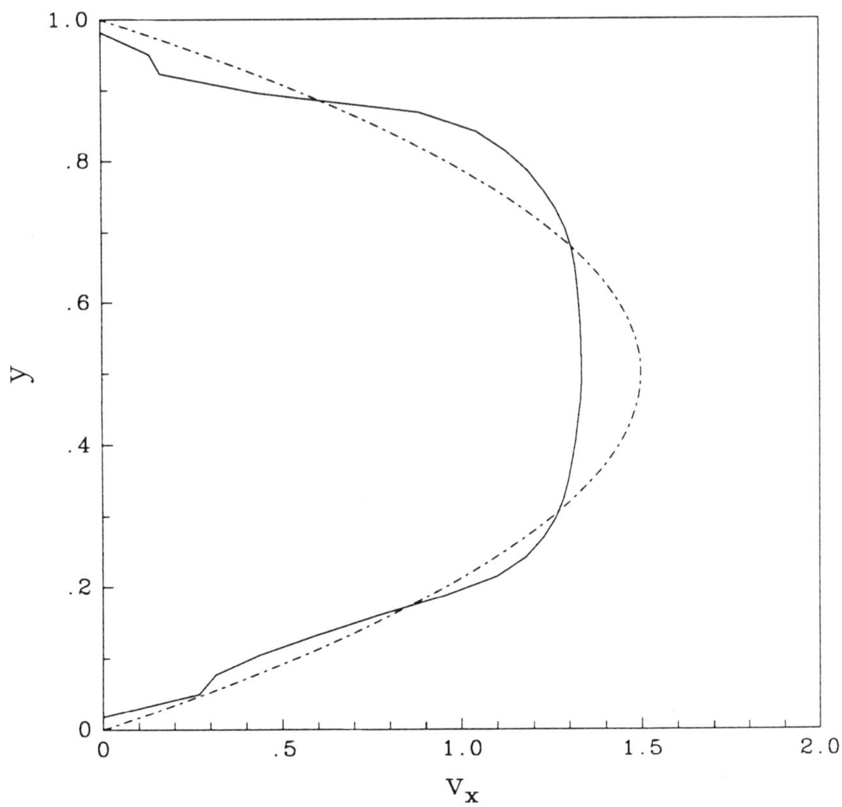

Figure 26.11: *The time-averaged particle velocity is plotted as a function of the separation between the walls for* $\alpha_A = 0.4$. *The dashed curve is the parabolic Poiseuille profile. The time average was performed for* $800 < t < 1200$.

Our simulations on the other hand show an opposite phase slip—the observed particle flux is 1.18 times that of an equivalent Newtonian fluid with a uniform concentration profile.

Finally, the shear-induced migration idea of Leighton and Acrivos has been incorporated into a continuum model for predicting suspension behavior by Phillips et al. (1991). They use an equation of the form (26.11) to model the stress, but include terms describing the particle flux due to stress and shear-rate gradients in the particle conservation equation. A simple correlation is used for the concentration dependence of the suspension viscosity. The model predictions for the velocity profiles are quite reasonable. However, the concentration profiles, while having a maximum at the centerline and a minimum at the walls as seen experimentally, have qualitatively and quantitatively the wrong shape. In particular, the predicted concentration profiles have a sharp cusp at the

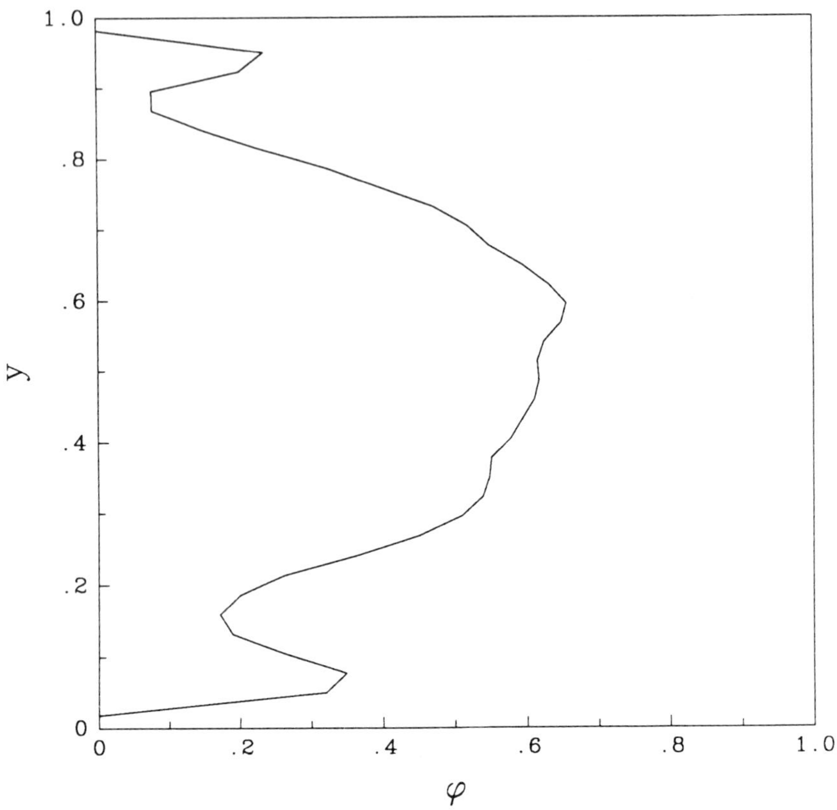

Figure 26.12: *The time-averaged particle concentration is plotted as a function of the separation between the walls for* $\alpha_A = 0.4$. *The spikes near the walls are particles stuck to the walls through the lubrication forces. The time average was performed for* $800 < t < 1200$.

centerline with a discontinuous, infinite, slope. By symmetry, the true concentration profile must be flat with zero slope at the centerline.

This theory fails because it is "local" in that the particle flux at a point is proportional to the shear rate $\dot{\gamma}$ at the same point. Because the shear rate vanishes at the centerline, the concentration gradient must become infinite in order to achieve a balance. However, as Figure 26.13 shows the particle velocity fluctuations at the centerline do not vanish and a simple kinetic theory-like argument would suggest that the particle diffusivity should not be zero at the centerline, even though the shear rate vanishes there. It seems that a different, perhaps non-local, theory is needed to capture this effect and bring the modeling into accord with the experiments. Furthermore, the current theory predicts a behavior that is independent of the particle to gap size. This is also not in accord with simulation or experiment, for as $a/H \to 0$ the profiles should all

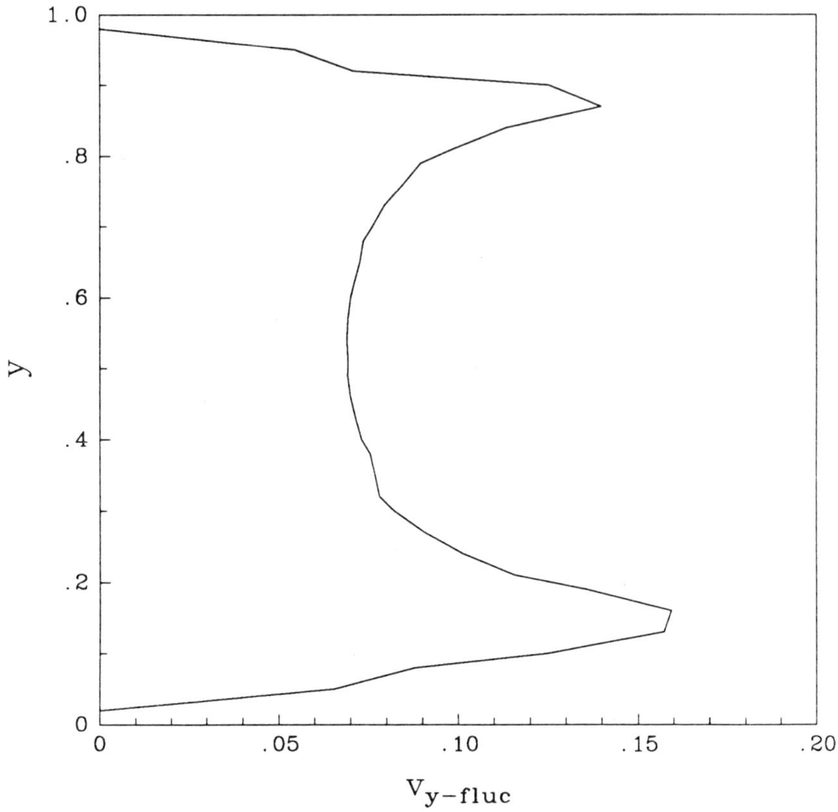

Figure 26.13: *The time-averaged root-mean-square fluctuation in the y-component of the particle velocity is plotted as a function of position between the walls. Note at the centerline the velocity fluctuation is not zero and there would thus be a diffusive motion of the particles even though the shear rate is zero.*

return to their Newtonian-like form (apart from a region of order the particle size near the walls where the finite size will be manifest). Thus, the intrinsic length scale ratio a/H must appear in the theory.

26.6 ELECTRORHEOLOGICAL FLUIDS

The last example of Stokesian dynamics is the simulation of electrorheological (ER) fluids. ER fluids are suspensions of particles in a viscous non-conducting oil with a particle-fluid dielectric mismatch. With an applied electric field this dielectric mismatch creates strong polarization forces that cause the particles to form chains aligned with the electric field. The changes in the suspension microstructure and the electrostatic forces among the particles combine to increase the effective viscosity of an ER fluid by several orders of magnitude (Carlson et al. 1990; Marshall et al. 1989; Bonnecaze 1991). Furthermore, and this is the true utility of an ER fluid, the viscosity can be *controlled* by the magnitude of the applied electric field. At low field strengths an ER fluid is a low viscosity Newtonian fluid, while at high field strengths it turns into a solid with measurable static and dynamic yield stresses. With a tunable viscosity a variety of active damping and other devices can be envisioned.

Typical suspended materials in an ER fluid are 1-100μm size particles of cornstarch, silica, or even zeolites, while the suspending fluids are nonconducting solvents such as silica oil, chlorohydrocarbons, or corn oil. To minimize settling problems the fluids are blended so that the particles are close to neutrally buoyant. The zero field viscosity of the suspension with particle loadings ranging from 0.05 - 0.50 by volume fraction are up to an order of magnitude greater than the suspending fluid viscosity. The effective viscosity of the ER suspension can be 100,000 times greater for electric field strengths of about 1 kV/mm perpendicular to the direction of flow. The particle Reynolds number is less than 0.1 (maximum shear rates are about 0.1 sec^{-1}), and the fluid Reynolds number is small enough so inertia is negligible. Also, the particle Péclet number is extremely large, in excess of 10^6 usually, so Brownian forces are negligible. Indeed the thermal forces must be small otherwise the Brownian motion would disrupt the chains of particles that are important to the ER response.

The key to observing the ER response is the induced electrostatic polarization forces due to the dielectric mismatch between the particles and the fluid. When an uncharged particle is placed in an electric field, it develops an induced dipole if the surrounding medium has a different dielectric constant. This dipole is further enhanced by the presence of other particles since they intensify the local electric field experienced by any one particle. The dipole, of course, can be thought of as a concentration of positive charge at one end of the particle and negative charge at the other end. With this simple picture and using the interactions of two particles as a paradigm, we can qualitatively explain why ER fluids form chains of particles aligned with the electric field.

Two particles whose line of centers is parallel to the electric field will experience an attractive force between the positive and negative charges at the surfaces nearest and furthest from each other and repulsion between charges of the same sign. There is then a net attraction of the particles along their line of centers so that they move toward each other. If we apply the same considerations to particles whose line of centers is perpendicular to the electric field, we find there is a net repulsion moving the particles away from each other along their line of centers. For particles whose line of centers is neither parallel nor perpendicular to the electric field, the particles may experience both attraction and repulsion causing the pair to translate and rotate into alignment with the electric field. Thus, the forces due to polarization interactions cause the formation of particle chains aligned with the electric field, and the strength of these chains strongly affects the viscosity of an ER fluid.

The operating conditions—time scales, magnitude of forces, relevant parameters, etc.—for ER fluids can be estimated by simple scaling arguments. An external electric field E_0 induces a dipole moment in the particles of magnitude $S \sim \epsilon a^3 \beta E_0$, where a is the particle size, ϵ the dielectric constant of the suspending fluid and $\beta = (\epsilon_p - \epsilon)/(\epsilon_p + 2\epsilon)$ gives the dielectric mismatch between the particles, ϵ_p, and the fluid. (The particles generally do not have a permanent dipole moment as do, for example, ferrofluids, where the permanent magnetic dipole moments give rise to their unique properties (Rosensweig 1985).) A dipole in a uniform electric field experiences no force. In an inhomogeneous field, however, a dipole will experience a force and thus move, with its motion being resisted by the viscous fluid. The inhomogeneous local field is created by the presence of the other particles, each producing the electric field of a dipole. The electric force on a dipole scales as $F_e \sim S \cdot \nabla E \sim \epsilon a^2 \beta^2 E_0^2$. It is important to note the scaling with the square of the electric field, a direct result of the induced dipole moments. The time scale for chain formation comes from a comparison of the viscous force $F_v \sim \eta a U \sim \eta a^2/t$ to the above electrostatic force, giving $t \sim \eta/\epsilon(\beta E)^2$, which for an applied field of 1 kV/mm can be as small as one millesecond. Finally, when a steady shear flow of shear rate $\dot{\gamma}$ is applied the relevant dimensionless parameter is the ratio of the viscous to electrostatic forces called the Mason number, $Ma = \eta\dot{\gamma}/2\epsilon(\beta E)^2$. $Ma \to \infty$ corresponds to hydrodynamic domination and $Ma \to 0$ electrostatic domination. Note, that both the time scale for chain formation and the Mason number are independent of the particle size.

To simulate an ER suspension we need simply take the evolution equation (26.4), neglect the effects of Brownian motion and use the electrostatic interactions among the particles for $\mathbf{F}^P$:

$$\Delta \mathbf{x} = \{\mathbf{U}^\infty + \mathbf{R}_{FU}^{-1} \cdot [Ma^{-1}\mathbf{F}^P + \mathbf{R}_{FE} : \mathbf{E}^\infty]\}\Delta t. \qquad (26.17)$$

Here, Ma appears as the natural dimensionless parameter characterizing the ratio of the hydrodynamic to the electrostatic forces. The stress is determined from (26.7), without, of course, $\mathbf{S}^B$. All that remains is to determine the proper form for the interparticle electrostatic forces.

The exact electrostatic interparticle forces in an ER suspension can be computed from the solution of the many-body potential problem and the integration of the Maxwell stresses over each particle. This is practically impossible, however, for anything more than two spherical particles. There is an alternative method that avoids the need for detailed knowledge of the local potential field that is quite simple and elegant—using the electrostatic energy of the system. Since the electrostatic interparticle force is conservative, the force on any particle is the negative of the gradient of the electrostatic energy with respect to the particle's position, and this energy can be written in terms of a geometric system capacitance tensor.

Recall that the energy of a system of charged particles in the absence of an electric field is one-half the sum of the product of the particle charges and their local potentials (Jackson 1975, Landau and Lifshitz 1984). For charge-free particles in an electric field, the electrostatic energy density is analogously given by,

$$\mathcal{U} = \frac{1}{2V} \sum_{\alpha} \mathbf{S}_E^{\alpha} \cdot \mathbf{E} = \frac{1}{2V} \mathbf{S}_E \cdot \mathcal{E}. \tag{26.18}$$

Here $\mathcal{U}$ is the energy per unit volume for a system of N particles in a volume V. The induced dipole on particle α is $\mathbf{S}_E^{\alpha}$ and $\mathbf{E}$ is the macroscopic applied electric field. The slightly more compact form on the RHS has $\mathbf{S}_E$ as the vector containing the $3N$ components of all the particle dipoles and $\mathcal{E}$ as the complementary $3N$ electric field vector. Please note that in (26.18) $\mathbf{S}_E$ and $\mathbf{E}$ refer to electrostatics and not hydrodynamics.

Since Laplace's equation and the boundary conditions are linear in the electric field, there exists a relationship among the charge, dipole, particle potential and external electric field completely analogous to the grand resistance tensor $\mathcal{R}$ introduced in Section 26.2. For charge-free particles in an external electric field, this "grand capacitance" tensor can be solved to give the particle dipoles:

$$\mathbf{S}_E = \hat{\mathbf{C}}(\mathbf{x}) \cdot \mathcal{E}, \tag{26.19}$$

where,

$$\hat{\mathbf{C}}(\mathbf{x}) = \mathbf{C}_{S\phi} \cdot (\mathbf{C}_{q\phi})^{-1} \cdot \mathbf{C}_{qE} - \mathbf{C}_{SE}. \tag{26.20}$$

Here $\hat{\mathbf{C}}(\mathbf{x})$ is the dipole-electric field capacitance tensor for charge-free particles and the $\mathbf{x}$ reminds us that it (and all the capacitance tensors) are functions of the instantaneous particle positions like the hydrodynamic resistance tensors. The capacitance tensors are also a function of the particle-to-fluid dielectric constant ratio. The small capacitance tensors

$\mathbf{C}_{q\phi}$, etc. give the coupling between the charge and potentials, just as $\mathbf{R}_{FU}$ couples forces and velocities in the resistance tensor.

The electrostatic energy density is now,

$$\mathcal{U} = \frac{1}{2V}\mathcal{E} \cdot \hat{\mathbf{C}}(\mathbf{x}) \cdot \mathcal{E}, \tag{26.21}$$

and then from classical physics the force on particle α is

$$\mathbf{F}^{\alpha} = -\frac{\partial V\mathcal{U}(\mathbf{x})}{\partial \mathbf{x}_{\alpha}} = -\frac{1}{2}\mathcal{E} \cdot \frac{\partial \hat{\mathbf{C}}(\mathbf{x})}{\partial \mathbf{x}_{\alpha}} \cdot \mathcal{E}, \tag{26.22}$$

and similarly the torque on the particle is

$$\mathbf{T}^{\alpha} = -\frac{\partial V\mathcal{U}(\mathbf{x})}{\partial \psi_{\alpha}} = -\frac{1}{2}\mathcal{E} \cdot \frac{\partial \hat{\mathbf{C}}(\mathbf{x})}{\partial \psi_{\alpha}} \cdot \mathcal{E}, \tag{26.23}$$

where $\mathbf{x}_{\alpha}$ and ψ_{α} are the particle's position and Euler angle vectors, respectively. Here we see clearly that there is no torque on a spherical particle since its Euler angle is irrelevant to its dipole strength and the system electrostatic energy. Also note that both the energy and the interparticle force scale with the square of the electric field as expected from our simple scaling analysis. The capacitance tensor can be constructed in a manner completely analogous to the resistance tensor, and the procedure gives excellent results for effective conductivities, electrostatic forces, etc. for spherical particles (Bonnecaze 1991, Bonnecaze and Brady 1990, 1991a).

The above method has been applied to simulate a monolayer of 25 spherical particles at an areal fraction $\alpha_A = 0.4$ for a variety of Mason numbers and a dielectric constant ratio of 4 (Bonnecaze 1991, Bonnecaze and Brady 1991b,c). An initial test of the simulation was to see if particle chaining was observed and to compare the predicted chaining time to experimental observations. The spheres were placed randomly within the periodic cell monolayer and allowed to order under the action of an applied electric field at $Ma = 10^{-3}$. Since there is no shear flow, the choice of Ma is arbitrary in so far as the time scales as $\dot{\gamma}^{-1} = Ma^{-1}\eta/(2\epsilon(\beta E)^2)$ in the simulation. Figure 26.14 shows the structure obtained after a dimensionless time of 0.1. The drawn square is the outline of the periodic cell and the electric field points from the bottom to the top. From our simulation we can estimate the chaining time in an ER fluid. The dimensional chaining time is the simulation chaining time multiplied by $Ma^{-1}\eta/2\epsilon(\beta E)^2$. Marshall et al. used an ER fluid with $\eta \approx 0.01$ Pa-s, $\epsilon = 8 \times 8.854 \times 10^{-2}$ farad/m, and an electric field of about 1kV/mm. For a simulation chaining time of 0.1, the dimensional chaining time is about

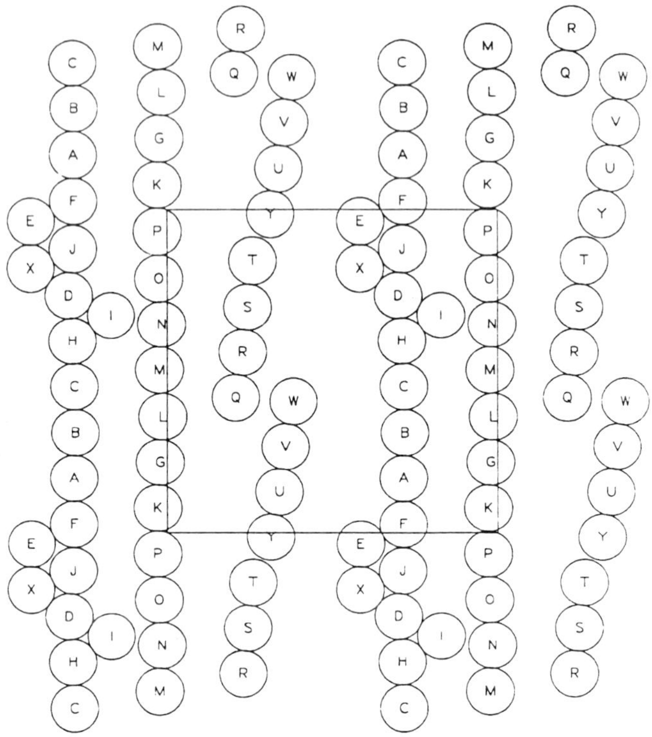

Figure 26.14: *A "snapshot" of the suspension microstructure for sphere ordering with no shear flow at* $t = 0.1$*. Note the spanning chains parallel to the electric field.*

0.03 seconds. Marshall et al. report that the ER suspension "sets up" in less than 0.1 seconds which is consistent with our simulation results. The microstructure in (26.14) is not yet at steady state. Steady state was achieved in a dimensionless time of order 2, with the chains fully aligned with the direction of the electric field.

Dynamic simulations were performed of an infinite, unbounded monolayer of spherical particles with a particle to the fluid dielectric constant ratio of 4. The periodic cell was square in shape, and there were 25 particles per cell. The simulations were performed with and without the addition of the near-field electrostatic interactions, though near-field hydrodynamic interactions were always included since they are necessary to prevent particle overlap. The simulation times, T, were typically 10.0 (non-dimensionalized by the shear rate) but were sometimes less. In all cases the reported average of the effective viscosities had reached steady-state values. The summary of the effective viscosities are presented in Figure 26.15. Simulations were also performed on 49 particles in a square

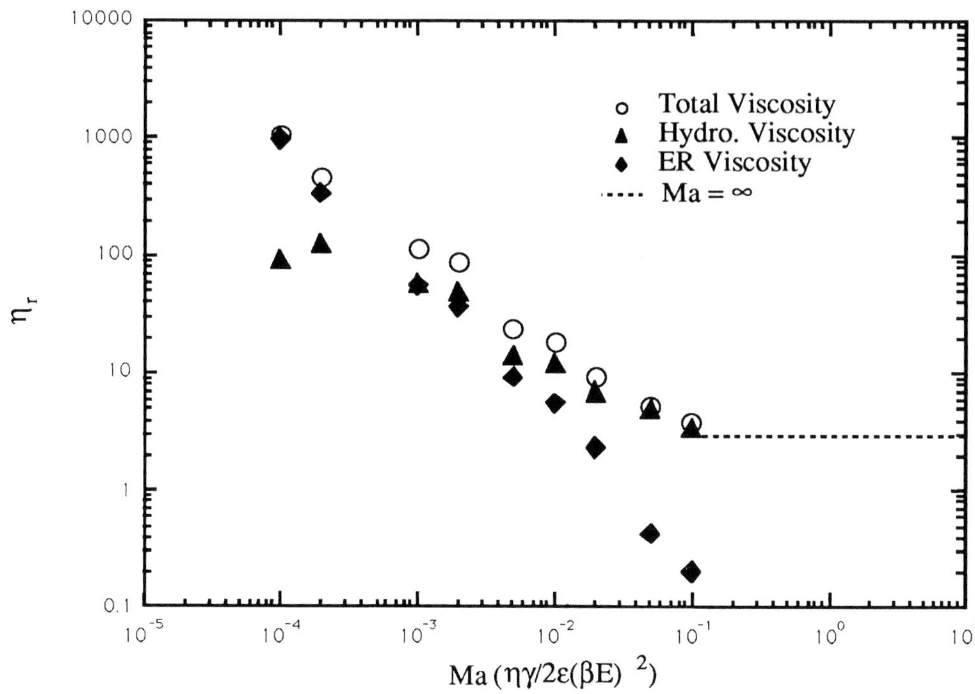

Figure 26.15: *The total, hydrodynamic and ER relative viscosities without the addition of near-field electrostatic interactions are shown as a function of the Mason number for an areal fraction* $\alpha_A = 0.4$. *The dotted line corresponds to the no electric field viscosity.*

periodic cell; there is little difference in the predicted effective viscosities between 25 and 49 particles.

The total viscosity increases with decreasing Ma with and without the addition of near-field electrostatic interactions. In fact the viscosity scales as the inverse of Ma for small Ma, or, equivalently, the viscosity scales with the square of the applied electric field as observed experimentally. The hydrodynamic viscosity, the contribution due to the resistance to straining of the particles, increases with decreasing Ma up until about $Ma = 2 \times 10^{-4}$ where it appears to reach a plateau value of about 100. The ER viscosity steadily increases with decreasing Ma over the entire range presented here. At high Ma, when the viscous forces dominate, the viscosity is almost entirely the hydrodynamic contribution. Note the total viscosity is approximately equal to the $Ma = \infty$ limit (pure shear, no electrostatics) at about $Ma = 10^{-1}$. As the Ma decreases, the ER viscosity contribution to the total viscosity steadily increases and eventually dominates at $Ma = 2 \times 10^{-4}$. The difference in the effective viscosities computed with and without the near-field electrostatic interactions is only about 20% at the smallest Ma of 1×10^{-4}. At this Ma the hydrodynamic viscosity is about the same independent of the addition of near-field electrostatic interactions, so the entire increase is due to the increased electrostatic interactions and not to any fundamental changes to the microstructure. This rather small difference is due to the modest dielectric constant ratio used here, which is typical of current ER fluids. It is expected that for higher dielectric constant ratios, the addition of the near-field interactions would become more important, especially since the interactions are singular for $\epsilon_p/\epsilon = \infty$.

If the viscosities in Figure 26.15 are converted to shear stresses by multiplying by the Mason number, then the resulting shear stress-shear rate behavior is well described by a Bigham plastic model:

$$\tau = \tau_B + \eta_B \dot{\gamma}. \tag{26.24}$$

The shear stress approaches a constant value or Bingham yield stress, τ_B, at small Ma or shear rates. At high Ma or shear rates the viscosity increases linearly, as a Newtonian fluid with viscosity η_B. Here the shear stress is non-dimensionalized by $2\epsilon(\beta E)^2$, and we see the Bingham yield stress scales with the square of the electric field as observed by Marshall et al. Actually, the Bigham model is only approximate in that η_B is really a function of $\dot{\gamma}$, rather than a constant as in the Bigham model.

Figure 26.16 compares the simulation results, with and without near-field electrostatic interactions, to the experimental results of Marshall et al. Their ER fluid consists of hydrated poly (metha-crylate) particles in a chlorinated hydrocarbon oil, and the effective viscosities were measured over a large range of Ma numbers. To compare the simulation results to their experimental results, the monolayer areal fraction is converted to a volume fraction by multiplying α_A by two-thirds. This "two-thirds" rule is correct in the dilute limit where the Einstein viscosity correction is $\frac{5}{2}\alpha_s$ in three-dimensions and $\frac{5}{3}\alpha_A$ in a monolayer. It is approximately correct in the limit of closest packing where two-thirds of the maximum packing areal fraction of a two-dimensional lattice is very close to the maximum packing volume fraction for the three-dimensional analog lattice. The simulation data then corresponds to an approximate volume fraction of $\alpha_s = 0.27$. The closest volume fraction from the experimental data reported was $\alpha_s = 0.23$.

The comparison between the two is extremely good at both extremes of the Ma number. At the intermediate values of Ma, the simulation viscosities very slightly underpredict the experimental viscosities. The experimental suspensions, even in the absence of the electric field, have a yield stress at low shear rates due to some other interparticle colloidal

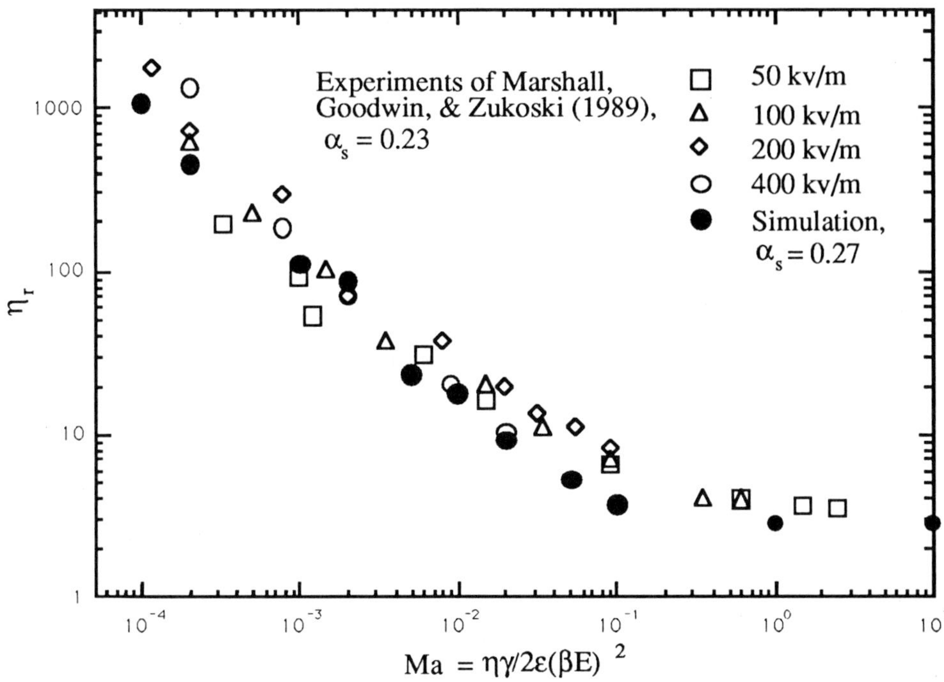

Figure 26.16: *Comparison of the simulation results for the steady shear viscosity of an ER fluid to the experimental results of Marshall et al. illustrated with the open symbols and the applied field used. The equivalent volume fraction of the monolayer from the "two-thirds" rule is 0.27, and the experimental volume fraction is 0.23.*

forces. At low Ma where the electrostatic forces dominate these colloidal forces are unimportant. At the intermediate Ma numbers these colloidal forces might be as important as the electrostatic forces in altering the microstructure and changing the hydrodynamic viscosity as well as making their own contribution to the viscosity. In any event the comparison is exceptionally good, and it appears that the simulation captures both the qualitative and quantitative features of an ER suspension.

We have also used the simulation technique to explore further the nature of and develop a model for the dynamic yield stress τ_B. These model simulations give static and dynamic yield stresses in very good agreement with experiment (Bonnecaze and Brady 1991c).

26.7 CONCLUSIONS

In the preceding sections I have presented the Stokesian dynamics method and applied it to four studies: 1) the structure, diffusion and rheology of concentrated colloidal suspensions, 2) the hydrodynamic transport properties of random and ordered dispersions of prolate spheriods, 3) pressure driven channel flow of neutrally buoyant suspensions of nonBrownian hard spheres, and 4) simulations of dielectric particles forming an electrorheological fluid. The good agreement bewteen simulation and experiment for the evolution of the viscosity with Péclet number for Brownian suspensions, the extensional viscosities of rod suspensions, the concentration profiles in channel flow and the shear viscosities and yield stresses in ER fluids indicates that Stokesian dynamics is capable of excellent *quantitative* (as well as qualitative) predictive ability in a wide array of problems. Further, we have, in many cases for the first time, a detailed understanding of the mechanisms causing such behavior.

The work presented in this chapter provides only a few examples of the ability of dynamic simulation to elucidate fundamental aspects of the physics of suspensions. The methods presented can be modified to include polydisperse systems and virtually any form of interparticle force is easily incorporated into the dynamic simulation. The current state of the computations are such that for 27 particles in three dimensions with Brownian motion for 50,000 time steps, which represents one (α_s, Pe) point in the simulations discussed above, requires approximately 5 hours on an IBM RISC/6000 530 or similar system. For 64 particles the time requirements increase by a factor of 10. (There is a roughly N^3 scaling.) The point is that fully three dimensional simulations with 27 or 64 particles can be routinely performed on the current generation of desktop workstations. Thus, the range of problems of both a fundamental and practical nature that can be addressed through Stokesian dynamics is quite extensive and quite accessible.

Finally, I have presented very little comparison with theory, becuase there are currently very few theories for the phenomena discussed here. Clearly, there is a need for further theory, experiment and simulation of particulate flows.

ACKNOWLEDGEMENTS

Portions of this work were done in collaboration with Georges Bossis, Ronald Phillips, Thanh Phung, Roger Bonnecaze and Prabhu Nott whose help was indispensable. This work was supported in part by grants CBT-8451597, CBT-8696067, INT-8413695, CTS-9020646 from the National Science Foundation, by the Camille and Henry Dreyfus Foundation, by the Office of Naval research through grant N00014-90-J-1945 and by the San Diego Supercomputer Center.

REFERENCES

Ackerson, B. J., "Shear Induced Order and Shear Processing of Model Hard Sphere Suspensions," *J. Rheol.*, **34**(4), 553-590 (1989).

Ackerson, B.J., "Shear Induced Order in Equilibrium Colloidal Liquids," *Physica A*, **174**,15-30 (1991).

Allen, M.P. and D.J. Tildesley, *Computer Simulation of Liquids*, Oxford University Press, Oxford (1989).

Batchelor, G.K., "The Stress Generated in a Non-dilute Suspension of Elongated Particles by Pure Straining Motion," *J. Fluid Mech.*, **46**, 813-829 (1971).

Batchelor, G.K., "Developments in Microhydrodynamics," in *Theoretical and Applied Mechanics*, W. T. Koiter, ed. North-Holland, Amsterdam, 33-55 (1976).

Beenakker, C.W.J. and P. Mazur, "Diffusion in a Concentrated Suspension II." *Physica A*, **126**, 349-370 (1984).

Bibbo, M.A., S M. Dinh and R.C. Armstrong, "Shear Flow Properties of Semiconcentrated Fiber Suspensions," *J. Rheol.*, **29**, 905-929 (1985).

Boersma, W.H., J. Laven and H. N. Stein, "Shear Thickening (Dilatancy) in Concentrated Dispersions," *AIChE J.*, **36**, 321-332 (1990).

Bonnecaze, R.T., *Macroscopic Properties of Electrically Interacting Suspensions*, Ph.D. Thesis, California Institute of Technology (1991).

Bonnecaze, R.T. and J.F. Brady, "A method for determining the effective conductivity of dispersions of particles," *Proc. Roy. Soc. A*, **430**, 285-313 (1990).

Bonnecaze, R.T. and J.F. Brady, "The Effective Conductivity of Random Suspensions of Spherical Particles," *Proc. Roy. Soc. A*, **432**, 445-465 (1991a).

Bonnecaze, R.T. and J.F. Brady, "Dynamic Simulation of an Electrorheological Fluid," *J. Chem. Phys.* (to appear), (1991b).

Bonnecaze, R.T. and J.F. Brady, "Yield Stresses in Electrorheological Fluids," *J. Rheol.*, (to appear), (1991c).

Bossis, G. and J.F Brady, "Self-diffusion of Brownian Particles in Concentrated Suspensions under Shear," *J. Chem. Phys.*, **87**, 5437-5448 (1987).

Bossis, G. and J.F. Brady, "The Rheology of Brownian Suspension," *J. Chem. Phys.*, **91**, 1866-1874 (1989).

Brady, J.F. and G. Bossis, "Stokesian Dynamics," *Ann. Rev. Fluid Mech.*, **20**, 111-157 (1988).

Brady, J.F., R.J. Phillips, J. Lester and G. Bossis, "Dynamic Simulation of Hydrodynamically Interacting Suspensions," *J. Fluid Mech.*, **195**,

257-280 (1988).

Campbell, C.S., "Rapid Granular Flows," *Ann. Rev. Fluid Mech.*, **22**, 57-92 (1990).

Carlson, J.D., A.F. Sprecher and H. Conrad, eds., "Electrorheological Fluids," *Porc. 2nd. Intl. Conf. ER Fluids*, Technomic Pub. Co., (1990).

Claeys, I.L., *Hydrodynamic Transport Properties of Suspensions of Non-Brownian Prolate Spheriods*, Ph.D. Thesis, California Institute of Technology, (1991).

Claeys, I.L. and J.F. Brady, "Lubrication Singularities of the Grand Resistance Tensor for Two Arbitrary Particles," *Physico Chemical Hydrodynamics* **11**, 261-293 (1989).

Claeys, I.L. and J F. Brady, "The Dynamics of Prolate Spheroids in Stokes Flow. Part I: Finite Number of Particles in an Unbounded Fluid," *J. Fluid Mech.* (submitted), (1991a).

Claeys, I.L. and J.F. Brady, "The Dynamics of Prolate Spheroids in Stokes Flow. Part II: Statistically Homogeneous Dispersions," *J. Fluid Mech.* (submitted), (1991b).

Claeys, I.L. and J.F. Brady, "The Dynamics of Prolate Spheroids in Stokes Flow. Part III: Hydrodynamic Transport Properties of Crystalline Dispersions," *J. Fluid Mech.* (submitted), (1991c).

Durlofsky, L.J., *Topics in Fluid Mechanics: I. Flow between Finite Rotating Disks. II. Simulation of hydrodynamically interacting particles in Stokes flow*, Ph.D. Thesis, Massachusetts Institute of Technology, (1986).

Durlofsky, L.J. and J.F. Brady, "Dynamic Simulation of Bounded Suspensions of Hydrodynamically Interacting Particles," *J. Fluid Mech.*, **200**, 39-47 (1989).

Durlofsky, L.J., J F. Brady and G. Bossis, "Dynamic Simulation of Hydrodynamically Interacting Particles," *J. Fluid Mech.*, **180**, 21-49 (1987).

Eckstein, E.C., D.G. Bailey and A.H. Shapiro, "Self-Diffusion of Particles in Shear Flow of a Suspension," *J. Fluid Mech.*, **79**, 191-208 (1977).

Heyes, D.M., "Rheology of Molecular Liquids and Concentrated Suspensions by Microscopic Dynamic Simulations," *J. nonNewt. Fluid Mech.*, **27**, 47-85 (1988).

Hoffman, R.L., "Discontinuous and Dilatant Viscosity Behavior in Concentrated Suspensons. I. Observation of a Flow Instability," *Trans. Soc. Rheol.*, **16**, 155-173 (1972).

Jackson, J.D., "Classical Electrodynamics," 2nd Edition, Wiley, New York., (1975).

Jefferey, D.J. and A. Acrivos, "The Rheological Properties of Suspensions of Rigid Particles," *AIChE J.*, **22**, 417-432 (1976).

Karnis, A., H.L. Goldsmith, and S.G. Mason, "The Kinetics of Flowing Dispersions. I. Concentrated Suspensions of Rigid Particles," *J. Colloid Interface Sci.*, **22**, 531-553 (1966).

Koh, C.J., *Experimental and Theoretical Studies on Two-Phase Flows*, Ph.D. Thesis, California Institute of Technology, (1991).

Koh C.J., P. Hookham and L.G. Leal, "An Experimental Investigation of Concentrated Suspension Flows in a Rectangular Channel," *J. Fluid Mech.*, (submitted) (1991).

Landau, L.D. and E.M. Lifshitz, *Electrodynamics of Continuous Media*, Vol. 8, Pergamon Press, (1984).

Laun, H.M., "Rheological Properties of Aqueous Polymer Dispersions," *Angew. Makromo. Chem.*, **123/124**, 335-359 (1988).

Leighton, D., personal communication, (1991).

Leighton, D. and A. Acrivos, "Measurement of Shear Induced Self- Diffusion in Concentrated Suspensions of Spheres," *J. Fluid Mech.*, **177**, 109-131 (1987a).

Leighton, D. and A. Acrivos, "The Shear-Induced Migration of Particles in Concentrated Suspensions," *J. Fluid Mech.*, **181**, 415-439 (1987b).

Marshall, L., J.W. Goodwin and C.F. Zukoski, "The Effects of Electric Fields on the Rheology of Non-Aqueous Concentrated Suspensions," *J. Chem. Soc. Faraday I*, **85**, 2785-2795 (1989).

Mewis, J and A B. Metzner, "The Rheological Properties of Suspensions of Fibres in Newtonian Fluids Subjected to Extensional Deformations," *J. Fluid Mech.*, **62**, 593-600 (1974).

O'Brien, R.W., "A Method for the Calculation of the Effective Transport Properties of Suspensions of Interacting Particles," *J. Fluid Mech.*, **56**, 401-427 (1979).

Ottewill, R.H. and N.St.J. Williams, "Study of Particle Motion in Concentrated Dispersions by Tracer Diffusion," *Nature*, **325**, 232-234 (1987).

Pätzold, R., "Die Abhängigkeit des Fliessverhaltens Konzentrierter Kugelsuspensionen von der Strömungsform: Ein Vergleich der Viskosität in Scher- und Dehnströmungen," *Rheol. Acta.*, **19**, 322-344 (1980).

Phillips, R.J., J.F. Brady and G. Bossis, "Hydrodynamic Transport Properties of Hard-Sphere Dispersions. I. Suspensions of Freely Mobile Particles," *Phys. Fluids*, **31**, 3462-3472 (1988).

Phillips, R.J., R.C. Armstrong, R.A. Brown, A.L. Graham and J.R. Abbot, "A Constitutive Equation for Concentrated Suspensions that Accounts for Shear-Induced Particle Migration," preprint, (1991).

Powell, R.L., "Rheology of Suspensions of Rodlike Particles," *J. Stat. Phys.*, **62**, 1073-1094 (1991).

Pusey, P.N. and W. van Megen, "Measurement of the Short-Time Self-Diffusivity of Particles in Concentrated Suspension. Evidence for Many-particle Hydrodynamic Interactions," *J. de Phys.*, **44**, 285-291 (1983).

Rallison, J.M. and E.J. Hinch, "The Effect of Particle Interactions on Dynamic Light Scattering from a Dilute Suspension," *J. Fluid Mech.*, **167**, 131-168 (1986).

Rosenweig, R.E., *Ferrohydrodynamics*, Cambridge University Press, (1985).

van der Werff, J. C. and C. G. de Kruif, "Hard-sphere Colloidal Dispersions: The Scaling of Rheological Properties with Particle Size, Volume Fraction, and Shear Rate," *J. Rheol.*, **33**(3), 421-454 (1989).

van Megan, W., S.M. Underwood and I. Snook, "Tracer Diffusion in Concentrated Colloidal Dispersions," *J. Chem. Phys.*, **85**, 4065-4072 (1986).

Weider, T., U. Stottut, W. Loose and S. Hess, "Order in Fluids: Shear-Induced Anisotropy in Dense Fluids of Spherical Particles and in Gases of Rotating Molecules," *Physica A*, **174**, 1-14 (1991).

Zuzovksy, M., P.M. Adler, and H. Brenner, "Spatially Periodic Suspensions of Convex particles in Linear Shear Flows. III. Dilute Arrays of Spheres Suspended in Newtonian Fluids," *Phys. Fluids*, **26**, 1714-1723 (1983).

CHAPTER 27

NUMERICAL SIMULATION OF SUSPENSION FLOW ON HIGH PERFORMANCE COMPUTERS

S. Kim
Y. O. Fuentes

The determination of the motion of microscopic objects through a viscous fluid is a fundamental problem that is encountered in many areas of industrial importance, as well as in the physical and biological sciences (Brownian motion, sedimentation, and electrophoresis). In manufacturing processes we may include particle-induced device contamination (Kim and Lawrence, 1988), colloid stability (Zeichner and Schowalter, 1977), suspension and polymer rheology (Bird, *et al.*, 1987), hindered sedimentation, filtration, as well as diverse particle fractionation techniques. Materials-processing (polymers, liquid crystals (Dinh and Armstrong, 1984) and ceramics) can also involve motion of particles in a viscous medium. Some aspects of biotechnology, such as efficient drug design and *de novo* protein architecture have also been cast in this framework (McCammon, *et al.*, 1976).

At the microscopic level in these systems, inertia is essentially negligible; the particle-based Reynolds number is typically less than 10^{-4}. Therefore, the evolution of these systems is governed by a force and torque balance for each submerged object, α, which we may write generally as (Brenner, 1972)

$$\boldsymbol{F}_{\alpha}^{\text{Hyd}} + \boldsymbol{F}_{\alpha}^{\text{Det}} + \boldsymbol{F}_{\alpha}^{\text{Br}} = \boldsymbol{0}\ . \tag{27.1}$$

Here $\boldsymbol{F}_{\alpha}^{\text{Hyd}}$, $\boldsymbol{F}_{\alpha}^{\text{Br}}$ and $\boldsymbol{F}_{\alpha}^{\text{Det}}$ are respectively the hydrodynamic drag, the Brownian force, and the net sum of all deterministic forces. The equation for the balance of angular moment is

$$\boldsymbol{T}_\alpha^{\mathrm{Hyd}} + \boldsymbol{T}_\alpha^{\mathrm{Det}} + \boldsymbol{T}_\alpha^{\mathrm{Br}} = \boldsymbol{0} \ , \tag{27.2}$$

where $\boldsymbol{T}_\alpha^{\mathrm{Hyd}}$, $\boldsymbol{T}_\alpha^{\mathrm{Br}}$ and $\boldsymbol{T}_\alpha^{\mathrm{Det}}$ are respectively the hydrodynamic torque, the Brownian torque, and the net sum of all deterministic torques.

These deterministic forces and torques are usually further decomposed into those due to *external* agents such as gravity and externally applied electromagnetic fields, and *internal contributions* such as those from colloidal forces between the particles. (See also the more complete discussion in Chapter 26).

For objects submerged in a liquid such as water at room temperature, of size greater than about a micron, Brownian forces are negligible in comparison to the deterministic forces of interest. The force and torque balances then yield the matrix relation (Brenner, 1972)

$$\begin{pmatrix} \boldsymbol{U}_\alpha \\ \boldsymbol{\omega}_\alpha \end{pmatrix} = \begin{pmatrix} \boldsymbol{a}_{\alpha\beta} & \boldsymbol{b}_{\alpha\beta}^{\dagger} \\ \boldsymbol{b}_{\alpha\beta} & \boldsymbol{c}_{\alpha\beta} \end{pmatrix} \begin{pmatrix} \boldsymbol{F}_\beta^{\mathrm{Hyd}} \\ \boldsymbol{T}_\beta^{\mathrm{Hyd}} \end{pmatrix} = - \begin{pmatrix} \boldsymbol{a}_{\alpha\beta} & \boldsymbol{b}_{\beta\alpha}^{\dagger} \\ \boldsymbol{b}_{\alpha\beta} & \boldsymbol{c}_{\alpha\beta} \end{pmatrix} \begin{pmatrix} \boldsymbol{F}_\beta^{\mathrm{Det}} \\ \boldsymbol{T}_\beta^{\mathrm{Det}} \end{pmatrix} . \tag{27.3}$$

The multiplication between matrix elements is actually tensor-vector dot products and we employ the Einstein summation convention so that β repeated in the above equation implies summation over all β. In this manner, the evolution of the state of the system (particle positions and velocities) can be determined given the so-called *mobility tensors* $\boldsymbol{a}_{\alpha\beta}$, $\boldsymbol{b}_{\alpha\beta}$ and $\boldsymbol{c}_{\alpha\beta}$ relating particle motions and hydrodynamic forces and torques.

For smaller objects, Brownian forces are important and the analysis becomes more involved. The force balance then becomes

$$\boldsymbol{F}^{\mathrm{Hyd}} = -\boldsymbol{F}^{\mathrm{Det}} - \boldsymbol{F}^{\mathrm{Br}} \ , \tag{27.4}$$

and the governing equation is a *stochastic* differential equation. Alternatively, if the system is described by a distribution function $\psi(\boldsymbol{Q}, t)$ giving the probability of system configuration $\boldsymbol{Q}$, a *Fokker-Planck equation* (Bird, *et al.*, 1987),

$$\frac{\partial}{\partial \boldsymbol{Q}_\alpha}(\dot{\boldsymbol{Q}}_\alpha \psi) = 0 \tag{27.5}$$

is obtained, with velocities $\dot{\boldsymbol{Q}}_\alpha$ given by either $\boldsymbol{U}_\alpha$ or $\boldsymbol{\omega}_\alpha \times (\boldsymbol{x} - \boldsymbol{r}_\alpha)$. These in turn are related to the hydrodynamic force and torque by Eq. 17.3 as before. The Brownian force is obtained from

$$\boldsymbol{F}_\alpha^{\mathrm{Br}} = -kT \frac{\partial}{\partial \boldsymbol{Q}_\alpha} \ln \psi \ , \tag{27.6}$$

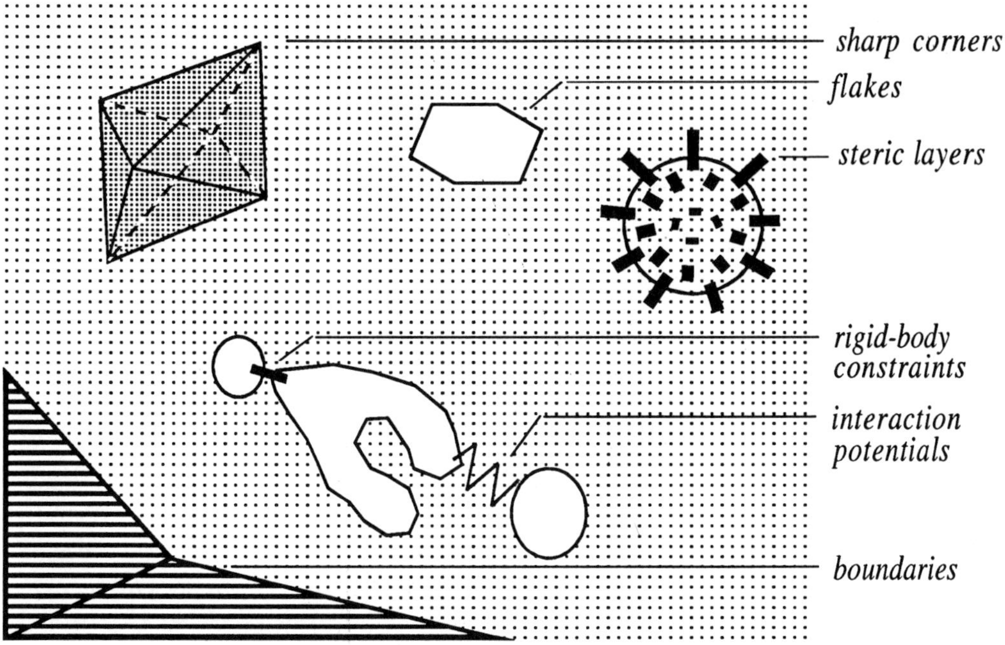

FIGURE 27–1 Hydrodynamic interactions between complex shapes in a viscous fluid, with the presence of bounding walls.

while for an axisymmetric particle with symmetry axis $\boldsymbol{u}_\alpha$, the Brownian torque is obtained from (Brenner, 1972)

$$\boldsymbol{T}_\alpha^{\mathrm{Br}} = -kT\boldsymbol{u}_\alpha \times \frac{\partial}{\partial \boldsymbol{u}_\alpha} \ln \psi \ . \tag{27.7}$$

For the class of problems of interest in this paper, the SDE and FPE approaches are equivalent, and the transformation from one to the other may be found in standard references on stochastic processes (Gardiner, 1985). The main point is that (for both SDE and FPE approaches) the statistical solution for system evolution requires the mobility relations between particle motions and hydrodynamic forces and torques.

Our challenge then is to determine and understand the relation between particle motions and forces in the highly complex geometries described at the beginning of this paper (see Figure 1). Of course, for simple limiting cases such relations are well-known. The use of Stokes law to calculate the sedimentation velocity of an isolated sphere is a familiar example. However, in the problems mentioned earlier, there are severe complexities due to sharp changes in surface curvatures. Furthermore, there are strong hydrodynamic couplings between: particles in concentrated suspensions; between particle and boundaries in deposition problems; and between subunits that act as semi-independent entities in proteins. Without proper attention to algorithm development, the computational requirements quickly exceed the capability of the most powerful computers currently available.

This is not to say that these problems cannot be addressed using the more powerful computers that will most certainly become available in the near future. Indeed, we will devote the rest of this paper in devising a strategy for solving such problems on massively-parallel computers consisting of a large number of high-speed processors networked together. Our focus on parallel processing seems to be appropriate, given the revolutionary changes taking place in the development of state-of-the-art computer architecture. Albeit with some simplifications, one can state that this *revolution* will be dramatically different from the *evolution* of computers over the past two decades. For example, a mathematical model implemented as a numerical code some twenty years ago would run with perhaps minor revisions, on a modern computer at a greatly-enhanced speed. However, we are nearing the end of this type of improvement. The performance of a single processor is now nearing the limitations imposed by the speed of light and semiconductor linewidths. With optimal technology, the speed of a single processor may perhaps be enhanced by a factor of 10. Clearly, the answer lies in parallel processing, for that is the one computational dimension which has yet to be fully exploited. Of course, some algorithms are more amenable to parallel processing than others and the distinction between *sequential* and *parallel* structure becomes paramount. Our challenge then is to reformulate the mathematical model so as to eliminate sequential computational structures from our computational algorithms. We expect the philosophical approach espoused in this chapter to become a common theme in the near future.

27.1 The Computational Strategy

We eliminate spatial methods, such as finite-elements and finite differences from further consideration. The solution of three-dimensional flow problems with such complex, moving boundaries seem to be beyond the scope of the most powerful computers available in the foreseeable future. Boundary integral formulations of the governing Stokes equations for an incompressible Newtonian fluid of viscosity μ,

$$-\nabla p + \mu \nabla^2 \mathbf{v} = 0 , \qquad \nabla \cdot \mathbf{v} = 0 , \tag{27.8}$$

reduce three-dimensional flow problems to the following two dimensional integral equation,

$$\mathbf{v}(\boldsymbol{x}) = \frac{1}{8\pi\mu} \oint_S \boldsymbol{f}(\boldsymbol{\xi}) \cdot \mathcal{G}(\boldsymbol{x}-\boldsymbol{\xi}) \, dS(\boldsymbol{\xi}) + \frac{1}{2} \oint_S \boldsymbol{K}(\boldsymbol{x}-\boldsymbol{\xi}) \cdot \mathbf{v}_s(\boldsymbol{\xi}) \, dS(\boldsymbol{\xi}) \, . \tag{27.9}$$

In this integral velocity representation (due to Lorentz (Lorentz, 1896)) the unknowns correspond to the surface tractions $\boldsymbol{f} = \boldsymbol{\sigma} \cdot \boldsymbol{n}$ and surface velocities $\mathbf{v}_s$. Here, $\mathcal{G}$ denotes the Oseen tensor, the fundamental solution or Green's function of the Stokes equations, which is given explicitly by

$$\mathcal{G}_{ij}(\boldsymbol{x}) = \frac{1}{|\boldsymbol{x}|}\delta_{ij} + \frac{1}{|\boldsymbol{x}|^3}x_i x_j \ . \tag{27.10}$$

The two integral terms on the RHS of the integral representation are known as the single layer and double layer potentials (Odqvist, 1930), in analogy with the corresponding terminology in electrostatics. The kernel of the double layer term is given by

$$\boldsymbol{K}(\boldsymbol{x} - \boldsymbol{\xi}) = -2\boldsymbol{n}(\boldsymbol{\xi}) \cdot \boldsymbol{\Sigma}(\boldsymbol{x} - \boldsymbol{\xi}) \ , \tag{27.11}$$

where $\boldsymbol{\Sigma}(\boldsymbol{x})$ is the stress field of $\mathcal{G}(\boldsymbol{x})/8\pi\mu$, or more explicitly,

$$\boldsymbol{\Sigma}(\boldsymbol{x}) = -\frac{3}{4\pi}\frac{\boldsymbol{xxx}}{|\boldsymbol{x}|^5} \ . \tag{27.12}$$

In Eq.(11) we introduce the factor of 2 in the definition of $\boldsymbol{K}$ so that the eigenvalues of $\boldsymbol{K}$ will be between ± 1. When the boundary velocities are provided, the integral equation becomes a *Fredholm equation of the first kind* for the tractions.

The reduction in dimensionality represents a significant advance as far as numerical solution of the Stokes equation is concerned (Youngren and Acrivos, 1975). However, a direct application to the problems of interest is not feasible in practice, because the boundary integral equations lead to large, dense systems of linear equations. Even with the most powerful supercomputers, accurate solutions are possible for only the simplest geometries and numerical difficulties are encountered with increasing geometric complexity or system size (Tran-Cong and Phan-Thien, 1989). Indeed, we can see this from the fact that the single layer integral operator is *compact*, and in infinite-dimensional Hilbert spaces, the inverse of a compact operator is unbounded. Thus in the sense of Hadamard, Fredholm equations of the first kind are *ill-posed*, since very different inputs are mapped by the compact operator to very similar outputs. In practice, this ill-posedness is not apparent unless very fine meshes are used, which in turn may or may not be necessary, depending on the length scales in the problem at hand. We now consider an alternate integral representation, which ultimately yields a *Fredholm equation of the second kind* and a highly-parallel algorithm.

27.2 Mathematical Background

The flow past an assembly of particles in bounded and unbounded domains can be represented by the alternate form (see Kim and Karrila, 1991)

$$\mathbf{v}(\boldsymbol{x}) = -\boldsymbol{F}^{\mathrm{Hyd}}\cdot\frac{\mathcal{G}(\boldsymbol{x})}{8\pi\mu} - \boldsymbol{T}^{\mathrm{Hyd}}\times\frac{\mathcal{R}(\boldsymbol{x})}{8\pi\mu} + \oint_S \boldsymbol{K}(\boldsymbol{x}-\boldsymbol{\xi})\cdot\boldsymbol{\varphi}(\boldsymbol{\xi})\, dS(\boldsymbol{\xi}) \,, \quad (27.13)$$

in which $\boldsymbol{F}^{\mathrm{Hyd}}$ and $\boldsymbol{T}^{\mathrm{Hyd}}$ are the hydrodynamic forces and torques on the particles,

$$\boldsymbol{T}^{\mathrm{Hyd}} \times \mathcal{R}(\boldsymbol{x}) = \boldsymbol{T}^{\mathrm{Hyd}} \times \frac{(\boldsymbol{x}-\boldsymbol{r}_\alpha)}{|\boldsymbol{x}-\boldsymbol{x}_c|^3} \,, \quad (27.14)$$

is the field of a rotlet or point torque, $\boldsymbol{\varphi}(\boldsymbol{\xi})$ represents the unknown double layer density (which is to be found) and the surface integration is performed over the union of all surfaces bounding the fluid. The double layer distribution alone cannot represent a net force or torque on the particle, just as a double layer distribution in electrostatics cannot represent a net charge; this deficiency is remedied by adding the fields of the point force and point torque, $-\boldsymbol{F}^{\mathrm{Hyd}}\cdot\mathcal{G}(\boldsymbol{x})/8\pi\mu$ and $-\boldsymbol{T}^{\mathrm{Hyd}}\times\mathcal{R}(\boldsymbol{x})$ respectively, a procedure known as completion, hence the name *Completed Double Layer Boundary Integral Equation Method.* Note that this introduces 6 new unknowns per particle surface. However, as we would expect from the Fredholm alternative, the null space of $\mathcal{K}$ is nontrivial; in fact, $\mathcal{N}(\mathcal{K}) = \mathcal{R}(\mathcal{K}^*)^\perp$ and the 6 null solutions per particle surface are precisely the six independent rigid body motions of the particle surface. The removal of this degeneracy provides exactly the right number of new equations, thus providing a unique solution.

Evaluating the preceding equation at the surface of the particles, we obtain the following equations for the unknowns ($\boldsymbol{\varphi}$, $\boldsymbol{F}^{\mathrm{Hyd}}$ and $\boldsymbol{T}^{\mathrm{Hyd}}$), in terms of the given boundary conditions, usually some rigid-body motion for each particle. Explicitly, if for the particle of interest, the rigid-body motion is $\boldsymbol{U}+\boldsymbol{\omega}\times\boldsymbol{x}$, then we have

$$\boldsymbol{U}+\boldsymbol{\omega}\times\boldsymbol{x} = -\boldsymbol{F}^{\mathrm{Hyd}}\cdot\frac{\mathcal{G}(\boldsymbol{x})}{8\pi\mu} - \boldsymbol{T}^{\mathrm{Hyd}}\times\frac{\mathcal{R}(\boldsymbol{x})}{8\pi\mu} + \boldsymbol{\varphi}(\boldsymbol{x}) + \mathcal{K}(\boldsymbol{\varphi}) \,, \qquad \boldsymbol{x}\in S \,. \quad (27.15)$$

where we have used the operator notation for the last term,

$$\mathcal{K}(\boldsymbol{\varphi}) = \oint_S \boldsymbol{K}(\boldsymbol{x}-\boldsymbol{\xi})\cdot\boldsymbol{\varphi}(\boldsymbol{\xi})\, dS(\boldsymbol{\xi}) \,.$$

Note that on the surface $\boldsymbol{\varphi}$ also appears outside the integral, thus yielding a *second-kind* equation. The velocity produced by a double layer distribution suffers a jump of strength $2\boldsymbol{\varphi}(\boldsymbol{\xi})$ across the surface at $\boldsymbol{\xi}$, and the term appearing outside the integral follows from replacing the limit of the surface integral, as the surface is approached, by the integral evaluated on the surface, the difference being exactly $\varphi(\xi)$ or one-half of the total jump.

The advantage of this method is that upon discretization, the linear system is well-conditioned, with very large elements on the diagonal — indeed, the ratio of off-diagonal to diagonal elements scales with the mesh size of the boundary integral method. Thus while very fine scale discretizations inevitably lead to increased computational costs, this loss is partially offset by the increasing disparity between the diagonal and off-diagonal elements. Furthermore, the form of the second-kind equation suggests an iterative solution; such algorithms are readily cast as a parallel algorithm. In the matrix-vector multiplication, the elements in the product vector can be calculated independently with the work load balanced evenly over all processors. In fact, each row of the matrix may be handled by a separate processor, so that this algorithm is the oft-cited example of parallel processing with very fine *granularity* (see Figure 2).

However, a direct application of iterative methods using the diagonal elements will not work in general. The convergence of iterative methods of the form

$$x_{n+1} = Mx_n + b \tag{27.16}$$

depends on the eigenvalues of M. The precise statement is that the iterative equation will converge to the solution if and only if all eigenvalues of M have norm less than one. If this condition is met, the rate of convergence is dictated by the largest eigenvalue. With each successive iteration, the "distance" of approximate solution x_n from the actual solution will diminish by at least a factor equal to the norm of this dominant eigenvalue.

The double layer operator $\mathcal{K}$ has eigenvalues of the form (Odqvist, 1930) (Karrila and Kim, 1989)

$$\lambda = \frac{E^{(i)} - E^{(o)}}{E^{(i)} + E^{(o)}} \tag{27.17}$$

where $E^{(i)}$ and $E^{(o)}$ are the energy dissipation rates inside and outside the surfaces, of the velocity field produced by the eigenfunction acting as a double layer density. So the eigenvalues of $\mathcal{K}$ lie on the real line between -1 and 1 (inclusive), with the end points of the spectrum corresponding to a zero velocity field on one side of the particle surface.

For the case when S is the surface of a single isolated sphere, the spectrum of $\mathcal{K}$ has been determined analytically (Karrila, Fuentes and Kim, 1989). The explicit result for the eigenvalues are:

$$\text{Branch 1:} \quad (n \geq 1) \qquad \lambda_n = \frac{-3}{2n+1} \tag{27.18}$$

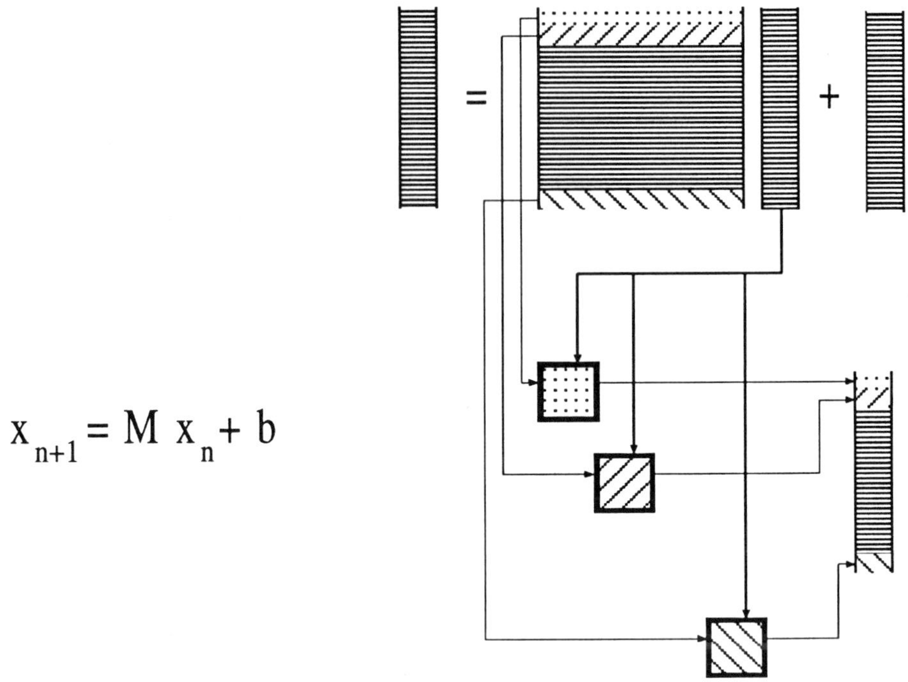

FIGURE 27–2 The parallel algorithm for iterative solution of a very large linear system $x = Mx + b$.

$$\text{Branch 2:} \quad (n \geq 1) \qquad \lambda_n = \frac{-3}{(2n-1)(2n+1)} \tag{27.19}$$

$$\text{Branch 3:} \quad (n \geq 0) \qquad \lambda_n = \frac{3}{2n+1} \tag{27.20}$$

with the degeneracy of $2n+1$ at the index n for each branch. The iterations typically converge to three significant figures after 5 iterations, because the spectral radius of the deflated system is 3/5.

The essence of the computational problem is that any given completion procedure may increase or decrease the spectral radius. We have developed (Karrila, Fuentes and Kim, 1989) the completion procedure for the double layer which also simultaneously shifts all eigenvalues $\lambda = -1$ (*i.e.* the eigenvalues associated with the eigenspace corresponding to the null space of $1 + \mathcal{K}$) to the origin, by a procedure known as Wielandt's deflation. In general, the term deflation refers to algorithms for moving eigenvalues to the origin. Wielandt's deflation uses only the eigenvectors of the operator, and not those of the adjoint. (For our problem, at $\lambda = -1$, the eigenfunctions of $\mathcal{K}$ are known analytically since these are the rigid body motion null solutions, while the eigenfunctions of $\mathcal{K}^*$ are unknown). The major disadvantage of Wielandt's deflation is that the eigenvectors of the deflated system are unknown. Consequently, Wielandt's deflation is usually a *se-*

quential operation, with deflation interlaced with eigenvector calculations. However, by using an orthonormal basis, the entire eigenspace can be deflated in parallel (Karrila, Fuentes and Kim, 1989). The following example with matrices illustrates the essential ideas behind the method. To execute Wielandt's deflation for a single particle, we complete the double layer and obtain a unique solution by requiring

$$-\boldsymbol{U}_\alpha = \sum_{i=1}^{3} -(U_\alpha)_i \boldsymbol{\delta}_i = \sum_{i=1}^{3} \boldsymbol{\varphi}^{(i,\alpha)} \int_{S_\alpha} \boldsymbol{\varphi}^{(i,\alpha)} \cdot \boldsymbol{\varphi}\, dS \,, \qquad (27.21)$$

$$-\boldsymbol{\omega}_\alpha \times (\boldsymbol{x} - \boldsymbol{r}_\alpha) = \sum_{i=4}^{6} \boldsymbol{\varphi}^{(i,\alpha)} \int_{S_\alpha} \boldsymbol{\varphi}^{(i,\alpha)} \cdot \boldsymbol{\varphi}\, dS \,, \qquad (27.22)$$

where $\{\boldsymbol{\varphi}^{(i,\alpha)}\}$, $i = 1, \ldots, 6$, $\alpha = 1, \ldots, N$ form an *orthonormal* basis for the null space. On each particle surface (fix α) the first three null functions are, except for constant scale factors, simply the unit coordinate vectors, while the last three are rotations about the coordinate axes, again with constant scale factors (the moment of inertia per unit mass, about those axes). We substitute the expression for $\boldsymbol{U}_\alpha$ and $\boldsymbol{\omega}_\alpha$ into the boundary integral equations, Eq. 17.15 In operator notation, we see that $\mathcal{K}$ is replaced by

$$\mathcal{K} + \sum \boldsymbol{\varphi}^{(i,\alpha)} \langle\, \boldsymbol{\varphi}^{(i,\alpha)}, \bullet \,\rangle \,. \qquad (27.23)$$

Each $\boldsymbol{\varphi}^{(i,\alpha)}$ is still an eigenfunction, but now its eigenvalue is 0. Generalizing this process to all particle surfaces, we shift all $6N$ eigenvalues of $\mathcal{K}$ at -1 to 0. This can be visualized most readily in the Jordan canonical form of the system. In the following illustration from (Karrila, Fuentes and Kim, 1989) the addition of the term,

$$\sum_i \boldsymbol{\varphi}^{(i,\alpha)} \langle\, \boldsymbol{\varphi}^{(i,\alpha)}, \bullet \,\rangle$$

can be described pictorially as follows:

$$\begin{pmatrix} -1 & 0 & 0 & 0 & \ldots & 0 \\ 0 & -1 & 0 & 0 & \ldots & 0 \\ 0 & 0 & -1 & 0 & \ldots & 0 \\ 0 & 0 & 0 & \lambda_7 & \ldots & 0 \\ \vdots & \vdots & \vdots & \vdots & & \vdots \\ 0 & 0 & 0 & 0 & \ldots & 1 \end{pmatrix} + \sum_i \boldsymbol{\varphi}^{(i,\alpha)} \langle\, \boldsymbol{\varphi}^{(i,\alpha)}, \bullet \,\rangle$$

$$= \begin{pmatrix} 0 & 0 & 0 & X & \ldots & X \\ 0 & 0 & 0 & X & \ldots & X \\ 0 & 0 & 0 & X & \ldots & X \\ 0 & 0 & 0 & \lambda_7 & \ldots & 0 \\ \vdots & \vdots & \vdots & \vdots & & \vdots \\ 0 & 0 & 0 & 0 & \ldots & 1 \end{pmatrix} \qquad (27.24)$$

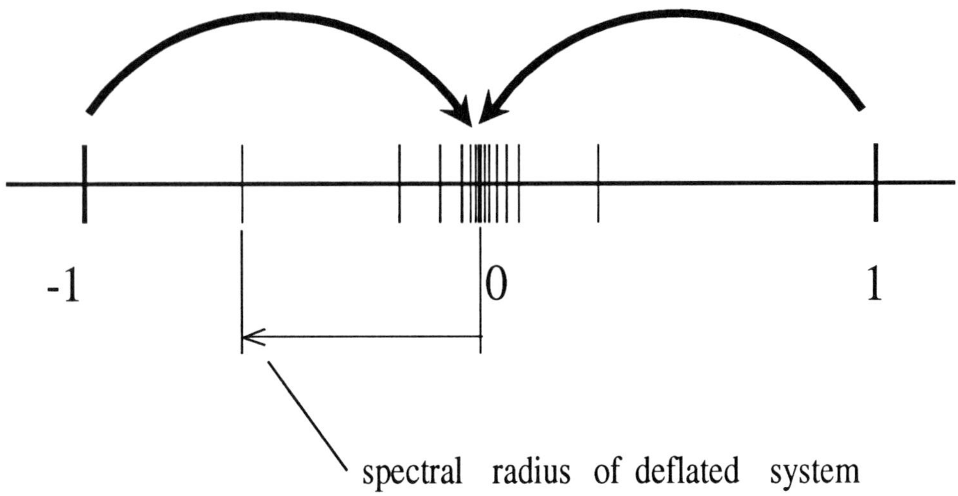

FIGURE 27–3 The discrete spectrum of the compact operator, $\mathcal{K}$. The origin is a limit point.

Wielandt deflation destroys the Jordan form but *the matrix remains block triangular.*

The 3 by 3 blocks on the upper left corner of these matrices represent schematically the $6N$ by $6N$ blocks of the actual system of equations. This entire block gets mapped to zero, *only if an orthonormal basis is used for the deflated eigenspace.* It is important to do so, since then the deflated matrix will be *triangular* and the eigenvalues may be read off the diagonal. Note that the eigenvalues of a block-triangular matrix are given by those of the diagonal blocks and this implies that the remaining eigenvalues have not been shifted. Finally, we used the Jordan canonical form to simplify the illustration, but the result applies to the original system, since similarity transformations applied to obtain the Jordan form preserve eigenvalues. Furthermore, the parallel computational structure is maintained since we deflate the entire eigenspace of dimension $6N$ simultaneously with all processors performing the same number of floating point operations.

We now exploit the fact that the double layer integral operator is compact (Ramkrishna and Amundson, 1985) (Friedman, 1982) and use a central result from the theory of compact operators, namely that the set of eigenvalues (spectrum) of compact operators is discrete, with at most one limit point, which if present, can be only at the origin. A typical spectrum for the double layer operator $\mathcal{K}$ is shown in Figure 27-3. Thus, when all eigenvalues -1 are deflated, we may employ relaxation methods to solve the system, since an appropriate shift of the post-deflation spectrum yields a spectral radius of less than one and the iterations must converge. However, an even more efficient iterative scheme is obtained by first deflating the eigenvalues at $\lambda = 1$, as shown in Figure 27-3. This is possible since the eigenfunctions of $\mathcal{K}^*$ are also known analytically at $\lambda = 1$ (Karrila, Fuentes and Kim, 1989).

The explicit procedure for deflating the eigensystem at $\lambda = 1$ is as follows. Let $\tilde{\varphi}$ be an eigenvector for the eigenvalue λ of the adjoint, denoted by $\mathcal{K}^*$. For the discretized system,

$$\mathcal{K}^*(\tilde{\varphi}) = \lambda\tilde{\varphi} \qquad \text{or} \qquad [\mathcal{K}^*]^{-1}(\tilde{\varphi}) = \lambda^{-1}\tilde{\varphi} \,. \tag{27.25}$$

Then

$$(\mathcal{K} + q\tilde{\varphi}\tilde{\varphi}^t)\mathcal{K}^{-1} = 1 + q\left([\mathcal{K}^*]^{-1}\tilde{\varphi}\tilde{\varphi}^t\right)^t = 1 + \frac{q}{\lambda}\tilde{\varphi}\tilde{\varphi}^t \tag{27.26}$$

so that

$$\mathcal{K}^{-1} = (\mathcal{K} + q\tilde{\varphi}\tilde{\varphi}^t)^{-1}(1 + \frac{q}{\lambda}\tilde{\varphi}\tilde{\varphi}^t) \tag{27.27}$$

where q is an arbitary parameter which we shall set presently. Therefore, the problem $\mathcal{K}(x) = b$, has the solution

$$x = (\mathcal{K} + q\tilde{\varphi}\tilde{\varphi}^t)^{-1}(1 + \frac{q}{\lambda}\tilde{\varphi}\tilde{\varphi}^t)b \tag{27.28}$$

so that the same solution is obtained from the system,

$$\mathcal{K}(x) + q\tilde{\varphi}\,\langle\, \tilde{\varphi}, x \,\rangle = b + \frac{q}{\lambda}\tilde{\varphi}\,\langle\, \tilde{\varphi}, b \,\rangle \;. \tag{27.29}$$

If we set $q^{-1} = -|\tilde{\varphi}|^2$, then the eigenvalue $\lambda = 1$ in the original problem is now at 0 in the modified system of equations, while all other eigenvalues remain fixed.

In the Jordan canonical form, the 'before' and 'after' pictures of the deflation of $\lambda = 1$ are as follows:

$$[\mathcal{K}] = \begin{pmatrix} 0 & 0 & 0 & X & \ldots & X \\ 0 & 0 & 0 & X & \ldots & X \\ 0 & 0 & 0 & X & \ldots & X \\ 0 & 0 & 0 & \lambda_7 & \ldots & 0 \\ \vdots & \vdots & \vdots & \vdots & & \vdots \\ 0 & 0 & 0 & 0 & \ldots & 1 \end{pmatrix}$$

$$\rightarrow \quad \left[\mathcal{K} - |\tilde{\varphi}|^{-2}\tilde{\varphi}\tilde{\varphi}^t\right] = \begin{pmatrix} 0 & 0 & 0 & X & \ldots & Y \\ 0 & 0 & 0 & X & \ldots & Y \\ 0 & 0 & 0 & X & \ldots & Y \\ 0 & 0 & 0 & \lambda_7 & \ldots & Y \\ \vdots & \vdots & \vdots & \vdots & & \vdots \\ 0 & 0 & 0 & 0 & \ldots & 0 \end{pmatrix} \,.$$

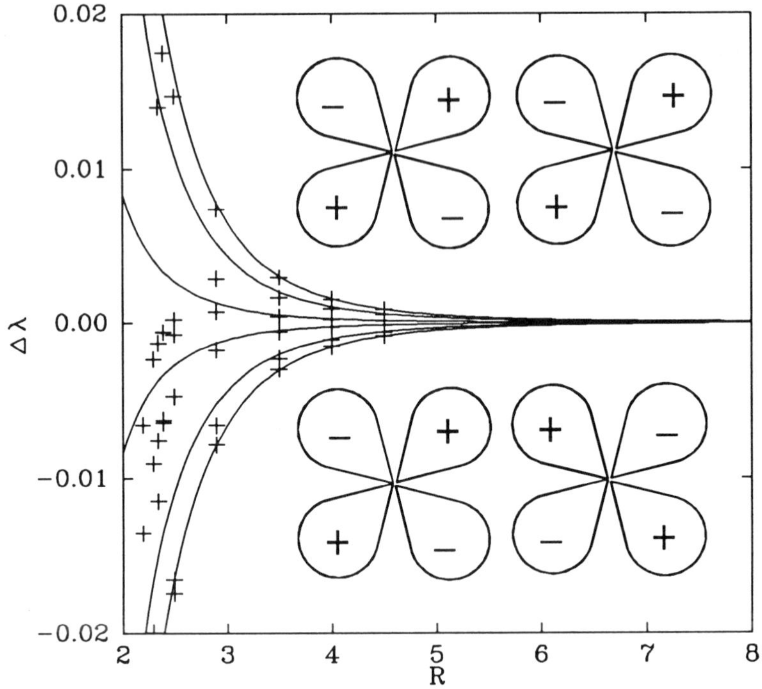

FIGURE 27–4 Hydrodynamic interaction induced splitting of the *outermost* eigenvalues of the two-sphere system, as a function of center-center separation r.

The placement of Y's in the last column indicates that these elements are the ones that are modified.

For a suspension of spheres, the spectrum is perturbed only slightly by the presence of hydrodynamic interactions. This point is illustrated in Figure 27-4 which is taken from Kim and Karrila (1991). We see that the interactions induce a split in the spectrum, analogous to the situation with energy levels in quantum mechanics. Here however, the spectral radius undergoes only a minor perturbation, even for relatively close surfaces, because the eigenfunctions (particle double layer densities) interact rather weakly as force-free and torque-free disturbances.

27.3 Tractions and Stresslets

Although the double layer density φ is not a physical quantity, the solution procedure furnishes directly the most important hydrodynamic relations, *viz.* the relation between hydrodynamic force, $\boldsymbol{F}$, the net hydrodynamic torque $\boldsymbol{T}$, and the particle motions. Note that we have the capability of solving *mobility* problems directly, since $\boldsymbol{F}^{\mathrm{Hyd}}$ and $\boldsymbol{T}^{\mathrm{Hyd}}$ appear directly in the boundary equations. The more common procedure of first solving the resistance problem, in which the motions are given and the force and torque are found, and then inverting to obtain the mobilities, becomes cumbersome and inefficent in many-body systems.

In the following discussion, we consider direct methods for calculating other quantities of interest, *viz.* the symmetric force dipole or stresslet, used in the calculation of the bulk stress in the suspension, as well as surface tractions for particles in rigid body motion.

27.3.1 Stresslets for Bulk Stress Calculations

There are other moments of the surface traction which are of interest in suspension rheology. For example, the bulk stress ($\boldsymbol{\sigma}^{\mathrm{eff}}$) in a suspension may be expressed as a sum of solvent ($\boldsymbol{\sigma}^s$) and particle contributions, (Batchelor, 1970)

$$\boldsymbol{\sigma}^{\mathrm{eff}} = \boldsymbol{\sigma}^s + \langle n\boldsymbol{S} \rangle \tag{27.30}$$

where $\boldsymbol{S}$ is the symmetric Stokes dipole or stresslet on each particle, defined by

$$\boldsymbol{S} = \frac{1}{2}\int_{S_p} [(\boldsymbol{\sigma}\cdot\boldsymbol{n})\boldsymbol{\xi} + \boldsymbol{\xi}(\boldsymbol{\sigma}\cdot\boldsymbol{n})] \; dS \; - \; \frac{\boldsymbol{\delta}}{3}\int_{S_p} (\boldsymbol{\sigma}\cdot\boldsymbol{n})\cdot\boldsymbol{\xi} \; dS \; . \tag{27.31}$$

The stresslet can be obtained exactly from the double layer density by the relation,

$$\boldsymbol{S} = -2\mu \oint_S (\boldsymbol{n}\boldsymbol{\varphi} + \boldsymbol{\varphi}\boldsymbol{n}) \; dS(\boldsymbol{\xi}) \; , \tag{27.32}$$

so that the numerical implementation of this work to suspension rheology is also possible.

27.3.2 Tractions in Rigid Body Motion

For the special case of rigid body motion, it is possible to obtain the tractions directly from the double layer density, without significant post-processing costs (Karrila and Kim, 1989). Using the Lorentz reciprocal theorem (Brenner, 1964), the drag on a particle in arbitrary ambient Stokes flow $\mathbf{v}^{\infty}$ may be expressed as

$$\boldsymbol{e}_i \cdot \boldsymbol{F}^{\mathrm{Hyd}} = \langle \, (\boldsymbol{\sigma}^{\mathrm{RBM}} \cdot \boldsymbol{n})^{(i)}, \mathbf{v}^{\infty} \, \rangle \tag{27.33}$$

where $(\boldsymbol{\sigma}^{\mathrm{RBM}} \cdot \boldsymbol{n})^{(i)}$ is the traction produced when the particle is in rigid body translation in the i-th coordinate direction.

If we interpret the preceding inner product as a linear functional mapping $\mathbf{v}^{\infty}$ to $\boldsymbol{F}^{\mathrm{Hyd}}$ then according to the Riesz representation theorem,

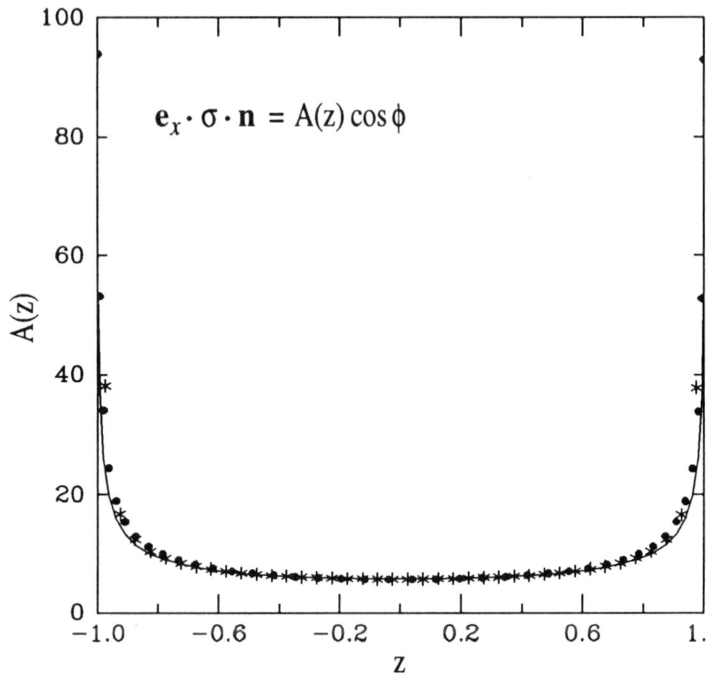

FIGURE 27–5 The x-component of the traction, $\boldsymbol{\sigma} \cdot \boldsymbol{n} = A(z)\cos\phi$ on a prolate spheroid of aspect ratio 10 in transverse (x-direction) translation. The solid line is the analytical result.

$(\boldsymbol{\sigma}^{\mathrm{RBM}} \cdot \boldsymbol{n})^{(i)}$ is the unique "vector" that performs this role. However, the hydrodynamic force calculations in our algorithm are always of the form,

$$\boldsymbol{e}_i \cdot \boldsymbol{F}^{\mathrm{Hyd}} = \langle\, \boldsymbol{C}, \boldsymbol{\varphi} \,\rangle \tag{27.34}$$

with $\boldsymbol{\varphi} = \boldsymbol{A}^{-1}(\mathbf{v}^{\infty})$ so that

$$\boldsymbol{e}_i \cdot \boldsymbol{F}^{\mathrm{Hyd}} = \langle\, \boldsymbol{C}, \boldsymbol{A}^{-1}(\mathbf{v}^{\infty}) \,\rangle = \langle\, (\boldsymbol{A}^{-1})^*(\boldsymbol{C}), \mathbf{v}^{\infty} \,\rangle \tag{27.35}$$

from which we infer $(\boldsymbol{\sigma}^{\mathrm{RBM}} \cdot \boldsymbol{n})^{(i)} = (\boldsymbol{A}^{-1})^*(\boldsymbol{C})$.

In Figure 27-5, we show the agreement between the numerical implementation of this procedure and the analytical result (solid line) for the traction on a prolate spheroid. The two discrete curves indicate slightly different discretizations.

27.4 The Parallel Algorithm

Our method possesses certain attractive features for the solution of particulate flow problems in complex geometries. In many problems, the iterative solutions converges so rapidly, that system construction represents a significant fraction of the overall computational time. But that also can

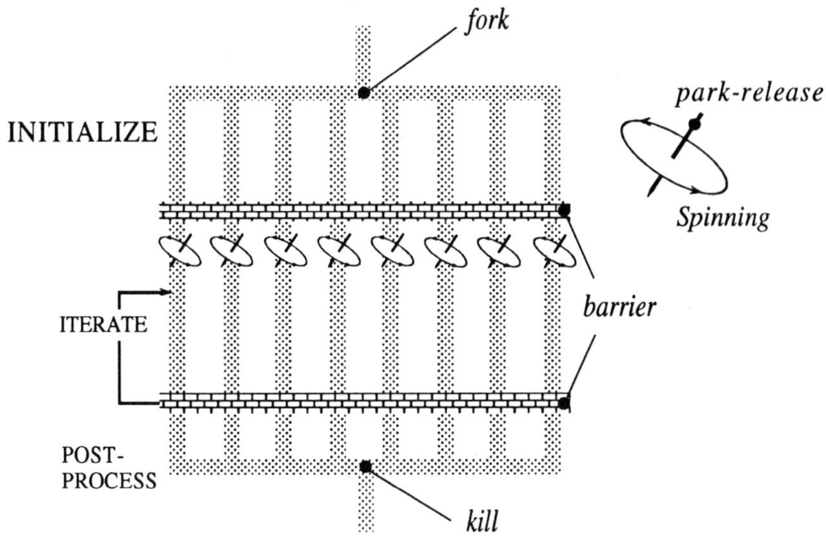

FIGURE 27–6 Parallel microprogramming showing the location of the fork and barriers for synchronization, for an 8-processor computer.

be done in parallel so that the overall algorithm remains parallel. Figure 27-6 shows schematically, how our program runs on the Sequent Symmetry. Since processor loads are evenly balanced, spinning (idle processors requesting more work) within a task is negligible. Between tasks, spinning is suppressed by parking the extra processors. We have achieved 96% efficiencies (see Table) on a Sequent Symmetry with 20 processors (Intel 80386 processors with Weitek 1167 Floating Point Accelerators and shared common memory) without the use of machine-specific trickery to improve performance. The spheres are discretized with 320 boundary elements, so that for N_p spheres, the matrix M is $960N_p$ by $960N_p$. The computation time per iteration *with the matrix recalculated at each iteration to reduce memory requirements* is tabulated in Table 1.

27.4.1 Communication Bottlenecks and Block Asynchronous Iterations

Overcoming communication bottlenecks is perhaps the greatest challenge when scaling up from current parallel computing technology to the machines of the future. This point is illustrated in Figure 27-7. The crossbar and bus-based communication schemes are not scalable. If we try to put thousands of processors, we end up with too many wires (in the cross bar) or traffic jams (bus contention in the bus-based architecture). Meshes and hypercube interconnect (also shown), are examples of scalable architecture.

	Number of Processors		
N_p	1	8	16
1	649.68	83.89	43.55
2	1571.14	201.97	104.66
4	4146.77	530.66	273.78
8	12243.04	1559.13	797.17

Table 27.1 Benchmarks on the Sequent Symmetry: computation times (seconds) per iteration *vs.* N_p, number of 320-element spheres, and number of processors (1, 8 and 16).

While the number of connections only grows as the log of the number of processors, we still face a problem with communication. Information exchange between distant processors requires multiple hops, *i.e.*, there is a time delay. Clearly, algorithms development must be in the direction of reducing net communication between processors.

In Figure 27-8, we show schematically our current investigations into reduction of communication between processors. In suspension simulation, the boundary elements on the same particle are strongly coupled, while those on different particles are less so, as we would expect on physical grounds. This implies that Gauss-Seidel iterative strategies can be employed. For each particle, we use a processor with its own memory. Very rapid internal communications facilitate iterative improvements for each particle. Interactions between particles are obtained in a slower cycle, but the number of such cycles are greatly reduced.

27.5 Summary and Conclusions

We call the new method the *Completed Double Layer Boundary Integral Equation Method* (CDL-BIEM) and its essential computational structure is that of an iterative solution falling in the general class of relaxation methods. The construction of the iterative scheme was not haphazard or accidental, but follows from a series of new mathematical theorems for integral representations for viscous flows (Kim and Karrila, 1991). An important feature of this method is that the system of equations remains well-conditioned with increasing size. Furthermore, on conventional, single CPU computers, computation time increases only as the square of the number of unknowns (and thus also as the square of the number of particles). An even more attractive feature is that the algorithm is parallel, so

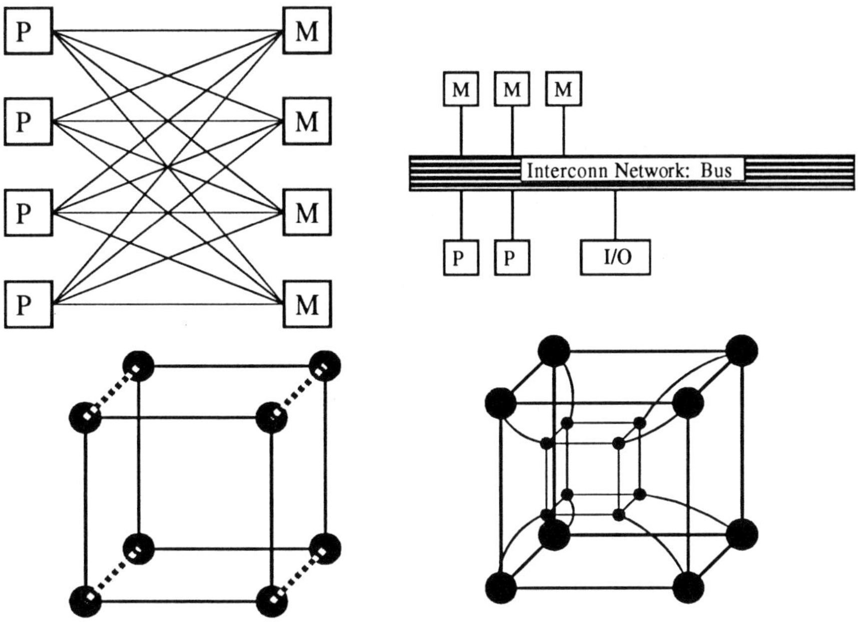

FIGURE 27–7 Interconnect schemes: cross bar, bus-based, and hypercube message passing.

that the computation time t decreases with the number of processors.

There are a number of promising applications of the new method, the common theme being the the study of many-body interactions and complex geometries with fine scale discretization (Pakdel and Kim, 1991). Examples of challenging problems in materials processing in which a better treatment of many-body interactions is needed include stability and rheology of concentrated colloidal suspensions, structural stability of colloidal flocs and aggregates, dynamics of rod-like fibers such as those found in liquid crystals, and buildup and destruction of structure in electro-rheological fluids. In addition to these direct applications in modeling, more work is needed on well-defined and controlled numerical tests of two fundamental concepts of suspension rheology: *screening* and *pairwise additivity*. The idea that beyond a certain distance, two test particles are screened by intervening particles is certainly an intuitively appealing concept which greatly simplifies the development of theory. Theoretical developments are also greatly simplified if many-body interactions can be treated by simply summing over all interactions between pairs. In both cases, our new method provides a tool for establishing the validity and ascertaining the limits of these important concepts. Preliminary results may be found in Fuentes and Kim (1991).

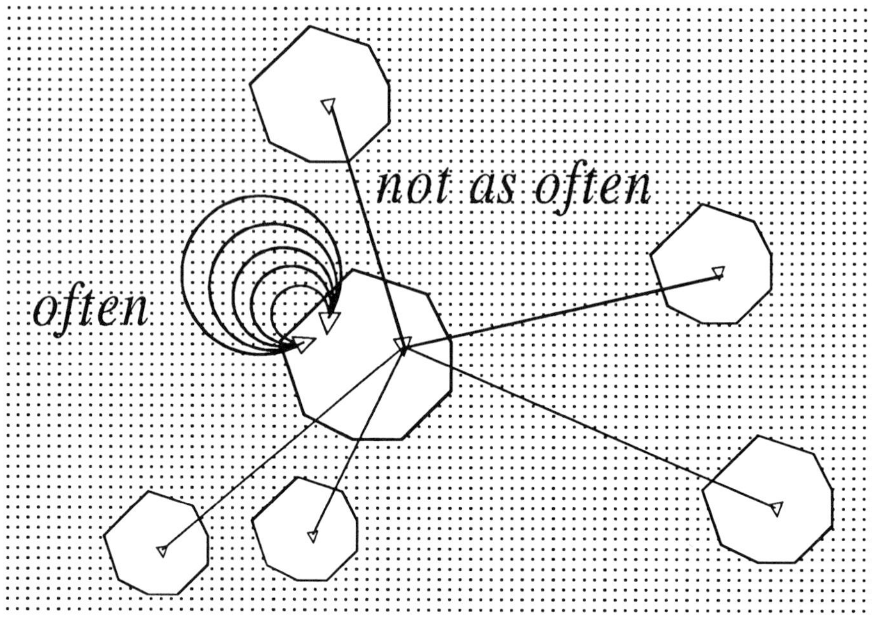

FIGURE 27–8 Communication bottlenecks as the motivation for asynchronous iterative strategy.

The suspension fluid mechanics research community is on the verge of a dramatic breakthrough in computational capabilities. As part of the *high-performance computing initiative*, the goal is to produce TeraFLOP (10^{12} floating point operations per second) parallel machines before the end of this decade, and thus create new paradigms in computing. Among the applications of this new era in computatational engineering, we envision computer simulations that aid our efforts to link suspension microstructure to macroscopic phenomena associated with materials processing, and the prediction of multiphase material properties. However, only *parallel algorithms* will be able to take participate in this revolution.

Acknowledgments

This chapter is based upon work supported by the National Science Foundation grant CTS-9015377 and the American Chemical Society Petroleum Research Fund. The Sequent Symmetry was purchased with the National Science Foundation grant DCR-8521228 to the Computer Sciences Department, for Topaz — A Laboratory for Research in Parallel Computing. We thank Profs. Mary Vernon and Mark Hill of the UW-Madison Computer Sciences Department for helpful comments on parallel computing.

REFERENCES

Batchelor, G. K., "The Stress System in a Suspension of Force-Free Particles," *J. Fluid Mech.*, Vol. 41, 1970, pp. 545-570.

Bird, R. B., Curtiss, C. F., Armstrong, R. C. and Hassager, O., *Dynamics of Polymeric Liquids*, Vol. 2, 2d ed., John Wiley & Sons, New York, 1987.

Brenner, H., "The Stokes Resistance of an Arbitrary Particle — IV. Arbitrary Fields of Flow," *Chem. Engng. Sci.*, Vol. 19, 1964, pp. 703-727.

Brenner, H., "Suspension Rheology," *Progress in Heat and Mass Transfer*, Vol. 5, 1972, pp. 89-129.

Dinh, S. M. and Armstrong, R. C., "A Rheological Equation of State for Semiconcentrated Fiber Suspensions," *J. Rheology*, Vol. 28, 1984, pp. 207-227.

Friedman, A., *Foundations of Modern Analysis*, Dover, New York, 1982.

Fuentes, Y. O. and Kim, S., "Foundations of Parallel Computational Microhydrodynamics: Communication Scheduling Strategies," *Rheology Research Center Report*, No. 127, 1991, pp. 1-50.

Gardiner, C. W., *Handbook of Stochastic Methods for Physics, Chemistry and the Natural Sciences*, 3d ed., Springer-Verlag, New York, 1985.

Karrila, S. J., *Linear Operator Theory Applied to Fast Computational Strategies for Hydrodynamic Interactions in Viscous Flows*, Ph.D. Dissertation, University of Wisconsin - Madison, 1988.

Karrila, S. J. and Kim, S., "Integral Equations of the Second Kind for Stokes Flow: Direct Solution for Physical Variables Removal of Inherent Accuracy Limitations," *Chem. Engng. Commun.*, Vol. 82, 1989, pp. 123-161.

Karrila, S. J., Fuentes, Y. O. and Kim, S., "Parallel Computational Strategies for Hydrodynamic Interactions between Rigid Particles of Arbitrary Shape in a Viscous Fluid," *J. Rheology*, Vol. 33, 1989, pp. 913-947.

Kim, S. and Karrila, S. J., *Microhydrodynamics: Principles and Selected Applications*, Butterworth–Heinemann, Boston, 1991.

Kim, S. and Lawrence, C. J., "Suspension Mechanics for Particle Contamination Control," *Chem. Engng. Sci.*, Vol. 43, 1988, pp. 991-1016.

Lorentz, H. A., "A General Theorem Concerning the Motion of a Viscous Fluid and a Few Consequences Derived from It," *Versl. Konigl. Akad. Wetensch. Amst.*, Vol. 5, 1896, p.168 (see also Collected Papers, Vol. 4, pp. 7-14, Martinus Nijhoff, The Hague, 1937).

McCammon, J. A., Gelin, B. R., Karplus, M. and Wolynes, P. G., "The Hinge-Bending Mode in Lysozyme," *Nature*, Vol. 262, 1976, pp. 325-326.

Odqvist, F.K.G., "Über die Randwertaufgaben der Hydrodynamik zäher Flüssigkeiten (On the Boundary Value Problems in Hydrodynamics of Viscous Fluids)," *Math. Z.*, Vol. 32, 1930, pp. 329-375.

Pakdel, P. and Kim, S., "Mobility and Stresslet Functions of Particles with Rough Surfaces: a Numerical Study," *J. Rheology*, Vol. 135, 1991, pp. 797-823.

Phan-Thien, N., Tullock, D. and Kim, S., "Completed Double Layer in Half–Space: a Boundary Integral Method," *Computational Mechanics*, 1991, forthcoming.

Power, H. and Miranda, G., "Second Kind Integral Equation Formulation of Stokes' Flows Past a Particle of Arbitrary Shape," *SIAM J. Appl. Math.*, Vol. 47, No. 4, 1987, pp. 689-698.

Ramkrishna, D. and Amundson, N. R., *Linear Operator Methods in Chemical Engineering*, Prentice-Hall, Englewood Cliffs, N.J., 1985.

Tran-Cong, T. and Phan-Thien, N., "Stokes Problems in Multiparticle Systems: A Numerical Method for Arbitrary Flows," *Physics of Fluids*, Vol. A1, No. 3, 1989, pp. 453-461.

Youngren, G. K. and Acrivos, A., "Stokes Flow Past a Particle of Arbitrary Shape: A Numerical Method of Solution," *J. Fluid Mech.*, Vol. 69, 1975, pp. 377-403.

Zeichner, G. and Schowalter, W. R., "Use of Trajectory Analysis to Study Stability of Colloidal Dispersions in Flow Fields," *A.I.Ch.E. J.*, Vol. 23, 1977, pp. 243-254.

CHAPTER 28

NUMERICAL SIMULATION OF THE MOTION OF PARTICLES AT LARGE REYNOLDS NUMBERS

A. S. Sangani
A. Prosperetti

28.1 INTRODUCTION

Given the very widespread occurrence of multi-phase flows in industry (oil, chemical, nuclear and conventional energy, and others) and Nature (e.g. breaking waves, volcanos, geysers), the difficulties that are encountered in their understanding and modelling impose a significant handicap and expense in our dealing with them. A typical example is the current impossibility to predict flow-regime transitions, e.g. from bubbly to slug flow. Yet, the two regimes behave very differently in their pumping requirements and structure loading, with the latter one generating significant vibrations that can seriously jeopardize the integrity of industrial plants.

A large number of mathematical models – some of which are documented in the previous chapters – have been proposed for the description of these flows. By necessity, these models are based on an averaged description in which no attempt is made to follow the motion of the individual interfaces. Rather, the equations are written in terms of suitably averaged quantities (see e.g. Chapter 16). Just as in turbulence modelling, however, such an approach is bound to give rise to an incomplete set of equations. It may be safely stated that today the central problem in the modelling of multi-phase flows is the closure of the system of equations produced by averaging. Even aside from concerns of physical realism, the closure issue is crucial as most of the existing two-phase models are inherently unconditionally unstable and therefore it is in no way obvious to what extent their numerical solutions are comparable with experiment or offer predictive design tools.

The closure problem can be tackled analytically in few and severely constrained cases such as small volume fractions or nearly parallel separated flows. Progress beyond these limited situations however usually requires recourse to intuitive physical arguments or *ad hoc* approximations the validity of which is difficult to assess. Typical in this respect is the recent and important work of Wallis (1989, 1991) who undertook to examine in detail the role of inertial forces in the disperse flow of rigid spheres in an inviscid fluid. He is forced to introduce coefficients that are impossible to calculate accurately analytically and that can only be estimated with unknown accuracy.

In this situation direct numerical simulation appears to be a very attractive alternative for the exploration of multi-phase flows. It should be stressed that these simulations are not an end in themselves, but are to be considered as a tool for a better understanding of the physical mechanisms and for the development of reliable averaged equations useful in applications. Ideally, one might try to derive averaged equations that approximate the average results computed from the direct simulations. While, then, in this sense, the philosophy is comparable to that of physical experimentation, in principle at least, this approach enjoys several advantages over the latter. The amount of data produced is orders of magnitude more detailed and, most importantly, different physical effects that operate simultaneously in nature can be manipulated and studied separately. While these advantages are not yet entirely realized due to the limitations of existing computers, it is already possible to obtain interesting results that would not be available otherwise. Furthermore, it is important to start developing now the mathematical techniques and conceptual strategies that will bear their fullest fruits as computers become more powerful.

Chapters 26 and 27 describe numerical simulation techniques useful for granular flows and low Reynolds number disperse flows. Here we consider the opposite limit of very large Reynolds number for which inertial coupling is dominant and viscous effects are small (but not negligible). We shall summarize a method that we have found very useful for this class of problems, and we shall describe a number of results for linear oscillatory flows and for convective flows. For a more detailed description of the mathematical method the reader is referred to the original papers (Sangani and Acrivos, 1982; Sangani and Yao, 1988; Sangani and Didwania, 1991; Sangani *et al.*, 1991). A related approach which is also useful in the presence of boundaries is given in Kim and Prosperetti (1991).

For simplicity, we shall confine ourselves to rigid spherical particles with a free-slip boundary condition. Our model is thus a rough approximation to a bubbly liquid in which surface tension effects are strong enough to prevent an appreciable deformation of the bubbles and pressure variations slow and gentle enough that the radius can be considered constant. The applicability of the method is however more general as shown in Sangani (1991) and Sangani *et al.* (1991). Furthermore, it may be expected that our results would provide some insight about inertial interaction effects also relevant for the case of deformable bubbles or solid particles at the surface of which the no-slip condition applies.

28.2 METHOD OF SIMULATION

In the limit of high particle Reynolds number of present concern the strongest mechanism of interaction among the particles is inertial. Viscous effects are small and can be included approximately building on a potential flow solution as will be described later. The calculation of the velocity potential accounting for the interactions among the particles is therefore the most important element in the simulation and is summarized here to the extent necessary for an understanding of the basic approach. Complete details can be found in Sangani and Yao (1988).

For the sake of simplicity we take the particles to be equal rigid spheres with radius a and volume σ. We use units such that the fluid density is unity. The particles' density is denoted by ρ_*, their mass by $m = \rho_* \sigma$, and their volume fraction by

$$\alpha_g = \frac{4}{3}\pi a^3 n = \sigma n, \qquad (28-1)$$

where n is the particles' number density. The greek index $\gamma = 1, 2, \ldots, N$ labels the particles and the generic particle is denoted by the index α.

28.2.1 Velocity Potential

To express the interaction among the particles it is very convenient to use simultaneously different representations of the velocity potential. In the neighborhood of the generic particle α we use the standard representation

$$\phi = \sum_{l=0}^{\infty} \sum_{m=0}^{l} \left\{ \left[C_{lm}^{\alpha} + D_{lm}^{\alpha} \left(\frac{a}{r}\right)^{2l+1} \right] Y_l^m + \left[\tilde{C}_{lm}^{\alpha} + \tilde{D}_{lm}^{\alpha} \left(\frac{a}{r}\right)^{2l+1} \right] \tilde{Y}_l^m \right\}, \qquad (28-2)$$

where the solid harmonics Y_l^m, $\tilde{Y}_l^m$ are the product of r^l times the surface harmonics

$$Y_l^m(\theta, \varphi) = r^l P_l^m(\cos \theta) \cos m\varphi, \qquad (28-3)$$

$$\tilde{Y}_l^m(\theta, \varphi) = r^l P_l^m(\cos \theta) \sin m\varphi. \qquad (28-4)$$

Although not expressly indicated by the notation, the origin of the spherical coordinate system adopted here is at the center $\mathbf{x}^\alpha$ of the particle. Using this expansion it is a simple matter to impose the kinematic boundary condition

$$\mathbf{n}^\alpha \cdot \mathbf{v}^\alpha = \mathbf{n}^\alpha \cdot \nabla\phi, \qquad (28-5)$$

where $\mathbf{v}^\alpha$ is the particle's velocity and $\mathbf{n}^\alpha$ the outward unit normal. The result is

$$v_1^\alpha = C_{10}^\alpha - 2D_{10}^\alpha, \qquad (28-6)$$

with similar expressions for v_2^α and v_3^α, while, for $l > 1$,

$$l\, C_{lm}^\alpha - (l+1)\, D_{lm}^\alpha = 0, \qquad (28-7)$$

with the same relation satisfied by $\tilde{C}_{lm}^\alpha$ and $\tilde{D}_{lm}^\alpha$. With $\mathbf{v}^\alpha$ available from the equation of motion, to be discussed below, these conditions establish one set of relations between the C's and the D's. The other relations necessary to close the problem arise from the requirement that the potential also satisfy the kinematic boundary conditions on the other particles.

The local representation (28-2) is not very convenient to account for the interaction among the particles. For this purpose we use an expansion of the potential in terms of fundamental solutions of the Laplace equation and their derivatives. For the case of a finite number of particles we write

$$\phi = \sum_{\gamma=1}^{N} \sum_{n=1}^{\infty} \sum_{m=0}^{n} \partial_1^{n-m} \left[A_{nm}^\gamma \Delta_m + \tilde{A}_{nm}^\gamma \tilde{\Delta}_m \right] \frac{1}{|\mathbf{x} - \mathbf{x}^\gamma|}, \qquad (28-8)$$

where, to obtain a more compact expression, we have changed coordinates from $\mathbf{x} \equiv (x_1, x_2, x_3)$ to $\mathbf{x} \equiv (x_1, \xi, \eta)$, with ξ and η defined by

$$\xi = x_2 + ix_3, \qquad \eta = x_2 - ix_3, \qquad (28-9)$$

and we have set

$$\Delta_m = \partial_\xi^m + \partial_\eta^m, \qquad \tilde{\Delta}_m = i\left(\partial_\xi^m - \partial_\eta^m\right), \qquad (28-10)$$

with the notation

$$\partial_1 \equiv \frac{\partial}{\partial x_1}, \quad \partial_\xi \equiv \frac{\partial}{\partial \xi}, \quad \partial_\eta \equiv \frac{\partial}{\partial \eta}. \tag{28 – 11}$$

Singularity methods for boundary-value problems are well-known (see e.g. Lamb, 1932; Chwang and Wu, 1974, 1975; Weinbaum *et al.*, 1990). In the present case a simple derivation of the representation (28-8) is to write Green's identity for ϕ and carry out a Taylor series expansion around the particle's centers using the fact that the potential satisfies Laplace's equation. It may be noted that an arbitrary solution of Laplace's equation can also be added to the right-hand side of (28-8) to represent, for instance, an imposed flow.

Explicit relations between the coefficients A of (28-8) and the coefficients C and D of (28-2) can be found in Sangani and Yao (1988). These relations, together with (28-6) and (28-7), permit a complete solution of the potential problem. From a computational viewpoint it is important to note that, while the coefficients D^α_{nm} are directly proportional to the corresponding A^α_{nm}, the C^α_{nm} equal an infinite linear combination of A^γ_{nm}'s the coefficients of which are given by derivatives of the fundamental solutions $1/|\mathbf{x} - \mathbf{x}^\gamma|$.

Conceptually, the method proceeds in the same way for the case of an infinite number of particles. Since no techniques exist to handle directly the case of an infinite disordered arrangement of particles, a widely used artifice consists in arranging N particles at random in a cubic cell, and then to fill up the space with copies of this cell. Clearly, this arrangement of particles can be viewed as consisting of the superposition of N interpenetrating infinite cubic lattices. The idea underlying this approximation is that, in practice, there is a screening effect that renders the details of the effect of distant particles on a given particle unimportant. Hence, provided the artificial correlation length resulting from the periodic distribution is sufficiently large, the difference with a truly infinite random medium is negligible.

An important technical difference with the case of a finite number of particles is that the simple-minded idea of extending the summation in (28-8) to an infinite number of terms fails because the series thus obtained does not converge. The underlying reason is that the functions $1/|\mathbf{x}-\mathbf{x}^\gamma|$ are not appropriate fundamental solutions of Laplace's equation in the periodic case as they fail to be orthogonal to the solutions of the adjoint homogeneous problem. Hasimoto (1959) has shown that, for a single infinite lattice, the appropriate fundamental solution is given by

$$S_1(\mathbf{x}) = \frac{1}{\pi\tau} \sum_{\mathbf{k}\neq\mathbf{0}} \frac{1}{k^2} \exp[-2\pi i(\mathbf{k}\cdot\mathbf{x})], \qquad (28-12)$$

where the **k**'s are the vectors in the reciprocal lattice and τ is the volume of the fundamental cell. Other representations of this function more useful for its manipulation and calculation are given in Sangani *et al.* (1991) and Hasimoto (1959). For the case in which the fundamental cell contains N particles, we thus write in place of (28-8),

$$\phi = \sum_{\gamma=1}^{N} \sum_{n=1}^{\infty} \sum_{m=0}^{n} \partial_1^{n-m} \left[A_{nm}^{\gamma} \Delta_m + \tilde{A}_{nm}^{\gamma} \tilde{\Delta}_m \right] S_1(\mathbf{x} - \mathbf{x}^{\gamma}). \qquad (28-13)$$

As before, one then connects the coefficients A and $\tilde{A}$ to those of the local solid harmonic expansions (28-2) near each sphere of the fundamental cell, thereby solving the potential problem. Mathematically, the relations between the coefficients have the same structure described above for a finite number of particles.

28.2.2 Equation of Motion of the Particles

We write the equation of motion of the particles as

$$m\frac{d\mathbf{v}^{\alpha}}{dt} = -\int_{S^{\alpha}} p\mathbf{n}^{\alpha} dS^{\alpha} + \mathbf{f}^{\alpha}, \qquad (28-14)$$

where $\mathbf{n}^{\alpha}$ is the unit normal directed out of the particle into the fluid, p the pressure field in the fluid, and $\mathbf{f}^{\alpha}$ denotes forces of other nature (e.g., the particles' weight, viscous resistance, electrical forces, etc.) acting on the particle α. From Bernoulli's integral, up to an inconsequential function of time,

$$p = -\frac{d\phi}{dt} + \frac{1}{2}\mathbf{u}\cdot\mathbf{u} - \mathbf{g}\cdot\mathbf{x}, \qquad (28-15)$$

where d/dt denotes the material derivative, $\mathbf{u} = \nabla\phi$ the fluid's velocity field, and $\mathbf{g}$ the body force. Upon substitution into (28-14), use of standard theorems for the differentiation of material integrals, and other manipulations (see e.g. Oğuz and Prosperetti, 1990), one finds

$$m\frac{d\mathbf{v}^{\alpha}}{dt} + \frac{d\mathbf{I}^{\alpha}}{dt} = \int_{S^{\alpha}} \left[\frac{1}{2}(\mathbf{u}\cdot\mathbf{u})\mathbf{n}^{\alpha} - (\mathbf{n}^{\alpha}\cdot\mathbf{v}^{\alpha})\mathbf{u}\right] dS^{\alpha} - \sigma\mathbf{g} + \mathbf{f}^{\alpha}. \quad (28-16)$$

In this equation we have introduced the impulse associated to the α-th particle defined as usual by

$$\mathbf{I}^{\alpha} = -\int_{S^{\alpha}} \phi\, \mathbf{n}^{\alpha}\, dS^{\alpha}. \tag{28-17}$$

From the local representation (28-2) of the potential one readily finds

$$I_1^{\alpha} = -\sigma\,(C_{10}^{\alpha} + D_{10}^{\alpha}), \tag{28-18}$$

with similar relations for the other two components.

The integrals in the right-hand side of (28-16) are readily calculated using the local expansion (28-2) for the velocity potential. For example

$$\begin{aligned}\int_{S^{\alpha}} \mathbf{u}\cdot\mathbf{u}\, n_1\, dS^{\alpha} &= \sum_{n=1}^{\infty}\sum_{m=0}^{n} \Lambda_{nm}\left[\frac{1}{a^2}\left(F_{nm}^{\alpha}F_{n+1,m}^{\alpha} + \tilde{F}_{nm}^{\alpha}\tilde{F}_{n+1,m}^{\alpha}\right)\right.\\ &\quad \left. + \frac{1}{n(n+2)}\left(F_{nm}^{\alpha\prime}F_{n+1,m}^{\alpha\prime} + \tilde{F}_{nm}^{\alpha\prime}\tilde{F}_{n+1,m}^{\alpha\prime}\right)\right], \end{aligned} \tag{28-19}$$

where

$$\Lambda_{nm} = 4\pi\,\frac{n\,(n+2)\,(n+m+1)!}{(2n+1)\,(2n+3)\,(n-m)!}\,(1+\delta_{mo}), \tag{28-20}$$

and

$$F_{nm} = r^n\left[C_{nm}^{\alpha} + D_{nm}^{\alpha}\left(\frac{a}{r}\right)^{2n+1}\right]_{r=a}, \qquad F_{nm}^{\alpha\prime} = \left.\frac{dF_{nm}}{dr}\right|_{r=a}, \tag{28-21}$$

with $\tilde{F}_{nm}^{\alpha}, \tilde{F}_{nm}^{\alpha\prime}$ given by similar expressions with C and D replaced by $\tilde{C}$ and $\tilde{D}$.

28.2.3 Numerical Aspects

The numerical solution of the problem described in the previous section requires a truncation of the series appearing in Eqs. (28-2), (28-8), (28-13) to some highest value $n_{Max} = N_s$ of the order of singularity retained. For the oscillatory flow case described in the next section we have studied the effect of truncation and found a surprisingly fast convergence. Errors are less than a few percent even for $N_s = 1$, and rapidly decrease with increasing N_s. In the final calculations we have taken $N_s = 5$, raised to 6 for the largest volume fractions considered. We estimate that the errors introduced by this truncation are less than 0.1%. A similar very strong convergence occurs with increasing number of particles in the fundamental cell used for the simulation of random distributions (Sangani *et al.*, 1991) and we have mostly used $N = 16$ particles in the simulations described below.

The particle distribution is obtained by subjecting each particle of an initially ordered distribution to several thousand random displacements. The degree of randomness of the final result and the effect of the spurious correlation length introduced by the artificial periodicity can be estimated by comparing the radial distribution function $g(r)$ (giving the particle number density at a distance r from a given particle divided by the average particle number density) computed numerically with the available solution of the Percus-Yevick equation (Throop and Bearman, 1965). A typical result for $\alpha_g = 0.3$ is shown in Figure 1, where the line is the exact result and the circles have been obtained with 19 random configurations of 16 particles in the fundamental cell. For $\alpha_g = 0.3$, the cell size is approximately 3 particle diameters. It is seen that the distributions that we use approximate fairly well the exact results, particularly at the small separations that are of greatest importance for the particle interactions.

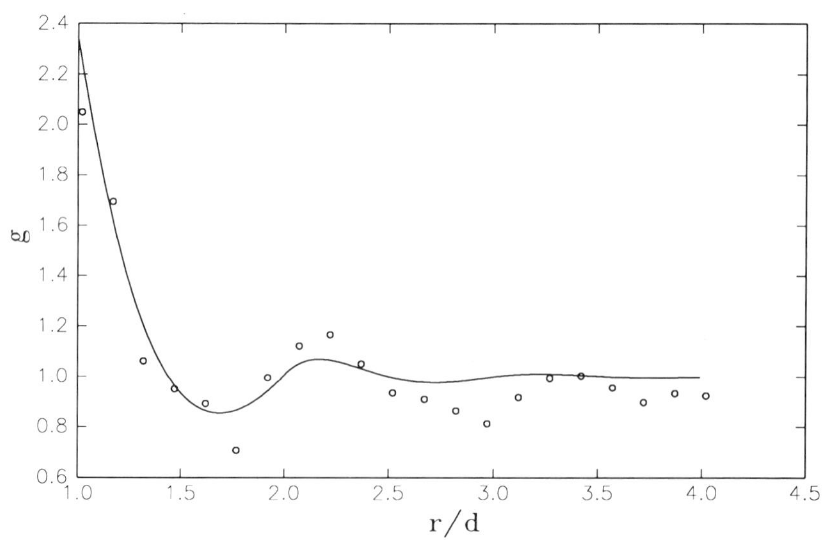

Figure 28-1. *Solution of the Percus-Yevick equation for the radial distribution function* g *compared with the present numerical results for 16 particles and 19 configurations (circles) with* α_g *= 0.3.* r *is the distance from the test particle and* d *is the diameter of the particles.*

For the determination of the potential in each configuration, one has to calculate the coefficients C^{γ}_{nm} in terms of the A^{γ}_{nm}. Setting up the matrix for this system requires the evaluation of a number of derivatives of the fundamental solutions of Laplace's equation of order N_s^2 for $N(N-1)/2$ bubble pairs. For $N_s = 5$ and $N = 16$, on the Cornell Theory Center supercomputer this task requires a time of the order of 15 seconds per configuration. The next step is to solve the linear system, which typically requires several seconds. Thus one potential calculation only requires about 20 seconds.

28.3 ADDED MASS FOR OSCILLATORY OR IMPULSIVE MOTION

Consider a small-amplitude oscillatory motion proportional to $\exp(i\omega t)$. With the neglect of the quadratic terms in the right-hand side, (28-16) reduces to

$$m\mathbf{v}^{\alpha} + \mathbf{I}^{\alpha} = \frac{-\sigma \mathbf{g} + \mathbf{f}^{\alpha}}{i\omega} \equiv \mathbf{G}^{\alpha}. \qquad (28-22)$$

If $\mathbf{G}^{\alpha}$ is prescribed, it is clear from (28-6) and (28-18) that this relation establishes a connection between C^{α}_{1j} and D^{α}_{1j} ($j = 1, 2, 3$) as had been anticipated earlier after Eq. (28-7).

One can get a similar expression in the case in which the motion of the particles and the fluid is impulsively started from rest. Upon integration of (28-16) from $t = 0$- to $t = 0+$ and neglect of the surface term in the right-hand side that, being continuous, cannot contribute to the integral, we find

$$m\mathbf{v}^{\alpha} + \mathbf{I}^{\alpha} = \int_{0-}^{0+} (-\sigma \mathbf{g} + \mathbf{f}^{\alpha})\, dt \equiv \mathbf{G}^{\alpha}. \qquad (28-23)$$

Other than for the two different definitions of $\mathbf{G}^{\alpha}$, this equation is identical to (28-23) so that we can deal with both problems at the same time.

In the present section we shall take $\mathbf{G}^{\alpha}$ to be independent of α. This would be the case if the system is subject to a gravitational field $\mathbf{g}$ and the only contribution to $\mathbf{f}^{\alpha}$ arises from the particles' weight. Another such situation would occur when $\mathbf{g}$ is a uniform macroscopic pressure gradient with an associated length scale much larger than the dimensions of the system under consideration and $\mathbf{f}^{\alpha} = 0$.

To define the added mass coefficient C_a we express the average force

$< \mathbf{F} >$ acting on the particles as

$$< \mathbf{F} >= \frac{1}{2} C_a \sigma < \dot{\mathbf{U}} - \dot{\mathbf{v}} > + \sigma < \dot{\mathbf{U}} > - \sigma \mathbf{g} + \mathbf{f}, \qquad (28-24)$$

where $< \mathbf{U} >$ is the mixture velocity and dots indicate time differentiation. Comparison with (28-23) or (28-22) shows then that

$$C_a < \mathbf{v} - \mathbf{U} >= 2 \left(< \mathbf{U} > - \frac{< \mathbf{I} >}{\sigma} \right). \qquad (28-25)$$

In a well-known paper van Wijngaarden (1976) solved the impulsive version of this problem in the dilute limit for massless spheres and found the result

$$C_a = 1 + 2.76 \alpha_g + O(\alpha_g^2). \qquad (28-26)$$

Kok (1989) and Biesheuvel and Spoelstra (1989) used an alternative definition and, in the dilute limit, found the different result

$$C_a = 1 + 3.32 \alpha_g + O(\alpha_g^2). \qquad (28-27)$$

As discussed by Sangani *et al.* (1991), the origin of the difference lies in the fact that C_a depends on the velocity distribution of the particles after the application of the impulse. This fact is made plausible – if not quite proved – by Eq. (28-23) which implies that a specific distribution of particle velocities requires a specific statistical distribution of $\mathbf{G}^\alpha$'s. Since the forces causing the motion must be different to give rise to different velocity distributions, different values of C_a will then follow. In van Wijngaarden's case, the final velocity distribution is determined implicitly by assuming $\mathbf{G}^\alpha$ to be the same for every particle. By the method of its derivation, on the other hand, the other result (28-27) assumes a uniform velocity distribution in the final state. Allowing the velocity distribution to be determined by the forces rather than taking it as given has a more direct appeal on physical intuition and is the route that we follow. It may be noted, however, that the final velocities of all particles after the application of the impulse are equal if they are arranged on a single periodic lattice. Similarly, for the case of very massive particles ($m \to \infty$), the final velocity vanishes and is therefore also uniform. In these cases, therefore, the two approaches to the definition of C_a lead to the same conclusion, only dependent on the geometry of the system (e.g., random or regular arrays). Our results would then correspond to an extension of (28-27) to the case of finite volume fraction.

Figure 28-2 shows C_a as a function of the particle volume fraction α_g for massless particles, $\rho_* = 0$. It is interesting to note that the results for random arrays are practically indistinguishable from those for body-centered cubic arrays or those obtained from the simple cell theory by Zuber(1964) which gives

$$C_a = \frac{1 + 2\alpha_g}{1 - \alpha_g}. \qquad (28-28)$$

It is shown in Sangani *et al.* (1991) that (28-28) is exact for cubic arrays up to order $\alpha_g^{10/3}$ and it is seen here that it remains a good approximation also at large volume fractions. It may be concluded that C_a is relatively insensitive to the details of the particle distribution.

The dependence of C_a on the particle density ρ_* is also very weak, as shown in Figure 28-3. The difference between the values of C_a for $\rho_* = 0$ and ∞ when α_g equals 0.3 is less than 8%. The difference is even smaller for α_g equal to 0.1 and 0.5. Since different values of ρ_* imply different particle velocities, this finding suggests that C_a is also only weakly dependent on the velocity distribution. Thus, the differences between the two definitions that lead to (28-26) and (28-27) remain small at all volume fractions.

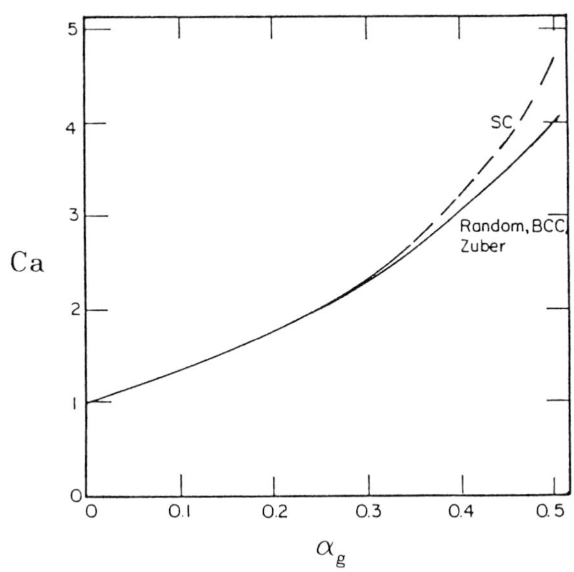

Figure 28-2. *Added mass coefficient as a function of α_g. The results for random arrays, body-centered cubic arrays, and cell theory of Zuber (1964) (28-28) are all represented by the solid line while those for the simple cubic arrays are represented by the dashed line.*

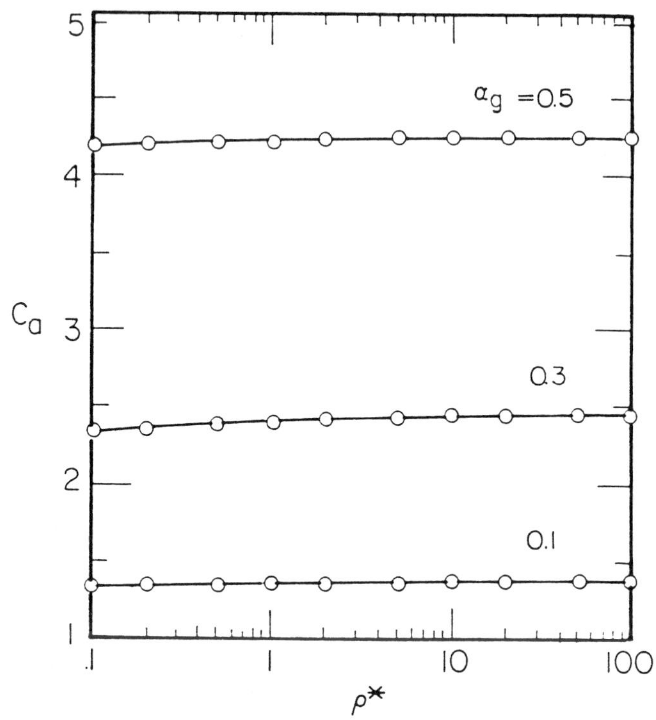

Figure 28-3. *The added mass coefficient as a function of the ratio of the particles' density to the liquid density and the volume fraction of the disperse phase.*

The results of Figure 28-2 have been obtained by using 16 particles in the fundamental cell and the averages are over 12-15 configurations. The results of Figure 28-3 were instead generated by using only one configuration, the same for all values of ρ_*, and averaging over the particles of that configuration. It may be remarked that C_a as defined by (28-25) is actually a second-rank tensor unless the configuration is perfectly isotropic, which in our calculations is true only approximately. In order to calculate all the components of C_a, for each configuration, we take **G** in three mutually orthogonal directions. We find that the off-diagonal elements of C_a are typically one to two orders of magnitude smaller than the diagonal ones, which themselves exhibit differences of a similar magnitude from each other.

An interesting question that direct numerical simulations permit one to examine is the magnitude of the fluctuations about the mean values that can be encountered. For this calculation we write an equation similar to (28-25) for each particle, rather than for averaged quantities. Figure 28-4 shows the fraction of the number of times that the value of C_a^α for an individual particle defined this way was found to lie in a specified small interval. The circles are obtained from a histogram of the computed values and the line is a smooth interpolation through these values. The maximum on the curve corresponds to $C_a^\alpha \simeq 2.44$ corresponding roughly to the average value for $\alpha_g = 0.3$. This average value is different from the value of 2.3 shown in Figure 28-2 which was obtained from Eq. (28-25) by averaging the velocities of all the bubbles first. It is seen here that the particles actually experience the average value of C_a only about 70% of the times. In the other realizations the value can be very different, in this case as low as 2.13 (with a 13% probability) or as high as 3.82 (with a 3% probability). This large spread of the results is very likely representative of what may be encountered in an actual experiment and perhaps should discourage any attempt to a determination of C_a to a greater precision than Zuber's formula (28-28). These results have been obtained with 19 configurations and 16 particles for a total of 304 values of C_a.

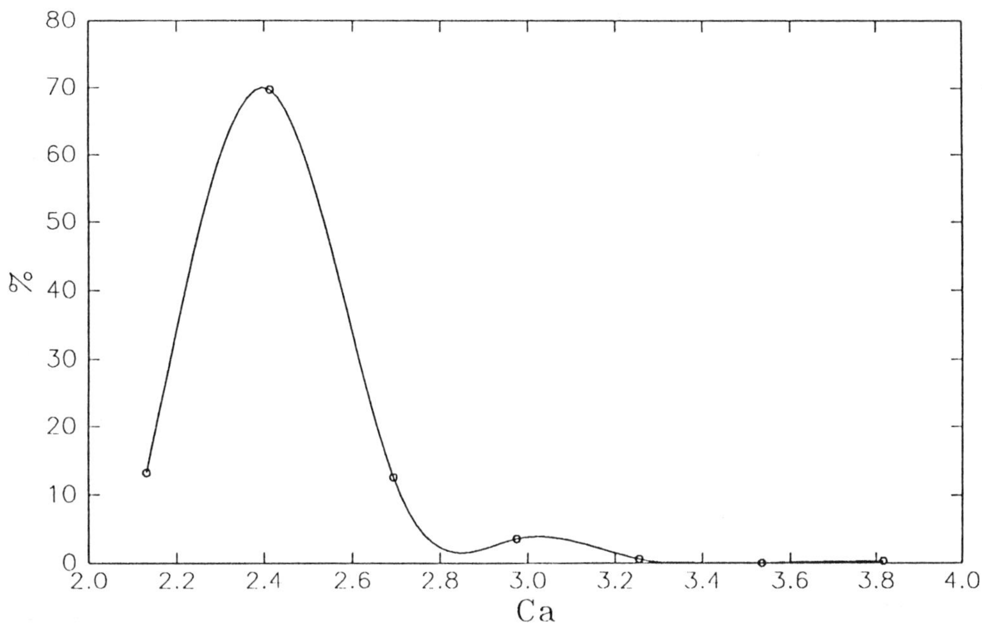

Figure 28-4. *Fraction of the number of times that the value of* C_a^α *for an individual particle was found to lie in a specified small interval for 19 simulations with* $\alpha_g = 0.3$ *and 16 particles in the unit cell.*

The variance of the particles' velocity, defined by

$$Var = \frac{< \mathbf{v} \cdot \mathbf{v} > - < v >^2}{< v >^2}, \qquad (28-29)$$

is shown in Figure 28-5 as a function of α_g, again for $\rho_* = 0$. It has a broad maximum corresponding to a standard deviation around $< v >$ of the order of 15%. In this figure the dashed straight line on the left is the dilute analytic approximation $Var = 0.275\,\alpha_g + O(\alpha_g^2)$ (Sangani *et al.*, 1991), while the broken part of the curve beyond $\alpha_g = 0.5$ is an extrapolation to 0 in correspondence of the maximum packing of a random collection of spheres. These results were obtained with 15 configurations for each α_g and 16 particles in the unit cell.

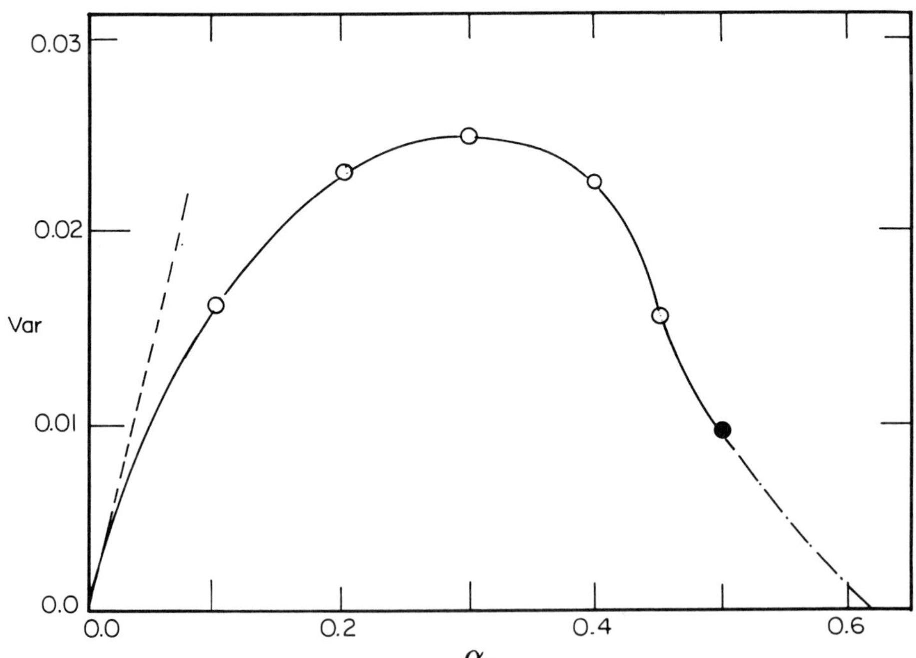

Figure 28-5. *Variance of the particle velocity defined by (28-29) as a function of the disperse phase volume fraction for* $\rho_* = 0$. *The dashed straight line on the left is the analytic approximation valid in the dilute limit, while the dashed part of the curve beyond* $\alpha_g = 0.5$ *is an extrapolation to 0 at the maximum packing of a random collection of spheres.*

28.4 VISCOSITY

The range of problems that can be investigated without the introduction of drag forces is rather limited and it is therefore important to be able to account – if in an approximate way – for viscous effects.

In large Reynolds number flows past free surfaces on which the tangential stress vanishes but not the relative tangential velocity, vorticity generation is weak and the effect of viscosity is confined to a thin layer adjacent to the surface (see e.g. Batchelor, 1967). The boundary-layer analysis of Moore (1963) has shown that, in these conditions, the correction to the velocity as derived from the potential approximation is uniformly small everywhere in the limit of $Re \to \infty$. This analysis cannot be employed, however, to estimate the viscous drag on individual bubbles to leading order, $O(Re^{-1})$, as the boundary layer approximation breaks down near the rear stagnation point so that the contribution to the drag from the viscous correction to the pressure in the wake behind the bubble is difficult to evaluate. Tam (1981) has addressed this problem but his results are inconclusive. For the case of steady translation the difficulty was side-stepped by Levich (1962) who argued that, since the velocity field is only slightly perturbed from the inviscid potential approximation, the total viscous dissipation can be estimated to leading order simply from the potential flow solution. Thus, if $\mathbf{F}_v^\alpha$ is the viscous force acting on a bubble moving with the velocity $\mathbf{v}^\alpha$ in a fluid at rest at infinity, he writes

$$\begin{aligned} \dot{E}_v = -\mathbf{F}_v^\alpha \cdot \mathbf{v}^\alpha &= \frac{\mu}{2} \int_V \left(\frac{\partial u_i}{\partial x_j} + \frac{\partial u_j}{\partial x_i} \right)^2 dV \\ &\simeq -\mu \int_{S^\alpha} n_i \, \partial_i \left(\partial_j \phi \, \partial_j \phi \right) dS^\alpha, \end{aligned} \qquad (28\text{-}30)$$

where μ is the liquid's viscosity. Setting $\phi = \frac{1}{2} \mathbf{v}^\alpha \cdot \nabla r^{-1}$ one readily finds

$$\mathbf{F}_v^\alpha = -12\pi \, \mu \, a \, \mathbf{v}^\alpha. \qquad (28-31)$$

In Levich's calculation the translational velocity was constant. The same result, however, was proven to hold also for a time-dependent velocity and small displacements in Prosperetti (1977). Recently, Kang and Leal (1988) proved that actually (28-31) gives correctly the instantaneous force on the bubble whatever the time dependence of the translational velocity. In other words, the memory effects typical of the parabolic problem that governs the diffusion of vorticity only appear at higher order in an inverse-Reynolds-number expansion.

It is not immediately clear how to extend the previous results to the case of many bubbles (or particles). The rigorous approach would be to follow the method of Kang and Leal, which however presents some technical difficulties that do not arise in the case of a single bubble because certain integrals involving the vorticity distribution vanish by symmetry. A commonly used method appeals to the analytical dynamics formalism of Rayleigh's dissipation function according to which

$$\mathbf{F}_v^\alpha = \frac{1}{2}\nabla_{\mathbf{v}^\alpha} \dot{E}_v, \qquad (28-32)$$

with $\dot{E}_v$ given by (28-30). The justification offered for this procedure is essentially that this formula does lead to (28-31) in the case of a single bubble.

While this is the approach we shall also follow here, it must be pointed out that there are reasons why this procedure is unsatisfactory. In the first place, the very existence of a Rayleigh dissipation function and its identification with $\dot{E}_v$ rests on the hypothesis that the viscous force is linear in the velocity and only depends on its instantaneous value. While this can be verified a posteriori from (28-31) for the special and very symmetrical case of an isolated particle, there is no proof that the property holds more generally. Secondly, if the flow velocity undergoes a correction of order $Re^{-1/2}$ in a boundary layer with a thickness of the same order, it would seem in principle that the kinetic energy of the flow should undergo a correction of order Re^{-1}. In a Lagrangian approach, this would introduce viscous effects in other terms in addition to those arising from Rayleigh's dissipation function. Again, this does not happen for an isolated particle, but no proof exists for the general case. Because of these uncertainties, it is certainly gratifying that, for the case of small-amplitude oscillatory motion, the results found from (28-32) coincide with those obtained by an exact analysis briefly described in the next section.

28.5 VISCOUS EFFECTS IN OSCILLATORY MOTION

For the case of small-amplitude (linear) oscillatory motion, it is possible to rigorously evaluate the effect of viscosity in the limit in which the viscous boundary layer adjacent to the surface of the particles is thin. The details of this calculation are given in Sangani *et al.* (1991). Here we only provide a broad outline sufficient to understand the general approach.

The equations to be solved are the linearized Navier-Stokes equations

$$\nabla \cdot \mathbf{u} = 0, \qquad \frac{\partial \mathbf{u}}{\partial t} = -\nabla p + \nu \nabla^2 \mathbf{u}, \tag{28-33}$$

where ν is the kinematic viscosity, in the limit in which the quantity

$$\mu_* = \frac{\nu}{\omega a^2}, \tag{28-34}$$

is small. Now we supplement the kinematic boundary condition (28-5) by the requirement that the tangential stress along the interface vanishes. Since we still assume the particles to be rigid, we cannot impose any condition on the normal stresses.

The starting point for the solution of the problem is the representation of the velocity field in terms of a scalar and a vector potential, the latter one decomposed into a toroidal and a poloidal part

$$\mathbf{u} = \nabla\phi + \sum_{\gamma=1}^{N} \nabla \times \{\nabla \times [(\mathbf{x} - \mathbf{x}^\gamma)\Psi^\gamma] + (\mathbf{x} - \mathbf{x}^\gamma)\chi^\gamma\}. \tag{28-35}$$

The scalar potential ϕ satisfies the Laplace equation as in the inviscid case, while the toroidal and poloidal components satisfy

$$\nabla^2 \Psi^\alpha = \frac{i}{a^2 \mu_*}\Psi^\alpha, \qquad \nabla^2 \chi^\alpha = \frac{i}{a^2 \mu_*}\chi^\alpha. \tag{28-36}$$

These equations show that, in the limit of small μ_*, the vorticity decays exponentially over a length scale of the order of $a\sqrt{\mu_*} << a$ so that, approximately (to order μ_* included, as can be shown), the vorticity fields of different particles do not interact. Hence, the interaction among the particles is through the irrotational component of the velocity field as in the inviscid case. This does not mean, of course, that there is no effect of viscosity because now the boundary conditions must be imposed on the complete velocity field (28-35) rather than the potential part alone. As will be clear in a moment, the end result is effectively a new boundary condition on the potential ϕ that is different from the inviscid case. Since this potential satisfies the Laplace equation, the viscous correction to the field around one bubble is felt by all the other ones as well.

For rigid spherical particles with free-slip, zero tangential stress boundary conditions, it is found that the toroidal field χ^α is of order $\mu_*^{3/2}$ and can therefore be neglected to the present level of approximation (Sangani *et al.*, 1991). As is clear from the field equations (28-36), the poloidal field Ψ^α can be expanded in series of surface harmonics, with each radial term dependent on a constant that arises from the integration of the radial part of the equation (the other constant is determined by the requirement that Ψ^α vanishes at infinity). The expression for the velocity field near the generic particle obtained in this way can then be substituted into the kinematic and dynamic boundary conditions, which are then projected onto the surface harmonic components. Each pair of the resulting relations contains three constants, two from ϕ and one from Ψ^α. The latter one can be eliminated with the result

$$n\,C^\alpha_{nm} - (n+1)\,D^\alpha_{nm} \;=\; -\,2\,(n+1)(2n+1)\,\mu_*\,D^\alpha_{nm} + O(\mu_*^{3/2}). \quad (28-37)$$

This equation replaces the previous one (28-7) applicable to the inviscid case. At this point the poloidal component has effectively been eliminated and it is therefore clear that the same solution procedure described in the inviscid case can be followed for the calculation of the coefficients C^α_{nm}, D^α_{nm}, etc.

To identify the viscous contribution to the force we look for solutions of the equations for A^α_{nm}, C^α_{nm}, D^α_{nm}, etc. in the form

$$A^\alpha_{nm} = A^{0\alpha}_{nm} + \mu_* A^{1\alpha}_{nm} + \ldots, \qquad (28-38)$$

etc. The zero order solution is the same as the one described in section 28.2. From the next order contribution one finds the following result for the viscous force acting on particle α

$$F^\alpha_{v1} = 4\pi\mu D^{1\alpha}_{10}, \qquad (28-39)$$

with similar expressions for the other components.

Once the viscous force on each particle is calculated in this way, we may proceed to calculate the average viscous drag coefficient that, on the basis of (28-31), we define by

$$<\mathbf{F}_v> = -\,12\,\pi\,\mu\,a\,C_d\,<\mathbf{v}-\mathbf{U}>. \qquad (28-40)$$

This definition ensures that $C_d \rightarrow 1$ for well-separated dilute arrays. Numerical results for C_d are shown in Figure 28-6 for massless spheres. The solid line is for a random distribution, the dashed line for the body-centered cubic array, and the dash-and-dot line for the simple cubic array. The dilute-limit approximation

$$C_d \simeq \frac{1}{(1-\alpha_g)^2}, \qquad (28-41)$$

derived for cubic arrays (Sangani *et al.*, 1991) deviates less than 10% from the numerical results for $\alpha_g \leq 0.5$.

As already mentioned in the previous section, the results of this calculation are identical to those given by the dissipation function method. This is interesting not only in view of the uncertainties about that method, but also computationally. Indeed, the evaluation of the viscous force from (28-32) is computationally much more demanding than the scheme described in this section.

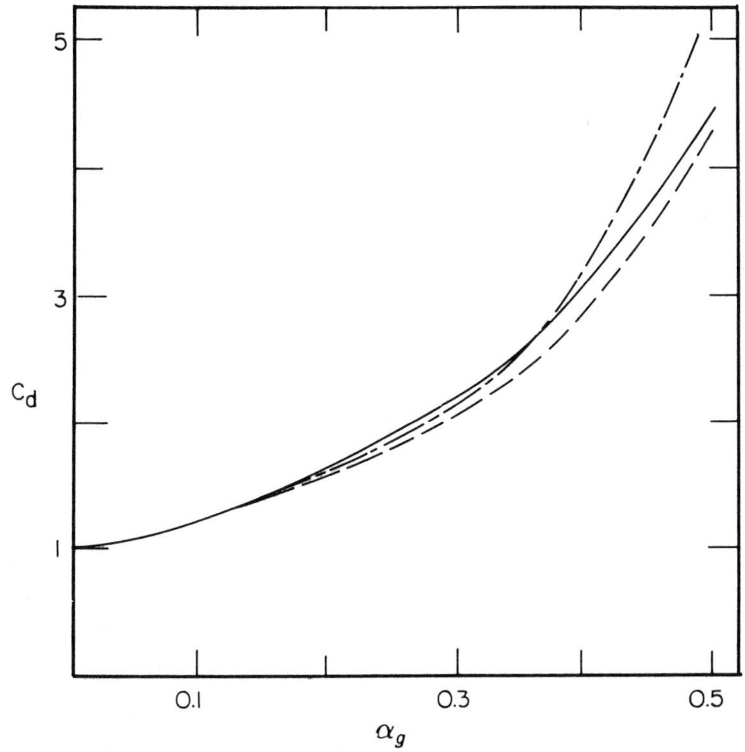

Figure 28-6. *Average viscous drag coefficient of massless spheres in oscillatory flow for a random distribution (solid line), a body-centered cubic array (dashed line), and a simple cubic array (dash-and-dot line).*

28.6 BUOYANT RISE OF SPHERES

The examples studied in the previous sections were particularly simple in that the initial microstructure was preserved. In most cases of interest, however, the microstructure evolves as a consequence of the flow with effects of the utmost importance on the flow itself. As an example of these situations we now consider the buoyant ascent of rigid massless particles ("bubbles") under the action of a constant body force. The flow is simulated in the frame of reference in which the total volume flow rate is zero. In addition to the general evolution of the flow, of particular interest are several average quantities such as the mean impulse (or, equivalently, the added mass coefficient) and the disperse-phase pressure. In the case of a slightly inhomogeneous system these average quantities could be used in conjunction with averaged equations to study e.g. the propagation of kinematic waves in bubbly liquids.

We consider N randomly placed spheres within the basic unit cell at time $t = 0$ given some initial velocity distribution and follow their motion as described by Eq. (28-16). The bubble mass is taken to vanish and the force $\mathbf{f}^\alpha$ is taken to be the viscous force $\mathbf{F}_v^\alpha$ on the α-th bubble. As before, we are interested in the limit of large Reynolds number and therefore we use the potential flow approximation for the evaluation of all the terms of (28-16) except $\mathbf{F}_v^\alpha$ which we evaluate from (28-32) or, equivalently, (28-39).

For pairs of bubbles, the Bernoulli interaction is repulsive when the line joining the bubbles is oriented within 35° of the body force while attraction prevails for other orientations (Kok, 1989). In view of this inertial effect, there is a large portion of phase space in which bubbles attract each other and eventually collide. If the simulation is to proceed for any appreciable length of time, it is therefore necessary to model this process. We have formulated a collision model (described in detail in Sangani and Didwania, 1991) based on the assumption that liquid momentum and kinetic energy are conserved during collisions. For slip-free surfaces this is reasonable as the viscous boundary layers are thin and the duration of the interaction between colliding bubbles very short. The experimental results of Kok (1989) – who found that in a relatively clean liquid the trajectories of two bubbles closely follow the potential theory predictions and the collision is nearly instantaneous – support this approach.

As already mentioned before, the convergence of the numerical results with the number of singularities retained in expressions such as (28-13) is very fast and therefore all the simulations were carried out retaining terms up to quadrupoles. The computational time on the Cornell Theory Center supercomputer is about 1.2 sec per time step for the case of 16 bubbles in the unit cell. Further details on these calculations can be found in Sangani and Didwania (1991).

We use the particle radius a as a length scale and the ascent velocity of a single bubble, $v_\infty = a^2 g/9\nu$, as a velocity scale. The fundamental time is thus a/v_∞. The Reynolds number defined in terms of the previous quantities is given by

$$Re = \frac{a^3 g}{9\nu^2}, \qquad (28-42)$$

while the Froude number $Fr = v_\infty^2/ag$ equals $Re/9$. In the results to be described the volume fraction is $\alpha_g = 0.1$ and $Re = 200$.

We show in Figure 28-7 the configuration of 16 particles in the fundamental cell at several times. The initially random configuration is seen to evolve into a configuration in which all the bubbles within the unit cell arrange themselves essentially in a horizontal plane so that a very significant amount of anisotropy develops in the flow. The formation of this structure is due to the inertial effects mentioned previously. The same tendency has been found experimentally in a two-dimensional system by Fortes *et al.* (1987). Since that experiment was conducted with solid particles exhibiting strong wakes, our results suggest that the structure arises independently of the wakes.

The robustness of the planar structure with respect to variations in bubble sizes and other conditions is evidently an important matter. We have run simulations with a polydisperse system in which the particle radius was given by $a(1+\zeta)$, with ζ a random variable uniformly distributed between -0.2 and 0.2. The corresponding spread in the terminal velocity of ascent is approximately a factor of 2, so that one would expect a considerable tendency for the larger bubbles to escape the planar formations. The numerical simulations shown in Figure 28-8 (in which all the bubbles are shown as having the same radius for convenience) surprisingly show that the planar structure appears nevertheless to form and be robust.

Finally, for the simulation of Figure 28-7, we show in Figure 28-9 the variation with time of the average velocity, of the average added mass coefficient

$$C_a = 2\frac{< I_1 >}{\sigma}, \qquad (28-43)$$

of two viscous force coefficients defined respectively by

$$C_{d1} = \frac{< F_{1v} >}{12\pi a\mu < v_1 >}, \qquad C_{d2} = \frac{\dot{E}_d}{12\pi a\mu < v_1 >^2}, \qquad (28-44)$$

and of a kinetic energy coefficient given by

$$E = \frac{\frac{1}{2} < \mathbf{I}\cdot\mathbf{v} >}{< \mathbf{v} >^2}. \qquad (28-45)$$

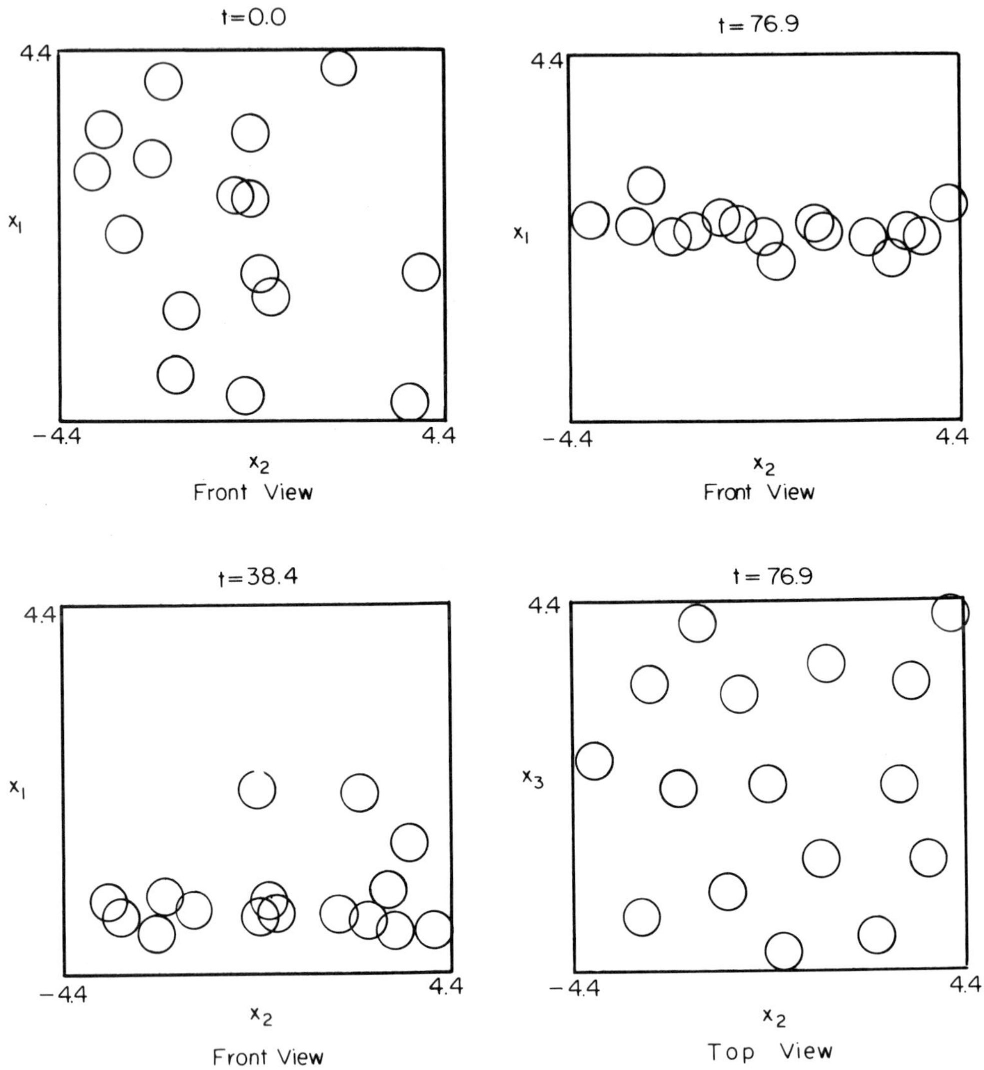

Figure 28-7. *Numerical simulation of the buoyant rise of 16 equal spheres. Note how the initial random configuration (upper left panel) gradually gives rise to a planar structure orthogonal to the direction of ascent (right two panels). The particle Reynolds number is 200 and the volume fraction 10%.*

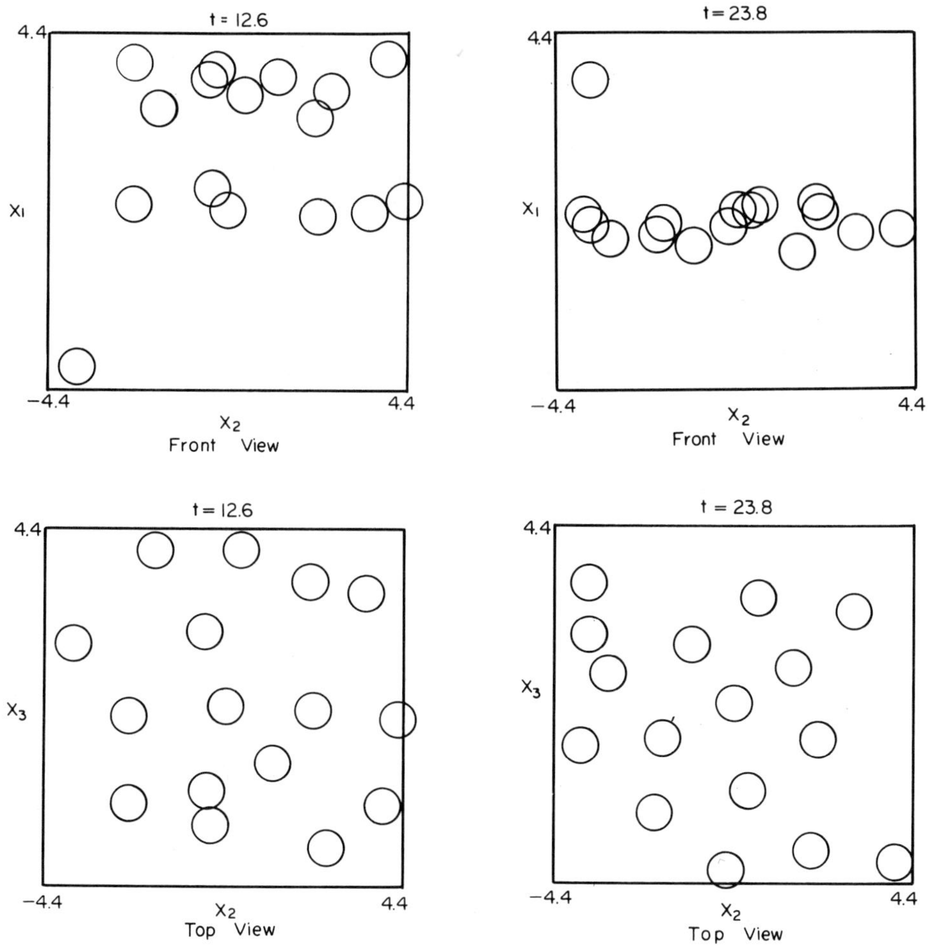

Figure 28-8. *Same as the previous figure, but for a polydisperse mixture. The spheres are shown with the same radius for convenience. Note that a planar structure forms in this case also.*

It is seen that, while the values of C_a and C_{d1} are initially close to those found for random arrays in Figures 28-3 and 28-6, the formation of the planar structures causes a large increase in these quantities.

It must be recognized that it is difficult to judge the full significance of our results due to the fact that they have been obtained for a periodic system. For instance, these simulations cannot tell us whether the plane structure would tend to grow indefinitely in the horizontal direction or would break up. Further work is needed to answer this and other questions. Nevertheless, the extremely strong influence of the Bernoulli attraction brought out by the simulations is quite interesting.

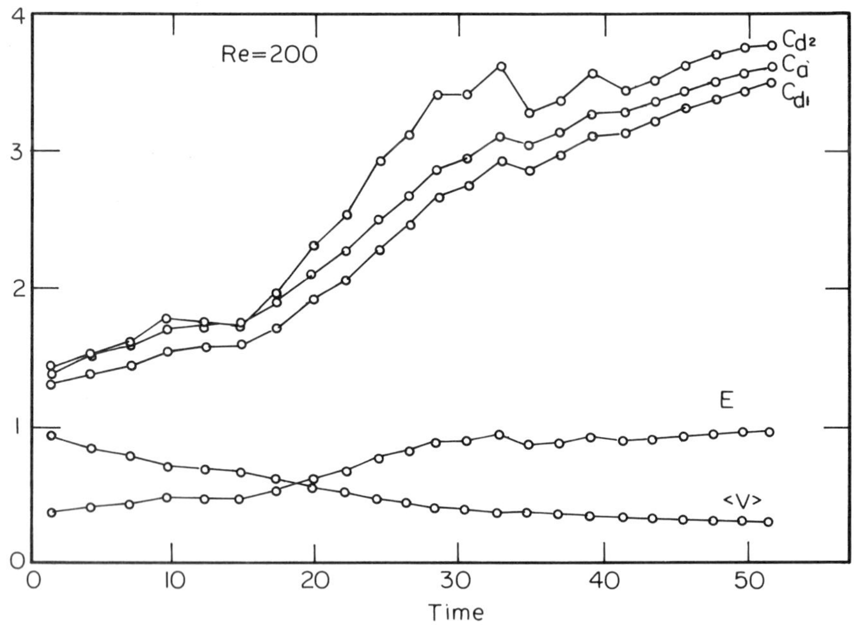

Figure 28-9. *Time evolution of several average quantities for the simulation of Figure 28-7.*

28.7 CONCLUSIONS

We have described a fast and efficient method for the numerical simulation of the motion of particles in a fluid at high Reynolds number. The approach has been illustrated with a number of relatively simple examples which shed considerable light on several issues that arise in the mathematical modelling of multi-phase flow.

We have considered examples of small-amplitude oscillatory motion and we have found that the added mass coefficient is represented very closely by the cell-theory result of Zuber for several different particle arrangements. This is significant because the available asymptotic analytical results valid at small volume fractions suggest bigger differences and some dependence on the precise assumptions made in the calculation of this quantity.

We have also shown simulations of the buoyant rise of particles. We have found a very strong and robust tendency toward the formation of stable structures of the disperse phase that may well be one of the most significant features of disperse two-phase flow responsible for the formation of kinematic waves and flow regime transitions. This tendency – which almost certainly lies at the basis of the well-known ill-posedness of most averaged equations – in nature is countered by turbulence and other flow disturbances (e.g. wakes) that we have not included in our model. It might be thought that the transition from disperse bubbly to slug flow is ultimately the result of the competition between these two effects.

If the formation of a microstructure is of such importance in multiphase flow, an averaged-equation description must be able to account for this effect. This obvious remark raises some troubling questions. In the case considered here, for example, volume-averaged equations based on the fundamental cell would clearly be quite inadequate as the average value of α_g remains constant throughout the simulation. Clearly, averaging over much smaller length scales (comparable with the particle diameter) would be necessary. But, if this is the case, is the fundamental premise of volume averaging – the separation of scales – a reasonable assumption? At the very least, if short-range effects are important, it would seem that models should not be limited to first-order spatial derivatives but should include higher-order ones as well.

The work described in this chapter only scratches the surface of the issues and possibilities opened by the direct numerical simulation of disperse multi-phase flow at high Reynolds numbers. We hope to have shown that this is a fruitful line of research already with the presently available computational resources. One may expect significant further progress as more powerful computers become available.

ACKNOWLEDGEMENTS

The authors wish to thank D.Z. Zhang for some help with the numerical calculations. They are also grateful for the use of the supercomputer facilities at the Cornell Theory Center.

This work has been supported by the Department of Energy, Office of Basic Energy Sciences, under grants DE-FG02-89ER14043 and DE-FG02-90ER14136. A.S. also acknowledges support from the National Science Foundation under Grant No. CBT-8800451.

REFERENCES

Batchelor, G.K., *An Introduction to Fluid Dynamics.* (Cambridge University Press, Cambridge, 1967), pp. 364-367.

Biesheuvel, A. and Spoelstra, S, "The added mass coefficient of a dispersion of spherical gas bubbles in liquid," *International Journal of Multiphase Flow*, Vol. 15, 1989, pp. 911-924.

Chwang, A.T., and Wu, T.Y., "Hydromechanics of low-Reynolds-number flow. Part 1. Rotation of axisymmetric prolate bodies," *Journal of Fluid Mechanics*, Vol. 63, 1974, pp. 607-622.

Chwang, A.T., and Wu, T.Y., "Hydromechanics of low-Reynolds-number flow. Part 2. Singularity method for Stokes flow," *Journal of Fluid Mechanics*, Vol. 67, 1975, pp. 787-815.

Fortes, A., Joseph, D.D., and Lundgren, T., "Nonlinear mechanics of fluidization of beds of spherical particles," *Journal of Fluid Mechanics*, Vol. 177, 1987, pp. 467-483.

Hasimoto, H., "On the periodic fundamental solutions of the Stokes equations and their application to the viscous flow past a cubic array of spheres," *Journal of Fluid Mechanics*, Vol. 5, 1959, pp. 317-328.

Kang, I.S., and Leal, L.G., "The drag coefficient for a spherical bubble in a uniform streaming flow," *The Physics of Fluids*, Vol. 31, 1988, 233-237.

Kim, H.S., and Prosperetti, A., "Numerical simulation of the motion of rigid spheres in potential flow," submitted to *SIAM Journal on Applied Mathematics*, 1991.

Kok, J.B.W., "Dynamics of gas bubbles moving through liquid," Doctoral Thesis, University of Twente, Enschede, The Netherlands, 1989.

Lamb, H., *Hydrodynamics*, 6th edition. (Cambridge University Press, Cambridge, 1932), Chapter V.

Levich, V.G., *Physicochemical Hydrodynamics.* (Prentice-Hall, Englewood Cliffs, 1962), p. 445.

Moore, D.W., "The boundary layer on a spherical gas bubble," *Journal of Fluid Mechanics*, Vol. 16, 1963, pp. 161-176.

Oğuz, H.N., and Prosperetti, A., "A generalization of the impulse and virial theorems with an application to bubble oscillations," *Journal of Fluid Mechanics*, Vol. 218, 1990, pp. 143-162.

Prosperetti, A., "On the stability of spherically symmetric flows," *Atti della Accademia Nazionale dei Lincei, Rendiconti della Classe di Scienze fisiche, matematiche e naturali*, Vol. 62, 1977, pp. 196-203.

Sangani, A.S.,"A pairwise interaction theory for determining the linear acoustic properties of dilute bubbly liquids," *Journal of Fluid Mechanics*, in press, 1991.

Sangani, A.S., and Acrivos, A., "Slow flow through a periodic array of spheres," *International Journal of Multiphase Flow*, Vol. 8, 1982, pp. 343-360.

Sangani, A.S., and Didwania A.K., "Dynamic simulations and average properties for buoyant bubbles rising through a liquid," in preparation, 1991.

Sangani, A.S., and Yao, C.,"Bulk thermal conductivity of composites with spherical inclusions," *Journal of Applied Physics*, Vol. 63, 1988, pp. 1334-1341.

Sangani, A.S., Zhang, D.Z., and Prosperetti, A., "The added mass, Bassett, and viscous drag coefficients in non-dilute bubbly liquids undergoing small-amplitude oscillatory motion," *The Physics of Fluids A*, in press, 1991.

Tam, P.D., "The unsteady drag on a spherical bubble at large Reynolds numbers," *Applied Scientific Research*, Vol. 38, 1981, pp. 245-254.

Throop, G.J., and Bearman, R.J. "Numerical solutions of the Percus-

Yevick equation for hard-sphere potential," *Journal of Chemical Physics*, Vol. 42, 1965, pp. 2408-2411.

van Wijngaarden, L., "Hydrodynamic interaction between gas bubbles in liquid," *Journal of Fluid Mechanics*, Vol. 77, 1976, pp. 27-44.

Wallis, G.B., "Inertial coupling in two-phase flow: Macroscopic properties of suspensions in an inviscid fluid," in *Multiphase Science and Technology*, Hewitt, G.F., Delhaye, J.M., and Zuber, N., Eds. (Hemisphere, Washington, 1989), Vol. 5, pp. 239-361.

Wallis, G.B., "Two-phase potential flow," in *Proceedings of the Ninth Symposium on Energy Engineering Sciences*, Frederick, D., Ed. (Argonne National Laboratory, Argonne, CONF-9105116, 1991), pp. 32-39.

Weinbaum, S., Ganatos, P., and Yan, Z.Y., "Numerical multipole and boundary integral techniques in Stokes flow," *Annual Review of Fluid Mechanics*, Vol. 22, 1990, 275-316.

Zuber, N., "On the dispersed two-phase flow in the laminar flow regime," *Chemical Engineering Science*, Vol. 19, 1964, 897-917.

INDEX